# Advances in Intelligent Systems and Computing

Volume 530

**Series editor**

Janusz Kacprzyk, Polish Academy of Sciences, Warsaw, Poland
e-mail: kacprzyk@ibspan.waw.pl

*About this Series*

The series "Advances in Intelligent Systems and Computing" contains publications on theory, applications, and design methods of Intelligent Systems and Intelligent Computing. Virtually all disciplines such as engineering, natural sciences, computer and information science, ICT, economics, business, e-commerce, environment, healthcare, life science are covered. The list of topics spans all the areas of modern intelligent systems and computing.

The publications within "Advances in Intelligent Systems and Computing" are primarily textbooks and proceedings of important conferences, symposia and congresses. They cover significant recent developments in the field, both of a foundational and applicable character. An important characteristic feature of the series is the short publication time and world-wide distribution. This permits a rapid and broad dissemination of research results.



More information about this series at http://www.springer.com/series/11156

Juan Manuel Corchado Rodriguez
Sushmita Mitra · Sabu M. Thampi
El-Sayed El-Alfy
Editors

# Intelligent Systems Technologies and Applications 2016

*Editors*
Juan Manuel Corchado Rodriguez
Department of Computer Science,
School of Science
University of Salamanca
Salamanca
Spain

Sushmita Mitra
Machine Intelligence Unit
Indian Statistical Institute
Kolkata, West Bengal
India

Sabu M. Thampi
Indian Institute of Information Technology
and Management, Kerala (IIITM-K)
Thiruvananthapuram, Kerala
India

El-Sayed El-Alfy
King Fahd University of Petroleum
and Minerals
Dhahran
Saudi Arabia

ISSN 2194-5357 ISSN 2194-5365 (electronic)
Advances in Intelligent Systems and Computing
ISBN 978-3-319-47951-4 ISBN 978-3-319-47952-1 (eBook)
DOI 10.1007/978-3-319-47952-1

Library of Congress Control Number: 2016953657



Printed on acid-free paper

This Springer imprint is published by Springer Nature
The registered company is Springer International Publishing AG
The registered company address is: Gewerbestrasse 11, 6330 Cham, Switzerland

# Preface

Intelligent systems refer broadly to computer embedded or controlled systems, machines and devices that possess a certain degree of intelligence with the capacity to gather and analyze data and communicate with other systems. There is a growing interest in developing intelligent technologies that enable users to accomplish complex tasks in different environments with relative ease. The International Symposium on Intelligent Systems Technologies and Applications (ISTA) aims to bring together researchers in related fields to explore and discuss various aspects of intelligent systems technologies and their applications. The second edition, ISTA'16 was hosted by The LNM Institute of Information Technology (LNMIIT), Jaipur (Rajasthan), India, during September 21–24, 2016. ISTA'16 was co-located with the First International Conference on Applied Soft computing and Communication Networks (ACN'16). This edition of the symposium was organised in association with Indian Unit for Pattern Recognition and Artificial Intelligence (IUPRAI).

We received 210 submissions and after rigorous review process, 80 papers were selected for publication. All the papers were evaluated on the basis of their significance, novelty, and technical quality. Each paper was rigorously reviewed by the members of the program committee. The contributions have the flavors of various aspects in the related fields and hence, it would be interesting to see the diversity of ideas that the authors came up with.

There is a long list of people who volunteered their time and energy to put together the conference and who warrant acknowledgment. We would like to thank the authors of all the submitted papers, especially the accepted ones, and all the participants who made the symposium a successful event. Thanks to all members of the Technical Program Committee, and the external reviewers, for their hard work in evaluating and discussing papers. The EDAS conference system proved very helpful during the submission, review, and editing phases. Our most sincere thanks go to all keynote and tutorial speakers who shared with us their expertise and knowledge

We are grateful to the General Chairs and members of the Steering Committee for their support. Our most sincere thanks go to all keynote and tutorial speakers who shared with us their expertise and knowledge. Special thanks to members of the organizing committee for their time and effort in organizing the conference. We thank The LNM Institute of Information Technology (LNMIIT), Jaipur for hosting the event.

We wish to express our thanks to Thomas Ditzinger, Senior Editor, Engineering/Applied Sciences Springer-Verlag for his help and cooperation.

Salamanca, Spain — Juan Manuel Corchado Rodriguez
Kolkata, India — Sushmita Mitra
Thiruvananthapuram, India — Sabu M. Thampi
Dhahran, Saudi Arabia — El-Sayed El-Alfy

# Organization

## Chief Patron

Lakshmi N. Mittal, Chairman, LNMIIT

## Patron

S.S. Gokhale, Director, LNMIIT

## General Chairs

Sushmita Mitra, Indian Statistical Institute, Kolkata, India
Junichi Suzuki, University of Massachusetts Boston, USA

## ACN'16 Steering Committee

Ngoc Thanh Nguyen, Wroclaw University of Technology, Poland
Janusz Kacprzyk, Polish Academy of Sciences, Poland
Sankar Kumar Pal, Indian Statistical Institute, Kolkata, India
Hans-Jürgen Zimmermann, RWTH Aachen University, Aachen, Germany
Nikhil R. Pal, Indian Statistical Institute, Kolkata, India
Sabu M. Thampi, IIITM-K, India
Mario Koeppen, Kyushu Institute of Technology, Japan
Michal Wozniak, Wroclaw University, Warsaw, Poland
Zoran Bojkovic, University of Belgrade, Serbia

Oge Marques, Florida Atlantic University (FAU), Boca Raton, Florida, USA
Ranjan Gangopadhyay, LNMIIT Jaipur, India
Nabendu Chaki, University of Calcutta, India
Abdennour El Rhalibi, Liverpool John Moores University, UK
Salah Bourennane, Ecole Centrale Marseille, France
Selwyn Piramuthu, University of Florida, USA
Peter Mueller, IBM Zurich Research Laboratory, Switzerland
Robin Doss, School of Information Technology, Deakin University, Australia
Md Zakirul Alam Bhuiyan, Temple University, USA
Axel Sikora, University of Applied Sciences Offenburg, Germany
Ryan Ko, University of Waikato, New Zealand
Sri Krishnan, Ryerson University, Toronto, Canada
El-Sayed El-Alfy, King Fahd University of Petroleum and Minerals, Saudi Arabia
Junichi Suzuki, University of Massachusetts Boston, USA
Parag Kulkarni, iknowlation Research Labs Pvt Ltd, and EKLaT Research, India
Narsi Bolloju, LNMIIT Jaipur, India
Sakthi Balan, LNMIIT Jaipur, India

## Organizing Chairs

Raghuvir Tomar, LNMIIT
Ravi Prakash Gorthi, LNMIIT

## Organising Secretaries

Sandeep Saini, LNMIIT
Kusum Lata, LNMIIT
Subrat Dash, LNMIIT

## Event Management Chair

Soumitra Debnath, LNMIIT

## Publicity Co-chair

Santosh Shah, LNMIIT

## TPC Chairs

Juan Manuel Corchado Rodriguez, University of Salamanca, Spain
El-Sayed M. El-Alfy, King Fahd University of Petroleum and Minerals, Saudi Arabia

## TPC Members/Additional Reviewers

Taneli Riihonen, Aalto University School of Electrical Engineering, Finland
Anouar Abtoy, Abdelmalek Essaâdi University, Morocco
Oskars Ozolins, Acreo Swedish ICT, Sweden
Atilla Elçi, Aksaray University, Turkey
Ambra Molesini, Alma Mater Studiorum—Università di Bologna, Italy
Waail Al-waely, Al-Mustafa University College, Iraq
Vishwas Lakkundi, Altiux Innovations, India
Deepti Mehrotra, AMITY School of Engineering and Technology, India
Jaynendra Kumar Rai, Amity School of Engineering and Technology, India
Rakesh Nagaraj, Amrita School of Engineering, India
Shriram K Vasudevan, Amrita University, India
Amudha J, Amrita Vishwa Vidyapeetham, India
GA Shanmugha Sundaram, Amrita Vishwa Vidyapeetham University, India
Algirdas Pakštas, AP Solutions, United Kingdom
Eun-Sung Jung, Argonne National Laboratory, USA
Valentina Balas, Aurel Vlaicu University of Arad, Romania
Shanmugapriya D., Avinashilingam Institute for Higher Education for Women, India
Nisheeth Joshi, Banasthali University, India
Mike Jackson, Birmingham City University, United Kingdom
Vaclav Satek, Brno University of Technology, Czech Republic
Elad Schiller, Chalmers University of Technology, Sweden
Yih-Jiun Lee, Chinese Culture University, Taiwan
Yuji Iwahori, Chubu University, Japan
Mukesh Taneja, Cisco Systems, India
Filippo Vella, CNR, Italian National Research Council, Italy
Ciza Thomas, College of Engineering Trivandrum, India
Ivo Bukovsky, Czech Technical University in Prague, Czech Republic
Deepak Singh, Dayalbagh Educational Institute, India
Suma V., Dayananda Sagar College of Engineering, VTU, India
Sasan Adibi, Deakin University, Australia
Tushar Ratanpara, Dharmsinh desai University, India
Saibal Pal, DRDO, India
Salah Bourennane, Ecole Centrale Marseille, France

N. Lakhoua, ENIT, Tunisia
Adil Kenzi, ENSAF, Morocco
Monica Chis, Frequentis AG, Romania
Akihiro Fujihara, Fukui University of Technology, Japan
Zhaoyu Wang, Georgia Institute of Technology, USA
Monika Gupta, GGSIPU, India
G.P. Sajeev, Government Engineering College, India
Anton Fuchs, Graz University of Technology, Austria
Kalman Graffi, Heinrich Heine University Düsseldorf, Germany
John Strassner, Huawei, USA
Abdelmajid Khelil, Huawei European Research Center, Germany
Anthony Lo, Huawei Technologies Sweden AB, Sweden
Agnese Augello, ICAR-CNR, Italian National Research Council, Italy
Kiril Alexiev, IICT-Bulgarian Academy of Sciences, Bulgaria
Björn Schuller, Imperial College London, UK
Mahendra Mallick, Independent Consultant, USA
Seshan Srirangarajan, Indian Institute of Technology Delhi, India
Ravibabu Mulaveesala, Indian Institute of Technology Ropar, India
Kaushal Shukla, Indian Institute of Technology, Banaras Hindu University, India
Sreedharan Pillai Sreelal, Indian Space Research Organization, India
Joel Rodrigues, Instituto de Telecomunicações, Portugal
Chi-Hung Hwang, Instrument Technology Research Center, Taiwan
Md Mozasser Rahman, International Islamic University Malaysia, Malaysia
Nahrul Khair Alang Md Rashid, International Islamic University Malaysia, Malaysia
Mohd Ramzi Mohd Hussain, International Islamic University Malaysia, Malaysia
Kambiz Badie, Iran Telecom Research Center, Iran
Amir Hosein Jafari, Iran University of Science and Technology, Iran
Engin Yesil, Istanbul Technical University, Turkey
Lorenzo Mossucca, Istituto Superiore Mario Boella, Italy
Vivek Sehgal, Jaypee University of Information Technology, India
Binod Kumar, JSPM's Jayawant Institute of Computer Applications, Pune, India
Qiang Wu, Juniper Networks, USA
Fathima Rawoof, KS School of Engineering and Management, Bangalore, India
Hideyuki Sawada, Kagawa University, Japan
Ismail Altas, Karadeniz Technical University, Turkey
Kenneth Nwizege, Ken Saro-Wiwa Polytechnic, Bori, Nigeria
Sasikumaran Sreedharan, King Khalid University, Saudi Arabia
Mario Collotta, Kore University of Enna, Italy
Maytham Safar, Kuwait University, Kuwait
Noriko Etani, Kyoto University, Japan
Kenichi Kourai, Kyushu Institute of Technology, Japan
Ernesto Exposito, LAAS-CNRS, Université de Toulouse, France
Philip Moore, Lanzhou University, P.R. China

Raveendranathan Kalathil Chellappan, LBS Institute of Technology for Women, India
Issam Kouatli, Lebanese American University, Lebanon
Grienggrai Rajchakit, Maejo University, Thailand
Ilka Miloucheva, Media Applications Research, Germany
Michael Lauer, Michael Lauer Information Technology, Germany
Kazuo Mori, Mie University, Japan
Su Fong Chien, MIMOS Berhad, Malaysia
Sheng-Shih Wang, Minghsin University of Science and Technology, Taiwan
Prasheel Suryawanshi, MIT Academy of Engineering, Pune, India
Sim-Hui Tee, Multimedia University, Malaysia
Huakang Li, Nanjing University of Posts and Telecommunications, P.R. China
Chong Han, Nanjing University of Posts and Telecommunications, P.R. China
Mustafa Jaber, Nant Vision Inc., USA
Shyan Ming Yuan, National Chiao Tung University, Taiwan
Yu-Ting Cheng, National Chiao Tung University, Taiwan
Guu-Chang Yang, National Chung Hsing University, Taiwan
Mantosh Biswas, National Institute of Technology-Kurukshetra, India
Dimitrios Stratogiannis, National Technical University of Athens, Greece
Anton Popov, National Technical University of Ukraine, Ukraine
I-Hsien Ting, National University of Kaohsiung, Taiwan
Rodolfo Oliveira, Nova University of Lisbon, Portugal
Stefanos Kollias, NTUA, Greece
Dongfang Zhao, Pacific Northwest National Laboratory, USA
Naveen Aggarwal, Panjab University, India
Manuel Roveri, Politecnico di Milano, Italy
Mihaela Albu, Politehnica University of Bucharest, Romania
Radu-Emil Precup, Politehnica University of Timisoara, Romania
Houcine Hassan, Polytechnic University of Valencia, Spain
Kandasamy Selvaradjou, Pondicherry Engineering College, India
Ravi Subban, Pondicherry University, Pondicherry, India
Ninoslav Marina, Princeton University, USA
Siddhartha Bhattacharyya, RCC Institute of Information Technology, India
Branko Ristic, RMIT University, Australia
Kumar Rajamani, Robert Bosch Engineering and Business Solutions Limited, India
Ali Yavari, Royal Melbourne Institute of Technology—RMIT, Australia
Mr. A.F.M. Sajidul Qadir, Samsung R&D Institute-Bangladesh, Bangladesh
Anderson Santana de Oliveira, SAP Labs, France
Mahendra Dixit, SDMCET, India
Ljiljana Trajkovi, Simon Fraser University, Canada
J. Mailen Kootsey, Simulation Resources, Inc., USA
Yilun Shang, Singapore University of Technology and Design, Singapore
Chau Yuen, Singapore University of Technology and Design, Singapore
Gwo-Jiun Horng, Southern Taiwan University of Science and Technology, Taiwan
Manjunath Aradhya, Sri Jayachamarajendra College of Engineering, India

Shajith Ali, SSN College of Engineering, Chennai, India
Anthony Tsetse, State University of New York, USA
Sanqing Hu, Stevens Institute of Technology, USA
Peng Zhang, Stony Brook University, USA
Rashid Ali, Taif University, Saudi Arabia
Meng-Shiuan Pan, Tamkang University, Taiwan
Chien-Fu Cheng, Tamkang University, Taiwan
Sunil Kumar Kopparapu, Tata Consultancy Services, India
Peyman Arebi, Technical and Vocational University, Iran
Dan Dobrea, Technical University "Gh. Asachi", Romania
Jose Delgado, Technical University of Lisbon, Portugal
Eitan Yaakobi, Technion, Israel
Angelos Michalas, Technological Education Institute of Western Macedonia, Greece
Grammati Pantziou, Technological Educational Institution of Athens, Greece
Biju Issac, Teesside University, Middlesbrough, UK
Stephane Maag, TELECOM SudParis, France
Eduard Babulak, The Institute of Technology and Business in Ceske Budejovice, Czech Republic
Haijun Zhang, The University of British Columbia, Canada
Hiroo Wakaumi, Tokyo Metropolitan College of Industrial Technology, Japan
Minoru Uehara, Toyo University, Japan
Ruben Casado, TreeLogic, Spain
Qurban Memon, United Arab Emirates University, UAE
Jose Molina, Universidad Carlos III de Madrid, Spain
Vinay Kumar, Universidad Carlos III de Madrid, Spain
Jose Luis Vazquez-Poletti, Universidad Complutense de Madrid, Spain
Juan Corchado, Universidad de Salamaca, Spain
Gregorio Romero, Universidad Politecnica de Madrid, Spain
Antonio LaTorre, Universidad Politécnica de Madrid, Spain
Luis Teixeira, Universidade Catolica Portuguesa, Portugal
Eraclito Argolo, Universidade Federal do Maranhão, Brazil
Marco Anisetti, Università degli Studi di Milano, Italy
Angelo Genovese, Università degli Studi di Milano, Italy
Roberto Sassi, Università degli Studi di Milano, Italy
Ruggero Donida Labati, Università degli Studi di Milano, Italy
Giovanni Livraga, Università degli Studi di Milano, Italy
Paolo Crippa, Università Politecnica delle Marche, Italy
Nemuel Pah, Universitas Surabaya, Indonesia
Kushsairy Kadir, Universiti Kuala Lumpur British Malaysian Institute, Malaysia
Hua Nong Ting, Universiti Malaya, Malaysia
Ku Nurul Fazira Ku Azir, Universiti Malaysia Perlis, Malaysia
Farrah Wong, Universiti Malaysia Sabah, Malaysia
Asrul Izam Azmi, Universiti Teknologi Malaysia, Malaysia
Norliza Noor, Universiti Teknologi Malaysia, Malaysia

Musa Mailah, Universiti Teknologi Malaysia, Malaysia
Rudzidatul Dziyauddin, Universiti Teknologi Malaysia, Malaysia
Siti Zura A. Jalil, Universiti Teknologi Malaysia, Malaysia
Salman Yussof, Universiti Tenaga Nasional, Malaysia
Ku Ruhana Ku-Mahamud, Universiti Utara Malaysia, Malaysia
Nhien-An Le-Khac, University College Dublin, Ireland
Hector Menendez, University College London, Spain
Alberto Nuñez, University Complutense of Madrid, Spain
Eduardo Fernández, University Miguel Hernández, Spain
Mariofanna Milanova, University of Arkansas at Little Rock, USA
Iouliia Skliarova, University of Aveiro, Portugal
Luís Alexandre, University of Beira Interior, Portugal
Amad Mourad, University of Bejaia, Algeria
Robert Hendley, University of Birmingham, UK
Mohand Lagha, University of Blida 1, Algeria
Francine Krief, University of Bordeaux, France
Otthein Herzog, University of Bremen, Germany
Kester Quist-Aphetsi, University of Brest France, France
Angkoon Phinyomark, University of Calgary, Canada
Yuanzhang Xiao, University of California, Los Angeles, USA
Marilia Curado, University of Coimbra, Portugal
Vasos Vassiliou, University of Cyprus, Cyprus
Chen Xu, University of Delaware, USA
Mr. Chiranjib Sur, University of Florida, USA
Abdallah Makhoul, University of Franche-Comté, France
Na Helian, University of Hertfordshire, UK
Hamed Vahdat-Nejad, University of Isfahan, Iran
Emilio Jiménez Macías, University of La Rioja, Spain
Simon Fong, University of Macau, Macao
Carl Debono, University of Malta, Malta
Kenneth Camilleri, University of Malta, Malta
Davide Carneiro, University of Minho, Portugal
Jorge Bernal Bernabé, University of Murcia, Spain
Adel Sharaf, University of New Brunswick, Canada
Jun He, University of New Brunswick, Canada
Sandeep Reddivari, University of North Florida, USA
Salvatore Vitabile, University of Palermo, Italy
Alain Lambert, University of Paris Sud, France
Andrea Ricci, University of Parma, Italy
Sotiris Kotsiantis, University of Patras, Greece
Sotiris Karachontzitis, University of Patras, Greece
Ioannis Moscholios, University of Peloponnese, Greece
Francesco Marcelloni, University of Pisa, Italy
Maurizio Naldi, University of Rome “Tor Vergata”, Italy
Massimo Cafaro, University of Salento, Italy

Ernesto d'Avanzo, University of Salerno, Italy
Angel Barriga, University of Seville, Spain
Shireen Panchoo, University of Technology, Mauritius
Mohamed Ba khouya, University of Technology of Belfort Montbeliard, France
Feiping Nie, University of Texas at Arlington, USA
Georgios Kambourakis, University of the Aegean, Greece
Massimiliano Comisso, University of Trieste, Italy
Tharrenos Bratitsis, University of Western Macedonia, Greece
Ahmet Turkmen, University of Wisconsin Sout, USA
Luis Sequeira, University of Zaragoza, Spain
Mikulas Alexik, University of Zilina, Slovakia
Traian Rebedea, University Politehnica of Bucharest, Romania
Feliksas Kuliesius, Vilnius University, Lithuania
Paramartha Dutta, Visva-Bharati University, India
Chitti Babu B., VSB-Technical University of Ostrava, Czech Republic
Krzysztof Kulpa, Warsaw University of Technology, Poland
Koushik Majumder, West Bengal University of Technology, India
Viorel Negru, West University of Timisoara, Romania
Grzegorz Debita, Wroclaw University of Technology, Poland
Nihan Kahraman, Yildiz Technical University, Turkey

# Contents

## Part II Networks/Distributed Systems

# Part I
# Image Processing and Artificial Vision

# A Color Image Segmentation Scheme for Extracting Foreground from Images with Unconstrained Lighting Conditions

Niyas S, Reshma P and Sabu M Thampi

Indian Institute of Information Technology and Management- Kerala, india
e-mail: {niyas.s,reshma.nair}@iiitmk.ac.in

**Abstract** Segmentation plays a functional role in most of the image processing operations. In applications like object recognition systems, the efficiency of segmentation must be assured. Most of the existing segmentation techniques have failed to filter shadows and reflections from the image and the computation time required is marginally high to use in real time applications. This paper proposes a novel method for an unsupervised segmentation of foreground objects from a non-uniform image background. With this approach, false detections due to shadows, reflections from light sources and other noise components can be avoided at a fair level. The algorithm works on an adaptive thresholding, followed by a series of morphological operations in low resolution downsampled image and hence, the computational overhead can be minimized to a desired level. The segmentation mask thus obtained is then upsampled and applied to the full resolution image. So the proposed technique is best suited for batch segmentation of high-resolution images.

**Keywords** Thresholding . Morphological operation . Upsampling . Downsampling.

## 1 Introduction

Image segmentation is a crucial process in image analysis and computer vision applications. Image segmentation splits images into a number of disjoint sections such that the pixels in each section have high similarity and pixels among different sections are highly divergent. Since the detection of the foreground area of an image is an important task in image analysis, researchers are in search of accurate segmentation algorithms that consumes less time. Image segmentation is frequently used as the pre-processing step in feature extraction, pattern recognition, object recognition, image classification and image compression [1]. While considering



J.M. Corchado Rodriguez et al. (eds.), *Intelligent Systems Technologies and Applications 2016*, Advances in Intelligent Systems and Computing 530,
DOI 10.1007/978-3-319-47952-1_1

an object recognition system, the primary task is the accurate extraction of the foreground area of the whole image. Various features can be extracted from this foreground area and further classification is based on the extracted features. If the segmentation is inefficient, relevant features cannot be extracted from the region of interest and may lead to false predictions.

Image Segmentation can be widely classified into supervised and unsupervised segmentation [2-3] methods. Supervised segmentation algorithms use prior knowledge by using a training set of images. However, in unsupervised algorithms, the segmentation process depends on parameters from the test image itself. Adoption of a particular algorithm among various supervised and unsupervised techniques depends on various factors like image type, nature of foreground and background, target application and computation time. Segmentation using Otsu's [4] thresholding is an example for unsupervised segmentation while Markov Random Field [5] based segmentation belongs to the supervised approach.

Unsupervised image segmentation methods can be further classified into thresholding-based, edge-based and region-based segmentation [1]. The thresholding-based segmentation [6] finds a threshold from a gray scale or color histogram of the image and this threshold acts as the barrier to segment the image into foreground and background areas. Edge-based segmentation [7] is suitable for boundary detecting applications such as text recognition. In region-based segmentation, the process starts with a few seed pixels and these seed points merge with the neighboring pixels with similar property around the seed pixel area. This process repeats until every pixel in the image gets scanned.

In the proposed work, the main objective is to develop an efficient segmentation algorithm that can perform well with color images with shadows and reflections from light sources due to non-uniform lighting conditions. The segmented output should be free from background region and noise, and can be used in object recognition applications [8]. Edge-based segmentation approach often fails to detect complex object boundaries, when the image is distorted by shadows or reflection noise. The efficiency of region based segmentation relies on the selection of appropriate seed points, and may end in erroneous results, if the selected seed points are incorrect. Existing threshold based techniques are simple and the computation time required is low compared to other unsupervised segmentation methods. However, the thresholding should be adaptive and should remove image background, shadows and reflection noise from the image

This article proposes an accurate threshold-based image segmentation technique for color images. In this system, the input image gets initially filtered by an adaptive median filter [9]. The filtered image is then downsampled to a lower resolution, and a thresholding is applied to segment the foreground area. The thresholding is based on certain parameters and these parameters help to remove shadows and high intensity light reflections from the image. The mask obtained after thresholding might contains noise elements and these are eliminated by applying a series of morphological operations. The mask thus obtained is then upsampled to the original resolution and is used to segment the foreground area of the image.

The proposed technique is intended for application in object recognition systems, where images need to be segmented prior to classification stage. Here the segmentation mask is generated in the lower resolution image, and the processing time can be reduced to a greater extend and thousands of images can be segmented within a short duration of time. Also the segmentation efficiency is much better since the algorithm removes shadows and reflections from the system. The article is organized into following sections: Section 2 briefly describes some related works on unsupervised image segmentation. In Section 3, the methodology of the proposed work is explained. Discussion about the experimental results is conducted in section 4. Finally, concluding remarks are drawn in Section 5.

## 2 Literature Review

Segmentation results become vulnerable in real world cases due to the impact of reflections from the light sources, non-uniform background. Image segmentation using edge detection methods fails to get the exact border in blurred images and images with complex edges especially in unconstrained illumination conditions. Region based segmentation techniques consume more time and segmentation accuracy cannot be guaranteed in segmenting multi-colored objects. Image thresholding [10] is considered as one of the simple methods to segment an image. Although, the operation is simple, choosing the optimal threshold value is a critical task. This is most commonly used in images where the contrast between foreground and background pixels is high. Most of the threshold-based image segmentation methods are not suitable for images with illumination variations.

Reviews of various segmentation techniques like edge based, threshold, region based, clustering and neural network are explained in the articles [11,12]. Different segmentation methods have been proposed based on active contour models [13-18]. This strategy is particularly suitable for modeling and extracting complex shape contours. The active contour based segmentation is especially suited for the segmentation of inhomogeneous images. In region growing method [19-20] pixels with comparable properties are aggregated to form a region. Several modified region-based segmentation techniques [21-24] have been evolved to improve the segmentation efficiency.

Otsu is an old, but effective method used for segmenting gray level images. Here the image is segmented via histogram-based thresholding. The optimal threshold is evaluated on the basis of maximum between-class variance and minimum within-class variance. Even though the method shows satisfactory results in various images, it becomes unusable, when the difference of gray-level distribution between objects and background is modest. Several popular modifications of Otsu's methods are used in various applications. Methods based on Log-Normal and Gamma distribution models are explained in an article by A. ElZaart et al. [25]. In Otsu methods based on Log-Normal distribution and Gamma distribution, different models for determining maximum between-cluster variance are used.

Another method [26] proposed by Q. Chen et al., discusses an improved Otsu image segmentation along with a fast recursive realization method by determining probabilities of diagonal quadrants in 2D histogram. Article [27] proposes a modified Otsu's thresholding along with firefly algorithm for segmenting images with lower contrast levels. But the algorithm efficiency is not satisfactory in removing shadows from the image.

Watershed transform is a kind of image thresholding based on mathematical morphological operations, which decomposes an image into several similar and non-overlapping regions [28-31]. The approach uses region based thresholding by analyzing peaks and valleys in the image intensity. Standard watershed transform and its various modifications are widely used in both grayscale and color image segmentations [32-34]. The papers [35-37] analyze the drawbacks of the classical watershed segmentation and a new watershed algorithm proposed, based on a reconstruction of the morphological gradient. Here morphological opening and closing operations are used to reconstruct the gradient image, removes noise and avoids over-segmentation. Even though the segmentation results are outstanding in images with proper white balance, this algorithm is not advisable for classifying real-time images with shadows and reflections.

In digital image applications, clustering technique [38] is another widely used method to segment regions of interest. K-means [39] is a broadly utilized model-based, basic partitioned clustering technique which attempts to find a user-specified 'K' number of clusters. While using K-means algorithm in image segmentation [40-45], it searches for the final clusters values based on predetermined initial centers of pixel intensities. Improper initialization leads to generation of poor final centers that induce errors in segmented results.

The main objective of the proposed method is to segment the exact foreground area in the image even if shadows and reflection noises are present. Existing thresholding methods like Otsu's segmentation are inadequate in removing shadows from the image. Since Watershed approaches use regional peaks for segmentation, the accuracy will be much dependent on the lighting conditions and hence such methods cannot be used in images with unconstrained lighting conditions. Clustering techniques can work well with high contrast images. However, the computation overhead of such methods is too high to be used in the batch segmentation of high resolution images. The proposed method uses an advanced thresholding approach along with appropriate mathematical morphological operations to extract the exact foreground area from the image.

## 3 Proposed Algorithm

The Proposed system aims at developing an efficient segmentation system for real world color images with minimal computational overhead.. The subsequent steps of the algorithm are shown in Fig.1.

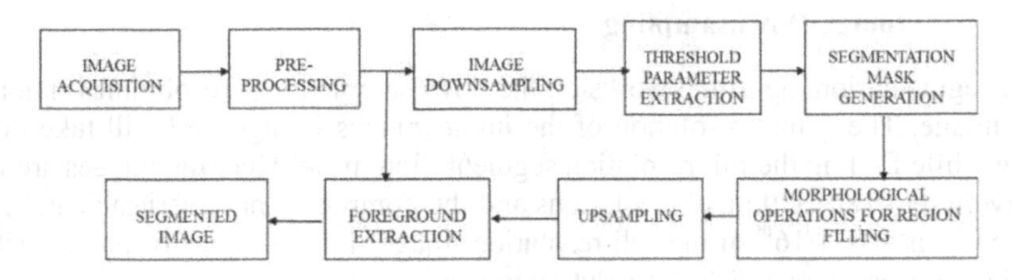


Fig 1: Proposed System Workflow

### 3.1 Image Acquisition and Pre-processing

The images were captured by a 5 MP webcam with 1024x1024 pixel resolution and 24-bit color depth. For creating the database of images, different objects were placed either on a nearly black background area or on a white surface. Image background can be a table top or anything with a nearly uniform texture. The images were captured in unconstrained lighting conditions and many images seemed to be affected by impulse noise [46], shadows and reflections of light sources.

A filtering process is used to remove impulse noise textures and tiny unwanted objects from the image. Adaptive Median Filtering (AMF) [47-49] is applied to remove impulse noise from the image. Since the input is a color image, AMF need to be applied to the individual color planes and then combined together, so as to result in the noise free color image. Before segmenting the foreground, a background color detection process is used to check whether the object is placed on a white surface or dark surface. This is calculated by finding the average pixel intensity among the border pixels of the image using equation (1),

$$B = \begin{cases} Black & ; \quad if\ P_{avg} < 50 \\ White & ; \quad if\ P_{avg} > 150 \\ Bad\ quality & ; \quad otherwise \end{cases} \tag{1}$$

where $P_{avg}$ is the average pixel intensity of the border pixels of the grayscale image. If the value of $P_{avg}$ is on the lower side of gray level, the image can be treated as the one with black background and if it is on the higher side of the grayscale, it is considered as a white background image. The segmentation result may not be good if $P_{avg}$ lies in the middle range of gray intensity scale. Segmentation efficiency appears to be good when the object is placed in nearly white or black backgrounds.

### 3.2 Image Downsampling

The segmentation algorithm works on the downsampled low resolution version of the image. The actual resolution of the input images is high and will take much time while finding the full resolution segmentation mask. Here the images are first converted to 180x320 pixel resolutions and the computational overhead can be reduced to nearly 1/16$^{th}$ of the full resolution image. Further steps of the algorithm will be processed on this low resolution image.

### 3.3 Extraction of Threshold Parameters

The primary objective of the algorithm is to filter shadows and reflections (from light sources) from the background area. Complex modeling of reflection and shadows are avoided here and a simple way to detect most of the noisy pixels with minimum time, is proposed. Firstly, individual color planes: Red, Green and Blue, get separated and two parameters are calculated at every pixel position of the image. The parameter *Dxy*, represents the average of the difference of pixels in different color planes at the location (x,y) and is obtained as

$$D_{xy} = \frac{|(i_r - i_g)| + |(i_r - i_b)| + |(i_g - i_b)|}{3} \quad (2)$$

where $i_r$, $i_g$, and $i_b$ are the intensity values of red, green and blue color plane at position (x,y). Another parameter *Sxy* , the average of the sum of individual color pixels at the location (x,y) is obtained by

$$S_{xy} = \frac{(i_r + i_g + i_b)}{3} \quad (3)$$

From the earlier background color detection phase, the image can be classified either into white background or black background. Let us first consider a white background image. The effect of shadows in this image might be higher than that of a black background image. Normally the pixels in the shadow region are closer to the gray-level axis of the RGB color space in Fig.2.

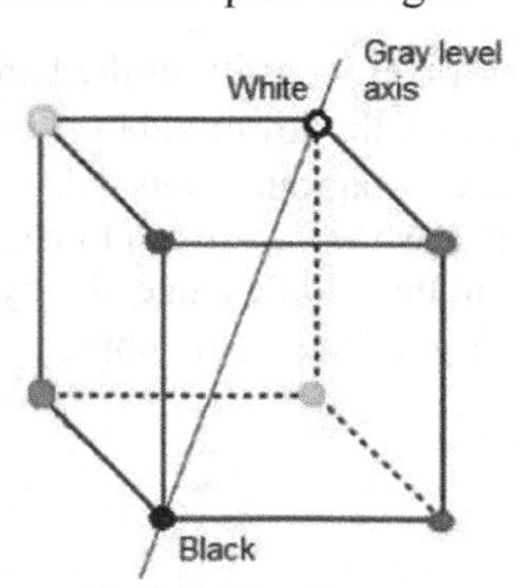


Fig.2. RGB color space

The exact gray levels in RGB color space are shown in the following Table 1.

Table 1. Intensity of color components in RGB color space

| Gray level Intensity | Intensity of color components in RGB Space | | |
|---|---|---|---|
| | R | G | B |
| 1 | 1 | 1 | 1 |
| 2 | 2 | 2 | 2 |
| 3 | 3 | 3 | 3 |
| . . . | . . | . . | . . |
| 255 | 255 | 255 | 255 |

The color component values of ideal gray pixels in a color image are shown in Table 1. Since the pixels in the shadow region belong to the gray level, they can be easily detected by checking the difference of individual color components.

### 3.4 Segmentation-Mask Generation

By finding the pixels nearer to the gray level axis, most of the pixels in the shadow region can be identified and removed. For a white background image, the mask, required to segment object is obtained as

$$M = \begin{cases} 0 \ ; \ if\ D_{xy} < T1\ and\ S_{xy} > T2 \\ 1 \ ; \quad otherwise \end{cases} \quad (4)$$

Where $M$ is the mask and $T1$ & $T2$ are the thresholds used. The threshold values can be adjusted to the optimum values by evaluating some sample images in the dataset. These values are selected in such a way that the pixels nearer to the pure black region are preserved while white background, shadows and reflections are removed. Based on our calculation we set $T1 = 30$ and $T2 = 70$. Since the reflection region and background lies in the high intensity range, it will also be removed by the above thresholding process. After this step the pixels of the Mask, $M$ in the object region is '1' and the background region has '0' value. While considering the images with black background, the distortion due to shadows is comparatively low. So, Otsu's thresholding is applied to create the mask $M$, with binary '1' in the object region and '0' in the background portion.

### 3.5 Morphological operations for region filling

The ultimate aim of the proposed algorithm is to segment ROI from raw images and this segmented image shall be used for recognition and classification purposes. Even after the preceding thresholding operation, the resultant binary mask *M*, may suffer from holes in the object region and may have small noise pixels in the background and are as shown in Fig.3(b).

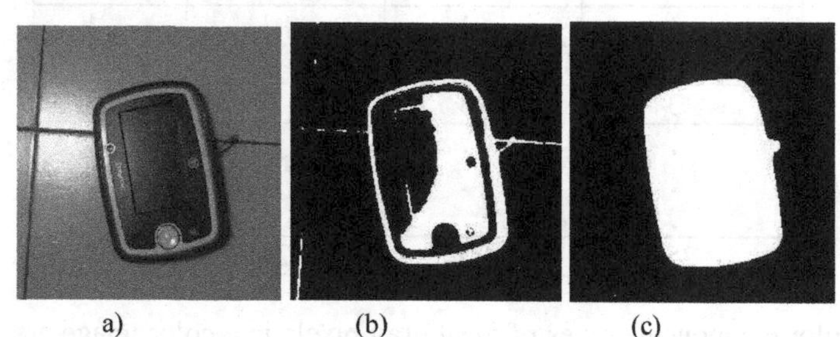

a) (b) (c)
Fig. 3 (a) Input image (b) Mask obtained after initial thresholding, (c) Mask obtained after morphological operations

The issue can be solved by using proper mathematical morphological operations [1]. Morphological operations depend only on the relative arrangement of pixel values and are particularly, suited for processing binary images. Due to the filtering operations used in the section 3.3, pixels in the range of gray axis will be removed and may result in holes in the object region. For getting the complete object region, a region filling operation is applied and the small white noisy pixels are removed by morphological opening of the mask '*M*' with a disc shaped structural element '*S*', having a radius of 3 pixels. Morphological opening is the erosion followed by a dilation, using the same structuring element for both operations. It can be represented as

$$M \circ S = (M \ominus S) \oplus S \tag{5}$$

where $\ominus$ and $\oplus$ denote erosion and dilation respectively. The Mask after this operation is as shown in the Fig.3(c)

### 3.6 Mask Upsampling

All the above segmentation steps were processed on the downsampled image and the mask thus obtained has only the low resolution. The Mask should be upsampled to the original resolution in order to operate on the actual resolution image.

### 3.7 Foreground area Extraction using Mask

After getting the full resolution mask, the object area from the input image is extracted by using the following expression.

$$O_{xy} = \begin{cases} 0 & ; \quad if\ M_{xy} = 0 \\ I_{xy} & ; \quad if\ M_{xy} = 1 \end{cases} \tag{6}$$

where '$I$' is the Median filtered input image, '$O$' is the Output image and $M$ is the mask obtained by the proposed method. '$x$' and '$y$' are the pixel coordinates in the images. The result after segmentation is shown in the Fig.4.

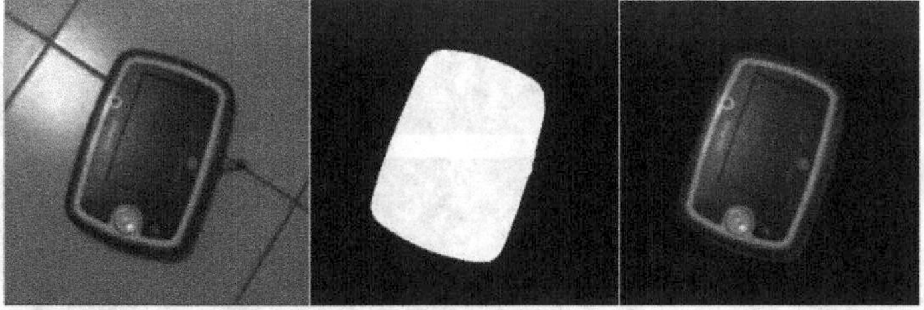

Fig.4 (a) Input image (b) Final Mask (c) Segmented Output

## 4 Experimental results

The proposed segmentation method aims for extracting object areas from images taken in unconstrained light conditions. Here we used a set of images taken by a digital webcam of 5MP resolution. The test images have a resolution of 1024x1024-pixel resolution and 24-bit color depth. Here MATLAB R2013b is used as the software tool to implement the proposed algorithm. All testing processes were executed on a system with Intel i5 processor and 4GB RAM. The results of proposed segmentation algorithms are compared with some traditional unsupervised methods like Active Contour based, Otsu's thresholding and K-means segmentation. The segmentation results of some sample images are shown in Fig.5.

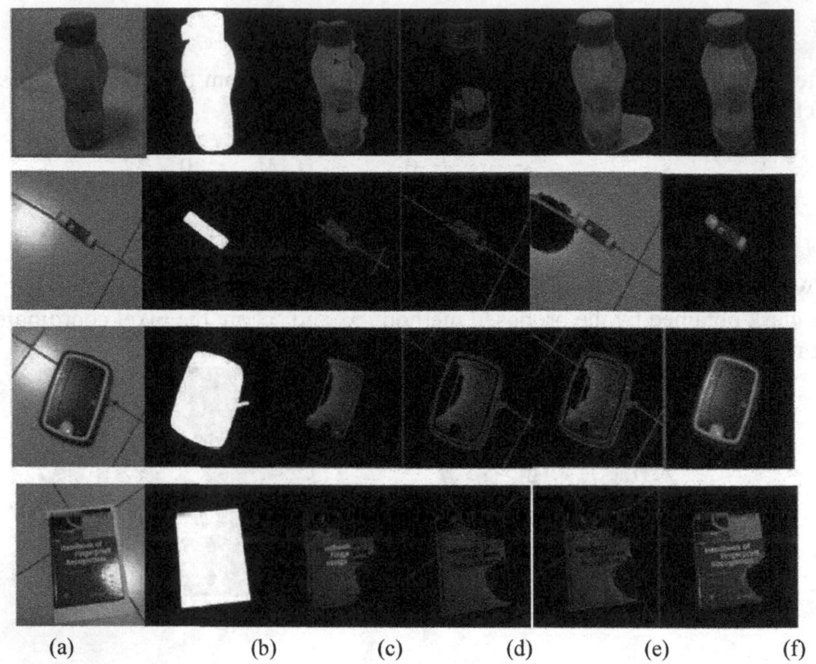

(a) (b) (c) (d) (e) (f)

Fig.5. Segmentation results (a)Input image (b)Ground Truth (c )Active contour (d)K-means (e)Otsu's segmentation (f)Proposed method

Segmentation efficiency is analyzed on the basis of object-based measures [50]. Measures were focused on two aspects: The first one is a pixel to pixel similarity measurement of segmented image and ground truth image. The second one evaluates how effectively the object boundary is detected. The proposed segmentation approach has been tested with 50 images in various working conditions and the result shows considerable improvement in comparison with the performance of existing segmentation methods. The computation time required is slightly higher than the Otsu's method. However, the exact foreground or background extraction is possible with this approach which would be further helpful in the recognition processes.

### 4.1 Pixel based Object Measures

In this analysis the ground truth image and the segmented output are compared and the following parameters were evaluated. Let $P_g$ and $N_g$ refers to the positive and negative pixel classes in the ground truth image respectively. Similarly, $P_s$ and $N_s$ are the positive and negative pixels in Segmented result.

True Positive (TP): Pixels that are detected as an object and also labeled as so on the ground truth image.

$$\text{TP} = P_s \cap P_g \tag{7}$$

True Negative (TN): Pixels that are detected as a background and also labeled as so on the ground truth image.

$$\text{TN} = N_s \cap N_g \tag{8}$$

False Positive (FP): Pixels that are detected as an object and labeled as background on the ground truth image.

$$\text{FP} = P_s \cap N_g \tag{9}$$

False Negative (FN): Pixels that are detected as background and labeled as an object region in the ground truth image.

$$\text{FN} = N_s \cap P_g \tag{10}$$

From the above parameters, the following performance measures were calculated.

$$Precision = \frac{TP}{TP+FN} \tag{11}$$

$$Recall = \frac{TP}{TP+FP} \tag{12}$$

Even though Precision and Recall [51] are robust measures to represent the accuracy, harmonic mean of these parameters gives a general tradeoff between them. This parameter is termed as F- measure and is given as

$$F = 2 * \frac{(Precision*Recall)}{(Precision+Recall)} \tag{13}$$

## 4.2 Boundary based Object Measures

Boundary based measures help to check how well the system identifies the boundary of the foreground object. For finding these measures, Canny edge detection [52] is applied on both the ground truth and segmented image and all the parameters: Precision, Recall and F-measure are again calculated by comparing the boundaries of ground truth and segmented result.

The following table shows the performance comparison of the proposed method with the existing segmentation techniques.

Table 2: Performance Analysis of various Segmentation Approaches

| Measure Representation | | Active Contour | K-means | Otsu's Method | Proposed Method |
|---|---|---|---|---|---|
| Pixel-based Measures | Precision | 0.93 | 0.68 | 0.85 | 0.98 |
| | Recall | 0.65 | 0.85 | 0.60 | 0.96 |
| | F-Score | 0.76 | 0.84 | 0.68 | 0.97 |
| Boundary-based Measures | Precision | 0.12 | 0.05 | 0.07 | 0.79 |
| | Recall | 0.19 | 0.18 | 0.29 | 0.81 |
| | F-Score | 0.16 | 0.09 | 0.11 | 0.80 |
| Average computation time (in Seconds) | | 9.28 | 93.56 | 1.74 | 2.31 |

From Table 2, It can be concluded that the proposed method performs with better segmentation accuracy both in pixel-wise and boundary based approach. Since the work aims at a fast and accurate segmentation, the time complexity should also be considered. The computation time taken for various segmentation methods along with the proposed approach is given below: -

Table 3: Analysis of computation time for various segmentation techniques on sample images

| Input Image | The computation time (in Seconds) for segmentation methods | | | |
|---|---|---|---|---|
| | Active contour based Segmentation | K- means segmentation | Otsu's thresholding | Proposed method |
| Bottle | 9.21 | 100.78 | 1.32 | 2.08 |
| Glue | 7.9 | 85.07 | 1.37 | 1.73 |
| Tab | 9.53 | 111.62 | 1.54 | 2.68 |
| Text | 9.58 | 69.79 | 2.18 | 2.79 |

From Table 2 and 3 it can be seen that computation time required for the proposed method is much less than that of Active contour and K-means segmentation. In comparison with a modified Otsu's method [27], the computation time is slightly high for the proposed algorithm. Otsu's segmentation techniques are faster since they operate directly on gray level histogram while the proposed method use pixel by pixel information for thresholding and hence requires a little more execution time. However, the segmentation accuracy (refer table 2) of the proposed method is much better than that of Otsu's method.

## 5 Conclusion and Future work

The proposed method is suitable for the segmentation of large number of images within a short duration of time. The approach is an unsupervised one and is devoid of training time. From the experimental results, it can be seen that the segmentation accuracy is considerably good and the object boundaries are much clear and accurate, when compared to some of existing techniques in the related area. The segmented results are also free from the effects of shadows and reflections from various light sources. Hence the segmentation approach can be efficiently used for various image processing applications. Also, the result analysis shows that computational overhead is reasonably low when compared to other similar algorithms. By incorporating modifications in selecting optimal threshold values, the proposed system can be further improved. Also the proposed segmentation can be extended as a part of an efficient object recognition system.

**Acknowledgment** We would like to accord our sincere gratitude to the support provided by Indian Institute of Information Technology and Management, Kerala (IIITM-K). This research work is funded by Centre for Disability Studies (CDS), Kerala and we acknowledge the support provided by them.

## References

1. R. C. Gonzalez, et al.: Digital Image Processing. 3rd edition, Prentice Hall, ISBN 9780131687288, 2008.

2. C. Wang and B. Yang.: An unsupervised object-level image segmentation method based on foreground and background priors, 2016 IEEE Southwest Symposium on Image Analysis and Interpretation (SSIAI), Santa Fe, NM, 2016, pp. 141-144.

3. Xiaomu Song and Guoliang Fan.: A study of supervised, semi-supervised and unsupervised multiscale Bayesian image segmentation. Circuits and Systems, 2002. MWSCAS-2002. The 2002 45th Midwest Symposium on, 2002, pp. II-371-II-374 vol.2.

4. Otsu, N.: A threshold selection method from gray level histogram, IEEE Trans. Syst. Man Cybern., 1979, 9, (1), pp. 62–66

5. T. Sziranyi and J. Zerubia.: Markov random field image segmentation using cellular neural network.IEEE Transactions on Circuits and Systems I.Fundamental Theory and Applications, vol. 44, no. 1, pp. 86-89, Jan 1997

6. S. Zhu, X. Xia, Q. Zhang and K. Belloulata.: An Image Segmentation Algorithm in Image Processing Based on Threshold Segmentation.Signal-Image Technologies and Internet-Based System, 2007. SITIS '07. Third International IEEE Conference on, Shanghai, 2007, pp. 673-678.

7. R. Thendral, A. Suhasini and N. Senthil.: A comparative analysis of edge and color based segmentation for orange fruit recognition.Communications and Signal Processing (ICCSP), 2014 International Conference on, Melmaruvathur, 2014, pp. 463-466

8. Z. Ren, S. Gao, L. T. Chia and I. W. H. Tsang.: Region-Based Saliency Detection and Its Application in Object Recognition.IEEE Transactions on Circuits and Systems for Video Technology,May 2014 vol. 24, no. 5, pp. 769-779,

9. Md. Imrul Jubair, M. M. Rahman, S. Ashfaqueuddin and I. Masud Ziko.: An enhanced decision based adaptive median filtering technique to remove Salt and Pepper noise in digital images. Computer and Information Technology (ICCIT), 2011 14th International Conference on, Dhaka, 2011, pp. 428-433.

10. Liang Chen, Lei Guo and Ning Yang Yaqin Du.: Multi-level image thresholding. based on histogram voting. 2nd International Congress on Image and Signal Processing, CISP '09., Tianjin, 2009

11. Ashraf A. Aly1, Safaai Bin Deris2, Nazar Zaki3.: Research Review for Digital Image Segmentation techniquesInternational Journal of Computer Science & Information Technology (IJCSIT) Vol 3, No 5, Oct 2011

12. Arti Taneja; Priya Ranjan; Amit Ujjlayan.: A performance study of image segmentation techniques Reliability, Infocom Technologies and Optimization (ICRITO) (Trends and Future Directions), 4th International Conference, 2015

13. Kass M,Witkin A,Terzopoulos D.: Snake:active contour models. Proc.Of 1st Intern Conf on Computer Vision, London,1987,321～331

14. G. Wan, X. Huang and M. Wang.: An Improved Active Contours Model Based on Morphology for Image Segmentation. Image and Signal Processing, 2009. CISP '09. 2nd International Congress on, Tianjin, 2009, pp. 1-5

15. B. Wu and Y. Yang.: Local-and global-statistics-based active contour model for image segmentation. Mathematical Problems in Engineering, vol. 2012

16. S. Kim, Y. Kim, D. Lee and S. Park.: Active contour segmentation using level set function with enhanced image from prior intensity. *2015 37th Annual International Conference of the IEEE Engineering in Medicine and Biology Society (EMBC),* Milan, 2015, pp. 3069-3072.

17. T. Duc Bui, C. Ahn and J. Shin.: Fast localised active contour for inhomogeneous image segmentation. *IET Image Processing*, vol. 10, no. 6, pp. 483-494, 6 2016.

18. J. Moinar, A. I. Szucs, C. Molnar and P. Horvath.: Active contours for selective object segmentation. *2016 IEEE Winter Conference on Applications of Computer Vision (WACV),* Lake Placid, NY, 2016, pp. 1-9.
19. Thyagarajan, H. Bohlmann and H. Abut.: Image coding based on segmentation using region growing. Acoustics, Speech, and Signal Processing. IEEE International Conference on ICASSP '87., 1987, pp. 752-755
20.Jun Tang.: A color image segmentation algorithm based on region growing. Computer Engineering and Technology (ICCET), 2010 2nd International Conference on, Chengdu, 2010, pp. V6-634-V6-637
21. X. Yu and J. Yla-Jaaski.: A new algorithm for image segmentation based on region growing and edge detection. Circuits and Systems, 1991., IEEE International Sympoisum on, 1991, pp. 516-519 vol.1
22. Ahlem Melouah.: Comparison of Automatic Seed Generation Methods for Breast Tumor Detection Using Region Growing Technique. Computer Science and Its Applications, Volume 456 of the series IFIP Advances in Information and Communication Technology. pp 119-128
23. S. Mukherjee and S. T. Acton.: Region Based Segmentation in Presence of Intensity Inhomogeneity Using Legendre Polynomials. *IEEE Signal Processing Letters*, vol. 22, no. 3, March 2015, pp. 298-302
24. P. K. Jain and S. Susan.: An adaptive single seed based region growing algorithm for color image segmentation. *2013 Annual IEEE India Conference (INDICON)*, Mumbai, 2013, pp. 1-6.
25. D H Al Saeed, A. Bouridane, A. ElZaart, and R. Sammouda.: Two modified Otsu image segmentation methods based on Lognormal and Gamma distribution models. Information Technology and e-Services (ICITeS), 2012 International Conference on, Sousse, 2012, pp. 1-5.
26. Q. Chen, L. Zhao, J. Lu, G. Kuang, N. Wang and Y. Jiang.: Modified two-dimensional Otsu image segmentation algorithm and fast realization. IET Image Processing, vol. 6, no. 4, , June 2012, pp. 426-433
27. C. Zhou, L. Tian, H. Zhao and K. Zhao.: A method of Two-Dimensional Otsu image threshold segmentation based on improved Firefly Algorithm. *Cyber Technology in Automation, Control, and Intelligent Systems (CYBER), 2015 IEEE International Conference on,* Shenyang, 2015, pp. 1420-1424.
28. Serge Beucher and Christian Lantuéj.: Uses of watersheds in contour detection. Workshop on image processing, real-time edge and motion detection/estimation, Rennes, France (1979)
29. L Vincent and P Soille.: Watersheds in digital spaces: an efficient algorithm based on immersion simulations. IEEE Transactions on Pattern Analysis and Machine Intelligence, vol. 13, no. 6, Jun 1991, pp. 583-598
30. Serge Beucher and Fernand Meyer.: The morphological approach to segmentation: the watershed transformation. Mathematical Morphology in Image Processing (Ed. E. R. Dougherty), pages 433–481 (1993).

31. Norberto Malpica, Juan E Ortufio, Andres Santos.: A multichannel watershed-based algorithm for supervised texture segmentation. Pattern Recognition Letters, 2003, 24 (9-10): 1545-1554

32. M. H. Rahman and M. R. Islam.: Segmentation of color image using adaptive thresholding and masking with watershed algorithm. Informatics, Electronics & Vision (ICIEV), 2013 International Conference on, Dhaka, 2013, pp. 1-6

33. A. Shiji and N. Hamada: Color image segmentation method using watershed algorithm and contour information. Image Processing, 1999. ICIP 99. Proceedings. 1999 International Conference on, Kobe, 1999, pp. 305-309 vol.4

34. G. M. Zhang, M. M. Zhou, J. Chu and J. Miao.: Labeling watershed algorithm based on morphological reconstruction in color space. Haptic Audio Visual Environments and Games (HAVE), 2011 IEEE International Workshop on, Hebei, 2011, pp. 51-55

35. Qinghua Ji and Ronggang Shi.: A novel method of image segmentation using watershed transformation. Computer Science and Network Technology (ICCSNT), 2011 International Conference on, Harbin, 2011, pp. 1590-1594

36. B. Han.: Watershed Segmentation Algorithm Based on Morphological Gradient Reconstruction. Information Science and Control Engineering (ICISCE), 2015 2nd International Conference on, Shanghai, 2015, pp. 533-536

37. Y. Chen and J. Chen.: A watershed segmentation algorithm based on ridge detection and rapid region merging. *Signal Processing, Communications and Computing (ICSPCC), 2014 IEEE International Conference on,* Guilin, 2014, pp. 420-424.

38. S. Chebbout and H. F. Merouani.: Comparative Study of Clustering Based Colour Image Segmentation Techniques. *Signal Image Technology and Internet Based Systems (SITIS), 2012 Eighth International Conference on,* Naples, 2012, pp. 839-844.

39. J. Xie and S. Jiang.: A Simple and Fast Algorithm for Global K-means Clustering. Education Technology and Computer Science (ETCS), 2010 Second International Workshop on, Wuhan, 2010, pp. 36-40

40. S. Vij, S. Sharma and C. Marwaha.: Performance evaluation of color image segmentation using K means clustering and watershed technique. Computing, Communications and Networking Technologies (ICCCNT), 2013. Fourth International Conference on, Tiruchengode, 2013, pp. 1-4

41. N. A. Mat Isa, S. A. Salamah and U. K. Ngah.: Adaptive fuzzy moving K-means clustering algorithm for image segmentation. in IEEE Transactions on Consumer Electronics, vol. 55, no. 4, November 2009, pp. 2145-2153

42. Hui Xiong, Junjie Wu.: Kmeans Clustering versus Validation Measures: A Data Distribution Perspective, 2006

43. Jimmy Nagau, Jean-Luc Henry. L.: An optimal global method for classification of color pixels. International Conference on Complex, Intelligent and Software Intensive Systems 2010

44. Feng Ge, Song Wang, Tiecheng Liu.: New benchmark for image segmentation evaluation. Journal of Electronic Imaging 16(3), 033011 (Jul–Sep 2007)
45. Ran Jin,Chunhai Kou,Ruijuan Liu,Yefeng Li.: A Color Image Segmentation Method Based on Improved K-Means Clustering Algorithm. International Conference on Information Engineering and Applications (IEA) 2012, Lecture Notes in Electrical Engineering 217
46. C. Y. Lien, C. C. Huang, P. Y. Chen and Y. F. Lin, "An Efficient Denoising Architecture for Removal of Impulse Noise in Images," in IEEE Transactions on Computers, vol. 62, no. 4, pp. 631-643, April 2013.
doi: 10.1109/TC.2011.256
47. R. Bernstein.: Adaptive nonlinear filters for simultaneous removal of different kinds of noise in images. IEEE Transactions on Circuits and Systems, vol. 34, no. 11, Nov 1987, pp. 1275-1291
48. Weibo Yu, Yanhui, Liming Zheng, Keping Liu.: Research of Improved Adaptive Median Filter Algorithm. Proceedings of the 2015 International Conference on Electrical and Information Technologies for Rail Transportation Volume 378 of the series Lecture Notes in Electrical Engineering. pp 27-34
49. K. Manglem Singh and P. K. Bora.: Adaptive vector median filter for removal impulses from color images. Circuits and Systems, 2003. ISCAS '03. Proceedings of the 2003 International Symposium on, 2003, pp. II-396-II-399 vol.2
50. J. Pont-Tuset and F. Marques.: Supervised Evaluation of Image Segmentation and Object Proposal Techniques. IEEE Transactions on Pattern Analysis and Machine Intelligence, vol. 38, no. 7, July 1 2016, pp. 1465-1478
51. T. C. W. Landgrebe, P. Paclik and R. P. W. Duin.: Precision-recall operating characteristic (P-ROC) curves in imprecise environments.18th International Conference on Pattern Recognition (ICPR'06), Hong Kong, 2006, pp. 123-127.
52. J. Canny.: Computational Approach to Edge Detection. IEEE Transactions on Pattern Analysis and Machine Intelligence, vol. PAMI-8, no. 6, Nov. 1986, pp. 679-698

# Automatic Diagnosis of Breast Cancer using Thermographic Color Analysis and SVM Classifier

Asmita T. Wakankar [1], G. R. Suresh [2]

[1] Sathyabama University, Chennai, India
asmita_wakankar@yahoo.com
[2] Easwari Engineering College, Chennai, India
sureshgr@rediffmail.com

**Abstract.** Breast cancer is the commonly found cancer in women. Studies show that the detection at the earliest can bring down the mortality rate. Infrared Breast thermography uses the temperature changes in breast to arrive at diagnosis. Due to increased cell activity, the tumor and the surrounding areas has higher temperature emitting higher infrared radiations. These radiations are captured by thermal camera and indicated in pseudo colored image. Each colour of thermogram is related to specific range of temperature. The breast thermogram interpretation is primarily based on colour analysis and asymmetry analysis of thermograms visually and subjectively. This study presents analysis of breast thermograms based on segmentation of region of interest which is extracted as hot region followed by colour analysis. The area and contours of the hottest regions in the breast images are used to indicate abnormalities. These features are further given to ANN classifier for automated analysis. The results are compared with doctor's diagnosis to confirm that infra-red thermography is a reliable diagnostic tool in breast cancer identification.

**Keywords:** Breast cancer, Thermography, Segmentation, Level Set, ANN

## 1 Introduction

Breast cancer is a type of cancer caused by breast tissue either the inner lining of milk ducts or the lobules that supply the ducts with milk.[1] Breast cancer is caused by combination of multiple factors like inheritance, tissue composition, carcinogens, immunity levels, hormones etc. Currently, the most common methods for detecting the breast diseases are Mammography, Doppler Ultrasonography, Magnetic Resonance Imaging (MRI), Computed Tomography Laser Mammography (CTLM), Positron Emission Mammography (PEM).[2] However theses imaging techniques only provide the anatomical structure information of tumor lacking functional information. Infrared Thermography is functional imaging technique that can detect cancerous tissue indicating cancer infection, inflammation, surface lesions and more. [2] All objects in the universe emit infrared radiations which is a function of their tempera-



J.M. Corchado Rodriguez et al. (eds.), *Intelligent Systems Technologies and Applications 2016*, Advances in Intelligent Systems and Computing 530,
DOI 10.1007/978-3-319-47952-1_2

ture. [4]The objects with higher temperature emit more intense and shorter wavelength infrared radiations. Infrared cameras are used to detect this radiation, which is in the range of 0.9-14 μm and produce image of that radiation, called Thermograms. In case of cancer, once a normal cell begins to transform, it's DNA is changed to allow for the onset of uncoordinated growth. To maintain the rapid growth of these precancerous and cancerous cells, nutrients are supplied by the cells by discharging the chemicals. [3]This keeps existing blood vessels open, awaken inactive ones and generate new ones. This process is called as 'Neoangiogenesis'. [5]This chemical and blood vessel activity in both pre-cancerous tissue and the developing cancer area is always higher. This in turn increases local tissue surface temperatures leading to detection by infrared imaging. These thermal signs indicate a pre-cancer stage that can't be detected by physical examination, or other types of structural imaging technique. Thermography is radiation free, painless and non-invasive technique. It is the best option for screening of young women, pregnant women or women with fibrocystic, large, dense breasts or women with metallic implants.

## 2 Literature Review

Though extensive work has taken place in the area of thermographic analysis for breast cancer detection, considering the scope of this paper, we have focussed only on the work that uses segmentation techniques like K means, Fuzzy C means and Level Set method. N. Golestani et al 2014 suggested use of K means, Fuzzy C means and Level set method for detection of hottest region in breast thermogram. Abnormality is detected by measuring shape, size and borders of hot spot in an image. [9] Lavanya A. used particle swarm optimization for threshold level determination and K means clustering algorithm to detect hottest region in IR thermogram. Using fractal measures the cancerous and non cancerous cases are identified from segmented region.[14] Authors have implemented K means and Fuzzy C means algorithms earlier for finding the hot spot in breast thermogram.[2] S.S.Suganthi et al implemented segmentation algorithm using anisotropic diffusion filter which smoothens intra region preserving sharp boundaries. The ROI are segmented by level set function based on improved edge information.[15] Prabha S. et al employed Reaction diffusion based level set method to segment the breast using edge map as stopping boundary generated by total variation diffusion filter. Wavelet based structural texture features are calculated for asymmetry analysis. The results are compared against ground truth. [16] Srinivasan s. et al worked on the segmentation of front breast tissues from breast thermograms using modified phase based distance regularized level set method. Local phase information is used as edge indicator for level set function. Region based statistics and overlap measures are computed to compare and validate the segmented ROI against ground truth. [17]

## 3 System Block Diagram

Early methods of breast thermogram interpretation were solely based on subjective criteria. [6] The doctors look for changes in colour and vascular patterns in the thermograms to detect the abnormality. As the results depend upon experience and skill set of an individual, the variability between diagnoses is very high. To generate more accurate, consistent and automated diagnosis, the proposed method recommends the use of colour analysis as a first stage. Further, few statistical parameters like entropy, skewness and kurtosis are calculated to arrive at abnormality detection. The system block diagram is as shown in the Fig.1

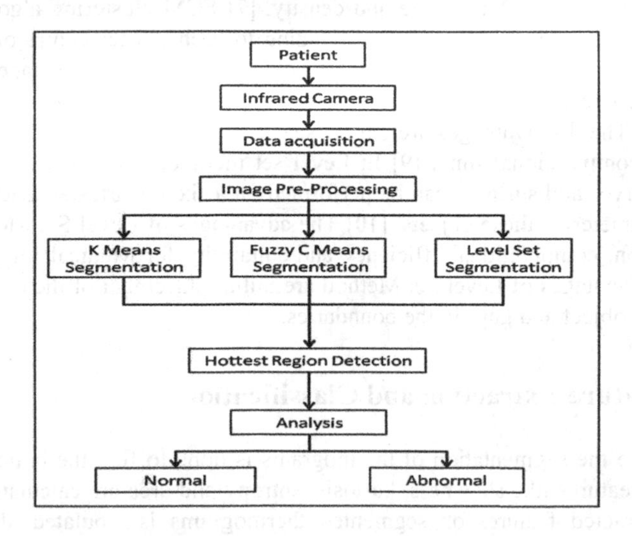


**Fig. 1.** System Block Diagram

The thermograms are captured at Breast Cancer Clinic using FLIR E30 Infrared camera considering the Pre-Themographic imaging instructions. [2] Total 27 samples are taken including normal and abnormal cases. Few more thermograms are taken from the online thermogram database, Ann Arbor Thermography, (http://aathermography.com) Some image preprocessing steps like background removal and resizing are carried out to remove the undesired body portion. [6] Three methods are used for segmentation of hot spot namely K-Means Clustering, Fuzzy C Means, and Level set method. The statistical parameters indicating abnormality are calculated which are further given to Support Vector Machine (SVM) for classification purpose and finally the results are compared with doctor's diagnosis.

## 4 Segmentation

The colour bar present on the thermograms gives idea of the temperature distribution indicating the coolest part as blue; intermediate temperature as yellow and red and the warmest part of the image as white. [2] Colour segmentation of the thermal images using clustering algorithms is done to find hottest regions. There are several techniques that can be applied to colour image segmentation. In this study, three techniques namely K-means clustering, fuzzy C means (FCM) clustering and level set are used for colour segmentation of breast infrared images. K means and Fuzzy C means techniques are based on the least square errors while the level set method is based on partial differential equations. K-means clustering computes the distances between the inputs and centers, and assigns inputs to the nearest center. This method is easy to implement, relatively efficient and computationally faster algorithm. But, the clusters are sensitive to initial assignment of centroids. The algorithm has a limitation for clusters of different size and density. [7] FCM clustering algorithm assigns membership to each data point corresponding to each cluster centre on the basis of distance between the cluster centre and the data point.[8] The advantages of FCM are it gives best result for overlapped data set and comparatively better than k-means algorithm. The disadvantages are cluster sensitivity to initial assignment of centroids and long computational time. [9] In Level set method, numerical computations involving curves and surfaces can be performed on a fixed Cartesian grid without having to parameterize these objects. [10] The advantages of Level Set Method are easy initialization, computational efficiency and suitability for Medical Image Segmentation. Disadvantages of Level Set Method are initial placement of the contour, embedding of the object and gaps in the boundaries.

# 5 Feature Extraction and Classification

Once the segmentation of thermograms is done to find the hottest region, the statistical features like skewness, kurtosis, entropy and area are calculated. The result of the extracted features on segmented thermograms is tabulated along with the graphical representation indicating the better performance of Level Set method.

Classification algorithms categorize the data based on the various properties of images. They work in two stages: training and testing. In the training phase, specific image properties are separated and using these, training class, is prepared. During the testing phase, these feature-space categories are used to classify images. There are different algorithms for image classifications. In this work, Support Vector Machine (SVM) is used for image classification. [11] SVM classifies data by finding the best hyperplane that separates all data points in two distinct groups. SVM method is effective in high dimensional spaces and is memory effective. SVM has a limitation when the number of features is much greater than the number of samples. [12]

## 5.1 Evaluation of Classifier(SVM) Performance

To evaluate the classifier performance, three parameters namely sensitivity, specificity and accuracy are calculated. Confusion matrix helps to visualize the outcome of an algorithm.

1. Sensitivity / true positive rate:

In case of medical diagnosis, sensitivity is the proportion of people who have the disease and are tested positive. Mathematically, this can be expressed as:

$$\text{sensitivity} = \frac{\text{number of true positives}}{\text{number of true positives} + \text{number of false negatives}}$$

2. Specificity / true negative rate:

Specificity is the proportion of people who are healthy and are correctly diagnosed. Mathematically, this can also be written as:

$$\text{specificity} = \frac{\text{number of true negatives}}{\text{number of true negatives} + \text{number of false positives}}$$

3. Accuracy / Error Rate:

It is the percent of correct classifications. It is the proportion of the total number of predictions that are correct. [29]

$$\text{accuracy} = \frac{\text{number of true positives} + \text{number of true negatives}}{\text{number of true positives} + \text{false positives} + \text{false negatives} + \text{true negatives}}$$

## 6 Results & Discussion:

In this work, a total of 34 thermograms (27 captured images and 7 online images) were studied. Thermograms of various cases like normal subject, inflammatory cancer patient and patient with fibrocystic change are used to test the system. By using three segmentation techniques K means, FCM and level set method, the hottest region for each case is identified. [14] The results of three different approaches are compared. Following session gives the details of the segmentation, feature extraction and analysis. A comparison of segmentation results between k-means, fuzzy c-means, and level set for a Normal case is shown in figure 2. The segmented region by K means and FCM are yellow and green and is symmetric for both the breasts. The hottest spot window is blank showing that it is a normal case.

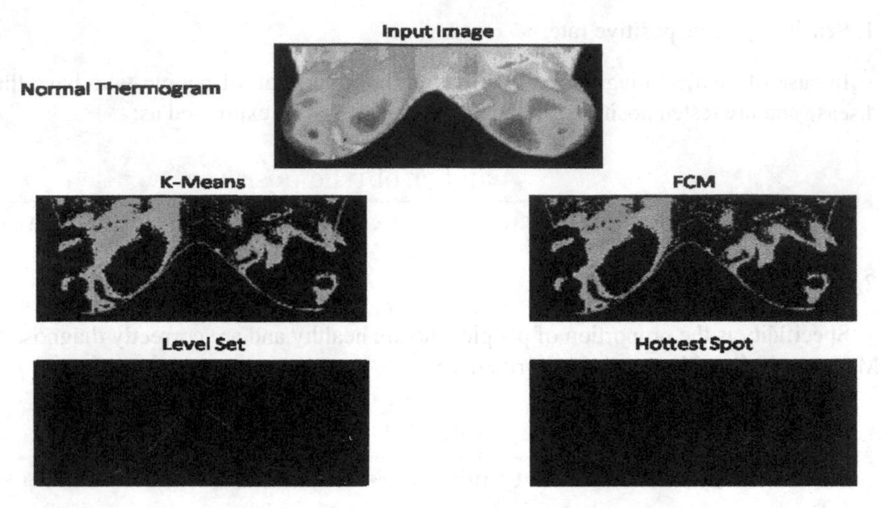


**Fig. 2.** Hot Region Segmentation of Normal Thermogram

Figure 3 shows segmentation results of all the three methods, for a fibrocystic case. The hottest regions shown by red and orange colour are separated for both the breasts. The area of hottest region is more in the left breast as compared to right breast indicating asymmetry. Also, the Hot spot window shows the presence of small white patches. Thus we can say that there is an abnormality in the left breast.

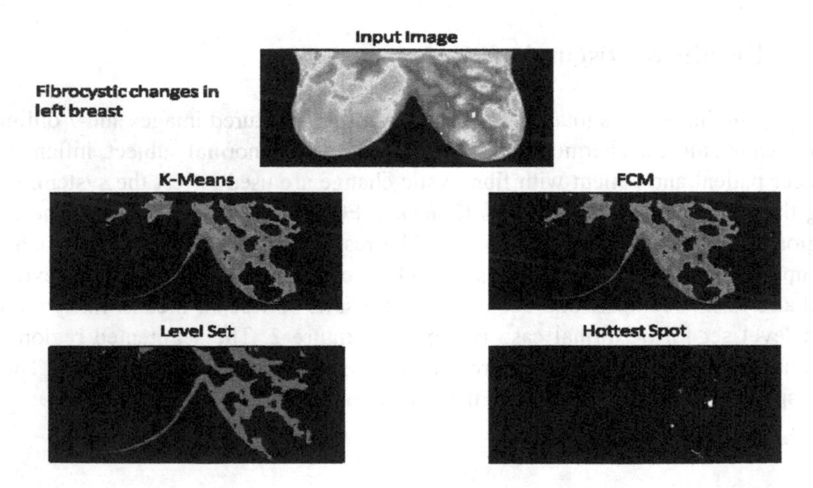


**Fig. 3.** Hot Region Segmentation of Fibrocystic Changes in Breast Thermogram

Segmentation results for an inflammatory breast cancer case are shown in Figure 4. The hottest spot window shows the presence of white patch in the right breast. So we can notice the asymmetry and elevated temperature for right breast indicating the abnormality.

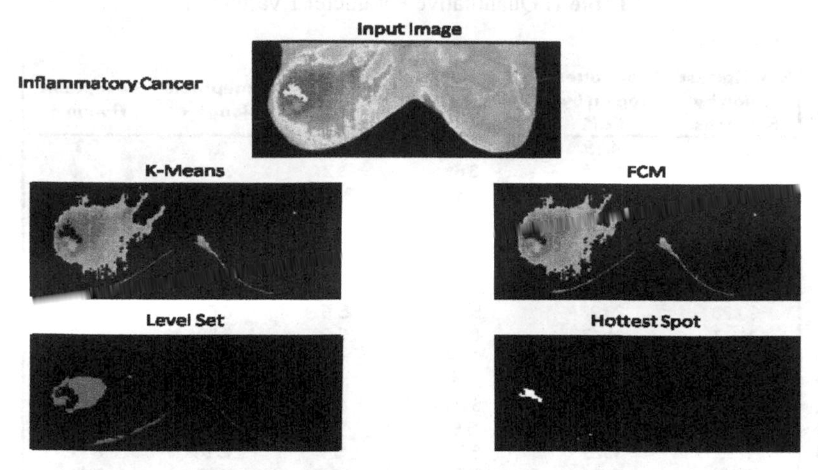


**Fig. 4.** Hot Region Segmentation of Inflammatory Cancer Patient Thermogram

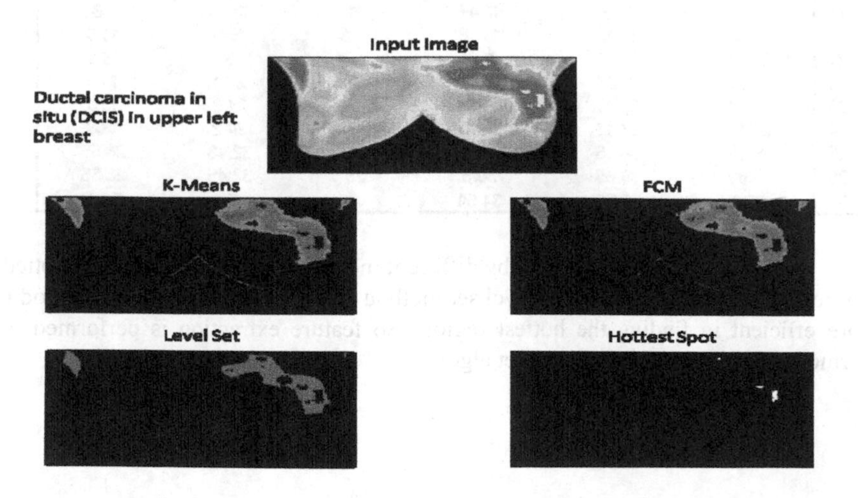


**Fig. 5.** Hot Region Segmentation of DCIS Patient Thermogram

Figure 5 shows a comparison of segmentation results between k means, fuzzy c means, and level set, for a DCIS in left breast cancer case.

To compare the performance of segmentation algorithms, area of hottest region segmented by three methods is calculated. The results are compared with the actual area of abnormality in thermograms, calculated using the doctors diagnosis. Table 1 shows the Quantitative Parameter Evaluation for 27 patients.

**Table 1.** Quantitative Parameter Evaluation

| Sr. No. | % of Hottest region by K-means | % of Hottest region by FCM | % of Hottest region by Level Set | Actual % of Hottest region | Acceptable Range + | Acceptable Range - |
|---|---|---|---|---|---|---|
| 1 | 2.64 | 4.29 | 4.71 | 3 | 3.3 | 2.7 |
| 2 | 10.4 | 10.39 | 3.61 | 3.2 | 3.52 | 2.88 |
| 3 | 21.28 | 21.54 | 7.58 | 3.32 | 3.652 | 2.988 |
| 4 | 14.93 | 14.93 | 6.63 | 4 | 4.4 | 3.6 |
| 5 | 10.92 | 10.91 | 3.95 | 4 | 4.4 | 3.6 |
| 6 | 5.55 | 4.48 | 4.4 | 4.1 | 4.51 | 3.69 |
| 7 | 5.22 | 5.81 | 4.9 | 4.5 | 4.95 | 4.05 |
| 8 | 23.29 | 23.28 | 5.69 | 5.5 | 6.05 | 4.95 |
| 9 | 14.47 | 14.47 | 6.17 | 5.8 | 6.38 | 5.22 |
| 10 | 9.98 | 9.97 | 5.48 | 6 | 6.6 | 5.4 |
| 11 | 9.95 | 9.71 | 9.36 | 8.6 | 9.46 | 7.74 |
| 12 | 9.91 | 9.91 | 9.08 | 8.5 | 9.35 | 7.65 |
| 13 | 10.08 | 9.43 | 9.6 | 8.8 | 9.68 | 7.92 |
| 14 | 11.22 | 11.29 | 9.72 | 9.1 | 10.01 | 8.19 |
| 15 | 16.2 | 17.34 | 11.61 | 10.5 | 11.55 | 9.45 |
| 16 | 12.02 | 12.01 | 11.54 | 10.8 | 11.88 | 9.72 |
| 17 | 12.5 | 12.5 | 12.49 | 11.5 | 12.65 | 10.35 |
| 18 | 19.35 | 19.33 | 10.18 | 11.5 | 12.65 | 10.35 |
| 19 | 12.5 | 12.5 | 12.47 | 11.5 | 12.65 | 10.35 |
| 20 | 37.19 | 37.12 | 15.48 | 15 | 16.5 | 13.5 |
| 21 | 21.93 | 21.27 | 21.24 | 21 | 23.1 | 18.9 |
| 22 | 21.93 | 21.27 | 21.24 | 21 | 23.1 | 18.9 |
| 23 | 23.54 | 23.12 | 23 | 22 | 24.2 | 19.8 |
| 24 | 27.51 | 27.47 | 23.41 | 22 | 24.2 | 19.8 |
| 25 | 36.89 | 36.85 | 30.71 | 29.5 | 32.45 | 26.55 |
| 26 | 36.89 | 36.85 | 30.9 | 29.5 | 32.45 | 26.55 |
| 27 | 35.8 | 35.55 | 34.54 | 32 | 35.2 | 28.8 |

A graph of segmented area by different methods versus actual area is plotted. Figure 6 shows that the results of level set method are superior and thus the method is more efficient in finding the hottest region. So feature extraction is performed on thermograms segmented by Level Set algorithm.

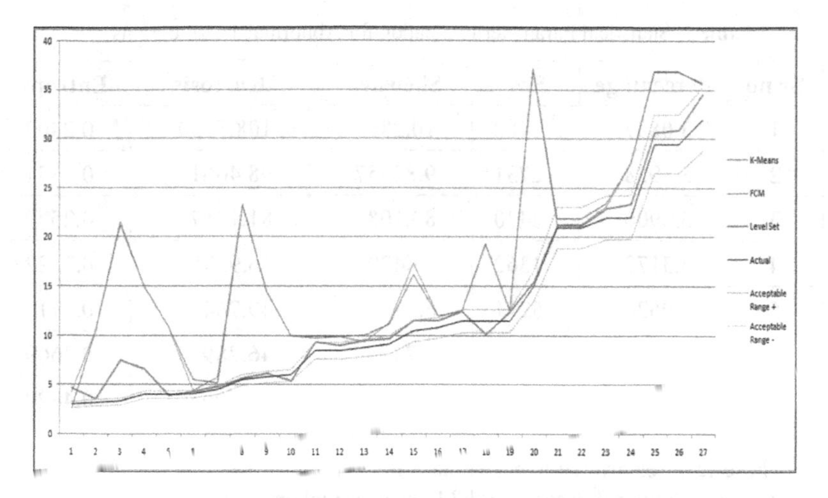


**Fig. 6.** Comparison of Segmentation Algorithms

A set of features like Skewness, Kurtosis, Entropy, Size and Area of hottest region is calculated. HOS based feature extraction includes the feature Entropy which describes the temperature distribution in the ROI. The Statistical Parameter Evaluation of normal thermograms for 7 volunteers is shown in table 2. It is observed that the parameters percentage, size of the hottest region and entropy are having lesser value for the Normal Thermograms and the parameters skewness and kurtosis are having higher value.

**Table 2.** Statistical Parameter Evaluation for Normal Thermograms

| Sr no | Percentage | Size | Skewness | Kurtosis | Entropy |
|---|---|---|---|---|---|
| 1 | 0.3533 | 89 | 26.6588 | 910.3917 | 0.2679 |
| 2 | 0.4769 | 639 | 29.5497 | 947.6723 | 0.1489 |
| 3 | 1.2917 | 452 | 15.1652 | 231.9787 | 0.1403 |
| 4 | 1.5567 | 471 | 15.855 | 265.4904 | 0.2634 |
| 5 | 10.08 | 721 | 13.7705 | 196.4618 | 0.199 |
| 6 | 1.972 | 2456 | 12.424 | 156.5434 | 0.1918 |
| 7 | 2.3006 | 873 | 12.6953 | 167.1503 | 0.1965 |

Table 3 gives the details for abnormal thermograms. It is observed that the parameters percentage, size of the hottest region and entropy are having higher value for the Abnormal Thermograms and the parameters skewness and kurtosis are having lesser value.

**Table 3.** Statistical Parameter Evaluation for Abnormal Thermograms

| Sr no | Percentage | Size | Skewness | Kurtosis | Entropy |
|---|---|---|---|---|---|
| 1 | 3.0837 | 1480 | 10.2861 | 108.5159 | 0.2402 |
| 2 | 3.1958 | 2731 | 9.82557 | 98.4661 | 0.2351 |
| 3 | 3.8906 | 2430 | 8.9108 | 81.1057 | 0.2579 |
| 4 | 4.3172 | 1362 | 8.6429 | 76.9598 | 0.3322 |
| 5 | 5.9626 | 5211 | 6.97 | 49.7647 | 0.371 |
| 6 | 6.3851 | 2363 | 6.7477 | 46.7892 | 0.3669 |
| 7 | 7.4077 | 5136 | 6.2694 | 40.5905 | 0.4428 |

These features are fed to SVM classifier which categorizes the thermograms into Normal and Abnormal cases. Total 34 cases are taken as case study, in which 10 cases are used for training of classifier and 24 cases are used for testing purpose. Out of 24 cases, 18 were abnormal and 6 were normal cases. The system correctly identified 16 abnormal cases and all 6 normal cases. The above values are shown in the form confusion matrix as in table 4. The two nonconclusive cases were online images and had issues related to image capturing.

**Table 4.** Confusion Matrix

| | **Disease Present** | **Disease Absent** |
|---|---|---|
| **Test Positive** | TP=16 | FP=0 |
| **Test Negative** | FN=2 | TN=6 |

To evaluate the classifier performance, the following three parameters Sensitivity, Specificity and Accuracy are calculated. Table 5 shows these values.

**Table 5.** Classifier Performance

| Sr.No. | Parameters | Value (%) |
|---|---|---|
| 1. | Sensitivity | 88.8 |
| 2. | Specificity | 100 |
| 3. | Accuracy | 91.6 |

## 7 Conclusion

Infra-red Thermography in medicine is a non-invasive and a non-ionizing method for understanding the internal system, especially for a cancerous tumor, as it relates temperature to physiological parameters such as metabolic rate. [6] This work presented an approach that deals with Automatic Analysis of Breast Thermograms using colour segmentation and feature extraction for abnormality detection. The hottest region of breast thermograms was detected using three image segmentation techniques: k-means, fuzzy c-means (FCM) and level set. Experimental results have shown that the level set method has a better performance as against other methods as it could highlight hottest region more precisely. Some useful features like skewness, kurtosis, area and entropy were extracted from them and given to SVM for classification into normal and abnormal cases. The classifier performed well with the accuracy of 91 %. Proposed method based on Colour Analysis proves to be an effective way for classification of Breast Thermograms helping medical professionals to save their time without sacrificing accuracy. An extension of developed methodology is to quantify degradation of cancer in terms of percentage.

### References

1. Ganesh Sharma, Rahul Dave, Jyostna Sanadya, Piyush Sharma and K.K.Sharma: Various Types and management of Breast Cancer: An Overview, Journal of Advanced Pharmaceutical Technology and Research, pp 109-126,1(2) (2010)
2. Shahari Sheetal, Asmita Wakankar: Color Analysis of thermograms for Breast Cancer Detection: Proceedings IEEE International Conference on Industrial Instrumentation and Control (2015)
3. Harold H. Szu, Charles Hsu, Philip Hoekstra, Jerry Beeney: Biomedical Wellness Standoff Screening by Unsupervised Learning: Proceedings SPIE 7343, Independent Component Analyses, Wavelets, Neural Networks, Biosystems and Nanoengineering VII, 734319 (2009)
4. Hossein Zadeh, Imran Kazerouni, Javad Haddaadnia: Diagnosis of Breast Cancer and Clustering Technique using Thermal Indicators Exposed to Infrared Images, Journal of American Science, Vol 7(9), pp 281-288 (2011)
5. Pragati Kapoor, Seema Patni, Dr. S.V.A.V Prasad: Image Segmentation and Asymmetry Analysis of Breast Thermograms for Tumor Detection, International Journal of Computer Applications, Volume 50, No 9 (2012)
6. Asmita Wakankar, G. R. Suresh, Akshata Ghugare: Automatic Diagnosis of Breast Abnormality Using Digital IR Camera, IEEE International Conference on Electronic Systems, Signal Processing and Computing Technologies (2014)
7. Saad N.H.,N.Mohonma Sabri,A.F.Hasa,Azuwa Ali, Hariyanti Mahd Saleh: Defect Segmentation of Semiconductor Wafer Image Using K Means Clustering, Applied Mechanics & Materials, Vol 815, pp 374-379 (2015)
8. Anubha, R.B. Dubey: A Review on MRI Image Segmentation Techniques, International Journal of Advanced Research in Electronics and Communication Engineering, Vol 4, Issue 5 (2015)
9. N. Golestani, M. Tavakol, E.Y.K.Ng: Level set Method for Segmentation of Infrared Breast Thermograms, EXCLI Journal, pp 241-251(2014)

10. Chuming Li, Rui Huang et al: A Level Set Method for Image Segmentation in the Presence of Intensity Inhomogenitis with Application to MRI, IEEE Transactions in Image Processing, Vol 20, No 7 (2011)
11. U. Rajendra Acharya, E.Y.K.Ng, Jen Hong Tan, S. Vinitha Sree: Thermography based Breast Cancer Detection Using Texture Features and Support Vector Machine, Journal of Medical Systems, Vol 36, pp 1503-1510 (2012)
12. Kieran Jay Edwards, Mahamed Gaber: Adopted Data Mining Methods, Studies in Big Data, pp 31-42 (2014)
13. Nijad Al Najdawi, Mariam Biltawi, Sara Tedmori: Mammogram Image Visual Enhancement, mass segmentation & Classification, Applied Soft Computing, Vol 35, pp 175-185 (2015)
14. A. Lavanya: An Approach to Identify Lesion in Infrared Breast Thermography Images using Segmentation and Fractal Analysis, International Journal of Biomedical Engineering and Technology, Vol 19, No 3 (2015)
15. S.S.Suganthi, S.Ramakrishnan: Anisotropic Diffusion Filter Based Edge Enhancement for Segmentation of Breast Thermogram Using Level sets, Elsevier, Biomedical Signal Processing and Control, Vol 10 (2014)
16. Prabha S, Anandh K.R, Sujatha C.M, S.Ramakrishnan: Total Variation Based Edge Enhancement for Levelset Segmentation and Asymmetry Analysis in Breast Thermograms, IEEE, pp 6438- 6441 (2014)
17. S.S.Srinivasan, R. Swaminathan: Segmentation of Breast Tissues in Infrared Images Using Modified Phase Based Level Sets, Biomedical Informatics and Technology, Vol 404, pp 161-174 (2014)

# Enhancement of Dental Digital X-Ray Images based On the Image Quality

**Hema P Menon, B Rajeshwari**
**Department of Computer Science and Engineering,**
**Amrita School of Engineering, Coimbatore.**
**Amrita Vishwa Vidyapeetham, Amrita University, INDIA.**

**p_hema@cb.amrita.edu** , **rajiinspire@gmail.com**

*Abstract*: Medical Image Enhancement has made revolution in medical field, in improving the image quality helping doctors in their analysis. Among the various modalities available, the Digital X-rays have been extensively utilized in the medical world of imaging, especially in Dentistry, as it is reliable and affordable. The output scan pictures are examined by practitioners for scrutiny and clarification of tiny setbacks. A technology which is automated with the help of computers to examine the X-Ray images would be of great help to practitioners in their diagnosis. Enhancing the visual quality of the image becomes the prerequisite for such an automation process. The image quality being a subjective measure, the choice of the methods used for enhancement depends on the image under concern and the related application. This work aims at developing a system that automates the process of image enhancement using methods like Histogram Equalization(HE), Gamma Correction(GC),and Log Transform(LT). The decision of the enhancement parameters and the method used is chosen, with the help of the image statistics (like mean, variance, and standard deviation). This proposed system also ranks the algorithms in the order of their visual quality and thus the best possible enhanced output image can be used for further processing. Such an approach would give the practitioners flexibility in choosing the enhanced output of their choice.

*Keywords*: Image Enhancement, Dental Images, X-Ray Images, Log Transform, Histogram Equalization, Gamma Correction and Entropy.



J.M. Corchado Rodriguez et al. (eds.), *Intelligent Systems Technologies and Applications 2016*, Advances in Intelligent Systems and Computing 530,
DOI 10.1007/978-3-319-47952-1_3

# 1. INTRODUCTION

Image enhancement has a great influence in vision applications and also takes a major role in it. The main objective of enhancement is to elevate the structural aspect of an object without causing any downgrade in the input image and to improve the visual qualities of the image in order to make it appropriate for a given application. The aim of enhancement is to improve the perception of the details and information in the images [3].It improves the view of the image thereby providing a processed input for further processing. In general, in case of images an enhancement is performed to improve its contrast and brightness, so that the objects of interest are clearly and distinctly visible from the background.

There are different modalities in medical imaging such as Computed Tomography (CT), Magnetic Resonance Imaging (MRI), Positron Emission Tomography (PET), X-ray, PET-CT, and Biomarkers etc. Among these, Digital X-rays have been extensively used for image acquisition in the field of medical imaging. This work focuses on the Dental X-Ray images, which are used by dentists for the analysis of the Tooth structure and anomalies present. Human clarification is predominantly based on the knowledge and the experience of the person. The acquired image quality depends on various factors like the imaging sensors, lighting present, noise and the area/ object being imaged [5]. Making analysis from an image that has poor visual quality is often a difficult and time consuming task. Additionally, if there are a large number of X-rays to be examined, as in case of structure analysis, then the process may be time consuming. In all such cases an automated tool that could help in the analysis process would be very useful. Such automation requires segmentation, feature extraction and recognition of the regions in the image under consideration. For these processes to be accurate one aspect is the quality of the input image. To ensure that the image is of good quality, an analysis of the image statistics, like mean and standard deviation, which gives the details of the brightness and contrast of the input image, would be useful. In this work one such approach for automating the enhancement process by considering the image statistics has been carried out. The organization of the paper is as follows: Section 2 discusses the review of literature in enhancement of dental X-Ray images. The proposed enhancement system has been elaborated in section 3 and the results and analysis are discussed in section 4. Section 5 gives the conclusion driven from this work.

## 2. BACKGROUND STUDY

Image Enhancement has been a topic of research in the field of image processing. A huge amount of literature is available pertaining to this in general. But with respect to enhancement of Digital X-Ray images and Dental image especially, the availability of literature is scarce. Hence in this work focus has been given to enhancement of dental x-ray images. This section discusses the existing literature available for enhancing such x-ray images.

Yeong-Taeg Kim, [1] conducted a study on contrast enhancement using Bi-Histogram equalization. This work focused on overcoming the drawbacks of Histogram Equalization. The input image is decomposed into sub images. Histogram Equalization is performed for the sub images. The equalized images are bounded in such a way that they occur around the mean. Foisal et al. [2] has discussed a method for Image Enhancement Based on Non-Linear Technique. The method uses Logarithmic Transform Coefficient Histogram Matching for enhancing images with dark shadows which is caused due to the restricted dynamic range. Fan et al. [3] has conducted a study on improved image contrast enhancement in multiple-peak images. The study is based on histogram equalization. This study put forth a new technique which works well for the images with multiple peaks and this mainly focuses on the enrichment of the contrast of the given input image. M. Sundaram et al. [4] has studied the histogram based contrast enhancement for mammogram images. This is used to detect micro calcification of mammograms. Histogram Modified Contrast Limited Adaptive Histogram Equalization (HM-CLAHE) is used which controls the level of enhancement in compared with histogram equalization. G. N. Sarage and Dr Sagar Jambhorkar, [5] studied the enhancement of chest x-ray images. Filtering techniques like mean and median filtering are applied to improve the quality of the images. These filters reduce the noise in the given image. D Cheng and Yingtao Zhang, [6] has performed a study in order to detect Over-Enhancement of images. Over – enhancement causes edge-loss, changes the texture of the image and makes the image appear unrealistic. The cause for over-enrichment is analysed and the over-enriched portions are identified accurately in this work. The level of over-enhancement is assessed using quantitative methods. Ikhsan et al., [7] has discussed a new method for contrast enhancement of retinal images that would help in diabetes screening system. In their work three techniques for improving image contrast namely, Histogram Equalization (HE), Contrast Limited Adaptive Histogram Equalization (CLAHE) and Gamma Correction (GC) had been used. Ritika and Sandeep Kaur, [8] studied the contrast enhancement techniques for images and proposed a technique based on mathematical morphology analysis to enhance image contrast. Here HE and CLAHE have been analysed to reduce the low contrast in the images. In CLAHE, local enhancement of the image is done without amplifying the unnecessary noises. Datta et al. [9] conducted a study to identify a contrast enhancement method of retinal images in diabetic screening system. This work enhances the image using various techniques and ensures s the brightness and quality of the image is preserved. Performance parameters like SSIM (Structure Similarity Index Measurement) and AAMBE (Average Absolute Mean Brightness Error) are applied to preserve quality of the image while enhancing. Of late the use of Partial Differential Equations (PDE) for enhancement and restoration of medical images [10] is gaining prominence. Huang et al. [11] studied the efficient contrast enhancement. It was achieved by adaptive gamma correction with weighting distribution. It provides image enhancement by modifying the histograms. Murahira and Akira Taguchi

[12] explain a method for contrast enhancement that analyses grey-levels in the image using a histogram and compares it with conventional histogram. Abdullah-Al-Wadud et al. [13] studied contrast enhancement using dynamic histogram equalization (DHE). This divides the image using local minima and equalization is performed on individual portions thereby reducing the side effects of conventional method. Stark [14] worked on image contrast enhancement by generalising the histogram equalization. It is performed using various forms of cumulative functions. Hadhoud, [15] studied enhancement of images using adaptive filters and Homomorphic processing. Both low and high adaptive filters are used and they are in combined in a domain with homomorphism and performs contrast enhancement. Debashis Sen et al. [16] performed a study on contrast enhancement by automatic exact histogram specification. It performs quantitative evaluation of the results. Mehdizadeh and Dolatyer [17] have analysed the use of Adaptive HE on dental X-Rays. Ahmad et. Al., [18] have in their work compared different ways of Equalizing the Histogram for enhancing the quality of dental X-Ray images. Turgay Celik et.al. [19] proposed an adaptive algorithm for image equalization which enhances the contrast of the given image automatically. They used Gaussian mixture model and analysed the gray level distribution of the image.

## 2.1 Findings and Observations:

Very few researchers have ventured into the enhancement and analysis of dental X-Ray images. In most of the work existing for enhancing the X-Ray image quality, the Histogram Equalization method has been adopted. This method is found to give good results when modelled adaptively. For enhancing the contrast and brightness of images methods like Log-Transform and Gamma Correction have also been used. From the nature of these methods, it is obvious that they can also be used for enhancing dental images. It would be more effective if the selection of the enhancement method is made based on the quality statistics of the input image. In this regard there is scarcity of literature.

## 3. PROPOSED SYSTEM

To enhance the visual quality of the Dental X-ray image the contrast and the brightness (visual quality) has been enhanced in this work by taking into account the input image statistics. Three algorithms namely Histogram Equalization (HE), Gamma Correction (GC) and Log Transform (LT) have been analyzed for 40 dental X-Ray images. The Image statistics used are the mean, variance and standard deviation. The algorithm is ranked based on the statistics obtained.

The aim here is to choose an appropriate enhancement algorithm for the given image and also to decide the optimum value for enhancement parameters from the statistics of the image in case of Gamma correction. This method improves the quality of the input image based on the amount of enhancement needed for that image. Figure 1 shows the overall system design.

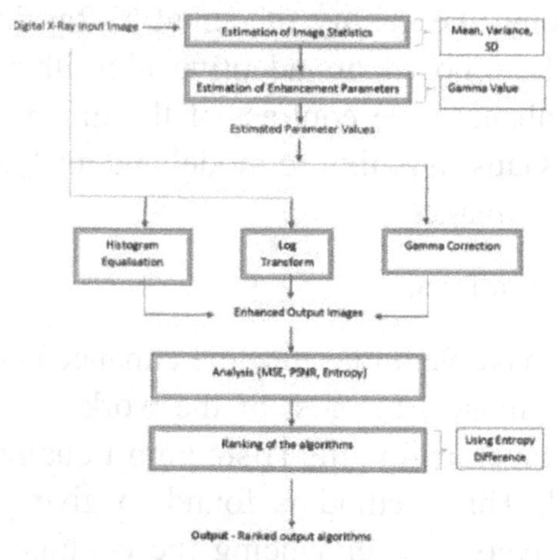


Figure. 1. Shows the system architecture representing the flow of input data into the system and output from the system.

The input image is enhanced using the specified algorithms. For Gamma Correction the value for the Gamma (ϒ) parameter is decided based on the mean and standard deviation of the input image.

If input image has poor contrast/brightness the Gamma > 1

If input image has high contrast/ brightness, then Gamma <1

Else If image is of Good Contrast then Gamma = 1 (No enhancement is needed).

These statistics are inferred from the image histogram. The mean tells about the image brightness and the standard deviation gives the information regarding the image contrast. The enhanced output images obtained using the three algorithms are then compared for finding out the best enhanced output, by analysing the entropy of the input and the output images. The results obtained and the ranking procedure is discussed in detail the next session.

## 4. RESULTS AND ANALYSIS

The results obtained for a sample data set image is shown in figure 2. For the 40 images considered the Gamma correction is performed for gamma values of 0.3, 0.45, 0.6, 0.75 and 0.9.

The statistics obtained from the input and output images of figure 2 are shown in table 1 and table 2 respectively.

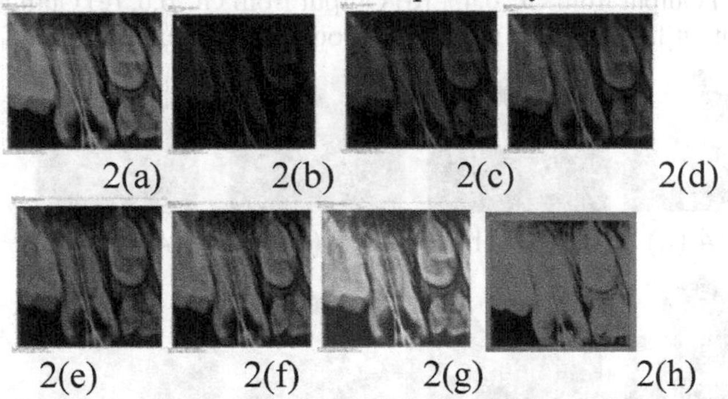


Figure 2: (a) Input image1, (b) Output from Gamma Correction with gamma value 0.3 (GC 0.3), (c) Output from GC 0.45, (d) Output from GC 0.6, (e)Output from GC 0.75, (f) Output from GC 0.9, (g) Output from HE, (h) Output from LT.

Table. 1. Image Statistics of the Input image given in figure 2(a).

| No of rows | No of columns | Mean | Variance | Standard Deviation | Entropy |
|---|---|---|---|---|---|
| 3300 | 2550 | 239.3891 | 50.4773 | 7.1047 | 6.2115 |

Table. 2. Image statistics for the output images obtained (2(b) – 2(h)) using HE, LT and GC.

| | GC_0.3 | GC_0.45 | GC_0.6 | GC_0.75 | Gc_0.9 | LT | HE |
|---|---|---|---|---|---|---|---|
| MEAN | 247.1122 | 243.9598 | 241.9565 | 240.6961 | 239.8948 | 245.6629 | 236.9495 |
| VARIANCE | 37.4095 | 43.3556 | 46.6463 | 48.5363 | 49.7119 | 40.0432 | 53.4065 |
| STANDARD DEVIATION | 6.1163 | 6.5845 | 6.8298 | 6.9668 | 7.0507 | 6.3279 | 7.3079 |
| ENTROPY | 4.0117 | 5.0793 | 5.6072 | 5.9188 | 6.0979 | 4.9952 | 6.1031 |

Figure 3, 4 and 5 shows the input and the corresponding outputs obtained for three different images in the data set considered.

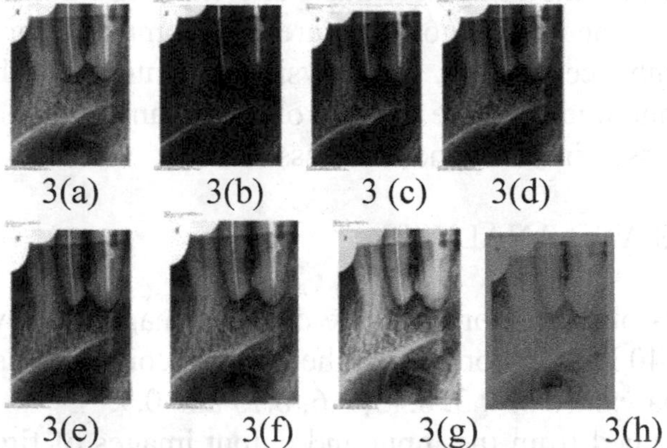


Figure 3: (a) Input image2, (b) Output from Gamma Correction with gamma value 0.3 (GC 0.3), (c) Output from GC 0.45, (d) Output from GC 0.6, (e) Output from GC 0.75, (f) Output from GC 0.9, (g) Output from HE, (h) Output from LT.

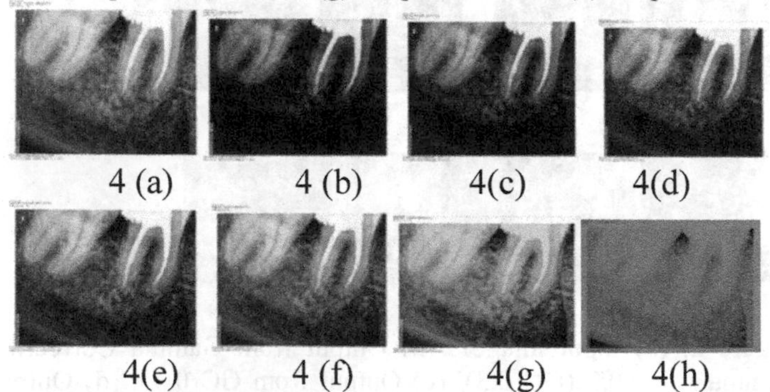


Figure 4: (a) Input image3, (b) Output from Gamma Correction with gamma value 0.3 (GC 0.3), (c) Output from GC 0.45, (d) Output from GC 0.6, (e) Output from GC 0.75, (f) Output from GC 0.9, (g) Output from HE, (h) Output from LT.

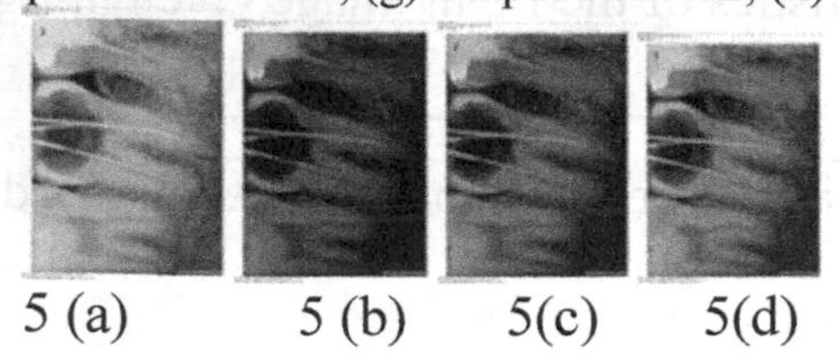

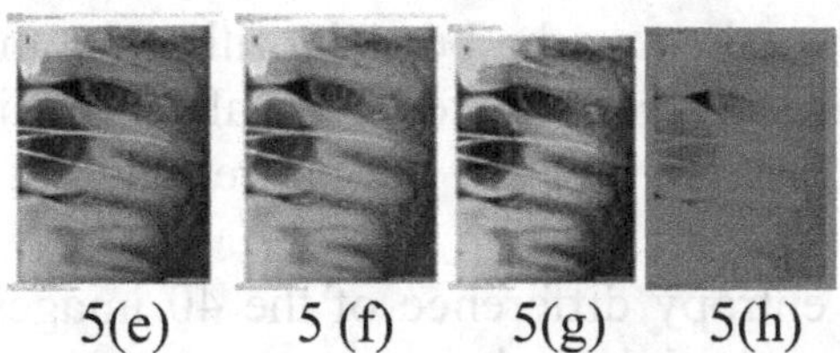


Figure 5: (a) Input image4, (b) Output from Gamma Correction with gamma value 0.3 (GC 0.3), (c) Output from GC 0.45, (d) Output from GC 0.6, (e) Output from GC 0.75, (f) Output from GC 0.9, (g) Output from HE, (h) Output from LT.

The entropy of the input image is compared with the entropy of output obtained from each algorithm. The algorithms are then ranked based on the entropy differene between the input and the output image. The algorithm with lower entropy difference has been ranked as the best , followed by the rest in increasing order of the entropy differences and the results are tabulated in table 3.

Table. 3. Ranking of the algorithms for the input images in figures2(a), 3(a), 4(a) and 5(a) Based on Entropy Difference Value (EDV).

| | Ranking for Image in figure 2(a) | | Ranking for image in figure3(a) | | Ranking for image in figure4(a) | | Ranking for image in figure5(a) | |
|---|---|---|---|---|---|---|---|---|
| Rank | Algorithm | EDV | Algorithm | EDV | Algorithm | EDV | Algorithm | EDV |
| 1 | HE | 0.1084 | GC_0.9 | 0.1364 | GC_0.9 | 0.1066 | GC_0.9 | 0.03183 |
| 2 | GC_0.9 | 0.1136 | HE | 0.1790 | GC_0.75 | 0.3037 | GC_0.75 | 0.1003 |
| 3 | GC_0.75 | 0.2927 | GC_0.75 | 0.3874 | HE | 0.3086 | HE | 0.1695 |
| 4 | GC_0.6 | 0.6043 | GC_0.6 | 0.7923 | GC_0.6 | 0.6188 | GC_0.6 | 0.2167 |
| 5 | GC_0.45 | 1.1323 | LT | 1.3004 | GC_0.45 | 1.1883 | GC_0.45 | 0.4430 |
| 6 | LT | 1.2164 | GC_0.45 | 1.5027 | LT | 1.5928 | GC_0.3 | 1.0679 |
| 7 | GC_0.3 | 2.1997 | GC_0.3 | 2.8181 | GC_0.3 | 2.2957 | LT | 1.9934 |

INFERENCE:

- Visual observation of the results obtained for the images also reveals that for the input image in figure 2(a) the Histogram Equalization (HE) gave the best result. Similarly, for the images in figures 3, 4 and 5 the Gamma Correction with gamma value of 0.9 was found to be better than the other methods.
- The behaviour of the algorithm changes with the quality of the input image considered.

For all the 40 images in the data set the entropy difference have been calculated. To assess the behaviour of the algorithm on the different types of images in the data set considered the standard deviation of the entropy difference for the algorithms has been computed. The algorithm for which the deviation is less can be considered as the one that is giving the similar enhanced result (better enhanced images) for most of the images in the data set and hence can be taken as the best algorithm for the given data set. Table 4 shows entropy difference values obtained for few images and the Scatter Plot for the entire data set is shown in figure 6.

Table.4. shows the entropy difference of the 40 images in the data set and the overall standard deviation.

| Data Set | Methodology Used | | | | | | |
|---|---|---|---|---|---|---|---|
| | GC_0.3 | GC_0.45 | GC 0.6 | GC_0.75 | GC_0.9 | LT | HE |
| 1 | 2.2 | 1.132 | 0.604 | 0.293 | 0.114 | 1.216 | 0.108 |
| 2 | 2.296 | 1.188 | 0.619 | 0.304 | 0.107 | 1.593 | 0.309 |
| --- | --- | --- | --- | --- | --- | --- | --- |
| --- | --- | --- | --- | --- | --- | --- | --- |
| 39 | 1.036 | 0.609 | 0.415 | 0.23 | 0.097 | 1.89 | 0.256 |
| 40 | 0.614 | 0.247 | 0.124 | 0.035 | 0.021 | 2.163 | 0.216 |
| SD | 0.713 | 0.394 | 0.216 | 0.11 | 0.041 | 0.456 | 0.056 |

From the Table.4 the minimum standard deviation of the entropies of all the 40 images from the data set is 0.041 corresponding to Gamma Correction with gamma value 0.9(GC_0.9).

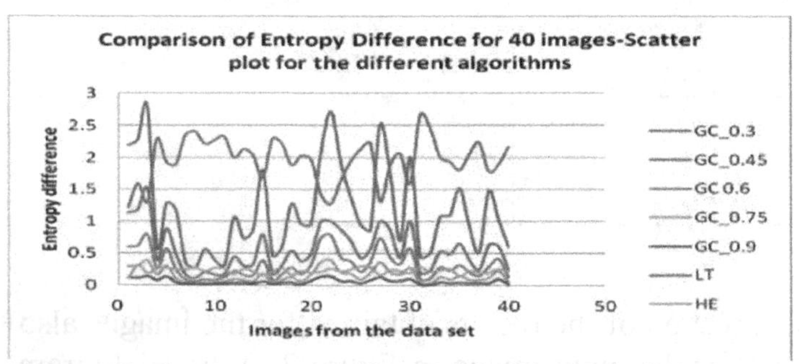


Figure.6. Scatter graph (for Table.4.) depicting the entropy difference of each algorithm for all 40 images.

Scatter graph shown in Figure.6 also shows that GC_0.9 has the minimum entropy difference for most of the input images from the data set. This shows that the Gamma Correction with gamma value 0.9 is the most suitable enhancement algorithm for this dataset. Table 5 gives details of the number of images in the data set for which a particular algorithm was ranked as $1^{st}$, $2^{nd}$, $3^{rd}$, etc.

This analysis may vary, if a different data set is considered.

Table.5. shows the inference obtained from the survey from the survey of the dataset.

| Algorithm | Rank1 | Rank2 | Rank3 | Rank4 | Rank5 | Rank6 | Rank7 |
|---|---|---|---|---|---|---|---|
| GC_0.3 | 0 | 0 | 0 | 0 | 0 | 31 | 9 |
| GC_0.45 | 0 | 0 | 0 | 7 | 30 | 3 | 0 |
| GC_0.6 | 0 | 0 | 19 | 21 | 0 | 0 | 0 |
| GC_0.75 | 0 | 32 | 8 | 0 | 0 | 0 | 0 |
| GC_0.9 | 39 | 1 | 0 | 0 | 0 | 0 | 0 |
| HE | 1 | 7 | 13 | 12 | 7 | 0 | 0 |
| LT | 0 | 0 | 0 | 0 | 3 | 6 | 31 |

## 5. CONCLUSION

In this work a method for finding the best possible enhancement algorithm for a given image and dataset has been proposed. There are lot of image enhancement methods available but it has been observed that most of the techniques cannot be generalised for different types of images. The behaviour of each algorithm depends on the quality of the input image. The enhancement procedure selected must be based on the input image statistics. Here an automated system was developed that automates the process of image enhancement, in such a way that the decision of the enhancement parameters and the method is done, with the help of the image statistics (like mean, variance, entropy and standard deviation). The system also ranks the algorithms so that the end user can have a choice, as image quality has always been a subjective measure. This could be used as a pre-processing for any image processing application.


## 6. ACKNOWLEDGEMENT

We would like to acknowledge Karthik Sundar and Shivnesh Kumar for their help in the execution of this work.


## REFERENCES


[1] Yeong-Taeg Kim, "Contrast Enhancement Using Brightness Preserving Bi-Histogram Equalization," *IEEE Transactions on Consumer Electronics*, Vol. 43, No. 1, 1997.

[2] Md. Foisal Hossain, Mohammad Reza Alsharif, and Katsumi Yamashita, "Medical Image Enhancement Based on Non-Linear Technique and Logarithmic Transform Coefficient Histogram Matching," *International Conference on Complex Medical Engineering, IEEE*/ICME, 2010.

[3] Fan Yang and Jin Wu, "An Improved Image Contrast Enhancement in Multiple-Peak Images Based On Histogram Equalization," International Conference on Computer Design and Applications (ICCDA), Vol. 1, 2010.

[4] M. Sundaram, K. Ramar, N. Arumugam, and G. Prabin, “Histogram Based Contrast Enhancement for Mammogram Images,” International Conference on Signal Processing, Communication, Computing and Networking Technologies (ICSCCN), IEEE, 2011.

[5] G.N.Sarage and Dr Sagar Jambhorkar, “Enhancement of Chest X-Ray Images Using Filtering Techniques,” International Journal of Advanced Research in Computer Science and Software Engineering, Vol. 2, Issue 5, 2012.

[6] H. D. Cheng and Yingtao Zhang, “Detecting of Contrast Over-Enhancement,” *International Conference on Image Processing (ICIP), IEEE,* 2012.

[7] Ili Ayuni Mohd Ikhsan, Aini Hussain, Mohd Asyraf Zulkifley, Nooritawati Md. Tahir and Aouache Mustapha, “An Analysis of X-Ray Image Enhancement Methods for Vertebral Bone Segmentation,” *International Colloquium on Signal Processing & its Applications (CSPA), IEEE*, 2014.

[8] Ritika and Sandeep Kaur, “Contrast Enhancement Techniques for Images– A Visual Analysis,” *International Journal of Computer Applications,* Vol. 64, No. 17, 2013.

[9] N. S. Datta, P. Saha, H. S. Dutta, D. Sarkar, S. Biswas and P. Sarkar, “A New Contrast Enhancement Method of Retinal Images in Diabetic Screening System,” *International Conference on Recent Trends in Information Systems (ReTIS), IEEE,* 2015.

[10] Laxmi Laxman, V. Kamalaveni and K. A. Narayanankutty, “Comparative Study On Image Restoration Techniques Using the Partial Differential Equation and Filters,” *International Journal of Engineering Research and Technology*, Vol.2, Issue 7, Jul 2013.

[11] Shih-Chia Huang, Fan-Chieh Cheng, And Yi-Sheng Chiu, “Efficient Contrast Enhancement Using Adaptive Gamma Correction with Weighting Distribution,” *IEEE Transactions On Image Processing*, Vol. 22, No. 3, Mar 2013.

[12] Kota Murahira and Akira Taguchi, “A Novel Contrast Enhancement Method Using Differential Gray-Levels Histogram”, *International Symposium on Intelligent Signal Processing and Communication Systems (ISPACS),* Dec 7-9, 2011.

[13] M. Abdullah-Al-Wadud, Md. Hasanul Kabir, M. Ali Akber Dewan, and Oksam Chae, "A Dynamic Histogram Equalization for Image Contrast Enhancement," *IEEE Transactions on Consumer Electronics,* Vol .53 , Issue 2 ,May 2007.

[14] J. Alex Stark, "Adaptive Image Contrast Enhancement Using Generalizations of Histogram Equalization," *IEEE Transactions On Image Processing*, Vol. 9, No. 5, May 2000.

[15] Mohiy M. Hadhoud, "Image Contrast Enhancement Using Homomorphic Processing and Adaptive Filters,"*16 th National Radio Science Conference, Nrsc'98*, Am Shams University, Egypt,Feb. 23-25, 1999.

[16] Debashis Sen and Sankar K. Pal, "Automatic Exact Histogram Specification for Contrast Enhancement and Visual System Based Quantitative Evaluation", *IEEE Transactions On Image Processing,* Vol. 20, No. 5, May 2011.

[17] M.Mehdizadeh and S.Dolatyar, "Study of Effect of Adaptive Histogram Equalization on Image Quality in DigitalPreapical Image in Pre Apex Area", *Research Journal of Biological Science"* ,pp: 922- 924, vol: 4, issue: 8, 2009.

[18] Siti Arpah Ahmad, Mohd Nasir Taib, Noor Elaiza Abdul Khalid, and Haslina Taib, "An Analysis of Image Enhancement Techniques for Dental X-ray Image Interpretation", *International Journal of Machine Learning and Computing*, Vol. 2, No. 3, June 2012.

[19] Turgay Celik and Tardi Tjahjadi, "Automatic Image Equalization and Contrast Enhancement using Gaussian Mixture Modelling", *IEEE Transactions On Image Processing*, Vol.21, Issue.5, Jan 2012.

# MRI/CT IMAGE FUSION USING GABOR TEXTURE FEATURES

Hema P Menon[1], K A Narayanankutty[2]
[1]Departmentment of Computer Science and Engineering
[2]Department of Electrical Communications and Engineering
Amrita School of Engineering, Coimbatore
Amrita Vishwa Vidyapeetham, Amrita University, India
[1]p_hema@cb.amrita.edu, [2]ka_narayanankutty@yahoo.com

***Abstract***

Image fusion has been extensively used in the field of medical imaging by medical practitioners for analysis of images. The aim of image fusion is tocombine information from different images in the output fused image without adding artefacts. The output has to contain all information form the individual images without introducing artifacts. In images that contains more textural properties, it will be more effective in terms of fusion, if we include all the textures contained in the corresponding individual images. Keeping the above objective in mind, we propose the use of Gabor filter for analysing the texture, because under this method the filter parameters can be tunned depending upon the textures in the corresponding images. The fusion is performed on the individual textural components ot the two input images and then all the fused texture images are combined together to get the final fused image.To this the fused residual image obtained by combining the residue of the two images can be added to increase the information content. This approach was tested on MRI and CT images considering both mono-modal and multi-modal cases and the results are promising.

**Keywords:** Image Fusion, Gabor Filters, Texture Analysis, Magnetic Resonance Imaging (MRI) images, Computed Tomography (CT) images, Fusion Factor, Fusion Symmetry, Renyi Entropy.



J.M. Corchado Rodriguez et al. (eds.), *Intelligent Systems Technologies and Applications 2016*, Advances in Intelligent Systems and Computing 530,
DOI 10.1007/978-3-319-47952-1_4

# 1. INTRODUCTION

In medical imaging different information pertaining to the same body part is obtained when using different modalities. For crucial diagnosis generally images from more than one modality (Multi-modal) is often used by doctors. In cases such as analysis of effect of medicine on the disease progression or the study of the pre and post operative images, the images acquired using the same modality (Mono-modal) at different instances of time is used. In either case the inconvenience caused to the doctors in looking at two different images for the analysis is the main reason for fusion gaining more prominence in the medical field. The medical images can be acquired using different imaging modalities like Magnetic Resonance Imaging (MRI), Computed Tomography (CT), Positron Emission Tomography (PET), Single Photon Emission Computed Tomography (SPECT), Functional MRI (fMRI) and X-Ray imaging. In this work fusion of both the Multi-Modal (CT-MRI) and Mono-modal (CT-CT, MRI-MRI) images have been considered.

Image Fusion is the process of producing single composite image through combining information from multiple images. The single image thus produced will contain the relevant information from the inputs and will be immensely useful for medical practitioners for analysis and diagnosis of Mono-modal and Multi-modal images. Major focus of many existing fusion techniques is to retain the coarse information or the edge information of the input images in the fused output image. In case of medical images the salient features in an image depends on the imaging modality used. The edge, perceptual contrast and the texture of the images are very vital for diagnosis, and transferring all these components from the source images to the fused output is very essential. In almost all applications, the main goal of image fusion is to produce an enhanced new image which retains all the important details in the source images with a better visual quality. Observations on medical (especially MRI) images revels that the image has immense textural details in it.

In this work the textural features of an image have been exploited for fusion purpose. The image texture can be extracted using many approaches. The use of Gabor Filters is a commonly adopted method for texture analysis. Gabor filter is a tunable band pass linear filter and the details of the filter design and the proposed fusion scheme is given in section 3. Section 4 discusses the results obtained and the quantitative analysis done. The conclusion of the work done is given in section 5. The following section 2 gives the background study on image fusion.

## 2. Background Study on Image Fusion

In medical field, depending on the acquisition method and the quality of the images acquired, a wide variety of applications arise [1]. The significance of fusion in the field of medical imaging has increased drastically in the recent years. Image fusion can be performed in Spatial Domain or in Frequency domain [2, 3, 4 and 6]. Various features can be used for fusion like texture, region, shape, intensity value and edges. The traditional methods for fusion include the Principal Component Analysis [2 and 7] and the Intensity Hue Saturation based fusion of images [8]. Discrete Wavelet Transform (DWT) [5, 9] and Pyramidal Decomposition [10] have been widely used for image fusion in the past few years. Recently, image fusion approaches based on Partial Differential Equation (PDE) have been are gaining prominence [11]. Naidu et.al, [12] have in their work explored the effectiveness of combining the PCA with the DWT to improve the image quality.

The image fusion techniques can be grouped into pixel based [12], feature based [13] and decision based [14] methods. For medical images pixel-level fusion methods [12] are generally used. In feature based fusion scheme the salient features from the source images, like the pixel intensities, edges or texture are used for fusion. Decision level fusion works on the basis of certain decision rules that are applied on pixels or the features extracted prior to fusing them. The advance in image fusion was then with the idea of decomposing the image into its various spectral components. The foundation to this approach was the use of pyramidal decomposition. There are various methods used for pyramidal decomposition: Steerable Pyramid [17], Laplacian pyramid [18 and 19], Gradient Pyramid, Filter Subtract

Decimate Pyramid [20], Ratio pyramid and Morphological pyramid [21]. All these methods follow the same process; the difference is in the selection of the filter for the decomposition process.

The advent of Wavelets was a major breakthrough in the field of signal and image processing. The review of literature pertaining to this is vast [22, 23, 24 and 25]. Image fusion using wavelets has been used widely in remote sensing [26] and medical imaging applications by many researchers. In medical imaging this has been used successfully in the fusion of multimodal images [27], especially the CT and MRI images [28]. Wavelet based image fusion techniques involves decomposition of the source images into its corresponding frequency sub-bands using the 2D DWT, and then fusing the coefficients of each band using the primary selection rules. Then apply the IDWT to get back the fused image. DWT is generally used for multi-resolution analysis, wherein the image is decomposed into many levels, the number of which depends on the application. Among the different type of wavelets available, Curvelets and Slantlets have been used immensely for CT-MRI image fusion in the recent past [29][30]. It can be observed from the review of literature that the focus has been on trying to preserve the maximum relevant information from the source images in the fused output image. Hence, most fusion schemes bank on the multi-resolution analysis, but have been giving more emphasis on any one of the features in the source images. The most recent advance in image fusion is the use of the different components (features) present in the source images for fusion, which can be clubbed with multi-resolution analysis.

## 3. Image Fusion Using Gabor Filters

In images that contains more textural properties, it will be more effective in terms of fusion, if all textures contained in the corresponding individual images are included in the fusion process. Advantage of Gabor filter is that the parameters can be tunned depending upon the textures in the corresponding images.

### 3.1 Gabor Filter Kernel

Gabor filter is a tunable band-pass-linear filter. Gabor filter kernal is a Gaussian weighted sinusoidal. The filter can be tuned by manipulating the frequency and the orientation parameters of the sinusoidal. The spatial spread of the Gaussian helps to model the time scale. For extracting texture features from an image, a bank of Gabor filters with different frequencies and orientations has to be designed. The general equation for 2D Gabor filter kernel is given by the following equation

$$f(x, y, \omega, \theta, \sigma_x, \sigma_y) = 1/2\pi\sigma_x\sigma_y (\exp[(-\tfrac{1}{2})((x/\sigma_x)^2 + (y/\sigma_y)^2) + j\omega (x\cos\theta + y\sin\theta)]$$

Where **σ** is the spatial spread, **ω** is the frequency, **θ** is the orientation.

The Gabor filter bank used is tuned for 4 frequencies and 9 orientations with equal spatial spread along x and y axis ($\sigma_x = \sigma_y = 2$). The frequencies used are 4, 8, 16 and 32. The orientations for the Gabor filter bank kernels have been chosen as $0^o$, $22.5^o$, $45^o$, $67.5^o$, $90^o$, $112.5^o$, $135^o$, $157.5^o$ and $180^o$. The filter bank used consisting of 4 banks of 9 filters each decomposes the image into its 36 textural components.

## 3.2 Gabor Filter Image Fusion Scheme and its significance

The individual texture components of the two images are fused together as shown in figure 1a. The textures are extracted from both the input images using the filter bank shown in the figure 2. The filter bank used is shown in figure 3. The Texture fusion process can be represented as given in equation 1:

$$Tex\_Fuse_I = \sum_{i=1}^{36} \left(\frac{1}{2}\right) (Gt_i(S) + Gt_i(T)) \quad (1)$$

Where, $Gt_i(S)$ and $Gt_i(T)$ are the $i^{th}$ Gabor-textural component of Source and Target image S and T respectively.

To this the final residue of the Gabor filtered images are added to retain the complete information. Residual image (ResImg) is an image generated by subtracting the original image (I) from the Gabor filtered image (Gf) as given by equation 2.

$$ResImg = I(x, y) - Gf(x, y) \quad (2)$$

The process of generating the residual images is shown in figure 1b.

$$Res\_Fuse_I = \frac{1}{2} (ResImg(S) + ResImg(T)) \quad (3)$$

Where, $ResImg(S)$ and $ResImg(T)$ are the Residual images of Source and Target images S and T respectively.

The final fused image can be expressed using the equation 4.

$$\text{FinalFuse}_I = (\text{Tex\_Fuse}_I + \text{ResFuse}_I) \quad (4)$$

Depending on the image texture the response of each of the filters in the filter bank varies. This can be estimated by analyzing the statistical properties of the filter magnitudes.

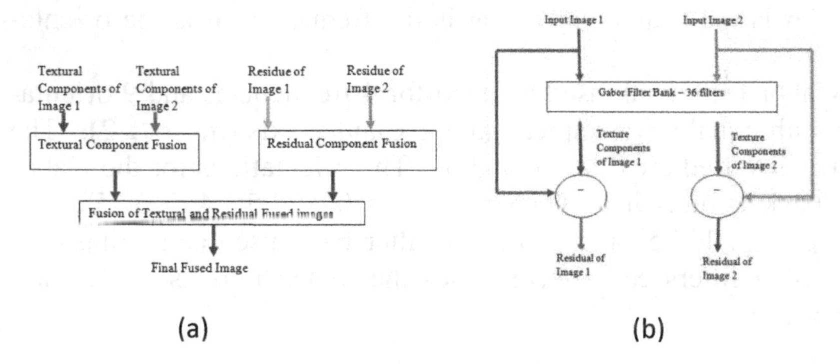


**Figure 1: a)Overall System Design for Textural Analysis Image Fusion and b) Generation of Residual Images**

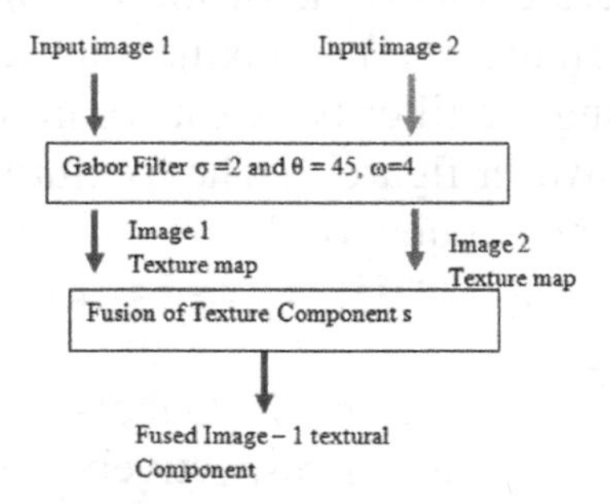


**Figure 2: Textural Component Fusion Scheme**

## 3.3 Results and Discussions

Experiments were conducted on MRI, CT and X-Ray images. For each image 36 textural components are generated and then they are individually fused to get 36 texture fused images.

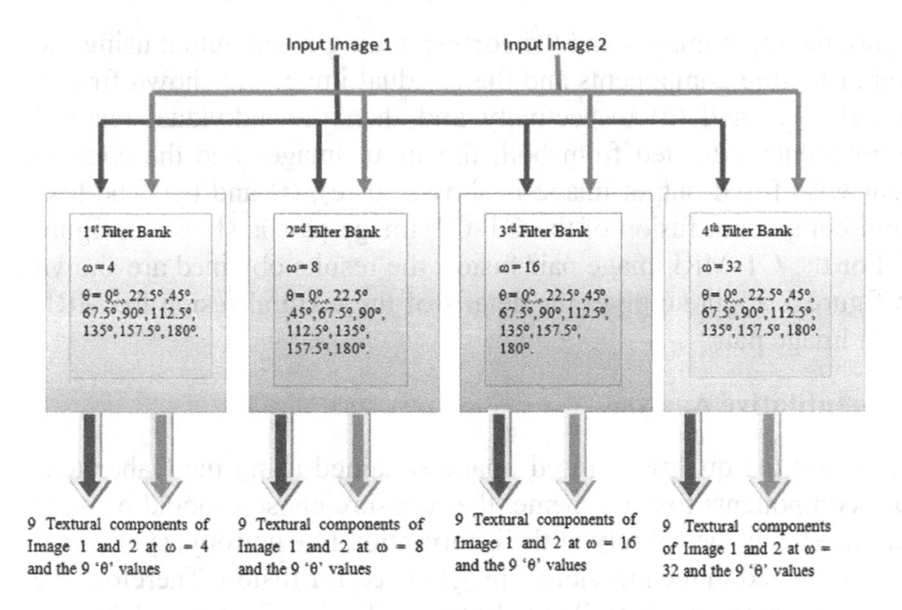


**Figure 3: Gabor Filter Bank Design with 36 Filters (4 banks with 9 filters each)**

***Significance of the proposed scheme***

36 textural components each from the source and the target images are extracted using the designed Gabor filter bank. Then, corresponding textural components of both the images are fused together to get 36 fused textural components. Since the Gabor filters are pre-tuned to certain orientations and frequencies empirically, there can be cases wherein the input image under consideration might have textural features that could not be captured by the designed filter bank. It is also noted that MRI images have minute textural variations in them. In order to retain theses missing features in the fused image, residual images from the source and the target images are fused separately and then added to the final fused output. This is found to be effective in ensuring structural quality of the fused image.

Since the numbers of outputs generated are too many, as a sample only the Gabor textural components at 22.5°, 90°, 45° and 135° orientations have been shown here for all the images discussed. In all figure the input images and the corresponding fused output using the Gabor texture components and the residual image are shown first in (a), (b), (c) and (d) respectively and then the individual textural components extracted from both the input images and the component wise fused output image is shown in (e), (f) and (g). The textural component fusion of the CT-CT image pair is shown in figure 4. For the CT-MRI image pair fusion the results obtained are shown in figure 5. Figure 6 gives the details of the textural fusion for MRI-MRI image pair.

## 4 Quantitative Analysis

To assess the quality of fused image obtained using the Gabor texture components fusion scheme, the measure chosen should be such that it reflects the entirety of the information like entropy, symmetry, etc. contained in the individual images used for fusion. Therefore we are using measures like Renyi Entropy, Fusion Factor and Fusion Symmetry for analysing the fused images in this work. The value of Renyi entropy measure obtained should be high for a good fusion technique. Larger Fusion Factor indicates that more information has been transferred from input images into the fused output image. The smaller the Fusion Symmetry value, better the result. The values obtained for each image type is tabulated in table 1 for analysis.

**Table 1: Comparison of Renyi Entropy measure for CT-CT, CT-MRI MRI-MRI images**

| Measures Used | Data Set type | | |
|---|---|---|---|
| | CT-CT | CT-MRI | MRI-MRI |
| Renyi Entropy | 2.2699 | 5.1384 | 2.6532 |
| Fusion Factor | 4.7977 | 4.2753 | 3.2360 |
| Fusion Symmetry | 0.0136 | 0.00096 | 0.0148 |

**CT-CT Image pair Textural Component Fusion**

a) b) c) d)

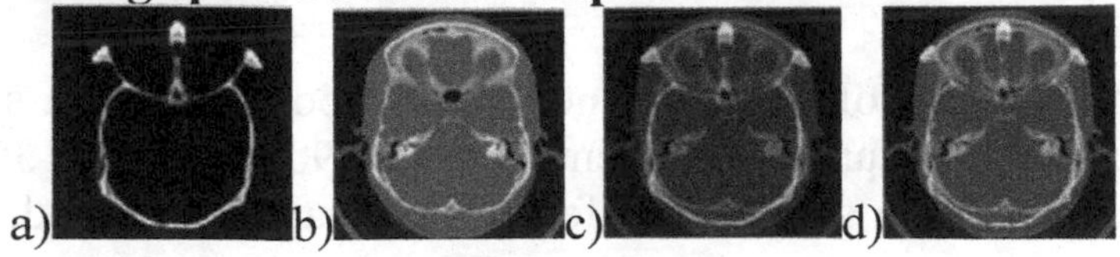

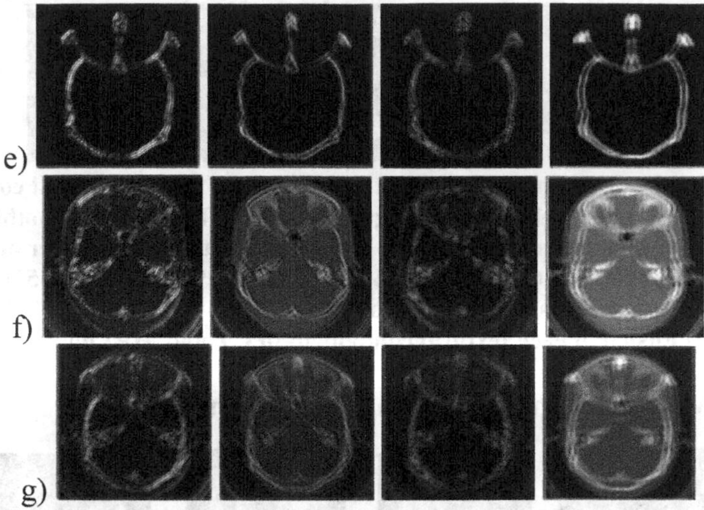


**Figure 4: CT – CT image pair fusion using Gabor textural components and residue. (a) Input CT image 1 (b) Input CT image 2 (c) Combining all the fused textural components (d) Final fused output image after adding the residue (e) Gabor textural components at 22.5°, 90°, 45° and 135° of CT image 1 (f) Gabor textural components at 22.5°, 90°, 45° and 135° of CT image 2 (g) Fusion of each textural component of CT - CT image pair.**

## CT-MRI Image Pair - Textural Component fusion

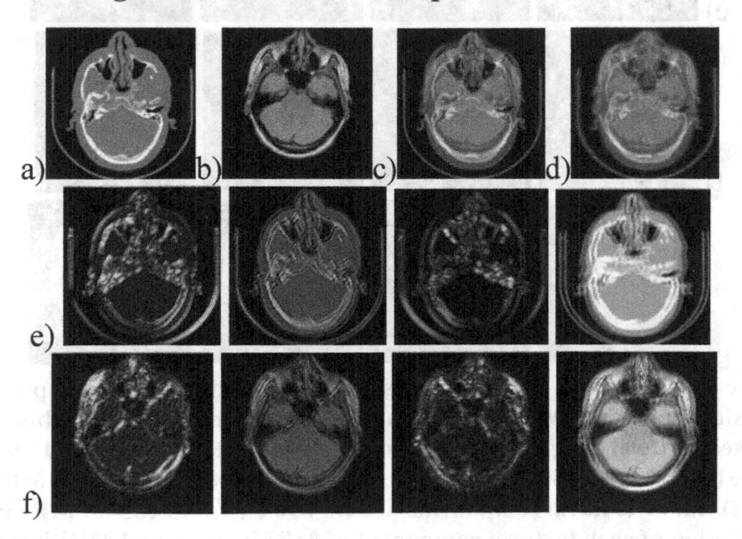

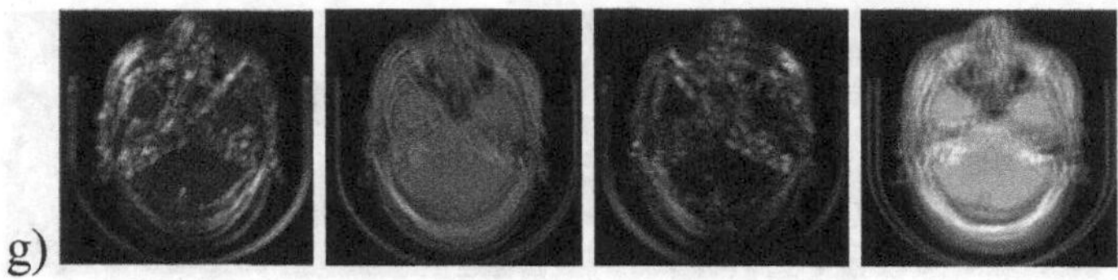


**Figure 5: CT – MRI Image Pair – Set 2 fusion using Gabor textural components and residue. (a) Input CT image (b) Input MRI image (c) Combining all the fused textural components (d) Final fused output image after adding the residue (e) Gabor textural components at 22.5°, 90°, 45° and 135° of CT image (f) Gabor textural components at 22.5°, 90°, 45° and 135° of MRI image (g) Fusion of each textural component of CT and MRI image.**

## MRI – MRI image pair Textural Component Fusion

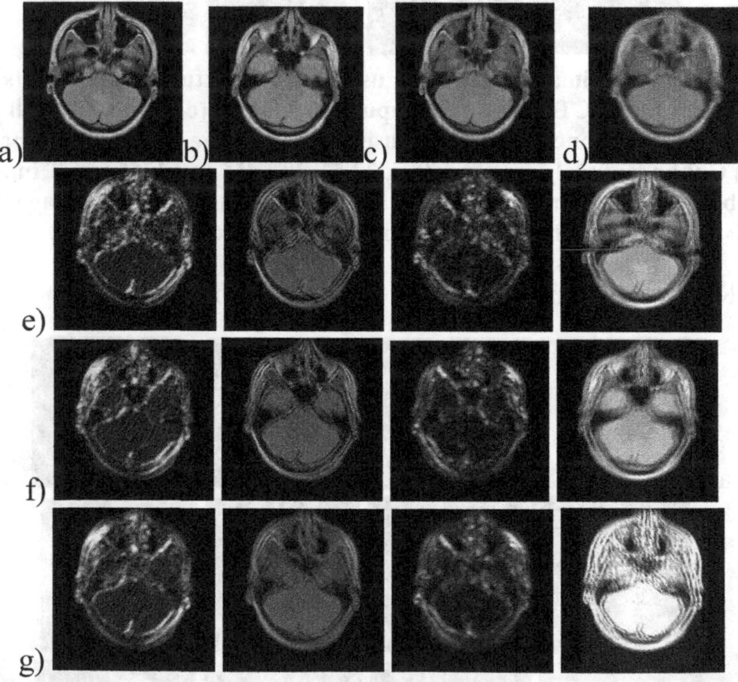


**Figure 6: MRI – MRI Image Pair fusion using Gabor textural components and residue. (a) Input MRI image 1 (b) Input MRI image 2 (c) Combining all the fused textural components (d) Final fused output image after adding the residue (e) Gabor textural components at 22.5°, 90°, 45° and 135° of MRI image 1 (f) Gabor textural components at 22.5°, 90°, 45° and 135° of MRI image 2 (g) Fusion of each textural component of MRI image 1 and MRI image 2.**

### 4.1 Findings and Inferences

From the table it can be observed that the values obtained for the quality measure indicate that fusing the images using texture at every level increases the information content in the fused output. It can be inferred that use of all the texture content in the input images for fusion is the most effective method for transferring maximum information from input images to the fused output image. This can be achieved by using properly tuned Bank of Gabor filters as established through our analysis. This fusion scheme was found to be more effective in case of MRI – CT image fusion.

## 5. CONCLUSION

In this work the applicability of use of Gabor features for fusion of MRI and CT mono-modal and multi-modal case has been assessed. These images contain lot of textural information In case of medical images to extract the textural component we suggest the use of Gabor filters as they are tunable filters and can be used for extracting irregular texture. The use of coarse information from the image and the image residue also is useful in improving the quality of the fused image. For MRI images, the use of texture components extracted using Gabor and the residue could be preferred.

REFERENCES

[1] Flusser, J., Sroubek, F., and Zitov, B. (2007), "Image Fusion: Principles, Methods and Applications", *Lecture Notes, Tutorial European Signal Processing Conference 2007.*

[2] Yang, J., Ma, Y., Yao, W., and Lu, W. T. (2008), "Spatial Domain and Frequency Domain Integrated Approach to Fusion Multi focus Images," *The International Archives of the Photogrammetry, Remote Sensing and Spatial Information Sciences*, Vol. 37, Part.B7.

[3] Metwalli, M.R., Nasr, A.H., Allah, O.S.F., and El-Rabaie, S.(2009), "Image fusion based on principal component analysis and high-pass filter", *International Conference on Computer Engineering & Systems, IEEE*, 2009, pp. 63-70.

[4] Hariharan, H., Gribok, A., Abidi, M.A., and Koschan, A. (2006), "Image fusion and enhancement via empirical mode decomposition", *Journal of Pattern Recognition Research,*

Vol.1, No.1, pp. 16-32.

[5] Nikolov, S., Hill, P., Bull, D., and Canagarajah, N. (2001), “Wavelets for image fusion”, *International Conference on Wavelets in signal and image analysis, Springer*, 2001, pp. 213-241.

[6] Li, W., and Zhang, Q. (2008), “Study on data fusion methods with optimal information preservation between spectral and spatial based on high resolution imagery”, *The International Archives of the Photogrammetry, Remote Sensing and Spatial Information Sciences*, Vol. 36, pp. 1227-1232.

[7] Chiang, J.L. (2014), “Knowledge-based principal component analysis for image fusion”, *International Journal of Applied Mathematics & Information Sciences*, Vol. 8, No. 1, pp. 223-230.

[8] Dou, W., and Chen, Y. (2008), “An improved IHS image fusion method with high spectral fidelity”, *The International Archives of the Photogrammetry, Remote Sensing and Spatial Information Sciences*, Vol. *37*, pp. 1253-1256.

[9] Amolins, K., Zhang,Y., and Dare, P. (2007), “Wavelet based image fusion techniques - An introduction, review and comparison”, *Journal of Photogrammetry and Remote Sensing*, Vol. 62, No. 4, pp. 249-263.

[10] Burt, P.J., and Adelson, E.H. (1983), “Laplacian pyramid as a compact image code,” *IEEE Transactions on Communications*, Vol. 31, No. 4, pp. 532-540.

[11] Socolinsky, D.A., and Wolff, L.B. (2002), “Multispectral Image Visualization Through First-Order Fusion”, *IEEE Transactions on Image Processing*, Vol. 11, No. 8, pp. 923-931.

[12] Naidu, V.P.S., and Raol, J.R. (2008), “Pixel-level image fusion using Wavelets and Principal Component Analysis”, *Defence Science Journal*, Vol. 58, No. 3, pp. 338-352.

[13] Calhoun, V. D., and Adali, T. (2009), “Feature-based fusion of medical imaging data”, *IEEE Transactions on Information Technology in Biomedicine,* Vol. 13, No. 5, pp. 711-720.

[14] Luo, B., Khan, M.M., Bienvenu, T., Chanussot, J., and Zhang, L. (2013), “Decision-based fusion for pansharpening

of remote sensing images", *Geoscience and Remote Sensing Letters, IEEE*, 2013, Vol. 10, No. 1, pp. 19-23.

[15] Naidu, V.P.S., and Raol, J.R. (2008) , "Fusion of out of Focus Images Using Principal Component Analysis and Spatial Frequency", *Journal of Aerospace Sciences and Technologies*, Vol. 60, No. 3, pp. 216-225.

[16] Zhang, Y. (2004), "Understanding image fusion", *Journal of Photogrammetric engineering and remote sensing*, Vol. 70, No. 6, pp. 657-661.

[17] Liu, Z., Tsukada, K., Hanasaki, K., Ho, Y.K., and Dai, Y.P. (2001), "Image Fusion by using Steerable Pyramids", *Pattern Recognition Letters*, Vol. 22, No. 9, pp. 929-939.

[18] Choudhary, B. K., Sinha, N. K., and Shanker, P. (2012), "Pyramid Method in Image Processing", *Journal of Information Systems and Communication*, Vol. 3, No. 1, pp.269-273.

[19] Wang, W., and Chang, F. (2011), "A Multi-focus Image Fusion Method Based on Laplacian Pyramid", *Journal of Computers*, Vol. 6, No. 12, pp.2559-2566 .

[20] Anderson, H. (1987), "A filter-subtract-decimate hierarchical pyramid signal analyzing and synthesizing technique," *U.S. Patent 718 104*, 1987.

[21] Laporterie, F., and Flouzat, G. (2003), "The morphological pyramid concept as a tool for multi-resolution data fusion in remote sensing", *Journal of Integrated computer-aided engineering*, Vol. 10, No. 1, pp. 63-79.

[22] Pajares, G., and Manuel de la Cruz, (2004), "A Wavelet based image fusion tutorial", *Pattern Recognition*, Vol. 37, No. 9, pp. 1855-1872.

[23] Li, H., Manjunath, B. S., and Mitra, S.K. (1995), "Multisensor image fusion using the wavelet transform", *Journal of Graphical models and image processing*, Vol. 57, No. 3, pp. 235-245.

[24] Heng Ma, Chuanying Jia and Shuang Liu, (2005) "Multisource Image Fusion Based on Wavelet Transform", *International Journal of Information Technology*, Vol. 11, No. 7, pp. 81-91.

[25] Chipman, L.J., Orr, T.M., and Lewis, L.N. (1995), "Wavelets and Image Fusion", *IEEE Transactions on Image Proc-*

*essing*, Vol. 3, pp. 248-251.

[26] Moigne, J.L., and Cromp, R.F. (1996), “The use of Wavelets for remote sensing image registration and fusion”, *Technical Report TR-96-171, NASA*, 1996.

[27] Wang, A., Sun, H., and Guan, Y. (2006), “The application of Wavelet Transform on Multimodal Image Fusion”, *IEEE International Conference on Networking, Sensing and Control, (ICNSC),* 2006, pp. 270-274.

[28] Yang, Y., Park, D.S., Huang, S., Fang, Z., and Wang, Z. (2009), “Wavelet based approach for fusing Computed tomography and Magnetic Resonance Images”, *Control and Decision Conference* (CCDC’09), Guilin, China, June 2009, pp. 5770-5774.

[29] Arathi T and Latha Parameswaran, “An image fusion technique using Slantlet transform and phase congruency for MRI/CT”, *International Journal of Biomedical Engineering and Technology,* Vol. 13, Issue 1, pp. 87-103, 2013.

[30] Sruthy, S. , Latha Parameswaran, and Ajeesh P. Sasi. "Image Fusion Technique using DT-CWT", *IEEE International Multi-Conference on automation, computing, control, communication & compressed sensing* (iMac4S), Kottayam, pp. 160-164, 22-23 March, 2013.

# Face recognition in videos using Gabor filters

S. V. Tathe, A. S. Narote and S. P. Narote

**Abstract** Advancement in computer technology has made possible to evoke new video processing applications in field of biometric recognition. Applications include face detection and recognition integrated to surveillance systems, gesture analysis etc. The first step in any face analysis systems is near real-time detection of face in sequential frames containing face and complex objects in background. In this paper a system is proposed for human face detection and recognition in videos. Efforts are made to minimize processing time for detection and recognition process. To reduce human intervention and increase overall system efficiency the system is segregated into three stages- motion detection, face detection and recognition. Motion detection reduces the search area and processing complexity of systems. Face detection is achieved in near real-time with use of haar features and recognition using gabor feature matching.

*Keywords:* Motion Detection, Face detection, Face Recognition, Haar, Gabor.

## 1 Introduction

Cameras are vital part of security systems, used to acquire information from environment. Visual information contains most of the information in the field of view of camera. Video cameras are installed for security reasons in areas that requires continuous monitoring for avoiding criminal activities. Cameras serve as observers

S. V. Tathe
Research Student, BSCOER, Pune, e-mail: swapniltathe7@gmail.com

A. S. Narote
S. K. N. College of Engineering, Pune, e-mail: a.narote@rediffmail.com

S. P. Narote
M. E. S. College of Engineering, Pune e-mail: snarote@rediffmail.com



J.M. Corchado Rodriguez et al. (eds.), *Intelligent Systems Technologies and Applications 2016*, Advances in Intelligent Systems and Computing 530,
DOI 10.1007/978-3-319-47952-1_5

to monitor movements and behavior of individuals or objects and preserve the data for future analysis if required.

There has been rapid growth in research and development of video sensors and analyzing technologies. It includes video, audio, thermal, vibration and various other sensors for civilian and military applications. The role of monitoring system is to detect and track objects, analyze movements and respond accordingly. Some systems are equipped with heterogeneous sensors with small scale processing to meet real-time constraints. Video analysis has become more important and need of time.

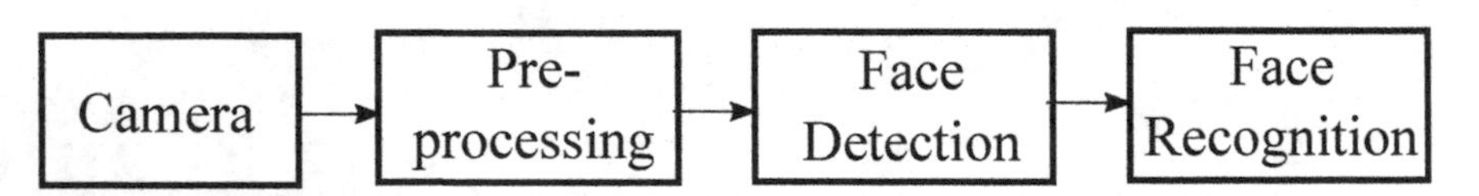


**Fig. 1** Proposed System

The system proposed here is intended to detect the person coming in the view of camera and recognize identity to restrict the movements of unauthorized person. The first stage captures the video. Pre-processing algorithms are employed to remove the unwanted artifacts like noise present in the image. The second stage detects faces in frames. Next stage is to identifying the detected human being in the camera view.

The paper is organized as follows: Section 1 gives the brief idea of proposed system. Section 2 briefly provides state-of-art work done in the field of human detection and recognition. Section 3 presents the techniques in details used in proposed system. In Section 4 results of proposed system are presented and discussed. Section 5 discusses the conclusions and issues related to proposed system. It also provides guidelines and direction to future work.

## 2 Related Work

Automatic face detection is the first step in many applications like face recognition, face tracking, surveillance, HCI, face retrieval, etc.The color information in images is sensitive to the intensity. Skin pixel values for $r$ plane is in the range 0.36 to 0.456 and $g$ value in 0.28 to 0.363. In [1] combined smoothed 2-D histogram and Gaussian model is used for skin pixel detection to overcome individual limitations to increase accuracy to 90%. In *HSV* model pixel is classified as skin pixel if it has values $0 \leq H \leq 50$ and $0.20 \leq S \leq 0.68$ [2]. Chae et al. [3] constructed statistical color model using Baye's probability and classified pixels to skin and non-skin with artificial neural network. In [4] combination of two models *HSV* and $YC_bC_r$ is used with classification range of $Cb \leq 125$ or $Cb \geq 160$; $Cr \leq 100$ or $Cr \geq 135$; $26 < Hue < 220$. Skin color based method is sensitive to illumination variation and fails if background contains skin color like objects. Complex methods uses depth features to detect face based on geometry and texture pattern. Face is detected using edge maps and skin color thresholding in [5]. Dhivakar et al. [6] used $YC_bC_r$ space for skin pixel detection and Viola-Jones [7] method to verify correct detection. Mehta et al. [8] proposed LPA and LBP methods to extract textural features with LDA dimensionality reduction and SVM classification. Park et al.[9] used $3 \times 3$ block rank patterns of gradient magnitude images for face detection. Lai et al. [10] used logarithmic difference edge maps to overcome illumination variation with face verification ability of 95% true positive rate. Viola and Jones [7] proposed real-time AdaBoost algorithm for classification of rectangular facial features. Xiao et al.[11] used advanced haar features to compensate for pose variation. In [12] Viola Jones face detector is combined with color depth data to reduce number of false positives. Haar features are robust to illumination changes and has proven method for real-time face detection.

The first stage in face recognition is face detection for extracting discriminant features and second stage for classification of these features to find similar face. The methods available are classified into component-based and appearance-based methods [13]. The component-based methods detects the facial components like eyes, eyebrows, nose and mouth which is a complex task. The appearance based methods takes into consideration the entire image for processing and detects the texture patterns. These methods are simpler as compared with the component-based approaches.

Meshgini et al. [13] proposed a method with combination of Gabor wavelets for feature extraction, linear discriminant analysis for dimensionality reduction and support vector machine for classification. Gabor wavelets extracts facial features characterized by spatial frequency, locality and orientation to overcome changes due to illumination, facial expression and pose. Many researchers have made use of PCA-LDA to reduce size of feature vector [14, 15]. Neural network is used for classification of these reduced vectors [16, 17]. Gangwar et al. [18] introduced Local Gabor Rank Pattern method that uses both magnitude and phase response of filters to recognize face. In [19] Histogram of Co-occurrence Gabor Phase Patterns are introduced for face recognition that has better recognition rate than state-of-art methods. In [20] Gabor discriminant features are extracted using Fisherface to reduce feature vector with good recognition rate. Table 1 gives state of art study of Gabor filter method in the field of face recognition.

## 3 Methodology

Automatic face detection is the very first step in many applications like face recognition, face retrieval, face tracking, surveillance, HCI, etc. The human face is difficult to detect due to changes in facial expression, beard, moustache, glasses, scales, lightening conditions and orientation of face.

**Table 1** Literature Survey Summary

| SrNo. | Author | Year | Feature Extraction | Feature Selection | Classification | Remarks |
|---|---|---|---|---|---|---|
| 1 | Abhishree et al. [21] | 2015 | Gabor | Binary Particle Swarm | Euclidean Distance | Robust system. Improved speed of recognition. |
| 2 | Sekhon et al. [17] | 2015 | Database Face Value | - | Neural Network | Back propagation network takes time to train a system. |
| 3 | Chelali et al. [16] | 2014 | Gabor & DWT | - | Neural Network | High training and recognition speed. High recognition rate |
| 4 | Juan et al. [22] | 2013 | Gabor | Non-linear quantization | SVM | Private face verification system on server without any interaction |
| 5 | Ragul et al.[14] | 2013 | Gabor | PCA & LDA | Neural Network & SVM | Reduces databse size & robust recognition. |
| 6 | Nguyen et al. [15] | 2012 | Non-linear Gabor | PCA-LDA | Hamming Distance | Efficient due to use of non-linear features |
| 7 | Lin et al. [23] | 2012 | Gabor Wavelet | Orthogonal locality preserving projection | nearest neighbor classifier | Better recognition rate with added system complexity. |
| 8 | Serrano et al. [24] | 2011 | Gabor Filter | - | Neural Network | Recognition rate can be increased with 6 frequencies and $\sigma = 1$ |
| 9 | Tudor Barbu [25] | 2010 | 2D Gabor | - | Minimum average distance classifier | High recognition rate. |
| 10 | Wang et al. [26] | 2009 | Gabor | 2D PCA | SVM | Robust method. SVM is better than NN. |

## *3.1 Haar Feature Based Face Detection*

Haar face detection is popular method for real-time applications proposed by Viola and Jones [7]. The detection technique represents shape of an object in terms of wavelet coefficients [27]. The four variance based Haar features are shown in Figure 2. To speed up the process of haar feature computation integral image representation is used [28, 29]. Variance of random variable $X$ is given as follows:

$$Var(X) = E(X^2) - \mu^2 \tag{1}$$

where $E(X^2)$ is expected value of squared of $X$ and $\mu$ is expected value of $X$.

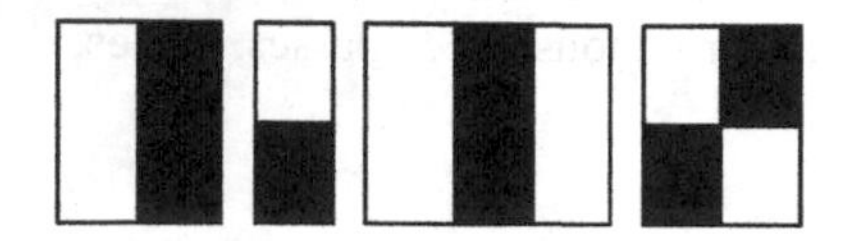

**Fig. 2** Haar basis feature

The value of a rectangle feature is computed as the difference between sum of variance values in white region and sum of variance values in dark region. These features are computed using integral image given by Eq. 2 and Eq. 3 respectively [30].

$$I(x,y) = \sum_{x=1}^{m}\sum_{y=1}^{n} f(x,y) \tag{2}$$

$$I^2(x,y) = \sum_{x=1}^{m}\sum_{y=1}^{n} f^2(x,y) \tag{3}$$

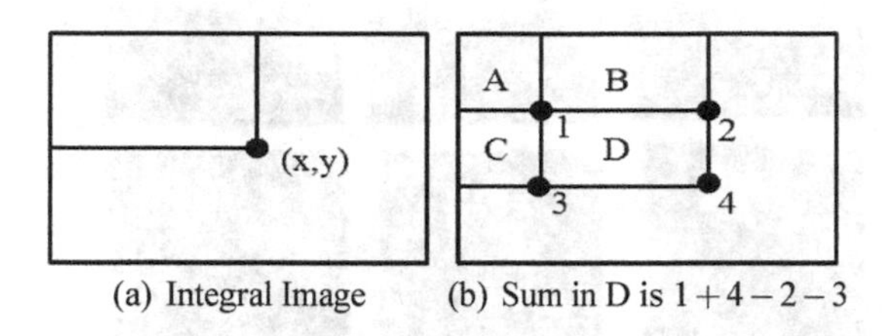


**Fig. 3** Integral image for haar features

Integral image is shown in Fig. 3. The value of the integral image at any location is the sum of the pixels values to left and above it. The sum of pixels within region $D$ can be computed as $1+4-2-3$. The feature values are computed using equations below.

$$\mu = \frac{1}{N}(I_1 + I_4 - I_2 - I_3) \tag{4}$$

$$E\left(f(x,y)^2\right) = \frac{1}{N}(I_1^2 + I_4^2 - I_2^2 - I_3^2) \tag{5}$$

$$Var(f(x,y)) = E\left((x,y)^2\right) - \mu^2 \tag{6}$$

where $N$ is number of elements within region $D$.

Each feature is given to Haar classifier to detect face region. Figure 4 shows face detection using Haar feature extraction.

Haar based method is capable of detecting multiple faces in image. This method is not robust to face pose variation. It shows good results in illumination variations.

## 3.2 Face Identification using Gabor Features

Gabor filter has response similar to bandpass filter. The characteristics of Gabor filter has made it popular for texture representation and discrimination. In spatial domain, 2D Gabor filter is viewed as Gaussian kernel function modulated by a complex

**Fig. 4** Face detection using Haar Features

sinusoidal plane wave. Gabor filter enhances the objects in image that has frequency and orientation similar to filter properties. Gabor filter is represented by [13]:

$$\Psi_{\omega,\theta}(x,y) = \frac{1}{2\pi\sigma^2} exp\left(-\frac{x'^2+y'^2}{2\sigma^2}\right) exp\left(j\omega x'\right) \tag{7}$$

$$x' = xcos\theta + ysin\theta, y' = -xsin\theta + ycos\theta \tag{8}$$

where $\omega$ is angular frequency of complex sinusoidal plane wave, $\theta$ is orientation of the Gabor filter and $\sigma$ represents sharpness of the Gaussian function along $x$ and $y$ directions. The relationship between $\sigma$ and $\omega$ is given as $\sigma \approx \frac{\pi}{\omega}$.

A set of Gabor filters with different frequencies and orientations which form a Gabor filter bank is used to extract useful features from an face image. The Gabor filter bank with 5 frequencies and 8 orientations is shown in Fig. 5. Eqs. 9 and 10 presents values for different scales and orientations.

$$\omega_u = \frac{\pi}{2} \times \sqrt{2}^u, u = 0, 1, \ldots, 4 \tag{9}$$

$$\theta_v = \frac{\pi}{8} \times v, v = 0, 1, \ldots, 7 \tag{10}$$

The face image is convolved with Gabor filter(Eq. 11) to extract features with different scales and orientations.

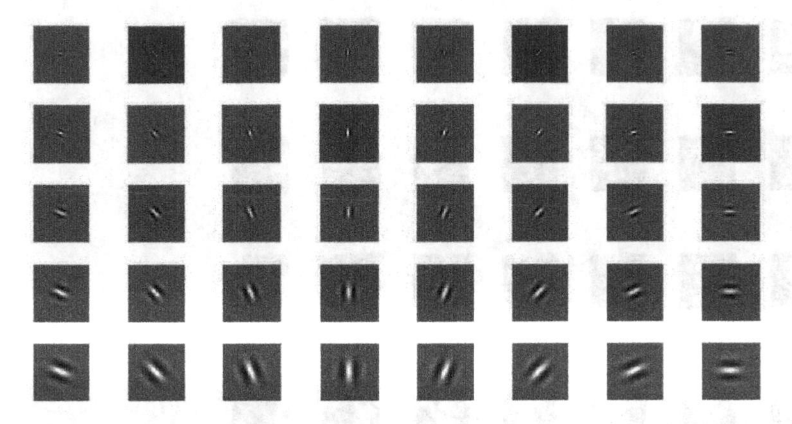

**Fig. 5** Gabor filters at 5 scales and 8 orientations

$$G_{u,v}(x,y) = I(x,y) * \Psi_{\omega_u,\theta_v}(x,y) \tag{11}$$

where $G_{u,v}(x,y)$ is convolution result for Gabor filter at scale $u$ and orientation $v$. Fig. 7 shows magnitude and Fig. 8 shows real part of the convolution results of a face image (Fig. 6) with Gabor filters in Fig. 5.

**Fig. 6** Face Image

Every image $I(x,y)$ is represented as set of Gabor coefficients $G_{u,v}(x,y)|\ u = 0,1,\ldots,4;\ v = 0,1,\ldots,7$. A feature vector is constructed by reducing dimensionality reduction. The size of the resulting vector is large i.e it contains large number of features.

In recognition step the unknown face is convolved with the Gabor filter. This feature vector is then compared with all the face vectors in training set (database). The Euclidean is calculated to find distance of query image with the images in database. The vector which has minimum euclidean distance is said to have best match. The recognition time is more due to large number of features in vectors.

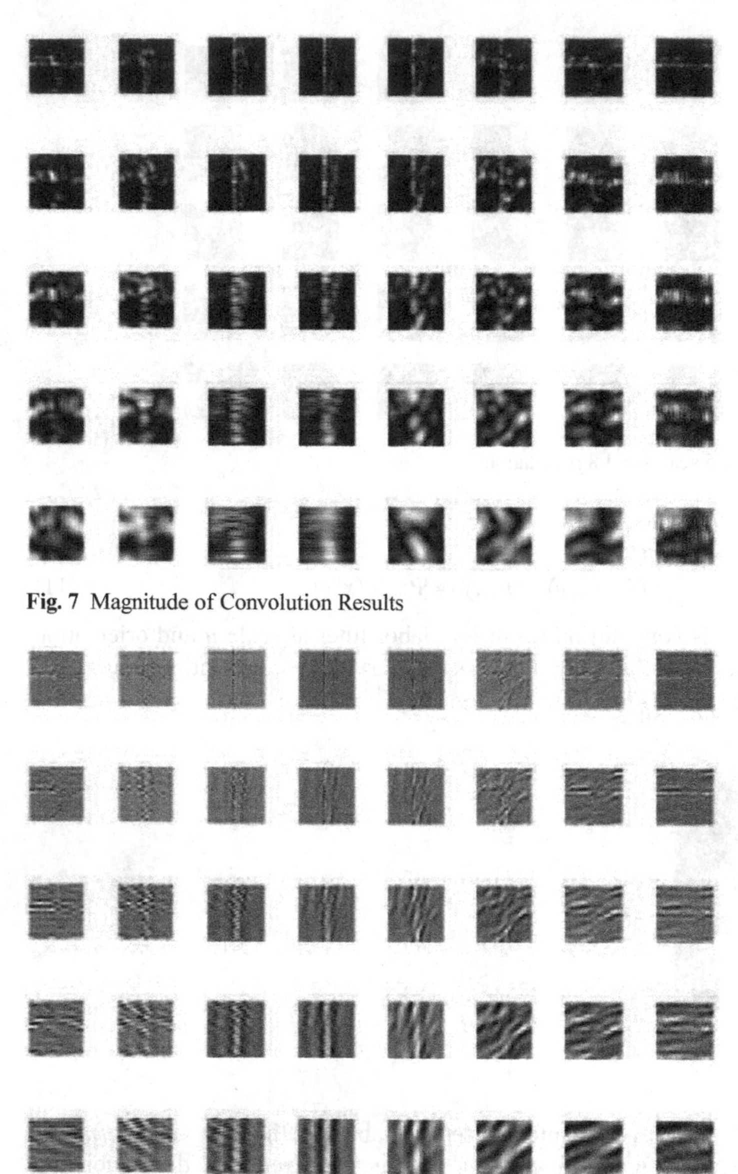

**Fig. 7** Magnitude of Convolution Results

**Fig. 8** Real Part of Convolution Results

# 4 Results

The algorithms are applied on video sequences and following results are obtained (Tab. 2). It is observed that haar based face detection algorithm requires less computation time and has good detection rate for frontal face. Recognition time required is large due to large number of features in feature vector. Out of total 6535 frames of all 21 videos face was detected in 4740 frames. Out of 4740 faces correct recognition was obtained in 1844 frames. The Gabor filter is capable of representing face efficiently with pose variations.

**Table 2** Performance Evaluation

| Video | Recognition Time | No. Frames | of Face Detection | True Detection | False Detection |
|---|---|---|---|---|---|
| 1 | 28.5168 | 237 | 71 | 37 | 34 |
| 2 | 40.03501 | 329 | 99 | 31 | 68 |
| 3 | 88.74552 | 257 | 223 | 98 | 125 |
| 4 | 90.22506 | 339 | 225 | 138 | 87 |
| 5 | 122.0239 | 448 | 305 | 7 | 298 |
| 6 | 112.9269 | 438 | 283 | 4 | 279 |
| 7 | 112.8361 | 353 | 283 | 96 | 187 |
| 8 | 112.7831 | 404 | 283 | 195 | 88 |
| 9 | 77.04045 | 198 | 191 | 2 | 189 |
| 10 | 81.87446 | 248 | 206 | 19 | 187 |
| 11 | 31.17757 | 78 | 78 | 65 | 13 |
| 12 | 50.01316 | 128 | 124 | 74 | 50 |
| 13 | 106.4006 | 324 | 252 | 224 | 28 |
| 14 | 102.756 | 353 | 252 | 216 | 36 |
| 15 | 72.21552 | 258 | 176 | 82 | 94 |
| 16 | 78.93236 | 328 | 191 | 95 | 96 |
| 17 | 99.17065 | 346 | 238 | 108 | 130 |
| 18 | 162.7538 | 426 | 392 | 289 | 103 |
| 19 | 161.4159 | 318 | 392 | 48 | 344 |
| 20 | 162.8405 | 388 | 392 | 16 | 376 |
| 21 | 34.63272 | 337 | 84 | 0 | 84 |

## 5 Conclusions

The proposed system will detect the facial features of human being and verify the identity of a person. Haar based face detection is efficient face detection algorithm. It is used for real-time applications. Gabor face recognition is better than many other methods available in the field. It compares large number of features hence requires more time for recognition. Reducing feature set may reduce computation time with small decrease in accuracy of system. With use of discriminant feature representation the recognition rate can be increased with less computation time. Further improvement in computational complexity will reduce evaluation time that can help to use this system in real time applications.

## References


1. W. R. Tan, C. S. Chan, P. Yogarajah, and J. Condell, "A fusion approach for efficient human skin detection," *IEEE Transactions on Industrial Informatics*, vol. 8, no. 1, pp. 138–147, February 2012.
2. I. Hemdan, S. Karungaru, and K. Terada, "Facial features-based method for human tracking," in *17th Korea-Japan Joint Workshop on Frontiers of Computer Vision*, 2011, pp. 1–4.
3. Y. N. Chae, J. Chung, and H. S. Yang, "Color filtering-based efficient face detection," in *19th International Conference on Pattern Recognition*, Dec 2008, pp. 1–4.
4. M. H. Yap, H. Ugail, R. Zwiggelaar, and B. Rajoub, "Facial image processing for facial analysis," in *IEEE International Carnahan Conference on Security Technology*, Oct 2010, pp. 198–204.
5. R. Sarkar, S. Bakshi, and P. K. Sa, "A real-time model for multiple human face tracking from low-resolution surveillance videos," *Procedia Technology*, vol. 6, pp. 1004 – 1010, 2012.
6. B. Dhivakar, C. Sridevi, S. Selvakumar, and P. Guhan, "Face detection and recognition using skin color," in *3rd International Conference on Signal Processing, Communication and Networking*, March 2015, pp. 1–7.
7. P. Viola and M. J. Jones, "Robust real-time face detection," *International Journal of Computer Vision*, vol. 57, no. 2, pp. 137–154, May 2004.
8. R. Mehta, J. Yuan, and K. Egiazarian, "Face recognition using scale-adaptive directional and textural features," *Pattern Recognition*, vol. 47, no. 5, pp. 1846–1858, May 2014.
9. K. Park, R. Park, and Y. Kim, "Face detection using the 3 × 3 block rank patterns of gradient magnitude images and a geometrical face model," in *IEEE International Conference on Consumer Electronics*, 2011, pp. 793–794.
10. Z.-R. Lai, D.-Q. Dai, C.-X. Ren, and K.-K. Huang, "Multiscale logarithm difference edgemaps for face recognition against varying lighting conditions," *IEEE Transactions on Image Processing*, vol. 24, no. 6, pp. 1735–1747, June 2015.
11. R. Xiao, M. Li, and H. Zhang, "Robust multipose face detection in images," *IEEE Transactions On Circuits And Systems For Video Technology*, vol. 14, no. 1, pp. 31–41, 2004.
12. L. Nanni, A. Lumini, F. Dominio, and P. Zanuttigh, "Effective and precise face detection based on color and depth data," *Applied Computing and Informatics*, vol. 10, pp. 1–13, 2014.
13. S. Meshgini, A. Aghagolzadeh, and H. Seyedarabi, "Face recognition using gabor-based direct linear discriminant analysis and support vector machine," *Computers and Electrical Engineering*, vol. 39, no. 910, p. 727745, 2013.
14. G. Ragul, C. MageshKumar, R. Thiyagarajan, and R. Mohan, "Comparative study of statistical models and classifiers in face recognition," in *International Conference on Information Communication and Embedded Systems*, Feb 2013, pp. 623–628.
15. K. Nguyen, S. Sridharan, S. Denman, and C. Fookes, "Feature-domain super-resolution framework for gabor-based face and iris recognition," in *IEEE Conference on Computer Vision and Pattern Recognition*, June 2012, pp. 2642–2649.
16. F. Z. CHELALI and A. DJERADI, "Face recognition system using neural network with gabor and discrete wavelet transform parameterization," in *International Conference of Soft Computing and Pattern Recognition*, 2014, pp. 17–24.
17. A. Sekhon and P. Agarwal, "Face recognition using back propagation neural network technique," in *International Conference on Advances in Computer Engineering and Applications*, 2015, pp. 226–230.
18. A. Gangwar and A. Joshi, "Local gabor rank pattern (lgrp): A novel descriptor for face representation and recognition," in *IEEE International Workshop on Information Forensics and Security (WIFS)*, Nov 2015, pp. 1–6.
19. C. Wang, Z. Chai, and Z. Sun, "Face recognition using histogram of co-occurrence gabor phase patterns," in *2013 IEEE International Conference on Image Processing*, Sept 2013, pp. 2777–2781.

20. X. Y. Jing, H. Chang, S. Li, Y. F. Yao, Q. Liu, L. S. Bian, J. Y. Man, and C. Wang, "Face recognition based on a gabor-2dfisherface approach with selecting 2d gabor principal components and discriminant vectors," in *3rd International Conference on Genetic and Evolutionary Computing*, Oct 2009, pp. 565–568.
21. T. M. Abhishree, J. Latha, and K. M. S. Ramchandran, "Face recognition using gabor filter based extraction with anisotropic diffusion as a pre-processing technique," in *International Conference of Advanced Computing Technologies and Applications*, 2015, pp. 312–321.
22. J. R. Troncoso-Pastoriza, D. Gonzlez-Jimnez, and F. Prez-Gonzlez, "Fully private noninteractive face verification," *IEEE Transactions on Information Forensics and Security*, vol. 8, no. 7, pp. 1101–1114, July 2013.
23. G. Lin and M. Xie, "A face recognition algorithm using gabor wavelet and orthogonal locality preserving projection," in *International Conference on Computational Problem-Solving*, Oct 2012, pp. 320–324.
24. ngel Serrano, I. M. de Diego, C. Conde, and E. Cabello, "Analysis of variance of gabor filter banks parameters for optimal face recognition," *Pattern Recognition Letters*, vol. 32, no. 15, pp. 1998 – 2008, 2011.
25. T. Barbu, "Gabor filter-based face recognition technique," *Proceedings of the Romanian Academy*, vol. 11, no. 3, pp. 277–283, 2010.
26. X. M. Wang, C. Huang, G. Y. Ni, and J. g. Liu, "Face recognition based on face gabor image and svm," in *2nd International Congress on Image and Signal Processing*, Oct 2009, pp. 1–4.
27. S. Gundimada and V. Asari, "Face detection technique based on rotation invariant wavelet features," in *Proceedings of International Conference on Information Technology: Coding and Computing*, vol. 2, April 2004, pp. 157–158.
28. L. Zhang and Y. Liang, "A fast method of face detection in video images," *2nd International Conference on Advanced Computer Control ICACC*, vol. 4, pp. 490–494, 2010.
29. P. K. Anumol, B. Jose, L. D. Dinu, J. John, and G. Sabarinath, "Implementation and optimization of embedded face detection system," in *Proceedings of 2011 International Conference on Signal Processing, Communication, Computing and Networking Technologies*, 2011, pp. 250–253.
30. C. N. Khac, J. H. Park, and H. Jung, "Face detection using variance based haar-like feature and svm," *World Academy of Science, Engineering and Technology*, vol. 60, pp. 165–168, 2009.

# Convolutional Neural Networks based Method for Improving Facial Expression Recognition

Tarik A Rashid

## Abstract

*Recognizing facial expressions via algorithms has been a problematic mission among researchers from fields of science. Numerous methods of emotion recognition were previously proposed based on one scheme using one data set or using the data set as it is collected to evaluate the system without performing extra preprocessing steps such as data balancing process that is needed to enhance the generalization and increase the accuracy of the system. In this paper, a technique for recognizing facial expressions using different imbalanced data sets of facial expression is presented. The data is preprocessed, then, balanced, next, a technique for extracting significant features of face is implemented. Finally, the significant features are used as inputs to a classifier model. Four main classifier models are selected, namely; Decision Tree (DT), Multi-Layer Perceptron (MLP) and Convolutional Neural Network (CNN). The Convolutional Neural Network is determined to produce the best recognition accuracy.*

***Keyword: Facial Behaviors Recognition, Convolutional Neural Networks, Human Computer Interaction***

## 1. Introduction

Conveying strong and communicative feelings can be best carried out by a person via various manners, namely; physique, feeling, attitude, movement, natural sings and etc. Clearly, there are some vital corporal organs that constitute the face of a person. These are nose, mouth, ears, and eyes. Through these corporal organs a person is facilitated to sense taste, whiff, perceive, view and distinguish scenes such as objects and places. Additionally, different signs can be outputted through some vital feelings. On the whole, interactions between human beings can be categorized into vocalized and non-vocalized action messages. Obviously, vocalized action messages are fallen into vocal movements, whereas non vocal action messages are fallen into engagements of body and functional responses. Experimentally speaking, the emotion of a listener in relation to whether he enjoys or hates what he gets is only seven percentage, this is subjected to the words orally articulated, and thirty eight percentage on vocal utterances. It is also indicated that facial expressions inspire this emotion to fifty five percentage [1]. It is commonly known that human beings' social communications can be determined highly by their direct person to person' facial expressions. It was suggested by some of the top logicians in [2] that discrepancies of nations' culture are regarded as key rea-



J.M. Corchado Rodriguez et al. (eds.), *Intelligent Systems Technologies and Applications 2016*, Advances in Intelligent Systems and Computing 530,
DOI 10.1007/978-3-319-47952-1_6

sons for determining discrepancies of facial expressions. Based on Darwin's studies, it was concluded that facial behavior can be categorized into wide-ranging sets. In 1872, he endorsed in his evolutionary notion the most common facial behaviors. Darwin's footsteps were followed by other logicians, as they equally revealed common reactions and facial behaviors, despite the fact that each logician acclaimed an exclusive hypothetical basis for his expectation. For the most part, over the last twenty years, the area of human computer interaction or HCI has technologically been advanced and played a major part in developing the field of computer science by engendering various practices and applications that connect human beings behavior with intelligence that can be build inside computer devices. Accordingly, developing this sort of relation has fascinated many researchers and scholars. One could say that certainty of comprehending the capabilities of computers to identify different facial expression is the most imperative reason or motive for the rapid development in this filed.

This research uses convolutional neural network to classify facial emotions and it adds three major contributions such as: 1) implementing an approach that can recognize facial expression in static images. 2) Two thoroughly facial data sets are used to evaluate the suggested technique. 3) Also, increasing recognition accuracy via using various extracted feature sets. The rest of this work of research can be arranged in the following sections: Previous research works are outlined in section two, then, the proposed method is explained in section three. Afterwards, analytical outcomes are established in detail, and finally, the key points are outlined.

## 2. Related Works

In 2005, a new method is led in [3] for detecting feelings via frontal facial images. They separated the face for the purpose of feature extraction into feature sections namely: eyes, mouth, and secondary. They extracted features via geometrical matching and shaping data for every section. They utilized fuzzy filtering based approach in their work, for color pictures along with analysis of graphical representation for extracting the face section. Depending on virtually face modeling via using graphical representations, the elements of the face are extracted. The Linear Matrix Inequality optimization method is acted as a fuzzy classifier for classifying pictures. Five classes of emotional feelings are used for the classification task. In 2010, an LPT visual descriptor technique is implemented for feature extraction in facial pictures, then, inference system based Neuro-Fuzzy approach is implemented for classification to decide among five facial expressions [4]. In [5], in 2010, a recognition system for recognizing gestures of human is developed. The system was mainly based on a classification technique explained by the inventor of AddBoost for detecting the face of a human. Classification task is conducted through neural networks on various gesture pictures. In 2012, in [6] both recognizing gestures of hand and face as two classifiers via the multimodal method are cautiously considered. Therefore, another classifier is used to produce each classifier's outcome which is considered as a subsequent expression. Features of face are extracted via Ekmanand ideal model. Bayesian model is used for categorizing gestures of hand and face in the data set. Bayesian model is then accepting features from both systems and performing task.

In this research paper, a technique for recognizing facial expressions with different imbalanced data sets of facial expression is presented, the data is pre-processed and balanced using SMOTE, then, an extraction technique for facial feature is used. Finally, three classification models are used namely; Decision Tree, MLP and Convolutional Neural Network.

## 3. Framework

Figure 1. Illustrates the suggested framework of emotion recognition. Detail descriptions of the suggested framework are described below:-

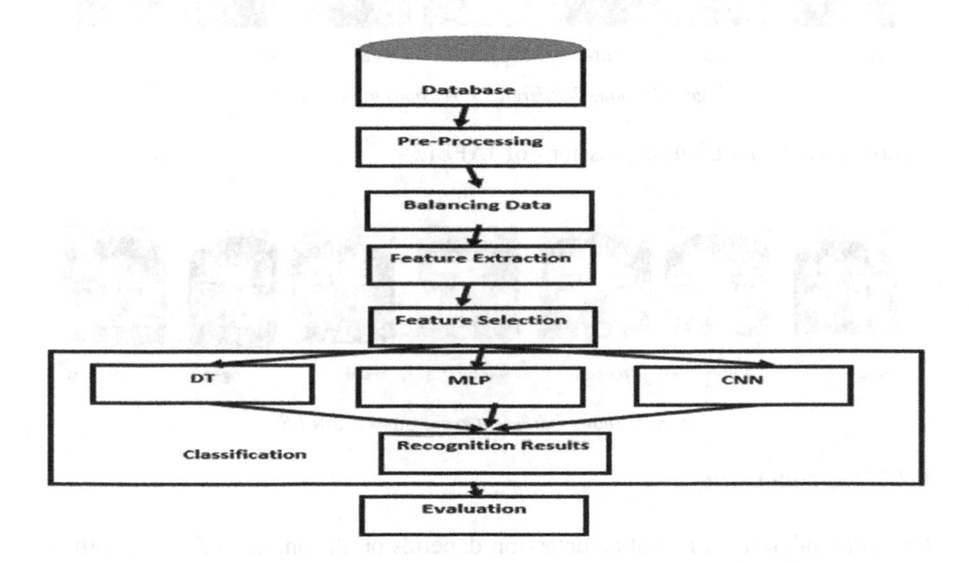


*Figure 1: The emotion recognition system structure*

### 3.1 Facial Database

Basically, the step of collecting suitable images is considered as the core phase in any recognition system, simply because the outcomes of recognition system are depended greatly on the explanation of pictures. This work of research has considered a number of databases and included a number of measures for evaluating these databases. Examples of these measuring points are: database size, spectacles presence; color of skin, the presence of hair in face (beard and moustache). The databases of JAFFE and Bosphorus are considered as the most suitable databases in this work (See Table 1).

*Table 1: Facial Database Specifications*

| Criteria | Bosphorus | JAFFE |
|---|---|---|
| **Subjects** | 105 | 10 |
| **Samples** | 558 | 213 |
| **No. of Expressions** | 7 | 7 |
| **Color** | Color | Grayscale |
| **Culture** | Different Culture | Japanese |

Figure 2, shows 7 various expressions in Bosphorus.

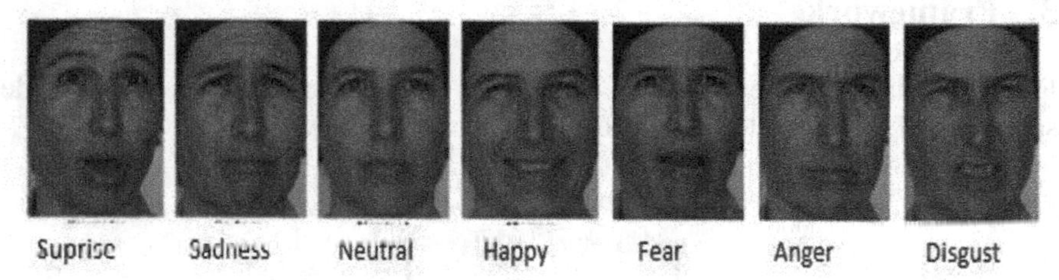


*Figure 2: Shows 7 different expressions in Bosphorus.*

Figure 3, shows 7 various expressions of JAFFE.

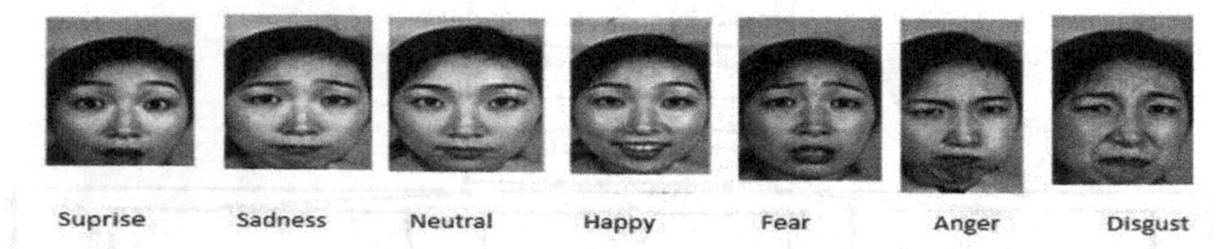


*Figure 3: Shows 7 different expressions in JAFFE*

## 3.2 Preprocessing

The recognition rate of emotion detection depends on the amount of pixels within images, for that reason, the size of all input images must be identical in the emotion recognition system, and above all, during the recognition stage. It is also important to note that the size of images must be constant for each of the databases. It was found that that the JAFFE database does not require to perform any sifting on its images, this is because all the pictures have a fixed size 256 X 256 pixel. On the other hand, for Bosphorus, a conversion procedure is needed to consistently convert all the pictures, thus, every image is converted to 470 X 384 pixel.

## 3.3 Balancing Data

Imbalanced dataset is one of the major problems which acts as a hinder issue for generalization. A dataset is considered to be an imbalanced dataset if classification labels are not spread similarly. There are several techniques that have ability to tackle problems concerned with imbalanced data, these techniques are namely; Down-sizing, resampling and recognition-based [9] and Synthetic Minority Over-sampling TEchnique (SMOTE) [10]. In this research work, SMOTE is used to balance the dataset. SMOTE generates synthetic instances from the minority class so that to rise the number and approximately balance the majority class. This method is taking k nearest neighbor for each instance in minority class and over-samples the class through generating synthetic instances. Basically, synthetic instances can be produced via finding the difference between both an instance and the nearest neighbor of that instance. The technique will help a decision area for the minority class to be generalized and this will increase accuracy rate of the whole system [10]. Let us assume that there is an instance $in_c$ (9, 6) to be considered and let $in_n$ (6, 4) be its nearest neighbor. Thus, the SMOTE process can be expressed as follows:-

$$
\begin{aligned}
& for\, in_c(fea_{1,1}, fea_{1,2}) \\
& fea_{1,1} = 9,\ fea_{1,2} = 6 \\
& Diff_1 = fea_{1,2} - fea_{1,1} \\
& = -3 \\
& for\, in_n(fea_{2,1}, fea_{2,2}) \\
& fea_{2,1} = 6,\ fea_{2,2} = 4 \\
& Diff_2 = fea_{2,2} - fea_{2,1} \\
& = -2 \\
& Sothenewinstnaceis: \\
& in_{new} = in_c + rand(0-1) * (Diff_1, Diff_2) \\
& = (9,6) + rand(0-1) * (-3,-2)
\end{aligned} \tag{1}
$$

The output between 0 and 1 is produced via using the *rand* operation. The pseudo code (which is taken from reference [10]) will give a wider view to the SMOTE technique. Table 2 shows facial databases after applying the SMOTE Process. The number of instances in each dataset is increased. Three different class models are used based on the number of expression. These are; seven classes, six classes and five classes.

*Table 2: Facial Database Specifications*

| Classes | Bosphorus | JAFFE |
|---|---|---|
| seven classes: all classes are included | 1494 | 1576 |
| six classes: all classes are included, except for fear class | 1508 | 1680 |
| five classes: all classes are included, except for fear and sadness classes | 1482 | 1672 |

## 3.4 Facial Feature Extraction

The feature extraction approaches of previous works were mainly linked to key facial features that have great impact on recognition of key parts of the face. Ultimately, these parts are normally regarded as parts of interest. The basic idea of the feature extraction technique is to extract pictured features from main facial parts

(1-mouth 2-eyes, 3-secondary). Accordingly, symmetrical data can be filtered from these three parts. This process can be carried out via two separate stages namely extracting and detecting facial features. In this work, Luxand Face SDK library is used. It is considered as a suitable library in the area of face recognition, simply because it has numerous physical appearance such as including a direct integration into applications, and backing up many different programming languages [11, 12]. In this work, only 66 points or features were detected via the above mentioned library. Only 27 features are found important than others. Figures 4 and 5 show selected features in both databases via the library.

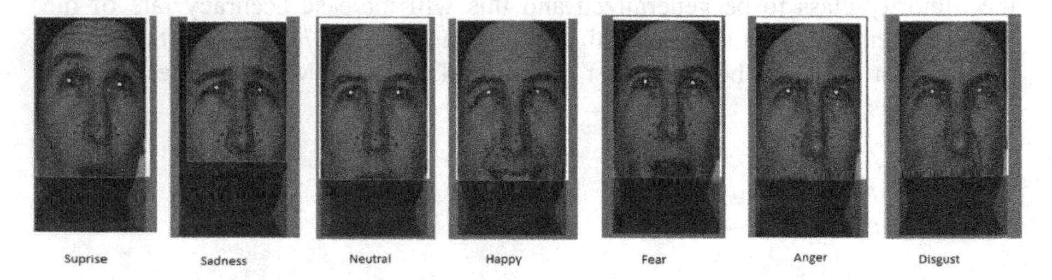


*Figure 4: Features found by the library in Bosphorus Database*

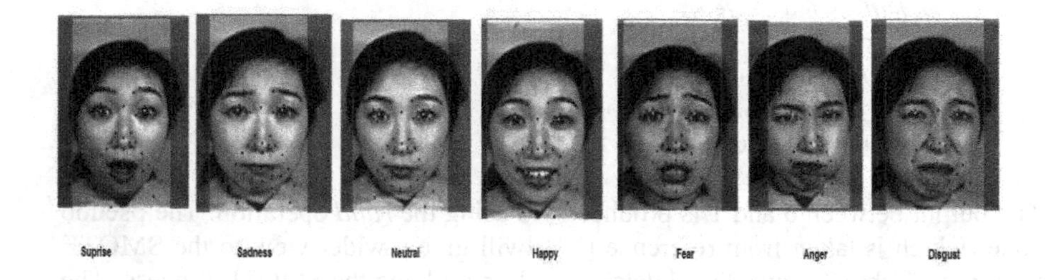


*Figure 5: Features found by the library in JAFEE Database*

Additionally, finding the final facial feature is the second step. This step can be realized through the void that exists in between two similar selected features. Consequently, the void can be determined via the Euclidian Distance approach. Since images are going to be used for experiments, the measurements are going to be in pixels. If *p (x1, y1)* and *q (x2, y2)* then the distance is given by:-

$$D(p,q) = \sqrt{(x_1 - x_2)^2 + (y_1 - y_2)^2} \tag{2}$$

Recognizing facial features from a number of combinations that may well be derived from these points is another vital technique which needs further research and examinations. Thus, the foremost ordinarily used features are designated [3, 13, 14, 12].

## 3.5 Feature Selection

It is clear that amalgamating a number of identical features of a face (identical features on both sides of the face) could be considered as one of the strategies for discovering helpful features. Features like the breadth of each eye could be similar. Still, the classification accuracy can be decreased by having near duplicate features. A classic response for this can be via merging these features. To do so, the typical of each pair of duplicate features can be computed. Accordingly, the number of features can be reduced from 15 features to only 10 features, having said that the classification accuracy can be decreased favorably.

## 3.6 Classification

For the experiments, different models have been chosen based on the expressions (shown in Table 2) included in each model described below.

### *3.6.1 Multi-Layer Perceptron*

In this research work, the MLP for the classification of emotions is experimented. The structure of MLP involves three or more layers (input, hidden or processing and output). Each layer in the network has several neurons. Neurons are connected via a link known as a synapse or a weight connection. Each weight connection carries a particular weight value. The number of neurons in each layer is set to unique values [14, 15].

### *3.6.2 Decision Tree*

Decision Tree is a technique that assigns the value of a class to an instance subject to the attributes values of that instance [16]. Basically, the decision tree defines the learning algorithm. To improve human readability for any learned trees, sets of 'if –then' rules can be considered. The technique is used to tackle problems such as medical diagnosing, forecasting loan risks, text classification, web page detection, etc. The structure of decision tree can be shown in Figure 6. The tree involved nodes and leaves. The value of the class can be identified by a leaf, whereas, a node can test an attribute [16, 17]. The details of the C 4.5 algorithm and its parameters are explained in the following sub sections.

#### *A. C4.5 theory and algorithm*

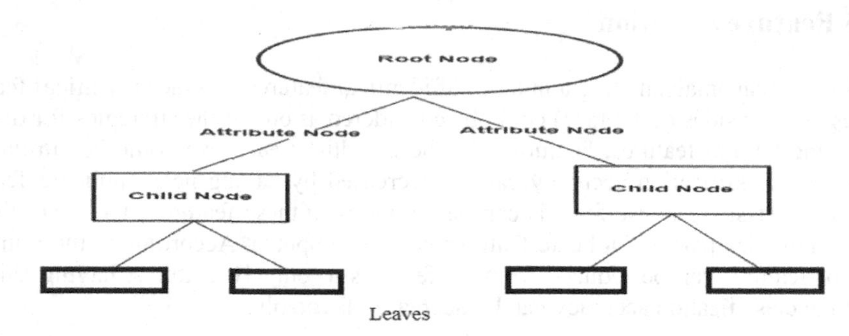


*Figure 6: Structure of c4.5 tree*

C4.5 technique was first coined by Quinlan in 1993. It is basically dependent on the concepts of Shannon theories, which are known as Entropy of Shannon. The technique can determine about information that can be generated via an occurrence [16, 17].

### B. Shannon entropy

Supposing there is $q$ instance, and distribution of possibility is represented by $P = (p1, p2,..., pn)$. The information that contained within this distribution is known as entropy. The entropy can be expressed as follows [16, 17]:

$$Entropy(p) = -\sum_{i=1}^{n} p_i \times \log(p_i) \tag{3}$$

### C. The gain information G (p, T)

Entropy is very valuable and significant not only for determining the classes mixing degree but also is used to determine the tree positions. Gain of a test $T$ attribute and a position $P$ at the current node can be expressed by [16, 17]:-

$$Gain(p,T) = Entropy(p) - \sum_{j=1}^{n} (p_j \times Entropy(p_j)) \tag{4}$$

The attribute $T$ can have some possible values represented by $(p_j)$. This is a way for attributes ranking and constructing the decision tree [16, 17].

### 3.6.3 Convolution Neural Network

CNN is considered as another category of standard neural network [18-20]. The network first was coined by Fukushima in 1980 and then developed by LeCun et al. It is constructive to integrate previous knowledge into structural design of the network so that to obtain better generalization. CNNs usually can spend less time on pre-processing. They intend at using spatial information among data elements. The main properties of CNNs are as follows: CNNs have local receptive fields; CNNs do contain shared weights; and they also have spatial sub sampling to guarantee degree of distortion invariance, scale, and shift. CNNs have two main concepts; Sparse Connectivity (SC) and Shared Weights (SW). They are described as follows:-

a) SC: This can be expressed by feeding inputs into neurons in *m* hidden layer from a subset of neurons at *m-1* layer that holds a spatial neighboring receptive fields. Thus, the behavior of receptive fields can be generated and therefore, guarantees that 'Filters' yield a robust response to local patterns at inputs [18-20]. This process can be expressed in Figure 7, a.

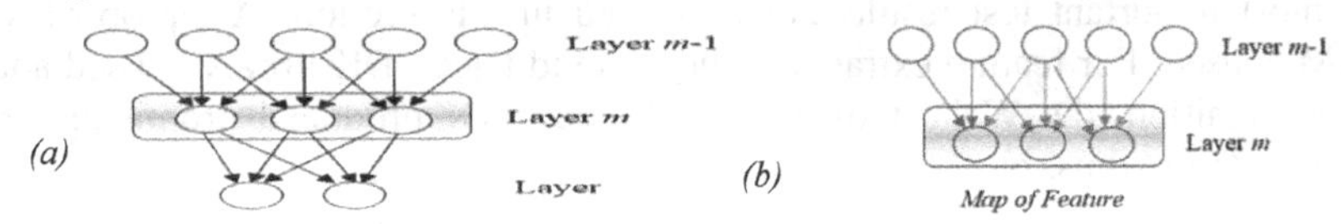


*Figure 7: (a), Shows served inputs into neurons and (b), shows map of feature*

b) SW: In CNNs, filters *'hi'* can be replicated. So, this means that identical bias vector and weight matrix are shared by these replicated filters which would form a map of feature that can be applied through the whole input data. The process of weight replication does the following actions [18-20]: increases effectiveness of learning; decreases estimated parameters; and it helps the network to learn location invariant feature detection.

Figure 7, (b), shows three hidden neurons that are part of the same map of feature. Clearly, weights with the identical color are sharing the same constraints. CNNs use particular operations that help detect input image features with high level, these operations are convolution and subsampling [18-20]:-

a) Convolution: This is the supreme operation in the network. The repeated application of a function across regions of the whole data set produces the map of feature, this means, input data convolution via a linear filter is used, a bias term is added and then non-linear functions are applied. Assuming if there is the $k^{th}$ map of feature at a $h^k_{ij}$, given layer, thus, filters of this map of feature can be decided by $W_k$ , $b_k$, weights and bias respectively, then the map of feature map $h^k$ can be calculated via the following non-linear equation [18-20].

$$h^k_{ij} = \tanh\left(\left(w^k * x\right)_{ij} + b_k\right) \tag{5}$$

It is desired to have each hidden layer poised by multi-map of feature $\{h^k_{ij}, k = 0.....K\}$, so that to create a wealthier representation of the data. *W*, is the hidden layer weights which are represented with 4D tensor in which features for every mixture of map of feature destination, map of feature source, vertical position source, and horizontal position source are contained. *B* is the biases which are represented as vectors in which one single element for each map of feature destination is contained [18-20].

b) Sub Sampling: Max pooling is widely used for this process. The map of feature is divided by max pooling into a group of sub regions. These sub regions are not overlapping and the maximum value can be formed in each sub region. The maximum value will help reduce computation at the upper layers. Besides, it also be responsible for some translation invariance procedures.

## 3.7 Simulations and Evaluations

Although several tests have been conducted throughout this research work, but, the most important test results are described in this section. A laptop of 4 GB RAM is used. For feature extraction, the Luxand Face SDK library is used and for the recognition, the Weka tool is used. Since CNN produced promising results among others, thus, only CNN's parameters are explained such as; input features (IF), Output Class (C), Number of Hidden Layer Neurons (HN) , Learning Rate (LR), Input Layer Dropout Rate (IDR), Hidden Layers Dropout Rate (HDR), Maximum Iterations, Weight Penalty (WP), see Table 3. Table 4, demonstrates brief experimental outcomes for the used models DT, MLP and CNN. The test results of the JAFFE database for 7 classes model are 89.29, 89.88, 94.56 respectively, whereas the test results of an equivalent model of the Bosphorus database are 76.71, 80.02, 82.79 respectively. A major reason behind this result is that the JAFFE database provider prefers to exclude the 'Fear' class since they believe that the expressions in the pictures are not clear. Also, some scientific literature claims that 'fear' may be processed differently from the other basic facial expressions. It is also obvious that the classification accuracy of the recognition rate increases as the number of labels decreases, the reason behind this is that the feature values will be less intrusive of each other. Thus, the classes will be highly segregated.

*Table 3: CNN Classifier Parameters for Bosphorus Database*

| DB | IF | O | HN | L | IDR | HDR | WP | MI |
|---|---|---|---|---|---|---|---|---|
| **Bosphorus** | F1-F15 | 1,2,3,4,5,6,7 | 100 | 0.9 | 0.09 | 0.5 | 1.00E-08 | 1000 |
| | F1-F15 | 1,2,4,5,6,7 | 100 | 0.9 | 0.09 | 0.5 | 1.00E-08 | 900 |
| | F1-F15 | 1,2,4,5,7 | 90 | 0.9 | 0.001 | 0.4 | 1.00E-08 | 700 |
| **JAFFE** | F1-F15 | 1,2,3,4,5,6,7 | 100 | 0.7 | 0.09 | 0.09 | 1.00E-10 | 2000 |
| | F1-F15 | 1,2,4,5,6,7 | 80 | 0.8 | 0.09 | 0.01 | 1.00E-04 | 800 |
| | F1-F15 | 1,2,4,5,7 | 60 | 0.9 | 0.09 | 0.01 | 1.00E-06 | 1500 |

*Table: 4 demonstrates Classification Outcomes*

| Database Sets | Models | DT | MLP | CNN | Ref [12] |
|---|---|---|---|---|---|
| **JAFFE** | 7 Class | 89.2986 | 89.8898 | 94.5631 | 65.1160 |
| | 6 Class | 91.7566 | 97.2684 | 97.6844 | 77.7778 |
| | 5 Class | 94.5772 | 97.8967 | 98.8887 | 90.0000 |
| **Bosphorus** | 7 Class | 76.7109 | 80.026 | 82.7902 | 78.3784 |
| | 6 Class | 79.5750 | 82.5455 | 89.0005 | 85.5670 |
| | 5 Class | 89.6321 | 91.7162 | 90.5762 | 89.5349 |

In addition, the lowest recognition rate as mentioned above was improved by merging duplicate features. The models of 7 class using DT for both JAFFE and Bosphorus databases provided the least results. However, the models of 5 class using CNN on both databases presented the best results. The fifth column in Table 4, shows that CNN outperformed the model in [12].

## 4. Conclusion

In this paper, the area of facial behavior recognition using computer processes is considered to establish emotional interactions between humans and computers. Two types of unbalanced data of facial expression are used. The balancing process is carried out using SMOTE technique to improve the generalization and increase the accuracy results of the system, next, features are decreased using feature extraction technique. Different classification approaches are examined to tackle fundamental problems such as using various database sets, using various sub sets per database, and using various structural models. The Convolutional Neural Network is determined to produce the best recognition accuracy.

## REFERENCES

1. Pantic, M.: *Facial Expression Analysis by Computational Intelligence Techniques*, Ph.D. Thesis, Faculty Electrical Engineering, Mathematics and Computer Science, Delft University of Technology, Delft, Nederlands, (2001).
2. Ekman, P.: *Universal Facial Expressions of Emotions*, California Mental Health Research Digest, (1940), 8(4), 151-158.
3. Hwan, M. Joo, H., Park, B.: *Emotion Detection Algorithm Using Frontal Face Image*, 12th International Conference on Computer Applications in Shipbuilding, (2005); KINTEX, Gyeonggi-Do, Korea.
4. Chatterjee, S., Shi, H.: *A Novel Neuro Fuzzy Approach to Human Emotion Determination*, International Conference on Digital Image Computing: Techniques and Applications, Sydney, Australia, (2010), 282-287.
5. Raheja, L. Kumar, U.: *Human Facial Expression Detection from Detected in Captured Image Using Back Propagation Neural Network*, International Journal of Computer Science and Information Technologies, (2010), 2(1), 116-123.
6. Metri , P., Ghorpade, J., Butalia, A.: *Facial Emotion Recognition Using Context Based Multimodal Approach*, International Journal of Emerging Sciences, (2012), 2 (1), 171-182.
7. Savran, A., Alyüz, N., Dibeklioğlu, H., Çeliktutan , O. Gökberk, B. Sankur, B. Akarun, L.: *Bosphorus Database for 3D Face Analysis*, First European Workshop on Biometrics and Identity Management, Roskilde, Denmark, (2008), 47-56.
8. Lyons, J., Akamatsu, S., Kamachi, M., Gyoba, J.: *Coding Facial Expressions with Gabor Wavelets*, Proceedings of 3rd IEEE International Con-

ference on Automatic Face and Gesture Recognition, Nara Japan, IEEE Computer Society, (1998), 200-205.
9. Japkowicz, N.: *Learning from Imbalanced Data Sets: A Comparison of Various Strategies*, AAAI Press, (2000), 10-15.
10. Nitesh V. Chawla, K.W.B., Lawrence O. Hall, W. Philip Kegelmeyer, *SMOTE: Synthetic Minority Over-sampling Technique*, Journal of Artificial Intelligence Research, (2002), 16, 321–357.
11. Luxand Inc., Luxand FaseSDK, *Detect and Recognize Faces with Luxand FaceSDK*, Available from: http://www.luxand.com/facesdk/, (Accessed: 7 May 2012).
12. Ahmed, H., Rashid, T., Sidiq, A.: *Face Behavior Recognition through Support Vector Machines*, International Journal of Advanced Computer Science and Applications, (2016), 7(1), 101-108.
13. Khandait, P. Thool, C. Khandait, D.: *Automatic Facial Feature Extraction and Expression Recognition Based on Neural Network*, International Journal of Advanced Computer Science and Applications, (2011), 2(1), 113-118.
14. Siddiqi, H. Lee, S., Lee, K., Mehmood, A. Truc, H.: *Hierarchical Recognition Scheme for Human Facial Expression Recognition Systems*, Sensors (2013), 13 (12), 16682-16713.
15. Engelbrecht, P.: *Computational Intelligence: An Introduction*, 2nd Edition, John Wiley & Sons, Ltd, Chichester, England, (2007).
16. HSSINA, B., et al.: *A comparative study of decision tree ID3 and* C4.5, International Journal of Advanced Computer Science and Applications. (2014), 4(2), 13-19.
17. Galathiya, A., A. Ganatra, and C. Bhensdadia: *Classification with an improved Decision Tree Algorithm*. International Journal of Computer Applications, (2012), 46.
18. George, N.: *Deep Neural Network Toolkit & Event Spotting in Video using DNN features*, master thesis, department of computer science and engineering, Indian institute of technology madras, (2015).
19. LISA Lab: *DeepLearning 0.1 Documentation, Convolutional Neural Networks*, Retrieved October, 20, 2015, from http://deeplearning.net/tutorial/lenet.html, (2015).
20. Stutz, D: *Understanding Convolutional Neural Networks*, (Seminar Report, Fakultät für Mathematik, Informatik und Naturwissenschaften Lehr- und Forschungsgebiet Informatik VIII Computer Vision, (2014).

# Composition of DCT-SVD Image Watermarking and Advanced Encryption Standard Technique for Still Image

Sudhanshu Suhas Gonge[1,1], Ashok Ghatol[2],

{sudhanshu1984gonge, vc_2005}@rediffmail.com

**Abstract.** Nowadays, multimedia technology is developing rapidly. It provides a best platform for multiple media like image, audio, video, text, etc.Due to continuous improvement in technology it has attracted large number of user and made presentation of information through user friendly media. Internet technology helps the user in transmission of data in various media format. In this research, data is considered as digital image. However, there is need to provide security and copyright protection to digital image. There are many security algorithms like blowfish, data encryption standards, advanced encryption standard, RSA,RC5,Cast-128, triple data encryption standard,IDEA,etc.The copyright protection can be provided to image by using digital image watermarking techniques. These techniques broadly classified into two broad categories i.e. transform domain technique and spatial domain technique. In this research work, combination of transform domain technique i.e. combined discrete cosine transform with singular value decomposition used for digital image watermarking and 256 bit key advanced encryption algorithm for security of digital image against various attacks, like Gaussian, cropping, salt pepper noise, median attack ,jpeg compression attack, etc is discussed.

**Keywords:** Digital image watermarking, Frequency domain technique, DCT, SVD, Security techniques, AES.

## 1 Introduction

Recently, multimedia uses the combination of text, image, graphic, animation, audio & video. With the help of digital communication & internet technology, it can be easily transfer from source to destination [1]. Media broadly classified into two categories i.e. static media and time varying media. Static media is further classified

[1] Research Scholar, Faculty of Engineering &Technology, Santa Gadge Baba Amravati University, Amravati.

[2] Former Vice-Chancellor, Dr. Babasaheb Ambedkar Technological University, Lonere, Maharashtra, India.

Email :{ sudhanshu1984gonge, vc_2005}@rediffmail.com.





J.M. Corchado Rodriguez et al. (eds.), *Intelligent Systems Technologies and Applications 2016*, Advances in Intelligent Systems and Computing 530,
DOI 10.1007/978-3-319-47952-1_7

into text, image, graphics, etc. Time varying media is classified into sound movies, and animations [1-4]. Classification of media is as shown in following Fig.1.

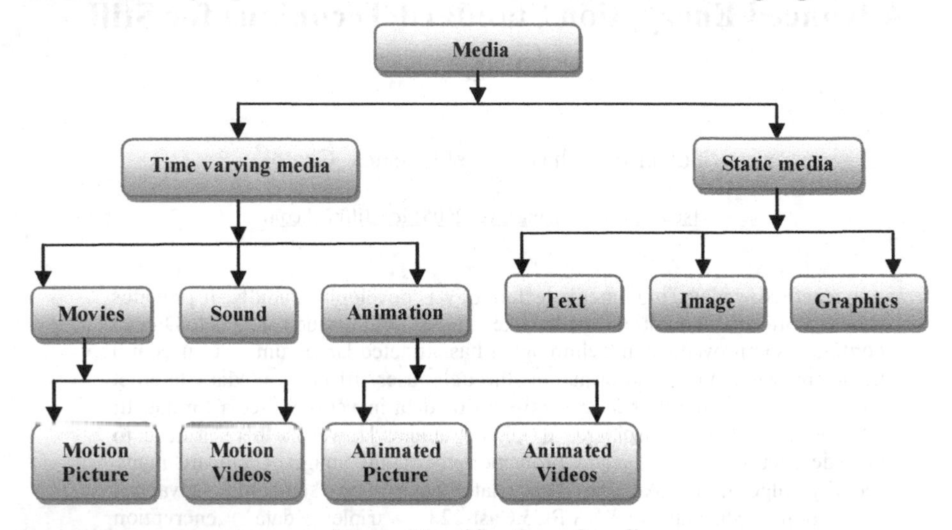


**Fig.1.** Classification of media.

Following digital communication channel is used for transmission of digital data using media shown in Fig.2.

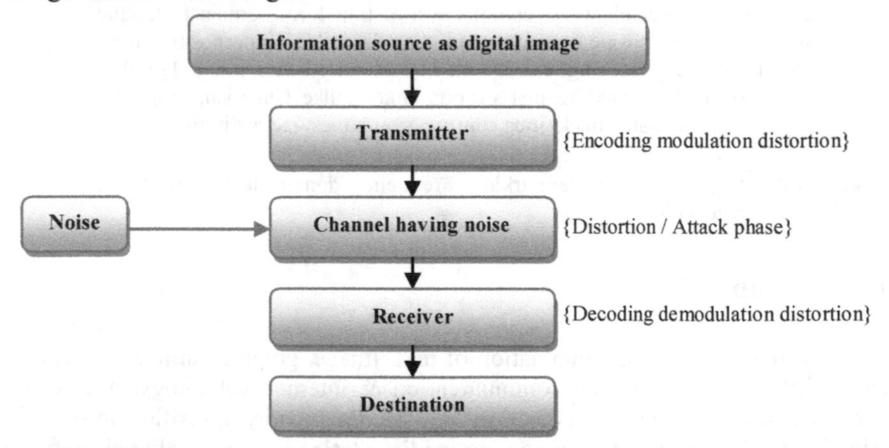


**Fig. 2.** Digital communication channel used for transmission of digital data using media.

This digital media can be easily distributed, modified and duplicated. Since, there is need of security and copy right protection [2-6]. Digital image watermarking has various properties of hiding watermark schemes for copyright protection:-

a) Invisible.
b) Robust.
c) Indefectibility.
d) Unambiguous.

In [1-7], Advanced encryption standard algorithm using 256 bit key has various properties like:-

a) Authentication.
b) Integrity.
c) Non-repudation.
d) Access control.
e) Availability.

# 2 Proposed DCT-SVD Digital Image Watermarking Algorithm

This algorithm is classified into three basic parts:-

a) Embedding phase: - In this phase, watermark image is used as watermark signal to embed in covered image for authorization, copyright protection, to avoid duplication, etc [4-20].
b) Attack phase: - In this phase, the watermarked covered image is transmitted through channel. As channel has noise, it may distort the image. However, there are some intruders who are always watching the data flow in the channel may attack intentionally or non- intentionally [4-10]. In this research work, intentional attacks are used for experiment purpose which is mention in abstract above [4-20].
c) Extraction phase:-In this phase, after receiving watermarked image, watermark image is extracted [4-20]. The metrics like similarity i.e. robustness in terms of normalized cross- correlation coefficient, imperceptibility in terms of peak signal to noise ratio is calculated of extracted watermark image[11-17].
d) The purpose of using DCT is for controlling the frequency location of the pixels quantization distortion. It provides good JPEG compression ratio [17].
e) The purpose of using SVD is to reduce noise from the image and has good compact energy coefficient pixels of image [17-20].

## 2.1 Working of DCT-SVD Watermark Embedding & AES Algorithm

Following are the steps used for DCT-SVD digital image watermarking:-
Step 1:- Select the covered image for watermarking purpose.
Step 2:- Take watermark image for embedding purpose.
Step 3:- Apply 2-D discrete cosine transform on covered image.
Step 4:- Apply singular value decomposition transform on DCT processed image.
Step 5:-Generate pseudo random number sequence i.e.w_0 and w_1which are differ from each other for embedding watermark image bit by using the attribute of component 'I' which having grater complexity block.
Step 6:- Apply the equation, if watermark bit is 1 then,

$$I_w = I + \alpha * W\_1 = U_{w\_1} * I_{w\_1} * V^T_{w\_1} \quad (1)$$

Otherwise,

$$I_w = I + \alpha * W\_0 = U_{w\_0} * I_{w\_0} * V^T_{w\_0} \quad (2)$$

Step 7:- With the help of step 5 & step 6, modify the components of 'U' and 'V'.
Step 8:-Apply inverse SVD transform on image after modifying the 'U' & 'V' component of the image.
Step 9:-After taking inverse of 2-D discrete cosine transform, combined DCT-SVD watermarked image is obtained.
Step 10:- Perform advanced encryption standard operation on DCT-SVD watermarked image using 256 bit key.
Step 11:- Finally, combination of DCT-SVD digital watermarked & AES encrypted image is obtained. Following fig.3 explains working of combination of DCT-SVD watermark embedding process and AES algorithm.

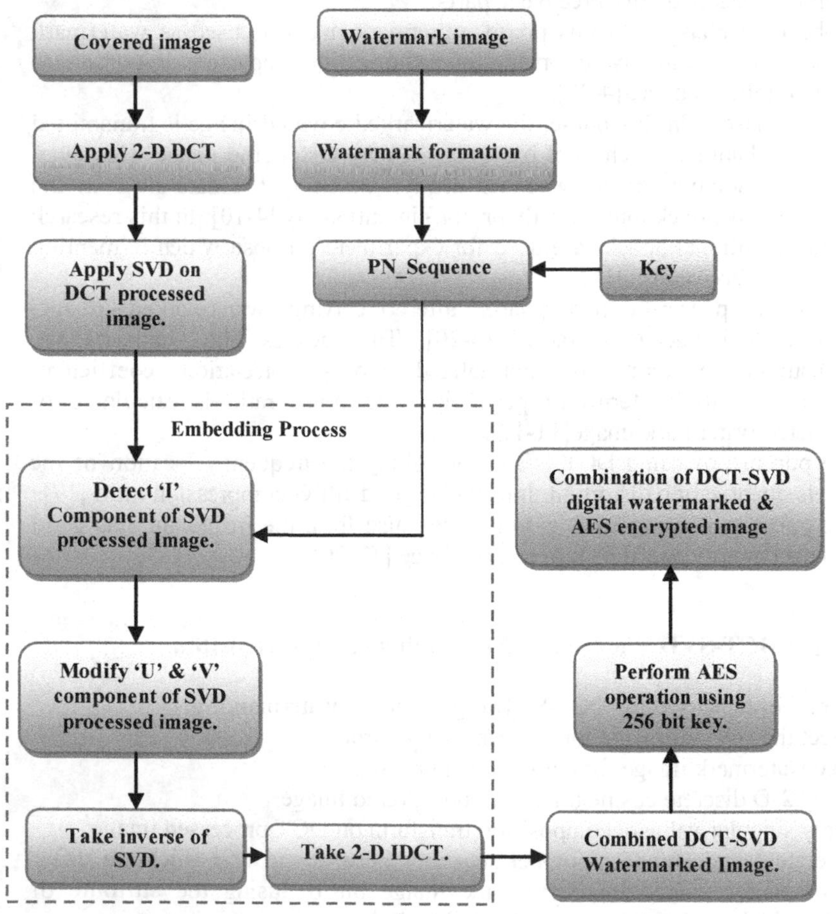


**Fig. 3.** Flow diagram of working of combination of DCT-SVD watermark embedding process and AES algorithm using 265 bit key.

## 2.2 Working of DCT-SVD Watermark Extraction & AES Decryption Algorithm

Step 1:- Read, received combined watermarked & AES encrypted image.

Step 2:- Perform AES decryption operation on combined DCT-SVD watermarked & AES encrypted image using 256 bit key.
Step 3:- After AES decryption process, decrypted combined DCT-SVD watermarked image is obtained.
Step 4:-Apply 2-D DCT on received decrypted combined DCT-SVD watermarked image.
Step 5:- Apply singular value decomposition transformed on DCT processed image.
Step 6:-The complexity of the block is determined by calculating non-zero co-efficient present in 'I' component of each block.
Step 7:-To create again pseudo random number sequence i.e.w_0 and w_1, the same packet of seed pixel are used which were used for embedding process.
Step 8:- Extract the watermark bit '1' if, the correlation with w_0 is less than w_1, if not extracted bit of watermark is consider as 0.
Step 9:-Similarity between original watermark and extracted watermark is calculated, by reconstruction of watermark with the help of extracted watermark bits from image.
The working of AES decryption of DCT-SVD watermarked image & DCT-SVD extraction process of watermark is explained in following fig.4.

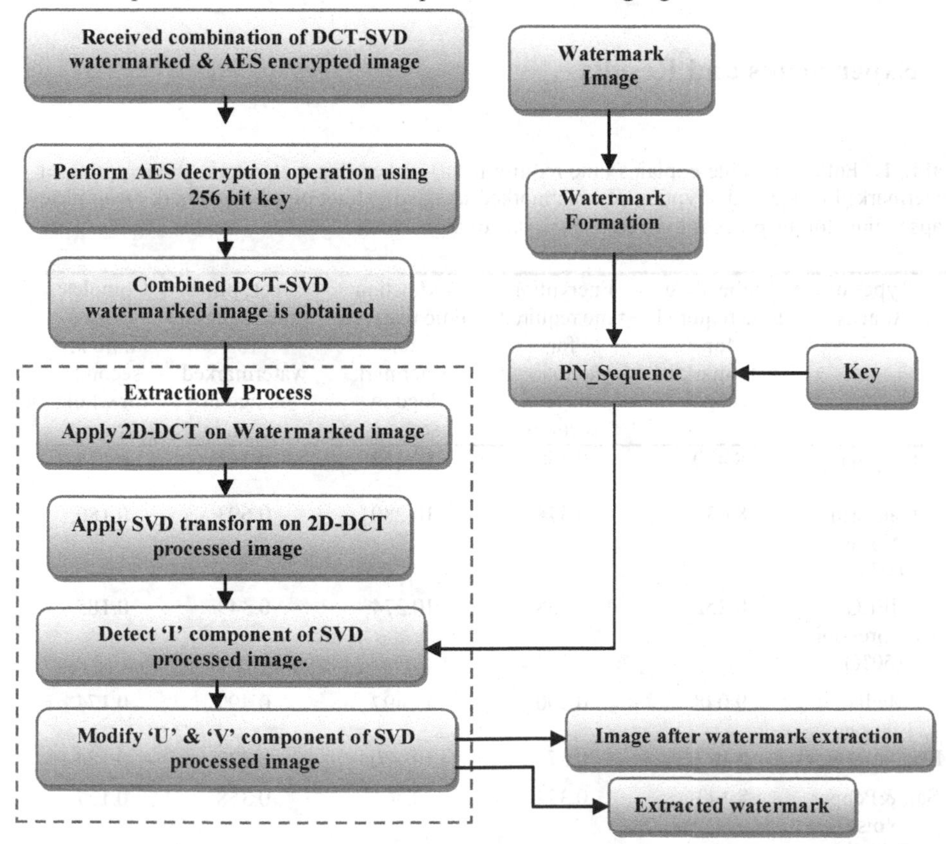


**Fig.4.** Flow diagram of working of 265 bit key AES decryption of DCT-SVD watermarked image & DCT-SVD extraction process of watermark.

## 3 Working of Advanced Encryption & Decryption Standards

Following Fig.5 explains the advanced encryption process of DCT-SVD watermarked image and advanced decryption process of DCT-SVD watermarked image [1-7].

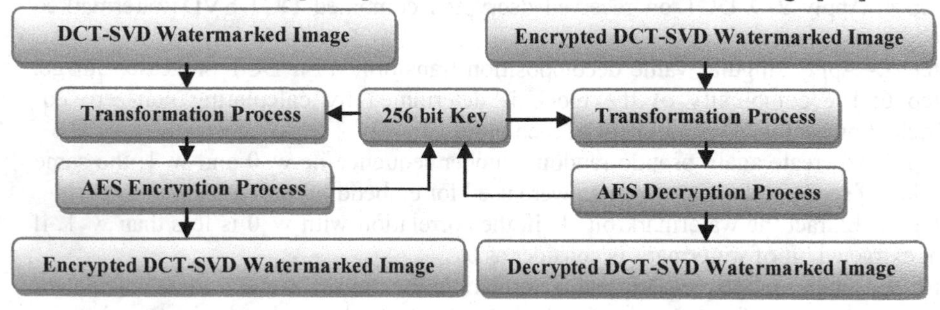


**Fig.5** Flow diagram of working of 265 bit key AES encryption & decryption process of watermarked image.

## 4 Experiments and Results

**Table 1.** Following table explains time require for a) embedding watermark b) encryption of watermarked image c) decryption of watermarked image d)extraction of watermark e)complete elapsed time for the process for a embedding factor alpha =0.5.

| Types of Attacks | Embedding time required for watermark in seconds | Encryption time required for watermarked image in seconds | Extraction time required for watermark logo in seconds | Decryption time required for watermarked image in seconds | Complete elapsed time in seconds for whole process |
|---|---|---|---|---|---|
| Cropping | 8.470 | 0.358 | 10.452 | 0.343 | 0.174 |
| Gaussian Noise (0.04 dB) | 8.034 | 0.374 | 10.280 | 0.693 | 0.180 |
| JPEG Compression (50%) | 8.252 | 0.358 | 10.374 | 0.343 | 0.185 |
| Median Filtering | 9.048 | 0.390 | 11.497 | 0.499 | 0.174 |
| Rotation (45°) | 8.127 | 0.374 | 10.670 | 0.327 | 0.184 |
| Salt &Pepper Noise (0.04 db) | 8.143 | 0.374 | 10.561 | 0.358 | 0.177 |
| Without any Attack | 8.533 | 0.358 | 10.046 | 0.358 | 0.174 |

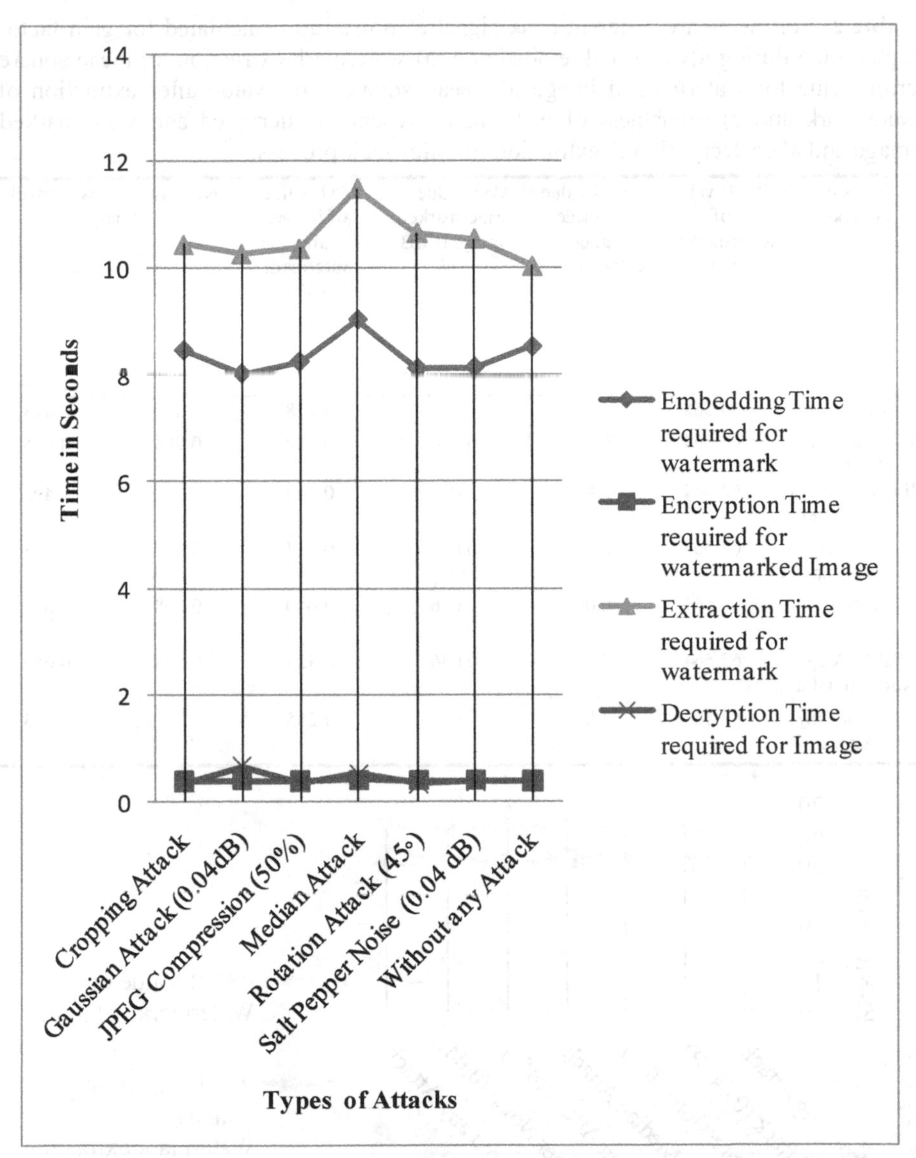


**Fig.6.** Graphical representation of different types of attacks versus time required for embedding ,encryption, decryption and extraction process for gain factor alpha=0.5

From above Fig.6 it is clear that the time required for embedding process required is less as compared to extraction process of watermark. It is also been observed that encryption time required for Gaussian attack and median attack is less as compare to decryption time whereas, cropping attack, jpeg compression attack, rotation attack & salt pepper noise attack with density of 0.04dB requires more time for encryption process as compare to decryption process. It is observed that the encryption time required is same as that of decryption time when attack is not applied.

**Table 2.** Following are different peak signal to noise ratio calculated for gain factor alpha = 0.5 during a)watermark embedding ,b) watermark extraction, c) mean square error value for watermarked image ,d) mean square error value after extraction of watermark and e) robustness of watermark present in encrypted and watermarked image and after decryption & extraction of watermark process.

| Types of Attacks | PSNR value of watermarked image in dB | PSNR value of image after extraction in dB | MSE value of watermarked image in dB | MSE value of image after watermark extraction in dB | NCC value of watermark Image | |
|---|---|---|---|---|---|---|
| | | | | | Before Encryption & watermarked image | After decryption & extraction of watermark |
| Cropping | 56.568 | 54.745 | 0.143 | 0.218 | 1 | 0.443 |
| Gaussian Noise (0.04 dB) | 62.564 | 50.494 | 0.036 | 0.580 | 0.964 | 0.065 |
| JPEG Compress -ion (50%) | 62.564 | 53.882 | 0,036 | 0.265 | 1 | 0.493 |
| Median Filtering | 62.564 | 52.401 | 0.036 | 0.374 | 0.999 | 0.043 |
| Rotation (45°) | 62.564 | 50.068 | 0.036 | 0.640 | 0.008 | 0.090 |
| Salt &Pepper Noise (0.04 db) | 62.564 | 53.037 | 0.036 | 0.323 | 0.928 | 0.040 |
| Without any Attack | 62.564 | 53.882 | 0.036 | 0.265 | 1 | 0.493 |

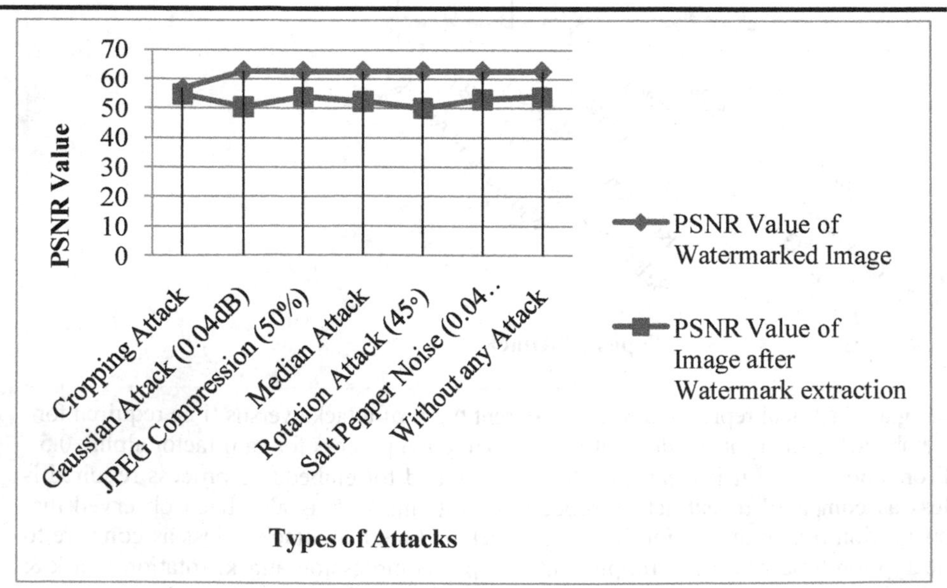


**Fig.7.** Graphical representation of different types of attacks versus PSNR value of watermarked image and PSNR value after watermark extraction.

Fig.7 explain the peak signal to the noise ratio of watermarked image and watermark image after extraction for gain factor (alpha=0.5).It is also been observed that psnr value is less for cropping attack as compare to other attacks. The psnr value of DCT-SVD watermarked image is same for all attacks except cropping attack. The psnr value of image after extraction of watermark varies & psnr value for cropping attack is high as compare to other.

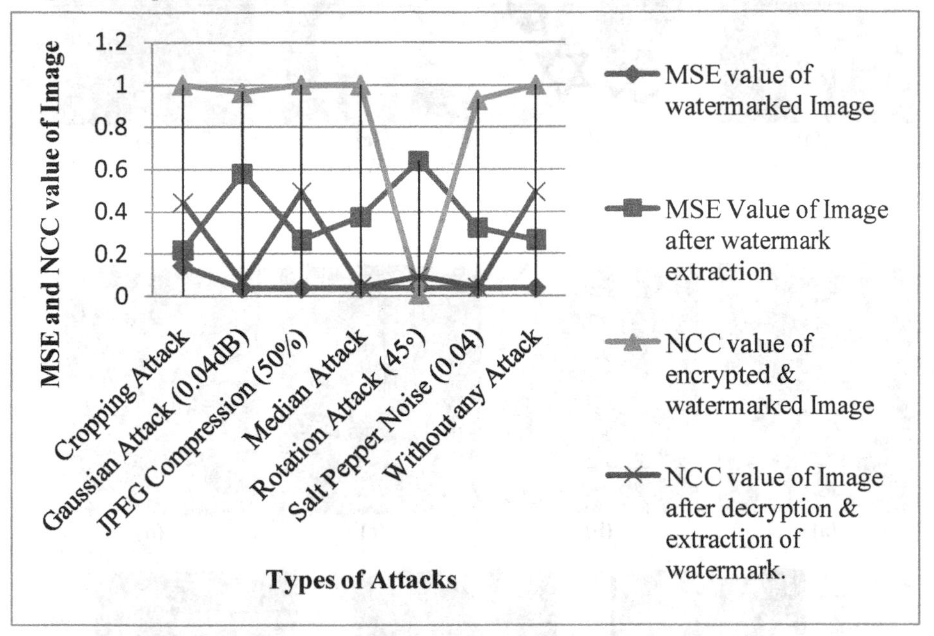


**Fig.8.** Graphical representation of different types of attacks versus MSE value & NCC value for gain factor (alpha=0.5).

Fig.8 explain the graphical representation of mean square error value & normalized cross co-relation co-efficient i.e. robustness of watermark against different types of attack. It is also observed that mean square error value of watermarked image is high after applying cropping attack as compare to other attacks. The mean square error value for DCT-SVD watermarked image for all attacks is same i.e. (0.036dB) as mention in table-2.The value of mean square error of image after extraction of watermark value get increased by 6 to 17 time more as compare to MSE value watermarked image after applying attack. Robustness of watermark is calculated by normalized cross-correlation co-efficient parameter.NCC value of watermark is 1 for AES encrypted & DCT-SVD watermarked image even after applying cropping attack & jpeg compression attack as well as for without applying any attack. However, even after AES 256 bit key encryption, there is an impact of attacks on encrypted watermarked image like Gaussian noise attack with an intensity 0.04 dB, median filter attack, salt & pepper noise attack with an intensity of 0.04 dB, rotation attack by angle of $45^0$ .It is observed that rotation attack has most powerful impact on combined AES encrypted & DCT-SVD watermarked image. It is also observed that, it fails to maintain the quality of image & robustness of watermark after extraction From above graph & Table.2, it can be seen that the watermark can't recovered properly even after successful completion of decryption process as shown in following Fig.9,Fig.10,Fig.11and Fig.12.

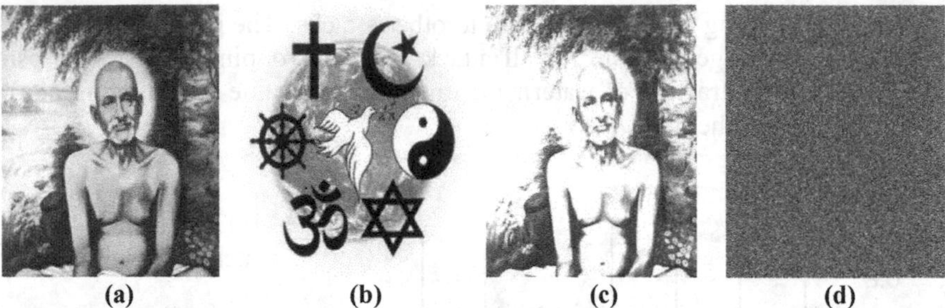

(a) (b) (c) (d)

**Fig.9** **(a)** Cover Image of Shri Sant Gajanan Maharaj. **(b)** Watermark logo used for embedding. **(c)**Watermarked image of Shri Sant Gajanan Maharaj. **(d)** Combination of DCT-SVD watermarked & AES 256 bit key encryption image of Shri Sant Gajanan Maharaj for gain factor (alpha=0.5).

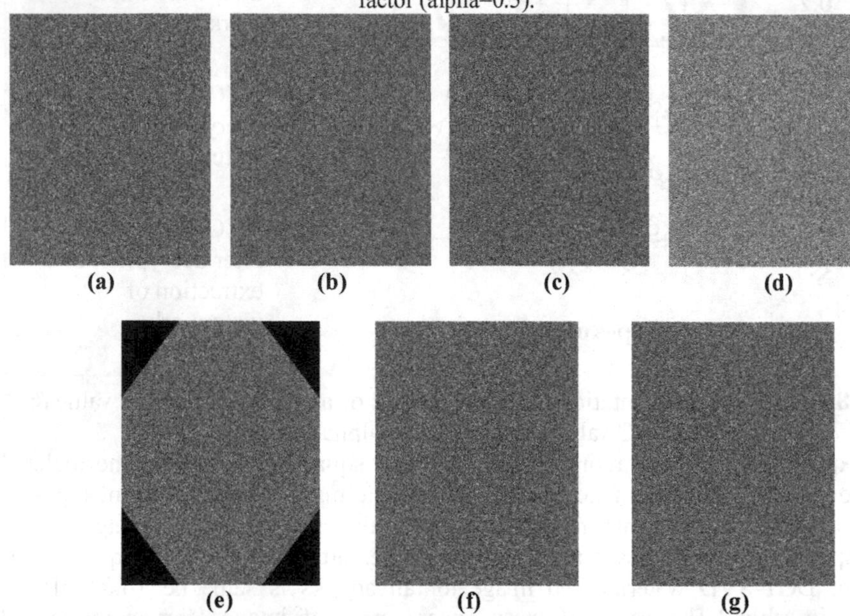

(a) (b) (c) (d)

(e) (f) (g)

**Fig.10** **(a)** Cropping attack. **(b)** Gaussian noise attack with intensity of 0.04 dB. **(c)** JPEG compression attack up to 50%. **(d)** Median filter attack. **(e)** Rotation attack with 45°. **(f)** Salt and pepper noise attack with intensity of 0.04dB. **(g)** Without any attack. On Combination of DCT-SVD watermarked & 256 bit key AES encryption image of Shri Sant Gajanan Maharaj for gain factor (alpha=0.5).

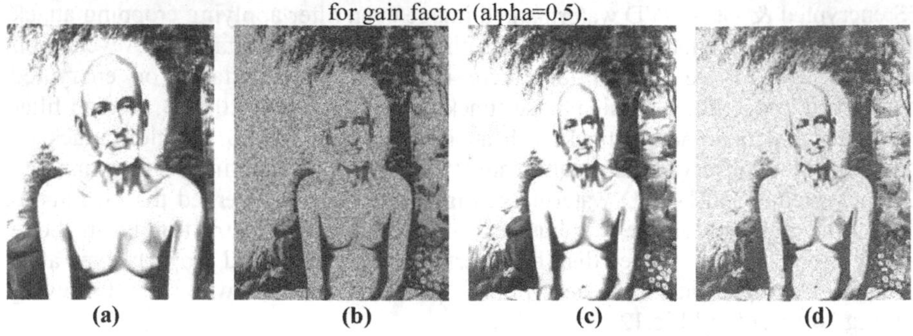

(a) (b) (c) (d)

**Fig.11** Decrypted watermarked images of Shri Sant Gajanan Maharaj for gain factor (alpha=0.5) after applying attacks like **(a)** Cropping attack. **(b)** Gaussian noise attack with intensity of 0.04 dB. **(c)** JPEG compression attack up to 50%. **(d)** Median filter attack. **(e)** Rotation attack with 45°. **(f)** Salt and pepper noise attack with intensity of 0.04dB. **(g)** Without any attack.

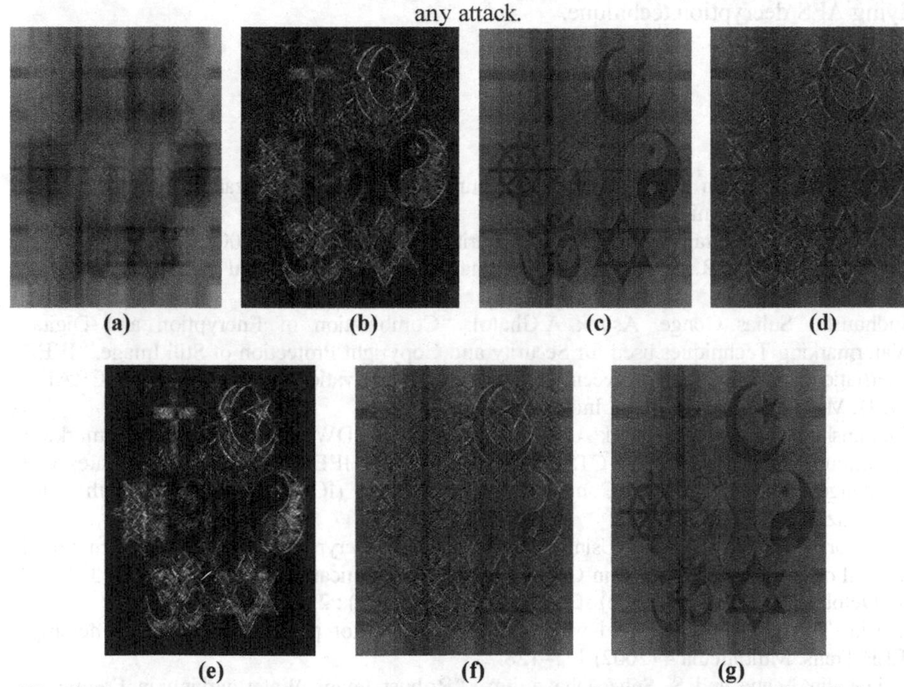


**Fig.12** Extracted watermark logo from decrypted watermarked images of Shri Sant Gajanan Maharaj for gain factor (alpha=0.5) after applying attacks like **(a)** Cropping attack. **(b)** Gaussian noise attack with intensity of 0.04 dB. **(c)** JPEG compression attack up to 50%. **(d)** Median filter attack. **(e)** Rotation attack with 45°. **(f)** Salt and pepper noise attack with intensity of 0.04dB. **(g)** Without any attack.

# 6 Conclusion

In this research paper, integration of watermarking and encryption technique is explained. The paper explains the composition of DCT-SVD watermarking & advanced encryption standard technique is explained using 256 bit key. It also explains the impact of cropping attack, Gaussian noise attack, jpeg compression attack, salt & pepper noise attack, median filter attack, rotation attacks and without applying attack is explained in detail. To analysis the research work the parameter like imperceptibility i.e. PSNR value of image, NCC value i.e. robustness of image, MSE i.e. mean square error is consider The research work discussed about the rotation attack has more impact as compared to other attack for embedding gain factor (alpha=0.5).It also tells about mean square error value for cropping attack is more as compare to other while, it remains same for other attacked DCT-SVD watermarked image. The work explains robustness of the watermark is maintained maximum up to 49% shown in table 2. The peak signal to the noise ratio of DCT-SVD watermarked image have less value as compared to the peak signal to noise ratio value after extraction of watermark from image. The Robustness of the watermark is maintained even after encryption but it fails to recover while performing extraction process after applying AES decryption technique.

## References

1. Data Communication and Networking, Fourth Edition by-Forouzan Copyright © the McGraw-Hill Companies, Inc.
2. W.Stallings, Cryptography and Network Security. NJ: Prentice Hall, 2003.
3. R. C. Gonzalez and R. E. Woods, "Digital Image Processing", Second edition,pp. 411-514, 2004.
4. Sudhanshu Suhas Gonge, Ashok A.Ghatol, "Combination of Encryption and Digital Watermarking Techniques used for Security and Copyright Protection of Still Image." IEEE International Conference on Recent Advanced and Innovations in Engineering (ICRAIE-2014), May 09-11, 2014, Jaipur, India.
5. Sudhanshu Suhas Gonge, Ashok A. Ghatol , "Combined DWT-DCT Digital Watermarking Technique Software Used for CTS of Bank." in 2014 IEEE Conference on Issues and Challenges in Intelligent Computing Techniques (ICICT-2014), Feb 7th and 8th.Ghaziabad,U.P.,India .
6. Kaladharan N, "Unique Key Using Encryption and Decryption of Image." International Journal of Advanced Research in Computer and Communication Engineering Vol. 3, Issue 10, October 2014, ISSN (Online) : 2278-1021,ISSN (Print) : 2319-5940.
7. R. Liu, T. Tan, An SVD-based watermarking scheme for protecting rightful ownership, IEEE Trans. Multimedia 4 (2002) 121–128.
8. G. Dayalin Leena and S. Selva Dhayanithy, "Robust Image Watermarking in Frequency Domain", Proceedings of International Journal of Innovation and Applied Studies ISSN 2028-9324 Vol. 2 No. 4 Apr. 2013, pp. 582-587.
9. Haowen Yan, Jonathan Li, Hong Wen ,"A key points bases blind watermarking approach for vector geospatial data", Proceedings of Elsevier Journal of Computers, Environment and Urban Systems, 2012, Volume 35, Issue 6, pp. 485–492.
10. A. Kannammal, K. Pavithra, S. Subha Rani, "Double Watermarking of Dicom Medical Images using Wavelet Decomposition Technique", Proceedings of European Journal of Scientific Research, 2012, Vol.70, No. 1, pp. 46-55.
11. Feng Wen-ge and Liu Lei,"SVD and DWT Zero-bit Watermarking Algorithm". 2nd International Asia Conference on Informatics in Control, Automation and Robotics, 2010, 2nd March, 2011

12. Liwei Chen, Mingfu Li. "An Effective Blind Watermark Algorithm Based on DCT." Proceedings of the 7th World Congress on Intelligent Control and Automation June 25 - 27, 2008, Chongqing, China.
13. Aleksandra Shnayderman, Alexander Gusev, and Ahmet M. Eskicioglu, "An SVD-Based Gray scale Image Quality Measure for Local and Global Assessment." IEEE TRANSACTIONS ON IMAGE PROCESSING, VOL. 15, NO. 2, FEBRUARY 2006.
14. Kuo-Liang Chung *, Wei-Ning Yang , Yong-Huai Huang,Shih-Tung Wu, Yu-Chiao Hsu, "On SVD-based watermarking algorithm." Applied Mathematics and Computation 188 (2007) 54–57.
15. Cheng-qun Yin, Li, An-qiang Lv and Li Qu, "Color Image Watermarking Algorithm Based on DWT-SVD." Proceedings of the IEEE International Conference on Automation and Logistics August 18 - 21, 2007, Jinan, China.
16. Juan R. Hernández, Fernando Perez Gonzalez, Martin Amado. "DCT-Domain Watermarking Techniques for Still Images: Detector Performance Analysis and a New Structure." IEEE Transaction on Image Processing, vol. 9, no.1, January2000.
17. Lin, S. and C. Chin, 2000. "A Robust DCT-based Watermarking for Copyright Protection," IEEE Trans. Consumer Electronics, 46(3): 415-421.
18. Nikolaidis, A. and I. Pitas, 2003. "Asymptotically optimal detection for additive watermarking in the DCT and DWT domains," IEEE Trans. Image Processing, 2(10): 563-571.
19. Golub & Van Loan – Matrix Computations; 3rd Edition, 1996.
20. Golub & Kahan – Calculating the Singular Values and Pseudo-Inverse of a Matrix; SIAM Journal for Numerical Analysis; Vol. 2, Issue 2; 1965.

# Performance Analysis of Human Detection and Tracking System in Changing Illumination

M M Sardeshmukh; Mahesh Kolte; Vaishali Joshi

***Abstract***—Detection and tracking of a human in a video is useful in many applications such as video surveillance, content retrieval, patent monitoring etc. This is the first step in many complex computer vision algorithms like human activity recognition, behavior understanding and emotion recognition. Changing illumination and background are the main challenges in object/human detection and tracking. We have proposed and compared the performance of two algorithms in this paper. One continuously update the background to make it adaptive to illumination changes and other use depth information with RGB. It is observed that use of depth information makes the algorithm faster and robust against varying illumination and changing background. This can help researchers working in the computer vision to select the proper method of object detection.

***Key words***: Object detection , machine vision, machine learning, image database, computer vision

## I Introduction

Over the past decade, our lives have been influenced by computers and worldwide tremendously. Computers extend our possibilities to communicate by performing repetitive and data-intensive computational tasks fundamentally. Because of recent technological advances, video data has become more and more accessible and now became very useful in our everyday life. Today, our commonly used consumer hardware, such as notebooks, mobile phones, and digital photo cameras are allowing us to create videos, while faster internet access and growing storage capacities enable us to publish directly or share videos with others. However, even though the importance of video data is increasing, the possibilities to analyze it in an automated fashion are still limited. In any surveillance application the first step is detection and tracking the human or interested object. This task becomes difficult in changing illumination. The background required to be updated continuously which increases time complexity. In this paper we proposed two approaches of human/object detection and tracking, one uses updating background continuously and other makes the use of depth information along with RGB for object detection and tracking. The comparison of these two approaches is presented in this paper

This work was supported in part by BCUD SPPU Pune



J.M. Corchado Rodriguez et al. (eds.), *Intelligent Systems Technologies and Applications 2016*, Advances in Intelligent Systems and Computing 530,
DOI 10.1007/978-3-319-47952-1_8

which will help the researchers to select the appropriate method for their work.

## 2. Background

The first step in any video surveillance application is Segmentation. For detecting the object knowledge of background, is essential. Any changes in illumination and viewpoint make this task critical. Differentiating the objects that are of similar color as of background turns more difficult. Depth information can be used along with color information to make the segmentation process robust and less complex. The traditional approaches for activity recognition in videos assume well-structured environments and they fail to operate in largely unattended way under adverse and uncertain conditions from those on which they have been trained. The other drawback of current methods is the fact that they focus on narrow domains by using specific concept detectors such as human faces, cars and buildings.

## 3. Methodology

3.1 Object detection and tracking in changing illumination and background
In many of the practical applications the illumination condition is not static whereas it keeps on changing. Detection and tracking of the object in this dynamic condition becomes difficult and challenging. This model detects and tracks the object in varying illumination. This method consists of continuously updating background for object detection and tracking. The foreground is extracted from the video sequence by learning background continuously by checking all pixel values from previous frames and subtracting it from test frame.

The system is executed in three steps

1. Continuously updating background modeling algorithm.
2. Object feature extraction.
3. Classification.

With background modeling algorithm, we can remove the constant pixel values means remove pixels whose values remain constant over some frames. Algorithm successfully removes noise, shadow effects and robust to the illumination change.

Real time features are extracted from the frames extracted from the video sequence. The features include the height, width, height width ratio, centroid. For a pedestrian, the periodic hand and leg movement is unique, and this is the feature that is used to classify the human from all other classes.

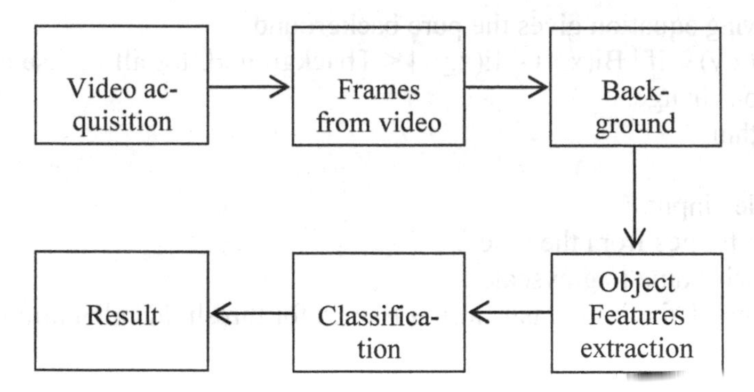


The real-time features that are acquired from the feature extraction stage are a base to classify the detected object into various categories such as pedestrian, group of pedestrian, vehicles, etc. Background subtraction is particularly a commonly used technique for motion segmentation in static scenes. It attempts to detect moving regions by subtracting the current image pixel-by-pixel from a reference background image that is created by averaging images over time in an initialization period. Some background subtraction techniques have many problems, especially when used outdoors. For instance, when the sun is covered by clouds, the change in illumination is considerable. In this framework, we have implemented continuously updating background modeling algorithm that ensures the removal of illumination change during the day, dynamic scene changes. The following figure shows the simple diagram for background subtraction.

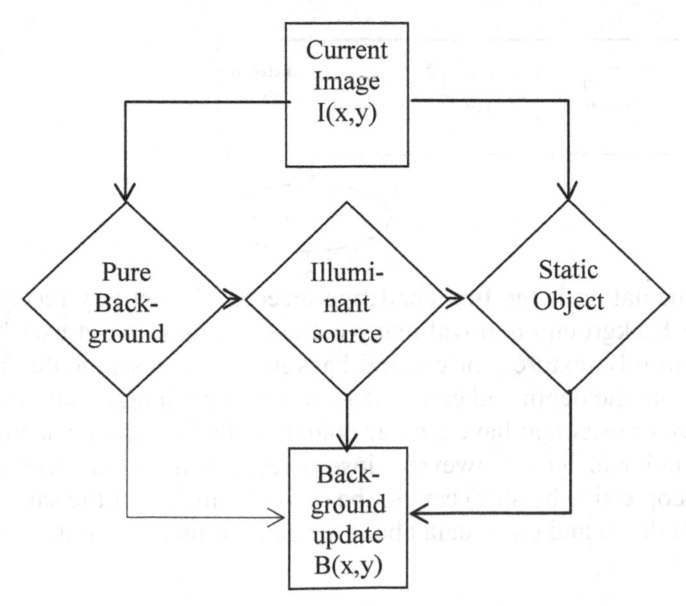

The following equation gives the pure background.
B(x,y) = I(x,y) if | Bi(x,y) – Ii(x,y) | < Tbackground, for all i ; else check illumination change.
b) Algorithm
Start.
Read a video input.
Extract the frames from the video.
Convert each frame to grayscale.
Consider initial six frames as learning frames for threshold calculation.
Calculate threshold values for background.
Consider next ten frames for object detection.
Compare incoming frame pixels with threshold values.
If more than 40% of the pixels are changed then it is temporary illumination change, consider next frame else extract the foreground object from the entire frame.
Deduce boundary for foreground object.

c) Parameters of the model
1. Video type : AVI consist of moving objects like Humans and vehicles
2. Illumination conditions: Varying (outdoor scenes)
3. No of videos: 30
4. Frames per video 30 to 40

3.2 Human/ Object Detection using RGB –D Information

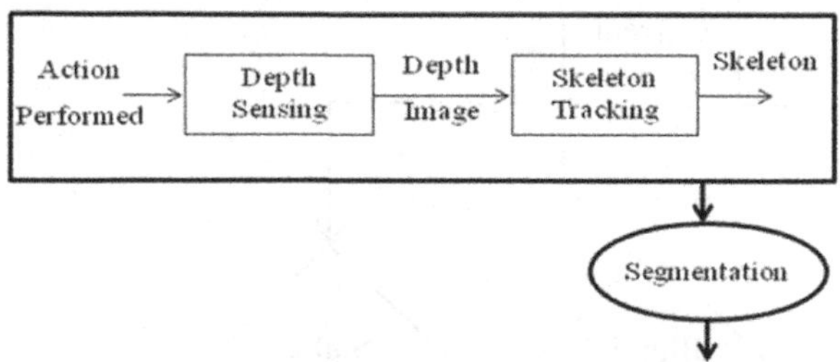


Proper segmentation is the first challenge faced in the activity recognition project. For background removal many methods require that a user have a known, uniformly textured, or colored background. For segmentation, we are considering the depth and color information of the image, into account. To recognize objects that have similar colors depth data can be segmented easier than color images. However, close objects with similar depths (e.g. two close people side by side) can not be easily identified at the same time. Having both depth and color data about the same frame is a quite common thing. Moreover, it is possible to recognize better the objects in the site by exploiting both geometries (depth) information and the color cues. By doing segmentation, we are generating a silhouette image. The steps included in segmentation are as follows:

1) We have done preprocessing on depth image that is used for reduction of noise that is coming from surround objects such as wall
2) After preprocessing we have done binarization of a picture.

$$Bim(i,j)=\begin{cases}1 & if\ Pim(i,j)>1\\ 0 & if\ Pim(i,j)<1\end{cases}$$

Where, Pim = preprocessed image, Bim = binary image
3) We have obtained locations of standing persons using the sum of columns and rows.

$$Location_{PersonY\ axis} = i\ if\ Sum(i)_{horizontal} > T_H$$

We have selected first and last value of in Locationpersons y-axis as the y-axis limits for bounding box. Similar process is done for Locationpersons x-axis.
4) By using location on x and y-axis we have found the width and height of bounding box.

$$Width_{boundingbox1} = L_2 - L_1$$

where L2= location(2) person x-axis and L1=location(1) person x-axis

$$Width_{boundingbox2} = L_4 - L_3$$

where L4= location(4) person x-axis and L3=location(3) person x-axis

$$Width_{boundingbox2} = L_4 - L_3$$

where L2= location(2) person y-axis and L1=location(1) person yx-axis
5) Later we have only selected the pixel locations that are present in bounding box boundaries decided by location of height and width-

$$I=\begin{cases}Bim(i,j) & if\ L_{1x}<i<L_{2x}\ \&\&\ L_{1y}<i<L_{2y}\\ 0 & elsewhere\end{cases}$$

Where, I = segmented image (i,j)
L1x = locationperson x-axis (1)
L2x = locationperson x-axis (2)
L1y = locationperson y-axis (1)
L2y = locationperson y-axis (2)

This gives us the segmented human/object present in the video that is moving. This proposed method is less complex and more accurate as we have not used the continuous updating of the background. Due to the use of depth information along with RGB obtained from Kinect sensor the segmentation becomes independent of illumination condition.

b) Parameters of the model

1. Dataset : MMS 3-D HAR database
2. Illumination conditions: varying
3. Background information: no
3. No of videos: 110
4. Frames per video: 100
5. Type of video: RGB and depth

## 4.Results

Experimentation carried out with model I i.e. object detection and tracking in varying illumination and background shows the results at different stages in detection and tracking of moving object in the video. This simulation is done on total thirty videos recorded by RGB camera in outdoor environment.

a) Result for video containing moving human

Table (1): Segmentation Result for Background Subtraction using Continuous Updation of Background.

a) Input Video

b) Sample extracted frame

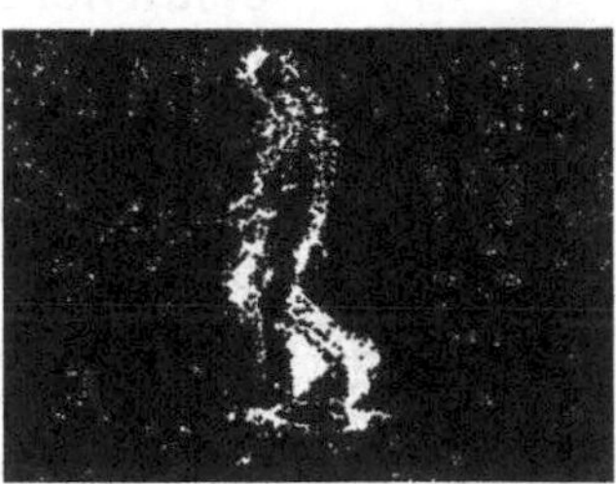

c) Detected object/ human d) Noise removal

e) Detected human after noise removal

Experimentation on model II i.e. human activity recognition with RGB-D data is carried out on the MMS 3-D HAR dataset comprising total hundred and ten videos for ten activities performed by eleven individuals in four lighting conditions. Results obtained at various stages are given below.

a) Dataset

MMS 3-D HAR dataset consist of total hundred and ten videos of ten human-human activities recorded by eleven individuals in four lighting conditions. Table 4.2.1 shows all these activities and the recorded information

Table : Dataset Categories with Depth and Skeleton Information

| Action Type | Color | Depth | Skeleton |
|---|---|---|---|
| Approach | | | |
| Depart | | | |
| Handshake | | | |
| Kick | | | |
| Lifting | | | |
| Patting | | | |
| Pointing | | | |
| Punch | | | |
| Push | | | |
| Salam | | | |

b) Segmentation

The first step in the activity recognition is segmentation. Table shows the detection of the silhouette in the different illumination condition.

Table : Segmentation using RGB-D Data in Varying Illumination

| Illumination condition | Color | Depth | Silhouette |
|---|---|---|---|
| L0 | | | |
| L1 | | | |
| L2 | | | |
| L3 | | | |

From the results, it is observed that there is no effect of changing illumination on the extracted silhouette this is due to the use of depth information shown in the table along with RGB information. This makes the segmentation less complex, less time consuming and more accurate. Time required for segmentation in RGB-D case is around 3 sec whereas if background is updated continuously to tackle the problem of varying illumination it takes around 25 sec for segmentation. Therefore it is difficult to maintain the frame repetition rate required in video processing to make it real time.

Comparison of segmentation algorithm based on time complexity:

The computational and experimental models are compared for the time required for segmentation. Model 1 and model 2 tested with the videos recorded in varying illumination condition. For model 1 the videos are captured by sony pc camera, whereas for model 2 videos are captured by Kinect Microsoft sensor.

It is observed that the time required for model 1 is more as compared to model 2. The very important point which can be noted here is in case of model 2 even the illumination condition is varying the time required for execution is very less as compared to other models.

Table 3 : Comparison of segmentation algorithm based on time complexity

| Sr. No. | Object Detection Algorithm | No. of Frames in video | Time Elapsed in sec | Lighting Condition |
|---|---|---|---|---|
| 01 | Background Subtraction using continuous updating | 40 | 25.62 | Changing |
| 02 | Background Subtraction using RGB-D data | 100 | 3 | Changing |

## 5.Conclusion

In model II depth and colour information is used in segmentation because of which segmentation is accurate and also independent of illumination condition. So, it is possible to detect human/ object from the recorded video with varying illumination conditions, which makes algorithm robust. Continuous updating the background is not required. Hence, algorithm becomes less time complex and the possibility of real time implementation increases.

## References

Han, Sang Uk, et. al. "Empirical Assessment of a RGB-D Sensor on Motion Capture and Action Recognition for Construction Worker Monitoring." Visualization in Engineering Springer Open Journal, Vol. 1, 2013, pp. 1-13.

Victor Escorcia, Mara A. Dvila, Mani Golparvar-Fard and Juan Carlos Niebles, "Automated Vision-based Recognition of Construction Worker Actions for Building Interior Construction Operations using RGBD Cameras", Construction Research Congress 2012, 2012 pp. 879-888.

Fanellom Sean Ryan et al. "One-shot Learning for Real-time Action Recognition," Pattern recognition and Image Analysis, Springer Berlin Heidelberg, Vol.2, 2013, pp. 31-40.

Yamato J., Ohya J., Ishii, K., "Recognizing Human Action in Time-Sequential Images using hidden Markov Model," Proceedings of IEEE Society Conference on Computer Vision and Pattern Recognition, 1992, pp. 379-385.

Natarajan P., Nevatia R., "Coupled Hidden Semi Markov Models for Activity Recognition", Workshop on Motion and Video Computing, 2007, pp.1-10.
Li W., Zhang Z., Liu Z, "Action Recognition Based on a Bag of 3D Points, Proceeding of IEEE Computer Society Conference on Computer Vision and Pattern Recognition , Vol.2 , 2010, pp. 13-18.
Poppe Ronald, "A Survey on Vision-Based Human Action Recognition", Image and Vision Computing Journal IEEE, Vol. 28(6) , 2010, pp. 976-990.
Thomas B. Moeslund, Adrian Hilton, Volker Kruger, "A Survey of Advances in Vision-based Human Motion Capture and Analysis", Computer Vision and Image Understanding (CVIU) Elsevier Journal, Vol. 104 (2-3), 2006, pp. 90-126..
Michael S. Ryoo, Jake , K. Aggarwal, "Semantic Representation and Recognition of Continued and Recursive Human Activities", International Journal of Computer Vision (IJCV), Vol. 82 (1), 2009, pp. 124-132.
Abhinav Gupta, Aniruddha Kembhavi, Larry S. Davis, "Observing Human Object Interactions using Spatial and Functional Compatibility for Recognition", IEEE Transactions on Pattern Analysis and Machine Intelligence (PAMI), Vol. 31 (10), 2009, pp. 1775-1789.

# Comparative Analysis of Segmentation Algorithms Using Threshold and K-Mean Clustering

S. S. Savkare, A. S. Narote and S. P. Narote

**Abstract** Worldwide many parasitic diseases infect human being and cause deaths due to misdiagnosis. These parasites infect Red Blood Cells (RBCs) from blood stream. Diagnosis of these diseases is carried out by observing thick and thin blood smears under the microscope. In this paper segmentation of blood cells from microscopic blood images using K-Mean clustering and Otsu's threshold is compared. Segmentation is important innovation for identification of parasitic diseases. Number of blood cells is an essential count. Preprocessing is carried out for noise reduction and enhancement of blood cells. Preprocessing is fusion of background removal and contrast stretch techniques. Preprocessed image is given for segmentation. In segmentation separation of overlapping blood cells is done by watershed transform. Segmentation using K-Mean clustering is more suitable for segmentation of microscopic blood images.

## 1 Introduction

In biomedical engineering microscopic blood image analysis is important for diagnosis of parasitic diseases. Currently diagnosis is carried out by examining thin or thick blood smear under the microscope. These blood smears are Giemsa stained because it colorizes blood cells and parasites are highlighted in purple colors [4, 15, 17]. Images are first given to preprocessing step. In preprocessing background is removed so noise reduction is not required. Blood cells are enhanced using

S. S. Savkare
JSPM Narhe Technical Campus, Pune, e-mail: swati_savkare@yahoo.com

A. S. Narote
S. K. N. College of Engineering, Pune, e-mail: a.narote@rediffmail.com

S. P. Narote
MES College of Engineering, Pune, e-mail: snarote@rediffmail.com

 
J.M. Corchado Rodriguez et al. (eds.), *Intelligent Systems Technologies and Applications 2016*, Advances in Intelligent Systems and Computing 530,
DOI 10.1007/978-3-319-47952-1_9

contrast stretch. This preprocessed image is given to the K-Mean clustering for cluster formation. Separation of overlapping blood cells is done using watershed transform. Segmentation is carried out using Otsu's threshold. The results of these two segmentation techniques are compared. K-Mean clustering is able to segments blood cells from low stained images. Threshold technique is unable to segment blood cells from low stained images. K-Mean clustering technique is unsupervised.

The remainder of the paper is organized as follows: section 2 briefly reviews important researches on blood cell segmentation, section 3 describes preprocessing technique, segmentation of blood cells using Otsu's threshold and K-Mean clustering, separation of overlapping cells using watershed transform, section 4 discuss and compare results of both segmentation techniques. Finally, concluding remarks are given based on segmentation results in section 5.

## 2 Literature Review

This section reviews techniques used for blood cell segmentation from microscopic blood images and results. Images used for processing are acquired by connecting high resolution camera to the microscope. Researchers used stained thin blood slides for image processing [5, 13]. Thick blood smears are used for quantitative analysis and thin blood smears are used for morphology.

Arco et al. [3], proposed a method for malaria parasite enumeration in which histogram, threshold, and morphological operations are utilized. Image becomes much bright and dark pixels more dark after applying Gaussian filter. Image is segmented using adaptive threshold. In adaptive threshold average intensity value is calculated. To separate WBC or parasite morphological opening is used. Result is evaluated and compared with manual count. Adaptive histogram equalization and threshold improve the results. Morphological operations can be performed before and after segmentation for improvement in results. Separation of connected component is done by Run Length Encoding and tagging each pixels. Average accuracy 96.46% is obtained by this proposed technique.

Purnama et al. [14], proposed a technique to diagnosis malaria and avoid dependency on experts. Proposed technique has three steps : preprocessing, feature extraction, and classification. Database images are captured using thick blood film images, which are cropped and resized into 64×64 pixels. For algorithm development, classification and parameter adjustment training set is utilized. Features such as mean, kurtosis, standard deviation, entropy and skewness are extracted from the red-green-blue channel color histogram, hue channel of the HSV histogram, and hue channel of HSI histogram for training purpose. Classification method uses genetic programming. Genetic programming works by generating and running thousands of programs and choose the most effective programs to use. Classification rate of two class classifier is 94.82% for not parasite, 95.00% for parasite and six classes has 88.50% for not parasite, 80.33% for trophozoit, 73.33% for schizont, 77.50% for gametocyte stage.

For segmentation of microscopic blood images most of the researchers prefer K-Mean clustering [1, 2, 10, 12, 18] and thresholding [3, 6, 15, 16] techniques. For separation of overlapped blood cells researchers prefered template matching [7] or watershed transform [9, 11].

## 3 Methodology

The blood cell segmentation technique is described in this section. The processing is carried on Giemsa stained thin blood images. The steps followed are image acquisition, preprocessing and segmentation. Images are captured utilizing high resolution camera connected to the microscope. Total 50 images are used for testing of algorithm. Some database is available on online library [4]. For blood cell segmentation staining is important because it demarcates blood cells and highlights parasites present in blood cells. Acquired images are further processed for preprocessing.

### *3.1 Pre-Processing*

Aim of preprocessing is to reduce noise of an input image and to enhance area of interest. Here blood cells are important to segment. Preprocessing technique implemented utilizes fusion of background removal and contrast stretch. For background removal pixel intensity representing background should be known. These values are calculated by gray-scale histogram. Intensities of all background pixels are set to 255 so background becomes white in color. Once background is removed, noise reduction does not required. Contrast stretch expands and compress piece of intensity level. Range of intensity representing blood cells is decreased, so cells become darker. Pixels which are stained but they are part of background are increased in intensity, so they become brighter. Figure 1 shows gray-scale histogram before and after preprocessing technique. Figure 2 shows output of preprocessing technique. Implemented preprocessing algorithm enhance successfully low contrast and blurred image (Fig. 2 c and d)

### *3.2 Otsu's Threshold*

Preprocessed image obtained is given to segmentation algorithm. Most of the researchers utilized Otsu's threshold for segmentation of microscopic blood images. Global threshold calculation is shown in Equation 1 [8].

$$g(x,y) = \begin{cases} 1 & if(x,y) \geq T \\ 0 & \text{otherwise} \end{cases} \tag{1}$$

Global threshold calculate threshold globally which does not consider local parameter of an image. Otsu's threshold calculate threshold value automatically. It maximizes between class variance of two classes. If $K$ is threshold any value then pixels having intensity below $K$ are class 1 and above $K$ are class 2. To calculate between class variance (equation 2), probability of class 1 (equation 3), probability of class 2 (equation 4), Mean intensity of class 1 ($\mu_1$) and Mean intensity of class 2 ($\mu_2$) need to be calculated as shown below [8].

$$\sigma_B^2 = P_1 P_2 (\mu_1 - \mu_2)^2 \tag{2}$$

$$P_1 = \sum_{i=0}^{k} p_i \tag{3}$$

$$P_2 = \sum_{i=k+1}^{L} p_i \tag{4}$$

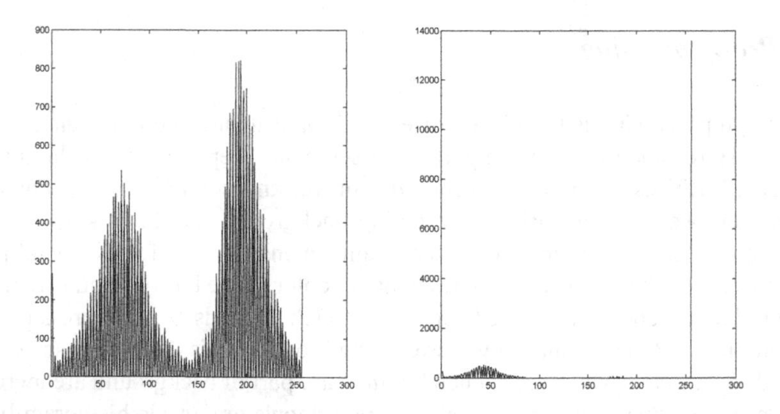

Fig. 1 (a,c) Original Image; (b,d) Preprocessed Grayscale Image.

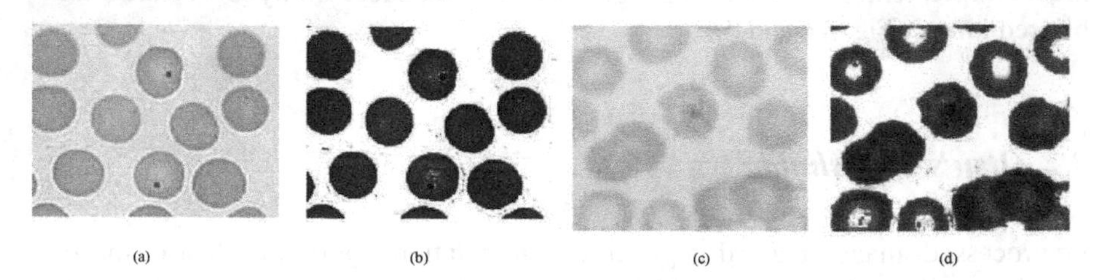


Fig. 2 (a,c) Original Image; (b,d) Preprocessed Grayscale Image.

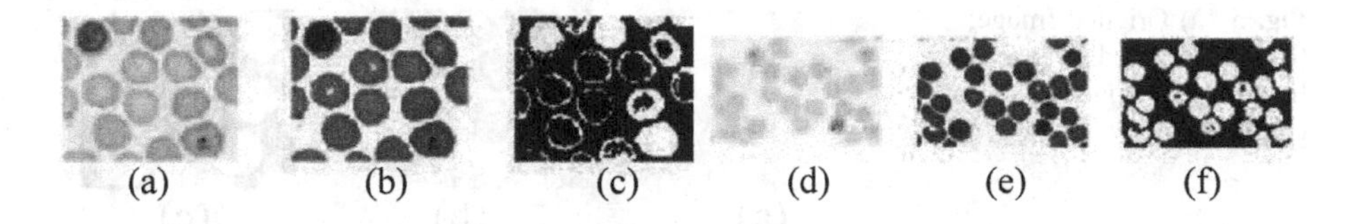


**Fig. 3** (a, d) Original image, (b,e) Preprocessed Grayscale image, (c,f) Output of Otsu's Threshold.

**Fig. 4** (a) Original image, (b) Binary Image of Cells, (c) Separated Overlapped Blood Cells using Watershed Transform.

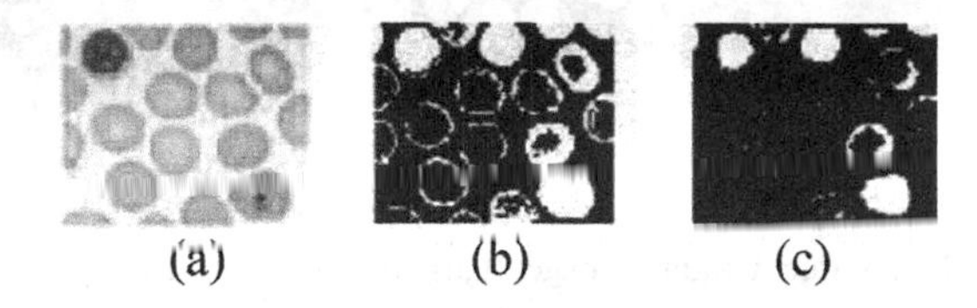


$$\mu_1(k) = \sum_{i=0}^{k} \frac{iP_i}{P_1(k)} \tag{5}$$

$$\mu_2(k) = \sum_{i=k+1}^{L} \frac{iP_i}{P_2(k)} \tag{6}$$

Figure 3 shows results of segmentation using Otsu's threshold. Segmented images are given for post-processing which includes removal of unwanted objects and separation of overlapping blood cells. A watershed transform on distance transform of binary image is applied to get ridge lines. These ridge lines are superimposed to get separated blood cells. Results of separation algorithm is shown in figure 4. Total number of blood cell count is compared with manual count.

### *3.3 K-Mean Clustering*

k-means is unsupervised learning algorithms that segments image into $k$ clusters. The researchers prefer K-Mean clustering for segmentation of microscopic blood images. Preprocessed images are given as an input. Here number of clusters are considered k=2, one for foreground (blood cells) and other for background. Algorithm of K-Mean clustering is as follows:

1. Consider number of clusters k=2.
2. At random take two centroid having maximum difference.
3. Assign each data points based on euclidean distance which is calculated as shown in following equation. Data points are assigned to closest centroid.

$$J = \sum_{1}^{k} \sum_{n \in S_j} (|x_n - \mu_j|)^2 \tag{7}$$

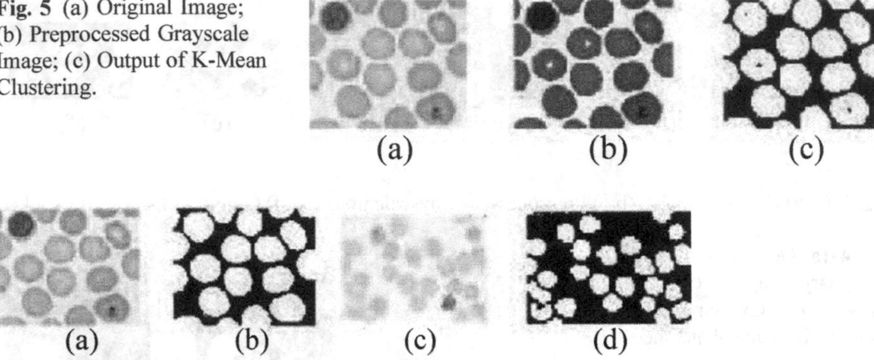


**Fig. 5** (a) Original Image; (b) Preprocessed Grayscale Image; (c) Output of K-Mean Clustering.

**Fig. 6** (a,c) Original Image; (b,d) Post-processed Binary Image of Cells.

where $x_n$ is data points and $\mu_j$ is geometric centroid.

4. Now reassign centroid based on average value of data points in group 1 and group 2.
5. Based on new centroid reassign data points to closest centroid.

Figure 5 shows segmentation results sing K-Mean clustering. Segmented binary image of blood cells is post-processed for unwanted objects removal using morphological operation. Watershed transform is used for separation of overlapping blood cells. Total number of blood cells are compared with manual count. Figure 6 shows post-processed results on output of K-Mean clustering.

## 4 Results

Preprocessed images are given to segmentation for extraction of blood cells from an image. Results are compared for 50 images. Segmentation using Otsu's threshold could not extract blood cells which are low stained or blurred. Figure 3 (c) shows that some blood cells shown in original image are not extracted in final image. Due to this blood cells count is reduced. K-Mean clustering is able to extract all blood cells in an image even though they are low stained or blurred. Table 1 shows blood count in 50 database images by manual counting, using Otsu's threshold and K-Mean clustering techniques. This count shows that Otsu's threshold fails to segment low stained blood cells.

**Table 1** Comarision Table of Segmentation Techniques

| Number of Database Images | Number of Blood Cells | | |
|---|---|---|---|
| | Manual | Otsu's Threshold | K-Mean Clustering |
| 50 | 940 | 846 | 921 |

## 5 Conclusions

Preprocessing algorithm is successfully tested on blurred and low stained image. It reduces noise in an image and increase contrast of blood cells. Preprocessing technique is depends on accurate calculation of threshold valued from grayscale histogram. Segmentation results of K-Mean clustering and Otsu's threshold is compared with manual count. Algorithms are tested on 50 images. In 50 images total number of blood cells are 940, 921, and 846 manually, by K-Mean and Otsu's threshold respectively. Otsu's threshold can not segments low stained blood cells, so number of count is low compared to manual count. Segmentation accuracy is 98% and 90% respectively for K-Mean clustering and Otsu's threshold. Results show that K-Mean clustering is suitable for segmentation of microscopic blood images.

## Acknowledgment

The authors thank Dr.N.L. Dhande, Consulting Pathologist, Pune, India, for supporting the research with microscopic thin blood images.

## References

1. Aimi Salihah Abdul-Nasir, Mohd Yusoff Mashor, and Zeehaida Mohamed. Colour image segmentation approach for detection of malaria parasites using various colour models and k-means clustering. pages 41–55, 2012.
2. AS Abdul Nasir, MY Mashor, and Z Mohamed. Segmentation based approach for detection of malaria parasites using moving k-means clustering. In *Biomedical Engineering and Sciences (IECBES), 2012 IEEE EMBS Conference on*, pages 653–658. IEEE, 2012.
3. J.E. Arco, J.M. Górriz, J. Ramírez, I. Álvarez, and C.G. Puntonet. Digital image analysis for automatic enumeration of malaria parasites using morphological operations. *Expert Syst. Appl.*, 42(6):3041–3047, April 2015.
4. DPX. Laboratory identification of parasite of public health concern, image library and diagnostic procedure.
5. J. Gatc, F. Maspiyanti, D. Sarwinda, and A.M. Arymurthy. Plasmodium parasite detection on red blood cell image for the diagnosis of malaria using double thresholding. In *Advanced Computer Science and Information Systems (ICACSIS), 2013 International Conference on*, pages 381–385, Sept 2013.
6. S. Ghosh and A. Ghosh. Content based retrival of malaria positive images from a clinical database. In *Image Information Processing (ICIIP), 2013 IEEE Second International Conference on*, pages 313–318, Dec 2013.
7. Diaz Gloria, Gonzalez Fabio A., and Romero Eduardo. A semi-automatic method for quantification and classification of erythrocytes infected with malaria parasites in microscopic images. *Journal of Biomedical Informatics*, 42(2):296–307, 2009.
8. Rafael C Gonzalez. *Digital image processing*. Pearson Education India, 2009.
9. N.A. Khan, H. Pervaz, A.K. Latif, A. Musharraf, and Saniya. Unsupervised identification of malaria parasites using computer vision. In *Computer Science and Software Engineering (JCSSE), 2014 11th International Joint Conference on*, pages 263–267, May 2014.

10. J. Laosai and K. Chamnongthai. Acute leukemia classification by using svm and k-means clustering. In *Electrical Engineering Congress (iEECON), 2014 International*, pages 1–4. IEEE, March 2014.
11. P. Lepcha, W. Srisukkham, L. Zhang, and A. Hossain. Red blood based disease screening using marker controlled watershed segmentation and post-processing. In *Software, Knowledge, Information Management and Applications (SKIMA), 2014 8th International Conference on*, pages 1–7. IEEE, Dec 2014.
12. Clara Mosquera-Lopez and Sos Agaian. Iterative local color normalization using fuzzy image clustering. In *SPIE Defense, Security, and Sensing*, pages 875518–875518. International Society for Optics and Photonics, 2013.
13. M.C. Mushabe, R. Dendere, and T.S. Douglas. Automated detection of malaria in giemsa-stained thin blood smears. In *Engineering in Medicine and Biology Society (EMBC), 2013 35th Annual International Conference of the IEEE*, pages 3698–3701, July 2013.
14. I.K.E. Purnama, F.Z. Rahmanti, and M.H. Purnomo. Malaria parasite identification on thick blood film using genetic programming. In *Instrumentation, Communications, Information Technology, and Biomedical Engineering (ICICI-BME), 2013 3rd International Conference on*, pages 194–198, Nov 2013.
15. SS Savkare and SP Narote. Automated system for malaria parasite identification. In *Communication, Information Computing Technology (ICCICT), 2015 International Conference on*, pages 1–4, Jan 2015.
16. Nazlibilek Sedat, Karacor Deniz, Ercan Tuncay, Sazli Murat Husnu, Kalender Osman, and Ege Yavuz. Automatic segmentation, counting, size determination and classification of white blood cells. *Measurement*, 55:58–65, 2014.
17. F Boray Tek, Andrew G Dempster, and Izzet Kal. Parasite detection and identification for automated thin blood film malaria diagnosis. *Computer Vision and Image Understanding*, 114:21–32, 2010.
18. Man Yan, Jianyong Cai, Jiexing Gao, and Lili Luo. K-means cluster algorithm based on color image enhancement for cell segmentation. In *Biomedical Engineering and Informatics (BMEI), 2012 5th International Conference on*, pages 295–299, Oct 2012.

## Semi-Supervised FCM and SVM in Co-Training Framework for the Classification of Hyperspectral Images

Prem Shankar Singh Aydav[1], Sonjharia Minz[2]
[1]DTE UP, [2]JNU New Delhi
{premit2007,sona.minz}@gmail.com

**Abstract** Collection of labeled samples is very hard, time-taking and costly for the Remote sensing community. Hyperpectral image classification faces various problems due to availability of few numbers of labeled samples. In the recent years, semi-supervised classification methods are used in many ways to solve the problem of labeled samples for the hyperspectral image classification. In this Article, semi supervised fuzzy c-means (FCM) and support vector machine (SVM) are used in co-training framework for the hyperspectral image classification. The proposed technique assumes the spectral bands as first view and extracted spatial features as second view for the co-training process. The experiments have been performed on hyperspectral image data set show that proposed technique is effective than traditional co-training technique.

**Keywords:** Co-training; Fuzzy C-Means; Support Vector Machine; Hyperspectral image classification.

### 1 Introduction

Hyperspectral image classification faces various problems due to limited number of labeled samples, spatial relationship, high dimensionality and noisy data. Many supervised classifications techniques like SVM [1], extreme learning machine [2], multinomial logistic regression [3] etc. have been used for the hyperspectral image classification. The existence of spatial relationship among pixels is one of the main characteristics of hyperspectral images. To improve classification accuracy, the spatial information has been integrated in many ways [4]. Spatial feature extraction and post processing methods are two main approaches to include spatial information [5]. Morphology based feature extraction [6], Co-occurrence based features [7], and Gabor filters based features [8], are leading methods to extract spatial features from hyperspectral remote sensing images.

Collection of labled samples for Hyperspectral image classification is a big problem for both machine learning and remote sensing communities. The supervised classification techniques are not able to achieve good classification accuracy with few numbers of labeled samples. In machine learning community, various semi-supervised classification methods are proposed which learn the classifier with the help of both labeled and unlabeled training samples [9]. To solve the problem of labeled samples in the area of remote sensing, semi-supervised technique have modified and explored in different way for the Hyperspectral image classification



J.M. Corchado Rodriguez et al. (eds.), *Intelligent Systems Technologies and Applications 2016*, Advances in Intelligent Systems and Computing 530,
DOI 10.1007/978-3-319-47952-1_10

[10]. The semi-supervised learning techniques either modify the objective functions of traditional classifier or train the traditional classifier in iterative manner to utilize unlabeled samples.

The objective function of SVM has been modified to include unlabeled samples for the semi-supervised classification of hyperspectral images [11]. The objective function of neural network has been also modified by adding a regulizer term for the semi-supervised classification of hyper spectral images [12]. On the other hand, self-labeled semi-supervised techniques utilize unlabeled samples by training the supervised classifier in iterative manner without any modifications in their objective functions [13]. Self-learning technique has been also applied successfully which utilize spatial information to train a supervised classifier in iterative manner for the semi-supervised classification of the hyperspectral images [14]. One of the major problems of the hyperspectral images that it contains more than hundred spectral bands and when spatial features are extracted the dimensions increases even more. The training of classifiers on high dimensional data set with few numbers of training samples is difficult, and this problem is known as curse of dimensionality. To reduce this effect, we have used co-training semi-supervised method in which spectral band are taken as first view, and spatial features are treated as second view. The supervised classifier is trained with limited data in each view may not be able to extract the exact distribution of classes present in data set that's why clustering technique have been integrated with co-training to help the learn supervised classifier in each spatial and spectral view. To evaluate effectiveness of proposed framework, SVM is used as supervised classifier and Semi-supervised kernel fuzzy c-means (SKFCM) is used as clustering technique.

The main purpose of this Article is to integrate semi supervised clustering technique in co-training process to learn ensemble of supervised classifier on the spatial and spectral view.

Rest of the Article is organized as follows: Section 2 describes Semi-supervised kernel fuzzy c-means, Support Vector machine and spatial feature extraction. Section 3 describes the proposed co-training technique, Section 4 describes experimental details and Section 4 describes the conclusion and future directions.

## 2 **Basic Concepts**

In this paper, we have used SKFCM for the clustering, SVM as supervised classifier and 2-D Gabor filter for the spectral feature extraction. Following are the brief introduction about these techniques:

### 2. 1 Semi-Supervised Fuzzy C-Means

The clustering technique is integrated in our proposed framework to learn most confident samples in spatial and spectral view respectively. The fuzzy c-means is a partition based clustering algorithm which group every sample to each cluster-through assigning a membership value [15].The main objective of fuzzy c-means is to minimize the overall distance of each sample from each cluster:

$$J_m = \min \sum_{i=1}^{c} \sum_{k=1}^{n} (\mu_{ik})^m \|X_k - V_i\|^2 \quad (1)$$

$$\text{s.t}0 \leq \mu_{ij} \leq 1, \sum_{j=1}^{c} \mu_{ij} = 1 \quad (2)$$

where $c$ is the number of clusters, $n$ the number of samples in dataset, $m$ the fuzzifier value. Let the $\mu_{ik}$ denotes the membership value of the $k^{th}$ samples from the $i^{th}$ cluster, $V_i$ denotes the $i^{th}$ cluster center.
For nonlinear clustering fuzzy c-means have been modified with following objectives function with kernel metric:

$$J_k = \min 2 * \sum_{i=1}^{c} \sum_{k=1}^{n} (\mu_{ik})^m \left(1 - K(X_k, V_i)\right) \quad (3)$$

$$\text{s.t.}0 \leq \mu_{ij} \leq 1, \sum_{j=1}^{c} \mu_{ij} = 1 \quad (4)$$

where $K(X, Y)$ denotes the kernel function.
The performance of traditional fuzzy c-means can be improved through incorporating some labeled sample information. In paper [16] Bensaid, Hall et al.1996 have proposed a semi-supervised fuzzy c means by changing the clustering process. Their technique is used here in kernel fuzzy c-means for semi-supervised clustering. In clustering process, the cluster center is updated by both labeled and unlabeled samples. During clustering process, the membership values of unlabeled samples are updated, and the membership values of labeled data remain unchanged. Let $X$ denotes set of labeled and unlabeled data and $U$ denotes the membership matrix of labeled and unlabeled samples in which membership of labeled samples is known.
The cluster center and membership values of unlabeled samples can updated as following way:

$$V_i(semi) = \frac{\sum_{j=1}^{j=nl} (\mu_{ij,l})^m \left(1 - K(X_{j,l}, V_i)\right) X_{j,l} + \sum_{j=1}^{j=nu} (\mu_{ij,u})^m (1 - K(X_{j,u}, V_i)) X_{j,u}}{\sum_{j=1}^{j=nl} (\mu_{ij,l})^m K(X_{j,l}, V_i)) + \sum_{j=1}^{j=nu} (\mu_{ij,u})^m K(X_{j,u}, V_i)} \quad (5)$$

$$\mu_{ij,u} = \frac{1}{\sum_{k=1}^{k=c} \left( \frac{(1 - K(X_{j,u}, V_i))}{(1 - K(X_{j,u}, V_k))} \right)^{\frac{1}{m-1}}} \quad (6)$$

where $X_{j,u}$ is the unlabeled samples, $X_{j,l}$ is the labeled samples, $\mu_{ij,u}$ is the fuzzy membership value of unlabeled samples $X_k$, $\mu_{ij,l}$ is the membership value of labeled samples $X'_j$ to the $i^{th}$ cluster, $nl$ denotes number of labeled samples and $nu$ denotes the number of unlabeled samples.

## 2.2 Support Vector Machine

Support vector machine is supervised classifier which tries to maximize the margin between hyper planes and minimize the training error [17]. It has produced a very robust result for the classification of hyper spectral images. SVM can be used for both linear and nonlinear data sets. For nonlinear separable data sets kernel functions are used to transform datasets in higher dimensions.
The objective of support venter machine is to minimize following equations.

$$\min_{w,b}\left(\frac{1}{2}\|w\|^2 + c\sum_{i=1}^{n} \epsilon_i\right) \tag{7}$$

$$st.\ y_i(w^t x + b) \geq 1 - \epsilon_i \quad \forall_i = 1, \ldots, n \tag{8}$$
$$\epsilon_i \geq 0\ \forall_i = 1, \ldots, n$$

It can by Lagrange multiplier and the decision function is given by:

$$f(x) = sgn\left[\sum_{i=1}^{l} \alpha_i\, y_i x + b\right] \tag{9}$$

Where $y_i$ denote support vecter and $\alpha_i$ is non-zero lagrange multipliers corresponding to$y_i$ .

## 2.3 Spatial Feature Extraction

In proposed method, the spatial features have been taken as the second view for the co-training process. Several spatial feature extraction techniques are developed for the extraction of texture features but Gabor filter are used frequently for the remote sensing image texture extraction [18]. "A Gabor function consists of a sinusoidal plane wave of certain frequency and orientation modulated by a Gaussian envelope". It can be defined as:

$$G_{ij}^{real}(x,y) = \frac{1}{2\pi\sigma_i^2}\exp\left(-\frac{x^2+y^2}{2\sigma_i^2}\right) \times \cos\left(2\pi(f_i\, x\cos\theta_j + f_i\, y\sin\theta_j)\right) \tag{10}$$

$$G_{ij}^{imag}(x,y) = \frac{1}{2\pi\sigma_i^2}\exp\left(-\frac{x^2+y^2}{2\sigma_i^2}\right) \times \sin(2\pi(f_i\, x\cos\theta_j + f_i\, y\sin\theta_j)) \tag{11}$$

Where $f_i$ is the central frequency, $\theta_j$ is the orientation and $x$, $y$ are the coordinates. Let $i$ and $j$ represents the index for scale and orientation respectively. These pair filters are joined for the real value outputs and dropping the imaginary value. To extract features from images a set of Gabor filters are considered with different frequency and orientation. Let $I$ (x,y) is the image pixel and $G$ is the filter with scale $i$ and orientation $j$ the filter image pixel is obtain by:

$$F_{ij}(x,y) = G_{ij}(x,y) * I(x,y) \tag{12}$$

The set of texture for pixels is obtain by:

$$p_{textue}(x, y) = [F_{11}(x, y), \ldots, F_{m,n}(x, y),] \quad (13)$$

For the hyper spectral remote sensing image, the filter bank can applied on each band or on the few principal components of hyper spectral sensing image.

## 3 Proposed Co-training Approach

Co-training semi-supervised technique is a self-labeled technique in which two classifiers help each other to learn the classification model with limited number of labeled and a pool of unlabeled samples [19]. Co-training usually applied on the data sets which contain two disjoint set of independent features. In co-training framework, the unlabeled samples predicted by one classifier are used to train other classifier in iterative manner. It is assumed that labeled predicted by one classifier may be informative for another classifier. The success of co-training algorithms depends on the diversity exits in both views. It is assumed that spectral and spatial features will provide two set of independent view. In our proposed framework, the classifier learned with spectral view is treated as first classifier and classifier learned with spatial features is treated as second classifier.

Co-training framework starts with few labeled samples, so there is possibility that the supervised classifier will not able to extract the exact underlying distribution of classes present in the data sets in their respective view. Recently in machine learning community few techniques have been proposed which uses clustering technique to learn the classifier in self-learning manner [20]. These techniques have been applied successfully in remote sensing image classification with self-learning framework [21, 22].

The clustering techniques are known to fetch meaningful information from unlabeled data. To increase the performance of traditional co-training, we have integrated semi-supervised clustering technique with co-training process. In this paper, the SKFCM is used which learn the cluster centers from both unlabeled and labeled samples. In proposed framework, it is considered that the cluster number is equivalent to the number of classes present in hyperspectral remote sensing data. The proposed frame work start with training of supervised classifier and learning clustering hypothesis in each spectral and spectral view. The unlabeled pixels classified with high confidence by both clustering and classification technique in spectral view is added to labeled pool of spatial view and vice versa.

Let $L$ is the labeled pixels set and $U$ is the unlabeled pixels set. The proposed technique starts with training of a supervised classifier $f$ on both spectral and spatial view with initial labeled pixels $L$. It also learns the cluster centers in each spectral and spatial view with set of labeled and unlabeled pixels. After that, the trained classifier in each view are used to classify the unlabeled pixels in their respective views. The samples classified with high confident in spectral view are added to spatial view and vice versa. This process is continues till some predefined iterations. The confidence in each spatial and spectral view can be calculated by product rule:

$$conf\big(x_i(j)\big) = f(x_{ij}) * \mu_{ij} \tag{14}$$

Where $\mu_{ij}$ denotes the membership values of $i^{th}$ pixel to the $j^{th}$ class acquired by clustering technique, $f(x_{ij})$ denotes the confidence of $i^{th}$ pixel to the $j^{th}$ class acquired by supervised classifier and $conf\big(x_i(j)\big)$ is the overall confidence acquired by both clustering and classifier technique. The whole method is summarized in the algorithm1:

**Algorithm 1- SFCM and SVM in co-training framework on spatial and spectral view**

**Input:** A supervised classifier *f*, clustering technique *c*, labeled data *L*, Unlabeled data pool *U*, Max Iteration *maxitr*, Number of classes' k, Spectral and spatial view, *L1=L, L2=L.*

**Output:** *An ensemble of two classifiers*

***Repeat***

1. Learn supervised classifier *f1* with labeled data *L1* on spectral view.
2. Learn the cluster center with unlabeled data *U* and labeled data *L1* and estimate membership values on spectral view.
3. Classify unlabeled data pool *U* with trained classifier *f1* on spectral view.
4. Get the confidence of unlabeled samples by using equation (14) and select $t_i$ most confident samples from each class *i* on spectral view.
5. Add most confident samples set $(T1 = \cup_{i=1}^{k} t_i)$ ) in labeled pool of spatial view $(L2 = L2 \cup T1\ )$ and remove those samples from unlabeled pool $(U = U - T1)$.
6. If (*U==0*) then return to step 13.
7. Learn classifier *f2* with labeled data *L2* on spatial view.
8. Learn the cluster center with unlabeled data *U* and labeled data *L2* and estimate membership values on spatial view.
9. Classify unlabeled data pool *U2* with trained classifier *f2* on spatial view.
10. Get the confidence of unlabeled samples by using equation (14) and select $t_i$ most confident samples from each class on spatial view.
11. Add most confident samples set $(T2 = \cup_{i=1}^{k} t_i)$ in spectral view labeled data pool $(L1 = L1 \cup T2\ )$ and remove those samples from unlabeled pool $(U = U - T2)$.

***Until*** *maxitr* or *U*==0

12. Learn supervised classifier *f1* with labeled data *L1* on spectral view.
13. Learn classifier *f2* with labeled data *L2* on spatial view.

14. Return ensemble of classifiers *f1, f2* trained on spectral and spatial view respectively.

## 4 Experimental Results and Analysis

The performance of proposed technique has been shown with the experiment on the Botswana hyperspectral image data set. The data set has 30 meter spatial resolution and collected by NASA at the location of Okavango Delta, Botswana [23].The experiment was performed on the 11 major classes of data sets. The data set contains 241 spectral bands of 10-nm resolution from rage of 400-2500nm.The original data set also contain water absorption and noisy bands which are removed in the experiments. The experimental output of the proposed technique is compared with traditional co-training semi- supervised technique. In the experiment, co-training is abbreviated as COSVM and proposed technique is abbreviated as COSVM_ SKFCM.

### 4.1 Design of the Experiment

To perform the training of the classifier in co-training fashion the spectral bands of the data set are treated as first view and spatial features extracted from the 2-d Gabor filter are treated as second view. To extract Gabor filter based features 10 principal components of spectral band have been considered. On each principal component the 2-d Gabor filter is applied by considering four directions (0,$\pi$/4, $\pi$/2, 3$\pi$/4) and four frequency scale (0.5,0.25,0.125,0.0625). The "9x9" neighbouring window is considered for the extraction of spatial features from each pixels in the data set. The window size has been decided through experiments. Ten random set of samples have been chosen in which each set contain 4 labeled samples per class and 1100 unlabeled samples. The parameters of SVM with Gaussian kernel are optimized through cross-validation. In KSFCM fuzzifier value is set as 2 and Gaussian kernel parameter is obtained through cross-validation. The 10 iterations of co-training have been performed. Two pixels which are most confident from each class are put in to labeled samples set from unlabeled data pool in each iteration. To evaluate the performance of proposed technique the test is performed on unlabeled data set and overall classification accuracy (OA) with kappa coefficient (KC) has been acquired.

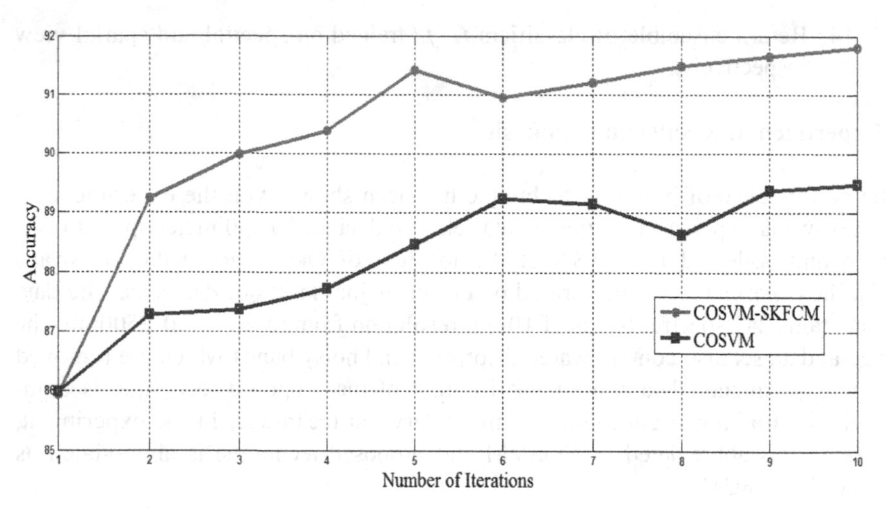


Figure 1 Accuracy of Botswana Dataset

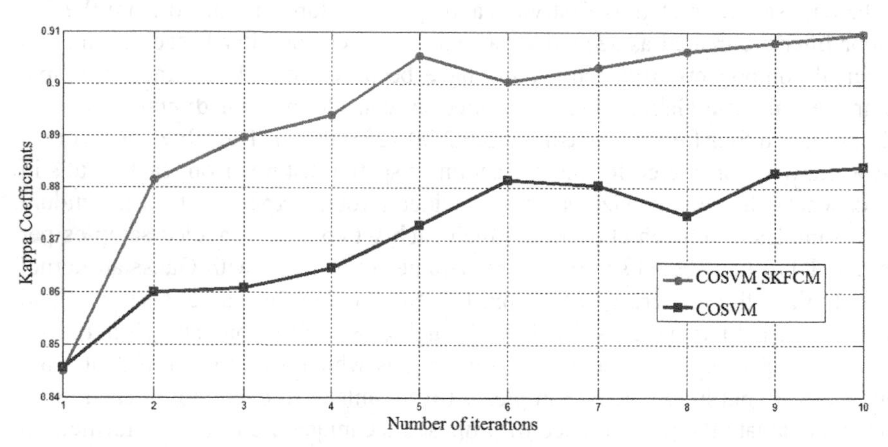


Figure 2 Kappa coefficient of Botswana Dataset

## 4.2 Analysis of Results

In the figure 1, the average accuracy obtained by both techniques in iterations is shown. The analysis of this graph indicates that the accuracy has improved by both techniques by adding most confident samples in iterations. The analysis of graph also shows that the rate of improvement in accuracy of proposed technique COSVM_ SKFCM is greater than traditional cotraining semi-supervised technique. After tenth iterations COSVM_KSFCM have achieved **91.7970%** accuracy while COSVM has achieved 89.4820.The proposed technique COSVM_ SKFCM has achieved 2.31498% more average accuracy than COSVM. In the figure 2, the kappa coefficient is presented as the function of number of iteration number. The analysis of the graph shows that COSVM_ SKFCM has achieved higher kappa coefficient in each iterations. In the final iteration kappa coefficient obtained by proposed technique is **0.9096** which is greater than kappa static achieved by COSVM (0.8841). The average accuracy obtained by each class is shown in Table 1. The analysis of the results shows that the proposed technique has achieved better accuracy than old co-training method for the most of classes. The COSVM_ SKFCM has produced better accuracy for classes 4,5,6,78,10,11. For the class 1, 3 the accuracy achieved by both techniques are almost same.

Table 1 Class accuracy of the Botswana data sets

| **Class** | **COSVM** | **COSVM_ SKFCM** |
|---|---|---|
| 1.Water | **99.9254** | 99.8510 |
| 2.Floodplain grasses | **97.1672** | 93.3549 |
| 3. Floodplain grasses | **96.5916** | 96.2547 |
| 4.Reeds | 89.1487 | **91.7099** |
| 5.Reparain | 64.9262 | **75.3769** |
| 6.Fires Car | 98.9901 | **99.3809** |
| 7.Island Interior | 94.9006 | **95.8935** |
| 8.Acacia woodlands | 82.7175 | **88.0887** |
| 9.Acacia shrub lands | **84.2306** | 82.6599 |
| 10.Acacia grasslands | 96.7543 | **98.9607** |
| 11.mixed mopane | 93.7721 | **96.4887** |
| **Overall Accuracy** | 89.4820 | **91.7970** |
| **Kappa Coefficients** | 0.8841 | **0.9096** |

## 5 Conclusions

This Article presents a co-training framework with semi-supervised clustering and SVM for the semi-supervised spatial-spectral classification of hyperspectral images. The clustering technique helps in the training of supervised classifiers to each spectral and spatial view. The analysis of experimental results shows that the developed technique is better than traditional co-training algorithm. For the spatial feature extraction, only Gabor Filter are used but in future, the performance can be evaluated on other technique like morphological features, Co-occurrence based features, etc. The proposed frame work is wrapper model which can integrate any clustering technique.

## References

1. Melgani, F., & Bruzzone, L. (2004). "Classification of hyperspectral remote sensing images with support vector machines". IEEE Transactions on Geoscience and Remote Sensing, 42(8), 1778-1790.
2. Heras, D. B., Argüello, F., & Quesada-Barriuso, P. (2014). "Exploring ELM-based spatial–spectral classification of hyperspectral images." International Journal of Remote Sensing, 35(2), 401-423.
3. Li, J., Bioucas-Dias, J. M., & Plaza, A. (2010). "Semisupervised hyperspectral image segmentation using multinomial logistic regression with active learning. Geoscience and Remote Sensing", IEEE Transactions on, 48(11), 4085-4098.
4. Li, M., Zang, S., Zhang, B., Li, S., & Wu, C. (2014). "A Review of Remote Sensing Image Classification Techniques: the Role of Spatio-contextual Information". European Journal of Remote Sensing, 47, 389-411.
5. Huang, X., Lu, Q., Zhang, L., & Plaza, A. (2014). "New Postprocessing Methods for Remote Sensing Image Classification: A Systematic Study". IEEE Transactions on Geoscience and Remote Sensing, 52(11), 7140 – 7159.
6. Fauvel, M., Tarabalka, Y., Benediktsson, J. A., Chanussot, J., & Tilton, J. C. (2013). "Advances in spectral-spatial classification of hyperspectral images." Proceedings of the IEEE, 101(3), 652-675.
7. Soh, L. K., & Tsatsoulis, C. (1999). "Texture analysis of SAR sea ice imagery using gray level co-occurrence matrices." IEEE Transactions on Geoscience and Remote Sensing, 37(2), 780-795.
8. Rajadell, O., Garcia-Sevilla, P., & Pla, F. (2013). Spectral–spatial pixel characterization using Gabor filters for hyperspectral image classification.Geoscience and Remote Sensing Letters, IEEE, 10(4), 860-864.
9. Schwenker, F., & Trentin, E. (2014). Pattern classification and clustering: A review of partially supervised learning approaches. Pattern Recognition Letters, 37, 4-14.
10. Persello, C., & Bruzzone, L. (2014). "Active and semisupervised learning for the classification of remote sensing images". IEEE Transactions on Geoscience and Remote Sensing, 52(11), 6937 – 6956.
11. Chi, M., & Bruzzone, L. (2007). "Semisupervised classification of hyperspectral images by SVMs optimized in the primal." Geoscience and Remote Sensing, IEEE Transactions on, 45(6), 1870-1880.
12. Ratle, F., Camps-Valls, G., & Weston, J. (2010). Semisupervised neural networks for efficient hyperspectral image classification. IEEE Transactions on Geoscience and Remote Sensing, 48(5), 2271-2282.

13. Triguero, I., García, S., & Herrera, F. (2013). Self-labeled techniques for semi-supervised learning: taxonomy, software and empirical study. Knowledge and Information Systems, 1-40.
14. Dópido, I., Li, J., Marpu, P. R., Plaza, A., Dias, J. B., & Benediktsson, J. A. (2013). "Semi- supervised self-learning for hyperspectral image classification." IEEE Trans. Geosci. Remote Sens, 51(7), 4032-4044.
15. Bezdek, J. C., Ehrlich, R., & Full, W. (1984). "FCM: The fuzzy c-means clustering algorithm." Computers & Geosciences, 10(2), 191-203.
16. Bensaid, A. M., Hall, L. O., Bezdek, J. C., & Clarke, L. P. (1996). "Partially supervised clustering for image segmentation." Pattern Recognition, 29(5), 859-871.
17. Vladimir N. Vapnik, "Statistical Learning Theory" New York wiley, 1998.
18. Jain, A. K., & Farrokhnia, F. (1990, November). "Unsupervised texture segmentation using Gabor filters". In Systems, Man and Cybernetics, 1990. Conference Proceedings, IEEE International Conference on (pp. 14-19). IEEE.
19. Blum, A., & Mitchell, T. (1998, July). "Combining labeled and unlabeled data with co-training." In Proceedings of the eleventh annual conference on Computational learning theory (pp. 92-100). ACM.
20. Gan, H., Sang, N., Huang, R., Tong, X., & Dan, Z. (2013). "Using clustering analysis to improve semi-supervised classification." Neurocomputing, 101, 290-298.
21. Aydav, P. S. S., & Minz, S. (2015, January). Exploring Self-learning for spatial-spectral classification of remote sensing images. In Computer Communication and Informatics (ICCCI), 2015 International Conference on (pp. 1-6). IEEE.
22. Aydav, P. S. S., & Minz, S. (2015). Modified Self-Learning with Clustering for the Classification of Remote Sensing Images. *Procedia Computer Science, 58*, 97-104.
23. http://www.ehu.es/ccwintco/index.php/Hyperspectral_Remote_Sensing_Scenes.

# Segmentation of Thermal Images Using Thresholding-Based Methods for Detection of Malignant Tumours

Shazia Shaikh, Hanumant Gite, Ramesh R. Manza, K. V. Kale and Nazneen Akhter

**Abstract.** Segmentation methods are useful for dividing an image into regions or segments that may be meaningful and serve some purpose. This study aims at application of thresholding-based segmentation methods on thermal images of skin cancer and analysis of the obtained results for identification of those algorithms that are superior in performance for identification of malignancy and for extracting the ROI (Region Of Interest). Global thresholding methods were applied on the skin cancer thermal images for a comparative study. While the MinError(I) method gave the most desirable results, Huang, Intermodes, IsoData, Li, Mean, Moments, Otsu, Percentile and Shanbhag were effective and consistent in performance as well. Other methods like MaxEntropy, Minimum, RenyiEntropy, Triangle and Yen could not yield expected results for segmentation and hence could not be considered useful. The investigations performed, provided some useful insights for segmentation approaches that may be most suitable for thermal images and consequent ROI extraction of malignant lesions.

Shazia Shaikh
Department of Computer Science & Information Technology, Dr. Babasaheb Ambedkar Marathwada University, Aurangabad (Maharashtra) 431001 India.
e-mail: shazia_shaikh07@yahoo.com
Hanumant Gite
Department of Computer Science & Information Technology, Dr. Babasaheb Ambedkar Marathwada University, Aurangabad (Maharashtra) 431001 India.
e-mail: hanumantgitecsit@gmail.com
Ramesh R. Manza
Department of Computer Science & Information Technology, Dr. Babasaheb Ambedkar Marathwada University, Aurangabad (Maharashtra) 431001 India.
e-mail: manzaramesh@gmail.com
K. V. Kale
Department of Computer Science & Information Technology, Dr. Babasaheb Ambedkar Marathwada University, Aurangabad (Maharashtra) 431001 India.
e-mail: kvkale91@gmail.com
Nazneen Akhter
Maulana Azad College of Arts, Science and Commerce, Dr. Rafiq Zakaria Campus, Aurangabad (Maharashtra) 431001 India.
e-mail: getnazneen@gmail.com



J.M. Corchado Rodriguez et al. (eds.), *Intelligent Systems Technologies and Applications 2016*, Advances in Intelligent Systems and Computing 530,
DOI 10.1007/978-3-319-47952-1_11

# 1 Introduction

Segmentation techniques are popularly applied in medical image processing for separation of objects of interests from their backgrounds. Various studies [1-7] have experimented and applied threshold-based segmentation techniques on thermal images for successful disease identification, most of which have been done for breast cancer identification. But there seem to be no attempts reported towards application of these techniques towards lesion extraction of skin. In this work, segmentation methods based on global thresholding have been applied on thermal images of skin cancer to identify their suitability in segmentation of malignant lesions from the healthy skin. The segmented images were further processed for ROI extraction of the regions that the lesions were occupying. The tool chosen for this study is ImageJ 1.48V which is a menu-driven and open source application and the source codes are readily available publicly in the form of plug-ins. Provisions have also been made for modifying and reusing these codes. The results highlight some algorithms that are more suitable in segmenting thermal images as compared to the other algorithms. The algorithms were rated based on some performance factors and the ROI extraction was performed from the segmented images obtained from the best rated algorithms.

# 2 Background

As the cells are the basic building blocks of our body, malfunctioning in their lifecycle due to some causes may lead to formation of tumours that are developed as a result of formation of new unneeded cells and also due to presence of cells that must expire [8]. This causes cancer which is a non-contagious but a fatal disease. Cancer can affect almost all organs of human body. Cancer is characterized by stages - from the time it begins to the level of development (size), how far it has spread in the body, and the damage it has caused. Therefore, it can be said that the sooner the cancer is detected, the higher are the chances for successful treatment and cure. The tumour formation in the body can be of two types- benign and malignant. Tumours may form when the balanced cell lifecycle is disturbed [9]. The malignant tumour is a confirmation of cancer while the benign tumour is not a cause of great concern as it is usually not dangerous [10]. Factors like obesity, exposure to carcinogenic chemicals, sunlight exposure, excessive drinking etc. maybe some of the causes of tumour formation. As the lesions grow, they start to get clearly visible and their characteristics are easily identifiable. In case of suspicion, through a screening method, the smallest possible malignancy in its early stage is detected [11].

Since the era of Hippocrates, health and body temperature have been associated with each other [12, 13]. Infrared radiations given out by human body were first discovered by Herschel in the 1800s [13]. Investigations based on study of temperature profiles of lesions are interesting as the malignant tumours of body tend to be warmer and exhibit higher temperatures from the healthy skin around owing to increase in the metabolic activity taking place inside those tumours [14]. The analysis of temperature of skin surface forms the basis of medical thermography. The foundation of Infrared Thermography (IRT) is the measurement of radiant thermal energy distribution which is emitted from a target [15]. The thermal measurement of skin on each of its points can be represented on the corresponding pixel in a thermal image of the region using false colours [16]. Figure 1 shows some examples of use of thermal imaging in identifying abnormalities through temperature profiling of body parts. The invention of thermal imaging systems in 1950, gave a direction to the research community for further study and analysis of diseases through the perspective of considering the thermal characteristics that the body displays when affected by those diseases. The earliest reports of application for thermography for medical research dates back to 1940s when the first thermograms used were made from experimental imaging equipment that was made using a liquid nitrogen cooled indium antimonide detector [17]. The advancements in thermal imaging systems till date have led to availability of extremely sophisticated thermal imaging systems that are highly sensitive and provide great amount of detailing.

For the detection of diseases like diabetic neuropathy, carcinoma of breast, and vascular disorders at the periphery, thermography has exhibited success [1] and it has also been employed for identification of problems related to dermatology, imaging of the brain, heart, gynaecology and other medical conditions [18].

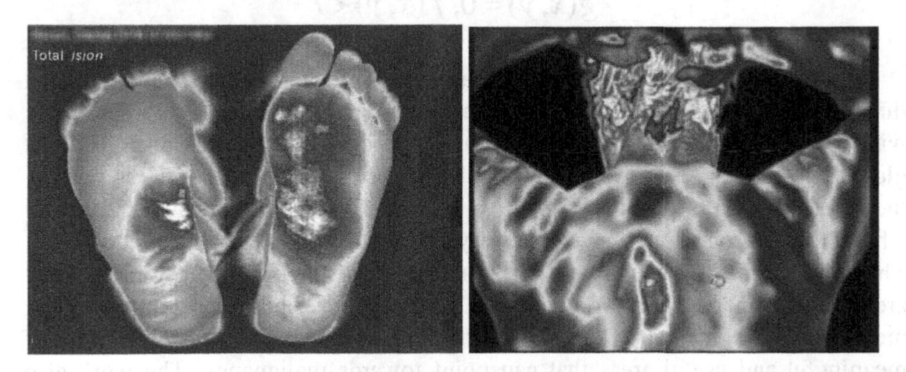

**Fig. 1** Thermal imaging for medical diagnostic purposes, Image Courtesy [19, 20]

The identification of the ROI is the first task of a thermography based analysis and processing system [1]. Thermal images do not have shape and specific limits and therefore, ROI extraction from these types of images is challenging [21]. The softwares that are commercially available, employ regular shapes like rectangles, squares, circles, etc for defining ROIs which is a poor representation of some anatomical regions [1] and it leads to inefficient extraction of ROIs as these anatomical regions are usually complex and possess quite irregular morphologies. Hence there is need for further exploration of image processing techniques that can be useful for correct extraction of lesions from their thermal regions. Segmentation techniques are popularly applied in medical image processing for separation of objects of interests from their backgrounds.

In image segmentation, an image is divided into set of regions based on the similar features shared by pixels in that image. One of the objectives of this process is to segregate an image into sections for performing some more image processing tasks on the segmented regions [22]. Identification of image sections based on the pixels having similarity in terms of intensity values forms the basis of thresholding-based methods. Rigid objects on a contrast background are obtained as a result of applying threshold based segmentation techniques [1]. The image is first converted into its grey-level version and the final segmented image is binary. The binary format is suitable to select the meaningful objects from the segmented image. A thresholded image can be represented as g(x,y) and the thermal image is a function f(x,y) and their relation is given as below.

$$g(x, y) = 1, f(x, y) > T$$

Or

$$g(x, y) = 0, f(x, y) < T$$

A computationally fast and simple approach of segmentation is the global thresholding method which is based on the assumption that image has a histogram with two modes which leads to procurement of a binary image as a result [23]. In global thresholding, T is used for all the image pixels as a single value indicating the threshold. On the contrary, adaptive or local thresholding methods are useful when the background has uneven intensity distribution and a varying contrast [23]. The adaptive method works with different values of threshold for different local areas of the image. This work aims to implement various global thresholding methods for segmentation of skin cancer thermal images for extraction of meaningful and useful areas that can point towards malignancy. The work also aims to assess the performance of all algorithms to identify the ones which are most effective in providing the expected results and meaningful extraction of ROIs.

## 3 Related Work

A basic step in analysis of thermal images is image segmentation [1]. This section discusses the methods and approaches that were applied to thermal images of patients for the purpose of image segmentation for diagnosis of diseases. For segmentation of breast thermal images S. Motta et al. [2] applied global variance by Otsu's threshold to subtract the background with the final aim of segmentation of both breasts as symmetrically as possible and were satisfied with the results of their study. With the aim of mass detection in thermal images, R., M. et al. [3] applied thresholding based methods like grey threshold, RGB technique, and K-means technique and pointed out specifically that the RGB technique suffered from the limitation of data loss. Dayakshini, D. et al. [4] proposed a novel method of breast segmentation called Projection Profile Analysis for identifying the boundaries in the breast thermograms. For edge detection of the breast, they successfully used the Sobel operator while adaptive thresholding was applied for changing the ROI's color to white and reported that the overall results of their approach were satisfactory. For processing breast images by automated segmentation, De Oliveira et al. [5] applied various image processing steps like threshold, clustering, identification and improvement of corners on 328 lateral thermal images of breasts. A thresholding method was applied by them to separate the background and patient body in two modes. 96% and 97% of mean values of accuracy and sensitivity were obtained by them. Duarte A. et al. [1] developed an application using Matlab for characterizing thermal images that facilitates selection of any ROI without considering its geometry and provides options for its further optimizations. However, the thresholding algorithm implemented by them is not specified. Barcelos E. et al. [6] performed a study on 28 subjects who were athletes for the purpose of preventing muscle injury. They proposed automated recognition of flexible ROIs that have shape of body by optimizing the Otsu's method of segmentation and further performing progressive analyses of thermograms. The proposed segmentation approach outperformed the other two automatic segmentation methods of mean and Otsu. On the contrary, Amin Habibi et al. [24] reported ACM technique as showcasing better precision in extraction of breast shape as compared to the thresholding methods.

## 4 Methodology

This section describes the tool chosen for study, information about the dataset and the methods applied as part of study.

### 4.1 Tool, Plugins

For this study, the image processing tool used is ImageJ 1.48V as it is open source. ImageJ offers multitudes of plug-ins (which can be developed by a user with the help of its in-built editor and java compiler) and macros for achieving extensive range of image processing tasks.

### 4.2 Dataset

This work was undertaken using five thermal images of skin cancer that were provided by Dr. Javier Gonzalez Contreras [25] a prominent researcher. The subjects, whose thermal images have been used, were already diagnosed with skin cancer through other available diagnostic techniques. In future, the public availability of standard databases of skin cancer thermal images would be a vital means of further research in this area.

### 4.3 Methods

Segmentation methods are widely used to extract ROIs from images. In case of thermal images, there is a smooth variance of colours throughout the captured image. This aspect reflects the absence of sharp changes in thermal images and also highlights the characteristic of heat propagation across any biological body or an object. For this reason, the task of ROI extraction becomes challenging. The main aim of this work was to perform thresholding based segmentation of skin cancer thermal images to analyze the methods that are most suitable and close to desired segmentation results.

ImageJ provides global thresholding algorithms by means of the plugin named Auto Threshold. This plugin offers 16 algorithms for global thresholding. They are Default, Huang, Intermodes, IsoData, Li, MaxEntropy, Mean, MinError(I), Minimum, Moments, Otsu, Percentile, RenyiEntropy, Shanbhag, Triangle and Yen. The results obtained are segmented binary images. With the use of Try All option, a montage containing results of all the 16 algorithms was obtained which is very useful for a visual analysis and comparison of the most suitable result. Figure 2 shows the colour thermal images of a sample where the malignancy can be observed at the area of the nose and its surrounding. In this work, the colour thermal images were first converted to 8-bit gray scale images (as can be observed in the sample shown in figure 3) as the application of the algorithms directly on the colour thermal images gave undesirable results.

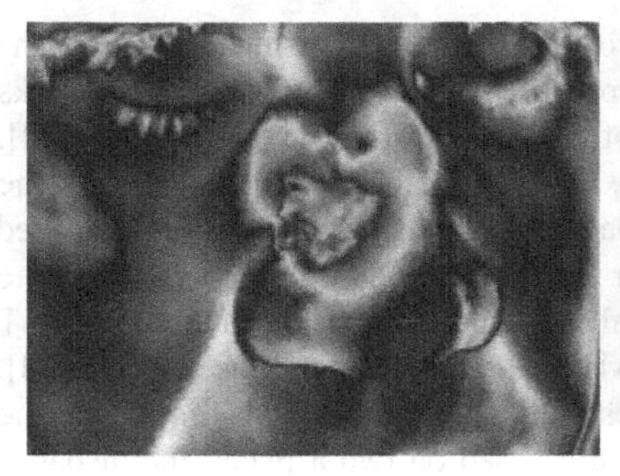

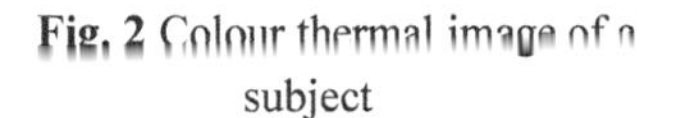
**Fig. 2** Colour thermal image of a subject

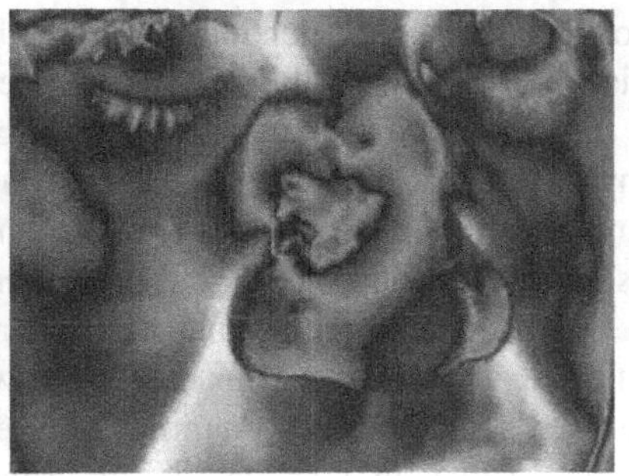

**Fig. 3** Gray scale conversion of colour thermal image

The algorithms that fall under global thresholding are histogram-based and they give results of 8-bit and 16-bit images in a binary form. Also known as iterative Intermeans, the default method is the modified form of IsoData algorithm [26] and is the default method of global thresholding that is used in ImageJ. This method segments objects from their background by calculating a mean value that is equal to threshold T or below T. The iterations continue to find mean and increment T until it is greater than the total mean [27]. The Default and IsoData are similar methods in the plug-in. The Huang method applies the Huang's fuzzy thresholding method by using the Shannon's entropy function for image segmentation. The Huang's method minimizes the value of fuzziness of an image and defines a fuzzy range to calculate the value of threshold [28]. The Intermodes method works with a bimodal histogram that is levelled by means of iterations by a running average of value 3. This method iterates until only two local maxima remain. The addition of these maxima and half of this addition is the threshold value T [29]. The Li's method [30] is an iterative version of the minimum cross entropy thresholding [31] that is computationally quick and derives the threshold by decreasing the cross entropy using one-point iteration approach. This method is based on image thresholding using grey-level histograms. The MaxEntropy method implements the method proposed by Kapur et al. [32] which selects a threshold from the grey level histogram of a picture using the concept of entropy. The working of Mean method is based on application of mean of the grey levels as the value of T [33]. The method of MinError(I) is the iterative version[34] of the Minimum Error Thresholding method [35] which is an efficient approach to the problem of minimum error thresholding. The Minimum method is based on the assumption of a bimodal histogram that is levelled iteratively with a running average size of 3, until only two local maxima remain [29]. Moments method is the Tsai's method [36] to automatic thresholding using the principle of preservation of moment from the input image to the output image that gives meaningful grey levels to the classes. The Otsu's method of segmentation is an unsupervised method of automatic threshold selection that looks for that value of T, which decreases the variance between two classes [37]. The percentile method by Doyle W. [38] works on the assumption that the fraction of the foreground pixels is 0.5 [34]. RenyiEntropy is homogenous to the MaxEntropy method though it applies the Renyi's entropy method [34]. The Shanbhag method is an entropy-based modified version of the method proposed by Kapur et. al. [32]. The triangle method is the extended version of the original Triangle algorithm and it calculates the value of T by finding out the max peak side from where the data extends the farthest [34]. The Yen's method is a multilevel thresholding algorithm that finds out the threshold values by consideration of criterion based on two aspects- the disparity between the thresholded images and original ones and also the count of bits required for the representation of the thresholded image [39].

# 5 Results and Discussion

The results of the segmentation algorithms that were applied, have been assigned performance factors based on visual interpretation. Table 1 describes those factors and their interpretation along with the ratings to be assigned to the algorithms based on their performance factors.

**Table 1** Performance factors for analysis of segmentation algorithms and their interpretation

| Performance Factor | Interpretation | Ratings based on performance factor |
|---|---|---|
| Effective and Consistent | The method is successful in segmenting the image as required for lesion identification. It is possible to extract the lesion. | 1 |
| Inconsistent | The method is producing varying results. Lesion can be identified in some cases and cannot be identified in some cases | 2 |
| Ineffective | The method is not successful in segmenting the images as required for lesion identification. Lesion remains unseen and therefore cannot be extracted. | 3 |

In thermal images, the changes in temperature are indicated with changes in colour. The thermal changes in biological bodies, when seen through thermal images depict some degree of smoothness during the traversal from one colour frequency to another and there is lack of sudden and sharp change. Hence, the thresholding algorithms applied over thermal images have different performances as compared to their performances with other types of images. In this work, the algorithms Default, Huang, IsoData, Mean, MinError(I), Otsu, Percentile, and Shanbhag have been found close to delivering desired results with a fair amount of consistency. The areas with higher temperature were segmented from the surrounding area of lower temperature effectively with the above mentioned methods. Though it was semi-automatic, the ROI extraction of all samples was most conveniently done after segmentation using the MinError(I) method. The methods Intermodes, Li, and Moments have been found as inconsistent in expected segmentation. With these three methods, the segmented warmer areas were approximations of the expected results. The most ineffective and undesired results were delivered by MaxEntropy, Minimum, RenyiEntropy, Triangle and Yen as no traces of warm segments could be seen in their segmentation results. The results after applying the sixteen algorithms on one of the samples can be seen in table 2 and table 3 along with the performance factors assigned to each of them. The ROIs extracted from other samples after segmentation using the MinError(I) method can be seen in table 4. Table 5 lists out the performance factor allotment to each of the algorithms in tabulated and summarised form based on their application to the dataset.

**Table 2** Results of algorithms 1-8 from the 16 thresholding algorithms and their performance factor based rating

| Algorithm | Result of Thresholding | Rating based on performance factor |
|---|---|---|
| Default |  | 1 |
| Huang |  | 1 |
| Intermodes |  | 2 |
| IsoData |  | 1 |
| Li |  | 2 |
| MaxEntropy | | 3 |
| Mean |  | 1 |
| MinError(I) |  | 1 |

**Table 3** Results of algorithms 9-16 from the 16 thresholding algorithms and their performance factor based rating

| Algorithm | Result of Thresholding | Rating based on performance factor |
|---|---|---|
| Minimum |  | 3 |
| Moments |  | 2 |
| Otsu |  | 1 |
| Percentile |  | 1 |
| RenyiEntropy | | 3 |
| Shanbhag |  | 1 |
| Triangle | | 3 |
| Yen |  | 3 |

**Table 4** ROI extraction of samples after segmentation using the MinError(I) method

| Sample | Result of MinError(I) | ROI extraction |
|---|---|---|
| 1. |  | 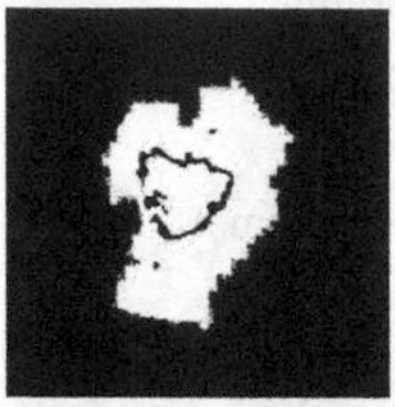 |
| 2. | 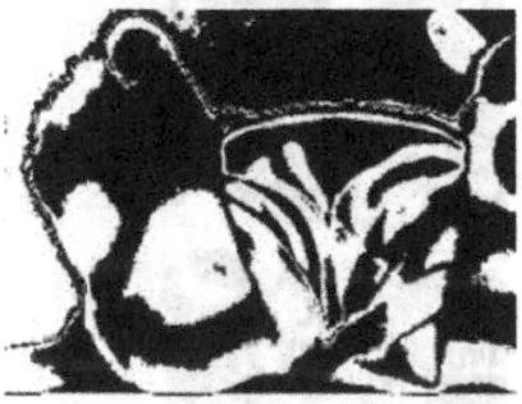 | 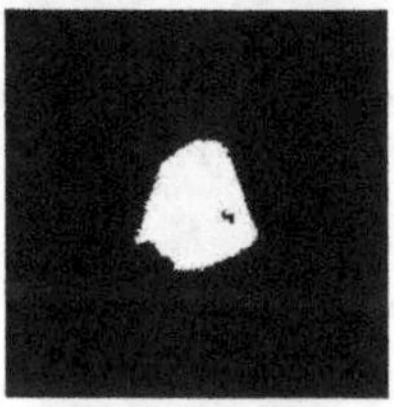 |
| 3. |  | 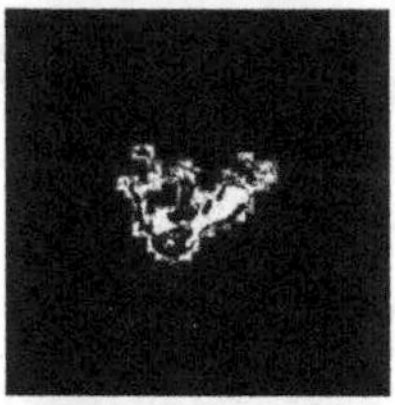 |
| 4. |  | 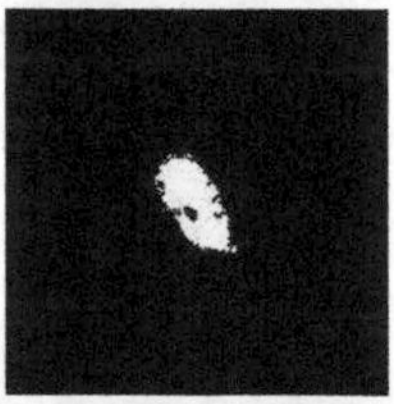 |
| 5. |  |  |

**Table 5** Summary of 16 thresholding algorithms indicating their rating based on performance factor allotted to them.

| Algorithm | Rating based on performance factor allotted to the algorithm |
|---|---|
| Default | |
| Huang | |
| IsoData | |
| Mean | 1 |
| MinError(I) | |
| Otsu | |
| Percentile Shanbhag | |
| Intermodes | |
| Li | 2 |
| Moments | |
| MaxEntropy | |
| Minimum | |
| RenyiEntropy | 3 |
| Triangle | |
| Yen | |

## 6 Conclusion and Future Work

Segmentation techniques have been widely acknowledged and used for analysis of medical images since they perform fairly at separating region of interest from the background data. In this work, some of the thresholding based segmentation methods proved effective in giving the desired result. Hence they can be termed as useful methods that may be optimized further and considered to be included into future automated systems that are capable of ROI extraction, image analysis and processing. Furthermore, the availability of an extensive dataset with thermal images of malignant and non-malignant skin lesions would be a vital means to validate these results and enable the study to further extend towards classification of tumours into malignant and non-malignant types.

## Acknowledgement

The authors are very thankful to UGC, Delhi for provision of grants for carrying out this research work under the MANF Scheme and BSR fellowship. The authors would also like to thank Dr. Javier Gonzalez Contreras (PhD, Researcher, and

Professor at the Autonomous University of San Luis Potosi, Mexico) for providing the skin cancer thermal images that were used for this work.

## References

1. Duarte, A., Carrão, L., Espanha, M., Viana, T., Freitas, D., Bártolo, P., Faria, P., Almeida, H.: Segmentation Algorithms for Thermal Images. Procedia Technology. 16, 1560-1569 (2014).
2. S. Motta, L., Conci, A., Lima, R., Diniz, E., Luís, S.: Automatic segmentation on thermograms in order to aid diagnosis and 2D modeling. Proceedings of 10th Workshop em Informática Médica. pp. 1610-1619 (2010).
3. R., M.Thamarai, M.: A Survey of segmentation in mass detection algorithm for mammography and thermography. International Journal of Advanced Electrical and Electronics Engineering (IJAEEE). 1, (2012).
4. Dayakshini, D., Kamath, S., Prasad, K., Rajagopal, K.: Segmentation of Breast Thermogram Images for the Detection of Breast Cancer – A Projection Profile Approach. Journal of Image and Graphics. 3, (2015).
5. P. S., J., Conci, A., Pérez, M., Andaluz, V.: Segmentation of infrared images: A new technology for early detection of breast diseases. De Oliveira, J.P.S., Conci, A., Pérez, M.G. and Andaluz, V.H., 2015, March. Segmentation of infrared images: A new technology for early detection of breast diseases. In Industrial Technology (ICIT), 2015 IEEE International Conference on. pp. 1765-1771. IEEE (2015).
6. Barcelos, E., Caminhas, W., Ribeiro, E., Pimenta, E., Palhares, R.: A Combined Method for Segmentation and Registration for an Advanced and Progressive Evaluation of Thermal Images. Sensors. 14, 21950-21967 (2014).
7. Mahajan, P., Madhe, S.: Morphological Feature Extraction of Thermal Images for Thyroid Detection. Technovision-2014: 1st International Conference. pp. 11-14. International Journal of Electronics Communication and Computer Engineering (2015).
8. Cancer: MedlinePlus, https://www.nlm.nih.gov/medlineplus/cancer.html.
9. Tumor: MedlinePlus Medical Encyclopedia, https://www.nlm.nih.gov/medlineplus/ency/article/001310.htm.
10. Benign and Malignant Tumors: What is the Difference?, https://thetruthaboutcancer.com/benign-malignant-tumors-difference/.
11. Mohd. Azhari, E., Mohd. Hatta, M., Zaw Htike, Z., Lei Win, S.: Tumor detection in medical imaging: a Survey. International Journal of Advanced Information Technology. 4, 21-30 (2014).
12. Fernández-Cuevas, I., Bouzas Marins, J., Arnáiz Lastras, J., Gómez Carmona, P., Piñonosa Cano, S., García-Concepción, M., Sillero-Quintana, M.: Classification of factors influencing the use of infrared thermography in humans: A review. Infrared Physics & Technology. 71, 28-55 (2015).
13. Ring, E.: The historical development of temperature measurement in medicine. Infrared Physics & Technology. 49, 297-301 (2007).
14. Pirtini Çetingül, M.Herman, C.: Quantification of the thermal signature of a melanoma lesion. International Journal of Thermal Sciences. 50, 421-431 (2011).

15. Kylili, A., Fokaides, P., Christou, P., Kalogirou, S.: Infrared thermography (IRT) applications for building diagnostics: A review. Applied Energy. 134, 531-549 (2014).
16. Mitra, S.Uma Shankar, B.: Medical image analysis for cancer management in natural computing framework. Information Sciences. 306, 111-131 (2015).
17. Ring, E., Hartmann, J., Ammer, K., Thomas, R., Land, D., W. Hand, J.: Infrared and Microwave Medical Thermometry. Experimental Methods in the Physical Sciences. 43, 393-448 (2010).
18. Lahiri, B., Bagavathiappan, S., Jayakumar, T., Philip, J.: Medical applications of infrared thermography: A review. Infrared Physics & Technology. 55, 221-235 (2012).
19. Other applications, medical thermography, infrared screening | Infrarood borstscreening medische thermografie, http://www.medicalthermography.com/other-applications-medical-thermography/.
20. Body Thermal Imaging | Arizona Cancer Screening | Body and Breast Thermal Imaging Tempe, http://arizonacancerscreening.com/thermal-imaging/full-body-thermal-imaging-phoenix.
21. Zhou, Q., Li, Z., Aggarwal, J.: Boundary extraction in thermal images by edge map. Other Applied Computing 2004 - Proceedings of the 2004 ACM Symposium on Applied Computin. pp. 254-258. , Nicosia (2016).
22. G. Sujji, E., Y.V.S., L., G. Jiji, W.: MRI Brain Image Segmentation based on Thresholding. International Journal of Advanced Computer Research. 3, 97-101 (2013).
23. Tavares, J.Jorge, R.: Computational vision and medical image processing. Springer, Dordrecht (2011).
24. Habibi, A. Shamsi, M.: A Novel Color Reduction Based Image Segmentation Technique For Detection Of Cancerous Region in Breast Thermograms. Ciência e Natura. 37, 380-387 (2015).
25. Gonzalez, F., Martinez C, C., Rodriguez R, R., Machuca E, K., Segura U, V., Moncada, B.: Thermal signature of melanoma and non-melanoma skin cancers. 11th International Conference on Quantitative InfraRed Thermography (2012).
26. Riddler, T.: Picture Thresholding Using an Iterative Selection Method. IEEE Transactions on Systems, Man, and Cybernetics. 8, 630-632 (1978).
27. Auto Threshold - ImageJ, http://fiji.sc/Auto_Threshold.
28. Huang, L. Wang, M.: Image thresholding by minimizing the measures of fuzziness. Pattern Recognition. 28, 41-51 (1995).
29. Prewitt, J. Mendelsohn, M.: THE ANALYSIS OF CELL IMAGES*. Annals of the New York Academy of Sciences. 128, 1035-1053 (2006).
30. Li, C. Tam, P.: An iterative algorithm for minimum cross entropy thresholding. Pattern Recognition Letters. 19, 771-776 (1998).
31. Li, C. Lee, C.: Minimum cross entropy thresholding. Pattern Recognition. 26, 617-625 (1993).
32. Kapur, J., Sahoo, P., Wong, A.: A new method for gray-level picture thresholding using the entropy of the histogram. Computer Vision, Graphics, and Image Processing. 29, 273-285 (1985).
33. Glasbey, C.: An Analysis of Histogram-Based Thresholding Algorithms. CVGIP: Graphical Models and Image Processing. 55, 532-537 (1993).

34. Auto Threshold - ImageJ, http://imagej.net/Auto_Threshold.
35. Kittler, J.Illingworth, J.: Minimum error thresholding. Pattern Recognition. 19, 41-47 (1986).
36. Tsai, W.: Moment-preserving thresolding: A new approach. Computer Vision, Graphics, and Image Processing. 29, 377-393 (1985).
37. Otsu,Nobuyuki.,: A Threshold Selection Method from Gray-Level Histograms. IEEE Transactions on Systems, Man, and Cybernetics. 9, 62-66 (1979).
38. Doyle, W.: Operations Useful for Similarity-Invariant Pattern Recognition. Journal of the ACM. 9, 259-267 (1962).
39. Jui-Cheng Yen, Fu-Juay Chang, Shyang Chang,: A new criterion for automatic multilevel thresholding. IEEE Transactions on Image Processing. 4, 370-378 (1995).

# Recognition of Handwritten Benzene Structure with Support Vector Machine and Logistic Regression a Comparative Study

**Shrikant Mapari,**

Assistant Professor, SICSR, Symbiosis International University, Pune (MH), India.
Shrikant.mapari@sicsr.ac.in, mshreek@gmail.com.

**Dr. Ajaykumar Dani**

Professor, G H Raisoni Institute of Engineering & Technology (GHRIET), Pune(MH), India.
ardani123@gmail.com

**Abstract**

A chemical reaction is represented on a paper by chemical expression which can contain chemical structures, symbols and chemical bonds. If handwritten chemical structures, symbols and chemical bonds can be automatically recognized from the image of Handwritten Chemical Expression (HCE) then it is possible to automatically recognize HCE. In this paper we have proposed an approach to automatically recognize benzene structure which is the most widely used chemical compound in aromatic chemical reactions. The proposed approach can recognize benzene structure from the image of HCE. We have developed two classifiers to classify the benzene structure from HCE. The first classifier is based on Support Vector Machine (SVM) and the second classifier is based on logistic regression. The comparative study of the both classification technique is also presented in this paper. The outcome of comparison shows that both classifiers have accuracy of more than 97%. The result analysis shows that classification technique based on SVM classification performs better than classification technique using logistic regression.

**Key Words**: Handwritten chemical expression, recognition, classification, support vector machine, logistic regression, SURF descriptor, classifiers, contours.



J.M. Corchado Rodriguez et al. (eds.), *Intelligent Systems Technologies and Applications 2016*, Advances in Intelligent Systems and Computing 530,
DOI 10.1007/978-3-319-47952-1_12

# 1. Introduction

The chemical reactions containing benzene compounds are called as aromatic reaction. The benzene in an aromatic reaction is represented by a standard graphical shape when this reaction is expressed on paper in either handwritten or printed format. A handwritten benzene structure on a paper written by pen or pencil is a form of hand drawn graphical symbol. The recognition of such hand drawn graphical symbol from its scanned image is challenging task due to the free style of drawing [1]. The benzene structure is represented by using geometric shapes like hexagon, lines and circle. The recognition of handwritten Benzene structure means recognition of these hand drawn geometric shapes. The major hurdle in recognition of handwritten benzene structure is the shape distortions introduced by irregular nature of handwriting. The shape distortions can reduce the uniformity and regularity of shapes. There are some existing systems which recognize the hand drawn benzene structures. However these systems have been developed for online handwritten recognition i.e. the characters or graphical symbols written by using different types of electronic devices. Some of these existing systems have used predefined geometric structures to draw benzene structures. In this paper we had developed a system which automatically recognizes handwritten benzene structure from the scanned images of HCE written on paper by pen (i.e. ink) or pencil (Offline hand written characters or symbols). Automatic identification of benzene structure is an important step in the recognition of complete HCE and subsequent identification of HCE from a statement written in natural language by ink or pencil.

Our proposed system accepts a scanned image of handwritten chemical expression or statements as input and recognizes the benzene structure contained in it. As a component of proposed system we have developed two classifiers. This system has been developed with two different classifiers which automatically assign class label "Y" (Benzene structure) or "N" (other chemical symbol) to chemical symbols of HCE. The output of the system is an image which shows the recognized benzene structure with rectangle around it and molecular formula of benzene.

There are four processing stages in the proposed system. These processing stages are pre-processing, segmentation, feature extraction and classification. In the first stage the input image is pre-processed with different image processing function to obtain sharpen binary image. This pre-processed image is segmented in next step to separate the handwritten chemical structure, symbols and text. After segmentation, the features of image segments are identified and feature vector is created using SURF descriptor (explained in section 3.4.1). This feature vector is used to train the developed classifiers based on SVM and logistic regression.

The rest of the paper is organized as follows. In section 2 work done for recognition of handwritten benzene structure is discussed. The proposed the proposed system explained in section 3. The experimental results and analysis are presented in section 4. Finally we conclude in section 5.

## 2. Existing Work

The chemical expressions in aromatic chemical reactions are expressed using chemical structures and bonds [2]. These chemical structures are drawn by using graphical symbols. The recognition of graphical symbols and shapes is the one of the emerging area of research since last decade. An approach to measure the similarities in the shapes and use these similarities to recognize the objects has been developed by Belongie et al. [3]. The identification of online handwritten chemical structures and symbols is challenging task due to special structures of chemical expression, broken characters and free handwriting styles [4]. A method to recognize online handwritten chemical structure and symbols is proposed in [4]. This method uses eleven (11) dimensional localized normalized feature vectors along with Hidden Markov Model (HMM). A model has been proposed by Remel et al. in [5] to recognize graphical entities from online chemical formulas. This model is based on text localization module and structural representation of image. The proposed model separates the graphical entities from the text. It may not recognize the molecular formula represented by graphical entities completely. A double stage classifier based on Support Vector Machine (SVM) and HMM has been proposed by Zhang et al. in [7] to recognize online handwritten chemical structures and symbols. A System has been developed by Tang et al. in [8] to recognize the online progressive handwritten chemical formulas. This system accepts the inputs from pen based and touch devices where the strokes of digital ink are used for recognition task.

There few more research papers which deal with recognition of printed chemical structures and symbols. A novel approach has been proposed by Park Jungkap et al. in [6] to extract chemical structure from digital raster images of scientific literature. The system developed in this work is named as ChemReader. This system converts these structures into system readable format. This system is developed for printed chemical structure and is not tested with handwritten chemical structure.

The above discussion related to current work being carried out for the recognition of handwritten chemical structures and symbols shows that the most of the reported work is carried out for online mode (i.e. when HCE is written using some type of electronic device). There are some research papers which deal with recognition of printed chemical structures. The printed chemical structures have regularity in their size and shape which makes the recognition task much simpler as compared to hand written structures. The recognition of offline handwritten chemical symbols is challenging task as the recognition has to be done on scanned images. As already stated the main challenge in offline handwritten character recognition is the irregularity in size and shapes of free hand drawn chemical structures by multiple users. There is not much of research work done in the area of recognition of chemical structures which are written by hand using pen or pencil. In this paper we have proposed a system to recognize a benzene structures from scanned images of HCE. In this system we have proposed two classifiers, which classify the benzene structures from other chemical symbols. These two classifiers are based on SVM and logistic regression.

# 3. The Proposed System

The proposed system accepts scanned image of HCE as an input and recognizes the benzene structure from it and highlights molecular formula of benzene in the output. The major processing steps in this system are pre-processing, segmentation, feature extraction and classification and recognition. The process flow of our system is shown in Figure 2 which is similar to any standard image processing system. The benzene structure is drawn with using geometric shapes like hexagon, lines and circle as shown in Figure 1. When these shapes are drawn with free handwriting different types of irregularities are introduced due to different writing styles of individuals. This introduces elements of complexity while recognizing them. The three different ways of representing benzene structures on paper are shown in Figure 1.

*Figure 1. Benzene Representation with hexagons lines and circle*

It can be seen from these representations of benzene that there are three correct representations of benzene structures. Any one of these three representations can be used in HCE. Two correct representations are hexagon and lines insides it showing alternate double bonds. The third correct representation is a hexagon and circle inside it showing benzene ring. When these structures have drawn with free handwriting there can be irregularities in drawing these shapes and recognition of such irregular shapes is a challenging task. Our proposed system has smartly handled this challenge and result shows that it has successfully recognize the handwritten benzene structures from image of HCE.

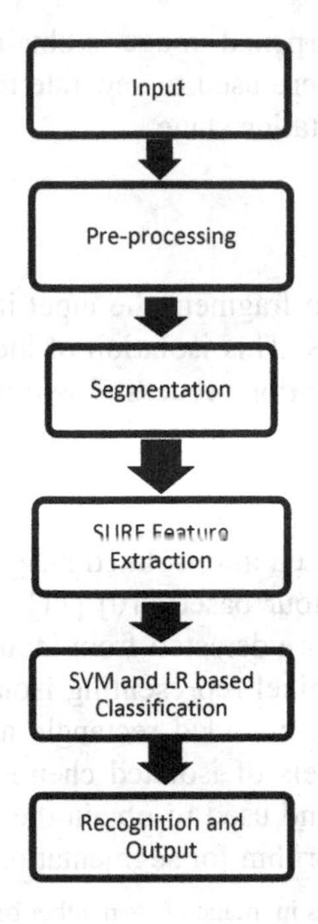


*Figure 2. Architecture of proposed model*

The processing steps of our proposed model are described in next following paragraphs.

## 3.1 Input

The scanned images of HCE are input for the proposed system. In order to test this system data is collected as scanned images of HCE from different people and sources. The chemical expressions of aromatic chemical reactions were identified from books of organic chemistry. These identified expressions were provided to different people and were asked to write them in their own handwriting on paper using pencil or pen. In addition to these images of HCE from Chemistry notes of students and teachers were collected for aromatic chemical reactions. These notes were written using pen on paper. These handwritten chemical expressions (HCE) were scanned with digital scanner and converted to the Portable Network Graphics (PNG) format images. These images of HCE were used as input for our system.

## 3.2 Pre-processing

The scanned input images of HCE were in color mode. This mode is indicated by RGB combination. These colored images have to be converted to binary image for segmentation purposes. In this stage of pre-processing the input image is converted to binary image. Then sharpened image is obtained by applying Laplacian filter. These sharpened images were used to separate the chemical structures and inorganic symbols (IS) in segmentation stage.

## 3.3 Segmentation

The segmentation is carried out to fragment the input image of HCE into isolated chemical structures, bonds and IS. This isolation of individual chemical symbols helps the classification and recognition process. The input to segmentation process is sharpened binary image. The HCE image is made up of continuous group of pixels called as image objects. These image objects were separated from background by applying distance transform [9] on the input image. The separated foreground image object was fragmented into isolated image of chemical structure and inorganic symbol (IS) using contour based [10] [11] segmentation technique. In this technique different contours are detected from input image of HCE. The contour is the group of continuous pixel representing isolated chemical structure or IS. These detected contours have bounded rectangle associated with them. This rectangle contains the image pixels of isolated chemical structures. The area of bounded rectangle is calculated and used to obtain the isolated image segment of chemical structure or IS. The algorithm for segmentation is as follows:

| | |
|---|---|
| Assumption | Let R = number of rows in image , C = number of column in image<br>Img= matrix of size R,C to store image<br>ImgGray= matrix representing gray scale image<br>Seg= Array of image segments |
| Step 1 | Img= ReadImage(File Path) |
| Step 2 | ImgGray=ConvertToGray(Img) |
| Step 3 | Apply Laplacian filter on image ImgGray |
| Step 4 | Apply distance transform to ImgGray |
| Step 5 | Let contours is Contours array<br>contours=findcontours(ImgGray) |
| Step 6 | For I = 0 to size of contours<br>brect= boundedrectangle (contours[I])<br>Seg[I]= ImgGray(bract) |
| Step 7 | Next I |
| Step 8 | Seg is the segment array contains all segments. |
| Step 9 | End |

Algorithm1.

The above Algorithm1 accepts input image of HCE and generates image segment array as an output. This image segment array is used as input for feature extraction stage.

## 3.4 Feature Extraction

In this stage segmented image is used as an input and the corresponding feature vector is obtained. The key point descriptor known as SURF descriptor [12] is used to obtain feature vector in our system. This feature vector is used for classification of benzene structure from HCE. The brief introduction about this descriptor is given in next section.

### 3.4.1 SURF Descriptor

Speeded- Up Robust Feature (SURF) descriptor is a key point descriptor with the best performance for pattern recognition. This descriptor detects key points of image. The SURF descriptor was proposed by Bay et al. [12]. It has robust descriptor features and widely used for image recognition tasks. This SURF algorithm has two steps, in first step, the interesting points of image called as key points are detected by using Hessian Matrix [13] and basic Laplacian based detector. In second step it generates the descriptor for these key points by using the distribution of Haar-wavelet responses within neighborhoods of these points. It extracts 64 values of floating points which was termed as feature vector and used for pattern recognition.

The SURF descriptor is used to find the key points from the images of handwritten benzene structures. This descriptor is used because of its scale and rotation invariant properties [12]. This invariant nature of SURF towards scale and rotation helps to recognize benzene structure from images of any size and orientation. The feature vector obtained by using SURF is used in next stage for classification of the image segment into classes of benzene or inorganic symbol. The classification is explained in next section.

## 3.5 Classification and Recognition

In classification step the image segments are classified in two classes. The first class represents the image segments which possibly represents benzene structure. This class has been labeled as BZ class. The second class is labeled as IS class and represents the image segments containing inorganic chemical symbols. The classifier accepts the image segment array and feature vector generated by SURF descriptor as an input and classifies these image segments into BZ class or IS class.

We had developed two different classifiers using SVM and logistic regression to classify these image segments into its respective class.

### 3.5.1 SVM classifier

Support vector machine is the one of the most powerful classifier for the problems in pattern recognition and machine learning. SVM was introduced by Vapnik[14][15] . SVM is widely used to develop the applications for recognition of handwritten character [16], cursive handwriting [17] and handwritten digit recognition [18]. In area of handwritten recognition SVM has reported better recognition rate than any other techniques. The working of SVM is based on kernel based learning [19]. In implementation of SVM three types of kernel are used which are named as liner, polynomial and Radial Basic Function (RBF).

In this paper we have developed a classifier using RBF kernel based SVM. The Table1 shows the parameters and its values used for creating the SVM classifier.

| Parameter Name | Values |
|---|---|
| Gama | 0.15 |
| C | 1.35 |
| Epsilon | 0.000001 |
| Max Iteration Count | 1000 |

Table1. List of Parameters for SVM Classifier

## 3.5.2 Logistic Regression based classifier

There are many classification systems which are based on logistic regression (LR) [20] technique. LR is widely used in data mining problems [20] as well as for pattern recognition and classification problems in medical domain such as cancer detection [21] [22]. The Logistic regression based classifier developed in this paper accepts the feature vectors as input and classifies these inputs into two classes. This classifier accepts the SURF descriptor as an input and generates the respective class label as an output.

These two classifiers have been trained with training samples and the performance of these two classifiers has been tested.

## 3.5.3 Recognition

Once the class of image segment is identified, the image segments with class label as BZ class are considered for recognition. In the output of recognition stage, the image segments labeled as BZ are shown within red rectangle and molecular formula of benzene as $C_6H_6$ inside the rectangle.

# 4 Experimental Results and Analysis

We have identified 15 different aromatic chemistry reactions from text book of chemistry and these reactions were hand drawn by 40 different people of different ages and belonging to different domains. Then these handwritten expressions are scanned and converted in PNG image format. In addition to it aromatic chemistry reactions from handwritten chemistry notes of students and chemistry teachers were collected. These students and teachers were selected at random and the images of HCE were captured from their handwritten notes. Overall 350 different images of HCE were collected. These images contain 4035 different chemical structures and inorganic symbols. There were 475 benzene structures among these 4035 chemical structures and remaining structures were inorganic symbols (IS).

This data set was then divided in three parts i.e. training set, testing set and validation set. Training set contained 60% of data, testing set contained 20% of data and validation data set contained remaining 20% of data.

The input images were preprocessed and sharpened binary images were obtained. These binary images were segmented by applying segmentation algorithm described in section 3.3. This algorithm uses distance transform [9] and contour based segmentation technique [10] [11] to obtain isolated image segments. Then key points are detected by using SURF descriptor [12] and feature vector is created. The benzene structure with SURF detected key points (indicated by set of small dark black circles) can be seen in Figure 3. The Figure 4 shows the key points detected by SURF detector for inorganic symbols.

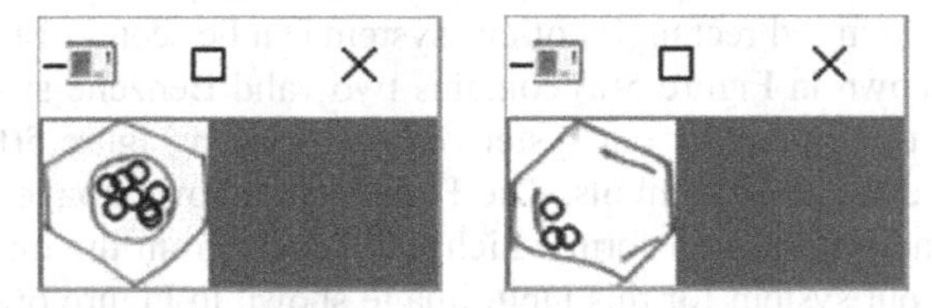

Figure 3. SURF Key descriptors drawn on Benzene structures

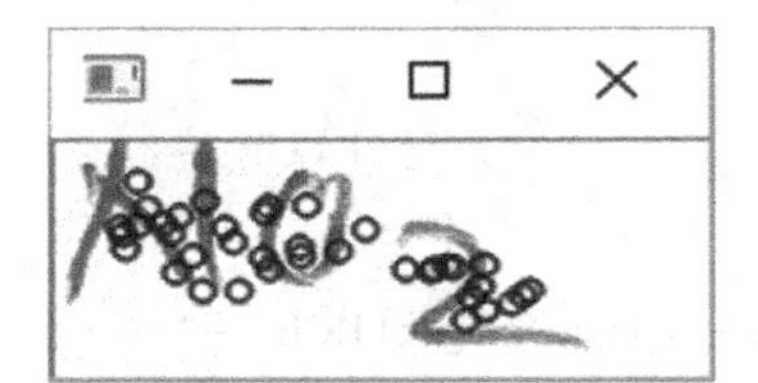

Figure 4. SURF Key descriptors drawn on Inorganic symbols

The feature vector is created from the key points detected for different image segments (Figure 3 and 4) and is used as feature vector for classification.

In the classification stage the image segment is classified into BZ class (possible Benzene structure) and IS class (possible inorganic structure). The classifiers based on SVM and LR was trained using training data set. They were tested using testing and validation data set. The performance comparison of these classifiers can be seen in Table2.

| Sr.No. | Class Label | Data Samples used | Correctly Recognized Samples by SVM Classifier | Correctly Recognized Samples by LR Classifier | Recognition % for SVM Classifier | Recognition % for SVM Classifier |
|---|---|---|---|---|---|---|
| 1 | BZ Class | 475 | 469 | 463 | 98.74 | 97.48 |
| 2 | IS Class | 3560 | 3510 | 3501 | 98.59 | 98.35 |

Table2. Performance Comparison of SVM and LR Classifier

It can be seen from Table2 that there is no major difference in the performance of two classifiers. The accuracy of both the classifiers is around 98%. However the performance of SVM classifier is slightly better than LR classifier.

Above results are not compared with those obtained in other cited papers as all these papers which have been cited deal with graphical shapes created using either electronic devices or printers. As such results are not directly comparable.

The snapshot of input images to our system and recognized Benzene symbol as output (enclosed in red rectangle) of our system can be seen in Figures 5 to 9. The input image shown in Figure 5(a) contains two valid Benzene symbols. The Benzene symbols recognized by our system can be seen in Figure 5(b). This reaction contained valid Benzene symbols. The Figure 6(a) shows image of reaction containing benzene symbol in a form which is different from the one in Figure 5(a). The output of our system for this input image shown in Figure 6(a) can be seen in Figure 6(b).

Figure 5(a). Input image of HCE.

Figure 5(b). Output showing recognized benzene structure.

Figure 6(a). Input image of HCE.

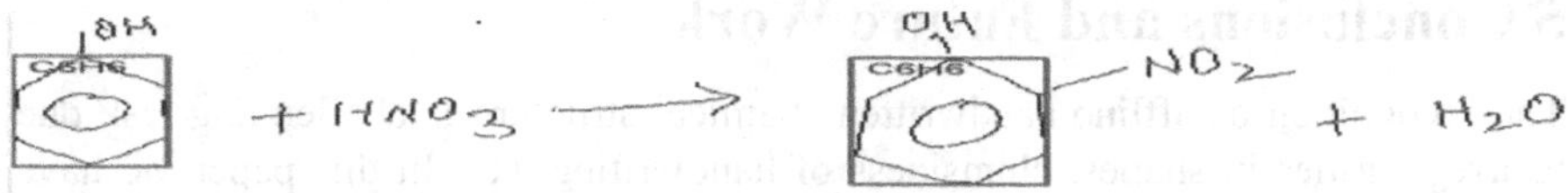

Figure 6(b). Output showing recognized benzene structure.

It can be seen from Figures 7(a) and 7(b) that the developed system can successfully handle irregularities in shapes of benzene structures (due to differences the nature of hand writing). The Figure 7(a) shows the input image of a reaction which contains three irregular Benzene structures. It can be seen in Figure 7(b) that the benzene structures with irregular shapes were correctly recognized. In Figures 8 and 9, reactions with invalid Benzene structure can be seen. The first Benzene structures in both of these reactions are invalid. In Figure 8, it is only hexagon and in Figure 9 it is hexagon with single line. It can be seen in Figures 8 and 9, that these symbols are not identified as valid Benzene symbols by our system. A valid Benzene symbol must have hexagon and three alternate bond lines or ring (circle) inside the hexagon. It can be seen that a hexagon or hexagon with one or two lines inside it are not recognized as valid benzene structure.

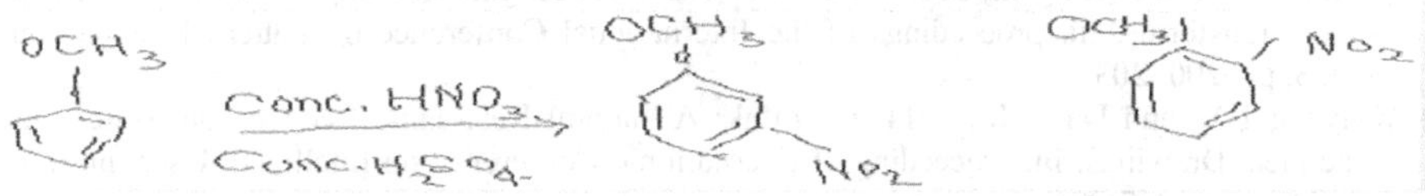

Figure 7(a). Input image of HCE with irregular shape benzene.

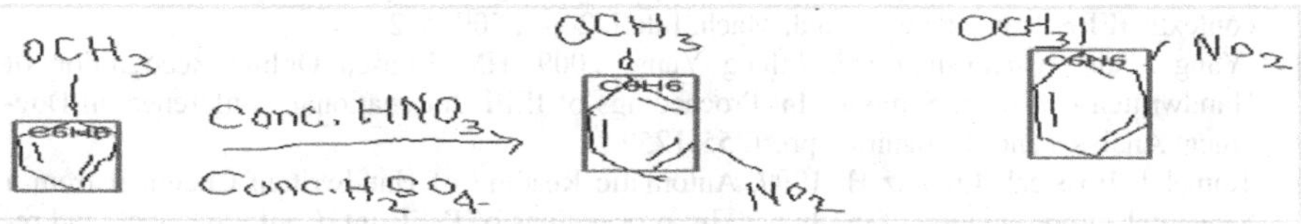

Figure 7(b). Output showing recognized benzene structure.

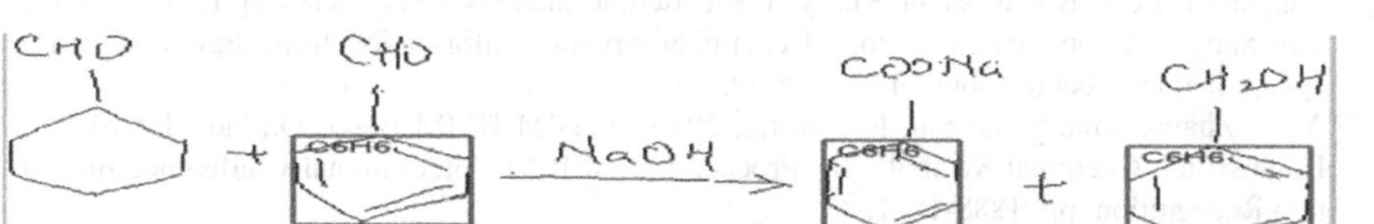

Figure 8. System does not recognize benzene structure drawn as hexagon.

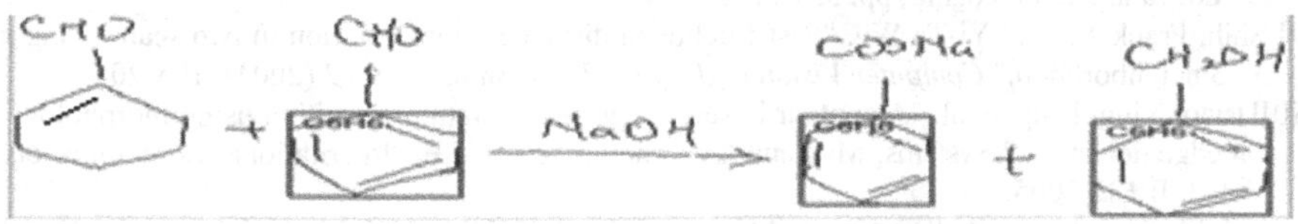

Figure 9. System does not recognize benzene structure drawn as hexagon and single line.

# 5 Conclusions and Future Work

The recognition of offline handwritten chemical structure is challenging task due to irregularities in shapes, clumsiness of handwriting, etc. In this paper we have developed a system to recognize handwritten chemical structures from scanned images of HCE. In the proposed system two classifiers based on SVM and LR for classification of image segments have been developed. The comparative study of both classifiers shows that there is not much difference between performances of these two classifiers. The experimental results have reported the accuracy around 98% for both these classifiers. The SVM based classifier has slightly higher accuracy as compared to LR classifier.

In the future work it is proposed to extend this system to recognize inorganic symbol. Another proposed extension is to handle the complex structures used in aromatic chemistry.

# References

[1] Tabbone S., Wendling L., 2002. Technical symbols recognition using the two-dimensional radon transforms. In: proceedings of the International Conference on Pattern Recognition, vol. 3, pp. 200–203

[2]Ouyang T.Y. and Davis R., 2011. ChemInk: A Natural Real-Time Recognition System for Chemical Drawings. In: proceeding of International Conference on Intelligent User Interfaces, ACM, pp. 267-276.

[3]Belongie S., Malik J., Puzicha J., 2002. Shape matching and object recognition using shape contexts. IEEE Trans. Pattern Anal. Mach. Intell. 24(4), 509–522.

[4] Yang Zhang, Guangshun Shi, Jufeng Yang, 2009. HMM-based Online Recognition of Handwritten Chemical Symbols. In: Proceedings of IEEE International Conference on Document Analysis and Recognition pp. 1255-1259.

[5] Ramel J, BossierJ, Emptoz H, 1999. Automatic Reading of Handwritten Chemical from a Structural Representation of the Image, In: proceedings of IEEE Int. Conf. doc. anal. and recognit. pp.83-86.

[6] Jungkap Park, Gus R Rosania, Kerby A Shedden, Mandee Nguyen, Naesung Lyu and Kazuhiro Saitou, Automated extraction of chemical structure information from digital raster images, Chemistry central Journal , 3:4. 2009.

[7] Yang Zhang, Guangshun Shi, Kai Wang, 2010. A SVM-HMM Based Online Classifier for Handwritten Chemical Symbols. In: Proceedings of IEEE International Conference on Pattern Recognition, pp. 1888-1891.

[8] Peng Tang, Siu Cheung Hui, Chi-Wing Fu, 2013a. A Progressive Structural Analysis Approach for Handwritten Chemical Formula Recognition, In: proceedings of IEEE Int. Conf. on doc. anal. and recognit., pp. 359-363.

[9] Shih, Frank Y., and Yi-Ta Wu. "Fast Euclidean distance transformation in two scans using a 3× 3 neighborhood." *Computer Vision and Image Understanding* 93.2 (2004): 195-205.

[10]Hsiao, Ying-Tung, et al. "A contour based image segmentation algorithm using morphological edge detection." Systems, Man and Cybernetics, 2005 IEEE International Conference on. Vol. 3. IEEE, 2005.

[11] Malik, Jitendra, et al. "Contour and texture analysis for image segmentation." International journal of computer vision 43.1 (2001): 7-27.

[12] Bay, Herbert, Tinne Tuytelaars, and Luc Van Gool. "Surf: Speeded up robust features." *Computer vision–ECCV 2006*. Springer Berlin Heidelberg, 2006. 404-417.

[13] Mikolajczyk, K., Schmid, C.: Indexing based on scale invariant interest points. In: ICCV. Volume 1. (2001) 525 – 531.

[14] Cortes, Corinna, and Vladimir Vapnik. "Support-vector networks." *Machine learning* 20.3 (1995): 273-297.

[15] Vapnik V.,The Nature of Statistical learning Theory.,Springer,1995.

[16] Nasien, Dewi, Habibollah Haron, and Siti Sophiayati Yuhaniz. "Support vector machine (SVM) for english handwritten character recognition." *2010 Second International Conference on Computer Engineering and Applications*. IEEE, 2010.

[17] Camastra, Francesco. "A SVM-based cursive character recognizer." *Pattern Recognition* 40.12 (2007): 3721-3727.

[18] Bellili, Abdel, Michel Gilloux, and Patrick Gallinari. "An hybrid MLP-SVM handwritten digit recognizer." *Document Analysis and Recognition, 2001. Proceedings. Sixth International Conference on*. IEEE, 2001.

[19] Müller, Klaus-Robert, et al. "An introduction to kernel-based learning algorithms." *Neural Networks, IEEE Transactions on* 12.2 (2001): 181-201.

[20] Chen, W., Chen, Y., Mao, Y., & Guo, B. (2013, August). Density-based logistic regression. In *Proceedings of the 19th ACM SIGKDD international conference on Knowledge discovery and data mining* (pp. 140-148). ACM.

[21] Samanta , B., G. L. Bird, M. Kuijpers, R. A. Zimmerman, G. P. Jarvik, G. Wernovsky, R. R. Clancy, D. J.Licht, J. W. Gaynor, and C. Nataraj, Prediction of periventricular leukomalacia. Part I: Selection of hemodynamic features using logistic regression and decision tree algorithms. Artificial Intelligence in Medicine 46 (3) (2009): 201-215.

[22] Zhou, X., K. Y. Liu, and S. T. C. Wong, Cancer classification and prediction using logistic regression with Bayesian gene selection. Journal of Biomedical Informatics 37(4) (2004): 249-259.

# Image And Pixel Based Scheme For Bleeding Detection In Wireless Capsule Endoscopy Images

Vani V. and K.V.Mahendra Prashanth

**Abstract** Bleeding detection techniques that are widely used in digital image analysis can be categorized in 3 main types: image based, pixel based and patch based. For computer-aided diagnosis of bleeding detection in Wireless Capsule Endoscopy (WCE), the most efficient choice among these remains still a problem. In this work, different types of Gastro intestinal bleeding problems:Angiodysplasia, Vascular ecstasia and Vascular lesions detected through WCE are discussed. Effective image processing techniques for bleeding detection in WCE employing both image based and pixel based techniques have been presented. The quantitative analysis of the parameters such as accuracy,sensitivity and specificity shows that YIQ and HSV are suitable color models; while LAB color model incurs low value of sensitivity. Statistical based measurements achieves higher accuracy and specificity with better computation speed up as compared to other models. Classification using K-Nearest Neighbor is deployed to verify the performance. The results obtained are compared and evaluated through the confusion matrix.

## 1 Introduction

Capsule endoscopy has been proved as a standard well-established non-invasive technique for identification of obscure Gastro Intestinal (GI) bleeding. The standard method for reading capsule endoscopy images involves watching streams of 50,000 capsule endoscopy images; it is time consuming; requiring about 17-60 minutes of average reading time[1]. An important challenge in the area of capsule endoscopy is

Vani V.
SJB Institute of Technology, Visvesvaraya Technological University,Bengaluru 560060, India, e-mail: vaniv81@gmail.com

K.V.Mahendra Prashanth
SJB Institute of Technology, Visvesvaraya Technological University,Bengaluru 560060, India e-mail: kvmprashanth@sjbit.edu.in

 
J.M. Corchado Rodriguez et al. (eds.), *Intelligent Systems Technologies and Applications 2016*, Advances in Intelligent Systems and Computing 530,
DOI 10.1007/978-3-319-47952-1_13

identification and detection of obscure GI bleeding. Suspected Blood Identification system (SBIS) developed by the capsule manufacture has reported to provide false alarm rates with sensitivity in the range of 30%-80% [2].

Vascular Ectasias, Vascular lesions and Angiodysplasia are the major bleeding disorders in the GI tract[3]. Angiodysplasia is the second leading cause of bleeding in lower GI tract in patients older than 60 years; and the most common vascular abnormality of GI tract [4]. Vascular ecstasias [4] is a rare but significant cause of lower GI bleeding in elderly of age group 60 years.

The three classical methods for bleeding detection are based on image, pixel and patch. Image based detection use various features of image such as statistical features and color similarity for bleeding detection [6, 7]. Since the image based features works on the whole image, it is much faster but provides poor performance. While pixel based detection relies on bleeding detection pixel wise based on neural network, thresholding and color features [8],[9],[10]. This method provides better classification but high computational cost. The patch based method provides a trade-off between performance and speed. In this method the images are divided into patches/blocks through methods such as LBP (Local Binary Pattern) and the most informative patches are further used for classification [11, 12].

This paper describes the design and implementation of image based and pixel based methodology to detect bleeding in WCE images. The image based techniques employed are color model, histogram and statistical based measurements. The pixel based measurements employed is the color range ratio through thresholding technique. The study on bleeding detection techniques reveals the various possibilities through which bleeding can be detected; yet there exists various trade-offs in metrics such as sensitivity and specificity. Hence an effort to provide a quantitative comparison of these methods have been employed in this paper; thus providing a simple and efficient choice of these methods.The experiments were carried out on several WCE images consisting of normal images and various types of bleeding images collected from Endoatlas image database [13].

## 2 Image Based - Color Model

Since medical experts differentiate bleeding from non-bleeding based on color information; choosing an appropriate color model is very critical; that investigates which color model would be efficient for the computer program to extract the color information. This study aims at investigation of commonly used colors (RGB, CIELAB and YIQ) to enable the usage of computationally and conceptual simple bleeding detection methods [15].

It is observed that second component of CIE Lab color space and YIQ provides better identification of bleeding region [16, 17]. Further HSI color model proved to be an ideal tool for further image processing and classification of bleeding anamoly; since HSI color space is much closer to human perception than any other domain. Hue is a color attribute which describes the pure color. Saturation is a relative color purity i.e lack of white color. Intensity is the brightness. Maximum intensity corresponds to pure white and minimum intensity corresponds to pure black.

The YIQ model is designed originally to separate chrominance from luminance. Y component contains luminance or intensity component, while I and Q components contain color information [18]. The conversion equation is quite simple to implement. The RGB to YIQ color conversion were done as follows:

$$Y = 0.3 \times R + 0.59 \times G + 0.11 \times B; \tag{1}$$

$$I = 0.6 \times R - 0.28 \times G - 0.32 \times B; \tag{2}$$

$$Q = 0.21 \times R - 0.52 \times G + 0.31 \times B; \tag{3}$$

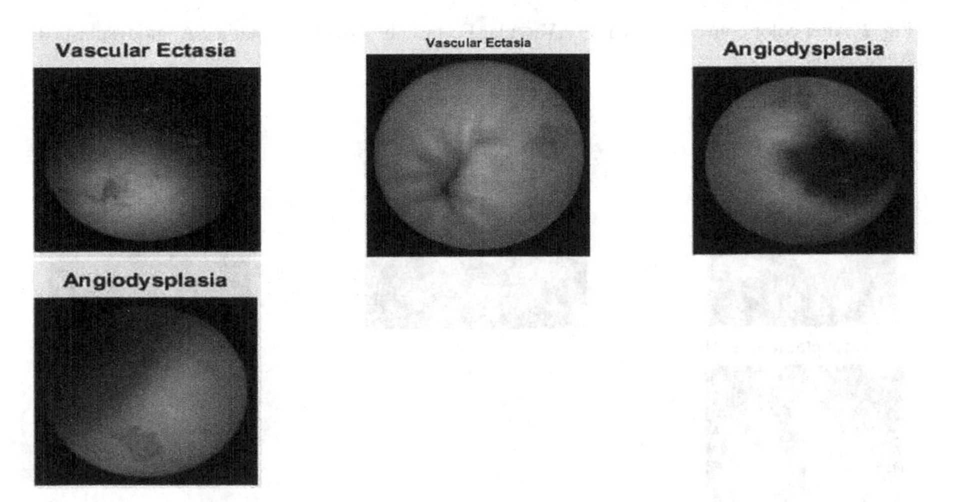


**Fig. 1** Original Images a.Vascular Ecstasia b. Vascular Ecstatia c.Angiodysplasia d. Angiodysplasia [13]

The original WCE images of size 150 * 150 pixels and resolution 75(horizontal) * 75(vertical) dpi obtained from Endoatlas image database [13] for bleeding disorders such as Vascular ecstasia and Angiodysplasia are shown in Fig 1. Fig 2 shows the same images after color conversion to YIQ color space. It is evident that bleeding region is more prominent in YIQ color space as compared to the original images shown in Fig 1 [13].

CIELAB color space is a color opponent space; L represents the lightness and A and B represents the color opponent dimensions. Color conversion from RGB to CIELAB color space for the second component of LAB color space is shown in Fig 3; indicating that A channel has the tendency in separating the non-informative regions from informative(i.e bleeding) region [14]

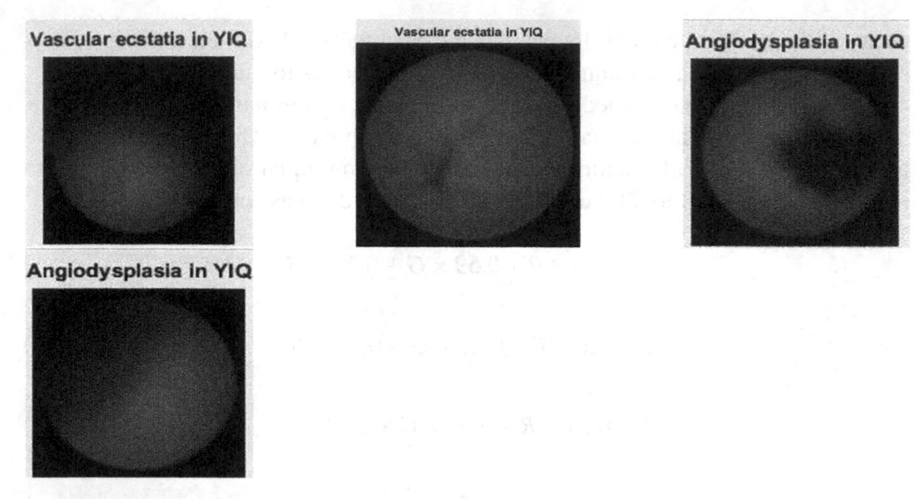


**Fig. 2** After color conversion to YIQ a.Vascular Ecstasia b. Vascular Ecstatia c.Angiodysplasia d. Angiodysplasia

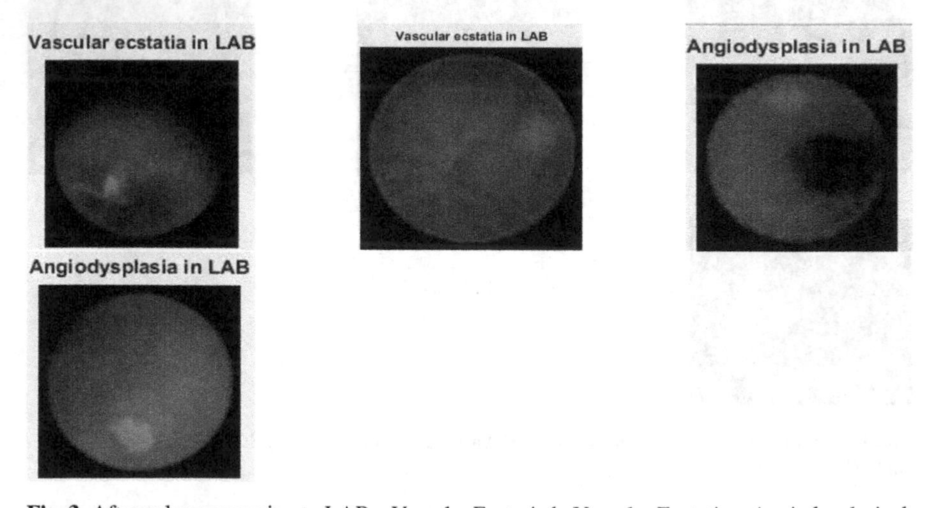


**Fig. 3** After color conversion to LAB a.Vascular Ecstasia b. Vascular Ecstatia c.Angiodysplasia d. Angiodysplasia

To convert RGB to LAB color space, it is important to determine an absolute color space for RGB since RGB is not an absolute color space. The standard technique to convert RGB to standard color space XYZ is shown in equations(4)-(6)

$$[X] = [[-3.240479\ -1.537150\ -0.498535] \times [R]] \tag{4}$$

$$[Y] = [[-0.969256\ 1.875992\ 0.041556] \times [G]] \tag{5}$$

$$[Z] = [[0.055648\ -0.204043\ 1.057311] \times [B]] \quad (6)$$

The color conversion from XYZ to LAB color space can be done as shown in equations (7)-(9)

$$CIEL = \left(116 \times \left(\frac{X}{X'}\right) - 16\right) \quad (7)$$

$$CIEA = 500 \times \left(\left(\frac{X}{X'}\right) - \left(\frac{Y}{Y'}\right)\right) \quad (8)$$

$$CIEB = 200 \times \left(\left(\frac{Y}{Y'}\right) - \left(\frac{Z}{Z'}\right)\right) \quad (9)$$

where X',Y'and Z' are tristimulus values of LAB color space

## 3 Pixel Based Color Range Ratio

The color descriptor based on range ratio color of RGB pixel values was evaluated. The range ratio were evaluated for every pixel [19]. The goal is to segregate the bleeding pixel from non bleeding pixel through the pixel value. Segregation of bleeding pixels through purity of bleeding color (R=255;G=0;B=0) failed to identify many bleeding regions as shown in Fig 4. Hence every pixel was investigated to develop a threshold value for bleeding pixel. Two variations of the color range ratio detection methods are proposed; first method for exact blood color detection and second method for more tolerant identification of colors close to bleeding. The investigation required bleeding pixel to have the range of values for R,G,B as follows:$((R \leq 177\&R \geq 215)\&(G \leq 99\&G \geq 56)\&(B \leq 52\&B \geq 20))$. The results obtained through color range ratio are obtained as shown in Fig.4. The experimental evaluation leads to satisfactory results.

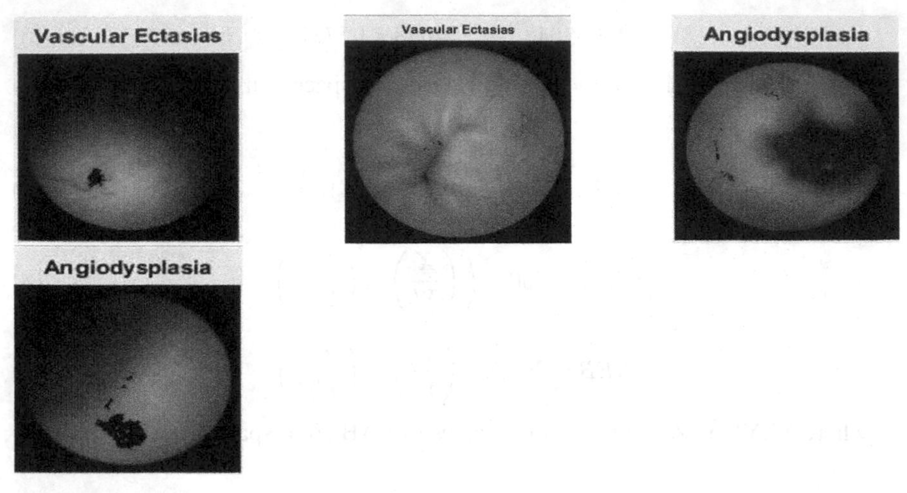


**Fig. 4** Pixel based bleeding detection using range ratio color a.Vascular Ecstasia b. Vascular Ecstatia c.Angiodysplasia d. Angiodysplasia

## 4 Pixel Based - Color Similarity Measurement

A color similarity technique to measure the similarity between a pixel in RGB to a known typical bleeding pixel is employed. Lower the color similarity, lesser is the similarity between the two pixels; higher the color similarity, higher is the similarity between two pixels[25].

The color similarity is estimated through closeness of pixel values between P(R,G,B) and P0(R0,G0,B0) where P(R,G,B) is the color vector of the pixel being tested and C0(R0,G0,B0) is the known typical bleeding pixel. The similarity of color vector depends on amplitude and direction similarity. The amplitude similarity between P(R,G,B) and P0(R0,G0,B0) is measured by absolute value of difference $|P| - |P0|$ . The coefficient of amplitude similarity is obtained as given in equation (10)

$$n = 1 - \frac{(|P| - |P0)|}{|P0|}; \tag{10}$$

$$|P| = \sqrt{R^2 \times G^2 \times B^2}; \tag{11}$$

$$|P0| = \sqrt{R0^2 \times G0^2 \times B0^2}; \tag{12}$$

Equation (10) shows that when $|P| = |P0|$, coefficient of amplitude similarity n is equal to 1. Higher the value of n, higher is the similarity and vice versa.

The direction similarity between the pixels shown in Fig. 5 is measured by the angle $\alpha$ computed as

$$\cos\alpha = \frac{P \times P0}{|P|/times|P0|} = \frac{R \times R0 + G \times G0 + B \times B0}{\left(\sqrt{R^2 + G^2 + B^2}\right) \times \sqrt{(R0^2 + G0^2 + B0^2)}} \tag{13}$$

Smaller the value of $\alpha$, larger is the function value of $cos\alpha$ and more similar are the directions of color vectors. The pixel value is calculated as bleeding pixel, non-bleeding pixel or suspected bleeding based on following function.

$$\begin{aligned} Output(X) = \{Bleeding; if G(X) \& C(X) = 1 \\ Non-Bleeding; if G(X) \| C(X) = 0 \\ SuspectedBleeding; others \end{aligned}$$

where & represents logical AND operation and || represents logical OR operation.

G(X) is the gray intensity similarity coefficient and C(X) is the chroma similarity coefficient respectively.

$$\begin{aligned} G(X) = \{1 if d(X) \geq 0 \\ 0 if d(X) < 0 \end{aligned}$$

d(X) is the classifying function given as $d(X) = n - n_{min}$
where $n_{min}$ is the threshold value of similarity co-efficient chosen as 0.96

$$\begin{aligned} C(X) = \{1 if d_C(X) \geq 0 \\ 0 if d_C(X) < 0 \end{aligned}$$

where $d_C(X) = \cos\alpha - \cos\alpha_{min}$

where $\cos\alpha_{min} = 0.96$

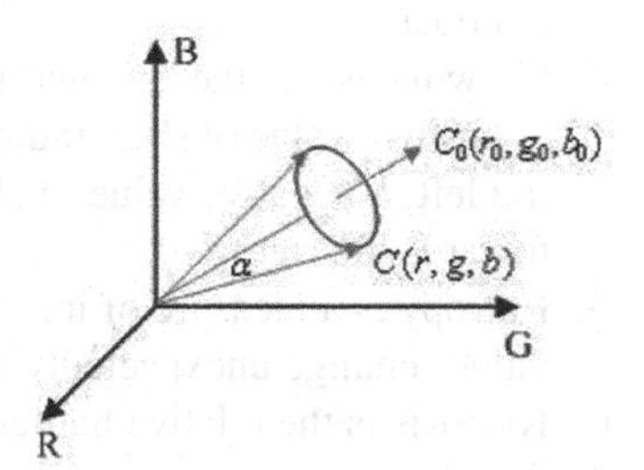


**Fig. 5** Color vector similarity measurement [25]

The results obtained through color similarity measurement shown in Fig 6 shows that possibilities of suspected bleeding is more in darker areas; hence increasing the number of false positive cases, thereby decreasing specificity .Thus decreasing the ability of the system to correctly identify the non-blood frame.

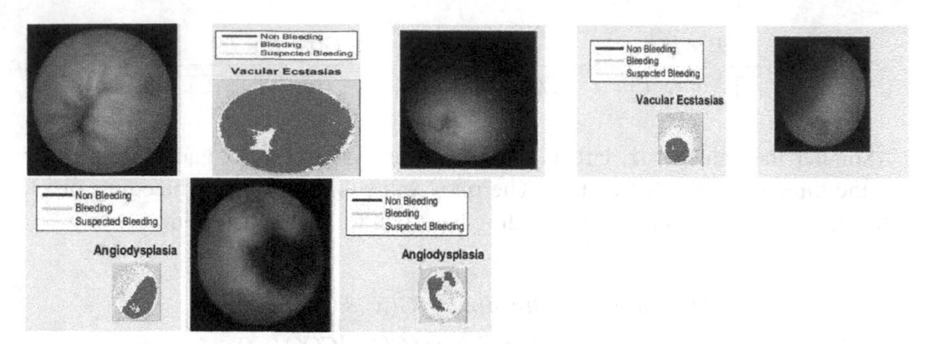


**Fig. 6** Color similarity based bleeding detection a.Vascular Ecstasia b. Vascular Ecstatia c.Angiodysplasia d. Angiodysplasia

## 5 Statistical Measures for RGB Color Plane

A very effective technique to search for most similar images out of huge database is Content Based Image Retrieval (CBIR) system [21]; which proves to be beneficial for WCE due to its huge image database. Hence statistical features obtained from the colour histogram are tested to detect and delineate bleeding regions from set of capsule endoscopy images.

Colour is an important cue for medical experts to identify different regions in an image. Hence it is suitable to employ colour features for extraction. In the colour images, colour distribution of each plane is modelled by its individual histogram. Statistical based measures like mean, variance, standard deviation, entropy and skew of first order colour histograms are statistical indicators that categorize various properties of images.

1. Mean of an image provides the average brightness of an image
2. Variance of an image provides information on distribution of the pixel intensities around mean intensity value. Higher value of variance indicates higher contrast
3. Standard Deviation is the square root of variance. It provides information about contrast
4. Skew measures the asymmetry about the mean in the intensity level distribution. A positive value of skew indicates histogram distribution has a longer tail towards the left; a negative value of skew indicates histogram distribution has longer tail towards the right
5. Entropy is a measure of information content or uncertainity. Images where pixel values change unexpectedly, have large entropy values
6. Kurtosis is the relative flatness or peakness of a distribution compared to normal distribution
7. Moment is the distribution of random variable about its mean. Higher order moments are related to spread and shape of the distribution

The above statistical parameters were applied on the R/G intensity ratio domain to determine the features for bleeding. The behavior of each statistical parameter were investigated for bleeding and non-bleeding images.

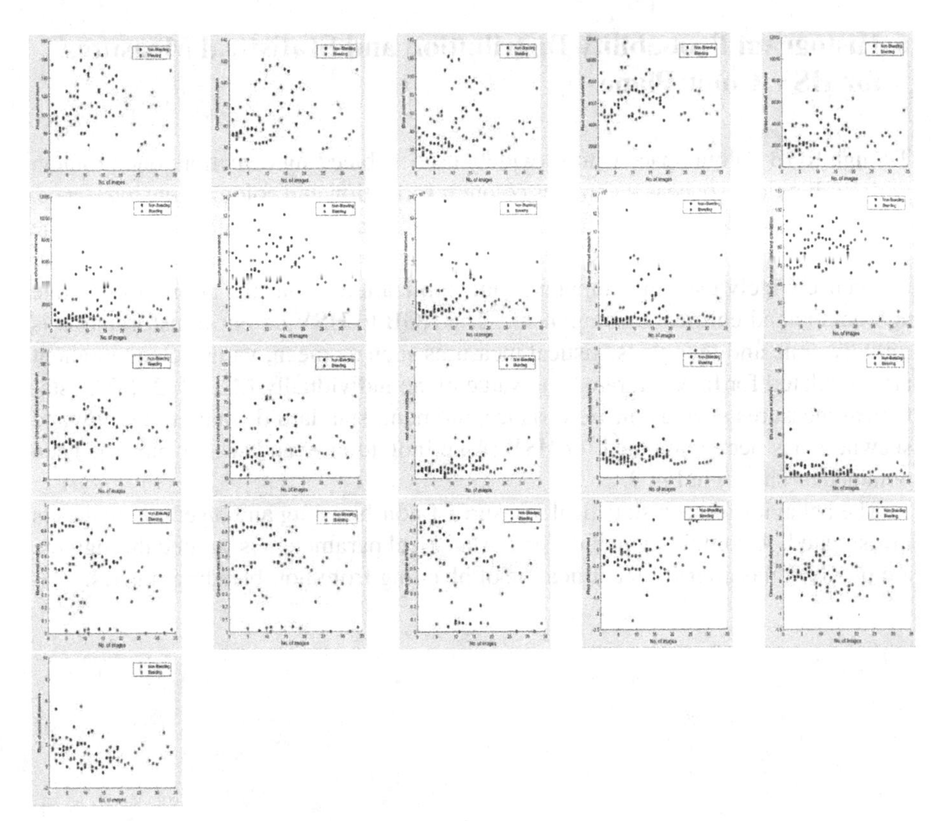

**Fig. 7** Scatter plot of bleeding and non-bleeding images in RGB domain (a)Mean of Red channel (b)Mean of Green channel (c)Mean of Blue channel (d)Variance of red channel (e)Variance of Green channel (f)Variance of Blue channel (g)Moment of Red channel (h)Moment of Green channel (i)Moment of Blue channel (j)Standard deviation of Red channel (k)Standard deviation of Green channel (l)Standard deviation of Blue channel (m)Kurtosis of Red channel (n)Kurtosis of Green channel (o)Kurtosis of Blue channel (p)Entropy of Red channel (q)Entropy of Green channel (r)Entropy of Blue channel (s)Skewness of Red channel (t)Skewness of Green channel (u)Skewness of Blue channel(Red scatter plots indicate bleeding images;Blue scatter plots indicate non-bleeding images)

The scatter plots show that only selective feature combinations produce differentiable cases of bleeding and non-bleeding.

The investigation of statistics in Fig 7 reveals that presence of bleeding region lowers the value of mean, moment, variance and standard deviation distribution as compared to non-bleeding images. However there is a certain degree of overlap between the two classes. Certain statistical measures shows relatively higher value for entropy, kurtosis and skew; measured for bleeding disorders such as colon carcinoma and colonic polyps with bleeding.

## 6 Histogram Probability Distribution and Statistical Measures for HSV Color Plane

Though RGB colour space is most widely used in bleeding detection; one problem with RGB colour space is it contains colour information and colour intensity. Hence, it is difficult to identify the colour from an individual RGB colour component. The HSV colour scheme is much closer to conceptual human understanding of colours and hence widely used in computer vision application. It has the ability to separate achromatic and chromatic components. The RGB to HSV colour transformation is initially done and then the statistical measures such as mean, variance and moment are calculated for hue, saturated and value plane individually [22, 23, 24]. The statistical measures such as mean, variance, moment, standard deviation, entropy and skewness has been computed for HSV plane individually and scatter plot has been obtained as shown in Fig 8.

The behavior of each statistical measure of non-bleeding and bleeding images is investigated. A combination of different statistical parameters is studied through the aid of KNN classifier for identification of bleeding from non-bleeding images.

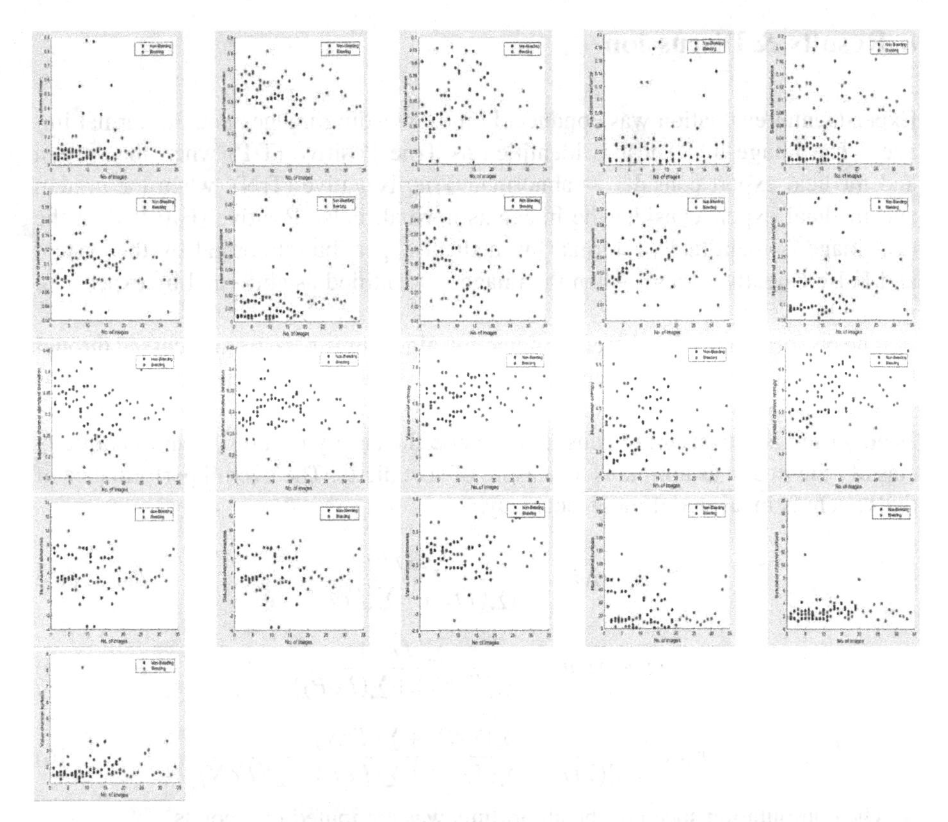

**Fig. 8** Scatter plot of bleeding and non-bleeding images in HSV domain (a)Mean of hue channel (b)Mean of saturated channel (c)Mean of value channel (d)Variance of hue channel (e)Variance of saturated channel (f)Variance of value channel (g)Moment of hue channel (h)Moment of saturated channel (i)Moment of value channel (j)Standard deviation of hue channel (k)Standard deviation of saturated channel (l)Standard deviation of value channel (m)Entropy of hue channel (n)Entropy of saturated channel (o)Entropy of value channel (p)Skewness of hue channel (q)Skewness of saturated channel (r)Skewness of value channel (s)Kurtosis of hue channel (t)Kurtosis of saturated channel (u) Kurtosis of value channel(Red scatter plots indicate bleeding images;Blue scatter plots indicate non-bleeding images)

# 7 K Nearest Neighbor KNN Classifier

The KNN classifier based on Leave-One-Out cross validation technique for k=1 was adopted for classification of bleeding and non-bleeding images and performance was compared. KNN classifier was used for all different combination of statistical features calculated in the HSV and RGB plane. The results obtained through KNN classifiers is tabulated in Table 1.

# 8 Results & Discussion

Experimental evaluation was conducted on 61 bleeding images and 38 normal images. The image under test is identified as True Positive (TrP) when the system and medical expert consider as abnormal; True Negative (TrN): when the system and medical expert consider the image as normal; False Positive (FaP): when the test image is identified as normal by medical expert but abnormal by the system and False Negative (FaN): when the image is identified as abnormal by expert and normal by system.

The performance of the bleeding detection algorithms have been accessed through following parameters: sensitivity, specificity and accuracy which are computed as shown in equations (17-19). The ability of the system to correctly identify the image as abnormal is a measure of sensitivity; while the ability of the system to correctly classify the image as normal is a measure of specificity. The overall performance of the system is measured through accuracy.

$$sensitivity = \frac{\Sigma(TrP)}{(\Sigma(TrP) + \Sigma(FaN))} \tag{14}$$

$$specificity = \frac{\Sigma(TrN)}{(\Sigma(TrN) + \Sigma(FaP))} \tag{15}$$

$$accuracy = \frac{\Sigma(TrP) + \Sigma(TrN)}{\Sigma(TrP) + \Sigma(TrN) + \Sigma(FaP) + \Sigma(FaN)} \tag{16}$$

The computation speed of the algorithms was computed in seconds.

**Table 1** Comparative Evaluation of Performance

| Method | Algorithm | Accuracy(%) | Sensitivity(%) | Specificity(%) | Computation time(seconds) |
|---|---|---|---|---|---|
| Image Based | YIQ Color model | 70.7 | 92.85 | 43.9 | 30.743 |
| Image Based | LAB Color model | 57.57 | 49.09 | 68.18 | 30.743 |
| Image Based | Statistical measures(RGB)* | 80.8 | 88.52 | 68.42 | 27.875 |
| Image Based | Statistical measures(HSV)* | 82.82 | 90.3 | 66.67 | 30.6875 |
| Image Based | Statistical measures(HSV)** | 78.78 | 83.87 | 70.3 | 27.0313 |
| Image Based | Statistical measures(HSV)*** | 80.80 | 90.3 | 64.86 | 13.0938 |
| Pixel Based | Color range ratio | 73.73 | 77.61 | 65.6 | 16.743 |
| Pixel Based | Color Similarity | 52.52 | 91.89 | 36 | 233.862 |

(*indicates all statistical measures **indicates entropy,skew,kurtosis ***mean,variance and moment)

The graphical representation of comparative evaluation is shown in Fig 9 which depicts a precise graphical comparison among the various methods.

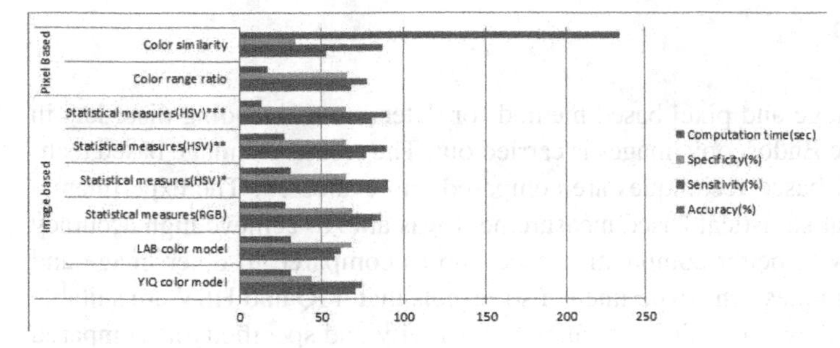


**Fig. 9** Graphical representation of comparative evaluation performance (*indicates all statistical measures **indicates entropy,skew,kurtosis only ***mean,variance and moment only)

Statistical based measurement for HSV color space for all statistical measures provides superior accuracy of 89.89% with superior sensitivity of 90.3%; while statistical based measurement for RGB color space achieves the best specificity of 68.42%. As observed, pixel based measurement required an appreciable computation time; whereas statistical based measurements consumed very less computation time.

However, having a higher value of sensitivity or specificity remains a matter of debate. In bleeding detection physician may prefer to have more false positive hit; but more number of false positives can also be annoying. The low value of sensitivity for LAB color model is due to the more number of false negatives; as the system was unable to recognize faint traces of bleeding in the images.

The performance metrics are represented with the help of a confusion matrix as shown in Table 2.

**Table 2** Confusion Matrix

| Algorithm | #TP | #TN | #FP | #FN |
|---|---|---|---|---|
| YIQ Color model | 52 | 18 | 23 | 4 |
| LAB Color model | 27 | 30 | 14 | 28 |
| Statistical measures(RGB)* | 54 | 26 | 12 | 7 |
| Statistical measures(HSV)* | 56 | 24 | 13 | 6 |
| Statistical measures(HSV)** | 52 | 26 | 11 | 10 |
| Statistical measures(HSV)*** | 56 | 24 | 13 | 6 |
| Color range ratio | 52 | 21 | 11 | 15 |
| Color similarity | 34 | 18 | 32 | 3 |

(*indicates all statistical measures **indicates entropy,skew,kurtosis only ***mean,variance and moment only)

## 9 Conclusion

In this paper, image and pixel based method for detection of bleeding disorders in Wireless Capsule Endoscopy images is carried out. The results of image based techniques and pixel based techniques are compared and evaluated. The experimental results shows that statistical based measurement was able to achieve high accuracy and specificity with better computation speed up as compared to other image and pixel based techniques. The experiment also reveals that YIQ and HSV are suitable colour model as they show higher accuracy, sensitivity and specificity as compared to LAB color space.

To summarize image based methods failed to identify small bleeding regions; thereby resulted in poor performance but faster computation time. Pixel based methods suffered from higher trade-off between sensitivity and specificity and large computation time.

The future work is aimed at patch based bleeding detection techniques and to reduce the trade-off between sensitivity and specificity. Further, research needs to be carried out to improve the overall accuracy, sensitivity and specificity through use of better features and classifiers.

**Acknowledgements** We would like to thank Dr.Anand Dotihal , Gastroenterologist, Bangalore for his valuable medical guidance on the different bleeding disorders in numerous images.

## References

1. Stein, Adam C., et al. "A Rapid and Accurate Method to Detect Active Small Bowel Gastrointestinal Bleeding on Video Capsule Endoscopy." Digestive diseases and sciences 59.10 (2014): 2503-2507.
2. Choi, Hyuk Soon, et al. "The sensitivity of suspected blood indicator (SBI) according to the background color and passage velocity of capsule endoscopy." JOURNAL OF GASTROENTEROLOGY AND HEPATOLOGY. Vol. 25. COMMERCE PLACE, 350 MAIN ST, MALDEN 02148, MA USA: WILEY-BLACKWELL, 2010.
3. Hara, Amy K., et al. "Small Bowel: Preliminary Comparison of Capsule Endoscopy with Barium Study and CT 1." Radiology 230.1 (2004): 260-265.
4. Gunjan, Deepak, et al. "Small bowel bleeding: a comprehensive review." Gastroenterology report (2014): gou025.
5. Nguyen, Hien, Connie Le, and Hanh Nguyen. "Gastric Antral Vascular Ectasia (Watermelon Stomach)–An Enigmatic and Often-Overlooked Cause of Gastrointestinal Bleeding in the Elderly." Issues 2016 (2016).
6. Ghosh, T., et al. "An automatic bleeding detection scheme in wireless capsule endoscopy based on statistical features in hue space." Computer and Information Technology (ICCIT), 2014 17th International Conference on. IEEE, 2014.
7. Guobing, P. A. N., X. U. Fang, and C. H. E. N. Jiaoliao. "A novel algorithm for color similarity measurement and the application for bleeding detection in WCE." International Journal of Image, Graphics and Signal Processing 3.5 (2011): 1.
8. Pan, Guobing, et al. "Bleeding detection in wireless capsule endoscopy based on probabilistic neural network." Journal of medical systems 35.6 (2011): 1477-1484.

9. Bourbakis, N., Sokratis Makrogiannis, and Despina Kavraki. "A neural network-based detection of bleeding in sequences of WCE images." Bioinformatics and Bioengineering, 2005. BIBE 2005. Fifth IEEE Symposium on. IEEE, 2005.
10. Poh, Chee Khun, et al. "Multi-level local feature classification for bleeding detection in wireless capsule endoscopy images." Cybernetics and Intelligent Systems (CIS), 2010 IEEE Conference on. IEEE, 2010.
11. Lau, Phooi Yee, and Paulo Lobato Correia. "Detection of bleeding patterns in WCE video using multiple features." Engineering in Medicine and Biology Society, 2007. EMBS 2007. 29th Annual International Conference of the IEEE. IEEE, 2007.
12. Fu, Yanan, Mrinal Mandal, and Gencheng Guo. "Bleeding region detection in WCE images based on color features and neural network." Circuits and Systems (MWSCAS), 2011 IEEE 54th International Midwest Symposium on. IEEE, 2011.
13. Atlas of Gastrointestinal Endoscopy. 1996 [online]. Available: http://www.endoatlas.com/index.html
14. Hunter Labs (1996). "Hunter Lab Color Scale". Insight on Color 8 9 (August 1-15, 1996). Reston, VA, USA: Hunter Associates Laboratories
15. Sharma, Gaurav, and H. Joel Trussell. "Digital color imaging." Image Processing, IEEE Transactions on 6.7 (1997): 901-932
16. J. Schanda, Colorimetry: Understanding the CIE system: Wiley. com, 2007
17. Szczypiski, Piotr, et al. "Texture and color based image segmentation and pathology detection in capsule endoscopy videos." Computer methods and programs in biomedicine 113.1 (2014): 396-411
18. Hughes, John F., et al. Computer graphics: principles and practice. Pearson Education, 2013
19. Al-Rahayfeh, Amer A., and Abdelshakour A. Abuzneid. "Detection of bleeding in wireless capsule endoscopy images using range ratio color." arXiv preprint arXiv:1005.5439 (2010).
20. Ghosh, T., et al. "An automatic bleeding detection scheme in wireless capsule endoscopy based on histogram of an RGB-indexed image."Engineering in Medicine and Biology Society (EMBC), 2014 36th Annual International Conference of the IEEE. IEEE, 2014
21. Sergyan, S., Color histogram features based image classification in content-based image retrieval systems In: Applied Machine Intelligence and Informatics, 2008. SAMI 2008. 6th International Symposium on, pp. 221224, 2008
22. Ghosh, T., et al. "An automatic bleeding detection scheme in wireless capsule endoscopy based on statistical features in hue space." Computer and Information Technology (ICCIT), 2014 17th International Conference on. IEEE, 2014
23. Shah, Subodh K., et al. "Classification of bleeding images in wireless capsule endoscopy using HSI color domain and region segmentation." URI-NE ASEE 2007 Conference. 2007
24. Ghosh, Tonmoy, et al. "A statistical feature based novel method to detect bleeding in wireless capsule endoscopy images." Informatics, Electronics & Vision (ICIEV), 2014 International Conference on. IEEE, 2014.
25. Guobing, P. A. N., X. U. Fang, and C. H. E. N. Jiaoliao. "A novel algorithm for color similarity measurement and the application for bleeding detection in WCE." International Journal of Image, Graphics and Signal Processing 3.5 (2011): 1.

# Leaf Recognition Algorithm for Retrieving Medicinal Information

**D Venkataraman, Siddharth Narasimhan, Shankar N, S Varun Sidharth, Hari Prasath D**

***Abstract*** --- India has a vast history of using plants as a source of medicines. This science is termed as Ayurveda. But, sadly somewhere in the race of keeping up with medicinal science and technology, India as a country has lost its track in the field of Ayurveda. Researchers and medicinal practitioners today, in spite of knowing that allopathic medicines are made using certain plant extracts, are oblivious about the medicinal the properties of plants. This paper aims at eradicating this problem, and hence strives to help potential users make better use of plants with medicinal properties. The dataset consists of 300 images of different types of leaves. The classification of the leaves is done with the help of a decision tree. Our system is an easy to use application which is fast in execution too. The objective of doing this paper is to develop an application for leaf recognition for retrieving the medicinal properties of plants. The recognition of leaves is done by extracting the features of the leaves from the images. The primary stakeholders involved with this project are researchers, medical practitioners and people with a keen interest in botany. We believe that this application will be an important part of the mentioned stakeholders' daily lives. The primary purpose that this paper serves is to solve the problem of not knowing the useful properties of many plants.

***Abbreviations Used:***

PNN - Probabilistic Neural Network, PCA - Principal Component Analysis,

RGB – Red Green Blue

D Venkataraman [1] email: d_venkat@cb.amrita.edu

Siddharth Narasimhan [1] email: siddharthn1@hotmail.com

Shankar N [1] email: shankar12243@gmail.com

S Varun Sidharth [1] email: varunsvs@gmail.com

Hari Prasath D [1] email: jayanhari555@gmail.com

[1] Department of Computer Science and Engineering, Amrita School of Engineering, Coimbatore, Amrita Vishwa Vidyapeetham, Amrita University, India





J.M. Corchado Rodriguez et al. (eds.), *Intelligent Systems Technologies and Applications 2016*, Advances in Intelligent Systems and Computing 530,
DOI 10.1007/978-3-319-47952-1_14

# 1. Introduction

### 1.1 Plant classification using PNN [1]:

Plants exist everywhere on the Earth. Many of these plants carry vital information which can be used in a variety of different purposes, mainly medicinal purposes. Hence, it is extremely important to create a repository which will consist of the medicinal properties of plants. The first step in doing so is to create an artificial intelligence system which will automatically identify and classify the plants. Classification of a plant using the leaf image is always considered to be a better option when compared with cell and molecule biology methods. Sampling the leaves and photographing them are less expensive and convenient to implement. Once this is done, the leaf image can be stored in a computer, and then the leaf recognition system will automatically classify them. The paper that we have studied and cited above tries to extract the features without any interference by the user. The paper proposes the main features of image pre-processing there are various steps involved in it. Initially the fed RGB image is converted into Binary image after this the smoothing of image is done, after which the smoothed image is filtered in Laplacian filter the last step is Boundary enhancement this is final step is also done effectively. The feature extraction is classified into two based on the features they poses. They are some basic geometric features which include diameter, length, width, leaf area, leaf perimeter, form factor, perimeter to diameter ratio, aspect ratio. The important concepts used in the paper we referred [1] are Principal component analysis and probabilistic neural network.

### 1.2 Extracting vein features [2]:

This paper, which deals with the extraction of vein features, helped us understand that each leaf has a specific vein pattern, which can be used to classify the plant it belongs to. The cited paper tells us that as an inherent trait, the veins of a leaf has all the information that is needed in order to classify the plant. It is one of the most definite feature of a leaf which helps in plant classification. Hence, if vein patterns are one of the features being considered to classify the plant, then the system becomes more accurate. But, having said that, it is a complex process to carry out. According to this paper [2], their approach is capable of extracting the vein patterns much more efficiently than other existing methods, in order to classify the plant. Studies have shown, the approach mentioned in this paper [2] can also reduce the execution time as compared to a direct neural network approach.

### 1.3 Principal Component Analysis [3]:

This paper helped us get an insight into how PCA is used in the process of feature extraction. In the cited paper [3], they have considered 7 features which are orthogonolized using PCA. By studying the cited paper [3], we got a clear understanding about how PCA can be put to effective use in our project. It deals with mapping the image in the form of pixels, in the form of principal components. The principal components are orthogonal because they are eigenvectors of the covariance matrix, which is symmetric.

### 1.4 Image scale and rotation [4]:

This paper addresses the issue of aligning multiple images under transformations like rotation, translation and scaling. The method described in the referred paper [4] utilizes the shift invariance property of the bispectrum which negates the translation component effect. From the bispectrum, only the information of phase is kept, so that we get a better result even with irregular changes of lighting in the image. Then, from the remaining spectrum, the scaling and the rotation parameters are estimated. The examples cited in the referred paper [4] tell us that the method is quite durable against the problem of noise in the image too. Basically, there are two kinds of approaches available to estimate the scaling, rotation and translation parameters between two images: feature based and featureless approach. The feature based approach utilizes vital features like edges, comers, contours.

# 2. Algorithms Used

### 2.1 Canny Edge Detection Algorithm:

The Canny Edge Detection algorithm is a multi-staged algorithm which correctly detects all the edges in an image.

The result of Canny edge detection algorithm is better than other edge detection algorithms because:

1. Non Maximum Suppression - Edges candidates which are not dominant in their neighbourhood aren't considered to be edges.

2. Hysteresis Process - While moving along the candidates, given a candidate which is in the neighbourhood of an edge the threshold is lower.

These 2 steps reduce the number of "False" edges and hence create a better starting point for farther process like Hough Transformation.

### 2.2 Gaussian Filter:

Since all edge detection results are easily affected by image noise, it is essential to filter out the noise to prevent false detection caused by noise. To smooth the image, a Gaussian filter is applied to convolve with the image. This step will slightly smooth the image to reduce the effects of obvious noise on the edge detector. The kernel of the Gaussian filter is shown below.

$$\mathbf{B} = \frac{1}{159}\begin{bmatrix} 2 & 4 & 5 & 4 & 2 \\ 4 & 9 & 12 & 9 & 4 \\ 5 & 12 & 15 & 12 & 5 \\ 4 & 9 & 12 & 9 & 4 \\ 2 & 4 & 5 & 4 & 2 \end{bmatrix} * \mathbf{A}.$$

A is the matrix representation of input image.

B is the output matrix after Gaussian filter.

# 3. Proposed System Architecture

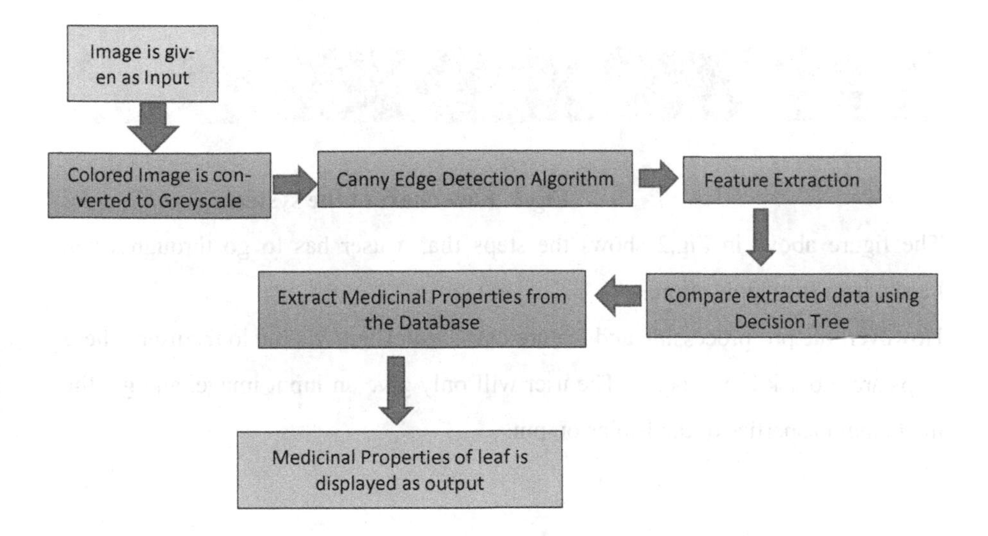


**Fig. 1: Proposed architecture diagram of the system**

The image of a leaf is given as the input, from where the image is pre-processed. This involves conversion to greyscale and applying the edge detection algorithm. Then the features of the leaf are extracted from the image. These features are represented in the form of a numerical value. These values are compared with already stored ideal values for each of the leaves. This comparison of values helps to correctly identify a leaf. Once identified, its medicinal properties are retrieved, and given as output.

# 4. Flowchart

The figure shown below shows the flowchart of the system.

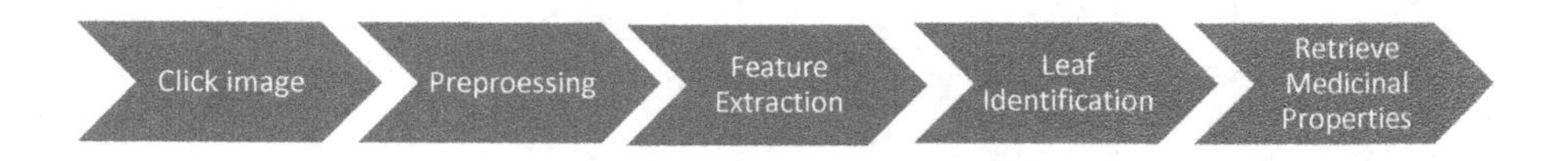


**Fig.2: Flowchart of the system**

The figure above in Fig.2 shows the steps that a user has to go through when he/she uses this application.

However, the pre-processing and feature extraction is not visible to the user. These steps are a black box process. The user will only give an input image, and get the medicinal properties of the leaf as output.

# 5. Internal Working of the System

As previously stated, the system works by extracting features from a particular leaf and checking the obtained values with the initially stored ideal values for the same leaf.

## 5.1 Feature Set

1. **Length of leaf:**

   This is the distance between the two ends of the main vein of the leaf. This is then used to calculate the aspect ratio of the leaf.

2. **Breadth of leaf:**

   This is the distance from the left most point in a leaf to the right most point in the leaf. This can be calculated by drawing lines that are orthogonal to the main vein of the leaf. The longest line which can be drawn is the breadth of the leaf.

3. **Aspect Ratio**

   The aspect ratio of a leaf is the ratio of the length to its breadth.

4. **Diameter**

   The diameter of a leaf is the maximum distance between any two points which lie inside the area covered by the leaf.

5. **Shape of leaf (Convex Hull)**

   Using the Convex Hull algorithm we can find the coordinates of the points under which the entire area of the leaf is covered. Using these coordinates the application will distinguish one leaf from another.

6. **Leaf perimeter**

   The number of pixels that are encompassed within a leaf is nothing but the perimeter of the leaf.

7. **Rectangularity (R)**

   The process includes drawing a rectangle outside the image, so that the image just fits in the rectangle. R= LpWp/A, where Lp is length, A is leaf area and Wp is breadth.

### 5.2 Using Features to classify leaves

The first feature that the application considers is the aspect ratio of the leaf. This is calculated suing the length and the breadth of the leaf. The ratio of the length to the breadth gives the aspect ratio. Most of the leaves which are visually clearly distinct can be separated and classified using aspect ratio only. Or it will help in eliminating the wrong ones.
Another feature of a leaf is its rectangularity. If the aspect ratio of two leaves is almost same, then the algorithm takes into consideration the rectangularity to distinguish them. The rectangularity is based on a certain formula.
The shape of a leaf can also be used to distinguish different leaves from each other. This is done using Convex Hull method. For every leaf a set of points is gotten as output. Using this, the area of the leaf is found out. The area as well as the points help in distinguishing different leaves. Given four distinct leaves, convex hull method will distinguish each leaf based on the points. The values are compared with the ideal values that are stored in a database. The name of the plant is gotten as output.

### 5.3 Dataset

The data set consists of 300 leaves of different varieties. The images in the dataset are taken in different orientations (horizontally and vertically). It comprises of a range of different varieties of leaves that we have collected locally. This dataset is trained using decision tree training. Some rules that were followed while creating the dataset are:

1. The background of all the images in the dataset is white in color.

2. The angle at which each image is clicked is same for all the images.

3. The distance of the camera from the leaf surface has been maintained constant for all the images.

These rules help in keeping uniformity in the dataset. This will reduce the errors in the classification of leaves. Also, it ensures that the processing time is not too high.

Leaves from the same plant, but of different sizes will ideally have same aspect ratio. Another feature in the application ensures that the user can upload an image of the leaf from the gallery itself, instead of clicking a picture.

This ensures that the angle at which the image is taken is always constant.

This helps in confirming that the aspect ratio of leaves from the same plant is constant.

### 5.4 Classification using decision tree

We have used the concept of decision trees to train the dataset and classify the leaves. During the initial stages of our application development we tried to do the classification using PNN, but later, we moved to decision tree. This is because the processing time for classification using decision tree is much lesser compared to that using PNN. This affects the efficiency of the application directly.

Decision tree is an approach for predictive modelling used in data mining and machine learning, by which the data in data sets can be classified into classes.

By using decision trees, the system eliminates half of the cases at each step. This is extremely beneficial because in a system like this, where the dataset is huge in number, the processing takes a chink of the total time. Hence, it has to be minimized wherever possible.

The following figure shows the decision tree for our proposed system, which is used for the classification of the leaves. This image shows the classification for Indian Mulberry leaf and Tulsi leaf.

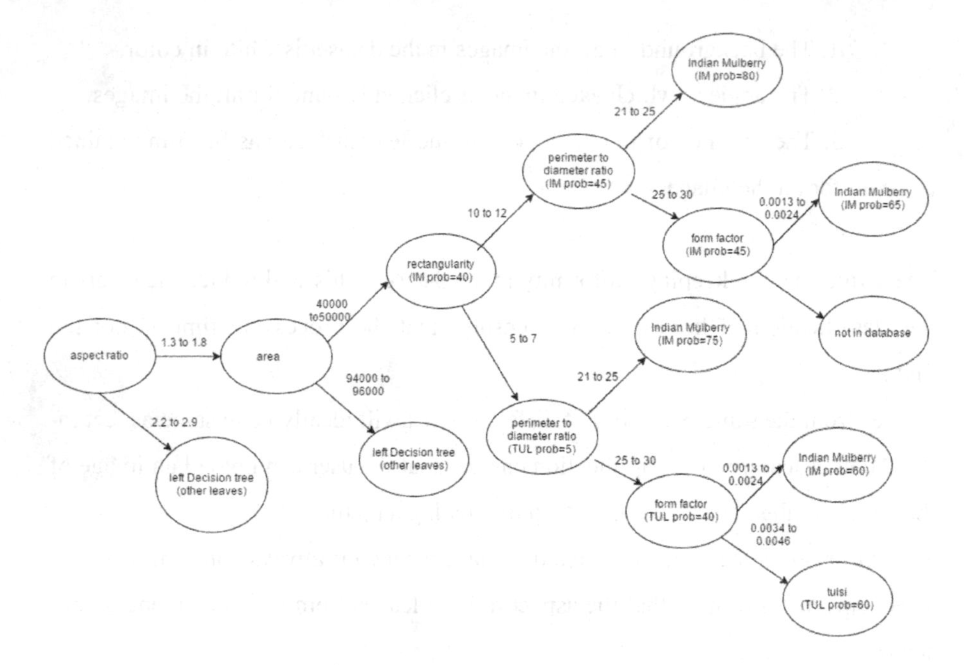


**Fig.3: Decision tree for proposed system showing classification of Tulsi and Indian Mulberry leaf**

IM prob – Probability that the input image is Indian Mulberry

TUL prob – Probability that the input image is Tulsi

### 5.4.1 Efficiency of using decision tree for classification of leaves.

The efficiency of using decision tree is quite high. This is because at every step of classification, half of the dataset which is being classified is being eliminated. The size of the classification set gets reduced to half of what it was in the previous step.

### 5.4.2 Using decision tree over back propagation neural networks (BPNN) for classification

Decision tree classification techniques have been used for a wide range of classification problems and becoming an increasingly important tool for classification of remotely sensed data. These techniques have substantial advantages for land use classification problems because of there flexibility, nonparametric nature, and ability to handle nonlinear relations between features and classes. It is shown by a number of studies that neural classifiers depends on a range of user defined factors, that ultimately limits their use. Training a decision tree classifier is much faster and these classifiers are easy to read and interpret as compared to a neural classifier which is a "black box".

## 5.5 Working

The screenshots in Fig.4 below show the image of a leaf getting pre-processed. It shows the conversion of the image to greyscale, and then Canny Edge Detection algorithm being applied to it.

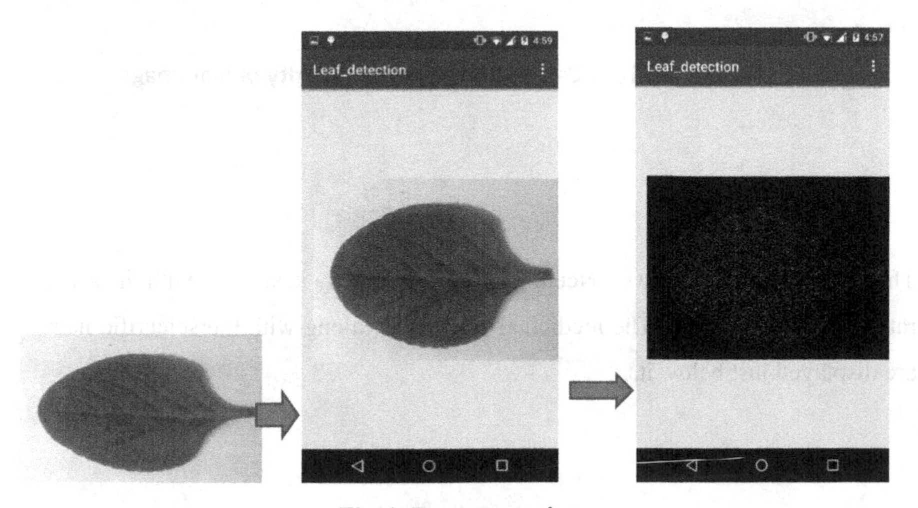


**Fig.4: Pre-precessing**

The calculation of features like rectangularity, form factor, perimeter to diameter ratio are not displayed in the application to the user. These are calculated in the back-end and are used in the process of leaf classification. The below shown screenshot in Fig.5 shows one such instance where the rectangularity is being calculated by the application in the back-end.

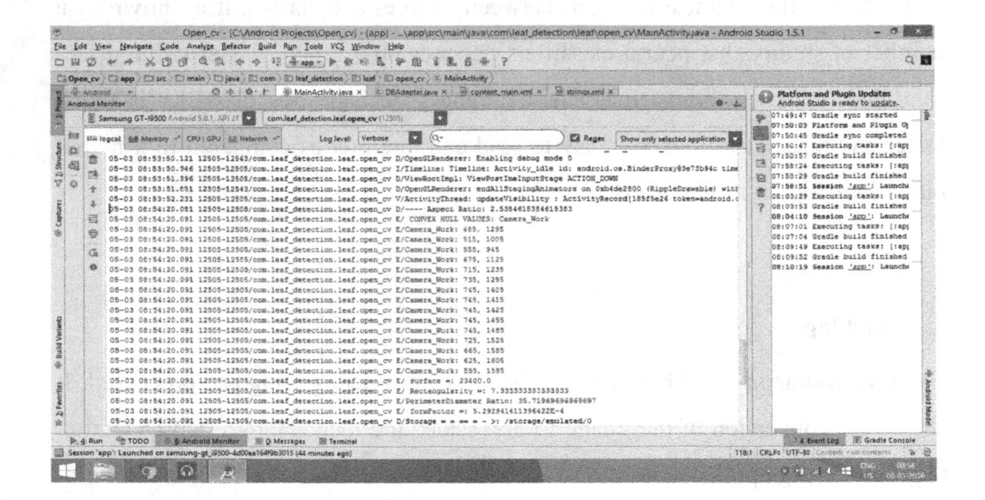

**Fig.5: Calculation of rectangularity of leaf image**

The following figure shows a Neem leaf being correctly identified, with the aspect ratio getting calculated. The medicinal properties, along with the scientific name are displayed just below it.

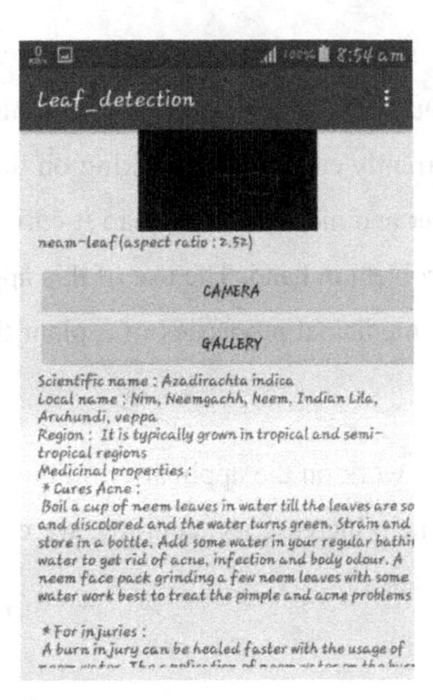


**Fig. 6: Medicinal properties of Neem leaf displayed**

### 5.6 Comparison with other similar works:

When compared with other similar papers which deal with the recognition and classification of leaves, this paper does not outshine the others in the classification process. However, it does not perform any worse than the others. And at the same time, this paper also has an added component of retrieving the medicinal properties of the plant, which is a unique feature that has not been dealt by any other paper in the past.

# 6. Conclusion

The proposed system makes use of image processing to identify a particular leaf and thus retrieve the medicinal properties of the respective plant. The application can automatically recognize different plants with the help of leaf images loaded from digital cameras. The features of the leaf are extracted. The leaf is correctly identified and the user gets the medicinal properties of the plant. The user gets all the information that they are looking for in text format, which is displayed in a very friendly interface. Our system is an easy to use application which is fast in execution too. We are currently considering working on this application in the future, so as to make it better and more efficient than it currently is. The application works well to solve the problem in hand. The use of this application will definitely help in finding the useful medicinal properties of a plant that are unknown to the user.

We believe that with more work on the application in the future and adding some more features which will help increase its accuracy, we can make this application a very useful one for the users.

# 7. References

[1] Stephen Gang Wu, “A Leaf Recognition Algorithm for Plant Classification Using Probabilistic Neural Network”, IEEE International Symposium on Signal Processing and Information Technology, pp.1-16, 2007.

[2] H. Fu and Z. Chi, "Combined Thresholding and Neural Network Approach For Vein Pattern Extraction From Leaf Images," IEE Proceedings- Vision, Image and Signal Processing, Vol. 153, No. 6, December 2006.

[3] Jonathon Shlens, “A Tutorial on Principal Component Analysis”, [Online]. Available: http://www.cs.cmu.edu/helaw/papers/pca.pdf

[4] Ching-Yung Lin, “Rotation, Scale, and Translation Resilient Watermarking for Images”, “IEEE Trans Image Processing”, Vol.10, No. 5, 2001.

[5] http://mathematica.stackexchange.com/questions/21059/how-can-i-detect-a-rectangular-region-of-interest-in-a-picture, “Details on Rectangularity in Image Processing”

[6] http://docs.opencv.org/2.4/doc/tutorials/imgproc/imgtrans/canny_detector/canny_detector.html, "Canny Edge Detection Algorithm in OpenCV"

[7] http://mathworld.wolfram.com/ConvexHull.html, "Understanding Convex Hull Algorithm"

[8] http://springerplus.springeropen.com/articles/10.1186/2193-1801-2-660,"Image Processing For Plant Classification"

# *Camouflaged Target Detection and tracking using thermal infrared and visible spectrum imaging*

*Miss. Supriya Mangale and Mrs. Madhuri Khambete*

***Abstract***— **This paper describes the new robust Thermo-Visible moving object detection system under different scenarios such as camouflaged, glass, snow, similar object and background color or temperature etc. Background subtraction is performed separately in Thermal infrared and Visible spectrum imaging modality by formation of mean background frame and use of global thresholding,then using connected component theory and fusion rule, moving objects are detected and blob based tracked in adverse situations also.**

***Keywords***- Visible Video, Thermal Video, Mean-background, Global-thresholding, Fusion rule etc.

## 1. INTRODUCTION

In the today's world of technology, Video surveillance has become an active topic of research. As safety and security at different public places, Military purposes are nowadays important issues to be discussed.

Generally, the sensor technology used for detection and tracking of moving objects is visible imaging .The reason behind it is, visible images are rich with information, containing strong edges, color and other features (textures) with low noise. However, using only visible image information may not be enough in challenging situations such as camouflaged, varying illumination condition or adverse environmental conditions such as rain, smoke, fog etc. Under such conditions, the acquired visible images are very noisy.

Now, the features of thermal-infrared images are very distinctive compared to visible-light spectrum images. The gray level value of the objects in thermal image represents the temperature and radiated heat, and is independent from lighting conditions. Hence objects with varying illumination conditions or in camouflaged conditions can be easily detected by thermal-infrared images. On the other hand, the spatial resolution and sensitivity of thermal-infrared images is very low compared to visible spectrum images. Hence thermal infrared images are noisy with low image quality. As well as Objects with similar background and foreground temperature are difficult to detect using Thermal imaging. Above all the detection through glass is not possible with thermal-infrared cameras.

Hence under challenging situations, single modality sensors such as only thermal-infrared or only visible spectrum cannot meet the requirement. So in-order to develop robust and reliable moving object detection system, it is necessary to fuse useful information from both the modalities.

The information obtained from visible spectrum sensor is complementary to the information obtained from thermal-infrared sensors.

---

Miss Supriya Mangale
Research Scholar at College of Engineering, Pune411052, supriya.mangale@cumminscollege.in
Dr.Mrs. Madhuri Khambete
MKSSS's Cummins College of Engineering ,Karvenagar Pune 411052,principal@cumminscollege.org



J.M. Corchado Rodriguez et al. (eds.), *Intelligent Systems Technologies and Applications 2016*, Advances in Intelligent Systems and Computing 530,
DOI 10.1007/978-3-319-47952-1_15

## 2. LITERATURE REVIEW

In background subtraction technique, background model of scene is build, then deviation of incoming frame from it , is calculated. Moving object is detected by significant change between image region and background model. We have assumed static background in our proposed method.
Frame differencing of temporally adjacent frames is basic method proposed by Jain and Nagel [2],then approximate median ,Running gausssian were being studied. However it was found by T. Bouwmans, F. El Baf, B. Vachon [20]that single gaussian was incorrect model to analyse outdoor scenes.Hence a mixture of Gaussians was used to model the pixel color by stauffer and Grimson[3].-The wallflower database is one of the standard datasets for evaluating Background Subtraction(BGS) algorithms for visible spectrum Imaging. We found that on Light Switch and Camouflage database the Mixture of Gaussian method failed.

In each Background subtraction algorithm ,selection of threshold (Global or local) is very critical.Moving object detection system developed by Q. Zhu, M. C. Yeh, K. T. Cheng and S. Avidan [9], A. Rakotomamonjy ,M. Bertozzi, A. Broggi, M. Del Rose, M. Felisa [14],Q. J. Wang, and R. B. Zhang [10] use Histogram of Gradients (HOG) for feature detection along with Support Vector Machine(SVM) classifier . Stocker A[7] and S.P.N. Singh ,P.J.Csonka, K.J.Waldron [8]Optical flow techniques are commonly used for motion detection .

Wei Li, Dequan Zheng, Tiejun Zhao, Mengda Yang [11] developed Moving object detection in thermal imaging using HOG features with SVM classifiers. Marco San-Biagio Marco Crocco Marco Cristani[12] used Recursive Segmentation (Using adaptive thresholding) is use to detect Region Of Interest(ROI's) in Thermal Image. The disadvantage of Thermal infrared imaging is, they are noisy and have poor texture. The behavior of Visible and thermal imaging under different conditions are shown in **Fig.1**

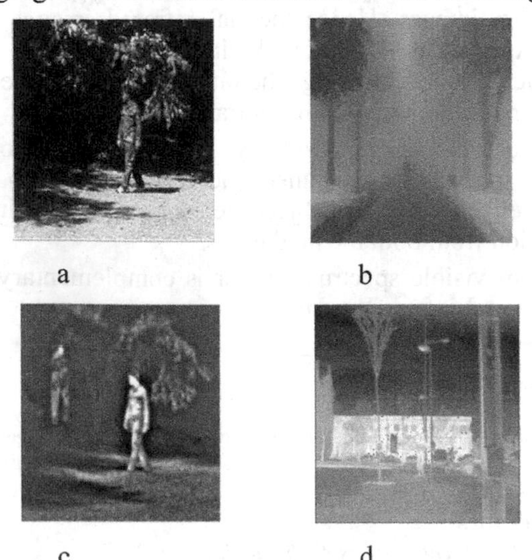


**Fig. 1**: Visible and Thermal Imaging

In **Fig. 1,Fig. a** shows Visible Imaging in good environmental condition. **Fig.b** shows failure in capture of Fog situation by visible camera. **Fig.c**Thermal imaging with different foreground and background temperature. **Fig.d** shows Failure of thermal camera when object and background temperature are same.

M.Talha and Stolkin[24], the moving objects are tracked using visible or thermal modality. One modality whichever has less Bhattacharya coefficient or less correlation with previous frame is selected at a time. Finally tracking is done using Particle filter. Drawback of selection of one modality at a time can be explained using from Fig (5) ,Row 3 column 1 and Row 4 column 1,we see vehicle is traced in visible modality and two persons are traced in thermal modality. Selection of one modality is better only when object cannot be detected by one modality at all such as glass where thermal totally fails or heavy camouflaged condition. Moreover particle filter is very complex and requires more time for processing.

The advantage of using thermal and visible image sensors together are, the color information is provided by visible imaging and temperature information by thermal infrared imaging of the same scene.[1]

In our proposed method we have performed detection and tracking of objects using both the modalities. We have taken into account different scenarios such as snow, glass, camouflaged condition, similar object and background temperature or color etc.

## 3. PROPOSED METHOD

In our method we have used the concept of formation of mean background frame. A mean background frame is formed for both the modalities separately. In this algorithm a background is recorded without any foreground present in that area. The advantage of mean background frame is it takes into account slightly varying illumination changes, waving of trees etc and forms robust background visible frame. The mean background frame in thermal imaging considers only temperature difference between foreground and background frames as it is illumination invariant while it considers illumination changes in visible imaging. We create a mean background frame of visible and thermal database as follows,

Series of N preceding frames $F$(x,y) are averaged to form B background frame at time instant t,

$$IM_b = \frac{1}{m*n} \sum_{i=1}^{N} \sum_{x=1}^{m} \sum_{y=1}^{n} F(x, y, t-1) \quad (1)$$

Where m*n is frame size

- For visible frames ,the background subtraction is done using standard deviation rule(std)

$IM_{aa} = IM_c - IM_b$ (2)

$std_f = std\ (IM_{aa})$ (3)

$IM_{bb} = IM_b - IM_c$ (4)

$std_b = std\ (IM_{bb})$ (5)

if $std_f > std_b$

  $IMv = IM_{aa}$ (6)

else

  $IMv = IM_{bb}$

end

Where $IM_c$ is current frame, $IM_b$ mean background frame and $IM_v$ is Visible Image after subtraction. Standard deviation indicates maximum presence of foreground after background subtraction.
This is needed as sometimes major region is covered by background and minor by foreground and vice-versa so subtraction may not result in selection of actual or appropriate foreground. Such is not case of thermal images due to good contrast.
Hence for thermal images $IM_t = IM_c - IM_b$ (7)
Where $IM_t$ is Background subtracted Thermal image.

After mean background subtraction, we have used global thresholding using Otsu's algorithm for binarization of gray level images. It separates foreground moving objects from background which are known as blobs. Foreground pixels are denoted as white pixels and background pixels as black. A threshold is said to be globally optimal if the number of misclassified pixels is minimum. For each frame of thermal and visible we get different optimal threshold level k* which compensates the disadvantages of mean filtering in which threshold is not function time. Moreover Global thresholding considers bimodality of histogram which is not considered again in mean filtering.
Here principle of image Segmentation is extended for background subtraction of moving objects. The regions are usually segmented by identifying common properties. The simplest property that pixels in a visible region can share is its intensity and temperature in thermal images.
The major problem of thresholding in visible imaging is that we consider only pixel intensity not any relationship between the pixels. Hence the pixels identified by thresholding process may not be contiguous.
Due to this pixels may get included or excluded in foreground or background region increasing the false positive rates or false negative rates respectively.
The above problem can be solved by thermal imaging as thermal image provides good contrast which helps to identify relationship between pixels of similar temperatures. Hence background and foreground objects can be easily segmented.
If pixels of foreground and background have same temperatures then they might not be easily segmented using thermal imaging but can be segmented by visible imaging due to difference in pixel intensities of foreground and background.
The advantage of binarization of Visible and Thermal sensors using global thresholding is visible imaging provides pixel intensity information (color, texture or features etc) and thermal imaging relationship of pixel intensities (good contrast) due to similar temperatures to build a Robust Moving Object Detection systems.
In this correspondence, our discussion will be limited to threshold selection using Otsu's global thresholding algorithm where only the gray-level histogram suffices without other a priori knowledge.

Moreover **Otsu**'s Global thresholding method [23] is Non-Parametric and unsupervised method of automatic global threshold selection.

Otsu's algorithm is basically clustering-based image thresholding method. It binaries the gray level image. The algorithm is based on assumption that image consists of two classes of pixels-foreground and background pixels i.e is bi-modal histogram,optimum threshold is calculated in such a way that it separates this two classes by maximizing inter-class varaiance or minimizing intra-class variance.
The basic relation is

$$\sigma_T^2 = \sigma_B^2 + \sigma_W^2 \tag{8}$$

Where $\sigma_W^2$ within-class variance, $\sigma_B^2$ is between-class variance, and $\sigma_T^2$ is total variance of levels.
Our Goal is to search for a threshold k that maximizes the criterion measures in thermal and visible imaging.The discrimant criterion measures is used to calculated optimum threshold k*

$$\eta(k) = \sigma_B^2(k)/\sigma_T^2 \quad (9)$$

In order to maximize $\eta$, maximize $\sigma_B^2$ as $\sigma_T^2$ is independent of $k$.
The optimal or threshold k* is

$$\sigma_B^2(k*) = \max_{1\leq k<L} \sigma_B^2(k) \quad (10)$$

In our case $\sigma_B^2(k*) = \max_{1\leq k<L} \sigma_{VB}^2(k) + \max_{1\leq k<L} \sigma_{TB}^2(k)$ (11)

As η * is invarian to affine transformation,it is considered as distinct measure .

Hence $0\leq \eta * \leq 1$ (12)

The lower bound (zero) is in images having a single constant gray level, and the upper bound (unity) is in two-valued images.
Moreover Global thresholding binaries the frames which directly provides foreground as white pixels and background as black pixels.

After performing blob detection of visible and thermal frames using Otsu's algorithm, Depending upon the video dataset, if distance between cameras and object is large, threshold used for selection of moving object connected components in the frame is small and if distance between cameras and object is small the threshold is large. Connected components show the presence of moving object in frame. Moreover it is used to remove the unwanted noise.
Addition of binary thermal and visible frames is carried to get fused frame.

$IM=IM_v+ IM_t$ or IM=IMv || IMt (13)

This is logical 'OR' operation used to extract and fused the useful information from both the modalities .It helps to get additional information about the moving object.
After performing background subtraction or detection of moving objects, moving object tracking is done using blob analysis. Centroids, area of each blobs present in a frame are calculated and a Rectangular bounding box are inserted around the detected blobs. Pixel co-ordinates along with width and height are calculated of each bounding box.
The advantage of tracking moving objects present in fused frame is, the object which may lose its track in one modality can be traced by other modality.
So using complementary information from both the modalities as well as through global thresholding and blob analysis, we have developed a Robust Moving object detection and tracking system.

## 4. EVALUATION PARAMETERS

I. Let TP=true positives, TN=true negatives,
FP=false positives,FN=false negatives

Sensitivity=True Positive Rate(TPR)=TP/(TP+FN) (14)

Specificity=TrueNegativeRate(TPR)=TN/(TN+FP) (15)

Precision=TP/(TP+FP) (16)

a. Precision should be as high as possible.
b. Precision depends on TPR and FAR.

II. Percentage Error: Percentage error compares estimated value with exact value. It gives absolute difference between exact value and estimated value as a percentage of exact value.

If **'p'** are number of true targets detected by our method and **'P'** are number of true targets
provided by Groundtruth then

$$\%\text{Error}=\left|\frac{p-P}{P}\right| \times 100 \quad (17)$$

Ideally error should be zero and practically as low as possible.

**4.1 Results**: We have worked on different videos from INO's Video Analytics Datasets, found at URL: www.ino.ca/Video-Analytics-Dataset as well as on OCTVBS datasets. INO have also provided ground truth with each video. We have also implemented the object detection and tracking on Niche database ,created using a FLIR E-30 series thermal imaging camera with 320 × 240 pixel resolution, mounted side by side with a standard, color camera. We have Compared proposed method with visible and Thermal imaging separately and also with[6].

**Table1:** Comparison of proposed method with different existing methods on INO_ Trees and Runners Video Dataset.

| Sr.NO | Frame no. | Precision of Only Visible (V) | Precision of Only Thermal(T) | Precision of VT-Fused by MOG | Precision Of VT using Proposed Method |
|---|---|---|---|---|---|
| 1. | 341 | 0.3177 | 0.6731 | 0.8961 | 0.8906 |
| 2. | 361 | 0.4083 | 0.6796 | 0.8787 | 0.8214 |
| 3. | 381 | 0.4259 | 0.6796 | 0.8787 | **0.9166** |
| 4. | 401 | 0.3390 | 0.7734 | 0.9215 | **0.9933** |
| 5. | 421 | 0.3390 | 0.7807 | 0.9215 | **0.9923** |
| 6. | 441 | 0.4393 | 0.7366 | 0.8291 | **0.9900** |
| 7. | 461 | 0.3854 | 0.5551 | 0.7823 | **0.9496** |

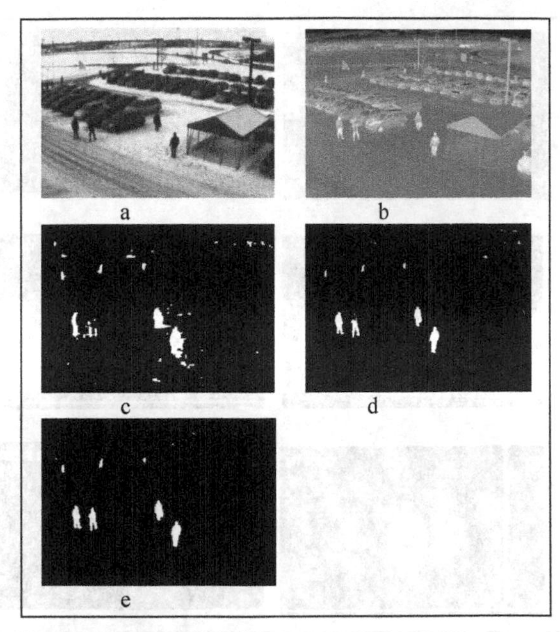


**Fig.2:** This is Frame no-851 of INO_Parking Snow Video Dataset, **Fig.a** shows Visible image and **Fig.b** Thermal Image. **Fig.c** shows Thermo-Visible Fused Image by Mixtures of Gaussian, **Fig.d** shows Thermo-Visible fused Image by Global Thresholding. **Fig.e** shows Ground Truth image.

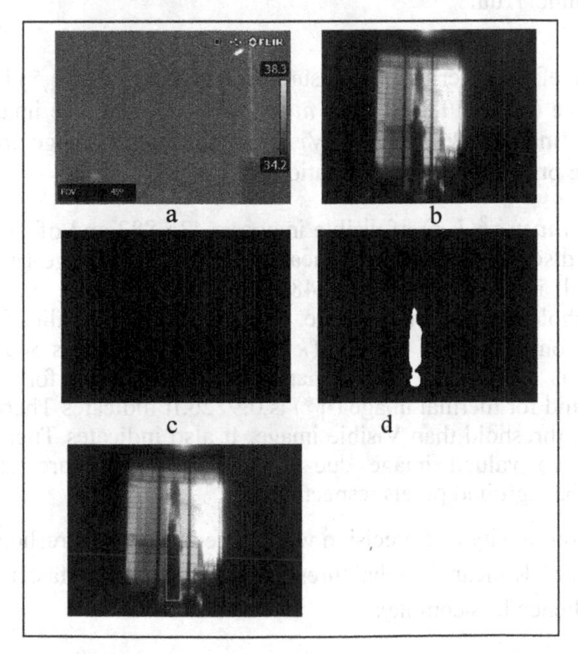


Fig.3: This Video shows Failure of Thermal camera for tracking of objects behind Glass.Row1 shows the Thermal and Visible Frame. Row 2 shows object detection in Thermal and Visible frame by proposed method and Row3 Fused Tracked results by Proposed Method.

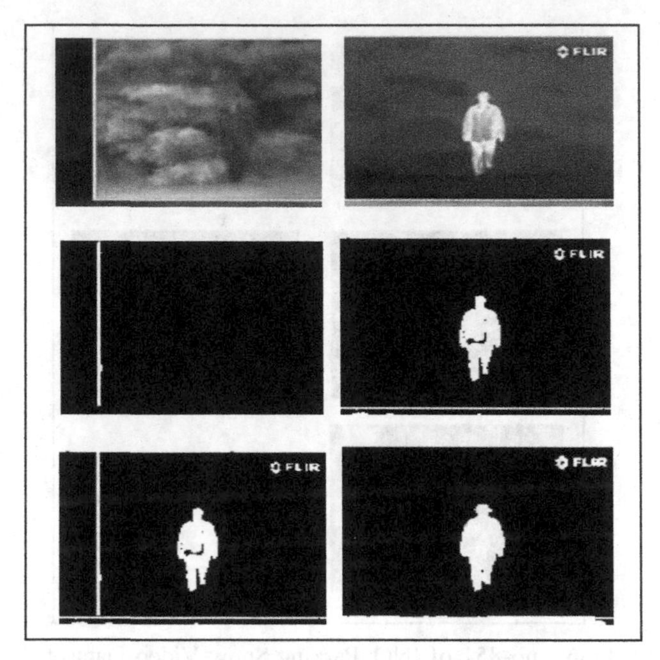


**Fig.4**: Extremely Camouflage Condition tracked by FLIR where Visible fails and Thermal works, Row1 shows Visible and Thermal image. Row2 shows Globally Thresholded Visible and Thermal Image. Row3 shows fused image by proposed method and Ground Truth

From **Fig.2**, The effective criterion measure for visible image ($\eta_v$*) is 0.8874 and for thermal image ($\eta_t$*) is 0.8132. The $max\ \sigma_B^2(k*)$ of visible image is 322.70 and of Thermal Image is 264.09.Both Visible andThermal Image are effectively threshold. Hence provides better information

From **Fig.3,** the $max\ \sigma_B^2(k*)$ of visible image is 435.883 and of Thermal Image is 34.7054.The discriminant criterion measure for visible image ($\eta_v$*) is 0.7934 and for thermal image ($\eta_t$*) is 0.4448.It indicates Visible Image is more effectively threshold than Thermal image. **Fig.3,** Thermal modality fails to detect through glass. From **Fig.4**, the $max\ \sigma_B^2(k*)$ of visible image is 549.9240 and of Thermal Image is 1973.3.The discriminant criterion measure for visible image ($\eta_v$*) is 0.7934 and for thermal image ($\eta_t$*) is 0.9226.It indicates Thermal Image is more effectively threshold than Visible image. It also indicates Thermal Image is approximately two valued image due to similar temperature of almost all foreground and background pixels respectively.

If we compare complexity and precision with respect to other thresholding methods like Fuzzy-K means. Global thresholding works much faster and gives same accuracy, hence less complex.

**Table 2** : Comparison of Elapsed time between proposed method and Mixtures of Gaussian for different INO-Video dataset.

| Sr.No | DATASETS | Elapsed Time by MOG | Elapsed Time by Proposed method |
|---|---|---|---|
| 1. | INO-MAIN ENTRANCE DATASET | 48.6532sec | 37.104010 sec |
| 2. | INO-COAT DEPOSIT DATASET | 499.5432 sec | 388.686089 sec |
| 3. | INO-CLOSE PERSON DATASET | 25.6577 sec | 17.380562 se |
| 4. | INO-PARKING SNOW DATASET | 587.081047 sec | 393.854919 sec |
| 5. | INO-GROUP FIGHT DATASET | 555.7865sec | 430.299108 sec |

**Table 3** : Comparison of Precision between proposed method and Mixtures of Gaussian for different INO-Video dataset.

| INO-MAIN ENTRANCE DATASET | | | | | | | |
|---|---|---|---|---|---|---|---|
| FRAME NOS | 201 | 226 | 251 | 376 | 451 | 501 | 551 |
| PRO METHOD | **0.936** | **0.875** | **0.971** | **0.980** | **0.999** | **0.919** | **0.997** |
| MOG | 0.783 | 0.860 | 0.531 | 0.727 | 0.890 | 0.641 | 0.908 |
| INO-COAT DEPOSIT DATASET | | | | | | | |
| FRAME NOS | 220 | 320 | 620 | 720 | 1423 | 1620 | 1720 |
| PRO METHOD | **0.872** | **0.940** | **0.740** | **0.629** | **0.333** | **0.508** | **0.544** |
| MOG | 0.8853 | 0.8836 | 0.113 | 0.479 | 0.0171 | 0.512 | 0.490 |

| INO-CLOSE PERSON DATASET | | | | | | | |
|---|---|---|---|---|---|---|---|
| FRAME NOS | 146 | 161 | 181 | 196 | 201 | 226 | 236 |
| PRO METHOD | **0.982** | **0.989** | **0.968** | **0.964** | **0.951** | **0.941** | **0.921** |
| MOG | 0.722 | 0.693 | 0.677 | 0.663 | 0.695 | 0.594 | 0.607 |
| INO-PARKING SNOW DATASET | | | | | | | |
| FRAME NOS | 751 | 851 | 1651 | 2251 | 2351 | 2551 | 2741 |
| PRO METHOD | **0.808** | **0.928** | **0.821** | **0.836** | **0.787** | **0.780** | **0.793** |
| MOG | 0.516 | 0.553 | 0.671 | 0.387 | 0.289 | 0.382 | 0.311 |
| INO-GROUP FIGHT DATASET | | | | | | | |
| FRAME NOS | 378 | 528 | 678 | 1028 | 1178 | 1278 | 1328 |
| PRO METHOD | **0.898** | **0.877** | **0.968** | **0.849** | **0.660** | **0.604** | **0.696** |
| MOG | 0.548 | 0.491 | 0.588 | 0.457 | 0.248 | 0.499 | 0.513 |

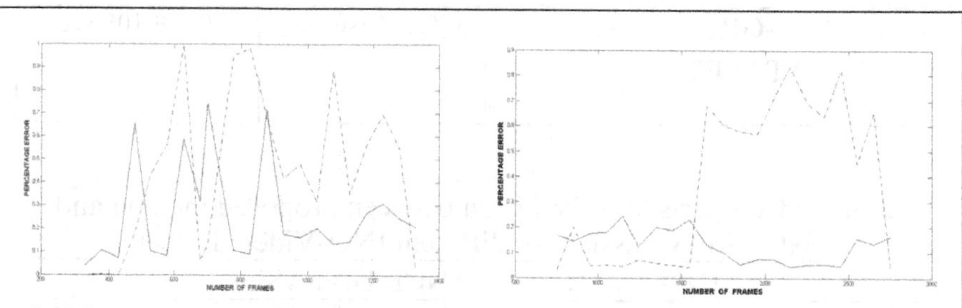


**Graph: Row1 column1** Indicates comparison of **Percentage-error** with respect to Number of Frames of INO-Group fight dataset .Dotted line shows error by Mixtures of Guassian and continuos line ,error by proposed method.
**Row1 column2** Indicates comparison of **Percentage-error** with respect to Number of Frames of INO- Parking Snow dataset .dotted line shows error by Mixtures of Guassian and continuos line ,error by proposed method.

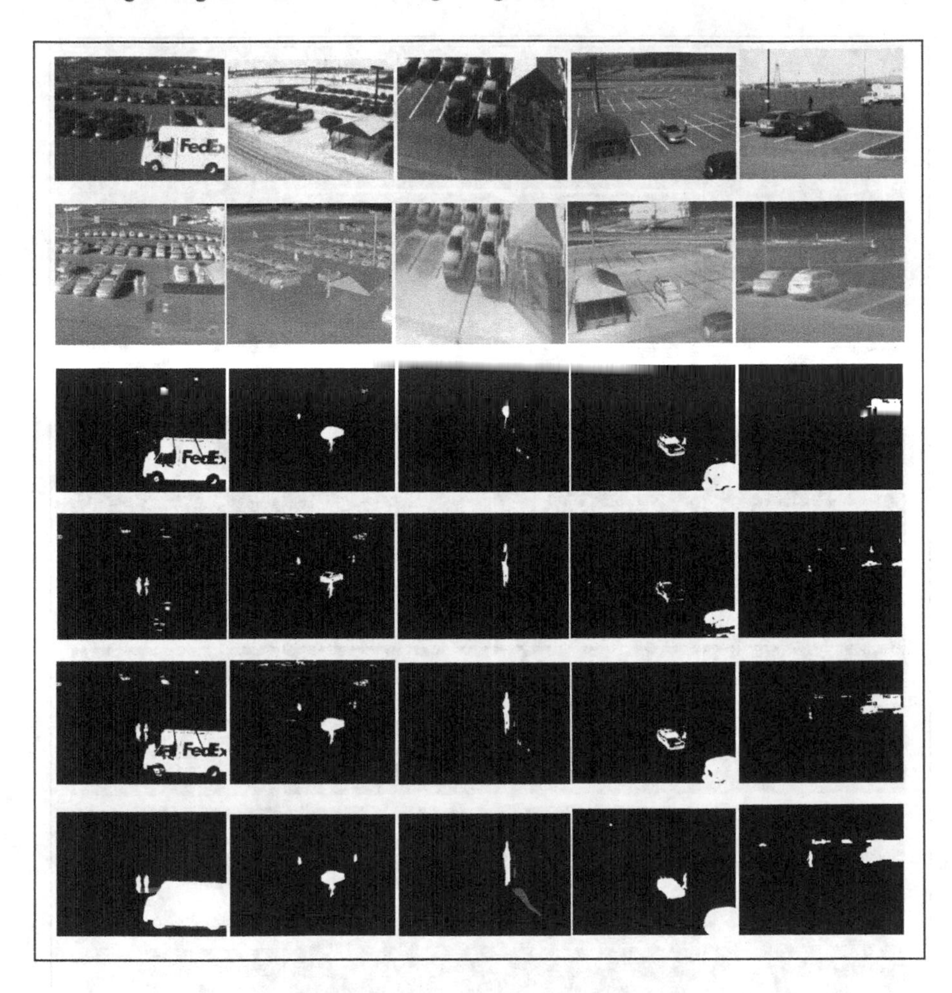


**Fig5:**Moving Object Detection in different INO-Datasets with our Proposed method. Row 1 and Row 2, visible and Thermal frame nos-320,2051,201,320,551 of INO-Group Fight, Parking Snow, Close person, Coat Deposit, Main Entrance Video datasets respectively.Row3 shows results by using Only Visible frames. Row 4 shows results by only Thermal frames. Row5 Result by Proposed method. Row 6 Ground Truth Results.

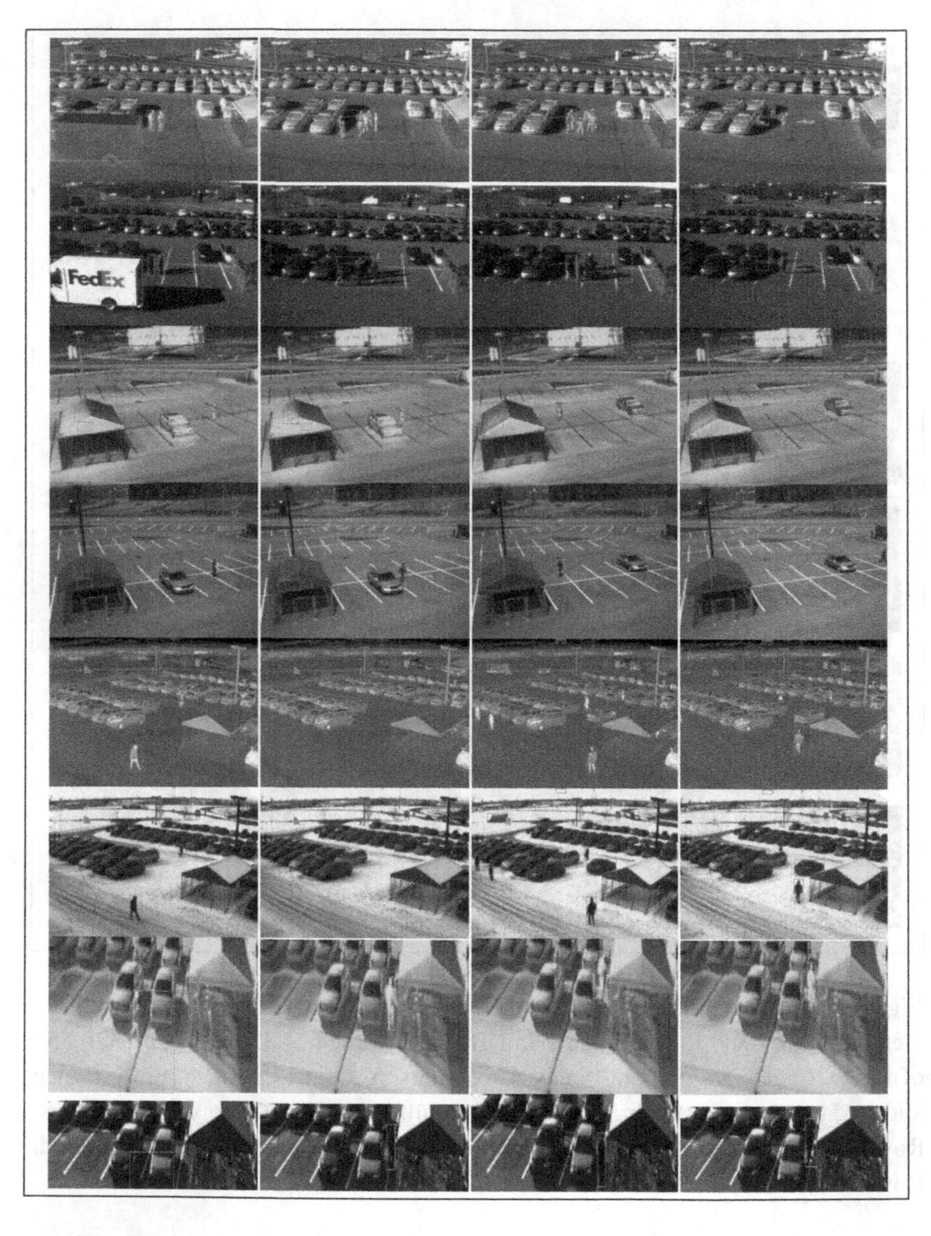


**Fig.6:** Tracking results using Proposed method for different frames of different INO-Dataset.Here Objects tracked in Thermal and Video datasets are shown. Row1 and Row2 shows INO-Group Fight Datasets Thermal and Visible Frame Nos-692,950,1099,1265 respectively. Row3 and Row4 shows INO-Coat Deposit Datasets Thermal and Visible Frame Nos-352,422,1610,1723 respectively. Row5 and Row6 shows INO-Parking Snow atasets Thermal and Visible Frame Nos-311,574,1997,2027 respectively. Row7 and Row8 shows INO-Close person Datasets Thermal and Visible Frame Nos-175,201,220,239 respectively.

5. **CONCLUSION:** We have developed a reliable Moving Object detection system using fusion of Thermal-Infrared and Visible Imaging. It is found that proposed method provides better precision over other existing methods. Moreover it is less complex and works faster than other methods.

This new algorithm exhibits several useful properties:

- Combine use of Visible and Thermal modality nullifies the drawbacks of both the methods.
- Simple Global-Thresholding technique and formation of mean background frame provides selected threshold for each frame to be function of time and also bimodality of histogram.
- It maximizes between class variances and minimizes within class variances. Hence depending upon modality, object classification becomes easier.
- The above methods can be used in adverse situations like snow, glass, camouflage etc.
- The objects which are not detected in one modality are easily detected by other modality. hence it proves that using single modality is insufficient for moving object detection.
- Tracking using simple blob analysis make object tracking simpler than other complex trackers such as particle filter or kalman filter.
- Precision and percentage of error is are better than Mixtures of Gaussian techniques.
- This paper lacks in shadow removal by fusion techniques.

## 6. REFERENCES

[1] Ciaran O C 2005 Thermal Infrared and Visible Spectrum fusion for Multi-modal Video Analysis.,Transfer Report.

[2] Jain R and Nagel H 1979. On the analysis of accumulative difference pictures from image sequences of real world scenes. IEEE Trans. Patt. Analy. Mach. Intell. 1, 2, 206–214.

[3] Stauffer C. and Grimson, W. 2000. Learning patterns of activity using real time tracking. IEEE Trans.Patt. Analy. Mach. Intell. 22, 8, 747–767.

[4] Schultz D , Burgard W, Dieter F, and Armin B. C 2001 Tracking multiple moving targets with a mobile robot using particle filters and statistical data association. In IEEE International Conference on Robotics and Automation,vol 2.pp.1665-1670.

[5] Carine Hue, Jean-Pierre Le Cadre, and Patrick P′erez.2001 A Particle filter to track multiple objects". In IEEE Workshop on Multi-Object Tracking,pp.61-68.

[6] Mangale S and Khambete M.2015,Moving Object detection using visible Spectrum Imaging and Thermal Imaging, In IEEE Internation Conferenec on Industrial Instrumentation and Control..

[7] Stocker A 2002 An improved 2D optical flow sensor for motion segmentation. In IEEE International Symposium on Circuits and Systems,vol.2,pp.332-335.

[8] S.P.N. Singh, P.J. Csonka, K.J. Waldron 2006 Optical flow aided motion estimation for legged locomotion. InIEEE International Conference on Intelligent Robots and Systems pp.1738-1743.

[9] Q. Zhu, M. C. Yeh, K. T. Cheng and S. Avidan 2006 Fast human detection using a cascade of histograms of oriented gradient. In IEEE Computer Society Conference on Computer Vision and pattern Recognition(CVPR), vol. 2: 1491-1498.

[10] Q. J. Wang, and R. B. Zhang 2008 LPP-HOG: A New Local Image Descriptor for Fast Human Detection.In IEEE International Symposium on Knowledge Acquisition and Modeling Workshop.pp.640-643.

[11] Wei Li, Dequan Zheng, Tiejun Zhao, Mengda Yang 2012 An Effective Approach pedestrian Detection in Thermal Imagery.In IEEE International Conference on Natural Computation.pp.325-329.

[12] Marco San-Biagio Marco Crocco Marco Cristani 2012 Recursive Segmentation based on higher order statistics in thermal imaging pedestrian detection.In In IEEE International Symposium on Communication Control and Signal Processing pp.1-4.

[13] H.Torresan, B.Turgeon, C.Ibarra: Castanedo, P.Hebert, X.Maldague 2004 Advance Surveillance systems: Combining Video and Thermal Imagery for Pedestrian Detection. In Proc. of SPIE Thermosense XXVI, SPIE,vol.5405, pp.506-515.

[14] M. Bertozzi, A. Broggi, M. Del Rose, M. Felisa, A. Rakotomamonjy and F. Suard 2007 A Pedestrian Detector Using Histograms of Oriented Gradients and a Support Vector machine Classifier" In IEEE Intelligent Transportation systems Conference .

[15] Jan Thomanek, Holger Lietz, Gerd Wanielik 2010 Pixel-based data fusion for a better Object detection in automotive application. In IEEE International Conference on Intelligent Computing and Intelligent Systems (ICIS),vol.(:2),285-390.

[16] Ju han and Bir Bhanu 2007 Fusion of color and infrared video for moving human detection. In ELSEVIER Pattern Recognition 40,1771-1784.

[17] Siyue Chen, Wenjie Zhu and Henry Leung 2008 Thermo-Visible video fusion using probabilistic graphical model for human tracking.In IEEE International Symposium on Circuits and Systems.978-1-4244-1684-4..

[18] Eun-Jin Choi and Dong-Jo Park 2010 Human Detection Using Image Fusion of Thermal and Visible Image with New Joint Bilateral Filter" In 5th IEEE International Conference on Computer Science and Convergence Information Technology(ICCIT).882-885.

[19] E.Thomas Gilmore III , Preston D.Frazier and M.F.Chouikha 2009 Improved Human detection using Image Fusion. In IEEE ICRA Workshop on People Detection and Tracking Kobe, Japan.

[20] T. Bouwmans, F. El Baf, B. Vachon 2008 Background Modeling using Mixture of Gaussians for Foreground Detection - A Survey. Recent Patents on Computer Science, Bentham.Science Publishers,1(3),pp.219-237.

[21] Stauffer C. and Grimson, W. 1999--Adaptive background mixture models for real-time tracking.In IEEE Computer Society Conference on Computer Vision and Pattern Recognition vol-2 i-xviii.

[22] Alper Y,Omar J ,Mubarak Shah 2006,Object Tracking,ACM Computing Surveys,vol.38,No.4,Article 13.

[23] Otsu N 1979 A Threshold Selection Method from Gray-Level Histograms,. InIEEE Trans. Systems, Man, and Cybernetics, Vol. SMC-9.

[24] Talha M and Stolkin 2014 Particle Filter Tracking of Camouflaged Targets by Adaptive Fusion of Thermal and Visible Spectra Camera Data. In IEEE Sensors Journal, Vol.14 NO.1

# Heuristic Approach for Face Recognition using Artificial Bee Colony Optimization

Astha Gupta and Lavika Goel

**Abstract** Artificial Bee Colony (ABC) algorithm is inspired by the intelligent behavior of the bees to optimize their search for food resources. It is a lately developed algorithm in Swarm Intelligence (SI) that outperforms many of the established and widely used algorithms like Genetic Algorithm (GA) and Particle Swarm Optimization (PSO) under SI. ABC is being applied in diverse areas to improve performance. Many hybrids of ABC have evolved over the years to overcome its weaknesses and better suit applications. In this paper ABC is being applied to the field of Face Recognition, which remains largely unexplored in context of ABC algorithm. The paper describes the challenges and methodology used to adapt ABC to Face Recognition. In this paper, features are extracted by first applying Gabor Filter. On the features obtained, PCA (Principal Component Analysis) is applied to reduce their dimensionality. A modified version of ABC is then used on the feature vectors to search for best match to test image in the given database.

## 1 Introduction

Swarm intelligence is a decentralized and self-organized behavior of a natural or an artificial system. This concept was introduced in 1898 by Gerardo Beni and Jing Wang, under Artificial Intelligence in context of cellular robotic systems. SI systems consist of a population of simple agents (also called as boids) which interact with one another in order to share local information. The merit of SI algorithms lies in the simplicity of the rules that each agent follows without any centralized structure to dictate their behavior, thus involving some level of randomness.

---

Astha Gupta, M.Sc.(Tech.) Information Systems
BITS Pilani, Pilani, Rajasthan e-mail: astha736@gmail.com

Lavika Goel, Assistant Professor
BITS Pilani, Pilani, Rajasthan e-mail: lavika.goel@pilani.bits-pilani.ac.in

 
J.M. Corchado Rodriguez et al. (eds.), *Intelligent Systems Technologies and Applications 2016*, Advances in Intelligent Systems and Computing 530,
DOI 10.1007/978-3-319-47952-1_16

Such interactions lead to globally intelligent behavior without agents worrying about achieving it. Examples of algorithms under SI are Ant Colony, Bird Flocking, Herding, Particle Swarm, Genetic Algorithm, Bacterial Growth etc. Survey on SI by Mishra et al. [1] and Keerthi et al. [2] provides further detail on different algorithms under SI. As ABC belongs to the class of SI algorithms, it shares same nature and advantages as stated above.

ABC has been applied to many different areas like Software Testing, Neural Network etc. In this paper ABC is being applied to Face Recognition. Face Recognition in itself is a huge domain with variety of algorithms. It appears to be a simple task for humans and rarely do we consider the complexity of the problem. Human ability to detect faces under various conditions within seconds has always been intriguing. In case of computing machines, the accuracy and the run time has lagged far behind that of a human's ability. Despite numerous efforts recognition accuracy continues to remains low, about 20 times lower than recognition by fingerprint analysis and matching according to Rawlinson et al. [3]. Therefore, a lot of work is required towards obtaining an efficient solution in this domain.

Apart from the Introduction this paper has 7 sections. Section 2 and 3 describe ABC and Face Recognition briefly. Section 4 states the problem statement and explains the challenges faced in detail. Section 5 contains the methodology used in the paper. Section 6 discusses the results and section 7 gives conclusion and future work. List of references has been provided at the end in section 8.

## 2 Artificial Bee Colony Algorithm

ABC is one of the algorithms under SI, which was proposed by Karaboga in 2005. It was used by him to optimize numerical problems. Paper by Karaboga et al. [4] provides a deep insight into the algorithm. ABC is inspired by foraging behavior of honey bees specifically based on model proposed by Tereshko and Loengarov in 2005 [5].

ABC algorithm emulates the procedure by which bees share information in order to maximize their honey collection and minimize their efforts. In the algorithm food sources are locations where nectar is available. Optimization is achieved by sharing information and diving responsibility amongst themselves. They organize themselves in three groups.

1. ***Employed Bees***: Goes to the food sources to get nectar.
2. ***Scout Bees***: Explore different food sources through random search.
3. ***Onlooker Bees***: Wait at the dance area, for scout bees to come back and do 'waggle' dance which gives information about food sources.

In the beginning, food sources are initialized. As these sources start to deplete, Scout Bees search for new sources. Scout Bees after completion of their search come back to bee hive and report at the dance area. The strength and inclination of the dance is the indicator of the quality and the location of food sources. This information along with prior information is used to select new food sources which will be used by Employed Bees. When food sources deplete, the cycle is repeated. In ABC food sources are searched in a multidimensional space. It provides a procedure for selecting solution based on fitness of new foods position. ABC has been thoroughly explored and experimented with Karaboga et al. [6,7] and Bolaji et al. [8] provide an extensive survey in the algorithm as well as its variants and their application in different fields, from its applications in bioinformatics, clustering and scheduling to various hybrids that have been proposed in past years such as GA, ACO etc. ABC algorithm has a home page [9] listing various developments, work and contribution for reference.

## 3 Face Recognition

Face Recognition can be classified into a general class of pattern recognition problem. Hence, analogous components are required for facial recognition over a large database. These components can be obtained by extracting features of faces and using them for recognition. An essential requirement for recognition is that the features must be generated for every image and recognition must be robust when subjected to variations in the conditions. Detection of face and recognition have their own challenges, therefore for a Face Recognition system can be thought of three step procedure as shown in Figure 1.

**Fig. 1**Flow chart describing three steps involved in a Face Recognition system

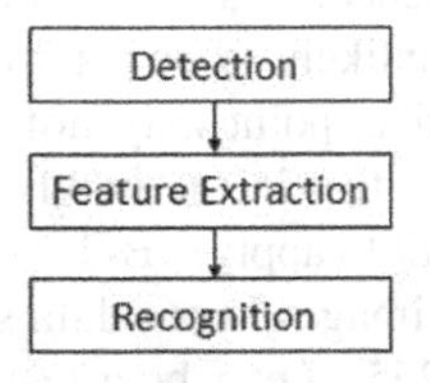


An overview of different techniques used in domain of Face Recognition is given in “Volterra Kernel based Face Recognition using artificial bee colony optimization” [10].

Face Recognition has many applications such as surveillance, indexing and search in public or private records, biometric authentication etc. Face Recognition is quite useful and has gained popularity over the years, but is not yet widely used due to low recognition rate. Poor performance can be attributed to the below listed problems:

- ***Occlusion***: Presence of elements such as beards, glasses or hats introduce variability.
- ***Expression***: Facial gestures can hamper with detection
- ***Imaging conditions***: Different cameras and lighting condition

Earlier works in ABC for Face Recognition mostly involve Face Detection and Feature Extraction. Simerpreet Kaur et al. [11] proposes use of ABC for Face Detection. Employing ABC for Feature Selection has been proposed in many papers, not specifically to domain of Face Recognition but Image Processing in general. There are significant works combining ABC and appearance based model but there are very few instances for algorithms using ABC in model based approach. One such work in this domain is modifying Adaptive Appearance Model with help of Artificial Bee Colony [12]. Applications of ABC in Face Recognition is novel and has a lot of room for exploration.

## 4 Problem Statement and Challenges

***Problem Statement***: Given a test image and pool of images (different subjects with some changes such as angle of the face with respect to camera, different pose and facial expressions), find the subject that matches the best in the database.

***Challenges***: Nature of algorithm poses the biggest challenge. In ABC, the optimization of search is generally done in a continuous space (as far as surveyed). Solutions are generated by modifying previous ones with some randomness and adding them to a set. In the case of a continuous search space, all the points in the set will be considered as possible solution but as per mentioned problem statement, only a data point that represents an image (from the given pool of images) can be considered as a possible solution. Hence, if a new solution is produced by randomly modifying existing ones, it is unlikely to be a part of solution set, since values of all attributes in the generated data point may not correspond to any image in the given pool. Thus, it can be said that in the newly generated points most have no meaning which makes it difficult to apply ABC.

Let's say there are only two images in our data set, represented by $\boldsymbol{a}$ and $\boldsymbol{b}$, with each attribute in the range [0, 255]. Let $\boldsymbol{x}$ be a test image which needs to detected as subject $\boldsymbol{a}$ or$\boldsymbol{b}$. As mentioned earlier ABC generates new solution from the existing ones, which at the beginning of the algorithm are initialized randomly. Let x1 be a point generated by modifying random initial points.

***Pool of Images: a*** = [125, 160, 217], ***b*** = [90, 100, 120]
***Test Image: x*** = [125, 177, 217]
***Generated Point: x1*** = [124, 165, 216]

If we output x1 directly, then it will have no meaning since it is neither equal to **a** nor **b**. This shows that most of the solutions generated will have no meaning because it is very difficult to generate solutions that can exactly match any of the pool images. Hence, a function is required that can map solutions generated to most similar data point in the pool of image. Mapping function is detailed further in the paper.

# 5 Methodology

The technique adopted in this paper is to use Gabor Filter and PCA to generate features for the test and training data set, similar to approach applied by Gupta et al. [13]. The best fit solution found during ABC algorithm run is marked as output for a given input image. This approach is also depicted in Figure 2 below in as a 5 step process.

**Fig. 2** Overall flow of the algorithm.

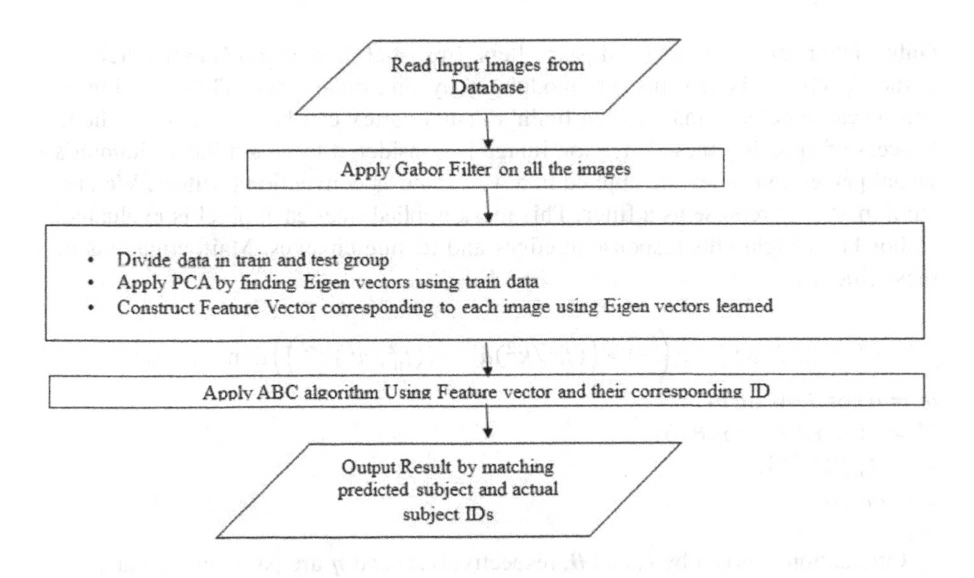


Standard ORL database distributed by the Olivetti Research Laboratory, Cambridge, U.K. has been used. Database consists of 40 subjects each having 10 different images in different conditions such as illumination, pose and facial expression. 400 distinct images with size 92 x 112 pixels and 8-bit grey levels are present in dataset [14]. Some of the images from the ORL database are shown below in Figure 3.

**Fig. 3** OLR database representation

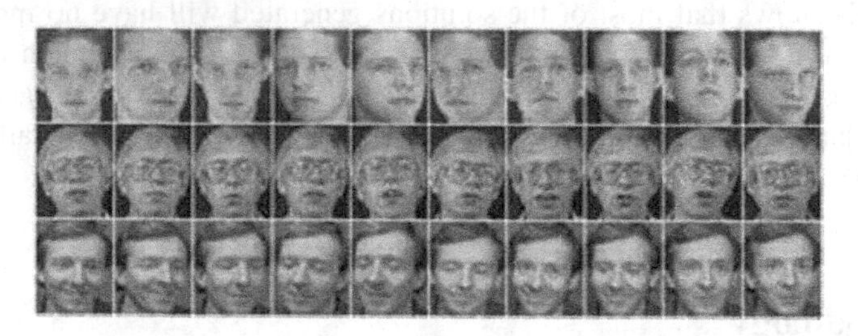

Sections below provides details about step for Gabor Filter, PCA and Implementation of ABC algorithm. First two sections are used for generating feature vectors and section 5.3 explains the ABC algorithm steps with flow chats.

## *5.1 Gabor Filters*

Gabor Filters are complex band pass filters. In spatial domain 2D Gabor Filter is a kernel function whose values are modulated by sinusoidal wave. They are significant because cells in mammalian brain's visual cortex can be modelled by them. Process of applying these filters on image is considered to be similar to human's visual perception. They are applied in a way same as conventional filters. We create a mask that represents a filter. This mask applied over each pixel is evaluated. Gabor Filter highlights response at edges and texture changes. Main equations of these filters are:

$$\varphi(a,b) = (f_u^2/\pi K n)\exp\left(-1 * \left((f_u^2/k^2)a'^2 + (f_u^2/n^2)b'^2\right)\right)\exp(j2\pi f_u a')(1)$$

$$a' = a\cos\theta + b\sin\theta(2)$$

$$b' = -a\sin\theta + b\cos\theta\ (3)$$

$$f_u = f_m/2^{u/2}\ (4)$$

$$\theta = v\pi/8 \quad (5)$$

Orientation is given by $f_u$ and $\theta$, respectively. $\varkappa$ and $\eta$ are parameters that evaluates the ratio of central frequency to size of Gaussian envelope. Commonly used values are $\varkappa = \eta = \sqrt{2}$ and $f_m = 0.25$ (***m*** is for maximum). A filter bank of five different scales and eight different orientations is created. Where $u = 0,1\ldots\ldots p-1$ and $v = 0,1\ldots,r-1$, where $p = 5$ and $r = 8$. Figure 4 displays Gabor Filter for different combinations of scale and orientation.

**Fig. 4** Filter bank for 40 different combinations of scale and orientation

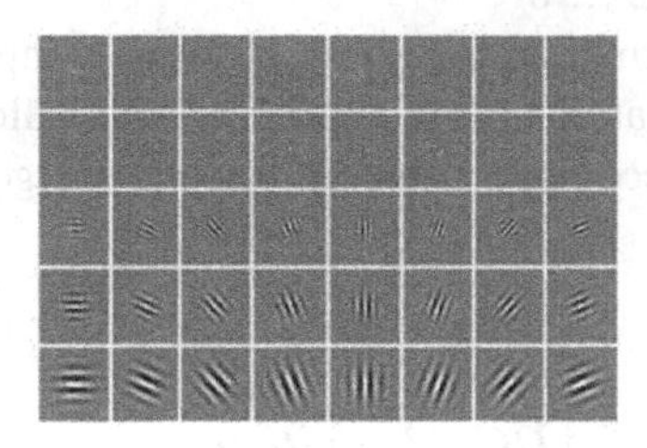

## *5.2 Principal Component Analysis*

Principal Component Analysis (PCA) is an extensively used and very popular method for feature reduction in Face Recognition. PCA has been used by Bahurupi et al. [15] to create a subspace on which a new data point (image) is projected and classification is done by Euclidean distance. Abdullah et al. [16] provides an insight into PCA algorithm and different optimizations that can be used to overcome its drawbacks. Faces are represented as Eigen Faces which are linear combination of Eigenvectors. Eigenvectors are derived from covariance matrix of training data. Number of Eigen Faces will be equal to no of images in dataset. Eigen Vectors represent new spaces in which faces are represented.

1. Let ***I*** be the training set
   $I = \{i_1, i_2 \ldots . i_p\}$
   Where ***p*** is the number of images in data set
2. Represent data set in a 2D Matrix with dimension ***a x p***, where a is equal to number of pixels in an image. Therefore, each column is representing an image in dataset where each element is a pixel value of the image
3. Calculate average face for a given data set
   $$\mu = {}^{1}/_{p} \sum_{r=1}^{p} i_r$$
4. Calculate covariance matrix
   $$c = \frac{1}{p} \sum_{r=1}^{p} (i_r - \mu)(i_r - \mu)'$$
5. Calculate Eigen Vectors and Eigen Values from the covariance matrix
   $cv = lv$
   Where ***l*** is Eigenvalues and ***v***is Eigenvectors for convolution matrix
6. Calculate Eigen Faces using Eigen Vectors
   $U = (i_n - \mu) * v$
   n = 1,2…p , U are called Eigen Faces due to ghost like appearance of face.
7. Calculate features using Eigen Faces
   $P_n = U^T * (i_n - \mu)$

$P_n$ represents the weights for each image corresponding to Eigenvectors in $c$ for an image. $n = 1, 2 \ldots p$

While testing, normalize with help of mean computed in step 3. Use steps 6 and 7 to calculate features for test images. After calculating features for both test and train data, recognize subjects for test images. This is done through ABC algorithm.

## 5.3 Implementation of ABC Algorithm

A merit of ABC that is fundamental to the implementation is the ability to optimize in high dimensional space and solve a complex problem. In contrast to conventional algorithms which are prone to divergence and local extrema, ABC avoids such scenarios by involving randomness to an extent.

As a pre-processing step Gabor Filter and PCA is applied (explained in previous sections) and feature vectors are generated for both train and test set. On this feature vector set, ABC algorithm is executed for recognition. The main steps of the algorithm are briefly illustrated.

Initialize: Initialize all the parameters for example
  Number of Employed Bees (**E**, no of solution/ no of food sources)
  Number of times main loop should run (**cLimit**)
  Number of times a food source can stay in the solution space (**limit**)
REPEAT
  Employed Bees Phase: Going to food source and extracting nectar
  Onlooker Bees Phase: Takes decision regarding fitness of a solution
  Scout Bees Phase: Discover new sources
UNTIL (requirements are met or **cLimit** reached?)

Initially, three parameters ***E***, ***cLimit*** and ***limit*** are set by the user. These parameters can later be tweaked to adjust the performance of the algorithm. The number of Employed Bees is set to a constant ***E***, which can also be considered to be number of solutions or number of food sources. Variable ***cLimit*** serves the purpose of an upper limit on the number of times main loop can be iterated. Variable ***limit*** is used to define the number of iterations after which the algorithm considers a food source to be exhausted and conduct search for new food sources. Figure 5, 6 and 7 detail the three phases of ABC algorithm.

In employed bee phase of each of the iterations, best solutions are chosen to be included in the solution set and rest are removed. In the first iteration, Scout Bees have no information about new food sources, thus the solution set will be composed of randomly initialized Employed Bees. As the algorithm progresses with the implementation food sources are removed from the solution set (for multiple reasons), and new solution set is formed each time considering both current solution set (Employed Bees) and solutions discussed by Scout Bees. The number of solutions in the solution set cannot exceed $E$ at any given point of time.

**Fig. 5** Employed Bee Phase

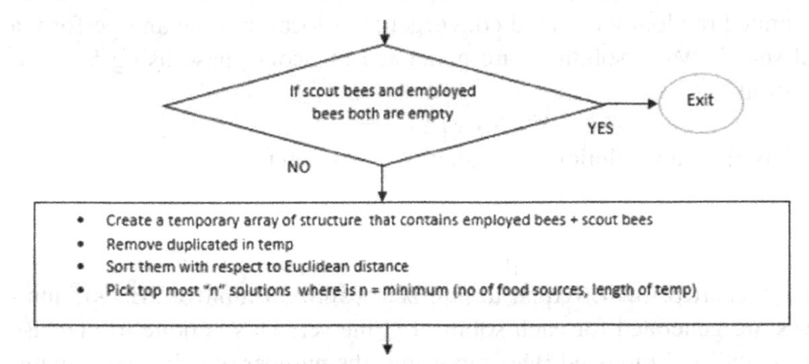


The onlooker bee phase keeps a check on the solution set. It removes exhausted sources. In this step an upper bound on the Euclidean distance between food source and test image is maintained, such that no image in the solution set exceeds the upper bound.

**Fig. 6** Onlooker Bees Phase

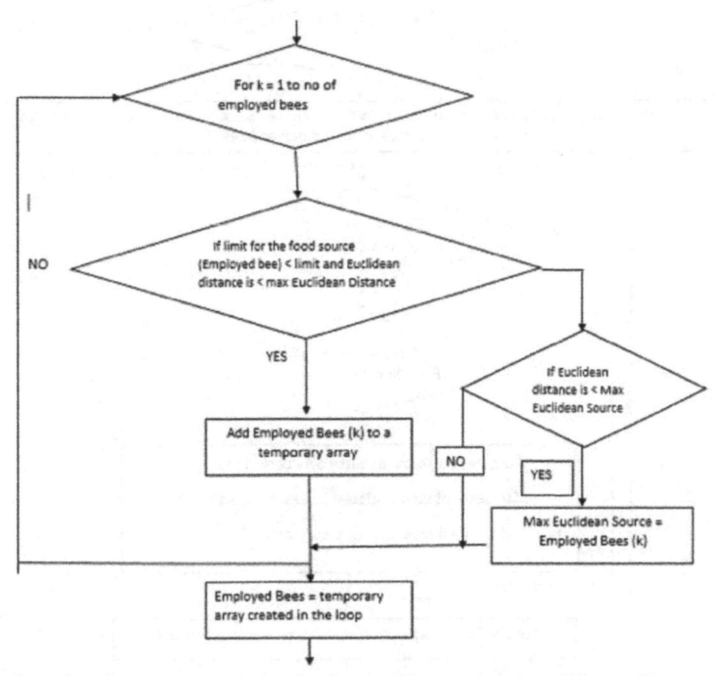


A solution that has exceeded the limit variable or has Euclidean distance greater than upper bound is removed from the solution set. If the Euclidean distance of removed solution is lesser than the existing upper bound, then that distance is updated as the new upper bound. The upper bound of Euclidean distance is stored in a variable (Max Euclidean Source).

Scout bee phase is the last phase in a given iteration. In this phase, sources that were exhausted in the onlooker phase are replaced with new solutions. New solutions are generated randomly to avoid convergence at local minima and perform a better global search. More solutions are generated in Scout phase using 5% randomness for local search.

$$X' = X_i + r\,(X_i - X^*)$$

Where X' is the new solution generated, Xi is a solution in the solution set around which new solutions have to be generated, X* is the test image for which algorithm is being run and is the randomness that we want to include in the procedure of creating new solution. In this paper results are reported for r= 5%.

As per Flow chart below ***E*** (equal to number of initial Employed Bees), number of sources are generated for each solution in the set. This is done to increase local search capability. Employed Bees represents the number of solutions remaining in the set. Number of Food sources is the number of Employed Bees initially fixed (***E***), this is a parameter that remains constant though the computation.

**Fig. 7** Onlooker Bee Phase

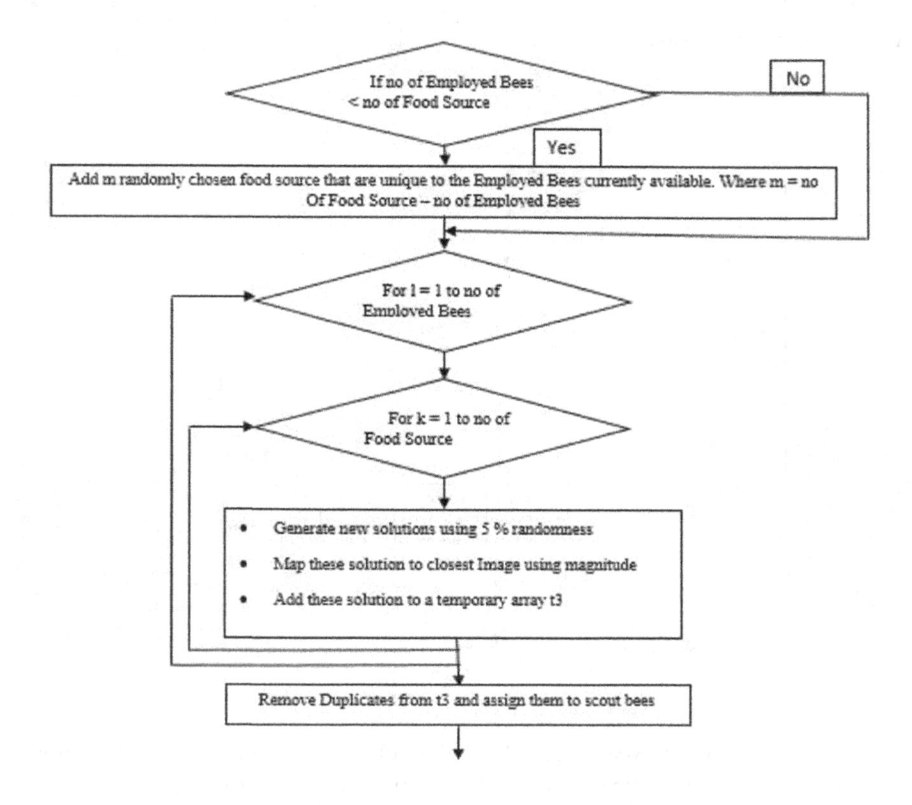

The mapping function as mentioned in section 2 is evaluated based on the similarity of the generated solution to an image in the provided database. Most similar image is chosen as an output of the mapping function. Similarity is measured by the closeness of a source's absolute value of the feature vector with respect to absolute value of the images in the data set. All the images are sorted based on absolute value and stored as a pre-processing step to look up the closest image for a given input source. Euclidean distance is used as a fitness function to choose the best solution in the solution set. The best solution has the least Euclidean distance from the given test image.

$$Fitness\ (X_i, X^*) = (X_i - X^*)^{1/2}$$

## 6 Results and Discussions

Performance of the algorithm was analyzed by measuring the accuracy. Accuracy is percentage of images in which subject was detected correctly, given certain number of test images. Accuracy with different values of the three parameters are calculated in order to find the best values that gives maximum efficiency. As described in section 3 we have three parameters, ***cLimit***, number of Food Sources (***E***) and ***limit***.

**Fig. 8** Graph for accuracy when cLimit = 15, for limit = 5 and 3.

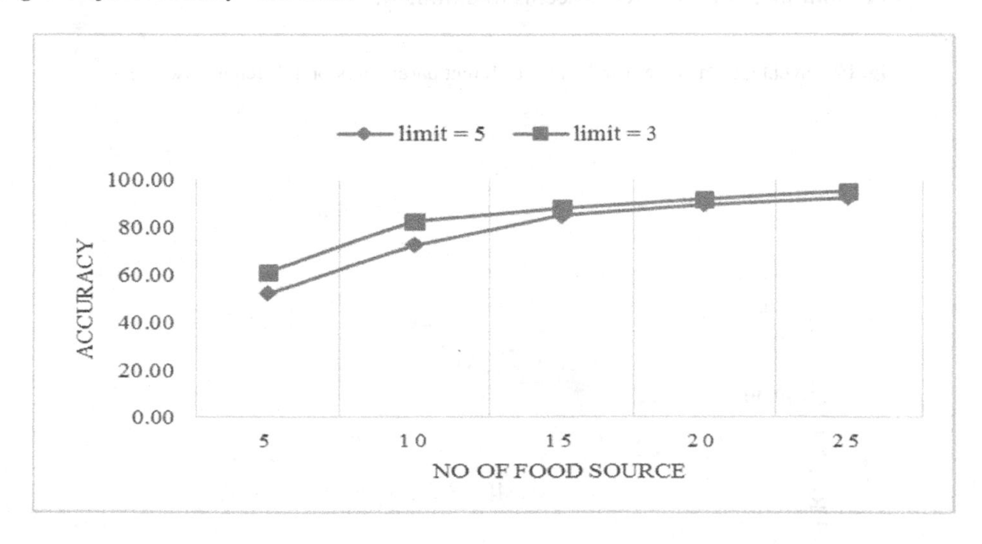

**Fig. 9** Graph for accuracy when cLimit = 20, for limit = 5 and 3.

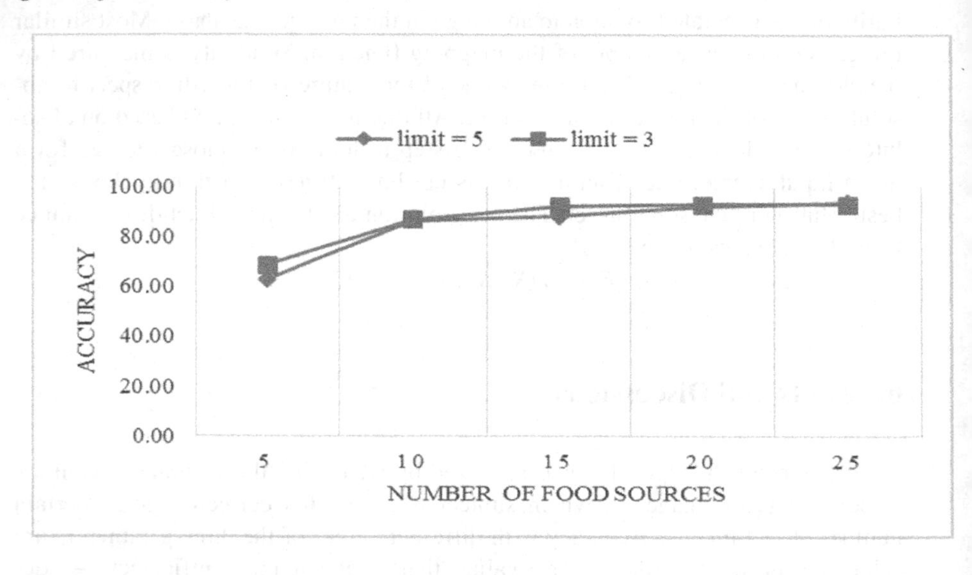


Figure 8 & 9 clearly demonstrates for ***limit*** = 3 accuracy seems to be better than ***limit*** = 5, especially for lower values of food sources. As number of food sources or ***cLimit*** increases difference seems to diminish.

**Fig. 10** Percentage Database searched for different parametersfor 4 different case.

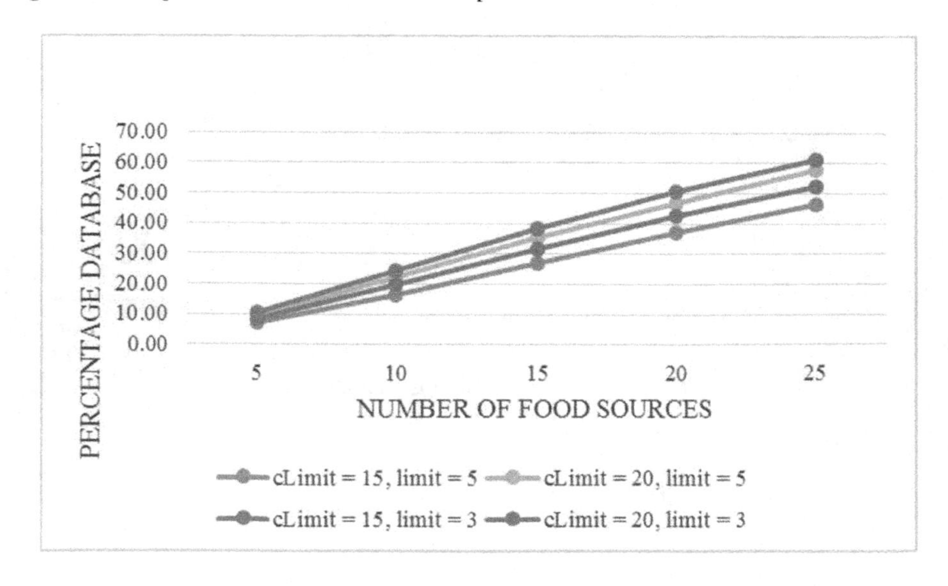

**Fig. 11** Accuracy when ***limit*** = 3, for varying number of food sources and ***cLimit***.

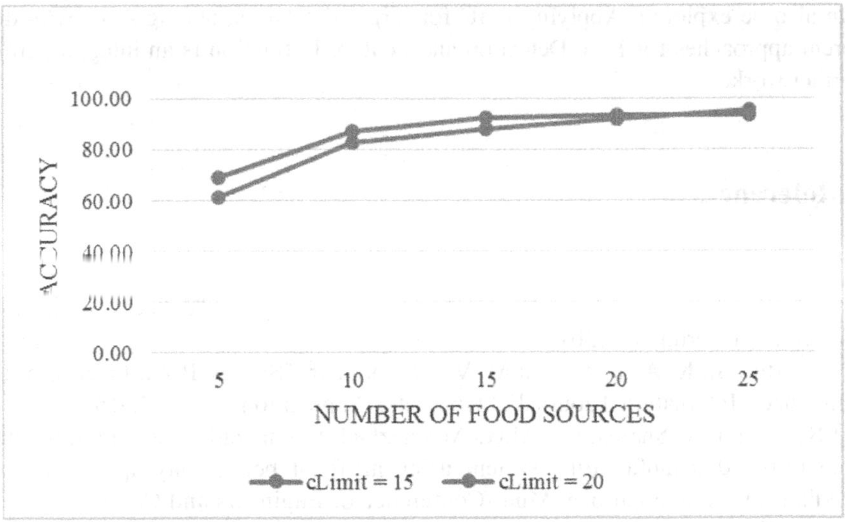

Amount of database searched for different parameters is demonstrated in Figure 10. The highest accuracy during experiment is 95.83%, which was achieved with 52.5% database search by adjusting ***cLimit*** = 15, **limit**= 3 and ***number of food sources*** = 25. Figure 11 give accuracy for ***limit*** = 3 and different ***cLimit***.

## 7 Conclusion and Future Work

ABC algorithm is one of the very efficient algorithms for optimization that can be employed easily in complex domains. In this paper it has been applied to optimizing search for the right subject for a given image. The advantage here is unlike traditional methods, we don't have to search the whole database and go through every image, only by going through few images we can search the database. Previously, knowledge of ABC has been applied in Face Detection and Feature Extraction but this is the first work where ABC has been applied in recognition phase (as far as seen). ABC proves to be helpful in searching database efficiently. In ABC, bees work independently with little communication amongst them, therefore this behavior has natural parallelism and will be quite scalable in parallel domain with complex and multi-dimensional data.

Future work in ABC with Face Recognition pertaining to this paper is to use an implementation of ABC for parallel computation, in a Face Recognition system. Abu et al. Surveyed ABC and its applications, where one of the major contributors were parallel and grid computing [17]. Using hybrids in place of just ABC such as Genetic Algorithm (GA) [18] and Particle Swarm Optimization (PSO) [19, 20] can also be explored. Applying ABC for larger databases, testing ABC with different approaches for Face Detection and Feature Extraction is an integral part of further work.

# 9 References

[1] Mishra, Er AK, Dr MN Das, and Dr TC Panda. "Swarm intelligence optimization: editorial survey." International Journal of Emerging Technology and Advanced Engineering 3.1 (2013).
[2] Keerthi, S., K. Ashwini, and M. V. Vijaykumar. "Survey Paper on Swarm Intelligence." International Journal of Computer Applications 115.5 (2015).
[3] S. Ajorlou, I. Shams, and M.G. Aryanezhad. Optimization of a multiproduct conwip-based manufacturing system using artificial bee colony approach. Proceedings of the International Multi-Conference of Engineers and Computer Scientists, 2, 2011
[4] Sagar Tiwari, SamtaGajbhiye,"Algorithm of Swarm Intelligence Using Data Clustering", International Journal of Computer Science and Information Technologies, Vol. 4 (4) , 2013, Page no 549 - 552
[5] Karaboga, Dervis. "Artificial bee colony al-gorithm."scholarpedia 5.3 (2010): 6915.
[6] Karaboga, Dervis, and BahriyeBasturk. " A Powerful and Efficient Algorithm for Numerical Function Optimization: Artificial Bee Colony (ABC) Algorithm.", Journal of Global Optimization (2007), 12 Apr. 2007
[7] Karaboga, Dervis, BeyzaGorkemli, CelalOzturk, and NurhanKaraboga. "A Comprehensive Survey: Artificial Bee Colony (ABC) Algorithm and Applications." Artificial Intelligence Review 42.1 (2012),11 March 2012
[8] Bolaji, AsajuLa'Aro, and AhamadTajudinKhader. "Artificial Bee Colony Algoritm, Its Variants and Application: A Survey." Journal of Theoretical and Applied Information Technology, Vol. 47, Issue 2, 20 Jan. 2013,Pages 434-59.
[9] Karaboga, D., and B. B. Akay. "Arti-ficial bee colony (ABC) algorithm homepage." Intelligent Systems Research Group, Department of Computer Engineering, Erciyes University, Turkiye(2009).
[10] Chakrabarty, Ankush, Harsh Jain, and Amitava Chatterjee. "Volterra Kernel Based Face Recognition Using Artificial Bee Colonyoptimization." Engineering Applications of Artificial Intelligencem, Vol.26, Issue 3, March 2013, Pages 1107–1114
[11] Simerpreet Kaur, RupinderKaur,"An Approach to Detect and Recognize Face using Swarm Intelligence and Gabor Filter",International Journal of Advanced Research in Computer Science and Software Engineering,Volume 4, Issue 6, June 2014

[12] Mohammed Hasan Abdulameer,"A Modified Active Appearance Model Based on an Adaptive Artificial Bee Colony",Hindawi Publishing Corporation Scientific World Journal, 2014
[13] Gupta, Daya, LavikaGoel, and Abhishek Abhishek. "An Efficient Biogeography Based Face Recognition Algorithm." 2nd International Conference on Advances in Computer Science and Engineering (CSE 2013). Atlantis Press, 2013.
[14] "Popular Face Data Sets in Matlab Format." Popular Face Data Sets in Matlab Format. AT&T Laboratories Cambridge, n.d. Web. 27 May 2016.
[15] Bahurupi, Saurabh P., and D. S. Chaudhari. "Principal component analysis for face recognition." International Journal of Engineering and Advanced Technology (IJEAT) ISSN (2012): 2249-8958.APA
[16] Abdullah, Manal, MajdaWazzan, and Sahar Bo-saeed. "Optimizing face recognition using PCA."arXiv preprint arXiv:1206.1515 (2012).
[17] Abu-Mouti, Fahad S., and Mohamed E. El-Hawary. "Overview of Artificial Bee Colony (ABC) algorithm and its applications." Systems Conference (SysCon), 2012 IEEE International. IEEE, 2012.
[18] Yuan, Yanhua, and Yuanguo Zhu. "A hybrid artificial bee colony optimization algorithm."Natural Computation (ICNC), 2014 10th International Conference on. IEEE, 2014.
[19] Hu, Wenxin, Ye Wang, and Jun Zheng. "Research on warehouse allocation problem based on the Artificial Bee Colony inspired particle swarm optimization (ABC-PSO) algo-rithm." Computational Intelligence and Design (ISCID), 2012 Fifth International Symposium on. Vol. 1. IEEE, 2012.
[20] Li, Mengwei, HaibinDuan, and Dalong Shi. "Hybrid Artificial Bee Colony and Particle Swarm Optimization Approach to Protein Secondary Structure Prediction." Intelligent Control and Automation (WCICA), 2012 10th World Congress on. IEEE, 2012.

# ILTDS: Intelligent Lung Tumor Detection System on CT Images

Kamil Dimililer Yoney Kirsal Ever and Buse Ugur

**Abstract** Cancer detection and research on early detection solutions play life sustaining role for human health. Computed Tomography images are widely used in radiotherapy planning. Computed Tomography images provide electronic densities of tissues of interest, which are mandatory. For certain target delineation, the good spatial resolution and soft/hard tissues contrast are needed. Also, Computed Tomography techniques are preferred compared to X-Ray and magnetic resonance imaging images. Image processing techniques have started to become popular in use of Computed Tomography images. Artificial neural networks propose a quite different approach to problem solving and known as the sixth generation of computing. In this study, two phases are proposed. For first phase, image pre-processing, image erosion, median filtering, thresholding and feature extraction of image processing techniques are applied on Computed Tomography images in detail. In second phase, an intelligent image processing system using back propagation neural networks is applied to detect lung tumors.

**Key words:** Intelligent Systems, Digital Image Processing, Artificial Neural Networks, CT Lung Cancer Images, Thresholding, Feature Extraction,

---

Kamil Dimililer
Electrical and Electronic Engineering Department, Faculty of Engineering, Near East University, Nicosia, North Cyprus e-mail: kamil.dimililer@neu.edu.tr

Yoney Kirsal Ever
Software Engineering Department, Faculty of Engineering, Near East University, Nicosia, North Cyprus e-mail: yoneykirsal.ever@neu.edu.tr

Buse Ugur
Biomedical Engineering Department, Faculty of Engineering, Near East University, Nicosia, North Cyprus e-mail: buse.ugur@neu.edu.tr

 
J.M. Corchado Rodriguez et al. (eds.), *Intelligent Systems Technologies and Applications 2016*, Advances in Intelligent Systems and Computing 530,
DOI 10.1007/978-3-319-47952-1_17

## 1 Introduction

Last three decades, cancer is one of the most common reasons for fatal injuries, that the death rate of cancer is the highest compared to the other diseases caused by. Researches showed that in 2012, cancer had an outstanding cause of death all around the world and elucidated 8.2 million deaths [4]. American Cancer Society (ACS) [3] stated that in 20 years, lung cancer mortality rates in men has dropped nearly 30 percent where as, mortality rate in women has started to regress in last 10 years.

Lung cancer is one of the common cancer types which is found in both men and women all over the world that has very low survival rate after the diagnosis. Survival from lung cancer is directly proportional in conjunction with its growth at its detection time. However, people do have a higher possibility of survival if the cancer can be detected at early stages [5]. Early detection is a main factor to prevent tumor growth in the body. Especially in lung cancer early treatment stages can increase the chance of survival.

Lung cancer also known as carcinoma is multiplying and growing of abnormal cells. Lung cancer is grouped into two, which are known as non small cell and small cell lung cancer. They are categorized according their cellular characteristics [6]. Non-small cell lung cancer has three subtypes: Carcinoma, Aden carcinoma and Squamous cell carcinomas [6]. As for the stages, there are four stages of carcinoma; I, II, III and IV, where staging is relying on tumor size, tumor and lymph node location.

Chest Radiography (x-ray), Computed Tomography (CT), Magnetic Resonance Imaging (MRI scan) and Sputum Cytology are techniques used to diagnose lung cancer [1], [2]. New technologies are greatly needed to diagnose lung cancer in its early stages, because these techniques could be expensive and time consuming in later stages. Image processing techniques are widely used in different medical fields to guide medical staff for detection of cancer in earlier stages [1]. Since the researches showed that, the detection of disease that is observed on time, survival rate of lung cancer patients have been increased from 14% to 49% within last five years [2].

Image processing techniques have been used frequently in various medical fields. Additionally, artificial neural networks propose a quite different approach to problem solving and known as the sixth generation of computing [7]. Digital image processing is widely used in the field of medical imaging [10,11,12]. For such tasks, image processing techniques were used with the aid of artificial intelligence tools such as back propagation neural network in order to achieve the optimum and most accurate results [12]. In this research two phases naming, lung tumor detection system (LTDS), and intelligent lung tumor detection system (ILTDS) are designed. The aim of this paper is to perform image processing techniques, as well as artificial neural networks and their interrelated analysis methods to Health care, especially to the carcinoma patients. Most of medical researchers appraised the analysis of sputum cells for early diagnosis of lung cancer. Ultimate research relay on quantitative information for instance, the shape, size and the ratio of the affected cells [7].

## 2 Proposed Image Processing System

In this section lung tumor detection by using Image processing techniques is proposed. Proposed system's steps will be explained in detail.

### 2.1 Image Acquisition

First step is to started with taking a collection of computed tomography (CT) scan images. The main advantage of using CT images is to have low noise and better clarity, compared to X-Ray and MRI images. NIH/NCI Lung Image Database Consortium (LIDC) is an on-line CT image dataset available for the researchers in the field of digital image processing. The acquired images are in raw form. Figure 1 shows an original CT image for a lung.

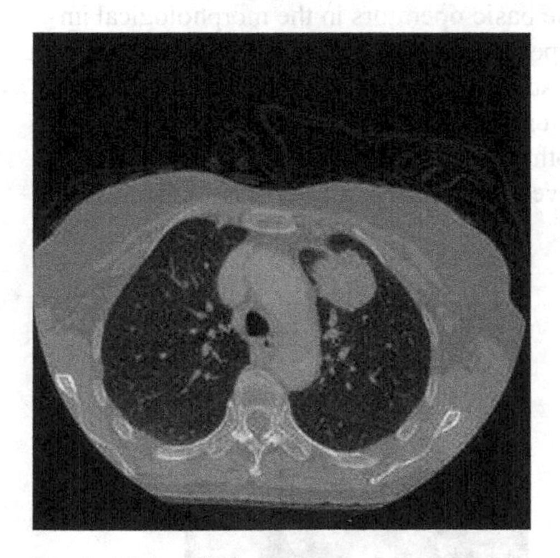

**Fig. 1** Original CT image

### 2.2 Image Pre-processing

After image acquisition, all the images have gone a few preprocessing steps such as grayscale conversion, normalization, noise removal, and binary image.

### 2.2.1 Gray Scale Conversion

This step involves conversion of original DICOM images which are in RGB format to grey colour. It converts RGB images to grayscale by removing the tint and saturation information while maintaining the luminance.

### 2.2.2 Normalization

Acquired image is converted from `512 x 512` to `256 x 256` pixels values. This size gives sufficient information while the processing time is low.

### 2.2.3 Noise Removal

For removing the noise, erosion and `5 x 5` median filter [14] are applied to the system respectively. Erosion is one of the basic operators in the morphological image processing. The main effect of the operator on a binary image is to erode away the confines of sites of foreground pixels such as white pixels. In order to reduce the noise within the lung images median filter is preferred to be applied, because this type of filter preserves details and smooth non-impulsive noise. If the aim is to simultaneously minimise noise and conserve edges, a median filter is more impressive [8].

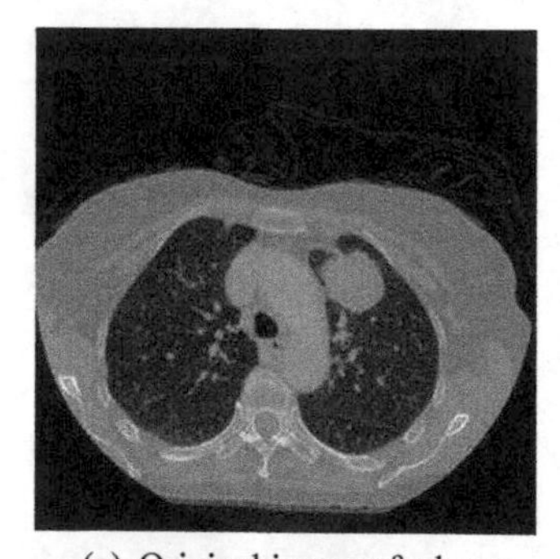

(a) Original image of a lung

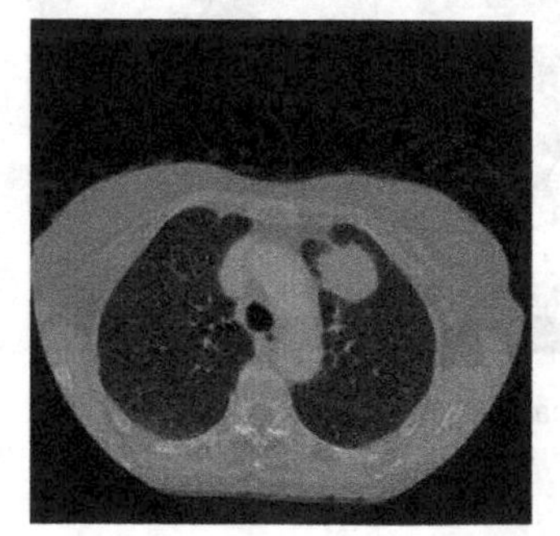

(b) Erosion applied

**Fig. 2** Original CT image with Erosion

### 2.2.4 Binary Image

Grayscale image is converted to binary image having pixels 1's (black) and 0's (white).

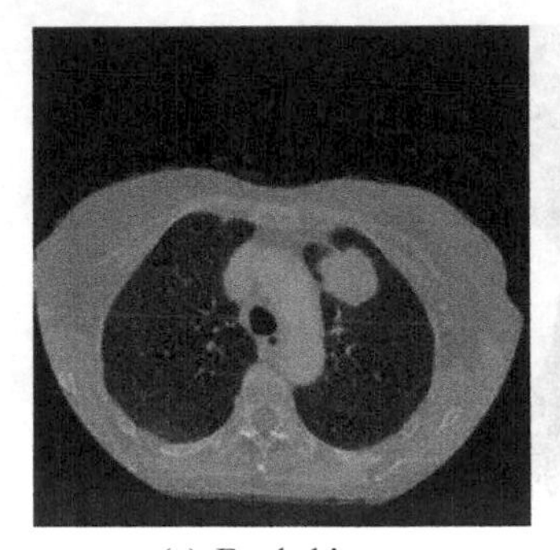

(a) Eroded image

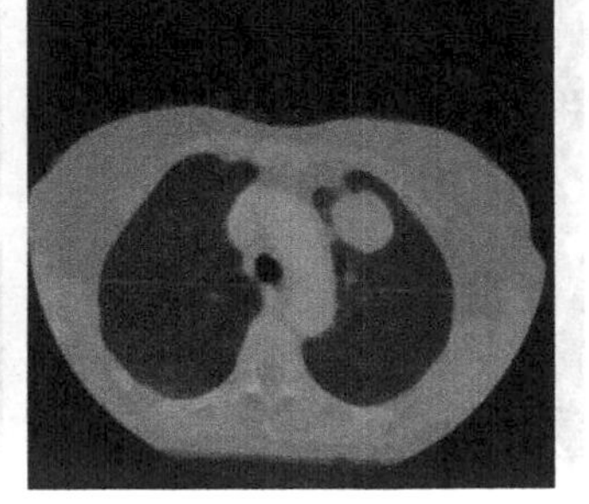

(b) Median filter applied

**Fig. 3** Median filtered applied to eroded image

## *2.3 Image Thresholding*

Threshold value is a specific value that turns a grayscale image into a binary image due to thresholding method. The main idea of this method is when multiple levels are selected, determine the threshold value or values.Nowadays, thresholding methods are developed especially for computed tomography (CT) images. Thresholding method is the easiest method for image segmentation. The proposed system used one threshold value i.e. `T`. As stated in [13], image properties are classified according to field of applications. Considering these applications, the images in the proposed system are divided into three, namely `low brightness, LB (T1)`, `medium brightness,MB (T2)`, and `high brightness,HB (T3)` where ranges are defined as `0-85,86-170,171-255` respectively. In binary CT image, if the T1 is less than the percentage of white pixels, then full lung is affected. Considering segmented binary image set, if the percentage of white pixels are higher than the T2 and T3, then the right lung and left lung are affected respectively. Because of this, low brightness images are chosen.

Figure 4(a) shows a median filter applied image, where as figure 4(b) represents threshold segmentation is applied. This is the area with the consistency values higher than the defined threshold. High consistency areas mostly involves of cancerous tissue. Through the threshold segmentation, the location of cancerous tissue is determined.

# 3 Feature Extraction

Image feature extraction stage plays an important role in this project, since image processing techniques that use algorithms and techniques to localize and eliminate various desired portions or shapes (features) of an image.

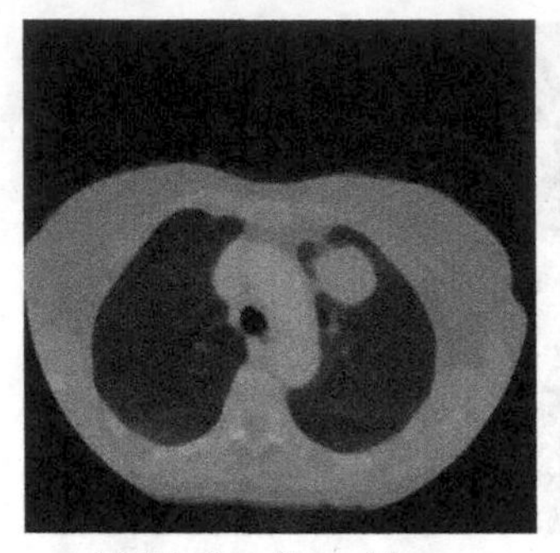

(a) Median filtered image

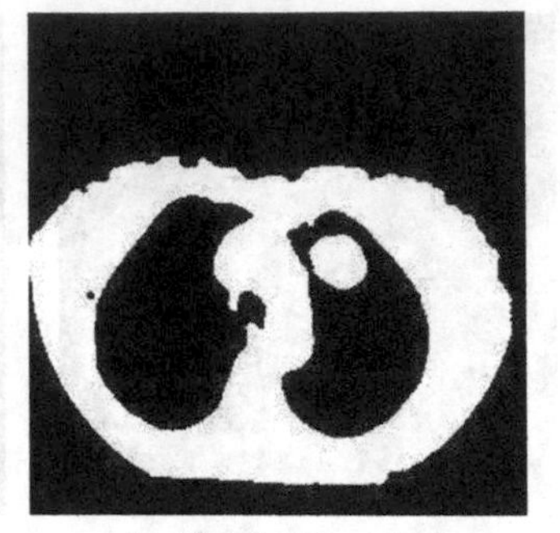

(b) Threshold applied

**Fig. 4** Threshold applied to Median filtered image

After performing segmentation on a lung region, the features can be achieved from it and the diagnosis rule can be designed to detect the cancer nodules in the lungs. In search for better diagnosis, the false detection of cancer nodules resulted in segmentation [9].

Two methods to predict the probability of carcinoma presence are defined as follow. First approach is binarization and the second is image subtraction method.

## *3.1 Binarization*

One of the techniques of image processing is binarization which converts an image grey level to a black and white image. It is one of the pre processing method that is used. Two colours are used for a binary image that are black (0) and white(1), which are also known as bi-level or two-level images.

In normal lung images, number of white pixels much less than the number of black pixels. Therefore when it start to count black pixels for normal and abnormal images,an average can be obtained that can be used as a threshold. As stated in [10], if the number of black pixels are greater than the threshold, it indicates that image is normal, otherwise abnormal image is considered.

## *3.2 Image Subtraction*

In this step, images that have tumor cells are clarified by using filtering and thresholding. Then tumor cells are removed from images. For removing tumor cells from images, objects are eliminated under specific pixel values.

In last step, in order to create tumor cells alone, difference between filtered image and image that tumor cells removed are taken into account.

These two approaches are combined and applied successfully. Figure 5 shows these steps. Following equation shows extraction rate for the above approaches.

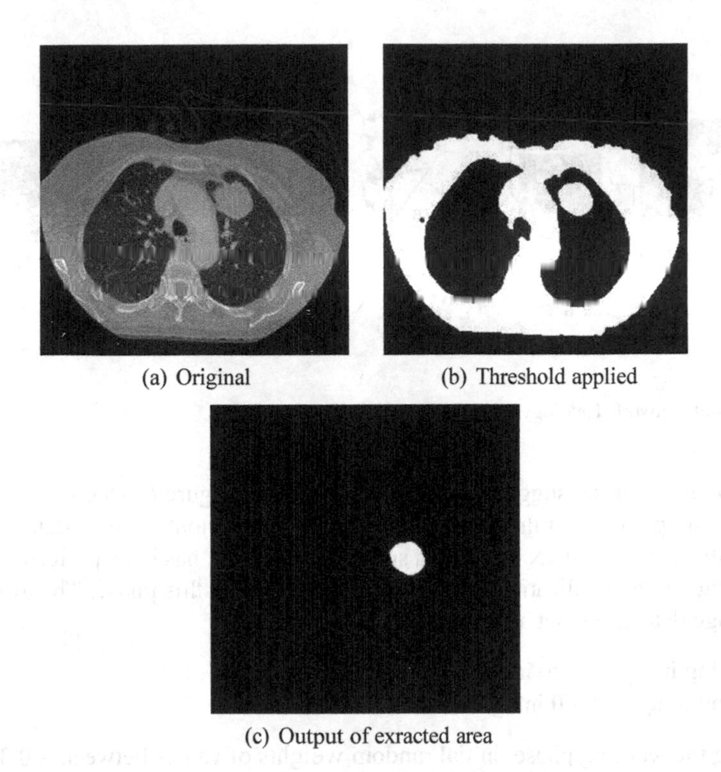

(a) Original (b) Threshold applied

(c) Output of exracted area

**Fig. 5** Feature Extraction

$$ER_{LTDS} = \frac{I_{CE}}{I_{TOT}} x100, \quad (1)$$

where ER is extraction rate, $I_{CE}$ is correctly extracted image and $I_{TOT}$ is total number of test images.

## 4 Neural Network Phase

A conventional, 3-layer back propagation neural network is used in Intelligent Lung Tumor Detection System (ILTDS) with 64 x 64 input images, having 4096 input neurons, 25 hidden neurons and 2 output neurons classifying the organs with tumors and without tumors. Organs are classified in binary coding as [1 0] for the Organ with Tumor and [0 1] for the Organ without Tumor. The sigmoid activation function has been preferred to be used for activating purposes of the neurons in both hidden and output layers.

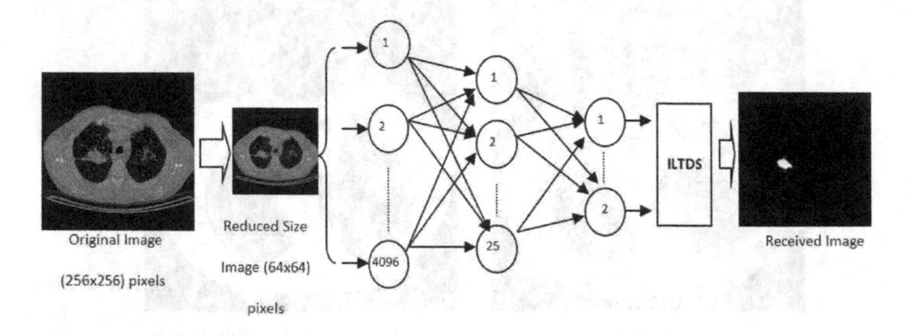


**Fig. 6** Neural Network Topology of ILTDS

The topology of the suggested network is shown in figure 6. Due to the implementation simplicity, and the availability of sufficient "input target" database for training, the use of a BPNN which is a supervised learner, has been preferred.

Training and generalization (testing) are comprised in this phase. The available Lung image database is set as follows:

1. Training image set: 20 images
2. Testing image set: 60 images

During the learning phase, initial random weights of values between −0.35 and 0.35 were used. These random weights of values are determined after several experiments [15]. In order to reach the required minimal error value and considerable learning, the learning rate and the momentum rate were adjusted during various experiments. An error value of 0.013 was considered as sufficient for this application. Table 1 shows the obtained final parameters of the trained network.

**Table 1** NEURAL NETWORK FINAL PARAMETERS

| Input Layer Nodes(ILN) | 4096 |
|---|---|
| Hidden Layer Nodes(HLN) | 25 |
| Output Layer Nodes(OLN) | 2 |
| Learning Rate | 0.004 |
| Momentum Rate | 0.7 |
| Minimum Error | 0.013 |
| Iterations | 1722 |
| Training Time | 40.88 seconds |

$$RR_{ILTDS} = \frac{I_{CR}}{I_{TOT}} x100, \tag{2}$$

where RR is recognition rate, $I_{CR}$ is correctly recognised image and $I_{TOT}$ is total number of test images.

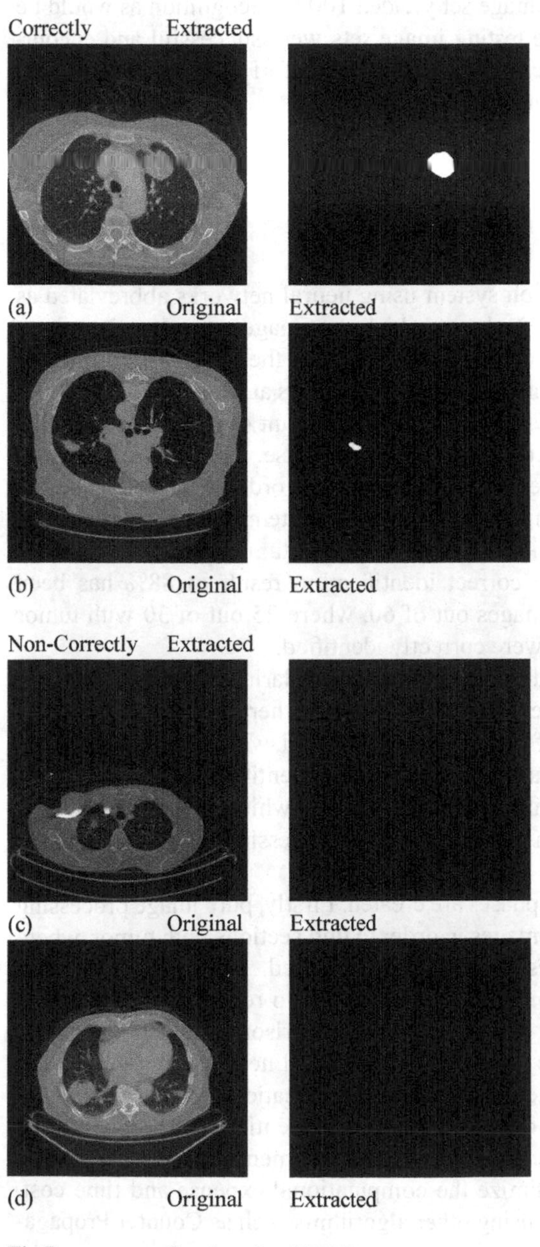


**Fig 7** Results of ILTDS

These results were obtained using a 8196 MB RAM and 16 CPUs of Windows 2012R2 64 bits OS Virtual machine that has a 1 Gbits connection speed to a super computer that has IBM Blade HS21 XM Intel Xeon 2.33 GHz CPU with 1280 core units. Through this application, a novel intelligent lung tumor detection system that consists of flexibility, robustness, and speed have been demonstrated. Tumor identification results using the training image set yielded 100 % recognition as would be expected. ILTDS results using the testing image sets were successful and encouraging. An overall correct identification of 54 images out of the available 60 lung images yielded a 90% correct identification.

## 5 Conclusions

In this paper, a lung tumor detection system using neural networks abbreviated as ILTDS, is presented. ILTDS uses 2 phases which are image preprocessing phase and neural network implementation phase used to detect the lungs with tumors or without tumors. Image preprocessing phase in ILTDS aims at providing meaningful representations of lung patterns while reducing the amount of data, thus reducing both the computational and time costs. After training phase, a non- linear relationship between input and output patterns has been created in order to detect the tumors within the images. A successful implementation of the system, ILTDS, has been implemented, to identify the lungs with tumors and without tumors.

Considering LTDS, an overall correct identification result of 88% has been achieved, in total number of 53 images out of 60, where 25 out of 30 with tumor and 28 out of 30 with non-tumor, were correctly identified.

This helps the radiologists and doctors physical similarities between these tumors. However, ILTDS can also be trained to recognize other tumor types.

An overall identification rate of 90% has been achieved, where 26 out of 30 with tumor and 28 out of 30 with non-tumor, were correctly identified.

The neural network training time was 40.88 seconds, which is considered to be encouraging. The run time for both phases (image preprocessing and neural network generalization) was 0.01 seconds.

Summarising the research, two phases are created. Firstly, pure image processing techniques are applied to original images in order to find sections with tumor, where as in second phase, original images are preferred to be used.

Future work will include the modification of ILTDS to recognize the brain tumor images as well as lung tumor images. Also a comparison of threshold applied images with original ones will be implemented in neural network phase, in order to improve validity of the suggested work. Some modifications will be applied on the back propagation neural networks to increase the identification rate using an improved database. Additionally, modifications to the segmentation process may be carried out in order to further optimize the computational expense and time cost. Finally, neural network arbitration using other algorithms such as Counter Propagation Neural Networks(CPNN), in addition to the back propagation model, will be implemented and compared.

## References


1. Mokhled S. AL-TARAWNEH, "Lung Cancer Detection Using Image Processing Techniques", 2012
2. Chaudhary, A. and Singh, S.S., "Lung Cancer Detection on CT Images by Using Image Processing", 2012
3. American Cancer Society, Inc. No.011482, 2012.
4. World health organization Cancer: www.who.int/gho/ncd/mortality_morbidity/cancer/en
5. Ilya Levner, Hong Zhangm ,"Classification driven Watershed segmentation", *IEEE TRANSACTIONS ON IMAGE PROCESSING*, vol.16, NO.5, MAY 2007
6. Matthais Rath, "Cancer", Cellular Health Series, February 2001
7. Almas Pathan, Bairu.K.saptalkar, "Detection and Classification of Lung Cancer Using Artificial Neural Network", *International Journal on Advanced Computer Engineering and Communication Technology*, vol.1 Issue.1.
8. Md. Badrul Alam Miah , Mohammad Abu Yousuf, "Detection of Lung Cancer from CT Image Using Image Processing and Neural Network", *2nd International Conference on Electrical Engineering and Information & Communication Technology (ICEEICT)* 2015 , Jahangirnagar University, Dhaka-1342, Bangladesh, 21-23 May 2015.
9. Disha Sharma, Gangadeep Jindal, "Identifying Lung Cancer Using Image Processing Technique", *International Conference of Computational Techniques and Artificial Intelligence*, 2011
10. Ada, Ranjeet Kaur, "Feature extraction and principal component analysis for lung Cancer Detection in ct images", March 2013.
11. Dimililer K (2012) "Neural network implementation for image compression of x-rays", *Electronics World* 118 (1911):26-29
12. Dimililer K (2013) "Back propagation neural network implementation for medical image compression", *Journal of Appl. Math.* 2013, Available:http://dx.doi.org/10.1155/2013/453098
13. Khashman A., Dimililer K (2005) "Comparison Criteria for Optimum Image Compression", *The International Conference on "Computer as a Tool"*, 2005, Available:http://dx.doi.org/10.1109/EURCON.2005.1630100
14. Roger Bourne, *Fundamentals of Digital Imaging in Medicine* by Springer-Verlag London Limited, 2010.
15. Khashman A., Sekeroglu B. and Dimililer K., "Intelligent Identification System for Deformed Banknotes", *WSEAS Transactions on Signal Processing*, ISSN 1790-5022, Issue 3, Vol. 1, 2005.

# Blink Analysis using Eye gaze tracker

Amudha J, S. Roja Reddy, Y. Supraja Reddy

***Abstract.*** An involuntary action of opening and closing the eye is called blinking. In the proposed work, blink analysis has been performed on different persons performing various tasks. The experimental suite for this analysis is based on the eye gaze coordinate data obtained from commercial eye gaze tracker. The raw data is processed through a FSM(Finite State Machine) modeled to detect the opening and closing state of an eye. The blink rate of a person varies, while performing tasks like talking, resting and reading operations. The results indicate that a person tend to blink more while talking when compared to reading and resting. An important observation from analysis is that the person tends to blink more if he/she is stressed.

**Keywords:** Blink detection; Eye tribe; Eye movement; Finite state machine

## 1 Introduction

The process of measuring either the point of gaze (where one is looking) or the motion of an eye relative to the head is called Eye tracking. Eye positions and eye movement are measured using the device known as Eye tracker.

Blink of an eye can be detected using various methods such as extracting eye position from video images [15], using coils or using electro-oculogram [2]. There are three different types of eye blinks: Spontaneous, Reflex and Voluntary. Spontaneous blinks are those which occur regularly as a normal process like breathing unlike reflex blinks that occurs when there is an external disturbance. Voluntary blinks on the other hand happen according to the person's will. On an average a healthy adult should have a blink rate of 17blinks/min.

Blink analysis is useful in various research domains such as Visual Systems, detecting disorders like Computer Vision Syndrome [1], Human Computer Interaction[6], Product Design. This is used to develop applications which works

---

Amudha J
Dept of Computer Science & Engineering, Amrita School of Engineering, Bengaluru, e-mail: j_amudha@blr.amrita.edu
S. Roja Reddy
Dept of Computer Science & Engineering, Amrita School of Engineering, Bengaluru, e-mail: lakshmareddysareddy@gmail.com
Y. Supraja Reddy
Dept of Computer Science & Engineering, Amrita School of Engineering, Bengaluru, e-mail: suprajareddy506@gmail.com

 
J.M. Corchado Rodriguez et al. (eds.), *Intelligent Systems Technologies and Applications 2016*, Advances in Intelligent Systems and Computing 530,
DOI 10.1007/978-3-319-47952-1_18

based on blink of any eye such as controlling a mouse cursor. These types of applications can assist severely disabled persons in operating computers easily.

It also finds its application like measuring stress levels of a person, recognition of activities, emotions etc.

## 2 Literature Survey

Based on various works carried out in the areas of eye tracking research, the eye tracking applications can be categorized into one of the following five areas like Computer Vision Syndrome, Emotions and Activities, Biometric, Drowsiness of driver, and Human Computer Interaction. The figure below depicts the broad classification of the literature survey.

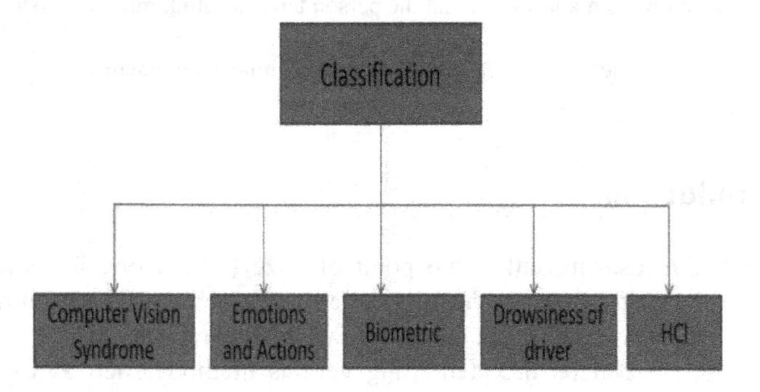


**Fig. 1.** Classification of Literature survey

The technique to detect if an individual is affected by Computer Vision Syndrome (CVS) is outlined in [1]. In [1] video based blink detection method is used to detect the blinks. The acquired data is used in analyzing the difference between number of blinks of a person affected with computer vision syndrome and a normal person. A real time eye fatigue detection system is developed to identify whether a person is affected by CVS.

The work involving eye movement analysis is presented in [2]. The works discussed on a new sensing medium for activity recognition. Eye movement data was taken using an EOG(Electro-oculogram) system. In this, the algorithms were designed and evaluated for detecting three different eye movement characteristics such as saccades or jolts, fixations (focusing), and blinks. It also suggested a method for evaluating repetitive patterns of eye movements.

In [3], there was a special emphasis laid upon a system that uses eye blink patterns as biometric. Firstly, user blinks are recorded according to the cadence of any specific song and then these patterns are stored in a database. With the help of computer vision, blinks of a person are detected. These are then compared to blinked cadence of a specific song to detect the song and also the user who performed the blink.

In [4], the majority of work is focused upon a system which monitors driver's state to evaluate his drowsiness, fatigue, and distraction accurately. It is done by analyzing both the driving patterns and images of the driver. Also an alarm is triggered at an appropriate time depending upon the state of the driver which is derived as per the proposed logic.

In [5], a survey on various methods which detect drowsiness or unconsciousness of the driver is presented. It classifies into three different methods: Physiological signals, Using In-vehicle sensors and Computer Vision based techniques.

In [6], the work deals mainly with developing a low-cost aid for physically handicapped persons to communicate with computer. The idea here is to use Electro-oculogram (EOG) signals which are generated by the eye movements, to control the computer. This helps physically challenged people to interact with computer easily.

Similar to [6], in [7], it is all about using eye movements to control the mouse cursor on a screen. Additionally each eye blink corresponds to a single mouse click.

In [8], an analysis was done to explore the temporal connection between smiles and eye blinks. Using video sequences and active appearance models algorithm, eye blinks were detected.

In [9], a different approach based on hidden markov models and Support Vector Machines is proposed for blink detection. It has a higher accuracy rate than an eye tracker

In [10], a new approach is presented for detecting blinks. A combination of boosted classifiers and Lucas—Kanade tracking is used to track face and eyes which is in turn used for detecting eye-blinks. This method is evaluated using video-sequences.

The literature reports quite number of methods to detect blink, however quite a few has been reported on blink detection using eye gaze data obtained from the commercial eye tracker to analyze on various task activities and to identify whether he is stressed.

## 3 Experimental Design

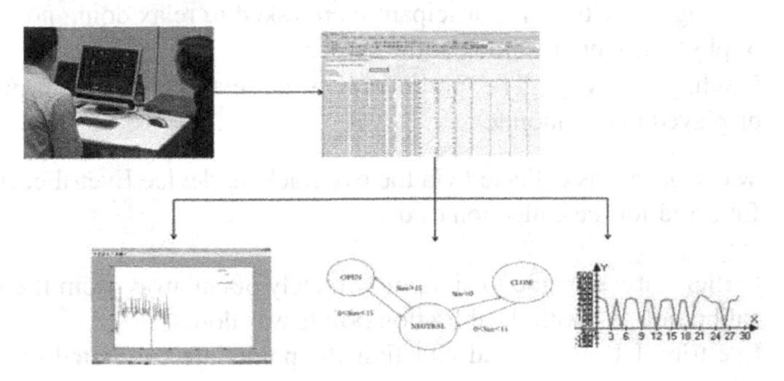


**Fig. 2.** Fig 2: System Architecture

Eyetribe is an eye gaze tracker through which we have performed the experiment of collecting the eye gaze data from various persons.

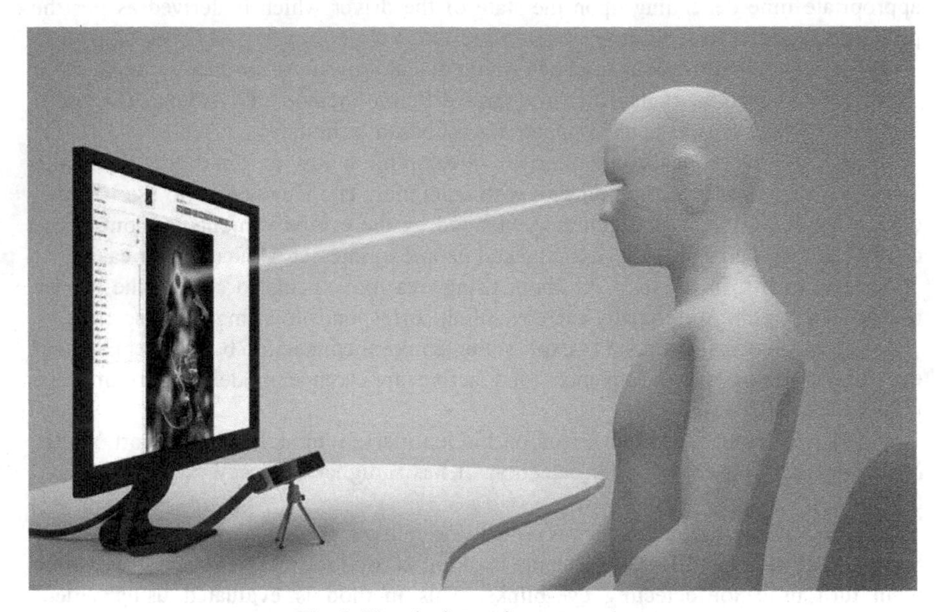

**Fig. 3.** User in front of an eye tracker

We have taken data from various persons of different ages ranging from 7-20, both genders, while they were performing different activities such as Talking, Resting and Reading.

- Talking Activity: The participant was continuously engaged in talk with the observant and was asked to look at the screen while talking.
- Resting Activity: The participant were asked to relax doing nothing mentally or physically but just look at the screen.
- Reading Activity: The participant were made to read a pdf document displayed in the monitor.

The raw eye gaze was collected via the eye tracking device Eyetribe. Below are the steps we followed for the collection of data:

- Participant was made to sit approximately 60cm away from the screen and a calibration test with 9 calibration points was done.
- Eye tribe UI is launched and first the points are calibrated to calculate the average error on screen.
- Once the calibration is successful, data recording gets initiated and the recorded data is exported into a txt file.

- The Gaze coordinates and pupil size are extracted from the raw data obtained from eye tribe excluding other irrelevant parameters.
- The activity graphs of gaze patterns while performing various activities were estimated.
- The pupil size obtained were analyzed using FSM to estimate
- The average size of pupil when it is open/close state.

*A. Graphs*

Figure 4.1, 4.2, 4.3 depicts the graphs of a normal person performing tasks like Talking, Restıng and Reading activities.

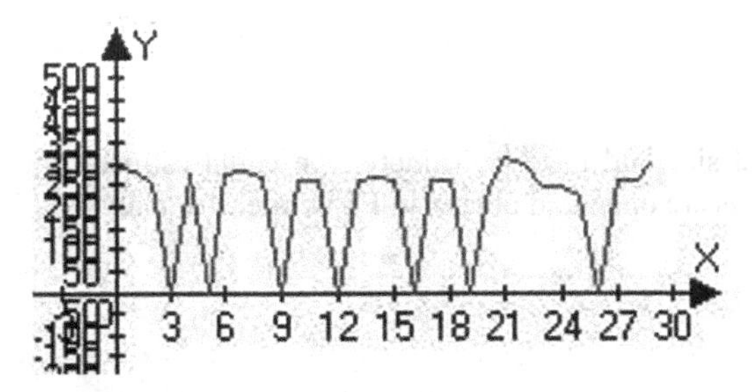


**Fig 4.1:** Talking Activity

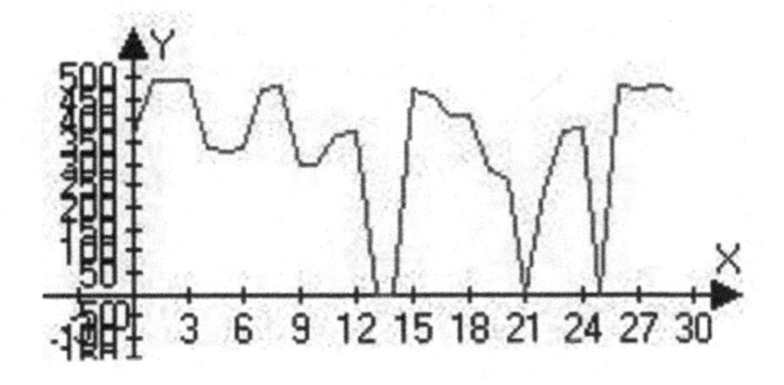


**Fig 4.2:** Resting Activity

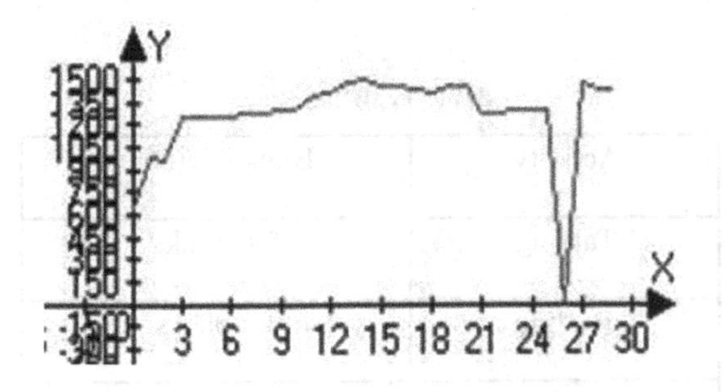


**Fig 4.3:** Reading Activity

The Graphs are drawn based on the eyegaze coordinates that are collected from Eyetribe. Along the X-axis denotes the time whereas Y-axis denotes the gaze point i.e. the point where the person is looking at. The Eyetribe raw data's coordinate value is zero, indicating probably a blink. So, when the graph touches the X-axis it is a blink.

*B. FSM*

Finite State Machine (FSM) is simply a state machine or a mathematical model used to represent computer programs and sequential logic circuits.

Using Pupil size: Identifying opening and closing states. Using gaze-coordinates: Identify blink (when coordinates are zero).

Based on Pupil size and gaze coordinates we could estimate the average size of pupil when the eyes are open and obtain an FSM after the analysis.

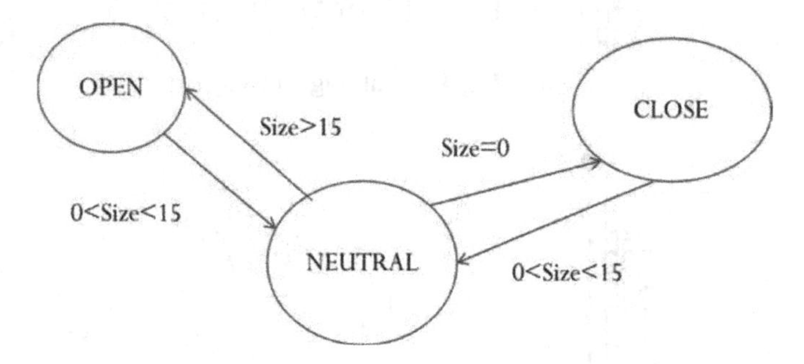


**Fig 5:** Finite State Machine

*C. Results*

The experimental suite was conducted for a period of 30sec. The average blink rate for each activity is shown in Table 1.

**Table 1.** Blink Rate

| Activity | Blink rate(30s) |
|---|---|
| Talking | 5-6 blinks |
| Resting | 4-5 blinks |
| Reading | 1-2 blinks |

Number of blinks are in the following descending order talking > resting > reading. It is observed that blinkrate of a person is about 10 to 15 times per minute and this is taken as a threshold value. The blinkrate is more than the threshold when a person is talking. When the person is reading, his blinkrate will be lesser than the threshold. When he/she is resting, blinkrate is in the threshold range. The other important observation is that when a person is stressed he tends to blink more.

Based on our analysis we built a simple application to measure the stress levels of a person. We have found a threshold value for the blink rate to find out if a person is stressed or not. If the person's blink rate is more than the threshold value we conclude that the person is stressed. The Figure 7 indicates the online analysis of the application which identifies whether the participant is stressed.

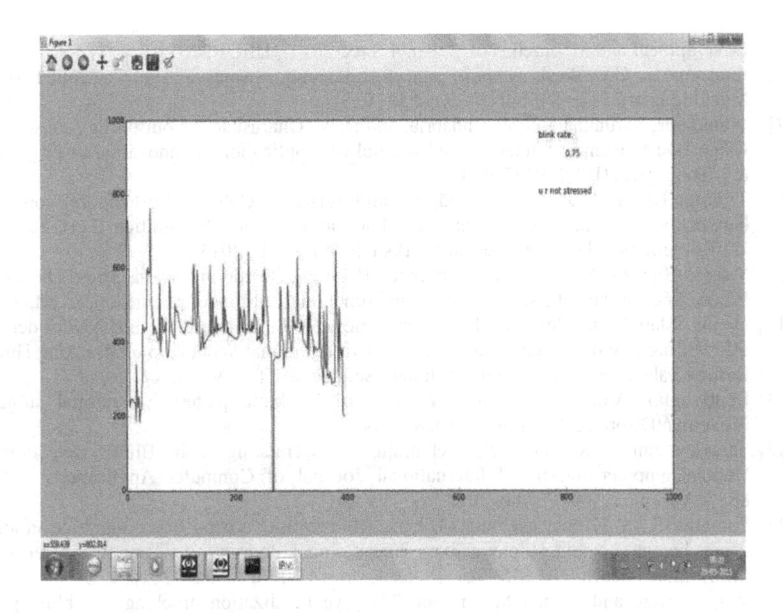


**Fig 7:** Sample result of the application

# 4 Conclusion and Future scope

The research work aimed at analyzing the eye blinks of a person while performing various activities and to develop a simple application to measure stress levels. It does not require any kind of prior learning.

The scope of this project can be further extended by building an application to detect various other activities by using multi modal inputs like video, audio etc. It can be even done by using eye blink patterns as the parameter.

In a more ambitious outlook, it can be extended to driver's drowsiness detection system.

## References

[1] Divjak, Matjaz, and Horst Bischof, “Eye Blink Based Fatigue Detection for Prevention of Computer Vision Syndrome”, In MVA (pp. 350-353).
[2] Bulling, Andreas, et al. "Eye movement analysis for activity recognition using electrooculography." IEEE transactions on pattern analysis and machine intelligence 33.4 (2011): 741-753.
[3] Westeyn, Tracy, et al. "Biometric identification using song-based blink patterns." Proc. HCI Int'l'05 (2005).
[4] Kim, Jinkwon, et al. "Driver's Drowsiness Warning System Based on Analyzing Driving Patterns and Facial Images." 23rd International Technical Conference on the Enhanced Safety of Vehicles (ESV). No. 13-0158. 2013.
[5] Mathiarasi, N., and T. Sureshkumar. "A Survey on Driver's Drowsiness and Unconsciousness Detection Methodologies." International Journal of Engineering Development and Research. Vol. 2. No. 4 (Dec 2014). IJEDR, 2014..
[6] Pradeep, S. G., Akesh Govada, and Kendaganna Swamy. "Eye Controlled Human Machine Interface (e-VISION)." Eye 2.5 (2013).
[7] Wankhede, Shrunkhala, S. A. Chhabria, and R. V. Dharaskar. "'Controlling Mouse Cursor Using Eye Movement." International Journal of Application or Innovation in Engineering & Management (IJAIEM) (2013).
[8] Trutoiu, Laura C., Jessica K. Hodgins, and JeffreyF. Cohn. "The temporal connection between smiles and blinks.", Automatic Face and Gesture Recognition (FG), 2013 10th IEEE International Conference and Workshops on. IEEE, 2013.
[9] Marcos-Ramiro, Alvaro, et al., “Automatic Blinking Detection towards Stress Discovery”, Proceedings of the 16th International Conferenceon Multimodal Interaction. ACM, 2014.
[10] Divjak, Matjaž, and Horst Bischof. "Real-timevideo-based eye blink analysis for detection oflow blink-rate during computer use." First International Workshop on Tracking Humans forthe Evaluation of their Motion in Image Sequences (THEMIS 2008). 2008.
[11] Bentivoglio, Anna Rita, et al. "Analysis of blinkrate patterns in normal subjects." MovementDisorders 12.6 (1997): 1028-1034.
[12] Saravanakumar, S., and N. Selvaraju. "EyeTracking and Blink Detection for HumanComputer Interface." International Journal of Computer Applications (0975 – 8887) Volume 2 – No.2, May 2010
[13] Sumathi, S., S. K. Srivatsa, and M. Uma Maheswari. "Vision based game development using human computer interaction." International Journal of Computer Science and Information Security, IJCSIS, Vol. 7, No. 1, pp. 147-153, January 2010, USA
[14] Wu, Junwen, and Mohan M. Trivedi. "An eye localization, tracking and blink pattern recognition system: Algorithm and evaluation." ACM Transactions on Multimedia Computing, Communications, and Applications (TOMM) 6.2 (2010): 8
[15] Amudha J, Madhura S, Parinitha Reddy, Kavitha, Hitha Nandakumar,” An Android based Mobile Eye Gaze Point Estimation System for studying the Visual Perception in Children with Autism”, International Conference on Computational Intelligence in Data Mining (ICCIDM'14), Veera Surendra Sai Technology, Odisha, Burla Dec 20-21st 2014
[16] Hitha NandaKumar, Amudha J,” A Comparative analysis of a Neural-based Remote Eye Gaze Tracker, International Conference on Embedded Application (ICES2014), 3-5th July 2014,ASE, Coimbatore, IEEExplore, Scopus Indexed
[17] Khilari, Rupal. "Iris tracking and blink detection for human-computer interaction using a low resolution webcam." Proceedings of the Seventh Indian Conference on Computer Vision, Graphics and Image Processing (pp. 456-463) ACM,2010.

# A Survey of Brain MRI Image Segmentation Methods and the Issues Involved

Reshma Hiralal, Hema P Menon
Department of Computer Science and Engineering
Amrita School of Engineering, Coimbatore
Amrita Vishwa Vidyapeetham, Amrita University, India
reshma.hiralal@gmail.com, p_hema@cb.amrita.edu

*Abstract*

***This paper presents a survey on the existing methods for segmentation of brain MRI images. Segmentation of brain MRI images has been widely used as a preprocessing, for projects that involve analysis and automation, in the field of medical image processing. MRI image segmentation is a challenging task because of the similarity between different tissue structures in the brain image. Also the number of homogeneous regions present in an image varies with the image slice and orientation. The selection of an appropriate method for segmentation therefore depends on the image characteristics. This study has been done in the perspective of enabling the selection of a segmentation method for MRI brain images. The survey has been categorized based on the techniques used in segmentation.***

***Keywords: Human Brain, Magnetic Resonance Image (MRI), Segmentation, Thresholding, Clustering, Region Growin, Neural Networks, Deformable Models.***

## 1. Introduction

Image processing is being extensively used in the field of medical imaging for analysis of images. Image analysis usually requires segmentation of the image (under consideration) into homogeneous



J.M. Corchado Rodriguez et al. (eds.), *Intelligent Systems Technologies and Applications 2016*, Advances in Intelligent Systems and Computing 530,
DOI 10.1007/978-3-319-47952-1_19

regions for feature extraction and analysis. Medical images are acquired using various modalities like the Computed tomography (CT), Magnetic Resonance Image (MRI), X-radiation (X-ray), Polyethylene terephthalate (PET), Functional MRI (fMRI) to mention the commonly used ones. This study concentrates on the processing of images obtained using MRI. The Magnetic Resonance Image (MRI) produces images using radio wave energy and magnetic field. It is a non-invasive imaging technique that is used for human body structures [1]. MRIs are capable of producing information which cannot be produced by any other technologies such as X-ray, CT-scan and ultrasound. CT gives details of the bony structures and MRI provides tissue level details. The MRI images of the human brain are acquired in typically three orthogonal orientations. They are called as axial, coronal and sagittal image slices, depending on the orientation (views) in which it has been obtained. The human brain is composed many parts of which for automatic segmentation only a few are considered, depending on the necessity. The most commonly researched parts for segmentation are: Gray matter (GM), Cerebrospinal Fluid (CSF) and white matter (WM). In more detailed analysis the parts that have been segmented are the tumor regions [2, 3].

Segmentation is the process of extracting or classifying the required brain parts. The part to be segmented depends on the problem at hand. Segmentation is the primary step of image analysis. Image segmentation can be broadly classified into Basic and Advanced methods. At a finer level they can be further divided into the following categories: 1) Thresholding 2) edge based method, 3) region based methods, 3) clustering methods. 4) Morphology based method, 5) neural networks, 6) deformable models and 7) genetic algorithm

The survey given in this paper is also based on these categories and relevance to use in brain MRI image. Segmentation process is also influenced by the presence of artifacts in the images under consideration. Human brain MRI images have various degrees of defects such as: Intensity inhomogeneity or Intensity Non Uniformity (INU) and Partial Volume Effect (PVE). The INU is occurred due to non –uniformity in data acquisition, which results in shading of MRI, images (human brain parts). Region based methods are highly affected by the intensity inhomogeneity [4, 5]. Partial volume effect is referred to as the effect wherein several types of tissues may be contained in a single image voxel [5, 6]. In such cases the Edge based methods may not have very much use due to the complexity of the structure of tissue [4]. To analyze these issues a thorough study of the existing segmentation algorithms is needed and hence has been taken as a topic of study in this paper. The organization of the paper is as follows:

Section II gives the review of literature on the existing segmentation techniques that are used in brain MRI image segmentation. The observations and findings from the literature have been discussed in section III, followed by the conclusions in section IV.

## 2. Review of Literature

This section presents a survey of the methods that have been used in the recent past for segmentation of brain MRI images. A survey of the segmentation methods has been done by many researches in different perspectives [5, 1, 7, 2, 3, 8]. A lot of literature is available for tumor segmentation [2, 3]. Kasiri et.al [1] conducted a comparative study of three different software tools which are used to segment the brain tissues based on both quantitative and qualitative aspects. The software's are SP M8 (Statistical Parametric Mapping), FSL 4.1, and Brain Suit 9.1 was used for segmenting the brain parts into white matter, grey CSF. They concluded that BrainSuit classifies WM and GM, show robust in performance tissue classification. SPM8 classifies Skull-striping and tissue segmentation and distinguishing CSF from other tissues

Liu et.al [2] has presented a survey on the segmentation methods that are used for Tumour detection. Yang et.al [3] gives a survey on the brain tumour segmentation methods such as atlas based segmentation, clustering based segmentation, continues deformable model based segmentation. A quantitative analysis of the results using standard evaluation parameters has also been discussed. The paper concludes that no one single segmentation method can be used for segmentation of all brain parts. The study is categorized based on the segmentation technique being implemented. It also gives details of the type of the brain parts that have been segmented using a particular technique.

## 2.1. Segmentation Process and Classification

Segmentation is the process of splitting images into segments according to the property specified according to which each region is to be extracted. The fundamental process of image analysis, understanding, interpretation and recognition, is image segmentation. There are different algorithms for segmenting brain MRI images. These segmentation algorithms can be broadly classified into basic and advanced methods based on the techniques that are used.

The brain images are generally classified into mainly three parts: White Matter (WM), Gray Matter (GM) and the Cerebrospinal Fluid (CSF).

### 2.2. *Basic Segmentation Method*

#### 2.2.1. Threshold Based Segmentation Methods.

Threshold based segmentation is the process of segmenting images based on the threshold value. The threshold value could be hard or soft. This method works well in cases where there is a large variation in the intensity between the pixel values. Thresholding is sensitive to noise and intensity in-homogeneities, when the image is a low contrast image it produces scattered output rather than connected regions. This affects the threshold selected and hence the segmented output. Thresholding method of image segmentation was used by different researchers in different variation and used in combination with other methods such as [10, 11, 12, 13, 14]

Evelin et.al [10] segmented brain MRI tissues into Grey matter, White matter and Cerebrospinal Fluids using thresholding based segmentation. A single fixed threshold was used in this case and hence only two classes of regions could be generated. The author reports that thresholding cannot be used for multi-channel images, but is an effective method for segmentation of Tumor regions. Aja-Fern´andez et.al [11] proposed a soft computing method which is fully automatic. They compared the results of proposed soft computing results with hard thresholding, k-means, fuzzy c-means (FCM), and thresholding. Further, concluded that soft computing is more efficient and faster because it is fully automatic, so manual selection of thresholds can be avoided. Since the operations are conducted in special level the system will be robust to noise and artifacts. This method can also be used in the segmentation of Ultrasound images and Radiography images.

### 2.2.2. Region Growing

Region growing works by grouping of pixels with similar intensities. It starts with a pixel or a group of pixel and grows to the neighbouring pixels with the same intensity. Region growing grows until it finds a pixel with does not satisfy the required pixel intensity. The major advantage is that region growing helps in extracting regions with the same intensity and it also extracts connected components. The region growing based segmentation method segments, regions with same intensity values. Region growing can be useful only if the region to be segmented contains maximum intensity variation. The main disadvantage of region growing is that it requires manual choosing of seed points.

Deng et.al [15] proposed a system which overcomes the difficulty in selecting the seed point the system used an adaptive region growing method based on variance and gradient inside and along boundary curves. This system gives an optimized segmented contour, for the detection of tumour and other abnormality in brain MRI. Since the results obtained are optimal this method can be used for clinical purposes in diagnosing diseases. Zabir et.al [16] has developed a system to segment and detect Glioma (which is a malignant tumour) from brain MRI image using region growing and level sets evaluation. Xiang et.al [17] suggested a hybrid 3D segmentation of brain MRI images which uses fuzzy region growing, mathematical morphological operators, and edge detection using Sobel 3D edge detector to segment white matter and the whole brain. Thus, from the results shown it is clear that the hybrid model is much more accurate and efficient in the 3D segmentation of the results. The hybrid system produces better result than the individual algorithms. Alia et.al [18] segmented brain parts of sclerosis lesions brain MRI images using a new clustering algorithm based on Harmony Search and Fuzzy C-means algorithm. This algorithm improves the standard HS algorithm to automatically evolve the appropriate number of clusters. The paper concludes that the proposed algorithm is able to find the appropriate number of naturally occurring regions in MRI images.

*2.2.3. Morphological segmentation.*

Morphological operators were used by [12, 17, 19]. This can be used in combination with other methods to get accurate and good results. Nandi [19] detected brain tumour from brain MRI images using morphological operators. The system is a hybrid combination of thresholding, watershed algorithm and morphological operators. Results obtained clearly suggest that a morphological operator gives a clear segmentation of tumour cells rather than k-means clustering. Further, they conclude that more work can be done to classify the tumour as malignant or benign tumour. Roger Hult [12] segmented cortex from MRI slices based on Grey level morphology on brain MRI imagcs. The system uses a histogram based method and hence finds the threshold value for the segmentation of brain parts from the non-brain parts. Binary and morphological operators are used in the segmentation of the brain parts of the brain MRI. They have suggested that the algorithm can be used in the segmentation of coronal MRI data.

*2.3.Advance Segmentations Methods*

### 2.3.1. Cluster Based Segmentation

Clustering is the process of classification of pixel values into different classes, without training or knowing previous information of the data. It clusters pixels with the same intensity or probability into same class. Clustering based segmentation is of different types basically they k-means, FCM. Clustering methods were used in detecting brain tumours. It has been observed that a combination of one or more basic and advanced segmentation enables the tumour detection more efficient. Such work, by combining multiple methods for tumour detection has seen reported by many researchers [20, 21, [22]. Qurat-ul et.al [20] proposed a system to segment brain tumour which detect the tumour region using naives Bayes classification and segment the detected region using k-mean clustering and boundary detection. The system was capable of achieving accuracy of 99% in detecting the tumour affected area.

The classification of regions of the brain image into WM, GM, and CSF uses different clustering algorithms has been reported by various researchers [4, 8, 13, 23, 25, 32]. Jun and Cheng [4] segmented WM, GM, CSF based on adaptive k-means clustering method and statistical Expectation-Maximization (EM) segmentation. From the results obtained they conclude that adaptive k-means clustering works better than the EM segmentation. They conclude that the system is capable of generating DE MR Images. Agarwal et.al [24] segmented brain MRI based on level sets segmentation and bias-field corrected fuzzy c-means. Further, concludes that the hybrid model produces an output which is better than the one obtained from the conventional level set and c-means clustering and this method can be used by the Radiologists and suggested that the proposed system suitable in detecting tumours development.

Zhidong et.al [26] proposed an automatic method for the 3D segmentation of brain MRI images using fuzzy connectedness. The paper focuses on two areas they are accurate extraction of brain parts and second focus is on the automatic seed selection. The proposed system accurately selects the seed point and extracts the MRI brain parts.

2.3.2. Neural Networks based segmentation.

Neural networks are used for the segmentation of GM, WM, and CSF by many researchers [9, 27, 28, 39]. Hussain et.al [9] detected Tumour and Edema in the human brain from MRI. They have claimed a segmentation accuracy of 99%, 82% and 99 % for segmentation of WM, GM, CSF tissues respectively. Tumour and edema tissues are segmented at a rate of 98% and 93% mean accuracy rate respectively. Hence, the paper concluded that proposed system can be used in segmenting tumour and edema with a high efficiency.

Segmentation using Neural Networks has also been used widely for detection of brain tumour [30, 31, 32, 33]. Sumithra and sexena [30] presented a neural network based system of classifying MRI images. The system consists of namely 3 different steps they are feature extraction, dimensionality reduction and classification. The classifier classifies the MRI images as benign, malignant and normal tissues and the classification accuracy of the proposed system is given as 73%.

### 2.3.3. Deformable Models

WM, GM, and CSF were segmented from MRI brain images using deformable models also [34, 35]. Anami et.al [34] using level set based approach. Modified fuzzy C means (MFCM) are used in the initial segmentation. Thus obtained result is given to levelsets methods. The results obtained from such a combinational system are more accurate than the individual MFCM and level sets. The time complexity of the combinational segmentation system is less than the levelsets method. With the accuracy level of segmentation this method can be used for further investigation of the Radiologist in diagnosing abnormalities. Chenling et.al [35] proposed a system using the AntTree algorithm. Further, they concluded that the improved AntTree algorithm is characterized as fast, robust, accurate and time complexity of the proposed system is less than the one obtained from k-means and FCM algorithm. Shaker and Zadeh [36] used level sets and atlas based segmentation for segmenting Hippocampus, Amygdala and Entorhinal cortex from the brain MRI image. GM is segmented from the brain MRI using atlas based segmentation and level sets are applied to the GM to produce the Hippocampus, Amygdala and Cortex.

Soleimani and Vincheh [37] proposed a system to detect the tumour present in brain MRI images using ant colony optimization. The system also improved the performance of the ant colony algorithm and the paper concludes that the suggested improved algorithm gives a more accurate result than the one with conventional algorithm. Karim [38] proposed a system which detects the organs in risk from a brain MRI. They used atlas based segmentation method and snakes to detect the organs which are at risk. They have also used canny edge detector to detect the edges and segment them. The use of deformable models made it accurate to segment the parts with accurate precision. The paper concludes that this model can be used in finding abnormal organs in human brain and this can be very beneficial for doctors in diagnosing. Zabir et.al [39] detected glioma tumour cells from human brain MRI. The segmentation and recognition process of Glioma cells are done by region growing and level sets method.

Juhi and Kumar [40] proposed a system which automatically segments brain tissues using random walker and active contours.

The boundary box algorithm makes a box around the tumour affected area and further segments the parts. They have compared the two and concluded that random walker is the best method with fast segmentation and provide accurate results.

## 3. *FINDINGS AND OBSERVATIONS*:

I. It has been observed from the survey that the Thresholding has been a widely used method for segmentation and is found to work well in cases where the intensity differences between the regions are more. This method is generally used for tumor detection from brain MRI images. From the above study it is evident that most of the research has been done for the segmentation, brain MRI images into White Matter, Gray Matter, and Cerebrospinal Fluid.

II. In cluster based methods, the use of FCM is most commonly followed by researchers for the segmentation of brain parts. Level sets can be used for segmenting intricate brain parts, but the issues with this approach are the seed pixel selection and the computational time. Another method that is generally used is the Atlas Based Segmentation. But this requires a proper Atlas to be available for all cases.

III. From this survey, it has been found that there is sacristy of literature in terms of segmentation of all regions of the brain MRI image. Brain image consists of more than 300 regions, of which the existing segmentation techniques are able to segment only a few (approximately 3-10). The maximum number of regions that can be segmented using any single segmentation algorithm is 3 to 5. There is no single segmentation algorithm that can extract every part present in the brain image. In applications like generic Atlas creation from the brain MRI images all parts have to be segmented and labeled. Hence, for this we need to use different algorithms or hybrid methods.

The table 1 gives the summary of the survey conducted.

Table 1: Segmentation methods and their segmented regions.

| Segmentation Method | Author's name | Segmented Region | Methodology used |
|---|---|---|---|
| Thresholding | Evelin et.al [10] | GM,WM,CSF | Single value thresholding |

| | | | |
|---|---|---|---|
| | Dawngliana et.al's [14] | Tumour | Multilevel thresholding, morphological operators and level sets |
| Region Growing | Zabir et.al [16] | Glioma | Region growing and level sets |
| Morphological | Nandi [19] | Tumour | Thresholding, watershed algorithm and morphological operators |
| | Roger Hult [12] | Cortex | Histogram based method, Binary and morphological operators, |
| Cluster Based Segmentation | Qurat-ul et.al [20] | Tumour | Naives Bayes classification, k-mean clustering, boundary detection |
| | Singh et.al [21] | Tumour | FCM and LSM, level set segmentation and fuzzy c-means clustering |
| | Jun and Cheng [4] | WM,GM,CSF | Adaptive k-means clustering method and statistical Expectation-Maximization (EM) segmentation |
| | Kong et.al [23] | WM,GM,CSF | Super voxel-level segmentation |
| | Agarwal et.al [24] | WM,GM,CSF | On level sets, bias-field corrected fuzzy c-means |
| | Liu and Guo [25] | WM,GM,CSF | Wavelets and k-means clustering method |
| Neural Networks. | Talebi et.al [27] | WM,GM,CSF | MLP feed-forward neural network, FCM |
| | Amin and Megeed [32] | Tumour | PCA and WMEM |
| Deformable Models | Anami et.al [34] | WM,GM,CSF | Modified fuzzy C means, level sets method |
| | Chenling et.al [35] | WM,GM,CSF | Anttree algorithm |
| | Shaker and Zadeh [36] | Hippocampus, Amygdala and Entorhinal cortex | Level sets and atlas based segmentation |
| | Soleimani and Vincheh [37] | Tumour | Ant colony optimization |
| | Karim [38] | Tumour | Atlas based segmentation |
| | Zabir et.al [39] | Glioma | Region growing and level sets method |
| Genetic Algorithm | Tohka [6] | GM,WM,CSF | FCM and genetic algorithm. |

# 4. CONCLUSION

In this paper review of existing literature on segmentation of brain MRI images has been discussed.

Hence, a thorough understanding of image under consideration and the segmentation methods is necessary for proper segmentation of the image. We hope that this review will help, researchers have an understanding of the algorithms that can be used for segmentation. This study shows that there is immense scope for further research in the field of segmentation of brain MRI images.

## REFERENCE

[1] Keyvan Kasiri, Mohammad Javad Dehghani, Kanran Kazemi, Mohammad Sadegh Helfroush, Shaghayegh Kafshgari, " Comparison Evaluation Of Three Brain MRI Segmentation Methods In Software Tools", 17th Iranian conference of Biomedical Engineering (ICBME), pp- 1-4, 2010.

[2] Jin Liu, Min Li, Jianxin Wang, Fangxiang Wu, Tianming Liu, and Yi Pan, "A Survey of MRI-Based Brain Tumour Segmentation Methods", Tsinghua Science And Technology, Volume 19, 2014

[3] Hongzhe Yang, Lihui Zhao, Songyuan Tang, Yongtian Wang, " Survey On Brain Tumour Segmentation Methods", IEEE International Conference On Medical Imaging Physics And Engineering, pp-140-145,2013.

[4] Jun Yung, Sung-Cheng Huang, "Methods for evaluation of different MRI segmentation approaches", Nuclear Science Symposium, vol 3 pp. 2053-2059,1998.

[5] Rong Xu, Limin Luo and Jun Ohya. "Segmentation of Brain MRI", Advances in Brain Imaging, Dr. Vikas Chaudhary (Ed.), ISBN: 978-953-307-955-4, 2012

[6] Jussi Tohka, "Partial volume effect modelling for segmentation and tissue classification of brain magnetic resonance images: A review", World Journal Of Radiology, vol-6(11), pp-855-864, 2014

[7] Mohammed Sabbih Hamound Al-Tammimi, Ghazali Sulong, "Tumor Brain Detection through MR Images: A Review of Literature", Journal of Theoretical and Applied Information Technology, vol 62, pp. 387-403, 2014.

[8] Jun Xiao, Yifan Tong , "Research of Brain MRI Image Segmentation Algorithm Based on FCM and SVM", The 26th CHINESE Control and decision conference, pp 1712-1716, 2014.

[9] S.Javeed Hussain, A.Satyasavithri, P.V.Sree Devi, "Segmentation Of Brain MRI With Statistical And 2D Wavelets Feature By Using Neural Networks", 3rs International Conference On Trendz In Information Science And Computing, pp-154-158, 2011.

[10] G. Evelin Sujji, Y.V.S. Lakshmi, G. Wiselin Jiji, "MRI Brain Image Segmentation based on Thresholding ", International Journal of Advanced Computer Research, Volume-3, pp-97-101, 2013
[11] Santiago Aja-Fern´andez, Gonzalo Vegas-S´anchez-Ferrero, Miguel A. Mart´ın Fern´andez, "Soft thresholding for medical image segmentation", 32nd Annual International Conference of the IEEE EMBS, pp- 4752-4755, 2010.
[12] Roger Hult, "Grey-Level Morphology Based Segmentation Of MRI Of Human Cortex", 11th International Conference On Image Analysis And Processing, pp-578-583, 2001.
[13] Xinzeng Wang,Weitao Li, Xuena Wang, Zhiyu Qian, "Segmentation Of Scalp, Skull, CSF, Grey Matter And White Matter In MRI Of Mouse Brain", 3rd International Conference On Biomedical Engineering And Informatics, pp-16-18, 2010.
[14] Malsawn Dawngliana, Daizy Deb, Mousum Handique, Sudita Roy, " Automatic Brain Tumour Segmentation In MRI; Hybridized Multilevel Thresholding And Level Set" International Symposium On Advanced Computing And Communication, pp-219-223, 2015
[15] Wankai Deng, Wei Xiao,He Deng,Jianuguo Liu, "MRI brain tumor segmentation with region growing method based on the gradients and variances along and inside of the boundary curve", 3rd international conference on biomedical engineering and informatics (BMEI), vol 1, pp-393-396, 2010
[16] I. Zabir, S. Paul, M. A. Rayhan, T. Sarker, S. A. Fattah, C. Shahnaz, "Automatic brain tumor detection and segmentation from multi-modal MRI images based on region growing and level set evolution", IEEE International WIE Conference on Electrical and Computer Engineering, pp- 503-506, 2015
[17] Zhang Xiang, Zhang Dazhi, Tian Jinwen, Liu Jian, " A Hybrid Method For 3d Segmentation Of MRI Brain Images", 6th International Conference On Signal Processing, Vol-1, pp- 608-611, 2002.
[18] Osama Moh'd Alia, Rajeswari Mandava, Mohd Ezane Aziz, "A Hybrid Harmony Search Algorithm to MRI Brain Segmentation" 9th IEEE Int. Conf. on Cognitive Informatics, pp-712-721, 2010

[19] Anupurba Nandi, "Detection Of Human Brain Tumour Using MRI Images Segmentation And Morphological Operators", IEEE International Conference On Computer Graphics, Vision And Information Security, pp-55-60, 2015

[20] Qurat-ul Ain, Irfan Mehmood, Naqi, M. Syed., Arfan Jaffar, M, "Bayesian Classification Using DCT Features for Brain Tumour Detection", 14th international conference, pp-340-349, 2010

[21] Pratibha Singh, H.S. Bhadauria, Annapurna Singh, "Automatic Brain MRI Image Segmentation using FCM and LSM", 2014 3rd International Conference on Reliability, Infocom Technologies and Optimization (ICRITO) (Trends and Future Directions), pp-8-10, 2014

[22] Dr. M. Karnan, T. Logheshwari, "Improved Implementation of Brain MRI image Segmentation using Ant Colony System", 2010 IEEE International Conference on Computational Intelligence and Computing Research (ICCIC), pp-1-4, 2010.

[23] Youyong kong, Yue Deng, and Qionghai Dai, "Discriminative clustering and feature selection for brain MRI segmentation", IEEE Signal Processing Letters, vol 22, pp-573-577, 2014.

[24] Pankhuri Agarwal, Sandeep Kumar, Rahul Singh, Prateek Agarwal, Mahua Bhattacharya, "A combination of bias-field corrected fuzzy c-means and level set approach for brain MRI image segmentation", 2015 Second International Conference on Soft Computing and Machine Intelligence, pp-85-88, 2015.

[25] Jianwei Liu, Lei Guo, "A New Brain MRI Image Segmentation Strategy Based on Wavelet Transform and K-means Clustering", 2015 IEEE International Conference on Signal Processing, Communications and Computing (ICSPCC), pp-1-4, 2015.

[26] Liu Zhidong, Lin Jiangli Zou Yuanwen, Chen Ke, Yin Guangfu," Automatic 3D Segmentation Of MRI Brain Images Based On Fuzzy Connectedness", 2nd International Conference On Bioinformatics And Biomedical Engineering, pp-2561-2564, 2008

[27] Maryam Talebi Rostami, Jamal Ghasemi, Reza Ghaderi, "Neural network for enhancement of FCM based brain MRI segmentation", 13th Iranian conference on Fuzzy Systems, 2013.

[28] Maryam Talebi Rostami, Reza Ghaderi, Mehdi Ezoji, Jamal Ghasemi, "Brain MRI Segmentation Using the Mixture of FCM and RBF Neural Networks", 8th Iranian Conference on Machine Vision and Image Processing, pp-425-429, 2013.
[29] K. J. Shanthi, M. Sasi Kumar and C. Kesavadas, "Neural Network Model for Automatic Segmentation of Brain MRI", 7th International Conference on System Simulation and Scientific Computing, pp-1125-1128, 2008.
[30] Sumitra, Saxene, "Brain Tumour Detection and Classification Using Back Propagation Neural Networks", I. J. Image, Graphics and Signal Processing. 45-50, vol-2,2013
[31] Dipali M. Joshi, Dr.N. K. Rana, V. M. Misra, "Classification of Brain Cancer Using Artificial Neural Network", 2nd International Conference on Electronic Computer Technology (ICECT), pp-112-115, 2010.
[32] Safaa.E.Amin, M. A. Megeed, "Brain Tumour Diagnosis Systems Based on Artificial Neural Networks and Segmentation using MRI", The 8th International Conference on INFOmatics and systems, pp-119-124, 2012.
[33] D. Bhuvana and P. Bhagavathi Sivakumar, "Brain Tumor Detection and Classification in MRI Images using Probabilistic Neural Networks", In Proceedings of the Second International Conference on Emerging Research in Computing, Information, Communication and Applications (ERCICA-14), pp- 796-801, 2014.
[34] Basavaraj S Anami, Prakash H Unki, "A Combined Fuzzy And Level Sets Based Approach For Brain MRI Image Segmentation", Fourth National Conference On Computer Vision, Pattern Recognition And Graphics, pp-1-4,2013
[35] Li Chenling, Zeng Wenhua, Zhuang Jiahe, "An Improved AntTree Algorithm for MRI Brain Segmentation", Proceedings of 2008 IEEE International Symposium on IT in Medicine and Education, pp-679-683, 2008.
[36] Matineh Shaker, Hamid Soltanian-Zadeh, "Automatic Segmentation Of Brain Structures From MRI Integrating Atlas-Based Labeling and Level Set Method", Canadian Conference on Electrical and Computer Engineering, pp- 1755 – 1758, 2008.

[37] Vahid Soleimani, Farnoosh Heidari Vincheh, "Improving Ant Colony Optimization for Brain MRI Image Segmentation and Brain Tumour Diagnosis", 2013 First Iranian conference on pattern recognition and image analysis (PRIA), pp-1-6, 2013.

[38] Boudahla Mohammed Karim, "Atlas And Snakes Based Segmentation Of Organs At Risk In Radiotherapy In Head MRIs" Third IEEE International Conference In Information Science And Technology, pp-356-363, 2014

[39] Ishmam Zabir, Sudip Paul, Md. Abu Rayhan, Tanmoy Sarker, Shaikh Anowarul Fattah, and Celia Shahnaz, "Automatic Brain Tumor Detection and Segmentation from Multi-Modal MRI Images Based on Region Growing and Level Set Evolution", IEEE International WIE Conference on Electrical and Computer Engineering, pp-503-506, 2015.

[40] Juhi P.S, Kumar S.S, " Bounding Box Based Automatic Segmentation Of Brain Tumours Using Random Walker And Active Contours From Brain MRI", International Conference On Control, Instrumentation, Communication And Computational Technologies, pp-700-704,2014.

# Scene Understanding in Images

Athira S, Manjusha R and Latha Parameswaran

**Abstract** Scene understanding targets on the automatic identification of thoughts, opinion, emotions, and sentiment of the scene with polarity. The sole aim of scene understanding is to build a system which infer and understand the image or a video just like how humans do. In the paper, we propose two algorithms- Eigenfaces and Bezier Curve based algorithms for scene understanding in images. The work focuses on a group of people and thus, targets to perceive the sentiment of the group. The proposed algorithm consist of three different phases. In the first phase, face detection is performed. In the second phase, sentiment of each person in the image is identified and are combined to identify the overall sentiment in the third phase. Experimental results show Bezier curve approach gives better performance than Eigenfaces approach in recognizing the sentiments in multiple faces.

## 1 Introduction

Most of the work to date on sentiment analysis focuses more on textual sentiment analysis. When people accelerated the use of videos and images to share their views and opinions [5], the importance of scene understanding increased. Many researches

Athira S
Athira S, M.Tech. Scholar, Dept. of Computer Science and Engineering, Amrita School of Engineering, Coimbatore, Amrita Vishwa Vidyapeetham, Amrita University, India.
e-mail: athira.sakunthala@gmail.com

Manjusha R
Manjusha R, Assistant Professor, Dept. of Computer Science and Engineering, Amrita School of Engineering, Coimbatore, Amrita Vishwa Vidyapeetham, Amrita University, India.
e-mail: r_manjusha@cb.amrita.edu

Latha Parameswaran
Latha Parameswaran, Professor, Dept. of Computer Science and Engineering, Amrita School of Engineering, Coimbatore, Amrita Vishwa Vidyapeetham, Amrita University, India.
e-mail: p_latha@cb.amrita.edu



J.M. Corchado Rodriguez et al. (eds.), *Intelligent Systems Technologies and Applications 2016*, Advances in Intelligent Systems and Computing 530,
DOI 10.1007/978-3-319-47952-1_20

claim that linguistic part of the message contribute only 7% to the overall impression of the message, the para-linguistic part contributes 38% and the remaining 55% is contributed by the mimics and body language. So, developing a system that uses the visual features is a promising way of understanding the given scene.

Some works in scene understanding uses the 3D layout of the entire scene in order to exploit the geometric and appearance information of the scene. It also uses the semantic and conceptual information. Generally, it deals with learning the given image to figure out what actually is happening in the scene. The proposed algorithm mainly focuses on finding the sentiments of a group of people.

Scene understanding [16] goes behind the pixel details to understand the underlying scene. It is of three types- pixel based scene understanding, region based and window based scene understanding. In pixel based scene understanding, each pixel is tagged with a class label. The objects and backgrounds are classified into different classes. Only one single category is assigned to each pixel. In region based scene understanding, each region is divided to coherent regions, and every region is marked with a class label. Each region is tagged with its class label [13]. In window based scene understanding, the objects within the window is analyzed and is tagged with a class label.The visual information in the image is used to find the polarity-positive or negative, of the scene. The low-level features in the visual information plays a vital role to find the positivism and negativism in a scene which refers to a group of people.

The paper is catalogued as follows: Sect. 2 explains the related work which discusses the different area and approaches used in scene understanding. Sect. 3 addresses an overview of the proposed work. The subsections in Sect. 3 explain each phase of the work in detail. Finally, Sect. 4 shows the result and the comparison of two algorithms in learning the sentiments in the scene.

## 2 Related Work

Scene understanding in images, with the aid of other algorithms, supports to learn the correlation between the essential components in the image, thus predicting its sentiment. For face detection, Ijaz Khan et.al, 2013 [1] proposed an efficient method which combined both Viola Jones and Skin color pixel detection algorithm. In paper [12], the authors proposed a face detection methodology where gabor wavelets filters are used.

Ren C. Luo et al.,2012 [6] proposed an integrated system that can track multiple faces and analyze the atmosphere of the scene. They developed the system for human robot interaction. In the work, Feature vector based algorithm and Differential Active Appearance Model Features based Approach were combined to find the sentiments of the scene. In paper [11], the authors found the sentiments in Online Spanish Videos with the help of audio, visual and textual features within the video. They collected videos from the social media website YouTube. In paper [13], the objects in the scene are annotated with its class label. Jianbo Yuan et al., 2013 [3]

**Fig. 1** Training Phase of the system

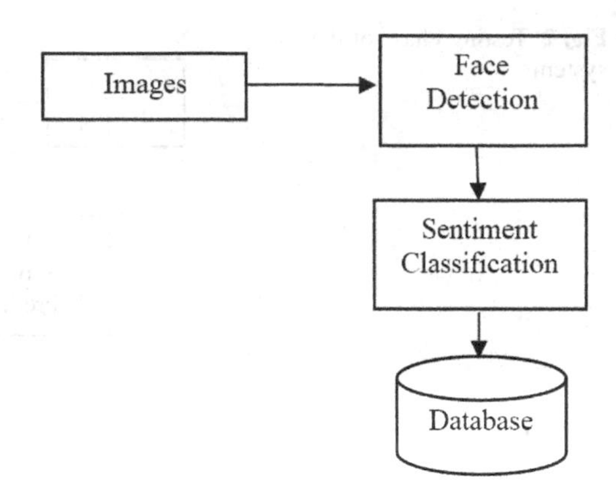


used midlevel attributes to find the sentiment of the scene. They considered images of games, tours, disasters etc and its sentiment was calculated. They also included images of humans and used eigenfaces approach to find the emotions of each face and later combined the emotions to tell the total sentiment of the image [5]. In paper [4], the authors used Action Units to find the emotion of individuals. Another approach as proposed by Yong-Hwan Lee et al., 2013 [17] used Bezier curve to find the emotion in the face. The control points were used to draw the Bezier curve for eye and mouth regions. The number of control points required for the curve depends on its type. Cubic Bezier curve requires four control points, Quadratic Bezier curve needs three control points, etc [10] [14]. In paper [8], the authors proposed a new algorithm to spot expressions in video. Variuos other approaches like neural network based methods are also used to recognize the emotion in a given face. PCA [18] reduces the dimensionality of the image and the result was given as input to neural network [2].

## 3 Proposed Work

The research work is explained through different phases- face detection, sentiment identification and classification, and overall sentiment of the total group.Fig. 1 shows the training phase of the system and Fig. 2 shows the testing phase of the system. Fig. 3 shows the result of face detection. The paper mainly discusses the performance of two algorithms- Eigen face and Bezier curve, in analyzing a scene with the help of sentiments.

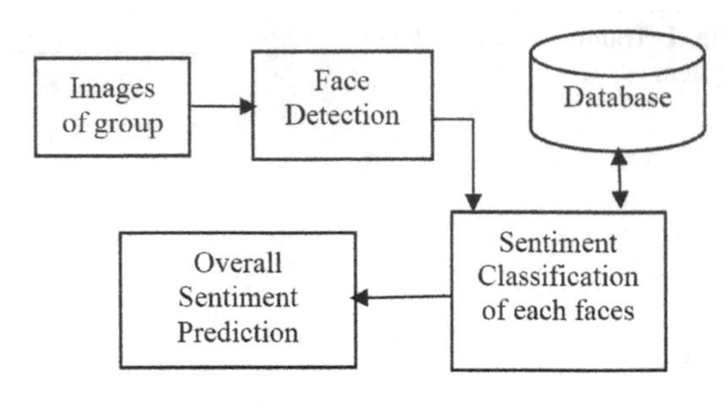


**Fig. 2** Testing Phase of the system

**Fig. 3** Face detection in group photo

## *3.1 Sentiment Classification*

Polarity of the image plays a vital part in the work. Here, two sentiments namely positive and negative has been considered. In this work, two algorithms- Eigenfaces and Bezier curve, are used to find the sentiment in the face, which is later used for sentiment calculation of the entire scene.

### 3.1.1 Eigenfaces Approach

In the algorithm, Eigenface [7] [15] for each sentiment is determined. On the given input image, principal components are calculated and weights are assigned to it. The steps for generating eigenfaces are as follows: s A set Z having m face images is obtained in the first step. Each image is transformed into a vector of size N and placed into the set.

$$S = \{\Gamma_1, \Gamma_2, \ldots, \Gamma_M\} \tag{1}$$

After it is transformed into a vector, the mean image is calculated. The mean image is

$$\Psi = \frac{1}{M} \sum_{a=1}^{M} \Gamma_(a) \tag{2}$$

Calculation is done between the input image and the mean image to find their difference.

$$\Theta_i = \Gamma_i - \Psi \tag{3}$$

Principal components are determined using the covariance matrix. These eigen vectors are used to assign weight to each image. For an input image, the faces are

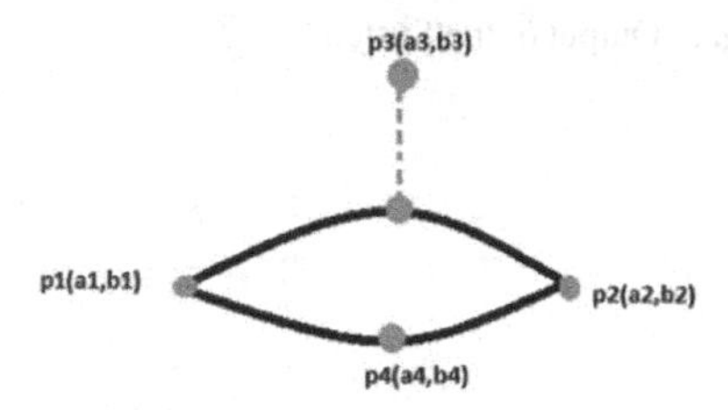


**Fig. 4** Eye boundary modelling using Bezier curve

transformed into its corresponding eigenface. The weight of the input is compared with that of the training images to check, to which sentiment, the image is closer. The sentiment class of the input image is the class to which the distance was minimal. Both Euclidean distance and Canberra distance has been used for finding the matching score.

### 3.1.2 Bezier Curve Approach

This algorithm uses contour points, also called control points, to generate the Bezier curve that passes through the first and last control points. The left eye, right eye and mouth detected using Viola Jones algorithm is taken as input to the algorithm. Here Bezier curve is drawn over the eye and mouth region for sentiment classification. A simple Bezier curve drawn over these control points approximates the original curve of the eye and mouth.

Eye boundaries are modelled using Bezier curves [19]. In this work, quadratic Bezier curve has been used. It uses two parabolas- one for upper eyelid and the other for lower eyelid. It is assumed that only the upper eyelids come into action whenever eyes are opened or closed wherein the control point in the lower eyclids will be in the lower boundary of eye itself. Fig. 4 represents the Bezier model of eye. p1, p2, p3 and p4 form the control points. For the lower eyelid, the control point lies in the eyelid. The control point p3 of upper Bezier curve is moved upward from its initial position until the luminance gradient flow begins to decrease [5]. The Bezier curve is defined by

$$x = (1-t)^2 a_1 + 2t(1-t)(a_3 - a_1) + t^2 a_2 \tag{4}$$

$$y = (1-t)^2 b_1 + 2t(1-t)(a_3 - b_1) + t^2 b_2 \tag{5}$$

The input eye image is converted into binary image, from which two extreme horizontal endpoints are selected as control points. In the work, Bezier curve is calculated for the input image and is compared with the average Bezier curve of each sentiment. Euclidean and Canberra distance measure has been used for finding the distance between the two.

**Fig. 5** Output of the System

**Fig. 6** Output of the System

### *3.2 Sentiment Prediction*

The sentiment of each face in an image is identified, and are combined to identify the sentiment of the total scene. If the sentiment is positive, a score of +1 is assigned and -1 otherwise. If the total score is a positive value, the overall sentiment of the scene is considered as positive otherwise negative.

## 4 Experiment Results

The system has been trained using 200 images including both positive and negative sentiment. The training images are collected from different databases - FEI, JAFFE [9] and some from google images. A total of 200 images are used in testing and these images include only group photographs. The system uses the same set of training and testing images for both the algorithms. Fig. 5 and Fig. 6 shows the output of the system when Bezier curve algorithm is applied on the test input image. The message box within the image shows the overall sentiment of the scene. Fig. 7 and Fig. 8 shows the output of the system along with test image on the left side of the figure, when Eigenface algorithm is used. In Fig. 7, all the faces are tagged with their corresponding sentiment and the overall polarity is shown in the message box within the image.

**Fig. 7** Input and Output of the System

**Fig. 8** Input and Output of the System

## 4.1 *Sentiment Classification using Eigenfaces*

In Eigenfaces algorithm, both Canberra distance and Euclidean distance measures are used to figure out which distance measure gives better results. Table 1 shows the confusion matrix of the Eigenfaces algorithm in sentiment prediction using Canberra distance. The confusion matrix shows that out of 110 positive images, only

**Table 1** Confusion Matrix using Canberra distance

| Sentiment | Positive | Negative | Total |
|---|---|---|---|
| **Positive** | 32 | 78 | 110 |
| **Negative** | 63 | 27 | 90 |
| **Total** | 95 | 105 | 200 |

32 are correctly classified as positive and the remaining 78 images are incorrectly classified. Likewise, out of 90 negative images, 27 of them are correctly classified and remaining images are misclassified. This gives a low accuracy of 29.5%. Table 2 represents the performance metrics of the algorithm using Canberra distance. Table 3 shows the confusion matrix of the Eigenfaces algorithm using Euclidean distance. The confusion matrix indicates that out of 110 positive images, 70 of them are correctly classified as positive and the remaining 40 images are incorrectly clas-

**Table 2** Performance measure using Canberra distance

| Measures | Value (%) |
|---|---|
| Accuracy | 29.5 |
| Misclassification Rate | 70.5 |
| TPR/Sensitivity/ Recall | 29.9 |
| Precision | 33.68 |
| F1 score | 31.22 |

sified as negative. Likewise, out of 90 negative images, 56 of them are classified correctly and 34 images are misclassified. The system gives an accuracy of 63% when Euclidean distance was used as the distance measure. Table 4 represents the performance metrics of the algorithm. Results explains that Eigenface algorithm

**Table 3** Confusion Matrix using Euclidean distance

| Sentiment | Positive | Negative | Total |
|---|---|---|---|
| **Positive** | 70 | 40 | 110 |
| **Negative** | 34 | 56 | 90 |
| **Total** | 104 | 96 | 200 |

**Table 4** Performance measure using Euclidean distance

| Measures | Value (%) |
|---|---|
| Accuracy | 63 |
| Misclassification Rate | 37 |
| TPR/Sensitivity/ Recall | 63.63 |
| Precision | 67.3 |
| F1 score | 65.42 |

gives better accuracy when Euclidean distance is used as the distance measure.

## *4.2 Sentiment Classification using Bezier curve*

Here, both Canberra and Euclidean distance are used for comparing the accuracy of the system. Table 5 shows the confusion matrix of Bezier curve algorithm using

Euclidean distance. Table 6 shows the performance metrics of the system using

**Table 5** Confusion Matrix using Euclidean distance

| Sentiment | Positive | Negative | Total |
|---|---|---|---|
| **Positive** | 84 | 26 | 110 |
| **Negative** | 23 | 67 | 90 |
| **Total** | 107 | 93 | 200 |

**Table 6** Performance measure using Euclidean distance

| Measures | Value (%) |
|---|---|
| Accuracy | 75.5 |
| Misclassification Rate | 24.5 |
| TPR/Sensitivity/ Recall | 76.36 |
| Precision | 78.5 |
| F1 score | 77.41 |

Euclidean distance measure. Table 7 shows the confusion matrix of Bezier curve system when Canberra distance is used as distance measure.

Table 8 shows the performance metrics of the system when Canberra distance is

**Table 7** Confusion Matrix using Euclidean distance

| Sentiment | Positive | Negative | Total |
|---|---|---|---|
| **Positive** | 97 | 13 | 110 |
| **Negative** | 12 | 78 | 90 |
| **Total** | 109 | 91 | 200 |

used. From the above findings, we conclude that Bezier curve gives better accuracy when Canberra distance measure is used. The performance of the algorithms suggest Bezier curve as a suitable algorithm for analyzing the sentiments in a scene; which gives an accuracy of about 87.5%. Table 9 shows the accuracy of the algorithms in understanding a group of people.

**Table 8** Performance measure using Euclidean distance

| Metric | Value (%) |
|---|---|
| Accuracy | 87.5 |
| Misclassification Rate | 12.5 |
| TPR/Sensitivity/ Recall | 88.18 |
| Precision | 88.99 |
| F1 score | 88.58 |

**Table 9** Comparison based Accuracy

| Algorithm Used | Euclidean | Canberra |
|---|---|---|
| Eigenfaces | 63 | 29.5 |
| Bezier Curve | 75.5 | 87.5 |

# 5 Conclusion

In this paper, we propose two algorithms for understanding the sentiments of a group people and also compared their performance in finding the sentiments of the scene. Through the observed result, we show that Bezier curve approach is advisable to identify the sentiments in a scene. The algorithm can be also be applied in live videos to obtain its sentiment and can be used in applications like classrooms, auditoriums, etc. In classrooms, the sentiment of the students towards the class can be identified and can be improvised depending on the sentiment identified. The system can be improved if audio, visual and textual features in the video are integrated.

The system is implemented in MATLAB R2013a and it executes in reasonable time.The overall performance can be improved by adding more positive and negative images in the training set.

# References

1. Ijaz Khan, Hadi Abdullah and Mohd Shamian Bin Zainal (2013) Efficient eyes and mouth detection algorithm using combination of Viola Jones and Skin color pixel detection, International Journal of Engineering and Applied Sciences.
2. J. Zhao and G. Kearney (1996) Classifying facial emotions by back propagation neural networks with fuzzy inputs. International Conference on Neural Information Processing,1:454-457
3. Jianbo Yuan, Quanzeng You, Sean Mcdonough, Jiebo Luo(2013), Sentribute: Image Sentiment Analysis from a Mid-level Perspective, Association for Computing Machinery, doi:10.1145/2502069.2502079
4. KaalaiSelvi R, Kavitha P, Shunmuganathan K (2014), Automatic Emotion Recognition in video, International Conference on Green Computing Communication and Electrical Engineering, p1-5, doi:10.1109/ICGCCEE.2014.6921398

5. Li-Jia Li, Richard Socher, Li Fei-Fei(2009), Towards Total Scene Understanding: Classification, Annotation and Segmentation in an Automatic Framework, Computer Vision and Pattern Recognition(CVPR),http://vision.stanford.edu/projects/totalscene/
6. Luo, R., Lin, P., Chang, L. (2012) Confidence fusion based emotion recognition of multiple persons for human-robot interaction. International Conference on Intelligent Robots and Systems (ICIRS), p 4590-4595
7. M. A. Turk , A. P. Pentland (1991), Face recognition using eigenfaces, Computer Vision and Pattern Recognition, IEEE Computer Society Conference
8. Matthew Shreve, Jesse Brizzi, Sergiy Fefilatyev, Timur Luguev, Dmitry Goldgof and Sudeep Sarkar (2014) Automatic expression spotting in videos. Image and Vision Computing, Elseiver
9. Michael Lyons, Miyuki Kamachi, and Jiro Gyoba, Facial Expression Database: Japanese Female Facial Expression Database, http://www.kasrl.org/jaffe.html
10. Navleen Kour, Madhu Bahl(2014) Emotion Extraction in Color Images using Hybrid Gaussian and Bezier Curve Approach, International Journal of Application or Innovation in Engineering and Management. Available via http://www.ijaiem.org/Volume3Issue9/IJAIEM-2014-09-25-60.pdf
11. Perez Rosas, Veronica, Rada Mihalcea, Louis-Philippe Morency(2013) Multimodal Sentiment Analysis Of Spanish Online Videos", IEEE Intelligent Systems,28(3):38-45
12. R. Karthika, Parameswaran Latha, B.K., P., and L.P., S. (2016) Study of Gabor wavelet for face recognition invariant to pose and orientation. Proceedings of the International Conference on Soft Computing Systems, Advances in Intelligent Systems and Computing, 397: 501-509
13. S. L. Nair, Manjusha R and Parameswaran latha (2015) A survey on context based image annotation techniques. International Journal of Applied Engineering Research,10:29845-29856
14. Sarfraz, M., M. R. Asim, A. Masood (2010), Capturing outlines using cubic Bezier curves, Information and Communication Technologies: From Theory to Applications.539 - 540,doi:10.1109/ICTTA.2004.1307870
15. Thuseethan, Kuhanesan S (2014) Eigenface Based Recognition of Emotion Variant Faces. Computer Engineering and Intelligent Systems,5(7)
16. Xiao J., Russell, B. C., Hays J., Ehinger, K. A., Oliva, A., Torralba(2013), Basic Level Scene Understanding: from labels to structure and beyond, doi:10.1145/2407746.2407782
17. Yong-Hwan Lee, Woori Han, Youngseop Kim (2013) Emotion Recognition from Facial Expression Analysis using Bezier Curve Fitting, 16th International Conference on Network Based Information Systems,doi: 10.1109/NBiS.2013.39
18. Yu-Gang Jiang, Baohan Xu, Xiangyang Xue(2014), Predicting Emotions in User Generated Videos, Association for the Advancement of Artificial Intelligence, Canada, July
19. Z.Hammal, A. Caplier(2004) Eyes and eyebrows parametric models for automatic segmentation, 6th IEEE Southwest Symposium on Image Analysis and Interpretation, p 138-141.

# Part II
# Networks/Distributed Systems

# Flexible Extensible Middleware Framework for Remote Triggered Wireless Sensor Network Lab

**Ramesh Guntha, Sangeeth K and Maneesha Ramesh**[1]

**Abstract** Wireless sensor networks (WSN) are also being actively researched and learned as part of graduate and undergraduate technical disciplines as they are being applied in variety of fields such as landslide detection, smart home monitoring etc. It is essential for the students to learn the WSN concepts through experimentation. But it needs a lot of time, hardware resources and technical expertise to design, program, setup, and maintain these experiments. A shared remote triggered (RT) WSN test-bed which can be accessed, controlled, monitored through internet would be of great help for many students to learn these concepts quickly without any setup at their end. Our RT WSN test-bed consists of 11 experiments, each deigned to explore a particular WSN concept. Each experiment consisting of a WSN setup, a web interface for user input and results, and middleware layer to communicate between web and WSN [1, 2, and 3]. In this paper we present the architecture of our flexible and extensible middleware framework; a single code base supporting all the experiments which can be easily extended for any other WSN experiment just by configuring. We also illustrate the success of our system through the analysis of various usage statistics.

## 1 Introduction and Related Work

WSN systems are being used for a wide range of applications such as landslide monitoring, smart home monitoring, monitoring ocean and atmospheric pollutions etc., and hence it's being researched and learned as part of undergraduate and graduate engineering courses. As is the case with any complex engineering subject, learning through experimentation in the lab is a must for the deeper understanding of the WSN concepts. Setting up such a WSN lab requires careful study of the critical concepts of WSN, designing the experiment for each concept in such a way that it helps the student to understand, explore and learn that concept quickly, formulating the related study material, developing required animations and other multi-media material to augment the study material, purchasing various wireless equipment and setting up the WSN system for the experiment, writing the embedded code to

[1] Amrita Center for Wireless Networks and Applications, Amrita School of Engineering, Amritapuri Campus, Amrita Vishwa Vidyapeetham University, e-mail: {rameshg, sangeethk, maneesha }@ am.amrita.edu



J.M. Corchado Rodriguez et al. (eds.), *Intelligent Systems Technologies and Applications 2016*, Advances in Intelligent Systems and Computing 530,
DOI 10.1007/978-3-319-47952-1_21

program the WSN equipment to run the experiment, writing the code for the web layer to accept student input and to provide the experiment results, and finally writing the code to program the middleware to help communication between the web and WSN layers. After setting up an experiment like this it has to be maintained to make sure any hardware, power, and software failures gets identified quickly and fixed immediately to keep the experiment always available for students. Thus it takes a lot of time, resources and equipment to setup an experiment and maintain it and hence it is not affordable for all the institutes to provide such advanced laboratory facilities. A remote triggered WSN lab can help a lot of students across the globe to learn various WSN concepts without going through all the effort to setup and maintain the experiment. Our RT WSN lab offers 11 experiments to help students learn all the critical WSN concepts like sending and receiving data, range vs. power, clustering, data acquisition, duty cycle etc. [1, 2, and 3]. Various test beds like Indriya, Motelab, Twist etc uses remote reconfiguration mechanism for their sensor network test bed. In Indriya [4], the input to the web layer is the uploaded application exe plus MIG generated Java class files which defines the structure of messages that application dumps to the USB port. In Motelab [5], the user uploads executables and class file. Using MySQL database, the user details and program details are stored. It has a DBLogger which starts at the beginning of every job. In Twist [6], the web layer input for remote reconfiguration is the executable. The user uploads the executable and gets the data logged results.

The middleware of our RT WSN system is extensible and flexible as it is designed as a set of configurable modules that can be assembled in any sequence. It has various modules for reading/writing to various devices like WSN Gateway and Multimeter. The COM ports, baudrate, read/write patterns, delimiters, output xml tags are configurable for each experiment. The read write modules can be configured to be invoked from web layer, based on delay timer, and/or in an automatic sequence. For a given experiment there could be any number of read/write operations going to multiple combinations of WSN gateways and Multimeters. Such modularity and configurability allows a single middleware codebase to cater to all the WSN experiments and it can be easily extended for any future experiments with little or no code changes. In this paper we present the architecture and design details of the RT WSN Lab's middleware.

In the rest of the paper we will present the brief overview of experiments and lab setup, architecture, design, usage data, and finally end with conclusions.

## 2 Experiments and Lab Setup

We offer 11 unique experiments in our RT WSN lab to enable the students to learn various important WSN concepts. The experiments are designed to explore basic NesC programming, sending and receiving data among the wireless motes, range

Vs connectivity and power, duty cycle Vs power consumption, sensor data acquisition, data collection frequency Vs power consumption, wireless propagation, wireless sensor network, WSN data acquisition, clustering algorithms, and time synchronization concepts.

Fig. 1 shows the layout of WSN equipment for a few experiments along with the cameras to show the live feed to users. The lab consists of 11 experiments connected to 6 computers spread across the room. The equipment include over 80 wireless motes, around 8 to 10 Multimeters, power equipment, and connecting wires.

Table 1 shows the read-write configurations of the experiments. Some experiments are reading writing to WSN gateways only, while the others are writing to WSN gateways and reading from a combination of WSN gateways and Multimeters. Some experiments are even reading and writing to multiple Multimeters. While the writing to WSN gateway is triggered by the web URL in all the experiments, the reading from WSN and Multimeters is triggered by web URL for some experiments, whereas for the others it is triggered automatically after the writing is done. For some experiments the reading stops after the first set of values, whereas for some others the reading continues for ever, and for some other experiments the reading stops after certain number of values.

Table 2 shows the data acquisition configurations for all the experiments. The experiments which are reading form WSN gateways get multiple data elements separated by start and end tag delimiters, whereas the experiments reading from Multimeters get only single current value and hence do not have delimiters. The data from Multimeters is modified by a multiplication factor and also formatted into the xml document by the specified tags.

Thus these tables shows both the differences and commonalities among the experiments, these commonalities are drawn out as various modules and configurations so that single codebase can be used to serve as middleware for all the experiments.

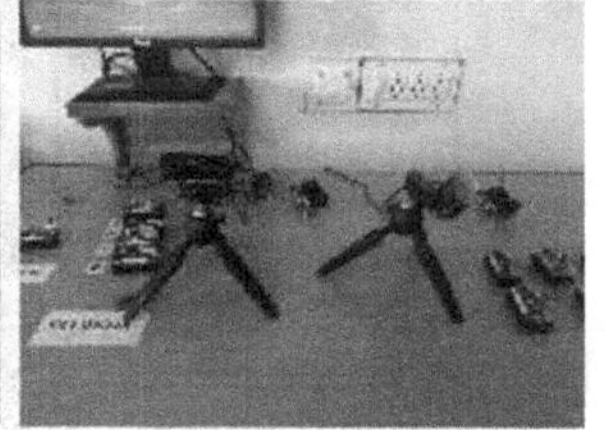

**Fig. 1.** WSN lab setup of few experiments

**Table 1.** Experiments and Read-Write configurations

| S. No | Experiment Name | WSN | Multimeter (Number) | Input - Output | Output Trigger | Comments |
|---|---|---|---|---|---|---|
| 1 | NesC Programming | Y | | WSN-N/A | N/A | Visual output only |

| | | | | | | |
|---|---|---|---|---|---|---|
| 2 | Send & Receive | Y | | WSN-N/A | N/A | Visual output only |
| 3 | Range Vs Connectivity Vs Antenna Power | Y | | WSN-WSN | URL | |
| 4 | Duty Cycle Vs Power Consumption | Y | Y (1) | WSN-Multimeter | Automatic With Delay | Multimeter communication is triggered automatically after input to WSN after a delay of 3 seconds |
| 5 | Sensor Data Acquisition | Y | | WSN-WSN | URL | |
| 6 | Data collection frequency Vs Tx Vs Power Consumption | Y | Y (1) | WSN-Multimeter | URL | |
| 7 | Wireless Propagation | Y | | WSN-WSN | URL | |
| 8 | Wireless Sensor Network | Y | Y (3) | WSN-WSN & Multimeter | URL | One output is from WSN, 2nd one is from a Multimeter, 3rd is from all Multimeters. |
| 9 | WSN Data Acquisition | Y | Y (3) | WSN-WSN & Multimeter | URL | One output is from WSN, rest three are from Multimeters |
| 10 | Clustering Algorithms | Y | | WSN-WSN | URL | |
| 11 | Time Synchronization | Y | | WSN-WSN | URL | It executes the data flow for 10 times and then terminates |

**Table 2.** Experiments and Data acquisition configurations

| S. No | Experiment Name | Start tag | End tag | Value multiplier | Output tag |
|---|---|---|---|---|---|
| 1 | NesC Programming | | | | |
| 2 | Send & Receive | | | | |
| 3 | Range Vs Connectivity Vs Antenna Power | Y | Y | | |
| 4 | Duty Cycle Vs Power Consumption | | | Y | Y |
| 5 | Sensor Data Acquisition | Y | Y | | |
| 6 | Data collection frequency Vs Tx Vs Power Consumption | | | Y | Y |
| 7 | Wireless Propagation | Y | Y | | |
| 8 | Wireless Sensor Network | Y | Y | Y | Y |
| 9 | WSN Data Acquisition | Y | Y | Y | Y |
| 10 | Clustering Algorithms | Y | Y | | |
| 11 | Time Synchronization | Y | Y | | |

## 3 Middleware Architecture

The RT WSN lab's middleware facilitates the communication between the web layer and WSN system (Fig. 2). The middleware runs on NodeJS server; a high performance, event driven, and extensible, heavily adopted, and open source application server. The ExpressJs package is used to route the incoming web commands. The configuration module reads the JSON configuration from the file and helps in find the configuration details of the experiment. The COM port manager is responsible in creating, flushing, and closing the COM ports needed for each experiment. Input and Output modules are responsible for writing and reading data to the devices and flushing unread data from the buffers. The WSN and Multimeter modules are responsible for communicating with the respective devices. The Data processor module is responsible for parsing the data from devices, modifying, and formatting the data as required by the web layer. The Core module is the complex heart of the system which coordinates with all the other modules to achieve the communication between web and WSN layers.

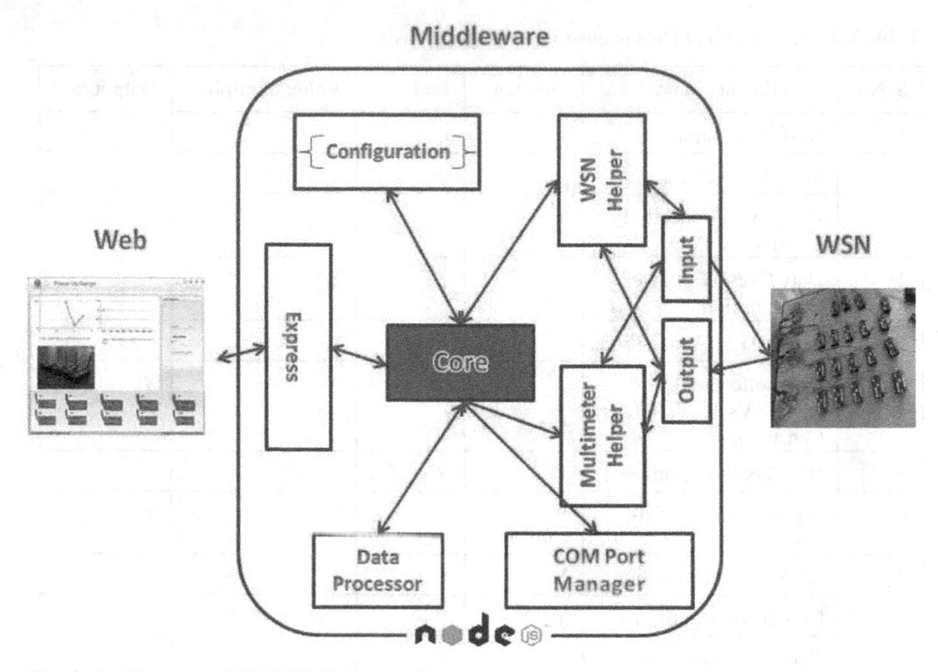


**Fig. 2.** Architecture of RT WSN lab's middleware

# 4 Middleware Design

In this section we present the configuration and workflow of the Core module of the middleware

Fig. 3 shows the configuration for the experiment ' Wireless Sensor Network ' as an example. It can be noted that the experiment handles four URLs from the web layer. The corresponding configuration for each URL is listed as separate JSON block. The first URL corresponds to the write instruction to WSN gateway, communicating the user commands to the WSN system. The second URL reads the results from the WSN gateway. The third and fourth URLs give commands to Multimeter and get the reading from the same. The corresponding COM ports, baudrate, start tag, end tag, idle time out duration, output xml tag, device type etc., configurations can be noted at each of the configuration block. This illustrates the flexibility and extensibility of the middleware as it can cater to many verities of experiments just by adapting the configurations.

```
/9/": {
  "urlPrefixes":["/9/wireless_senor_network/input/",
    "/9/wireless_senor_network/output/",
    "/9/wireless_senor_network/leaf_power/output",
    "/9/wireless_senor_network/leaf_cluster_sink_power/output"],
  "/9/wireless_senor_network/input/":{
    "readWrite":"WRITE",
    "deviceType":"WSN",
    "baudRate":57600,
    "serialPort": "COM4",
    "idleTimeoutSecs:":120,
    "outputDataStartTag":"/",
    "outputDataEndTag";"$"
  },
  "/9/wireless_senor_network/output":{
    "readWrite":"READ",
    "deviceType":"WSN",
    "baudRate":57600,
    "serialPort": "COM4",
    "outputTagName":"wsn_output"
  },
  "/9/wireless_senor_network/leaf_power/output":{
    "readWrite":"WRITE_READ",
    "deviceType":"MultiMeter",
    "baudRate":19200,
    "outputTagName":"leaf_power",
    "serialPort": "COM12",
    "commands": [
      ":FUNC 'CURR:DC';:CURR:DC:DIG 7;:CURR:DC:NPLC 1.000000;:CUR
      ":FUNC 'CURR:DC';:CURR:DC:AVER:TCON REP;:CURR:DC:AVER:COUN
      ":DATA?"
    ],
    "outputValueMultiplier":3
```


**Fig. 3.** Configuration details for an experiment which communicates with both WSN gateway and Multimeter

Fig. 4 illustrates the algorithm of the most important module of the middleware, the Core module. As soon as the server starts, it reads the configurations from Config.json file and stores in cache. As the request from user comes in, the URL is matched with the urlPrefixes attribute in the configuration object and matched experiment and the subsection is found. The related comports are prepared for the usage. Based on the instruction type, either write, read or write-read operations are performed either on WSN gateway or on the Multimeter. The read operation simply reads the accumulated data from the cache so far and returns to the user. Whereas the write operation creates the data event listeners, performs appropriate write operations on the respective devices. The event listeners would be responsible in capturing and processing the output data from WSN system. In case of Multimeter, it needs to be configured first with a series of commands before requesting the data. These commands must be executed with a certain delay to avoid hanging of the system. In case of write-read, first the write operation is performed, if there are multiple COM ports involved, then there should be a delay of 1 second between communicating with each of the COM ports to avoid system hang.

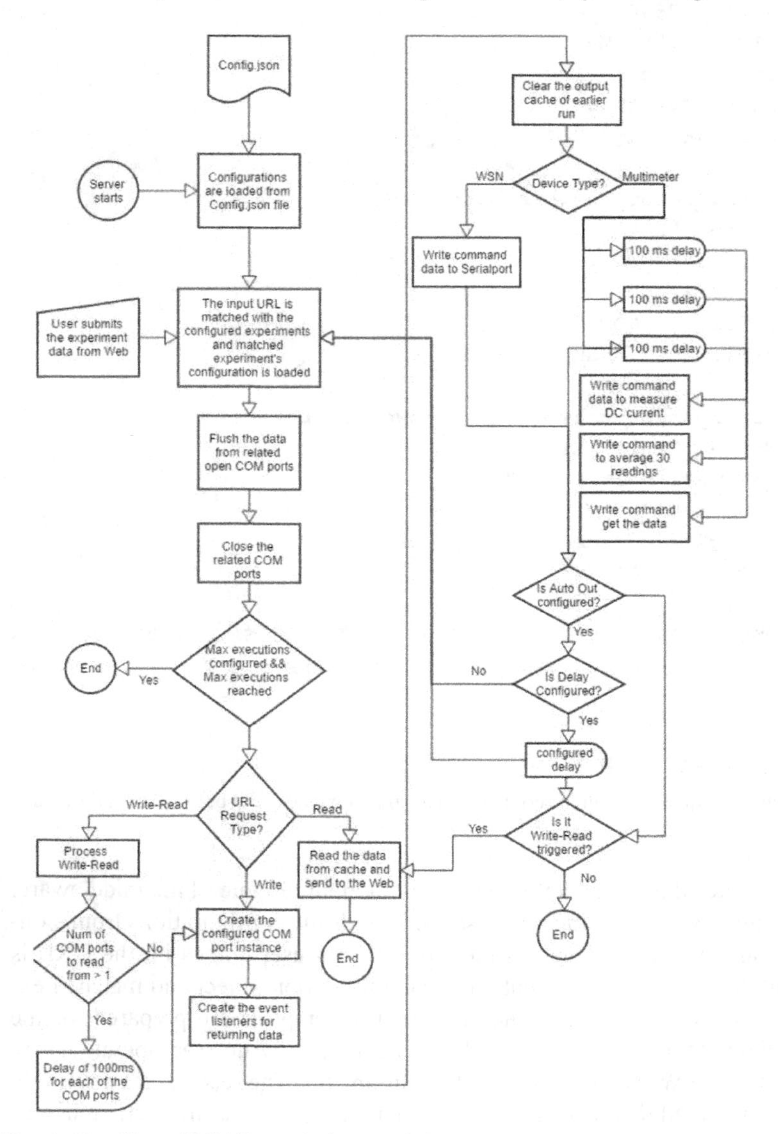


**Fig. 4.** Algorithm of Middleware's Core module

Fig. 5 illustrates the algorithm of the data processing logic of the core module. As the data comes in from the WSN system, it's sent for processing. The processor first looks to see the configured start and end tags for the experiments write operation, then parses each chunk out of the data stream based on these tags and stores the chunks in a cache, which will be eventually read out during the read request. The processor then sends the data through the value multiplier code based on the configuration, after that based on the output tag configuration the data is formatted into xml tags, which are sent to the user upon issuing the read request.

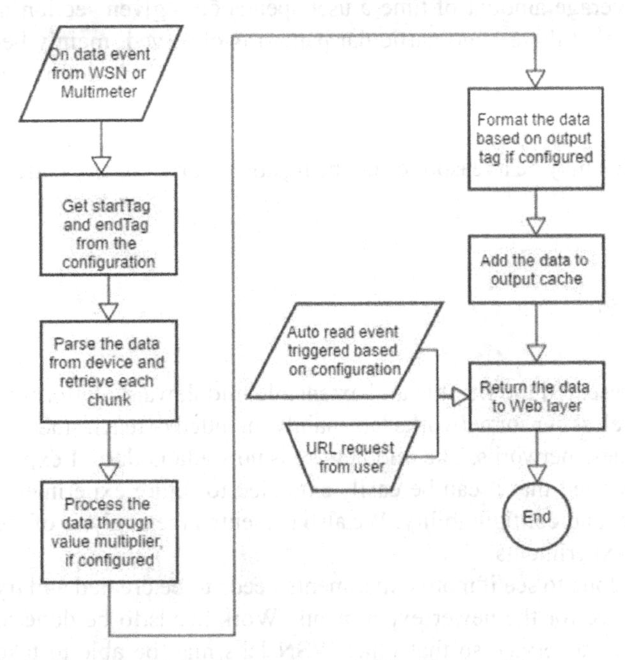


**Fig. 5.** Algorithm of data processing part of the Core module

## 5 Usage Data Analysis

The RT WSN Lab has been operational for the past couple of years. Every day several students schedule experiments in the available time-slots. Each experiment is divided into few sections to facilitate faster learning. The students are presented with the background theory of the experiment to make them familiarize with the concept, then procedure of the experiment is explained, a self-evaluation is presented to assess how well the student understood the theoretical concepts and then the student is led to experiments and related result pages. We have analyzed a year worth of usage data and the analysis is presented below.

Fig. 6 presents the number of views of each section for all the experiments. It can be noted that user behavior is more or less consistent across all the experiments, a very high viewership during the beginning sections of experiments and it quickly reduces to below 1/6th of the viewership as the user proceeds to the assignment, reference and feedback sections. This can be analyzed as lack of sustained interest for the students to go through the entire course of the experiment, further studies needed to be done to see what changes can be made the process to increase the viewership across the experiment process.

Fig. 7 presents the average amount of time a user spends on a given section in seconds. Here it's noted that there is no particular pattern is observed, mainly because of repeated users, as the users get more familiar with theoretical sections of the experiment, they tend to spend more time in assignment sections. And some concepts may be more difficult to grasp or more interesting compared to others to different students, and this may be a reason for not being able to observe any pattern in this statistic.

## 6 Conclusions

In this paper we have presented the flexible and extensible middleware platform of our remote triggered wireless sensor networks lab, mainly intended to teach students various concepts of wireless networks. The middleware is now adapted to 11 experiments and it is demonstrated that it can be easily extended to future experiments because of its modularity and configurability. We also presented the analysis of the usage metrics of all the experiments.

Analysis needs to be done to see if more experiments needs to be created and try and extend the middleware for the newer experiments. Work needs to be done to expose the middleware as a service so that other WSN labs may be able to take advantage of this middleware.

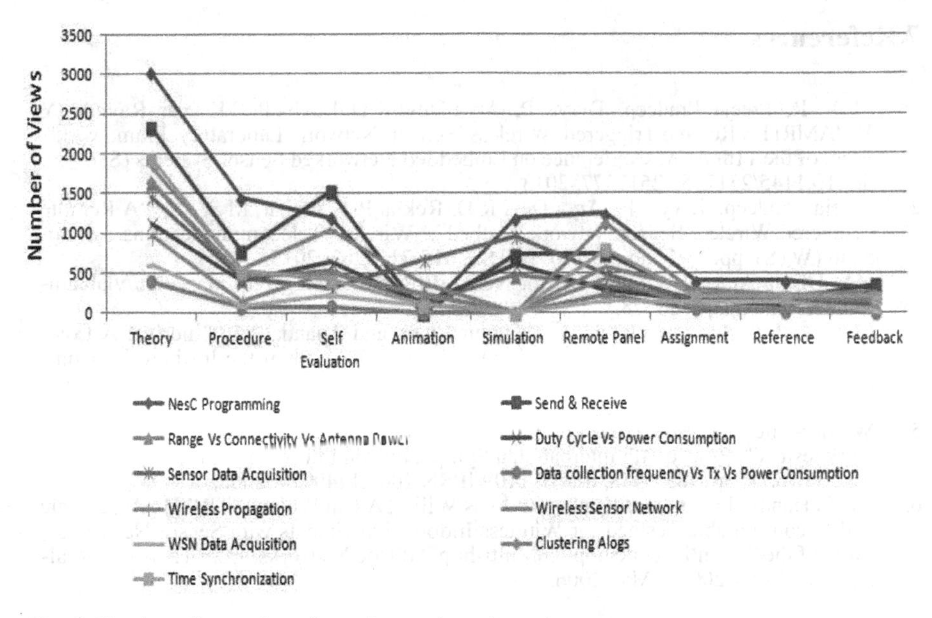


**Fig. 6.** Number of page views for each experiment's section

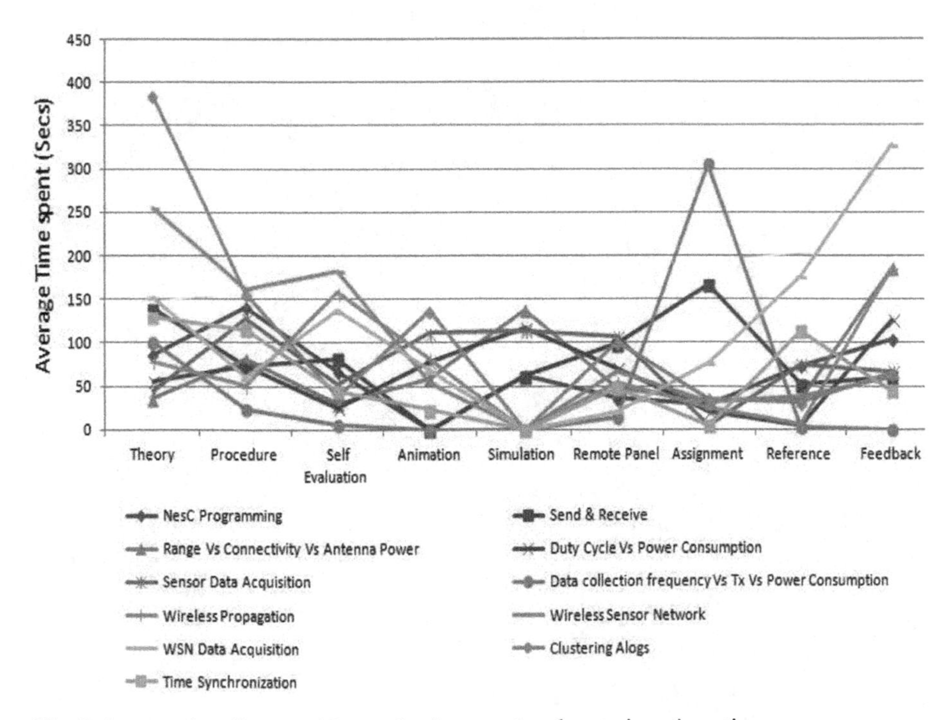


**Fig. 7.** Average time (in seconds) spent by the user at each experiment's section

# 7 References


1. M V R, Preeja Pradeep, Divya P, Arya Devi R D, Rekha P, S Kumar, Rayudu Y V, "AMRITA Remote Triggered Wireless Sensor Network Laboratory Framework", Proc. of the 11th ACM Conference on Embedded Networked Sensor Systems (Sensys), doi:10.1145/2517351.2517377, 2013
2. Preeja Pradeep, Divya P, Arya Devi R D, Rekha P, S Kumar, M V R, "A Remote Triggered Wireless Sensor Network Testbed", Wireless Telecommunications Symposium (WTS), pp. 1 - 7, doi:10.1109/WTS. 2015.7117262, 2015.
3. VALUE @ Amrita: Wireless Sensor Network Remote Triggered Lab, http://vlab.amrita.edu/index.php?sub=78.
4. Manjunath Dodda venkatappa, MunChoon Chan, and Ananda A.L, "Indriya: A Low-Cost, 3D Wireless Sensor Network Testbed", Lecture Notes of the Institute for Computer Sciences, Social Informatics and Telecommunications Engineering, vol. 90, pp. 302-316, 2012
5. Werner-Allen, G., Swieskowski, P. and Welsh, M., "MoteLab: a wireless sensor network testbed", Proc. of 4th International Symposium on Information Processing in Sensor Networks, pp. 483 - 488, doi:10.1109/IPSN.2005.1440979, 2005.
6. Vlado Handziski, Andreas Kopke, Andreas Willig, Adam Wolisz, "TWIST: A Scalable and Reconfigurable Testbed for Wireless Indoor Experiments with Sensor Networks", Proc. of the 2nd Intl. Workshop on Multi-hop Ad Hoc Networks: from Theory to Reality, (RealMAN 2006), May 2006.

# Energy Efficient Deflate (EEDeflate) Compression for Energy Conservation in Wireless Sensor Network

**Pramod Ganjewar**

Research Scholar, Sathyabama University,Chennai, Tamilnadu, India,

MIT Academy of Engineering, Alandi(D.), Pune, Maharashtra, India.
pdganjewar@comp.maepune.ac.in

**Barani S.**

Department of E & C,

Sathyabama University, Chennai, Tamilnadu, India,
baraniselvaraj77@gmail.com

**Sanjeev J. Wagh**

Department of Information Technology,

Government College of Engineering, Karad, Maharashtra, India.

sjwagh1@yahoo.co.in

**Abstract**

*WSN comprises of sensor nodes distributed spatially to accumulate and transmit measurements from environment through radio communication. It utilizes energy for all its functionality (sensing, processing, and transmission) but energy utilization in case of transmission is more. Data compression can work effectively for reducing the amount of data to be transmitted to the sink in WSN. The proposed compression algorithm i.e. Energy Efficient Deflate (EEDeflate) along with fuzzy logic works effectively to prolong the life of Wireless Sensor Network. EEDeflate algorithm saves 7% to 10% of energy in comparison with the data transmitted without compression. It also achieves better compression ratio of average 22% more than Huffman and 8% more than Deflate compression algorithm. This improvements in terms of compression efficiency allows saving energy and therefore extends the life of the sensor network.*

Keywords

Wireless Sensor Networks, Radio Communication, Energy Consumption, Data Transmission, Data Compression, Fuzzy Logic.

## 1. Introduction

Wireless Sensor Networks are gaining lots of attention with advancements in the technology. WSN is a network of many self-organized



J.M. Corchado Rodriguez et al. (eds.), *Intelligent Systems Technologies and Applications 2016*, Advances in Intelligent Systems and Computing 530,
DOI 10.1007/978-3-319-47952-1_22

small tiny nodes, to perform the functions of sensing and processing for gathering the environmental information. In the wireless sensor network tiny nodes are provided with battery resource. The data transmission is most energy consuming task performed by sensor node [1] i.e. more energy is consumed in transmission. This energy need to be conserved or minimized.
Now a days the issue of energy conservation in WSN has taken a verge and many solutions with Data-driven approaches are proposed. But here data compression is used for energy conservation. At one end encoding is applied and the data is presented at another end in encrypted form, which can be easily recovered by the process of decryption. After compressing lesser bits [2][3] are required for representing the original value, which can be applied in wireless sensor network before sending the data to sink. Deflate is lossless compression algorithm which utilizes the capabilities of LZ77 and Huffman for performing compression. The Energy Efficient Deflate (EEDeflate) algorithm is proposed for improving the lifetime of wireless sensor network by minimizing the energy consumption, packet delay, enriching the compression ratios and throughput of network. The rest of the paper is arranged as Related Work, Proposed Work, discussion and Conclusion followed by acknowledgement.

## 2. Related Work

Jonathan et al. [4] proposed a fast and efficient lossless adaptive compression scheme (FELACS) for WSN. This is faster as it generates coding in less time, it utilizes low memory, it has low complexity and effective in handling packet losses.

Tommy Szalapski and Sanjay Madria [5] presents Tinypack compression, which is a set of energy-efficient methods. These methods achieves high compression ratios, reduce the latency, storage and bandwidth usability. Here Huffman compression is used, which in turn helps in consumption of bandwidth in more precise way.

Mohamed Abdelaal and Oliver Theel [6] introduced the innovative and adaptive data compression technique using Fuzzy transform for decreasing the bandwidth, memory and energy. Using this 90% of power can be saved with good compression ratio.

Emad et al. [7] proposed a approach for minimizing the power consumption spent in the communication which takes place between sensor and sink. The approach is “Model-based Clustering” approach which works on the principles stated by mixture-model based clustering.

According to Xi Deng and Yuanyuan yang [8] suggestions compression is not always capable of performing the task of reducing packet delay in WSN rather it is a combination of both network and hardware conformations. They had proposed the adaptive approach by compressing the data whenever it is required based on online decisions. Massimo et al. [9] presented the adaptation schemes for the lossless compression algorithm i.e. Lossless Entropy Compression (LEC) by applying the concept for rotating the prefix–free tables.

Mohammad et al. [10] proposed second mode of Deflate algorithm which is appropriate for real-time applications and thus the improvement is done by adding the advantage of modifying the table adaptively depending on the input provided.

Danny Harnik et al. [11] provides the implementation of deflate by making it fast by utilizing the fast version of LZ77 (LZ4 package) and Huffman (Zlib).

Wu Weimin et al. [12] worked on the drawbacks of Deflate i.e. memory overflow and presented the solution to this by exchanging or replacing the Huffman encoding with Marchov chain instead of changing with the static and dynamic Huffman.

Ganjewar et. al [19] proposed threshold based data reduction technique. They proved that considerable amount of energy can be saved using TBDRT data reduction technique in WSN.

## 3. Proposed Work

As per the requirement of the application, WSN can be deployed for monitoring the surrounding. The sensed data is transmitted by sensor nodes to the sink by consuming more amount of energy as compared to other to the energy consumed for other operations like sensing and processing. According to review done in this field, to transmit a single bit of data 0.4μJ of energy is required. Transmission in WSN is quite costly in terms of energy. Data compression can be used as one of the

major solution to reduce the energy, which is require for transmission. This will help in reducing the data to be transmitted, makes less energy usage for sending that data.
Proposed work includes the combination of compression algorithm and fuzzy logic applied after compression. Performance analysis is done by calculating energy consumption (in Joules), compression ratio, end-to-end delay (in seconds) and throughput (in kbps).

*A. Metrics taken for performance analysis*

*1. Energy Consumption*

Consumed energy is calculated as a difference of, initial energy and balance final energy of the network. The % energy consumption is measured w.r.t. its initial energy. The % energy consumed by the network is calculated with respect to the initial energy of the network. The initial energy of network is the average of initial energy of all the sensor nodes in the network, similarly final balance energy of the network is the average balance energy left in the all the sensor nodes of the network.

% Energy Consumed = (Initial Energy – Final Energy) / Initial Energy *100.

Avg. Energy Consumed = Sum of % energy consumed by all nodes / No. of nodes.

Energy is calculated as $E_d = I * V * T$

Where $E_d$ – Energy to decrease I – Total Current Usage
V – Voltage Supply T – Duration

Remaining Energy is calculated as $E_r = E_r - E_d$

Where $E_r$ – Remaining Energy of Node & $E_d$ – Energy to Decrease

% Energy Consumed = ((Initial Energy – Final Energy) * 100 ) / Initial Energy

Avg. Energy Consumed = Energy Consumed by all the nodes / No. of Nodes.

*2. Compression Ratio:*

Compression ratio is computed by the following formula

Compression Ratio = [ 1 - (packet size / packets received) ] * 100.

Higher values of compression ratio represents the ability of compression algorithm is high.

*3. Throughput:*

Throughput is calculated as the total number of packets delivered per second, i.e. the total number of messages delivered per second.

$$\text{Throughput} = (\text{Total bytes received} \ * 8) / 1000$$

*4. End To End Packet Delay:*

It is used to calculate the time between packet origination and packet reaching time at the destination, if transmission is successful. The delay occurs for finding of route, queuing packet retransmission etc.

$$\text{Packet delay} = \text{Receive time at destination} - \text{Transmit time at source}$$

*B. Proposed Architecture*

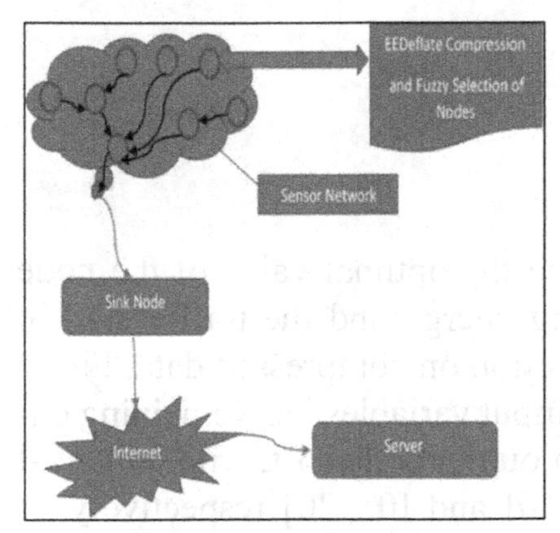


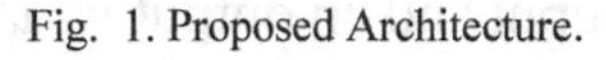
Fig. 1. Proposed Architecture.

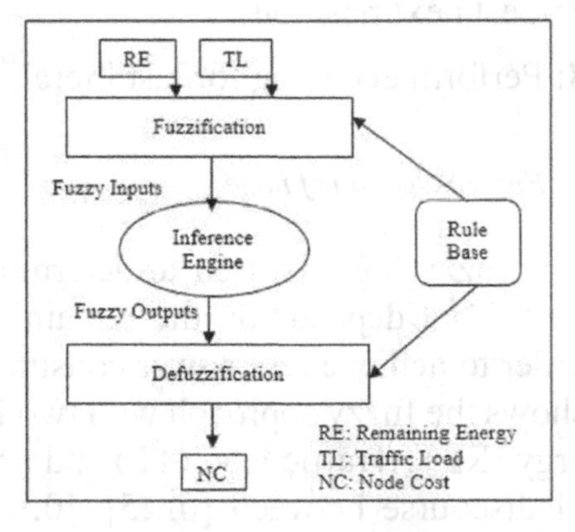


Fig. 2. Fuzzy structure with two inputs (RE, TL) and one output (node cost).

Fast Deflate is proposed using the fast version of LZ77 i.e. LZ4, has capability of performing the compression at very high speed [11]. But it is widely used at machine level, here it is used for compressing sensor data which is generally more repetitive as compared to text data.

EEDeflate works by dividing the data into blocks with specified size and by finding the matching in the block, will takes place by using the hashing. But it gives rise to two issues. First, specify the boundary limit for compression and second, it should work for the remaining data. EEDeflate is more efficient than Huffman and also added fuzzy logic concept [13] for achieving the less power consumption. The data is forwarded to the nodes, which are capable of it, which is determined by calculating the node cost of the nodes coming in the path. Steps of working of fuzzy logic are as follows

1: Defining Hash table.

2: Finding match by searching using hash.

3: Encode literal length.

4: Copy literals.

5: if next match is found then

a) Encode offset  b) Start counting till Min match verified.

c) Encode Match length  d) Test end of chunk

6: Fill table

7: Test next position

8: Perform encoding for last literal.

*C. Fuzzy Selection of node*

The fuzzy logic is used to determine the optimal value of the node cost. That depends on the remaining energy and the traffic load in order to achieve less power consumption on compressed data. Fig. 2 shows the fuzzy approach with two input variables i.e. Remaining energy (RE) , Traffic load (TL) and an output node cost, with universal of discourse between [0…5], [0…10] and [0…20] respectively. It uses five membership functions for each input and an output variable. For the fuzzy approach, the fuzzified values are processed by the inference engine, which consists of a rule base and various methods to inference the rules. The rule base is simply a series of IF-THEN rules that relate the input fuzzy variables and the output variable using linguistic variables each of which is described by fuzzy set and fuzzy implication operator AND.

*D. Working of Fuzzy Logic*

In the path of transmission many nodes are included. It is more energy saving, if we send data to node with maximum energy to avoid packet drops. Thus it can be reliably done by determining the cost of the flow. Using fuzzy logic node cost is determined from remaining energy and Traffic load of the particular flow, after that Perform Fuzzification to determine input values of fuzzy inputs for remaining energy and Traffic load by representing them with fuzzy sets. Calculate the outputs for fuzzy inputs from the rule base i.e. through Inference engine. Then perform the Defuzzification which includes the calculation of Node cost by using the output of rule base and center of output member function.

*E. Experimental Setup*

The simulation of WSN is done as per the assumptions in Table 1.

*F. Experimental Results*

Performance of the system is analyzed on the basis of parameters like Energy Consumption, Compression Ratio, Throughput, End-to-End Delay. The results of the Huffman, Deflate and EEDeflate algorithm are compared with the results of normal transmission. Graphical representation of simulation are shown in Fig. 2,3,4,5 and 6.

## 4. Discussion

Implementation of Huffman, Deflate and EEDeflate algorithms proves that considerable amount of energy can be saved using data compression in WSN. EEDeflate saves more energy.

| Sl. No. | Parameter | Value |
|---|---|---|
| 1. | Topological Area | 1500 x1500m |
| 2. | Number of nodes | 20 to 200 |
| 3. | Topology | Random |
| 4. | Network type | Multihop |
| 5. | Routing protocol | AODV |
| 6. | Packet size | 64 KB |
| 7. | Initial Energy | 30 J |
| 8. | Initial voltage | 3.0 Volts |
| 9. | Simulation time | 25 seconds |
| 9. | Low battery threshold | 0.1% of IE |
| 10. | High battery threshold | 0.15 % of IE |

Table 1 : Simulation Parameter

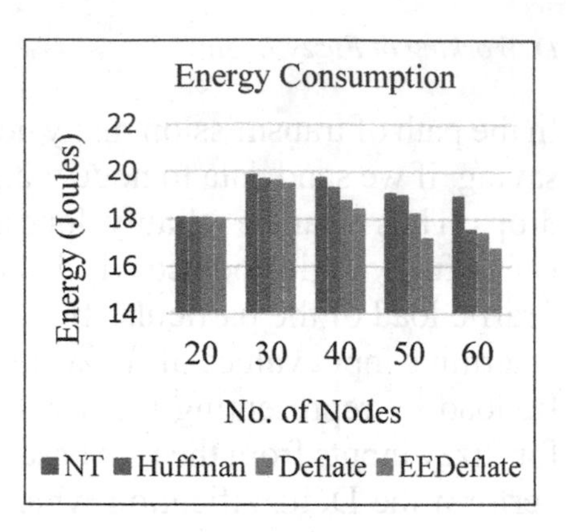


Fig. 3. Energy Consumption Using NT, Huffman, Deflate and EEDeflate.

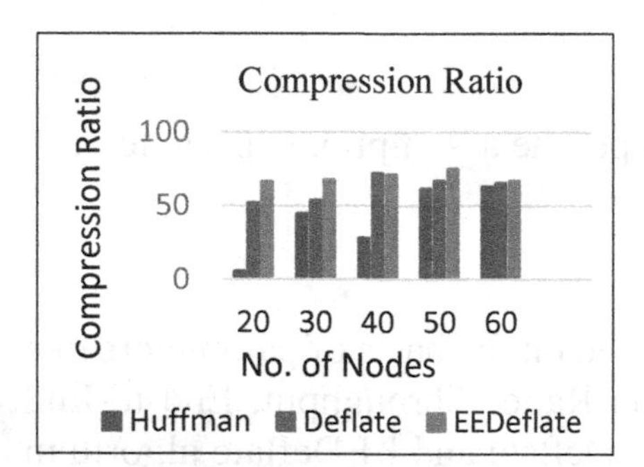


Fig. 4 Compression Ratio Using Huffman, Deflate and EEDeflate.

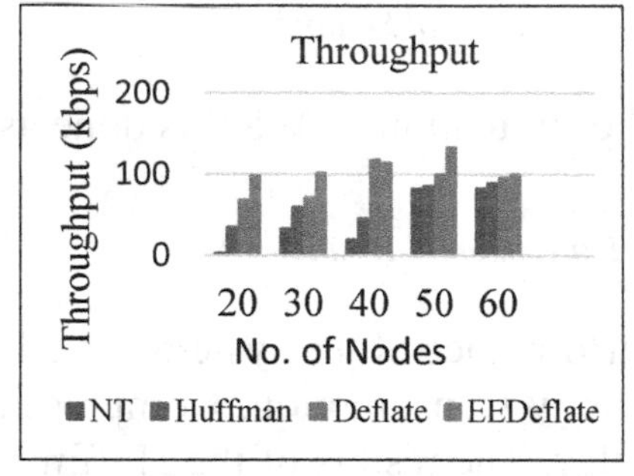


Fig. 5. Throughput Using NT, Huffman, Deflate and EEDeflate.

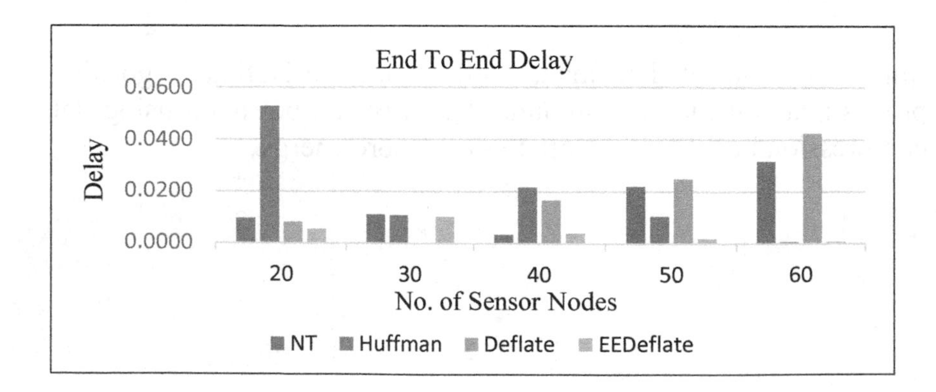


Fig. 6 End To End Delay Using NT, Huffman,. Deflate and EEDeflate

## 5. Conclusion

The Energy Efficient Deflate (EEDeflate) algorithm is efficient than the fast version of Deflate and Huffman algorithm. It is applied for reducing the data to be transmitted in WSN, this in turn reduce the energy consumed for transmission. Along with energy conservation, it focuses on compression ratios, end-to-end delay and throughput. EEDeflate saves 7% to 10% of energy as compared to normal data transmission without compression. It achieves better compression ratio up to 22% - 25% more than Huffman and 8% - 10% more than Deflate compression. This improvements in terms of compression efficiency allows saving energy and therefore extends life of the network.

**Acknowledgments**

Authors would like to appreciate the support and guidance of Dr. B. P. Patil and Dr. S. S. Sonawane. They are also thankful to MIT AOE, Alandi(D.) for providing their resources required for this work.

References

[1] Jennifer Yick, Biswanath Mukheerjee, and Dipak Ghosal,"Wireless Sensor Network Survey", Elsevier, Computer Networks ,2008, pp. 2292-2330.

[2] M.A. Razzaque, Chris Bleakley, Simon Dobson, "Compression in Wireless Sensor Networks: A Survey and Comparative Evaluation", ACM Transactions on Sensor Networks, Vol. 10, No. 1, Article 5, Publication date: November 2013.

[3] Tossaporn Srisooksai , Kamol Keamarungsi , Poonlap Lamsrichan , Kiyomichi Araki ,"Practical data compression in wireless sensor networks: A survey" ,Elsevier,Journal of Network and Computer Applications,35,37–59,2012.

[4] Jonathan Gana Kolo, S.Anandan Shanmugan, David Wee Gin Lim, Li-Minn Ang, "Fast and Efficient Lossless Adaptive Compression Scheme for Wireless Sensor Networks", Elsevier, Computer and Electrical Engineering, June 2014.

[5] Tommy Szalapski · Sanjay Madria," On compressing data in wireless sensor networks for energy efficiency and real time delivery", Springer, Distrib Parallel Databases 31,151–182, 2013.

[6] Mohamed Abdelaal and Oliver Theel," An Efficient and Adaptive Data Compression Technique for Energy Conservation in Wireless Sensor Networks", IEEE Conference on Wireless Sensors, December 2013.

[7] Emad M. Abdelmoghith, and Hussein T. Mouftah," A Data Mining Approach to Energy Efficiency in Wireless Sensor Networks", IEEE 24th International Symposium on Personal, Indoor and Mobile Radio Communications: Mobile and Wireless Networks,2013.

[8] Xi Deng, and Yuanyuan Yang," Online Adaptive Compression in Delay Sensitive Wireless Sensor Networks", IEEE Transaction on Computers, Vol. 61, No. 10, October 2012.

[9] Massimo Vecchio, Raffaele Giaffreda, and Francesco Marcelloni, "Adaptive Lossless Entropy Compressors for Tiny IoT Devices", IEEE Transcations on Wireless Communications, Vol. 13, No. 2, Februrary 2014.

[10] Mohammad Tahghighi, Mahsa Mousavi, and Pejman Khadivi, "Hardware Implementation of Novel Adaptive Version of De flate Compression Algorithm", Proceedings of ICEE 2010, May 11-13, 2010.

[11] Danny Harnik, Ety Khaitzin, Dmitry Sotnikov, and Shai Taharlev, "Fast Implementation of Deflate", IEEE Data Compression Conference, IEEE Computer Society, 2014.

[12] Wu Weimin, Guo Huijiang , Hu Yi , Fan Jingbao, and Wang Huan, "Improvable Deflate Algorithm", 978-1-4244-1718-6/08/$25.00 ©2008 IEEE.

[13] Imad S. AlShawi, Lianshan Yan, and Wei Pan, "Lifetime Enhancement in Wireless Sensor Networks Using Fuzzy Approach and A-Star Algorithm", IEEE Sensor Journal, Vol. 12, No. 10, October 2012.

[14] Ranganathan Vidhyapriya1 and Ponnusamy Vanathi," Energy Efficient Data Compression in Wireless Sensor Networks", International Arab Journal of Information Technology, Vol. 6, No. 3, July 2009.

[15] Tommy Szalapski, Sanjay Madria, and Mark Linderman, "TinyPack XML: Real Time XML compression for Wireless Sensor Networks", IEEE Wireless communications and Networking Conference: Service,Applications and Business, 2012.

[16] S. Renugadevi and P.S. Nithya Darsini, "Huffman and Lempel-Ziv based Data Compression Algorithms for Wireless Sensor Networks", Proceedings of the 2013 International Conference on Pattern Recognition, Informatics and Mobile Engineering(PRIME),Februrary 21-22,2013.

[17] Jaafar Kh. Alsalaet, Saleh I. Najem and Abduladhem A. Ali(SMIEEE), "Vibration Data Compression in Wireless sensor Network",2012.

[18] Mo yuanbin, Qui yubing, Liu jizhong, Ling Yanxia,"A Data Compression Algorithm based on Adaptive Huffman code for Wireless Sensor Networks",2011 Fourth International Conference on Intelligence Computation Technology and Automation, 2011.

[19] Ganjewar Pramod, Sanjeev J. Wagh and S. Barani. "Threshold based data reduction technique (TBDRT) for minimization of energy consumption in WSN". 2015 International Conference on Energy Systems and Applications. IEEE, 2015.

# Secure and Efficient User Authentication Using Modified Otway Rees Protocol in Distributed Networks

Krishna Prakash[1] , Balachandra Muniyal[2], Vasundhara Acharya[3], Akshaya Kulal[4]

[1,2,3,4] Dept.of Information and CommunicationTechnology
Manipal Institute of Technology
Manipal University, Karnataka 576104

Email: kkp_prakash@yahoo.com

**Abstract** Authentication protocol is used for authenticating the two communicating entities in order to build a secure communication channel between them to exchange messages. This authentication is built using exchange of keys. The existing authentication protocol has flaws of mutual authentication and tried to overcome the existing flaw. The paper combines the Otway Rees protocol with a new protocol termed as CHAP (Challenge handshake authentication protocol). It authenticates the communicating entities by using a shared key. Authors also compare symmetric algorithms and choose the best algorithm for encryption.

**Keywords:** CHAP, Otway Rees Protocol, Symmetric encryption.

## 1 Introduction

As the requirement of secure systems came in to existence the authentication protocol arose. It is being used in various fields such e-commerce, banking applications. Otway Rees is one of the authentication protocol used for security of the networks like internet [1]. It is used for proving the identity of the individuals communicating over the network. It is used for preventing the replay attack or eavesdropping. But the recent studies on Otway's Rees Protocol found that it has the problem of mutual authentication. In this Alice can identify Bob but Bob cannot identify Alice where Alice and Bob are sender and receiver. Hence we overcome this problem using a combination of Otway Rees with CHAP [5, 6]. CHAP is defined as a one-way authentication method and it can also be used for two way authentication [4]. In order to achieve the two way authentication a three way handshake is established.



J.M. Corchado Rodriguez et al. (eds.), *Intelligent Systems Technologies and Applications 2016*, Advances in Intelligent Systems and Computing 530,
DOI 10.1007/978-3-319-47952-1_23

## 2 Background

Otway and Rees proposed a shared-key authentication protocol which involves two principals and an authentication server (1987).The protocol is very attractive as it provides timeliness and has smaller number of messages in it. Moreover, it makes no use of synchronized clocks, and is easily implemented as two nested remote procedure calls. When a new key exchange protocol is introduced in a wide area network, security people strive hard in order to ensure that there are no loop holes or flaws in protocol implementation and specification design is done in such a way that it becomes difficult for a hacker to break the security of information transmitted over the network. When selecting a secure key exchange protocol, it is required that the protocol key will never be sent unscrambled in an environment beyond the control, and at the end of the protocol it must also ensure that participating entities are able to obtain that session key without any modification or interference of malicious users.

But in the real time scenario malicious users are able to manipulate the session key without breaking in to the system and even the participating entities remain unaware of situation of session key being reused. The Otway-Rees Protocol makes it possible to distribute a session key $K_{AB}$ (Key between Alice and Bob) created by the trusted server S to two principals A and B. This key will encrypt the information transmitted between these two principals. Here the principal refers to the two participating entities. Sharing this secret key makes it possible to establish a private network which creates a safe tunnel for transmission of information. As well, this protocol helps in authenticating the principals to ensure the integrity of messages and that the key has been correctly distributed to the correct principals. This prevents the key from reaching the malicious user such as those of a hacker who is hijacking a session or performs a man-in-the-middle attack. The actual protocol has one way authentication and the paper proposes an approach to have two way authentications and overcome the flaws of existing Otway Rees protocol.

## 3 Existing Otway Rees PROTOCOL

In this protocol [2], there are two participating entities namely Alice and Bob at source and destination respectively. Trent plays the role of trusted third party. The protocol is as follows:

### 3.1 Otway Rees Protocol

1. Alice ⟶ Bob: M, Alice, Bob $\{R_A,$ M, Alice, Bob$\}$ $K_{AT}$;

2. Bob ⟶ Trent: M, Alice, Bob $\{R_A,$M, Alice, Bob$\}K_{AT}$ $\{R_B\}K_{BT}$,$\{$M, Alice, Bob$\}K_{BT}$;

3. Trent ⟶ Bob: M,$\{R_A,K_{AB}\}$ $K_{AT}$, $\{$ $R_B,K_{AB}\}$ $K_{BT}$ ;

4. Bob ⟶ Alice: M,{$R_A$,$K_{AB}$} $K_{AT}$;

M is called as the run identifier for Alice and Bob in order to keep the track of run between them .$R_A$ and $R_B$ are the nonce of Alice and Bob respectively. $K_{AT}$ is the shared key between Alice and Trent. $K_{BT}$ is the shared key between Bob and Trent. At the end of the protocol they establish a shared session key $K_{AB}$. In the first message, Alice sends to Bob, the run identifier , encrypts and sends the nonce with shared key of the Alice with Trent. In the second message, Bob sends to Trent the run identifier with identities of Alice and Bob and also sends the nonce and associated identities encrypted with shared key with Trent respectively. In the third message, Trent after receiving the message from Bob is able to retrieve the random number Alice, $R_A$ using its shared key $K_{AT}$ and random number Bob, $R_B$ using its shared key $K_{BT}$. It generates the session key $K_{AB}$. With this information it generates the message above and sends it to Bob. In the last message, Bob receives the message from Trent .It removes the last encrypted part with his shared key, decrypts this sub-message with his key $K_{BT}$, retrieves the session key $K_{AB}$, and sends the remaining part of the message to Alice. Alice is also able to retrieve the session key $K_{AB}$, based on the last part of message send by Bob.

But this protocol had a flaw which was discovered by Boyd and Mao. In this attack Malice begins a conversation with Bob acting as Alice. He then intercepts the message with Bob to Trent. He modifies the identity with his own and does not touch the Bob's first cipher chunk and replaces Bob's second cipher chunk {M,Alice,Bob}$K_{BT}$ with and old chunk {M,Alice,Bob}$K_{BT}$ which he has identified in the previous conversation with Bob. After sending message to Trent, the Trent thinks that the two participating entities are Malice and Bob. But Bob thinks that he is having an conversation with Alice. Bob uses the session key which he thinks that he has shared with Alice but in reality it is shared with Malice. Bob shares the confidential information without being aware that it is Malice. The messages of are as follows:

### 3.2 Attack on Otway Rees protocol.

1. Malice("Alice") ⟶ Bob : M, Alice, Bob {$R_M$, M, Malice, Bob}$K_{MT}$;
2. Bob ⟶ Malice("Trent "): M, Alice, Bob {$R_M$,M, Malice, Bob}$K_{MT}$, {$R_B$}$K_{BT}$,{M, Alice, Bob}$K_{BT}$;

2'.Malice("Bob") ⟶ Trent: M, Malice, Bob {$R_M$,M, Malice, Bob}$K_{MT}$, {$R_B$}$K_{BT}$,{M,Malice,Bob}$K_{BT}$;

3. Trent ⟶ Bob: M,{$R_M$ ,$K_{MB}$} $K_{MT}$, { $R_B$,$K_{MB}$} $K_{BT}$ ;
4. Bob ⟶ Malice("Alice"): M,{$R_M$,$K_{MB}$} $K_{MT}$;

## 4 Proposed Model

In this model, an approach to integrate the CHAP Protocol[3] with Otway Rees in order to overcome the problem of mutual authentication is proposed by using the one way authentication property of CHAP Protocol. In the existing Otway Rees protocol, Alice is able to identify Bob but Bob is not ensured if it is communicating with Alice only. We use CHAP Protocol to establish a secure session between Alice and Bob. This prevents any hacker from misusing the system by masquerading as the participating entities.

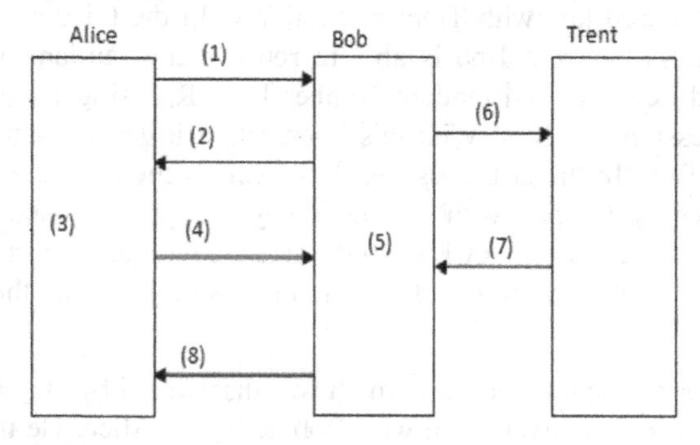


**Fig. 1.** Proposed Model for Modified Otway Rees Protocol

### 4.1 Modified Otway Rees Protocol Messages

(1)Alice ⟶ Bob: M, Alice, Bob $\{R_A$, M, Alice, Bob$\}$ $K_{AT}$
(2)Bob ⟶ Alice: $R_B$
(3)Alice computes: Ha=Hash ($K_{AB}$, $R_B$)
(4)Alice ⟶ Bob: $H_a$

(5) Bob computes the Hash value again and compares it with Ha sent by Alice. Bob computes Hash ($K_{AB}$,$R_B$)=Ha. The Process proceeds only if the two hash values are equal.

(6)Bob ⟶ Trent: M, Alice, Bob $\{R_A$, M, Alice, Bob$\}$ $K_{AT}$, $\{R_B$, M, Alice, Bob$\}$ $K_{BT}$
(7)Trent ⟶ Bob: M,$\{R_A$,$K_{AB}\}$ $K_{AT}$, $\{R_B$,$K_{AB}\}$ $K_{BT}$
(8) Bob ⟶ Alice M,$\{R_A$,$K_{AB}\}$ $K_{AT}$

In the first message, Alice sends to Bob the run identifier,encrypts and sends the nonce with shared key of the Alice with Trent. In the second message, Bob sends to Alice his nonce. The CHAP Protocol is used for mutual authentication. In the third message, Alice computes the hash value of shared key between Alice and Bob and nonce of Bob($K_{AB}$, $R_B$).In the fourth message, Alice sends to Bob the hash value computed in the previous stage. In the fifth message, Bob computes the Hash value again Hash ($K_{AB}$,$R_B$) and compares it with Ha sent by Alice. Process proceeds only if the two hash values are equal. In the sixth message Bob sends to Trent the run identifier with identities of Alice and Bob and also sends the nonce and associated identities encrypted with shared key with Trent respectively. In the seventh message, Trent after receiving the message from Bob is able to retrieve the random number Alice, Ra using its shared key $K_{AT}$ and random number Bob, $R_B$ using its shared key $K_{BT}$. It generates the session key $K_{AB}$. With this information it generates the message above and sends it to Bob .In the eighth message, Bob receives the message from Trent .It removes the last encrypted part with his shared key, decrypts this sub-message with his key $K_{BT}$, retrieves the session key $K_{AB}$, and sends the remaining part of the message to Alice. Alice is also able to retrieve the session key $K_{AB}$, based on the last part of message send by Bob at last stage.

Using all of the above messages Alice and Bob which are two communicating entities are able to communicate with each other and achieve mutual authentication. They are able to establish a secure tunnel for communication.

# 5 Results

## 5.1 Performance Analysis of three different symmetric encryption algorithms

In this section, the performance of three different symmetric encryption algorithms to perform the encryption are discussed. The performance test is carried out using the eclipse IDE for execution of the algorithms. The parameter used to measure the performance is time of execution. The execution time is in milliseconds. From below table, it is identified that the Advanced Encryption Standard (AES) is the algorithm that has an upper hand over the other two algorithms.

**Table 1.** Performance analysis

| Algorithm | Time(in milliseconds) |
|---|---|
| 1.AES | 1.364 |
| 2.Blowfish | 12 |
| 3.DES | 1.447 |

## 5.2 Discussion

The AES is an alternative for DES algorithm has faster executing capability. Various parameters are exchanged between communicating entities of the model using symmetric encryption technique. The usage of AES in the above case will result in the better performance of the system.

## 6 Conclusion and Future Scope

In this paper, a modified Otway Rees protocol is discussed for achieving mutual authentication. For mutual authentication, the integration of Otway Rees protocol with a new protocol termed as CHAP is done. The limitation of original Otway Rees protocol is addressed using the introduction of CHAP to handle bidirectional authentication. The performance analysis of different symmetric encryption algorithms are carried using execution time. The future scope of the current work may be addressing the same issues of the protocol in asymmetric key cryptosystem. The user authentication issues can also be researched using public key certificate as research topic.

## References

1. Li Chen, "Improved Otway Rees Protocol and its Formal Verification" IEEE Transl. China, pp. 498-501,2008
2. Wenbo Mao Modern Cryptography: Theory and practice, Prentice Hall PTR, 2003.
3. F Paul.Syverson,C Paul.van Oorschot."On Unifying Some Cryptographic Protocol Logics",Proceedings of the IEEE Computer Society Symposium in Security and Privacy in Los Alamitos,1994.
4. D OTWAY, REES. "Efficient and timely mutual authentication", Operating Systems Review, 1987,2(1):8 - 10.
5. W simpson,Request for Comments 1994,PPP Challenge Handshake Authentication Protocol (CHAP),Network Working Group,California,1996.
6. G.Zorn,Request for Comments:2759:Microsoft PPP CHAP Extentions-Version 2,Network Working Group,Microsoft Corporation,2000.

# Enhancing Group Search Optimization with Node Similarities for Detecting Communities

**Nidhi Arora and Hema Banati**

**Abstract** Recent research in nature based optimization algorithms is directed towards analyzing domain specific enhancements for improving results optimality. This paper proposes an Enhanced Group Search Optimization (E-GSO) algorithm, a variant of the nature based Group Search Optimization (GSO) algorithm to detect communities in complex networks with better modularity and convergence. E-GSO enhances GSO by merging node similarities in its basic optimization process for fixing co-occurrences of highly similar nodes. This leads to avoidance of random variations on fixed node positions, enabling faster convergence to communities with higher modularity values. The communities are thus evolved in an unsupervised manner using an optimized search space. The experimental results established using real/synthetic network datasets support the effectiveness of the proposed E-GSO algorithm.

**Keywords:** Community, Group Search Optimization, Node Similarity, Complex Networks.

## 1 Introduction

Representing social relations in the form of complex networks were introduced by Moreno's in its sociometry theory [18] and had seen vide applicability in varied disciplines since then. Sociometry is a field of study to quantify the relationships/interactions among entities of a system to build social interaction networks or graphs and apply graph theoretical concepts for their analysis. Entities in these networks are represented as nodes and their relationships are represented as edges.

Presence of densely connected hidden communities of nodes is inherent in such networks due to behavioural interdependence. Communities represent groups of

1

Nidhi Arora
Kalindi College, University of Delhi, Delhi, India. e-mail:nidhiarora@kalindi.du.ac.in

Hema Banti
Dyal Singh College, University of Delhi, Delhi, India. e-mail:hemabanati@dsc.du.ac.in

 
J.M. Corchado Rodriguez et al. (eds.), *Intelligent Systems Technologies and Applications 2016*, Advances in Intelligent Systems and Computing 530,
DOI 10.1007/978-3-319-47952-1_24

of people/entities sharing common interests which can be intuitively related to high level of collaborations with in the group (dense edge connections) as compared to collaborations outside their group. Detecting communities can reveal various hidden mechanisms prevailing through networks and can support various strategic decisions. Communities need to be fetched from a network by applying suitable community detection (CD) algorithm which aims at partitioning the nodes of the network into densely connected subsets/clusters. A more formal method of finding communities [23] is based on dividing the nodes of a network in various subsets: the summations of overall in degrees (edges with in community) of all the nodes in all the subsets are maximized and their respective summations of out degrees (edges outside community) are minimized. Various quality metrics which measure the strength of in degrees vs. out degrees for community subsets are also proposed in the literature [27] Finding communities by maximizing in degrees/minimizing out degrees is defined to be NP hard [26] making heuristic algorithms a natural way of solution to CD problem.

Girvan and Newman [11] first proposed heuristic divisive algorithm based on an edge betweenness metric. At each level the edge with the highest edge betweenness is removed. The drawback of this approach is the absence of stopping criteria. Later Newman [19] proposed a heuristic agglomerative "Fast algorithm" to evolve communities by optimizing a modularity metric and reported communities automatically having highest modularity value. Clauset, Newman and Moore (CNM) algorithm [8] improves the "Fast algorithm" by stopping at a point where change in the modularity shows negative values. Some other heuristic based algorithms are [5], [6], [25].

Recently various nature based soft computing heuristic approaches (evolutionary/swarm) have been applied in the CD problem domain by various researchers. These include Memetic algorithm [10], Genetic algorithm [1]; Differential Evolution [28], ABC algorithm [13] to name a few. These approaches have been able to provide optimal/near optimal solutions for communities to a considerable extent as compared to basic heuristic approaches. To improve the optimality of the results, various domain specific enhancements and modification in the basic nature based optimization approaches have also been tested recently. These include Modified PSO [12], Modified GSO [3], Enhanced Firefly [2] and others [4], [7], [21]. Hence experimentation with newer modifications /enhancements in the basic evolutionary algorithms to heuristically reduce searching space and improve convergence/optimality provides a ground for further research.

This paper proposes a nature based CD algorithm called Enhanced Group Search Optimization (E-GSO) algorithm by enhancing Group search Optimization (GSO) algorithm with node similarities. E-GSO experiments with a novel heuristic to incorporate node similarities in the evolutionary/ swam heuristics to detect optimal communities. The intuition for enhancement is to reduce random evolutions and improve convergence. GSO algorithm is chosen for evolutions due to its efficiency to optimize a quality function using small population by applying efficient group food searching strategies of animals. With the proposed enhancement, the E-GSO algorithm is able to provide community solutions with better optimality and convergence as compared to GSO /Mod-GSO and many other CD approaches.

The following section presents the basic GSO algorithm followed by proposed E-GSO algorithm.

## 2 Basic Strategies: Group Search Optimization

Group Search Optimization (GSO) algorithm [14] is a nature based algorithm based on the swarm food searching behaviour of animals. The optimization process is divided into three main phases: Initialization, Evolving and Reporting. Animals are randomly placed in an n-dimensional searching space during initialization phase by using randomly initialized n-dimensional position vectors. The fitness of an animal is dependent on the optimality of its position vector in the search space. Animals are allotted jobs as per their fitness as follows: Producer (fittest), 80 percent Scroungers (average fit) and 20 percent Rangers (least fit). All animals have head angles to define their optical search area. In the evolving phase, Producer, the leader of the group scans for three better locations (in terms of fitness) in a searching bout as described in [14]. On finding a better fit location, producer relocates to improve its fitness. Scroungers, the followers copy producer by doing a real point crossover with the producer's. Rangers, the random movers randomly move to new locations in the search space. The improved members merge again for the selection of new producer/ scroungers/rangers and repeat next cycle of food searching. The producer vector is reported as output in the reporting phase after the completion of searching bouts.

## 3 Node Similarity based Group Search Optimization

The proposed Enhanced Group Search Optimization algorithm (E-GSO) is a discrete variation of the swarm based Group Search Optimization algorithm [14]. It incorporates node similarities in the evolutionary process of GSO to reduce the amount of search space movements while optimizing modularity [19] to detect communities. Modularity measures the difference in the actual strength of with in community edges of detected communities w.r.t. a random model. A higher modularity value signifies dense communities. E-GSO uses ZLZ node similarity index to measure node similarities.

### 3.1 Node Similarity Index

Node similarity index is a metric which measures the level of resemblance or closeness between a pair of nodes in a network. Among many node similarity indexes [20] Zhou-Lü-Zhang (ZLZ) node similarity index [30] is an efficient index. It considers two nodes to be similar if they share more number of common neighbours as shown in equation 1. For any pair of connected nodes 'i' and 'j' in a network N, ZLZ similarity index value $Z_{ij}$ is calculated by performing summations over the inverse of the degrees of all common neighbouring nodes of 'i' and 'j'.

$$Z_{ij} = \begin{cases} \sum_{X \in Nei(i) \cap Nei(j)} \frac{1}{K(X)} & \text{if } i, j \text{ are connected} \\ 0 & \text{otherwise} \end{cases} \quad (1)$$

Zhou et al. [30] conducted a comparative analysis (Table 1) amongst various node similarity metrics, on two real networks (Karate (RN1) and Football (RN2)) and two artificial networks (AN1-1000 nodes and AN2-5000 nodes). The study which was done on the basis of CPU time and NMI levels (a metric for measuring level of similarity of extracted communities in comparison to real communities) established the efficiency of ZLZ similarity metric over other similarity metrics. The potential of revealing the most similar communities in the least time motivated us to use ZLZ similarity metric with in GSO evolution for detecting communities, in the paper.

**Table 1.** Node similarity metrics comparative analysis

| Metric | NMI Levels | | | | Metric's CPU time(in ms) | | | |
|---|---|---|---|---|---|---|---|---|
| | RN1 | RN2 | AN 1 | AN2 | RN1 | RN2 | AN1 | AN2 |
| Jaccard | 0.906 | 0.885 | 0.970 | 0.914 | 20 | 30 | 955 | 8786 |
| CN | 1 | 0.864 | 1 | 0.997 | 16 | 16 | 514 | 1975 |
| Salton | 1 | 0.922 | 0.992 | 0.983 | 4 | 4 | 542 | 2395 |
| Sorenson | 0.914 | 0.938 | 0.979 | 0.937 | 5 | 5 | 542 | 2371 |
| ZLZ | 1 | 0.946 | 1 | 1 | 5 | 5 | 505 | 1950 |

### 3.2 The proposed algorithm

The proposed E-GSO community detection algorithm calculates node similarities initially for all the nodes and stores them in a similarity vector. A threshold $\varepsilon$ is used to fix the level of acceptable similarity value for nodes to fix their co-occurrence in the similarity vector. This similarity vector is then used in the evolution phases of GSO algorithm. Pseudocode of the E-GSO algorithm is given in fig. 1.

**Fig. 1.** E-GSO Pseudocode

***Input***: *- Network N (n nodes, m edges), Population size= p, No. of generations= max, Similarity threshold= ε.*

***Processing*** *:*

Step 1. Generate a similarity vector for storing highly similar nodes of network N which have similarity level >= threshold ε.

Step 2. Initialization of p individuals to create first generation of population:

a) Create p number of empty individual vectors of size n each: every i[th] position in a vector represents a node in network N.

b) Repeat for each i[th] position in the all P individual vectors:

- If the i[th] node position has valid value in similarity vector then assign it to i[th] position in the position vector V: V[i] =similarity vector[i].
- If the node position has "NF" in similarity vector then use any random neighbour from neighbour set of i[th] node : V[i]=random node (neighbouring nodes[i])

c) Form clusters for all p individuals and calculate fitness.

Step 3. Evolve population for max number of generations:

a) Select Producer, Scroungers and Rangers according to fitness.

b) Let Producer generate three new individuals varying least similar node positions using optical scan and neighbouring nodes.

c) Scroungers improvise themselves by copying Producer vector using crossovers.

d) Rangers generate new individuals randomly.

e) Calculate fitness of all new individuals and merge the best fitted individuals to form next generation.

f) If maximum generations reached than go to step 4 else go to step 3(a).

Step 4. Report the clusters of best fit individual at max generation as final best communities to the output.

***Output:*** *Optimal node communities partitions for network N.*

The main four phases (steps) of E-GSO are described below.

*INPUT*: Network N (n, m): {n is a no. of nodes, m is the no. of edges}, size of population=p, similarity threshold=ε, number of maximum generation=max.

**Step 1.** *Node Similarity Calculation*: ZLZ similarity measure is calculated for each pair of connected nodes using equation 1. Fig. 2 presents the python pseudocode for implementing ZLZ similarity metric. The function takes a network N and similarity threshold ε as an input parameter and returns an instance of a similarity vector. The common neighbours of any two connected nodes are generated by using intersection (&).The degrees of the all the common neighbours are used in the ZLZ equation to quantify the similarity among the two connected nodes. A similarity vector of size n (for n nodes) is created where each index represents a node and will be assigned the number of its highest similar neighbouring node with similarity >= threshold ε otherwise "NF" (not found) is assigned.

Following functions are used in the pseudocode: function neighbours (i) returns the list of connected nodes of a node 'i', function degree (i) returns the degree of a node 'i', function append(x) is used with a vector/array to append an element 'x' in the array and function sort() is used to sort a table on a particular key which stores multiple values/keys in a row.

**Fig. 2.** Python based pseudocode for ZLZ node similarity calculation

```
1.    function similar_nodes (Network N, similarity threshold ϵ ):
2.    { node_list= N.nodes()
3.      similarity_vector = []
4.      for i in node_list:
5.       { table=[] ; N1= g.neighbors (i)
6.        for j in N1:
7.        {  N2=g.neighbors (j); total=0
8.           N3=list(set(N1)&set(N2))
9.           for node in N3:
10.          {  total=total+1/float ( g.degree(node)) }
11.          table.append (j, total) }
12.       table.sort (key=lambda y : y[1], reverse=True)
13.       if table[0][1]>=ϵ than similarity_vector.append (table[0][0])
14.       else  similarity_vector.append ("NF") }
15.   return similarity_vector}
```

**Step 2.** *Initialization*: The population having ‘p’ individuals is generated by using locus based adjacency representation [1, 2].

**Fig. 3.** Locus based adjacency representation and clustering

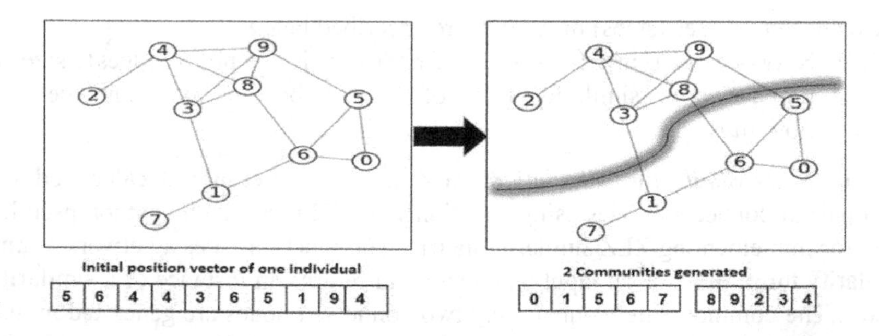


The representation detects number of disjoint communities without fixing them a-priori. For an ‘n’ node network N, each individual is represented using n-dimensional position array $\underset{V}{\rightarrow}$, where each of the $i^{th}$ position and its value represent respective nodes of the network. The initialization strategy is safe as v[i] =j implies node i has a connection with node j in the network as shown in fig. 3. An additional similarity vector look up is performed before assigning values to each index in this array. If similarity vector’s position contains a valid entry, it is given to the position array else if it contains “NF”, value in the position array is assigned randomly from its neighbours.

*Clustering and Selecting*: A linear time iteration through each of the p individual arrays puts nodes into disjoint communities and their modularity values are calculated. Fig. 2 presents the identification of communities in a sample 10 node graph. An initial position array/vector of size 10 (as nodes in the network=10) is shown. A value j on position i represent an edge between them and hence added to the same community. If any one of i or j is present in an already created community than the other is appended to it else a new community set is created to contain them. The process divides the single individual’s vector into multiple community vectors. The clustered community vectors are arranged in decreasing order of their modularity fitness. Out of these, the vector having the best modularity is chosen as producer, next best 80 percent are chosen as scroungers and rest are chosen as rangers.

**Step 3.** *Evolving*: To evolve the current population, evolutionary strategy of GSO is applied for ‘max’ number of generations. This phase is further divided into Producing, Scrounging/Ranging and Merging.

- *Producing*: In this phase producer vector optimizes its fitness by varying its position vector and generating three new position vectors. The new position vectors are generated keeping similarity table values for highly similar nodes and applying evolutionary strategies of GSO [14] to remaining nodes positions. The values generated by the equations are brought in acceptable range of connected nodes by performing a simple modulus with number of connected nodes to generate an index for a lookup in the list of connected nodes.
- *Scrounging and Ranging*: Scroungers optimize by performing single point crossover with the producer [3] instead of performing real point crossover. Rangers perform random variations in their position vectors to generate random position vectors.
- *Merging*: The best optimized ‘p’ position vectors are merged again to generate new set of individuals while discarding rest.

**Step 4.** *Reporting*: The final producer’s communities are reported as output.

## 4 Experimental Analysis

The proposed E-GSO algorithm is implemented in python 2.7 on Intel core I3, 2.40 GHz, 4GB RAM. Initial population size= 48, number of maximum generations=200 and the results are averaged over 20 runs. Python implementation of modularity [19] and NMI metrics [9] are used to compare the quality of E-GSO's generated communities with other algorithms.

*Data Set:* E-GSO was executed on 4 real worlds and 1 artificial GN Benchmark dataset. Real world datasets include: Karate Club Network (34 nodes,78 edges)[29], American College Football Network (115 nodes,613 edges) [11], Dolphin Network (62 nodes, 159 edges)[17], Political Books Network (105 nodes, 441 edges) [15]. Artificial datasets include eleven variations of classic GN Benchmark network( 128 nodes with average degree 16) [11] which are artificially generated using LFR software [16]. The value of mixing parameter $\mu$ for each network is varied from 0.1 to 0.6 (gap 0.5) to allow for change in the the strength of connections inside and outside the communities.

*Algorithms Compared* :The results generated on varied real world and artificial datasets are compared with four non evolutionary :-CNM [8], Blondel Multilevel [6], RB [25] , LPA[24] and four evolutionary algorithms :- GSO[14]; Mod-GSO [3] ; Firefly [2] ; Moga-Net[22] .

## *4.1 Results Analysis*

The actual nodes interconnections in the Karate dataset and the actual communities present are shown in fig. 4.

**Fig. 4.** Friendship network of Karate Club and its actual communities at modularity=0.3582

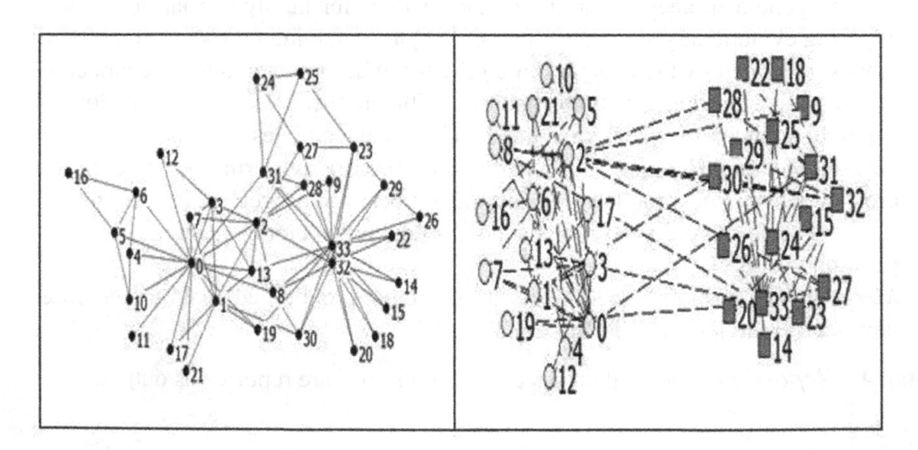

*Parameter tuning of similarity threshold* $'\varepsilon'$ : To tune the value of ZLZ similarity threshold $'\varepsilon'$ , experiments are conducted on different datasets dataset. The results for Karate club dataset are shown below. ZLZ similarity metric calculations for each node is done using function *similar_nodes ()* shown in fig. 1. Table 2 depicts 34 nodes of karate club dataset and their two most similar neighbouring nodes.

To find the optimal value for $'\varepsilon'$ such that only highly similar nodes are fixed together in the position vectors, a set of experiments are performed for values of $\varepsilon$ ranging from 0.5 to 1.0. For each value of threshold, the algorithm was executed 10 times. Table 3 depicts sample (3 runs) for 0.5 to 1.0 values of $\varepsilon$ in karate club dataset.

**Table 2.** Two most similar neighbours: Karate dataset.

| Node | Two most similar nodes with similarity values | Node | Two most similar nodes with similarity values | Node | Two most similar nodes with similarity values |
|---|---|---|---|---|---|
| 0 | 1(2.05), 3(1.16) | 12 | 0(0.16),3(0.06) | 24 | 31(0.33),25(0.16) |
| 1 | 0( 2.05),2( 0.68) | 13 | 0(0.37),2(0.34) | 25 | 31(0.33),24(0.16) |
| 2 | 0(0.93),1( 0.68) | 14 | 33(0.08),32(0.05) | 26 | 33(0.25),29(0.05) |
| 3 | 0(1.16),2(0 .62) | 15 | 33(0.08),32(0.05) | 27 | 33(0.2),23(0.05) |
| 4 | 0(0.58),10(0.06) | 16 | 5(0.25),6(0.25) | 28 | 33(0.16),31(0.05) |
| 5 | 0(0.58),6(0.56) | 17 | 0(0.11),1(0.06) | 29 | 33(0.78),32(0.25) |
| 6 | 0(0.58),5(0.56) | 18 | 33(0.08),32(0.05) | 30 | 33(0.28),32(0.25) |
| 7 | 0(0.37),2(0.34) | 19 | 0(0.11),1(0.06) | 31 | 33(0.41),24(0.33) |
| 8 | 32(0.40),33(.33) | 20 | 33(0.08),32(0.05) | 32 | 33(3.56),8(0.40) |
| 9 | 33(0),2(0) | 21 | 0(0.11),1(0.06) | 33 | 32(3.56),29(0.78) |
| 10 | 0(0.58),4(0.06) | 22 | 33(0.08),32(0.05) | | |
| 11 | 0(0),Nil | 23 | 33(0.58),32(0.30) | | |

Table 3 shows the number of node positions fixed in the similarity vector. The positions in the vectors which are not fixed are represented using "NF" and are evolved using GSO's evolution mechanism. The initial and final best modularities ($Q_{best}$) for each value of $\varepsilon$ are also shown.

**Table 3.** Parameter tuning for similarity threshold parameter $\varepsilon$

| $\epsilon$ | Fixed nodes | Similarity Vector as per $\epsilon$ | Initial $Q_{best}$ | Final $Q_{best}$ |
|---|---|---|---|---|
| 0.5 | 12 | [1, 0, 0, 0, 0, 0, 0, "NF", "NF", "NF", 0, "NF", "NF", "NF", "NF", "NF", "NF", "NF", "N"", "NF", "NF", "NF", "NF", 33, "NF", "NF", "NF", "NF", "NF", 33, "NF"", "NF", 33, 32] | 0.3799<br>0.3874<br>0.3759 | 0.3897<br>0.3897<br>0.3897 |
| 0.6 | 7 | [1, 0, 0, 0, "NF", "NF", "NF", "NF", "NF", "NF", "NF", "NF", "NF", "NF", "NF", "NF", "NF", "NF", "NF", "NF", "NF", "NF", "NF", "NF", "NF", "NF", "NF", "NF", "NF", 33, "NF", "NF", 33, 32] | 0.3973<br>0.4020<br>0.4030 | 0.4197<br>0.4197<br>0.4197 |
| 0.7 | 7 | [1, 0, 0, 0, "NF", "NF", "NF", "NF", "NF", "NF", "NF", "NF", "NF", "NF", "NF", "NF", "NF", "NF", "NF", "NF", "NF", "NF", "NF", "NF", "NF", "NF", "NF", "NF", "NF", 33, "NF", "NF", 33, 32] | 0.4022<br>0.3809<br>0.3990 | 0.4197<br>0.4197<br>0.4197 |

| | | | | |
|---|---|---|---|---|
| 0.8 | 6 | [1, 0, 0, 0, "NF", "NF", "NF", "NF", "NF", "NF", "NF", "NF", "NF", "NF", "NF", "NF", "NF", "NF", "NF", "NF", "NF", "NF", "NF", "NF", "NF", "NF", "NF", "NF", "NF", "NF", "NF", 33, 32] | 0.4022<br>0.4172<br>0.3889 | 0.4197<br>0.4188<br>0.4188 |
| 0.9 | 6 | [1, 0, 0, 0, "NF", "NF", "NF", "NF", "NF", "NF", "NF", "NF", "NF", "NF", "NF", "NF", "NF", "NF", "NF", "NF", "NF", "NF", "NF", "NF", "NF", "NF", "NF", "NF", "NF", "NF", "NF", 33, 32] | 0.3865<br>0.4188<br>0.3943 | 0.4197<br>0.4197<br>0.4197 |
| 1.0 | 5 | [1, 0, "NF", 0, "NF", "NF", "NF", "NF", "NF", "NF", "NF", "NF", "NF", "NF", "NF", "NF", "NF", "NF", "NF", "NF", "NF", "NF", "NF", "NF", "NF", "NF", "NF", "NF", "NF", "NF", "NF", 33, 32] | 0.3800<br>0.394<br>0.4030 | 0.4197<br>0.4188<br>0.4197 |

The final best modularities are best for threshold 0.6, 0.7 and 0.9. However the variations in the initial best modularities are observed to be minimum for $\varepsilon$ value 0.6. This indicates a more optimized initial population leading to faster convergence in this case. Similarly for other datasets the minimum variations in the initial and final best modularities were observed to be for threshold value 0.6, hence the value of $\varepsilon$ is fixed at 0.6 for E-GSO.

E-GSO results (modularities and communities) in the best cases are compared with four non-evolutionary community detection algorithms as shown in table 4 and table 5 respectively. The modularity values obtained by E-GSO in the datasets were found to be much better than the modularities reported by other algorithms. The number of communities detected were also more than the actual communities and comparable to other algorithms.

**Table 4.** Modularity values generated by various CD algorithms

| Method | Karate (Actual=0.3582) | Polbooks (Actual=0.4149) | Football (Actual=0.554) | Dolphin (Actual=0.3735) |
|---|---|---|---|---|
| E-GSO | 0.4197 | 0.5271 | 0.6045 | 0.5285 |
| CNM | 0.3806 | 0.5019 | 0.5497 | 0.4954 |
| Multilevel | 0.4188 | 0.5204 | 0.6045 | 0.5185 |
| RB | 0.4020 | 0.5228 | 0.6005 | 0.5247 |
| LPA | 0.3719 | 0.4983 | 0.6044 | 0.3238 |

**Table 5.** Number of communities generated by various CD algorithms

| Algorithms | Karate (Actual=2) | Polbooks (Actual=3) | Football (Actual=12) | Dolphin (Actual=2) |
|---|---|---|---|---|
| E-GSO | 4 | 5 | 10 | 5 |
| CNM | 3 | 4 | 6 | 4 |
| Multilevel | 4 | 4 | 10 | 5 |
| RB | 3 | 6 | 12 | 5 |

| LPA | 4 | 3 | 10 | 4 |
|---|---|---|---|---|

A detailed comparative analysis of the results of E-GSO for 20 independent executions with four evolutionary CD algorithms is done in table 6 and table 7. The results indicate that the best and the average modularity values of E-GSO are better than the listed algorithms in almost all the datasets. It is worth here to note that the results of E-GSO were higher than GSO/Mod-GSO for 200 iterations (Mod-GSO achieved comparative results on 400 iterations) indicating better convergence and optimality of E-GSO. It is thus observed that the similarity driven node position fixation results in lesser standard deviations and convergence rate for E-GSO.

**Table 6.** Modularity variations: Karate and Dolphin datasets.

| Algorithm | Karate | | | Dolphin | | |
|---|---|---|---|---|---|---|
| | $Mod_{best}$ | $Mod_{avg}$ | Stdev | $Mod_{best}$ | $Mod_{avg}$ | Stdev |
| E-GSO | 0.4197 | 0.4197 | 0 | 0.5285 | 0.5150 | .21e-2 |
| Mod-GSO | 0.4197 | 0.4031 | 0.2e-2 | 0.5277 | 0.5156 | .32e-2 |
| GSO | 0.3845 | 0.3810 | 0.3e-1 | 0.4298 | 0.4013 | 0.2e-1 |
| Firefly | 0.4185 | 0.4174 | 0.1e-2 | 0.5151 | 0.5149 | 0.2e-3 |
| Moga-Net | 0.4159 | 0.3945 | 0.8e-2 | 0.5031 | 0.4584 | 0.1e-1 |

**Table 7.** Modularity variations: Football and Polbooks datasets.

| Algorithm | Football | | | Polbooks | | |
|---|---|---|---|---|---|---|
| | $Mod_{best}$ | $Mod_{avg}$ | Stdev | $Mod_{best}$ | $Mod_{avg}$ | Stdev |
| E-GSO | 0.6045 | 0.6010 | .2e-2 | 0.5271 | 0.5228 | .37e-2 |
| Mod-GSO | 0.5900 | 0.5800 | 0.2e-2 | 0.5255 | 0.5213 | 0.3e-2 |
| GSO | 0.4276 | 0.3890 | 0.3e-2 | 0.4447 | 0.4234 | 0.4e-1 |
| Firefly | 0.6011 | 0.5974 | 0.3e-2 | 0.5185 | 0.5180 | 0.5e-3 |
| Moga-Net | 0.4325 | 0.3906 | 0.1e-1 | 0.4993 | 0.4618 | 0.12e-1 |

Communities in Karate club dataset evolved by GSO and E-GSO are depicted using different geometrical shapes in fig. 5. It is evident that the actual communities shown in figure 4 contained hidden communities within, which were not revealed and reported initially. The two communities represented by circles and squares in actual dataset (fig. 4) are bifurcated further to four communities by E-GSO shown using circles/squares and triangles/checked squares respectively in fig. 5. It is evident that GSO algorithm could only reveal three communities with modularity value 0.3845, missing on significant hidden patterns for better modular bifurcations. E-GSO however revealed better modular partitions of nodes leading to increase in modularity values.

**Fig. 5.** Communities detected by GSO and E-GSO

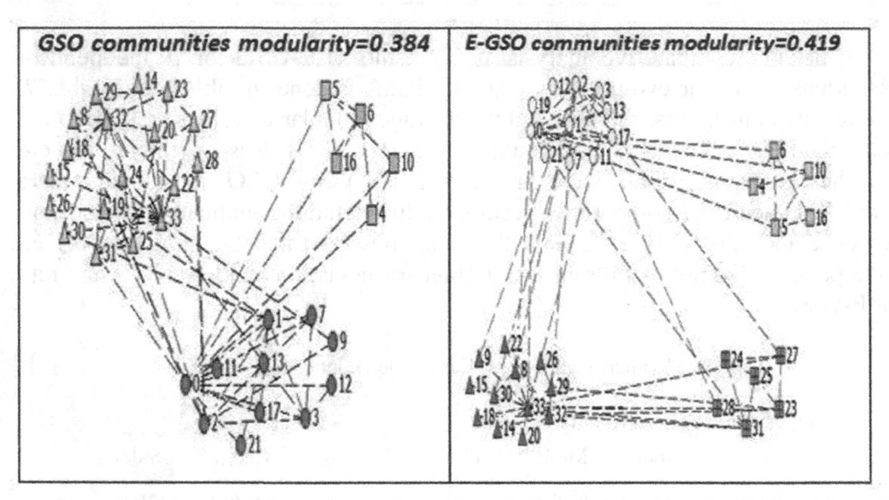


Experiments are also performed on 11 variations of the classic GN benchmark dataset to evaluate and compare the efficiency of E-GSO with standard algorithms.

**Fig. 6.** NMI levels on Artificial GN Benchmark dataset

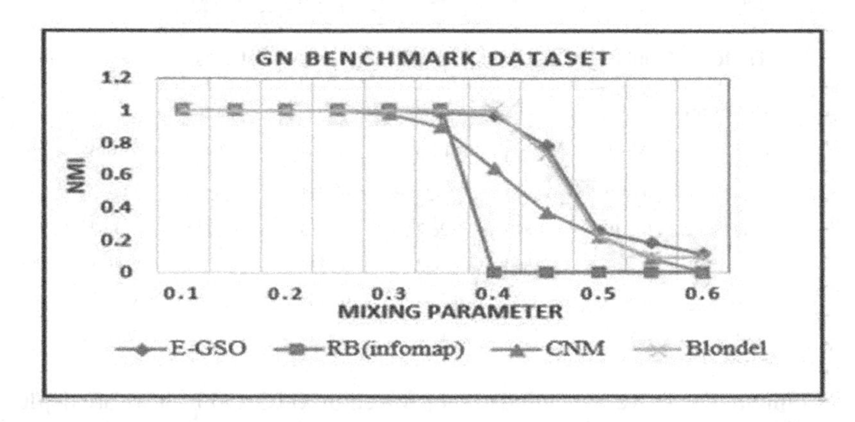


True community structures in artificially generated datasets can be used for comparison using NMI (Normalised Mutual Information) (Danon et al. 2005) similarity metrics. A high NMI value indicates high similarity of generated communities with actual communities. Fig. 6 above reveals that E-GSO is able to evolve communities with high NMI similarity values.

## 5 Conclusion

This paper proposes an enhanced E-GSO algorithm for detecting communities with better convergence and modularities, in complex networks. The algorithm achieves this by incorporating node similarities in the optimization process of GSO algorithm. This fixes the co-occurrence of highly similar nodes in the position vectors of the population. The group food searching optimization of GSO is subsequently applied to evolve the node positions with similarities less than a threshold (obtained after experimental analysis). A reduction in the number of random variations and optimization of search space is hence achieved. The efficiency of the algorithm was verified by comparing its performance with four evolutionary and non-evolutionary each, community detection (CD) algorithms using four real world and eleven artificial datasets. The results revealed a visible improvement in the optimality and convergence speed of E-GSO as compared to many CD algorithms including basic GSO and Mod-GSO. E-GSO reported 0.4197, 0.5271, 0.6045 and 0.5285 modularity values in Karate Club, Political Books ,Football and Dolphin datasets respectively which are higher than most of the compared CD algorithms. Detection of accurate communities with high NMI values in artificial datasets was also observed. The algorithm thus shows a potential in the community detection problem.

## References

1. Agrawal, R.: Bi-objective community detection in networks using genetic algorithm. In: Contemporary Computing (pp. 5-15). Springer Berlin Heidelberg (2011)
2. Amiri, B., Hossain, L., Crawford, J.W., Wigand, R.T.: Community detection in complex networks: Multi–objective enhanced firefly algorithm. Knowledge-Based Systems. 46, 1-11 (2013)
3. Banati H., Arora N.: Modelling Evolutionary Group Search Optimization Approach for Community Detection in Social networks. In: Proceedings of the Third Int. Symposium on Women in Computing and Informatics. pp. 109-117, ACM (2015a)
4. Banati H., Arora N.: TL-GSO: - A hybrid approach to mine communities from social networks. In: IEEE International Conference on Research in Computational Intelligence and Communication Networks, pp. 145-150, IEEE (2015b)
5. Barber, M.J., Clark, J.W.: Detecting network communities by propagating labels under constraints. Physical Review E. 80(2), 026129 (2009).
6. Blondel, V.D., Guillaume, J.L., Lambiotte, R., Lefebvre, E.: Fast unfolding of communities in large networks. Journal of statistical mechanics: theory and experiment. 2008 (10), P10008 (2008)
7. Cao, C., Ni, Q., Zhai, Y.: A novel community detection method based on discrete particle swarm optimization algorithms in complex networks. In: Evolutionary Computation (CEC), 2015 IEEE Congress on. pp. 171-178, IEEE (2015, May).
8. Clauset, A., Newman, M. E., Moore, C.: Finding community structure in very large networks. Physical review E. 70(6), 066111 (2004).
9. Danon, L., Díaz-Guilera, A., Duch, J., Arenas, A.: Comparing community structure identification. Journal of Stastical Mechanics: Theory and experiment. 2005 (09), P09008 (2005).

10. Gach, O., Hao, J. K.: A memetic algorithm for community detection in complex networks. In: Parallel Problem Solving from Nature-PPSN XII . pp. 327-336. Springer Berlin Heidelberg (2012)
11. Girvan, M., Newman, M. E.: Community structure in social and biological networks. In: Proceedings of the national academy of sciences. 99(12), 7821-7826 (2002)
12. Gong, M., Cai, Q., Chen, X., Ma, L.: Complex network clustering by multiobjective discrete particle swarm optimization based on decomposition. Evolutionary Computation. IEEE Transactions on. 18(1), 82-97 (2014)
13. Hafez, A. I., Zawbaa, H. M., Hassanien, A. E., Fahmy, A.A.: Networks Community detection using artificial bee colony swarm optimization. In: Proceedings of the Fifth International Conference on Innovations in Bio-Inspired Computing and Applications IBICA 2014 . pp. 229-239. Springer International Publishing (2014)
14. He, S., Wu, Q. H., & Saunders, J. R.: Group search optimizer: an optimization algorithm inspired by animal searching behavior. Evolutionary Computation. IEEE Transactions on. 13(5), 973-990 (2009)
15. Krebs, V.: http://www.orgnet.com/cases.html (2008)
16. Lancichinetti, A., Fortunato, S., Radicchi, F.: Benchmark graphs for testing community detection algorithms. Physical review E. 78(4), 046110 (2008)
17. Lusseau, D., Schneider, K., Boisseau, O. J., Haase, P., Slooten, E., Dawson, S.M.: The bottlenose dolphin community of Doubtful Sound features a large proportion of long-lasting associations. Behavioral Ecology and Sociobiology. 54(4), 396-405 (2003)
18. Moreno, J.L.: Who shall survive? Foundations of Sociometry, group psychotherapy and socio-drama (1953)
19. Newman, M. E.: Fast algorithm for detecting community structure in networks. Physical review E. 69(6), 066133 (2004)
20. Pan, Y., Li, D.H., Liu, J.G., Liang, J.Z.: Detecting community structure in complex networks via node similarity. Physica A: Statistical Mechanics and its Applications. 389(14), 2849-2857 (2010)
21. Pizzuti, C.: Boosting the detection of modular community structure with genetic algorithms and local search. In: Proceedings of the 27th Annual ACM Symposium on Applied Computing. pp. 226-231. ACM (2012a)
22. Pizzuti, C.: A multiobjective genetic algorithm to find communities in complex networks. Evolutionary Computation. IEEE Transactions on. 16(3), 418-430 (2012b)
23. Radicchi, F., Castellano, C., Cecconi, F., Loreto, V., Parisi, D. Defining and identifying communities in networks. Proceedings of the National Academy of Sciences of the United States of America. 101(9), 2658-2663 (2004)
24. Raghavan, U. N., Albert, R., Kumara, S.: Near linear time algorithm to detect community structures in large-scale networks. Physical Review E. 76(3), 036106 (2007)
25. Rosvall, M., Bergstrom, C. T.: Maps of random walks on complex networks reveal community structure. Proceedings of the National Academy of Sciences. 105(4), 1118-1123 (2008).
26. Schaeffer, S.E.: Graph clustering. Computer Science Review. 1(1), 27-64 (2007)
27. Shi, C., Yu, P. S., Yan, Z., Huang, Y., Wang, B.: Comparison and selection of objective functions in multiobjective community detection. Computational Intelligence. 30(3), 562-582 (2014)
28. Wang, G., Zhang, X., Jia, G., Ren, X.: Application of algorithm used in community detection of complex network. International Journal of Future Generation Communication and Networking. 6(4), 219-230 (2013)
29. Zachary, W. W.: An information flow model for conflict and fission in small groups. Journal of anthropological research. 452-473 (1977)
30. Zhou, T., Lü, L., Zhang, Y.C.: Predicting missing links via local information. The European Physical Journal B. 71(4), 623-630 (2009)

# Performance Tuning Approach for Cloud Environment

Gunjan Lal, Tanya Goel, Varun Tanwar, and Rajeev Tiwari

Department of Computer Science
UPES
gunjan.vk@gmail.com, tgtanya_goel@yahoo.co.in, varuntanwar1@gmail.com, rajeev.tiwari@ddn.upes.ac.in/errajeev.tiwari@gmail.com

**Abstract.** In a cloud environment, the workload that has to be maintained using virtualization is limited by the available hardware resources. A Virtual Machine(VM) allocation policy has a crucial role in the cloud computing life cycle. There are different techniques that can be used for these allocation and scheduling processes, which can impact the performance and working of the cloud environment. In this paper, analysis and implemention of the existing allocation techniques has been performed using Cloud Sim. We have further used an approach for performance tuning of the light load cloud environment,by sorting tasks and VMs on basis of a heuristics function. Then have shown performance tunning on basis of parameters like execution time, make span and throughput of allocated tasks.

**Keywords:** CloudSim; cloud computing; algorithm; performance; virtualization; task allocation policy

## 1 Introduction

According to National Institute of Standards and Technology (NIST) [1], Cloud computing is a model for enabling ubiquitous, convenient, on-demand network access to a shared pool of configurable computing resources (e.g., networks, servers, storage, applications, and services) that can be rapidly provisioned and released with minimal management effort or service provider interaction. The applications and tasks hosted in a cloud computing model have a complex composition, deployment and configuration needs. To model such a Cloud environment, a simulation toolkit, known as CloudSim [2] is used. For task allocation simulation, through CloudSim, we can study how the Cloudlets (the application service or a task that has to be executed on the cloud) are allocated to the Virtual Machines (A virtualized version of actual physical hardware that can be provisioned to the consumer for executing his tasks).[3]



J.M. Corchado Rodriguez et al. (eds.), *Intelligent Systems Technologies and Applications 2016*, Advances in Intelligent Systems and Computing 530,
DOI 10.1007/978-3-319-47952-1_25

### 1.1 What is Task Allocation Algorithm?

Task allocation [4] is the method of process organization when workloads related to one task (cloudlets) are spread among different organizational units (Virtual Machines) that perform own portions of the work. In a cloud model, the hardware resources are ideally allocated to the virtual machines based on the applications to be deployed on them. The policies used to allocate cloudlets to specific virtual machines so that the load can be managed dynamically in the cloud are known as Task Allocation Policies and Algorithms.

### 1.2 Why is optimization of Task Allocation required?

Task Allocation Policies and Algorithms are used to map specific cloudlets to VMs so as to tune the performance of the cloud environment. The performance parameters may include execution time of the cloudlets, makespan, throughput of the system and utilization ratio of the hardware resources. [5] Optimization of Task Allocation results in increased performance of the cloud environment by decreasing the time taken for execution of each cloudlet along with increasing the throughput of the system and the utilization of the underlying hardware resources.

## 2 Literature Review

### 2.1 First-come-first-serve (FCFS)[6,7]

It is the easiest algorithm for task allocation. Processes are allotted on the basis of their time of arrival in the ready queue. As it is a non-preemptive algorithm, when a process has a CPU allotted to itself, it is executed to completion.

FCFS is more predictable than other algorithms since it provides a fair sense of allotment and execution. However, it is poor since the waiting time and response time of some processes may be large. In CloudSim, FCFS is the default technique used for mapping cloudlets to virtual machines.
**Advantages of FCFS scheduling:**

- Easy to understand and operate.
- Useful when transactions are not voluminous.
- Avoids obsolescence and deterioration.

**Disadvantages of FCFS scheduling:**

- Waiting time of processes may increase rapidly.
- Does not consider priority of processes.

### 2.2 Shortest Job First (SJF) algorithm [7,8]

is also known as Shortest-Process-Next (SPN) algorithm. In this scheduling, when a system (virtual machine) is available, it is assigned to the task (cloudlet) that has smallest length among the waiting cloudlets.
Shortest Job First may be:

- Non-preemptive algorithm
- Preemptive (Shortest Remaining Time First) algorithm

**Advantage:**

- The SJF algorithm is suitable for batch jobs in which the execution time is known beforehand.
- Since this type of scheduling results in the lowest average time for a set of tasks, therefore it is optimal.

**Disadvantage:** SJF scheduling requires precise knowledge of the execution time of the task.

### 2.3 Hungarian:

The Hungarian algorithm[9] follows combinatorial optimization that solves the scheduling problem in polynomial time. It was developed in 1955 by Harold Kuhn. Later, James Munkres reviewed the algorithm in 1957 and found that it is (strongly) polynomial. The time complexity of the algorithm was found to be $O(n^4)$, however it was optimized to achieve a running time of $O(n^3)$. Hungarian Algorithm solves matching problems by taking a cost matrix as input and reducing it. In CloudSim, Hungarian Method is used for binding the cloudlet to a particular virtual machine with an objective of attaining the least cost for the entire task. Therefore, each cloudlet can be executed well. [10]

## 3 Problem Formulation

In a cloud environment, the hardware resources are provided as virtual machines on which the tasks, i.e., cloudlets are executed. The Task Allocation Algorithms allocate cloudlets or applications to the virtual machines based on their respective configurations and architectures, so that the load can be managed dynamically and efficiently in the cloud. Optimization of these techniques results in performance tuning of the cloud environment by affecting parameters such as execution time of the cloudlets, makespan, and throughput of the system.[11] [12]

### 3.1 Open Issues

According to recent surveys and studies [11] [12], a cloud environment's performance can be tested on various parameters, on the basis of the primary focus of the service provider.

- The rate of power consumption and the total power consumed are the main areas where an energy efficient system is desired.
- Infrastructure response time helps in dynamic provisioning of the resources.
- The makespan achieved by a technique, when minimized, reduces the power consumed by the complete system.

- Also, the difference between the makespan and execution time when minimized, increases the utilization ratio of that resource. Execution time is thus, inversely proportional to the execution time of that resource.
- The price/performance ratio helps estimate the cost and further design the service catalogue.
- Parameters like audit ability and applicability of Secure Socket Layer (SSL) increase the security measures of the environment.

### 3.2 Problem Statement

Performance Tuning of Cloud Environment using CloudSim is a paper which aims at analyzing and implementing the existing Task Allocation techniques using CloudSim, and further proposes a new technique for the same. The goal of the paper is to tune the performance of the cloud environment based on parameters such as execution time of the cloudlets, throughput and make span of tasks allocated.

### 3.3 Methodology

Over the course of the paper, we have studied and analyzed various task allocation techniques, namely FCFS, SJF and Hungarian. We have also studied the package and class structure of the cloud simulation tool, CloudSim. The techniques have also been implemented on CloudSim and their performance parameters have been compared. Further, a new task allocation technique has also been proposed in this paper which has later been tested for performance on CloudSim.
For the implementation of our goal, we have considered the following performance parameters:

- Execution time of the cloudlets
- Makespan
- Throughput of the system

All the techniques have been tested with the same Cloudlet and VM configurations and compared for performance and the results have been represented graphically.

## 4 Tools Used

### 4.1 Cloud Sim [2]

CloudSim is defined as an extensible simulation toolkit that delivers a technique for modeling and simulation of a Cloud environment. It eases the creation and management of virtual machines and also helps to simulate the datacenter. The chief aim of CloudSim is to offer an extensible simulation framework that aids modeling and simulation of Cloud infrastructures and applications.[13] It provides the packages and classes[14] for DataCenter, DataCentreBroaker, VirtualMachine, CloudLet, VMAllocationPolicy etc.

### 4.2 Eclipse

Eclipse is a platform or an Integrated Development Environment (IDE) which has been intended for web and application development. Composed most part in Java, the platform does not give a considerable measure of end-client usefulness, however, it supports development of integrated features based on plug-ins in various languages including C, C++, JavaScript, Fortran, PHP, Perl, Python, Prolog, Ruby Scala, R, Groovy, Clojure, etc.[15]

## 5 Implementation and Output

In this paper, we proposed a new task allocation technique.

### 5.1 Proposed Solution and its Overview

The proposed technique aims to tune the performance of the cloud environment by reducing the execution time of the cloudlets on VMs
For the implementation of our goal, we have considered the following performance parameters:

- Execution time of the cloudlets

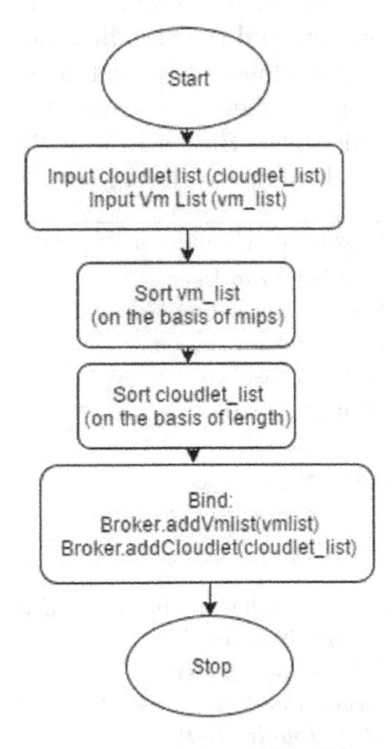


**Fig. 1:** Proposed algorithm Flowchart

- Makespan
- Throughput of the system

The techniques that have been studied so far, namely FCFS, SJF and Hungarian perform task allocation in a cloud environment such that high performance can be achieved.

The proposed technique basically works by allocating a VM of higher configuration to a more demanding cloudlet and a VM of lower configuration to a lighter application. A heuristic function is created, which first arranges cloudlets in decreasing order of the length of the cloudlet. Similarly, the list of VMs are arranged in the decreasing order of their processing power. After the sorting is complete, the cloudlet is assigned to the VM with the corresponding index number in the list of VMs. This ensures proper utilization of the resources as well as reduces the execution time of the application tasks.

### 5.2 Flowchart

Flowchart of proposed technique is shown is Figure 1

### 5.3 Algorithm

Algorithm1 takes a list of VMs and a list of Cloudlets as input from the user binds the cloudlets to the VMs according to the execution time of the cloudlets. It is performed in 3 steps: A, B and C. Part A performs sorting on the VMs list in decreasing order of their computing power i.e. MIPS (Million Instructions Per Second) using Selection Sort and the cloudlets are swapped in Line 7. Part B performs sorting on the list of cloudlets in decreasing order of their execution time using Selection Sort. The cloudlets are swapped in Line 15. Part C executes by submitting the list of VMs to the Broker in Line 20 and the list of Cloudlets in Line 21, thereby binding cloudlets to the VMs.

**Algorithm 1** Proposed Resource Allocation Algorithm

1: **Input:**vm_list and cloudlet_list from User
2: **Output:**Cloudlets bind to Vm according to their execution time
3: **A:** Sort VMs in decreasing order on the basis of their MIPS
4: **for** $i$ in (0 to $vm\_list.length()$) **do**
5:     **for** $j$ in (1 to $vm\_list.length()$) **do**
6:         **if** ($vm\_list[i].mips < vm\_list[j].mips$) **then**
7:             Swap($vm\_list[i], vm\_list[j]$)
8:         **end if**
9:     **end for**
10: **end for**
11: **B:**Sort Cloudlets in decreasing order on the basis of their execution time
12: **for** $i$ in (0 to $cloudlet_list.length()$) **do**
13:     **for** $j$ in (1 to $cloudlet_list.length()$) **do**
14:         **if** ($cloudlet\_list[i].ExecutionTime < cloudlet\_list[i].ExecutionTime$) **then**
15:             Swap($cloudlet\_list[i], cloudlet\_list[j]$)
16:         **end if**
17:     **end for**
18: **end for**
19: **C:** Bind Cloudlet to VM
20: $Broker.submitVmList(vm\_list)$
21: $Broker.submitCloudletList(cloudlet\_list)$

```
Initialising...
Starting CloudSim version 3.0
Datacenter_0 is starting...
Broker is starting...
Entities started.
0.0: Broker: Cloud Resource List received with 1 resource(s)
0.0: Broker: Trying to Create VM #0 in Datacenter_0
0.0: Broker: Trying to Create VM #2 in Datacenter_0
0.0: Broker: Trying to Create VM #1 in Datacenter_0
0.1: Broker: VM #0 has been created in Datacenter #2, Host #0
0.1: Broker: VM #2 has been created in Datacenter #2, Host #0
0.1: Broker: VM #1 has been created in Datacenter #2, Host #0
0.1: Broker: Sending cloudlet 2 to VM #0
0.1: Broker: Sending cloudlet 0 to VM #2
0.1: Broker: Sending cloudlet 1 to VM #1
1250.1: Broker: Cloudlet 0 received
1333.4333333333332: Broker: Cloudlet 1 received
1400.0973333333332: Broker: Cloudlet 2 received
1400.0973333333332: Broker: All Cloudlets executed. Finishing...
1400.0973333333332: Broker: Destroying VM #0
1400.0973333333332: Broker: Destroying VM #2
1400.0973333333332: Broker: Destroying VM #1
Broker is shutting down...
Simulation: No more future events
CloudInformationService: Notify all CloudSim entities for shutting down.
Datacenter_0 is shutting down...
Broker is shutting down...
Simulation completed.
Simulation completed.

========== OUTPUT ==========
Cloudlet ID    STATUS    Data center ID    VM ID    Time    Start Time    Finish Time
    0         SUCCESS         2              2      1250       0.1         1250.1
    1         SUCCESS         2              1      1333.33      0.1         1333.43
    2         SUCCESS         2              0      1400       0.1         1400.1
Proposed Algorithm  Executed!
```

**Fig. 2:** Output screen of proposed approach

### 5.4 Output

The output screen of proposed approach, is shown in Figure 2.

## 6 Performance Testing

In this paper, we have studied and analyzed First-Come-First-Serve, Shortest Job First, and Hungarian Matching techniques. Further, a new task allocation technique has also been proposed in this paper which has later been tested for performance on CloudSim. The following performance parameters have been taken into account:

- Execution time of the cloudlets
- Make span
- Speedup
- Throughput of the system

All the techniques have been tested with the same Cloudlet and VM configurations and compared for performance and the results have been represented graphically. [16]

### 6.1 Configurations of VMs and Cloudlets

The Virtual Machines and the Cloudlets that have been used for experimentation have the configurations [16] as shown in Table 1 and Table 2.

**Table 1:** VM Configuration

| VMid | **0** | **1** | **2** |
|---|---|---|---|
| MIPS | 250 | 150 | 200 |
| Size | 10000 | 10000 | 10000 |
| RAM | 512 | 512 | 512 |
| Bw | 1000 | 1000 | 1000 |
| Pes | 1000 | 1000 | 1000 |
| VMM | Xen | Xen | Xen |

**Table 2:** Cloudlet Configuration

| Cloudletid | **0** | **1** | **2** |
|---|---|---|---|
| Pes | 1 | 1 | 1 |
| Length | 250000 | 350000 | 200000 |
| File Size | 300 | 300 | 300 |
| Output Size | 300 | 300 | 300 |

### 6.2 Execution Time

Execution time is the total time taken by a cloudlet to finish its task on the allotted virtual machine. Per cloudlet execution time of each technique is tested, and shown in Figure 3

### 6.3 Makespan

Time difference between the start and finish of sequence of cloudlets is known as Makespan. The makespan of each technique has been shown in Figure 4

### 6.4 Speed Up Percentage

Speed Up Percentage is the percentage by which the proposed technique runs faster than conventional techniques. The speedup percentage of proposed technique with respect to FCFS, SJF and Hungarian is shown in Table 3.

### 6.5 Throughput

Throughput is the number of processes executed per unit time (in seconds) by a technique. The throughput of each technique is shown in Table 4

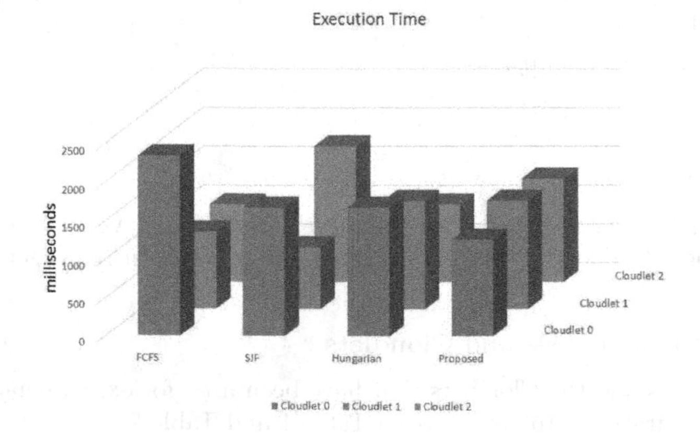


**Fig. 3:** Execution Time of Techniques on each cloudlet

**Table 3:** Speed-up percentage of each technique with respect to proposed technique

| | FCFS | SJF | Hungarian |
|---|---|---|---|
| Speed up of our Approach w.r.t | 39% | 20% | 16% |

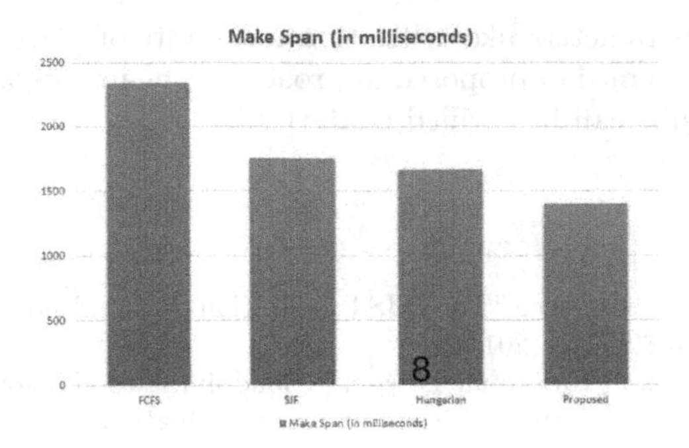


**Fig. 4:** Makespan of each technique

**Table 4:** Throughput of each technique

| FCFS | SJF | Hungarian | Our Approach |
|---|---|---|---|
| 1.2857 | 1.7142 | 1.8000 | 2.1428 |

## 7 Conclusion and Future Work

In this paper, we have studied and analyzed various task scheduling techniques, such as FCFS, SJF and Hungarian, for a Cloud environment. There performance parameters like throughput, make span and speedup is computed on a light load cloud environment. The implementation environment has 3 virtual machines and 3 cloud lets of varying configurations. We have taken a new optimized approach for allocations. It is implemented on same environment and comparison on performance parameters is given as below:

### 7.1 Execution Time

The average execution time per VM of the proposed technique is lesser than the conventional techniques, as shown in Table 5

**Table 5:** Percentage of lesser average execution time per VM for our approach w.r.t. peer techniques.

| | **FCFS** | **SJF** | **Hungarian** |
|---|---|---|---|
| **Our Approach** | 8.07% faster | 5.53% faster | 2.04% faster |

### 7.2 Makespan

The makespan of the proposed technique is lesser than the conventional techniques, by the percentages shown in Table 6

**Table 6:** Percentage decrement in make span of Our Approach w.r.t. peer techniques.

| | **FCFS** | **SJF** | **Hungarian** |
|---|---|---|---|
| **Our Approach** | 39.99% lesser | 20% lesser | 16% lesser |

In near future parameters like utilization ratio, rate of power consumption and total power consumed of proposed approach can be investigated w.r.t peer techniques and results can be verified analytically also.

## References

1. Mell, P. M., and T. Grance. "The NIST definition of cloud computing", NIST special publication 800-145 (2011)
2. Beloglazov, Anton, and Rajkumar Buyya. "CloudSim: a toolkit for modeling and simulation of cloud computing environments and evaluation of resource provisioning algorithms" 2010 Wiley Online Library (wileyonlinelibrary.com) (2010)
3. Perla Ravi Theja, and SK. Khadar Babu. "Resource Optimization for Dynamic Cloud Computing Environment: A Survey" International Journal of Applied Engineering Research, ISSN 0973-4562 Volume 9, Number 24 (2014)
4. Jingsheng Lei, et al. "Cloud Task and Virtual Machine Allocation Strategy in Cloud Computing Environment" Network Computing and Information Security, Second International Conference, NCIS Proceedings (2012).
5. Sandeep Tayal, Tasks Scheduling Optimization for the cloud computing systems, (IJAEST) International Journal Of Advanced Engineering Sciences and Technologies, Volume-5, No.2, (2011).
6. Yahyapour Ramin and Schwiegelshohn Uwe. "Analysis of First-Come-First-Serve Parallel Job Scheduling" Proceedings of the Annual ACM-SIAM Symposium on Discrete Algorithms (1998)
7. S.Rekha, R.Santhosh Kumar. "MJHP - Job scheduling Algorithm for Cloud Environment", International Journal of Technical Research and Applications e-ISSN: 2320-8163 (2014)
8. Hong C, Caesar M, Godfrey P. "Finishing flows quickly with preemptive scheduling" SIGCOMM Comput Commun Rev. (2012).
9. H. W. Kunh, "The Hungarian Method for the Assignment Problem", Naval Research Logistics Quarterly, vol.1, pp. 83 -97 (1955)
10. Kushang Parikh, et al. "Virtual Machine Allocation Policy in Cloud Computing Using CloudSim in Java" International Journal of Grid Distribution Computing (2015)
11. Vijindra, and Sudhir Shenai. "Survey on Scheduling Issues in Cloud Computing." Procedia Engineering 38 (2012)
12. P. T. Endo, Resource allocation for distributed cloud: Concept and Research challenges, IEE, pp. 42-46
13. S.M. Ranbhise, and K.K.Joshi. "Simulation and Analysis of Cloud Environment" International Journal of Advanced Research in Computer Science & Technology (2014)
14. Buyya, Rajkumar, Rajiv Ranjan, and Rodrigo N. Calheiros. "Modeling and Simulation of Scalable Cloud Computing Environments and the CloudSim Toolkit: Challenges and Opportunities." 2009 International Conference on High Performance Computing & Simulation (2009)
15. Desrivieres, J., and J. Wiegand. "Eclipse: A Platform for Integrating Development Tools." IBM Syst. J. IBM Systems Journal 43, no. 2 (2004): 371-83. doi:10.1147/sj.432.0371.
16. S. Majumdar, Resource Management on cloud: Handling uncertainties in Parameters and Policies, CSI Communication, (2011), pp.16-19.

# Enhanced User Authentication Model in Cloud Computing Security

**Prof. (Ms.) Kimaya Ambekar**
Assistant Professor, K. J. Somaiya Institute of Management Studies & Research, kimaya.ambekar@somaiya.edu

**Prof. (Dr.) Kamatchi R.**
Professor, Amity University, rkamatchiiyer@gmail.com

**Abstract**
The rate of technological advancement in the globe has increased rapidly in the last decade. There is a fair rate of enhancement in the various areas like Information Technology, Communication Technology and also on the area of its application like virtualization and utility computing. These all advancement has led to the conceptualization of Cloud Computing. Cloud computing is nothing but the variety of services on pay-as-per-use model. The increased security breaches are the main hindrance for increased use of cloud computing in business sector. There are various security measures are available to provide a personalized security framework based on the business needs. This paper proposes a completely new security model using VPN to provide a secured authentication to the users. The first section discusses on the various characteristics of cloud computing with its extended support to the business model. The final section proposes an advanced security model using VPN and analyses its impact on the cloud computing system.

**Keywords:** Key words: Virtual Private Network, Cloud Computing, Virtualization

## 1 Introduction

The services offered by cloud service provider on rental basis can be accessed over the internet remotely. Companies invest a huge amount as a Capex on software and equipments. According to 80:20 rule, organizations invest only 20% of the capital on the core applications for business. To increase the Opex(Operational expenditure) and decrease Capex(Capital Expenditure) Cloud computing is the best choice [1]. Almost all business sectors have adopted the cloud computing as a prime technology for the simple facts like resource sharing and optimal resources utilization, worldwide access and many more.



J.M. Corchado Rodriguez et al. (eds.), *Intelligent Systems Technologies and Applications 2016*, Advances in Intelligent Systems and Computing 530,
DOI 10.1007/978-3-319-47952-1_26

On the basis of the way of deployment, cloud computing can be divided into 4 major categories: [2]

**2.1. Public cloud:** It is a type of cloud in which services are owned by the CSP and provided to client free or using pay as you go model. Example: Google App is a public cloud in which Google provide services free for some data usage but after that it charges pay as you go model.

**2.2. Private cloud:** In this type of cloud, services are dedicated to a particular organization on site or off site. And it is maintained either by organization or by CSP or by third party.

**2.3. Hybrid cloud:** It can be a combination of more than two deployment models (i.e. public, private, community). It gives advantages of all the models it combined. Example: An organization using private cloud for data storage and networks and using Gmail as mailing server.

**2.4. Community cloud:** This is useful for the organizations whose goal or mission is common. Such organizations can collaboratively use the services given by CSP. These services can be managed by third party or CSP. Example: Educational or healthcare institutes can come together to use private cloud which can be shared by all associated organizations

Cloud computing mainly offers hardware, platform and software. Hardware services include servers, routers, communication channels etc. Platform includes framework, database etc and software include minimal customizable or non-customizable software. Other than these main services many CSPs are also provide database, monitoring, identity etc as a service independently.

## 2 Cloud Security

Cloud computing brought about revolutionary changes in the usage of ICT applications. As it eases the working along with data, applications and/or infrastructure access, it also threatens the users with various drawbacks. Security can be considered as one of the major shortcoming of the cloud computing. Many researchers are already talked a lot about various facets of cloud security. The Security issues in the cloud computing can be categorized on various factors. Some of them can be listed as follows[3][4] :

- **Data Security**

  Data is the most essential asset for any organization. The security of data in transit or in storage, Data theft or loss, Data modification, data segregation, data leaks, deletion of data etc can be various types of threats in data related security in cloud computing.[5]

- **Identity and Access Management:**

  Authentication, authorization and access control can be seen as a subset of identity and access management framework for any cloud solution. Intruder can impersonate a legal user by hacking the authentication details. When penetrating in the cloud, he/she can create a havoc using the user's right.

Also intruder can elevate the privileges which can be a dreadful scenario.

- **Network Security**

  Network configuration and communication are the essentials for cloud computing. Hence, securing the intermediate network is a vital criterion in overall security aspect of cloud computing. Resource sharing, virtualization, Internal and external attackers, both can create security threats to cloud environment. Usage of proper protocols, firewalls, encryption can help in enhancing network security

- **Virtualization related threats**

  Virtualization is the heart of the cloud services. Hypervisor helps to create a logical unit for every user In cloud environment. Managing these virtual machines for security reason is a tedious job. Identification and management of these virtual machines are important when user saves the data on cloud. Sharing the resources may lead to data leakage problem where sensitive information can be disclosed to other unauthorized person. Attacks like cross-VM attack are also very common because the memory resources are mostly shared.

- **Governance, Compliance & Legal issues**

  Governance talks about the user control over the data and also on the security policies. User may lose the control on his/her own data, locality of the data, data redundancy etc. Due to lack of standards over the protocol or data storage, users cannot migrate to other cloud service provider. Sometimes, integrating two or more software for the same purpose from two different CSPs can also create a problem.

  Compliance is how the user can access and assess the cloud. It majorly talks about the SLA (Service Level Agreement) related threats. Auditing the logs for security reasons are the necessity for healthy secured cloud services.

  Legal issues are related law and judicial requirement, may be of organization or a state/ county. Location of the store data is the most essential concern.

## 3 Authentication in Cloud Computing

Authentication is the most basic but most imperative process in any security related processes. This is a practice in which system identifies the individual or user, usually on the basis on username and associated password. Authentication is the first step towards security. It shows whether user is what he/she claims to be or not. [6] This process can reduce the unauthorized and inappropriate access of services. Authorization and identity management are two auxiliary processes which also help in tightening security. Identity management is the process in which user's identity is mapped for the roles and access privileges on objects/resources. Authorization is a process in which users are provided rights/permissions to access a particular object/resource.

## 3.1 Recent trends in Authentication:

Recent researches have shown a huge amount of work done in these three mentioned processes [7]. Various techniques have been followed to construct authentication process more rigorous and stringent. They can be listed as follows:

### 3.1.1. PIN/Password based authentication:

This is the simplest and easiest way for authentication. In this, legitimate user needs to enter a Personal Identification Number i.e. PIN (e.g. Bank ATM/ Online banking) or a password which can be alphanumeric. After entering the same the server will validate the request and then the services will be granted. These PIN/Passwords are generally kept secret to avoid unauthorized access[8].

### 3.1.2. One-Time Password based authentication(two factor/ three factor/Multifactor):

One Time Password (OTP) can be considered as one more level of security.

#### 3.1.2.1. Two Factor Authentication:

Two factor authentications is a mechanism in which a user enters a userId and Password. Then he/she receives a message on the mobile which will have a special password. User may get a list of passwords or there can be some applications that can provide an OTP which will be synced with the provider. After entering the OTP user will be authorized to access the services. For extra security, this OTP can be encrypted by the server using user's public key and decrypted by user using private key. [9]. This mechanism is resilient and impressive in replay or phishing attacks. It can also prevent Man-in-the-middle (MITM) attacks.

#### 3.1.2.2. Three or Multi-Factor Authentication

In three factor authentication, user uses a smart card to get recognized by the system and then uses its credentials i.e. user id & password for authentication. This can also be supported by OTP or pre-scanned biometric templates for extra security.[10] [11]

### 3.1.3. Single Sign On(SSO):

In cloud computing, one cloud service provider may provide multiple services and one user (may be an individual or an organization) can opt for one or multiple services. In normal approach, user needs to be authenticated every time when he/she asks for a new service. Using Single Sign On technique, one central server will be created which will maintain the authentication details of every user for different applications. User, no longer have to use authentication details again and again for the same cloud service provider. E.g. Google appEngine uses SSO for their applications. [12] Researchers are moving towards a technique in which authentication for multiple services from different cloud service provider will be provided by SSO. [13]

**3.1.4. Encryption**

Encryption is very powerful technique which is used in security mechanisms.

**3.1.4.1. Public-Key Authentication**

During the process of authentication an intruder can learn the passwords by number of ways. In this mechanism, a pair of key is generated instead of writing password. Private Key will be known only by the user and a public key will be known by all. User will create a signature using its private key and server will decrypt with users public key for verification. The main problem with this method is that the user have to input the big paraphrase while authentication. To ease this problem, there are many authentication agents like Pageant of PuTTY are available. These agents ask signature at the starting of the session/ machine and later these agents automatically produces signatures. This type of scheme is used as manner by which the problems of non-reputability, authenticity and confidentiality can be solved. [14][15]

Digital signature can also be looked as an authentication technique. This also uses the idea of public key authentication. This is generally used for validation purpose.

**3.1.4.2. Symmetric-Key Authentication**

There is a slight variation in public key authentication method. In this scheme user uses only one key. User shares the same key with the server. When authenticating, user sends a user name and password with a randomly generated number/OTP encrypted using shared secret key. At the server end, it decrypts using user's shared key. If it matches, user is authenticated. This also can be used as two- way authentication systems.[16][17]

**3.1.5. Biometrics based authentication**

Other than external factors, users can use what they are having to authenticate themselves using physiological characteristics. The information related to the user is accumulated by digitized measurements of user's characteristics. At the time of authentication, server extracts important information from the scanned data of the user and tries to match with archived data. If the data matches then user is authenticated. There can be multiple types of biometric techniques available. They can be fingerprints, iris, retina, face, voice etc [13].

# 4 Conventional model:

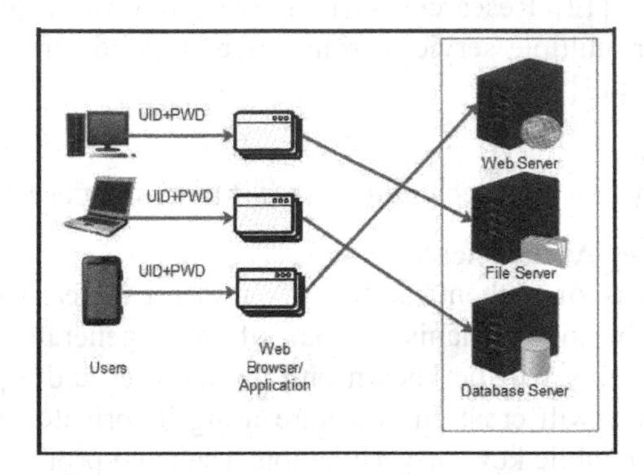


**Fig.1.** The conventional Model in cloud security

In the conventional cloud scenario, services are hosted on one or multiple servers. As we know the architecture of cloud computing, virtualization can be seen as a core of it. Using hypervisors, virtual machines will be configured. Services will reside on these virtual machines and those machines will ultimately reside on servers. Each service will have an associated IP and corresponding name. Whenever user requests for any service, the request will be redirected to the appropriate IP. In such scenario, when user needs to be authenticated, he/she sends user name and password to the server (IP address of Service/server). The server verifies the same for an appropriate match and allows the user to get the access of the same service. [19]

In the conventional model, if an intruder enters in to the network then the IP address of the service and eventually of server will be exposed. At the same time an intruder can get the source and destination IPs which he/she can use for further destruction. User Ids and (hashed) password may also be compromised. When the user credentials get exposed, the server becomes more prone for further destruction. This destruction can go from just passive attacks like eavesdropping to active attacks like data manipulation, elevation of privileges etc. This will also attract attacks on reliability and availability like DOS or DDOS attacks. This can also give rise to attacks like MIMT (Man in the middle), Hijacking (in this host server can be hijacked which create maximum damage), access control to authorized users etc.

# 5 Proposed model:

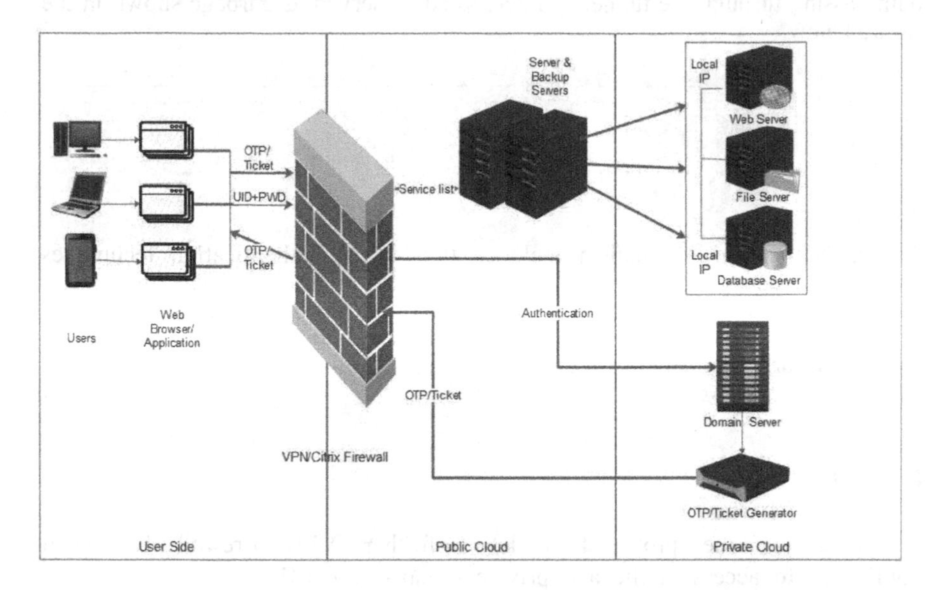


**Fig.2.** The Proposed Model

The Proposed model is divided in three different areas.

- There will be a User Side, Public Cloud Side and Private Cloud Side. This scenario would be more successful with a firewall setup. There will be a VPN Firewall which sits in between User and Cloud Host. [20]
- When a user connects itself to the Internet, a live IP is assigned to the machine. User requests a service using a browser or any equivalent application, first step for authentication will be username and password. User provides username and associated password. A domain server which is responsible for creating and authenticating users will be placed in private cloud for extra security.
- The user credentials will be passed by VPN firewall to the domain server for authentication. If the username and password is authenticated then the OTP (One Time Password)/Ticket generator will be activated. It will generate OTP/Ticket for the session and send it back to the user through the firewall. User will receive the OTP/Ticket which it will send again to the server for authentication.
- One server which contains a list of services and the actual services will be placed in the servers in the private cloud environment. These servers will also have backup servers in the case of failures.
- When the user is authenticated, user will be shown the list of services based on its roll or authorization. When user selects any service present in the list, the server fetches the local IP of the server where the requested service is placed and redirect user automatically.

The change in the firewall prevents insider as well as outsider attacks since the data passing through the tunnel, if intercepted, observed as garbage shown in the figure below.

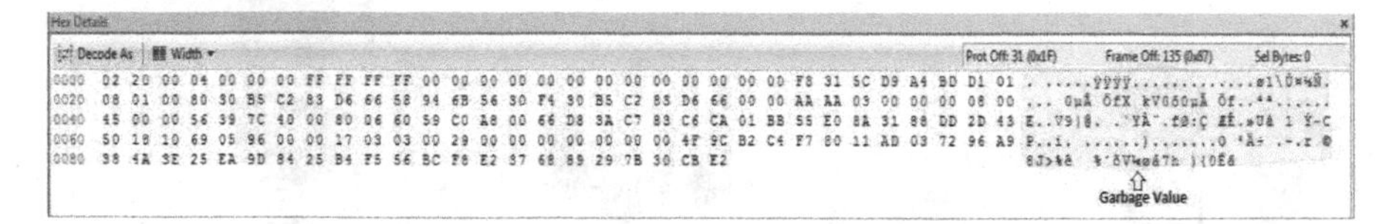


The authentication mechanism includes two factor authentication techniques which can reduce attacks like relay, man-in-the-middle etc. The two factor authentication adds an extra layer of security for cloud computing. The main servers where the services will be placed will be residing on private cloud which also add security quotient in the given scenario.

# 6 Testing:

For testing the proposed model SoftEther VPN firewall has been configured to access public and private cloud servers.[21]

To facilitate public cloud, a server was created which will list all the services authorized for the user.

To enable private cloud, four different servers at different locations have been created for the performance analysis and security perspective. All the Servers having the static IP and domain name. The Servers are as follows

| Service Type | Operating System Used |
|---|---|
| Terminal Service | Windows 2008 R2 |
| SSH/Telnet Service | Redhat Linux |
| FTP/SFTP | Ubuntu |
| HTTP/HTTPS | Windows 2008 R2(IIS enabled) |

Also three users with credentials have formed and configured them in VPN firewall.

- User1 is allowed to use Terminal Service and Web application/Web service (i.e. HTTP/HTTPS & WS)
- User 2 is allowed to use SSH/Telnet Service
- User 3 is allowed to use File transfer Service (i.e. FTP/SFTP Service)

With the help of softEther VPN Client, users can connect and authenticate themselves to the server. After authentication, user will be assigned a dynamic IP. According to their authorization, user will see the list of services.

Using Microsoft Network Monitor, we have analyzed the packets which are going through the VPN Tunnel. By evaluating network data and deciphering the network protocols we found that it was garbage data. From this we can confirm that using VPN firewall the data and network security is enhanced.

Due to limited resources only three users have created for testing purpose but this surely can be tested on larger scale too.

## 7 Performance Analysis

Quality of Service(QoS) is very essential for any networking application. We can analyse the network performance on various parameters. Metrics on which a network performace is analyzed can be response time, throughput, error rate, bit per second rate, availability, bandwidth etc. As a sample evealuation measure, we have taken response time as a parameter for the performance analysis.We have analysed convetional model(without VPN firewall) and proposed model(with VPN firewall) using Solarwind Orion. Solarwind Orion helped to collect raw data from the network for various services explained above. We have calculated the 95$^{th}$ Percentile of average response time for all the services on both the cases(without and with VPN).

- **Case1: Terminal Services Average Response Time:**
  Average Response time for Terminal Services with VPN: 961.66 ms
  Average Response time for Terminal Services without VPN: 987.55ms

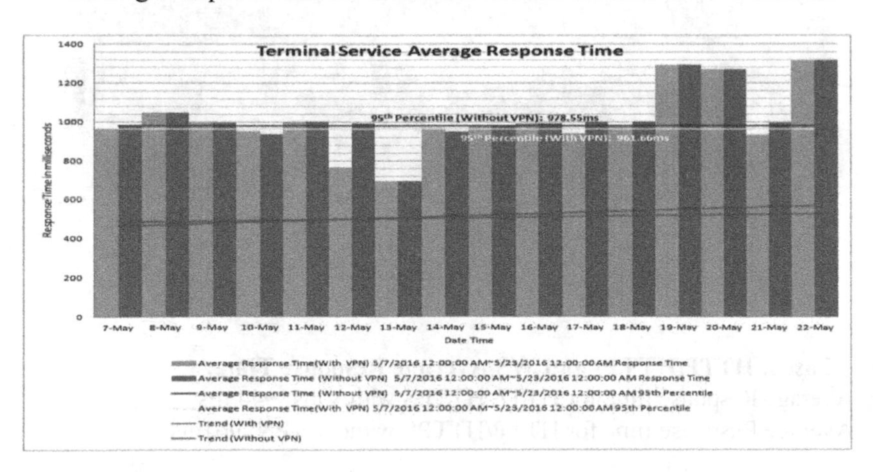


- **Case2: SSH/Telnet Average Response Time:**
  Average Response time for SSH/Telnet with VPN: 921ms
  Average Response time for SSH/Telnet without VPN: 942.4 ms

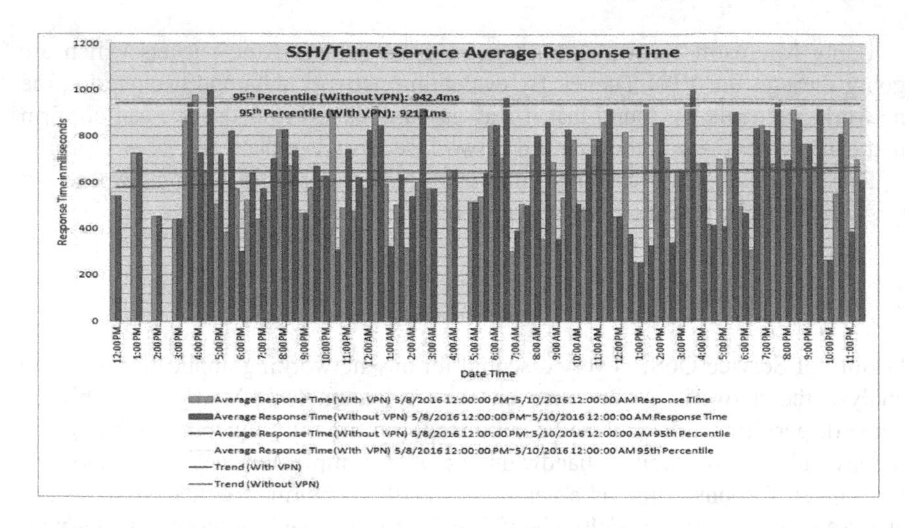


- **Case3: FTP/SFTP Service Average Response Time:**
  Average Response time for FTP/SFTP with VPN: 10.64 ms
  Average Response time for FTP/SFTP without VPN: 14.84 ms

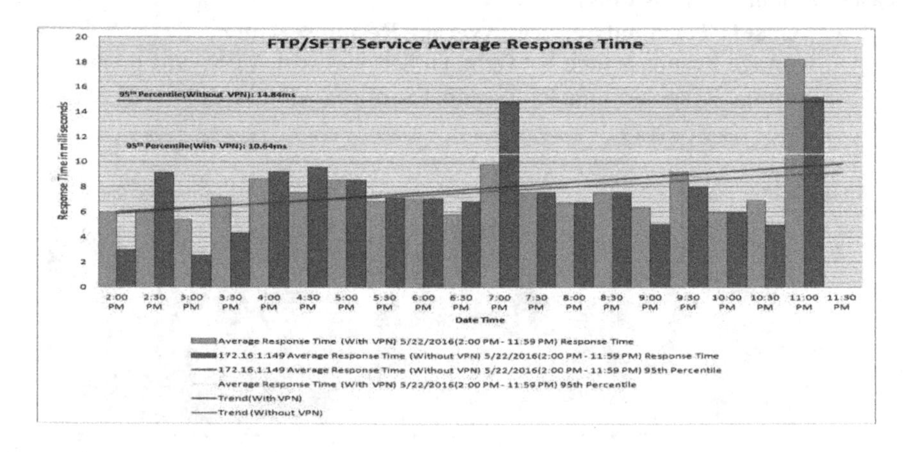


- **Case4: HTTP/HTTPs and WS Average Response Time:**
  Average Response time for HTTP/HTTPS with VPN: 944 ms
  Average Response time for HTTP/HTTPS without VPN: 961ms

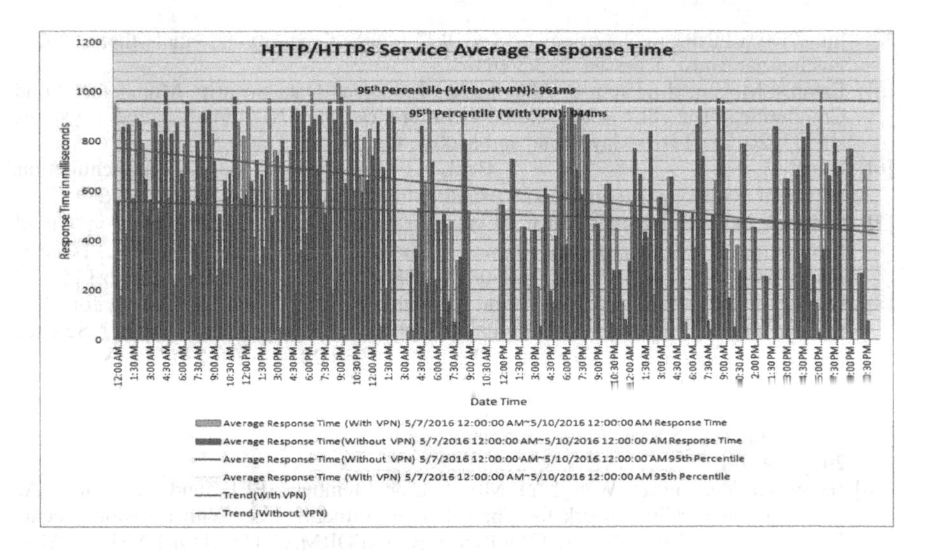


## 8 Conclusion:

Cloud computing can be seen as a new era of computing. This has come with loads of benefits and few drawbacks mainly like security. Security is considered as a major hindrance for the complete adoption of cloud technology in various business sectors. For enhancing security in cloud, authentication plays a major role. From all the above analysis, we can see that VPN firewall gives better response time than conventional firewall. Thus the first part of the proposed model can be seen more productive than the conventional model. At the same time, we can see that the packets on the network can be seen as a garbage value, if sniffed. This has given better security for cloud computing. This kind of security enhancement can lead to prevalent usage of cloud computing in the business sector to achieve its complete utilization.

## References:

[1] Ben Kepes, Cloudonomics: The Economics of Cloud Computing, https://support.rackspace.com/white-paper/cloudonomics-the-economics-of-cloud-computing/, accessed on 30/4/2016

[2] Nelson Mimura Gonzalez et al, A framework for authentication and authorization credentials in cloud computing, 2013 12th IEEE International Conference on Trust, Security and Privacy in Computing and Communications, 978-0-7695-5022-0/13 $26.00 © 2013 IEEE, DOI 10.1109/TrustCom.2013.63

[3] Eric Chabrow, 10 Realms of Cloud Security Services, Computer Security Alliance Foresees Security as a Service, http://www.bankinfosecurity.com/10-realms-cloud-security-services-a-4097/op-1, accessed on 1/5/2016

[4] AditiTyagi , 7 Types Of Security Essentials That Cloud Hosting Demands,

http://www.businesscomputingworld.co.uk/7-types-of-security-essentials-that-cloud-hosting-demands/, accessed on 2/5/16
[5] EmanM.Mohamed, Hatem S. Abdelkader, Enhanced Data Security Model for Cloud Computing, The 8th International Conference on INFOrmatics and Systems (INFOS2012) - 14-16 May Cloud and Mobile Computing Track
[6] Vangie Beal, authentication, http://www.webopedia.com/TERM/A/authentication.html, accessed on 3/5/16
[7] S.Ziyad and S.Rehman, Critical Review of Authentication Mechanisms in Cloud Computing, IJCSI International Journal of Computer Science Issues, Vol. 11, Issue 3, No 1, May 2014, ISSN (Print): 1694-0814 | ISSN (Online): 1694-0784, pg no 145
[8] DeepaPanse, P. Haritha, Multi-factor Authentication in Cloud Computing for Data Storage Security, International Journal of Advanced Research in Computer Science and Software Engineering, Volume 4, Issue 8, August 2014 ISSN: 2277 128X
[9] GeetanjaliChoudhury, JainulAbudin, Modified Secure Two Way Authentication System in Cloud Computing Using Encrypted One Time Password, (IJCSIT) International Journal of Computer Science and Information Technologies, Vol. 5 (3) , 2014, 4077-4080
[10] Jiangshan Yu, Guilin Wang, Yi Mu, Senior Member, IEEE, and Wei Gao, An Efficient Generic Framework for Three-Factor Authentication With Provably Secure Instantiation, IEEE TRANSACTIONS ON INFORMATION FORENSICS AND SECURITY, VOL. 9, NO. 12, DECEMBER 2014, pg no 2302
[11] AmlanJyoti Choudhury1, Pardeep Kumar1, Mangal Sain1, Hyotaek Lim2, Hoon Jae-Lee2, A Strong User Authentication Framework for Cloud Computing, 2011 IEEE Asia -Pacific Services Computing Conference,978-0-7695-4624-7/11 $26.00 © 2011 IEEE DOI 10.1109/APSCC.2011.14
[12] Margaret Rouse, single sign-on (SSO), http://searchsecurity.techtarget.com/definition/single-sign-on, accessed on 2/5/2016
[13] ArvindMeniya, HarikrishnanJethva, Single-Sign-On (SSO) across open cloud computing federation, International Journal of Engineering Research and Application(IJERA), ISSN: 2248-9622, Vol 2, Issue 1, Jan-Feb 2012, pp-891-895
[14] Using public keys for SSH authentication, Chapter 8, http://the.earth.li/~sgtatham/putty/0.55/htmldoc/Chapter8.html accessed on 01/05/2016
[15] Karamjit Singh, IshaKharbanda, Role of Public Key Infrastructure in Cloud Computing,ISSN: 2278 - 7844, ijair.jctjournals
[16] Tom Roeder, Symmetric-Key Cryptography, http://www.cs.cornell.edu/courses/cs5430/2010sp/TL03.symmetric.html
[17] e-Authentication, http://www.infosec.gov.hk/english/itpro/e_auth_method.html accessed on 01/05/2016
[18] ZeeshanJavaid, Imran Ijaz, Secure User Authentication in Cloud Computing, 9778-1-4799-2622-0/13/$31.00 © 2013 IEEE
[19] Dayananda M S, Ashwin Kumar, Architecture for inter-cloud services using IPsec VPN, 2012 Second International Conference on Advanced Computing & Communication Technologies,978-0-7695-4640-7/12 $26.00 © 2012 IEEE
[20] SoftEther VPN Manual, https://www.softether.org/4-docs/1-manual, Accessed on 09/05/2016

# A new discrete imperialist competitive algorithm for QoS-aware service composition in cloud computing

Fateh Seghir[1], Abdellah Khababa[1] Jaafer Gaber[2], Abderrahim Chariete[2], and Pascal Lorenz[3]

[1] University of Ferhat Abbas Sétif-1, Campus El Bez. Sétif 19000, Algeria
seghir.fateh@gmail.com & khababa_abdlh@yahoo.fr
[2] University of Technology Belfort-Montbeliard, 90010 Belfort Cedex, France
jaafar.gaber@utbm.fr & charieteabderrahim@gmail.com
[3] University of Haute Alsace, 68008 Colmar, France
pascal.lorenz@uha.fr

**Abstract.** In this paper, an effective Discrete Imperialist Competitive Algorithm (DICA) is proposed to solve the QoS-aware cloud service composition problem, which is known as a non-polynomial combinatorial problem. To improve the global exploration ability of DICA, as inspired by the solution search equation of Artificial Bee Colony (ABC) algorithm, a new discrete assimilation policy process is proposed, and differently from the assimilation strategy of the original ICA, colonies moved toward their imperialists by integrating information of other colonies in the moving process. To enhance the local exploitation of DICA and to accelerate the convergence of our algorithm, the proposed assimilation process is also applied among imperialists. The performance of the proposed DICA is evaluated by comparing DICA with other recent algorithms, and the obtained results show the effectiveness of our DICA.

**Keywords:** Cloud Service Composition, Quality of Service (QoS), Optimization, Imperialist Competitive Algorithm (ICA)

## 1 Introduction

The **Q**oS-aware **C**loud **S**ervice **C**omposition (QCSC) problem is very important in cloud computing and service oriented architecture, which is a combinatorial multi-objective optimization problem [1, 2]. The QCSC involves the selection of atomic cloud services to construct a composite service in order to solve a complex goal by simultaneously optimizing some objectives, where each objective is related to an attribute's **Q**uality of **S**ervice (QoS) of the selected cloud services, such as cost, response time, availability and throughput... etc.

Traditionally, the QCSC problem has been solved using Mixed/Integer linear-Programming (MILP) approaches [3, 4]. The MILP methods are very strong and can obtain optimal solutions for the solved problems, but these approaches suffer from the problem of scalability entailed by the huge search space of cloud



J.M. Corchado Rodriguez et al. (eds.), *Intelligent Systems Technologies and Applications 2016*, Advances in Intelligent Systems and Computing 530,
DOI 10.1007/978-3-319-47952-1_27

services on the net, which limits the MILP methods to a certain extent. Recently, and with a view to addressing this issue, a new type of methods, namely optimization methods have been proven to be more effective to solve the QCSC problem. This kind of approaches can get near-optimal global solutions with a reasonable response time, so, it has aroused the interest of academia in the past few years. Various studies using optimization approaches have been proposed to solve the QCSC problem, e.g. Particle Swarm Optimization (PSO) [2], Ant Colony optimization (AC) [5], Genetic Algorithm (GA) [6]... etc.

In recent years, an evolutionary algorithm, the Imperialist Competitive Algorithm (ICA), has been put forward by Esmaeil and Lucas [7], which is inspired by the sociopolitical process of imperialistic competition. It has proven to be more effective to solve some academic and engineering optimization problems. An overview is provided in [8]. Referring to the literature reviews, there are only two works about the ICA for solving the QCSC problem proposed by A. Jula et al in [9,10]. In [9], a hybrid approach using ICA and an improved gravitational attraction search is presented to solve the QCSC problem, In the said study only two QoS criteria are considered in the optimized objective function: Execution time and Execution fee. In [10], a Classified Search Space Imperialist Competitive Algorithm (CSSICA) is proposed to tackle the QCSC problem. Due to the huge search space of QCSC's problem, the cloud service repository is classified into three categories using the PROCLUS clustering algorithm, which is based on service time values of all provided single services. The clustering step is extremely helpful in the ICA part of CSSICA insofar as it permits more appropriate solutions to be obtained for the QCSC problem. Compared with other evolutionary algorithms, ICA has attracted the attention of many researchers in the past few years, given that it has a superior performance in many applications [8]. However, the classical ICA presents the population stagnation phenomena caused by the imperialists' absorption policy [11]. Moreover, it is developed to resolve the continuous optimization problems, so, developing a new discrete version of ICA to solve the QCSC problem is our motivation in this study.

In this study, a new **D**iscrete **I**mperialist **C**ompletive **A**lgorithm (DICA) is proposed to solve the QCSC problem through improving the global exploration search of the original ICA and adjusting it to solve the QCSC's combinatorial problem. First, the initial population is randomly generated and classified into two classes: imperialists and countries. Next, to improve the global exploration ability of our DICA, a novel discrete assimilation policy process is proposed. Subsequently, with a view to enhancing the local exploitation of DICA and to accelerate the convergence of our algorithm, the proposed absorption process is also performed among imperialists. Finally, the experimental results show that the DICA is more effective and can obtain near-optimal global solution with an acceptable CPU-time compared to other recent algorithms in the literature.

The remainder of this paper is organized as follows: In section 2 the QCSC problem is described after a succinct introduction of the canonical algorithm ICA. In section 3, the proposed algorithm is explained in details. In section 4, our experimental results are shown. Finally, the conclusion and the future works are given in section 5.

# 2 ICA and the QCSC problem formulation

## 2.1 The imperialist competitive algorithm (ICA)

ICA is a novel evolutionary search algorithm that simulates the process of sociopolitical behavior of countries [7], which has been applied to solve several optimization problems [8]. At the beginning of the algorithm, an initial population of countries is randomly generated, after that each country which represents a unique solution to the solved problem is calculated according to its cost function. Next, some of the best countries are chosen in population as imperialists, whereas the other countries are considered as colonies of theses imperialists. The colonies are dispersed among the mentioned imperialists to construct empires by using a simple assignment procedure. The number of assigned colonies for an imperialist is defined according to its power, where the best imperialist has greater number of colonies compared to weak imperialist. After distributing all colonies among imperialists, each colony starts to move toward its related imperialist. This process of moving is called the ***assimilation or absorption process***. To prevent the population falling into local optimum, like the mutation operator in GA, ICA has a ***revolution process***, in which some of colonies for each empire are affected by making a change of their sociopolitical characteristics. During the movement of colonies, it may be the case that a colony gets the best position (best cost) compared to its relevant imperialist. So a swapping of positions between that colony and its imperialist is applied. This process of exchanging positions between imperialists and their colonies in ICA is referred to as the ***updating positions process***. The ***imperialistic competition process*** augments the survival chance of powerful empires through possession of other colonies whilst it leads to the collapse of the weakest empires through loss of their colonies. During the competition process, the larger empires experience an increase in their power with a concomitant decrease in the power of the weaker ones, and when the latter lose all their colonies, the ***collapsing process*** is invoked by eliminating these empires. All mentioned processes are repeated for each evolution (i.e. iteration) until the ending criterion of the ICA is satisfied.

## 2.2 The QCSC problem formulation

The following notations are used in the rest of this paper:

- $ACSC = \{T_1, T_2, \ldots, T_n\}$, denotes an abstract cloud service composition, where $T_{i(i=1,2,\ldots n)}$ are the elementary tasks that comprise it and its size is $n$ abstract services. For each task $T_i$ , a set $S_i = (CS_i^1, CS_i^2, \ldots, CS_i^m)$ of $m$ concrete services having the same functionalities with different $QoS$ values is associated to it, where $CS_i^j$ represents the $j$th cloud service in the set $S_i$.

- The set $QoS = \{q_1, q_2, \ldots q_r\}$ of $r$ proprieties for each cloud service $CS$ can be categorized into two sub-sets: positive and negative QoS parameters, denoted as $QoS^+$ and $QoS^-$ respectively, where for each positive QoS, larger values indicate higher performance (e.g. throughput) while for negative QoS, smaller values indicate higher performance (e.g. response time).
- The set $GCst = \{Cst_{q_1}, Cst_{q_2}, \ldots, Cst_{q_k}\}$ represents the global constraints defined by the end-user, where $Cst_{q_t}$ with $t \leq k$ and $k \leq r$ is the global QoS constraint over the QoS criterion $q_t$. eg. The response time of the optimal composite cloud service must be less than 20 ms (i.e. $Cst_{Res\ Time} = 20\ ms$).
- For each attribute $q_t \in QoS$, a related user's preference $w_{q_t}$ is given, which is representing the weight value of $q_t$, with $w_{q_t} \in [0, 1]$ and $\sum_{t=1}^{r} w_{q_t} = 1$.

Since several QoS attributes must be optimized simultaneously, the objective of QCSC aims to find the best **C**omposite **C**loud **S**ervice $CCS = (CS_1^{j_1}, CS_2^{j_2}, \ldots, CS_n^{j_n})$ which can maximize the overall positive QoS attributes and minimize the negative QoS attributes by selecting each concrete cloud service $CS_i^{j_i}$ ($j_i \in [1, m]$) associated to each task $T_i$ from the set $S_i$, where these selected services satisfy the global QoS constraints $GCst$. From such a perspective, the QCSC can be formulated as a multi-objective optimization problem. In our study the multi-objective QCSC problem is converted to a single-objective optimization one by using the Simple Additive Weighting (SAW) [12] method which can be written as follow:

$$\begin{aligned} &\max : Score(CCS) = \textstyle\sum_{t=1}^{r} Q(CCS)_{q_t} * w_{q_t} \\ &s.t : \forall t = 1...k \begin{cases} agg(CCS)_{q_t} \leq Cst_{q_t} \text{ if } q_t \in QoS^- \\ agg(CCS)_{q_t} \geq Cst_{q_t} \text{ if } q_t \in QoS^+ \end{cases} \end{aligned} \tag{1}$$

where:
- $Score(CCS)$ is the global score value of the concrete $CCS$.
- $Q(CCS)_{q_t}$ is the normalized aggregate QoS of the $CCS$ in the $t$th attribute.
- $r$ indicates the number of QoS attributes and $k$ is the number of constraints.
- $w_{q_t}$ is the weight value for the $q_t$, which is specified by the end-user.
- $agg(CCS)_{q_t}$ is the aggregate QoS of $CCS$ in the $t$th attribute, which will be defined next in this subsection.
- $Cst_{q_t} \in GCst$ is the user QoS specified constraint related to the $t$th attribute.

The normalized aggregate QoS of the $CCS$ (i.e. $Q(CCS)_{q_t}$) for the negative and positive QoS attributes are given in (Eqs. 2 and 3) respectively:

$$Q(CCS)_{q_t} = \begin{cases} \frac{agg(q_t^{max}) - agg(CCS)_{q_t}}{agg(q_t^{max}) - agg(q_t^{min})} & \text{if } agg(q_t^{max}) \neq agg(q_t^{min}) \\ 1 & \text{if } agg(q_t^{max}) = agg(q_t^{min}) \end{cases} \tag{2}$$

$$Q(CCS)_{q_t} = \begin{cases} \frac{agg(CCS)_{q_t} - agg(q_t^{min})}{agg(q_t^{max}) - agg(q_t^{min})} & \text{if } agg(q_t^{max}) \neq agg(q_t^{min}) \\ 1 & \text{if } agg(q_t^{max}) = agg(q_t^{min}) \end{cases} \tag{3}$$

Where

- $agg(q_t^{max})$ and $agg(q_t^{min})$ are evaluated by aggregating the maximum/minimum values of $q_t$ for each set $S_i$ corresponding to the tasks $T_i$ of $CCS$.

The overall aggregate QoS of each $CCS$ in the $t$th attribute (i.e. $agg(CCS)_{q_t}$) is evaluated according to the type of interconnection structures among the atomic cloud services of $CCS$ (i.e. sequential, parallel, conditional and loop models) as shown in Fig. 1. Table. 1 illustrates the aggregate formulas for some positive and negative QoS criteria [2–4]. As far as simplicity is concerned, only the sequential model is taken into account, where other type of models can be converted into sequential ones by using techniques cited in existing studies[4].

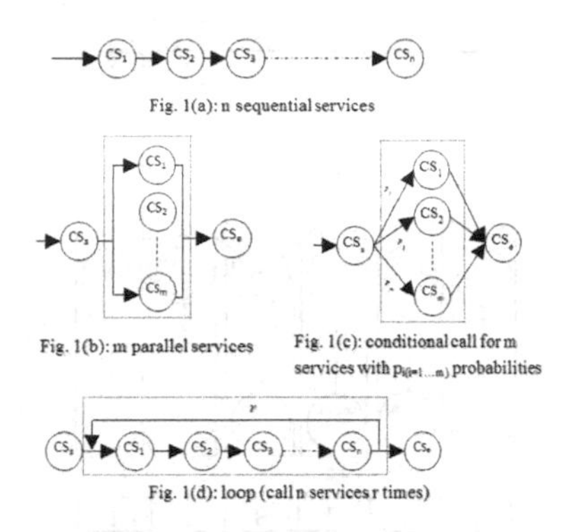

Fig. 1(a): n sequential services
Fig. 1(b): m parallel services
Fig. 1(c): conditional call for m services with $p_{i(i=1...m)}$ probabilities
Fig. 1(d): loop (call n services r times)


**Fig. 1.** Composition models

**Table 1.** Aggregation formulas applied for some QoS attributes

| Parameter ($q_t$) | Type of interconnection structures among cloud services | | | |
|---|---|---|---|---|
| | $n$ sequential services | $m$ parallel services | call service $cs_i$ with $p_i$ probability | call a service $cs_i$ $r$ times |
| $q_t$ = Price | $\sum_{i=1}^{n} q^j_{Price,i}$ | $\sum_{i=1}^{m} q^j_{Price,i}$ | $\sum_{i=1}^{m} q^j_{Price,i} * p_i$ | $r * q^j_{Price,i}$ |
| $q_t$ = Response time | $\sum_{i=1}^{n} q^j_{Time,i}$ | $Max(q^j_{Time,i})$ | $\sum_{i=1}^{m} q^j_{Time,i} * p_i$ | $r * q^j_{Time,i}$ |
| $q_t$ = Availability | $\prod_{i=1}^{n} q^j_{Ava,i}$ | $\prod_{i=1}^{m} q^j_{Ava,i}$ | $\sum_{i=1}^{m} q^j_{Ava,i} * p_i$ | $(q^j_{Ava,i})^r$ |
| $q_t$ = Reliability | $\prod_{i=1}^{n} q^j_{Rel,i}$ | $\prod_{i=1}^{m} q^j_{Rel,i}$ | $\sum_{i=1}^{m} q^j_{Rel,i} * p_i$ | $(q^j_{Rel,i})^r$ |
| $q_t$ = Throughput | $Min_{i=1}^{n}(q^j_{Thr,i})$ | $Min_{i=1}^{m}(q^j_{Thr,i})$ | $Min_{i=1}^{m}(q^j_{Thr,i} * p_i)$ | $q^j_{Thr,i}$ |

For the conditional call of $m$ services $cs_{i(i=1...m)}$ with $p_i$ probabilities: $\sum_{i=1}^{m} p_i = 1$.

## 3 The proposed algorithm (DICA)

In this section, the components of our proposed DICA for solving the QCSC problem are explained in details.

### 3.1 Initialization of empires (initial population)

In the population, each individual representing a solution for the QCSC is called: a country. A country is expressed by an array of integers. Each integer ($j\{j = 1 \ldots m\}$) in the array presents the selected service ($CS_i^j$) from its related concrete services' set ($S_{i\{i=1 \ldots n\}}$). Fig. 2, illustrates an example of a country representing a concrete $CCS$ of six abstract cloud services. In this study an initial population (i.e. initial countries) of size $PopSize$ is generated randomly. In the generated

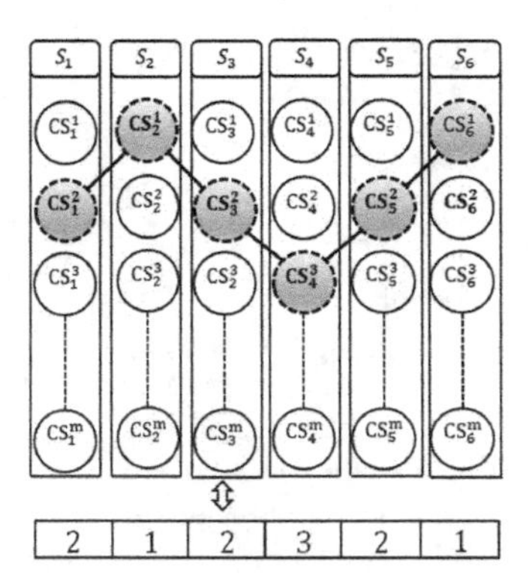


**Fig. 2.** Array integer encoding of solution

population, each country ($cont$) can be evaluated as follow:

$$Cost(cont) = \begin{cases} 0.5 + 0.5 * Score(cont) & \text{if cont is a feasible solution} \\ 0.5 * Score(cont) - Pn(cont) & \text{if cont is an infeasible solution} \end{cases} \tag{4}$$

where, $Score(cont)$ is the score value of $cont$, which is refer to the sum formula given in Eq. 1 and $Pn(cont)$ is introduced in the $Cost$ function of $cont$ to guarantee the feasibility of countries by giving the feasible solutions more cost value than the infeasible ones. The defined penalty function in our paper is given in Eq. 5.

$$Pn(cont) = \begin{cases} 0 & \text{if all constraints are verified} \\ \sum_{t=1}^{k} Csts_{q_t}(cont)^2 * p_{q_t} & \text{If at least one constraint is not verified.} \end{cases} \tag{5}$$

Where

- $p_{q_{t(t=1\ldots k)}} \in [0, 1]$ is the penalty coefficient with $\sum_{t=1}^{k} p_{q_t}=1$
- $Csts_{q_t}(cont)$ is the amount value of the violation biased by $cont$ for the criterion $q_t$ which is calculated as given in Eq. 6.

$$Csts_{q_t}(cont) = \begin{cases} \frac{max(0, Cst_{q_t} - agg(cont)_{q_t})}{Cst_{q_t}} & \text{if } q_t \in QoS^+ \\ \frac{max(0, agg(cont)_{q_t} - Cst_{q_t})}{Cst_{q_t}} & \text{if } q_t \in QoS^- \end{cases} \tag{6}$$

In Eq. 6, $agg(cont)_{q_t}$ is the value of the aggregate QoS of *cont* for the QoS criterion $q_t$ where $Cst_{q_t}$ is the positive or negative constraint value for this attribute.

After evaluating all the countries of the initial population, the best countries of size *PopImp* are selected to be the **imperialists** and the other countries of size *PopCol* where (*PopCol* = *PopSize-PopImp*) represent the **colonies** of these imperialists which are distributed to those imperialists according to their powers. In order to define the initial number of colonies for each imperialist, the roulette wheel selection mechanism of the genetic algorithm is used to distribute the colonies to the imperialists. By way of consequence, the best imperialists have a greater number of colonies compared to the weaker imperialists. An **empire** is formed by one imperialist and its colonies. To allocate the colonies to the imperialists, the following pseudo-code simulating the roulette wheel selection mechanism is applied:

- **Step 1.**
  1. Calculate the normalized cost of each imperialist $imp_n$ as follow:

$$NC_{imp_n} = Cost(imp_n) - min(Cost(imp_s)) \tag{7}$$

  where $Cost(imp_n)$ is the cost of the $n$th imperialist given in Eq. 4 and $min(Cost(imp_s))$ denotes the minimum cost value of all imperialists.
  2. Forming the vector $P = [P_1, P_2, \ldots, P_{PopImp}]$ where $P_n = \frac{NC_{imp_n}}{\sum_{i=1}^{PopImp} NC_{imp_i}}$.
  3. Forming the vector $CP = [CP_1, CP_2, \ldots, CP_{PopImp}]$, where $CP_i$ is the cumulative power of each imperialist which is defined in Eq. 8.

$$\begin{cases} CP_1 = P_1 \\ CP_i = CP_{i-1} + P_i \quad i = 2 \ldots PopImp \end{cases} \tag{8}$$

- **Step 2.**
  **For** $k = 1 \ldots PopCol$
  1. Generate randomly the real number ($rand$) in the interval [0, 1].
  2. Assign the $k$th colony to the $i$th imperialist $imp_i$, such that $CP_{i-1} < rand \leq CP_i$.

  **End For**.

### 3.2 Discrete assimilation policy process

In the ICA algorithm, the assimilation process is modeled by moving colonies of empires toward their related imperialists, the reason for this moving is to increase the power of colonies by assimilating the culture and social-structure of their imperialists. Thus the whole algorithm converges at high speed, as result, the stagnation of population in a local optimum occurs with ease [11]. To improve the global search of colonies i.e. the exploration ability of our algorithm, we perform the moving of colonies in the research space by integrating information of other colonies in the moving process of these colonies toward their associated imperialists. Alternatively worded, colonies operate a simultaneous assimilation of some sociopolitical information from their imperialists and the neighbor colonies of their empires alike. This new process of moving is inspired by the solution search equation of Artificial Bee Colony (ABC) algorithm [13].

The assimilation policy of the original ICA is developed for continuous optimization problems, but the QCSC is a combinatorial problem. So, we must adjust the moving of colonies and make it adapt to solving our problem. The new assimilation process in the discrete situation is given in Eq. 9. Each colony assimilates some sociopolitical characteristics from a random selected colony in the corresponding empire (i.e. the third term on the right-hand side of Eq. 9), and the related imperialist (i.e. the second term on the right-hand side of Eq. 9).

$$NCol_{ij}^{e} = Col_{ij}^{e} + \lfloor \beta\mu(imp_{j}^{e} - Col_{ij}^{e}) + \theta(RCol_{kj}^{e} - Col_{ij}^{e}) \rfloor \tag{9}$$

In Eq. 9, $j$ is a dimension selected randomly from $n$dimensional vector $j \in [1 \ldots n]$, $n$ is the number of abstract services, $Col_{ij}^{e}$ denotes the $j$th element of the $i$th colony attached by the $e$th empire $e \in [1 \ldots SE]$, $SE$ is the current size of empires, $imp_{j}^{e}$ denotes the $j$th element of the $e$th imperialist, $RCol_{kj}^{e}$ denotes the $j$th element of the $k$th colony attached by the $e$th empire ($k$ is randomly selected from the colonies of empire $e$, such that $k \neq i$), $\beta$ is the assimilation coefficient with a value greater than 1 and close to 2, $\mu$ is a random number within the range $[0, 1]$ and $\theta$ is a random number between $[0, 1]$. The new index value $NCol_{ij}^{e}$ representing the selected service from the candidate set $S_j$ must be an integer number. So, in order to get an integer value for $NCol_{ij}^{e}$, the rounding down operation $\lfloor \; \rfloor$ is applied in Eq. 9. If $NCol_{ij}^{e}$ is less than 1 or greater than $m$ where $m$ is the number of concrete cloud services in the set $S_j$, then we will set the value of $NCol_{ij}^{e}$ as follows:

$$NCol_{ij}^{e} = \begin{cases} 1 & \text{if } NCol_{ij}^{e} < 1 \\ m & \text{if } NCol_{ij}^{e} > m \end{cases} \tag{10}$$

The position's cost value of the assimilated colony (i.e. $NCol_{i}^{e}$) will be evaluated using the cost function given in Eq. 4. Like the greedy selection strategy of the ABC algorithm [13], this new colony will replace its old position (i.e. $Col_{i}^{e}$) if its cost value is better than its previous cost; otherwise the assimilated colony will take its former position. So, the difference between our proposed assimilation and the classical assimilation policy process lies in the fact that the colonies' movement will be applied if the new position of the moved colony is ascertained to give more power to its empire. The applied greedy selection strategy between the old and new positions of colonies is described in Eq. 11.

$$\text{if } Cost(NCol_{i}^{e}) > Cost(Col_{i}^{e}) : Col_{i}^{e} = NCol_{i}^{e} \text{ and } Cost(Col_{i}^{e}) = Cost(NCol_{i}^{e}) \tag{11}$$

### 3.3 Moving of imperialists toward strongest imperialist

In the classic ICA algorithm, the assimilation policy process is performed only for colonies toward their imperialists. There is no interaction and exchanging of socio-structure among empires. But in the real world, weak imperialists try to improve their characteristics by assimilating some economic and sociopolitical characteristics from other empires and in particular from the strongest one. This imperialist assimilation policy increases the power of weak empires and increases their survival capability during the competition process, so, the local exploitation of the DICA is enhanced, as a result, the convergence of the proposed algorithm to the near-optimal global solution is accelerated. In this study the proposed discrete assimilation policy is performed for the $i$th imperialist ($Imp_i$) toward its strongest imperialist ($StImp$) as shown in Eq. 12 :

$$NImp_{ij} = Imp_{ij} + \lfloor \beta\mu(StImp_j - Imp_{ij}) + \theta(Imp_{kj} - Imp_{ij}) \rfloor \quad (12)$$

To accept the movement of each imperialist, the same previous greedy selection mechanism is employed between the old position of this imperialist and its new position.

### 3.4 Revolution

To prevent the population falling into a local optimum, which increases the exploration ability of the proposed algorithm, revolution operators are performed on some colonies which don't improve their sociopolitical characteristics (like language and culture) through a predetermined number of evolutions ($T$), i.e. colonies not absorbed by their imperialists. The proposed revolution operator in this study is as follows:

**Step 1.** For each colony ($Col_i$), assign to it an update iteration number ($Trial_i$) and let $Trial_i = 0$.

**Step 2.** In each evolution process of our DICA, set the $Trial_i$ for each $Col_i$ as follows: if the $Col_i$ updates its position, set the $Trial_i$ to zero; otherwise, increase it.

**Step 3.** Find the colony which does not updated its position after $T$ iterations ($Trial_i > T$ ), if there is more than one, select one randomly, and replace the selected colony with a new randomly generated one.

### 3.5 Empires competition process

Both, the power of the imperialist country and the power of its colonies are used to define the total power of an empire, which is evaluated by making an addition between the power cost of the imperialist country and a percentage of average power costs of its colonies. Eq. 13 shows the definition of the total power of an empire $Emp_n$.

$$TP_{Emp_n} = Cost(Imp_n) + \xi * mean(Cost(colonies\ of\ empire\ Emp_n)) \quad (13)$$

In Eq. 13, $TP_{Emp_n}$ is the total power of the $n$th empire, $\xi$ is a number between 0 and 1, the larger the value of $\xi$ the higher the effect of colonies on the power of $Emp_n$.

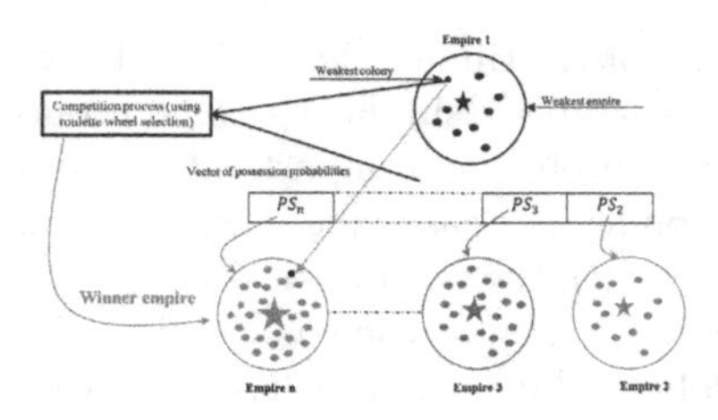


**Fig. 3.** Imperialistic competition.

In the imperialistic competition process, a competition between strong empires gets under way to secure some (usually one) of weak colonies from the weakest empire; this process of competition improves the power of the strong empires and decreases the power of the weak ones. When all the colonies of the weakest empire are attracted by the strong ones, the imperialist of this weak empire will be in turn possessed by the more powerful ones, by way of consequence, the weaker empires will collapse. Fig. 3 depicts this strategy of competition. It can be seen from this figure that the weakest colony of the weakest empire (i.e. *empire* 1) is under competition among the remaining empires (*empires* $2 \ldots n$). The winning empire in this competition is the (*empire* $n$). In order to find the victor empire which takes the weakest colony, we need to calculate the possession probability of each empire, which is defined as follows:

$$PS_{emp_n} = \left| \frac{NTP_{emp_n}}{\sum_{i=1}^{PopImp} NTP_{emp_i}} \right| \tag{14}$$

where, $PS_{emp_n}$ is the possession probability of the $n$th empire and $NTP_{emp_n}$ is its normalized total power which is described as follow:

$$NTP_{emp_n} = TP_{emp_n} - min(TP_{emp_i}) \tag{15}$$

where $TP_{emp_n}$ is the total power of the $n$th empire and $min(TP_{emp_i})$ denote the minimum total power value among all empires ($i = 1 \ldots PopImp$).

After evaluating the possession probabilities of all empires, the same mechanism of roulette wheel selection discussed in sub-section 3.1 is applied by using these probabilities in order to find the winning empire.

### 3.6 The ending criterion of DICA

Several stopping criteria can be used to end a meta-heuristic algorithm such as:

1. Fixing a specific time to stop the meta-heuristic algorithm;
2. A large number $MaxItr$ is defined as the maximum number of iterations;
3. The proposed DICA algorithm can also be terminated in the following fashion: when all empires are eliminated, except for the most powerful one, its near-global optimal solution represents the best solution returned by the DICA.

Pertaining to our proposed DICA, the second criterion is utilized to effect its termination.

### 3.7 The framework of the proposed algorithm

In the following, the running steps of our proposed DICA are detailed:

- **Step 1.** Initialize the algorithm's parameters: Population size ($PopSize$), number of imperialists ($PopImp$), colonies average cost coefficient ($\xi$), the assimilation coefficient ($\beta$), the trial number of revolution ($T$) and the maximum number of iteration ($MaxItr$).
- **Step 2.** Generate randomly the initial population ($Pop$), evaluate the cost value of each country in $Pop$ and create the initial empires as discussed in *subsection* 3.1.
- **Step 3.** If the DICA's ending criterion is reached, terminate our proposed algorithm and return the best solution; otherwise, perform steps 4 to 9.
- **Step 4.** In order to move colonies toward their associated imperialists, apply the proposed discrete assimilation process between each colony, a random selected colony in the corresponding empire and its related imperialist as discussed in *subsection* 3.2.
- **Step 5.** To accelerate the convergence of the proposed DICA and enhance the local exploitation of the DICA algorithm, apply the same discrete assimilation policy process for each imperialist toward its strongest imperialist as discussed in *subsection* 3.3.
- **Step 6.** Perform the revolution operator for colonies of empires as discussed in *subsection* 3.4.
- **Step 7.** For each colony, calculate the cost value of its new position, if it is better than its related imperialist, exchange positions between it and its relevant imperialist.
- **Step 8.** For each empire, calculate its total power, find the weakest colony of the weakest empire and make it under competition among imperialists as discussed in *subsection* 3.5.
- **Step 9.** Return to **Step 3**.

## 4 Experiments

Two data-sets taken into consideration to validate and illustrate the performance of our proposed algorithm to solve the QCSC problem.

- **QWS data-set**: it is a real data-set collected by E. Al-Masri et Q.H. Mahmoud [14], in which 2507 web service were provided. In this data-set, and for each web service, nine QoS criteria including throughput, availability, response time, ...ect are used to present the QWS's QoS model. In our study, four QoS parameters (i.e. the response time, the availability, the reliability and the throughput) are selected to perform the experiments on this data-set.

- **Random data-set**: in order to make sure that the experimental results obtained by our proposed algorithm are not biased for the used QWS data-set, where this QoS data-set contains only 2507 atomic services that is a small one in some extent, a large random data-set denoted as **QWR**, is generated consisting of simulating 20.000 atomic cloud services. For simplicity, each cloud service considers four QoS attributes: the price, the response time, the reliability, and the availability as the QoS model, where their random values in each experiment were generated randomly in the ranges [20,1500](ms), [2,15]($), [95,100](%) and [40,100](%) respectively.

In the following simulations, and for these two data-sets, it is easy to add or replace the used criteria in order to extend the QoS model without modifying the proposed service composition algorithm. The weights of the used attributes are initialized with the same value (e.g. for a QoS model with four attributes ($wq_{t(t=1\ldots4)} = 0.25$)) and by referring to the study in [5], the constraint violation ($Cst_{q_t}$) for each ($q_t$) is evaluated according to the following equation :

$$Cst_{q_t} = \begin{cases} \delta(agg(q_t^{max}) - agg(q_t^{min})) + agg(q_t^{min}) \text{ if } q_t \in QoS^+ \\ agg(q_t^{max}) - \delta(agg(q_t^{max}) - agg(q_t^{min})) \text{ if } q_t \in QoS^- \end{cases} \tag{16}$$

where ($\delta$) denotes the severity of the global QoS constraint for $q_t$ (heare, its value is set to 0.4).

A PC with an Intel(R) Core i3 CPU 1.9 GHz, 4 GB RAM, Windows 10 (64 bit) system and MATLAB R2013a is used to perform all experiments.

To validate the performance and the effectiveness of our DICA, we compared it with GA [6], PSO [2] and ICA [15]. Table 2 details the initial parameter setting of the compared algorithms. To end these algorithms, $MaxItr$ is set to 1000. Several experiments are used to verify the performance of DICA. Each

**Table 2.** The initial values of the algorithms' parameters

| $\text{Alg}_s$ | Parameters' values |
|---|---|
| GA[6] | $PopSize$=100, $P_c$=0.7 and $P_m$=0.01 |
| PSO[2] | $PopSize$=100, for $C_1$ , $C_2$ and the Inertia weight refer to [2]. |
| ICA[15] | $PopSize$=100, $PopImp$=10, $P_c$=0.7 , $P_m$=0.2 and $\xi$=0.05 |
| DICA | $PopSize$=100, $PopImp$=10, $\xi$=0.05, $\beta$=2 and $T$=100. |

experiment consists of simulating a sequential $ACSC$ with ($n$) abstract services, in which the relating set of concrete services for each abstract service has ($m$) concrete services. By varying the value of these parameters, the results were collected to be analyzed where each unique combination of these two parameters represents one simulation of the concerned experiment. The results obtained in our simulations are average and have been repeatedly reoccurring 20 times. For the above data-sets, Two scenarios are considered to set the values of $n$ and $m$ as illustrated in Table 3.

**Table 3.** The values of $n$ and $m$ for the two scenarios in WSQ and WSR data-sets

| Data-set | The first scenario (SN1) | The second scenario (SN2) |
|---|---|---|
| **QWS** | $m$=100 and $n$ differs from 5 to 25 with an increase of 5. | $n$=20 and $m$ differs from 25 to 125 with an increase of 25. |
| **QWR** | $m$=800 and $n$ differs from 5 to 25 with an increase of 5. | $n$=20 and $m$ differs from 200 to 1000 with an increase of 200. |

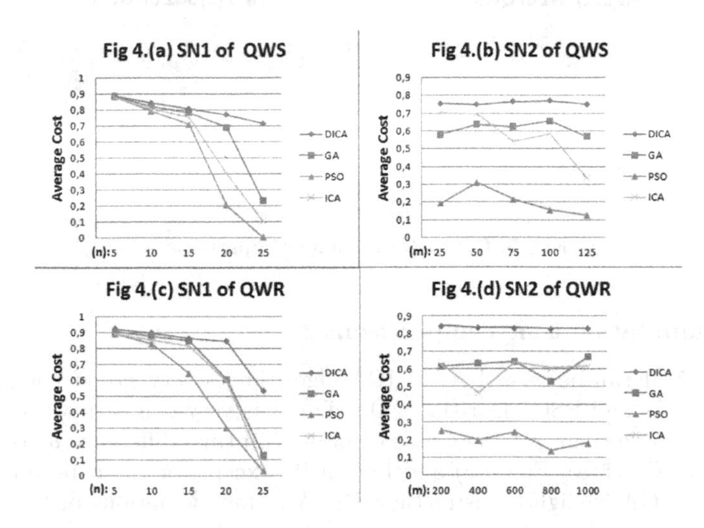


**Fig. 4.** Optimality comparisons.

## 4.1 Optimality comparisons

As shown in Fig. 4, the average best cost values of our DICA is better than the average best cost values of the other algorithms for the two scenarios on the both data-sests: the real and the random ones, i.e. fixing the number of concrete cloud services and incrementing the number of abstract cloud services as shown in Fig. 4(a)& Fig. 4(c) and vise-versa as seen in Fig. 4(b)& Fig. 4(d). It is also observed from this figure, that when the number of abstract services is with a large value, our DICA returns a feasible solution; whereas, the others return infeasible ones (i.e. in Fig. 4(a) & Fig. 4(c),and for $n$=25, the average cost value of DICA is greater than 0.5, by inverse, it is less than 0.5 for the other algorithms). Moreover, the DICA presents a good stability compared to the other approaches (i.e. In Fig. 4(b) & Fig. 4(d), when the number of abstract services is fixed with a value of 20 the variance of the DICA is not affected by the increasing of the number of concrete services, which is very small than the variances of GA,ICA and PSO approaches). From these results, the DICA is optimal than the other algorithms on the obtained near-optimal global solutions.

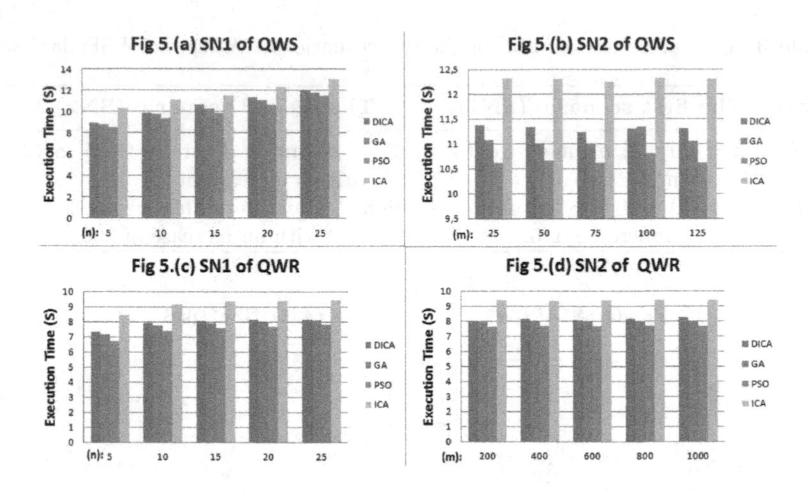


**Fig. 5.** Computation time comparisons.

### 4.2 Computation time comparaisons

From Fig. 5, it can be seen that the ICA algorithm is very slower compared to the DICA, GA and PSO algorithms. Our DICA is very competitive to the GA algorithm, where there is a negligible computation time difference among them as shown in Fig. 5(a), Fig. 5(c) and Fig. 5(d). Except for the experiment seen in Fig. 5(b), GA is slightly faster than DICA in four variations of the number of concrete services per five variations, but as can be observed from their related fitness values from Fig. 4(b) of the above optimality comparisons section, the average cost values of our DICA is more important than the average fitness values of GA. As a result, the DICA has a good tradeoff between the quality of near-optimal solution and its computation time compared to the GA algorithm. The PSO algorithm has lees execution time for the two scenarios on the WSQ and WSR data-sets, but with a very low fitness values for these scenarios.

Considering the previous comparisons results, we can conclude that our proposed algorithm is superior to the ICA, GA and PSO algorithms.

## 5 Conclusion and future works

In this study, a new Discrete Imperialist Competitive Algorithm (DICA) is proposed to solve the QoS-aware cloud service composition problem which is a non-polynomial combinatorial problem. To prevent the DICA from converging to the local optima, a novel discrete assimilation policy process is proposed which is inspired by solution search equation of Artificial Bee Colony algorithm (ABC). To improve the weak empires and furnish them with a greater chance of survival during the competition process, the proposed absorption process is also applied for imperialists and, by way of consequence; the local exploitation of our DICA is enhanced. Based on real-world and random datasets, our DICA is validated. Comparison of the results of our DICA with other recent algorithms in the literature demonstrates its superiority. In our study, the QoS attributes are assumed with fixed values. However, in real-situation, some QoS may be imprecise for many unexpected reasons, such as network connectivity and system congestion ...etc. So, our future work tends to improve DICA to deal with uncertain QoS.

## References


1. A. Jula, E. Sundararajan, Z. Othman, Cloud computing service composition: A systematic literature review. Expert Systems with Applications. 41(8), 3809–3824(2014)
2. F. Tao, D. Zhao, Y. Hu, and Z. Zhou, Resource Service Composition and Its Optimal-Selection Based on Particle Swarm Optimization in Manufacturing Grid System. IEEE Transactions on Industrial Informatics. 4(4), 315–327(2008)
3. D. Ardagna, B. Pernici, Adaptive Service Composition in Flexible Processes. IEEE Transactions on Software Engineering. 33(6),369–384(2007)
4. M. Alrifai , T. Risse, Combining Global Optimization with Local Selection for Efficient QoS-aware Service Composition . International World Wide Web Conference Committee 2009. Madrid. Spain. pp. 881–890.
5. Q. Wu, Q. Zhu, Transactional and QoS-aware dynamic service composition based on ant colony optimization. Future Generation Computer Systems. 29, 1112–1119(2013)
6. G. Canfora, M.D. Penta, R. Esposito, M.L. Villani, An Approach for QoS-aware Service Composition based on Genetic Algorithms. In: Proceedings of the conference on genetic and evolutionary computation. Springer–Berlin. 1069–75(2005)
7. G. Esmaeil and C. Lucas, Imperialist competitive algorithm: an algorithm for optimization inspired by imperialistic competition. Proceedings of the 2007 IEEE Congress on Evolutionary Computation. 4661–4667(2007)
8. S. Hosseini, A. Al Khaled, A survey on the Imperialist Competitive Algorithm metaheuristic: Implementation in engineering domain and directions for future research. Applied Soft Computing. 24, 1078–1094(2014)
9. A. Jula, Z. Othman, E. Sundararajan, A Hybrid Imperialist Competitive-Gravitational Attraction Search Algorithm to Optimize Cloud Service Composition. In: Memetic Computing (MC), IEEE Workshop. 37–43(2013)
10. A. Jula , Z. Othman , E. Sundararajan, Imperialist competitive algorithm with PROCLUS classifier for service time optimization in cloud computing service composition. Expert Systems with applications. 42, 135–145(2015)
11. C.H. Chen, W.H. Chen, Bare-bones imperialist competitive algorithm for a compensatory neural fuzzy controller. Neurocomputing. 173, 1519–1528(2016)
12. M. Zeleny, Multiple Criteria Decision Making. McGraw-Hill. New York. 1982
13. D. Karaboga, B. Basturk, A powerful and efficient algorithm for numerical function optimization: artificial bee colony (ABC). Journal of Global Optimization. 39, 459–471(2007)
14. E. Al-Masri, Q.H. Mahmooud, Investigating Web Services on the World Wide Web, 17th International Conference on World Wide Web (WWW), Beijing, April 2008, pp. 795–804.
15. B.M. Ivatloo, A. Rabiee, A. Soroudi, M. Ehsan, Imperialist competitive algorithm for solving non-convex dynamic economic power dispatch, Energy, 44 ,228–240(2012)

# Smart feeding in farming through IoT in silos

**Himanshu Agrawal, Javier Prieto, Carlos Ramos and Juan Manuel Corchado**

**Abstract** Smart farming practices are of utmost importance for any economy to foster its growth and development and tackle problems like hunger and food insecurity and ensure the well-being of its citizens. However, such practices usually require large investments that are not affordable for SMEs. Such is the case of expensive weighing machines for silos, while the range of possibilities of the Internet of Things (IoT) could intensively reduce these costs while connecting the data to intelligent Cloud services, such as smart feeding systems. The paper presents a novel IoT device and methodology to monitor quantity and quality of grains in silo by estimating the volume of grains at different time instants along with temperature and humidity in the silo. A smart feeding system, implemented via a virtual organization of agents, processes the data and regulates the grain provided to the animals. Experimental on-field measurements at a rabbit farm show the suitability of the proposed system to reduce waste as well as animal diseases and mortality.

---

Himanshu Agrawal, Javier Prieto and Juan Manuel Corchado
University of Salamanca, BISITE Research Group
C/ Espejo s/n, 37007, Salamanca (SPAIN)
e-mail: javierp@usal.es, corchado@usal.es

Himanshu Agrawal
Department of Mechanical Engineering, Indian Institute of Technology Jodhpur
Old residency road, Ratanada, Jodhpur 342011, Rajasthan (INDIA)
e-mail: ug201312013@iitj.ac.in

Carlos Ramos
Institute of Engineering, Polytechnic of Porto
Rua Dr. António Bernardino de Almeida 431, 4200-072 Porto (PORTUGAL)
e-mail: csr@sc.ipp.pt



J.M. Corchado Rodriguez et al. (eds.), *Intelligent Systems Technologies and Applications 2016*, Advances in Intelligent Systems and Computing 530,
DOI 10.1007/978-3-319-47952-1_28

## 1 Introduction

Smart farming techniques have been put in the forefront of many farmers and food produces as long as they can reduce costs, facilitate traceability, and increase security [1, 2, 3]. Many smart farming technologies require large investments in intelligent devices, however, the irruption of the Internet of Things (IoT) is increasingly reducing these investments. Such is the case of expensive weighing machines used to calculate the quantity of grain in silos. Silos are big structures that are used to store bulk quantity of materials like cement, grains, saw dust, food products, etc. (see Fig. 1). In agriculture, they are used to store grains and protect them from being attacked by insects, rodents, birds and also provide a medium to store large quantities of grain over long time with protection. The aim of this paper is to develop an affordable intelligent system able to measure the volume/weight of the grain in different silos, together with ambient conditions, and intelligently use these data to regulate feeding of animals.

But, what is the use of knowing the volume of grains in the silo at different time instants? The information on volume of grains plays a pivotal role in efficient management of stock. By proper monitoring the volume of grains currently present inside the silo, the farmers or industries can do a proper management of their available stock and plan accordingly when they have to refill the stock so that they are not at a risk of out of stock danger. The timely updates on the volume of grains will prevent them from losses incurred due to unavailability of stock and also improve food security and help to fight hunger, where one of the major causes of the hunger is due to improper management of grains after post-harvest. Also the temperature and humidity monitoring of the grains over time can help in ensuring the grain quality and it can alert the farmer of the degradation in the quality of grains, thus not only preventing losses for the farmer but also instrumental in tackling hunger and food insecurity due to poor quality of grains. Moreover, with the farmers and the owners of silo having updates on volume of grains in the silo through the continuous monitoring of the silo content, they are able to plan when they need to purchase the next stock according to the market availability of grains and the price of grains in the market.

In addition, one of the major causes of failure of silos filled with small grain size products is due to the distress in the structure which is caused as sometimes the silo is unable to take the loads of the product which has been filled inside it. Also sometimes failure occur after a little time after the outlet gate has been opened for discharge as during the discharge process the lateral pressure exerted by the stored grains on the walls of the silo exceeds the permissible level, since the pressure exerted by the grains during discharge is more than the pressure exerted by the stored grains and thus resulting in cracks in the walls or total sudden fall down of the structure. But once the user is aware of the volume of the grains inside the silo, the user will be able to prevent situations like overloading of the silo with grains, and prevent keeping silo filled with extra large quantities of grains for a long time and thus it can prevent failure of the silo due to these reasons.

Some approaches have dealt with the measuring of the volume of grains inside the silo, however, they present different drawbacks such as the requirement of constant contact between grains and sensor [4], or the limitation to liquid materials [5]. Moreover, these systems restrict their functionality to measuring the volume, while additional capabilities such as measuring ambient conditions, and intelligently feeding animals are considered within the virtual organization of agents presented in this paper.

This research paper deals with the development of an IoT device that can measure the volume of the grains inside the silo at different time instants and keep its user timely updated about the volume information along with the temperature and humidity inside the silo so that the user can properly and efficiently manage the quantity and quality of stock and prevent from out of stock risks, thus avoiding losses and providing efficient management. The IoT device sends the data via WiFi connection to a Cloud service that monitors silo and feeds animals. A virtual organization of agents governs the system, where there are intelligent agents in sub-organizations dedicated to collect data, fuse information, control feeding, or active alarms in the case a risky situation is detected [6, 7].

The rest of the paper is organized as follows: Section 2 summarizes related work regarding volume measuring at silos, Section 3 states the problem to be solved, Section 4 describes the deployed system, Section 5 discuss the experimental setup and experiments, and, finally, Section 6 draws the conclusions.

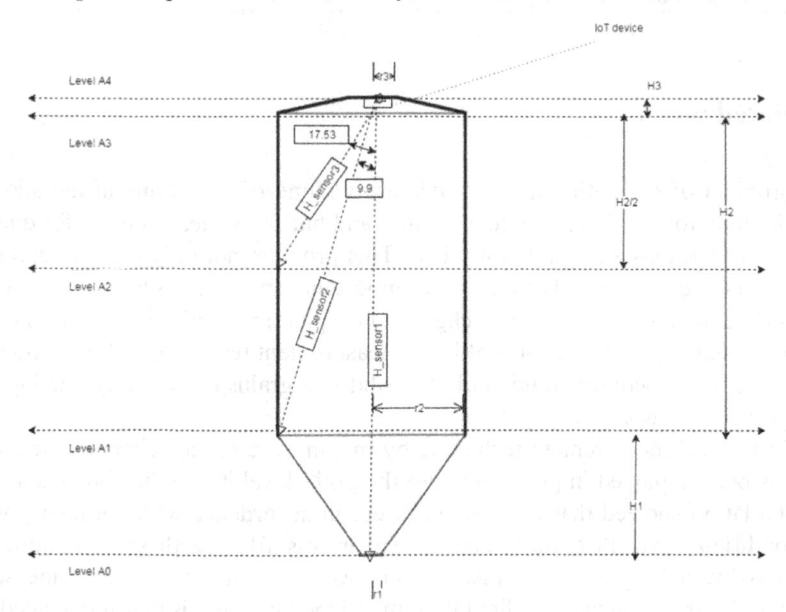


**Fig. 1.** Silo structure and notations.

**Table 1.** Notations

| Notation | Meaning |
|---|---|
| $H_{sensor1}$ | Distance of the obstacle measured from the GH-311 RT Ultrasonic sensor when the GH-311 RT Ultrasonic sensor mounted on HS-645 MG Servo motor makes an angle of 0 degrees with the vertical. |
| $H_{sensor2}$ | Distance of the obstacle measured from the GH-311 RT Ultrasonic sensor when the GH-311 RT Ultrasonic sensor mounted on HS-645 MG Servo motor makes an angle of 9.9 degrees with the vertical. |
| $H_{sensor3}$ | Distance of the obstacle measured from the GH-311 RT Ultrasonic sensor when the GH-311 RT Ultrasonic sensor mounted on HS-645 MG Servo motor makes an angle of 17.53 degrees with the vertical. |
| $\alpha_2$ | 9.9°= angle made by the GH-311 RT Ultrasonic sensor mounted on the rotating HS-645 MG Servo motor with the vertical |
| $\alpha_3$ | 17.53°= angle made by the GH-311 RT Ultrasonic sensor mounted on the rotating HS-645 MG Servo motor with the vertical |
| $r_1$ | Radius at level AO |
| $r_2$ | Radius at level A1 |
| $r_3$ | Radius at level A4 |
| $H_1$ | Height from level A0 to A1 |
| $H_2$ | Height from level A1 to A3 |
| $H_3$ | Height from level A3 to A4 |

## 2 Related work

The problem of estimating the approximate volume of the grains in the silo for quantity monitoring along with temperature and humidity measurement for quality monitoring has existed from a long time. This problem not only plays a vital role for farmers and industries but also is an important aspect of ensuring food security. Work related to this problem background of quantity monitoring inside silo has been done in the past in the form of level measurement techniques of the fluid present in the silo where this fluid could be solid like grains or liquid by making use of appropriate sensor.

Grain level measurement technique by making use of capacitance based sensors has been deployed in [4] to measure the grain level inside the silo. Their sensor simulation showed that the readings were in accordance with the theory with the condition being that the reference sensor was filled with grain completely (whereas by making use of ultrasonic here we can make non-contact measurements of the silo content). The limitation mentioned by them is that if one needs to know the grain content inside the silo by using the grain level measurement readings then it can come out to be faulty and this error is difficult to avoid for level measuring systems for solids like grain, powders, etc. as the grain accumulation inside the silo is not flat and it possesses a surface profile. This motivated a research to be carried out in the volume estimation of the grain content inside the silo.

A sensor based fluid level measurement system has been applied in [5] where the sensor design is based on a passive circuit comprising of inductor and capacitor and the substance should be able to be placed inside the capacitance plates. In a study on "liquid level sensor using ultrasonic lamb waves" [8], ultrasonic lamb waves are used to detect the presence of liquid as the wave characteristic changes when it comes in liquid contact. Liquid level measurement by using and applying technology of optical fibers has been applied in [9] which makes use of the measurement in the variations in amplitude where it is a function of the distance of the liquid.

Work related to the problem background of quality monitoring has been done through studying the temperature variations and moisture variations in the grains inside the silo by installing the sensors at different points inside the silo. The temperature variations in the headspace are larger than the temperature variations inside the grain mass and also the moisture change inside the grain mass is lower than the moisture change in the surface grain [10]. Also, when a large dry bulk of material is exposed at the surface to high relative humidity then the moisture diffusion rate through the material is extremely slow in practical conditions [11]. This motivated us to install the temperature and humidity sensor at the headspace in the silo rather than installing it at different locations inside the silo for temperature and humidity monitoring.

The quality monitoring plays an important role in detecting the presence of beetles in the grains inside the silo. In a study on the influence of relative humidity and thermal acclimation on the survival of adult grain beetles in cooled grain [12] it was found that the survival rate was shorter at 45% relative humidity than at 70% relative humidity. Also it was found that at 45% relative humidity all species survived when the grain temperature was at 13.5 °C but when the temperature was further brought down to 9 degrees, only S. granaries did survive 26 weeks. Also, in a study on the effects of temperature on the aerobic stability of wheat and corn silages [13], it was found that silage kept at 30 °C had the highest production of $CO_2$, largest pH increase and the largest count of yeast than the samples kept at 10, 20, and 40°C. The samples kept at 10 and 40 °C were found to be stable. So this motivated to measure temperature and humidity inside the silo to monitor grains over time.

## 3 Problem statement

Since the level of grains inside the vertical silo is not known and also the surface profile of the top portion of the grains inside the vertical silo is not known, it makes it difficult to estimate the quantity or volume of grains inside the silo at different time instants.

Without an effective system to measure the quantity or volume of grains inside the silo at different time instants, it hinders proper and efficient management of stock of grains thus, sometimes leading to scarcity of grains to the people leading to hunger and food insecurity and sometimes there are excess grains leading to losses to the farmers and industries and wastage of grains.

Without proper quality monitoring of grains in silo, the degradation in quality of grains could not be managed and thus it can lead to grain shortage, hunger, food insecurity and loss to the farmers.

In the following, we develop an IoT device that can give timely updates of the volume of grains, $V_{t,t \in \mathrm{N}}$, inside the silo at different time instants, $t \in \mathrm{N}$, along with the temperature, $T_{t,t \in \mathrm{N}}$, and humidity, $M_{t,t \in \mathrm{N}}$, data to monitor the quality of grain inside the silo. The IoT device shall be located at the top inside the silo and include an RF module that transmit data to an intelligent Cloud system, based on a virtual organization of agents, which decides whether to increase or reduce feeding of animals taking into account: consumed grain, temperature, humidity, animal age, animal weight, weight evolution, and the probability of disease.

*Assumption:* The surface profile of the top portion of the grains present in the silo is assumed conical.

There can be different cases depending at what level the grains are present. The different cases can be (see Fig. 2):

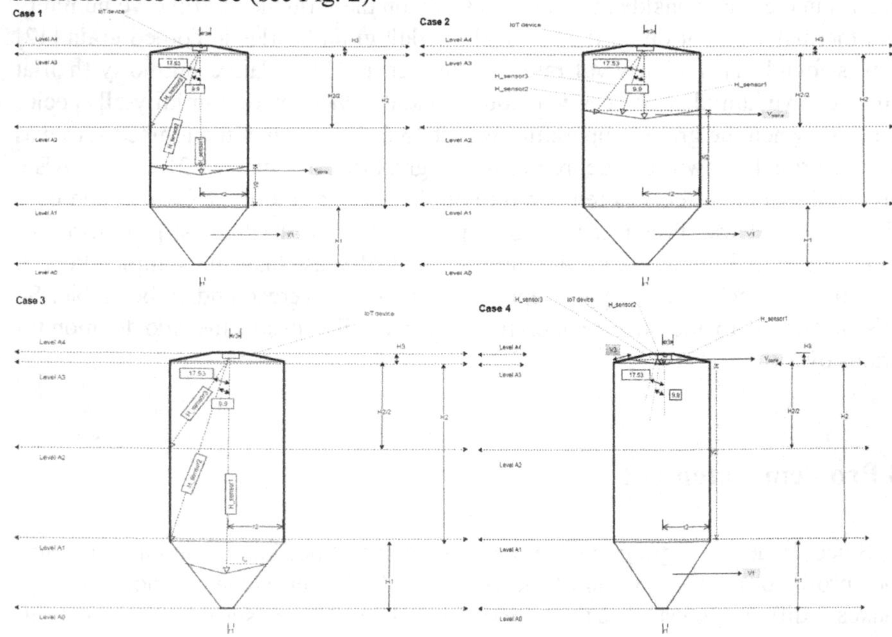


**Fig. 2.** Cases 1, 2, 3 and 4, respectively, can be distinguished depending on the grain level.

- Case 1: when the grain level is between A1 and A2. If this case holds, then, $H_3 + H_2/2 < H_{sensor1} < H_3 + H_2$, and $V_t = V_1 + V_2 - V_{cone}$.
- Case 2: when the grain level is between A2 and A3. If this case holds, then, $H_3 < H_{sensor1} < H_3 + H_2/2$, and $V_t = V_1 + V_2 - V_{cone}$.
- Case 3: when the grain level is below A1. If this case holds, then, $H_3 + H_2 < H_{sensor1} < H_3 + H_2 + H_1$, and calculation of $V_t$ is straightforward.
- Case 4: when the grain level is above A3. If this case holds, then, $0 < H_{sensor1} < H_3$, and $V_t = V_1 + V_2 + V_3 - V_{cone}$..

$H_{sensor1}$, $H_{sensor2}$ and $H_{sensor3}$ are the distances measured at the positions marked in Fig. 1, which correspond with angles 0, 9.9 and 17.53 degrees with the vertical, respectively. These angles were calculated based on the geometry of the silo. Then, the grain volume can be obtained, respectively, for the different cases once the values of $H_{sensor1}$, $H_{sensor2}$, and $H_{sensor3}$ are known.

## 4 System description

The IoT device contains an ultrasonic sensor mounted on a servo motor that rotates to measure the three distances marked in Fig. 1 between the top of the silo and the grain level. It also includes a temperature and humidity sensor that informs the user about temperature and humidity measurement data along with the volume of grains inside the silo at different time instants. The results are shown on a website and an Android Smartphone application. This information can help to eradicate the problems like scarcity of grains and wastage of grains caused due to the presence of unknown amount of grains or poor quality of grains inside the silo and thus help in efficient quantity and quality management of stock. Fig. 3 depicts the different modules that constitute the IoT device. It comprises:

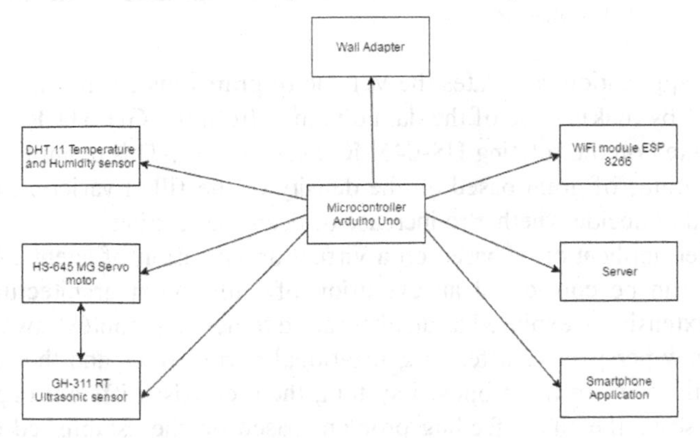


**Fig. 3.** Arduino Uno Microcontroller enables handling the different blocks that constitute the designed IoT device.

1. *Arduino Uno:* It is a microcontroller board based on ATmega328P that controls all other modules.
2. *GH-311 RT ultrasonic sensor:* This ultrasonic sensor is used in the device to measure the distance between the top of the silo and the grains at different angles. The sensing range of the sensor is 2-8000 mm.
3. *HS-645MG Servo motor*: The GH-311 RT ultrasonic sensor is mounted on the HS-645MG Servo motor so as to take the distance readings of the obstacle at 3 different angles namely 0, 9.9 and 17.53 degrees. The servo motor operates at a speed of 0.24s/60°.
4. *DHT 11 Temperature and Humidity sensor:* The DHT 11 sensor is used to measure the temperature and humidity inside the silo at different time instants. The DHT 11 sensor has a measurement range of 20-90% RH and 0-50 °C. The humidity accuracy is ±5% RH and the temperature accuracy is ±2 °C.
5. *WiFi Module - ESP8266:* The WiFi module is used in the IoT device to connect the microcontroller to a WiFi network and helps in transferring the data received from the 3 GH-311 RT sensor to the server via WiFi.
6. *Wall Adapter:* The wall adapter is used to convert the AC voltage from the power supply to the DC 5 volt supplied to the Arduino Uno.

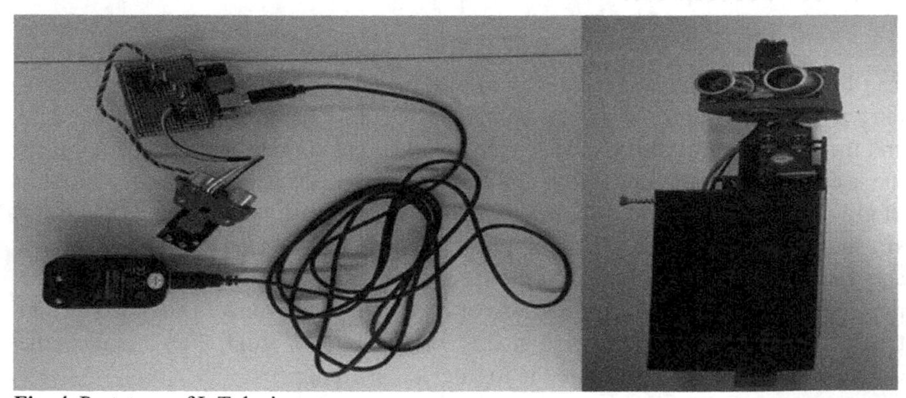

**Fig. 4.** Prototype of IoT device.

A server application calculates the volume of grains inside the silo at different time instants by making use of the data obtained from the GH-311 RT Ultrasonic sensor mounted on the rotating HS-645MG Servo motor. This quantity is translated into kilograms of grain based on the density of the filled variety. Finally, this value is used to decide whether to increase or decrease feeding.

The server application is based on a virtual organization of agents. Virtual organizations can be considered an evolution of multi-agent architectures, which have been extensively explored as an alternative to develop context-aware systems [14]. This new perspective offers organizational norms, roles and the services associated with them. In the proposed system, there coexist different organizations that jointly solve the smart feeding problem based on the established rules from case-based reasoning on past experiences [15] (e.g., it is known that low or too high feeding can derive in animal death): data gathering, context, rules establishment, decision making and feeding regulation organizations. The abovementioned variables influence the feeding rate: consumed grain, temperature, humidity, animal age, animal weight, weight evolution, and the probability of disease

Finally, the server application displays data (volume, temperature, humidity, etc.) inside the silo at different time instants on the web, and a Smartphone application fetches the data from the server and displays it to the user.

# 5 Experimental evaluation

In this section, we evaluate the case study of the smart feeding system working on a rabbit farm. We first describe the experiment and discuss the results afterwards.

## *5.1 Experimental setup*

The IoT device was fitted at the lid of the silo at the top such that the IoT device faces the grains. The silo was the model Aviporc 230/5, with diameter 2.3 m, height 8.15 and capacity 21.07 $m^3$ (approximately 12 tons of grain). The device was tested during 10 hours where the silo was filled with 6 tons of corn (approx. 10,5 $m^3$), and 1 ton was pulled out at intervals of 100 kilograms every hour.

The values of $H_{sensor1}$, $H_{sensor2}$, and $H_{sensor3}$ were determined with the help of the GH-311 RT Ultrasonic sensor mounted on the HS-645MG Servo motor, both of them were connected to the Arduino Uno. The HS-645MG Servo motor rotated from 0 to 25 degrees and the distance readings from the GH-311 RT Ultrasonic sensor were recorded when the GH-311 RT Ultrasonic sensor mounted on the HS-645MG Servo motor made angles of 0, 9.9 degrees and 17.53 degrees from the vertical of the silo.

According to measured $H_{sensor1}$, $H_{sensor2}$, and $H_{sensor3}$ values, the appropriate case can be determined and it will tell us where the grains are present at a time instant. With this information, the volume of the grains in the silo at different time instants can be known by using the geometry of the given silo.

The DHT 11 Temperature and Humidity sensor was also connected to the Arduino Uno. The DHT 11 sensor calculates the temperature and humidity inside the silo every 5 minutes.

The server application connects to the MySQL database and stores all the information on sensor's values like: Temperature, Humidity, $H_{sensor1}$, $H_{sensor2}$, $H_{sensor3}$ and volume of the grains in the silo in the database.

Finally, the Smartphone application connects to the server and fetches the data from the server's database about the volume of the grains present in the silo at any instant of time along with the values of temperature and humidity. The Smartphone application displays this data to the user.

## *5.2 Experimental results*

Fig. 5 shows the actual and measured values of grain during the mentioned 10-hours interval, together with measured temperature and humidity readings.

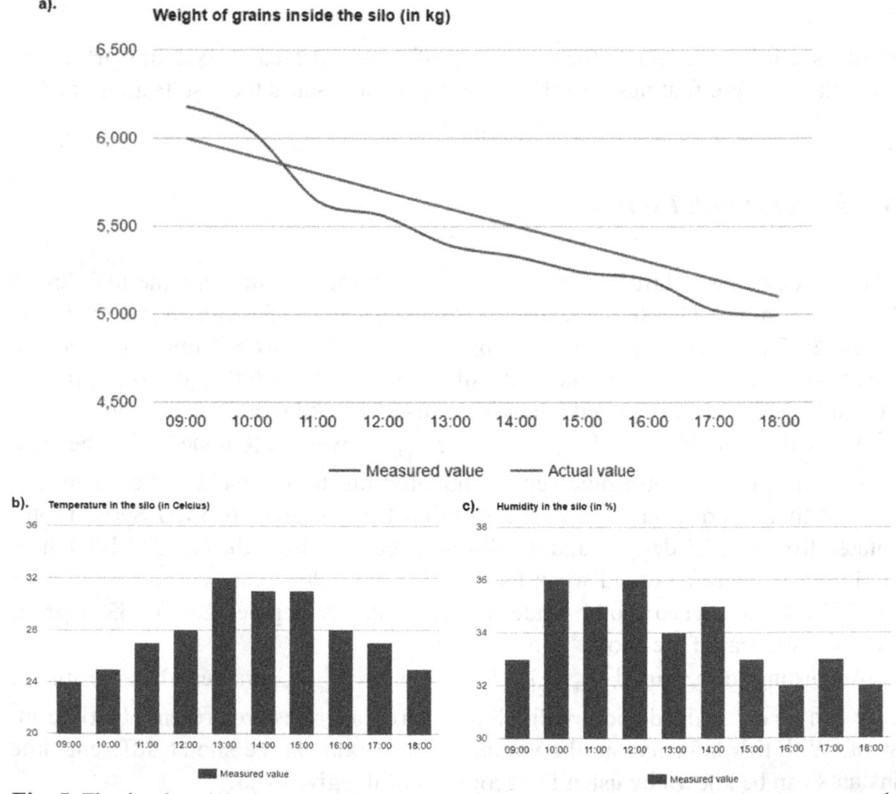


**Fig. 5.** The developed device smoothly estimates a) the volume of grain; b) the temperature; and c) the humidity in the silo, without the need of large investments for weighing machines.

The figure reflects the suitability of the proposed system where the error in weight estimation is below 4% in all the considered cases, thus, enabling the smart feeding system to reduce waste and animal mortality.

## 6 Conclusions

This paper has presented a novel approach for monitoring the quantity and quality of grains inside a silo as this plays a pivotal role in the field of agriculture and ensuring food security, thus tackling hunger and providing efficient and effective management of grain stock along with management of quality of grains. We have developed a novel IoT device based on ultrasonic sensors that measures the grain volume as well as ambient conditions, and sends this data to a Cloud application, based on a virtual organization of agents, that regulates feeding based on these data and case-based reasoning. The experimental results in a rabbit farm show that the proposed solution can be a very beneficial tool for the farmers, industry, individuals who own silo, etc. and help them in properly optimizing the available resources and in taking well informed decisions.

**Acknowledgments** This work has been partially supported by the European Commission (Seventh Framework Programme for Research and Technological Development) FP7-PEOPLE-2012-IRSES project EKRUCAmI (Europe-Korea Research on Ubiquitous Computing and Ambient Intelligence) under grant Ref. 318878.

## References

1. Stein, A.; Allard, D.; van de Kerkhof, B.; van Persie, M.; Noorbergen, H.; Schouten, L.; Ghauharali, R. (2015). Spatio-temporal Analysis of Remote Sensing and Field Measurements for Smart Farming, Procedia Environmental Sciences, 27:21-25.
2. Ryu, M.; Yun, J.; Miao, T.; Ahn, I.Y.; Choi S.C.; Kim, J. (2015). Design and implementation of a connected farm for smart farming system, IEEE Sensors 2015.
3. Andrewartha S.J.; Elliott N.G.; McCulloch J.W.; Frappell P.B. (2015). Aquaculture Sentinels: Smart-farming with Biosensor Equipped Stock, Journal of Aquaculture Research & Development 7(1):1-4.
4. İşiker, H.; Canbolat, H. (2009). Concept for a novel grain level measurement method in silos. Computers and electronics in agriculture 65(2):258-267.
5. Woodard, S.E.; Taylor, B.D. (2007). A wireless fluid-level measurement technique. Sensors and Actuators A: Physical 137(2):268-278.
6. Villarrubia, G., De Paz, J. F., Bajo, J., & Corchado, J. M. (2014). Ambient agents: embedded agents for remote control and monitoring using the PANGEA platform. Sensors, 14(8):13955-13979.
7. Tapia, D. I., Fraile, J. A., Rodríguez, S., Alonso, R. S., & Corchado, J. M. (2013). Integrating hardware agents into an enhanced multi-agent architecture for Ambient Intelligence systems. Information Sciences, 222:47-65.

8. Sakharov, V. E., Kuznetsov, S. A., Zaitsev, B. D., Kuznetsova, I. E., & Joshi, S. G. (2003). Liquid level sensor using ultrasonic Lamb waves. Ultrasonics, 41(4):319-322.
9. Vázquez, C., Gonzalo, A. B., Vargas, S., & Montalvo, J. (2004). Multi-sensor system using plastic optical fibers for intrinsically safe level measurements. Sensors and Actuators A: Physical, 116(1):22-32.
10. Jian, F., Jayas, D. S., & White, N. D. (2009). Temperature fluctuations and moisture migration in wheat stored for 15 months in a metal silo in Canada. Journal of stored products research, 45(2):82-90.
11. Pixton, S. W., & Griffiths, H. J. (1971). Diffusion of moisture through grain. Journal of Stored Products Research, 7(3):133-152.
12. Evans, D. E. (1983). The influence of relative humidity and thermal acclimation on the survival of adult grain beetles in cooled grain. Journal of Stored Products Research, 19(4):173-180.
13. Ashbell, G., Weinberg, Z. G., Hen, Y., & Filya, I. (2002). The effects of temperature on the aerobic stability of wheat and corn silages. Journal of Industrial Microbiology and Biotechnology, 28(5):261-263.
14. Villarrubia, G., De Paz, J. F., Bajo, J., & Corchado, J. M. (2014). Ambient agents: embedded agents for remote control and monitoring using the PANGEA platform. Sensors, 14(8):13955-13979.
15. Corchado, J. M., & Laza, R. (2003). Constructing deliberative agents with case-based reasoning technology. International Journal of Intelligent Systems, 18(12):1227-1241.

# The Use of Biometrics to Prevent Identity Theft

**Syed Rizvi[1], Cory Reger[2], and Aaron Zuchelli[3]**

**Abstract**

This paper investigates the identity theft and how to effectively use the biometric technology in order to prevent it from happening. Over the past few years, identity theft has become one of the most serious financial threats to corporations in the United States. Identity theft is when a thief steals sensitive information in order to engage in large financial transactions. When a thief successfully steals an identity, he or she may take out large amounts of loans and make purchases in the victim's name. Identity theft allows thieves to pose as the victim so that all of the thief's actions will be projected as actions from the victims. Thieves can damage corporations by compromising corporate credit cards as well as file documents to change the legal address of the victim's company. Throughout the past decade, corporations have been using variations of biometric technology to prevent identity theft. This paper presents a new scheme for corporate identity theft prevention using biometric technology. Specifically, we develop a biometric based authentication system consisting of encryption and decryption processes. To show the practicality of our proposed scheme, an attacker-centric threat mode is created.

Syed Rizvi,
Pennsylvania State University, Altoona PA, e-mail: srizvi@psu.edu

Cory Reger,
Pennsylvania State University, Altoona PA, e-mail: cjr5556@psu.edu

Aaron Zuchelli,
Pennsylvania State University, Altoona PA, e-mail: apz5024@psu.edu

 
J.M. Corchado Rodriguez et al. (eds.), *Intelligent Systems Technologies and Applications 2016*, Advances in Intelligent Systems and Computing 530,
DOI 10.1007/978-3-319-47952-1_29

## 1. Introduction

Identity theft is a crime that can happen to any victim at any time. They can have no idea it's happening, and even if it is happening it is incredibly hard to stop. Identity theft is one of the largest growing crimes in the world, and it affects 17.6 million people, or 7% of the US adult population in year 2014 [15]. The average victim of identity theft loses approximately $1,400, but the number varies greatly [15]. Once personal information is stolen, it can easily be passed on to someone else for fraudulent reasons. The methods of identity theft continue to grow, ranging from people stealing mail to the use of social engineering to steal personal data from data centers. According to the Bureau of Justice [15], there are three types of identity theft: unauthorized use of an existing account, unauthorized use of information to open an account, and misuse of personal information for fraudulent reasons.

According to the National Criminal Justice Reference Service, the most common type of identity theft is the theft of government documents for 34%, credit card theft served for 17%, bank fraud is about 8%, 6% employment related fraud and loan fraud at 4% [12]. With these statistics, 84% of the identity theft that occurred utilized a preexisting account. Unfortunately, even when you figure out your identity is being stolen it is not something that can be fixed immediately. Once you place a fraud alert, there are many steps you must do, such as freezing credit cards and ordering your credit report. After this, filing a police report is the best route of action. With just a police report or the fraud alert, the criminal who is committing the identity theft cannot be caught. But with both reports, the information can potentially be utilized to catch the criminal. Understanding how to prevent identity theft is the most important part in avoiding identity theft. Since information you put anywhere can be taken by someone else for fraudulent reasons, learning how to properly give out your personal information is a big step. Informing people about techniques that prevent identity theft is a key to avoiding the crime from occurring. Although personal identity theft seems very serious, corporate/business identity theft also occurs and can be more serious. Entire businesses can be completely ruined due to their accounts being compromised, and in some cases, there really is nothing to be done to fix the damage that was done.

In this paper, we present a new scheme for corporate identity theft prevention using biometric technology. Specifically, we develop a biometric based authentication system consisting of encryption and decryption processes. To show the practicality of our proposed scheme, both encryption and decryption algorithms are presented with an attacker-centric threat mode.

## 2. Related Works

***Authentication Requirements for Identity Theft:*** 13.Hoofnagle Chris Jay [13] refers the issue of credit authentication being a main cause of identity theft since the credit grantors could not properly authenticate who the borrowers actually were. Author claims that this issue is occurring due to the decline in public life, such as removing names from telephone books, city directories, and DMV databases. He argues that the lack of public information about ourselves has created this issue for ourselves because without the information readily available people are able to disguise who they truly are [13]. The second part of the paper talks about the FACTA (Fair and Accurate Credit and Transactions Act) processes its applications [11]. Multiple imposter's applications were studied, and although they contained some type of incorrect personal information, the credit grantors still chose to give the imposters access to their services. The third part of the article [13] refers to the implementation of strict liability for the credit grantors. This would ensure a financial cost for bad authentication processes, compensate victims more than the current system currently does, and would create new systems in detecting fraudulent accounts. In addition, this research provided twelve most common personal information items stolen for identity theft: 1. Credit card numbers 2. CW2 numbers (the back of credit cards) 3. Credit reports 4. Social Security (SIN) numbers 5. Driver's license numbers 6. ATM numbers 7. Telephone calling cards 8. Mortgage details 9. Date of birth 10. Passwords and PINs 11. Home addresses 12. Phone numbers.

The article also refers to social factors of identity theft. The level of understanding a person has of their role in protecting their privacy plays a big role in how that person goes about daily life using social engineering. People and institutions try to implement security mechanisms, but according to Applegate [14], social engineering is a "methodology that allows an attacker to bypass technical controls by attacking the human element in an organization" (i) 25% of Germans and 60% of Americans have shared their account passwords with a friend or family member. As result of this cultural attitude, 3% of Germans have experienced ID theft. (ii) 50% of Americans use family member names, important dates, nicknames, or pet's name as online accounts passwords. (iii) In all six countries, about 40% of consumers display their personal information in their social network profile and some of them use exactly the same information as their passwords. With statistics like this, it isn't very hard for information to be stolen.

***Identity Theft Prevention*:** Authors in [4] offer an in depth analysis of the damage and targeting results towards identity theft victims. In addition the article provides a unique technique for preventing identity theft. The article began with statistics provided by the Federal Trade Commission (FTC) involving the regions where identity theft has been most common. According to the article, the number of identity theft complaints has been rising significantly over the last couple of years.

The FTC data provided in the article suggests that identity theft crime has regional variation. The regions that had the most identity theft regions per 100,000 populations were District of Columbia, California, Arizona, Nevada, and Texas.

The article also includes results from a number of studies regarding identity thieves including the employment status of offenders, the relationship between the offenders and victims, the number of offenders, and the number of incidents for offenses of incidents. The study about the relationship between the offenders and victims suggests that the majority of attacks were completed without knowing the victim. The next study, employment status of offenders, suggested that 53% of offenders were unemployed whereas 41% were the employed. The last 6% were either unemployed or retired. Moving onto the number of offenders working together, the article states that there are additional publications that support the idea of identity thieves working together. However, the figure in the article showed that 64% of crimes are committed by single offenders, 33% involved two offenders, and 3% involved three offenders. The final study that we considered from the article was the number of incidents for offenses of incidents. The authors of the article conducted chi-square tests to show statistically significant differences between identity theft and all other combinations of offense type. The results of the study provide results for the statement that identity theft is greater than the percentage of increase in other theft oriented offenses.

The article concludes with three suggestions for preventing identity theft. The first suggestion is the use of biometrics. Biometrics would require the user to use some type of physical trait, such as a fingerprint, in order to authenticate himself to the system. A fingerprint scanner is the most widely used form of biometrics and is used for access to buildings, computer terminals, bank accounts, and more. However, biometrics has disadvantages such as being expensive, integration difficulty, and program effectiveness. Another disadvantage mentioned in the article was that biometrics has the ability to monitor employees, which could become an invasion of an individual's property. While biometrics has disadvantages, it may be useful if it helps prevent identity theft. The second suggestion was to educate the public. It is pretty logical that increased education of the public will allow them to be more aware of cyber space and help them identify any potential threats. The third and final suggestion was increased sentencing for offenders. Increased sentencing could reduce identity theft in that thieves will be scared of the punishment.

***Frameworks for Biometric Based Systems:*** With previous work suggesting biometric technology as a solution to identity theft, we found a framework for the creation of a biometric system. This journal article was written by Anthony Vetro and came from the Mitsubishi Electric research laboratory.

Vetro began his explanation by pointing out the issues involved with biometric technologies. Vetro identified the last of interoperability among sensors, enrollment processing models, database storage, and verification modules as one of the biggest problems with industrialized deployment of biometric technology.

According to the paper, "it is necessary to have a standardized minimum feature set as well as the ability to evaluate and report a set of performance metrics" to ensure the interoperability [9]. The paper offered four different biometric template protection schemes: secure sketch, fuzzy commitment, cancelable biometrics, and biometric comparison in the encrypted domain. Each protection scheme had a type of enrollment process followed by database storage, and then with a verification process. In the secure sketch scheme, the system computed the biometric sketch, stored the sketch in the database, and concluded with a test, which determined the biometric sketch's consistency. The fuzzy commitment, cancelable biometrics, and biometric comparison in the encrypted domain included keys. For example, in the fuzzy commitment, the first key is created by the user and is bonded with biometric and then code words with a mixed key and biometric. When the user tries to verify himself, he uses his fingerprint to generate a second key. The original key must be matched with the second key in order for the user to be verified [9].

The article continued with the standardization of frameworks. The paper pointed out that within the ISO/IECJTC1 suite of standards, biometric recognition systems are divided into five subsystems: Data capture, signal processing, store, comparison, and a decision [9]. According to the standards, a template protection system is made up of AD (Auxiliary Data) and PI (Pseudonymous Identifier) [9]. AD is "a portion of the enrollment information that may be used to reconstruct a PI during the biometric recognition process" [9]. The PI is "a portion of the enrollment information that uniquely represents the individual" [9]. The user's biometric information is stored in an AD database and a PI database. If a user were to try and provide a test biometric, the signal subsystem would combine the test biometric with the AD and generate an estimate of the PI [9]. If the PI matches the PI in the database, recognition is deemed positive.

***Standards for Biometric Protection:*** The main aspect of identity theft and biometric protection is the privacy of individuals. The ISO/IEC 24745 standard referenced in the paper suggested three critical privacy aspects that could be incorporated into each system. The presented aspects were irreversibility, unlink ability, and confidentiality. The use of irreversible signal processing transformations, data encryption, separating PI and AD storage, and biometric modalities were examples of ways for someone to incorporate the three privacy aspects into their template protection system [9]. Similar to privacy aspects, security aspects such as confidentiality, integrity, and revocability are important within the template. Secure templates will have ways to recover information after the attacker has compromised it. Since the terms privacy and security are difficult to differentiate, the paper recommended that the system creators ignores those terms and focuses on the specific aspects itself. The steps recommended to achieve security aspects were message authentication codes and digital signatures [7].

## 3. Proposed Biometric based Authentication System

Our proposed system has been designed for a corporate organization authenticating its employees. After researching the related works we concluded that the most important parts of a biometric system are the privacy and security aspects. The "Standardization of Biometric Template Protection" journal article by Shantuanu Rane provided common template structures for the most common biometric systems [9]. The template structures Rane provides are: the fuzzy commitment, cancelable biometric, and biometric comparison in the encrypted domain. Each of these systems either uses encryption, hashing, keys, or some type of method to ensure the security within the biometric system [9].

While most current technologies use one type of method, our proposed solution includes two-factor authentication along with a combination of schemes for users to completely authenticate themselves as well as keep their data within the system secure. In addition, our proposed system will include Asymmetric encryption of both the AD (Auxiliary Data) and PI (Pseudonymous Identifier) databases to provide confidentiality to the stored data. The AD database will consist of the password storage, and the PI database will consist of the fingerprint information. Each database will be encrypted so that if a hacker were to breach the database, the integrity and confidentiality of the information can still be kept. A brief description of the process for our proposed system beings with our two-factor authentication process that will consist of the creation of an initial password, a public key, and a private key. Once user A created his credentials, he will then scan an initial fingerprint that will be stored in the PI database. When user A goes to authenticate himself, user A will use his private key to verify the password. If the previous step is successful, then user A will then scan his fingerprint in an attempt to completely authenticate himself. If the password and fingerprint match, then the user will be authenticated.

### *3.1. Encryption Process*

An illustration of encryption process is shown in Fig.1. The first step of the process is for the user to create an initial password or phrase-key of mixed letters and symbols that only he or she knows. The user will also create a public key and a private key which will be discussed in the encryption portion of the paper. Creating a complex password will only improve the strength of the user's phrase-key. When creating a password the user should stay away from using words and try to be more random with what makes it up. One thing the user should keep in mind when creating a complex password is to use special characters combined with capital and lowercase letters and numbers. For our system, the user will be prompted with a set of requirements their password must abide by. These guidelines will consist of the following: the password must be between 8-20 characters long, it cannot contain the user's name, it must include an uppercase letter, a lowercase letter, a number, and a special character ( !, @, #, $, etc.) If the password created does not meet these requirements, it will be rejected by the system and the user will be prompted to make a new password meeting the requirements.

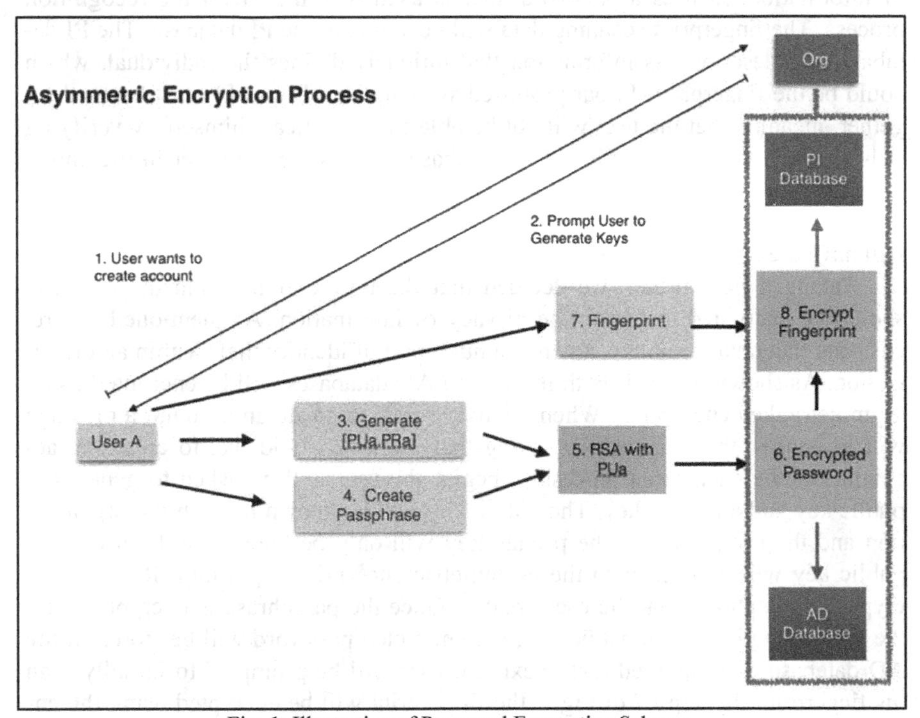


Fig. 1: Illustration of Proposed Encryption Scheme

The second step of the process is fingerprint scanning. Using a fingerprint scanner to further the security in a system is a good technique to follow. Everyone's fingerprint is basically an identification tag in a sense. It is unique to only one specific person. In other words, having a user's fingerprint as a password is almost the same as having a complex password. A fingerprint scanner works by placing a finger on a piece of glass with a camera behind it. The camera then takes a picture of the user's finger and compares it to the stored data (a database of fingerprints permitted.) When comparing the image taken with the image from a database it is looking for the same pattern of ridges and valleys. Since not everyone has the same pattern of ridges and valleys, that is what makes a fingerprint a good source of a password. By incorporating a fingerprint analysis stage in our biometric system, we are making our system that much more protected from intrusions.

The data entered by the user will be stored into two databases that are controlled by the organization. Both the password and fingerprint scanning data will be put into their own database. The password will be stored in what is called an AD database. As described in one of our related works, the AD database is a piece of information, such as a password, that is used with the PI for the recognition process. The fingerprint scanning data will be stored in the PI database. The PI database was described as information that uniquely defines the individual, which could be the fingerprint. In our proposed system, the AD and PI must be used together, meaning that the user will not be able to authenticate himself by verifying only the AD database or only the PI database. As described further in the paper, the user will authenticate himself by using the private password stored in the AD database. Once he verifies himself, the PI database will be activated and the user will have the opportunity to match the data stored in the PI database.

Taking a step further, we decided that the information within the database should be encrypted to ensure the privacy of information. As mentioned before, ensuring that data cannot be stolen is a huge part of identity theft within an organization. As shown in Fig.1, both the PI and AD databases will be encrypted using asymmetric key encryption. When the user wants to create an account, a message will be sent to the organization saying that the user would like to create an account. For the encryption process to begin, the user will be asked to generate a public key and a private key. The public key will be known by both the organization and the user, whereas the private key will only be known by the user. The public key will be used with the asymmetric encryption algorithm, RSA, to encrypt the passphrase that the user creates. Once the passphrase is encrypted with the RSA using the user's public key, the encrypted password will be stored in the AD database as a ciphered text. Next, the user will be prompted to initially scan his fingerprint. In step #7 of Fig.1, the fingerprint will be encrypted using the encrypted password and stored in the PI database. To provide more specific implementation details of the proposed encryption process, Algorithm A is designed.

**Algorithm A: Encryption Algorithm**

1. User A: ask Org C{Account} ; Org: ask User G{PU,PRa}
2. User A: G{PUa,PRa]
3. User A: {PWDa}
4. RSA[EPUa(PWDa)]
5. ST{EPUa(PWDa)} in AD
6. EPUa(PWDa) {EFINGERPRINTa}
7. ST{EFINGERPRINTa} in PI

### *3.2. Decryption Process*

The decryption process will begin with the user asking the organization for authentication, as shown in Fig.2. The organization will respond to the request by asking the user to provide the private key to the RSA decryption algorithm. While the user provides their private key to the decryption algorithm, the organization will retrieve the encrypted password that is stored in its AD database and applied to the RSA decryption algorithm as one of the inputs. If the decryption of the stored password with the private key entered by the user is not successful, the user will be denied access and will have to consult with the IT department. However, in the event of a successful decryption, our proposed system will assume the password owner as a partially legitimate user with the authenticated credentials. Subsequently, the user will be prompted to go into the next step in the decryption process. The next step in the process will require the user to scan their fingerprint. As the user scans their fingerprint, the organization will access the PI database and decrypt the stored fingerprint using the encrypted password. The decrypted stored fingerprint and scanned fingerprint will be compared. If the stored fingerprint matches the scanned fingerprint then the user will be authorized. If the stored fingerprint does not match the scanned fingerprint then the user will be denied access to the network and will have to consult with the IT department. To provide more specific implementation details of the proposed decryption process, Algorithm B is designed.

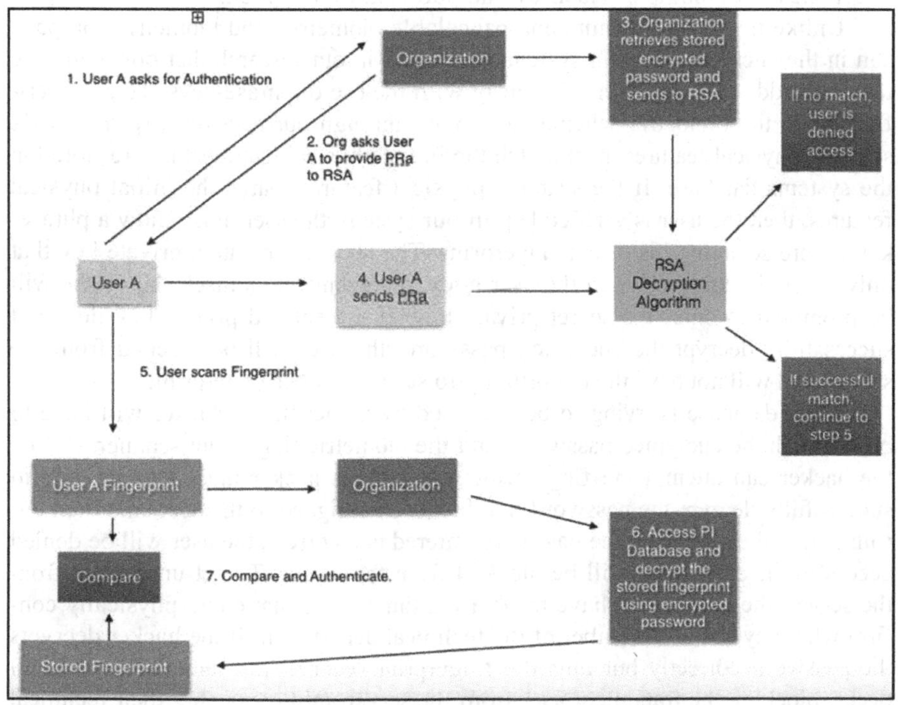


Fig. 2: Illustration of Proposed Decryption Scheme

**Algorithm B: Decryption Algorithm:**

0. User A:{PUa≠0 ; PRa≠0};org 1
1. User A: Auth
2. Org: S{EPUa(PWDa)} to RSA
3. User A: S{PRa} to RSA
4. DPRa[EPUa(PWDa)]
5. IF DPRa {EPUa(PWDa)} Cont. to #6 Else... Deny User
6. User A: Scan{FINGERPRINTb}
7. DEPUa(PWDa) [FINGERPRINTa]
8. Compare
9. If FINGERPRINTa = FINGERPRINTb... authenticate , Else... Deny User

### *3.3. How Proposed System is different from the Existing Solutions?*

After conducting our research, we concluded that many biometric templates utilize one of the following within their system: encryption, hashing, keys, multiple databases, etc. With that being said, we designed a system that has a combination between the use of multiple databases, and public key encryption.

Unlike the fuzzy commitment, cancelable biometric, and biometric comparison in the encrypted domain systems we found in our research that our biometric template adds an extra layer of security with the use of phrase-keys. To gain verification in the biometric schemes we found through our research [9], the user's scanned physical features must match the initial physical features that are stored in the systems database. If the scanned physical features match the initial physical features, then the user is verified [9]. In our system, the user must know a phrase-key before scanning his or her fingerprint. The user will create a private key that only he or she knows. When the user goes to authenticate himself, he or she will be prompted to enter the secret private key. If the entered private key does not successfully decrypt the encrypted password, the user will be blocked from the server, and will not have the opportunity to scan his or her fingerprint.

If the database is trying to be breached by a hacker, the hacker will have to breach both the encrypted password, and the biometric fingerprint scanner. Before the hacker can attempt the fingerprint scanner, the hacker must first be able to successfully decrypt the password that has been assigned to the account when enrollment in the system. If the password entered is incorrect, the user will be denied access to the server and will be blocked from the server. To get unblocked from the server, the user would have to be a legitimate user that could physically confirm who they are to a member of the technical department. If the hacker decrypts the password correctly but fails the fingerprint scanner, the same sequence will occur, blocking the fraudulent user from the server until contacting their technical department.

# 4. Performance Evaluation Using Case Studies

In this section, our goal is to show the practicality of our proposed scheme by creating an attacker-centric threat model as shown in Fig.3. Specifically, the model discusses (a) the motivations and goals of attackers for identity theft, (b) the attack vector which is typically used by the attackers for penetration, (c) the types of attacks used by the attackers to cause damage to individuals or organizations, and (d) the security defense of the proposed scheme.

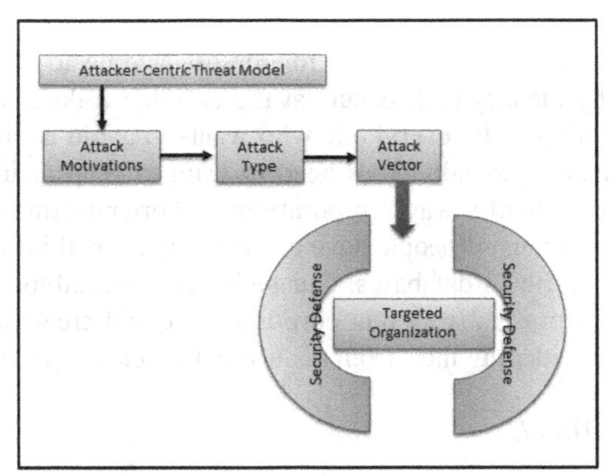


Fig.3: An Illustration of Attacker Centric Threat Model and Proposed Security Defense

## *4.1. Motives and Goals*

Unfortunately, identity theft is something that has happened for years, and is something that will continue to happen. The potential risk of identity theft continues to grow every day as technology continues to grow and advance. Whether a person at home just casually is using the internet to browse or a person working in cooperation, he/she putting valuable information out that can be used by someone to negatively affect that person. There are 12 main things that are stolen during the identity theft: 1. Credit card numbers 2. CW2 numbers (the back of credit cards) 3. Credit reports 4. Social Security (SIN) numbers 5. Driver's license numbers 6. ATM numbers 7. Telephone calling cards 8. Mortgage details 9. Date of birth 10. Passwords and PINs 11. Home addresses 12. Phone numbers [10] It was found that because of this data being stolen, over 2 billion dollars have been lost in just the United States alone in 2010 from identity theft related issues pertaining to the 12 things listed previously.

### 4.2. Attack Vector

Our personal information is obtainable in just about every way possible. We post our information on the internet, in the telephone book, and even our mail contains personal information. Attacks use many different techniques to obtain information to cause harm. Roughly 35% of information comes from cooperation and businesses, and the statistics are the following: 15% taken by a corrupt employee, 7% misuse of data taken from a transaction, 6% stolen from a company financial data. Approximately, 75% of identity theft happens in the public realm, with 5% from computer viruses, 3% from phishing, 1% information taken out of garbage, 15% by friends/relatives/acquaintances, and 30% from stolen or lost credit card/ check book [10].

With information readily available for almost anyone to obtain, it is pretty understandable why identity theft occurs at the rate that it does. Prevention techniques need to be adopted by everybody who wants to avoid having identity theft happen to them. People as individuals need to learn to keep their personal information protected in a smarter way. Corporations and organizations are a bit different because large amounts of people have all of their personal information embedded into their corporation's databases. Our technique to combine biometrics with the traditional password is a way that corporations could create a much stronger database and prevent identity theft from occurring to their cooperation.

### 4.3. Types of Attack

In order to gain an understanding of the range of different attacks, it is helpful to know where the threats originate. The Government Accountability Office (GAO) developed a "Threat Table" that identifies the threat, and then provides a description of each threat. The table contains nine threats as follows: Bot-network operators, Criminal Groups, Foreign intelligence services, Hackers, Insiders, Phishers, Spammers, Spyware/malware authors, and Terrorists [12]. A common theme in the descriptions of each threat was the use of spyware/malware to accomplish their objectives. Furthermore, a large amount of computer knowledge or a high skill level is no longer required to hack company databases. Hackers have the ability to download attack scripts from the internet and launch them against unsuspecting victims.

However, it is important to recognize that there are many less sophisticated techniques used by attackers, such as dumpster diving or mail theft. From a technology standpoint, the Center for Identity Management and Information Protection lists Credit/Debit Card Theft, Skimming, Pretexting, Man-in-the-Middle Attack, Pharming, Vishing, and various forms of Phishing Schemes [13]. Unfortunately, the list will continue to grow as attackers become more sophisticated. One of the best measures that can be used to thwart attackers is educating the public about safeguarding personal information.

### 4.4. Security Defense

Our proposed system, designed for a corporate organization aims to prevent attackers from gaining entry into databases by using biometric technology to authenticate its employees. In order to show the practicality of our proposal, refer back to the GAO Threat Table that identified an Insider as a potential threat. This person may be a disgruntled employee who, for example, may wish to gain access to the company Human Resources data base to steal information stored in employee data files. As an employee, the disgruntled worker knows that upon being hired, they must complete a comprehensive form that includes social security number, date of birth, dependent information, and bank details, including account number where the new hired employee would like to have his or her paycheck direct deposited. Companies that secure confidential information using traditional usernames and passwords run the risk of a data breach. Our biometric identification system is by far more secure because it uses additional controls by requiring fingerprint scanning to authenticate the user at point of entry to access company data files. Biometric recognition offers the following advantages: Users do not need to remember a password or carry an ID card, the person to be identified must be physically present at the point of identification, biometric systems are much harder to foil, and biometric traits cannot be stolen [14]. In our example, the insider has virtually no chance to access employee data files as fingerprint authentication is required for access.

The practicality of our system is further tested when considering the range of various attacks that companies potentially face. Specifically, companies are vulnerable to various phishing schemes, and the Center for Identity Management and Information Protection identifies Spear Phishing as a method attackers use against businesses. In this instance, spear phishers send emails to almost every employee of an organization stating they must send their username and password for verification purposes [13]. Company employees receiving the email who responded to the phishing scheme provided the attacker credentials that only gets them past the first step of our data access process, because our system login process require fingerprint authentication to access company databases.

## 5. Conclusion

After conducting several phases of research, we concluded the combination of security measures proposed in our solution can help reduce the risk of identity theft within an organization. As we found in our case studies, identity thieves usually carry out attacks for some type of personal or financial gain. With access to a corporate system, thieves can compromise financial information, important files, employee information, and more. With attacks from identity thieves towards corporations increasing, we proposed a biometric system for corporations to reduce the risk of identity theft. Our proposed solution was based off of several different templates that we found while researching related works. Throughout each phase of the creation of our proposed solution, we made key modifications in order to create a strong biometric system. When we were creating our proposed solution, we decided that adding a two factor authentication would prevent thieves from gaining access to sensitive information easily. The user's physical characteristic combined with the user's private key, makes it difficult for the intruders to gain possession of the credentials needed for authentication. For example, if the attacker gains possession of the private key but does not have the correct fingerprint, the attacker will not be able to gain access to system and impersonate the true employee. After we implemented the two-factor authentication aspect to our proposed solution, we utilized asymmetric encryption of the stored user password and fingerprint scan data. If someone were to hack the databases, the hacker will not be able to see the plaintext of the fingerprint data or the password that is stored in each database. With that being said, the hacker can only see the stored data's ciphertext, which will add an extra layer of security. For us, designing the encryption method and decryption method was the most difficult part. In our first version of the encryption method, we encrypted the user's password with the user's private key. After further consideration, we realized that if we encrypted the password with the user's private key, anybody who had the known public key would be able to easily make it to the fingerprint step. With that being said, our final solution encrypts the user's password with his public key so that only the user can decrypt the password with his private key. While we were researching biometric systems, we realized that the system can sometimes have fingerprint read errors. Once we realized the system could have read errors for fingerprint scans, we decided that the organization's IT department can assist with authenticating the user. In the event of the user not being able to successfully authenticate himself, the user can contact the IT department and get assistance. The use of the IT department when the system fails to succeed will ensure that the user is actually who they say they are. The combination of each security element in our solution prevents against intrusions from multiple angles.

## References

1. N. Piquero, M. Cohen and A. Piquero, 'How Much is the Public Willing to Pay to be Protected from Identity Theft?,' *Justice Quarterly*, vol. 28, no. 3, pp. 437-459, 2010.
2. K. Finklea, *Identity theft*. [Washington, D.C.]: Congressional Research Service, 2010, pp. 1-27.
3. H. Copes and L. Vieraitis, *Identity thieves*. Boston: Northeastern University Press, 2012.
4. Usa.gov, 'Identity Theft | USA.gov,' 2015. [Online]. Available: https://www.usa.gov/identity-theft.
5. K. Turville, J. Yearwood and C. Miller, 'Understanding Victims of Identity Theft: Preliminary Insights', in *Second Cybercrime and Trustworthy Computing Workshop*, 2010, pp. 60-68.

6. M. Shah and R. Okeke, 'A Framework for Internal Identity Theft Prevention in Retail Industry', in *European Intelligence and Security Information Conference*, 2011, pp. 366-371.
7. E.AÃ¯meur and D. Schonfeld, 'The ultimate invasion of privacy: Identity theft', in *Ninth Annual International Conference on Privacy, Security and Trust*, 2011.
8. E. Holm, 'Identity Crime: The Challenges in the regulation of identity crime,' in *International Conference on Cyber Security, Cyber Warfare, and Digital Forensic (CyberSec), 2012*, 2012, pp. 176-179.
9. S. Rane, 'Standardization of Biometric Template Protection,' *IEEE MultiMedia*, vol. 21, no. 4, pp. 94-99, 2014.
10. Hedayati, 'An analysis of identity theft: Motives, related frauds, techniques and prevention', *Journal of Law and Conflict Resolution*, vol. 4, pp. 1-12, 2012.
11. Fair and Accurate Credit Transactions Act of 2003, Pub. L. No. 108 - 159, 117 Stat. 1952 (2003)
12. Robert Shaler, Akhlesh Lakhtakia. "Acquisition of Sebaceous Fingerprint Topology Using Columnar Thin Films (CTF) on Forensically Relevant Substrates," NIJ-Sponsored, June 2013, NCJ 244195.
13. Hoofnagle, Chris Jay, Internalizing Identity Theft (Oct 1, 2009). 13 UCLA Journal of Law and Technology 1 (Fall 2009). Available at SSRN: http://ssrn.com/abstract=1585564
14. Scott D. Applegate, Major. 2009. Social Engineering: Hacking the Wetware!. Inf. Sec. J.: A Global Perspective 18, 1 (January 2009), 40-46. DOI=http://dx.doi.org/10.1080/19393550802623214
15. Harrell, E., and Langton, L. (2013). Victims of identity theft, 2012 (Report No. NCJ-243779). Washington, DC: Bureau of Justice Statistics. Available at: http://www.a51.nl/sites/default/files/pdf/vit14.pdf

# Implementation of Adaptive Framework and WS Ontology for Improving QoS in Recommendation of WS

Subbulakshmi S, Ramar K, Renjitha R, Sreedevi T U

**Abstract** With the advent of more users accessing internet for information retrieval, researchers are more focused in creating system for recommendation of web service(WS) which minimize the complexity of selection process and optimize the quality of recommendation. This paper implements a framework for recommendation of personalized WS coupled with the quality optimization, using the quality features available in WS Ontology. It helps users to acquire the best recommendation by consuming the contextual information and the quality of WS. Adaptive framework performs i) the retrieval of context information ii) calculation of similarity between users preferences and WS features, similarity between preferred WS with other WS specifications iii) collaboration of web service ratings provided by current user and other users. Finally, WS quality features are considered for computing the Quality of Service. The turnout of recommendation reveals the selection of highly reliable web services, as credibility is used for QoS predication.

**Keywords:** web services; quality factors; context information; ontology; optimization of QoS for WS.

## 1 Introduction

Web Services technology have paved way for the creation of efficient reusable components which are able to perform complex business services and provide useful

Subbulakshmi S, Renjitha R, Sreedevi T U
Department of Computer Science and Applications, Amrita School of Engineering, Amritapuri, Amrita Vishwa Vidhyapeetham, Amrita University, Kollam, Kerala, India , e-mail: subbulakshmis@am.amrita.edu, renji4646@gmail.com, sreedeviprasad79@gmail.com

Ramar K
Department of Computer Science and Engineering, Einstein College of Engineering, Tirunelveli, India , e-mail: kramar.einstein@gmail.com

 
J.M. Corchado Rodriguez et al. (eds.), *Intelligent Systems Technologies and Applications 2016*, Advances in Intelligent Systems and Computing 530,
DOI 10.1007/978-3-319-47952-1_30

high quality information to the end users. The details of the existing web services (WS) specified by the Web Service Providers are available in the registry of Web Services - UDDI (Universal, Description, Discovery, and Integration). The service requestor use the Web Service Description Language (WSDL) to access the web services as per their requirements. In the present era of information technology, user's demands are highly fluctuating. There comes the need for adaptive recommendation of the best web services, based on the context information of the users and the web service providers. The recommendation of the quality WS includes selection, discovery and filtering of the required services.

The selection and discovery of the WS based on the information in the UDDI and WSDL helps to meet the functional requirements of the user. The filtering of the WS based on the non-functional requirements is performed by considering the information available in WS-Policy specification of WS. WS-Policy specifications like WS-Security, WS-Reliable Messaging etc., are specified for each WS as per their non-functional properties. It is highly difficult to refer the WS-Policy of each WS separately in order make the best recommendation. To overcome the above issue, the Ontology of Web Service is created as a central repository which includes the hierarchical relationships between different WS and the non-functional properties of each WS.

The Adaptive framework for the Recommendation System proposes a new system which recommends a list of high quality WS based on both the functional and non- functional characteristics of the WS. The quality of the web service based on the functional characteristics is calculated based on the contextual information using the *collaborative filtering* (CR) and *content based search* (CB) between the users and web services. The quality of the web services based on the non-functional characteristics is calculated with the quality attributes of each web services retrieved from the ontology of web services and the previous users experiences using the credibility calculator of the web services.

The collaborative filtering is computed using the *Pearsons Correlation Coefficient* (PCC) between the ratings given for the web services by different users. To overcome the cold start problem in the above filtering the contextual based similarity search is computed. The 3-layer similarity search is computed using the *Cosine Similarity* between the contextual data of users and web services. It refers to the user's preferences-service profile, user's past references-service profile, service profile- other service profile. The creditability calculation of the web service starts with the *retrieval of the Quality of Service (QoS) values* of the web services from the ontology, and QoS values given by the experienced users. The *missing values* in the QoS of web services are filled up using *min-max normalization*. The credibility of the web service is calculated for each quality factor for that Web service. Then, the weighted QoS value for the web service is computed. Finally, web services with the highest QoS are recommended to users.

The rest of the paper is organized as follows. Section 2 has the details of the Literature Review of the existing recommendation system. Section 3 suggests an Adaptive Framework for the high quality Recommendation System. Section 4 explains the Experimental Results of the implemented system. Section 5 summarizes and Concludes with Future work.

## 2 Literature Review

The current system of content and collaborative based approach identifies the best web services based on the user preferences and other user's preferences for related WS. The Neighborhood based collaborative filtering methods were also designed to predict the list of recommended services. Memory based models were also used to predict the similarity between the users and web services based on their ratings preferences.

Content based systems were used to recommend WS, music or product to the users. They utilize a sequence of discrete characteristics of an item in order to recommend appropriate items with similar properties. The content based mail filtering [13] is an active mail-filter agent for an intelligent document processing system. Most recently the CB prediction [7] was used to measure the similarity between the user preference and content of services, also uses one more filtering method that takes similar user rating for the web service.

In 2001, Amazon proposed item-based collaborative filtering and they personalized the recommendation [15] for each of its users. In another system, they have combined item based collaborative filtering and demographics [5] for recommending the WS. This hybrid approach helps in avoiding problems like scalability and sparsity and it gives high performance than the previous recommendation systems. But this system does not always react to every user in the same manner, as the non-functional factors affect each user's environment.

A Personalized system [17] uses interactive query based reasoning to recommend the travel plan for the users by considering the preferences of the users, existing travel plans available and by avoiding the annoying and unwanted items in the travel plan. Importance was given to the interactive query with the customers mainly when the system fails to get best results. It performs case-based [16]] reasoning before making the preferred recommendations. The system produces the better plan, but has issues related to the irrelevant user interactive designs.

To overcome the cold start problem of the knowledge-based system, a Demographic recommendation system [18] uses the demographic information of the tourists and performs categorization with the help of different machine learning algorithms. Based on the classification [19] it is able to identify the best tourist attractions for the user. The system failed to produce accurate results as it uses only demographic information. It could produce better results, if the textual reviews and data are taken into consideration.

The new approach describes QoS method which is developed as a WS Composition in which QoS [12] is the main building block of precise recommendation approach. Here QoS data are divided into two categories and it uses intelligent strategies on those categories and produces an effective approach which is proved by a case study. It also includes the quality parameters like performance, availability, reliability, and stability, etc. In WS selection model [3], QoS consultant acts as mediator agent that gives priority to a user's preference for selecting QoS parameter and helps to select the best recommendation.

Our proposed system is an adaptive framework where content and collaborative based techniques are combined with the ontological based recommendation. It incorporates semantic knowledge along with non-functional quality factors of the web services.

# 3 Proposed Adaptive Framework Recommendation System

Selection of the optimized web service is main objective of the system as the current Internet era is flooded with numerous web services provided by various service providers. Selecting the highly reliable research paper is an important task for the people interested in doing research. We have taken the implementation of the system for the recommendation of web services providing best research papers, since ascertaining the best service provider is a challenging task. The architecture of our system is shown below:

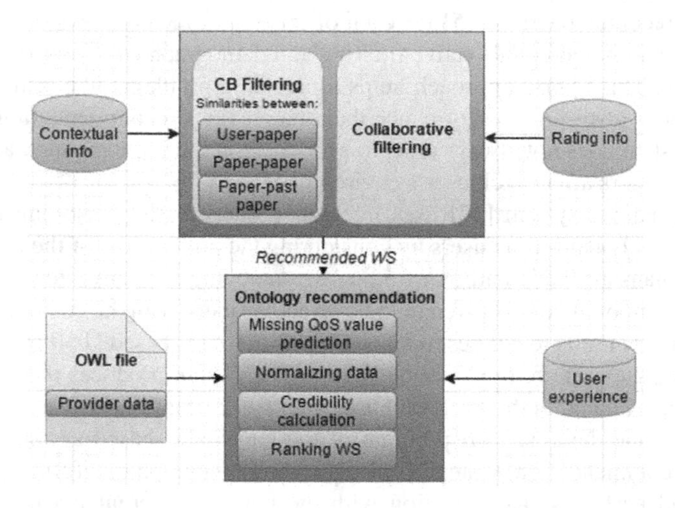


Fig. 1: System Architecture. ( IAFO-QRW)

## 3.1 *Content Based Method of Filtering*

Content Based filtering recommends services based on the contextual information related to the user preferences and WS profile information. In CB, the filtering process is done as a 3-layer similarity search. It uses the contextual data related to user profile, web service profile and past history of each user. Keyword selection gets an important role in the filtering method. For that the contextual information is identified from user preferences and WS profiles. Then the similarity between user and WS profile are calculated using Cosine Similarity algorithm. The formula for Cosine Similarity Algorithm is as follows:

$$Similarity = \frac{A.B}{||A|| * ||B||} = \frac{\sum_{i=1}^{n} A_i B_i}{\sqrt{\sum_{i=1}^{n} A_i^2}\sqrt{\sum_{i=1}^{n} B_i^2}}. \quad (1)$$

Vectors 'A' and 'B' are the values from WS profile, user profile and past history of users. The 3-layer similarity search is as follows:

### 3.1.1 User Profile with Paper Profile:

When the user want to find the recommendation related to a given paper, first the similarity between that user profile and paper profiles are calculated. For that, combination of current user's interested keywords set and the given paper's available keywords are computed against each of other papers in the paper profile. The papers with the highest similarity rates are selected.

Table 1: Sample data for User Profile - Paper Profile Similarity

| | Keywords (User Preferences) | | | | | Paper | Keywords in Paper Profile | | | | |
|---|---|---|---|---|---|---|---|---|---|---|---|
| User 1 | 1 | 1 | 1 | 0 | 0 | Paper 1 | 0 | 0 | 1 | 1 | 1 |
| | | | | | | Paper 2 | 1 | 1 | 1 | 0 | 0 |
| | | | | | | Paper 3 | 0 | 1 | 0 | 0 | 1 |
| | | | | | | Paper 4 | 1 | 0 | 0 | 1 | 0 |

In Table 1 user's preference is marked using binary encoding, where 1 means like and 0 means unlike for that keyword. The paper profile's binary encoding shows the presence (1) or absence (0) of a keyword.

Algorithm for measuring the similarity between User - Paper Profiles

1. U $\rightarrow$ keyword vector of the active user.
2. D$\rightarrow$ set of all papers(keywords) in paper-profile.
3. i$\rightarrow$ Keyword vector of a paper.
4. For all i in D
5. Cosine similarity $(U, D_i)$
6. Return the top N similar papers

### 3.1.2 Paper with Every Other Paper in Paper Profile:

The similarity between user's selected paper and each paper in the paper profile are calculated and the highest similar papers are taken for further computations.

Table 2: Sample Data for Paper - Other Paper Profile Similarity

| | Paper Profile | | | | |
|---|---|---|---|---|---|
| Paper 1 | 0 | 1 | 0 | 0 | 1 |
| Paper 2 | 1 | 0 | 1 | 0 | 0 |
| Paper 3 | 1 | 0 | 1 | 0 | 1 |
| Paper 4 | 0 | 1 | 1 | 1 | 0 |

### 3.1.3 Paper with Other Paper in Past-Paper Profile:

With the assumption that user's preference will be consistent over time, past user's searches are considered. They could have a greater influence in the current interest of the user. So, the active users past history is compared against paper-profile.

Table 3: Sample Data for Paper - Past Paper Similarity

| | Keywords in the Selected Paper | | | | | Paper | Keywords in Past Paper Profile | | | | |
|---|---|---|---|---|---|---|---|---|---|---|---|
| Paper 1 | 1 | 1 | 0 | 0 | 1 | Paper 2 | 1 | 1 | 0 | 0 | 1 |
| | | | | | | Paper 3 | 0 | 0 | 1 | 1 | 0 |
| | | | | | | Paper 5 | 0 | 0 | 1 | 0 | 0 |
| | | | | | | Paper 6 | 1 | 1 | 0 | 1 | 1 |

## 3.2 *Collaborative Method for Filtering:*

Collaborative approach recommends papers based on the peer users ratings for different papers. This method recommends top N list of papers based on correlation between the behaviors of the current user and other peer users. Pearsons correlation algorithm is used to measure the similarity of the rating between different users. Formula for Pearsons Correlation Algorithm is as follows:

$$PCC = \frac{N\sum xy - (\sum x)(\sum y)}{\sqrt{N\sum x^2 - (\sum x)^2}\sqrt{N\sum y^2 - (\sum y)^2}} \quad . \tag{2}$$

where N is the number of pairs of scores and x is given paper's users rating and y is the other paper's users rating.

Table 4: Sample Data used in Collaborative filtering.

| Users | Paper ratings | | | | |
|---|---|---|---|---|---|
| User 1 | 2 | 7 | 5 | 4 | 1 |
| User 2 | 9 | 0 | 6 | 5 | 3 |
| User 3 | 3 | 2 | 5 | 8 | 5 |
| User 4 | 4 | 1 | 4 | 7 | 4 |

After the computations of CB and CR approaches, we calculate the Relative Frequency of each service. It is calculated using the following formula:

$$Relative frequency(RF_i) = \frac{f_i}{\sum_{i=1}^{n} f_i} \quad . \tag{3}$$

where $f_i$ is sum of the similarity value and correlative value of the $i^{th}$ paper and n is the number of papers. Then the top N web services with the highest relative frequency values are identified and passed on to the ontology based filtering. Those services are considered as the best fit to the contextual information of the user requirements.

## 3.3 Ontology Based Filtering Method:

### 3.3.1 Ontology Creation for Web Services.

Ontology based filtering starts with the creation of ontology for web services. The QoS details of the web services retrieved from the WS Description, WS-Policy, and Service Level Agreement (SLA) are used for the creation of Ontology. Ontology is less ambiguous and contains controlled vocabulary of terms and theories. It reveals the semantic relationship between the web services.

Web Ontology Language (OWL) is used for creating the definition of web services, their relationships and properties. Our system mainly focuses on the non-functional quality properties of the WS. The quality factors are classified into two categories, dynamic and static. The values of dynamic factors are changed when the execution happens. The values of static quality factors are provided by the service providers.

The OWL file contains information about all WS in the form of classes, subclasses, properties and restrictions. Properties that we include are year of publication, quality factor values, such as availability, response time, experience, reputation. The hierarchy of the WS are created using Protege tool.

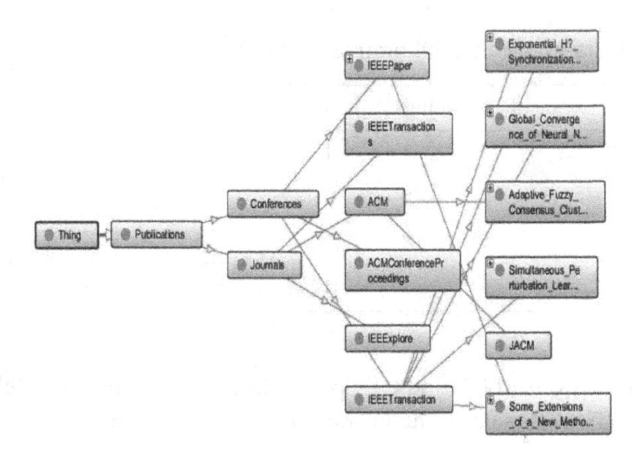


Fig. 2: Ontology hierarchy. ( IAFO-QRW)

The root of the tree hierarchy is *Thing* which holds all other classes. *Publications* is the sub class of Thing and it is the main division used to identify the quality of paper. The leaf nodes/sub-classes can contain the *research papers* according to the classification. Properties are included for papers and Publications as per requirements. They are specified as the model of restrictions in the OWL file. The class/sub class can have one or more restrictions and it reveals the behavior of the specific class. The overall view of the sample ontology created for web services is given below:

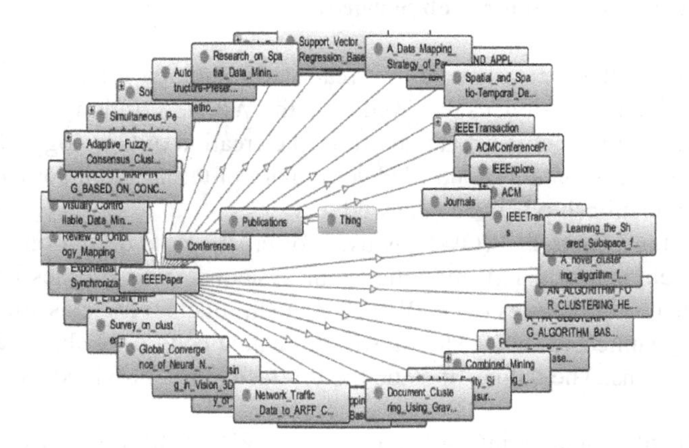


Fig. 3: Ontology for Web Services ( IAFO-QRW)

### 3.3.2 Missing Value Prediction

If the quality factors for any WS are missing in the OWL file, then those values are calculated as follows:

(a) *Response Time:* It depends on the user who is accessing and so it is calculated based on the location information of the user. The calculation is done with the *K nearest neighbor algorithm*, using the latitude and longitude values of the location of the user's IP address.

(b) *Availability:* It is the system active time and it is measured as,

$$Availability = \frac{UT}{(UT + DT)} \quad . \tag{4}$$

where $UT$ is the uptime and $DT$ is the downtime of WS.

(c) *Reputation:* It describes the satisfaction of users about the WS. Different users have different values for reputation. So the formula to calculate reputation is,

$$Reputation = \sum_{i=1}^{n} \frac{R}{N} \quad . \tag{5}$$

where $R$ is the end user's rating about a WS and $N$ is the number of times, the service have been graded.

### 3.3.3 Normalization of Values

QoS values for different quality factors given by the users and those provided by the service provider are normalized, in order to get the values in a specific range (0 , 1).

$$V' = \frac{v - minA}{maxA - minA}(new_{maxA} - new_{minA}) + new_{minA} \quad . \tag{6}$$

The above formula represents *min-max normalization* and *minA, maxA* denotes the minimum and maximum values in the set. The $(new_{maxA}, new_{minA})$ is the new range for normalization.

### 3.3.4 Credibility Calculation for the Quality Factors

The fidelity of WS recommendation always relies on the credibility of the service. It is calculated as the disparity between user execution values and provider given QoS values. The formula for credibility calculation is as follows:

$$Q_n = \begin{cases} \frac{1}{K}\sum_{k=1}^{K}(1 - \frac{|er_{nk-qn}|}{q_n}), & \text{if } er_{nk<2qn} \\ 0, & er_{nk\geq 2qn} \end{cases} \quad . \tag{7}$$

-

where K represents the number of execution results, $er_{nk}$ represents the $k^{th}$ execution result of the $n^{th}$ quality factor. $q_n$ is the QoS value given by the service provider for the $n^{th}$ quality factor.

#### 3.3.5 QoS Value for Each Web Service

It is calculated by assigning weightage for each QoS factor according to the priority or importance in the context of service selection using the given formula:

$$Q_{ws} = \frac{W_1Q_1 + W_2Q_2 + W_3Q_3 + .........}{N} \quad . \tag{8}$$

where $W_i$ and $Q_i$ are the weightage assigned and the creditability value for $i^{th}$ quality attribute of a particular web services, respectively.

### *3.4 Ranking of Web Services*

For ranking of the web services, we take into consideration both the functional and non-functional requirements of the users. The Web services that best comply the functional aspects are identified by using content and collaborative methods. The ontology based method, identify the high quality WS based on its non-functional properties. The overall score for each WS is calculated using the relative frequency method with the calculated similarity value, correlation value and QoS value of the WS. Then the final score is used for ranking the web services and the top web services are recommended to the users. The services are of high quality and best suits the user's requirements.

## 4 Experimental Results

Following are the experimental results of our adaptive framework for web services recommendation system. We have done experiments with the data set which includes the user's preferences related to the interested areas of research and web service profile information which provides contents of different research papers. We have used our system i) to identify the recommended papers with the conventional methods of collaborative filtering and content-based filtering and ii) add quality for those recommended web services by using the QoS factors. In order to find the performance, we compared the results of collaborative, content-based, combination of CB with CR and ontology based filtering. The recommendation accuracy was evaluated by examining the quality of top x rankings of Web services.

Conventional methods have their own limitations about performance and quality of service delivery. So in order to make the recommender system more powerful, QoS factors are included. The graph in fig. 4(a) shows the individual results of CB filtering done based on the paper selected by the user. It shows the similarity output as estimated in 3-layers. The combination of the 3-layers of CB results is calculated by taking the average of those similarities and it is shown in fig.4(b). The paper ID p2 shows the highest similarity value.

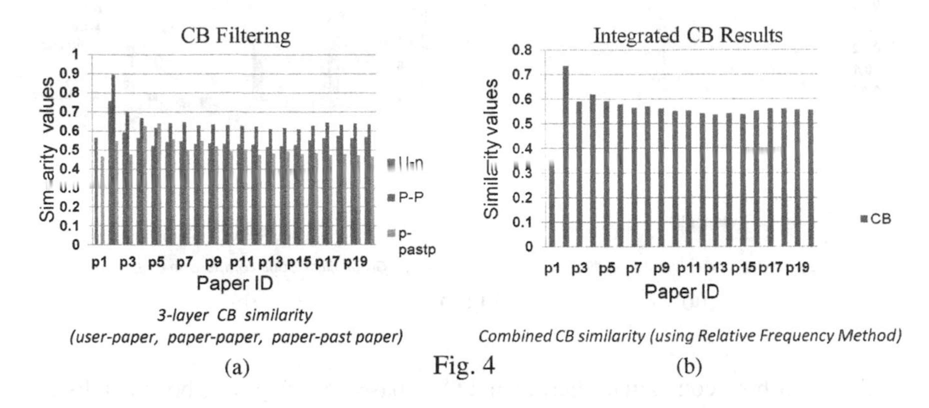


3-layer CB similarity (user-paper, paper-paper, paper-past paper)

Combined CB similarity (using Relative Frequency Method)

(a) Fig. 4 (b)

Then in collaborative filtering, ratings of different papers given by experienced users are computed against the selected paper using PCC (fig.5a). This method shows that the rating of paper ID with p7 has the highest value. The importance of the content based similarity has to be reflected in this result. So, first we show the comparative output of the CB and CR methods in Fig.5(b).

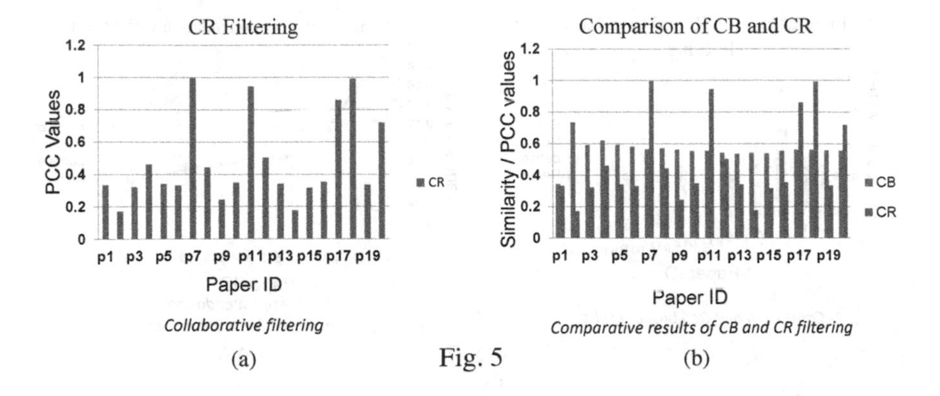


Collaborative filtering

Comparative results of CB and CR filtering

(a) Fig. 5 (b)

Then the individual results of the above methods are again integrated to get enhanced result with includes both user preferences and similarity of the paper selected. This is done by relative frequency method and its output is shown in fig.6(a). From those results, the top x papers are selected and the qualities of those papers are predicted by using ontology based filtering. The quality factor values for the web services given by the providers are available in the OWL file are retrieved and the credibility is calculated against the user experience matrix. The credibility for those quality factors of the web services is used to compute the QoS value for each web service. The comparison of the CB, collaborative, and QoS values is shown in fig.6(b).

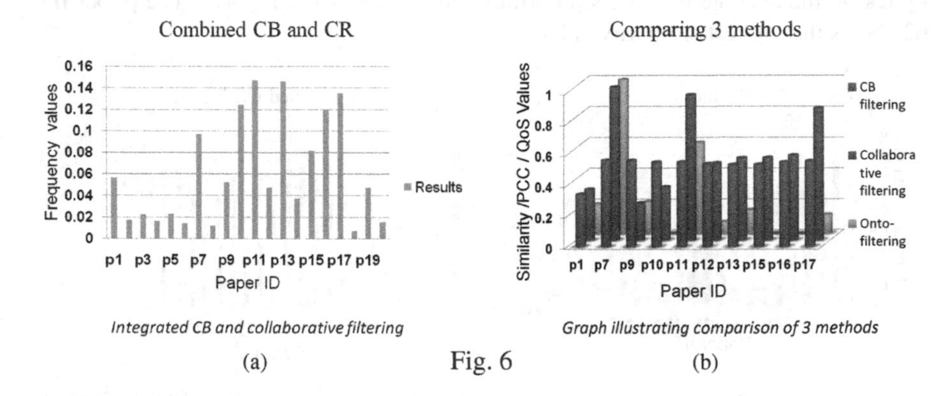


*Integrated CB and collaborative filtering* *Graph illustrating comparison of 3 methods*

(a) Fig. 6 (b)

To reflect both contextual filtering and QoS based filtering the above results are merged together as shown in the fig.7(a). The combined result selects the high quality papers with best fit the requirements of the user.

Finally, the ranking of those services is identified and they are included in the list of services recommended to the users. Thus the users would be able to get the best services from the huge volume of data available on the web. The ranked list of papers in the order of recommendation is shown in fig. 7(b).

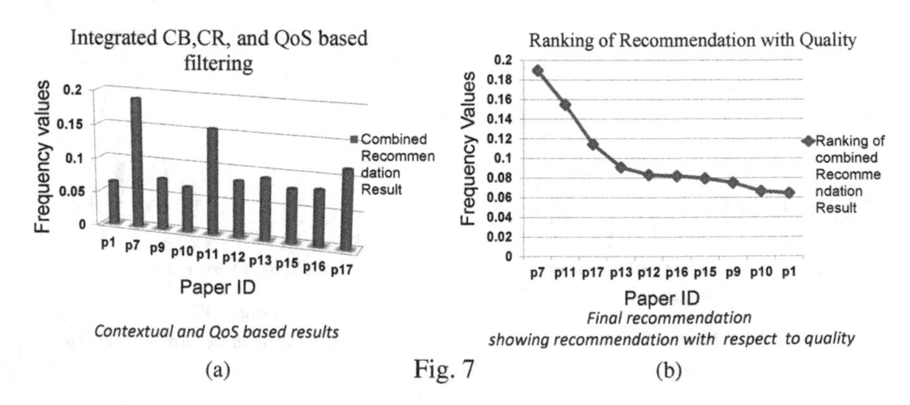


*Contextual and QoS based results* *Final recommendation showing recommendation with respect to quality*

(a) Fig. 7 (b)

The result analysis of the adaptive framework reveals the fact that even if a service fits the contextual information if it fails to satisfy the quality factor, those services cannot be considered for recommendation. Those services in real world would fail to satisfy user's expectation as they may not be reliable services.

## 5 Conclusion and Future Work

Current Web Service Information Retrieval systems desperately demands for the creation of new methodology for the retrieval of the web services, with high quality in par with their functional requirements. High quality WS are selected based on the non-functional properties (QoS). The adaptive framework presents QoS aware context based web service recommendation approach. The basic idea in our approach is to create a framework which predicts the set of web services that satisfies the contextual information related to the user requirements using similarity methods and collaborative filtering methods. The details of the web services from the WSDL, WS- specification are used for the creation of the web services ontology. The quality factor values of the web services retrieved from the ontology are used for prediction of QoS value for those services. The QoS values helps to identify the high quality web services. Finally, the relative frequency method is used to combine the results of similarity score, collaborative score and QoS values of the web services. The combined output is used to rank the web services and recommend the high quality services which best suits users preferences and importance. The results show a significant increase in the quality of predicting personalized web service in the recommendation system. The system helps the user find appropriate web service in a complex information search and decision task.

In our present system, the missing quality value is predicted only for Response Time, Reputation and Availability. The prediction of other quality factors like security, experience can be calculated in our future work. Moreover the functional properties could be included in the ontology in future, so that it would help to have semantic based filtering.

## References


1. Ahu Sieg,Bamshad Mobasher,Robin Burke: Improving the effectiveness of collaborative recommendation with ontology-based user profiles. Proceedings of the 1st International Workshop on Information Heterogeneity and Fusion in Recommender Systems, pages 39-46, 2010-09-26.
2. Deng-Neng CHEN, Yao-Chun CHIANG: Combining Personal Ontology and Collaborative Filtering to Design a Document Recommendation System. Journal of Service Science and Management, Vol.2, pages 322-328, 2013.
3. Divya Sachan,Saurabh Kumar Dixit,and Sandeep Kumar: QoS AWARE FORMALIZED MODEL FOR SEMANTIC WEB SERVICE SELECTION. International Journal of Web and Semantic Technology (IJWesT) Vol.5, No.4, October 2014.
4. E. Michael Maximilien, Munindar P. Singh: A Framework and Ontology for Dynamic Web Services Selection. IEEE Internet Computing, Volume:8 , Issue:5, Sept-Oct 2004 .
5. Jyoti Gupta: Performance Analysis of Recommendation System Based On Collaborative Filtering and Demographics. International Conference on Communication, Information and Computing Technology (ICCICT), Jan. 15-17, Page(s): 1-6, 2015.
6. Kyriakos Kritikos,Dimitris Plexousakis: Requirements for QoS-Based Web Service Description and Discovery. IEEE Transactions on Services Computing Vol.2, Issue: 4, 11 Sept. 2009.

7. Lina Yao and Quan Z. Sheng, Aviv Segev, Jian Yu: Recommending Web Services via Combining Collaborative Filtering with Content-based Features. IEEE 20th International Conference on Web Services, Page(s): 42-49, 2013.
8. Maringela Vanzin, Karin Becker, and Duncan Dubugras Alcoba Ruiz: Ontology-Based Filtering Mechanisms for Web Usage Patterns Retrieval. 6th International Conference, EC-Web, Copenhagen, Denmark, August 23-26, pp. 267277, 2005.
9. Nitai B. Silva, Ing-Ren Tsang, George D.C. Cavalcanti, and Ing-Jyh Tsang: A Graph-Based Friend Recommendation System Using Genetic Algorithm. IEEE Congress on Evolutionary Computation (CEC), pages 18-23 July 2010.
10. Sunita Tiwari,Saroj Kaushik: A Non Functional Properties Based Web Service Recommender System. International Conference on Computational Intelligence and Software Engineering (CISE), Page(s): 1 - 4, 10-12 Dec. 2010.
11. Wang Chun Hong, Yun Cheng, P.R.China: Design and Development on Internet-based Information Filtering System. International Conference on Educational and Network Technology (ICENT), 25-27 June 2010.
12. Wang Denghui, Huang Hao, Xie Changsheng: A Novel Web Service Composition Recommendation Approach Based on Reliable QoS. IEEE Eighth International Conference on Networking, Architecture and Storage (NAS), Page(s): 321 - 325,17-19 July 2013.
13. Yaqiong Wang, Junjie Wu, Zhiang Wu, Hua Yuan: Popular Items or Niche Items: Flexible Recommendation using Cosine Patterns. IEEE International Conference on Data Mining Workshop (ICDMW), Page(s): 205 - 212, 14-14 Dec. 2014.
14. Xi Chen, Zibin Zheng, Qi Yu, and Michael R. Lyu: Web Service Recommendation via Exploiting Location and QoS Information. IEEE Transactions on Parallel and Distributed Systems, Page(s): 1913 - 1924, July 2014.
15. Greg Linden, Brent Smith, and Jeremy York: Amazon.com Recommendations Item-to-Item Collaborative Filtering. Published by IEEE Computer Society on Page(s): 76 - 80, Feb. 2003.
16. Barry Smith,Case Based Recommendation, Springer Berlin Heidelberg,Pages 342-376, 2007.
17. Francesco Ricci,Fabio Del Missier: Personalized Product Recommendation through Interactive Query Management and Case-Based Reasoning, Springer-Verlag Berlin, Heidelberg, Pages 479-493, 2003.
18. Yuanyuan Wang,Stephen Chi-fai Chanand Grace Ngai: Applicability of Demographic Recommender System to Tourist Attractions - A CaseStudy on TripAdvisor. ACM International Conferences on Web Intelligence and Intelligent Agent Technology, Pages 97-101, 2012.
19. S. B. Kotsiantis, Zaharakis, Pintelas: Machine Learning: a review of classification and combining techniques, Springer Science+Business Media B.V. pp 159190, 10 November 2007.

# Cluster Based Approach to Cache Oblivious Average Filter Using RMI

**Manmeet Kaur, Akshay Girdhar and Sachin Bagga**

**Abstract** Parallel execution uses the power of multiple system simultaneously thus comes out to be an efficient approach to handle and process complex problems producing result in less execution time. Present paper represents implementation of a Cache-oblivious algorithm for de-noising of corrupted images using parallel processing approach. In present era, there is a need to work with large sized image. Sequential execution of any process will results in long time of execution ultimately degradation of performance. This paper focuses to implement the algorithm on distributed objects by Cluster using RMI and utilize the concept of multithreading to enhance the depth of distributed parallel technology.

## 1 Introduction

In distributed parallel processing, more than one processor or complete systems work in parallel manner for bringing outcome of a common problem. With advent of digital image capturing and graphic designing devices, it becomes a day to day activity to process images. Now- a-days images captured or designed are of bigger size. If one need to retouch or edit images then one have to deal with hundred of megabytes. When it comes about scientific, satellite or medical imaging the sea deepens up to terabytes of data. Fortunately, it is possible to grasp the advantage of distributed parallel technology for reducing processing time by running processes on network environment.
Distributed parallel processing uses divide and conquer method to solve a particular task i.e. single task is divided into number of sub-tasks and is given to individual processor or units where threads work to solve a sub task. In the end, results of all sub- tasks make the result of whole task when combined. Cluster is the name given to the distributed remote objects when they are combined and work in parallel fashion to solve a single problem. For communication and transmission of data between remotely distributed systems, out of various techniques RMI remote method invocation makes reliable connection and uses java Remote Method protocol which works on TCP/IP hence makes communication between remote objects possible [1]. Remote object referencing is done by RMI registry. [1]

[1] Manmeet Kaur
Guru Nanak Dev Engg. College, Ludhiana, Punjab, manmeet.yg@gmail.com
Akshay Girdhar, Sachin Bagga
Guru Nanak Dev Engg. College, Ludhiana, Punjab, akshay1975@gmail.com, sachin8510@gmail.com



J.M. Corchado Rodriguez et al. (eds.), *Intelligent Systems Technologies and Applications 2016*, Advances in Intelligent Systems and Computing 530,
DOI 10.1007/978-3-319-47952-1_31

## 2 Cluster Computing

Cluster computing is based on Master- Slave architecture, where number of distributed nodes i.e. system work for a common problem. Size of cluster can vary by increasing or decreasing the number of nodes. In it, if any of the nodes fails or stop working then the whole cluster is not affected and easily replaced by working node explicitly. In Cluster environment nodes are connected to each other through Local area network (LAN). Firstly, task is given to master system then master system distributes the task into number of sub task depending upon the number slave nodes available for processing. Then the sub task is given to individual slaves who work on its piece of work and send the result back to Master system. In the end, master system combines the result from all the slaves and gives the final output of given task [2].

## 3 Remote Method Invocation

For communication and data passing between remotely distributed objects, Remote Method Invocation (RMI) is used which provides network layer framework which act as intermediate [3]. RMI works by using fastest ways of establishing connection between client and server which is not even interrupted by Firewall.RMI can utilize any of the below written way-out to establish connection-

- Direct connection by using Socket
- HTTP POST request
- Standard HTTP Port 80 along with Common Gateway Interface (CGI)

Master system uses lookup() method to start communication with the remote systems. Every slave system will start RMI registry which will help master system to obtain reference to available remote slave systems.RMI uses Java Remote Method Protocol (JRMP) which works on TCP/IP layer. Connection provided by RMI is reliable.

## 4 Cache Oblivious Algorithm

Current technology demands for large capacity memories with fast access rate. As access time of memory depends upon cache memory and other features attached to cache [4], so utilization of available cache memories in better and efficient manner can be a meaningful option which leads to fulfillment of current demand other than purchasing large storage capacity devices.Eq.1 shows the relation of cache access time with effective access time-

$$Effective\ Access\ time = \mathrm{HTc} + 1 - \mathrm{H}\,(\mathrm{Tm} + \mathrm{Tc}) \quad (1)$$

Where Tc is time to access cache, H is hit ratio, and Tm is time to access memory. Further, Eq.2 shows dependency of cache access time-

$$Cache\ access\ time = Cache\ hit\ ratio + Cache\ miss\ rate \times Cache\ miss\ penality \quad (2)$$

Cache hit ratio and cache miss penalty is dependent on hardware. So if cache miss rate is reduced there is a chance of improvement of effective access time [5]. The algorithm which improves the cache access time by decreasing cache miss and is not dependent on cache size is known as Cache-Oblivious Algorithm.
when any application or component request for data, CPU look for particular data in cache level 1 i.e. L1, if it is not found, it is counted as 1 miss and the request is forwarded to higher Levels such as L2, L3 and Random Access Memory (RAM).Every time data not found it is counted as miss and if it is found in cache it is counted as Cache hit. More the CPU moves from one level to another, more number of cache misses.

## 5 Image Noise

Noise manipulates the actual pixel values of an image. The effected pixels get modified pixel values and rest of pixels retains their previous original values. Noise can be impulsive, additive or multiplicative in nature.

Suppose an original image is given by i (x,y), noise introduced is n(x,y) and distorted image at location (x,y) is $f(x,y)$ [6].
Then the image corrupted by additive noise is given by

$$f(x, y) = i(x, y) + \eta(x, y) \quad (3)$$

Similarly, if noise is multiplicative then the corrupted image will be written as

$$f(x, y) = i(x, y) * \eta(x, y) \quad (4)$$

Noises are divided into categories depending upon their nature and effects [7]. Some of the noises are shown in table 1 below [8]. Different filters work for the removal of particular noises [9]. In this paper, work is concentrated on Salt and pepper noise and the filter which works best for its removal.

Table 1. Noises with type

| Name of the Noise | Type |
|---|---|
| Salt and Pepper noise | Impulse Noise |
| Gaussian noise | Additive Noise |
| Speckle noise | Multiplicative Noise |
| Uniform noise | Additive Noise |
| Poisson noise | Additive Noise |

## 6 Average Filter

Average filter is the simplest Linear filters which removes noise by smoothing it [10]. Basically filters are matrix of [k x k] dimensions. Size of filter can be 3x3, 5x5and 11x11 and so on. For this work size of filter used is 3x3. Each cell of filter contains some value which is called its weight. In simple Average filter weight of each cell is 1[11] as shown in Fig.1.

**Fig. 1** 3 X 3 Simple average filter, where weight of each cell is 1

| 1 | 1 | 1 |
|---|---|---|
| 1 | 1 | 1 |
| 1 | 1 | 1 |

On a given image A[n x n], an average filter F[3 x 3] is applied, filters moves across image,

- Firstly in row 0 it moves from left to right manipulating value for A[0,0] to A[1,n]
- Row is incremented and above step is repeated for row 1.
- Process continues till whole image is covered.

If a(x, y) is current pixel of image on which average filter of size 3x3 is applied, then image pixel after applying filter is given by b(x , y) and is calculated as shown in Eq.5

$$b(x,y)=\{a(x-1)(y-1) + a(x-1)(y) + a(x-1)(y+1) + a(x)(y-1) + a(x)(y) + a(x)(y+1) + a(x+1)(y-1) + a(x+1)(y) + a(x+1)(y+1)\} / 9 \quad (5)$$

Where 9 is number of neighbors under filter

## 7 Literature Review

Singh Et al. (2014) in their paper "*Noise reduction in images using enhanced average filter*" explains that the most common noise which degrades the images is salt and pepper noise. It is impulsive in nature it means it changes the value of original image pixel to either maximum or minimum limit. Further, explained the different types of filters which can work in removal of salt and pepper noise. Out of which major emphasis is on Average filter. Mivule and Turner (2014) in their paper "*Applying moving average filtering for non-interactive differential privacy setting*" proposed a model which uses average filtering to implement differential privacy at various levels to finally achieve privatized and secured data set. Author further explain that average filter is the most used filter in field of image processing, having variety of applications. Ghosh and Mukhopadhyaya (2012) in their paper "*Cache Oblivious Algorithm of Average Filtering in Image Processing*" rises up the idea that algorithm for any task can be divided into two categories one is cache-dependent and other is cache-independent also known as cache oblivious. So after understanding the need to design a cache oblivious algorithm from literature studies, author has proposed a cache oblivious algorithm for average filtering. Using Cache grind, a tool of Val grind Simulator, author has proved that the proposed algorithm [5] has decreased the cache misses as compared to traditional algorithm of average filtering. Comparison is shown in table 2.

Table 2. Comparison of L2 cache misses in 10 lacs using Traditional and Cache Oblivious Average Filtering algorithm

| Image dimensions | Cache misses in Traditional Algorithm(Ct) | Cache misses in Cache Oblivious Algorithm(Co) | Difference Diff.=(Ct-Co) |
|---|---|---|---|
| 1000 * 1000 | 17.86 | 1.85 | 16.01 |
| 825 * 825 | 12.8 | 1.5 | 11.3 |
| 750 * 750 | 9.85 | 1.19 | 8.66 |
| 625 * 625 | 7.78 | 0.93 | 6.85 |
| 500 * 500 | 4.68 | 0.72 | 3.96 |

Kaur Et al. (2015) in their paper "*RMI Approach to Cluster Based Winograd's Variant of Strassen's Method*" demonstrates the use of RMI to make cluster taking Winograd's variant of Strassen as basic task. Author has compared the outcome of Socket based programming and Cluster based for the same data size i.e. matrix size. Table no.2 completely depicts the idea of author in terms of total execution time that RMI based cluster work in much better and efficient way than socket based technique.

Table 3. Comparison of RMI based with Socket Based in terms of Execution time

| **Nodes in Cluster** | **Image dimensions (Input)** | **Execution time taken using RMI (t1)** | **Execution time taken using Socket (t2)** | **Difference Diff.=(t2-t1)** |
|---|---|---|---|---|
| 4 | 256 * 256 | 0.262 sec | 1.369 sec | 1.107 sec |
| 8 | 256 * 256 | 0.0293 sec | 1.57 sec | 1.540 sec |
| 4 | 512 * 512 | 0.872 sec | 9.818 sec | 9.846 sec |
| 8 | 512 * 512 | 1.763 sec | 5.93 sec | 4.167 sec |
| 4 | 1024 * 1024 | 4.457 sec | 74 sec | 69.543 sec |
| 8 | 1024 * 1024 | 3.192 sec | 40.47 sec | 37.278 sec |

All of the papers are base paper for present work i.e. implementing cache-oblivious image de-noising technique on distributed objects making cluster. Further for communication between nodes of cluster a method is required, as Kaur Et al. (2015) has compared the working of RMI and Socket for Cluster and Results in table 3 has shown RMI is better than socket.

## 8 Algorithmic Approach

The proposed work implements the cache oblivious average filtering algorithm using multithreading on cluster which used Remote Method Invocation (RMI) approach. Using this approach large sized image can be filtered on multiple systems i.e. utilizing the computational power of more than one single system. The algorithm used is cache oblivious hence pixel value once fetched means once it enters cache memory, it is used again and again hence decreasing the cache misses and increasing overall execution time.
Cache oblivious average filtering algorithm completes its task in two steps

- Step 1: calculate the row-wise sum of neighbor including current pixel value, which are under 3 x 3 filter.
- Step 2: calculate the column-wise sum of neighbor including current pixel value and divide it by n i.e. number of pixels under filter.

**Fig. 2** Working of cache oblivious average filtering algorithm
i 3 X 3 input image
ii. Intermediate output after step1 iii. Final output after step 2

| M | N | O |
|---|---|---|
| P | Q | R |
| S | T | U |

i

| M+N | M+N+O | N+O |
|---|---|---|
| P+Q | P+Q+R | Q+R |
| S+T | S+T+U | T+U |

ii

| $\frac{(M+N)+(P+Q)}{4}$ | $\frac{(M+N+O)+(P+Q+R)}{6}$ | $\frac{(N+O)+(Q+R)}{4}$ |
|---|---|---|
| $\frac{(M+N)+(P+Q)+(S+T)}{6}$ | $\frac{(M+N+O)+(P+Q+R)+S+T+U}{9}$ | $\frac{(N+O)+(Q+R)+(T+U)}{6}$ |
| $\frac{(P+Q)+(S+T)}{4}$ | $\frac{(P+Q+R)+(S+T+U)}{6}$ | $\frac{(Q+R)+(T+U)}{4}$ |

iii

The time complexity of cache-oblivious average filtering algorithm is same as that of traditional algorithm but improved cache efficiency leads to less execution time as value of pixel once calculated is reused is the actual outcome of this algorithm [5].

## 9 Work Flow

The distributed master slave architecture uses Remote method invocation technique for communication between master and slave. To begin communication, it is necessary that master system have all the stubs from remote nodes attached to it i.e. slave systems. To generate stub every slave system use rmic (*remote method invocation compiler*). Further workflow is as follows:

1. All slave system start their bootstrap by running *start rmiregistry* on systems.
2. Image with noise is put in to master system. Master system coverts input image of m x n pixels to array of size m x n, where each element consist of grayscale value of pixel [12].
3. As large sized images are used in it. So for creating large array default assigned heap memory is not enough, there is a need to explicitly reserve heap memory. For this master system will run the java program in a specialized manner as shown in Eq. 6 below

   $$java - Xmx1200m\ Filename \quad (6)$$

   where 1200 is the heap memory space reserved explicitly
4. Master system creates sub-array from main array generating the workload for every slave attached in cluster. Division is done according to the number of slave nodes available in cluster as per the formula in Eq. 7

   $$\text{size of sub array} = \frac{\text{size of original array}}{\text{number of slave systems}} \quad (7)$$

5. Now 2d arrays are converted into 3d array say C[a][b][c] , where a is the number of 2d array whose dimension is b x c.
6. Master system uses multithreading for assigning sub tasks to the remote objects in parallel. [13].
7. Each slave system receives its sub-array on which it has to perform functions.
8. Every slave system performs average filtering as per cache oblivious algorithm on the data set i.e. sub array received by individual.
9. After performing filtration, every slave sends back the output generated individually. Master system maintains the check over thread using isAlive() function so that final result will be calculated only after all threads finishes task assigned to them
10. When master receives return sub-arrays from all slaves after completion of all threads, master combine them and makes single output 3d array. This 3d array is then converted to 2d array.
11. In final step, filtered noise-free image is regenerated from 2d array.

## 10 Benefits of Proposed Approach

RMI based cluster for cache oblivious average filtering algorithm provides a way-out to work with images in distributed environment to utilize the power of remotely connected systems and brings out lot of benefits as:

- **Array generation** – proposed approach converts image to 2-d array of equivalent size, where each element corresponds to grayscale value of particular pixel.
- **Decomposition of large sized images** –The proposed approach decompose the large sized images say satellite image, medical images etc into sub images.
- **Flexible Cluster Size** – Cluster created is flexible in nature means slave nodes can be easily added and removed from cluster.

- **Multithreading** – Master system starts new thread for every slave node. Hence multiple threads work in parallel fashion using multithreading.
- **RMI over Socket** – proposed approach uses RMI for creating cluster rather than Socket which is proved better than Socket programming [14].
- **Actual and virtual partitioning** – actual partitioning of image takes place at Master system and virtual partitioning of image is done at slave systems.
- **Compatibility** – the proposed clustered system is compatible with different operating platforms.
- **Load balancing and easy replacement** – in case any fault occurs with a node, then whole cluster is not become faulty. Rather than this the faulty node halts and the work load of faulty node is shifted to rest of proper working nodes. As system is flexible so faulty node can be easily replaced with working node.
- **Firewall no bar** – as RMI is used, so firewall either present on Master side or on slave side cannot interrupt in establishing connection.

## 11 Experimental Setup

For the implementation of present cluster, following configuration for experimental setup is required:

Table 4. Configuration of hardware equipments required for implementation of present work

| Component Name | Count | Configuration | OS Support | Remarks |
|---|---|---|---|---|
| Master System Or Client System | 1 | – Vendor :Genuine Intel<br>– System Type : x86_64<br>– RAM : 2 GB<br>– CPU :3000 MHz<br>– *L3 :3072K*<br>– *L2 :256K*<br>– *L1i & d :32K*<br>– NUMA node0 CPU(s) : 0,1<br>– Core Per Socket : 2<br>– Socket : 1<br>– CPU(s) : 2<br>– Thread per Core : 1 | Windows 7 | JAVA(RMI) Version:- 1.7.0_31 |
| LAN Adapter | 3 | – 1st card IP address - 192.168.13.10<br>– 2nd card IP address - 192.168.14.10<br>– 3rd card IP address - 192.168.15.10 | USB Ethernet 10/100 Base –T | |
| Slave systems Or Server Systems | 5 | **IP addresses**<br>– 192.168.13.51 to 192.168.13.52<br>– 192.168.14.51 to 192.168.14.52<br>– 192.168.15.51<br>– Vendor :Genuine Intel<br>– System Type : x86_64<br>– RAM : 2 GB<br>– CPU :3000 MHz | Windows 7 | JAVA(RMI) Version :- 1.7.0_31 |

| | | | |
|---|---|---|---|
| | | – *L3* :*3072K*<br>– *L2* :*256K*<br>– *L1i & d* :*32K*<br>– NUMA node0 CPU(s) : 0,1<br>– Core Per Socket : 2<br>– Socket : 1<br>– CPU(s) : 2<br>Thread per Core : 1 | |
| Ethernet Switch | 1 | Cisco switch with 24 ports | Non-blocking communication. |
| Cat 5 UTP Cable | | Cable upto 100 MB/s transmission rate | |

## 12 Performance Measurements

In order to validate the outcome of proposed clustered system performance metrics like Speedup, Efficiency, excessive parallel overhead is evaluated.

### 12.1 Speed Up

Comparison is done between uniprocessor system and multiprocessor system in terms of speedup. Speedup is calculated as shown in Eq.8.

$$Speed\ Up = \frac{computation\ time\ taken\ by\ uniprocessor\ system}{computation\ time\ taken\ by\ multiprocessor\ system} \tag{8}$$

### 12.2 Efficiency

It gives the measure of time for which the proposed cluster system is engaged in performing given task. Efficiency is calculated using formula shown in Eq. 9:

$$Efficiency = \frac{execution\ time\ taken\ by\ single\ system}{execution\ time\ taken\ X\ No.of\ systems\ in\ cluster} \tag{9}$$

### 12.3 Excessive Parallel Overhead

It calculates the extra time overhead which is consumed in management of parallel system other than actual execution. Excessive It is calculated as Eq. 10

$$Excessive\ Parallel\ overhead = N * C_t - St \tag{10}$$

Where N is count of systems in Cluster, Ct is time taken by single processor system and St is time taken by cluster.
Excessive Parallel Overhead of Cluster with different no. of nodes for images of different dimensions is shown in Fig.5.

For images of different dimensions Speed up, Efficiency and excessive Parallel overhead of cluster with different number of nodes is shown in table 5.

Table 5. Performance of cluster with various no. of nodes in term of Speed Up, Efficiency, Excessive parallel overhead

| No. of nodes | Image Dimensions | Memory for Heap Space | Execution time by cluster | Execution time by single system | Speed Up | Efficiency | Excessive parallel overhead |
|---|---|---|---|---|---|---|---|
| 2 | 2214 x2214 | Xmx1200m | 5.97 | 9.24 | 1.54 | 0.77 | 2.7 |
| 2 | 2268 x2268 | Xmx1200m | 6.20 | 9.58 | 1.54 | 0.77 | 2.8 |
| 2 | 3168 x3168 | Xmx1200m | 10.52 | 17 | 1.61 | 0.80 | 4.0 |
| 2 | 4230 x4230 | Xmx1200m | 18.76 | 30.26 | 1.61 | 0.80 | 7.2 |
| 2 | 5004 x5004 | Xmx1200m | 43.07 | 61.59 | 1.42 | 0.71 | 24.5 |
| 3 | 2214 x2214 | Xmx1200m | 6.85 | 9.24 | 1.34 | 0.45 | 11.3 |
| 3 | 2241 x2241 | Xmx1200m | 6.81 | 8.97 | 1.31 | 0.44 | 11.4 |
| 3 | 2268 x2268 | Xmx1200m | 6.89 | 9.58 | 1.39 | 0.46 | 11.0 |
| 3 | 4230 x4230 | Xmx1200m | 20.23 | 30.26 | 1.49 | 0.49 | 30.4 |
| 3 | 5004 x5004 | Xmx1200m | 28.29 | 61.59 | 2.17 | 0.72 | 23.2 |
| 4 | 2214 x2214 | Xmx1200m | 5.11 | 9.24 | 1.80 | 0.45 | 11.2 |
| 4 | 2268 x2268 | Xmx1200m | 5.63 | 9.58 | 1.70 | 0.42 | 12.9 |
| 4 | 3168 x3168 | Xmx1200m | 10.13 | 17.06 | 1.68 | 0.42 | 23.4 |
| 4 | 3276 x4230 | Xmx1200m | 10.70 | 27.62 | 2.58 | 0.64 | 15.1 |
| 4 | 5004 x5004 | Xmx1200m | 24.79 | 61.59 | 2.48 | 0.62 | 37.57 |
| 5 | 2250 x2250 | Xmx1200m | 5.77 | 8.72 | 1.51 | 0.30 | 20.1 |
| 5 | 2295 x2295 | Xmx1200m | 6.76 | 9.27 | 1.4 | 0.27 | 24.5 |
| 5 | 3195 x3195 | Xmx1200m | 10.94 | 16.85 | 1.54 | 0.30 | 37.8 |
| 5 | 4230 x4230 | Xmx1200m | 16.51 | 30.26 | 1.83 | 0.36 | 52.2 |
| 5 | 4995 x4995 | Xmx1200m | 45.14 | 70.81 | 1.56 | 0.31 | 154.8 |

Speed up of cluster with different no. of nodes is charted in the Fig. 3
Efficiency of cluster is shown in Fig. 4 and excessive parallel overhead is in Fig. 4.

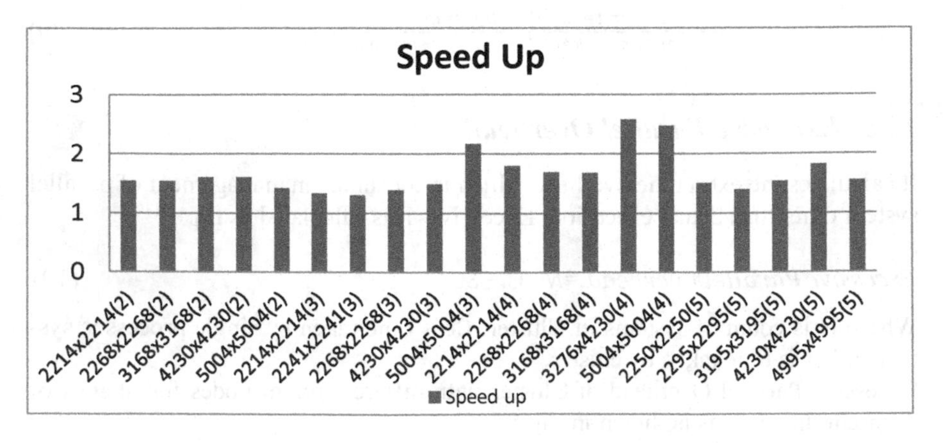


**Fig.3** Speed Up of cluster with different number of nodes

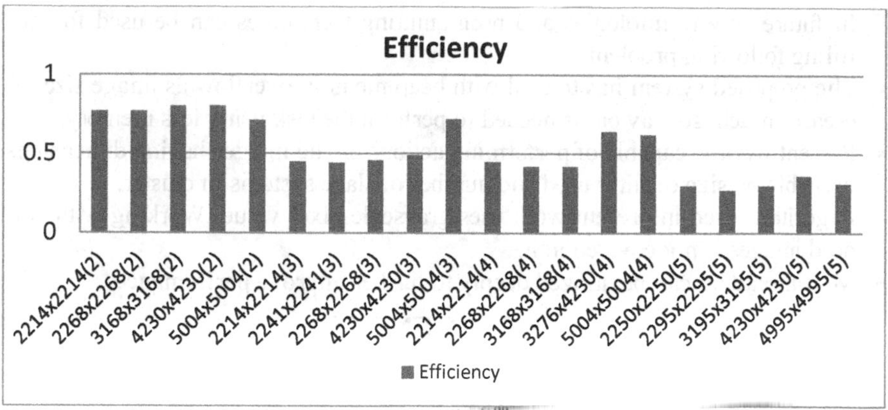


**Fig. 4** Efficiency of cluster with different number of nodes

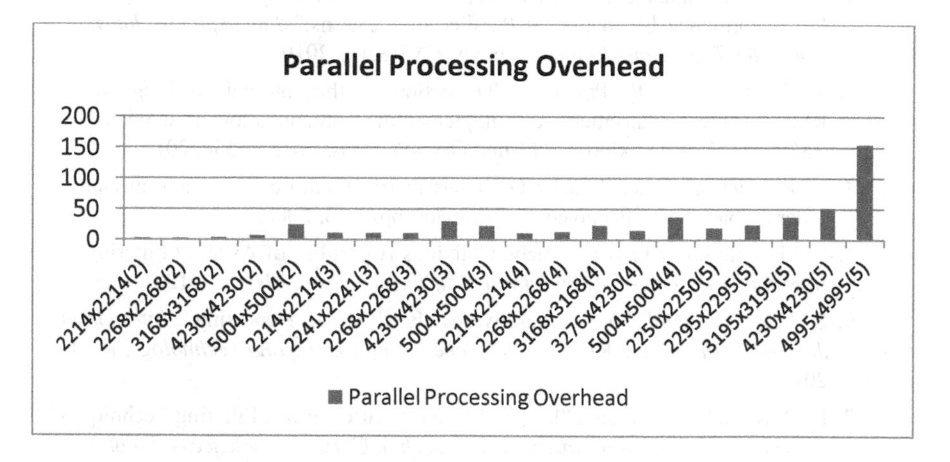


**Fig. 5** Parallel processing overhead of Cluster with Different No. of Nodes

## 13 Conclusion And Future Scope

From the above work, the conclusion derived is as follows:

- Cache oblivious average filtering algorithm for the removal of salt and pepper noise from image is working more effectively in parallel multithreaded cluster.
- Large sized problems of image processing can be easily and timely solved on multiple systems as compared to single system using cluster.
- The proposed architecture proves itself by consideration of performance metrics such as Speed up, efficiency and excessive parallel overhead.
- The partition size for decomposing image can be decided dynamically.
- Because of divide & conquer method workload of single system is distributed.
- NIC cards perfectly manage the network traffic, because of dedicated link.
- Sometimes excessive overhead increases as number of nodes increases in cluster because much of the time is spent in communication and transmission.
- Cache oblivious algorithm for average filtering is working on remotely distributed environment which increases the execution speed and output as compared to single system.

In future new technologies and programming techniques can be used for fulfilling following problems.

- The proposed system has to deal with heap memory overflow as image size increases much, so way out is needed to perform the task using less memory..
- Present work is capable of performing actions on the images having dimensions divisible by size of filter used and number of slave systems in cluster.
- Algorithm used in present work uses grayscale pixel value. Working with colored images is not covered in it.
- Multithreading can be utilized deeply to further improve performance.

# REFERENCES


1. P. S. Nivedita Joshi, "Remote Method Invocation – Usage &," *International Journal Of Engineering And Computer Science,* pp. 3136-3140, 2013.
2. G. S. Amit Chhabra, "A Cluster Based Parallel Computing Framework (CBPCF) for Performance Evaluation of Parallel Applications," *International Journal of Computer Theory and Engineering,* pp. 1793-8201, 2010.
3. V. K. D. a. H. B. Prajapati, "Dissection of the internal workings of the RMI,possible enhancements and implementing authentication in standard Java RMI," *Int. J. Networking and Virtual Organisations,* pp. 514-534, 2010.
4. C. Z. Michael Bader, "Cache Oblivious matrix multplication using an element ordering based on a Peono curve," *Elsevier,* pp. 1-13, 2006.
5. S. M. Mrityunjay Ghosh, "Cache Oblivious Algorithm of Average Filtering," in *International Conference on Informatics, Electronics & Vision*, 2012.
6. S. J. Jyotsna Patil, "A Comparative Study of Image Denoising," *International Journal of Innovative Research in Science, Engineering and Technology,* pp. 1-8, 2013.
7. E. D. K. Ankita Malhotra, "Image Denoising with Various Filtering Techniques," *International Journal of Advanced Research in Computer Science & Technology,* pp. 1-3, 2015.
8. D. J. A. Rohit Verma, "A Comparative Study of Various Types of Image Noise and Efficient Noise Removal Techniques," *International Journal of Advanced Research in Computer Science and Software Engineering,* pp. 616-622, 2013.
9. C. H. Youlian Zhu, "An improved median Filter for Image Noise Reduction," *Elsevier,* p. 8, 2012.
10. M. C. G. R. C. M. F. C. H. J. Mukesh C. Motwani, "Survey of Image Denoising Techniques," pp. 1-7, 2003.
11. K. V. Tarun Kumar, "A Theory Based on Conversion of RGB image to Gray image," *International Journal of Computer Applications,* pp. 1-7, 2010.
12. B. L. A. R. H. S.-H. Y. W.-J. H. Hantak Kwak, "Effects of Multithreading on Cache Performance," *IEEE* , pp. 176-184, 1999.
13. S. B. A. A. Harmanpreet Kaur, "RMI Approach to Cluster Based Winograd's Variant," in *IEEE*, 2015.
14. A. N. A. K. G. A. B. N. S. Harsh Prateek Singh, "Noise Reduction in Images using Enhanced Average Filter," in *International Journal of Computer Applications*, Gaziabad, 2014.
15. C. T. Kato Mivule, "Applying Moving Average filtering for Non- Interactive Differential Privacy Settings," *Elsevier,* pp. 1-7, 2014.

# INTERNET OF VEHICLES FOR INTELLIGENT TRANSPORTATION SYSTEM

Kundan Munjal[1,*] and Shilpa Verma

[1]Department of Computer Science and Engineering, Lovely Professional University, Jalandhar, Punjab – 144411,India

kundan.16806@lpu.co.in

[2]Department of Computer Science and Engineering , Pec University of Technology, Chandigarh,160012, India,

shilpaverma.pec@gmail.com

**Abstract.** Wireless communication between vehicles is a new era of communication that leads to intelligent transportation system. Internet of Things (IOT) is a new paradigm which is a combination of storing sensor data and computing the provided data to achieve useful information in the real world applications. The origin of IoT is Radio Frequency Identification (RFID) which is networked over the region to collect data, based on the concept of sensor data which is collected by RFID a user can derive important information after getting raw data from established network. In this paper we will discuss how Vehicles with sensors and actuators can absorb large amount of information from environment and provide this useful information to assist secure navigation pollution control and efficient traffic management we will also discuss various challenges architecture and cloud based implementation of Internet of vehicles

**Keywords:** Internet of Things; IoV; IoT; Big data analysis, VANET

## 1 Introduction

With WiFi and LTE Communication access, the communication is already evident but we need to move beyond the basic communication paradigm. Before IoV comes into role the architecture of VANET includes three types of communication as shown in figure. Vehicle to Vehicle Communication, Vehicle to Roadside Unit Communication and Vehicle to Infrastructure communication. All the vehicles in transportation system perform three acts. A vehicle can act as a sender, receiver, router. Inter Vehicle communications mainly concerned with broadcasting and multicasting of messages to a set of nodes or receivers and vehicles can broadcast messages after regular intervals. In vehicles to roadside unit communication a roadside unit (RSU) broadcast messages to all the vehicles in the network. It is not peer to peer broadcasting. It is a single hop broadcasting of messages for e.g. Speed limit warning message. Vehicle to Infrastructure Communication is multi hop but



J.M. Corchado Rodriguez et al. (eds.), *Intelligent Systems Technologies and Applications 2016*, Advances in Intelligent Systems and Computing 530,
DOI 10.1007/978-3-319-47952-1_32

unicast communication. Both roadside unit and vehicles act as a router and unicasts the message towards destination.
In Internet of vehicles as compare to Vanets vehicles are equipped with smart multi sensors, smarter technology for communication as a result large information of data collected and that information after processing and analysis will result in welfare of the society. As according to the vision of internet of things the computing theory need to be changed to some extent. Internet of things [2] demands:

a) Mutual understanding of state of its users and their appliances,
b) architecture and widely spread ubiquitous network that processes information,
c) various tools for analyzing the data.

P. Jaworski et. al had [17] suggested intelligent traffic clouds for urban traffic management system with two roles one as a service provider and other as a customer. All roles such as Agent based transportation (ATS), traffic strategy database and traffic strategy agent database has been encapsulated in system core and customer contains urban traffic management system and traffic contributors. Mario Gerla [15] highlighted urban sensing and efficient management of traffic and also analysed that by keeping information on vehicle and not on web we can make system cost efficient and proved the significance of using sensor information of local importance in vehicular cloud as compared to internet cloud. Lino Figueiredo et. al discussed various areas for intelligent transportation and also discussed the need of board computers, navigation systems and real time traffic and also focuses on fully automated notion initially for small number of applications thereafter increases gradually and lastly it also includes enhancement of model of roads, vehicles for efficient intelligent transportation system. In this paper we primarily focus on IoV implementation from data collection through sensors to analysis and at the end taking decisions based upon them and also proposed a model for achieving these objectives. In the first section we will discuss about various challenges in implementation of IoV followed by significance of using cloud and thereafter we will propose a model to achieve efficient ubiquitous computing from collection of data through sensors followed by processing and visualisation of analysed data.

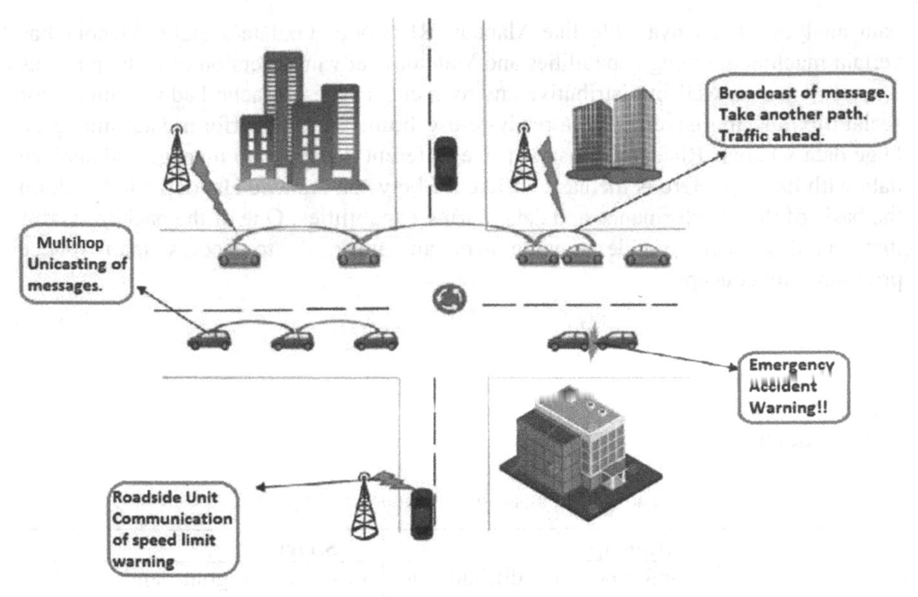


**Fig. 1.** VANET Architecture

# 2 CHALLENGES

## 2.1 HUGE STORAGE REQUIREMENT

Huge amount of data storage in petabytes is needed as large amount of information collected through sensors from large number of commercial vehicles. With advent of cloud based IoT implementation, every device on internet will send and receive information to and from cloud. It will be a good approach to move towards Platform as a Service (PaaS) that would enable for continuous expansion so that large amount of data can be handled efficiently. Object storage can also be considered that will not only be distributive but also it will be scalable in nature. Challenge of huge storage requirement can be met by Categorization and Archiving of data as it is necessary to ensure optimal use of resources by not loading them gratuitously.

## 2.2 EFFICIENT DATA ANALYSIS AND PROCESSING.

Collecting data through various sensors followed by processing and storing and then analyzing with accuracy in real time environment will be a great challenge for IoV Implementation. For this activity big data analytics is the important component which have been emerged with various computational software. Computational software like Apache Hadoop and Spark provide cost effective, fault tolerant and speedy solutions to the problems involving large datasets. There are other numerous open source big data analytics tools available like Mahout, RHadoop, Oxdata's H2O. Mahout has certain machine learning capabilities and Mahout is advance version of hadoop and as a result it works well in distributive environment. It uses Apache hadoop library for scalability and moreover it has a ready-to-use framework to perform data mining on huge data whereas Rhadoop consist of five different packages to manage and analyze data with hadoop. Here is the basic difference between Apache Hadoop and Spark on the basis of their performance and data sharing capabilities. One of the package is rmr that enables map provide programmers an approach to access map reduce programming concept.

### 2.3 SECURITY

Data Collected should be secure enough. Any change or lose of data might result into harming of all connected things and or involved in processing and analyzing the data.

**Table 1.** Comparison of Hadoop and Spark

| | **Hadoop** | **Spark** |
|---|---|---|
| **Difficulty** | MapReduce is difficult to Program and needs abstraction. | Easy to Program and do not need any abstraction. |
| **Data sharing** | **During Interactive operations** Slow in MapReduce due to replication, serialization, and disk IO. | **During Interactive operations** Data can be kept in memory for better execution times. |
| | **During Iterative operations,** Incurs substantial overheads due to data replication for iterative algorithms. | **During Iterative operations,** Distributed collection of objects which targets in-memory computing. |
| **Performance** | MapReduce does not leverage the memory of the Hadoop cluster to the maximum | Execute batch processing jobs about 10 to 100 times faster. |
| **Latency** | Distributed disk based (Slow) | Distributed memory based (Fast) Lower latency computations by caching the partial results across its distributed memory |

## 3 SIGNIFICANCE OF USING CLOUD BASED IMPLEMENTATION

Cloud Computing is dependent upon sharing resources instead of using local servers or own devices for computation. Cloud computing is standardized with various attributes like broad network access, rapid flexibility, calculated service and service at demand. These attributes differentiate the use of cloud from other services. Large number of services of cloud can be classified mainly into three categories as in figure 2. Infrastructure as a service (IaaS), Platform as a Service (PaaS) and Software as a Service (SaaS). IaaS provides virtual computing resources virtual here refers to third party service provider in terms of storage, server, hardware on behalf of its clients and it is highly scalable with respect to increase in demand. SaaS provide software as a service in which applications are hosted by the service provider so that huge customers can make maximum use of it through the internet. SaaS support various web services, service oriented architecture and Azax like development approaches. PaaS provides a platform where client can make use of various hardware and software tools needed for application development. The key advantage of using cloud computing is its capability that not only provides infrastructure [1] but also software as a service. According to R Buyya et. al [9] internet of things survey. IoT has two viewpoints one in Internet Centric and other is 'Thing' Centric. In former the data will be routed by the objects and it will involve various internet services role in play and in latter objects will act primarily. Use of cloud computing in implementing IoV will provide flexibility in terms of increased data storage, not only it will be cost efficient but also it will provide security of data with disaster recovery as an add-on. There are various cloud service providers available like Amazon web services (AWS). It consist of various attributes like AWS IoT DEVICE SDK that consist of various libraries for communication authentication and exchange of messages. It has its own Rule Engine that not only route the messages to AWS services but also alter messages based upon certain rules. Another popular cloud support is Microsoft Azure that provides large number of services including collection, storing, analysing and also visualizing both real time and historical.

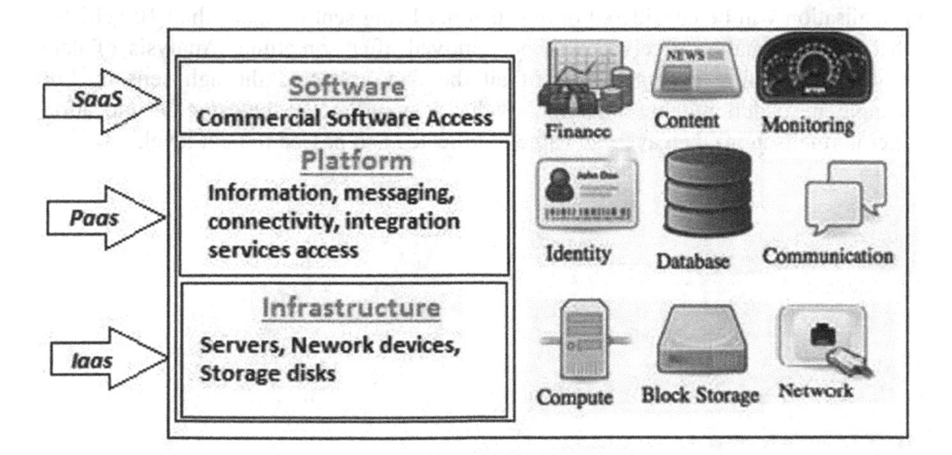


**Fig. 2. Cloud Computing Components**

# 4 PROPOSED MODEL

In real world things matter more as compare to ideas this section presents the IoV model. In IoV environment users will have access to various number of resources while moving from one place to other. Vehicles can communicate with each on the basis of established adhoc network. This communication may be Vehicle to RSU and V2Vehicle as well.

## 4.1 Role of sensors

Sensors will play an important role in collecting the information. We can have various sensors. Sensors for detecting pothole, sensing speed, sensing traffic jams and other sensing information. Every vehicle will participate in transferring the information and all information collected will be transferred to cloud.

## 4.2 Processing and analysing the data

The information collected will be structured and will be based on certain keywords. Processing of data involves identifying the type based upon keywords. For example, if data given by vehicle with keyword roaddept with its GPS location and sensor value of say 100 then the database relational table of roaddept will be consisting of two attributes one is GPS location and pothole sensor value. The analysing of data will take into account all the relation table roaddept values in account of same or nearby GPS location and then applying fitness function on all the sensing data values.

The values collected for GPS location and any sensor value will be normalized and then added into the table. After adding into table identical GPS values after normalisation will be considered only if that are being sent by more than 10 vehicles otherwise information received will be removed after sometime. Analysis of data involves calculating average value of all the data collected through sensors. For example, in roadddept table the pothole value is stored for all the same GPS location (after normalisation) then average value is collected and passed to next level.

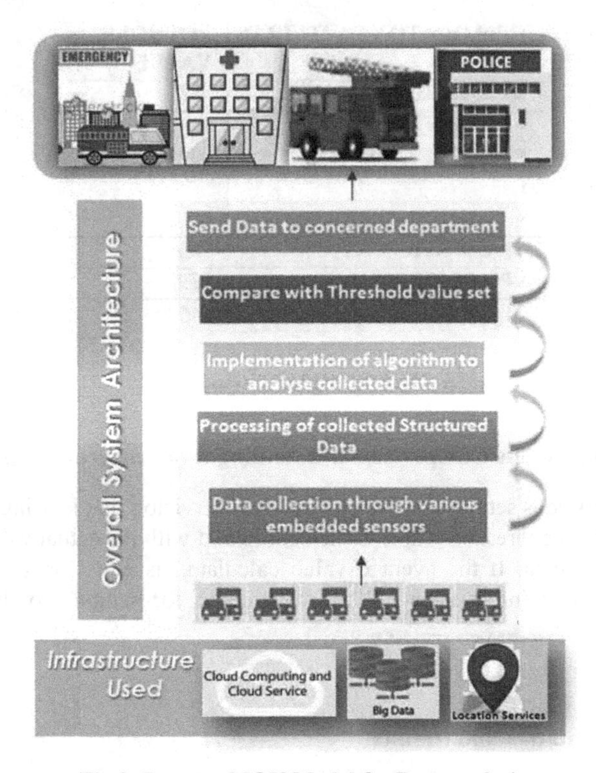


**Fig 3: Proposed IOV Model for Data analysis**

1. **Normalise GPS value w.r.t. location collected**
2. **Take all equal GPS values and store it in N**
3. **if (N>10)**
4. **'find avg of all values and store it in M'**
5. **if (thresholdvalue < M)**
6. **'send the information to concern department**

| GPSLOCATON CAPTURED | POTHOLE VALUE |
|---|---|
| GPSLOCATION "A" | 883 |
| GPSLOCATION "A" | 754 |
| GPSLOCATION "A" | 900 |
| : | : |
| : | : |
| GPSLOCATION "A" | 983 |
| GPSLOCATION "A" | 789 |
| AVG | 845.8 |

**Table 2.** *ROADDEPTT Database*

### 4.3 Threshold Value Comparison and sending it to concerned department

The threshold value is set for every object under supervision that is being given by the concerned dept. The threshold value is then compared with the actual value calculated after analysis of data. If the average value calculated is equal or greater than the threshold, then that information will be considered for sending to the concerned department.

### 4.4 Implementation and Results

Two firm wares have been established on the basis of OBU (On Board Unit) and RSU (Road Side Unit) are developed in 'C' Language using Contiki library functions along with some 'C' libraries. These firmwares have been compiled along with Contiki on motes, which are used in simulation to demonstrate the behaviour and functionality of OBU and RSU.

OBU: After executing OBU, reports like Accident, Road Condition etc. have been started broadcasting to allneighbours at random time interval. OBU also send level of severity with the help of:

```
static char* reports [ ] = {"Traffic", "Street light",
"Road condition", "Climate", "Accident"};
Static char* severity [ ] = {"Low", "Average", "High"};
```

RSU: Broadcast connection have been established using library provided Contiki OS and callback( ) function is called after receiving data from connection. When RSU receives a message from OBU it processes the message then increase the counter value of message it receives,Then RSU decides whether to disseminate the message or not based on the corresponding value of its threshold and reset the counter. Basically, when RSU receives a report from various OBU, it will check that "How may OBUs have sent same report?" and then counter will increase its value. This counter value is used to calculate particular threshold value which will be further compared using the threshold( ) user defined function.

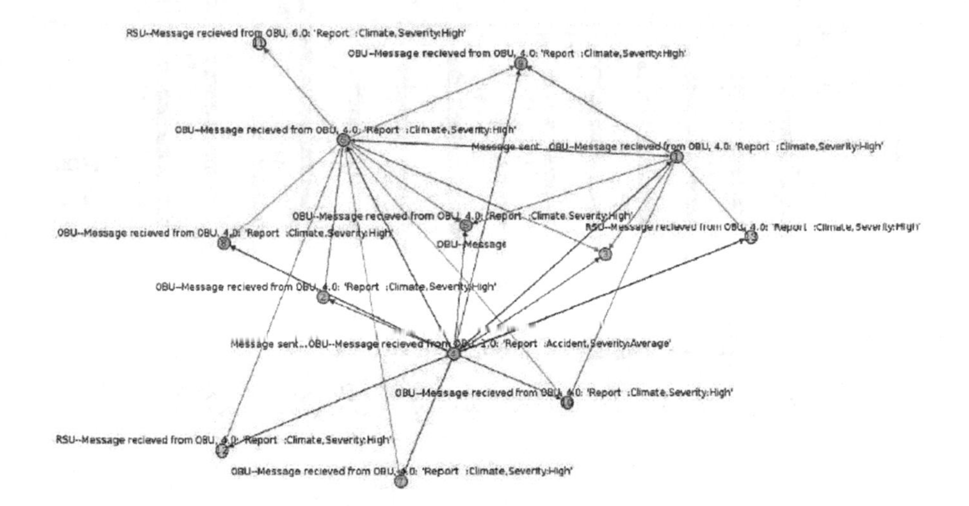

Fig. 3. Network Connection and communication

| Mote | ON | TX | RX |
|---|---|---|---|
| Sky_1 | 1.02% | 0.17% | 0.06% |
| Sky_2 | 1.14% | 0.22% | 0.08% |
| Sky_3 | 1.05% | 0.16% | 0.06% |
| Sky_4 | 1.09% | 0.17% | 0.08% |
| Sky_5 | 1.11% | 0.17% | 0.10% |
| Sky_6 | 1.08% | 0.16% | 0.09% |
| Sky_7 | 1.03% | 0.17% | 0.06% |
| Sky_8 | 1.12% | 0.23% | 0.07% |
| Sky_9 | 1.08% | 0.17% | 0.09% |
| Sky_10 | 1.07% | 0.16% | 0.08% |
| Z1_11 | 1.45% | 0.55% | 0.03% |
| Z1_12 | 1.26% | 0.39% | 0.04% |
| Z1_13 | 1.38% | 0.50% | 0.03% |
| Average | 1.14% | 0.25% | 0.07% |

Table 3. Power trace of sensors

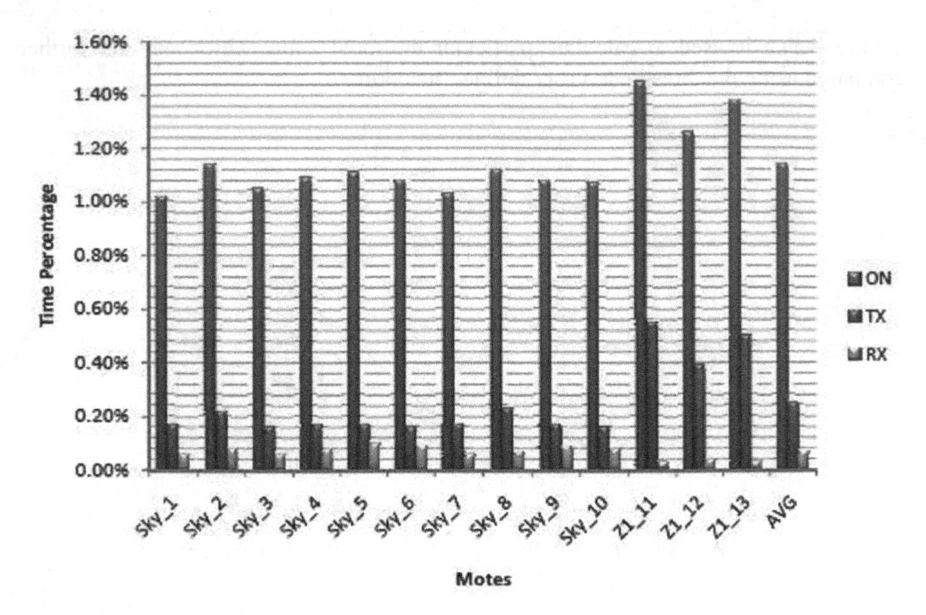


**Fig. 4. Power Trace chart of motes**

This chart shows total power consumption by the sensors. It is observed that RSUs (Z1_* motes) uses more power than OBUs as they process every received data. 'Tx' denotes the time which particular mote takes to transmit the data to other motes and 'Rx' shows the data about the time which a mote takes for receiving the data in whole. Rest time motes were Idle.

## 5 Conclusion and Future Work

The latest concept of Internet of things (IOT) is a state in which any object in this world whether living or non-living are provided with unique identity. These objects having identity are also provided with the ability to transfer information over the network without any interaction with human. Our first objective in the research is to identify the types of information collected by the vehicles. This identification will be based on future usage of information collected. This will lead to fixing of the Sensors in vehicles that will provide required information to be communicated. Second Objective is to design an efficient algorithm that analyse the collected structured data third Objective is to provide the analysis details to the proper channel so that necessary actions can be taken on time.

Future work involves including large number of connected objects so that enormous and different types of information can be collected, processed and analysed. Future work also involve efficient processing by getting exact values of GPS location so that we don't need to normalise and make it fast and time efficient.

## References

1. A. M. Ortiz, D. Hussein, S. Park, S. N. Han, and N. Crespi, ``The cluster between Internet of Things and social networks: Review and research challenges," IEEE Internet Things J., vol. 1, no. 3, pp. 206215, Jun. 2014.
2. A.P. Castellani, N. Bui, P. Casari, M. Rossi, Z. Shelby, M. Zorzi, Architecture and protocols for the Internet of Things: a case study, 2010, pp. 678–683.
3. B. Ahlgren, C. Dannewitz, C. Imbrenda, D. Kutscher, and B. Ohlman, "A Survey of Information-Centric Networking," IEEE Communications Magazine, vol. 50(7), pp. 26 – 36, July 2012.
4. C. Vecchiola, S. Pandey, and R. Buyya, "High-Performance Cloud Computing: A View of Scientific Applications", Proc. of the 10th International Symposium on Pervasive Systems, Algorithms and Networks (I-SPAN 2009), Kaohsiung, Taiwan, Dec. 2009
5. F. Wang, C. Herget, D. Zeng, "Developing and Improving Transportation Systems: The Structure and Operation of IEEE Intelligent Transportation Systems Society", IEEE Transactions on Intelligent Transportation Systems, Volume 6, Issue 3, Sept. 2005 Page(s):261 – 264.
6. Figueiredo, L.; Jesus, I.; Machado, J.A.T.; Ferreira, J.R.; Martins de Vehiclevalho, J.L., "Towards the development of intelligent transportation systems", Intelligent Transportation Systems, 2001. Proceedings. 2001 IEEE, 25-29 Aug. 2001 Page(s):1206 – 1211
7. G. Broll, E. Rukzio, M. Paolucci, M. Wagner, A. Schmidt, H. Hussmann,PERCI: pervasive service interaction with the internet of things, IEEE Internet Computing 13 (6) (2009) 74–81.
8. G. Dimitrakopoulos, P. Demestichas, "Intelligent Transportation Systems based on Cognitive Networking Principles", IEEE Vehicular Technology Magazine (VTM), March 2010.
9. J. Gubbi, R. Buyya, S. Marusic, M. Palaniswami "Internet of Things (IoT): A vision, architectural elements, and future directions", Future Generation Computer Systems 29 (2013) 1645–1660.
10. J. Gubbi, K. Krishnakumar, R. Buyya, M. Palaniswami, A cloud computing framework for data analytics in smart city applications, Technical Report No. CLOUDS-TR-2012-2A, Cloud Computing and Distributed Systems Laboratory, The University of Melbourne, 2012.
11. J. Sung, T. Sanchez Lopez, D. Kim, The EPC sensor network for RFID and WSN integration infrastructure, in: Proceedings of IEEE PerComW'07, White Plains, NY, USA, March 2007.
12. K. M. Alam, M. Saini, and A. El Saddik, ``tNote: A social network of vehicles under Internet of Things," in Internet of Vehicles Technologies and Services. Berlin, Germany: Springer-Verlag, 2014, pp. 227236.
13. K. Su, J. Li, and H. Fu, ``Smart city and the applications," in Proc. Int. Conf. Electron., Commun. Control (ICECC), Sep. 2011, pp. 1028103
14. L. Atzori, A. Iera, G. Morabito, The Internet of Things: a survey, Computer Networks 54 (2010) 2787–2805.

M. Bayly, M. Regan & S. Hosking, "Intelligent Transport System for Motorcycle Safety and Issues", European Journal of Scientific Research, ISSN 1450-216X Vol.28 No.4 (2009), pp.600- 611

15. M. Gerla, “Vehicular Cloud Computing,” in IEEE Med-Hoc-Net, June 2012.
16. N. Fernando, S. Loke, and W. Rahayu, “Mobile Cloud Computing: A Survey,” Elsevier Future Generation Computer Systems, vol. 29(1), pp. 84 – 106, July 2013.
17. P. Jaworski, T. Edwards, J. Moore, and K. Burnham, “Cloud computing concept for intelligent transportation systems,” in Proc. 14th Int. IEEE Conf. Intell. Transp. Syst. (ITSC), 2011, pp. 391–936.
18. R. Buyya, C.S. Yeo, S. Venugopal, J. Broberg, I. Brandic, Cloud computing and emerging IT platforms: vision, hype, and reality for delivering computing as the 5th utility, Future Generation Computer Systems 25 (2009) 599– 616.
19. X. Hu et al., ``Social drive: A crowdsourcing-based vehicular social networking system for green transportation," in Proc. 3rd ACM Int. Symp. Design Anal. Intell. Veh. Netw. Appl., 2013, pp. 8592.
20. Ying Leng, Lingshu Zhao. Novel design of intelligent internet-of-vehicles management system based on cloud-computing and internet-of-things, 2011 International Conference on Electronic & Mechanical Engineering and Information Technology. Harbin: IEEE, 2011:3190-3193.

# Handoff Schemes in Vehicular Ad-Hoc Network: A Comparative Study

**Prasanna Roy, Sadip Midya, Koushik Majumder**

**Abstract** Vehicles on road are increasing at a rapid pace recently. Passengers are also feeling the urge to get various other services while they are travelling. Thus, there is a necessity to bring about enhancements in the existing Intelligent Transport System (ITS). In VANET, handoff is required because a vehicle is mobile and move from one network region covered by an access point to another. During handoff the connection that has been established between the vehicle and the network should remain intact to maintain seamless connectivity. It is also necessary to bring about reduction in packet loss and handover delay. This paper provides a comparative study of some existing schemes that deal with handoff management in VANET. In addition to this we have carried out a qualitative analysis of these schemes to explore the new areas of research and the scopes for enhancement for providing better quality of services to the users.

**Keywords**: Handoff; VANET; packet loss; ad-hoc network; proactive caching.

## 1. Introduction

In Vehicular Ad-Hoc Networks (VANET), vehicles maintain connection to establish communication by forming an ad-hoc wireless network. A couple of communication modes are important in VANET which are Vehicle to Vehicle (V2V) and Vehicle to Infrastructure (V2I)[1]. Another variation of VANET communication is Hybrid Architecture in which the Vehicle to Vehicle and Vehicle to Infrastructure communication modes are combined [2, 3]. In VANET, the vehicles, which are the prime components, are highly mobile and keep on changing Point of Attachments (access points) too frequently. Thus, it becomes a challenge to manage handoff in such a network. Handoff or Handover is the process in which a vehicle stops receiving services from the old base station (access point) and begins receiving services from a new base station(access point). Thus Quality of Service in VANET depends on the delay caused during the handoff procedure.

Prasanna Roy
MAKAUT, BF-142, Sector 1 Salt Lake Road Kokata-700064,
prasanna.roy.durgapur@gmail.com

Sadip Midya
MAKAUT, BF-142, Sector 1 Salt Lake Road Kokata-700064,
sadip20@gmail.com

Koushik Majumder
MAKAUT, BF-142, Sector 1 Salt Lake Road Kokata-700064,
koushik@ieee.org



J.M. Corchado Rodriguez et al. (eds.), *Intelligent Systems Technologies and Applications 2016*, Advances in Intelligent Systems and Computing 530,
DOI 10.1007/978-3-319-47952-1_33

Primarily three components make up a VANET [4]. Firstly, vehicles are considered to be the nodes in VANET. Secondly, Infrastructure is made up of the Road Side Base Stations which form the backbone of VANET. Thirdly, Communication Channels which generally involve radio waves for signal transmission in VANET.VANET applications are classified into two categories, namely, real time application and non-real time applications. Real time applications are of high priority like video calls and GPS systems, while non-real time applications includes net browsing ,checking emails etc. The handoff procedures must be taken care of keeping in mind these kinds of applications [5].

In VANET, handoff management is carried out by the process of rerouting, in which a new path towards the destination is constructed. The mobile node's neighbors keep on changing as it keeps on moving. In order to ensure better handoff performance it is required to establish the path without delay through which the data is to be transmitted [6].There are various types of handoff techniques. Handoff can be horizontal or vertical depending on the network technology used. Horizontal handoff involves switching between Access Points belonging to the same network while the Vertical Handoff involves switching between Access Points belonging to different networks. Handoff can also be classified depending on the number of Access Points to which a mobile node is associated at a given time instant. Handoff can be hard or soft handoff. In hard handoff, also known as "break before make", the connection with the current access network must be broken by the Mobile Station before it can establish connection with the new one. In soft handoff a Mobile Station can remain connected with multiple networks while the handoff process. This handoff process is also known as "make before break".

This survey has been broken up into the following sections. Section 2 discusses some of the recent existing VANET handover protocols. A comparative study between the schemes is given in Section 3. Section 4 contains the conclusion and future scope.

## 2. Overview of Recent Handover Protocols for VANET

### *2.1 Mobility Handover Scheme*

A new mobility protocol is proposed in [7], where a new format of IPv6 addressing of the mobile unit (vehicles) is shown. In this protocol the upcoming target access point of the vehicle, to which it is to be handed off to is predicted by the vehicle's current access point. This prediction is done using an Angle of Arrival (AoA) technique which can determine the relative orientation of any vehicle or any stationary base station with respect to any standing infrastructure. The target access point is generally the one that is nearest to the current access point. In this scheme, the network layer handover is finished first. Then the handover at the data link layer takes place. As a result, the vehicle can still receive packets from the previous access point. This reduces the rate of packet loss significantly and continuous communication is maintained. VANET consists of vehicles, access routers and access points. The network communication is carried out using WiFi technology. Neighbor vehicles that are within one hop scope can carry out direct communication.

This scheme firstly, addresses the mobility handover of the vehicle within the same subnet. In this case the vehicle maintains its link layer connection with the previous access point so that it can still receive packets from the access point to which it was previously connected. During this time the access router under which both the target access point and source access point lies, is updated by the current access point about the switch of access points. The handoff process within a subnet is triggered by the source access point as shown in Fig.1

In the second scenario the mobility handover of the vehicle between two different subnets is being addressed. In this case the assistance of a vehicle connected to the target access point is required. When a vehicle is about to enter a new subnet covered by a different access router it transmits a pre-update message to the vehicle in front, which is already connected to the target access point. The vehicle in front then triggers the handoff procedure by communicating with its AR and informing about the arrival of a new vehicle. The Home Agent is also notified about the access point switch. The mobility handover delay in this case is shortened and cost is reduced because neighbor vehicle carries out the handover process. The Handover process in between subnets is shown in Fig. 2.

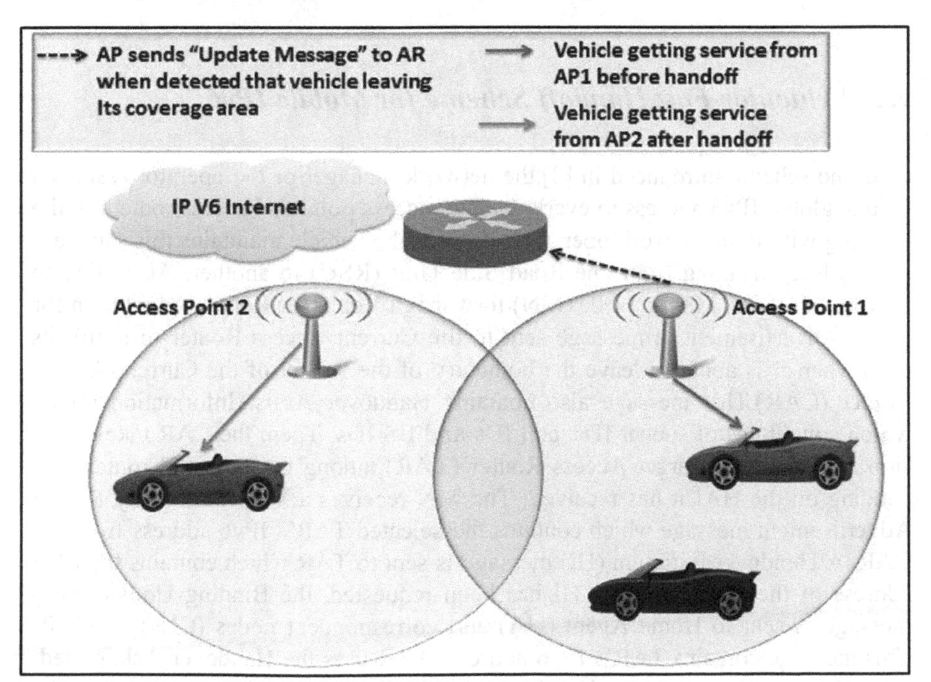


**Fig. 1** **Handoff Process within Subnets**

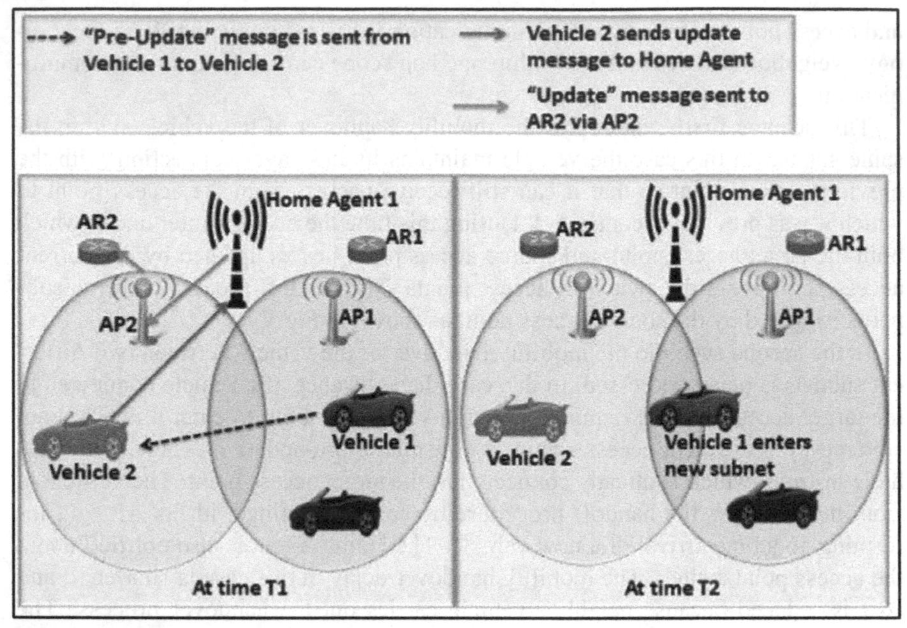


**Fig. 2 Handoff Process between Subnets**

## *2.2 Vehicular Fast Handoff Scheme for Mobile IPv6*

In the scheme introduced in [8],the network manager or the operator assigns a unique global IPv6 address to every vehicle, access point and access router. While roaming within the network operator's domain the vehicle maintains this same address while changing from one Road Side Unit (RSU) to another. According to this protocol, The Mobile Node (MN) forwards a Vehicular Router Solicitation for Proxy Advertisement (a message sent to the Current Access Router to notify its exit) when it is about to leave the boundary of the subnet of the Current Access Router (CAR).This message also contains Handover Assist Information (HAI) which constitutes of subnet IDs, cell IDs and BS IDs. Then, the CAR takes decision regarding the Target Access Router (TAR) among the potential routers depending on the HAI it has received. The MN receives a Vehicular Proxy Router Advertisement message which contains the selected TAR's IPv6 address from the CAR. A Handover Initiation (HI) message is sent to TAR which contains the IPv6 address of the MN. After the HI has been requested, the Binding Update (BU) message is sent to Home Agent (HA) and correspondent nodes (CNs) by CAR. This message contains TAR's IPv6 address. TAR uses the Handover Acknowledgement message as a reply to HI. Thus, a bidirectional tunnel is set up between the two Access Routers. Then TAR starts buffering those packets that have been forwarded by CAR and is meant for MN. Both CN and HA sends Binding Update Acknowledgement message as a reply to the Binding Update message. The MN gets attached to TAR's link layer connection after getting detached from CAR's link layer connection. A Neighbor Advertisement is sent out by MN to TAR. The packets that have been buffered are sent to the global IPv6 address of MN. MN and CN can now again participate in the end to end packet transmission process.

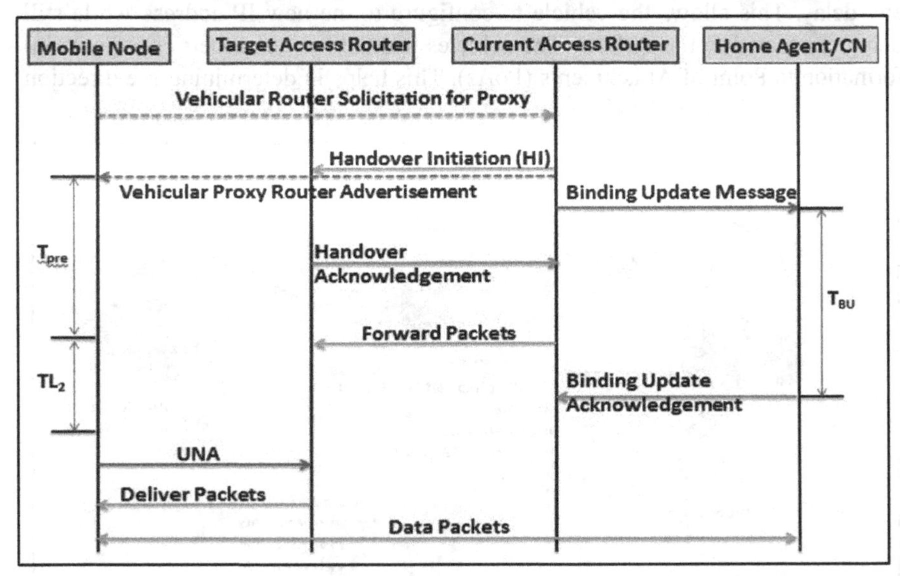


**Fig. 3 Sequence Diagram of VFMIPv6**

## *2.3 Proxy MIPv6 Handover Scheme*

A new scheme is proposed in [9] for enhancing the performance of the Proxy Mobile IPv6 Protocol (PMIPv6). Though this protocol is designed with an aim of providing uninterrupted Internet connection, some issues such as achieving seamless handoff still faces difficulty. Vehicles in VANET move with a very high speed resulting in very frequent handoffs. Adding to this, the process of handoff itself consumes significant amount of time. This might result in the increase in handoff latency and the packets might get dropped. Thus, the Early Binding Update Registration in PMIPv6 has been proposed in [9].

According to this protocol the Mobile Access Gateway has to maintain a pool of IP addresses. These addresses are assigned by an administrator. The Previous Mobile Access Gateway (PMAG) also contains a table that contains information about other neighboring Mobile Access Gateways. This table is used by the Previous Mobile Access Gateway (PMAG) to discover the New Mobile Access Gateway (NMAG) to which the vehicle is going to be attached. Whenever the PMAG detects that the vehicle is moving out of its range, it sends an Information Request message to the New Mobile Access Gateway. It is the duty of the new gateway to assign a unique IP address to the newly arriving vehicle. At the Local Mobility Anchor the vehicle's Binding Cache Entry is also updated. An Information Request Acknowledgement (IRA) is also sent to the vehicle by the NMAG without any delay. This allows the vehicle to configure to the new IP address while still being connected to the PMAG. The vehicles use GPS to send their coordinate information to Point of Attachments (PoAs). This helps in determining the direction in which the vehicle is moving.

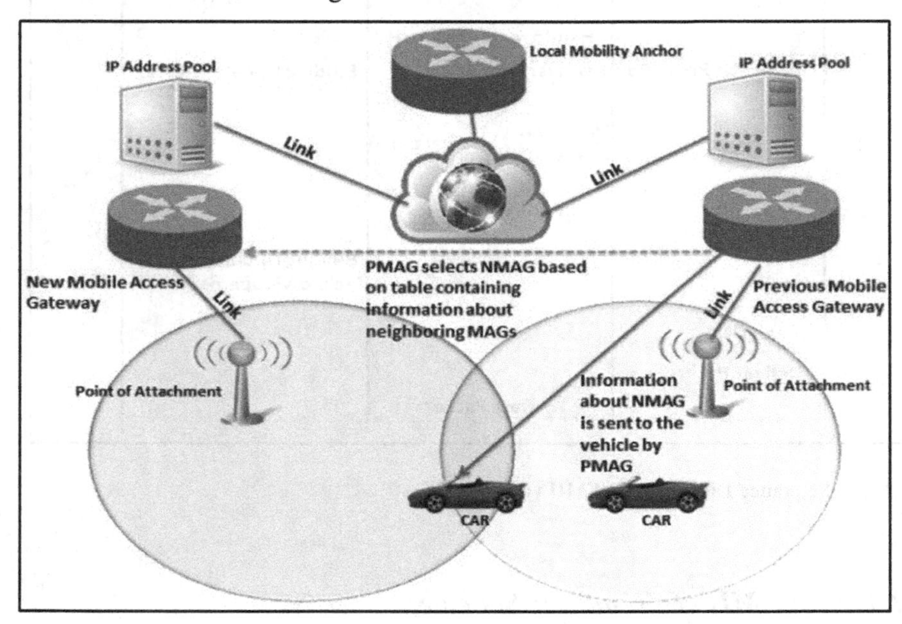


**Fig. 4 Early Binding Update Registration Scheme in PMIPv6**

## *2.4 Handoff using Multi-Way Proactive Caching*

According to the scheme in [10], a minimum number of Access Points are chosen as targets taking into consideration the neighbor graphs. When a vehicle is about to leave a particular network region covered by a RSU, it sends a disassociation message to its RSU. The RSU then sends all the packets meant for the MN(mobile node/vehicle)to all of its neighboring RSUs using multicast, since this scheme has no provision for determining the next RSU from beforehand. All the nRSU (new RSUs), receives the data packets meant for MN. The RSUs uses proactive caching of data packets meant for MN and starts transmitting the packets blindly. It is done since the nRSUs cannot determine whether the MN has entered its coverage area or not. This happens because this process does not have any association procedure. Now, it is not feasible for an RSU to go on transmitting without the knowledge of reception of packets by the MN. This would cause unnecessary packet loss. A solution to this problem is proposed here. Whenever the MN receives its first packet after handoff, it acknowledges it with a ACK message to the new RSU that sent it. The new RSU then transmits a Move-Notify message (IEEE 802.11f) to all its neighboring RSUs to inform them that the vehicle has entered its coverage area. The RSUs after receiving this Move-Notify message, stop transmitting the unnecessary packets and remove them from its cache.

Once the neighbors have been informed, the new Road Side Unit sends a message to the Gateway indicating that it is now serving as the current Road Side Unit. Every packet that the other Road Side Units receive from now on will be sent to this Road Side Unit that has been selected. An assumption is also made in this scheme that deployment of the Road Side Units in the highway is done in a fixed order. The scheme is explained in Fig.5.

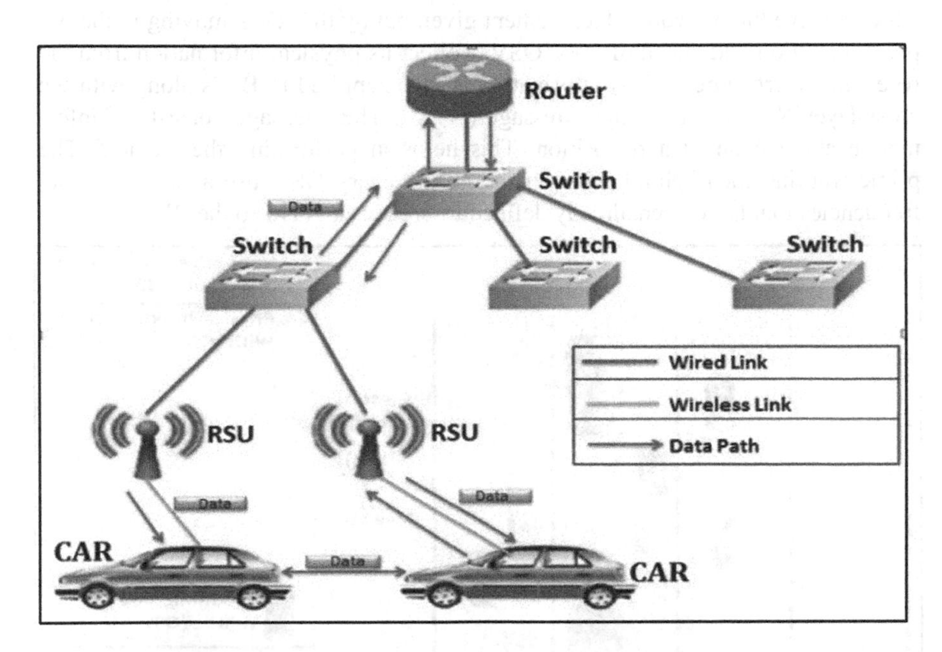


**Fig. 5 Wireless Access in Vehicular Environments (WAVE) Architecture**

## 2.5 Cross Layer Fast Handover

A newly designed wireless network technique known as WiMAX Mobile Multi-hop Relay (MMR) [11] is used in this scheme. This framework is most suitable for those vehicles that operate on high speed on highways. A Global Positioning System (GPS) is fitted in the vehicles to sense the location of the vehicle. According to this proposal, the larger vehicles like buses/trucks are used as relay nodes and are known by the name Relay Vehicles (RV). They provide mobile management and relay services to the neighboring vehicles present in a WiMAX network. The RV forms a cluster with other small vehicles that are being driven in its transmission range. The Oncoming Side Vehicles (OSVs) are relatively smaller vehicles that are being driven in the opposite lane of the road. The data packets from the vehicles are sent to the RV to which they are connected. RVs on receiving them send the packet to the internet.

Sometimes Broken Vehicles (BV) are also encountered in the networks which are out of the transmission range of RVs and seek connection to the RVs in front. OSVs receive the network advertisement given out by the RV's moving in the opposite direction and ahead of BVs. OSVs collect its physical information from the received advertisement. This information is then supplied to BV's along with the cross-layer Network Topology Message (NTM). This message consists of information about channel and position. This helps in performing the handoff. The prime working idea behind this Scheme is that every OSV use a set of channel frequencies that have been already defined to broadcast NTM to the BVs.

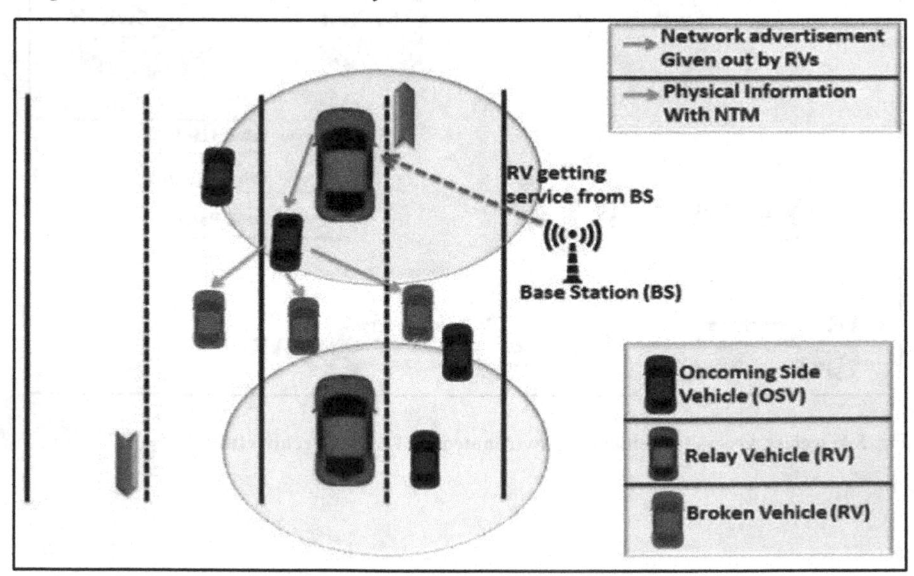


**Fig. 6** **Model Showing the Cross Layer Handover Process**

As soon as the BV's move in to the transmission area of the nearest RV, it can discover the channel of the OSVs to which it must listen. When the BV's receive the NTM, they carry out adjustments in the WiMAX adapter's channel frequency. To do this the BV compares its location with respect to that of the RV that has been targeted. It then scans only that RV that lies near to the BV. The handoff process is depicted in Fig. 6.

## *2.6 Handoff using IP Passing*

In general when a vehicle moves out of the current communication area of the serving Base Station (BS) and enters the boundary of the new base station then it gets the new IP address using the Dynamic Host Configuration Protocol (DHCP). This procedure is time consuming and can consume up to 100% of the vehicles connection time. However, according to the IP passing scheme [12] the concerned vehicle receives the new IP from the outgoing vehicles. In case the incoming vehicle is unable to get IP from other vehicles, the normal DHCP procedure is used to obtain the IP address. The following example gives an illustration of the scenario. Suppose a vehicle A leaves the coverage area of an AP, another vehicle B which is behind A can access the services of the same AP by reusing A's IP address. There are three major steps in the whole process. Firstly, IP information gathering. Secondly, IP passing from vehicle A to vehicle B. Finally, real time configuration of B's interface.

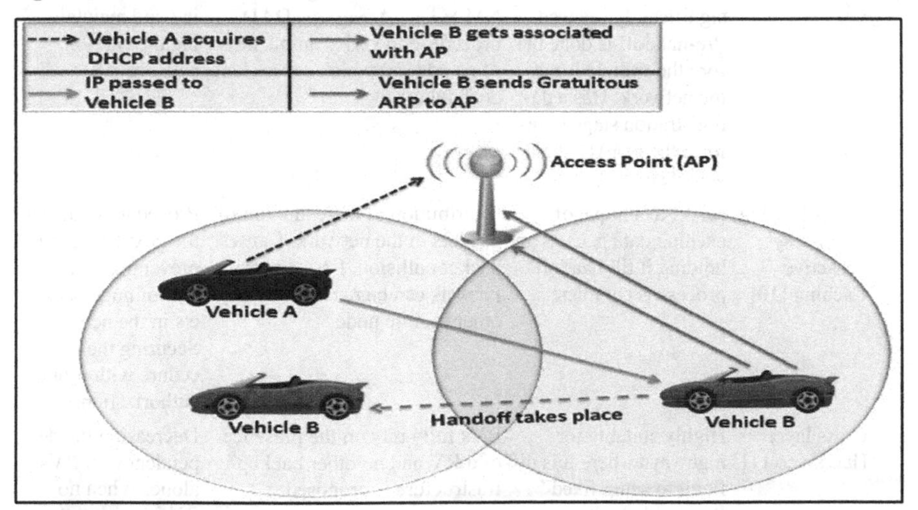


**Fig. 7 IP Passing Technique in VANET**

The steps that are involved in the IP passing have been illustrated in Figure 7 in detail. In first phase, a DHCP address is acquired by vehicle A. In second phase, the IP is passed to vehicle B. In phase three vehicle B gets associated with the AP. Finally, vehicle B broadcasts a Gratuitous ARP to AP.

**Table 1 Brief Overview of Advantages and Disadvantages along with Future Scope**

| Sl No. | Scheme | Advantage | Disadvantage | Future Scope/Area of Improvement |
|---|---|---|---|---|
| 1 | Mobility Handover Scheme[7] | Connection maintained all the time (even during handoff process). Reduction in packet loss. Reduction in delay of handover. No CoA required. | Heavily rely on the presence of a neighboring vehicle in the vicinity. High network overhead will occur during peak hours. | Handover management scheme to facilitate handover without requiring the assistance of another vehicle. |
| 2 | Vehicular Fast Handoff Scheme for Mobile IPv6[8] | Handover latency at the network layer is decreased by completely avoiding DAD. Incorporates pre-handoff procedure. | Inter-operator handoff is not demonstrated in this scheme. Not scalable. Assigning new addresses to vehicles is difficult. | Inter operator domain handover must be introduced. New addressing schemes should be included to make the network scalable. |
| 3 | Proxy MIPv6 Handover Scheme [9] | Ensures that the vehicle is always connected to the MAG. Packet loss is significantly reduced. Pre-handoff is done before the vehicle leaves the network. Has a deregistration stage to inform the PMAG about vehicle exit. | Care-of address has to be assigned to vehicle when it enters the transmission range of NMAG. CoA requires DAD procedure to verify uniqueness of an address which causes handoff delay | Elimination of overhead associated with assigning and maintaining the care-of address. |
| 4 | Handoff using Multi-way Proactive Caching [10] | Novel technique of caching data packets and holding it till handoff process is complete. | Distribution of huge amount of packets in the network. Causes packet collision. Less Secure. Packets can be received by any other mobile node. | Procedure to detect the next RSU, thus preventing multicast of huge packets in the network. Securing the procedure with proper authorization. |
| 5 | Cross layer Handover[11] | Highly suitable for highways where it is difficult to setup fixed Road Side Units. | BVs fully rely on the presence of a RV and no other backup infrastructure is proposed. | Decreasing the dependency on RVs alone. When no RV is present the Base Stations can take up the responsibility of assisting handoff or provid- |

| | | | | |
|---|---|---|---|---|
| | | | | ing the services. |
| 6 | Handoff using IP Passing [12] | No need to reconfigure the AP. Network overhead is reduced. Time spend in DHCP configuration reduced. | If there is no vehicle to pass on its IP then the IP needs to be acquired from DHCP, which takes up a lot of time. | It is better to design a network using unique home address throughout the network. |

## 3. Comparative Study between above Schemes

In [7], the handoff process in between subnet requires the assistance of the vehicle in front. If this vehicle is absent then the handoff process will get delayed. This problem can be solved by incorporating the solution that has been proposed in [8], where the vehicle before leaving the network sends a HAI message to the CAR. The CAR then performs the computation based on the parameters sent in the HAI and predicts the target access router. After that, it informs the vehicle using the vehicular proxy router advertisement. In [8], however the scheme is unable to address the scenario when the vehicle changes from the domain of one operator to another. The inter-operator handoff process has not been demonstrated clearly. This scheme lacks scalability. This can be improved by incorporating a proper infrastructure and detailed handoff procedure as in [9].

Another scheme in [10] uses proactive caching of data packets in the new RSU to which the vehicle is going to connect next. However in [10] there is no provision for determining the next access point/RSU to which the vehicle is to be connected from beforehand. This can be overcome by using the AoA technique discussed in [7] to determine the next AP or RSU so that packet loss is reduced significantly. The IP passing technique in [12] can be applied in scheme [9] to enhance the scheme. IP passing will avoid the complexity of the duplicate address detection procedure which is a must while using CoA in a network

In [11] if there is no Relay Vehicle present the Oncoming Side Vehicles (OSVs) and Broken Vehicles (BVs) are unable to carry out fast handoff. In such a case the Base Station should carry out the handoff process. A backup for relay vehicle must be provided. Moreover, a RV selection algorithm should be proposed if two or more RVs are present at the same time.

## 4. Conclusion and Future Scope of Work

The comparative analysis of the above scheme gives us a broader scope of future work on some crucial areas. An efficient handoff algorithm can be modeled if the scheme avoids dependency on other road vehicles for handoff. The handoff procedures in future must be triggered and managed solely by the infrastructure or the device that is providing the services, instead of the hosts. Deterministic learning algorithms can be included in VANET to make it more adaptive. The concept of caching is introduced in VANET in [10]. Handoff system designed using caching will reduce loss of data packets. In addition to this a standard handoff system must also cover the primary objectives of handoff such as reduced latency, reduced battery power consumption and load, and increased throughput of the overall network. The future focus should be to develop new solutions that will enhance the efficiency and performance of VANET.

**Acknowledgement:** The authors are grateful to West Bengal University of Technology TEQIP II program.

## References

1. Sharma, R., &Malhotra, J (2014). A Survey on Mobility Management Techniques in Vehicular Ad-hoc Network. *International Conference on Computing , Communication & Systems, 38-41*
2. Hartenstein, H., & Laberteaux, L. P. (2008). A tutorial survey on vehicular ad hoc networks. IEEE Communications magazine, 46(6), 164-171.
3. Toh, C. K. (2007, December). Future application scenarios for MANET-based intelligent transportation systems. In *Future generation communication and networking (fgcn 2007)* (Vol. 2, 414-417). IEEE.
4. Raw, R. S., Kumar, M., & Singh, N. (2013). Security challenges, issues and their solutions for VANET. *International Journal of Network Security & Its Applications*, *5*(5), 95.
5. Dinesh, M. D., PIIT, N. P., & Deshmukh, M. Challenges in Vehicle Ad Hoc Network (VANET). (2014, December)*International Journal of Engineering Technology, Management and Applied Sciences.* Volume 2 Issue 7
6. Wang, X., Le, D., & Yao, Y. (2015). A cross-layer mobility handover scheme for IPv6-based vehicular networks. *AEU-International Journal of Electronics and Communications*, *69*(10), 1514-1524.
7. Wang, X., & Qian, H. (2013). A mobility handover scheme for IPv6-based vehicular ad hoc networks. *Wireless personal communications*, *70*(4), 1841-1857.
8. Banda, L., Mzyece, M., & Noel, G. (2013, February). Fast handover management in IP-based vehicular networks. *In IEEE International Conference on Industrial Technology (ICIT), 2013* (1279-1284). IEEE.
9. Moravejosharieh, A., &Modares, H. (2014). A proxy MIPv6 handover scheme for vehicular ad-hoc networks. *Wireless personal communications*,*75*(1), 609-626.
10. Lee, H., Chung, Y. U., & Choi, Y. H. (2013, July). A seamless Handover Scheme for IEEE WAVE Networks based on multi-way Proactive Caching. In *Fifth International Conference on Ubiquitous and Future Networks (ICUFN), 2013* (356-361). IEEE.
11. Chiu, K. L., Hwang, R. H., & Chen, Y. S. (2009, June). A cross layer fast handover scheme in VANET. In *IEEE International Conference on Communications, 2009. ICC'09.*(1-5). IEEE.
12. Arnold, T., Lloyd, W., Zhao, J., & Cao, G. (2008, March). IP address passing for VANETs. In *Sixth Annual IEEE International Conference on Pervasive Computing and Communications, 2008. PerCom 2008.* (70-79). IEEE.

# Efficient Television ratings system with Commercial Feedback Applications

Aswin T.S, Kartik Mittal and Shriram K Vasudevan*
Dept. of Computer Science and Engineering,
Amrita School of Engineering,
Amrita Vishwa Vidyapeetham,
Amrita University, India.

**Abstract**

Everyone would have noticed several inconsistencies in TV ratings, with many popular programmes going off the air due to lack of sponsors, or certain shows getting undue patronage. Due to the absence of a uniform and transparent TV ratings system, several flaws exist in current rating systems, and they are open to manipulation. The purpose of our innovative invention is to mainly relay data to TV stations of what shows, programmes are watched by whom precisely without any chance for manipulation (with a responsibility of safeguarding one's private information and biometric data). Details of name or location are not asked, for sake of user's anonymity (a security perspective).

The proposed system is aimed at eliminating shortcomings of currently used ratings systems in the world that are bogged down by issues of transparency, insufficient data, and statistical approximation. Data pertaining to user (age and gender alone) and viewed statistics (shows, time viewed) is transmitted to the cable station, for purposes of 'feedback' i.e., showing vital statistics of the individual and the programme watched. This system will undoubtedly be one of the most reliable sources of feedback, compared to the existing systems [4].

**Keywords:** TRP (Target Rating Point) system, Broadcast Audience Research Council (BARC), Television Viewership in Thousands (TVT) system, Viewership statistics, Viewership transparency, TV programme rating system.



J.M. Corchado Rodriguez et al. (eds.), *Intelligent Systems Technologies and Applications 2016*, Advances in Intelligent Systems and Computing 530,
DOI 10.1007/978-3-319-47952-1_34

## Introduction

With advancements in technology today regarding feedback in entertainment and commercial presentation in the small screen, our proposed system proves that there is room for more accurate results that could be obtained. But this only possible if every TV manufacturer adopts our proposed standards uniformly, which requires minimal modification to be carried out in the respective hardware. Data of user's viewing statistics and user's statistics (specifically gender and age alone) is relayed to the cable station where data of all shows aired is compiled. Viewership statistics is then released to the general public as a region wise statistics, for purpose of transparency and for commercial interests (i.e., for advertisers to know where best they can invest their money to reach the target group).

Nielson Inc. in 1987 designed a system called the ***People Meter*** to store and relay data pertaining to age and gender of viewers and that method is in use even today, with demographic selection of people as a criteria [19]. The People Meter is a box which is attached to a TV set and the required data is transmitted wirelessly to the relay station. (Note: Our system makes use of the People Meter as an inbuilt box unit in the TV set.)

## Background Information

Although demographic selection of people is a good strategy to cater to all sections of the audience, and the best use of probability is employed, the reports on viewership data may not show actual precision despite having a large audience sample size.

The best way forward is undoubtedly to incorporate the system on a micro level (i.e. attach a People Meter in every TV set, as is proposed strategy) by having a component embedded in each TV set so as to include every sample viewer in the ratings statistics. In India, until recently, the Target Rating Point (TRP) system was employed as an audience measurement method. Following several protests over faulty implementation and manipulation of ratings, a new system, called Television Viewership in Thousands (TVT) system was introduced in July 2013, which was only different from the TRP system in having a larger sample size [7]. A new agency, the Broadcast Audience Research Council (BARC) was formed in May, 2015, to monitor TV ratings. The initial sampling is 22,000 Indian households within first two years, set to expand at the rate of 10,000 homes annually (with a budget of Rs. 280 crore) [1]. India has roughly around 160 million TV viewers currently (statistics of 77.5 million urban households and 76 million rural TV households, as of the 2011 census) [2]. According to a recent industry report by consultants KPMG India, India's television revenue grew to $7.9 billion in 2014 compared to $6.95 billion in 2013. The industry is projected to touch $16.25 billion by 2019, growing at a compound annual growth rate of 15.5 percent [3].

In all, India accounts for less than 10 per cent of the world's viewership. In 2013, the worldwide market revenue was estimated at 342 Billion Euros (roughly US $388 Billion as of 2015 exchange rates) [5]. Such an important worldwide industry ought to be regulated with more precision and accuracy. When a lot of money is getting involved in a system where lot of people are also involved, there is a necessity for precise and fool-proof measurement system which can be adopted as a global standard to ensure uniformity. [4]

The problem is that, even now, the focus is at household level data and not at the individual level, which could have been more beneficial. [8]

**Strategy on Cost Management**

The money spent worldwide on audience measurement can instead be given to TV manufacturers in lieu of a subsidy to include hardware components for the People Meter, embedding the biometric system in the TV remote. When made in bulk, pricing can eventually reduce the cost of every TV set manufactured in the world. The added advantage new being that every TV set would now be compatible with the new uniform TV measurement system.

**Architectural and construction details**

**Hardware Design**

The system is an electronic circuit system that can be embedded within the TV hardware. The circuit system consists of a PIC 4520 micro controller, a matrix keyboard, a timer and an LCD display. Initialized user data is stored in the inbuilt EEPROM in the

PIC. Consider the user's interaction with the TV unit by a remote control unit. Additional feature of biometric authentication may be used for the purpose of locking one's profile, so that only the designated user can access the profile.

The design of our intelligent TV is similar to latest TV units being sold in the markets, but with a few notable and appreciable exceptions.

Hardware composition is as follows:

1. An inbuilt chipset, to wirelessly (or through a cable) transmit the data pertaining to viewership, instead of installing it for selected users' systems (which is the usual practice).
2. Integrated timer, to start timing the viewership (with a gap of about 5-10 seconds, as channel flipping time)
3. The data stored will be in the form of RAM (Random Access Memory) so that stored data will be wiped once power supply is removed to the TV set. The reason is, with additional data storage of viewership data in the system (for multiple viewers, over a period of time), there is a possibility of working efficiency of the TV set lowering. Generated data is stored in RAM till the user switches the channel or decides to switch off the TV system, which is when the timer stops its action, transmits the generated data. Then the required action (changing the channel, or switching off the TV system) takes place. Data can also be stored within the TV set and transmitted at periodic intervals, (say, weekly). Storage of viewership can be supported in this case by installing memory component (of, say 100 MB) size.

The system itself need not store any viewership data. If user wants storage of viewership data for 'suggestion' of programmes, the uniform algorithms for prediction can be applied to the user's viewership data content and the results returned. Assuming the TV set takes maximum of 5 seconds to begin when powered on, the predictions can be retrieved by then, to minimize any waiting time. Consider a situation where user viewership can be monitored, like for children under 12, parental care, in schools also, since TV viewership may be for educational purposes.

4. A buffer storage of memory (internal memory of the system) of minimum of 4 users (with facility for increasing number of users) to store the users' age, gender and username (main aim: to differentiate between users with matching age, gender entries).
5. Biometric (fingerprint) scanner embedded in the TV remote and the required internal software to store users' biometric data and retrieve the profile on selection.[17][18]

**Technical description – Working Methodology**

Consider a household purchasing the TV set with our inbuilt system. The users will have to store their age, gender and unique identification (i.e. a unique username) on the TV set, which will be stored by our PIC micro controller (refer Fig 1 below). During each use, the user will have to select his/her user account and start viewing. For a user-friendly and 'quick' login experience, biometric system may be used, to identify the user, in an instant. Then the user selects desired channel. Once the desired channel is selected, a packet is being made, with the user's age, gender, channel and the timer segment. Once the channel is changed, or the system is turned off, the timer stops, and the packet is ready, to be transmitted to the 'relay station', where similar packets are received and viewership data is compiled and indexed accordingly. In our system, the Start/Stop buttons are meant for the manual start and end of viewing of the program, for the timer to begin and stop respectively. (The 'N' button from the circuit diagram below, has not been given any purpose in the program.) Once the viewership data is transmitted, the packet is 'reset' within the system, meaning that no viewership data remains in the system, for privacy concerns.

Moreover, no sensitive data reaches the 'relay station', so the user's anonymity is preserved. Additionally, a 'location' segment may be added by the relay station to the 'packet', if all viewership data is collected at one center (i.e. geo-specific data on viewership can also be obtained). In our system, the LCD display shows the 'packet'. This is the novelty we wish to claim credit for, the novelty of a system that can efficiently capture user's age, gender, channel and duration (i.e. viewership data) and efficiently, responsibly send for accurate data analysis. Transmission of the packet has not been implemented in our system, since that is not the novelty we wish to claim exclusivity for. User data can also be reset and changed at the user's convenience, with the 'reset' button. (Fig 2)

**Working of the system**

i.) Initialization: At time of setting up TV, biometric data(fingerprints) of all users is entered, each user's age and gender is stored (additionally, a 'username' field can be used to accommodate systems with users entering repeated age, gender combination). The user can use the biometric system, or can manually enter the user details in the TV with the help of the remote. For example, consider a sample profile:

| Username | Age | Gender |
|---|---|---|
| Dad | 40 | M |
| Mom | 40 | F |
| Kevin | 10 | M |
| Bob | 10 | M |

As mentioned above, the nickname is a unique identifier and distinguishes IDs with identical age and gender entries.

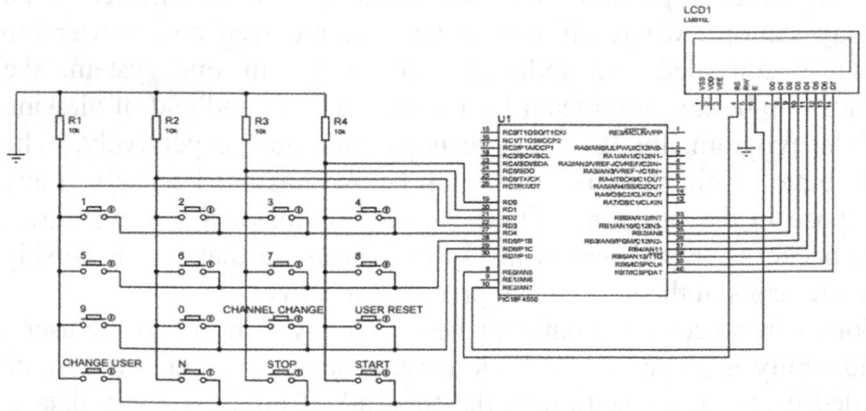


Fig. 1. The Architecture of the hardware Unit

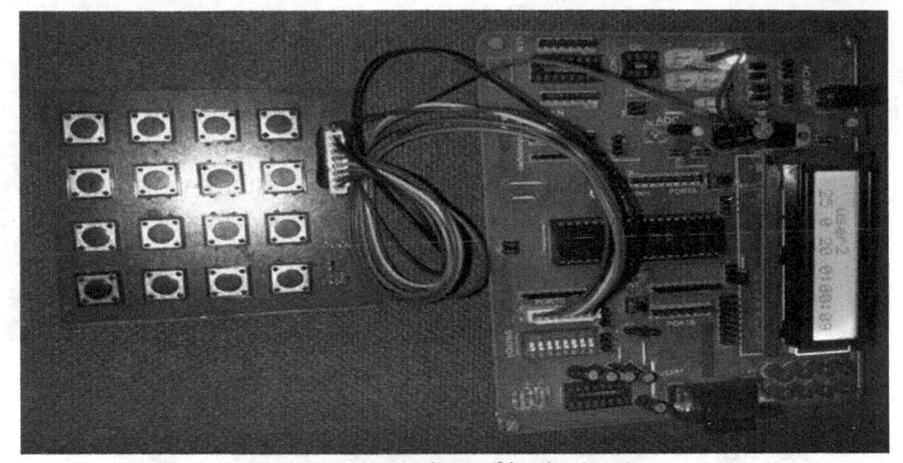

Fig. 2. A Snapshot of built circuit.

ii.) Working: The TV remote itself is an intelligent system, which recognizes the user (when the TV is powered on), selects the profile and intimates the TV system (through IR rays) that the particular user is operating the TV system (fig 2). Other facilities offered by the remote can be similar to a touchscreen phone, like time, alarm, number pad (for channel selection), additional security authentication like PIN number (according to user's convenience) may be added.

When recognized user starts TV, first he/she is identified, then as TV is working, timer notes program watched and duration of viewing, stores data in a buffer (neglecting channel flipping time, which may be assigned as less than 10 seconds). After a certain time interval, viewership statistics is returned to the transmitting satellite/cable station periodically, in the combination of several bits.

iii.) Transmission of data: From every 'regional' cable station, data can be transmitted to a central data collection unit (along with the extra variable of 'region') where the viewership data can be collected and displayed periodically. The main ideals of transparency, exclusivity, everyone's contribution may be noted. All internal security precautions are to be taken by the data collection units so that there is no possibility of tracing back any user/household.

iv.) Assimilation of data: A central data processing unit may be designed to receive all the data and map it appropriately.

For implementing our proposed system worldwide, a universal key of codes needs to be assigned for all the shows that are currently running in the world, that must be matched while viewing (recognizing) the TV show and assimilating data at the data collection units.

Assuming the data transfer is in the form of a series of bits, certain keys needs to be assigned universally for the age, gender entries as well. Design of keys may be done appropriately, to save space and follow the same universally, for sake of uniformity.

TV viewership is now supported by multiple network connected devices like smartphones, tablets, and laptops. Consider the software required to perform the functions of the intelligent TV system as a combination of biometric access, timer action, screen control functions and wireless transmission. (Refer Fig 4 below)

If the network connected devices are initialized too, then as a part of an authentication, biometric access may be checked when the user connects to the TV feed and selection of a profile can be intimated to the TV set, viewership data can be generated and transmitted to the TV set for further action.

**Claims**:

1. Continuous sampling of viewership: Since sampling 'unit' is inbuilt in our TV system, there is no fixed monitoring period of audience, making our approach more comprehensive and thorough.
2. Anonymous reporting of viewership: Only viewer's age, gender and viewership data is sent to the receiving station (no way of determining which household contributed this viewership data). Once sent, the viewership data is erased from the viewer's TV system (to ensure complete privacy of viewer).

The user's age and gender is asked only for more efficiency in TV viewership ratings. Unique Username for profile exists to maintain separate usernames.

3. Easy use system: New users can be initialized within seconds. Switching of profiles may also be supported in short time interval. Only user's gender and age, region is added to the viewership content before transmission. For e.g. M 23 NI could be added as a header to the viewership content transmitted (just a few integer bits more than the content itself), this code could imply that the viewership data is of a male, age 23 years, in North India. Region may be described as required.

**Advantages of our system over conventional ratings systems**

1. Anonymity of audience is lost in conventional systems, possibility of manipulation of ratings exist.
2. Selection of only few households (which is found in many programs, like the Nielsen Ratings that is prominently used in the United States ) implies that any findings of such surveys are an approximation, since entire sample size (i.e. entire TV viewing audience is not contributing their viewership data)
3. Selection of households, and not an individual's viewership implies that preferences to certain programs specific to gender, age is unimportant. While in the U.S the Nielsen ratings take into account the age (unknown if categorized or noted as it is), it remains to be seen if gender is taken into account. In India, only 'households' and not the individuals are taken into account.
4. Sampling in our system is continuous, for the entire time of TV viewership. In existing ratings systems, sampling is done for some pre-determined period of time, from a few weeks to a few months. For more accuracy, the sampling should be indefinite, which is one of the principles of our system's design.
5. Consider a programme that is well received among most people in a particular region and nowhere else. Current ratings systems would indicate very small interested audience and rate the system with low score, whereas our system can pinpoint geo-location of high number of viewers, so that sponsors can identify well received programmes. Broadcasters can telecast their

programmes flexibly so that large group at a particular region can view their desired programme. Several concerns have been raised about the efficiency of current ratings systems, due to well received programmes not being recognized by sponsors. [14][15][16]

6. Current operating rating systems, entities running them are not transparent, leading to integrity issues and potential manipulation of received data. [9][11][12][13] Since the entities, like Nielsen Inc. are registered as profit making entities, they are not bound to function transparently, share operational data.[10] Moreover, profit making entities would look to raise more profits in the future, by means of 'appeasement', catering to the subscriber. Our proposed system works on a transparent, non-profit model where the goal is that our system can be implemented for any region with complete accuracy.

**Potential commercial applications**

1. With complete data on world Television viewership, data analysis can be very significant for advertising, studying impacts, influences of programmes.
2. The viewership data can be segmented so advertisers can identify specific market segments that they would like to reach, for example: above 40 age group, in India.
3. Television channels can identify programmes that are receiving wide support and those are not performing well, etc.

**Results**

In this section, we evaluate our approach for the above mentioned system, where we have considered a datatset of 2000 entries, consisting of 4 regions, 5 channels and one week of viewership for age groups 5 to 80. The entries were generated with the help of a python script, and the output was then plotted using Microsoft Excel. A sample of 2 entries is tabulated below. Many combinations of data interpretation are possible, we have shown five graphs to highlight the same:

| Region | UserID | age | sex | channel | Duration in mins | time | Date |
|---|---|---|---|---|---|---|---|
| 3 | 2 | 39 | 1 | 2 | 17 | 16:42 | 25-03-16 |
| 1 | 3 | 8 | 1 | 5 | 11 | 20:44 | 22-03-16 |

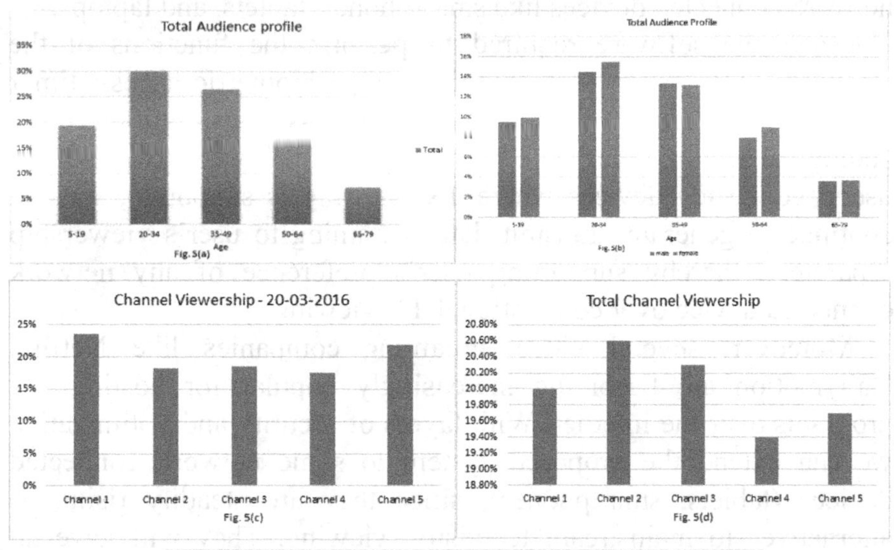


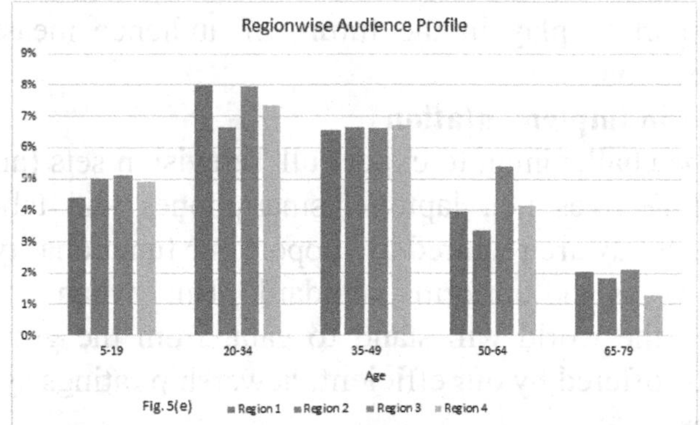


Fig. 5(a-e) Graphical analysis results

We inferred from the above graphs that majority of people who viewed the TV, were people in the range of (20-34) and majority were females. Channel 1 was most viewed on 20th March 2016, but channel 2 was viewed maximum times during the week.

**Further Scope**

As pointed above, TV viewership is now supported by multiple network connected devices like smartphones, tablets, and laptops.

Consider the software required to perform the functions of the intelligent TV system as a combination of biometric access, timer action, screen control functions and wireless transmission. If not adding the required hardware, installing the required software on all users' connected devices where TV viewing is supported, we can continue to generate, transmit data pertaining to user's viewership separately, thereby supporting user's preference of any network connected device over conventional TV viewing.

Moreover, several video streaming companies like Netflix, Dailymotion are becoming increasingly popular for hosting TV programs over the internet. With layers of security and optimization we can extend the proposed system to some network connected devices (tablets, smartphones), since they are steadily rising as alternatives to mainstream television viewing. They will have an important part to play in the future of audience measurement in world Television.

**Challenges in implementation**

It will be challenging, to expect all Television sets (and Network Connected Devices i.e., laptops, smartphones and tablets) to all contain the hardware required to support our functionality and adopt our system as a uniform standard. But when implemented universally, the world will stand to gain, from the growing list of applications offered by our efficient viewership ratings system [20].

**References**

[1] Business Standard. BARC's biggest achievement was getting the funding going. (2015, July 7). Retrieved from http://www.business-standard.com/article/companies/barc-s-

biggest-achievement-was-getting-the-funding-going-partho-dasgupta-115070700016_1.html
[2] The Economic Times. Meet BARC, a new TV rating system. (2015, May 30). Retrieved from http://economictimes.indiatimes.com/industry/media/entertainment/media/meet-barc-a-new-tv-ratings-system-to-create-a-new-set-of-winners-and-losers/articleshow/47477297.cms
[3] The Hollywood Reporter. India launches new TV ratings. (2015, April 30). Retrieved from http://www.hollywoodreporter.com/news/india-launches-new-tv-ratings-792527
[4] The Financial Express. All BARC, not bite. (2015, May 12). Retrieved from http://www.financialexpress.com/article/industry/companies/all-barc-no-bite/71492
[5] Business Standard. BARC to broaden TV rating system. (2016, March 31) Retrieved from http://www.business-standard.com/article/pti-stories/barc-to-broaden-tv-rating-system-to-include-laptops-phones-116033101079_1.html
[6] Business Standard. New TRP system to start early next year. (2013, April 18). Retrieved from http://www.business-standard.com/article/management/new-trp-system-to-start-early-next-year-113041801140_1.html
[7] The New Indian Express. Television viewership: TVT to replace TRP. (2013, July 26). Retrieved from http://www.newindianexpress.com/nation/Television-viewership-TVT-to-replace-TRP/2013/07/26/article1702592.ece
[8] Times Of India. TV ratings must be fair and researched: Prasad. (2016, March 30). Retrieved from http://timesofindia.indiatimes.com/city/mumbai/TV-ratings-must-be-fair-and-researched-Prasad/articleshow/51617481.cms
[9] New York Times. TV Ratings by Nielsen had errors for months. (2014, October 10). Retrieved from

http://www.nytimes.com/2014/10/11/business/media/tv-ratings-by-nielsen-had-errors-for-months.html?_r=0
[10] The Wall Street Journal. Nielsen Contests Rivals' Claims, Sparks Fly in Measurement Fight. (2015, September 30). Retrieved from http://www.wsj.com/articles/nielsen-contests-rivals-claims-sparks-fly-in-measurement-fight-1443633212
[11] The Wall Street Journal. Nielsen Isn't Leaving the Living Room. (2015, January 19). Retrieved from http://www.wsj.com/articles/nielsen-isnt-leaving-the-living-room-heard-on-the-street-1421688753
[12] Poynter. Nielsen method for TV ratings missing minorities, young people. (2013, October 16) Retrieved from http://www.poynter.org/2013/nielsen-method-for-tv-ratings-missing-minorities-young-people/225876/
[13] The Wall Street Journal. CNBC to Stop Using Nielsen for Ratings. (2015, Jan 6). Retrieved from http://www.wsj.com/articles/cnbc-to-stop-using-nielsen-for-ratings-1420520556
[14] Denny Meyera and Rob J. Hyndman. The accuracy of television network rating forecasts: The effects of data aggregation and alternative models. IOS Press. Model Assisted Statistics and Applications 1 (2005,2006) 147–155
[15] International Business Times. Why are TV shows cancelled so quickly? (2014, May 13) Retrieved from http://www.ibtimes.com/why-are-tv-shows-canceled-so-quickly-fans-mourn-death-surviving-jack-believe-others-1583307
[16] Business Insider. The 17 Best TV shows that were cancelled. ( 2014, October 17 ) Retrieved from http://www.businessinsider.in/The-17-Best-TV-Shows-That-Were-Canceled/articleshow/44855372.cms
[17] Device Configuration for Multiple Users Using Remote User Biometrics (US 20150205623) Patent granted to Apple Inc. in 2014.

[18] Electronic Device Operation Using Remote User Biometrics (US 20150206366) Patent granted to Apple Inc. in 2014.
[19] The Nielsen company website. Celebrating 25 years of the Nielsen People Meter. (2012, August 30) Retrieved from http://www.nielsen.com/in/en/insights/news/2012/celebrating-25-years-of-the-nielsen-people-meter.html
[20] Business Insider. The Future of TV Browsing Resides on iPad and Laptops. (2011, Sep 8). Retrieved from http://www.businessinsider.com/the-future-of-tv browsing-resides-on-ipad-and-laptops-2011-9?IR=T

# A Literature Survey on Malware and Online Advertisement Hidden Hazards

Priya Jyotiyana [1,*], Saurabh Maheshwari [2,*]
*Department of Computer Science and Engineering,
Government Women Engineering College, Ajmer, INDIA

[1]e-mail: priya7653@gmail.com
[2]e-mail: dr.msaurabh@gmail.com

**Abstract.** Malware is a malignant code that expands over the connected frameworks in system. Malvertising is a malicious action that can distribute malware in different forms through advertising. Malware is the key of advertising and generate the revenue and for various Internet organizations, extensive advertisement systems; for example, Google, Yahoo and Microsoft contribute a ton of effort to moderate malicious advertising from their advertise network systems. This paper specifically discusses various types of detection techniques; procedures and analysis techniques for detect the malware threat. Malware detection method used to detect or identify the malicious activities so that malware could not harm the user system. Moreover the study includes about malicious advertising. This paper will look at the strategies utilized by adware as a part of their endeavours to stay inhabitant on the framework and analyze the sorts of information being separated from the client's framework.

**Keywords:** malware classification; malware analysis; malware detection; malvertising

## 1 Introduction

The term malware originates from joining the two words malicious and software and to be utilized to demonstrate any undesirable programme as any unwanted malicious code included and changed and expelled from programming keeping in mind. Once when malware inject in system then it intentionally cause harm or change the functionally of system. A huge number of new malware appear and cause harms the functionality of system. Interestingly researchers create numerous new strategies to detect and remove the malware. Malware of computer system can be categorized into propagation, replication, corruption and self-execution. When malware inject in computer system cause harms the integrity, confidentiality and denial of service [1]. Malvertising is one of such activities where advertise can maliciously used by an attacker to distribute various forms of malware. It is serious consequences, because an attacker can purchase ad space to publish malicious ad on many popular website and cause harm the system silently. In addition user may be unaware of the fact of



J.M. Corchado Rodriguez et al. (eds.), *Intelligent Systems Technologies and Applications 2016*, Advances in Intelligent Systems and Computing 530,
DOI 10.1007/978-3-319-47952-1_35

malicious activity that they could meet the malicious content while browsing highly reputable website which may pace them silently at an even higher risk [2]. In present paper we examine the current circumstance with respect to adware and their finding strategies. The rest of the paper is structured as follows. Section II gives a well-known introduction about classification of malware i.e. virus, worm, adware, spy-ware and Trojan horse. Section III elaborates those malware analysis techniques. Section IV illustrates the various detection techniques. Section V gives a well-known theory of malvertising and how to inject malicious advertising in user system. We conclude the paper in Section VI.

## 2 Malware Taxonomy

As indicated by [3] malware can be ordered into three eras in light of their payload, empowering risk and spread system. In first era malware gives properties of infection which duplicates by some human activity, messages and document sharing. Second era malware offers the property of worms. These are hybrid in nature with a few elements of infection and Trojans which needn't bother with any human activity for replication. In third era malware gives the geographical region or association specific. Further we can see the classification of malware [4]:

- **Virus:** Virus Infection is a type of malware that oblige medium to stimulate. It spreads malware by joining any project and after that executes the code when user can perform any activity in framework. The host program work even after it has been influenced by virus infection. Virus can steal also some private data, unwanted data and can cause denial of service attack.
- **Worms:** The worm first going into the framework and spread its contamination through some liability and after that taking the advantage of data and exchange the record which permit it to stimulate autonomously.
- **Trojan horse:** A Trojan horse is a kind of malware that cover itself as an ordinary record. At first look it seems, to be totally safe. Once actuated, it can perform an extensive number of attacks on the host by popping up windows to harming the framework by erasing the documents and information, take private data, and so forth. Trojan horse can create the indirect accesses through which malevolent user can access the framework. At the point when malevolent user can access the framework, then it perform its noxious activities like taking private information, introducing malware, follow and checking user action, utilizing the PC as a part of botnet. In future, a Trojan horse might be prepared to do any type of or level of damage.
- **Spyware:** A spyware is a kind of malware that follow the user action without the client learning. It then transmits and assembles data to outsider i.e. attacker. The gathered information can include the website, browser and system information, etc which can visit by user. A spyware is a kind of malware that follow the user action without the client learning. It then transmits and assembles data to outsider i.e. attacker. The assembled data can incorporate the site which can visit by client,

program and framework data, and so on. Spyware can likewise control over the framework in a way like that of a Trojan horse.
- **Adware:** Adware is also called advertising supported programming which can perform its activity to display or download the advertisement to a user computer after the installation of malicious application or programming. As [2] an ad-injecting browser extension is one instance of adware, analysing all the malware activities of ad injecting extension also falls under the category of adware.

## 3 Review and Analysis of Malware

Malware can advance and perform their activity in quick way and measures to stop them have ended up troublesome that attackers utilize new signatures and encapsulate the activity which attackers perform so difficult to detect. Malware analysis types are as follows:
- **Static analysis:** It is additionally called behavioural investigation, which is utilized to examine and study about the malware behaviour. In static analysis, we analyse how malware interact with the environment, file tampered, services added, network association made, data captured, port opening, etc. Gathered information can reproduced and mapped and consolidate together to get complete picture of examination. It is the procedure of break down the execution of code without being really executing the record. Static examination declines less in breaking down vague malware that uses code complication technique. Researchers and scientists have been performing analysis aimed at static of malwares need to have decent information of low level computing paradigm and also with functional background. Programming analysers, debuggers and disassemblers can be distributed with as functioning devices in investigation with static malware. The actual encouraging point of static investigation over its dynamic partner is that the execution time overhead must be less or nonetheless [5].
- **Dynamic analysis**: is likewise called code examination, which is utilized to break down the code of malevolent programming. It is extremely hard to get the program of malware particularly executable record we have to examine the binary code. Malware code can be converted into its binary or assembly level by using debugger and decompiles. Dynamic examination may neglects to identify an activity which demonstrates the behavioural modifications in codes by various activate conditions over the span of its execution [5].
- **Hybrid analysis:** It is a combinational methodology of both static and element examination. Firstly it investigations about the signatures of malware code and afterward joins it with behaviour parameters for improvement of complete examination. Because of this methodology half and half investigation conquers the constraint of both static and element examination [6].

**Table 1.** Static analysis v/s dynamic analysis

| Method | Static analysis | Dynamic analysis |
|---|---|---|
| **Analysis** | Behavioural analysis | Code analysis |
| **unknown malware detection** | Easy detection | Hard to detect |
| **multipath malware analysis** | Easy analysis | Hard to analyse |
| **Speed and safety** | Fast and safe | Neither fast nor safe |
| **Code of execution** | Not executed | executed |

## 4 Techniques for Malware Detection

Malware identification methodologies are based on different techniques that are normal in PC programming investigation, i.e. Dynamic, static and hybrid. In malware recognition, it is about recognizing whether the code is malevolent or favourable [5].

- **Signature based techniques:** the vast majority of the antivirus devices depend on signature based identification procedures. The signatures are generated when the disassembled code of malware can examine. Different debuggers and dissemblers are accessible which can support in dissimulate the program. Dissimulate code is broke down and components can be extracted. These components can be separated in building the signatures of specific malware family [7]. Pattern matching and signature based detection techniques is most mainstream technique in malware detection [8].
- **Behaviour based technique:** the principle objective of this technique is to examine the behaviour of known or unknown malware. Various factors include in behaviour parameter such as types of attachment, source code destination address of malware, and other countable measurable components [7]. This procedure watch the conduct of project whether it is noxious or favourable. Since behaviour based procedure watch action of executable code, they are not vulnerable to the weakness of mark based ones. Essentially behaviour based strategies close whether a system is malignant or not by investigating what it does instead of what it says [8].
- **Heuristic based technique:** as we specified, some disadvantages of signature based and behaviour based procedures. Thus heuristic based identification technique is proposed to defeat these disservices. Heuristic based malware method analyse the behaviour of executable file by utilizing information mining and machine learning methods. Naive bayes were employed to classify malware and generous records [8].
- **Semantic based technique:** malware identifier can work by testing for marks which can likewise catch the syntactic attributes of machine level byte series of the malware. In syntactic methodology make a few indicators unprotected against code confusions, progressively utilized by malware correspondents that adjust

syntactic properties of malware byte grouping without influencing their execution behaviour. A semantic based system for investigation about malware identifier and demonstrating properties, for example, sound and fulfilment of these locators. Semantic based system follow the behaviour of malware and in addition infected program likewise checked and hide the irrelevant aspects of this behaviour [9].

# 5 Malvertising

The other software programme runs then Adware typically generate advertising such as pop-ups and banners. Some adware gather the information about activity which perform the users system and provide the information to the third party or attacker and as some instances is spyware disguised as adware. Adware perform the unwanted activity through malicious advertisement and cause harm the privacy, user's computer security software.

## 5.1 Different media and means for online advertising

Types of online medium Description [10]

- **Mobile Advertising:** Portable promoting is advertisement duplicate conveyed through remote cell phones, for instance, mobiles, tablet ,PCs ,highlight telephones, Portable publicizing may take the type of static media show advertisements, SMS (Short Message Service) or MMS (Multimedia Messaging Service) promotions, portable inspection advertisements, publicizing inside versatile sites, or advertisements inside portable applications.
- **Social media marketing:** Online networking advertising is business advancement led through social networking sites. Numerous groups spread their items by posting regular redesigns and giving unique offers through their online networking profiles.
- **Email Advertising:** Email circulation is promotion replica involving a whole email or a part of an email message. Email circulation might be spontaneous, in which case the sender may give the beneficiary a choice to quit future messages, or it might be sent with the beneficiary's earlier assent.
- **Banner Adverts:** Banner adverts are the most well-known kind of adware. It typically appears a little strip at the top point of the site. Client taps on banner advert, clients are coordinated to the site page of the advertisement. So for every click the proprietor of the host site will get an instalment from the advertisement.
- **Pop-up Adverts:** This notice opens another program in an alternate window. The client may or not trigger the advert. Case in point, they tap on a connection to go to an alternate site. The pop-up window expressly ought to be closed by clients with a specific end goal to end the advert.

- **In-text Adverts:** This advert is not like alternate sorts of adverts that we have looked as of not long ago. Because in this the text is adjusted. Keywords contain joins in the substance. On the off chance that the mouse goes over them the averts get to be noticeable.
- **Video Adverts:** This advert content grants notice strategies application from TV with the likelihood of the client communication. This advert has two distinct sorts. These are straight and non-direct linear advert is put into video content like business adverts are put into TV shows. The adverts temporarily take over from the video content. Non-direct adverts appear to be all the while with the video content. The substance may be hidden.

## 5.2 Delivery Vectors

Adware undertakings are presented on a structure in a variety of ways, however in an unmistakable and real to straight way. Most adware undertakings are obtained at first by examining the web or nearby some unimportant advancement maintained programming. The projects are infrequently introduced from a prominent site, but instead through social building standard advertisements, drive-by-downloads, and through shared systems with misdirecting filenames.

- **Social engineering banner ads.** Numerous sites use standard advertisement administrations where a promoting picture is put on their site. Tragically, a substantial number of these pennant advertisements are totally deceptive .Some banner promotions utilize a photo that mirrors a Windows message box with a sincere message misleading PC customers into tapping on the photo. When they tap on the fake message box they are redirected to various destinations that may begin the foundation of adware or further device the customer. For instance, some of these fake message boxes will express the customer's PC is polluted or have some other structure issue, for example, an off base clock. While tapping the fake message box, the customer is occupied to explain programming with conform the issue, when truth is told the customer was not contaminated or didn't have a misguided clock. The illustration appeared here originates from the Fast Click Ad Network [12].
- **Drive by downloads:** It is the activity of irritating a client to introduce a system as they check the web without the client really asking for the establishment of any project in any case. A drive by download is generally pretended through programmed website page revive or ActiveX control installers [11].
- **Automatic refresh**: Programmed page invigorate happens when a website page just diverts the program to an EXE bringing on an exchange brief to be shown.

This redirection can be accomplished in an assortment of habits together with utilizing basic hypertext markup language HTML or JavaScript [2].

Example 1: Using HTML and JavaScript of Page redirection

```
HTML
<meta http-equiv="refresh" content="0;url='http://127.0.0.1/example.exe">
JavaScript
<script>
location.href='http://127.0.0.1/example.exe'
</script>
```

— **Active X:** ActiveX is a game plan of headways made by Microsoft to permit undertakings to interoperate. Part of these advances is an ActiveX control. ActiveX controls are created as a development to static HTML content, giving more dynamic web satisfied. Exactly when seeing a site page that requirements an ActiveX control to summon dynamic substance, the project will induce the customer to download and present the ActiveX control. Most adware don't use this procedure to current component web satisfied, however quite to trap the customer into downloading and presenting the ActiveX manage, which is basically the adware program [14].

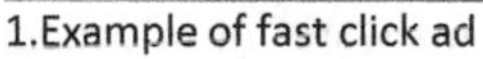
1.Example of fast click ad

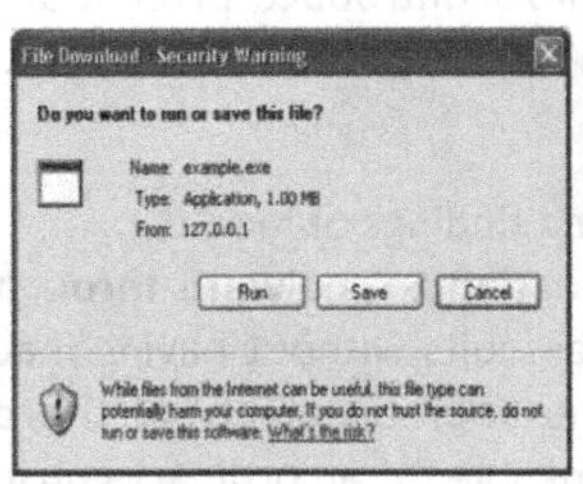


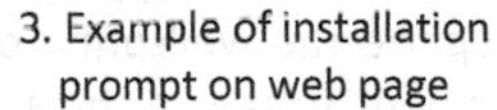
3. Example of installation prompt on web page

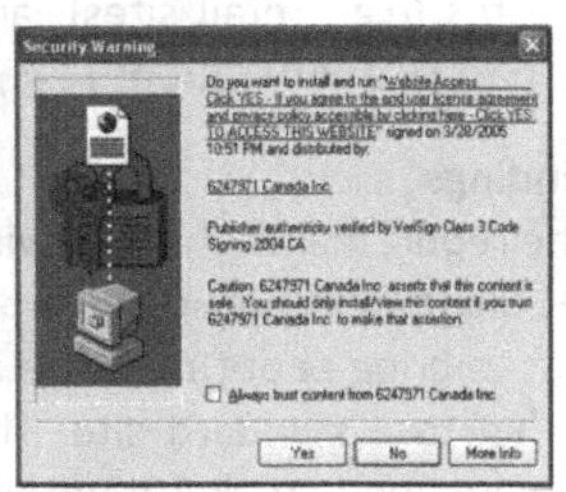


4.Example of an active x

**Fig. 1.** Examples of Delivery Vectors [13]

## 5.3 Way of Findings and Recommendations

The internet promoting industry can be minded confusing and hard to get it. In such a domain, deciding capable gatherings when things turn out badly can be troublesome [13].

**Problems occur due to**

The report makes the following problems:

— **Deceptive downloads:** misleading downloads attempt to pull in their casualties to download and introduce the particular programming that is malignant. The principal contrast from drive-by downloads is that attackers don't endeavour to and a weakness in the project modules to download and present a bit of malware, however rather they endeavour to trap the client into performing that approach purposefully. This happens by taking the client assume that here is some attractive substance on the visited web page [14].

- **Link Hijacking:** Link hijacking permits a notice to naturally divert clients to sites that they must not choose to visit. The commercials are merged into iframes, and the headway elevating scripts can't get to the Document Object Model (DOM) of the distributer's site page in light of the Same-Origin Policy (SOP) imperatives. Although, a malicious script contained in headway can involve the whole page to a preselected destination by setting Browser Object Model's (BOM) top Location variable. Along these lines, the casualty is diverted to a self-assertive area and not to the one she has at initially selected [14].
- **JavaScript Libraries for Ad Injection:** It is moderately direct for expansion designers to adapt their expansions through advertising combination. Like existing advertisement systems and advertising trades, there exists a flourishing advertise of advertising systems that give JavaScript advertisement combination libraries to expansion engineers to incorporate with their applications. These libraries introduce advertisements on pages by altering the DOM structure of the HTML and embedding extra HTML iframe's that contain the infused advertising content. The advertising provider may trigger the advertisement combination just on particular sites (e.g., retail sites), and/or introduce promotions when a client performs a particular operations (e.g., mouse drift on an item picture)[2].

**Findings**

The report makes the following findings of fact.

- **Customers hazard introduction to malware through regular action.** Customers can bring about malware assaults without having made any move other than going to a standard site. The unpredictability of the online advertising biological community makes it unthinkable for an ordinary consumer to withdraw from publicizing malware assaults, recognize the wellspring of the malware introduction, and figure out if the advertisement system or host site could have kept the assault.
- **The multifaceted nature of current web publicizing enhances blocks industry responsibility for malware assaults.** The internet promoting industry has developed in multifaceted nature to such a degree, to the point that every gathering can possibly guarantee it is not mindful when malware is conveyed to a client's PC through a notice. A standard online notice commonly experiences five or six middle people before being conveyed to a client's program, and the ad networks systems themselves infrequently convey the genuine commercial from their own servers. Much of the time, the proprietors of the host site went by a client doesn't recognize what notices will be appeared on their site.
- **Visits to standard sites can open customers to several obscure, or possibly hazardous, outsiders.** Subcommittee examination of a few well known sites observed that meeting even a standard site opens purchasers to several outsiders. Each of

those outsiders might be equipped for gathering data on the shopper and, in compelling situations, is a potential wellspring of malware.

- **Current frameworks may not make adequate motivators for internet promoting members to prevent consumer manhandle.** Since obligation regarding malware assaults and unseemly information gathering through online promotions is indistinct, internet publicizing members may not be completely incentivized to build up successful customer shields against misuse.

**Recommendations**

The Report makes the following recommendations [13].

- **Set up better practices and clearer standards to anticipate internet promoting manhandle.** Under the current administrative and authoritative structure, legitimate obligation regarding harms brought about through malvertising for the most part rests just with the fake performer being referred to. Since such on-screen characters are once in a while got and even less much of the time ready to pay harms, the damage brought about by vindictive notices is eventually conceived by purchasers who as a rule have done just visit a standard site. While buyers ought to be mindful so as to keep their working frameworks and projects overhauled to maintain a strategic distance from helplessness, advanced business substances, expansive and little, ought to find a way to diminish systemic vulnerabilities in their publicizing systems.
- **Clear up particular disallowed rehearses in internet publicizing to avoid manhandle and secure buyers.** Self-administrative bodies must try to create extensive security rules for anticipating web publicizing malware assaults. Without compelling self-control, consider issuing thorough directions to restrict misleading and out of line web publicizing hones that encourage or neglect to find a way to anticipate malware, intrusive treats, and unseemly information accumulation conveyed to Internet purchasers through online ads. More noteworthy specificity in disallowed or debilitated practices is required before the general security circumstance in the internet publicizing industry can move forward.

## 5.4 Methodology

In online advertising industry to understand some of the hazards it is important to comprehend two unique procedures: (1) how information is gathered on Web clients by outsiders and (2) when the use of data how advertises are delivered [13].

(a) Information is gathered on Web Clients by outsiders

- Cookies: The most essential capacity a treat serves is to distinguish a device. With a cookie, sites can know what number of significant machines—and, by augmentation, generally what number of special guests gone to their site. By permitting a site to recognize singular guests, treats can offer assistance sites give valuable administrations to guests. For instance, numerous anti-fraud provisions are cookie based, furthermore, most web "shopping cart" capacities require a cookie to affirm that the client who added one thing to their cart is the same client who has explored to an alternate part of the site.

— Information gathering and Advertising: Third parties can convey a treat since a few part of the host site draws upon substance from the third parties server. With regards to promoting, the outsider substance asked for by the host site is the ad itself. A call from the host site opens the entryway for a cookie to be put by the third parties whose substance was called for. Be that as it may, the third parties substance showed on the host site can be undetectable. Because the host site asked for some imaginary measure of substance from the third parties can now convey its cookie to the client's program too. Accordingly, information agents or different elements that convey no genuine substance to the host site can in any case convey treats by contracting with the host site to put a single pixel on their site.

— Cookie Controversies default setting: A program's default settings leave cookie active, since numerous considerate web capacities shoppers have generally expected are cookie based. A protection minded (and educated) client can maintain a strategic distance from all treat based following on the off chance that she so picks. Be that as it may, not very many Internet clients really adjust default program settings that organize shopper privacy. The default program setting along these lines has a huge effect in the utilization of cookie, and thusly how much information is assembled on Internet clients.

**(b) When the use of data how advertises are delivered.**

— **Efficient Process of Ad Delivery:** At the point when a client visits a site that uses ad network to convey its advertisements, the host site educates the client's program to contact the ad network. The advertisement system recovers whatever client cookie identifiers it can. Utilizing those identifiers, the ad network can get to its own database to see what other data about the client's history it has keeping in mind the end goal to distinguish the client's interests and demographic data. The promotion system can then choose which ad would be best to serve that specific client.

In spite of the fact that the ad network chooses which notice must be sent, it frequently does not convey the genuine ads. Rather, the advertisement system trains the client's program to contact a server assigned by the genuine publicist. The server that conveys the notice is most regularly called a substance content delivery network (CDN).

— **Role of Ad Tags in the Online Advertisement Delivery Process:** Another critical part of advertisement conveyance is the entangled way in which the client's program, the host site, the promotion system, and the publicist speak with each other. That correspondence is eventually accomplished through "ad labels," which are hypertext markup dialect (HTML) code sent between web advertising elements, which will eventually call up the right advertisement to be conveyed to a user. That HTML code passes on data about the advertisement space to be filled. The working of ad labels clarifies how web publicizing organizations can send notices to clients' programs without the advertising organizations quite realizing what that notice is. Advertisement labels are the messages that advise internet advisement organizations what advertisement to convey without really sending the ad itself between different organizations. Whenever a client visits a site, that host site sends an advertisement tag out to its system. That tag will contain some type of treat distinguishing proof so that the advertisement system will perceive the client.

The advertisement system's server will then quickly ring every single accessible data on the client and choose which commercial to convey. The promotion system will then send an advertisement tag back through the client's program, instructing it to recover the correct ad at a URL that the publicist has indicated. In any case, the real record at that URL can be carefully changed after that underlying quality control check so that when a client really experiences the promotion, a harmless and safe advertisement may have been changed into a vehicle for malware.

## 6 Conclusion and Future Work

Malware is representing a danger to user's PC frameworks as far as taking individual and private data, ruining or handicapping our security frameworks. This paper highlighting some current techniques joined by malware and adware detection and also gives the brief introduction of about the static, dynamic and hybrid malware analysis techniques. A few deficiencies in dynamic evaluation like distinct implementation way, major execution transparency so that static investigation more ideal than element. Adware typically generates advertising, such as pop-ups and banners, while the other software program is running and cause harm the computer system. It represents the danger for security and privacy of computer user.

In future perspective point of view the internet advertising genetic system is extremely unpredictable. Online sponsors do significantly more than only disappear content, realistic, or video commercials. Basic the work of online advertise are modern systems that can distinguish and target particular buyer group with important advertise, and also state of-the art security practices to screen the trustworthiness of these advertise delivery frameworks that lack of concern inside the web advertise industry may quick excessively careless security managements, making genuine vulnerabilities for Internet clients. Such vulnerabilities can develop more terrible without extra motivating forces for the most trained gatherings on the Internet to work with buyers and other partners to take compelling introductory methods.

## References

1. Imtithal A Saeed, Ali Selamat, Ali MA Abuagoub: *A survey on malware and malware detection systems*. International Journal of Computer Applications, 67(16), doi:10.5120/11480-7108.(2013)
2. Xinyu Xing, Wei Meng, Byoungyoung Lee, Udi Weinsberg, Anmol Sheth, Roberto Perdisci, and Wenke Lee. *Understanding malvertising through ad-injecting browser extensions*, 2015.
3. Andrew H Sung, Jianyun Xu, Patrick Chavez, and Srinivas Mukkamala. *Static analyzer of vicious executables*. In Computer Security Applications Conference, 2004. 20th Annual, pages 326–334. IEEE, 2004.
4. Robin sharp. *An introduction to malware*, spring 2012
5. Nitin Padriya and Nilay Mistry. *Review of behaviour malware analysis for android*. International Journal of Engineering and Innovative Technology (IJEIT) Volume, 2, 2008.
6. Muazzam Ahmed Siddiqui. *Data mining method for malware detection*.Proquest,2008
7. Ammar Ahmed Elhadi, Mohd Aizaini Maarof, and Ahmed Hamza Osman. *Malware detection based on hybrid signature behaviour application programming interface call graph*. American journal of applied Science, 9(3):283,2012
8. Zahra Bazrafshan, Hossein Hashemi, Seyed Mehdi Hazrati Fard, and Ali Hamzeh. *A survey on heuristic malware detection techniques*. In Information and Knowledge Technology (IKT), 2013 5th Conference on, pages 113–120. IEEE, 2013.
9. Mila Dalla Preda, Mihai Christodorescu, Somesh Jha, and Saumya Debray. *A semantics-based approach to malware detection*. ACM Transactions on Programming Languages and Systems, 30(5):1–54, 2008.
10. Antarieu, F. et al., 2010, *measuring the effectiveness of online advertising*. Study conducted by PwC for IAB France and the Syndicat de Régies Internet, Paris.
11. Apostolis Zarras, Alexandros Kapravelos, Gianluca Stringhini, Thorsten Holz, Christopher Kruegel, and Giovanni Vigna. *The dark alleys of Madison Avenue: Understanding malicious advertisements*. In Proceedings of the 2014 Conference on Internet Measurement Conference, pages 373–380. ACM, 2014.
12. John Aycock. *Spyware and Adware*, volume 50. Springer Science & Business Media, 2010.
13. Eric Chien. *Techniques of adware and spyware*. In the Proceedings of the Fifteenth Virus Bulletin Conference, Dublin Ireland, volume 47, 2005.
14. Zarras, Apostolis, et al. "*The dark alleys of Madison Avenue: Understanding malicious advertisements*." Proceedings of the 2014 Conference on Internet Measurement Conference. ACM, 2014.

# Design & Analysis of Clustering based Intrusion Detection Schemes for E-Governance

**Rajan Gupta[1], Sunil K. Muttoo[1] and Saibal K Pal[2]**

**Abstract** The problem of attacks on various networks and information systems is increasing. And with systems working in public domain like those involved under E-Governance are facing more problems than others. So there is a need to work on either designing an altogether different intrusion detection system or improvement of the existing schemes with better optimization techniques and easy experimental setup. The current study discusses the design of an Intrusion Detection Scheme based on traditional clustering schemes like K-Means and Fuzzy C-Means along with Meta-heuristic scheme like Particle Swarm Optimization. The experimental setup includes comparative analysis of these schemes based on a different metric called Classification Ratio and traditional metric like Detection Rate. The experiment is conducted on a regular Kyoto Data Set used by many researchers in past, however the features extracted from this data are selected based on their relevance to the E-Governance system. The results shows a better and higher classification ratio for the Fuzzy based clustering in conjunction with meta-heuristic schemes. The development and simulations are carried out using MATLAB.

## 1. Introduction

Threats in the digital world are rising continuously and one such area of concern is E-Governance. The digital data stacked by government on regular basis is growing by leaps and bounds and thus its security concerns are rising exponentially [5] [6]. Be it data transfer through the E-Governance network or storage of files and folders, the attackers are finding innovative ways to enter into the system and causing trouble to them [7][11].

---

[1] Department of Computer Science, Faculty of Mathematical Sciences, University of Delhi, North Campus, Delhi-110007, INDIA, e-mail: rgupta@cs.du.ac.in, guptarajan2000@gmail.com

[2] SAG Lab, Metcalfe House, DRDO, Delhi-110054, INDIA, e-mail: skptech@yahoo.com





J.M. Corchado Rodriguez et al. (eds.), *Intelligent Systems Technologies and Applications 2016*, Advances in Intelligent Systems and Computing 530,
DOI 10.1007/978-3-319-47952-1_36

There are various kinds of intrusions that are creating trouble to the E-Governance systems. For these anomalies, lots of intrusion detection systems are getting deployed within the E-Governance System. These intrusion detection systems may be used for the anomaly detection or misuse detection. But with time, attackers are also getting trained to break such detection systems and are making the system weak for anomalies based on transactional and storage capacities [1][3][9].

So, there is a need to continuously work on the development and evolution of different kinds of Intrusion Detection Systems and relate them to the specific patterns of the attacks in the E-Governance arena. That is why the current study discusses Intrusion detection schemes on the features which are relevant for data points related to E-Governance Information System. The identification of anomalies and intrusions are done using the data mining schemes like clustering. New age evolutionary and nature inspired schemes are also used to improve the working efficiency of existing algorithms. Also, various permutations and combinations in the experimental setup of the system are considered and most efficient ones are reported. From analytical point of view, a different metric is used for the comparison of performances of the various schemes.

## 2. Review of Literature

The work in the field of Intrusion Detection Schemes has been going on for long time but with different experimental setup, schemes and applications. This study will focus mainly on clustering scheme related literature so as to limit the scope of this study. The comparative analysis of K-Means and Fuzzy C-Means (FCM) have been presented and studied by various researchers in past however limited literature was found with respect to the combination of meta-heuristic schemes (esp. Particle Swarm Optimization (PSO)) along with the clustering techniques in context of features relevant to E-Governance attacks and the number of clusters formed during the implementation of various clustering techniques.

FCM was studied in combination with Quantum PSO by Wang [16] by considering 5 different types of clusters based on types of attacks. Attacks of Computer forensics were also studied by Wang [15] for which again the similar combination of FCM and PSO was used but the function from FCM was given as the objective function to PSO. Another study used this combination but with minimization of detection rate as the objective function [17]. Another study by Yuting [20] discussed about the PSO scheme addition to Mercer-Kernel based Fuzzy clustering for which PSO algorithm got its objective function from the membership function of fuzzy clustering only. The limitation with Kernel based clustering is that it works only on specific types of cluster which is not applicable in the current study or related to E-Governance. The PSO scheme was also used with K-means algorithms by few researchers like Xiao [18] and Munz [10]. While the superiority of

FCM is an established one over K-Means clustering, so the current experiment does not involve the combination of PSO with K-Means. But study by Munz certainly discussed the merits of using binary clusters in anomaly detection from where the current study adopts the usage of binary clustering as compared to attacks based clustering scheme.

# 3. Methodology

## *Clustering Techniques*

There are three types of techniques used for the experimental purpose – K-Means Clustering [8][12], Fuzzy C-Means (FCM) Clustering [13][19][21] and Optimization of FCM based on meta-heuristic scheme. K-Means algorithm is one of those simplest schemes which help in forming pre-defined data groups. It is an unsupervised type of learning algorithm for which the centers are decided later on while the number of clusters is decided beforehand. Once the cluster number is decided, the centers are allocated to the data set on random basis, keeping their distances away from each other and subsequent distance is calculated for various items. Nearer data items are grouped together with the clause of exclusivity where one item set belongs to one cluster only. On the other hand, FCM uses a membership function to assign nearness to every data points based on which clusters are formed. The cluster center has got nearness to every data item in the whole dataset and only the most near items to the center are clubbed into that particular cluster.

The center of the clusters formed can be optimized using various algorithms but the meta-heuristic algorithms have been used in this study to re-allocate the centers due to their inherent advantage of solving complex problems. Particle Swarm Optimization (PSO) [2][4] is one of the simplest techniques for optimization purpose and thus it has been used in conjunction with FCM to derive results. PSO works on the principle of food searching mechanism of swarm and takes into consideration both the local best and global best solutions. The velocities and positions are updated for every random search and wherever the best solution is achieved, those values are returned.

In this particular study, the objective function of detection rate is kept in consideration while calculating the swarm movements. Two features of the dataset are considered at a time on the basis of which the cluster center's value changes with every iteration and new clusters are formed. If the classification ratio and detection rate is achieved more than previous one, the new cluster centers are recorded. This process keeps on repeating until the change in cluster center stops.

The number of clusters are kept as two only in the experiment based on the nature of every item to be classified as intrusion or non-intrusion. Also the features on the basis of which clustering is performed are also kept in pair for easy calculations and accurate results. The concept of binary clusters also relies on the fact that data with unequal distribution of intrusions and non-intrusion can easily be accommodated and lesser number of cluster labeling leads to simpler analysis. For experimental purpose, the cluster formed with more number of data points or Bigger Size is considered as Intrusion Cluster and other one i.e. smaller in size of data points is labeled as Non-Intrusion cluster. In every iteration, the big cluster is optimized to find the number of intrusion more accurately while the small cluster is discarded in every iteration.

## *Data set & Implementation*

For experimental purpose, the Kyoto Dataset [14] has been used for finding the results. The choice of this dataset has been done due to the similarity in parameters required for Intrusion Detection in E-Governance system. The Kyoto dataset originally had 24 features out of which 14 are conventional. Out of these conventional features, six have been selected for this experiment in three pairs as they are closest related to intrusions taking place in E-Governance systems. The features are source port, destination port, source byte, destination byte, number of packets and their duration. Various types of attacks are covered by these features like flooding attack, data loss, SYN attack and the likes.

A total of 25000 data transactions were used for the experiment purpose with 6 different sizes of data set considerations. These sizes were 1000, 2000, 3000, 4000, 5000, and 10,000. All the implementation has been done using MATLAB. The results for K-Means, Fuzzy C-Means and FCM in Combination with Metaheuristic schemes are calculated on 6 different data sets using 3 different combinations of the parameters.

## *Evaluation Metric*

The evaluation parameter used for the comparison of various schemes is Classification Ratio. It is defined as the number of correctly detected data items divided by incorrect number of data items. It works on the assumption that every group will have at-least one correctly detected and one incorrectly detected item, respectively. For the optimization purpose, the objective function of maximization of classification ratio has been used. This classification ratio should be on the higher side which will portray that more number of observations are falling in the category of the correctly detected intrusions or anomalies. Higher the ratio, better is the intrusion detection scheme.

## 4. Experimental Results

This section will present the experimental results conducted on 6 different sizes of the data set on 3 different pairs of features selected from the dataset

**Table 1. Classification Ratio for three different clustering schemes across three types of clusters formed for 6 types of size for dataset**

| DATA SET | BIG CLUSTER | | | SMALL CLUSTER | | | OVERALL DATA | | |
|---|---|---|---|---|---|---|---|---|---|
| | T1 | T2 | T3 | T1 | T2 | T3 | T1 | T2 | T3 |
| **Source Port & Destination Port** | | | | | | | | | |
| 1000 | 3.12 | 3.12 | **3.83** | 7.33 | 7.33 | **8.32** | 3.65 | 3.65 | **4.52** |
| 2000 | 1.68 | 1.69 | **2.05** | **7.14** | 6.77 | 1.65 | 1.91 | **1.92** | 1.89 |
| 3000 | 1.52 | 1.54 | **2.08** | 5.85 | **5.95** | 1.45 | 1.73 | **1.77** | 1.74 |
| 4000 | 1.33 | 1.36 | **2.27** | 7.18 | **7.40** | 2.09 | 1.58 | 1.63 | **2.18** |
| 5000 | 1.34 | 1.37 | **2.36** | 6.45 | **6.53** | 2.22 | 1.60 | 1.64 | **2.29** |
| 10000 | 1.36 | 1.48 | **3.15** | 2.56 | **5.18** | 2.51 | 1.47 | 1.70 | **2.82** |
| **Source Bytes and Destination Bytes** | | | | | | | | | |
| 1000 | 2.32 | 2.41 | **17.67** | 10 | 21 | **89.67** | 2.34 | 2.47 | **22.81** |
| 2000 | 1.38 | 1.39 | **6.77** | 36 | 42 | **694** | 1.42 | 1.44 | **10.83** |
| 3000 | 1.26 | 1.26 | **5.27** | 59 | 59 | **1048** | 1.31 | 1.31 | **8.62** |
| 4000 | 1.08 | 1.08 | **4.71** | 63 | 68 | **629** | 1.11 | 1.11 | **8.12** |
| 5000 | 1.09 | 1.09 | **4.85** | 77 | 77 | 13.32 | 1.12 | 1.12 | **6.75** |
| 10000 | 1.67 | 2.32 | **5.39** | 653.5 | **1110** | 30.09 | 2.07 | 3.26 | **8.06** |
| **Number of Packets and their Duration** | | | | | | | | | |
| 1000 | 2.32 | 2.41 | **18.86** | 10 | **21** | 20.92 | 2.34 | 2.47 | **19.41** |
| 2000 | 1.38 | 1.39 | **9.73** | 36 | **42** | 15.63 | 1.42 | 1.44 | **11.50** |
| 3000 | 1.26 | 1.26 | **9.02** | 53 | **59** | 14.64 | 1.30 | 1.31 | **10.81** |
| 4000 | 1.08 | 1.08 | **8.10** | 62 | **68** | 16.23 | 1.11 | 1.11 | **10.59** |
| 5000 | 1.36 | 1.09 | **7.70** | 8.53 | 77 | 15.01 | 1.62 | 1.12 | **9.96** |
| 10000 | 1.43 | 2.10 | **7.28** | 3 | **496** | 13.88 | 1.56 | 2.87 | **9.14** |

Legend: T1 refers to Clustering Technique 1 which is K-Means only, T2 refers to Clustering Technique 2 which is Fuzzy C-Means only, and T3 refers to Clustering Technique 3 which is combination of PSO and Fuzzy C-Means [Best values are presented in bold]

The result as shown in Table 1 depicts the classification ratio obtained for different parameters and are categorized against different types of data set sizes chosen randomly from the dataset.

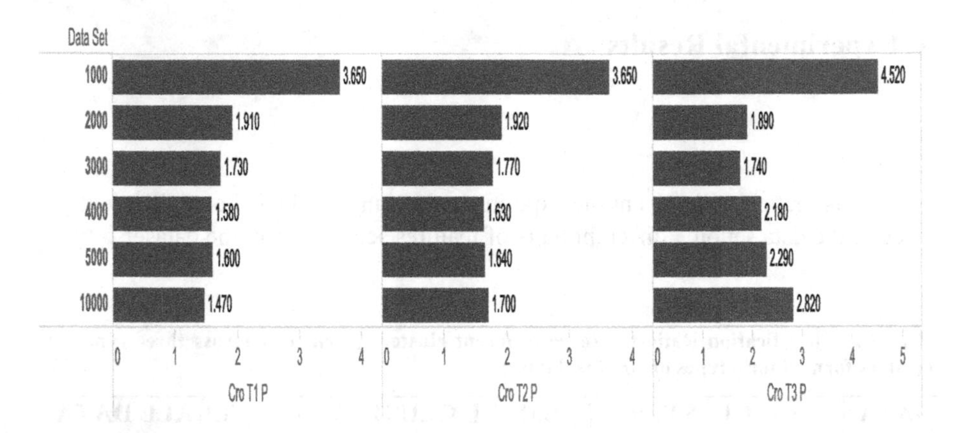


**Fig. 1. Classification Ratio for overall clusters for three techniques based on Ports as the system feature**

With respect to port as one of the system characteristic, it can be seen that for Big Cluster to which intrusion has been claimed, the classification ratio of technique 3 has been found higher in all the cases as compared to technique 1 and 2. This means that technique 3 has been effectively able to take out correct intrusions in this cluster as claimed when compared against other two techniques. However as seen for small cluster, which is considered to be a clean cluster, the classification ratio for the technique 3 goes down as compared to technique 1 and 2 where second technique has done well for the five of the six datasets. This means that second technique is good in grouping the correct non-intrusion data items. On the other hand, if we see on the overall basis, technique 3 emerges out to be best among the three techniques which suggest that third technique has better potential in grouping the data items as intrusions and non-intrusions. On the overall basis the classification ratios were higher for small data sets which decreased and got stabilized once the data size increased in case of ports as the system network parameter.

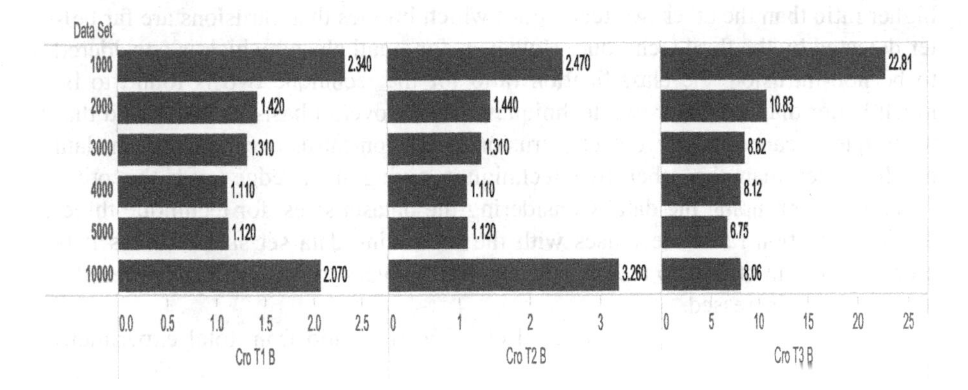


**Fig. 2. Classification Ratio for overall clusters for three techniques based on Bytes as the system feature**

With respect to Source Bytes and Destination Bytes as the parameter for the system network, it is found that for Big cluster, the third technique outperforms the other two techniques in all the cases by big margins. It clearly suggests that technique three has been able to find the intrusions more accurately than the other two techniques with lesser incorrect intrusions. However, for the small cluster, the classification ratio varies dramatically for all three techniques. The number of incorrect items is very low in case of bytes as network system characteristics. All the intrusions are correctly grouped by the three techniques and technique 3 again outperforms the other two in most of the cases of data sets. On the overall basis, the classification ratio seems to be higher for the smaller data set while it reduces as the size of dataset increases and stabilizes for larger datasets.

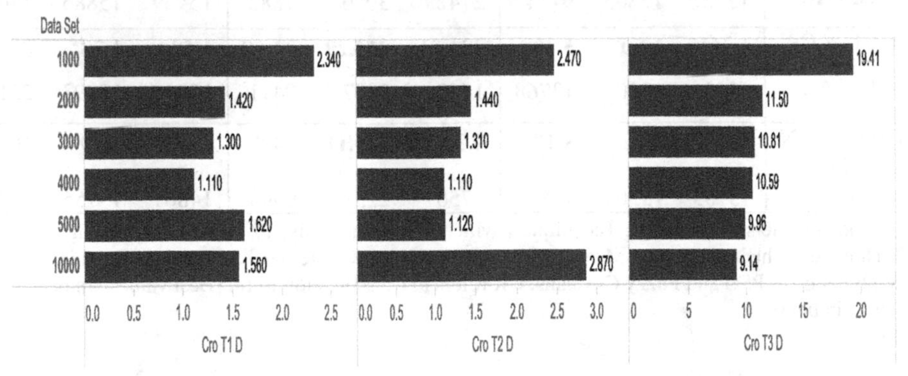


**Fig. 3. Classification Ratio for overall clusters for three techniques based on duration & number of packets as the system feature**

For the parameter related to the number of packets and their duration within the system network, it is found that for big cluster the technique 3 has substantially higher ratio than the other two techniques which implies that intrusions are far better detected by the third technique. However, for small cluster which is considered to be non-intrusion, the classification ratio for the technique two is found to be much better than the other two techniques. On the overall basis, it was found that technique 3 can correctly detect intrusions and non-intrusions items from data much higher than the other two techniques giving it an edge over the other schemes for grouping the data. Considering the dataset sizes, for technique three, the classification ration decreases with the increasing data set size whereas it is more initially in other two techniques and then reduces to get stabilized when the dataset size is increased.

As observed in Table 2, the overall classification ratio from total experiment size (25000 items) shows that for different parameters, different techniques performed well. However on the overall basis, technique three has been able to correctly identify the intrusions and non-intrusions from the total dataset with a higher classification ratio than the other two techniques. The absolute numbers also showcase the same fact that technique three is superior in judging the correct items to be intrusions or not. The classification ratio is reported to have large variations in small and big clusters due to their size i.e. number of data points getting segregated under small cluster and big cluster respectively.

**Table 2. Average of Classification Ratio for three different clustering schemes across three types of clusters formed and Total correct predictions overall**

| VARIABLE | BIG CLUSTER | | | SMALL CLUSTER | | | OVERALL DATA | | |
|---|---|---|---|---|---|---|---|---|---|
| | T1 | T2 | T3 | T1 | T2 | T3 | T1 | T2 | T3 |
| PORT CR | 1.43 | 1.49 | **2.60** | 4.30 | **6.05** | 2.18 | 1.62 | 1.74 | **2.40** |
| TOTAL | 12727 | 12865 | 9754 | 2742 | 3020 | 7882 | 15469 | 15885 | 17636 |
| BYTES CR | 1.36 | 1.50 | **5.40** | 221.71 | **355.29** | 34.60 | 1.52 | 1.78 | **8.22** |
| TOTAL | 13506 | 13506 | 13768 | 1552 | 2487 | 9411 | 15058 | 15993 | 23179 |
| DURATION CR | 1.35 | 1.46 | **8.12** | 4.32 | **250.11** | 14.85 | 1.47 | 1.70 | **10.11** |
| TOTAL | 13105 | 13504 | 12846 | 1756 | 2251 | 9903 | 14861 | 15755 | 22749 |

Legend: T1 refers to Clustering Technique 1 which is K-Means only, T2 refers to Clustering Technique 2 which is Fuzzy C-Means only, and T3 refers to Clustering Technique 3 which is combination of PSO and Fuzzy C-Means, CR refers to Classification Ratio [Best values are presented in bold]

## 5. Discussion

The results displayed in previous section shows the emergence of technique three on the basis of the classification ratio. Even the absolute value of correct classified items into intrusions and non-intrusions validate that the scheme three is more efficient than the other two schemes. This section discusses the possible explanation for this behavior. Then, the relevance of this analysis and scheme is discussed with respect to E-Governance. Finally the research contributions are highlighted at the end summing up the various insights obtained through this study.

Classification ratio was used to analyze the performance of three techniques. The first is clustering based on K-Means, second is clustering based on Fuzzy C-Means and third is clustering based on Fuzzy C-Means followed by optimization using meta-heuristic scheme. As per the definition of classification ratio, higher number of accurately classified items and lower number of inaccurately classified items will give better ratio. Since technique two and three are able to obtain higher ratios, it implies that both these techniques are capable of classifying the accurate items. On the other hand, one flip side of using this ratio could be a small number of accurate items and relatively smaller number of inaccurate items could also develop a higher ratio, but the absolute values obtained from the experiment reveals that the absolute number of correctly classified cases (Refer Table 2) are also better in techniques with higher classification ratio. This validates the results that ratio is a good indicator towards the assessment of three techniques and is not a misleading metric by having outlier values while conducting the experiment. Also classification ratio seems to be a simple and more logical metric to the government departments and officials working in the area of E-Governance as compared to other metrics like understanding derivation of ROC curve. So its utilization in this field of E-Governance will give more straightforward analysis to the team working in intrusion detection techniques.

With respect to the performance of three techniques, it can be said that K-Means generally gets trapped into the empty or sparse clusters leading to the unequal cluster sizing and formation of non-globular clusters. In the current experiment, the concept of only two clusters was used which would be classified as group of intrusions and a group of non-intrusions. The lower performance of K-Means can be attributed to the fact that initial centroids for the intrusion and non-intrusion clusters is difficult to judge and an evenly spread-out of data based on different parameters under study also led to difficulty in creating clearly demarcated two clusters where non-convex shape handling would have made things easier, which K-Means could not do.

For the scheme like Fuzzy C-Means based clustering, the concept of membership function makes it superior to the K-Means clustering. While K-Means requires exclusivity of the data in the clusters, the Fuzzy C-Means assigns membership to various data items and then forms clusters based on the nearness through this function. That is why the ratio in this scheme is found to be better than the K-Mean algorithm. However, the Fuzzy C-Means algorithm has got its own weakness like not achieving the optimum value of objective function (which could be classification ratio or a high detection rate or a low false alarm rate) as the distance is calculated on a static centroid value. Moving the centroid to optimum place can help in maximizing the objective function which any clustering scheme is not accustomed to. And here comes in the role of Meta-heuristic algorithms. For the experimental purpose, this study uses Particle Swarm Optimization as the nature inspired scheme to optimize the clusters. And in most of the cases, the algorithm successfully clusters the data set more accurately as compared to K-Means and Fuzzy C-Means. The inherent characteristic of meta-heuristic algorithms to solve complex problems in beneficial in this experiment as the items which are evenly spread out on the plane can be clustered more efficiently based on the objective function of maximizing classification ratio and detection rate.

From the E-Governance point of view, it can be seen that parameters like bytes, ports and duration of the packets can help in identifying the anomalies. For E-Governance services relating to lower number of transactions, the FCM-PSO scheme will work well. For very large number of transactions too this scheme will work well. However, in some cases where there are mid number of transactions performed in the E-Governance system, the scheme based on only FCM clustering can also be used.

## 6. Conclusion

The current study explored new ways to improve the clustering scheme for Intrusion Detection system and that too relevant to the E-Governance. The new Metric for Intrusion Detection Algorithm's Efficiency detection was used which is much simpler and easy to implement. The optimization of the clustering schemes brings in the possibility of using different meta-heuristic schemes on binary clustering in the field of Intrusion Detection. Also the different parameters used for Intrusion Detection can give a way forward to the technology team of E-Governance Systems for possible detection of attacks.

The future scope of the work includes usage of other clustering schemes and different meta-heuristic algorithms to frame even better Intrusion detection systems. Also the usage of Classification Ratio can be more frequent apart from traditional metrics by the researchers to formulate objective functions of their evolutionary schemes. The training and test set can be based on real life transactional data to find more patterns and insights.

## References

1. Anderson, J. P. Computer security threat monitoring and surveillance. Technical report, James P. Anderson Company, Fort Washington, Pennsylvania. February 1980.
2. Bai, Q. Analysis of particle swarm optimization algorithm. Computer and information science, (2010), 3(1), p180.
3. Denning, D. E. An intrusion-detection model. Software Engineering, IEEE Transactions, (1987), 13(2), 222-232.

4. Guolong, C.; Qingliang, C. &Wenzhong, G. A PSO-based approach to rule learning in network intrusion detection. In Fuzzy Information and Engineering, Springer Berlin Heidelberg. 2007. pp. 666-673.
5. Gupta, R.; Muttoo, S.K., & Pal, S.K. Review based security framework for E-Governance Services. *Chakravyuh* (2016), 11(1), 42-50.
6. Gupta, R.; Muttoo, S.K., & Pal, S.K. Analysis of Information Systems Security for e-Governance in India. *National Workshop on Cryptology* (2013), TSII, 17-25.
7. Gupta, R.; Pal, S.K., & Muttoo, S.K. Network Monitoring and Internet Traffic Surveillance System: Issues and Challenges in India. *Intelligent Systems Technologies and Applications*. Springer International Publishing, 2016. 57-65.
8. Jianliang, M.; Haikun, S. & Ling, B. The application on intrusion detection based on k-means cluster algorithm. In Information Technology and Applications, IFITA'09. International Forum on. Vol. 1, 2009. pp. 150-152 IEEE.
9. Liao, H. J.; Lin, C. H. R.; Lin, Y. C., & Tung, K. Y. Intrusion detection system: A comprehensive review. J. NetwComputAppl, (2013), 36(1), 16-24.
10. Münz, G.; Li, S. & Carle, G. Traffic anomaly detection using k-means clustering. In GI/ITG Workshop MMBnet. September 2007.
11. Muttoo, Sunil K., Rajan Gupta, and Saibal K. Pal. "Analysing Security Checkpoints for an Integrated Utility-Based Information System." *Emerging Research in Computing, Information, Communication and Applications*. Springer Singapore, 2016. 569-587.
12. Ravale, U.; Marathe, N. &Padiya, P. Feature Selection Based Hybrid Anomaly Intrusion Detection System Using K Means and RBF Kernel Function. Procedia Computer Science, (2015), 45, 428-435.
13. Ren, W.; Cao, J. & Wu, X. Application of network intrusion detection based on Fuzzy C-means clustering algorithm. In Intelligent Information Technology Application, 2009.IITA 2009.Third International Symposium on, Vol. 3, pp. 19-22. IEEE, 2009 November.
14. Song,J.; Takakura, H. & Okabe, Y. Description of Kyoto University Benchmark Data. Available at link: http://www.takakura.com/Kyoto_data/BenchmarkData-Description-v5.pdf. [Accessed on 15 March 2016].
15. Wang, D.; Han, B. & Huang, M. Application of fuzzy c-means clustering algorithm based on particle swarm optimization in computer forensics. Physics Procedia, (2012), 24, 1186-1191.
16. Wang, H.; Zhang, Y. & Li, D. Network intrusion detection based on hybrid Fuzzy C-mean clustering. In Fuzzy Systems and Knowledge Discovery (FSKD), 2010 Seventh International Conference on, Vol. 1, pp. 483-486.IEEE. 2010 August.
17. Wang, Y. Network Intrusion Detection Technology based on Improved C-means Clustering Algorithm. Journal of Networks, (2013), 8(11), 2541-2547.
18. Xiao, L.; Shao, Z. & Liu, G. K-means algorithm based on particle swarm optimization algorithm for anomaly intrusion detection. In Intelligent Control and Automation, 2006. WCICA 2006. The Sixth World Congress on Chicago. Vol. 2, pp. 5854-5858. IEEE. June 2006.
19. Xie, L.; Wang, Y.; Chen, L. &Yue, G. An anomaly detection method based on fuzzy C-means clustering algorithm. In The Second International Symposium on Networking and Network Security. (ISNNS) p. 89, April 2010.
20. Yuting, L. U. The Study on the Network Intrusion Detection Based on Improved Particle Swarm Optimization Algorithm. Int J Adv Comput Tech, (2013), 5(2), 17-23.
21. Zhang, H. & Zhang, X. Intrusion Detection Based on Improvement of Genetic Fuzzy C-Means Algorithm. In Advances in Information Technology and Industry Applications, Springer Berlin Heidelberg, 2012. pp. 339-346.

# Part III
# Intelligent Tools and Techniques

# Classroom Teaching Assessment Based on Student Emotions

Sahla K. S. and T. Senthil Kumar

**Abstract** Classroom teaching assessments are designed to give a useful feedback on the teaching-learning process as it is happening. The best classroom evaluations additionally serve as significant sources of data for instructors, helping them recognize what they taught well and what they have to deal with. In the paper, we propose a deep learning method for emotion analysis. This work focuses on students of a classroom and thus, understand their facial emotions. Methodology includes the preprocessing phase in which face detection is performed, LBP encoding and mapping LBPs are done using deep convolutional neural networks and finally emotion prediction.

## 1 Introduction

Facial emotions play an important role in everyday interaction of people with others. Feelings are as often as possible thought about the face, close by and body motions, in the voice, to express our emotions or enjoying. Late Psychology research has demonstrated that the people express their feelings fundamentally through outward appearances. Facial expression analysis a major part in sincerely rich man-machine communication (MMI) frameworks, since it utilizes all around acknowledged nonverbal signs to evaluate the users passionate state. There are enormous applications for emotion recognition systems. Some of them are used in biometrics, providing customer services, video recommender systems etc...

Sahla K. S.
Dept of Computer Science and Engineering, Amrita School of Engineering, Coimbatore, Amrita Vishwa Vidyapeetham, Amrita University, India e-mail: zelasahal@gmail.com

T. Senthil Kumar
Dept of Computer Science and Engineering, Amrita School of Engineering, Coimbatore, Amrita Vishwa Vidyapeetham, Amrita University, India e-mail: t_senthilkumar@cb.amrita.edu



J.M. Corchado Rodriguez et al. (eds.), *Intelligent Systems Technologies and Applications 2016*, Advances in Intelligent Systems and Computing 530,
DOI 10.1007/978-3-319-47952-1_37

With the enhancement in technologies computer vision became very popular in different areas of research. The facial detection, facial recognition, emotion analysis are some of the challenging areas. Automatic processing and analyzing of facial images is considered an important area. Here, facial identification and processing in still images and videos has much attention. Identifying the locations in images where the face is actually present is a fundamental issue. A great deal of work has been done in the zone of face localization basically concentrating on distinguishing proof part. There are numerous techniques for recognizing facial components, for example, eyes, nose, nostrils, eyebrows, mouth, lips, ears, and so on. These elements can be further used for recognition approach depending on systems. The same process is much more complicated in a real time system.

Basic classroom teaching assessments are simple, non-graded, anonymous and it is designed in a way to provide feedback on teaching assessments. Some of the popular methods of teaching assessments[10] like Background Knowledge Probe, Minute Paper tests, Muddiest Point, Whats the Principle?, Defining Features Matrix are depending on information learning of students. In a real-time scenario of classroom, students are having varying emotions. A continues assessment based on emotions can predict the overall class behavior like whether they are listening, laughing, sleeping, and some other activities.

In this paper, we propose a system using deep convolutional neural network technique for analyzing the classroom emotions. The classroom video get captured and undergoes key frame extraction, from which faces are identified. The convolution neural networks predicts the classes of emotion with highest probability as the resulted one. Based on the predicted emotion, an assessment will provide to the user for example; whether the students are happy so that class is interesting.

The rest of paper is organized as follows: Section II clarifies the related work which discusses the different area and approaches used in facial emotion detection. Section III elaborates the architecture of emotion recognition system. The subsections in Section III explain each phase of the work in detail. Section IV addresses the implementation part. Finally, Section V shows the results of proposed method in predicting emotions.

## 2 Related Work

Facial emotion recognition system has become a significant area of interest with an enormous number of applications that have emerged in this domain. The number of applications in which a users face is tracked by a video camera is growing exponentially. One can easily analyze the emotion so as to bring corresponding changes to the environment surrounds it.

## *2.1 Facial Image Processing*

For analyzing the emotions, face identification be the initial step. One of the challenge in this area is that recognizing only the face from inputted image or video.

Face identification by addressing the problems like illumination effects, occlusion is important. In prior work, Xu Zhao, Xiong Li, Zhe Wu, Yun Fu and Yuncai Liu[18] presented multiple cause vector quantization (MVCQ) a probabilistic generative model for identifying faces. The generated model use subcategories like eyes, mouth, nose and other features for face detection. Challenges like illumination, occlusion, pose and low resolution should be address through some robust method. In another piece of prior work, Huu-Tuan Nguyen and Alice Caplier[9] presented a Local Patterns of Gradients Whitened Principal Component Analysis (LPOG WPCA) which composed of WPCA at learning stage and K-NN for classification. The system calculate the ELBP value at every pixels and calculate average of pixels and its neughbors.

Other previous method for face detection was presented by Chao Xiong, Xiaowei Zhao, Danhang Tang, Karlekar Jayashree, Shuicheng Yan, and Tae-Kyun Kim[4] for detecting multiple faces and verifying occluded ones. The system was implemented with deep learning algorithms in CNN. Identifying faces in large scale unconstrained video data is essential in surveillance system. In prior work, Luoqi Liu, Li Zhang, Hairong Liu, and Shuicheng Yan[13] proposed an algorithm named Multitask Joint Sparse Reconstruction (MTJSR) in which information was inferred from the video by integrating extracted frame features.

The effectiveness of system depending on face location in image or video. Amir Benzaoui, Houcine Bourouba and Abdelhani Boukrouche[2] proposed a neural network approaches for efficient face identification in unconstrained images. In other work proposed by Raman Bhati, Sarika Jain, Nilesh Maltare, Durgesh Kumar Mishra[14] used Eigen faces in order to extract the feature vectors, neural network and BPNN. This approach generates more efficient result in terms of recognition and training time.

## *2.2 Facial Expression Synthesis*

Facial expressions help ones to make their behavior understandable to viewers and also support the mode of verbal communication easier. Several algorithms for facial expression analysis are there in the literature.

In prior work, Junghyun Ahn, Stephane Gobron, Quentin Silvestre and Daniel Thalmann[12] proposed a circumplex model of emotion. It defines an emotional space with different axis. In other work, Jasmina Novakovic, Milomir Minic and Alempije Veljovic[11] proposed a dimensionality reduction method PCA performed on feature set, so that retains those data which contribute more and ignoring others. In previous work, Damir Filko and Goran Martinovi c[5] proposed a system for emotion analysis with principle component analysis and neural networks. The system analyze the key features with PCA and thus easing the computer interactions. Another method for emotion analysis was proposed by Surbhi and Mr. Vishal Arora[15] in which analysis was done on captured images using extracting Eigen vectors and then fed as input to neural network for recognition part. In prior work, Ajay Reddy P K, Chidambaram M, D V Anurag, Karthik S, Ravi Teja K, Sri Harish. N and T. Senthil Kumar[16] proposed a system for video recommender in which emotion analysis is done by Facial Action Coding System. In order to satisfy the properties of real time system, in previous work Usha Mary Sharma, Kaustubh Bhattacharyya and Uzzal Sharma[17] proposed system with classifiers that performs more efficiently in analyzing emotions. K-Means, K-Nearest Neighbour, SVM, were some of the classifiers explained in it.

### *2.3 Learning Methodologies*

Recent studies show that, Convolutional Neural Network(CNN) become a predominant method in the field of image processing. CNN is applying in image classification problems like object recognition proposed by A. Krizhevsky, I. Sutskever, and G. E. Hinton[1], scene recognition system proposed by B. Zhou, A. Lapedriza, J. Xiao, A. Torralba, and A. Oliva[3], face verification methods explained in prior works by F. Schro, D. Kalenichenko, and J. Philbin. Facenet[6] and Y. Sun, X. Wang, and X. Tang[19], age and gender classification by G. Levi and T. Hassner[7], and more. In prior work, Gil Levi and Tal Hassner[8] represented a novel method for classifying emotions using CNN with LBP encodings. The system shows high accuracy rate for selected static images on multiple CNN architectures. Advantages in computation power, improvements in complexity of algorithms, availability of large data-sets, different modes for training models, various number of parameters, CNN is applying in various practical areas with high accuracy rate.

## 3 Proposed Model

### *3.1 Architecture of Emotion Recognition System*

The emotion recognition system is mainly for understanding the current state of human beings. Sometimes it be a way of communication between people. Here, the system is implemented in a classroom. Complete architecture of the proposed system is represented in Fig. 1. The Video Capturing is done inside the classroom by using two cameras. One for capturing the video of teacher and the second one is for capturing video of students. All the captured videos get stored into a database. After the Preprocessing steps, Emotion Prediction is done for both the students and teacher. Based on prediction, the system will provide an assessment about the class,

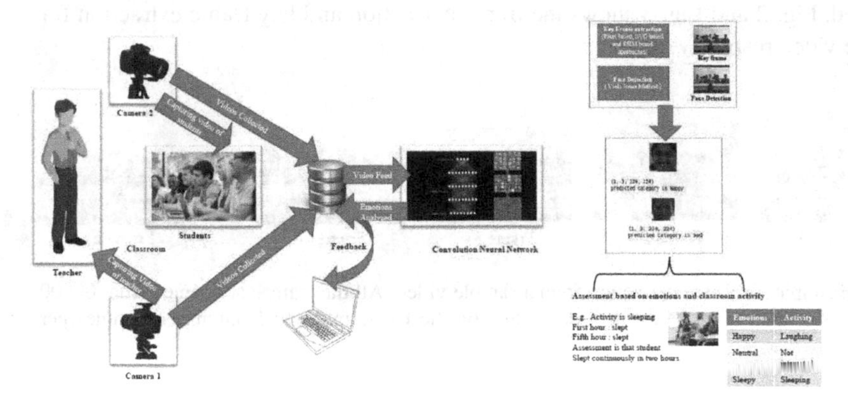


Fig 1 Architecture of the proposed system for Classroom Teaching Assessment Based on Student Emotions

whether the class is interesting or sleepy. The proposed system made an assumption that the students are not closely sitting in the classroom. It is because, the detection of faces and its features become very difficult in a class-room where students are closely sitting.

The methodology for proposed system includes Input Data Stage, Preprocessing, Feature Extraction and Emotion prediction. The experiment is done with both images and videos as the Input Data. Initial training of system is done with images and after that experiment done through captured videos. The main intension was to obtain high accuracy in prediction of emotions.

## *3.2 Preprocessing*

Before applying any image processing techniques to given input data, it is necessary to process the data in order to assure that it contains all the details implied by the method.

### 3.2.1 Key Frame Extraction

Frame extraction is the initial step, in which captured video get converted to simple frames. The process is done for getting the easiness in succeeding steps. Whereas Key Frame Extraction from the captured video is that, understanding whether there is a change in the scenery or not. A key-frame contains content a lot different from the previous ones. Here the similarity of each video frame is calculated by the intersection of color histograms. If the value is lower than the threshold, a screen-cut is detected. Fig. 2 and Fig. 3 shows the frame extraction and key frame extraction for sample video respectively.

**Fig. 2** Example for frame extraction from a sample video. All the frames are named from 00000 and the total number of frames will depending on the frame rate that is number of frames per second

**Fig. 3** Some of the Key Frames extracted from the sample video based on the similarity rate

### 3.2.2 Face Detection

Face detection is the process of detecting the faces in each frames. The efficiency is depending on effective detection of faces. Extracted key-frames get analyzed sequentially. The popular method Viola-Jones algorithm is considered for detecting the faces in the given image. The in-build class 'Haar feature extraction' in OpenCV is used for the purpose. Whenever a face get located, it will highlighted by drawing an ellipse/ circle around it. Initially every single RGB images get converted into gray scale and cropping them with region of face. Fig. 4 represents the sample input image which is a selected key frame from the video. Firstly the input get converted into the gray scale as show in second image and Haar feature extraction will perform on the gray scale image and draw an ellipse on detected faces as show in third figure. All the detected face get stored in the database by cropping with respect to face region.

## *3.3 CNN as classifier*

Multiple image representations is a successful method employed in previous works. Here, the same method is considered with CNN architectures. The cropped faces are

**Fig. 4** Different phases of face detection. a) Inputted key frame of sample video, b) Conversion of RGB Key frame into gray scale image and c) Haar feature face detection in the inputted key frame. Detected faces are show by drawing an ellipse

processed to obtain the inputs for convolution neural network. Each inputted image is processed as follows:

- LBP encoding method is applied to the pixels of each image using intensities of its neighbors. Considering central pixel as the threshold, LBP is generated by applying threshold on the generated neighbor's intensity values. Lower the threshold resulting to a pattern of 0 and higher the threshold results to 1. When neighbor contains 8 pixels, each encoding convert image intensity values to one of 256 LBP code values.
- The unordered code values are transformed to a metric space by Multi-Dimensional Scaling (MDS). Convolution operations are used as an averaging function on transformed points. Code-to-code dissimilarity scores between LBPs are calculated based on an approximation to the Earth Movers Distance. The similarity between the images intensities in code values are represent with estimated distance.
- CNN model is trained with the original RGB image and mapped code image to predict any of seven emotions. For efficient classification, CNN is applied to center crop of the inputted image. Weighted average over the output vectors are generated. The class with maximum average prediction be the final classification.

The cropped faces are given as input to the deep CNN. After the process of LBP encoding nad mapping of LBPs, CNN will generate the class of emotion. Fig. 5 shows the shows the predicted output along with the displayed image. CNN predicts emotion as'Happy', which is True Positive in nature.

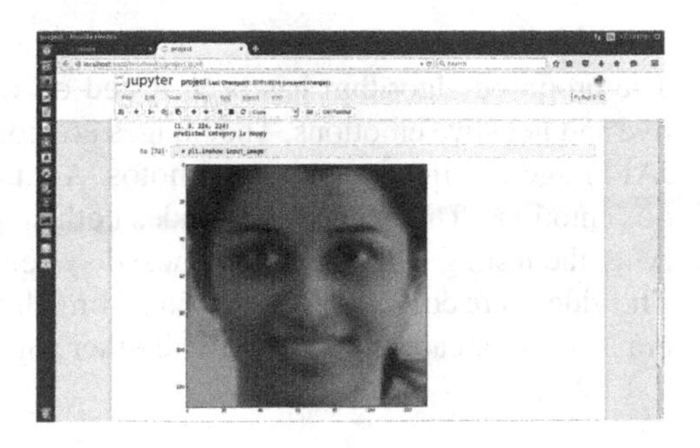


**Fig. 5** Predicted Emotion along with the input image

### *3.4 Predicted emotions and Mapping of classroom events*

After predicting the emotions of both students and teacher, the result get mapped to classroom events. The classroom events can be categorized as sleeping, laughing, watching outside, thinking etc The emotions for these activities can be either a mix of all emotions or respective single emotions. For example; the classroom event can be laughing then the emotions can be happy. An assessment can be made from the mapping like class is interesting. At the same way for a neutral face emotion the event can be thinking. So the assessment can be of either class is interesting or boring. The possible emotions, some events along with the action units are represented in the given Table 1. The importance gives a factor which is calculated out of 10. Greater value represents the fact that respective activity should consider more seriously.

## 4 Experiments

The preprocessing steps including key-frame extraction, face tracking and identification, cropping with region of face was done in OpenCV. Training and testing part was implemented in Caffe, an open source framework for deep learning in Convolution Neural Network.

Our test was done on captured videos of classroom. It includes the complete lecture of different periods. The videos were arranged together in a folder, from which each one undergoes preprocessing. The initial processing of key frame extraction was done using OpenCV. The generated frames were assembled together and face-identification was processed in those key frames. The detected face get cropped from the inputted frame, which was the next input to CNN architecture. Then assigning any of seven emotions: happy, sad, angry, surprise, fear, disgust and neutral to inputted image through CNN. Here the images contains head poses, both genders, age and other constraints.

## 5 Results

The proposed algorithm has been tested on different images including both positive and negative emotions. The images are collected from available databases CK, JAFFE, google images and class photos. A total 105 images are used in testing various emotions. The collection includes both single imaged photos and group photos. After the testing with images, proposed system get tested with the captured videos. The videos are collected by capturing from classroom using two cameras. One camera is used for capturing teacher and other for capturing the students.

**Table 1** Mapping of emotions to classroom events

| Activity | Emotions | Action Units | Importance |
|---|---|---|---|
| Watching outside  | Happy, sad, neutral or any other ones | 6 + 12 or 1 + 4 + 15 or 0 | 6 |
| Laughing  | Happy | 6 + 12 | 5 |
| Not interested / not paying attention  | Neutral / sleepy | 0, 43 | 7 |
| Thinking  | Neutral | 0 | 5 |
| Sleeping  | Sleeping | 43 | 6 |
| Listening  | Happy / Neutral | 6 + 12 or 0 | 10 |
| Talking | Happy / Sad | 6 + 12 or 1 + 4 + 15 | 4 |

1. Emotion Classification for images
   The proposed system has been tested with selected images from different databases, Google images and other classroom photos. The dataset includes both single and group images. The deep convolution neural network used to predict the emotion of respective inputs. Table 2 shows the confusion matrix of emotions in images using deep convolution neural network. The testing is performed for seven emotions namely: anger, disgust, happy, neutral, surprise, fear and sad. Evidently, the surprise emotion was never classified correctly. This may be due to challenging for system to classify as surprise that is features may not be extracting properly.

**Table 2** Confusion Matrix for images

| | Anger | Disgust | Happy | Neutral | Surprise | Fear | Sad |
|---|---|---|---|---|---|---|---|
| Anger | 33.3% | 40% | 0 | 6.6% | 0 | 20% | 0 |
| Disgust | 2% | 46.6% | 0 | 13.3% | 0 | 2% | 0 |
| Happy | 13.3% | 2% | 60% | 0 | 0 | 6.6% | 0 |
| Neutral | 13% | 13% | 0 | 46.6% | 6.6% | 20% | 0 |
| Surprise | 0 | 0 | 0 | 0 | 0 | 0 | 0 |
| Fear | 2% | 2% | 0 | 0 | 0 | 33.3% | 26.6% |
| Sad | 6.6% | 13% | 0 | 2% | 0 | 6.6% | 53.3% |

Another reason may be due to availability of only few images. The table shows that 60% happy is correctly classified and it is the highest accuracy as compared to others. 53.3% of Sad is correctly classified comes up in second position. In case of anger, the images get misclassified as disgust in about 40% which is more than the 33.35% of correctly classified anger. This may due to the similarity in features of both emotions.

2. Emotion Classification for video of students

The video was captured for an eight students in the classroom, who were sitting in a way that, there is enough gap between them. It undegoes the process of key frame extraction. For each extracted key frames, face detection undegoes. For each faces, emotion get detected through convolutional neural network. Table 3 shows the confusion matrix of emotions in selected key frames.

The sample input and predicted emotions are shown in Table 4. Since each frame

**Table 3** Tabulated Results of Emotion Detection

| Number of Trails | Correctly Predicted Emotions | Efficiency | Validation Accuracy | Validation on Error rate |
|---|---|---|---|---|
| 196 | 150 | 76.53 | 0.76 | 0.24 |
| 134 | 117 | 87.31 | 0.87 | 0.13 |

consists of eight students, total of 330 faces get detected from around 50 key frames. The detected faces undergoes emotion analysis in such a way that seven classes of emotions are considered. In first case out of 196 face, for 150 faces emotions get predicted correctly and in second case 117 emotions get predicted correctly. Thus system gives an verall efficiency of 80.9% in detecting various emotions and with a missclassification rate of 19.1%.

**Table 4** Predicted emotions to respective samples

| Output | Input |
|---|---|
| (1, 3, 224, 224) predicted category is Happy |  |
| (1, 3, 224, 224) predicted category is Sad |  |
| (1, 3, 224, 224) predicted category is Sad |  |
| (1, 3, 224, 224) predicted category is Fear |  |
| (1, 3, 224, 224) predicted category is Fear |  |

## 6 Conclusion

Emotion recognition system for classroom using deep learning is designed. Preprocessing methods on video made the system to work more efficiently. Here video analysis is performed on key frames of captured classroom videos. The identified faces from each key frames are given as input to the deep CNN architecture. Converting images to LBP codes made system more robust to illumination variations. Further, generated codes get mapped to a metric space. Finally, class with maximum weighted average get predicted by the multiple CNN architectures.

In this paper, emotions of students are considered and also the overall performance is calculated. As a future work, this system can be modified to provide a feedback on teaching. The work can be extended by mapping the student's emotion with classroom activities. This method of teaching assessment will help the teachers to improve their way of teaching.

## References


1. A. Krizhevsky, I. Sutskever, and G. E. Hinton (2012) Imagenet classification with deep convolutional neural networks. In Neural Inform. Process. Syst., pages 1097-1105
2. Amir Benzaoui, Houcine Bourouba and Abdelhani Boukrouche (2012) System for Automatic Faces Detection. Image Processing Theory, Tools and Applications, IEEE 354 - 358
3. B. Zhou, A. Lapedriza, J. Xiao, A. Torralba and A. Oliva (2014) Learning deep features for scene recognition using places database. In Neural Inform. Process. Syst., pages 487-495
4. Chao Xiong, Xiaowei Zhao, Danhang Tang, Karlekar Jayashree, Shuicheng Yan and Tae-Kyun Kim (2015) Conditional Convolutional Neural Network for Modality-aware Face Recognition. IEEE Int. Conf. on Computer Vision (ICCV), Santiago, Chile
5. Damir Filko and Goran Martinovi c (2013) Emotion Recognition System by a Neural Network Based Facial Expression Analysis, ATKAFF, 54(2), 263-272
6. F. Schro, D. Kalenichenko, and J. Philbin. Facenet (2015) A unified embedding for face recognition and clustering. In Proc. Conf. Comput. Vision Pattern Recognition, pages 815-823
7. G. Levi and T. Hassner (2015) Age and gender classification using convolutional neural networks. In Proc. Conf. Comput. Vision Pattern Recognition Workshops
8. Gil Levi and Tal Hassner (2015) Emotion Recognition in the Wild via Convolutional Neural Networks and Mapped Binary Patterns in Zhengyou Zhang; Phil Cohen; Dan Bohus; Radu Horaud & Helen Meng, ed., 'ICMI' , ACM, , pp. 503-510
9. Huu-Tuan Nguyen and Alice Caplier (2015) Local Patterns of Gradients for Face Recognition. Ieee Transactions on Information Forensics and Security, vol. 10, No. 8
10. https://cft.vanderbilt.edu/guides-sub-pages/cats/
11. Jasmina Novakovic, Milomir Minic and Alempije Veljovic (2011) Classification Accuracy of Neural Networks with PCA in Emotion Recognition, Theory and Applications of Mathematics & Computer Science, 1116
12. Junghyun Ahn, Stephane Gobron, Quentin Silvestre and Daniel Thalmann (2010) Asymmetrical Facial Expressions based on an Advanced Interpretation of Two-dimensional Russells Emotional Model, Engage, September 13-15
13. Luoqi Liu, Li Zhang, Hairong Liu, and Shuicheng Yan (2014) Toward Large-Population Face Identification in Unconstrained Videos. Ieee Transactions On Circuits and Systems for Video Technology, VOL. 24, NO. 11
14. Raman Bhati, Sarika Jain, Nilesh Maltare and Durgesh Kumar Mishra (2010) A Comparative Analysis of Different Neural Networks for Face Recognition Using Principal Component Analysis, Wavelets and Efficient Variable Learning Rate. Computer Vision & Communication Technology (ICCCT), IEEE Intl Conf, pp. 526-531
15. Surbhi and Mr. Vishal Arora (2013) The Facial expression detection from Human Facial Image by using neural network. International Journal of Application or Innovation in Engineering & Management, Volume 2, Issue 6, June
16. Ajay Reddy P K, Chidambaram M, D V Anurag, Karthik S, Ravi Teja K, Sri Harish. N and T. Senthil Kumar (2014) Video Recommender In Open/Closed Systems. International Journal of Research in Engineering and Technology, Volume: 03, Special Issue: 07, May
17. Usha Mary Sharma, Kaustubh Bhattacharyya and Uzzal Sharma (2014) Contemporary study on Face and Facial Expression Recognition System-A Review. International Journal of Advanced Research in Electrical, Electronics and Instrumentation Engineering, Vol. 3, Issue 4, April
18. Xu Zhao, Xiong Li, Zhe Wu, Yun Fu and Yuncai Liu (2013) Multiple Subcategories Parts-Based Representation for One Sample Face Identification. IEEE Transactions on Information Forensics and Security, VOL. 8, NO. 10, Oct
19. Y. Sun, X. Wang, and X. Tang (2014) Deeply learned face representations are sparse, selective, and robust. arXiv preprint arXiv:1412.1265

# Soft Computing Technique Based Online Identification and Control of Dynamical Systems

Rajesh Kumar, Smriti Srivastava, and J.R.P Gupta

Netaji Subhas Institute of Technology, Sector 3, Dwarka, New Delhi-110078
{rajeshmahindru23@gmail.com, smriti.nsit@gmail.com, jairamprasadgupta@gmail.com}

**Abstract.** This paper proposes a scheme for online identification and indirect adaptive control of dynamical systems based on intelligent radial basis function network (RBFN). The need to use intelligent control techniques arises as the conventional control methods like PID fails to perform when there is a non linearity in the system or system is affected by parameter variations and disturbance signals. In order to show the effectiveness of the proposed scheme, the mathematical models of the dynamical systems considered in this paper were assumed to be unknown. Since most real-world systems are highly complex and their precise mathematical descriptions are not available which further makes their control more difficult. These factors laid the foundation for the development of control schemes based on intelligent tools so that such systems can be controlled. One such scheme, based on RBFN, is presented in this paper. The key part of the scheme is the selection of inputs for the controller and in the proposed scheme; the inputs to the controller were taken to be the past values of plant's as well as of the controller's outputs along with the externally applied input. A separate RBFN identification model was also setup to operate in parallel with the controller and plant. Simulation study was performed on two dynamical systems and the results obtained show that the proposed scheme was able to provide the satisfactory online control and identification under the effects of both parameter variations and disturbance signals.

**Keywords:** Radial Basis Function Networks, Brushless DC motor, Identification and Adaptive Control, Water Bath System, Gradient Descent

## 1 Introduction

The basic difficulty with the conventional control methods like PID are their dependency on the knowledge about the mathematical model of the plant for tuning their own parameters [14]. Most of the time this information regarding the dynamics of the plant is partially known or unknown and hence they cannot be used. Even if the mathematical model of the system is known in advance, the problems of non linearity, variation in the operating point, disturbance signals etc in the system cannot be accurately handled by conventional methods. Also, when the parameters of the plant undergo change (because of any reason), PID controller parameters are required to be retuned. To deal with such uncertainties, various intelligent techniques were developed [5], [7], [12], [13], [15]. Processing the complex signals by ANN can be viewed as a curve-fitting operation in



J.M. Corchado Rodriguez et al. (eds.), *Intelligent Systems Technologies and Applications 2016*, Advances in Intelligent Systems and Computing 530,
DOI 10.1007/978-3-319-47952-1_38

the multidimensional space and this is the reason why many applications use ANN for realizing some highly complex non linear functions. This ability in ANN to approximate any continuous function comes from its sufficient large structure which contains appropriate amount of neurons that approximates any given unknown function in the form of their parameters values (which reaches to their desired values during the training) [2],[3]. ANN suffers from some problems that include slow learning, poor process interpretability and often stucking in the local minima during the learning procedure. A suitable alternative to the ANN is the non linear RBFN. Traditionally it has been in multidimensional space for interpolation [9],[11]. RBFN is a simple 3 layered network with non linear hidden layer. They offer several advantages over ANN like smaller extrapolation errors, higher reliability and faster convergence [8], [18]. The rest of the paper is outlined as follows: Section 2 describes the structure and mathematical model associated with the RBFN. In section 3, update equations for the parameters of RBFN identification model using the powerful technique of gradient descent are derived. Section 4 includes the derivation of update equations for RBFN controller parameters. In section 5 simulation study was done by considering two dynamical systems: Brushless dc motor and water bath system. The proposed scheme was also tested for robustness against the uncertainties like parameter variations and disturbance signals which affects the plant. Section 6 concludes the paper.

## 2 Radial Basis Function Network

The RBFN structure is shown in Fig. 1. The input vector X represents the input training samples $X = (x_1, x_2....x_m)$. The nodes in the hidden layer are known as radial centres. They have Gaussian radial basis function as their activation function. Each radial centre, $G_i$, represents one of the training samples and generates high output if present input training sample lies in its vicinity. This happens because of the mathematical definition of the Gaussian radial basis function. The weights connecting the inputs to the hidden layer nodes in RBFN have unity values. For SISO systems, output layer of RBFN contains only single neuron whose induced field (input) is a sum of weighted outputs obtained from the hidden layer nodes. The number of radial centres in the hidden layer of RBFN are taken to be q, where $q << m$ [1] and are functionally different from the neurons in MLFFNN. The adjustable output weight vector of RBFN is denoted by $W = [w_1, w_2, w_3....w_q)$. The output of RBFN is calculated as:

$$y_r(k) = \sum_{i=1}^{q} \phi_i(k) W_i(k) \tag{1}$$

where $y_r(k)$ represents the RBFN output at $k^{th}$ instant of time. The output of $i^{th}$ radial centre at any $k^{th}$ instant is found by the following expression:

$$\phi_i(k) = \phi \left\| X(k) - G_i(k) \right\| \tag{2}$$

where $\|X(k) - G_i(k)\|$ represents the Euclidean distance between $X(k)$ and $G_i(k)$ at the $k^{th}$ time instant. $\phi\,(.)$ is a Gaussian radial basis function which has the following

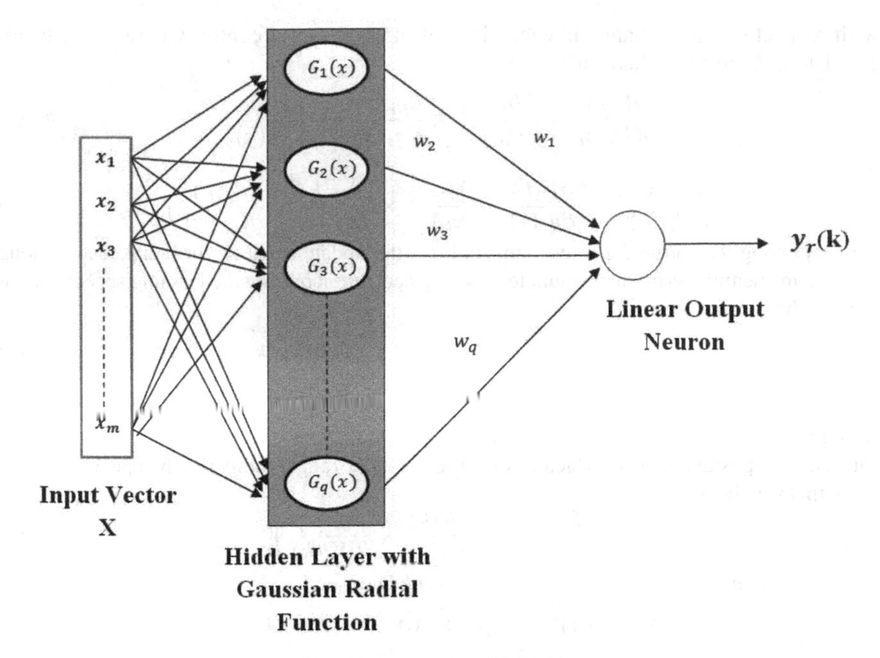


Fig. 1: Structure of RBFN

definition [1]:

$$\phi(t) = \exp\left(\frac{-t^2}{2\sigma^2}\right) \tag{3}$$

where $t = \|X(k) - G_i(k)\|$ and $\sigma = (\sigma_1, \sigma_2.....\sigma_q)$ are the widths of each radial centre.

## 3 Parameter Adjustment Rule for RBFN Identification model

The adjustable parameters of RBFN includes: radial centre, radial centre's width and the output weight vector. They are adjusted using training samples for reducing the cost function (which is generally taken to be an instantaneous mean square error (MSE)). To update these parameters, recursive update equations for each of them are obtained using gradient descent principle. The derivation is as follows: Let E(k) denotes the cost function value (instantaneous mean square error) at any $k^{th}$ time instant and is defined as follows:

$$E(k) = \frac{1}{2}(y_p(k) - y_r(k))^2 \tag{4}$$

where $y_p(k)$ is the plant's output whose dynamics was unknown and were identified using RBFN identification model and $y_r(k)$ denotes the output of RBFN at $k^{th}$ instant. Now, taking the derivative of $E(k)$ w.r.t $G_{ij}(k)$ will give the rate by which E(k) changes with respect to small change in each element of each radial centre, where $i = 1$ to $m$, $j = 1$ to $q$. Using the chain rule:

$$\frac{\partial E(k)}{\partial G_{ij}(k)} = \left(\frac{\partial E(k)}{\partial y_r(k)} \times \frac{\partial y_r(k)}{\partial \phi_j(k)} \times \frac{\partial \phi_j(k)}{\partial G_{ij}(k)}\right) \tag{5}$$

or

$$\frac{\partial E(k)}{\partial G_{ij}(k)} = \left(\frac{\partial E(k)}{\partial y_r(k)} \times \frac{\partial y_r(k)}{\partial \phi_j(k)} \times \frac{\partial \phi_j(k)}{\partial t_j(k)} \times \frac{\partial t_j(k)}{\partial G_{ij}(k)}\right) \tag{6}$$

On simplifying the partial derivative terms in Eq.6, the update equation for radial centres along with the momentum term (momentum term was added for increasing the learning speed without causing the instability) is:

$$G_{ij}\,(k+1) = G_{ij}\,(k) + \Delta G_{ij} + \alpha \Delta G_{ij}(k-1) \tag{7}$$

here $\Delta G_{ij} = \eta e_i(k) W_j(k) \frac{\phi_j(k)}{\sigma_j^2(k)} (X_j(k) - G_{ij}(k))$ and $\eta$ is the learning rate having value lying in $(0, 1)$ and $\alpha \Delta h_{ij}(k-1)$ represents the momentum term where, $\alpha$, known as momentum constant, is a positive number which lies between 0 and 1 range. Similarly, the update equation for output weights is:

$$\frac{\partial E(k)}{\partial W_j(k)} = \left(\frac{\partial E(k)}{\partial y_r(k)} \times \frac{\partial y_r(k)}{\partial W_j(k)}\right) \tag{8}$$

Thus, each element in $W = [w_1, w_2, w_3....w_q]$ is updated as:

$$W_j\,(k+1) = W_j\,(k) + \Delta W_j + \alpha \Delta W_j(k-1) \tag{9}$$

where $\Delta w_j = \eta e_i(k) \phi_j(k)$. Similarly, the update equation for widths of radial centres can be found as:

$$\frac{\partial E(k)}{\partial \sigma_j(k)} = \left(\frac{\partial E(k)}{\partial y_r(k)} \times \frac{\partial y_r(k)}{\partial \phi_j(k)} \times \frac{\partial \phi_j(k)}{\partial \sigma_j(k)}\right) \tag{10}$$

Each element in $\sigma = (\sigma_1, \sigma_2....\sigma_q)$ will be updated as:

$$\sigma_j(k+1) = \sigma_j(k) + \Delta\sigma + \alpha \Delta\sigma(k-1) \tag{11}$$

where $\Delta\sigma = \eta e_i(k) W_j(k) \frac{\phi_j(k) t_j^2(k)}{\sigma_j^3(k)}$

# 4 Update Equations for the Parameters of RBFN controller

Following the same procedure, the update equations for RBFN controller's parameters can be found as: Let $E(k) = 0.5(y_m(k) - y_p(k))^2$ where $y_m(k)$ denotes the desired output value. So:

$$\frac{\partial E(k)}{\partial G_{ij}(k)} = \left(\frac{\partial E(k)}{\partial y_p(k)} \times \frac{\partial y_p(k)}{\partial uc(k)} \times \frac{\partial uc(k)}{\partial \phi_j(k)} \times \frac{\partial \phi_j(k)}{\partial t_j(k)} \times \frac{\partial t_j(k)}{\partial G_{ij}(k)}\right) \tag{12}$$

where $\frac{\partial y_p(k)}{\partial uc(k)} = J(k)$ and is known as jacobian of plant and $uc(k) = RBFN$ controller output at the $k^{th}$ instant. The computation of J(k) required the knowledge of mathematical dynamics of the plant and since they were assumed to be unknown they were approximated by RBFN identification model which also provides this jacobian value. The rest of the partial derivatives in Eq. 12 are computed in a same fashion as done for the RBFN identification model case. Adaptive control involves direct and indirect control. Direct control only requires the value of error between the plant and desired output [10]. The stable laws that are applicable to direct control don't apply to RBFN and hence for that matter, indirect control is used. This method requires the knowledge of the model of the plant (for computing the jacobian value) for adjusting the parameters of RBFN controller.

## 5 Simulation Results

To show the effectiveness of the proposed scheme, two control problems were considered.

### 5.1 Brushless DC motor

The first control problem involves the control of a brushless DC motor [plant] whose model was taken from [6]. To work in the mat lab programming environment, the dynamics of the motor were discretized with sampling period T= 0.003sec. The resulting difference equation so obtained is given by:

$$y_p(k+1) = -y_p(k)(-2+150T) - y_p(k-1)(1-150T+380T^2) + 1100T^2r(k-1); \quad (13)$$

where $r(k)$ denotes the input voltage to the motor. The structure of the series-parallel identifica-

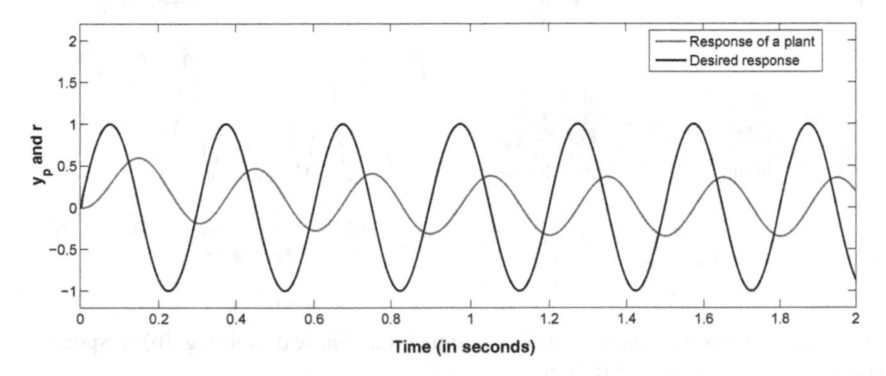


Fig. 2: Response of uncontrolled motor

tion model[10] based on RBFN was taken to be same as the unknown dynamics of plant and is given by:

$$y_r(k+1) = RBFN[y_p(k)] + 1100T^2r(k-1) \quad (14)$$

Where $y_r(k+1)$ denotes the RBFN output. The desired external input, $r(k)$, was taken to be $\sin(\frac{2\pi k}{250})$. Fig.2 shows the response of motor without the controller action. Fig.3 shows the general architecture of the proposed scheme for online identification and control of dynamical systems based on RBFN. As already mentioned earlier, RBFN based identification model provides the necessary past values of its output along with the jacobian value. The delay elements, $z^{-1}$, in the figure provides the delayed values of their input signals. The RBFN controller inputs in the proposed scheme includes: $r(k), y_p(k), y_p(k-n+1)$ and $uc(k-n+1)$ (controller past outputs), where n denotes order of the plant. This order, n, value may not be available and in that case it is usually taken to be 2. Both RBFN identifier and controller undergo the online training and hence will operate simultaneously during online identification and control. Fig.4a and Fig.4b shows the initial and final response of motor under the proposed a scheme action during the initial and final stages of online training. From the figure it can be concluded that parameters of RBFN controller were getting trained in a right direction and eventually reached to their desired values which made the plant's response equal to that of the desired response. The initial and final response of RBFN identifier are shown in Fig.5a and Fig.5b respectively. Here, also the learning

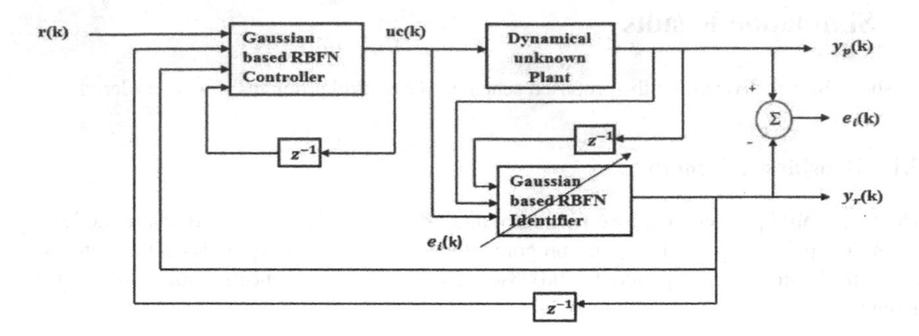


Fig. 3: General architecture of RBFN based identification and control scheme

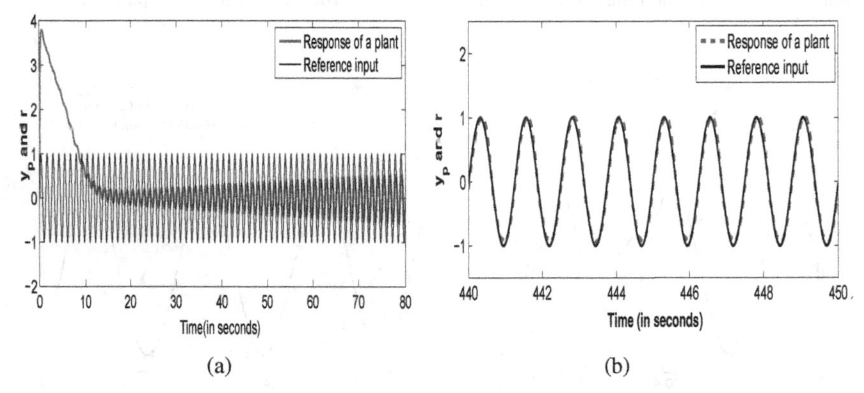


Fig. 4: (a) Motor response when controller is in its initial phase of training. (b) Response of motor during final stages of training.

of RBFN identifier is visible as its response became equal to that of plant after sufficient time of online training.

Now, the robustness of proposed scheme is evaluated by intentionally adding uncertainties like disturbance signal (at $k = 850^{th}$ instant) and varying the parameter (at $k = 500^{th}$ instant) in the system. From Fig.6, the effects of these uncertainties can be seen as a rise in the mean square error (MSE) value of controlled plant at the corresponding instants. The MSE, which was made zero earlier by the online training, shoots up again when these uncertainties entered in the system. But, it again went back to zero as RBFN controller and identification model's (identifier) parameters adjusted themselves so as to suppress the effects of these uncertainties.

## 5.2 Water Bath System

The second control system used is the water bath system and is taken from [17]. The differential equation describing the water bath system is given as

$$C\frac{dy_p(t)}{dt} = \frac{y_o - y_p(t)}{R} + h(t) \tag{15}$$

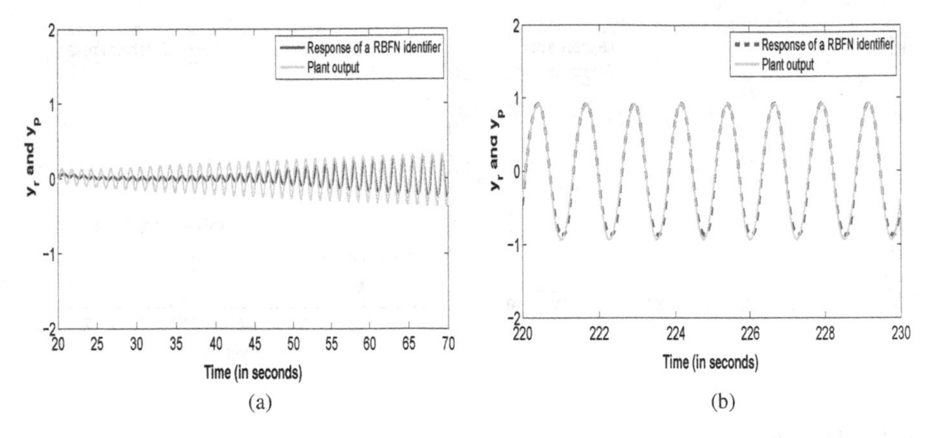


Fig. 5: (a) Initial stages of training identification model response. (b) Final identification model response during the end stages of training

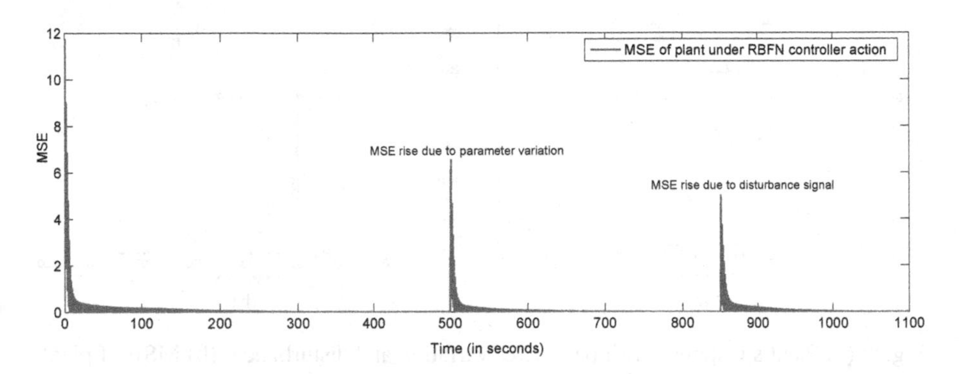


Fig. 6: MSE of Plant with parameter and disturbance signal during the online training

where $y_p(t)$ denotes temperature of water in the tank ($^oC$), $h(t)$ refers to the heat flowing into the water bath tank through the base heater (watts), symbol $y_o$ denotes the surroundings temperature and is assumed constant for simplicity. C and R denotes tank thermal capacity (joules$/^oC4$) and thermal resistance between surroundings and the tank border [4]. Both of them are assumed to be constant. So, we can write

$$x(t) = y(t) - y_o \tag{16}$$

On taking its derivative we can write

$$\frac{dx(t)}{dt} = -\frac{1}{RC}x(t) + \frac{1}{C}h(t) = -\alpha x(t) + \beta h(t) \tag{17}$$

where $\alpha = \frac{1}{RC}$ and $\beta = \frac{1}{C}$.

The discretized difference equation version of Eq.15 is:

$$y_p(k+1) = e^{-\alpha T}y_p(k) + \frac{[1-e^{-\alpha T}]\frac{\beta}{\alpha}}{1+e^{[0.5y_p(k)-40]}}u(k) + (1-e^{-\alpha T})y_o \tag{18}$$

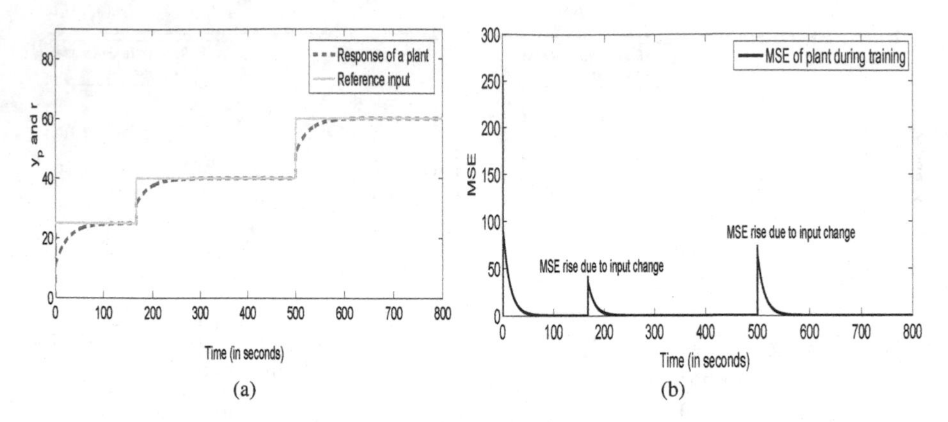


Fig. 7: (a) Plant's response under RBFN controller action. (b) MSE of plant during the training

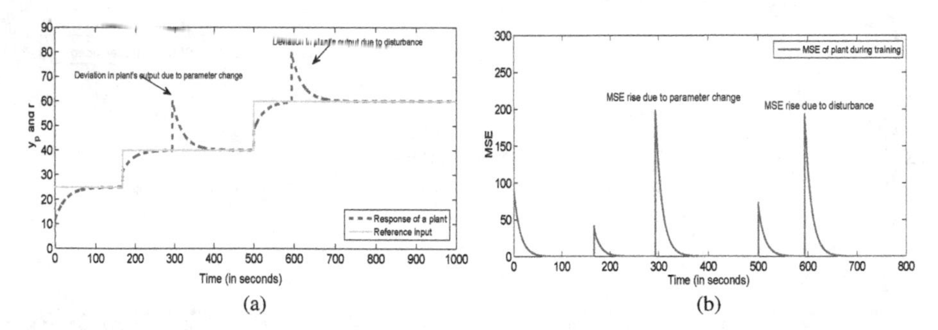


Fig. 8: (a) Plant's response with parameter variation and disturbance. (b) MSE of plant with parameter variation and disturbance.

The parameters values of above water bath system are: $\alpha = 1.0015e^{-4}$, $\beta = 8.67973e^{-3}$ and $y_o = 25(°C)$ which are computed from real water bath plant. The sampling period used for this system is $T = 30s$ [16]. Eq.18 represents a non linear water bath system's dynamics which was assumed to be unknown. It was identified during the online training by RBFN identifier. The plant under RBFN controller and identification model's action was tested for three set points:

$$r(k) = \begin{cases} 25(°C) \text{ if } k \leq 200 \\ 40(°C) \text{ if } 200 < x \leq 500 \\ 60(°C) \text{ if } 500 < x \leq 800 \end{cases} \tag{19}$$

From the Fig.7a and Fig.7b, it can be seen that as the online training progressed the controlled plant started to follow the desired set points. The MSE dropped towards zero as the training went on and became equals to zero. It then suddenly rises at those instants where set point value undergone change. But it again dropped to a zero due to the adjustments made by RBFN controller and identifier to their respective parameters during the online training. Fig.8a and Fig.8b shows the plots of the response of controlled plant and MSE respectively in the presence of parameter variation (occurred at k= $300^{th}$ instant) and disturbance signal (occurred at k= $600^{th}$ instant). In this case also the robustness characteristics of the proposed scheme can be seen. From the simulation study the suitability of RBFN as controller and identification model can be easily seen. Since it contains only 3 layers, the computation time elapsed during the updating of parameters (at each instant) of RBFN is very small as compared to ANN. RBFN have successfully handled the complexity of the unknown non linear dynamical system and shown its effectiveness against the parameter perturbation and disturbance signals.

# 6 Conclusion

This paper demonstrated the use of intelligent RBFN as an identifier and controller for online identification and adaptive control of the dynamical systems. The simulation study was done by taking two problems. The results so obtained reveal the effectiveness of the RBFN as a robust controller. The effects of parameter variation and disturbance signal were also got quickly compensated during the online control which shows the power of update equations for RBFN parameters. Since the plants dynamics were assumed to be unknown, the role of RBFN identification model was very crucial as it captured the dynamics of the plant during the online control and provides its past values (which serves as a portion of controller inputs) along with the value of rate of change of output of plant with change in the controller's output to the controller so that RBFN controller could adjusts its own parameters. From the results obtained in this paper it can be concluded that RBFN is able to effectively control and identify the complex plants. It can be applied to real time systems for real time control.

# References


1. Behera, L., Kar, I.: Intelligent Systems and control principles and applications. Oxford University Press, Inc. (2010)
2. Cybenko, G.: Approximation by superpositions of a sigmoidal function. Mathematics of control, signals and systems 2(4), 303–314 (1989)
3. Funahashi, K.I.: On the approximate realization of continuous mappings by neural networks. Neural networks 2(3), 183–192 (1989)
4. Gopal, M.: Digital Cont & State Var Met. Tata McGraw-Hill Education (2012)
5. He, W., David, A.O., Yin, Z., Sun, C.: Neural network control of a robotic manipulator with input deadzone and output constraint. IEEE Transactions on Systems, Man, and Cybernetics: Systems 46(6), 759–770 (2016)
6. Kang, L.L.K., Zhang, S.: Research and application of compound control based on rbf neural network and pid. In: Intelligent System and Knowledge Engineering, 2008. ISKE 2008. 3rd International Conference on. vol. 1, pp. 848–850. IEEE (2008)
7. Liu, Y.J., Li, J., Tong, S., Chen, C.P.: Neural network control-based adaptive learning design for nonlinear systems with full-state constraints (2016)
8. Lu, J., Hu, H., Bai, Y.: Generalized radial basis function neural network based on an improved dynamic particle swarm optimization and adaboost algorithm. Neurocomputing 152, 305–315 (2015)
9. Micchelli, C.A.: Interpolation of scattered data: distance matrices and conditionally positive definite functions. Constructive approximation 2(1), 11–22 (1986)
10. Narendra, K.S., Parthasarathy, K.: Identification and control of dynamical systems using neural networks. Neural Networks, IEEE Transactions on 1(1), 4–27 (1990)

11. Powell, M.J.: Radial basis functions for multivariable interpolation: a review. In: Algorithms for approximation. pp. 143–167. Clarendon Press (1987)
12. Singh, M., Srivastava, S., Gupta, J., Handmandlu, M.: Identification and control of a non-linear system using neural networks by extracting the system dynamics. IETE journal of research 53(1), 43–50 (2007)
13. Srivastava, S., Singh, M., Hanmandlu, M.: Control and identification of non-linear systems affected by noise using wavelet network. In: Computational intelligence and applications. pp. 51–56. Dynamic Publishers, Inc. (2002)
14. Srivastava, S., Singh, M., Madasu, V.K., Hanmandlu, M.: Choquet fuzzy integral based modeling of nonlinear system. Applied Soft Computing 8(2), 839–848 (2008)
15. Tan, M., Chong, S., Tang, T., Shukor, A.: Pid control of vertical pneumatic artificial muscle system. Proceedings of Mechanical Engineering Research Day 2016 2016, 206–207 (2016)
16. Tanomaru, J., Omatu, S.: Process control by on-line trained neural controllers. Industrial Electronics, IEEE Transactions on 39(6), 511–521 (1992)
17. Turki, M., Bouzaida, S., Sakly, A., M'Sahli, F.: Adaptive control of nonlinear system using neuro-fuzzy learning by pso algorithm. In: Electrotechnical Conference (MELECON), 2012 16th IEEE Mediterranean. pp. 519–523. IEEE (2012)
18. Wang, T., Gao, H., Qiu, J.: A combined adaptive neural network and nonlinear model predictive control for multirate networked industrial process control. IEEE Transactions on Neural Networks and Learning Systems 27(2), 416–425 (2016)

# Neuro-Fuzzy Approach for Dynamic Content Generation

Monali Tingane[1], Amol Bhagat[1], Priti Khodke[1], and Sadique Ali[1]

[1] Innovation and Entrepreneurship Development Center, Prof Ram Meghe College of Engineering and Management Badnera,
444701 Amravati, India
{ monali.tingane, amol.bhagat84, priti.khodke, softalis}@gmail.com

**Abstract.** E-Learning across all environments, whether it may be profit-making, educational or personal, it can be greatly used if the learning experience fulfilled both without delay and contextually. This paper presents the neuro-fuzzy based approach for dynamic content generation. Fast expansion of technology convert newly generated information into previous and invalidates it. This circumstance makes educators, trainers and academicians to use information suitably and efficiently. To achieve this objective, there must be several ways that is used by broad variety of educationists in a quick and efficient manner. These tools must permit generating the information and distributing it. In this paper, a web-based lively content generation system for educationists is presented. Neuro-fuzzy approach is used for developing this system. So that by considering performance of the learners the system will provide the content based on the knowledge level of that particular learner. It will help to individual learners to improve their knowledge level. Instructors can rapidly gather, bundle, and reorder web-based educational content, effortlessly import prepackaged content, and conduct their courses online. Learners study in an easily reached, adaptive, shared learning atmosphere.

**Keywords:** Dynamic content generation, e-learning system, intelligent tutoring system, learning management system, neuro-fuzzy system.

## 1 Introduction

Conveyance of knowledge, teaching or tutoring program with the help of electronic resources is defined as E-Learning. E-learning implicates the usage of a computer or electronic devices (e.g. a mobile phone) in particular manner to offer teaching, informative or educational material. In 21st century traditional education is certainly transmitted to web-based learning for institutes due to fast evolution of Internet. Teachers, academicians, mentors identify the online learning as a revolution that improves teaching, encourages lifetime learning understanding and influences each person. The fast development of e-learning organizations significantly supports and improves educational practices online by changing traditional learning activities and



J.M. Corchado Rodriguez et al. (eds.), *Intelligent Systems Technologies and Applications 2016*, Advances in Intelligent Systems and Computing 530,
DOI 10.1007/978-3-319-47952-1_39

presenting an innovative condition to students. In the online learning atmosphere, it is difficult for learners to choose the learning actions that greatly satisfy their requirements. The information overload problem is progressively simple in the huge data period. It is authoritative for an e-learning scheme to produce suitable contents to guide a student's appropriately. Now a day's requirement for man-made dynamic system is rising to become safer and additionally trustworthy as well as reliable. In the various fields and most importantly in e-learning area dynamic systems are very effective. E-Learning across all environments, whether it may be profit-making, educational or personal, it can be greatly used if the learning experience can be fulfilled both without delay (i.e. at the time when it is necessary) and contextually (i.e. personalized to the understanding, goals and individual preferences of the particular learner). Such learning necessities may vary from very precise learning desires e.g. how to bring out a specific action, to more unreserved learning goals, e.g. extend the understanding of the knowledge a particular discipline. Learning theories, like Andragogy [1] and Constructivism, sustain learning experiences, which focuses on individual learners motivations and encourage learner's answerability and engagement. Learner's needs are changed by time and requirement, so that system should be capable of fulfilling their needs.

The system which is adaptive to the every single learner condition, that is the system which facilitates learners with the learning material according to their requirement by producing the content dynamically, is referred as dynamic content generation system. Next age group educational environments are responding to these demands by attempting to sustain characteristics such as personalization, adaptivity and as per the demand learning item creation [2]. Adaptive interactive program is referred as the unique and important part in offering personalized that is "just-for-you" e-Learning. The advantage of this type of education is that it is dynamically appropriate to the learner's objectives, capability, prior knowledge, learning method, educational preferences and learning approach. Main complications with personalized e-Learning are that they typically access content which has been particularly developed for the personalized learning system. This is mainly the situation by means of Intelligent Tutoring Systems where, within the content itself personalization is fixed [3]. But in the dynamic systems the sequencing for the adaptivity is divided from the real teaching materials. This makes the chance by which educational material can be choose and put in the series that suits the learner, though, in such systems the educational material is, in over-all situations, still provided from an individual depository of educational sources. Dynamic content generation system will overcome the problems of e-learning system by providing the tutoring content to the learners dynamically by using neuro-fuzzy approach. In maximum technical practices, adaptive systems are come across that are considered with doubts in relation to development and constraints. These doubts can't be defined using static systems effectively, and hence, conventional control methodologies grounded on such prototypes are not likely to consequence in necessary outcome. In these situations, employing the soft computing practices remains a feasible option.

Fuzzy sets were proposed via Zadeh during year 1965 to symbolize/operate documents and material having dynamic doubts. Fuzzy knowledge is an active method used for handling challenging nonlinear procedures that are categorized by means of imprecise and undecided aspects. In such condition, fuzzy organizers are normally used in order to reach to the preferred presentation; the rule base of which is designed on the basis of the information of individual professionals. Still, this knowledge may not be adequate, for some multifaceted processes, and a number of methodologies [4], [5] estimated for formation of the IF–THEN instructions. These days, on behalf of this reason, the usage of neural networks (NNs) takes additional significance. Artificial neural systems can be measured as easy scientific simulations of brain-like structures and they task equivalent to scattered computing links. On the other hand, in distinction to conservative processers, which are planned to carry out precise job, the majority of neural networks essentially educated, or skilled. They can become skilled at different relations, novel functional dependencies and fresh configurations. Possibly the greatest significant benefit of neural networks is their adaptiveness [6]. Neural networks can spontaneously change their heaviness to enhance their achievement such as pattern recognizers, judgment producers, structure organizers, forecasters, etc. Adaptivity lets the neural network to execute effectively though the surroundings or the system presently examined differs with time. It will help in studying the learner's behavior by observing their actions and performance while studying. The grouping of fuzzy logic and Neural Networks can be used to resolve problems in the e-learning system by identifying and governing adaptive content generation systems. The important aspects designed for the application of Neural Networks in e-learning field are the features that they include for instance: education and simplification capacities, nonlinear planning, simultaneous calculations, and dynamism. Owing to this uniqueness, NNs turn out to be more widespread in regions like e-learning, artificial behavior, AI, governor, judgment creation, identification, robotics, etc. This system focuses on several factors including learners, teacher and admin, database for storing all the learning content and off course dynamic generation of the learning content means as per the learners need and knowledge level using neuro-fuzzy techniques. Proposed work gives the detailed description of the system.

The organization of this paper is given as. In Segment 2, orderly presentation of the literature review is given; which consist of the essence of the associated methodologies. Here, many diverse neural and fuzzy arrangements are studied for e-learning, and their parameters modernize procedures are specified. Also the detailed description for this neuro-fuzzy related work is given in table form. Likewise reviews regarding e-Learning systems, Intelligent tutoring systems, Electronic based content generation; Blended Learning based for e-Learning are given along with the table form description. Section 3 concentrates on the formulation of the recognized delinquent and it is devoted with proposed methodology; where the working of the neuro-fuzzy approach for dynamic content generation is stated. Experimental results are presented in Section 4. Section 5 summarizes the assumptions strained from the experimental results and provides the forthcoming directions.

## 2 Related Work

Quantifying and picturing student's interactive commitment in writing actions is proposed in [7]. They state this ability has providing a perfect groundwork to discover educational practices and their intellectual, emotional and communicative mechanisms. Furthermore, Google Docs has the benefit of supporting informal system combination and synchronous cooperative writing and it has been effectively useful in student project organization and shared writing practices. This paper mainly describes: 1) a new learning analytic (LA) system that combine interactive information of users writing, estimates the equal of commitment, and generates three types of conceptions, point-based, line-based and height-based visualizations; the paper furthermore studied 2) the concordance of the conditional commitment measures through associating these with member self-reports.

The construction and composition of an interactive video lecture platform is verified in [8] to compute student commitment using in-video quizzes. The use of in-video quizzes as an approach to diminish the deficiency of feedback presented to students and lecturers in videos and traditional lectures is also investigated. The results from this investigation prove that in-video quizzes were effective in producing an engaging and cooperative mode of content distribution. This work offers: by concentrating on student engagement an evaluation of the in-video quiz technique; study of student performance and interactions at the time of using in-video quizzes; and at the end, the design and development of an open source tool that simplifies access to lecture videos at the same time as gathering significant interaction data from learners for research purposes.

In E-learning, scholar's profile data shows a most important character in giving educational materials to the learners [9]. Presentation of the students can be upgraded with the help of adaptive e-learning; it is an important requisite for the learning. It tries to adapt the educational material, which fulfills the requirements of the learners. Therefore, receiving a dynamic Learning Object for matching the students made to order requirements is a significant matter. Two most important problems rise here in the previous e-learning: the identical educational resources is provided to every student without knowing there knowledge level under the "one size fit all" approach, and the information overload problem occurs due to massive amount of information availability.

The MTFeedback approach and its implementation are presented in [10]. The key objective of the MTFeedback approach is to repeatedly provide educators announcements related to collective educational events in an appropriate way throughout a lecture. It describes its instruments to spontaneously gather and exploit learner information in the teaching space. It prove in what manner the announcements can be revealed, in actual, to the instructors organizing a lecture in a reliable teaching space. This study describes, the announcements were produced through evaluating, in immediate, qualitative features of the information art details actuality constructed through the learners and linking learner records with the model of skilled information along with a set of mutual mistaken belief delivered through the instructor as portion of the proposal of the education jobs.

Resulting from earlier works, [11] offered and studied a recommender structure that facilities tutors, in a complete way, to choice suitable Learning entities from the available Learning Object Repositories (LORs) grounded on their enthusiastically rationalized ICT proficiency outline. The suggested system is offered and estimated in a two-layer demonstration coordinated through the generally acknowledged covered valuation procedure of adaptive educational systems, which has remain prolonged to

Recommender System in, a distinct exhibition and valuation of the "Teacher ICT Competence Profile Elicitation" Level and the "Learning Object Recommendation Generation" Level is executed. In [12], the aim is to generate innovative exercises from typescript. The effort offered now follows this final path: goal is to form apparatuses that repeatedly produce novel examination queries. The physical formation of MCQs, a frequently recycled manner of learner valuation, is problematic and timewasting. Therefore, the application of a programmed system might provide significant investments in educator period expended on exam making, sendoff additional period for jobs that have additional influence on learner education consequences. Generation of the distractors is the foremost objective of this paper. Systems that produce MCQs for vocabulary-training in linguistic education or in exact areas spread over diverse approaches to gain distractors. The consequences of the experimentations indicates that overall, the experimentation that practices diverse quantity and information resources to acquire the distractors achieved the greatest outcomes likewise determine that it is probable to spontaneously produce MCQs of satisfactory value for a underground language like Basque which can benefit instructors create MCQs for their examinations.

Web based educational tools like learning management systems (LMSs) is presented in [13]. When testing is a main concern for the educational processes then this method is exclusively significant for systematic and procedural sequences. Virtual and remote laboratories (VRLs) performed to insure this requirement in remote learning and to function as a didactical supplement for outdated direct courses. But, there is further a deficiency of: 1) conjunction and interoperability amongst both tools and 2) actual communication amongst learners while they work through VRLs, and/or inside an LMS; even however constructivist online learning atmospheres and VRLs previously occur. This methodology equipment this idea using two appreciated software presentations for e-learning and VRL improvement: Moodle and Easy Java Simulations (EJS).

A software suite is presented in [14] that allow instructors to implant the EQF into their course schemes and informative exercise of web-based and combined education. Though this set aimed to shelter every areas of advanced education (AE), its valuation taken a restricted scope because of the complications of employing advanced instructors for it. The European qualifications framework (EQF) is a European groups' endorsement and shared orientation system that associates the nations' recommendations systems and outlines. Thus, the EQF accompaniments and emphasizes present European flexibility apparatuses for instance Europass and ECTS. As unique type of S-CISE system is given in [15], the main purpose of the Online Classroom System are to make the communication experience in the physically dispersed educational atmosphere almost that in the old-style direct classroom, and additionally, to supplement the communication by provided that convinced additional instruments and functionalities to overwhelmed the weaknesses of the old-style face-to-face teaching space. There are two categorized of this system as asynchronous or synchronous. Asynchronous CISE (A-CISE) systems normally simplify the learning procedure with the help of consultation mass media like email, conversation panels, and expert learning management systems, which support learners and instructors to achieve education resources and trail development. But, they obligate restrictions on association and deficiency intelligence of closeness and earnestness, which is anywhere synchronous CISE (S-CISE) systems, is provided and has this compensations.

There is a large literature on e-learning, dating back over 20 years, with applications related to neuro-fuzzy, Dynamic content generation, and many areas other than computer vision as well, however here critically analyzed various research papers. The findings from this exhausting literature review are summarized here. Fuzzy technology is an efficient means aimed at dealing by way of difficult nonlinear procedures that are categorized by ill-defined and undecided influences. Uncertainties in the learning process cannot effectively be described by deterministic systems, because it varies as per individual learner's knowledge condition and knowledge level; therefore the systems which handle all the uncertainty in an effective manner are required. It is possible to use neuro-fuzzy approach for dynamic content generation in e-learning.

## 3 Proposed Neuro-Fuzzy Approach

To make web-based learning largely recognized and extensively used, web-based course quality necessarily superior than the old-style courses and courses have to be accessible everywhere. But, the stationary education resources that are offered online don't satisfy the necessities of educators and can't change straightforwardly as per the particular necessities. Hence, the software tool is required that will produce course materials according to the desires of instructors. So that e-learning is a web-based learning device or structure which allows learners to study anyplace and on anytime. Fuzzy set theoretical prototypes attempt to simulate social reasoning and the proficiency of managing ambiguity; nevertheless the neural networks prototypes tries to compete with the architecture and statistics illustration organizations of the human mind. To deal with the reality acknowledgment/ judgment creating complications, combination of the qualities of fuzzy set theory assures to offer, to an unlimited amount, additional smart structures (in relation of parallelism, error acceptance, adaptivity and ambiguity administration). Neural networks as well as fuzzy systems are trainable self-motivated system that approximate input output functions. They study from understanding with sample data and approximate function deprived of any mathematical model. A fuzzy system dynamicaly gathers and changes its fuzzy memory from illustrative numeric examples. On the other hand, neural networks instinctively produce and improve fuzzy instructions from exercise statistics. Fuzzy systems and neural networks correspondingly fluctuate in how they evaluate sample functions, what type of examples used and how they signify and stock these models. The objective of this work is to construct the design elements for a web-based e-learning system that generate the e-learning content dynamically using neuro-fuzzy approach. The proposed system is a neuro-fuzzy based dynamic content generation learning system that provides: content required for the learning process dynamically with the help of the neuro-fuzzy approach.

Four main modules are presents in dynamic content generation systems which include Tutor, Learner, Admin and Course Generation section. The structural design of the scheme is given in figure 1. An admin module contains two sub units that are

Instructor-Course Editor (ICE) and Teaching Materials Editor (TME). They can accomplish all Produce, Recover, Update, Remove procedures like adding new tutor by accepting there request, bring up to date tutor information, handing over courses to tutors, removing course/trainer statistics. Teaching materials are managed by the admin also with the help of adding fresh educational resources or removing them. Admin has a right to set the system preferences.

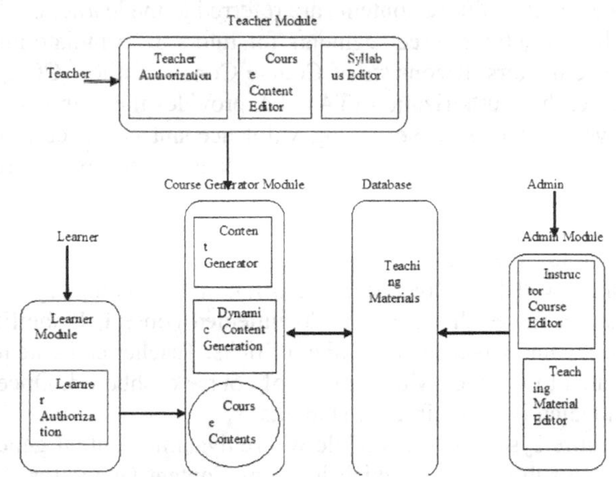


**Fig. 1.** Proposed System Architecture.

The teaching materials are stored in secure database. The communication with the students and teachers is established with textual content which is present into the database. All the details related to learners as well as tutor is stored in the database. It contains general information like name, address, contact no, e-mail-id, Personal details like date of birth, gender etc. Grouping of MySQL and relational database is used for storing purpose. To store the teaching materials MySQL database is chosen which is schema-free, document-based. This type of database is called as accessible and delivers extraordinary presentation, permits storage of huge amount of organized, semi- organized, and unorganized data. Remaining data is deposited into the general relational database.

The important element of the system from the time when it intervenes in all levels of the learning process is learner module. Learner module contains of two parts first is static and second is dynamic, in the static part all the information is static and hardly changes throughout a learning period and consists of: Identification of the learner such as name, surname, address, contact no, e-mail id, date of birth etc. Whereas in the dynamic part, information varies with learner's progress during the learning session, depends on the way followed by the learner to complete activities in relation with the followed learning goal and the developed abilities for their concepts. Learner every action is examined and saved in his learner model. Later this indicates at every step, the learner's knowledge level. The learner is evaluated on the basis of the selected learning goals. Learner himself/herself set the static part of the learner model, by a questionnaire, and the dynamic part is set by the system. This task contains of preparing for every particular learning goal the different individuals that better describe it. Each time the learner visit a proposed basic unit of the didactic plan, the dynamic part in his model is updating, considering the learners behavior. It is done for every idea of the education goal based on the existing information in the model. Learners have to fill one simple form for registering into the system. Only after that he/she will able to access the system. The person who is not carrying valid log-in ID and password is not able to use the system. Depending on the performance of the learner, which is obtained by analyzing their assignments and test paper results provided by the instructor, future contents are referred to the learners.

Teacher module structures three segments for tutors to formulate and bring about their virtual course matters. It consists of Course Content Editor (CCE) and Syllabus Editor (SE) and teacher authorization (TA). TA provides the authorized entry of the tutor into the system that is those having valid account can access to the system. Teachers login to the system by entering their user name and password. SE handles the management of course contents. This module supports teachers to make common information about their class, like essential education materials, fundamentals, course credits or agenda for the classes. Supplementary information for instance course explanation can as well be delivered. CCE offers a user boundary for the tutor to formulate course resources that contains documented content. It facilitates tutors in starting a different course by a small number of ticks. Teacher can able to view the list of courses allocated to him/her with the help of course dashboard. Once the course is created, it is immediately accessible for students.

Heart of the entire system is the module where dynamic content generation carried out. It consist three sub segments which includes Content Generator (CG), Dynamic content generation and Course View. CG has an authority of the formation of documentary course material. It receipts the essential statistics from Teacher Module wherever the course tutor fills the written statistics and creates demonstration style course constituents. In the previous dynamic content generation approach, the objective of dynamic courseware generation is to produce a personalized course by considering precise learning goals together with the original level of the students' knowledge. Here the difference is that the system with dynamic content generation observes and adapts to students' progress during the interaction of student with the system in the learning process. If the student performance is not matching our expectations, then the course can be re-planned and it may be dynamically. Final view of the created course is provided by CV, which will be presented to the student. This course view is updated, whenever the instructor updates the course materials. These CV modules perform as an association between the Instructor, Scholar segments and the folder. Produced course resources are deposited in the record and recovered at any time when they required.

First Learner need to make a registration on the system by filing a simple form only after it he/she can log into the system by entering valid user name and password. After log in properly they can access the course contents, choose their preferences, solve assignment, FAQs, test etc. They can make a request for particular teacher. Students can access process services provided by system. All there information that they have field at the time of registration as well as login and browsing details are saved into the database. Being the students they are the regular user of the system so the system will keep the track of there every action from log in to log.

First teacher will have filled the registration form as like students for logging in to the system. Initially teacher is registered as a student into the system then they have to send the request to the administrator; after accepting there request by the administrator teacher will get the right to create the course. User will have to mention weather he wants to be a student or instructor by choosing an appropriate option. After successful login teacher will have the ability to create course, assignments, FAQs, test etc. for the various courses. They can analyze the student performance and give them appropriate guidance to improve their knowledge. They have the ability to access various services provided by the system in the instructor domain. All there details that they have entered while filling the registration form is stored into the database. As teacher is the second user of the system after students there log in and logout details are also stored into the database.

## 4 Experimental Results and Discussion

The approach is implemented using PHP and database MySQL. The experimentation is carried out on Intel (R) Core (TM) i3-4200U CPU @ 1.60GHz 2.30 GHz processor. The RAM of the system used is 4GB and ROM is 500GB. The operating system is 64-bit and the processor is x64 installed on Windows 8. The experimentations are carried with several entries and some course contents.

This adaptivity plays an important role in personalized learning environment, which has the wide verity of learners each with different requirement and background. This technique is useful when the entire requirement of the learner is fulfill to learn the particular concept, that is, identify all the essential concepts, but not capable to understand the concept even after learning definite learning material. In such cases ITS does the actions are:
Providing explanation variants: The same information is presented in different way depends on understanding of the learner. The concept is presented in simplest form by explanation for all the concept points. The only problem with this approach is learner has to study more learning material than regular learning material for understanding of the same concept.
Provides additional learning material and task to perform which is not part of the course but help him to understand concept. Such additional material has to be provided by the instructor and associated with concept during domain knowledge building. Figure 2 provides the stepwise execution of proposed approach.

After login into the system students visit the courses as per there requirement. Where they can access the content solve the assignments, FAQs, tests etc. Then teacher will analyze their answer and gives feedback on it. The teacher and students collectively try to improve quality of the answer with the help of explanations, summaries, examples, hints, feedback and questions etc. Then teacher will analyze the total results and the find out whether the student will understand the particular concept or not. If the students will understands properly then it is ok; and if the performance is not up to the mark then the process will be repeated as shown below.

Name of the database created for this system is adaptiveits. This database contains total 120 tables. Following figure shows the list of tables in the database. It consist of the table for admin, user, instructor, there details, course materials have been stored into this database, tests, assignments, personal details, results survey and many more

things. Everything that is entered is stored into the adaptiveits until and unless it is deleted personally.
The comparative analysis of various learners result is provided to show that by providing the learners contents adaptively and making them to go through it again and again until their performance is not reaches up to the mark. Test results for various students in different attempts are analyzed. Table 1 shows the some student test result and the advice given by the instructor to improve the result. Here the marks of student in different attempts is given if the result is below 5 marks then the performance is considered as poor and instructed to start from beginner module, if it is above 5 but below 8 then it is considered as average and instructed to start from intermediate level and if it is above 8 then the performance is excellent no need of repetition.

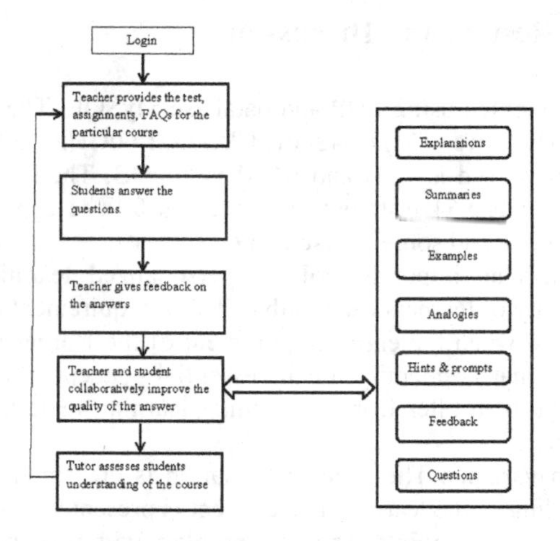


**Fig. 2.** Proposed System Stepwise Execution.

The typical proportion of right marks recorded for specific question is also calculated as shown in figure 3, which benefit instructor to examine the understanding of the learner for specific topic. The experiments are carried out with 10 students. Each student is instructed to solve the test and try its various attempts until his/her performance will be improved. All this details related to user and admin, also their test and survey details is stored in the database. From the experimental results it is found that the adaptive intelligent tutoring system will help to improve the student performance.

## 5 Conclusion

In this paper, using neuro-fuzzy approach a software tool is developed which is skilled in generating dynamic course content for e-learning. Condensed and high-performance fuzzy rule-base system can be produced by using the proposed approach is shown by result. Developed tool is practical, it can be developed to any type of courses effortlessly and has user-friendly boundary. This tool can be accessed from anyplace and anytime, because it is web-based. Out-of-date methods of e-learning are static that are not reliable and do not satisfying the learner's requirement. Without the

**Table 1.** Students test result details and dynamically generated course contents using proposed system.

| Login | Date and Time | Time Spend | Marks | Dynamically Generated Course Contents |
|---|---|---|---|---|
| student1 | 13/5/16 20:08 | 57s | 2/10 | Poor performance instructed to start from beginner module |
| student1 | 13/5/16 20:20 | 20s | 4/10 | Poor performance instructed to start from beginner module |
| student1 | 13/5/16 20:20 | 22s | 9/10 | Excellent performance |
| student2 | 13/5/16 20:22 | 26s | 4/10 | Poor performance instructed to start from beginner module |
| student2 | 13/5/16 20:23 | 1m 25s | 9/10 | Excellent performance |
| student6 | 13/5/16 20:35 | 24s | 3/10 | Poor performance instructed to start from beginner module |
| student6 | 13/5/16 20:34 | 40s | 6/10 | Average performance instructed to start from intermediate module |
| student6 | 13/5/16 20:36 | 23s | 10/10 | Excellent performance |
| student5 | 13/5/16 20:31 | 22s | 1/10 | Poor performance instructed to start from beginner module |
| student5 | 13/5/16 20:32 | 22s | 7/10 | Average performance instructed to start from intermediate module |

localhost/AdaptiveITS/mods/_standard/tests/results_all.php?tid=2

File Storage | Networking | Site map | Export Content | Links | Polls | Reading List | Gradebook | Web Search | Manage

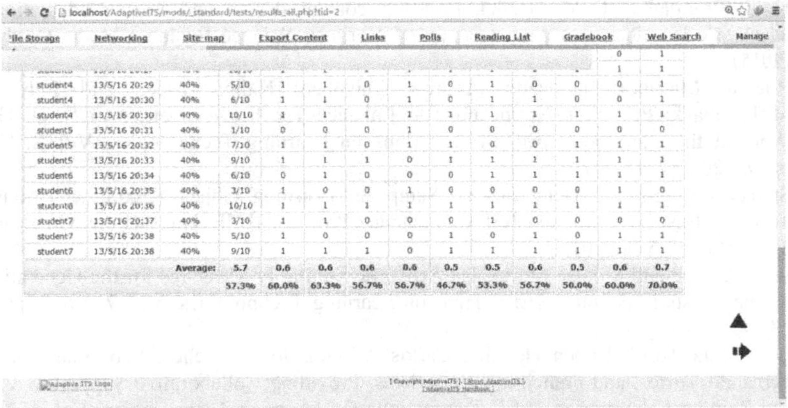

| | | | | | | | | | | | | | |
|---|---|---|---|---|---|---|---|---|---|---|---|---|---|
| student4 | 13/5/16 20:29 | 40% | 5/10 | 1 | 1 | 0 | 1 | 0 | 1 | 0 | 0 | 0 | 1 |
| student4 | 13/5/16 20:30 | 40% | 6/10 | 1 | 1 | 0 | 1 | 0 | 1 | 1 | 0 | 0 | 1 |
| student4 | 13/5/16 20:30 | 40% | 10/10 | 1 | 1 | 1 | 1 | 1 | 1 | 1 | 1 | 1 | 1 |
| student5 | 13/5/16 20:31 | 40% | 1/10 | 0 | 0 | 0 | 1 | 0 | 0 | 0 | 0 | 0 | 0 |
| student5 | 13/5/16 20:32 | 40% | 7/10 | 1 | 1 | 0 | 1 | 1 | 0 | 0 | 1 | 1 | 1 |
| student5 | 13/5/16 20:33 | 40% | 9/10 | 1 | 1 | 1 | 0 | 1 | 1 | 1 | 1 | 1 | 1 |
| student6 | 13/5/16 20:34 | 40% | 6/10 | 0 | 1 | 0 | 0 | 0 | 1 | 1 | 1 | 1 | 1 |
| student6 | 13/5/16 20:35 | 40% | 3/10 | 1 | 0 | 0 | 1 | 0 | 0 | 0 | 0 | 1 | 0 |
| student6 | 13/5/16 20:36 | 40% | 10/10 | 1 | 1 | 1 | 1 | 1 | 1 | 1 | 1 | 1 | 1 |
| student7 | 13/5/16 20:37 | 40% | 3/10 | 1 | 1 | 0 | 0 | 0 | 1 | 0 | 0 | 0 | 0 |
| student7 | 13/5/16 20:38 | 40% | 5/10 | 1 | 0 | 0 | 0 | 1 | 0 | 1 | 0 | 1 | 1 |
| student7 | 13/5/16 20:38 | 40% | 9/10 | 1 | 1 | 1 | 0 | 1 | 1 | 1 | 1 | 1 | 1 |
| | | Average: | 5.7 | 0.6 | 0.6 | 0.6 | 0.6 | 0.5 | 0.5 | 0.6 | 0.5 | 0.6 | 0.7 |
| | | | 57.3% | 60.0% | 63.3% | 56.7% | 56.7% | 46.7% | 53.3% | 56.7% | 50.0% | 60.0% | 70.0% |

[ Copyright AdaptiveITS ] [ About AdaptiveITS ]

**Fig. 3.** Student Analysis Displayed to Instructor for Overall Student Performance Analysis.

necessity for a professional, this software device makes it possible for teachers, instructors to make their personal courses and provided that courses to the students dynamically as per their requirement. Even though this tool in preparing the course materials and providing it to students; helps instructors, it has some boundaries. The major problem with this neuro-fuzzy method is to determine several predefined parameters heuristically. Wrong choice of these parameters may result in the problem of oscillation, though these parameters offer flexible adjustment between complexity and performance. The robustness and estimation of performance boundary are currently under investigation.

## References

1. Knowles, M.S. *Andragogy in Action*. San Francisco: Jossey-Bass. (1984).
2. Brusilovsky, Peter. Adaptive Hypermedia. In User Modelling and User-Adapted Interaction, 87- 110, Springer, (2001).
3. Conlan, O.; Wade, V.; Bruen, C.; Gargan, M. Multi-Model, Metadata Driven Approach to Adaptive Hypermedia Services for Personalized eLearning. Second International Conference on Adaptive Hypermedia and Adaptive Web-Based Systems, Malaga, Spain, (2002).
4. R. R. Yager and L. A. Zadeh, Eds., Fuzzy Sets, Neural Networks and Soft Computing. New York: Van Nostrand Reinhold, (1994).
5. J.-S. R. Jang, C. T. Sun, and E. Mizutani,Neuro-Fuzzy and Soft Computing. Englewood Cliffs, NJ: Prentice-Hall, ch. 17, (1997).
6. Konstantina Chrysafiadi and Maria Virvou, Fuzzy Logic for Adaptive Instruction in an E-learning Environment for Computer Programming, IEEE Trans on Fuzzy Systems, Vol. 23, No. 1, pp. 164-177, (2015).
7. Ming Liu, Rafael A. Calvo, Abelardo Pardo, and Andrew Martin, Measuring and Visualizing Students' Behavioral Engagement in Writing Activities, IEEE Transactions on learning technologies, Vol. 8, No. 2, pp.215-224, (2015).
8. Stephen Cummins, Alastair R. Beresford, and Andrew Rice, Investigating Engagement with In-Video Quiz Questions in a Programming Course, IEEE Transactions on Learning Technologies, (2015).
9. K. R. Premlatha, T. V. Geetha, Learning content design and learner adaptation for adaptive e-learning environment: a survey, Springer Science Business Media Dordrecht, (2015).
10. Roberto Martinez-Maldonado, Andrew Clayphan, Kalina Yacef, and Judy Kay, MTFeedback: Providing Notifications to Enhance Teacher Awareness of Small Group Work in the Classroom, IEEE Transactions on Learning Technologies, Vol. 8, No. 2, pp.187-200, (2015).
11. Stylianos Sergis and Demetrios G. Sampson, Learning Object Recommendations for Teachers Based On Elicited ICT Competence Profiles, IEEE Transactions On Learning Technologies, 5 Jan 2015.
12. Itziar Aldabe and Montse Maritxalar, Semantic Similarity Measures for the Generation of Science Tests in Basque", IEEE Trans on Learning Technologies, Vol. 7, No. 4, pp.375-387, (2014).
13. Luis de la Torre, Ruben Heradio, Carlos A. Jara, Jose Sanchez, Sebastian Dormido, Fernando Torres, and Francisco A. Candelas, Providing Collaborative Support to Virtual and Remote Laboratories, IEEE Transactions On Learning Technologies, Vol. 6, No. 4, pp.312-323, (2013).
14. Beatriz Florian-Gaviria, Christian Glahn, and Ramon Fabregat Gesa, A Software Suite for Efficient Use of the European Qualifications Framework in Online and Blended Courses, IEEE Trans on Learning Technologies, Vol. 6, No. 3, pp.283-296, (2013).
15. Linmi Tao and Meiqing Zhang, Understanding an Online Classroom System: Design and Implementation Based on a Model Blending Pedagogy and HCI, IEEE Trans on Human-Machine Systems, Vol. 43, No. 5, pp. 465-478, (2013).

# A Comprehensive Review on Software Reliability Growth Models utilizing Soft Computing Approaches

[1]Shailee Lohmor , [2] Dr. B B Sagar

[1]Research Scholar, Research and Development Centre, Bharathiar University, Coimbatore, India

[2]Assistant Professor, Birla Institute of Technology, Mesra- Ranchi, India

**Abstract** Software Reliability Engineering is an area that created from family history in the dependability controls of electrical, **auxiliaryAbstract** , and equipment building. Reliability models are the most prevailing devices in Programming Dependability Building for approximating, insidious, gauging, and assessing the unwavering quality of the product. In order to attain solutions to issues accurately, speedily and reasonably, a huge amount of soft computing approaches has been established. However, it is extremely difficult to discover among the capabilities which is the utmost one that can be exploited all over. These various soft computing approaches can able to give better prediction, dynamic behavior, and extraordinary performance of modelling capabilities. In this paper, we show a wide survey of existing delicate processing methodologies and after that diagnostically inspected the work which is finished by various analysts in the area of software reliability.

**Keywords:** Abstract Fuzzy logic (FL),Artificial Neural network (ANN), Genetic algorithm (GA), Artificial Bee Colony (ABC), Ant Colony Optimization (ACO), Cuckoo search Algorithm (CS), Simulated annealing (SA), Software Reliability Growth Model (SRGM).

## 1 Introduction

Software engineering is an inculcation, production of software with quality is the aim of software engineering based on these the software is distributed on time, in economics, and also fulfills its necessities [1]. Software plays an important role in various life-critical systems that frequently need an exceptionally rigorous guarantee process, however the utmost vital thing is software products development with higher quality and high reliability as quick as possible in time [2]. Reliability is



J.M. Corchado Rodriguez et al. (eds.), *Intelligent Systems Technologies and Applications 2016*, Advances in Intelligent Systems and Computing 530,
DOI 10.1007/978-3-319-47952-1_40

the significant factor in retrieving the quality of software, it is correlated to faults and mistakes. Software reliability is a significant factor which disturbs reliability of the system. It fluctuates from hardware reliability in that it redirects the design, precision, somewhat than developed perfection [3-7].

Software reliability growth is a significant problem in the software industry, meanwhile it can deliver serious information for the designers and testing staff during the testing stage [8]. The aim of SRGMs is to measure software reliability behavior and status, support to develop a more solid programming and conjecture when dependability has grown adequately to authorize item production [9]. The testing stage objective is to perceive and get rid of dormant software faults to ensure, as distant as likely, error free procedure of software at a specified time [10]. SRGMs are extensively utilized to approximation measurable software reliability measures. It specifies the common procedure of the requirement of the failure method on the influences mentioned [11]. The SRGM are usually distinct as stochastic counting procedures concerning the number of detected faults experienced in testing stage [12].

Any software system required to control smart reliability should endure intensive testing and debugging. These processes might be costly and time intensive and manager's still essentially correct info associated with however software system reliability grows [13]. SRGMs will approximate the amount of preliminary faults, software system reliability, intensity of failure, mean time-period between disasters. These models facilitate tracking and measure the expansion of reliability of the system software as software system is improved [14].

## 2 Soft-computing techniques

It is a combo of computing approaches that purpose to exploit the acceptance for inaccuracy and vagueness to attain robustness, tractability, and little solution price. Its principal components are FL, Chaos Theory, Neuro-computing, Evolutionary computing and Probabilistic computing. [15]. It is based on usual as healthy as artificial concepts and it is denoted as a computational brain [16]. In result, the soft computing role model is the human concentration. In this subsection we have chatted about the current soft computing order approaches as appeared in Figure 1.

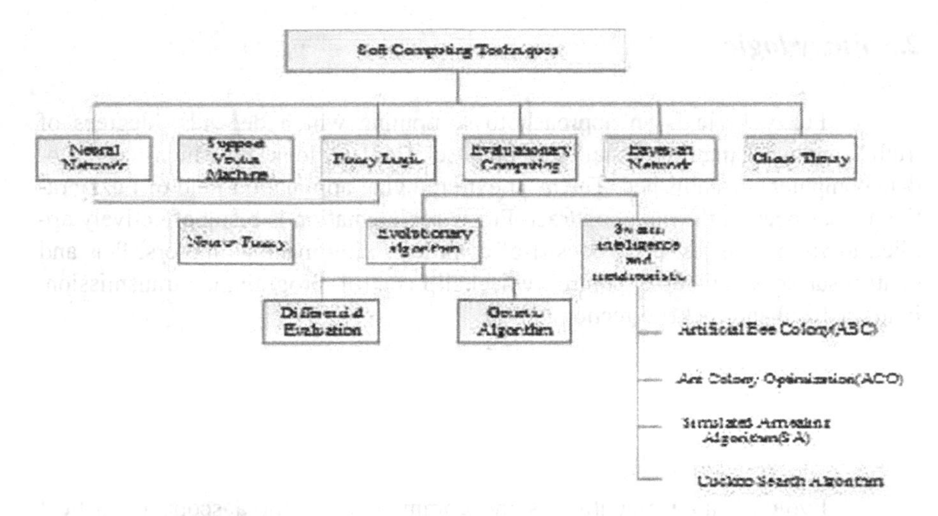


**Fig. .** 1: Soft Processing Technique

In Figure 1 we conversed about a portion of the soft processing approaches as neural systems, support vector machine (SVM), Fuzzy logic (FL), Evolutionary Computing, Bayesian Network and Chaos theory. And likewise some methodologies used in the game plan with the others as Neuro-Fluffy, mixture of Neural System and Fuzzy logic. The Evolution System is further isolated into Evolutionary method and Swarm Intelligence approaches.

## *2.1 Neural Network*

A neural network is a framework made out of numerous basic handling components working in parallel whose capacity is dictated by system structure, association qualities, and the processing implemented at computing nodes.

## *2.2 Support vector machine*

A Support vector machine (SVM) is a PC calculation that learns by illustration to allocate labels to objects. Case in point, an SVM can figure out how to recognize deceitful credit card activity by looking at hundreds or a huge number of fake and non-fake credit card action reports.

## 2.3 Fuzzy logic

Fuzzy logic is an approach to computing which depends "degrees of truth" somewhat than the usual "true or false" Boolean logic on which the up-to-date computer is established. The most extreme vital application field of Fuzzy rationale has been in the direction tract. Fuzzy regimentation is being effectively applied to various issues, these consist of complex flying machine motors, fans and control surfaces, helicopter control, wheel slip control, programmed transmission, industrialized and rocket direction.

## 2.4 Evolutionary Computing

Evolutionary Computing is the common name for a scope of critical thinking procedures in light of standards of organic advancement, for example, hereditary legacy and normal determination. It can likewise characterize as the stochastic pursuit and an optimization heuristic methodology decided from the exemplary progression hypothesis, which are refined on PCs in the real share of cases [17].

## 2.5 Bayesian Network

Bayesian Networks are graphical arrangement to characterize and reason around a dubious domain. The nodes in a Bayesian system characterize to an arrangement of arbitrary variables, from the area. An arrangement of circular segments unites sets of nodes, speaking to the immediate conditions between variables. The effect of management style on measurable proficiency, operational dangers, biotechnology, investigations of site convenience, consumer loyalty overviews, testing of web administrations and the social insurance frameworks are the different applications.

## 2.6 Chaos theory

A deterministic system is represented to be turbulent at whatever point its examination delicately relies on upon the startling circumstances. This equity indicates that two headings showing from two unmistakable near to early on conditions detached exponentially over the range of point. [19].

# 3 Reviews on soft processing approaches in Software Reliability Models

Soft processing is applied to reliability optimization, software fault diagnosis and in time sequence prediction during the software reliability analysis. In this part we will discuss about various soft computing approaches in the product dependability development model.

## *3.1 Reviews on Neural Network*

Neural systems are contracted model of essence neuronal framework, it is gigantically coordinated coursed preparing structure made up of exceedingly interconnected neural handling segments that be able to learn and along these lines have data and make it open for use. Parameter estimation of the official model and self-learning procedure the neural network has been connected to forecasting the future results. For expectation the feed forward network can be utilized.

Su, Yu-shen*et al.* [20] have proposed an AAN based method for software reliability estimation and modeling. They initially explained the ANN from the mathematical perspectives of software reliability modeling. They have shown how to apply ANN to forecast software reliability by design dissimilar components of ANN. Additionally, they used the ANN approach to construct a dynamic weighted combinational model (DWCM).

Karunanithi*et al.* [21, 22] predicted the increasing number of failures of design of initial ANN based on software reliability model. They utilized recurrent neural network, feed forward neural system, and Elman neural system in their writing and used execution time as the system. Reliability prediction in neural network can be worked in the following way, the network was trained by adjusting the neurons interconnection strength based on the software failure history.

Cai*et al.* [23] exhibited neural network based technique for software reliability predictions, utilized back propagation algorithm for data training. The execution of the methodology was assessed by utilizing an alternate number of inputs and hidden nodes. The neural network strategy took care of element information of software failures and various observed software failures. The outcome demonstrated that its execution relies on the way of the handled datasets.

Tian and Noore [24] projected a progressive neural system based strategy of prediction of software reliability, utilized numerous deferred into single output construction modeling. A genetic algorithm was utilized to universally advance the quantity of the delayed input neuron and the quantity of neuron in the concealed layer of the neural system construction modeling. Alteration of Levenberg–Marquardt calculation with Bayesian regularization was utilized to enhance the capacity to forecast software cumulative disappointment time. The outcome demonstrated that neural system building design greatly affects the execution of the system

Aljahdali*et al.* [25] investigated connections ANN models as an option way to deal with in order to determine these models by exploring the performance examination of four diverse connections ideal models for modelling the prediction of software reliability. The Elman recurrent neural systems were a powerful strategy for capacity expected due catching the dynamic conduct of the dataset.

Singh *et al.* [26] used feed forward neural system for gauging the software quality and associated back propagation algorithm to expect programming trustworthiness advancement design. The model had been connected to various failure data sets gathered from a few standard software ventures. The best portion of exhibiting strategy required, just the failure history as inputs and afterward added to its own inward model of failure procedure by utilizing back-propagation learning algorithm as a part of which the network weight was adjusted utilizing error spread back through the yield layer of the system.

Khoshgoftaar*et al.* [27] used the neural system for foreseeing the amount of flaws and exhibited a philosophy for displaying of static dependability. They chose a various regression quality model for the vital components of software complexity metrics gathered from an extensive business software framework. They prepared two neural systems one with the whole game plan of rule component and the other in the course of action of guideline component caused by various regression model determination lead us to a superior comprehension of the utility of neural network software quality models and the relationship between software intricacy measurements and software quality measurements.

Sitte [28] proposed neural system based software reliability desire model, with calibration of parametric models using some significant persistent measures. Both strategies were asserted to forecast as great or superior to the ordinary parametric models that have been utilized - with constrained results in this way. Every system connected its own particular consistency measure, hindering an immediate correlation. Neural Network expectations did not rely on upon earlier known models.

Sandeep Kumar Jain *et al.* [29] proposed a system to assess the unwavering quality of the product include segments using differing neural system designs. By then surveyed the shortcoming desire conduct in the game plan of portions over an aggregate execution time interim other than this the gauge of leaks was assessed in the complete programming. The prediction of the reliability of each component was considered from respective trained neural network architecture.

### 3.2 Reviews on Fuzzy logic

Fuzzy rationale is determined from fuzzy set theory handled with reasoning that is suitable rather than precisely found from conventional predicate rationale. A fuzzy model is a laying out between phonetic terms, associated with variables. Therefore the data to fuzzy model and yield from a fuzzy model can be either mathematical or phonemical.

Cai*et al.* [31] examined the improvement of fuzzy software reliability models set up of probabilistic software reliability models (PSRMs). It depends on the affirmation that software reliability was fuzzy in nature. The likelihood presumption for PSRMs was not sensible in view of the uniqueness of software.

Reformat [32] proposed a methodology prompting a multi procedure, information extraction and advancement of an exhaustive meta-model forecast framework. An issue of a restricted measure of information accessible for the development of models was also addressed. That was solved by building a few fuzzy-based models utilizing distinctive subsets of the information. In general, that prompts various models representing connections existing among information qualities.

AlJahdali*et al.* [33] scrutinized the use of Fuzzy logic on building SRGM to evaluate the predicted programming issues in the midst of the testing procedure. Proposed model involves an accumulation of direct sub-models, considering the Takagi-Sugeno approach and annexed beneficially using fuzzy participation capacities to speak to the foreseen programming disappointments as a component of chronicled measured disappointments. This purposed model gives a prevalent execution, displaying limit.

## 3.3 Reviews on Neuro-Fuzzy System

In the Neuro-fuzzy structure joins the Fuzzy rationale and neural systems. It can be generally used for software reliability modelling examination. Neuro-fuzzy depicts a technique for controlling neural systems with Fuzzy rationale.

KirtiTyagi*et al.* [35] proposed a model for surveying CBSS dependability, known as an adaptive Neuro fuzzy inference system (ANFIS) that relied on upon these two basic segments of the neural framework, delicate registering, and Fuzzy rationale. The CBSSs utilized to test the models were classroom-based tasks. Their outcomes demonstrated that the ANFIS enhanced the reliability estimation of the fuzzy inference system. A restriction of their proposed model was that while the four components for which the model was actualized were the most basic elements, there might be other pertinent variables that ought to be utilized.

## 3.4 Reviews on Genetic algorithm (GA)

A genetic algorithm is a modifying of machine learning in which gets its behavior from a similarity of the technique of improvement in nature. This is effected by the creation within a machine of a populace of the creatures portrayed by chromosomes. The strength of each chromosome is collected by evaluating it against a coal capacity. To reproduce the typical survival of the fittest methodology, best chromosomes exchange information to convey posterity chromosomes. Commonly, the technique is continued for endless to show signs of improvement fit game plan.

Liang *et al.* [36] proposed a genetic algorithm improving for the amount of postponing input neurons and the amount of neurons in the hid layer of the neural framework, predicting programming unwavering quality and disappointment time of programming. The advancement process decided the quantity of the delayed input neurons comparing to the past failure time data succession and the quantity of neurons in the shrouded layer. Once the ideal or close optimal neural system structural engineering was designed, change of Levenberg–Marquardt algorithm with Bayesian regularization was utilized as their Feed forward neural system training plan. That enhanced the capacity of the neural system to sum up well when the data were not known, mitigates the issue of over fitting in prior methodologies, and upgraded the algorithm exactness of next-step forecast of the total failure time.

Aljahdali*et al.* [37] investigated Genetic algorithms (GA) as an option way to deal with infer software reliability models. The gene-based methodology as one of the computational knowledge strategies. That was followed in predicating so as to anticipate faults in software reliability amid the software testing process utilizing software faults historical data. Additionally, a multi-objective genetic algorithm was connected to take care of the three issues recorded already by consolidating the conceivable reliance among progressive programming run and used outfit of creating so as to forecast models routines for evaluating the model(s) parameters with numerous and contending targets, through the system of GA improving.

Satya Prasad *et al.* [38] joined both imperfect troubleshooting and change-point issue into the product software reliability growth model (SRGM) in light of the definitely comprehended exponential circulation the parameter estimation was considered. Different NHPP SRGMs have been concentrated on with different suppositions.

## *3.5 Reviews on Genetic programming (GP)*

Genetic programming can be gained as a development of the genetic algorithm, a model for testing and picking the best choice among a plan of results, each spoke to by a string. Genetic programming utilizes above and beyond and makes the project or "function" the unit that is verified. Two methodologies are utilized to pick the fruitful system - crossbreeding and the competition or rivalry approach.

Costa *et al.* [39] modeled growth of software reliability with genetic programming. With sufficient information, the use of GP licenses to satisfy the backslide of basically any capacity, without the past learning about the behavior of the technique. In view of this, they expected that GP can be effectively utilized as a part of software reliability. This paper amplified past work and introduced consequences of two analyses that explore GP for programming unwavering quality demonstrating.

E. Oliveira *et al.* [40] enhanced software reliability models utilizing advanced strategies in view of genetic programming. Boosting Technique consolidates a few speculations of the training set to improve results. This methodology joined GP with Boosting systems. Boosting systems have been effectively used to enhance the execution of known techniques from the machine learning area. The main analysis utilized models taking into account time, which apply to various ventures. In the second trial, models taking into account test scope are investigated. The most essential change of this work is when considered models in view of time and got improved results by utilizing only one function set.

Y. Zhang *et al.* [41] anticipated MTBF disappointment information arrangement of software reliability by genetic programming algorithm. The transformative model of GP was then tacked and assessed by attribute criteria for some routinely used programming testing cases. Keeping in mind the end goal to take care of the issues of software reliability estimation and expectation, they chose meantime between Failures' (MTBF) time series of the software as the target time series for Genetic programming model, which can powerfully make reliability models in the perspective of various software failure time arrangement with no assumptions made for the software failure mechanics.Result demonstrated the higher forecast accuracy and better pertinence.

Eduardo Oliveira Costa *et al.* [42] presented another GP based approach, named ($\mu$+$\lambda$) GP. This calculation was displayed another technique to demonstrate the dependability of programming, named GP. The objective of the algorithm was to add to an assortment of the GP algorithm by changing the established GP. To assess this algorithm, two sorts of models: taking into account time and on scope were exhibited for trial results, which was constantly superior to anything established GP.

Zainab Al-Rahamneh*et al.* [43] proposed the use of Genetic programming (GP) as a transformative calculated way to deal with handle the software reliability modeling issue. Their investigations demonstrated that the proposed genetic-programming model better thought about ten different models found in literature.

## 3.6 Reviews on Artificial Bee Colony

In 2005, DervisKaraboga, cleared up another calculation, prodded by the insightful behavior of honey bees known as artificial bee colony. It is an upgraded instrument giving a populace based hunt procedure in which people called nourishments positions are balanced by the fake honey bees with time and honey bee's arrangement to see the spots of sustenance sources with high nectar entirety ultimately the one with most paramount nectar.

Tarun Kumar Sharma *et al.* [44] proposed a balanced interpretation of the ABC, the DABC (Dichotomous ABC), to improve its execution, to the extent joining to an individual perfect indicates and compensate the compelled measure of pursuit moves of one of a kind ABC. The thought was to move dichotomously in both bearings to produce another trial point. The execution of the ABC algorithm was dissected on five standard benchmark issues, furthermore, they investigated the relevance of the proposed calculation to evaluate the parameters of software reliability growth models (SRGM).

## 3.7 Reviews on Ant Colony Optimization

Ant Colony Optimization [45] is a technique which apply likelihood to take care of issues where the calculations are debasing with the assistance of diagrams to acquire effective ways.

LathaShanmugamet al. [46] talked about a framework evaluation strategy in view of the Ant Colony Algorithm. The result of the examination utilizing six common models exhibited that the calculation can be connected for assessing criterion. They accepted that the software framework was liable to fail arbitrarily because of software mistakes. At whatever point, there was a software failure, it was uprooted and accepted, that other mistakes were not presented

LathaShanmugamet al. [47] focused on the correlation and redesign of Ant Colony Optimization Methods for Modelling of Software Reliability. The Improved technique demonstrates noteworthy preferences in finding the integrity of fit for programming unwavering quality models, for example, limited and endless failure Poisson model and binomial models.

## 3.8 Reviews on Simulated annealing (SA) Algorithm

Simulated annealing (SA) is an iterative pursuit calculation roused by the hardening of metals [48]. Beginning with an introductory arrangement and equipped with sufficient irritation and assessment functions, the calculation performs a stochastic fragmentary inquiry of the state space.

Pai and Hong [61] associated Support vector machines (SVMs) for evaluating programming dependability where simulated annealing (SA) estimation was used to pick the parameters of the SVM model. This examination explained the possibility of the utilization of SVMs to estimate software reliability. Simulated annealing calculations (SA) were utilized to choose the parameters of an SVM model. Numerical illustrations taken from the past literature are utilized to exhibit the execution of software reliability prediction

### *3.9 Reviews on Cuckoo search Optimization*

This calculation subject to the commit brood parasitic conduct of some cuckoo species in gatherings with the Duty flight conduct of some organic product fliers and feathered creatures. Cuckoo search optimization is exceptionally powerful in getting better and acceptable answers to the issue of parameter estimation of Programming Unwavering quality Development Models. This calculation seeks methodology can professionally direct all through the issue look space and find great arrangements using less emphasis and lesser populaces.

NajlaAkram AL-Saati*et al.* [65] assessed parameters in perspective of the open disappointment information. Cuckoo look beats both PSO and ACO in discovering better parameters took a stab at using unclear datasets, yet more unfortunate if there ought to emerge an event of enhanced ACO. The Exponential, Power, S-Formed, and M-O models are pondered in this work. The hint system of the cuckoo can capably investigate all through the pursuit space on the issue and discover awesome game plans using less cycles and tinier populaces.

## 4 Performance Analysis

This section delivers a detailed comparison and performance analysis of the software reliability growth model using soft computing tips, and of several techniques within each category. In this review, we have compared different techniques based on the parameters like project data and its acquisition method, summary of the review. Table 1 shows the detailed comparative analysis of various soft computing techniques.

**Table . Table 1** Comparative analysis of different method

| S.no | Reference | Technology Used | Project data | Summary of review |
|---|---|---|---|---|
| 1 | [20] | Neural Network | DS1&DS2 | The use of ANN to give better prediction |
| 2 | [21] | Neural Network | DS-1K.Matsumoto (1988) DS-2J.D. Musa (1987) DS-3M. Ohba (1984) | Finds the use of feed forward neural systems as a software reliability growth model expectation. |
| 3 | [22] | Neural Network | DS from Yoshiro Tohma project | Resulting to preparing the neural system toward the end of a future testing and researching session. |
| 4 | [25] | Neural Network | Real time control project | The Elman repetitive NN is a vigorous methodology estimate dynamic behavior of the dataset. |
| 5 | [26] | Neural Network | Data gathered from three comparable systems (Kernel2, Kernel3andKernel4) and two | The splendid forecast results found in the neural framework models considered as a capable displaying |

| | | | | |
|---|---|---|---|---|
| | | | dissimilar systems (Kernel5andKernel6) | device for programming engineers. |
| 6 | [27] | Neural Network | Musa datasets S1 & SS3 | Neural Systems are proportionate or better example pointers. |
| 7 | [28] | Neural Network | Data collected from the local training set. | Evaluated the faults expectation and forecast of deficiencies is surveyed for the complete programming. |
| 8 | [32] | Fuzzy Logic | Real command & control project, Military and operating system, John D Musa, Bell lab. | Established models offer extraordinary performance modeling capabilities. |
| 9 | [33] | Fuzzy Logic | DS-1: Musa J D (1975) software reliability data DS-2: Pham H (2006) system software<br>Reliability | It doesn't require any due fuzzified systems autonomously and count time is decreased. |
| 10 | [36] | Genetic algorithm | Data from three projects. They are Military, Real Time Control and Operating System. | The consistency of the single AR model and collaborative of AR models trained by GA over the train and test data is concerned, the outfit of the models performed superior to the single model. |
| 11 | [37] | Genetic Algorithm | DS-1: Misra, P.N., 1983 Software reliability analysis. IBM Syst. DS-2: Pham, H., 1993. | This model is assessed as better than the substitute considers models with respect to all conditions are picked. |
| 12 | [38] | Genetic programming | DS-1John Musa at Bell Telephone Laboratories DS-2 failure data of a program called<br>Space | GP is a reasonable gadget to locate a scientific explanation to displaying software reliability. |

## 5 Future Enhancement

In this review diverse soft computing approaches are utilized as a part of various pattern with these models. We watched that Neural System philosophy is more revered by the experts in software reliability models. Genetic programming gives more precision than other delicate figuring approaches. Cuckoo Look, Recreated Tempering are used as a piece in this field, yet not all that broadly yet. To overcome the delay of release time and cost overrun for ongoing projects we can use the software reliability growth model (SRGM) using Adaptive Dolphin Echolocation (ADE) Algorithm. It is a very effective parameter estimation method for SRGMs and provide ten times better accuracy compare other heuristic algorithms.

## 6 Conclusion

In view of this paper, we explore some soft computing approaches, for better prediction, dynamic behavior, and extraordinary performance of modelling capabilities. Each algorithm is best in approach for parameter estimation of Programming Unwavering quality Development Models, reduced the issues raised streamlining and has a better rate. We have demonstrated the current soft computing approach in modelling of software reliability, with the dependence that it would assist as a kind of perspective to both gray and advanced, approaching scientists in this field, to reinforce their comprehension of flow patterns and help their forthcoming exploration predictions and bearings.

## References

1. Agarwal kk, Singh Y "Software Engineering.", New Age International Publisher, New Delhi, 2005.
2. Hsu, Chao-Jung, and Chin-Yu Huang."Optimal weighted combinational models for software reliability estimation and analysis." Reliability, IEEE Transactions, Vol. 63, No. 3, pp. 731-749.
3. Mohanty, Ramakanta, V. Ravi, and ManasRanjanPatra. "Software reliability prediction using group method of data handling." Springer, Berlin, Heidelberg on Rough Sets, Fuzzy Sets, Data Mining and Granular Computing, pp. 344-351, 2009.
4. Lakshmanan, Indhurani, and SubburajRamasamy. "An artificial neural-network approach to software reliability growth modeling." Procedia Computer Science Vol.57, pp. 695-702, 2015.
5. Kapur, P. K., and R. B. Garg"Cost, reliability optimum release policies for a software system with testing effort." Operations Research, Vol. 27, No. 2, pp. (109-116).
6. Musa JD "Software Reliability Engineering: More Reliable Software, Faster Development and Testing" McGraw-Hill, 1999.
7. Musa JD, Iannino A, Komodo K "Software Reliability: Measurement, Prediction and Application." McGraw-Hill, 1987.
8. Chiu K. C, Huang, Y. S, and Lee T. Z," A study of software reliability growth from the perspective of learning effects." Reliability Engineering & System Safety, Vol.93, No.10, pp. 1410-1421, 2008.
9. Chang Y. C, and Liu C. T, "A generalized JM model with applications to imperfect debugging in software reliability." Applied Mathematical Modelling, Vol.33, No.9, pp. 3578-3588, 2009.

10. Upadhyay R, and Johri P, “Review on Software Reliability Growth Models and Software Release Planning.” International Journal of Computer Applications, Vol.73, No.12, 2013.
11. CagatayCatal, “Software fault prediction: A literature review and current trends”, Expert systems with applications, Vol.38, No.4, pp. 4626-36, 2011.
12. Thanh-TrungPham, XavierDefago and Quyet-ThangHuynh, “Reliability prediction for component-based software systems: Dealing with concurrent and propagating errors”, Science of Computer Programming, Vol.97, pp. 426-57, 2015.
13. Zio E, “Reliability engineering: Old problems and new challenges”, Reliability Engineering and System Safety, Vol.94, No.2, pp. 125-41, 2009.
14. Ahmet Okutan and OlcayTanerYıldız, “Software defect prediction using Bayesian networks”, Empirical Software Engineering, Vol.19, No.1, pp. 154-181, 2014.
15. Bonissone P, Chen YT, Goebel, K, Khedkar P “Hybrid Soft Computing Systems: Industrial and Commercial Applications.” Proceedings of the IEEE, Vol.87, pp. 1641-1667, 1999.
16. Das, Santosh Kumar, Abhishek Kumar, Bappaditya Das, and A. P. Burnwal"On soft computing techniques in various areas." Int J Inform Tech ComputSci, Vol. 3, pp. 59-68.
17. Streichert F “Introduction to Evolutionary Algorithms.” Frankfurt Math Finance Workshop, University of Tubingen, Germany, 2002.
18. Sydenham, Peter H., and Richard Thorn, “Handbook of measuring system design”, Vol. 2
19.Boccaletti, S., Grebogi, C., Lai, Y.C., Mancini, H. And Maza, D, "The control of chaos: theory and applications." Physics reports Vol.329, No. 3, pp. 103-197.
20. Su Y. S, and Huang C. Y, “Neural-network-based approaches for software reliability estimation using dynamic weighted combinational models. “Journal of Systems and Software, Vol.80, No.4, pp. 606-615, 2007.
21. Karunanithi, Nachimuthu, Yashwant K. Malaiya, and Darrell Whitley. "Prediction of software reliability using neural networks." Software Reliability Engineering,International Symposium, IEEE, pp. 124-130.
22. Karunanithi, Nachimuthu, Darrell Whitley, and Yashwant K. Malaiya, "Using neural networks in reliability prediction." IEEE, Vol. 9, No. 4, pp. 53-59.
23. Cai KY, Cai L, Wang WD, Yu ZY, Zhang D, "On the neural network approach in software reliability modeling.", Journal of Systems and Software, Vol.58, No. 1, pp. 47-62.
24. Tian L, Noore A “Evolutionary neural network modeling for software cumulative failure time prediction.” Reliability Engineering and System Safety, Vol.87, pp. 45-51, 2005.
25. Aljahdali AS, Buragga KB “Employing four ANNs Paradigms for Software Reliability Prediction: an Analytical Study.” ICGST-AIML Journal, 2008.
26. Singh, Yogesh, and Pradeep Kumar "Prediction of software reliability using feed forward neural networks." Computational Intelligence and Software Engineering (CiSE), IEEE, pp. (1-5).
27. Khoshgoftaar TM, Szabo RM, Guasti PJ “Exploring the Behavior of Neural-network Software Quality Models.” Software Engg J, Vol.10, pp. 89-96, 1995
28. Sitte R “Comparison of Software Reliability Growth Predictions: Neural networks vs. parametric recalibration.” IEEE transactions on Reliability, Vol. 48, pp. 285- 291, 1999.
29Jain, Sandeep Kumar, and Manu Pratap Singh"Estimation for Faults Prediction from Component Based Software Design using Feed Forward Neural Networks.", 2013.
30. Aljahdali, Sultan, and Narayan C. Debnath,"Improved Software Reliability Prediction through Fuzzy Logic Modeling.", 2004.
31. Cai KY, Wen CY, Zhang ML “A critical review of software reliability modeling.” Reliability Engineering and System Safety, Vol. 32, pp. 357-371, 1991.
32. Reformat, Marek. "A fuzzy-based multimodal system for reasoning about the number of software defects.", International journal of intelligent systems, Vol.20, No. 11, pp. 1093-1115, 2005.
33. Aljahdali, Sultan "Development of software reliability growth models for industrial applications using fuzzy logic." Journal of Computer Science, Vol.7, No. 10, pp. 1574.

34. Chatterjee, S., Nigam, S., Singh, J.B. And Upadhyaya, L.N. "Application of fuzzy time series in prediction of time between failures & faults in software reliability assessment."Fuzzy Information and Engineering, Vol.3, No. 3, pp. 293-309, 2011.
35. Tyagi, Kirti, and Arun Sharma "An adaptive Neuro fuzzy model for estimating the reliability of component-based software systems."applied Computing and informatics, Vol.10, No. 1, pp. 38-51, 2014.
36. Tian L, Noore A "On-line prediction of software reliability using an evolutionary connectionist model." Journal of Systems and Software, Vol.77: pp. 173- 180, 2005.
37. Aljahdali, Sultan H., and Mohammed E. El-Telbany "Software reliability prediction using multi-objective genetic algorithm." Computer Systems and Applications, IEEE/ACS International Conference, pp. (293-300), 2009.
38.SatyaPrasad, R., O. NagaRaju, and R. R. LKantam. "SRGM with Imperfect Debugging by Genetic Algorithms."
39. Costa, Eduardo Oliveira, Silvia R. Vergilio, Aurora Pozo, and Gustavo Souza. "Modeling software reliability growth with genetic programming." Software Reliability Engineering,IEEE International Symposium, ISSRE, (pp. 10-pp) 2005.
40. Oliveira, E. O., Aurora Pozo, and Silvia Regina Vergilio, "Using boosting techniques to improve software reliability models based on genetic programming.", Tools with Artificial Intelligence, IEEE International Conference, ICTAI, pp. (653-650),2006.
41. Yongqiang, Zhang, and Chen Huashan. "Predicting for MTBF failure data series of software reliability by genetic programming algorithm." Intelligent Systems Design and Applications, ISDA'06. Sixth International Conference, Vol. 1, pp. (666-670), 2006.
42. Costa, Eduardo Oliveira, Aurora Trinidad Ramirez Pozo, and Silvia Regina Vergilio. "A genetic programming approach for software reliability modeling." Reliability,IEEE Transactions, Vol. 59, No. 1, pp. (222-230), 2010.
43. Al-Rahamneh, Zainab, Mohammad Reyalat, Alaa F. Sheta, SuliemanBani-Ahmad, and Saleh Al-Oqeili, "A New Software Reliability Growth Model: Genetic-Programming-Based Approach." Journal of Software Engineering and Applications, Vol.4, No. 8, pp. 476, 2011.
44. Sharma, Tarun Kumar, Millie Pant, and Ajith Abraham "Dichotomous search in ABC and its application in parameter estimation of software reliability growth models."Nature and Biologically Inspired Computing (NaBIC), Third World Congress, IEEE, (pp. 207-212), 2011.
45. Li, Weili, Qiaoyu Yin, and Xiaochen Zhang, "Continuous quantum ant colony optimization and its application to optimization and analysis of induction motor structure." Bio-Inspired Computing: Theories and Applications (BIC-
TA), Fifth International Conference, IEEE, (pp. 313-317), 2010.
46. Shanmugam, Latha, and Lilly Florence, "A Comparison of Parameter best Estimation Method for software reliability models."International Journal of Software Engineering & Applications, Vol. 3, No. 5, pp. 91, 2012.
47.Shanmugam, Latha, and Lilly Florence, "Enhancement and comparison of ant colony optimization for software reliability models." Journal of Computer Science, Vol. 9, No. 9, pp. 1232, 2013.
48. Kirkpatrick S, Gelatt C, Vecchi M "Optimization by simulated annealing." Science, Vol.220, pp. 498– 516, 1983.
49. Pai PF, Hong WC "Software reliability forecasting by Support vector machines with simulated vector machines with simulated annealing algorithms." The Journal of Systems and Software, Vol.79, pp. 747–755, 2006.
50. AL-Saati, Dr, NajlaAkram, and Marwa Abed-AlKareem, "The Use of Cuckoo Search in Estimatingthe Parameters of Software Reliability Growth Models." 2013.

# FEATURE EXTRACTION IN DENTAL RADIOGRAPHS IN HUMAN EXTRACTED AND PERMANENT DENTITION

Kanika Lakhani[1], Bhawna Minocha[2], Neeraj Gugnani[3]

[1] Research Scholar, Amity University, Noida, kanikalakhani@yahoo.co.in
[2] Assistant Professor, Amity University, Noida, bminocha@amity.edu
[3] Professor, DAV Dental College , Yamunanagar, drgugnani@gmail.com

**Abstract.** Feature extraction in dental images in the form of radiographs involves the identification of major defect areas. While analyzing complex radiograph images, one of the major problems stems from the types of defects present. Analysis with a large number of defects present generally requires a large amount of memory and computational power. Feature extraction applied over the radiographs, once the edge detection process is accomplished, derives combinations of the defects to get around the problems while still describing the problem areas with sufficient accuracy. The process has been implemented over a set of 20 of extracted human dentition for the identification of similar features to actualize the presence of defects in the dentition.

**Keywords:** Feature Extraction, Edge Detection, Dental radiographs, SOM

## 1 Introduction

Dental radiograph is an image that is recorded using X-ray radiation. Dental radiographs are actually X-ray radiation consisting of teeth, tooth-bones and soft tissues surrounding the oral cavity. Feature extraction, collection and analysis for dental clinical diagnostics is the chief requirement nowadays for dental science. In this realm of dental image analysis, most of the research done is crucial for the purpose of human biometric identification. Moving further with the realm is the diagnosis of dental diseases from radiographs that eases the job of a dentist.

Different branches in dentistry deal with different types of dental problems like Tooth Decay, Mouth Sores, Tooth Erosion, etc. Some of these problems have common symptoms that can be diagnosed easily. The idea is to detect these problems and be aware of the true picture before undergoing appropriate treatment. Furthermore, this can be used as a preliminary diagnostic aid by the dentists. The process to be followed has been depicted in the figure 1 below.

---

1



J.M. Corchado Rodriguez et al. (eds.), *Intelligent Systems Technologies and Applications 2016*, Advances in Intelligent Systems and Computing 530,
DOI 10.1007/978-3-319-47952-1_41

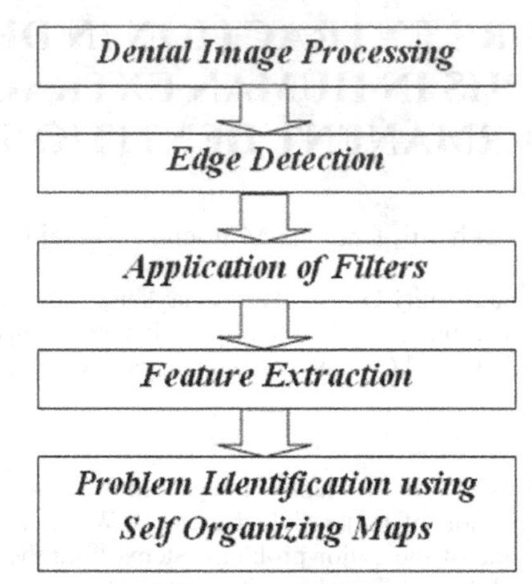


**Fig 1**. Steps for the identification of problems related to Permanent Dentition in Human Beings

The images retained from radiographic devices produce a relatively blurred image. A physician's or dentist's involvement in the diagnosis is of attern importance. It is not possible to detect fine details with ease in such radiographs. Latest technology with high contrast radiograph readers is available but that sounds heavily on a physician's pocket. This paper discusses an alternate technique for the same. It includes the processing of radiographs for identifying the exact location & depth of damage in affected tooth..

Edge detection process identifies and locates the lack of continuity, inequalities and varied orientations in an image. [11]This discontinuity describes the sudden changes in the pixel intensity. The discontinuities that occur in image intensity can either be Step edge, or Line edge. [8]These discontinuities are rare in real images because instant changes rarely occur.

The paper has been organized as follows. Section 2 discusses edge extraction using various edge extraction techniques. Section 3 shows the comparison of various techniques. Section 4 focuses on the observations & findings. Finally, the paper concludes with describing the scope for future work.

## 2 Conventional Edge Extraction Operators in Dental x-ray Images

**Sobel Operator:** Most of the edge detection method are based on the assumption that edges are found in the image where there is discontinuity. Based on this assumption the derivative is taken for image intensity value and the points are located where intensity derivatives have maximum value so as to locate the edges [4].

**Prewitt Operator:** [12] It is computationally less expensive and faster method for edge detection. It is only appropriate for noiseless and well contrasted images [6]. Prewitt approximation is applied on the derivatives of intensity function. Its results in edges where gradient of intensity function has maximum value. Prewitt operator detect two types of images: horizontal edges and vertical edges. The difference between the corresponding pixel intensities of the image results in edges. Derivative masks are used for edge detection technique. Prewitt operator generates two masks, one for detecting edges in horizontal direction and other for vertical direction.

**Canny Edge Extraction Algorithm for Dental X-Ray Images:** Canny edge extraction process is a multi stage algorithm that is popularly known as the optimal edge detector. The regions are represented by the local maxima which are marked as the edges in the gradient image. A non-maximal suppression is used to find the local maximum points in the gradient edge map. The weak edge areas are suppressed by double thresholding. The algorithm produces over segmented images from which none of the root features can be identified. Generally, the Canny edge extraction algorithm is assumed to provide optimum results, but is proving over segmented edge maps in the case of these dental x-ray images which have illumination variations, noise and different gradient angles.

## 3 Experimental Analysis of Edge Detection Techniques

The above techniques were implemented on a sample space of 20 teeth. The radiographs were taken for a set of 20 extracted teeth. Further, the radiographs were converted to jpeg format and the number of black and white pixels was calculated using Sobel operator, Prewitt Operator and Canny edge detection technique. Table1 represents the difference between the numbers of black and white pixels of each tooth in the sample space.

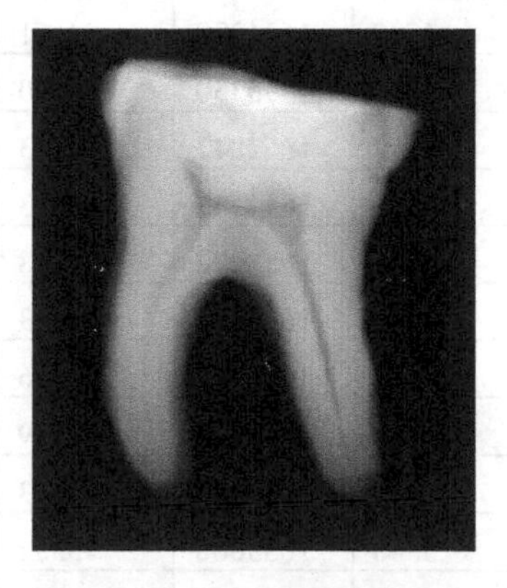

**Fig2.** Tooth No.6 Radiograph Image converted to jpeg format

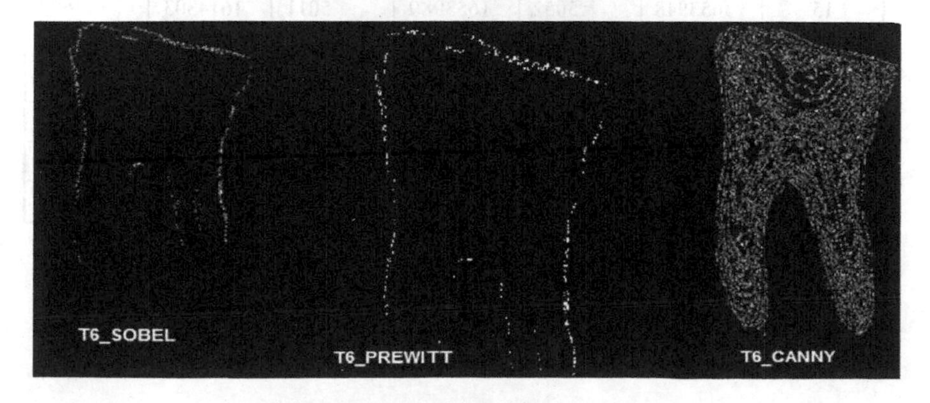


**Fig 3.** Comparison of Various Techniques on Tooth No.6

Figure 2 depicts the radiograph of tooth No. 6 selcted from the data set. Figure 3 shows the comparison of digital X-ray image of Tooth Sample No.6 converted using Sobel , Prewitt and Canny Techniques. When 3D plotted, as in Figure 4 and Figure 5, the curves on the graph were look-alike with not much difference. Thus, there occurs a need to furthur smoothen the images to get much clearer view of the image. Table 1 shows the count of pixels, i.e. number of black and white pixels present in the dental images after the application of various edge detection techniques.

Table1. Comparison of number of Pixels

| Tooth No | Sobel | | Prewitt | | Canny | |
|---|---|---|---|---|---|---|
| (.jpg) | no. of black pixels | no. of white pixels | no. of black pixels | no. of white pixels | no. of black pixels | no. of white pixels |
| 1 | 1654642 | 4358 | 1654643 | 4357 | 1602577 | 56423 |
| 2 | 1652873 | 6127 | 1652931 | 6069 | 1592284 | 66716 |
| 3 | 1655807 | 3193 | 1655805 | 3195 | 1599029 | 59971 |
| 4 | 1654293 | 4707 | 1654329 | 4671 | 1590427 | 68573 |
| 5 | 1653300 | 5700 | 1653350 | 5650 | 1607227 | 51773 |
| 6 | 1652825 | 6175 | 1652847 | 6153 | 1600715 | 58285 |
| 7 | 1654162 | 4838 | 1654147 | 4853 | 1618285 | 40715 |
| 8 | 1655245 | 3755 | 1655272 | 3728 | 1600768 | 582232 |
| 9 | 1655550 | 3450 | 1655557 | 3443 | 1597341 | 61659 |
| 10 | 1654731 | 4269 | 1654734 | 4266 | 1602566 | 56434 |
| 11 | 1654290 | 4710 | 1654271 | 4729 | 1593289 | 65711 |
| 12 | 1653874 | 5126 | 1653907 | 5093 | 1598133 | 60867 |
| 13 | 1653652 | 5348 | 1653668 | 5332 | 1621309 | 37691 |
| 14 | 1654329 | 4671 | 1654359 | 4641 | 1614303 | 44697 |
| 15 | 1653948 | 5052 | 1653989 | 5011 | 1614303 | 44697 |
| 16 | 1654502 | 4498 | 1654515 | 4485 | 1609869 | 49131 |
| 17 | 1653824 | 5176 | 1653793 | 5207 | 1623816 | 35184 |
| 18 | 1654397 | 4603 | 1654433 | 4567 | 1625303 | 33697 |
| 19 | 1652901 | 6099 | 1652917 | 6083 | 1621286 | 37714 |
| 20 | 1653490 | 5510 | 1653605 | 5395 | 1615613 | 43387 |

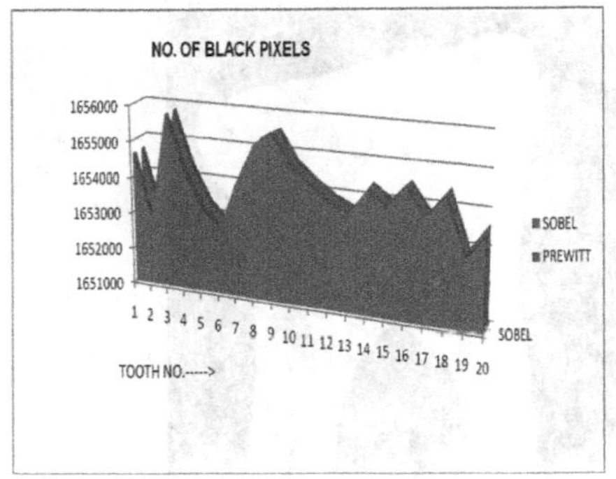


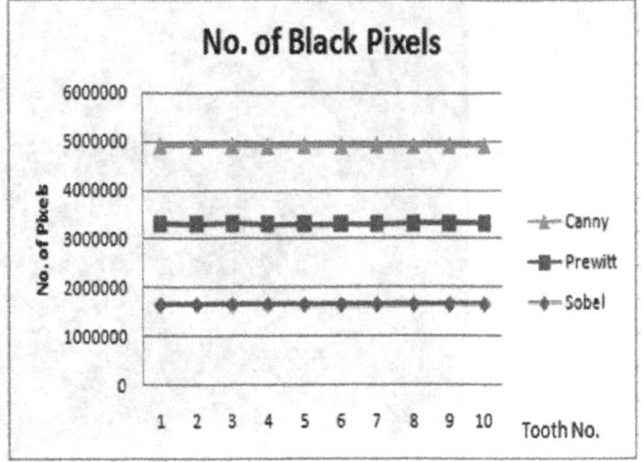


**Fig 4.** Comparative Graph of number of Black Pixels

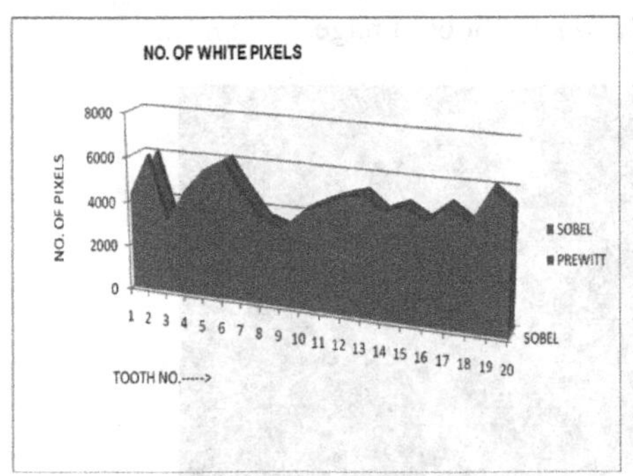


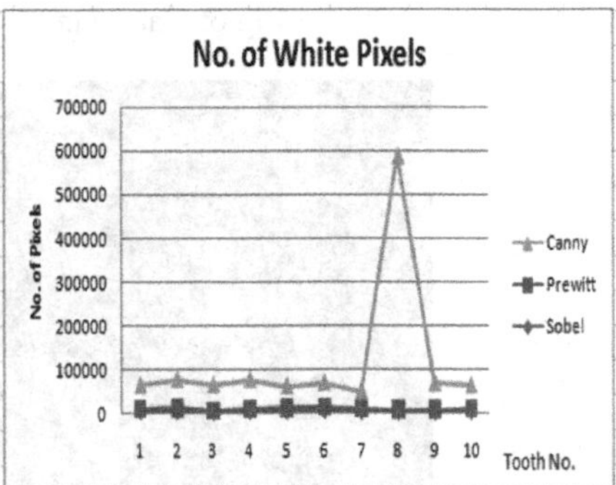


**Fig 5.** Comparative Graph of number of White Pixels

## 4 Implementation of Gaussian & Laplacian Filter on Dental Images

After converting the dental radiograph to Sobel and Prewitt images, somewhat similar results were obtained thereby generating the need for further application of filters for obtaining much clear view of the visual image.

When the number of white and black pixels were compared in both Sobel and Prewitt techniques, it was found that Sobel technique gives much better results than Prewitt images, though the difference was not much. Based on the assumption that a Sobel image may provide better results, Gaussian filter was applied on the Sobel image. Figure 4 shows the effect of Gussian filter on Tooth Sample No.6 converted to Sobel image. It seems that it would be much easier to implement feature extraction on the Gaussian image so formed.

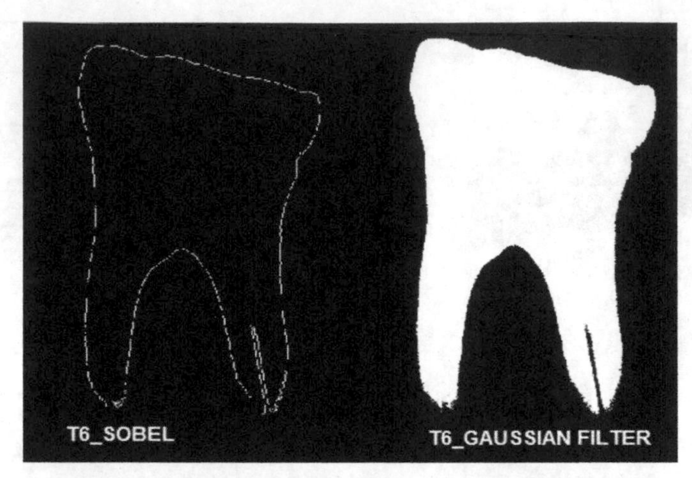


Fig 6. Gaussian Filter on T6_Sobel Image

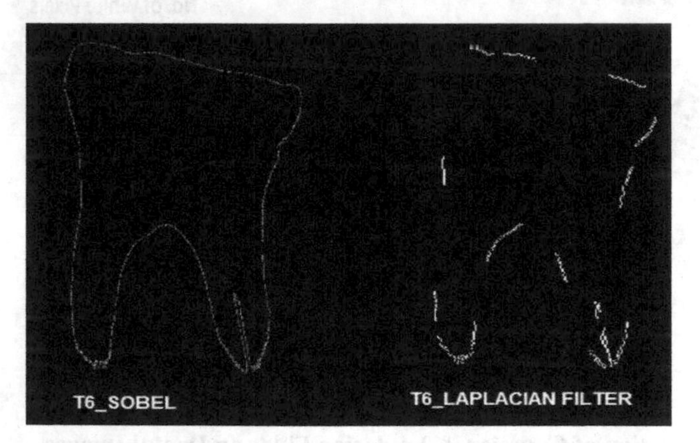


Fig7. Laplacian Filter applied on T6_Sobel Image

Since much better results were obtained when a smoothening filter, i.e. Gaussian filter was implemented on the Sobel converted radiograph, there may occur a need for finding the defects on the edges of the tooth than the surface. For this, the image needs to be sharpen. Laplacian filter was used on the Sobel image for further sharpening the image for the identification of defective edge in dental sample. Figure 5 depicts the effect of Laplacian filter on Sobel image.

## 5 Observation & Findings

From the above results, the challenges posed by various image processing edge extraction techniques on non-uniform dental images can be evaluated. Sobel and the Prewitt operators provide edges which are very nearer. In this paper, the issues obtained in using different image processing edge extraction methodologies on misaligned dental x-ray images have been discussed. The conventional edge extraction techniques ,i.e. Sobel and Prewitt, mentioned in this paper seems to be inadequate for successfully obtaining the edge features from dental x-ray images. When Gaussian filter was applied on the Sobel image, much clearer image was formed and thus feature extraction can now easily be applied over the Gaussian image so formed which can easily be compared with the normal tooth image for the identification of actual problem in the tooth. Also, It was found that Smoothening & Sharpening of images have a greater effect in disease diagnosis in medical images.

It can be concluded from the implemented data set that the features extracted, as in figure 6, from the proposed technique can describe the exact shape and the accuracy in classifying the dental image. The effect of pretreatment of the image so produced is not idyllic and the proposed technique is still relatively lacking perfection that still leaves a scope for further improvement.

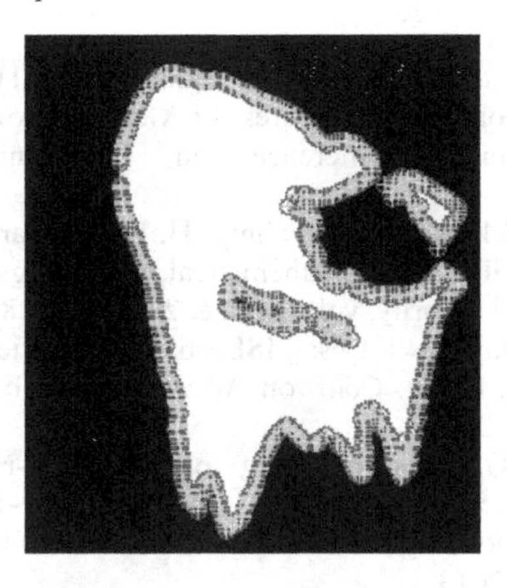

**Fig 6.** Feature extraction on Tooth No.6

## 6 Conclusion & Future Work

From the above results, the challenges posed by various image processing edge extraction techniques on non-uniform dental images can be evaluated. Sobel and the Prewitt operators provide edges which are not complete and sufficient. The Canny algorithm, known as the optimal edge extraction algorithm is producing over segmented edge map containing edges of unwanted noise areas also. Thus, a better algorithm which can make use of the linear character of the edges must be developed.

The issues obtained in this paper using different image processing edge extraction methodologies on misaligned dental x-ray images have been discussed. The conventional edge extraction techniques mentioned in this paper seems to be inadequate for successfully obtaining the edge features from dental x-ray images. Thus, study necessitates the emergence of an improved edge extraction algorithm over dental x-ray images.

Future work includes the implementation of Self Organizing Maps for the identification of problems related to permanent dentition. The extracted features in the radiographs shall be grouped together to form clusters and thus the classification of diseases shall be done.

## REFERENCES

[1] D. K. Sharma, Loveleen Gaur, Daniel Okunbor, "Image Compression and Feature Extraction with Neural Network", Proceedings of Academy the of Information and Management Sciences, 2007, Volume 11, No. 1, pp 33 - 37

[2] M.Sushma Sri, M. Narayana, Edge detection by using Look-up table. IJRET: International Journal of Research in Engineering and Technology, eISSN: 2319-1163 | pISSN: 2321-7308, Volume: 02 Issue: 10 | Oct-2013.

[3] S. Oprea, C. Marinescu , I. LiŇă, M. Jurianu, D. A. Visan, I. B. Cioc, " Image Processing Techniques used for Dental X-Ray Image Analysis" , Electronics Technology, ISSE 2008, pp 125-129.

[4] Azam Amini Harandi, Hossein Pourghassem[2011], "A Semi Automatic Algorithm Based on Morphology Features for Measuring of Root Canal Length", in Pro IEEE International Conference on Communication Software and Networks(ICCSN)

[5] E. H. Said, D. E. M. Nassar, G. Fahmy, H. H. Ammar. "Teeth segmentation in digitized dental X-ray films using mathematical morphology," IEEE Transactions on information forensic and security, vol. 1, Issue. 2, pp. 178-189, June. 2006.

[6] A J Solanki , K R Jain , N P Desai, ISEF based Identification of Dental Caries in Decayed Tooth, , Proc. of Int. Conf. on Advances in Information Technology and Mobile Communication 2013

[7] A. K . .lain, H. chen,'matching of dental x-ray images for human identification,Pattern Recognitin, vol. 37, Issue. 7, pp. 1519-1532

[8] Hema, Edge Extraction Algorithm using Linear Prediction Model on Dental X-ray Images, IJCA, Vol 100 2013

[9] Ziad M. Abood Edges Enhancement of Medical Color Images Using Add Images IOSR Journal of Research & Method in Education (IOSR-JRME) e-ISSN: 2320–7388,p-ISSN: 2320–737X Volume 2, Issue 4 (Jul. –Aug. 2013), PP 52-60

[10] Anita Patel, Pritesh Patel Analysis of Dental Image Processing IJERT Vol1, Issue 10

[11]Rohit Thanki, Deven Trivedi," Introduction of novel tooth for human identification based on dental image matching", International Journal of Emerging Technology and Advanced Engineering, ISSN 2250-2459, Volume 2, Issue 10, October 2012

[12] Huiqi Li, and Opas Chutatape, *"Automated Feature Extraction in Color Retinal Images by a Model Besed Approach"*, IEEE Transaction on Biomedical Engineering 2004, **51** (2), 246-254

# Multilayered Presentation Architecture in Intelligent eLearning Systems

**Uma Gopalakrishnan, Ramkumar N, P. Venkat Rangan and Balaji Hariharan**[1]

**Abstract** eLearning systems have become essential in enabling lectures from experts accessible to masses spread across distant geographic location. The digital representation of different components in the instructor's environment, such as the presentation screen and the instructor's video, in a remote location is often spatially disconnected. This results in missing out spatially related gestural cues such as pointing at the screen by the instructor. This paper discusses a method of creating a spatially connected, close to natural view of the instructor's environment by overlaying the instructor's video defined by his contour over the presentation screen content. This technique involves a calibration step required to match the camera perspectives of the different video components. We present a real-time and robust calibration technique allowing for automatic recalibration in a dynamic classroom scenario which involves changes in camera position and pan, tilt and zoom parameters.

**Keywords** – eLearning, Calibration, Feature-extraction, warping, homography

## 1 Introduction

Elearning systems are becoming increasingly popular in making expert lectures available to the masses. Most of the live eLearning solutions and video conferencing solutions such as AVIEW, Cisco WebEx, Skype, Elluminate etc.[3][4][5], give a representation of all the essential teaching elements such as instructor's video, teaching material in digital form, instructor's audio varying according to their interface design. While lecturing, the instructor makes use of gestural cues such as pointing to the teaching material. This gestural information can be lost in the remote classroom where the instructor is not physically present and the individual elements of the teaching environment, the instructor's video and the presen-

[1] Amrita Center for Wireless Networks and Applications, Amrita School of Engineering, Amritapuri Campus, Amrita Vishwa Vidyapeetham University, e-mail: {umag, ramkumar}@ am.amrita.edu, venkat@amrita.edu, balajih@am.amrita.edu



J.M. Corchado Rodriguez et al. (eds.), *Intelligent Systems Technologies and Applications 2016*, Advances in Intelligent Systems and Computing 530,
DOI 10.1007/978-3-319-47952-1_42

tation screen, are represented as spatially separate and uncorrelated digital video elements.

Our solution deals with making the viewing experience more natural by extracting the instructor's contour and overlaying on the contents of the presentation screen (screencast video) at the correct place preserving the spatial correlation. This creates a multilayered presentation at the remote location. In the local classroom, where the instructor is physically present, the instructor's video and the screencast are captured in different camera perspective. The instructor's body referred to as instructor contour in this paper is extracted from his/her video by eliminating the background. The instructor contour is then changed to match the perspective of the screencast. This spatial matching involves a calibration step which computes a 3x3 matrix transformation matrix used to warp the instructor contour to match the screencast perspective. This is done by displaying a predefined calibration image on the screen and finding points of matches in the images resulting from both the perspective.

Previous work in this area [1] used the SIFT algorithm [2] in this step to generate the transformation matrix, also known as the homography matrix. In this paper, we present improvements to the calibration method, making it robust and real-time using ArUco marker detection algorithm. ArUco marker detection is used to generate robust keypoints which are used to generate the homography matrix. In a dynamic classroom environment, the camera position and camera parameters such as pan, tilt and zoom need to be adjusted during a lecture session to capture a good view to be transmitted to the remote students. Whenever any such adjustments are done, the transformation matrix needs to be recalibrated in real-time without causing any interruptions to the lecture session. The performance analysis shows the auto recalibration is made real-time with the ArUco marker detection algorithm. We further analyse conditions to make the calibration step robust to changes in lighting conditions, camera distance and different levels of occlusion to the calibration image. Using ArUco markers improved the performance in these cases.

Rest of the paper is organized as follows. Section 2 describes the related work used in this paper. Section 3 explains system architechture. Section 4 describes the calibration step in detail using SIFT and ArUco. Section 5 presents the performance evaluation and a comparison of the two calibration methods and Section 6 gives a conclusion.

## 2 Related Work

The paper deals with enhancing the view a remote student gets of the teacher's environment. It is based on a previous work[1], which deals with instructor's contour extraction in real time using Microsoft Kinect's depth and colour sensors along with its libraries OpenNI2 and NITE for user map detection and mapping to the colour frame [6]. The method further heavily made use of homography matrix estimated from SIFT[2] feature matches. The SIFT algorithm, developed by David G. Lowe, describes detection of feature points referred to as keypoints in an image and computes a descriptor vector for each keypoint which is a list of 128 numbers describing the region around the keypoint. These descriptors are compared to find matches between two images.

The homography estimation was a initial calibration step after which the kinect position and another other parameters like pan, tilt and zoom need to be remain unchanged as real-time recalibration was not possible. This paper improves in the performance of the contour extraction and overlay method by using ArUco marker detection algorithm [7]. ArUco markers are predefined images used for camera calibration and marker detection in many real time augmented reality applications. In this paper we use this technique to find feature for robust matches in real-time.

## 3 System Architecture

To solve the spatial disconnect between the instructor's video and the screencast in the remote classroom, the instructor contour extraction and overlay method can be utilized. The instructor's video captures the instructor along with the background in the camera's view. From this video the instructor's body defined by the contour is detected and the background details removed such that only the instructor is seen in the video (referred to as the Instructor contour in this paper) which is then overlaid on the screencast at spatially correlated position. The perspective of the instructor's contour video in this process needs to be changed to match the screencast perspective. The following steps explain this process in detail and are illustrated in Fig. 1. And the resulting view in the remote location is seen in Fig. 2.

1. Calibration of Homography Matrix – Image registration

The instructor's video with the background details including the presentation screen is captured with a camera perspective other than the screencast video's perspective. To overlay the instructor's contour extracted from his/her video the mapping between both the perspectives is first calibrated. A predefined image is projected on the screen to compare the frames from both the videos and compute a 3x3 matrix called the homography matrix which defines the transition from one perspective to another. Section 3 explains this calibration process in more detail and describes the improved technique adopted for a real time elearning classroom scenario. Whenever there is a change in the perspectives which can occur due to changes camera position or camera parameters such as pan, tilt and zoom, the homography matrix needs to be calculated again. Real-time automatic recalibration is done every 't' seconds where t can be a user defined parameter which depends on the dynamic nature of the classroom.

2. Instructor contour extraction

The instructor's video is captured using Microsoft Kinect's camera. The depth sensors are used to identify the instructor mask. Open source libraries OpenNi2 and NiTe are used to extract the instructor mask which is compared to the color images to obtain the instructor's contour extracted frames.

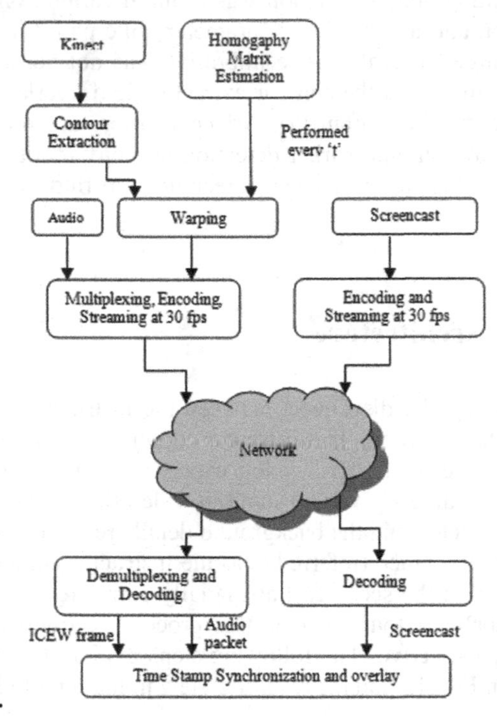


**Fig. 1** Flow chart.

3. Perspective mapping of the Instructor contour extracted frame

The instructor contour extracted (ICE) frame is warped by multiplying with the homography matrix computed during calibration. OpenCV libraries are used to perform this computation. This transforms the ICE frame to match the screencast frame's perspective. The resulting instructor contour extracted and warped (ICEW) frame can be directly overlaid on the screencast frame.

4. Encoding and transmitting

The high definition screencast frames are encoded and transmitted to the remote classroom at 10 times lower rate (3 fps) than the ICEW frames (30 fps) to obtain real time transmission without loss of quality. In our experiments this ratio good overlay results without loss of clarity in the screen content.

5. Decode and overlay

The ICEW frames and screencast frames are decoded at the remote classroom side and the ICEW frames are overlaid at a rate of 10:1 to maintaining the time synchronization.

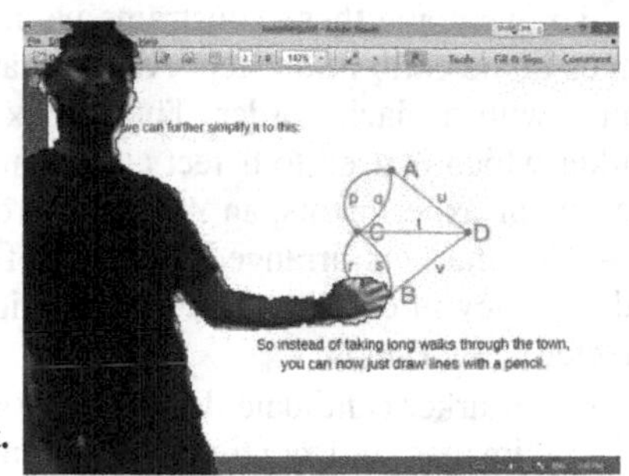


**Fig. 2** Instructor contour overlay on Screencast.

# 4 Calibration of Homography matrix

In the calibration process, a predefined image is selected and displayed on the screen. Our previous work in this area used the SIFT algorithm[2] to extract the keypoints from the images obtained from screencast and the kinect camera. The algorithm then computes and associates a descriptor vector with each of the key-points from both the image which describes a small neighborhood region around the keypoint. Matches between the keypoints from the kinect frame and the screencast frame are found by correlating the descriptor vectors.

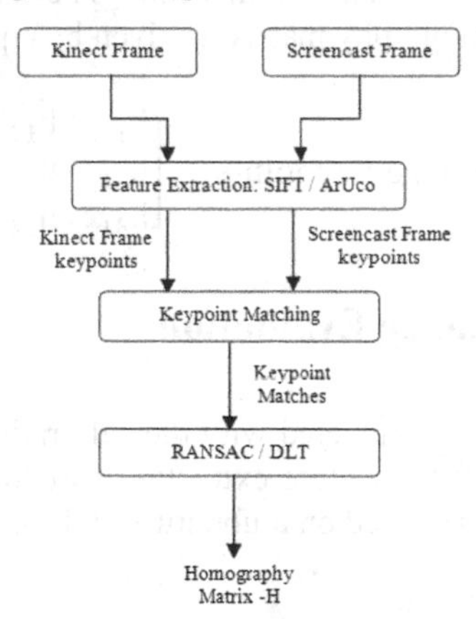


**Fig. 3.** Homography Estimation

In this method, once the initial calibration is done, camera cannot be moved afterwards, since the calibration is computation intensive and cannot be done real-time on the fly. Moreover, the image used for calibration fills the entire screen and the instructor needs to move out of the view of screen to detect sufficient number of keypoints. Occlusion results in errors. These factors can interrupt the flow of a classroom session. In addition, the kinect camera needs to be kept close to screen, limiting the area captured by the camera and proper lighting condition needs to be set such that all details are sufficiently captured.

To overcome these constrains we made use of ArUco marker detection techniques in the calibration step. ArUco markers are square markers made of a binary matrix with a black border. The matrix associates an identifier number with the marker which is used to detect and identify the presence of this marker in any image. In our experiments, an ArUco board is displayed on the screen, which is a set of ArUco markers arranged in a predefined manner. This is done for robustness and accuracy in conditions of illumination change and occlusion. The markers are detected in two steps:

- Marker candidate detection: A thresholding is performed on the grayscale images and contour extraction is done to identify approximate square shapes.
- Marker Identification: The inner regions of the selected contours are analysed for presence of valid markers belonging to a predefined dictionary. Here, a perspective transform is first done correct perspective changes brought by the camera and otsu's thresholding is applied on the resulting marker to split black and white bits.

The ArUco markers resulted in robust and real-time calibration of Homography matrix. The marker detection was possible in presence of large occlusion and changes in illumination conditions. During a classroom session, when the camera position or parameters such as pan, tilt and zoom needs to be adjusted, the recalibration of new homography matrix can be done on the fly. The next section elaborates these improvements in detail. The matching image points are used to estimate the homography matrix as given by equation 1.

$$\text{Screencast Image Keypoints} = \begin{bmatrix} h_{11} & h_{12} & h_{13} \\ h_{21} & h_{22} & h_{23} \\ h_{31} & h_{32} & h_{33} \end{bmatrix} * \text{Kinect Image Keypoints} \quad (1$$

# 5 Performance Evaluation

The tests were conducted with the calibration process implemented using openCV libraries for SIFT feature extraction and ArUco marker detection. The implementation was performed on a ubuntu 14.04 system with i7 processor and 8GB RAM.

## *5.1 Processing time:*

SIFT generated about 180 keypoint matches from the high detailed image. The keypoint generation in both the images and their descriptor matching took 1.42 seconds. In our scenario, this limits the calibration process to be an initial step before the start of a class. The camera position and parameters cannot be changed as the recalibration would disrupt the flow of the class. ArUco marker detection built for real-time application, took 11 ms to generate the matching marker keypoints from both the images, making real-time recalibration possible.

## *5.2 Robustness*

The correctness of the resulting matches and homography matrix were tested under three conditions:

1. Occlusion :

The marker images were partly occluded by the presence of other content on the screen or the instructor in front of the screen.

2. Large Camera Distance:

The Kinect was at first positioned with a minimum distance of 2 meters from the screen. This was minimum distance sufficient to capture the entire screen and the instructor besides the screen. The distance was increased to a larger distance to capture a wider view.

3. Lighting

The classroom environment tested with had diffused lights over the screen, in front and on sides. Different lighting scenarios were created the matches from both the algorithm tested. The worst lighting condition was when the lights above the screen were put on blurring the image details.

With all the above three conditions, Aruco marker worked with greater accuracy as the algorithm finds the definite marker corners without any ambiguity, as long as a marker square and its code is identified. In each scenario, the same image points are correctly extracted from both test images, without any false positives. When all the squares in the aruco marker board are identifiable in the algorithm, a correct match between all the 40 corners (from 8 squares in our test) can be detected. In our setup, correct matches were obtained with an occlusion of 80% of the marker board, a large camera distance of 4.6 meters and also the poorest lighting condition with lights above the screen on.

SIFT image calibration on the other hand worked with varying correctness in the presence of these undesirable external factors. A random high detail image gives a high number of correct matches between the images in optimal condition and progressively generates more false positives as the environment conditions are changed which lower the clarity of the image captured from kinect.

The SIFT algorithm when tested with the ArUco marker image as the calibration image, gave better matches with much reduced number of false positives, and generated enough matches for correct homography estimate in the varying environmental conditions. These results can be seen in Fig 4, 5 and 6. The figures show the correct matches resulting from ArUco marker detection. SIFT yielded approximately 180 matches using the high detailed image. A random 30 matches are shown in the figures. Under adverse external conditions we can see wrong matches which resulted in wrong homography matrix estimation using a random image(Fig 4b, 5b, 6b). Figures 4c, 5c, 6c shows the performance of SIFT with the ArUco marker image. Here lesser false positives are seen and resulted in correct homography estimation.

**Fig. 4.** Match at occlusion a)ArUco b)SIFT – random image c) SIFT – AruCo marker image

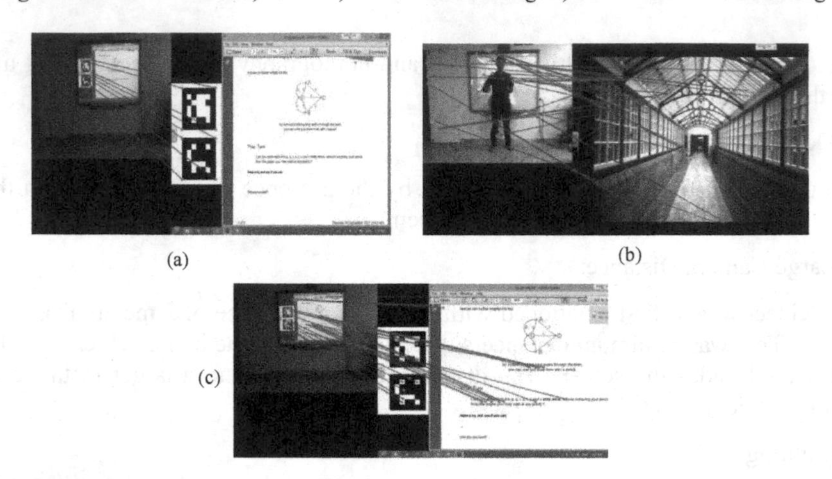

**Fig. 5.** Match at large distance a)ArUco b)SIFT – random image c) SIFT – AruCo marker image

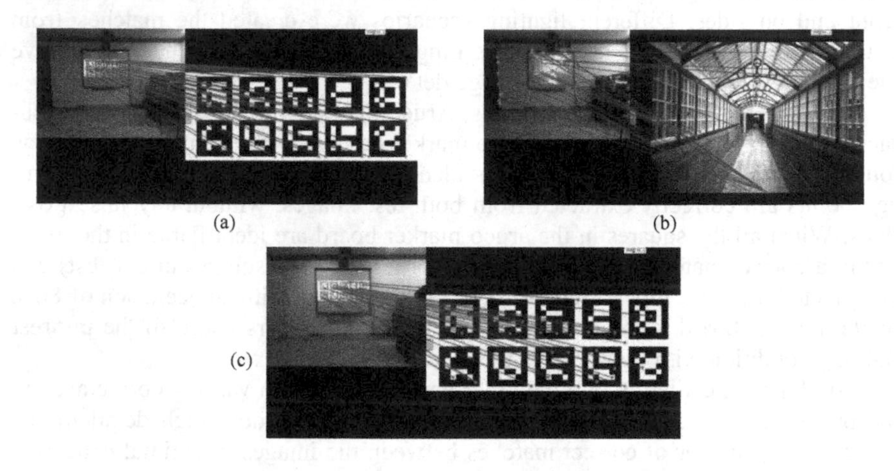

**Fig. 6.** Match at poor lighting a)ArUco b)SIFT – random image c) SIFT – AruCo marker image

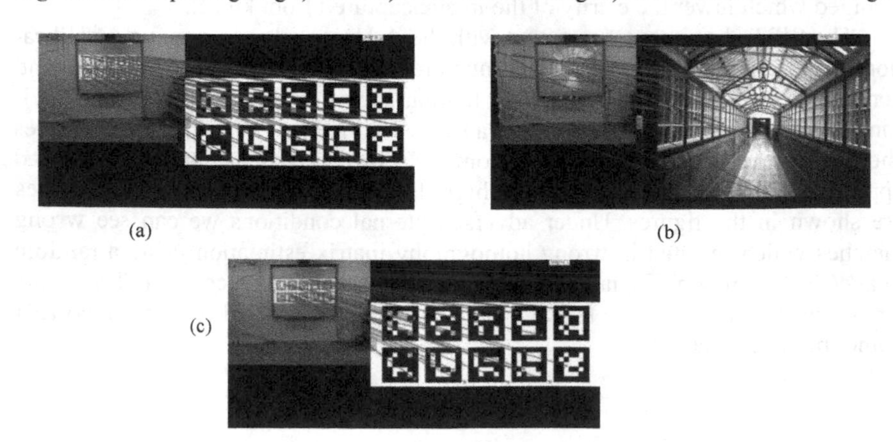

# 6 Conclusion

We discussed a method to represent the teaching environment in a close to natural way in a remote classroom in eLearning. This involved extracting the instructor's contour from his/her video and overlaying it on the screencast at a spatially correlated place. We presented a technique for real-time and robust calibration of images from the instructor's video and the screencast using ArUco marker detection algorithm. The real-time calibration enabled recalibration during an ongoing lecture with minimal intrusion, in the event of changes in camera position and camera parameters pan, tilt and zoom. The calibration step was made robust to changes in lighting conditions, camera distance from the screen and occlusion to the marker image.

# References

1. Ramkumar N, P. Venkat Rangan, G. Uma, B. Hariharan, "Instructor Contour Extraction and Overlay for Near Real Presence In E-Learning Systems", International Conference on Computer supported Education (2016)
2. Lowe, David G. "Distinctive image features from scale-invariant keypoints." International journal of computer vision 60.2 (2004): 91-110.
3. Bijlani, Kamal, Jayahari KR, and Ancy Mathew. "A-view: real-time collaborative multimedia e-learning." Proceedings of the third international ACM workshop on Multimedia technologies for distance learning. ACM, 2011.
4. WebEx, Cisco. "Cisco WebEx Meetings." Luettavissa: http://www. webex. com/products/web-conferencing. html. Luettu 18 (2014): 2014.
5. Karabulut, Aliye, and Ana Correia. "Skype, Elluminate, Adobe Connect, Ivisit: A comparison of web-based video conferencing systems for learning and teaching." Society for information technology & teacher education international conference. Vol. 2008. No. 1. 2008.
6. Cruz, Leandro, Djalma Lucio, and Luiz Velho. "Kinect and rgbd images: Challenges and applications." Graphics, Patterns and Images Tutorials (SIBGRAPI-T), 2012 25th SIBGRAPI Conference on. IEEE, 2012.
7. Garrido-Jurado, S., et al. "Automatic generation and detection of highly reliable fiducial markers under occlusion." Pattern Recognition 47.6 (2014): 2280-2292.

# 7 Conclusion

We [illegible] a method [illegible] the [illegible] [illegible] of natural [illegible] [illegible] [illegible] [illegible] the [illegible] [illegible] [illegible] [illegible] [illegible] [illegible] [illegible] [illegible] [illegible] and [illegible] calibration of images [illegible] [illegible] [illegible] [illegible] detection [illegible] [illegible] [illegible] [illegible] camera parameter [illegible] [illegible] [illegible] [illegible] of changes of lighting conditions [illegible] [illegible] [illegible] into the [illegible] image.

## References


1. [illegible] Computer Vision [illegible] (2016)
2. [illegible] (2004) [illegible]
3. [illegible] [illegible]
4. [illegible] [illegible]
5. [illegible] Society for [illegible] [illegible] 2005, No. 1 2005
6. [illegible] images: Challenges and [illegible] [illegible]
7. [illegible]

# Inverse Prediction of Critical Parameters in Orthogonal Cutting using Binary Genetic Algorithm

Ranjan Das[1]

**Abstract.** An inverse problem is solved for concurrently assessing the rake angle, the chip thickness ratio and the required cutting width in an orthogonal cutting tool, when subjected to a prescribed force constraint. The force components which can be obtained experimentally by mounting either suitable dynamometers or force transducers on a machine tool, are calculated here by solving a forward problem. Due to inherent complexities involved in the calculations of the gradients, genetic algorithm-based evolutionary optimization algorithm is used in the present study. The results of the inverse problem have been compared with those of the forward problem. It is observed that a good estimation of the unknowns is possible. The current study is projected to be of use to decide on the relevant cutting tool parameters and adjusting the cutting process in such a manner that the cutting tool works within the dynamic limits.

**Keywords:** orthogonal cutting, inverse problem, genetic algorithm, force.

## 1 Introduction

Orthogonal cutting process is one of the most common phenomena observed in any manufacturing processes in which the tool cutting edge remains perpendicular to the cutting velocity. Forces arising during the cutting process can be broadly classified as cutting force and shear force. These forces depend upon many factors such as the chip-thickness ratio, rake angle of the tool, friction coefficient at the chip-tool interface and the cutting width [1]. From the knowledge of these parameters, forces

---

[1] Ranjan Das
Department of Mechanical Engineering, IIT Ropar, Rupnagar, Punjab, 140001, India. E-mail: ranjandas@iitrpr.ac.in



J.M. Corchado Rodriguez et al. (eds.), *Intelligent Systems Technologies and Applications 2016*, Advances in Intelligent Systems and Computing 530,
DOI 10.1007/978-3-319-47952-1_43

can be experimentally calculated using suitable dynamometers, force transducers and from the amount of power consumption. Alternatively, they can also be calculated from suitable computational techniques. Problems of such kind in which the final objective is to obtain the final effect (which can be various forces) from the responsible causes (which can be the known parameters) are known as the direct (forward problems). On the other hand, there exists another category of problems where the final consequence is only known, but many parameters responsible for the observed/required effect are not known. The task is then to estimate the unknowns and the case thus becomes an inverse problem, which is mathematically ill-posed [2, 3]. The solution of inverse problems invariably requires either regularization or an optimization method.

Many studies dealing with forward problems demonstrating different methods to calculate forces in cutting operation. Some of these include, the strain-gauge type dynamometer [4], cutting force model [5], the finite element method (FEM) [6] and many other computational techniques [7]. In the area of manufacturing sciences, some studies dealing with inverse problems are also available. For instance, Huang and Lo [8] used steepest descent method to predict the surface heat flux satisfying a given temperature profile in a cutting tool. In orthogonal cutting operation, using the temperature data, the heat flux on the tool along with the heat transfer coefficient were estimated by Yvonnet et al. [9]. The objective function was defined by the calculated and the experimental temperature histories and the error was regularized using the Newton-Raphson method. For estimating the friction coefficient on integrated chip dam-bar trimming process, the Levenberg-Marquardt algorithm of optimization in combination with the FEM was used by Lin and Chen [10]. Recently, Luchesi et al. [11] have estimated the moving heat source in a machining process using the conjugate gradient method. Using an experimental setup, they have also shown the procedure of obtaining various temperatures and forces.

On the basis of above discussion, it is realized that so far the application of inverse problems in machining processes is concerned, most studies deal with the estimation of either one or two parameters only. Therefore, the present work is aimed that estimating three critical parameters for satisfying a particular force distribution in an orthogonal cutting process. The unknown parameters which have been estimated for meeting the objectives are the rake angle, the chip thickness ratio and the required width of the cut. The required force field has been calculated from a forward problem using various recognized values of the three parameters. Next, in the inverse optimization analysis, just the force field is considered known, and three parameters are assumed to be unknown, which have been simultaneously predicted to satisfy the force field. Although gradient-based techniques converge fast than the evolutionary optimization methods, but the latter are preferred when the objective function is complex, discontinuous and involve more variables. In the domain of inverse problems, currently the evolutionary principle-based algorithms are getting significant notice [12, 13]. On the basis of the above-mentioned argument and due to intrinsic complicatedness associated in the evaluation of gradients, in the present study, the Genetic Algorithm (GA) is implemented to serve the rationale of optimization.

## 2 Formulation

An orthogonal cutting operation (Fig. 1) has been considered with details as indicated in Table 1. In Fig. 1, $t$ and $t_c$ indicate the chip thickness and uncut chip thickness, respectively. These two parameters are again combined to represent the chip thickness ratio, $r$ according to $r = \frac{t}{t_c}$ that in turn relates to the rake and shear angles (i.e., $\alpha$ and $\phi$, respectively) according to $r = \frac{\sin\phi}{\cos(\phi - \alpha)}$ [14]. Proper adjustment of various machining parameters is an important task to avoid many problems such as formation of built-up edges, poor surface finish, excessive wear and many more. Using the data given in Table 1, other important parameters related to the machining can be computed. These are angle of shear plane $(\phi)$, frictional along with normal forces acting on the rake face ($F$ and $N$), angle of friction $(\beta)$, shearing and normal forces acting on the shear plane ($F_s$ and $N_s$). The procedure of calculating these parameters can be found in many texts on manufacturing [14, 15].

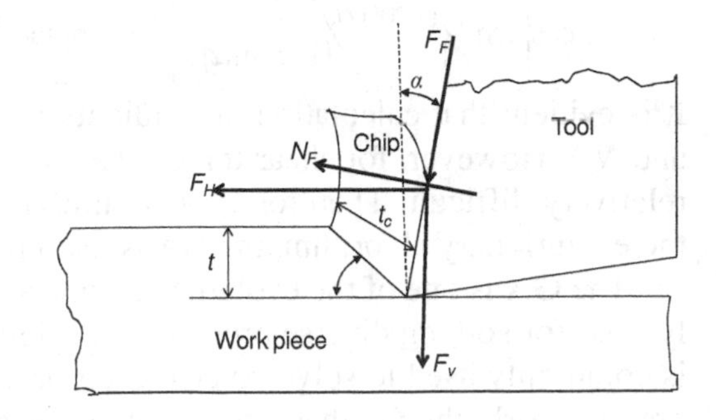


**Fig. 1** Cutting and friction force components in an orthogonal cutting process.

**Table 1** Details of various parameters considered in the cutting process.

| | |
|---|---|
| Rake angle ($\alpha$)* | 10° (= 0.1745 radians) |
| Chip thickness ratio ($r$) * | 0.3 |
| Horizontal element of cutting force ($F_H$) | 1290 N |
| Vertical element of cutting force ($F_V$) | 1650 N |
| Shear force along the shear plane ($F_S$) | 740.63 N |
| Normal force along the shear plane ($N_S$) | 1959.10 N |
| Width of cut ($b$) * | 0.006 m |

* *indicates the parameters which have been assumed to be unknowns and estimated by the inverse method.*

Next, a situation has been considered where the cutting forces ($F_H$ and $F_V$), the friction and normal forces ($F_F$ and $N_F$) and the shear forces ($F_s$ and $N_s$) are already known beforehand. However, a few critical factors responsible for the force distribution are unknown. These factors are the rake angle $(\alpha)$, the chip thickness ratio ($r$) and the required width of the cut ($b$). The aim is to correctly estimate three unknowns ($\alpha$, $r$ and $b$) sat isfying the given objective. In order to achieve the task, an objective function ($F$) is defined as below,

$$J=\left[F_F(n)-\tilde{F}_F\right]^2+\left[F_s(n)-\tilde{F}_s\right]^2+\left[N_F(n)-\tilde{N}_F\right]^2+\left[N_s(n)-\tilde{N}_s\right]^2 \tag{1}$$

where, $\tilde{F}_F$, $\tilde{N}_F$ $\tilde{F}_s$ and $\tilde{N}_s$ represent the exact values of the force components. Whereas, other terms are the functions of the matrix, $n\in[\alpha, r, b]$ containing the unknown coefficients and are regularized in an iterative process using the GA. In terms of the unknowns ($\alpha$, $r$ and $b$), the force components are expressed below [14],

$$F_F = F_H \sin\alpha + F_V \cos\alpha$$
$$N_F = F_H \cos\alpha - F_V \sin\alpha \, ;$$

$$F_s = F_H \cos\left[\tan^{-1}\left\{{(r\cos\alpha)}/{(1-r\sin\alpha)}\right\}\right] - F_V \sin\left[\tan^{-1}\left\{{(r\cos\alpha)}/{(1-r\sin\alpha)}\right\}\right] \tag{2}$$

$$N_s = F_V \cos\left[\tan^{-1}\left\{{(r\cos\alpha)}/{(1-r\sin\alpha)}\right\}\right] + F_H \sin\left[\tan^{-1}\left\{{(r\cos\alpha)}/{(1-r\sin\alpha)}\right\}\right]$$

It is evident that calculation of gradients is simple for cutting forces ($F_F$ and $N_F$). However, for shear forces ($Fs$ and $Ns$), gradient calculations are relatively difficult. Therefore, for minimizing the objective function ($F$), the evolutionary algorithm, the GA is used here.

The GA is one of the evolutionary optimization algorithms extensively used for solving diverse problems in science and engineering. The GA is commonly used to solve diverse categories of inverse problems. In the present work, the GA has been implemented for functional optimization of a set of promising unknown parameters ($\alpha$, $r$, $b$) minimizing the pertinent objective function, $J$. In the GA, generally various parameters to be estimated are encoded into binary strings which in turn create chromosomes. In the present work, a random chromosome structure involving three variables ($\alpha$, $r$, $b$) may be represented in the following manner,

$$\underset{(\alpha)}{1110110101} \qquad \underset{(r)}{0110111101} \qquad \underset{(b)}{1100110110} \tag{3}$$

Generally, the ranges of various parameters are different. In the present study, the searching ranges for various unknowns are considered as, $0.10\text{ rad.} \leq \alpha \leq 0.20\text{ rad.}$ $0.20 \leq r \leq 0.50$ and $0.001\text{ m} \leq b \leq 0.01\text{ m}$. For binary encoding, 10 randomly-generated bits are used which can generate $2^{10} = 1024$ in the range 0 to 1023 (i.e., $2^{10}$-1). In order to decode the rake angle, $\alpha$ from binary form to real value, the following rule has been adopted [16],

$$\alpha = 0.10 + \frac{(0.2-0.1)}{(2^{10}-1)} \times (\text{decimal correspondent of binary number}) \tag{4}$$

Similarly, other variables such as the chip thickness ratio along with the width of cut may be encoded as shown below [16],

$$r = 0.20 + \frac{(0.50-0.20)}{(2^{10}-1)} \times (\text{decimal correspondent of binary number}) \tag{5}$$

$$b = 0.001 + \frac{(0.01-0.001)}{(2^{10}-1)} \times (\text{decimal correspondent of binary number}) \tag{6}$$

After encoding, the strings are combined to create an individual of the GA. For example, according to Eqs. (4-6), a chromosome of 30 bits as indicated in Eq. (3) (i.e., 1110110101-0110111101-1100110110) corresponds to $\alpha = 0.19276$ rad., $r$ =0.33049 and $b$ =0.00823 m. A group of such solutions (individuals) is known as population. In this analysis, a population size of 30 has been taken. If the termination criterion is unsatisfied, then the solutions are reorganized by reproduction, crossover and mutation by succeeding generations of the GA. The crossover and mutation occurs depending upon the allocated probabilities and for this study, crossover and mutation probabilities of 0.80 and 0.03, respectively have been accounted for. During crossover, each pair of complete strings are selected randomly, followed by a random number generation between 0 and 1 that is compared against the crossover probability. If the random number generated is less than the crossover probability, then the crossover operation occurs among the strings pair as represented below [16],

| | | |
|---|---|---|
| Parents | **‘1010010101**-0010110’101-0101010101<br>**‘0101110010-**1100000’110-1100100110 | (7) |
| | *After crossover* | |
| Offspring | **‘0101110010-**1100000’101-0101010101<br>**‘1010010101**-0010110’110-1100100110 | (8) |

Once offsprings are generated, a random number is again generated within the range 0 and 1 and if it is less than the mutation probability, then the mutation occurs for that particular offspring. The mutation procedure involves flipping of a single bit of a string in the following manner,

| | | |
|---|---|---|
| *Before mutation* | **0101010011-**10011011‘1’1-1010110110 | (9) |
| *After mutation* | **0101010010-**10011011‘0’1-1010110110 | |

In the present study, an elitist strategy is adopted by replacing the worst member (solution) of the current generation with the best member (solution) of the preceding generation. For diverse class of inverse problems, the application the GA is can be found elsewhere [12, 16, 17]. Futhermore, several variants of GA along with its efficacy in multi-objective optimization problems have been also reported in many literatures [18, 19]. Consequently, further details of the GA are not elaborated here.

## 3 Results and discussion

In the current section, results and deliberations have been presented regarding the concurrent estimation of three parameters involving the rake angle, the chip thickness ratio along with the required width of cut for meeting a pre-defined force criterion. Table 2 presents the comparison between the estimated values obtained from the inverse method and the exact ones which were used in the forward method for calculating various force components. The respective search ranges for each variable are also indicated in the table caption. It is observed that the current estimations acquired from the inverse technique are in fine agreement with the corresponding actual parameters used in the forward method to obtain various components force.

Figure 2 presents the variation of the best value of the objective function ($J$) against number of generations of the GA. It is manifested from the figure that the value of the objective function gradually reduces from a high value to sufficiently low value in about 15 iterations and ahead of this, there is no significant change until the termination boundary of 50 iterations.

**Table 2**: Comparison of the exact and the GA-estimated parameters.
*Range*: $[0.1 \text{ rad.} \le \alpha \le 0.2 \text{ rad.}, 0.2 \le r \le 0.5,\ 0.001 \le b \le 0.01]$

| Parameter | Exact value (forward method) | GA-estimated value (inverse method) |
|---|---|---|
| Rake angle ($\alpha$) | $10^{o}$ (0.1746 radians) | $9.994^{o}$ (0.1745 radians) |
| Chip thickness ratio ($r$) | 0.3 | 0.3 |
| Width of cut ($b$) | 6 mm (0.006 m) | 7.167 mm (0.007167 m) |

In Fig. 3, the trend of the average value of the objective function, $J$ is considered against the number of the GA generations. Unlike the best value of the objective function where either a constant or continuously decreasing trend was discovered in Fig. 2, in this case, both increasing and decreasing along with nearly constant trends are observed. For instance, in Fig. 3, between 5th and 10th generation of the GA, such phenomenon can be easily observed. This is because the average objective function pertains to all members available within the population. Thus, even though the best value may either decrease or remain constant in two successive generations (iterations of the GA), that does not guarantee the average value also to follow the same trend. It is noticed that beyond 20 iterations, there is no noticeable change in both the best and the average values of the objective function, $J$. The observed value of the objective function was found to be less than, $O(J) < 10^{-3}$.

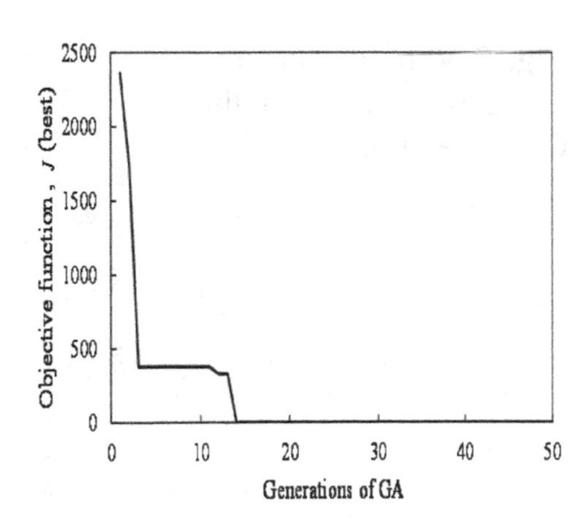


**Fig. 2** Iterative variation of the best value of the objective function

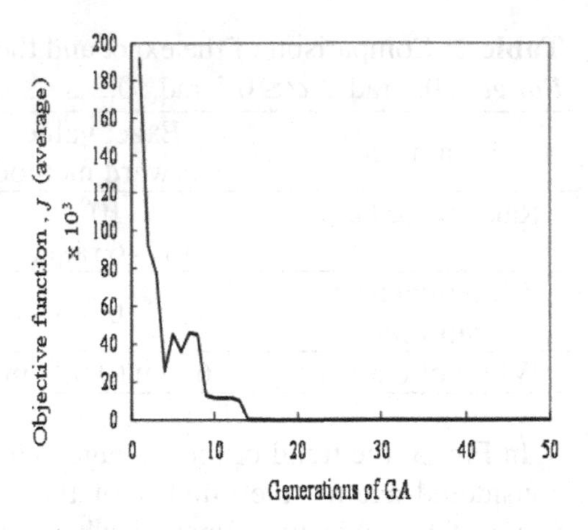


**Fig. 3** Iterative variation of the average value of the objective function

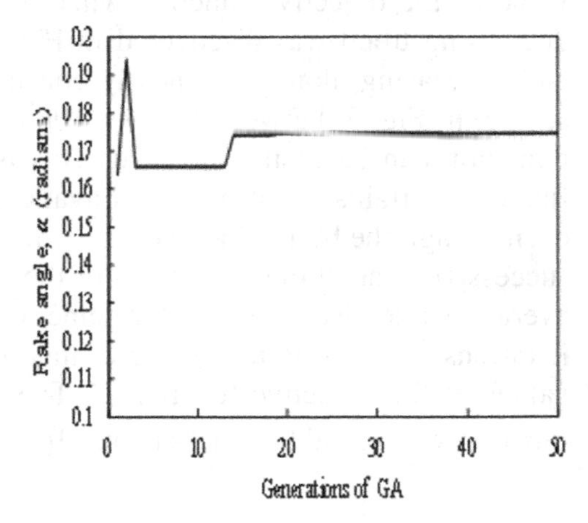


**Fig. 4** Variation of the rake angle with the GA generations

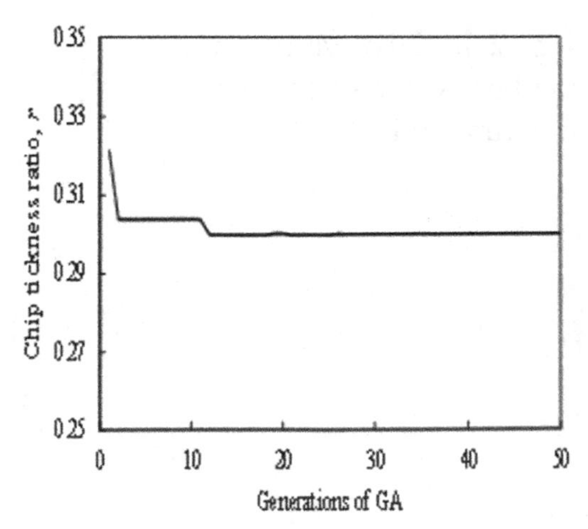


**Fig. 5** Variation of chip thickness ratio with the GA generations

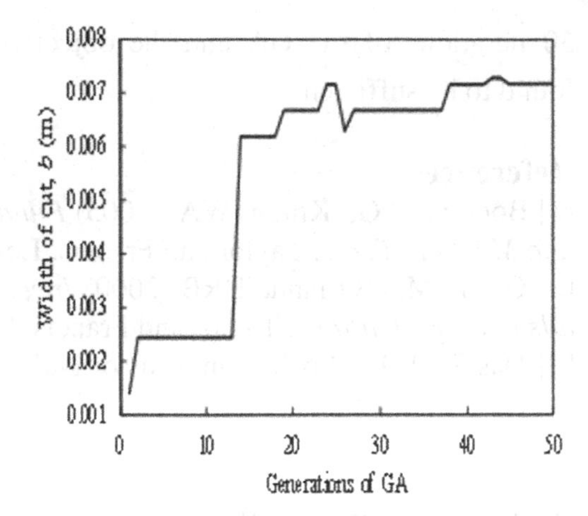


**Fig. 6** Variation of width of cut with generations of the GA

Figures 4-6 present the variation of three unknown parameters ($\alpha, r$ and b) during the inverse estimation process against number of generations (iterations) of the GA. It can be observed from Figs. 4 and 5 that the values of rake angle and chip thickness ratio $(\alpha, r)$ nearly remain unchanged beyond 20 iterations. However, even beyond 20 iterations, little variations are still observed to take place in the value of width of cut ($b$) as depicted in Fig. 6. But, it was revealed previously (in Figs. 2 and 3) that both the best along with the average values of the objective function, $J$ do not undergo significant variation beyond 20 iterations. This observation clearly reveals that sufficient flexibility occurs to select this parameter (i.e., width of cut) in satisfying a given force criterion. The present work can be further extended using the concept of swarm dynamics-based nature-inspired optimization algorithm [20].

## 4 Conclusion

An inverse optimization analysis is made for concurrently evaluating the rake angle, the chip thickness ratio and the required width of the cut in an orthogonal cutting process for meeting a pre-defined distribution of force criterion. The necessary forces are in turn obtained from a forward problem using some known values of the three unknowns, which are eventually evaluated from the inverse analysis using the GA. It is observed that the rake angle $(\alpha)$ and the chip thickness ratio ($r$) are estimated with an excellent accuracy. However, some flexibility and compromise can be made with respect to the required width of the cut ($b$). For optimization,

50 iterations of the GA and the objective function of $O(J) < 10^{-3}$ is found to be sufficient.

## References

[1] Boothroyd G, Knight WA (2005) *Fundamentals of Metal Machining and Machine Tools*. Taylor and Francis, London
[2] Ozisik MN, Orlande HRB (2000) *Inverse Heat Transfer: Fundamentals and Applications*. Taylor and Francis, London
[3] Das R (2012) Inverse analysis of Navier-Stokes equations using simplex search method. *Inverse Probl Sci Eng* 20: 445-462
[4] Wallace PW, Boothroyd G (1964) Tool forces and tool-chip friction in orthogonal machining. *J Mech Eng Sci* 6: 74-87
[5] Yang M, Park H (1991) The prediction of cutting force in ball-end milling. *Int J Mach Tool Manu* 31: 45-54
[6] Özel T, Altan T (2000) Determination of workpiece flow stress and friction at the chip–tool contact for high-speed cutting. *Int J Mach Tool Manu* 40: 133-152
[7] Bil H, Kılıç SE, Tekkaya AE (2004) A comparison of orthogonal cutting data from experiments with three different finite element models. *Int J Mach Tool Manu* 44: 933-944
[8] Huang CH, Lo HC (2005) A three dimensional inverse problem in predicting the heat fluxes distribution in the cutting tools. *Numer Heat Transfer A- Appl* 48: 1009-1034
[9] Yvonnet J, Umbrello D, Chinesta F, Micari F (2006) A simple inverse procedure to determine heat flux on the tool in orthogonal cutting. *Int J Mach Tool Manu* 46: 820-827
[10] Lin ZC, Chen CC (2007) Friction coefficient investigation on IC dam-bar trimming process by inverse method with experimental trimmed geometric profiles. *Int J Mach Tool Manu* 47: 44-52
[11] Luchesi VM, Coelho, RT (2012) An inverse method to estimate the moving heat source in machining process. *Appl Therm Eng* 45-46: 64-78
[12] Das R (2012) Application of genetic algorithm for unknown parameter estimations in cylindrical fin. *Appl Soft Comput* 12: 3369-3378
[13] Parwani AK, Talukdar P, Subbarao PMV (2012) Estimation of strength of source in a 2D participating media using the differential evolution algorithm. *J. Phys: Conf Ser* 369: 1-10
[14] Rao PN (2002) *Manufacturing Technology-Metal Cutting and Machine Tools*. McGraw-Hill, New York
[15] Kalpakjian S, Schmid SR (2007) *Manufacturing Processes for Engineering Materials*. Prentice Hall, New Jersey

[16] Das R, Mishra SC, Kumar TBP, Uppaluri R (2011) An inverse analysis for parameter estimation applied to a non-Fourier conduction-radiation problem. *Heat Transfer Eng* 32: 455–466
[17] Singla RK, Das R (2014) Application of decomposition method and inverse prediction of parameters in a moving fin. *Energy Convers Manage* 84: 268-281
[18] Nwaoha TC, Yang Z, Wang J, Bonsall S (2011) Application of genetic algorithm to risk-based maintenance operations of liquefied natural gas carrier systems. *Proc Inst Mech Eng E-J Process Mech Eng* 225: 40-52
[19] Singh K, Das R (2016) An experimental and multi-objective optimization study of a forced draft cooling tower with different fills. *Energy Convers Manage* 111: 417-430
[20] Shang Y, Bouffanais R (2014) Influence of the number of topologically interacting neighbors on swarm dynamics. *Scientific Reports*, 4: 4184-1-4184-7.

# Robust Control of Buck-Boost Converter in Energy Harvester: A Linear Disturbance Observer Approach

Aniket D. Gundecha, V. V. Gohokar, Kaliprasad A. Mahapatro and Prasheel V. Suryawanshi

**Abstract** An ingenious control of DC–DC buck-boost converter with uncertain dynamics is proposed in this paper. The proposed converter operates in buck-boost mode based on the uncertain input either from a photovoltaic cell (boost) or piezo-electric generator (buck). A linear disturbance observer is designed to alleviate the disturbances in load resistance and input source. The control is synthesized using sliding mode control. The stability of system is assured. The results are validated for a practical case of multi-energy harvesters.

## 1 Introduction

Energy harvesting plays a vital role in energizing low power devices. Energy harvesters are circuits which captures puny amount of energy from naturally available sources such as solar, thermal, wind, vibrations, acoustic and magnetic field energy [1]. Many practical systems are available that harvest energy from a single source. The challenge in such systems is to provide an uninterrupted power supply to the load affected by uncertain nature of these energy sources [2]. To improve the overall reliability and robustness of the system, energy from multiple sources should be harvested [3, 4, 5]. The most promising methods to harvest ambient power is solar power and vibration energy [6]. In many cases, solar energy and vibrations are not present continuously, hence both the sources are considered in this paper.

Aniket D. Gundecha
SSGM College of Engineering, Shegaon and MIT Academy of Engineering, Alandi (D), Pune
e-mail: adgundecha@entc.maepune.ac.in

V. V. Gohokar
Maharashtra Institute of Technology, Pune e-mail: vvgohokar@rediffmail.com

Kaliprasad A. Mahapatro · Prasheel V. Suryawanshi
MIT Academy of Engineering, Alandi (D), Pune



J.M. Corchado Rodriguez et al. (eds.), *Intelligent Systems Technologies and Applications 2016*, Advances in Intelligent Systems and Computing 530,
DOI 10.1007/978-3-319-47952-1_44

The disturbances from changes in load and variation in source such as solar panel and piezoelectric generator voltage will depend on the irradiance of sun and vibrations. To alleviate the effect of these disturbances, DC–DC buck-boost converter system with robust control is proposed in this paper. Passivity-based dynamical feedback controller is designed for converters of boost and buck-boost type [7]. Other control strategies are proposed such as adaptive and nonadaptive backstepping control approach [8], LQG [9], PID control [10] and fuzzy-logic [11]. A DC–DC converter is a nonlinear system, having unstable zero dynamics; sliding mode control (SMC) is derived to deal the contrary effects due to large variations in load [12, 13, 14]. The uncertainty in the load and input source is caused by various factors. The buck-boost converter output voltage is controlled through current. Hence sensors are required to sense the current, making the control complicated. A control technique which estimates uncertainties and disturbances can be a answer to the above mentioned problem. Several estimation techniques such as extended state observer [15, 16, 17, 18], observers for uncertain systems with uncertain inputs [19, 20] are applied to remove the disturbances. A novel control is proposed to handle mismatched uncertainties using extended disturbance observer [21], a nonlinear disturbance observer is designed in [22]. Combining sliding mode control with techniques that can estimate the uncertainties and disturbances is an attractive scheme.

In this paper, a linear disturbance observer based sliding mode control is developed to alleviate the effect of uncertainties on input and output side as shown in Fig 1. A reference current is designed to control the output voltage and linear disturbance observer is designed which can estimate the disturbances. The output voltage is controlled by a reference current and a linear disturbance observer based SMC is designed to track the reference current. The main outcomes of this paper are:

- Information of disturbance is not required
- Uncertainties are similar to the characteristics of natural energy sources
- Improvement in the performance without increase in control

The paper is organized as follows: The mathematical modelling of buck-boost in Section 2. Section 3 describes the linear disturbance observer based sliding mode control. Section 4 gives the analysis of stability. The performance is illustrated in Section 5 with discussion and conclusion in Section 6.

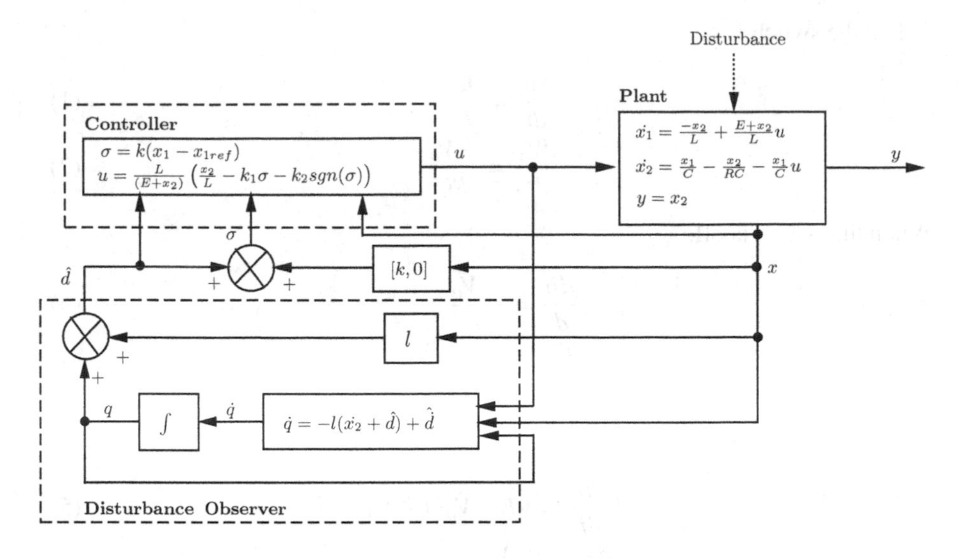


Fig. 1: Control Structure for Buck–Boost Converter

## 2 Modelling of Buck–Boost Converter System

A DC–DC buck–boost converter is shown in Fig. 2, where $E$ is the source voltage, $i_L$ is the inductance current, $C$ is the capacitance, $L$ is the inductance, $R$ is load resistance, $V_o$ is the voltage across capacitor and $\nu \in [0, 1]$ is the duty ratio to, control the PWM.

The dynamic model of converter is derived as follows.

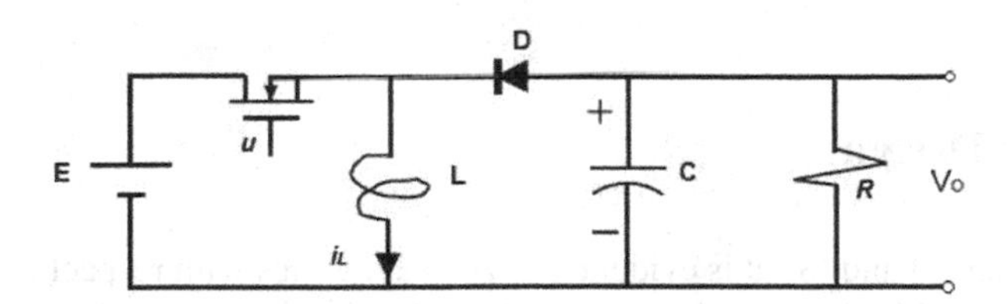


Fig. 2: Circuit diagram of buck–boost converter

When the switch is on,

$$\frac{di_L}{dt} = \frac{E}{L} \tag{1}$$

$$\frac{dV_o}{dt} = \frac{-V_o}{RC} \tag{2}$$

When the switch is off,

$$\frac{di_L}{dt} = \frac{-V_o}{L} \tag{3}$$

$$\frac{dV_o}{dt} = \frac{Ri_L - V_o}{RC} \tag{4}$$

The average system model is,

$$L\frac{di_L}{dt} = \nu E - V_o(1-\nu) \tag{5}$$

$$C\frac{dV_o}{dt} = \frac{-V_o}{R} - i_L(1-\nu) \tag{6}$$

From the above equations (5) and (6) it is clear that there are two states. Let $x_1 = i_L$ and $x_2 = V_o$. The final state space equations of Buck–Boost converter are given as,

$$\dot{x_1} = \frac{-x_2}{L} + \frac{E+x_2}{L}u \tag{7}$$

$$\dot{x_2} = \frac{x_1}{C} - \frac{x_2}{RC} - \frac{x_1}{C}u \tag{8}$$

where $u =$ control law to be designed.

## 3 Control Design

From equation (7) and (8), it is evident that zero dynamics with respect to the output voltage are unstable; hence voltage control method is not applicable [12]. Also the load resistance $R_L$ in the system can change due to variations in temperature. Since it is difficult to control output voltage directly, we need to design inductor current reference signal to track the output voltage. The inductor current is given as,

$$L\frac{di_L}{dt} = -V_o + Eu + V_o u \tag{9}$$

According to Voltage Second Balance of inductive voltage, $L\frac{di_L}{dt} = 0$

$$U_s = \frac{V_{\text{ref}}}{E + V_{\text{ref}}} \tag{10}$$

where, $U_s$ = steady state value of duty ratio. Similarly the voltage across capacitor is given as,

$$C\frac{dV_o}{dt} = \frac{-V_o}{R} + (1-u)i_L \tag{11}$$

According to Capacitor Charge Balance principle, $C\frac{dV_o}{dt} = 0$

$$I_{Ls} = \frac{V_{\text{ref}}}{(1-U_s)R} \tag{12}$$

where, $I_{Ls}$ = reference current. Substituting $U_s$ in equation (12),

$$I_{Ls} = \frac{V_{\text{ref}}}{E}\frac{(E+V_{\text{ref}})}{R} \tag{13}$$

Add and subtract $\frac{x_2}{R_oC}$ to equation (8) and assuming the nominal value $R_o$ which is known.

$$\dot{x}_2 = (1-u)\frac{x_1}{C} - \frac{x_2}{R_oC} + \frac{x_2}{C}\left(\frac{1}{R_o} - \frac{1}{R}\right) \tag{14}$$

Euation (14) can be rewritten as,

$$\dot{x}_2 = (1-u)\frac{x_1}{C} - \frac{x_2}{R_oC} + d \tag{15}$$

where $d$ = disturbance. The resistance value is:

$$R = \frac{R_oV_o}{V_o - R_odC} \tag{16}$$

Substituting (16) in (13) gives,

$$I_{Ls} = \left(\frac{E+V_{\text{ref}}}{E}\right)\left(\frac{V_{\text{ref}} - C\,d\,R_o}{R_o}\right) \tag{17}$$

From (17) the reference inductor current designed can be written as,

$$x_{1ref} = \left(\frac{E+V_{\text{ref}}}{E}\right)\left(\frac{V_{\text{ref}} - C\,\hat{d}\,R_o}{R_o}\right) \tag{18}$$

where, $\hat{d}$ is the estimated disturbance which will be derived in 3.1.

### 3.1 Linear Disturbance Observer

From equation (15) it is clear that there is disturbance which needs to be estimated; caused by load resistance changes and input voltage variations. From equation (16), the disturbance estimation due to load resistance changes and uncertainty in input voltage is designed based on a linear disturbance observer [21] which is designed as,

$$\hat{d} = q + lx_2 \tag{19}$$

$$\dot{q} = -l(x_2 + \hat{d}) + \hat{\dot{d}} \tag{20}$$

where $\hat{d}$ is the estimate of $d$, $q$ is auxiliary variable and $l$ is user chosen constant. The estimation error can be defined as,

$$\tilde{d} = d - \hat{d} \tag{21}$$

## *3.2 Sliding Mode Controller Design*

The objective is to design a robust control for eliminating disturbance and uncertainty in a DC–DC buck–boost converter. A sliding mode controller is proposed followed by a linear disturbance observer for estimating the disturbance [21]. The control structure for buck–boost converter system is shown in Fig. 1. The current and voltage are available, and the control signal will generate the PWM. A sliding surface is selected as,

$$\sigma = ke \tag{22}$$

where, $k > 0$, $e = x_1 - x_{1ref}$
Differentiating (22) gives,

$$\dot{\sigma} = k(\dot{x_1} - \dot{x_{1ref}}) \tag{23}$$

$$\dot{\sigma} = k\left(\frac{-x_2}{L} + \frac{E + x_2}{L}u + C\left(\frac{V_{ref} + E}{E}\right)\hat{\dot{d}}\right) \tag{24}$$

Selecting the control $u$ as

$$u = \frac{L}{(E + x_2)}\left(\frac{x_2}{L} - k_1\sigma - k_2 sign(\sigma)\right) \tag{25}$$

# 4 Closed loop Stability

In this section, the closed loop stability of the linear disturbance observer and DC–DC buck-boost converter to be controlled is analysed. The Lyapunov function is defined as,

$$V = \frac{1}{2}\sigma^2 \tag{26}$$

Taking derivative of the Lyapunov function defined above,

$$\dot{V} = \sigma\dot{\sigma} \tag{27}$$

$$\dot{\sigma} = k(\dot{x_1} - \dot{x_{1ref}}) \tag{28}$$

$$= k\left(\frac{-x_2}{L} + \frac{E + x_2}{L}u + C\left(\frac{V_{ref} + E}{E}\right)\dot{\hat{d}}\right) \tag{29}$$

Substituting $u$ from (25) in (29)

$$\dot{\sigma} = k\left(-k_l\sigma - k_s sgn(\sigma) + C\left(\frac{V_{ref} + E}{E}\right)\dot{\hat{d}}\right) \tag{30}$$

$$= k\left(-k_l\sigma - k_s sgn(\sigma) + C\left(\frac{V_{ref} + E}{E}\right)l\tilde{d}\right) \tag{31}$$

$$= k(-k_l\sigma - k_s sgn(\sigma) + \beta\tilde{d}) \tag{32}$$

where $\beta$ is a constsnt

$$\sigma\dot{\sigma} = k\sigma(-k_l\sigma^2 - k_s\sigma sgn(\sigma) + \beta\sigma\tilde{d}) \tag{33}$$

$$\leq k|\sigma|(-k_l|\sigma| - k_s|\sigma| + \beta|\sigma||\tilde{d}|) \tag{34}$$

The disturbance d is continous and satisfies,

$$|\tilde{d}| \leq \varepsilon \tag{35}$$

where $\varepsilon$ is a positive number.

It may be noted that, the knowledge of bound $\varepsilon$ is not required.

## 5 Results and Discussion

The performance of proposed linear disturbance observer based sliding mode control is tested for buck-boost applied to energy harvesting system. The output voltage from solar panel and piezoelectric generator is considered as input to the system. The system is also tested for dynamic changes in the load resistance. The different cases are tested for practical energy harvesting application systems. The simulation is carried out using MATLAB (ver 8.5) and Simulink (ver 8.5) considering practical conditions.

### *5.1 Case 1: Uncertain input (E)*

The buck-boost converter is tested for variations in the input voltage. Variable voltage of 6V to 9V from solar panel [23] and 11V to 13V from piezoelectric gener-

Table 1: *Parameter values of DC–DC buck-boost converter*

| Parameters | Meaning | Values |
|---|---|---|
| $E$ | Input Voltage | 8V–12V |
| $V_o$ | Output Voltage | 10V |
| $L$ | Inductance | 10mH |
| $C$ | Capacitance | 100$\mu$F |
| $R$ | Load Resistance | 60Ω–100Ω |
| $f_{sw}$ | Frequency | 25kHz |

ator [24] is given to the buck-boost converter system. The reference voltage considered is $V_{ref}$=10V and load resistance $R_o$=100Ω. The parameters of buck-boost converter are designed as in [26] and are given in Table 1. Fig 3a shows the variable input given to the system, Fig 3b shows the output voltage $V_o$ which is constant with respect to variations in input voltage $E$. Fig 3c is the control signal which is switching between boost and buck mode with respect to changes in the input source voltage. Fig 3d shows the inductor current.

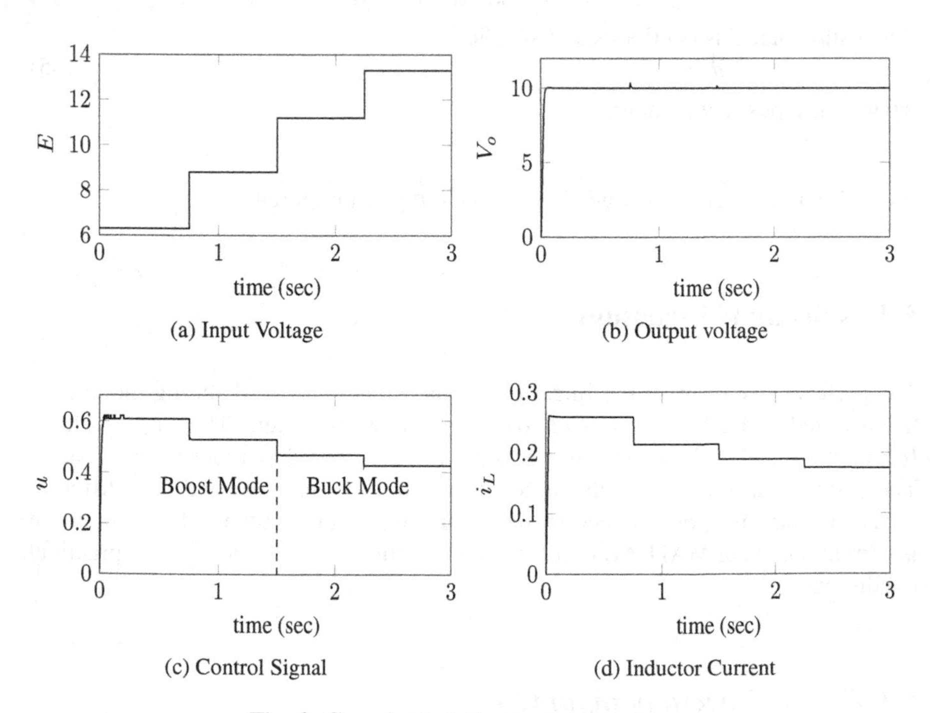


(a) Input Voltage

(b) Output voltage

(c) Control Signal

(d) Inductor Current

Fig. 3: Case 1: Variation in input voltage

### 5.2 Case 2: Variable resistance (R)

Fig 4 shows the performance of buck-boost converter for changes in load resistance. The load resistance may change due to various factors such as temperature, etc [25]. Disturbance in the form of variations in load resistance from 60Ω to 100Ω is inserted, reference voltage is set to $V_{ref}$ = 10V. Fig 4a shows the variation in inductor current. Fig 4b is the output voltage which is constant irrespective of uncertainties in load resistance. The control signal $u$ is below 0.5 which shows that it is a buck operation, since input voltage $E$=15V as shown in Fig 4c. The linear disturbance observer designed in 3.1 can estimate the changes in load resistance and tracks exactly the uncertainties in the load resistance as shown in Fig 4d.

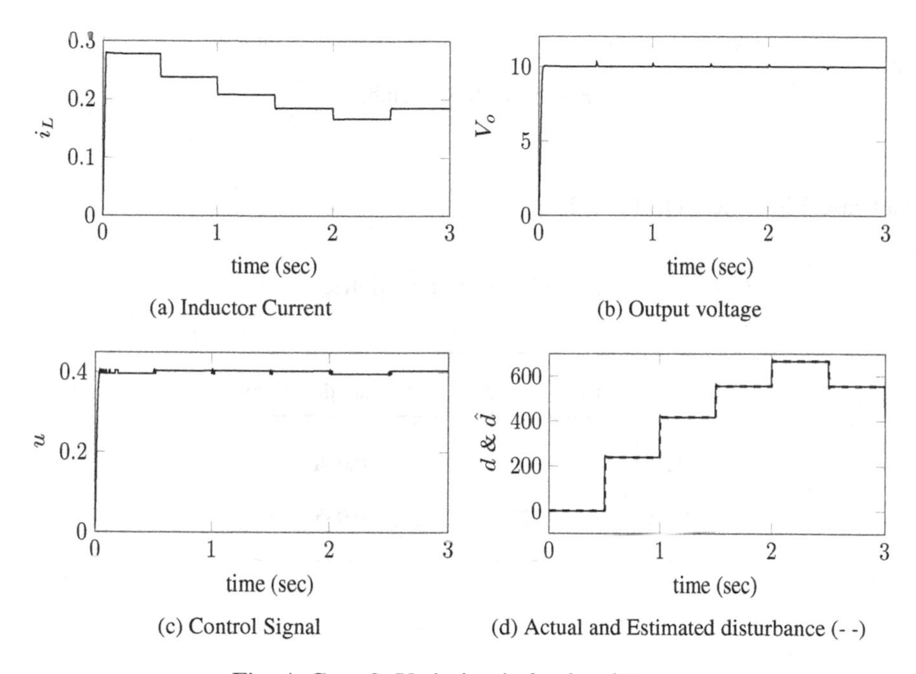


(a) Inductor Current

(b) Output voltage

(c) Control Signal

(d) Actual and Estimated disturbance (- -)

Fig. 4: Case 2: Variation in load resistance

The estimation of uncertainty is good, resulting in good tracking accuracy. From Fig 5a and Fig 5b it is proved that the ultimate boundedness of sliding variable is guaranteed; and the bounds can be lowered by appropriate choice of design parameters.

The result obtained in the above mentioned cases proves the robustness in the control algorithm designed. For case 1 of variable input voltage, actual characteristics of solar panel and piezoelectric generator are considered. The control signal accordingly switches between buck and boost mode. In case 2, load resistance was changed to create disturbance and accordingly the linear disturbance observer has track the changes accurately. In both the cases the control effort required is minimum, resulting in effective energy harvesting. The performance metrics for case:1

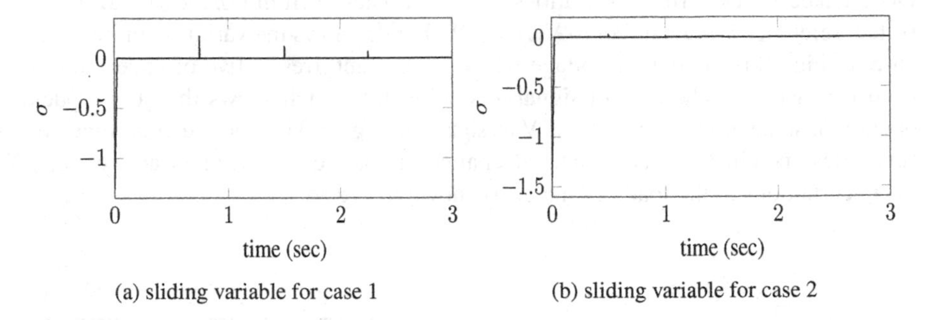

(a) sliding variable for case 1

(b) sliding variable for case 2

Fig. 5: Sliding variable

and case:2 is depicted in Table 2.

Table 2: Performance Metrics

| | Rise time (sec) | Settling time (sec) |
|---|---|---|
| Case:1 | 0.03 | 0.038 |
| Case:2 | 0.027 | 0.035 |

## 6 Conclusion

This paper addresses some of the important issues in buck-boost converters for multi-energy harvesting. The linear disturbance observer based sliding mode control, provides better tracking and rejection of disturbances such as variable load resistance and input voltage. The proposed control is tested for energy harvesting application by considering two energy sources i.e. solar and vibrations. Overall stability of system is proved, the boundness of the sliding variable and the output is guaranteed.

## References


1. G. Zhou, L. Huang, W. Li, and Z. Zhu.: Harvesting Ambient Environmental Energy for Wireless Sensor Networks: A Survey. Journal of Sensors (2014). doi: 10.1155/2014/815467
2. W. K. G. Seah, Z. A. Eu and H-P. Tan.: Wireless Sensor Networks Powered by Ambient Energy Harvesting (WSN-HEAP) Survey and Challenges. International Conference on Wireless Communication, Vehicular Technology, Information Theory and Aerospace & Electronic Systems Technology (2009). doi:10.1109/WIRELESSVITAE.2009.5172411
3. A. Schlichting, R. Tiwari and E. Garcia.: Passive multi-source energy harvesting schemes. Journal of Intelligent Material Systems and Structures (2012). doi:10.1177/1045389X12455723
4. S. Bandyopadhyay and A. P. Chandrakasan.: Platform Architecture for Solar, Thermal, and Vibration Energy Combining With MPPT and Single Inductor. IEEE Journal of Solid-State Circuits (2012). doi:0.1109/JSSC.2012.2197239
5. A. S. Weddell et al.: A Survey of Multi-Source Energy Harvesting Systems. Design, Automation & Test in Europe Conference & Exhibition (2013). doi: 10.7873/DATE.2013.190
6. S. Roundy.: Energy Scavenging for Wireless Sensor Nodes with a Focus on Vibration to Electricity Conversion. Ph. D. Dissertation, Dept. of EECS, UC Berkeley (2003).
7. H. S. Ramirez and R. Ortega.: Passivity-based controllers for the stabilization of DC-to-DC power converters. IEEE Conference on Decision and Control (1995). doi: 10.1109/CDC.1995.479122
8. H. El Fadil and F. Giri.: Backstepping Based Control of PWM DC-DC Boost Power Converters. IEEE International Symposium on Industrial Electronics (2007). doi: 10.1109/ISIE.2007.4374630
9. C. Chang.: Robust control of DC-DC converters: the buck converter. IEEE Conference on Power Electronics (1995). doi: 10.1109/PESC.1995.474951
10. R. D. Keyser and C. Ionescu.: A Comparative Study of Several Control Techniques Applied to a Boost Converter. IEEE $10^{th}$ Int Conf on Optimisation of Electrical and Electronic Equipment OPTIM pp. 71-78
11. P . R. Shiyas, S. Kumaravel and S. Ashok.: Fuzzy controlled dual input DC/DC converter for solar-PV/wind hybrid energy system. Electrical, Electronics and Computer Science (SCEECS) (2012). doi: 10.1109/SCEECS.2012.6184775
12. V. Utkin.: Sliding mode control of DC/DC converters. Journal of the Franklin Institute. vol. 350, no. 8, pp. 2146–2165 (2013). Elsevier
13. Y. He and F. L. Luo.: Sliding-mode control for dcdc converters with constant switching frequency. Electrical, IEE Proceedings - Control Theory and Applications (2006). doi: 10.1049/ip-cta:20050030
14. S. C. Tan, Y. M. Lai and C. K. Tse.: General Design Issues of Sliding-Mode Controllers in DCDC Converters . IEEE Transactions on Industrial Electronics (2008). doi: 10.1109/TIE.2007.909058
15. J. Han.: From PID to active disturbance rejection control. IEEE Trans. on Industrial Electronics. vol. 56, no. 3, pp. 900-906 (2009).
16. A. D. Gundecha, V. V. Gohokar, K. A. Mahapatro and P. V. Suryawanshi in *Control of DC-DC Converter in Presence of Uncertain Dynamics*, ed. by S. Berretti et al. Intelligent Systems Technologies and Applications, vol 384 (Advances in Intelligent Systems and Computing, 2015), pp. 315-326
17. S. E. Talole, J. P. Kolhe and S. B. Phadke.: Extended State Observer Based Control of Flexible Joint System with Experimental Validation. IEEE Transactions on Industrial Electronics, vol. 57, no. 4, pp. 1411–1419 (2010).
18. K. A. Mahapatro, A. D. Chavan, and P. V. Suryawanshi.: Analysis of Robustness for Industrial Motion Control using Extended State Observer with Experimental Validation. IEEE Conference on Industrial Instrumentation and Control (2015). doi: 10.1109/IIC.2015.7150586
19. Takahashi R. H. C and Peres P. L. D.: Unknown input observers for uncertain systems: a unifying approach and enhancements. (1996) doi: 10.1109/CDC.1996.572726

20. L. Jiang and QH. Wu.: Nonlinear adaptive control via sliding-mode state and perturbation observer. IEE Proceedings-Control Theory and Applications (2002). doi: 10.1049/ip-cta:20020470
21. D. Ginoya, P. D. Shendge and S. B. Phadke.: Sliding Mode Control for Mismatched Uncertain Systems Using an Extended Disturbance Observer. IEEE Transactions on Industrial Electronics. vol. 61, no. 4, pp. 1983–1992 (2014)
22. J. Wang, S. Li, J. Fan and Qi Li.: Nonlinear disturbance observer based sliding mode control for PWM-based DC-DC boost converter systems . The 27th Chinese Control and Decision Conference (2015). doi: 10.1109/CCDC.2015.7162338
23. SLMD600H10-$IXOLAR^{TM}$ High Efficiency SolarMD.
24. Volture, Piezoelectric Energy Harvestors. MIDE (2013)
25. J. L. Flores, A. H. Mndez, C. G. Rodrguez and H. S. Ramrez.: Robust Nonlinear Adaptive Control of a Boost Converter via Algebraic Parameter Identification. IEEE Transactions on Industrial Electronics. vol. 61, no. 8, pp. 4105–4114 (2014)
26. M. Green(2012) Design Calculations for Buck-Boost Converters. Texas Instruments. http://www.ti.com/lit/an/slva535a/slva535a.pdf

# Multi Objective PSO Tuned Fractional Order PID Control of Robotic Manipulator

Himanshu Chhabra[1],Vijay Mohan[1],Asha Rani[1],Vijander Singh[1]
[1] Instrumentation and Control Engineering Division, NSIT ,Sec-3 Dwarka University of Delhi, New Delhi, India,110078
him08miet@gmail.com,vijay13787@gmail.com,ashansit@gmail.com,vijaydee@gmail.com

**Abstract.** Designing of an efficient control strategy for robotic manipulator is a challenging task for control experts due to inherent nonlinearity and high coupling present in the system. The aim of this paper is to design precise tracking controller with minimum control effort for robotic manipulator. In order to fulfill the aforementioned purpose a fractional order PID controller is proposed. Multi objective particle swarm optimization (MOPSO) is used to optimize parameters value of FOPID controller. The integer order PID is also implemented for comparative study. Results show that the robustness of proposed controller towards trajectory tracking and uncertainty in parameters are superior over traditional PID controller.

**Keywords:** Robotic manipulator, PID, Fractional Order PID (FOPID), MOPSO

## 1 Introduction

Robotic manipulators are extensively utilized in various engineering applications such as process industries, nuclear plants, space and medical field, for the purpose of pick and place, path tracking and accurate positioning etc. To accomplish precise and effective control of end-effector of a robotic manipulator is essential. Various control schemes are implemented on robotic manipulator but traditional PID controller is still the primary choice for control engineers due to its simple realization, less cost and easy structure. In the case of non linear and uncertain systems, PID controller is not able to provide an efficient control due to the absence of more adjustable parameters.
In order to overcome the above shortcoming of PID controller **Padulbny [1]** proposed a new form of control scheme, which is the amalgamation of fractional calculus and conventional PID termed as Fractional order PID (FOPID) controller. Literature reveals that FOPID controller is implemented for different engineering applications by various researchers. **Dumlu and Koksal [2]** implemented FOPID controller for trajectory tracking control of a 3-DOF parallel manipulator. Mathematical method and pattern search technique are used to calculate the optimum values of FOPID parameters. The Simulation results reveal that FOPID not only improves transient response, but also reduces the steady state error in comparison to traditional PID controller. **Das *et.al* [3]** presented a performance comparison of different structures of



J.M. Corchado Rodriguez et al. (eds.), *Intelligent Systems Technologies and Applications 2016*, Advances in Intelligent Systems and Computing 530,
DOI 10.1007/978-3-319-47952-1_45

fractional order fuzzy PID (FOFPID) controllers applied to oscillatory fractional order processes with dead time for servo and regulatory problems. **Bingul and Karahan [4]** proposed particle swarm optimization (PSO) tuned FOPID for robot trajectory control. Performance of PSO tuned FOPID controller are compared with fuzzy logic controller (FLC) and PID. Results reveal the superiority of proposed controller.

In this article FOPID controller is proposed for trajectory tracking control of robotic manipulator. The objective of the work is to achieve precise control of robotic manipulator with minimum control effort. The two popular optimization algorithms Genetic Algorithm (GA) and Particle Swarm Optimization (PSO) provide certain advantages over other techniques i.e. ability to find global optima and gradient-free. PSO is preferred over GA because of its superiority in terms of their algorithm coefficients, robustness to population size and accuracy **[9]**.Multi objective PSO is used to estimate the optimized parameter values of FOPID. Classical PID is also designed for comparative study. The simulation results reveal the superiority of FOPID over its integer order counterpart.

Paper is organized as follows mathematical modeling of manipulator is explained in section 2, section 3 explains the implementation of FOPID control strategy, MOPSO is explained in section 4. The result and discussion are presented in section 5 and section 6 concludes the work.

## 2 Mathematical Modeling

The mechanical structure of a two degree of freedom robotic manipulator consist of two rigid links connected by means of revolute joints to guarantee human arm like mobility is shown in Fig.1.The mathematical model under consideration and all the parameter value are taken from **Ayala and Coelho [8]**.

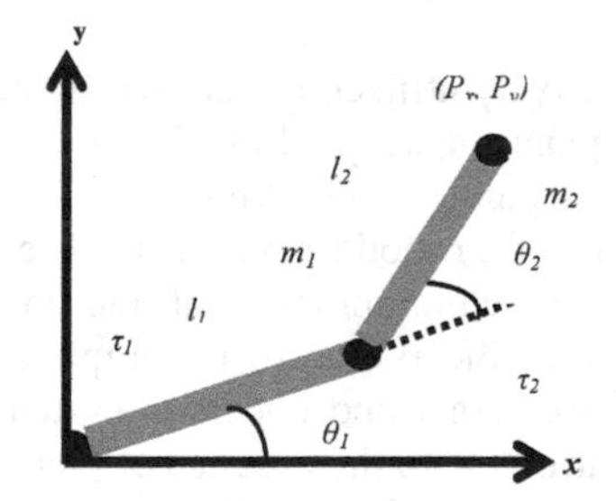


**Fig.1:** Mechanical model of two degree of freedom robotic manipulator

The dynamics of two link robotic manipulator is given as follows: **[5]**

$$\tau_1 = m_2 l_2^2 (\ddot{\theta}_1 + \ddot{\theta}_2) + (m_1 + m_2) l_1^2 \ddot{\theta}_1 + m_2 l_1 l_2 c_2 (2\ddot{\theta}_1 + \ddot{\theta}_2) - m_2 l_1 l_2 s_2 \dot{\theta}_2^2 - 2 m_2 l_1 l_2 s_2 \dot{\theta}_1 \dot{\theta}_2 + (m_1 + m_2) l_1 g c_1 + m_2 l_2 g c_{12} \tag{1}$$

$$\tau_2 = m_2 l_1 l_2 c_2 \ddot{\theta}_1 + m_2 l_1 g c_{12} + m_2 l_2^2 (\ddot{\theta}_1 + \ddot{\theta}_2) + m_2 l_1 l_2 s_2 \dot{\theta}_1^2 \tag{2}$$

Where $\theta_i, \dot{\theta}_i$ and $\ddot{\theta}_i$ are angular position, velocity and acceleration of the links. $\tau_i \in \Re^{n\times 1}$ is the torque vector for links. The subscript $i$ =1and 2 represents the link 1 and 2 parameters respectively. Arguments $c_1=\cos\theta_1, c_2=\cos\theta_2, c_{12}=\cos\cos(\theta_1+\theta_2), s_1=\sin\theta_1$ and $s_2=\sin\theta_2$ .

## 3 Control strategy

FOPID is designed by changing the integer order of differentiation and integration operator of conventional PID controller to non integer value, in time domain the description of manipulating variable is described by equation (3):**[1]**

$$\tau_i(t)=\left( K_{P_i}e_i(t)+K_{I_i}\frac{d^{-\lambda_i}}{dt^{-\lambda_i}}e_i(t)+K_{D_i}\frac{d^{\mu_i}}{dt^{\mu_i}}e_i(t)\right) \quad i=1,2 \tag{3}$$

Taking Laplace transform of above time domain equation (3) the transfer function can be given as:

$$TF=K_{P_i}+\frac{K_{I_i}}{s^{\lambda_i}}+K_{D_i}s^{\mu_i} \tag{4}$$

[$\lambda_i$, $\mu_i$] are additional parameters over PID controller, thus providing extra degree of freedom to the controller which enhances the performance of the system. Oustaloup's suggested the implementation of fractional operator '$s^a$' **[6],** which is a recursive distribution of $N_f$ number of zeros and poles with the following expression, characterized by higher order filter.

$$s^a=K_f\prod_{k=-N_f}^{N_f}\frac{s+z_k}{s+p_k} \tag{5}$$

Now, the gain zeros and poles of the filter are determined as:

$$K_f=(\omega_h)^a \tag{6}$$

$$z_k=\omega_l\left(\frac{\omega_h}{\omega_l}\right)^{\frac{k+N_f+\frac{1}{2}(1-a)}{2N_f+1}} \tag{7}$$

$$p_k=\omega_l\left(\frac{\omega_h}{\omega_l}\right)^{\frac{k+N_f+\frac{1}{2}(1+a)}{2N_f+1}} \tag{8}$$

Where, $a$ is the order in fractions, $N_f$ is the Oustaloup's constant and $2N_f+1$ is the order of the filter, and [$\omega_l$, $\omega_h$] is the lower and higher frequency.

## 4 Multi objective particle swarm optimization (MOPSO)

MOPSO is proposed by **Coello *et. al.* [7]** in 2004. It incorporates grid making and Pareto envelope, which is appropriate for generating compromised solution for conflicting objectives. The algorithm starts with initializing number of iterations (k), repository (r) and particles (i) in the search space. These particles are sharing information and moving towards global best on the basis of particles velocity and position as described in equation (9)-(10) respectively. Thus non-dominated solution is collected into a subset called repository, and every particle chooses its global best target, among members of this repository. For local best particle, domination based and probabilistic rules are utilized.

$$V_i^{k+1} = \beta V_i^k + a_1 r_1 \left( x_{lbest} - X_i^k \right) + a_2 r_2 \left( x_{gbest} - X_i^k \right) \tag{9}$$

$$X_i^{k+1} = X_i^k + V_i^{k+1} \tag{10}$$

Where $\beta$ is weight constant, $a_1$ and $a_2$ are acceleration coefficient, $r_1$ and $r_2$ are the random values between 0 and 1; $x_{lbest}$ and $x_{gbest}$ are the local and global best of the particle respectively. $X_i^k$ is the current position of $i^{th}$ particle at iteration $k$. $V_i^k$ is the velocity of $i^{th}$ particle at iteration $k$. In this work, the parameters of MOPSO are $k=100, i=20, r=10$ and the objective functions employed for this particular problem are defined as follows:

$$z_1 = \sum_{t=1}^{t_f} \left| \theta_{r_1}(t) - \theta_1(t) \right| + \sum_{t=1}^{t_f} \left| \theta_{r_2}(t) - \theta_2(t) \right| \tag{11}$$

$$z_2 = \sum_{t=1}^{t_f} \left| \tau_1(t) - \tau_1(t-1) \right| + \sum_{t=1}^{t_f} \left| \tau_2(t) - \tau_2(t-1) \right| \tag{12}$$

## 5 Result and Discussion

In this section, simulation results acquired for trajectory tracking and parametric uncertainty for FOPID and PID controllers are compared. For this work the trajectory selected for reference tracking was given by **Ayala and Coelho [8].** The search space for controller parameters are $Kp \in [0,300], Kd \in [0,50]$ and $Ki \in [0,50], \lambda \in [0,1], \mu \in [0,1]$. Fig.2 shows the obtained solution, repository and the selected solution for objectives $z_1$ and $z_2$ for FOPID controller. Corresponding parameters and integral absolute error values of FOPID and PID controller for link 1 and link 2 is given in Table.1. The trajectory tracking performance, controller output, variation of X and Y coordinate versus time and position error for both FOPID and PID controllers are shown in Fig.3. It is observed from the results that FOPID outperforms the PID controller.

**Table.1**: controller parameter, objective function and IAE value for controller

| Parameter for link1 | FOPID | PID | Parameter for link2 | FOPID | PID |
|---|---|---|---|---|---|
| $K_p$ | 254.011 | 184.86 | $K_p$ | 210.31 | 11.63 |
| $K_i$ | 17.0708 | 49.26 | $K_i$ | 20.8301 | 16.20 |
| $K_d$ | 13.9313 | 9.10 | $K_d$ | 22.7973 | 0.11 |
| $\lambda$ | 0.471938 | 1 | $\lambda$ | 0.373578 | 1 |
| $\mu$ | 0.946435 | 1 | $\mu$ | 0.522513 | 1 |
| $z_1$ | 16.4101 | 81.8172 | $z_1$ | - | - |
| $z_2$ | 13.1491 | 21.5917 | $z_2$ | - | - |
| IAE1 | 0.02054 | 0.009612 | IAE1 | 0.02816 | 0.1155 |

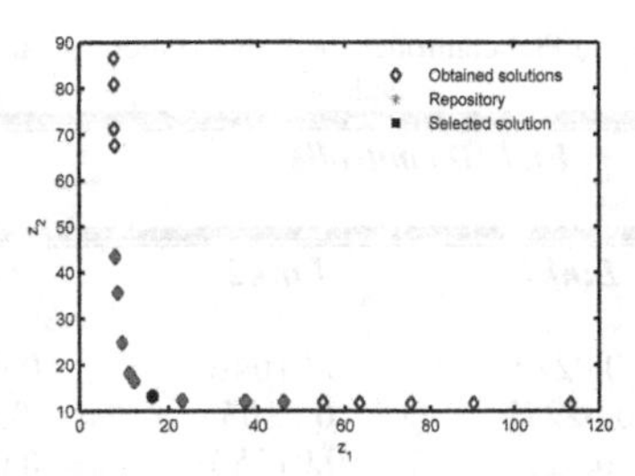


**Fig.2**: Pareto front for FOPID controller

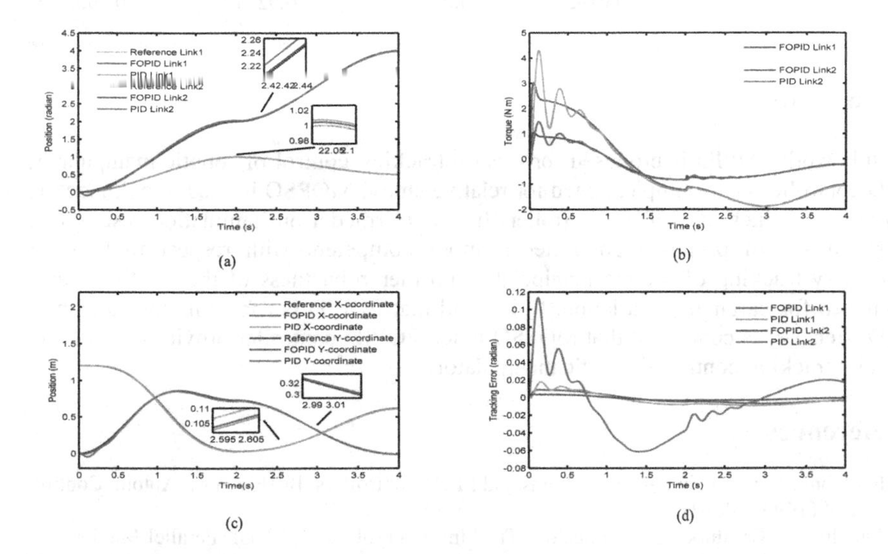


**Fig.3**: (a) Trajectory tracking performance, (b) Controller output, (c) Variation in X and Y- Coordinate versus time and (d) Tracking errors for FOPID and PID controller

## 5.1 Robustness Analysis

In this section, robustness of the proposed controller is tested under 10% increment and decrement in mass and length of the links. Table.2 shows the IAE variation for FOPID and PID controller in presences of parametric uncertainty. It is observed from Table.2, that IAE variation is very small for FOPID controller in comparison to PID which shows the effectiveness of proposed controller toward parametric uncertainty. Therefore FOPID controller is more robust than traditional PID controller.

**Table.2**: IAE value for FOPID and PID controller for 10% (a) increase and (b) decrease in parameter values

| *Uncertainty in system parameter (10%)* | *FOPID controller* | | *PID controller* | |
|---|---|---|---|---|
| | *Link1* | *Link2* | *Link1* | *Link2* |
| (a) Increases | | | | |
| $m_1$ & $m_2$ | 0.0226 | 0.01046 | 0.03103 | 0.1272 |
| $l_1$ & $l_2$ | 0.02272 | 0.0105 | 0.0312 | 0.1273 |
| $m_1$, $m_2$, $l_1$ & $l_2$ | 0.025 | 0.01155 | 0.03476 | 0.1394 |
| (b) Decrease | | | | |
| $m_1$ & $m_2$ | 0.01848 | 0.008561 | 0.02531 | 0.1037 |
| $l_1$ & $l_2$ | 0.01838 | 0.008528 | 0.02517 | 0.1032 |
| $m_1$, $m_2$, $l_1$ & $l_2$ | 0.01654 | 0.007675 | 0.02263 | 0.09286 |

## 6 Conclusion

In this work FOPID is proposed for precise tracking control of robotic manipulator. PID controller is also implemented for relative study. MOPSO is used to optimize the parameter values of FOPID controller. It is confirmed from simulation results that performance of proposed controller is more competent with respect to PID for trajectory tracking of robotic manipulator. Further robustness of the FOPID is also analyzed for parameters variation. It is found that FOPID is robust in comparison to PID. Hence it is concluded that MOPSO tuned FOPID controller provides robust and precise tracking control of robotic manipulator.

## References


1. Podlubny, I., : Fractional-order systems and $PI^{\lambda}D^{\mu}$-controllers. IEEE Trans. Autom. Control, vol.44 (1999) 208–214.
2. Dumlu, A., Erenturk, K., : Trajectory Tracking Control for a 3-DOF Parallel Manipulator Using Fractional-Order PID Control. IEEE Trans. on industrial electronics, vol.61 (2014) 3417-3426.
3. Das, S., Pan, I., Das, S., : Performance comparison of optimal fractional order hybrid fuzzy PID controllers for handling oscillatory fractional order processes with dead time. ISA Transactions vol.52 (2013) 550–566
4. Bingul, G., Karahan, O., : Fractional PID controllers tuned by evolutionary algorithms for robot trajectory control. Turk J Elec Eng & Comp Sci, vol.20 (2012) 1123-1136 .
5. Craig, J.J., : Introduction to Robotics Mechanics and Control. 3rd edn. Pearson Education, Inc (2005)
6. Oustaloup, A., Levron, F., Mathieu, B., & Nanot, F. M., : Frequency-band complex noninteger differentiator: characterization and synthesis. IEEE Transactions on Circuits and Systems-I: Fundamental Theory and Applications, vol.47 (2000) 25–39.
7. Coello Coello, C., Pulido, G., Lechuga, M., : Handling Multiple Objectives With Particle Swarm Optimization. IEEE Transactions on Evolutionary Computation, vol. 8 (2004) 256-279.
8. Ayala, H.V.H., Coelho, L.d.S., : Tuning of PID controller based on a multi objective genetic algorithm applied to a robotic manipulator. Expert Systems with Applications, vol.39 (2012) 8968-8974.
9. Duan,Y., Harley, R.G., Habetler, T.G., : Comparison of Particle Swarm Optimization and Genetic Algorithm in the Design of Permanent Magnet Motors. Power Electronics and Motion Control Conference IEEE (2009) 822-825.

# Book Recommender System using Fuzzy Linguistic Quantifier and Opinion Mining

Shahab Saquib Sohail, Jamshed Siddiqui, Rashid Ali

**Abstract.** The recommender systems are being used immensely to promote various services, products and facilities of daily life. Due to the success of this technology, the reliance of people on the recommendations of others is increasing with tremendous pace. One of the best and easiest ways to acquire the suggestions of the other like-minded and neighbor customers is to mine their opinions about the products and services. In this paper, we present a feature based opinion extraction and analysis from customers' online reviews for books. Ordered Weighted Aggregation (OWA), a well-known fuzzy averaging operator, is used to quantify the scores of the features. The linguistic quantifiers are applied over extracted features to ensure that the recommended books have the maximum coverage of these features. The results of the three linguistic quantifiers, 'at least half', 'most' and 'as many as possible' are compared based on the evaluation metric - precision@5. It is evident from the results that quantifier 'as many as possible' outperforms others in the aforementioned performance metric. The proposed approach will surely open a new chapter in designing the recommender systems to address the expectation of the users and their need of finding relevant books in a better way·

***Keywords:*** *book recommendation; opinion mining; soft computing; Ordered Weighted Aggregation (OWA); recommender system; feature extraction and analysis.*

## 1 Introduction

The advancement in technologies has provided the solution to various daily life problems with ease of access, less amount of time and minimal efforts. It has become easier to perform shopping, interacting with the people away from us, exploring the news across the globe and finding opinion of the public and experts just by sitting at home and using technologies-driven communication means like TV and the Internet, etc. One of the good aspects of the advancement in the technology is an easy-finding of the other's ideas and suggestion, and being benefited with their experiences. As, it is a common practice for human to rely upon experiences of their relatives and friends and follow their recommendations, these state-of-the-art technologies help them to reach the destination, i.e. selection of the right items they are seeking for [1]. One of the techniques which help the customer to find their desired item is recommender system



J.M. Corchado Rodriguez et al. (eds.), *Intelligent Systems Technologies and Applications 2016*, Advances in Intelligent Systems and Computing 530,
DOI 10.1007/978-3-319-47952-1_46

[2]. Recommender system makes use of the user's behavior and utilizes their neighbor, i.e. friends, relatives or a man with similar choices to recommend them their desired items [3, 4].

The opinion or reviews of a user for particular product is usually found on the various online shopping portals like amazon.com and ebay.com, etc. The opinions are usually unformatted and unstructured [5], hence, concluding a decision about a product from the available online customer's reviews is an interesting but tough job [6].

The online reviews from the customers, though about a specific product, sometimes found direction less and it is very difficult to conclude anything about the product for which reviews are provided. Thus extracting the reviews to get some features of the product is tedious task, although, is very important. Since, our concern is to recommend books, therefore, as a solution to this difficult job, we have applied human intelligence to extract features of the books from online customer's reviews available for various books at bestselling book sites like amazon.com, books.google.com and goodreads.com, etc. In this paper, we have categorized the features of a book on the basis of user's reviews, and score these features on the basis of lexicon of user's reviews. Depending upon positive or negative sense of the reviews and the intensity by which it is stated, we have scored the features.

Over these scores, OWA, an actively and frequently used soft computing method for aggregation operations, is applied [7]. With the help of OWA operator we can assign weights to the respective extracted features of the books. By using different linguistic quantifier we get different weights for obtained features, e.g 'most' is a linguistic quantifier, by using it those books are preferred in which 'most' of the features are found with a high score. Thus each book is assigned a value, and upon these values the books are sorted and ranked. The top books are recommended as a final recommendation for the users.

Rest of the paper is organized as follows. In section II, we have illustrated examples to give a background of opinion mining and OWA. Section III explains the combined approach for book recommendation composed of soft computing and opinion mining techniques. Experimental results and discussions are elaborated in section IV. We conclude in section V with some direction for future work.

## 2 Background

In this section we have described the related work and some basic idea about the terms and techniques which have been used in this paper to lay a foundation for readers to have some background knowledge of the work that may help them in understanding the problems better. First, we define opinion mining and give some related work to it followed by the definition of OWA and use of OWA in various fields of research.

### 2.1 Opinion Mining

Opinion mining is the field of data mining, mainly concerned with the extraction and analysis of the customers' opinions. It helps in drawing some meaningful conclusion from customer's reviews and feedbacks [8]. A comprehensive study of opinion mining is presented in [5]. An extensive work on opinion suggests that it is a combination of words and documents by means of which, sentiments of individuals or groups are expressed [9, 10, 11].

Opinion mining has emerged significantly in past few years for various applications such as product recommendation, business analytics, information retrieval, etc [12]. Like other area of research, opinion mining can play a vital role in recommendation technology. As user satisfaction remains always a challenge in recommender system [13, 14], the customer's opinion has become a base for exactly knowing the need of the customers. Hence, mining of customer's review has become a major focus for research in recent days [15]. Authors in [16] suggested explicit ratings and implicit opinions for accurate recommendation. Explicit and implicit user feedbacks have been used in evaluating recommender system by authors in [17, 18]. In [6] opinion mining based book recommendation technique has been proposed. The authors categorized the features of books and weight them according to their relevance in the eye of users which is calculated by their proposed mechanism.

The researchers have encountered two basic issues while mining customer's opinion; 1) expansion of opinion lexicon and 2) extraction of opinion target. [5, 12]. Opinion lexicon refers to list of opinion words which indicate the positivity or negativity of a particular word, such as excellent, great, good, bad, poor, etc. the lexicon helps in performing sentence level and document level tasks [8, 10, 21]. The lexicon is also effective for sentiment classification and summarization [19].

The authors in [19] described opinion mining and summarization. The authors have used lexicon-based method to interpret about a user's opinion sense, whether positive or negative. Related works on these topics include [20, 21, 22].

## 2.2 Ordered Weighted Aggregation (OWA)

Soft computing techniques have been used for various daily life problems, especially for the problems where uncertainty is concerned [23]. Ordered Weighted Averaging (OWA) is an averaging operator actively used in soft computing [7]. OWA has been used extensively in literature for various applications [24]. The researchers have used OWA for GIS applications [25, 26], book recommendation [27], exploring novel fuzzy queries for web searching [28], talent enhancement of sports person [29], etc. Mathematically, we give OWA as;

$$OWA(x_1, x_2, \ldots, x_n) = \sum_{k=1}^{n} w_k z_k \qquad \text{(I)}$$

Where $z_k$ is interpreted as; if the values of $x_i$ is re-ordered in descending order, it gives $z_1, z_2, z_n$, where $z_1 \geq z_2 \geq \ldots z_{n-1} \geq z_n$.

The weights '$W_k$' for OWA operator is calculated by using following equation [30].
$W_k = \{Q(k/m) - Q((k-1)/m)\}$, (II)
Where $k = 1, 2 \ldots m$.

Function $Q(r)$ for relative quantifier can be calculated as:

$$Q(r) = \begin{cases} 0 & \text{if } r<a \\ \frac{(r-a)}{(b-a)} & \text{if } a \leq r \leq b \quad \text{(III)} \\ 1 & \text{if } r>b \end{cases}$$

Q (0) = 0, $\exists r \; \varepsilon \; [0, 1]$ such that Q(r) =1, and a, b are parametric values such that a, b and r ε [0,1].

Example 1: for seven criteria, we have m = 7, it gives following weights; W(1) =0.0, W(2) =0.0, W(3) =0.0, W(4) =0.14285, W(5) =0.28571, W(6) =0.28571, W(7) =0.28571.

# 3 Book Recommendation

Our approach is composed of opinion mining and soft computing techniques. In this section, first, we discuss the opinion mining technique which is being used here, followed by illustration of soft computing technique. We use OWA here, thus respective steps involved in the recommendation process for implementing OWA and its quantifier, is explained with examples. Architecture for our recommender system (Fig. 1) is depicted and explained in this section as well.

## 3.1 Opinion Mining Technique

In the proposed technique for mining opinions, we have several steps to follow. The first step is to find the opinions of the users for books of a particular topic of interest. For collecting user's opinion on a particular book, we enter pre-defined queries, like, *"books on cloud computing"* to search engines. The results of the queries give links to the books on Search Engine Result Page (SERP). We store all these results in a book table data base. The book table information helps us in finding the online reviews and opinion of the user. These steps are demonstrated in Fig. 1, from 'queries-block' to 'opinion finding-block'. Opinion mining of the user's reviews is aimed to explore the features of the books and weight those features in accordance with the reviews of the users. Once we have obtained the user's opinion, we need to go through following major steps. 1) Feature extraction from opinions, 2) feature categorization, and 3) quantification to these features, i.e. assigning numerical values to features; we call it scoring the categorized features.

*1) Feature Extraction*: We have selected books from different area of computer science. For book of each area, we pass query like, "books on the 'area of computer science' ", e.g., the query, "books on cloud computing" is used for collecting the books on cloud computing. The names of the books that appear in top 100 links, with its content including Title of the book, who has authored the book, who is the publisher of the book, what is the edition, price and number of pages in the book, are stored. We have used Search Engine Optimization (SEO) tools for collecting all these information.

Again, we pass different queries like, "review of <book-name>", where <book-name> is obtained from book information table. It helps us in finding the opinion of the users of different books. We have selected more than 10 different area of computer science and each area has on average 40-50 different books in its top 100 links. A book may and may not have its review available on Internet. Some books have more than 200 reviews and some of them have no review, hence, approximately we have to work on $5x10^4$ reviews. Though we are not concerned with recommending all books but to give an idea to recommend top books on a specified topic, and hence we have illustrated all the effort with one book, as it is suffiient to demonstrate the procedure.

*2) Feature Categorization*: We have categorized seven features of books by examining book-reviews. The feature categorization is based on the commonality in user's reviews and emphasis of the users' feedback for a particular feature of the book in their respective reviews. The features are namely, multiple occurrences, Helpful content in the book, comprehensive material; availability in the market, Irrelevant content in the book, cost and user's rating respectively.

Multiple occurrences tell the multiple appearances of the same books on single query in SERP, 'Helpful content' means whether the content in the book is supportive or not. Comprehensive material implies sufficient illustration and comprehensive content in the book. Availability in the market tells the demand of the book and its selling trends, etc.

Irrelevant content is reciprocal to helpful content. Cost of a book also affect the user's choice for a particular book, thus we have categorized cost as a separate feature. User's rating signifies the overall importance of the book in the eyes of a user.

*3) Scoring the categorized features:* the reviews obtained from online review sites can be categorized as positive and negative reviews depending upon the positive and negative content contained in the reviews, respectively.

TABLE 1. POINT SCALE FOR REVIEW COMMENTS

| Examples of review words representing user's sentiments | Sense | Points scale (PS) |
|---|---|---|
| basic, Excellent, fantastic, great addition, invaluable, good job, useful, etc. | (+)ve | $0 \geq PS \leq 1$ |
| Disappointment, too costly, boring, failing, time pass, etc. | (-)ve | $-1 \geq PS \leq 0$ |
| Non availability of reviews | Neutral | PS=0 |

TABLE 2. FEATURES AND WEIGHTS

| Features | Weights Assigned | Weight Range |
|---|---|---|
| Occurrence | $S_{MO}$ | $0 \leq S_{MO} \leq 4$ |
| Helpfulness | $S_{HC}$ | $0 \leq S_{HC} \leq 4$ |
| Comprehensive material | $S_{CM}$ | $0 \leq S_{CM} \leq 3$ |
| Market availability | $S_{MA}$ | $0 \leq S_{MA} \leq 2$ |
| Irrelevant content | $S_{IC}$ | $-2 \leq S_{IC} \leq 0$ |
| Cost of books | $S_C$ | $-2 \leq S_C \leq 0$ |
| User's rating | $S_{UR}$ | $0 \leq S_{UR} \leq 1$ |

These reviews are quantified using -1 to 1 point scale. Table 1 shows the example of quantification of reviews. Further, these values are used as a base for scoring the categorized features. We denote respective scores as $S_{MO}$, $S_{HC}$, $S_{EM}$, $S_{MA}$, $S_{IC}$, $S_{C}$, $S_{UR}$ respectively. The different scores calculated for seven features are shown in table 2.

### 3.2 Soft Computing Technique

We have obtained different scores for categorized features of a book by using opinion mining technique, as demonstrated in the above section. Now, we implement averaging operator, OWA, to aggregate the assigned weights of all the features. We have considered all the seven features as 7 criteria for recommending the books using OWA operator. We use quantifier at least half, most and as many as possible for observing how many of the features have high value for a particular book, e.g. let us consider as many as possible, it implies '*as many as possible*' features that satisfy the user about concerned book.

Example 2: For calculating weights of the quantifier, 'as many as possible', we use parametric values as 0.5 and 1. As we have seven criteria, i.e. m = 7, it gives following weights; W(1) =0.0, W(2) =0.0, W(3) =0.0, W(4) =0.14285, W(5) =0.28571, W(6) =0.28571, W(7) =0.28571

Let us consider a review to understand how the scores are calculated and the way OWA with respective quantifiers are applied over the opinion scores.

" *I award this book 10 out of 10 for overall content, structure and identification of future research areas, this book has very nice introduction, very useful*"

Let we need to find the final value of a book whose code is given by 'CC.22' for the above review, the scores of the seven features are calculated as 1, 0.6, 0.4, 0.75, 0,0,0.5 respectively for features-multiple occurrences, Helpful content in the book, comprehensive material; availability in the market, Irrelevant content in the book, cost and user's rating respectively. The occurrence feature is found in the Google SERP. By observing the above review it is easily noticed that book has strong positive review thus gains 0.6 and 0.4 for content and material features. in the similar way all the reviews are collected. Depending upon these reviews final score for the book is recorded accordingly. For example 2, the OWA value comes out to be;

$$\text{OWA} = \sum_{k=1}^{n} w_k z_k$$

OWA (as many as possible) = 1*0 + 0.75*0 + 0.6*0 + 0.5*0.14285 + 0.4*0.28571 + 0*0.28571 + 0*0.28571

= 0.185709

Similarly, for finding values of OWA (at least half); we use parametric values as a=0 and b=0.5 for quantifier 'at least half'. the respective weights are; W(1) =0.28571, W(2) =0.28571, W(3) =0.28571, W(4) =0.14285, W(5) =0.0, W(6) =0.0, W(7) =0.0. And for quantifier 'most', we use parametric value as a=0.3 and b=0.8. The weights for quantifier is calculated as; W(1) =0.0, W(2) =0.0, W(3) =0.25714, W(4) =0.28571, W(5) =0.28571, W(6) =0.17142, W(7) =0.0.

# 4 Experiments and Results

As described in the section III, we get three different ranking by the proposed method, i.e. applying OWA over opinion mining technique with three different quantifiers. The detail scoring table for quantifier 'at least half' is shown in in table 3 for illustration. In the same way, the values of others quantifiers are obtained. These values serve as a base for books final ranking. The top 5 books' code and their respective OWA values for all three quantifiers are listed in table 4.

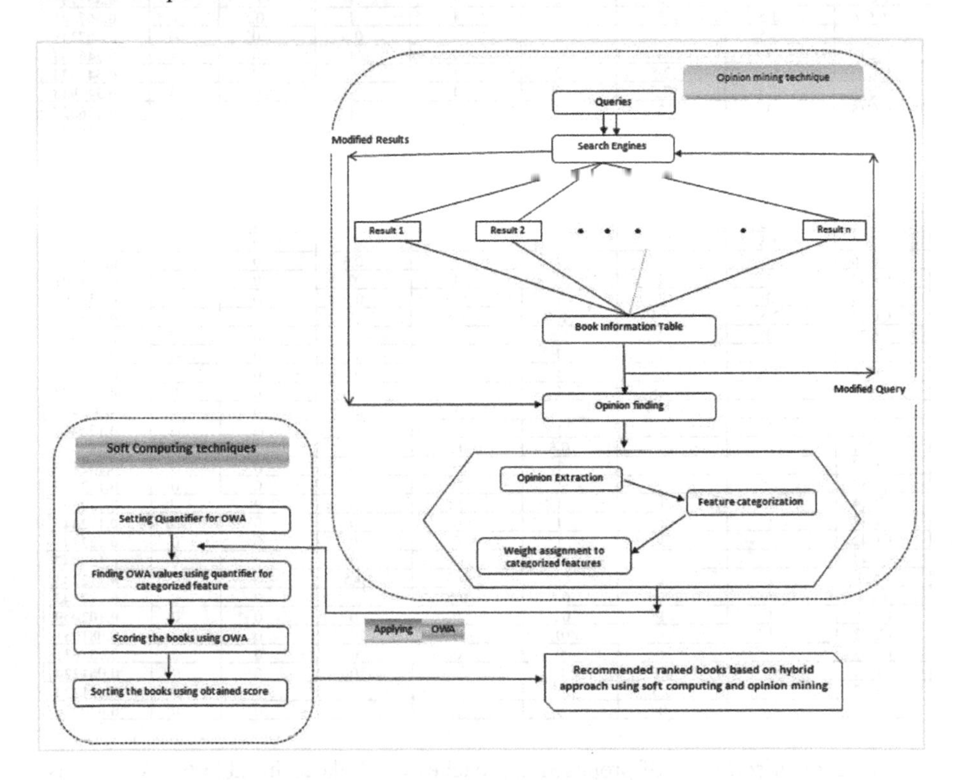


Fig 1. Architecture of the Recommender Technique

The real challenge here is to compare the ranking and find the best among them, as we do not have any true ranking or any bench mark of ranking for particular books on cloud computing. We have compared the three ranking with the ranking done in [mysore]. We have taken top 5 ranking of cloud computing books published in *'cloudzone'* [31] and *'saasaddict'* [32], both have published same ranking of books, hence we consider it as a base ranking. The base ranking of books is shown in table 5 [31, 32].

TABLE 3. DETAILS SCORE TABLE FOR QUANTIFIER 'AT LEAST HALF'

| Book code | Occurrence | Helpfulness | Comprehe-nsive material | Market availability | User's rating | Irrelevant content | Cost of books | Average |
|---|---|---|---|---|---|---|---|---|
| CC2 | 3.5 | 3.5 | 2.5 | 2 | 1 | -0.5 | - | **0.428564** |
| CC35 | 3.5 | 3.5 | 2.5 | 1 | 1 | 0 | 0 | **0.408156** |
| CC13 | 4 | 3 | 2 | 2 | 0 | 0 | - | **0.408156** |
| CC22 | 4 | 3 | 2 | 1.5 | 1 | 0 | 0 | **0.397952** |
| CC21 | 4 | 3 | 2 | 1 | 1 | -1.5 | - | **0.387749** |
| CC7 | 3.5 | 3 | 2 | 1 | 1 | -0.5 | - | **0.367341** |
| CC36 | 3.5 | 3 | 2 | 1 | 0.5 | 0 | -1 | **0.367341** |
| CC15 | 3 | 3 | 2 | 1 | 1 | 0 | - | **0.346933** |
| CC23 | 3.5 | 2.5 | 2 | 1 | 1 | 0 | -1 | **0.346933** |
| CC28 | 3.5 | 2.5 | 2 | 1 | 0.5 | 0 | 0 | **0.346933** |
| CC18 | 3 | 3 | 1.5 | 1 | 1 | 0 | 0 | **0.326525** |
| CC9 | 3 | 3 | 1.5 | 1 | 1 | 0 | -1 | **0.326525** |
| CC31 | 3.5 | 2 | 1.5 | 1 | 0 | 0 | 0 | **0.306117** |
| CC29 | 3 | 2.5 | 1.5 | 1 | 0.5 | 0 | 0 | **0.306117** |
| CC41 | 3 | 2.5 | 1.5 | 1 | 0.5 | 0 | - | **0.306117** |
| CC40 | 3 | 2 | 2 | 1 | 0 | 0 | 0 | **0.306117** |
| CC3 | 2.5 | 2 | 2 | 1.5 | 0 | 0 | -1 | **0.295913** |
| CC5 | 2.5 | 2.5 | 1.5 | 0.5 | 0 | 0 | 0 | **0.275506** |
| CC16 | 3 | 2 | 1 | 1 | 0.5 | 0 | 0 | **0.265301** |
| CC12 | 3 | 2 | 1 | 1 | 0 | 0 | - | **0.265301** |
| CC10 | 3 | 2 | 1 | 0.5 | 0.5 | 0 | - | **0.255098** |
| CC25 | 2.5 | 1.5 | 1.5 | 1.5 | 1 | 0 | 0 | **0.255097** |
| CC20 | 2 | 2 | 1.5 | 1 | 1 | 0 | -1 | **0.244894** |
| CC17 | 2 | 2 | 1 | 1 | 0.5 | 0 | -1 | **0.224486** |
| CC34 | 2.5 | 1.5 | 1 | 1 | 0 | 0 | 0 | **0.224486** |
| CC8 | 2 | 1.5 | 1 | 0.5 | 0.5 | 0 | 0 | **0.193874** |
| CC6 | 1.5 | 1.5 | 1 | 1 | 0.5 | 0 | - | **0.18367** |
| CC26 | 2 | 1 | 1 | 0.5 | 0 | -1 | -1 | **0.173466** |
| CC11 | 2 | 1 | 0.5 | 0.5 | 0 | 0 | 0 | **0.153059** |
| CC37 | 1.5 | 1 | 1 | 0 | 0 | 0 | 0 | **0.142855** |
| CC39 | 1.5 | 1 | 1 | 0 | 0 | 0 | 0 | **0.142855** |
| CC14 | 1 | 1 | 1 | 0.5 | 0 | 0 | -1 | **0.132651** |
| CC38 | 1 | 1 | 1 | 0 | 0 | 0 | 0 | **0.122447** |
| CC42 | 1 | 1 | 1 | 0 | 0 | 0 | 0 | **0.122447** |
| CC43 | 1 | 1 | 1 | 0 | 0 | 0 | 0 | **0.122447** |
| CC4 | 1 | 1 | 0.5 | 0.5 | 0.5 | -1 | - | **0.112243** |
| CC1 | 1 | 1 | 0.5 | 0.5 | 0 | 0 | 0 | **0.112243** |
| CC30 | 1 | 1 | 0.5 | 0 | 0 | 0 | 0 | **0.102039** |
| CC32 | 1 | 1 | 0 | 0 | 0 | 0 | 0 | **0.081631** |
| CC33 | 1 | 1 | 0 | 0 | 0 | 0 | 0 | **0.081631** |
| CC24 | 1 | 0.5 | 0 | 0 | 0 | 0 | 0 | **0.061224** |
| CC27 | 1 | 0.5 | 0 | 0 | 0 | 0 | 0 | **0.061224** |
| CC19 | 1 | 0 | 0 | 0 | 0 | 0 | 0 | **0.040816** |

The comparison table of proposed approaches is tabulated in table 6. A graph is shown in Fig. 2 for corresponding values demonstrate the performance of different approaches. It is evident from the Fig. 2 that the proposed approach which uses 'as many as possible' quantifier out performs other approaches. The ranking of books by best method with the details of books and authors are listed in table 7.

TABLE 4. THE TOP 5 BOOKS' CODE AND THEIR RESPECTIVE OWA VALUES

| Ranking | At least half | | Most | | As many as Possible | |
|---|---|---|---|---|---|---|
| | Code of the book | OWA | Code of the book | OWA | Code of the book | OWA |
| 1 | CC2 | 0.80712 | CC2 | 0.44854 | CC35 | 0.21428 |
| 2 | CC13 | 0.79998 | CC35 | 0.41426 | CC22 | 0.18570 |
| 3 | CC22 | 0.74284 | CC22 | 0.41139 | CC18 | 0.15713 |
| 4 | CC21 | 0.67141 | CC15 | 0.38568 | CC15 | 0.14999 |
| 5 | CC35 | 0.66427 | CC23 | 0.38568 | CC25 | 0.12856 |

TABLE 5. BASE RANKING OF BOOKS

| Rank position | 1 | 2 | 3 | 4 | 5 |
|---|---|---|---|---|---|
| Base ranking | CC22 | CC6 | CC7 | CC18 | CC15 |

TABLE 6. PRECISION@5 FOR DIFFERENT APPROACHES

| Approach | OWA (As many as Possible) | OWA (At least half) | OWA (Most) |
|---|---|---|---|
| Precision@5 | 0.6 | 0.2 | 0.2 |

TABLE 7. FINAL RECOMMENDED BOOKS

| Code | Title | Author | Publisher |
|---|---|---|---|
| CC 22 | Cloud Computing Explained: Implementation Handbook for Enterprises | John Rhoton | Recursive Press |
| CC 25 | Cloud Computing: Concepts, Technology & Architecture | Thomas Erl, Zaigham Mahmood, Ricardo Puttini | Prentice Hall |
| CC 35 | Microsoft Private Cloud Computing | Aidan Finn, Hans Vredevoort, Patrick Lownds, Damian Flynn | Sybex |
| CC 18 | Cloud Security and Privacy: An Enterprise Perspective on Risks and Compliance | Tim Mather, Subra Kumaraswamy, Shahed Latif | O'Reilly Media |
| CC 15 | Enterprise Cloud Computing: A Strategy Guide for Business and Technology Leaders | Andy Mulholland , Jon Pyke , Peter Fingar | Meghan-Kiffer Press |

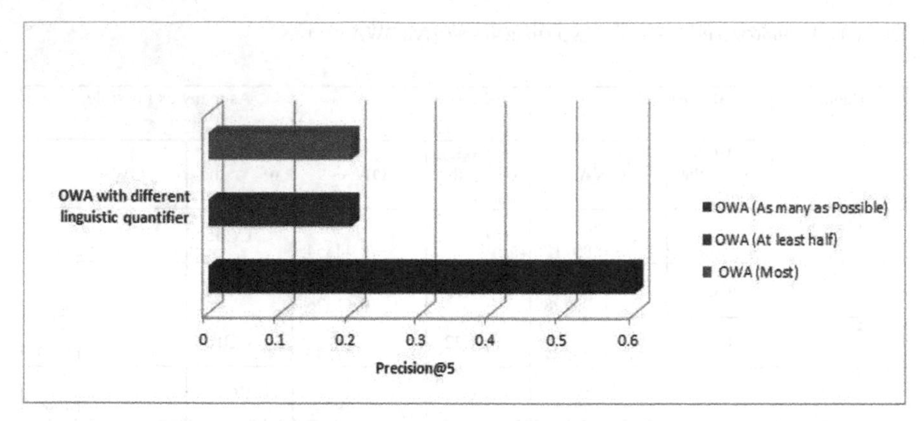


Fig. 2 Precision@5 for different approaches

# 5 Conclusion

The basic idea here is to present a recommender system for books by applying a combination of soft computing and opinion mining techniques. The OWA is used as a soft computing technique. As we have used OWA operator with three different quantifiers. The precision for ranking of books recommended by proposed approach using respective quantifiers are shown, and it turned out that the quantifier 'as many as possible' has greater value of precision.

The proposed approach can be implemented in various recommender systems where the recommendations are made by customer's opinion only and expected to be very useful. The combined approach, which we may call as a hybrid approach can enhance the recommendation process and also improves the quality of recommendations.

Further, these techniques can be integrated to implement it over a larger data set and to a variety of books for respective users. Also, we may have some explicit or implicit feedback from users directly to counter-check the performance of the system as well.

# References

1. J. Bobadilla, F. Ortega, A. Hernando, and A. Gutiérrez, "Knowledge-Based Systems Recommender systems survey," vol. 46, pp. 109–132, 2013.
2. G. Adomavicius and A. Tuzhilin, "Toward the Next Generation of Recommender Systems : A Survey of the State-of-the-Art and Possible Extensions," vol. 17, no. 6, pp. 734–749, 2005.
3. R. Burke, "Hybrid Web Recommender Systems," pp. 377–408, 2007.
4. R. Burke, A. Felfernig, and M. H. Göker, "Recommender Systems : An Overview," 1997.
5. B. Pang and L. Lee, "Opinion Mining and Sentiment Analysis," vol. 2, pp. 1–135, 2008.
6. S. S. Sohail, J. Siddiqui and R. Ali. "Book recommendation system using opinion mining technique." In Advances in Computing, Communications and Informatics (ICACCI), 2013 International Conference on, pp. 1609-1614. IEEE, 2013.
7. R. Yager, and Kacprzyk, J. eds., 2012. "The ordered weighted averaging operators: theory and applications," Springer Science & Business Media.

8. Y. Fang, L. Si, N. Somasundaram and Z. Yu. "Mining contrastive opinions on political texts using cross-perspective topic model." In Proceedings of the fifth ACM international conference on Web search and data mining, pp. 63-72. ACM, 2012.

9. V. Hatzivassiloglou and K. McKeown, "Predicting the semantic orientation of adjectives" EACL, pp-174-181, 1997.

10. S. Kim and E. Hovy,"Determining the sentiment of opinions," COLING, pp- 1367-1374, 2004.

11. B. Pang, L. Lee, and S. Vaithyanathan, "Thumbs up?: sentiment classification using machine learning techniques" In EMNLP, pp- 79-86, 2002.

12. B. Liu, M. Hu, and J. Cheng, "Opinion observer: Analyzing and comparing opinions on the web," WWW, pp-342-351, 2005.

13. G. Shani and A. Gunawardana. "Evaluating recommendation systems." In Recommender systems handbook, pp. 257-297. Springer US, 2011.

14. J. Beel, L. Stefan, G. Marcel, G. Bela, B. Corinna and N. Andreas, "Research paper recommender system evaluation: a quantitative literature survey." In Proceedings of the International Workshop on Reproducibility and Replication in Recommender Systems Evaluation, pp. 15-22. ACM, 2013.

15. X. Ding, B. Liu, and P. Yu, "A holistic lexicon-based approach to opinion mining," WSDM, pp-231-240, 2008.

16. H.Liu, , J. He, , T. Wang, , W. Song, and X. Du,"Combining user preferences and user opinions for accurate recommendation. Electronic Commerce Research and Applications", 12(1), pp.14-23, 2013.

17. S. S. Sohail, J. Siddiqui, and R. Ali. "User feedback scoring and evaluation of a product recommendation system." In Contemporary Computing (IC3), 2014 Seventh International Conference on, pp. 525-530. IEEE, 2014.

18. S. S. Sohail, J. Siddiqui, and R. Ali, "User Feedback Based Evaluation of a Product Recommendation System Using Rank Aggregation Method." In Advances in Intelligent Informatics, pp. 349-358. Springer International Publishing, 2015.

19. M. Hu and B. Liu, "Mining and summarizing customer reviews," KDD'04, 2004.

20. V. Hatzivassiloglou and J. Wiebe, "Effects of adjective orientation and gradability on sentence subjectivity," COLING, 2000.

21. P. Beineke, T. Hastie, C. Manning, and S. Vaithyanathan, "An Exploration of Sentiment Summarization," Proc. of the AAAI Spring Symposium on Exploring Attitude and Affect in Text: Theories and Applications, 2003.

22. N. Kaji and M. Kitsuregawa, "Automatic Construction of Polarity-Tagged Corpus from HTML Documents," COLING/ACL'06, 2006.

23. L. A. Zadeh, "Roles of soft computing and fuzzy logic in the conception, design and deployment of information/intelligent systems." In Computational intelligence: soft computing and fuzzy-neuro integration with applications, . Springer Berlin Heidelberg, 1998, pp. 1-9.

24. G Beliakov, A Pradera, T Calvo, "Aggregation Functions: A Guide for Practitioners, Springer, Heidelberg," Berlin, 2007.

25. J. Malczewski, "Ordered weighted averaging with fuzzy quantifiers: GIS-based multicriteria evaluation for land-use suitability analysis," International Journal of Applied Earth Observation and Geoinformation, vol. 8, no. 4, 2006, pp.270-277.

26. C.K. Makropoulos and D. Butler, "Spatial ordered weighted averaging: incorporating spatially variable attitude towards risk in spatial multi-criteria decision-making. Environmental Modelling & Software", vol. 21 no. 1, 2006, pp.69-84.

27. S. S. Sohail, J. Siddiqui, and R. Ali, "OWA based Book Recommendation Technique". Procedia Computer Science, vol. 62, 2015, pp.126-133.

28. M. M. S. Beg, "User feedback based enhancement in web search quality, Information Sciences," vol. 170, no. 2-4, 2005, pp. 153–172.

29. G. Ahamad, S. K. Naqvi and M. M. Beg. "An OWA-Based Model for Talent Enhancement in Cricket." International Journal of Intelligent Systems, 2015.

30. R. Yager, "On Ordered Weighted Averaging Aggregation Operators in Multicriteria Decision Making. IEEE Trans. Systems, Man and Cybernetics," vol. 18, no. 1, 1988, pp. 183-190.

31. https://dzone.com/articles/5-best-cloud-computing-books/ [accessed on 25-02-2016].

32. http://saasaddict.walkme.com/5-best-cloud-computing-books-you-should-check--out/ [accessed on 25-02-2016]

# Bio-inspired Model Classification of Squamous Cell Carcinoma in Cervical Cancer using SVM

**M.Anousouya Devi, S.Ravi, J.Vaishnavi, S.Punitha**

Department of Computer Science ,Pondicherry University, Pondicherry

**Abstract** Cervical cancer is a deadly cancer which occurs in women's of all age group without any pre-symptoms. This cancer can be detected at the earliest by the manual screening of Pap smear test and LCB test which suffers from high false positive rate and cost effectiveness. To overcome this disadvantage an automatic computerised system is used to enhance the efficiency and sensitivity in the detection of cervical cancer. There are three types of tissues in the cervix region of uterus as Columnar Epithelium (CE), Squamous epithelium (SE) and the Aceto white (AW) region. The AW region, when immersed in 5% acetic shows a change of white colour as abnormal cervical cells. In this paper, a bio-inspired model is used for the automatic detection of cervical cancer where a support vector machine is used to classify the squamous cell carcinoma which will produce more accurate results by reducing the false positive rates.

Key words: Cervical cancer, Pap smear test, LCB test, AW region, Squamous cell carcinoma, Support Vector Machine, Bio-inspired algorithm.

## I. Introduction

Cervical cancer is the second most common cancer that occurs in women of all age group. This cancer is very dangerous because it has high mortality rates as it cannot be detected at the earlier stage. This cancer occurs in the cervix region of the uterus which acts as a lid to the uterus. There are



J.M. Corchado Rodriguez et al. (eds.), *Intelligent Systems Technologies and Applications 2016*, Advances in Intelligent Systems and Computing 530,
DOI 10.1007/978-3-319-47952-1_47

no symptoms detected for the occurrence of this cancer in women. The Human Papilloma Virus (HPV) causes cervical cancer which has many gene types where the HPV16 and HPV 18 lead to cervical cancer. In 2016, there are nearly 1,685,210 cases in United States of America that diagnosis cervical cancer where 595,690 women will die because of this disease [1].

The mortality rate of cervical cancer in women has decreased the death rate in women because of the manual screening Pap test. This test must be carried out in all women where a doctor uses a spatula or a brush to collect the cervical cells from the cervix region. The collected sample cells are diagnosed [2] by a pathologist for the classification of normal and abnormal cells which will diagnose the presence of cervical cancer. The main drawback of this method is that there are only few experienced pathologist available to carry out this test efficiently and a pathologist can diagnose only 10 slides in a day which is time consuming and tends for human errors. The manual screening method is also cost effective.

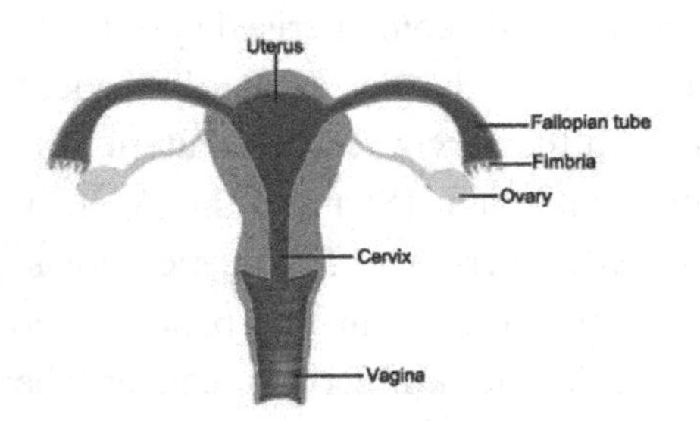


**Fig.1 . .Structure of Uterus**

The Liquid cytology based test (LCB) is another popular manual screening method where the cervical cells are immersed in 5% acetic acid which leads to a colour change of white region indicating the abnormal cells. This manual method suffers from time consumption [3] and the colour changes regions are not observed carefully. The prevention of cervical cancer can be made only by the early detection. The colour images of the cervix are taken by a camera with telephoto lens which is called as cervicography.

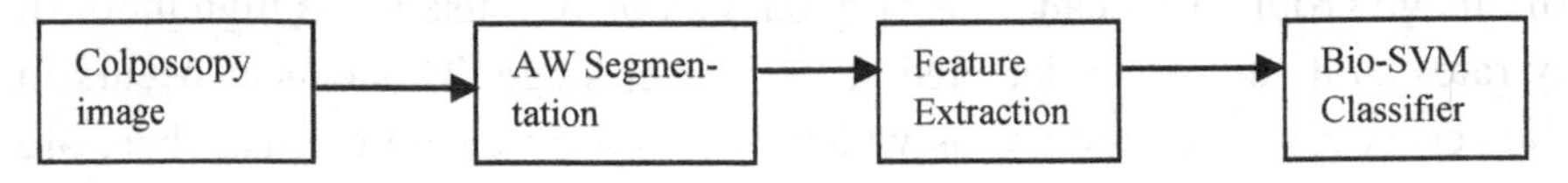


**Fig. 2** Cervical cancer Medical Image process

After the images are collected, the image is pre-processed by using a median filter in removing the debris and noise. The abnormal regions must be identified from the tissues which can be obtained from the Aceto-White (AW) region by segmenting the AW region using the K- means clustering algorithm where the Euclidean distance between the cells are calculated and segments the nucleus and cytoplasm where the abnormal cells are identified. The intensity and contrast of the colour pixels are calculated by Fourier transform. The classification of the cells as mild, normal and severe are done by using a bio-inspired algorithm with a support vector machine (SVM) by detecting the stages of cancer. The rest of the paper is organised as: section 2 briefs about the related work; section 3 gives a brief description about the AW segmentation process, section 4 describes the feature extraction process; section 5 explains the support vector machine classification with the bio inspired algorithm section 6 discusses the various methods used for the classification of cervical cancer with SVM features to be extracted from the colposcopy image and finally the paper is concluded in section 6.

## II. Related work

Holger Lange [4] proposes an automated method for detecting the Aceto-White region with multi-level region growing algorithm. The input image given is a colposcopy image which uses the RGB colour images for visual inspection in the metaplastic epithelium colour change of AW region into white. The regions of interest in the AW are obtained and classified with hue classifier. The multi-stage region growing is selected with Green channel saturation from the pixel values and segments the AW region of the cervix. Yeshwanth Srinivasan et al [5] propose a probabilistic method for the segmentation and classification of AW region in cervical cancer

The RGB colour image with high intensity pixels are extracted and the geometric features are extracted with a linear rotating structuring element (ROSE) to observe the colour changes in the pixel intensity value. The RGB colour space is converted into YCbCr model to extract the mosaic and non-mosaic region of the cervix. Yeshwanth et al [6] proposes a novel segmentation of AW region in the cervix and classification of the cervix lesion by the texture analysis. The RGB images are given as the input colposcopy image which is converted into YCbCr image to extract the image with high intensity values and cluster the pixel range with high and low values where the segmentation of AW region is determined with the set of cluster values. The texture features are extracted for the classification with mathematical morphological operations. Rajesh kumar et al [7] proposes a detection and classification method of biological interpretable features. The K-means segmentation algorithm is used to extract the features from the background image and K-nearest neighbourhood method is used as the classifier for differentiating the normal and cancer cells. Thus the proposed Bio-inspired model adds a weightage over the existing methods by using a fitness function evaluation which is not yet evaluated for detecting cervical cancer using a support vector machine classifier. The existing method lacks accuracy and specificity which can be overcome by the proposed method.

## III. Aceto-White Segmentation

There are three layers of tissues present in the cervix layer of the uterus as columnar epithelium (CE), Squamous epithelium (SE) and the Aceto white (AW) region. The application of acetic acid test [8] to AW region performs a visual colour change of white colour indicating the identification of abnormal cells in the manual process of LCB test where it suffers from dehydration of abnormal cells and the light intensity cannot be passed through large nuclear cell structure.

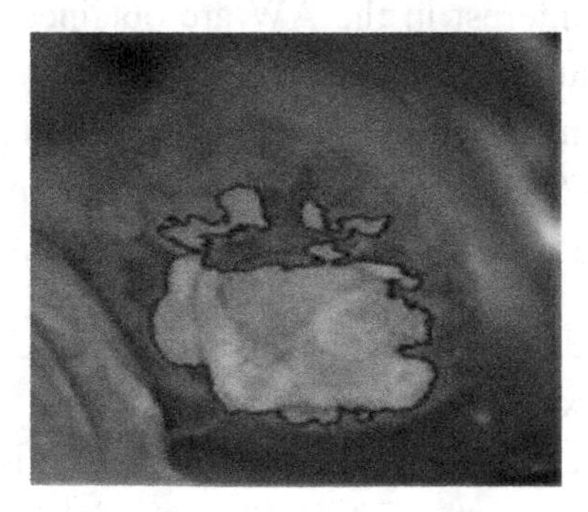

**Fig. 4.** *Aceto White regions Segmentation*

The identification of AW region is very important because of the indication level in the severity can be segmented and analysed from this region. The automated segmentation of AW region suffers from several drawbacks like the surface of the cervix layers is not flat and smooth. This makes it as a reason for the light intensity or colour feature extraction cannot [9] is located closer to the region and a clear perception of the image cannot be obtained. If the AW region is segmented and identified accurately then with the subjective comparison made with ground truth data, the results obtained will be more accurate than the other existing methods. In this proposed method, the colposcopy image [10] of the cervix is taken as the input, where the image of the cervix is extracted alone and cropped in order to concentrate only on the cervix layer and save the computational time. The first step is to remove [11] the unwanted or high reflectance of light from the images. The obtained cervix image will be in the RGB colour space format. The R, G, B high pixel intensity values are extracted separately. The K-means clustering algorithm [12] is used to calculate the Euclidean distance between the neighbourhood pixels and replaces by zero the unwanted high reflectance pixels. The pixels with the high intensity values are made into clusters and based on this cluster value the K value is determined. Thus the k value segments the AW region. The Euclidean distance between two pixels is calculated by the formula (1) given below:

$$D\,(a, b) \;=\; \sum \left| ai \right. \qquad (1)$$

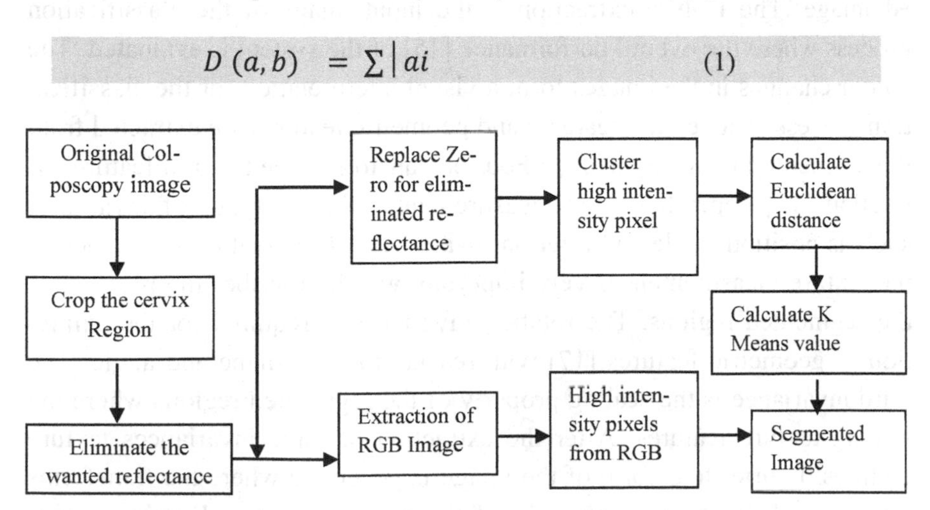


**Fig.5.** *Segmentation of K-Means Clustering algorithms in Aw Region*

The d(a,b) gives the Euclidean distance between the two pixels a and b where i value is the position of the pixels. The calculation of this distance helps to measure or identify the similarity in shape between the clusters. The K-means algorithm partition the regions into K clusters where $K < n$, where it assumes the attributes to form a vector space .The AW segmentation process with RGB images is very challenging task because of the un-uniform surface of the cervix which makes it difficult [13] for the identification of the regions. Figure 3 illustrates the colposcopy image of the AW region where the acetic acid test produces a change of white colour which is segmented for the classification of cervical cancer. The segmentation process of the K-Means clustering algorithm is explained in figure 4 where the block diagram explains the pixel clustering with high intensity values of R, G, B colour space and it is segmented by calculating the K value. The pixels with high [14] intensity are clustered and the nearest neighbour value is calculated by obtaining the Euclidean distance between the pixels. The value of k segments the image with respect to the white regions.

## IV. Fast Fourier Texture Transform

The segmentation and classification of cervical cancer with colposcopy images is dependent on the vital features to be extracted from the segmented image .The feature extraction is the input image of the classification process where the overall performance [15] of the system is evaluated. The colour changes in the images form a visual interpretation for the classification process. The texture features and geometric features are extracted from the segmented image by using a Fourier transform. The textural feature extraction [16] depends on the measurements of three types of invariance such as position, scale and rotation invariances. The position invariance of the texture measurement is very important which describes the position of the segmented regions. The rotation invariance is required for the extraction of geometric features [17] with respect to size, shape and angle. The third invariance is the scaling property of the segmented regions where the scaling texture features. After the extraction of three invariances texture features, Fourier transform of the image is generated where the transforms are grouped for the measurement of the values. The fast Fourier texture transform is used for the conversion of transform into pixel values by using the below equation.

$$FP = Pxy \tag{2}$$

The FP is the fast Fourier transform value which is converted into the pixels grouped value denoted by $P_{xy}$. The main advantage of using the fast Fourier texture transform is the clear quality of the image obtained by using a shift variance which makes the size of the transformed image equal to its original size. The extracted texture features with the position, rotating and scaling are represented in co-occurrence matrix which contains the pixel values of the high brightness intensity values between each pair of the pixels. The RGB colour images tend to have high brightness with the different colour space of R, G, B. Consider two pixels a, b the brightness between the pixels are measured by C1 and C2. The co-occurrence matrix is represented as given below

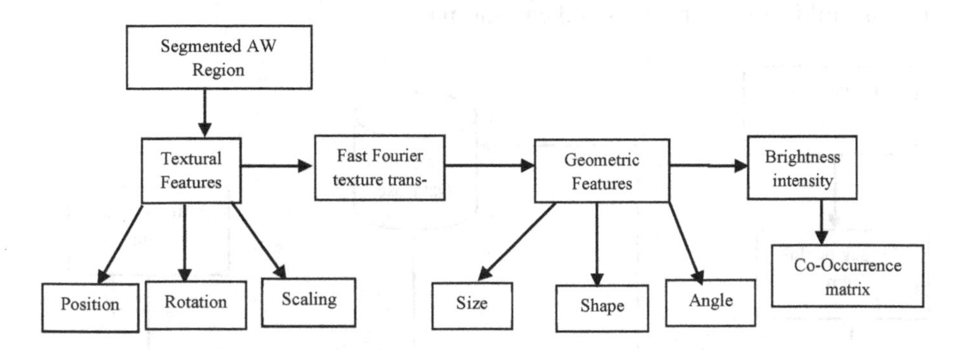


**Fig. 6.** *Fast Fourier texture transforms*

$$MC1C2 = \sum a = 1 \sum b = 1 \, (Pab = C1) \wedge (Pab = C2) \tag{3}$$

In figure 5, the Fast Fourier texture transform illustrates the types of the texture and geometric features extracted are represented with respect to the brightness intensity factor of the RGB images. The final pixel values are represented in terms of the co-occurrence matrix. The extracted features are given as the input for the Bio-support vector machine model for the classification of cervical cancer into normal, mild and severe stages which depends on the evaluation of fitness function value

## IV. Bio- inspired Support Vector Machines

The proposed classifier is a combination of the bio-inspired algorithm with the support vector machine model (SVM) which is termed as Bio-SVM. The texture and geometric features extracted from the segmented AW region is passed [18] to the Bio-SVM. The bio-inspired algorithm solves many complex problems and produces the optimal solution which is applied in support vector machine yields to produce more accurate results. In figure 7, the system architecture diagram of the [19] proposed is explained where the extracted textural and geometric features are given as input to the Bio-SVM where the features are initialized and the population value of the features is generated. The input images are a collection of colposcopy images called as cervigrams which is available at the national cancer library. The fitness function value is evaluated for each stage of categorization as mild, severe, normal and carcinoma.

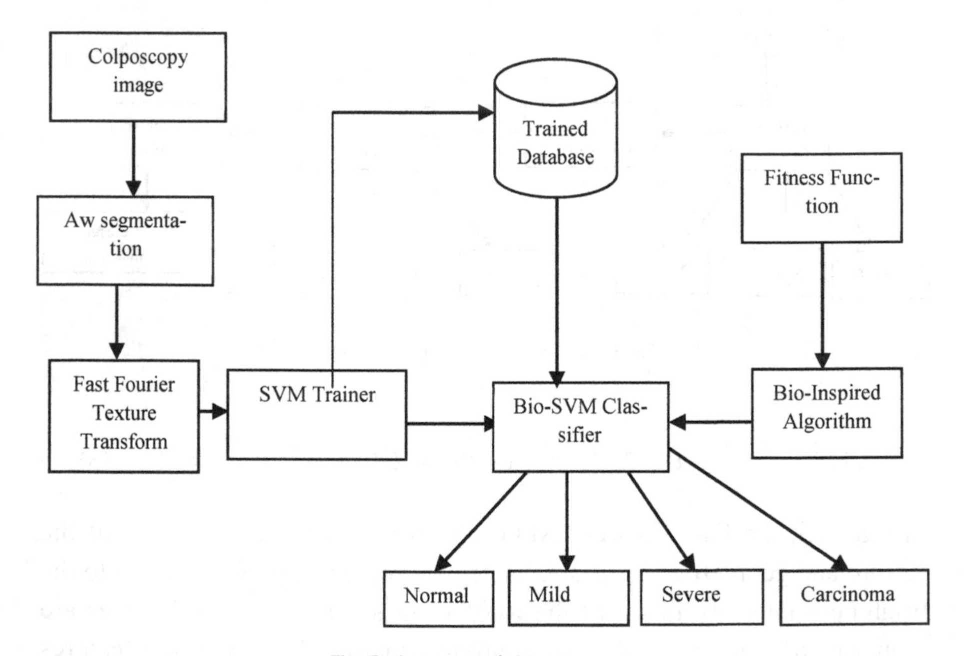


Fig. 7. System Architecture

The support vector machine uses a linear kernel model where the bio-inspired optimization computes a range of fitness function evaluation [20] for each classification of cervical cells. The database used is a trained set of data in order to make the process more efficient and effective. The training of data set for the SVM trainer is made iteratively with the manual calculated value from various experimental results. The values in the trained database is passed on to the SVM trainer for each input image to check the corresponding value of the severity classified as normal ,mild, severe and carcinoma .The trained dataset helps to save the computational time with respect to the analysis of value . The SVM trainer extracts the trained data set from the database and makes it easy for the Bio-SVM classification to classify the cervical cells as normal, mild, severe and carcinoma. The calculation of fitness function with the vector value range makes the classification with respect to each category. Thus this automatic detection overcomes all the drawbacks of the manual screening methods and will produce more accurate results than the existing methods. The support vector machine classification is used in many medical image applications [21] and yields accurate result and when this is combined with bio-inspired optimization algorithm will produce the optimal classification solution and detects the squamous carcinoma in cervical cancer and decreases the mortality rate in women.

## VI Discussion

In this paper, a bio-inspired model with support vector machine combination called Bio-SVM is proposed which is an automated method for the detection of cervical cancer which classifies the squamous carcinoma which is the deadly stage in women. Cervix is the opening mouth of the uterus. There are two parts in the cervix endocervix closest to the body of uterus and excocervix present next to vagina [22]. These two parts are covered by two types of cells called squamous cells and glandular cells respectively. The cervical cells are detected as precancerous cells called cervical intraepithelial neoplasia (CIN). There are two main types of cancer squamous cell carcinoma and adenocarcinoma. In this paper, the classification method for detecting the squamous cell carcinoma is proposed which occurs in the endocervix regions. Turid Torheim et al [23] proposes a dynamic contrast method for enhancing the cervical image contrast from MRI images and uses a support vector machine (SVM) classification with GLCM feature extraction for detecting cervical cancer. The SVM classification method produces more accurate results. Kun-Huang Chen et al [24] proposes a gene decision tree model for detecting cancer with particle swarm optimization. The proposed method is compared with all other classifiers where it is converted to SVM compatible mode and produces accurate results.

Table 1 illustrates the list of effective features to be extracted along with SVM classifier to produce accurate results. The extraction of features is the input parameter given to the classifier which process with a set of conditions and produces the results. Hence the selection of features must be valid which estimates the overall accuracy and performance of the system. The common features which can be extracted [25] from an image is size, shape, colour and texture. In this, the list of common features, the texture and colour plays a major role where the size and shape analyses of medical images cannot be made accurate because cervical cancer cells are shapeless and shape of the cells can only be assumed. The exact size of the image cannot be measured accurately.

So the features which can be represented for more clarity and accuracy are the texture and colour features. In the colposcopy image of the cervix, the texture feature extraction [26] is very challenging because the region of the cervix surface is unevenly distributed and varies from one surface to another. But the texture feature extraction will produce more accurate results than the other features because if extracted accurately, the prominence value can be measured accurately and does not provide any changes. Thus the Bio-SVM classifier uses the above mentioned features and yields more accurate results than other methods.

**Table 1.** *List of features extracted with Bio-SVM*

| S.No | Type of Feature | Feature description |
|---|---|---|
| 1 | Texture | Uniform or uneven surface of the region |
| 2 | Geometric | Size and shape |
| 3 | Rotation | The texture rotation invariance |
| 4 | Position | The location of the texture to be identified |
| 5 | Scaling | The colour changes with respect shading effects |

| | | |
|---|---|---|
| 6 | Contrast | The shades of image with darker and lighter regions identification |
| 7 | Intensity | The pixel differentiation value with colour |
| 8 | Correlation | The linearity measure of the image |
| 9 | Entropy | The measure of intensity within the region |

## VII. Conclusion and Future Work

The manual screening methods of cervical cancer suffers from high false positive rates and cost effective. There are only few experienced pathologist to perform this test which tends to produce manual errors to overcome the drawbacks in the manual screening process, a bio-inspired model with Support vector machine is used as a classifier to detect the Squamous Cell Carcinoma in Cervical Cancer [27] with colposcopy images which are in RGB colour space. The k-means clustering algorithm is used for the segmentation of AW region which changes into a white colour when immersed in 3-5 % acetic acid. The textural and geometric features are extracted from the segmented image by using a fast Fourier texture transform. The proposed bio-SVM classifier will produce more accurate results than the existing automated methods. In future, the proposed classifier will be implemented with colposcopy images of cervical cancer which will be classified into mild, normal and severe cells. This automatic detection and classification method will produce more accurate results.

REFERENCES

[1]http://www.cancer.org/acs/groups/content/@research/documents/document/acspc-047079.pdf
[2]Jantzen, J., Dounias, G., & Engineering, M. (n.d.). ANALYSIS OF PAP-SMEAR IMAGE DATA.
[3] Payne, N., Chilcott, J., & Mcgoogan, E. (2000). Liquid-based cytology in Standing Group on Health Technology Chair :, *4*(18).
[4] Holgersti-medicalcom, H. L. (n.d.). Automatic detection of multi-level acetowhite regions in RGB color images of the uterine cervix.
[5]Srinivasan, Y., Hernes, D., Tulpule, B., Yang, S., & Guo, J. (n.d.). A P robabilistic A pproach to S egmentation and C lassification of N eoplasia in U terine C ervix I mages U sing C olor and G eometric F eatures National Library of Medicine , Rockville , MD 20852 ;, *5747*, 995–1003.
[6]Srinivasan, Y., Gao, F., Mitra, S., & Nutter, B. (2006). Segmentation and classification of cervix lesions by pattern and texture analysis, *1*, 234–246.
[7]Kumar, R., Srivastava, R., & Srivastava, S. (2015). Detection and Classification of Cancer from Microscopic Biopsy Images Using Clinically Significant and Biologically Interpretable Features, *2015*.
[8]Cytoplasm and nucleus segmentation in cervical smear images using radiating GVF snake", Kuan Li et Al, 2011

[9]ZHANG, J.-W., ZHANG, S.-S., YANG, G.-H., HUANG, D.-C., ZHU, L., & GAO, D.-F. (2013). Adaptive Segmentation of Cervical Smear Image Based on GVF Snake Model. *Proceedings of the 2013 International Conference on Machine Learning and Cybernetics*, 14–17.
[10]Xue, Z., Long, L. R., Antani, S., & Thoma, G. R. (2010). Automatic extraction of mosaic patterns in uterine cervix images. *Proceedings - IEEE Symposium on Computer-Based Medical Systems*, 273–278
[11]Kumar, R. R., Kumar, V. A., Kumar, P. N. S., Sudhamony, S., & Ravindrakumar, R. (2011). Detection and removal of artifacts in cervical Cytology images using Support Vector Machine. *2011 IEEE International Symposium on IT in Medicine and Education*, *1*, 717–72
[12]Wang, Y., Crookes, D., Eldin, O. S., Wang, S., Hamilton, P., & Diamond, J. (2009). Assisted Diagnosis of Cervical Intraepithelial Neoplasia (CIN). *IEEE Journal of Selected Topics in Signal Processing*, *3*(1), 112–121. http://doi.org/10.1109/JSTSP.2008.2011157
[13]Chen, Y.-F., Huang, P.-C., Lin, K.-C., Lin, H.-H., Wang, L.-E., Cheng, C.-C., … Chiang, J. Y. (2014). Semi-automatic segmentation and classification of Pap smear cells. *IEEE Journal of Biomedical and Health Informatics*, *18*(1), 94–108. http://doi.org/10.1109/JBHI.2013.2250984
[14]Sahli, H., & Mihai, C. (2011). A Hybrid Approach for Pap-Smear Cell Nucleus, 174–183.
[15]Arteta, C., Lempitsky, V., Noble, J. A., & Zisserman, A. (2012). Learning to Detect Cells Using Non-overlapping Extremal Regions, (Figure 1), 348–356.
[16]Orozco-monteagudo, M., Taboada-crispi, A., & Sahli, H. (2013). in Pap-Smear Images, 17–24
[17]Lorenzo-ginori, J. V., & Curbelo-jardines, W. (2013). Cervical Cell Classification Using Features Related to Morphometry and Texture of Nuclei, 222–229.
[18]Zhang, J., & Liu, Y. (2004). Cervical Cancer Detection Using SVM Based, (2), 873–880.
[19]Cheng, C.-C., Hsieh, T.-Y., Taur, J.-S., & Chen, Y.-F. (2013). An automatic segmentation and classification framework for anti-nuclear antibody images. *Biomedical Engineering Online*, *12 Suppl 1*(Suppl 1), S5. http://doi.org/10.1186/1475-925X-12-S1-S5.
[20]Wang, W., Zhu, Y., Huang, X., Lopresti, D., Xue, Z., Long, R., Thoma, G. (2009). A Classifier Ensemble Based On Performance Level Estimation Department of Computer Science and Engineering , Lehigh University , Bethlehem , PA 18015 Communications Engineering Branch , National Library of Medicine , MD 20894. *Performance Evaluation*, 342–345
[21]Liu, Y., Zhou, J., & Chen, Y. (2008). Ensemble Classification for Cancer Data. *BioMedical Engineering and Informatics, 2008. BMEI 2008. International Conference on*, *1*, 269–273.
[22]http://www.cancer.org/cancer/cervicalcancer/detailedguide/cervical-cancer-what-is-cervical-cancer.
[23]Torheim, T., Malinen, E., Kvaal, K., Lyng, H., Indahl, U. G., Andersen, E. K. F., & Futsæther, C. M. (2014). Classi fi cation of Dynamic Contrast Enhanced MR Images of Cervical Cancers Using Texture Analysis and Support Vector Machines, *33*(8), 1648–1656.
[24]Chen, K., Wang, K., Tsai, M., Wang, K., Adrian, A. M., Cheng, W., … Chang, K. (2014). Gene selection for cancer identification : a decision tree model empowered by particle swarm optimization algorithm, 0–9.
[25]Paper, C., Mukhopadhyay, S., Technologies, N., Kanpur, T., Education, S., & Kanpur, T. (2016). Optical diagnosis of colon and cervical cancer by support vector machine, (May).
[26]Info, A. (2014). Australian Journal of Basic and Applied Sciences Selection of Optimal combinational features for identification of Cervical Cancer cells using Support Vector Machine, *8*(1), 583–589.
[27]Jusman, Y., Ng, S. C., Azuan, N., & Osman, A. (2014). Intelligent Screening Systems for Cervical Cancer, *2014*.

# Development of KBS for CAD modeling of a two wheeler IC Engine Connecting Rod: An approach

Jayakiran Reddy Esanakula; Cnv Sridhar; V Pandu Rangadu

**Abstract** The conventional CAD modeling methods of connecting rod are time-consuming because of the complex in geometry. A Little modification in shape or size of IC engine assembly will cause a considerable chain reaction in the geometry of connecting rod which leads to alter the CAD model because of various interrelated design issues. Consequently, the CAD model of the connecting rod needs to be altered so as to match the modification(s) of the engine. The advanced CAD modeling techniques such as parametric modeling technique offer the solutions to these issues. This paper introduces a knowledge-based system for CAD modeling of a two wheeler IC Engine Connecting Rod by using commercially available CAD package SolidWorks. An inference engine and relevant GUI are developed within the CAD software for assisting the design engineers. The developed system is an application of engineering which utilizes the reuse of the design knowledge.

**Keywords** Connecting Rod; Parametric Modeling; SolidWorks API; Macro; KBS;

## 1 Introduction

The connecting rod is a commonly used and highly dynamically loaded component for power transmission in IC Engine. The development of CAD model of connecting rod started almost three decades ago. However, the customers and the design industry are looking for the best, fast and easily modifiable CAD model. The CAD modeling of connecting rod includes the modeling of small end, big end or bearing end, I - beam etc. Fig. 1 shows a two wheeler connecting rod of an IC Engine. The strength and stability of the connecting rod lie in its I - beam. Fig. 2. shows the mid cross sectional view of I - beam. The thickness 't' of the I - beam is the key factor for the entire design of the connecting rod.

 
J.M. Corchado Rodriguez et al. (eds.), *Intelligent Systems Technologies and Applications 2016*, Advances in Intelligent Systems and Computing 530,
DOI 10.1007/978-3-319-47952-1_48

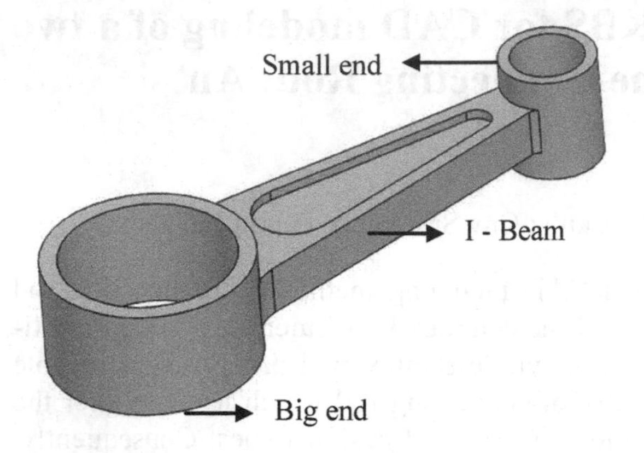


**Fig. 1.** Two wheeler connecting rod of an IC Engine

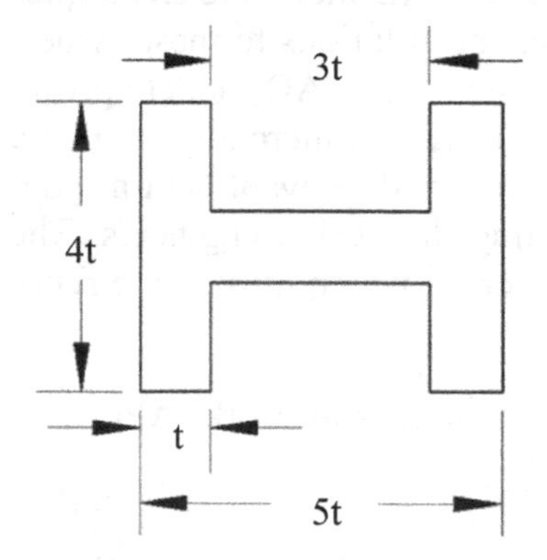


**Fig. 2.** The mid cross section view of I - beam of thickness 't'

Despite the fact that, the CAD modeling became as an unavoidable design practice in the modern design system, it is a time-consuming process because of shortage of skilled CAD modeling professionals [1]. Parametric modeling technique overcomes this issue as it is capable of regenerating the model based on the pre-specified task in less time in comparison with human professional [2]. Therefore, parametric modeling technique allows the CAD modeler to generate and regenerate the model with the help of parameters like the thickness of I - beam, the diameter of big end and small end etc. In addition, it is fast, proficient and interactive than conventional 3D modeling techniques. So, for developing the current knowledge-based system (KBS) for CAD modeling of a two wheeler IC engine connecting rod, parametric modeling may be useful as the connecting rod is comprised of repetitive geometry.

SolidWorks is a CAD software which supports 3D parametric modeling and uses VBA (Visual Basic Application) as a secondary development tool. SolidWorks API (Application Programming Interface) is a software development tool which facilitates to develop custom stand-alone executable files on windows platform. Moreover, SolidWorks API offers the modeling automation by make using of specific programming codes and functions of the software. With the help of Visual Basic (VB), API can access GUI (Graphical User Interface). Further, VB facilitates all Microsoft software to generate macros which are useful for executing the repeated tasks. A macro is a fragment of code which can store any operation done on SolidWorks screen and will reproduce the same operation later whenever requires. Macro file stores 3D coordinates of the drawn component on the screen. Almost all commercial CAD systems support macro file which is with different names: swb file of SolidWorks, program file of IDEAS, trail file of ProE, macro file of UG and script file of CATIA. Macro file of CATIA and SolidWorks are written in a VB code where as for UG, IDEAS, ProE are written in text files.

## 2 Literature Review

Prompt and painless solid modeling can be obtained by using macro programming and parametric modeling technique. Development of CAD model of the connecting rod can be attained with the help of these tools. These tools have the ability to simplifying the complexity of the CAD modeling by automating the frequently executing tasks using a predefined algorithm [3]. Wei et al. [4] suggested an approach for the generation of automated CAD model on SolidWorks platform. This approach is capable of reducing the CAD modeling time by reproducing the CAD model with altered dimensions. Similarly, Bo et al. [5] made an effort for the development of the standard parts library with SolidWorks in order to reduce CAD modeling time.

For automating the CAD modeling, Myung and Han [6] proposed a framework which parametrically models the machine tool. Parametric modeling technique not only simplifies the CAD modeling but also minimizes the modeling time with the support of macro code. Jayakiran et al. [2] proposed a knowledge-based system with parametric modeling technique and macro code for developing intelligent design system for bearing library construction and was demonstrated using SolidWorks software. Automation can be achieved in SolidWorks software using API, macro and VB. Using SolidWorks as platform for developing CAD model and VB as a development tool, Liu et al. [7], Luo et al. [8] and Dun et al. [9] has developed a system for modeling parts of automobile chassis, cylindrical cam and seed plate respectively.

Even though, current powerful parametric modeling tools and CAD packages are available to the designer or modeler, effective modeling is still lacking because of inefficient CAD modeling professionals [10,11]. In order to achieve quality CAD model, Company et al. [12] recommended some best modeling practices based on progressive refinement that can be reviewed with a series of coordinated rubrics. Additionally, Jorge et al. [13] proposed the strategies for design reusability using parametric CAD modeling.

To sum up, SolidWorks can reproduce the CAD model by using Visual Basic as the development tool. Automated CAD model can be achieved by interpreting geometric data of the part that is available in predefined text data form. Thus, automatic generation of 3D model can be obtained with increased design quality and efficiency. So, the developed CAD modeling system enable the skilled or unskilled user to generate the 3D model easily. This paper discusses the way to develop knowledge-based CAD modeling of connecting rod with VB and SolidWorks along with the macro code. The developed system is capable of automatic regeneration of 3D CAD model of connecting rod.

## 3 Methodology for Modeling and Automation

Currently, the majority of commercial CAD software makers are developing the software in such a way that the software should produce or reproduce the part model of any component based on the text inputs pertaining to the model along with the conventional approaches. For instance, for drawing a line on the screen, it is not required to give the inputs (starting and ending points) by making use of the mouse or pointer, they can also be given in the form of text (by entering the 3 dimensional coordinates of the starting and ending points) with the use of keyboard. Moreover, these text input can also be given through a computer program. Thus, the automation of any design system can be achieved with the help of the present CAD software and computer program. But, the programming can be done using specific programming languages like VBA etc. VBA is one of the easy and powerful programming language. The commercial CAD software SolidWorks is using the VBA as the programming editor. So, the authors have chosen the SolidWorks software for developing the current KBS for developing CAD modeling of connecting rod. For carrying out the design calculations a database containing the data related to the materials of connecting rod is required. Microsoft Access database management system has been chosen as database for this purpose. Incidentally, both VB and Access are developed by Microsoft Corporation, so, it is believed that, the data transfer error will not arise. The database has been developed in such a way that, all the relevant data for design calculation should be available in it and will be able to retrieve the same by the computer program whenever the demand arises for the current developing system. The list of materials of the connecting rod that are included in this database are Aluminium Alloy 6061, Gray Cast Iron, Titanium 6AL4V, ASTM A216 GRWCB, T6-2024, T651-7075 aluminum alloys etc. This database consists of all the details of these material properties pertaining to the design of connecting rod. Additionally, authors also used the macro code of the SolidWorks for automatic generation of the CAD model of the connecting rod. In order to executive the whole process in a trouble-free way, a knowledge-base has been developed with the knowledge and experience of connecting rod design experts. The knowledge base is an artificial intelligence tool and an organized repository which can represent and store concepts, facts, rules, data and assumptions about a particular subject. The knowledge to the knowledge-base can be attained from the domain experts. The domain expert can be an academician or industrial professional or both. The knowledge can be stored in the knowledge-base in such a way that a computer program can be able to retrieve the same whenever the system demands.

First and foremost, the developed KBS will calculate the thickness 't' of the I - beam by using all the convectional empirical formulas and material data from the database. Based on the thickness, the system will compute the other dimensions of the connecting rod like diameter of the big end, small end, distance between the ends, fillet radius etc. Later, the current KBS can generate the CAD model of the connecting rod with the help of these dimensions, parametric modeling technique and macro code. The detailed procedure of the development is given under.

### 3.1 Developing a generic 3D model

Initially, standard CAD model of connecting rod is developed manually with all features and constraints while recording its macro file. While generating CAD model, equations should be established between the related parameters of the connecting rod. The recorded macro file need to be edited to parameterize itself for incorporating into VB code of the SolidWorks API. Afterward, the same will be used for developing the CAD model of connecting rod.

### 3.2 Design calculations

Initially, the input data like the type of cooling, bore, stroke, maximum power, maximum torque, connecting rod material need to be collected from the user. Fig. 3 shows the graphical user interface (GUI) for two wheeler IC engine connecting rod, which was developed in SolidWorks using VB.

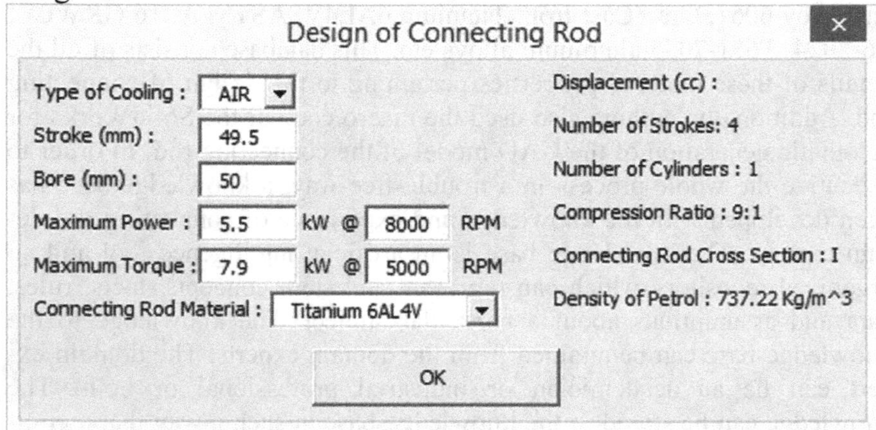


**Fig. 3.** GUI of two wheeler IC engine connecting rod

Based on the above input data, the thickness of the I-beam can be calculated with the standard empirical formulas and the data in the database.

### 3.3 Creating Graphical User Interface:

With the support of VB language, SolidWorks enables the user to create and develop their own GUI. Fig. 4 shows the developed connecting rod button. This button can be used for developing CAD model of connecting rod. At the time of developing the GUI, it is necessary to consider Microsoft ActiveX Data Objects 2.5 Library as reference. The GUI for the input was developed using standard VB tools like textbox, combo box, button etc. The GUI was developed in such a way that it should be able to collect all the required data from the user and should transfer the same to the developed logical algorithm for design calculations. All the design calculation can be done only after clicking the OK button. If the minimum required input data for design calculation is not available from the user then GUI will not allow the process to move further for design calculations.

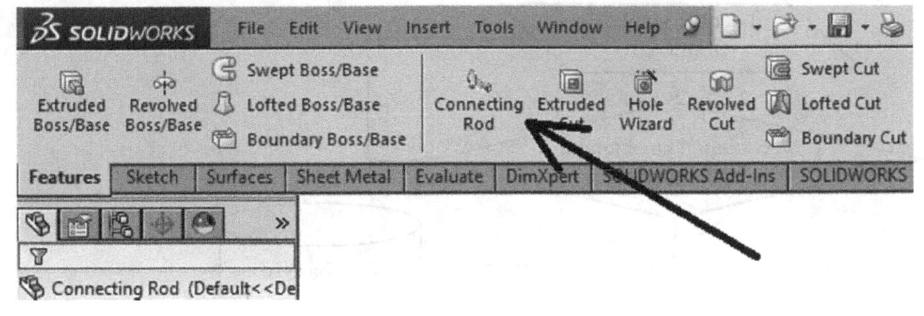


**Fig. 4.** Developed connecting rod button in the GUI of SolidWorks.

### *3.4 Program Development for design calculations*

As already discussed, the developed KBS can support logical algorithms for calculation. This feature helps the system in computing the design calculations. Manual design calculation is tedious and time-consuming to calculate the suitable dimensions using standard empirical formulas. In order to overcome this complexity, a computer program is developed for design calculations. This computer program will compute all design calculations and will finalize the geometrical dimensions of the connecting rod. The output data of this program will be sent to the SolidWorks software to generate the corresponding CAD model as per the calculated geometrical dimensions. Thus, design and modeling work is automated. Fig. 5 illustrates the flowchart of the design of the connecting rod that has been developed, followed and implemented in this work.

## 4 Results and Discussion

In this section, a sample result is presented that was generated by the current developed KBS. Table. 1 provides the sample input for the development of CAD model for two wheeler IC engine connecting rod. This sample input data is collected from one of the leading two wheeler connecting rod producing industry in southern part of India.

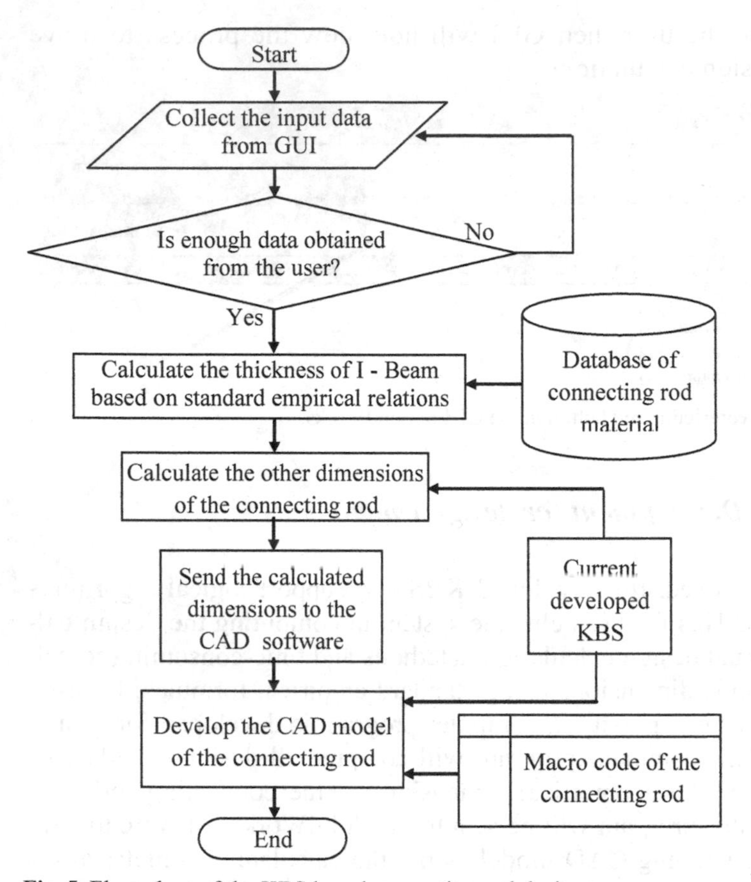


**Fig. 5.** Flow chart of the KBS based connecting rod design process.

Table 1: Sample input data for developing the CAD model of connecting rod.

| Parameter | Input data |
|---|---|
| Type of cooling | Air |
| Number of Strokes | 4 |
| Number of Cylinders | 1 |
| Bore (mm) | 50 |
| Stroke (mm) | 49.5 |
| Compression Ratio | 9:1 |
| Connecting rod cross section | I |
| Maximum Power | 5.5 kW @ 8000 RPM |
| Maximum Torque | 7.9 kW @ 5000 RPM |
| Density of Petrol | 737.22 kg/m$^3$ |
| Connecting Rod Material | Titanium 6AL 4V |

Using the above input data, the developed KBS calculated the thickness of the I-beam based on the standard empirical formulas and the data in the database. Upon calculations, the obtained thickness for the given input data is 2 mm. With this thickness, the developed KBS generated the CAD model of the connecting rod that is shown in Fig. 6. Based on comparison, it is found that the generated CAD model is very close in dimensions and material to the CAD model that was generated by the industrial CAD professionals. Moreover, it is observed that the time taken for generating this KBS based CAD model is 18 seconds in whole where as industry developed CAD model took 31 minutes for CAD modeling alone and 16 man-hours for design calculations.

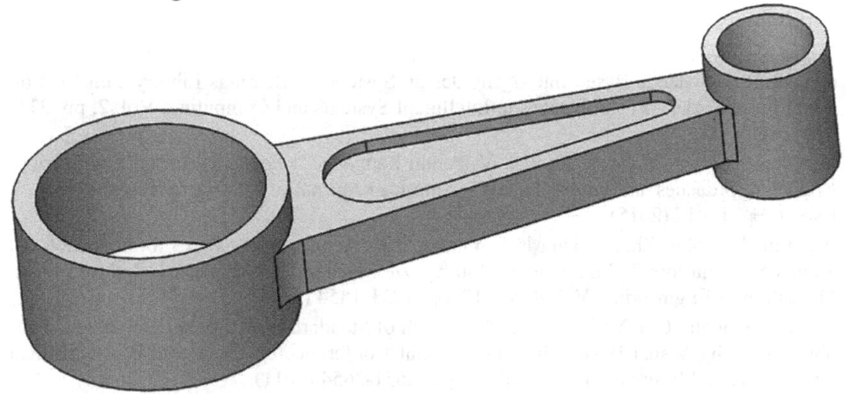

**Fig. 6.** CAD model of the connecting rod.

## 5 Conclusion

This paper presents an approach for development of KBS for CAD modeling of connecting rod by using knowledge-based parametric modeling technique. The basic thought behind the development of this approach is to reduce the CAD molding time. This developed system uses Visual Basic and SolidWorks redevelopment technologies. The procedure of system development is expounded with a reasonable example. This system is easy-to-use, trouble-free, handy and capable of improving design quality and efficiency. Results of this system satisfying the industrial standards. The system helped in dropping turn-around time by automating the process to eliminate manual operations. This system not only provides sophisticated functionality and trouble-free user interface but also make it simple for occasional or non-expert users. In addition, this system is capable of being reused since it is created in an object-oriented environment.

Though the presented and developed system is restricted to SolidWorks software, in future the same can be extended to the other CAD software for increasing the accessibility to all the users. The only addition work for achieving this is to change the macro file format of the SolidWorks software to the supporting format of the target CAD software.

## 6 References


[1] V. Naranje, S. Kumar, "A knowledge based system for automated design of deep drawing die for axisymmetric parts", Expert Systems with Applications, Vol. 41, pp. 1419–1431 (2014).

[2] Esanakula Jayakiran Reddy, C.N.V. Sridhar and V. Pandu Rangadu, "Research and Development of Knowledge Based Intelligent Design System for Bearings Library Construction Using SolidWorks API", Advances in Intelligent Systems and Computing, Vol. 2, pp. 311-319 (2016).

[3] Jayakiran Reddy E, C. N. V. Sridhar, V. Pandu Rangadu, "Knowledge Based Engineering: Notion, Approaches and Future Trends", American Journal of Intelligent Systems, Vol. 5, Issue 1, pp. 1 – 17 (2015).

[4] Wei Liu, Xionghui Zhou, Qiang Niu, Yunhao Ni, "A Convenient Part Library Based On SolidWorks Platform", International Journal of Mechanical, Aerospace, Industrial and Mechatronics Engineering, Vol. 8, No. 12, pp. 1851-1854 (2014).

[5] Bo Sun, Guangtai Qin, Yadong Fang, "Research of Standard Parts Library Construction for SolidWorks by Visual Basic", In: International Conference on Electronic & Mechanical Engineering and Information Technology, pp. 2651-2654 (2011).

[6] Sehyun Myung, Soonhung Han, "Knowledge-based parametric design of mechanical products based on configuration design method", Expert Systems with Applications, vol. 21, pp. 99–107 (2001).

[7] Liu G.T, G. Yan, H.M. Yang, "A Parametric Design Research on the Parts of Automobile Chassis Applied Solidworks", Computer Knowledge and Technology, Vol.7, No.1, pp. 245–249 (2011).

[8] Luo X.J, D.F. Liu, W. Yi, "Research of cylindrical cam CAD /CAM system based on SolidWorks", Modern Manufacturing Engineering, No.7, pp. 54–59 (2011).

[9] Dun G.Q, H.T. Chen, "SolidWorks API Methods of Modeling for Seed Plate Based on Visual Basic", Soybean Science, Vol.31, No.4, pp. 630–637 (2012).

[10] Bodein, Y., Rose, B., Caillaud, E., "Explicit Reference Modeling Methodology in Parametric CAD System" Computers in Industry, Vol. 65, No.1, pp. 136–147 (2014).

[11] Leahy K, "Promoting best practice design intent in 3D CAD for engineers through a task analysis", American Journal of Engineering Research, Vol. 2, No.5, pp. 71–77 (2013).

[12] Company P, Contero M, Otey J, Plumed R, "Approach for developing coordinated rubrics to convey quality criteria in MCAD training", Computer-Aided Design, Vol.63, pp. 101–117 (2015).

[13] Jorge D. Camba, Manuel Contero, Pedro Company, "Parametric CAD modeling: An analysis of strategies for design reusability", Computer-Aided Design, Vol. 74, pp. 18–31 (2016).

# Development of KBS for CAD modeling of Industrial Battery Stack and its Configuration: An approach

Jayakiran Reddy Esanakula, Cnv Sridhar and V Pandu Rangadu

**Abstract** The conventional CAD modeling methods of industrial battery stacks are time-consuming because of the complex in geometry and non-availability of geometry standards. A little modification in electrical power backup required for the customer will cause a considerable chain reaction in the geometry of industrial battery stacks which leads to alter the CAD model because of various interrelated design issues. The advanced CAD modeling techniques such as parametric modeling technique offer the solutions to these issues. This paper introduces a knowledge-based system for developing the CAD model of the industrial battery stack and its configuration and also to reduce the CAD modeling time. An inference engine and relevant GUI are developed within the CAD software for assisting the design engineers. The developed system is an application of engineering which utilizes the reuse of the design knowledge.

**Keywords** Battery stacks; Parametric Modeling; SolidWorks API;

## 1 Introduction

With the intention to sustain in the present volatile economy and to satisfy rapidly changing customer requirements, the companies are implementing the mass customization strategy. Implementing this strategy leads to certain complexities such as rise in the price of the product, necessity of more skilled labor when compare with the mass production strategy. It is obvious that the rise in the price of the product will reduce the sales because of the availability of alternate product in the present competitive market. So, with the aim of sustain in the market, the companies are looking for the alternate methods for designing the product. The traditional CAD/CAM tools are not capable enough to meet the needs of the present volatile market as they are developed to meet only the general needs of the market. So, they need to be customized for improving their effectiveness.

For achieving continuous production in modern industry, electrical power backup system is the most vital as the failure of electrical system in between the production will lead to production disaster. Hence,

 
J.M. Corchado Rodriguez et al. (eds.), *Intelligent Systems Technologies and Applications 2016*, Advances in Intelligent Systems and Computing 530,
DOI 10.1007/978-3-319-47952-1_49

electrical power backup system become as a essential for continuous production. The electrical power backup requirement of the industry is dependent on it production size. So, the battery manufacturers need to produce the battery and stack to suite the all needs of the customer. With the advancements in modern electrochemical technology attaining the variety of electrical power backup systems may not be difficult. At present, battery manufacturing companies are capable to produce the enough number of batteries within the stipulated time as per the customer requirement with the latest production rules and technology. Interestingly, the difficulty for the industrial battery manufactures is arising at manufacturing the stack and its assembly. The stack is a type of storage piece which mounts number of modules. Module is a set of identical batteries or individual battery cells. Stack is a assembly consisting of some number of modules and module is a assembly of some number of cells. These cells will be configured in a specific manner in series, parallel or a mixture of both to deliver the required electrical power output. Fig. 1 shows an industrial battery stack (IBS) arrangement for various needs. As the power requirement for the customer vary from one to another, the standard size of the stack to manufacture is not possible in the optimal production scenario. Hence, a dedicated stack need to be manufactured for every customer depending on their requirement. In other words, customization of the manufacturing process of the stack need to be done. Hence, this paper presents a preliminary approach for developing the knowledge-based system (KBS) for CAD modeling of industrial battery stack by customizing a CAD tool and configuring the same stack.

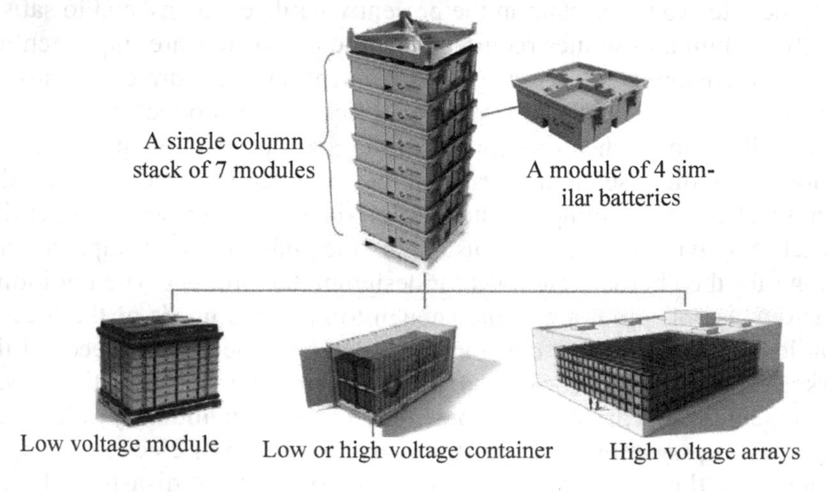


**Fig. 1.** An industrial battery stack arrangement for various needs.

Despite the fact that, the CAD modeling became as an unavoidable design practice in the modern design system, it is a time-consuming process because of shortage of skilled CAD modeling professionals [1]. Parametric modeling technique overcomes this issue as it is capable of regenerating the model based on the pre-specified task in less time in comparison with human professional [2]. Therefore, parametric modeling technique allows the CAD modeler to generate and regenerate the model with the help of parameters. In addition, it is fast, proficient and interactive than conventional 3D modeling techniques. So, for developing the current KBS, parametric modeling technique may be useful as the stack comprises of repetitive geometry.

SolidWorks is a CAD software which supports 3D parametric modeling and uses VBA (Visual Basic Application) as a secondary development tool. SolidWorks API (Application Programming Interface) is a software development tool which facilitates to develop custom stand-alone executable files on windows platform. Moreover, SolidWorks API offers the modeling automation by make using of specific programming codes and functions of the software. With the help of Visual Basic (VB), API can access GUI (Graphical User Interface). Further, VB facilitates all Microsoft software to generate macros which are useful for executing the repeated tasks. A macro is a fragment of code which can store any operation done on SolidWorks screen and will reproduce the same operation later whenever requires. Macro file stores 3D coordinates of the drawn component on the screen. Almost all commercial CAD systems support macro file which is with different names: swb file of SolidWorks, program file of IDEAS, trail file of ProE, macro file of UG and script file of CATIA. Macro file of CATIA and SolidWorks are written in a VB code where as for UG, IDEAS, ProE are written in text files.

## 2 Literature Review

Prompt and painless solid modeling can be obtained by using macro programming and parametric modeling technique. Development of CAD model of the industrial battery stack can be attained with the help of these tools. These tools have the ability in simplifying the complexity of the CAD modeling by automating the frequently executing tasks using a predefined algorithm [3]. Wei et al. [4] suggested an approach for the generation of automated CAD model on SolidWorks platform. This approach is capable of reducing the CAD modeling time by reproducing the CAD model with altered dimensions. Bo et al. [5] made an effort for the development of the standard parts library with SolidWorks. For automating the

CAD modeling, Myung and Han [6] proposed a framework which parametrically models the machine tool. Parametric modeling technique not only simplifies the CAD modeling but also minimizes the modeling time with the support of macro code. Jayakiran et al. [2] proposed a knowledge-based system with parametric modeling technique and macro code for developing intelligent design system for bearing library construction and was demonstrated using SolidWorks software. Automation can be achieved in SolidWorks software using Application Programming Interface (API), macro and VB. SolidWorks as platform for developing CAD model and Visual Basic (VB) as a development tool, Liu et al. [7], Luo et al. [8] and Dun et al. [9] has developed a system for modeling parts of automobile chassis, cylindrical cam and seed plate respectively.

Even though, current powerful parametric modeling tools and CAD packages are available to the designer, effective modeling is still lacking because of inefficient CAD modeling professionals [10,11]. For achieving quality CAD model, Company et al. [12] recommended some best modeling practices based on progressive refinement that can be reviewed with a series of coordinated rubrics. Additionally, Jorge et al. [13] proposed the strategies for design reusability using parametric CAD modeling.

To sum up, SolidWorks can reproduce the CAD model by using VB as the development tool. Automated CAD model can be achieved by interpreting geometric data of the part that is available in predefined text data form. Thus, automatic generation of 3D model can be obtained with increased design quality and efficiency. So, the developed CAD modeling system enables the skilled or unskilled user to generate the 3D model easily. This paper discusses the way to develop KBS for CAD modeling of industrial battery stack and its configuration with VB and SolidWorks along with the macro code. The intension of developing this KBS is to reduce the CAD modeling time. The developed system is capable of automatic regeneration of 3D CAD model of industrial battery stack.

## 3 Methodology for Modeling, Automation and Implementation

Currently, the majority of commercial CAD software produce the part model of a component based on the text input pertaining to the model along with the conventional approaches. These text input can also be given through a computer program. This feature of the CAD software can help in developing an automatic CAD modeling system. But, the programming can be done using specific programming languages like Visual Basic for Application (VBA) etc. VBA is one of the easy and powerful programming language. The commercial CAD software SolidWorks uses the VBA as the programming language. So, the authors have chosen the SolidWorks for developing the present KBS for CAD modeling system. Additionally, it is known that, SolidWorks is equipped with API to support the automation. Furthermore, authors also used the macro code of the SolidWorks for automatic generation of the CAD model. For the current KBS, a knowledge-base has been developed with the knowledge and experience of the domain experts. A Graphical User Interface (GUI) is developed for the system for collecting the input data from the user. Through this GUI, the user can send the query to the inference engine in the system understandable manner. The inference engine is a means which work based on the artificial intelligence system and uses logical rules of the knowledge-base for obtaining the solution for the posed query. This process would be iterative for obtaining the best result. In the present work, the inference engine uses the query to search the knowledge-base and generates the CAD model. A logical algorithm has been developed for carrying out the design calculation. This algorithm uses the knowledge-base which was developed earlier and follows a logical way for computing and finding the best dimensions of the product based on the input data. Because of the complexity of design calculations, it is tedious and time-consuming to manually calculate the suitable dimensions of the output product. In order to overcome this complexity, a logical computer program is developed for design calculations.

Based on the input data, the developed KBS will generate the individual CAD models of all the components of stack assembly (cells, sheet metal frames of module, connectors, terminals and channels). Then, the developed KBS will assemble the components together. Fig. 2 shows various components of the battery stack assembly. Then, The KBS decides the total number of cells required for attaining the customer required power output. For instance, if the total power output required is 240 volts; 600 Ah, then, the total number of cells of capacity 600Ah required are 120 of the each cell of 2 volts. Soon after deciding, the number of cells required, the number of stack column will be decided by the KBS by considering maximum allowable weight and maximum roof level. Later, those cells will be connected in a strategic way by the developed KBS with the various connectors like inter stock connector (ISC), inter module connector (IMC), inter cell connector vertical (ICC V), inter cell connector horizontal (ICC H). These connectors are used for connecting the cells within the module or between the modules or between the stack columns. Soon after completion of the battery connections, terminal plate assembly (TPA) will be connected and placed in the assembly in such a way that they should be always at the extreme ends of the stack column which are nearer to the usage in order to avoid short circuit and to maintain industrial safety standards. Later, this KBS can design and develop the base supporting channel for the stack frame for holding the whole weight of the entire assembly. This can be done with the help of input data like type of the channel and height of the channel. This KBS offers 3 variety of the channels namely, I channel, C channel and Penguin channel.

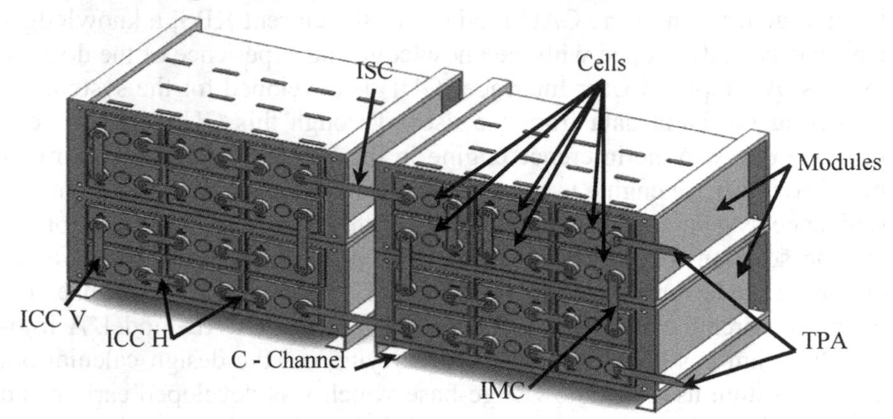


**Fig. 2.** Various components of two column stack with two module per column.

### *3.1 Developing a generic 3D model*

First and foremost, standard CAD model of industrial battery stack and it all components need to be developed manually with all features and constraints while recording its macro file. While generating CAD model, equations should be established between the related parameters of the stack assembly, especially equations development is required for sheet metal frame. The recorded macro file need to be edited to parameterize itself for incorporating into VB code of the SolidWorks API. Afterward, the same will be used for developing the CAD model of industrial battery stacks.

### *3.2 Design calculations*

Initially, the input data like battery capacity, available floor area, maximum roof level, maximum allowable weight, type of channel, height of the channel need to be collected from the user . Fig. 3 shows the GUI for developed KBS for generating CAD model of industrial battery stack, which was developed in SolidWorks using VB.

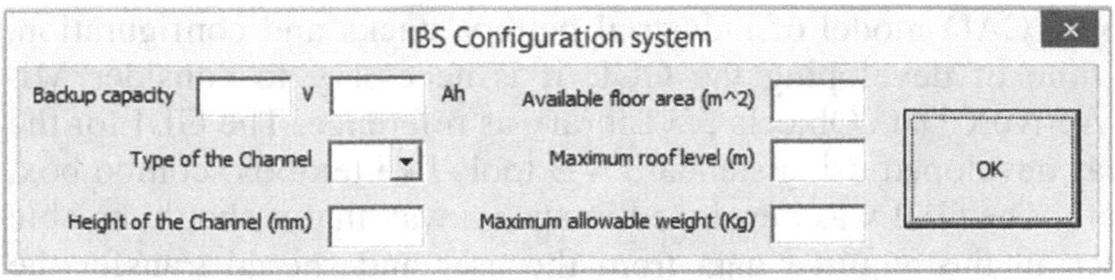

IBS Configuration system

Backup capacity V Ah

Type of the Channel

Height of the Channel (mm)

Available floor area (m^2)

Maximum roof level (m)

Maximum allowable weight (Kg)

OK

**Fig. 3.** GUI for developed KBS for generating CAD model of industrial battery stack.

Based on the input data, the developed KBS will decide the total number of cells needed for the customer. This task is crucial and simple to calculate. As discussed earlier, the total power output is the multiple of number of cells required with their capacity. Soon after, the size of connectors need to be calculated along with their geometry. Industrial battery stack requires variety of connectors which are discussed earlier. The length of the connector is dependent on its function, but the thickness and width is same for all the connectors. The length of each connector is dependent on the position where it is going to be placed. As, the connectors ICC H, ICC V are connected between the cells, the length is dependent on the geometry of the cell and the gap between the cell terminals. So, it can be calculated with the geometry of the cell. The length of the connector IMC is dependent on the geometry of the cell and thickness of the sheet metal frame of the module along with its geometry. The length of the IMC connector is also dependent on the sequence of the connector connections and on the input data like maximum roof level. In other words, the length of the IMC is the distance between the last terminal of the module and first terminal of the next module. The length of the ISC connector is also dependent on the sequence of the connector connections and on the available floor area. The length of the ISC is the distance between the last terminal of the stack column and the first terminal of the next stack column. Later, the geometry, shape and size of the TPA is a standard for every configuration as it is purely dependent on the total power output from the whole stack system. Finally, the thickness and the size of the channel can be calculated by considering the whole weight and floor area occupied by the single column stack assembly and type of the channel.

### *3.3 Creating Graphical User Interface:*

With the support of VB language, SolidWorks enable the user to create and develop their own GUI. Fig. 4 shows the developed button for industrial battery stack configuration system. This button can be used for developing CAD model of industrial battery stacks and configuration. At the time of developing the GUI, it is necessary to consider Microsoft ActiveX Data Objects 2.5 Library as reference. The GUI for the input was developed using standard VB tools like textbox, combo box, button etc. The GUI was developed in such a way that it should be able to collect all the required data from the user and should transfer the same to the developed logical algorithm for design calculations. All the design calculation can be done only after clicking the OK button. If the minimum required input data for design calculation is not available from the user then GUI will not allow the process to move further for design calculations and will be revert back for user attention.

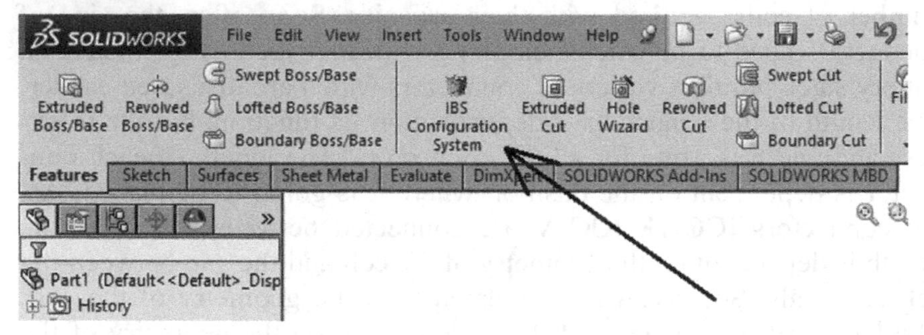


**Fig. 4.** Developed industrial battery stack configuration system button in the GUI of SolidWorks.

### *3.4 Program Development for design calculations*

As already discussed, this developed KBS can support logical algorithms for calculation. This feature helps the system in computing the design calculations. Manual design calculations is tedious and time-consuming to calculate the suitable dimensions using standard empirical formulas. In order to overcome this complexity, a computer program is developed for design calculations. This computer program will compute all design calculations and will finalize the geometrical dimensions of the industrial battery stack. The output data of this program will be sent to the SolidWorks software to generate the corresponding CAD model as per the calculated geometrical dimensions. Thus, design and modeling work is automated. Fig. 5 illustrates the flowchart of the development of KBS for CAD modeling of industrial battery stack and its configuration.

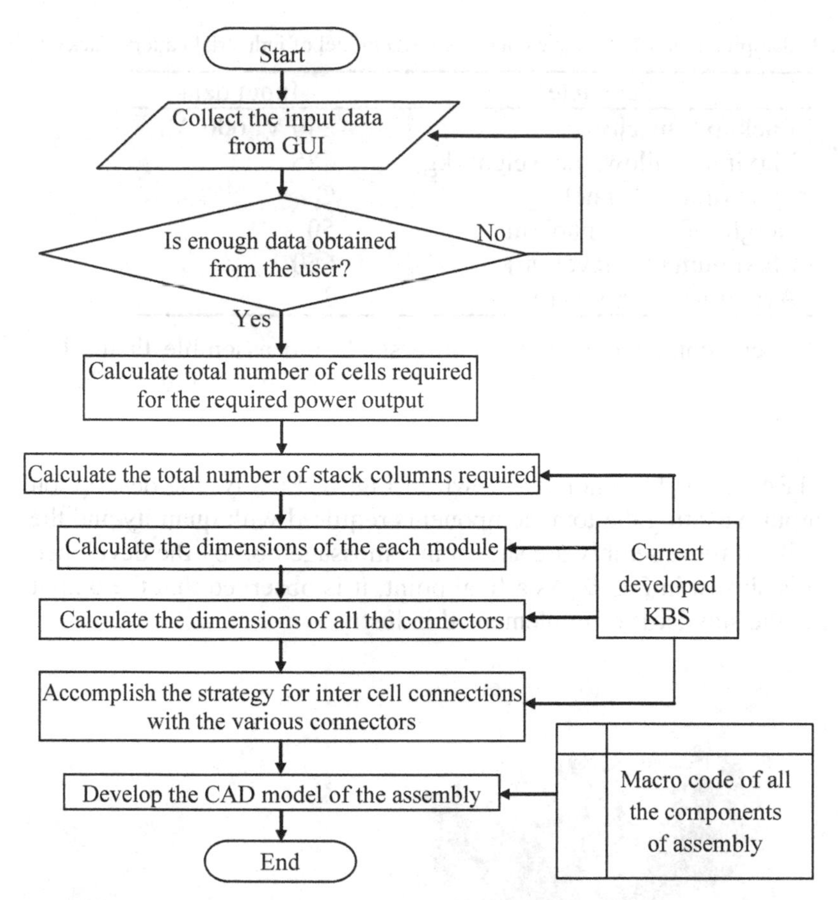


**Fig. 5.** Flow chart of development of KBS for CAD modeling of industrial battery stack and its configuration.

## 4 Implementation

The implementation of this developed system has been done in a battery manufacturing industry in southern part of Andhra Pradesh, India. The focus is to develop a KBS for industry battery stack and its configuration. The developed KBS can provide the CAD model as the output. From this output, the configuration of the stack and the variety of components required and their quantity along with dimensions can be known. Table. 1 provides the sample inputs for the development of KBS.

Using this input data, the developed KBS CAD modeling system calculated the required dimensions of the sheet metal frame of module, various connectors, TPA, C Channel based on the standard empirical formulas and industrial standards. Then, the developed KBS generates individual CAD

Table 1: Sample input data for developing the CAD model of industrial battery stacks

| Parameter | Input data |
|---|---|
| Backup Capacity | 240 V; 600 Ah |
| Maximum allowable weight (kg) | 485 |
| Type of the channel | C |
| Height of the channel (mm) | 50 |
| Maximum roof level (m) | 660 |
| Available floor area ($m^2$) | 3 |

model of every components of the entire stack and assemble them. The KBS developed output for the above input data is shown in Fig. 6. The individual generated CAD models of the components are shown in Fig. 7 to Fig. 9. The cell and the module are shown in the Fig. 7. The various connectors, TPA and C Channel are shown in Fig. 8. Finally, the description of the output narrating the total components required with quantity and the other details of the assembly are shown in a message box by the developed KBS and is shown in Fig. 9. As a final point, it is observed that the output is meeting the standards of implemented industry.

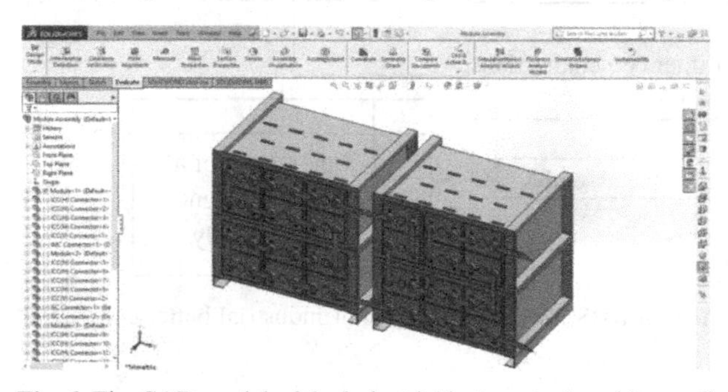

**Fig. 6.** The CAD model of the industrial battery stack and its configuration.

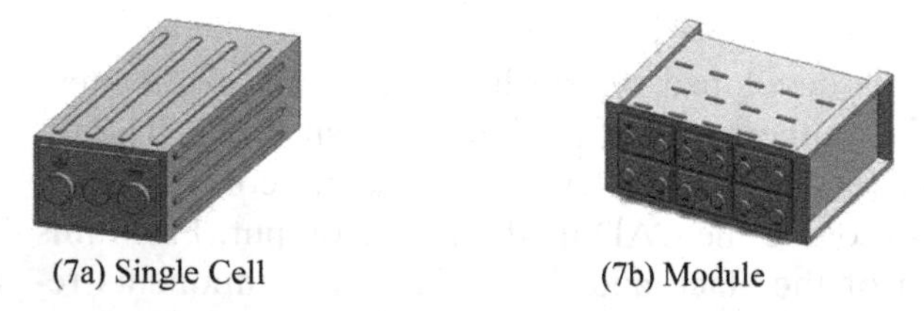

(7a) Single Cell (7b) Module

**Fig. 7.** The CAD model of the single cell and module containing six cells.

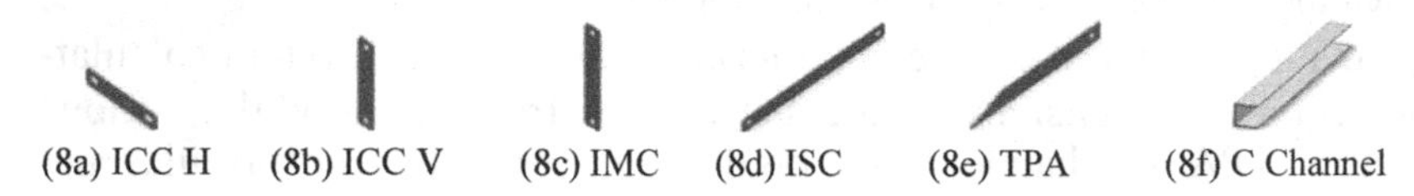

(8a) ICC H (8b) ICC V (8c) IMC (8d) ISC (8e) TPA (8f) C Channel

**Fig. 8.** The CAD model of various connectors.

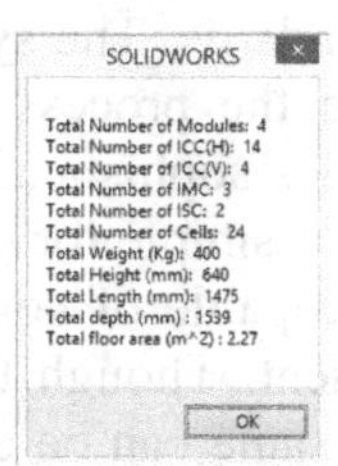


**Fig. 9.** Total quantity of components required and the other details of the assembly.

For checking the robustness of this KBS, some abnormal input data are imposed to the system and it proved itself as a stable and effective. Fig. 10 shows some of the other outputs that was generated with the abnormal data. For generating all the above outputs it took 19 sec to 37 sec on a computer with minimum hardware requirement of SolidWorks 2015. Hence, the developed system is proved as effective than human experts.

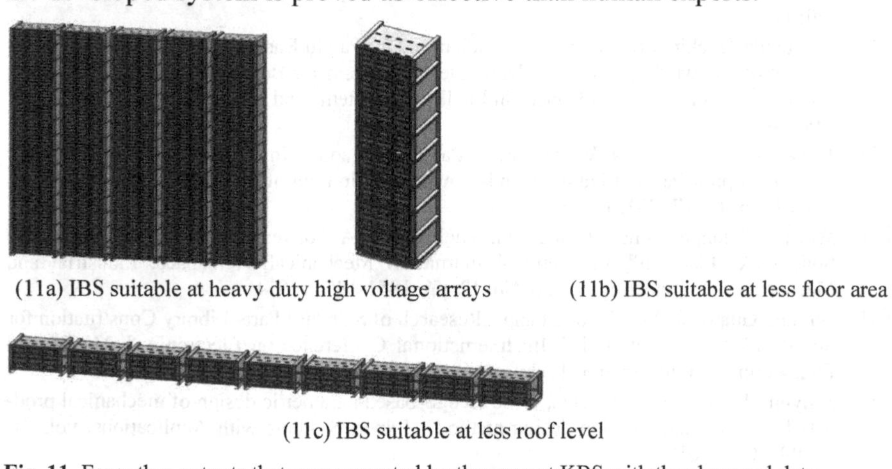

(11a) IBS suitable at heavy duty high voltage arrays

(11b) IBS suitable at less floor area

(11c) IBS suitable at less roof level

**Fig. 11.** Few other outputs that was generated by the present KBS with the abnormal data.

## 5 Conclusion

This paper presents an approach for development of KBS for CAD modeling of industrial battery stack and its configuration by using knowledge-based parametric modeling technique. The basic thought behind the development is to reduce the CAD molding time. This system uses VB and SolidWorks redevelopment technologies. The procedure of system development is expounded with a reasonable example. The system is easy-to-use, trouble-free, handy and capable of improving design quality and efficiency. Results of this system satisfying the industrial standards. The system helped in dropping turn-around time by automating the process to eliminate manual operations. This system not only provides sophisticated functionality and trouble-free user interface but also make it simple for occasional or non-expert users. In addition, this system is capable of being reused since it is created in an object-oriented environment. Though the system is restricted to SolidWorks software, in future the same can be extended to the other CAD software for increasing the accessibility to all the users. The only additional work is to change the macro file format of the SolidWorks software to the supporting format of the target CAD software.

## 6 References


[1] V. Naranje, S. Kumar, "A knowledge based system for automated design of deep drawing die for axisymmetric parts", Expert Systems with Applications, Vol. 41, pp. 1419–1431 (2014).

[2] Esanakula Jayakiran Reddy, C.N.V. Sridhar and V. Pandu Rangadu, "Research and Development of Knowledge Based Intelligent Design System for Bearings Library Construction Using SolidWorks API", Advances in Intelligent Systems and Computing, Vol. 2, pp. 311-319 (2016).

[3] Jayakiran Reddy E, C. N. V. Sridhar, V. Pandu Rangadu, "Knowledge Based Engineering: Notion, Approaches and Future Trends", American Journal of Intelligent Systems, Vol. 5, Issue 1, pp. 1 – 17 (2015).

[4] Wei Liu, Xionghui Zhou, Qiang Niu, Yunhao Ni, "A Convenient Part Library Based On SolidWorks Platform", International Journal of Mechanical, Aerospace, Industrial and Mechatronics Engineering, Vol. 8, No. 12, pp. 1851-1854 (2014).

[5] Bo Sun, Guangtai Qin, Yadong Fang, "Research of Standard Parts Library Construction for SolidWorks by Visual Basic", In: International Conference on Electronic & Mechanical Engineering and Information Technology, pp. 2651-2654 (2011).

[6] Sehyun Myung, Soonhung Han, "Knowledge-based parametric design of mechanical products based on configuration design method", Expert Systems with Applications, vol. 21, pp. 99–107 (2001).

[7] Liu G.T, G. Yan, H.M. Yang, "A Parametric Design Research on the Parts of Automobile Chassis Applied Solidworks", Computer Knowledge and Technology, Vol.7, No.1, pp. 245–249 (2011).

[8] Luo X.J, D.F. Liu, W. Yi, "Research of cylindrical cam CAD /CAM system based on SolidWorks", Modern Manufacturing Engineering, No.7, pp. 54–59 (2011).

[9] Dun G.Q, H.T. Chen, "SolidWorks API Methods of Modeling for Seed Plate Based on Visual Basic", Soybean Science, Vol.31, No.4, pp. 630–637 (2012).

[10] Bodein, Y., Rose, B., Caillaud, E., "Explicit Reference Modeling Methodology in Parametric CAD System" Computers in Industry, Vol. 65, No.1, pp. 136–147 (2014).

[11] Leahy K, "Promoting best practice design intent in 3D CAD for engineers through a task analysis", American Journal of Engineering Research, Vol. 2, No.5, pp. 71–77 (2013).

[12] Company P, Contero M, Otey J, Plumed R, "Approach for developing coordinated rubrics to convey quality criteria in MCAD training", Computer-Aided Design, Vol.63, pp. 101–117 (2015).

[13] Jorge D. Camba, Manuel Contero, Pedro Company, "Parametric CAD modeling: An analysis of strategies for design reusability", Computer-Aided Design, Vol. 74, pp. 18–31 (2016).

# Discrete Sliding Mode Control using Uncertainty and Disturbance Estimator

**Prasheel V. Suryawanshi, Pramod D. Shendge and Shrivijay B. Phadke**

**Abstract** This paper details design and validation of $\delta$-operator based discrete sliding mode control (DSMC) algorithm for uncertain dynamical systems. A unifying sliding condition is used and control is designed for model-following. The control law is synthesized by estimating states and uncertainties using UDE. The UDE used in combination with SMC makes it possible to use a smooth control without having to employ a smoothing approximation. The proposed control affords control over the magnitude of sliding width for a given sampling period. The stability of system is assured using Lyapunov criterion. The control design is validated on a benchmark motion control problem.

## 1 Introduction

The pervasive use of digital controllers has necessitated a need to synthesize discrete-time design for sliding mode control (SMC) [1, 2]. The digital implementation requires a finite sampling period that may cause instability along with chattering along the sliding surface. The discrete systems may become unstable with a large sampling period. Thus the discrete controllers at moderate sampling periods are of significance. In discrete-time, the control avoids instantaneous switching as it is computed at discrete instants of time. The system states can move towards the sliding plane but it is generally difficult for the control to maintain the states on the plane [3, 4]. The system can thus achieve only quasi-sliding. The magnitude of sliding width depends on the uncertainty and sampling interval [5]. A variety of discrete control laws for sliding mode [6, 7, 8, 9, 10, 11] have been proposed in the literature.

Prasheel V. Suryawanshi
MIT Academy of Engineering, Alandi (D), Pune, India and College of Engineering, Pune, India
e-mail: prasheels@gmail.com

Pramod D. Shendge and Shrivijay B. Phadke
College of Engineering, Pune, India



J.M. Corchado Rodriguez et al. (eds.), *Intelligent Systems Technologies and Applications 2016*, Advances in Intelligent Systems and Computing 530,
DOI 10.1007/978-3-319-47952-1_50

The delta-operator [12] can be used for synthesizing a controller in discrete-time domain [13, 14, 15]. The delta-operator offers attractive features like convergence to continuous part as sampling width decreases to zero and better representation of finite world length coefficient. The issue of matching conditions in discretized systems is addressed in [16] and unification of sliding condition in [17].
The traditional SMC also requires all system states to be available. The increased number of sensors for state measurement makes overall system complex and expensive. Observer design for systems with uncertainties, disturbances and noise, is a major problem. A sliding mode observer is designed in [18] that uses additional switching terms to counter the effects of uncertainties. The problem of state observation in presence of uncertainties of known bounds is dealt in [19], with an improvement in [20]. A combined state and perturbation observer for discrete-time systems [21] is an interesting approach.
A continuous-time SMC combined with UDE is extended to a discrete-time case in this paper. The robustness is assured by designing an observer that estimates states, disturbances and uncertainty. The major contributions of this work are:

(i) The UDE is extended to DSMC of uncertain dynamical systems. The control is synthesized by simultaneous estimation of states and uncertainty.
(ii) The SMC designs in continuous and discrete-time are unified. The UDE is able to mitigate the effects of quasi-sliding.
(iii) The ultimate boundedness of state estimation error, uncertainty estimation error and sliding variable is assured.
(iv) The design is validated experimentally on a laboratory set-up of motion control system.

The remaining of the paper is organized as: Section 2 details the problem formulation with necessary assumptions. The unifying sliding condition design is derived in section 3. Section 4 includes the model-following control along with uncertainty estimation. The observer design for estimation of uncertainty and states is explained next in section 5. The stability is proved in section 6. The robust performance is validated by an application to motion control in section 7 with results in section 8 and conclusion in section 9.

## 2 Problem Formulation

Consider a linear uncertain plant in continuous-time as,

$$\dot{x}(t) = A_c x(t) + B_c u(t) + F_c d(t) \tag{1}$$

where $A_{c\,(n\times n)} = A_{nc} + \Delta A_c$, $\quad B_{c\,(n\times 1)} = B_{nc} + \Delta B_c$, with '$nc$' as the nominal part of uncertain continuous time system, $x(t)$ is state, $u(t)$ is control input, and $d(t)$ is unknown disturbance with $\Delta A_c$ and $\Delta B_c$ are uncertainties.

**Assumption 1** *$A_c$ and $B_c$ is stabilizable and uncertainties $\Delta A_c$ and $\Delta B_c$ satisfy matching conditions given by,*

$$\Delta A_c = B_{nc}\Delta_a \qquad \Delta B_c = B_{nc}\Delta_b \qquad F_c = B_{nc}\Delta_f \tag{2}$$

The plant in (1) is discretized using $\delta$ operator [22] and the modified form given by [16]. The plant in (1) can now be written as,

$$\delta x_k = Ax_k + Bu_k + B(bu_k + E_k) \tag{3}$$

where, $E_k = \Delta A x_k + w_k$ and $w_k = (1 + \Delta_a B_{nc} T/2)\Delta_f d_k$ and the system parameters are given by,

$$A = \frac{e^{A_{nc}T} - I}{T}, \quad B = \frac{1}{T}\int_0^T \frac{e^{A_{nc}\tau} - I}{T} B_{nc} d\tau \tag{4}$$

$$b = \Delta_a B_{nc}(1 + \Delta_b)(T/2) + \Delta_b \tag{5}$$

$$\Delta A = \Delta_a[I + (A_{nc} + B_{nc}\Delta_a)T/2] \tag{6}$$

This system can be written as,

$$\delta x_k = Ax_k + Bu_k + Be_k \tag{7}$$

where $e_k$ is the lumped uncertainty.

**Remark 1** *Shift operator and z-transform, which forms the basis of most discrete time analysis are inappropriate with fast sampling, and have no continuous counterpart [22]. Better correspondence is obtained between continuous and discrete time, if the shift operator is replaced with a different operator, more like derivative.*

$$\delta x_k = \frac{x_{k+1} - x_k}{T} \tag{8}$$

*where $T$ is the sampling period*

**Remark 2** *The shift-operator ($q$) generally leads to simpler expressions and emphasizes the sequential nature of sampled signals. The delta-operator ($\delta$) leads to models that are more alike models in $\frac{d}{dt}$. Using $\delta$, any polynomial in $q$ of degree $n$ is equivalent to some polynomial in $\delta$ of degree $n$.*

**Assumption 2** *The disturbance vector $e_k$ is such that its derivatives up to $m^{th}$ order are finite.*

$$\|\delta^{(m)} e_k\| \le \mu \quad m \ge 0 \tag{9}$$

*where $\mu$ is a constant that is positive and $\delta^{(m)} e_k$ implies $m^{th}$ derivative of $e_k$.*

**Remark 3** *The assumption includes a varied class of disturbances. It may be noted that value of $\mu$ is not required for control design.*

## 3 Sliding Condition

Let $s_k$ be the value of sliding variable $\sigma$ at $k^{th}$ sampling instant and $\Delta s_k = s_{k+1} - s_k$. For sliding to occur,

$$s_{k+1}^2 < s_k^2 \tag{10}$$

Rearranging (10),

$$\left.\begin{array}{l} s_{k+1}^2 - s_k^2 < 0 \\ (s_{k+1} - s_k)(s_{k+1} - s_k + 2s_k) < 0 \end{array}\right\} \tag{11}$$

The equation (11) can be written as,

$$\Delta s_k^2 < -2s_k \Delta s_k \tag{12}$$

For sliding to occur, if $s_k > 0$, then $\Delta s_k < 0$ and if $s_k < 0$, then $\Delta s_k > 0$.
Using this logic in (12),

$$\left.\begin{array}{l} -2s_k < \Delta s_k < 0 \\ 2s_k < \quad \Delta s_k < 0 \end{array}\right\} \tag{13}$$

Multiplying the 2-equations in (13) by $s_k$ and $-s_k$ respectively,

$$-2s_k^2 < s_k \Delta s_k < 0 \tag{14}$$

As $\delta s_k = \frac{\Delta s_k}{T}$ the sliding condition (14) can be written as,

$$\frac{-2s_k^2}{T} < s_k \delta s_k < 0 \tag{15}$$

**Remark 4** *The condition (15) clearly shows that two conditions must be satisfied for sliding to occur in discrete time system. The condition (15) further collapses into the familiar $\sigma\dot{\sigma} < 0$, as the sampling time $T \to 0$.*

## 4 Model Following Control

Consider the discretized continuous system as in (7),

$$\delta x_k = Ax_k + Bu_k + Be_k \tag{16}$$

and the sliding surface as in [23],

$$s_k = B^T x_k + z_k \tag{17}$$

where,

$$\delta z_k = -B^T A_m x_k - B^T B_m u_{m_k}, \qquad z_0 = -B^T x_0 \tag{18}$$

The choice of $A_m$ and $B_m$ is made in such a way that,

$$\delta x_{m_k} = A_m x_{m_k} + B_m u_{m_k} \tag{19}$$

will have desired response, like the model in a model following system.

**Assumption 3** *The choice of model is such that, it satisfies the matching conditions $A - A_m = BL$ and $B_m = BM$, where $L$ and $M$ are known matrices of appropriate dimensions.*

From (17)

$$\begin{aligned}\delta s_k &= B^T \delta x_k + \delta z_k \\ &= B^T BLx_k - B^T BMu_{m_k} + B^T Bu_k + B^T Be_k\end{aligned} \tag{20}$$

Selecting

$$u_k = u_k^{eq} + u_k^n \tag{21}$$

The equivalent control ($u_k^{eq}$) can be derived as,

$$u_k^{eq} = -[Lx_k - Mu_{m_k}] - (B^T B)^{-1} Ks_k \tag{22}$$

where $K$ is a constant that is positive.
The sliding surface dynamics can be written as,

$$\delta s_k = (B^T B)u_k^n + (B^T B)e_k - Ks_k \tag{23}$$

**Remark 5** *Selecting $u_k^n = -\hat{e}_k$ and using $\tilde{e}_k = e_k - \hat{e}_k$ with $\hat{e}_k \cong e_k$, the dynamics in (23) can be written as, $\delta s_k = -Ks_k$ leading to $s_k \delta s_k = -Ks_k^2 < 0$. Further, if $K < \frac{2}{T}$ then discrete sliding condition $\frac{-2s_k^2}{T} < s_k \delta s_k < 0$ is satisfied for arbitrarily small values of $s_k$.*

## 4.1 Estimation of Uncertainty ($e_k$)

The UDE algorithm is based on the presumption that a signal can be approximated by using a filter of appropriate bandwidth. The lumped uncertainty $e(x,t)$ can be estimated as, $\hat{e}(x,t) = G_f(s)\,e(x,t)$, where $G_f(s)$ is a low-pass filter with an appropriate bandwidth and steady-state gain as unity.
Using (23),

$$e_k = (B^T B)^{-1}(\delta s_k + Ks_k) - u_k^n \tag{24}$$

Let $G_f(\gamma)$ be a digital filter with unity steady state gain, where '$\gamma$' is the transform variable as used in [22]. Therefore,

$$\hat{e}_k = e_k G_f(\gamma) \tag{25}$$

Now consider a filter given by,

$$G_f(\gamma) = \frac{1 - e^{\frac{-T}{\tau}}}{1 + T\gamma - e^{-T/\tau}} \tag{26}$$

which is digital equivalent of a continuous filter $G_f(s) = 1/(\tau s + 1)$, in $\delta$-domain. Therefore,

$$\hat{e}_k = [(B^T B)^{-1}(\delta s_k + K s_k) - u_k^n] G_f(\gamma) \tag{27}$$

$$= (B^T B)^{-1}(\delta s_k + K s_k)\frac{G_f(\gamma)}{1 - G_f(\gamma)} \tag{28}$$

$$= (B^T B)^{-1}\left(s_k + \frac{K}{\gamma} s_k\right)\left(\frac{1 - e^{-T/\tau}}{T}\right) \tag{29}$$

**Remark 6** *The accuracy of estimation depends on the order of filter. The error in estimation can be reduced, if a second-order filter in delta-form [22] is used as,*

$$G_f(\gamma) = \frac{\gamma \alpha \tau - \gamma e^{-T/\tau} + \alpha^2 \tau}{\tau(\gamma + \alpha)^2} \tag{30}$$

$$\alpha = \frac{1 - e^{-T/\tau}}{T} \tag{31}$$

## 5 Estimation of States and Uncertainty

The discrete plant and model as defined in (7) and (19) are rewritten as,

$$\left.\begin{aligned} \delta x_k &= A x_k + B u_k + B e_k \\ y_k &= C x_k \end{aligned}\right\} \tag{32}$$

$$\left.\begin{aligned} \delta x_{m_k} &= A_m x_{m_k} + B_m u_{m_k} \\ y_{m_k} &= C_m x_{m_k} \end{aligned}\right\} \tag{33}$$

An observer is defined as,

$$\left.\begin{aligned} \delta \hat{x}_k &= A \hat{x}_k + B u_k + B \hat{e}_k + J(y_k - \hat{y}_k) \\ \hat{y}_k &= C \hat{x}_k \end{aligned}\right\} \tag{34}$$

where, $\hat{x}, J, \hat{e}_k$ are observer state vector, observer gain matrix and estimate of the lumped uncertainty respectively.
The observation error $\tilde{x}_k = x_k - \hat{x}_k$ has an exponentially convergent dynamics,

$$\delta \tilde{x}_k = (A - JC)\tilde{x}_k + B \tilde{e}_k \tag{35}$$

where, $\tilde{e}_k = e_k - \hat{e}_k$.

**Assumption 4** *This assumption is necessary to guarantee the asymptotic stability of observer.*

1. *The pair $(A,B)$ is controllable*
2. *The pair $(A,C)$ is observable*
3. *The triplet $(A,C,B)$ has no invariant zeros, i.e. for all $\lambda \varepsilon C$*

$$Rank \begin{bmatrix} \lambda I - A & -B \\ C & 0 \end{bmatrix} = n+1$$

The state estimation error dynamics are derived from (32) and (34) as,

$$\begin{aligned} \delta \tilde{x}_k &= (A - JC)\tilde{x}_k + B\tilde{e}_k \\ y_k &= C\tilde{x}_k \end{aligned} \tag{36}$$

Using (25) and (26),

$$\hat{e}_k = e_k G_f(\gamma) = e_k \frac{1 - e^{-T/\tau}}{1 + T\gamma - e^{-T/\tau}} \tag{37}$$

where $\hat{e}_k$ is estimate of uncertainty and $G_f(\gamma)$ is a low-pass discrete filter. The lumped uncertainty is written using (34) as,

$$e_k = B^+(\delta \hat{x}_k - A\hat{x}_k - Bu_k) \tag{38}$$

$$= B^+[J(y_k - \hat{y}_k) + B\hat{e}_k] \tag{39}$$

$$= B^+ JC\tilde{x}_k + \hat{e}_k \tag{40}$$

Using (36) and (42),

$$\delta \hat{e}_k = \frac{1}{T}[B^+ JC\tilde{x}_k(1 - e^{-T/\tau})] \tag{41}$$

Subtracting both the sides of above equation from $\delta e_k$ and with Assumption 2,

$$\delta \tilde{e}_k = -\frac{1}{T}[B^+ JC\tilde{x}_k(1 - e^{-T/\tau})] + \delta e_k \tag{42}$$

Therefore, combining (36) and (42)

$$\begin{bmatrix} \delta \tilde{x}_k \\ \delta \tilde{e}_k \end{bmatrix} = \begin{bmatrix} (A - JC) & B \\ \frac{-B^+ JC}{T}(1 - e^{-T/\tau}) & 0 \end{bmatrix} \begin{bmatrix} \tilde{x}_k \\ \tilde{e}_k \end{bmatrix} + \begin{bmatrix} 0 \\ 1 \end{bmatrix} \delta e_k \tag{43}$$

The observer dynamics can be stabilized with appropriate choice of $J$ and $T$, if the pair $(A,C)$ is observable. Thus $\tilde{x}_k \to 0$ and $\tilde{e}_k \to 0$ under the Assumptions 2 and 4.

## 6 Stability

The error dynamics in (43) are written as,

$$\delta\tilde{d_k} = D\tilde{d_k} + E\,\delta e_k \tag{44}$$

where, $\delta\tilde{d_k} = [\,\delta\tilde{x}_k \;\; \delta\tilde{e}_k\,]\;\tilde{d_k} = [\tilde{x}_k \;\; \tilde{e}_k]$.
It is possible to select the gains of observer such that, the eigen values of $D$ can be arbitrarily placed in a circle with radius $1/T$ and center $(-1/T, 0)$. If the observer gains are selected such that all eigen values of $D$ have negative real parts, one can always find a positive definite matrix $P$ such that,

$$D^T P + PD = -Q \tag{45}$$

where $Q$ is a given positive definite matrix. Let $\lambda_d$ be the smallest eigen value of $Q$. Defining a Lyapunov function,

$$V(\tilde{d_k}) = \tilde{d_k}^T P\tilde{d_k} \tag{46}$$

and calculating $\delta V_1(\tilde{d_k})$ along (44),

$$\begin{aligned}\delta V(\tilde{d_k}) &= \tilde{d_k}^T(D^T P + PD)\tilde{d_k} + 2\tilde{d_k}^T PE\delta^r d_k \\ &\leq -\|\tilde{d_k}\|\,(\lambda_d\|\tilde{d_k}\| - 2\|PE\|\,\mu)\end{aligned} \tag{47}$$

Thus the estimation error $\|\tilde{d_k}\|$ is bounded by $\frac{2\|PE\|\mu}{\lambda_d}$. This implies,

$$\|\tilde{x}_k\| \leq \frac{2\,\|PE\|\,\mu}{\lambda_d} \qquad \text{and} \qquad \|\tilde{e}_k\| \leq \frac{2\,\|PE\|\,\mu}{\lambda_d} \tag{48}$$

**Remark 7** *The ratio* $\frac{\|PE\|}{\lambda_d}$ *depends on the eigen-values of D. The estimation error bounds can be lowered by selecting large values of observer gains, which in turn increases the noise sensitivity. Therefore, the choice of observer gains is a trade-off between quality of measurement and desired accuracy. The practical stability is thus proved in the sense of [24].*

## 7 Application : Motion Control

The motion control test bed [25] consists of a servo plant with computerized control. The test-bed comprises of two motors, one as disturbance and other as drive with position feedback through an encoder. The drive motor is connected to a drive disk through a timing belt. A second timing belt couples the drive disk to the speed reduction (SR) assembly and a third belt completes the drive train to the load disk. The drive and load have adjustable inertia, that is varied by weights. Backlash can be added by a mechanism in SR assembly. The load has a disturbance motor connected and is used to emulate viscous friction and disturbances at the plant output. A brake below the load disk is used to introduce coulomb friction.

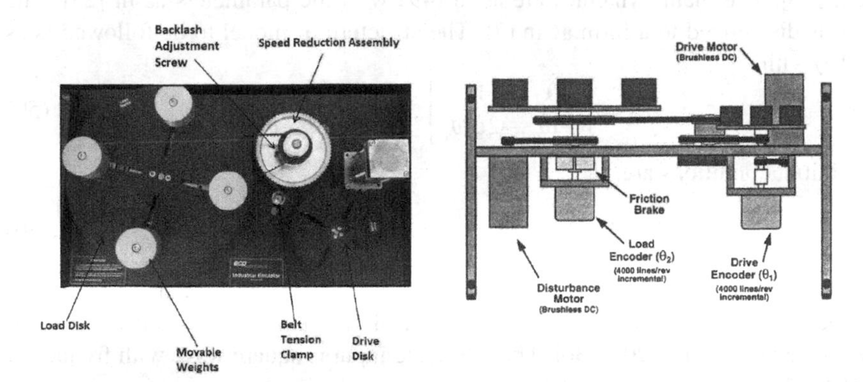


**Fig. 1** Motion Control Plant : ECP220

The motion control dynamics can be written as in [25],

$$J_r\ddot{\theta} + C_r\dot{\theta} = T_d \tag{49}$$

where $J_r$ is inertia at drive and $C_r$ is damping to drive. The parameter $T_d$ is the desired torque that is achieved by selecting appropriate control voltage ($u$) and hardware gain ($k_{hw}$). The plant dynamics can be modeled as,

$$\begin{bmatrix} \dot{x}_1 \\ \dot{x}_2 \end{bmatrix} = \begin{bmatrix} 0 & 1 \\ 0 & -\frac{C_r}{J_r} \end{bmatrix} \begin{bmatrix} x_1 \\ x_2 \end{bmatrix} + \begin{bmatrix} 0 \\ \frac{k_{hw}}{J_r} \end{bmatrix} u \tag{50}$$

where, $[x_1 \quad x_2]^T$ are states - position ($\theta$) and velocity ($\dot{\theta}$), $u$ is control signal in volts and $y$ is output position in degrees. The other parameters in (50) are,

$$C_r = C_1 + C_2\,(gr)^{-2} \tag{51}$$

$$gr = 6\,\frac{n_{pd}}{n_{pl}} \tag{52}$$

$$J_r = J_d + J_p\,(gr_{prime})^{-2} + J_l\,(gr)^{-2} \tag{53}$$

$$J_d = J_{dd} + m_{wd}\,(r_{wd})^2 + J_{wd_0} \tag{54}$$

$$J_p = J_{pd} + J_{pl} + J_{pbl} \tag{55}$$

$$gr_{prime} = \frac{n_{pd}}{12} \tag{56}$$

$$J_l = J_{dl} + m_{wl}\,(r_{wl})^2 + J_{wl_0} \tag{57}$$

$$J_{wl_0} = \frac{1}{2} m_{wl}\,(r_{wl_0})^2 \tag{58}$$

# 8 Results and Discussions

The control law is tested for model-following strategy on a industrial motion control plant [25]. The plant dynamics are as in (49) with the parameters as in [25]. The plant is discretized to a form as in (7). The structure of model to be followed is as in (19) with,

$$A_m = \begin{bmatrix} 0 & 1 \\ -\omega_n^2 & -2\zeta\omega_n \end{bmatrix}, \quad b_m = \begin{bmatrix} 0 \\ -\omega_n^2 \end{bmatrix} \tag{59}$$

The initial conditions are,

$$x(0) = [0 \quad 1]^T \qquad x_m(0) = [0 \quad 0]^T \tag{60}$$

The model parameters are $\zeta = 1$ and $\omega_n = 5$ in (59). The plant and model have a mismatch in initial conditions (60). The control gain is $K = 2$ and observer poles are located at $[-10 \ -20 \ -30]$. The reference input is square wave with frequency of 0.3 rad/sec and amplitude of 1.

## 8.1 *Case 1 : Nominal plant*

The accuracy of tracking and estimation of states is illustrated in Fig. 2.

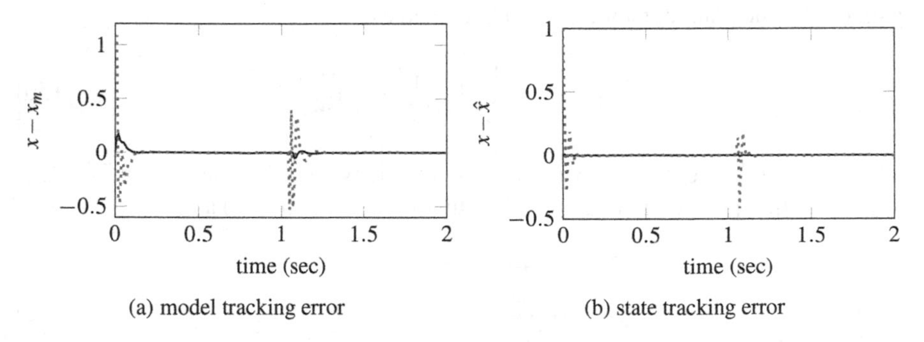


**Fig. 2** Estimation of states and uncertainty (10ms (solid) and 20 ms (dotted))

The control performance for model-following is shown in Fig. 3. The plant and model states is seen in Fig. 3a and 3b. The corresponding control effort (Fig. 3c) and sigma (Fig. 3d) are also shown. It is obvious that the tracking performance is good and the plant follows the desired model. The UDE is able to estimate the states as well as uncertainties and thus aids in robust performance. The sliding surface is also bounded and control effort is within limits to preclude saturation of actuator.

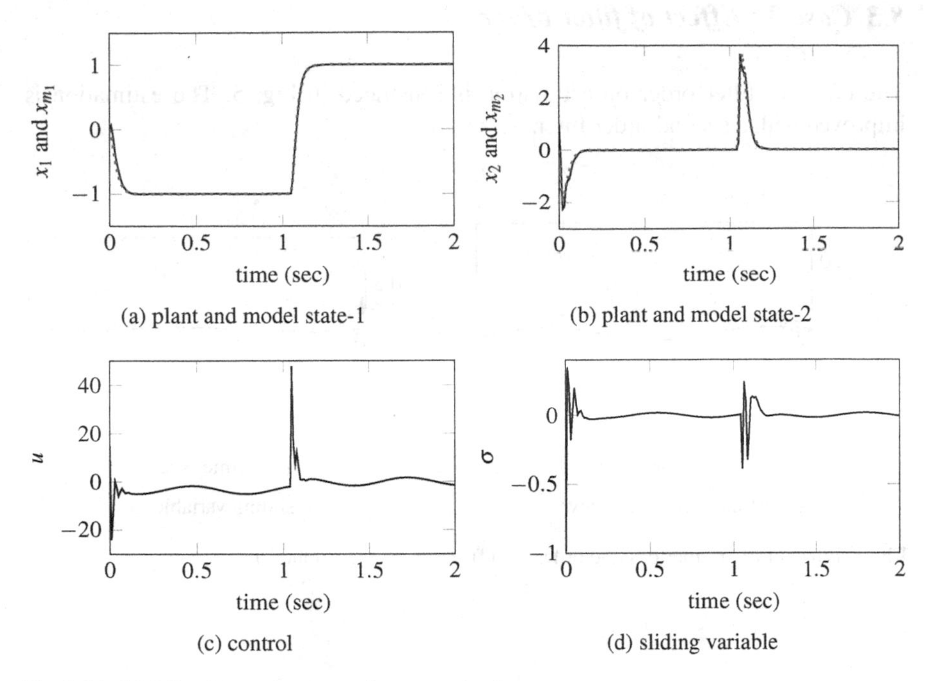


**Fig. 3** Model following performance for uncertain plant

## *8.2 Case 2 : Effect of sampling time*

The effect of sampling time is shown in Fig. 4. It is seen that the estimation and sliding width is consistent even with increase in sampling time.

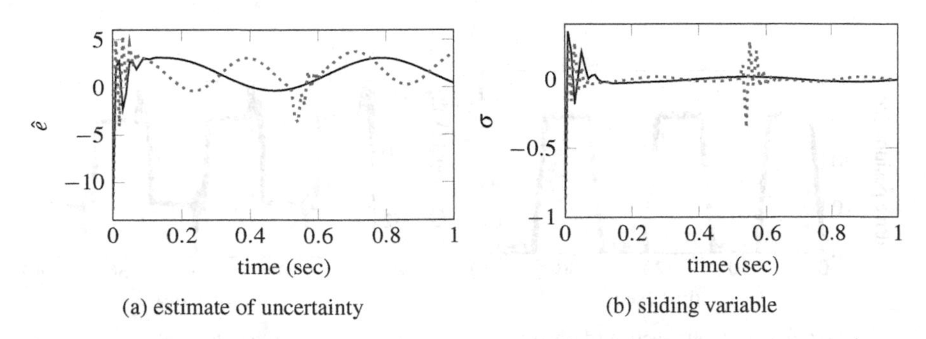


**Fig. 4** Effect of different sampling time (10ms (solid) and 20 ms (dotted))

### 8.3 Case 3 : Effect of filter order

The effect of filter order on estimation is illustrated in Fig. 5. The estimation is improved with a second order filter.

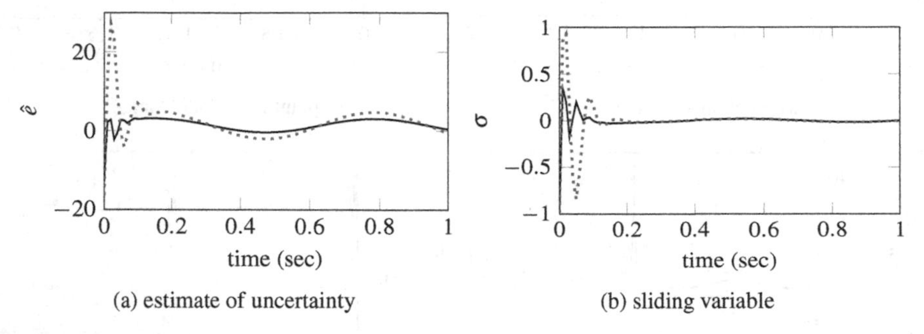


(a) estimate of uncertainty (b) sliding variable

**Fig. 5** Effect of filter order (second order (solid) and first order (dotted))

### 8.4 Case 4 : Experimental Validation

The set-up [25] is an electromechanical system, with brushless motors for disturbance generation and drive, and high resolution encoders. The control is tested for disturbances like backlash and coulomb friction. The control performance for a trajectory of ramp-type is shown in Fig. 6. The ramp is with a velocity of 20 deg/sec and amplitude of 30°.

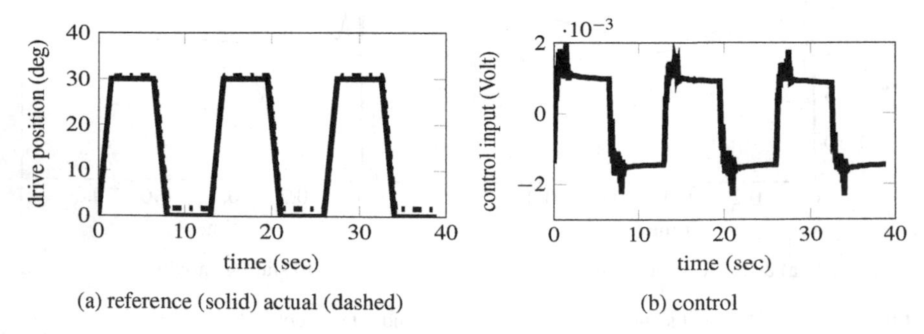


(a) reference (solid) actual (dashed) (b) control

**Fig. 6** Control performance for nominal plant

The tracking performance is also tested for 3-different trajectories as step, ramp, parabola with varied cycles and amplitude. The Table-1 shows cumulative results of model tracking error, state estimation error and unertainty estimation error.

**Table 1** RMS values of errors

| Trajectory | Model tracking | State estimation | Uncertainty estimation |
|---|---|---|---|
| Step | 0.1453 | 0.1679 | 0.3269 |
| Ramp | 0.1725 | 0.1699 | 0.2618 |
| Parabola | 0.1778 | 0.2067 | 0.3711 |

## 9 Conclusion

A SMC combined with UDE is extended to discrete-time case of an uncertain system. The control law is made implementable by designing an observer to give a robust controller-observer structure in discrete domain. A notable feature of the proposed design is that, it affords control over the magnitude of sliding width for a given sampling period. The UDE enables a reduction of quasi-sliding band and the use of second-order filter reduces the sliding width significantly. The use of $\delta$-operator in conjunction with a new sliding condition enables complete and seamless unification of the sliding-condition, control law and UDE. The ultimate boundedness is guaranteed for state estimation error, uncertainty estimation error and sliding variable and the bounds can be further lowered by appropriate selection of design parameters. The efficiency of design is confirmed on a motion control system.

## Acknowledgment

This work is supported by Board of Research in Nuclear Science, Department of Atomic Energy, Government of India, vide Ref. No. 2012/34/55/BRNS.

## References

1. Misawa, E. A.: Discrete time sliding mode control: The linear case. ASME Journal of Dynamic Systems, Measurement and Control, 119(1), 819–821 (1997).
2. Misawa, E. A.: Discrete-time sliding mode control for nonlinear systems with unmatched uncertainties and uncertain control vector. ASME Journal of Dynamic Systems, Measurement and Control, 119(3), 503–512 (1997).

3. Utkin, V. I.: Variable structure systems with sliding modes. IEEE Transactions on Automatic Control, Acc-22(2), 212-222 (1977).
4. Sarpturk, S. Z., Istefanopulos, Y., Kaynak, O.: On the stability of discrete-time sliding mode systems. IEEE Transactions on Automatic Control, 32(10), 930-932 (1987).
5. Furuta, K.: Sliding mode control of a discrete system. Systems & Control Letters, 14(2), 145–152 (1990).
6. Bartoszewicz, A.: Discrete-time quasi-sliding-mode control strategies. IEEE Transactions on Industrial Electronics, 45(4), 633–637 (1998).
7. Milosavljevic, C.: General conditions for the existence of a quasi-sliding mode on the switching hyperplane in discrete variable structure systems. Automation and Remote Control, 46(3), 307–314 (1985).
8. Chan, C. Y.: Discrete adaptive quasi-sliding mode control. International Journal of Control, 72(4), 365–373 (1999).
9. Gao, W., Wang, Y., Hamaifa, A.: Discrete time variable structure control systems. IEEE Transactions on Industrial Electronics, 42(2), 117–122 (1995).
10. Yu, X., Yu, S.: Discrete sliding mode control design with invariant sliding sectors. ASME Journal of Dynamic Systems, Measurement and Control, 122, 776–782 (2000).
11. Ramirez, H. S.: Nonlinear discrete variable structure systems in quasi-sliding mode. International Journal of Control, 54(5), 1171–1187 (1991).
12. Middleton, R. H., Goodwin, G. C.: Improved finite word length characteristics in digital control using delta operator. IEEE Transactions on Automatic Control, AC-31(11), 1015–1021 (1986).
13. Jabbari, F.: Lattice filters for RLS estimation of a delta operator-based model. IEEE Transactions on Automatic Control, 36(7), 869–875 (1991).
14. Collins, E. G.: A delta operator approach to discrete-time $H_\infty$ control. International Journal of Control, 72(4), 315–320 (1999).
15. Veselic, B., Perunicic-Drazenovic, B., Milosavljevic, C.: High-performance position control of induction motor using discrete-time sliding-mode control. IEEE Transactions on Industrial Electronics, 55(11), 3809–3817 (2008).
16. Tesfaye, A., Tomizuka, M.: Robust control of discretized continuous systems using the theory of sliding modes. International Journal of Control, 62(1), 209–226 (1995).
17. Ginoya, D. L., Shendge, P. D., Phadke, S. B.: Delta operator based extended disturbance observer and its applications. IEEE Transactions on Industrial Electronics, 62(9), 5817–5828 (2015).
18. Slotine, J. J. E., Hedrick, J. K., Misawa, E. A.: On sliding observers for nonlinear systems. SME Journal of Dynamic Systems, Measurement and Control, 109(1), 245–252 (1987).
19. Walcott, B. L., Zak, S. H.: Combined observer controller synthesis for uncertain dynamical systems with applications. IEEE Transactions on Systems, Man and Cybernetics, 18(1), 88–104 (1988).
20. Chen, W., Saif, M.: Novel sliding mode observer for a class of uncertain systems. In: American Control Conference, Minnesota, USA, June (2006).
21. Kwon, S. J., Chung, W. K.: A combined synthesis of state estimator and perturbation observer. ASME Journal of Dynamic Systems, Measurement and Control, 125(1), 19–26 (2003).
22. Middleton, R. H., Goodwin, G. C.: Digital Control and Estimation: A Unified Approach. Prentice Hall, New Jersy (1990).
23. Ackermann, J., Utkin, V. I.: Sliding mode control design based on Ackermann's formula. IEEE Transactions on Automatic Control, 43(2), 234–237 (1998).
24. Corless, M., Leitmann, G.: Continuous state feedback guaranteeing uniform ultimate boundedness for uncertain dynamic systems. IEEE Transactions on Automatic Control, 26(5), 1139–1144 (1981).
25. ECP220: User Manual: Model 220 Industrial Plant Emulator. Educational Control Products, Canada (2004).

# An Overview of Feature Based Opinion Mining

**Avinash Golande, Reeta Kamble, Sandhya Waghere**

*Abstract*— **Opinion mining is becoming a popular area in today's world but before the invention of web 2.0 people were only able to view the information but now they are also able to publish the information on Web in the form of comments and reviews. The user generated content forced organization to pay attention towards analyzing this content for better visualization of public's opinion. Opinion mining or Sentiment analysis is an autonomous text analysis and summarization system for reviews available on Web. Opinion mining aims for distinguishing the emotions and expressions expressed within the reviews, classifying them into positive or negative and summarizing into the form that is quickly understood by users. Feature based opinion mining performs fine-grain analysis by recognizing individual features of an object upon which user has expressed his/her opinion. This paper gives an idea of various methods proposed in the area of feature based opinion mining and also discusses the limitations of existing work and future direction in feature based opinion mining.**

***Keywords*— *Opinion mining, Feature extraction, Sentiment classification, Sentiment analysis, Opinion summarization***

## I. INTRODUCTION

Since the advancement of the technologies, many people use internet to express their feedback or opinion, as reviews, comments or question answers on forums, blogs and social websites, which increases the amount of user generated content on the internet. This user generated content can be very useful for both a user and an organization. For instance, tourists can check reviews and experiences published by other travelers on specific hotel on different tourism web sites before booking the hotel. For hotel organization, the reviews available on the web could be used to make surveys, focus groups in market research.

However, due to the large number of opinions or feedbacks published on web it is very difficult for users to analyze all web opinions. All reviews are in the form of plain text which is written in any natural language, therefore to get valuable information from those reviews we need help from other domains like Natural Language Processing (NLP) and Data Mining. To analyze and summarize the opinions that are expressed on web manually is a difficult task. Therefore, we require an automated sentiment analysis system which will provide us the feedback or opinion accurately.

Avinash Golande
Department of Computer Engineering, JSPM's RSCOE, Pune, India, avinash.golande@gmail.com
Reeta Kamble
Department of Computer Engineering, JSPM's RSCOE, Pune, India, reeta.kamble12@gmail.com
Sandhya Waghere
Department of IT Engineering, PCCOE, Pune, India, sandhya.waghere@gmail.com



J.M. Corchado Rodriguez et al. (eds.), *Intelligent Systems Technologies and Applications 2016*, Advances in Intelligent Systems and Computing 530,
DOI 10.1007/978-3-319-47952-1_51

Opinion mining or sentiment analysis involves an area of NLP, computational linguistics and text mining, and refers to a set of techniques that deals with data about opinions and tries to gain valuable information from them. Opinion mining has its applications in almost each and every promising area, from customer products, services, financial services, and healthcare to political elections and all social events.

**Motivation**

Customer's review contains very useful information which could be used for analysis purpose by the researchers. It is common behavior to consider other customer reviews about the product before purchasing it, even before the foundation of internet. In today's world people face the problem of handling huge amount of data or information, for eg. If we want to check opinion or reviews about an hotel we go through the internet and check review about that hotel but due to lot of reviews available it creates confusion in the mind of user which enhances the decision making process. Normally people try to purchase the best product form the market in cheapest price which satisfy their need.

From the e-commerce point of view, receiving customers' opinions can significantly improve its policies for the sake of increasing the income. Generally, each products contains thousands of opinions, therefore it is a difficult task for a customer to analyze all of them. Also, it could be a very tedious task to find opinions about some specific features about a product which is generally a need of a routine customer. Feature based opinion mining system examines opinions at feature level and produce comprehensive summary of opinions which help customers in the decision making.

The rest of the paper is structured as follows. In section II, we discuss basics of opinion mining and tasks of feature based opinion mining. Section III presents the outline of the related work, evaluation parameters. Section IV lists issues in feature based opinion mining. Finally, section V concludes the paper.

## II. OPINION MINING

Opinions are very important among all human activities as they highlight the behavior of the people. People always tend to know other peoples' opinions, whenever they need to make some decision. Nowadays, businesses and organizations always look for getting their customers' reviews about their goods and services. Each and every user wants to know the opinion of various other customers on a product before buying it, while in political election, public opinion about a specific political leader is important before making a voting decision.

An opinion is the personal view of an individual; it represents the individual's ideas, judgments, beliefs, assessments and evaluations about a specific topic, item or subject. One's opinion about a subject can either be positive or negative, which is referred as the semantic orientation or polarity or sentiment.

An opinion has three main elements, i.e.

- □ The opinion source: author of the review
- □ Target of the Opinion: object or its feature
- □ Opinion polarity: positive or negative

All of these elements are vital for opinion identification. "Opinion mining is the problem of recognizing the expressed opinion on a particular subject and determining the polarity of opinion". It provides broad view of the sentiments expressed via text, and to classify and summarize the opinions, it enables further processing of the data.

**A. Level of Opinion Mining**

In general, opinion mining has been explored mainly at three stages by researchers:

*1) Document level:* At document level opinion mining, whole opinion or review is classified into positive or negative. For example, consider a movie review; based on the opinion words present in the review we classify the movie reviews as positive or negative. The main issue of this level is that a whole review is expressed on a single subject. Thus, it is not applicable to reviews in which single review expresses opinion on multiple subjects.

*2) Sentence level:* At this level task of opinion mining is to classify opinion at sentence level. It categorizes every sentence into a positive, negative, or neutral opinion. Sentence which contains no opinion or unrelated words are considered as neutral opinion. The sentence level opinion mining systems may contain subjectivity classification as the pre-processing step. Subjectivity classification is the task of classifying sentences into the objective or subjective sentences. Objective sentences are the sentences that represent factual information and subjective sentences are the sentences that represent subjective opinions.

*3) Feature or Aspect level:* The document level as well as the sentence level analysis does not describe the exact liking of the people. Feature level opinion mining performs fine-grained analysis. It is also referred to as feature based or aspect based opinion mining [2]. Feature level analysis directly looks at the opinion itself instead of looking at language constructs like clauses, sentences or paragraphs. It is based on the fact that user may express his/her opinion on specific feature or aspect of an entity rather than entity itself. Feature or aspect of an entity upon which opinion is expressed is referred as target of an opinion.

An opinion whose target is not identified is of restricted use. Opinion targets are of much importance because single sentence may contain multiple targets from which each target has different opinion. For example, consider the sentence "Even though the music of the movie is not good, I like the story of the movie" obviously has a positive attitude but the sentence is not completely positive. The sentence is positive about the story, but due to its music quality it is negative. In various applications, opinion targets are represented by entities and/or their different aspects. Thus, the aim of this level is to identify polarities of entities and/or their aspects. Feature based opinion mining system provides a structured summary of opinions about entities and their aspects, and hence converts the unstructured data to structured data which can be further used for all kinds of qualitative and quantitative analysis.

**B. Opinion Mining Tasks**

Generally, Feature-based Opinion Mining studies consist of three tasks: Aspect identification, Sentiment classification and Summary generation.

1) *Aspect Identification:* The main goal of this task is to identify and extract relevant topics from the text which will be further used for summarization. In [3], Hu and Liu present a technique based in statistics and NLP. In their proposed system, syntax tree parsing and part-of-speech (POS) tagging are used to detect nouns and noun phrases (NP). Then, the most frequent nouns and NPs are identified by using frequent item set mining. Then using distinct linguistic rules, the discovered sets of nouns and NPs are filtered. The aspects of the entity that are made up of more than one word, usually represent real objects together and they also remove redundant aspects too. They also extract non-frequent features using an approach by finding nouns or NPs that appear near to opinion words with high frequency. This approach does not extract adjectives or any other kind of non-object aspects.
2) *Sentiment Classification:* The next task is sentiment classification which determines the semantic orientation of each aspect. Ding et al. [4] proposed system uses a lexicon and rule-based approach. Their method depends on opinion words, a list of positive and negative words contained in sentiment word dictionary. It is used to determine the semantic orientation of the words in the review. Distinct linguistic rules are put forward to consider other special words that change the orientation. These rules handles negative words ''no'' or ''not'' and also some general negation patterns. Though these rules may appear simple, but it is important to handle them with caution, as all rules or word will generate the different meaning each time they appear in the text. Ding et al. [4] developed the rules that use an aggregation score function to evaluate the semantic orientation of each feature in a sentence that contains multiple opinion words.

3) *Summary Generation:* The last task is summary generation, to represent processed results in a proper form which can be easily understood by users. Bing Liu [2], [5], defines a kind of summary which is called as

aspect-based opinion summary in which there are various bar charts showing number of positive and negative reviews about each and every aspect of a single entity. Liu et al [6] states that a set of selected products can be compared and described using the bar charts, which shows the set of all aspects of the chosen products in the chart. In this chart, each bar above or below the x-axis can be represented in two scales: (1) the percentage of positive or negative opinions on reviews and (2) the actual number of positive -or negative opinions normalized with the most number of opinions on any feature of any product.

## III. LITERATURE SURVEY

The research in the field of opinion mining has been increased since advancement in the field of NLP. K. Khan et al. [7] review the various advancement in opinion mining research. According to T sytsarau et al. [8] opinion mining methods can be classified in four approaches: Machine Learning, Dictionary, Statistical, and Semantic.

### A. Machine Learning Approach

The machine learning method consists following stages. First, a training dataset is acquired, that could be either annotated or not with sentiment labels. Second, each review is presented as a vector of features. Third, a classifier is trained to differentiate among sentiment labels by analyzing the relevant features. Finally, the trained classifier is used to identify sentiments for new reviews.

Zhai et al. 2011 [9] applied the machine learning method for opinion mining on Chinese language. Ngram feature extraction algorithm was used to extract the features from the labeled documents. The extracted features were used to represent document in vector form on which the training and classification steps were based. For the sentiment classification task Support Vector Machine (SVM) classifier was used. Spam reviews and comparative reviews are not considered in opinion mining task.

The performance of machine learning algorithms is very much dependent on the quantity and quality of training data, which is less compared to the extent of unlabeled data. The performance of this approach also depends on data selection or feature selection method and the choice of algorithm to some extent.

### B. Dictionary Approach

The Dictionary Approach uses a pre-built dictionary which defines semantic orientation of words, such as the SentiWordNet, which is the standard dictionary today. Existing opinion mining approaches use these dictionaries mainly for identifying semantic orientation of opinion words. Semantic orientation of a single sentence or review is generally calculated by averaging the semantic orientation values of individual words. For instance, most of the dictionary base methods aggregate the semantic orientation values for a sentence or whole review, and estimate the resultant polarity using simple rule-based algorithms.

Zhu et al. 2011 [10] uses Chinese sentiment lexicon Hownet for sentiment classification of restaurant reviews in Chinese language. It applies bootstrapping method for identification of features. A sentiment value of a review is computed by summing the sentiment values of all opinion words occurring in the review. The resultant semantic orientation value of a review shows its corresponding polarity, that is, greater than 0 for positive, equal to 0 for neutral and less than 0 for negative.

***C. Statistical Approach***

The Statistical Approach aims to obtain polarity values via the co-occurrence of adjectives in a corpus. Here, corpus-Specific dictionary is created to achieve adaptability. To solve the problem of unavailability of some words, this approach uses a very large corpus. It is also possible to use the complete set of indexed documents on the Web as the corpus for the dictionary construction Peter, Turney[11].

The existing statistical methods are based on the fact that similar opinion words mostly appear together in a corpus. Similarly, if two words usually appear together in the same context then they are possibly having the same semantic orientation. Thus the semantic orientation of a new word is determined by calculating the relative frequency of co-occurrence with other word, which invariantly preserves its polarity. To achieve this, Peter, Turney [11], [12] proposed to use the point-wise mutual information (PMI) criterion for statistical dependence, replacing probability values with the frequencies of term occurrence F(x) and co-occurrence F(x near y):

$$PMI(x, y) = log_2 = \frac{F(x\ near\ y)}{F(x)F(y)} \quad (1)$$

Semantic orientation for any word x is calculated as the difference between PMI values computed against positive words (words), e.g. “good”, and negative words (nWords), e.g. “bad” [11]:

$$PMI\text{-}IR(x) = \sum_{p \epsilon pWords} PMI(x,n) - \sum_{n \epsilon nWords} PMI(x,n) \quad (2)$$

Turney et al. considers the statistics of the web search engine of AltaVista to find out the co-occurrence frequencies (F).

**D. Semantic Approach**

The Semantic Approach depends on various principles for calculating the similarity between words and thus provides sentiment values directly. The basic principle of this approach is that semantically related words should obtain similar sentiment values [8].

WordNet is a dictionary that is used to determine sentiment polarities and defines different types of semantic relationships among words. WordNet can be used to determine the senses of words because some words can have multiple interpretations. The possibility to disambiguate senses of words using WordNet can serve as a way to include the context of these words into the opinion analysis task. Like statistical methods, two sets of seed words with positive and negative sentiments are used as a starting point for bootstrapping the construction of a dictionary.

Hu and Liu [2], [3], [6] used WordNet to acquire a list of sentiment words by iteratively expanding the initial set with synonyms and antonyms. However, this method has some shortcomings. This method does not deal with context dependent opinion words.

Ding, Liu and Yu [4] proposed a holistic lexicon based approach to handle the context dependent opinion words. This method applied special linguistic rules. This approach uses external information and indications in other sentences and other reviews, instead of looking at the current sentence alone. This method is improvement of the method used in [2].

M. Eirinaki et al. 2011 [13] presents a feature-based opinion mining technique which used HAC(High Adjective Count) to identify the features from opinion and proposed Max Opinion Score algorithm for sentiment classification. To identify the polarity of reviews, features extracted from title also considered separately.

The researchers of the opinion mining field have shown less concern to the domain of opinions. The accuracy of sentiment analysis methods highly depends on the domain of interest [14]. The majority of the work [2]-[4], [6], and [13] had been carried on Product domain.

M. Taylor at al. [15]-[16] has applied feature-based opinion mining techniques on hotels and restaurants reviews. Association rule mining was applied as feature extraction method. For sentiment classification holistic lexicon based approach [4] was used. However the algorithms were only able to extract 35% of the explicit features.

### E. Rule based method

Rule-based methods handle the previously listed challenges better as it is possible to build dedicated rules for anaphora resolution, processing negations and intensifiers, target identification [18], source identification, and meta-phor disambiguation [19]. However, in order to attain acceptable success rates it is necessary to restrict the domain of application as much as possible. Thus, their main disadvantage is the difficulty of building generic extraction patterns and lexicons to extract all sentiment-related expressions contained in the data and to assign them a relevant label in varying contexts. Some authors have proposed different solutions, such as expanding the affective lexicon with new entries based on semantic similarity [20] or linear programming [21].

Supervised machine learning methods make it possible to generate more interoperable models, but they require the availability of labeled data for training. The quality of the models learned strongly depends on the reliability of sentiment annotation, which is affected by raters' subjectivity as existing annotation guides are rare. Besides, this type of methods allows building models that are sometimes difficult to interpret and control, which is why hybrid (statistical and rule-based methods) are sometimes used [22] [23]. Also some authors propose methods that provide more control, e.g., [24] uses conditional random fields (CRF) tackling opinion source identification as a sequential tagging task, whereas [25] identifies the target of the opinion with CRF. In [27] a Bayesian network is also used to model pragmatic dependencies to improve agreement analysis in meetings.

The tendency is now to handle the drawbacks of each type of method using hybrid methods, which bring both the generalizable nature of machine-learning approaches and the in-depth modeling offered by semantic rules. For instance, [18] uses probabilistic models for disambiguation, and tools like Auto Slog provide supervised pattern trainers that have been used for opinion source identification [24], while [26] proposes merging unsupervised machine learning algorithms with supervised ones.

Different contexts and domains of application may intro-duce other specific challenges. It is thus important to keep in mind that the performance of such systems is strongly dependent on:

The type of data to be analyzed: the style and the language register of the writer/speaker, the quality of the syntactic structure of the data (tweets and oral transcription versus newspaper). For example, sentiment detection in social networks or transcriptions of speech must be able to cope with ill-formed linguistic structures [28] and other phenomena such as orthographic mistakes, emotions and other symbols [29].

The classes of sentiment that are considered: classification according to ten classes is indeed more difficult than for two classes. Performance also depends on the choice of classes being studied, e.g., discriminating between appreciation and judgment is generally more difficult than between positive and negative.

The quality of the ground-truth annotations: evaluating performance consists of analyzing differences between ground truth annotations and system decisions. In the field of sentiment, we cannot speak about actual human error, since the situation is more complex (subjectivity of the annotations, and imbalanced proportion of emotional and neutral contents [30]).

**F. Comparative Evaluation**

In this section we give an overview of different feature extraction and sentiment classification methods applied in feature based opinion mining as discussed earlier. Table I presents comparison of different feature extraction methods based on the parameter precision and recall. Table II presents comparison of different sentiment classification methods.

TABLE I. COMPARISON OF VARIOUS FEATURE EXTRACTION METHODS

| Paper | Domain | Language | Feature Extraction | | |
|---|---|---|---|---|---|
| | | | *Method* | *Precision* | *Recall* |
| Turney 2002 [11] | Automobile Movie | English | Pattern based extraction | - | - |
| Hu and liu 2006 [5] | Product | English | Association rule mining | 72 | 80 |
| M.Taylor 2013 [15] | Hotels | English | Association rule mining | 38 | 33 |
| Eirinaki 2012 [13] | Product | English | High Adjective Count algorithm | - | - |
| J. Zhu 2011 [10] | Hotels | Chinese | Multi-aspect bootstrapping | 69 | 56 |
| Ding, Liu, 2008 [4] | Product | English | Association rule mining | 72 | 80 |

TABLE II. COMPARISON OF VARIOUS SENTIMENT CLASSIFICATION METHODS

| Paper | Sentiment Classification | | | |
|---|---|---|---|---|
| | *Method* | *Accuracy* | *Precision* | *Recall* |
| Turney 2002 [11] | PMI-IR method | 74 | - | - |
| Hu and liu 2006 [5] | Rule based algorithm with WordNet | 84 | - | - |
| M.Taylor 2013 [15] | Holistic linguistic Method | - | 90 | 93 |
| Eirinaki 2012 [13] | Max opinion score algorithm using manual dictionary | 87 | - | - |
| Z.Zai 2011[9] | SVM | 83 | - | - |
| J. Zhu 2011 [10] | Dictionary based | 75 | - | - |
| Ding, Liu, 2008 [4] | Holistic linguistic Method | - | 92 | 91 |

# IV. CHALLENGES AND ISSUES

Despite numerous research efforts, feature based opinion mining studies and application still have limitations and margins for improvement.

- Limitations of natural language processing, such as context dependency, semantic relatedness and word sense ambiguity, have created feature based opinion mining challenging.

- Feature based opinion mining systems that uses machine learning approach are domain dependent which require manual labeling of data, a difficult task to accomplish.
- Opinion spamming has become an issue due to fake opinions in reviews and forum discussions given by the users which may affect the decision of genuine users.
- Domain dependency is another issue, because the target features specific to domain may have different meanings or interpretations when applied to different domain.
- As reviews are crawled from web, they are from different locale so it may consist of different human languages. Multilingual effect also makes opinion mining difficult.

## V. CONCLUSION

The paper discusses various methods of sentiment analysis and tasks of feature based opinion mining. We have stated some of the approaches of feature based opinion mining and compared various feature extraction and sentiment classification techniques. Majority of work done in opinion mining is at document level and sentence level. Most of them have paid little attention to opinion spam and domain dependency. Feature based opinion mining is still an open area of research due to the fact that we can improve the accuracy of feature based opinion mining system by adding spam opinion detection, comparative review detection and multilingual Review handling mechanisms. For spam opinion detection any machine learning methods can be applied and to evaluate opinion of foreign language machine translation techniques can be used.

## VI. REFERENCES

[1] Bing Liu. "Sentiment analysis and opinion mining." Synthesis Lectures on Human Language Technologies 5, Morgan and Claypool, pp. 1-167, 2012.

[2] Hu, Minqing, and Liu Bing. "Mining and summarizing customer reviews." In Proceedings of the tenth ACM SIGKDD international conference on Knowledge discovery and data mining, ACM, pp. 168-177, 2004.

[3] Hu, Minqing, and Liu Bing. "Mining opinion features in customer M. Taylor, Edison, Juan D. Velásquez, and Felipe Bravo-Marquez. "A reviews." In AAAI, vol. 4, pp. 755-760, 2004.

[4] Ding, Xiaowen, Bing Liu, and Philip S. Yu. "A holistic lexicon-based approach to opinion mining." In Proceedings of the 2008 International Conference on Web Search and Data Mining, ACM, pp. 231-240, 2008.

[5] Hu, Minqing, and Bing Liu. "Opinion extraction and summarization on the web." In AAAI, vol. 7, pp. 1621-1624, 2006.

[6] Liu, Bing, Minqing Hu, and Junsheng Cheng. "Opinion observer: analyzing and comparing opinions on the web." In Proceedings of the 14th international conference on World Wide Web, ACM, pp. 342-351, 2005.

[7] Khan, Khairullah, Baharum Baharudin, and Aurnagzeb Khan. "Mining Opinion Components from Unstructured Reviews: A Review." Journal of King Saud University-Computer and Information Sciences, 2014, doi: http://dx.doi.org/10.1016/j.jksuci.2014.03.009.

[8] Tsytsarau, Mikalai, and Themis Palpanas. "Survey on mining subjective data on the web." Data Mining and Knowledge Discovery vol. 24, pp. 478-514, 2012.

[9] Zhai, Zhongwu, Hua Xu, Bada Kang, and Peifa Jia. "Exploiting effective features for chinese sentiment classification." Expert Systems with Applications, vol. 38, pp. 9139-9146, 2011.

[10] Zhu, Jingbo, Huizhen Wang, Muhua Zhu, Benjamin K. Tsou, and Matthew Ma. "Aspect-based opinion polling from customer reviews." IEEE Transactions on Affective Computing, vol. 2, pp. 37-49, 2011.

[11] Turney, Peter D. "Thumbs up or thumbs down? Semantic orientation applied to unsupervised classification of reviews." In Proceedings of the 40th annual meeting on association for computational linguistics, Association for Computational Linguistics, pp. 417-424, 2002.

[12] Turney, Peter D., and Michael L. Littman. "Measuring praise and criticism: Inference of semantic orientation from association." ACM Transactions on Information Systems (TOIS), vol. 21, pp. 315-346, 2003.

[13] Eirinaki Magdalini, Shamita Pisal, and Japinder Singh. "Feature-based opinion mining and ranking." Journal of Computer and System Sciences, vol. 78, pp. 1175-1184, 2012.

[14] Cruz, Fermín L., José A. Troyano, Fernando Enríquez, F. Javier Ortega, and Carlos G. Vallejo. "'Long autonomy or long delay?' The importance of domain in opinion mining." Expert Systems with Applications, vol. 40, pp. 3174-3184, 2013.

[15] Marrese-Taylor, Edison, Juan D. Velásquez, Felipe Bravo-Marquez, and Yutaka Matsuo. "Identifying customer preferences about tourism products using an aspect-based opinion mining approach." Procedia Computer Science, vol. 22, pp. 182-191, 2013.

[16] Marrese-Taylor, Edison, Juan D. Velásquez, and Felipe Bravo-Marquez. "Opinion Zoom: A Modular Tool to Explore Tourism Opinions on the Web." In Web Intelligence (WI) and Intelligent Agent Technologies (IAT), IEEE/WIC/ACM International Joint Conferences, vol. 3, IEEE, pp. 261-264, 2013.

[17] novel deterministic approach for aspect-based opinion mining in tourism products reviews." Expert Systems with Applications, vol. 41, pp. 7764-7775, 2014.

[18] K. Bloom, N. Garg, and S. Argamon, "Extracting appraisal expressions," in Proc. Human Language Technol./North Amer. Assoc. Comput. Linguists, Apr. 2007, pp. 308–315.

[19] L. Zhang, "Exploration of affect sensing from speech and meta-phorical text," in Learning by Playing. Game-Based Education Sys-tem Design and Development. New York, NY, USA: Springer, 2009, 251–262.

[20] N. Malandrakis, A. Potamianos, E. Iosif, and S. Narayanan, "Emotiword: Affective lexicon creation with application to inter-action and multimedia data," in Computational Intelligence for Multimedia Understanding. New York, NY, USA: Springer, 2012, 30–41.

[21] Y. Choi and C. Cardie, "Adapting a polarity lexicon using integer linear programming for domain-specific sentiment classi-fication," in Proc. Conf. Empirical Methods Natural Language Pro-cess., 2009, pp. 590–598.

[22] A. Kaur and V. Gupta, "A survey on sentiment analysis and opinion mining techniques," J. Emerging Technol. Web Intell., vol. 5, no. 4, pp. 367–371, 2013.

[23] W. Medhat, A. Hassan, and H. Korashy, "Sentiment analysis algorithms and applications: A survey," Ain Shams Eng. J., vol. 5, no. 4, pp. 1093–1113, 2014.

[24] Y. Choi, C. Cardie, E. Riloff, and S. Patwardhan, "Identifying sources of opinions with conditional random fields and extrac-tion patterns," in Proc. Conf. Human Language Technol. Empirical Methods Natural Language Process., 2005, pp. 355–362.

[25] N. Jakob and I. Gurevych, "Extracting opinion targets in a single-and cross-domain setting with conditional random fields," in Proc. Conf. Empirical Methods Natural Language Process., 2010, 1035–1045.

[26] M.-T. Mart_ın-Valdivia, E. Mart_ınez-C_amara, J.-M. Perea-Ortega, and L. A. Urena~ Lopez,_ "Sentiment polarity detection in Spanish reviews combining supervised and unsupervised approaches," Expert Syst. Appl., vol. 40, pp. 3934–3942, 2013.

[27] M. Galley and K. McKeown, "Identifying agreement and dis-agreement in conversational speech: Use of Bayesian networks to model pragmatic dependencies," in Proc. 42nd Annu. Meeting Assoc. Comput. Linguistics, 2004, p. 669.

[28] K. Khan, B. Baharudin, A. Khan, and A. Ullah, "Mining opinion components from unstructured reviews: A review," J. King Saud Univ.-Comput. Inf. Sci., vol. 26, no. 11, pp. 258–275, 2014.

[29] M. A. Russell, Mining the Social Web. Newton, MA, USA: O'Reilly, 2011.

[30] Z. Callejas and R. Lopez_-Cozar,_ "Influence of contextual informa-tion in emotion annotation for spoken dialogue systems," Speech Commun., vol. 50, no. 5, pp. 416–433, 2008.

# Roadmap for Polarity Lexicon Learning and Resources: A Survey

**Swati Sanagar and Deepa Gupta**

**Abstract** Sentiment analysis opens door for understanding opinions conveyed in text data. Polarity lexicon acts as heart in sentiment analysis tasks. Polarity lexicon learning is explored using multiple techniques over years. This survey paper discuss polarity lexicon in two aspects. The first part is literature study which depicts from initial techniques of polarity lexicon creation to the very recent ones. The second part reveal facts about available open source polarity lexicon resources. Also, open research problems and future directions are unveiled. This informative survey is very useful for individuals entering in this arena.

**Keywords:** sentiment analysis; polarity lexicon; survey; transfer learning; sentiment lexicon.

## 1 Introduction

Today's internet era has provided easy access to huge online information. People around the world digitize their diverse opinions and experiences on internet. These text data experiences and opinions are analysed to extract valuable information using Sentiment analysis techniques. The word sentiment/opinion is individuals' view, understanding, or experience about some entity. Sentiment analysis is computational study of people's opinion and emotion about some event, topic, object, individual, or organization etc. This research primarily started since early 1990's when standard information retrieval was observed insufficient and research for more in-depth information [18] and individual's point of views

---

Swati Sanagar
Department of Computer Science and Engineering, Amrita School of Engineering, Amrita Vishwa Vidyapeetham, Amrita University, Bangalore, India. e-mail: swatisanagar@yahoo.co.in

Deepa Gupta
Department of Mathematics, Amrita School of Engineering, Amrita Vishwa Vidyapeetham, Amrita University, Bangalore, India. e-mail: g_deepa@blr.amrita.edu

 
J.M. Corchado Rodriguez et al. (eds.), *Intelligent Systems Technologies and Applications 2016*, Advances in Intelligent Systems and Computing 530,
DOI 10.1007/978-3-319-47952-1_52

[48] was explored. This initial research direction is explored for semantic orientation of adjectives [17] and also subjectivity [49] etc. In subsequent years, it received attention and is explored for product reputation mining [29], sentiment classification [35] etc. Sentiment analysis research has flourished over decade and it has unlocked numerous opportunities such as investor's opinion analysis for stock market [7] etc.

Consider following example from consumer review category.

- This APP is *amazing*!! *Very easy* to control and *user friendly*.
- *Waste of money*. This sweatshirt is *not well-made* at all.

The first example from Android Apps domain contain words/phrases '*amazing*', '*very easy*', and '*user friendly*' which indicate positivity. The second example from Clothes domain contain words/phrases '*waste of money*', '*not well-made*' which denote negativity. These are *opinion words* that convey individual's opinion and it is the only means to detect underlying sentiment in text data. These opinion words or phrases are compiled together to form lexicon. Polarity lexicon stores set of opinion words and their polarity orientation or scores. It is an important resource in sentiment analysis which avoids recreation efforts and help in faster processing. Most importantly, it is building block of most sentiment analysis applications. Sentiment analysis approaches are broadly categorised as machine learning/statistical, lexicon based, and hybrid. Machine learning/statistical approaches solve sentiment analysis as a regular text classification problem using machine learning algorithms. It makes use of syntactic and/or linguistic features such as part of speech patterns, pruning, n-grams etc. e.g. unigram, bigram features used with Naïve Bayes classifier [21] and enhanced with Support Vector Machine [39]. These features include opinion words as major contribution. Lexicon based approaches utilise and/or generate polarity lexicon which can be stored and reused for domain related sentiment analysis task. These approaches [26] are either corpus-based that use linguistic information or knowledge/dictionary based that use existing knowledge bases. Lexicon based approaches are focused on polarity lexicon learning. Hybrid approach is combination of machine learning/statistical and lexicon based approaches.

As per our knowledge this is first attempt of writing survey on polarity lexicon learning. This work is organised in five sections. Section 2 gives general polarity lexicon creation process. This survey paper discuss polarity lexicon learning in two ways, literature study in section 3 and facts about available open source polarity lexicon in section 4. The last section concludes with open research problems and future research direction.

## 2 General Process for Polarity Lexicon Learning

General steps of polarity lexicon learning are given in Fig. 1. The process starts from corpora selection and follows multiple steps.

- Corpora is unstructured text either labelled or unlabeled.
- Data pre-processing is used for data cleaning and may involve multiple steps such as tokenization, spell-check, removal of stop word, part-of-speech tagging, lemmatization, chunking, parsing, etc. Most of the times data pre-processing decision is taken by observing data.
- Opinion words/phrases are selected based on decision made for model. It may include n-grams, part-of-speech, syntactic patterns, collocations etc.
- Term weighting is performed for opinion words/phrases using different techniques that may include binary, frequency, or tf-idf based.
  Knowledge/Corpus based lexicon creation technique is/are applied to opinion word/phrase to assign weight based on lexical affinity technique or to assign polarity based on keyword spotting technique.

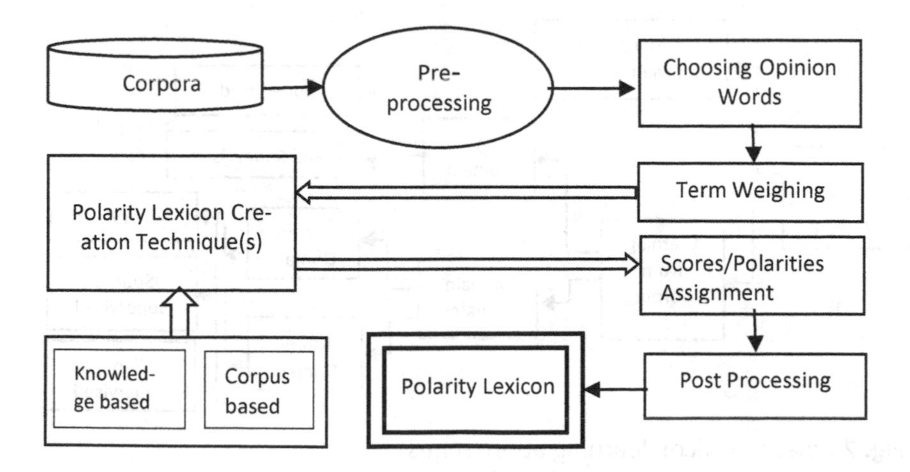


**Fig. 1** General procedure for polarity lexicon learning.

– Raw polarity lexicon may be post-processed for pruning and relevancy to get final polarity lexicon.

## 3 General Classification of Polarity Lexicon Learning Techniques

General classification of polarity lexicon learning approaches from our viewpoint are given in Fig. 2. Polarity lexicon learning approaches include learning from existing knowledge bases, learning using linguistic information from corpus, or learning manually. Manual approach is labour intensive, time consuming, and non-scalable which is inadequate for information era. Other two approaches are classified depending on usage of labelled data and method used for polarity lexicon learning. Literature study based on this is discussed in subsequent subsections.

### *3.1 Learning from Knowledge base/Dictionary*

Learning from knowledge bases/dictionaries approaches exploit linguistic resources such as language dictionaries, thesaurus, WordNet, and other lexicon resources. Polarity lexicon learns from single knowledge base or

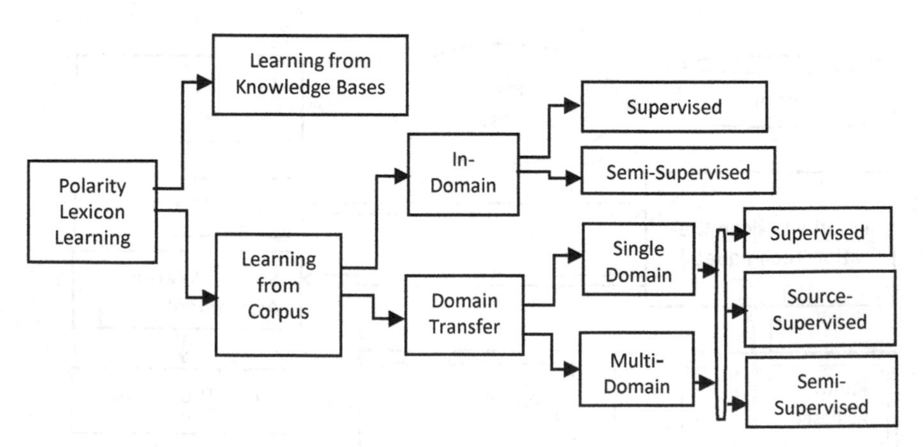


**Fig. 2** Polarity lexicon learning approaches

incorporate knowledge from multiple knowledge bases. Learning from single knowledge base observed using three major approaches that includes relation based, graph based, and hybrid.

One among early papers using this approach [19] take initial small set of labelled adjectives as opinion words. These words are used to expand polarity lexicon using WordNet. It adds synonyms to same polarity and antonyms to opposite polarity set. Adjectives are strong representatives of sentiment, but verbs and nouns also express sentiments. This fact is used to extend approach by considering verbs, nouns in addition to adjectives [22] as opinion words. WordNet-Affect [44] is created with similar criteria by including hyponymy, meronymy, holonymy relations using WordNet and other resources. First the core synsets are identified and then extended using relations. Basic assumption that synonym of positive word is also positive may not hold in every case. Some cross-verification mechanism is needed. Approach of finding synonym of polarity opinion word is extended [23] using probabilistic approach to confirm polarities of synonyms. Opinion word is assigned polarity based on its maximum closeness with polarity class across synsets. Initial set of polarity words is extended using synonym, antonym, hypernymy, hyponymy search [11] and used to generate feature vectors for polarity opinion words using glosses. Word feature vectors cosine normalised tf-idf score is assigned as polarity strength to word. Orientation of opinion words is considered subjective. Approach is further progressed [12] by considering subjectivity to assign positive, negative, and objective scores to each opinion word. Also, new words are added using WordNet graph navigation. Refined SentiWordnet is extended by adding objectivity along with subjectivity [13]. Positive, negative, and objective scores are assigned to each synsets available in WordNet by combining results from eight ternary classifiers. Labelled adjectives are used [1] to add more opinion words using different WordNet relations, POS-pruning, polarity overlapping etc. in multiple steps. Score is assigned to word depending on frequency of word in polarity class in multiple runs.

Graph based approaches often learn polarity lexicon using label propagation approach. Word is considered as node and relation as edge. Initially nodes are unlabelled, later on they are labelled with polarity. Synonym relation between adjectives is used as edge [20]. WordNet distance based measure is used for assigning semantic orientation to adjectives. It is used to calculate distance between adjectives using geodesic measure and in turn to calculate score for it. Also, connected syntactic components such as noun and verb are explored. Labelled polarity words are used to extract synonym graph from WordNet [38]. It is explored using three graph approaches Mincut, Randomized Mincut, and label propagation. Markov random walk model is used to create polarity lexicon [16] and two words are connected using WordNet synonym and hypernym relation. Polarity is assigned to words based on mean hitting time.

Hybrid approach is used Roget-like thesaurus to extract synonyms and antonyms from initial labelled opinion words. It used additional affix pattern rules for antonym generation [30] to increase polarity lexicon coverage. Two approaches are combined for large corpora [36]. Initial labelled word list is expanded using synonym, antonym relation from WordNet and conjunctions. Adjacency matrix is created for adjectives and synonym relation as edge weight. A constrained symmetric nonnegative matrix factorization (CSNMF) method is applied to matrix and an iterative process is used to cut graph of adjectives into positive and negative sets where each adjectives is assigned positive and negative score. Approach uses WordNet and corpora jointly for improved performance.

Usage of multiple knowledge bases is less explored in literature. Multiple knowledge bases such as General Inquirer, SentiWordnet, Subjectivity Clues and Moby thesaurus are combined using max rule, sum rule etc. [34] and evaluated on multiple domains.

Initial development of polarity lexicon learning from knowledge bases is observed mainly for relation based approach in 2000s. Subsequently graph based and hybrid approaches are evolved. Initial approaches assigned polarity to opinion words and later approaches assigned scores.

## *3.2 Learning from Corpus*

Learning from corpus approach utilize linguistic knowledge exploiting corpora and explore using statistical and semantic methods. It is divided into two main approaches in-domains learning and domain transfer learning. This approach learns opinion words and their orientation by unveiling facts from corpus.

### 3.2.1 In-Domain Learning

In-domain polarity lexicon learning uses domain corpora for training and for testing. It is supervised when labelled training data is used and semi-supervised when minimal labelled information is used.

Supervised polarity lexicon learning models provide highly accurate results. These models are mostly based on syntactic relation, semantic orientation etc. Supervised model [17] used knowledge of conjunction and morphological relation. This knowledge is combined in log linear model to check if two conjoined adjectives have same orientation using labelled adjectives. Many conjunction rules are used in addition to other rules [9] to extract orientation of word in context using initial labelled instances. Labelled opinion words are used [43] to build a phrase based polarity lexicon. Point wise Mutual Information (PMI) approach is used to calculate semantic orientation score of word using opinion word orientation. This is based on co-occurrence. Rule based approach based on dependency relation [37] used minimal labelled seed input. This double propagation approach extract opinion words and assign polarity using conjunction, negation etc. rules utilising other opinion words. Domain specific semi-supervised polarity lexicon creation framework [52] is applied to ten domains. Dependency relation and few labelled instances are used to design rules for polarity assignment.

Polarity lexicon creation approach is scaled to multiple domains [42] to improve scope. Scores are assigned using semantic orientation calculator using linguistic features such as negation, intensification etc. Multi-domain knowledge is used to construct polarity lexicon [46] to create domain dependent and domain independent polarity lexicon. Scores are assigned based on frequency proportionality in polarity context.

In-domain polarity learning is most obvious, highly accurate, and popular approach. Linguistic knowledge is explored using various perspective such as occurrence of conjunction, negation, intensification etc. and techniques such as rule based, co-occurrence, dependency relation etc.

### 3.2.2 Domain Transfer Learning

Domain transfer learning is based on storing knowledge while solving one problem and transferring knowledge to other domains. Transferred knowledge is learned either from single or multiple domains. Supervised transfer learning for multi-domain and single domain uses labelled data from both training i.e. source and target i.e. test domains. Source-supervised learning utilises labelled data only from source domain. Semi-supervised approach uses only minimal supervised information.

Supervised single domain knowledge transfer approach for POS feature ensemble and sample selection of target and source instances [51] are checked for closeness using principal component analysis. Based on this word scores of source domain are tuned to adapt to target domain. A source supervised structured correspondence learning model is used to identify frequently occurring features from source and target domains called pivot features [4]. These features are used to build correspondence with non-pivot features from source and target domains. Success of this method depends on choice of pivot features in both domains, based on which the algorithm learns a projection matrix that maps features from target domain into the feature space of source domain. Geodesic flow kernel adaptation algorithm used [15] source supervised approach. Domain independent opinion words available in labelled data of source domain and having similar word distribution as target domain are selected as Landmarks. Semi-supervised knowledge transfer approach for single domain is not observed.

Very few multi-domain knowledge transfer approaches are observed in literature. A supervised corpus based approach is explored by learning domain specific and domain independent lexicons [45]. An improved polarity lexicon is constructed by acquiring knowledge from multiple source domains which is transferred to multiple similar target domains. Polarity lexicon learned using multiple source supervised domains using various approaches [2] such as majority voting, weighted voting etc. is transferred to target domain. A semi-supervised multi-domain transfer learning approach [40] uses minimal seed words as labelled input and learns seed words and polarity lexicon from multiple source domains using iterative learning process. This process embeds Latent Semantic Analysis approach that assign score to opinion words. This process learn seed words applicable to group of domains belonging to a particular category. Learned seed words are used to create polarity lexicon for target domains. Also, source domain polarity lexicons are adapted to target domain.

A major advantage of domain transfer polarity lexicons is that it capture domain specific effects. Polarity lexicon customized to new domain can help avoid individual polarity lexicon construction from scratch.

Exploration of learning from knowledge bases is seen since early 2000. Learning from knowledge bases make easy and quick access to large number of sentiment words. Since most of these knowledge bases are generalized they end up creating generic polarity lexicon. They lack in domain orientations and do not provide solution to domain specific task. Corpus based approaches have gained popularity and importance, as they overcome this drawback and capture domain specific effects. Supervised corpus based approaches are highly accurate, but bottleneck for these techniques is availability of labelled data. This bring limitations when it comes to scaling up to multiple domains. Unsupervised approaches provide solution in this scenario. Although, scaling up also depends on other factors such as algorithm limitations. Domain transfer learning has capability to overcome these shortcomings. It show path towards development of polarity lexicon that preserves domain specific characteristics and still provide solution to multiple domains. This evolving futuristic approach is bringing revolution in polarity lexicon learning methods.

**Table 1** Open-source polarity lexicon resource

| Details → Lexicon ↓ | Size | Domain | Polarity | Method | Applicable | Corpora used |
|---|---|---|---|---|---|---|
| ANEW (1999)[5] | 1,040 | General | 9 point score | Manual | emotion and attention study | Brown Corpus |
| General Inquirer(GI) (2000)[14] | 11,788 | General | Categorical 182 tag categories | Manual | Content Analysis, social cognition, (social science) | Harvard , Lasswell dictionary |
| WordNet-Affect (2004)[41] | 2,874 SS[a] 4,787 WD[c] | General | + / - /neu/ ambiguous | Semi-automatic | Multi-Category SA[b] | Based on WordNet |
| Hu & Liu (2004)[19] | 6,800 | Social Media | +/- | Semi-automatic | Social Media data analysis | Social Media |
| MPQA (2005)[50] | 8,221 | General | + / - /neu Subjectivity | Semi-automatic | Subjectivity clues analysis | News documents |
| Micro-WNOp (2007)[28] | 1,105 SS | General | Pos/neg (0 to1) | Manual | General Purpose SA | WordNet |
| SentiWordnet 1.0(2006) 3.0(2010)[3] | 1,17,659 | General | (0 to 1) | Semi-automatic | General Purpose SA | Based on WordNet |
| LabMT (2011)[10] | 10,222 | Mixed | Rank | Semi-automatic | Ranking, time series analysis | Twitter postings |
| AFINN (2011)[33] | 2,477 | Micro-blogging | -5 to +5 | Manual | Micro-blogging SA | Twitter postings |
| NRC Word-Emotion(2011)[32] | 14,182 | General | 8 emotion, ( + / - ) | Manual | Multi-category SA | Macquarie Thesaurus |
| Warninger (2013)[47] | 13,915 | General | 9 point scale | Manual | emotion and attention SA | Anew, Categorical norms |
| Sentiment140 (2013)[31] | 62,468 U[d], 677,698 B[e], 480,010 P[f] | Micro-blogging | Context Frequency (+/-), score | Auto - matic | Twitter based SA | Twitter |
| NRC MD[g] Twitter (2014)[24] | 1,515 | Micro-blogging | ( -1 to1 ), ( 0 to 1 ) | Semi-automatic | Micro-blogging SA | Sentiment140 & Hashtag sentiment |
| NRC Laptop (2014)[25] | 26,577 U, 1,55,167 B | Consumer Product (Laptop) | Context Frequency (+/-), score | Auto - matic | Consumer product (Laptop) review SA | Amazon Consumer reviews |

| Details → Lexicon↓ | Size | Domain | Polarity | Method | Applicable | Corpora used |
|---|---|---|---|---|---|---|
| SenticNet (2014)[6] | 30,000 concepts | General | Affective scores -1 to 1 | Auto-matic | Concept level SA | WordNet-Affect, Open Mind |
| SentiSense (2014)[8] | 2,190 SS 5,496 WD | General | 14 emotion categories | Semi-au-tomatic | Intensity, emo-tion SA | WordNet 3.0 |
| Yelp Restau-rant(2014)[25] | 39,274 + 276,651 | Restaurant related | Frequency (+/-), score | Auto - matic | Restaurant review SA | Yelp Restaurant |
| Loughran McDona-ld Master(2015)[27] | 85,131 | Finance | 7 category & (+,-) | Semi-au-tomatic | Finance domain | 2of12inf list |

[a]Synsets, [b]Sentiment Analysis, [c]Words, [d]Unigram, [e]Bigram, [f]Pair of uni/bigram, [g]MaxDiff

## 4. Existing Open Source Polarity Lexicon Resources

Many polarity lexicon are created for different tasks representing different domains. Open source resources make research valuable, freely available, and receive wide acceptance from society. Countable, but widely useful open source polarity lexicons have been created over two decades. A brief informative description of major available open-source polarity lexicons for English language is given in Table 1. It describes key facts about polarity lexicon such as size, creation year, polarity measure used, application etc. In Table2 examples containing almost same opinion words are given for polarity lexicon listed in Table1. This describes variation in polarity/score assignment across the listed polarity lexicon. Every polarity lexicon have their own scales to assign polarity/score/category to words/phrases. Most of the examples are self-explanatory. Consider example from *LabMT* that stores rank from high as 1 and grows low. It store rank of word in different context such as word's rank in happiness index, word's rank in twitter frequency index etc. The word *happy* is ranked 4 indicating it as highly *happy word* compared to word *sad* which is ranked much low according to *happiness index. Warninger* lexicon is extended using earlier *Anew* polarity lexicon by size and adding emotional norm dif-

***Table 2*** *Examples from open-source polarity lexicon*

| Polarity Lexicon | Example and Description |
|---|---|
| ANEW | Word / Valence(M, SD)[h] / Arousal(M, SD) / Dominance(M, SD)<br>happy / 8.25, 1.39 / 7.00, 2.73 / 6.73, 2.28<br>sad / 1.61, 0.95 / 4.13, 2.38 / 3.45, 2.18 |
| General Inquirer (GI) | Happy: H4Lvd[i], positive, pleasure, emotion, related adjective: Joyous, pleased Sad: H4Lvd, negative passive, pain, emotion, related adjective: unhappy |
| WordNet-Affect | Affective category labels are emotion, behaviour, attitude, cognitive state<br>Word / Tag / Senses<br>joy / positive / joy, elated, gladden, gleefully<br>sad / negative / sadness, unhappy, sadden, deplorably |
| Hu & Liu | happy: positive; sad: negative |
| MPQA | happy: strong subjective, positive; sad: strong subjective, negative |
| Micro-WNOp | good: adjective, positive (1), negative(0) by 3 annotators;<br>ugly: adjective, positive(0,0), negative(0.75, 1) |
| SentiWordnet | Word / Positive / Objective / Negative / POS<br>happy / 0.875 / 0.125 / 0 / Adjective<br>sad / 0.125 / 0.125 / 0.75 / Adjective |
| LabMT | Word / Happiness / Twitter / GBks[j] / NYT[k] / Music Lyrics<br>happy / 4 / 65 / 1372 / 1313 / 375<br>sad / 10091 / 306 / 3579 / 3441 / 526 |
| AFINN | happy: positive, score(3); sad: negative, score(-2) |
| NRC Wd Emotion | happy: anticipation, joy, positive, trust; sad: not available |
| Warninger | Word / Valence / Word / Arousal / Word / Dominance<br>happiness / 8.48 / calm / 1.67 / uncontrollable / 2.18 |
| Sentiment140 | Word / PMI[l] score / Count in' +'context / Count in'- ' context<br>happy / 1.196 / 19174 / 6087<br>sad / -2.735 / 1442 / 23342 |
| NRC Laptop | Word/Phrase / PMI score / Count in' +'context / Count in'- ' context<br>happy / 1.121 / 3,646 / 268<br>sad / -1.342 / 74 / 64<br>best_laptop / 3.462 / 287 / 2 |

| Polarity Lexicon | Example and Description |
|---|---|
| NRC MD Twitter | happy : 0.953(0 to 1), 0.734(-1 to 1); sad : 0.219(0 to 1), -0.562(-1 to 1) |
| SenticNet | related concepts for celebrate_special_occasion are celebrate ( holiday, occasion, birthday, wedding, express appreciation)<br>Concept pleasantness attention sensitivity aptitude polarity<br>happy 0.894 0 0 0 0.298<br>sad - 0.919 0 0 0 -0.306<br>celebrate-special_occasion 0.93 0.724 0.0 0.0 0.551 |
| SentiSense | depression: Sadness category; exultation: Joy; adorable: Love |
| Yelp Restaurant | Word PMI score Count in' +'context Count in'- ' context<br>happy 0.825 15185 2118<br>sad -0.977 51 43 |
| LoughranMc-Doald Master | Happy : Positive added in 2009; closed : Negative added in 2009 |

[h](Mean, Standard deviation), [i]Harvard-4 & Lasswell dictionary, [j]Google Books, [k]New York Times, [l]Point-wise Mutual Information

ferentiation by gender, age etc. This comprises most of the available open source lexicon which we observed from our exhaustive search.

From available polarity lexicon, we observed various different facts and how open source polarity lexicon changed over time. Overall, the first decade from 1999 to 2010 has displayed initial and important polarity lexicon creation efforts, mostly carried out manually and semi-automatically. Trend of touching different areas of study is observed. First decade denotes focus on many different areas, but mainly based on general context. Current decade since 2011 represent information era where development of polarity lexicon is around social media contents. Moreover, usability of these polarity lexicon is not limited to study, but are vastly used for research, commercial, and social purposes. Many useful polarity lexicons are small in size and are having potential to expand. Many available polarity lexicon are created manually which brings limitation on extending it to other domains. Automated polarity lexicon creation need more exploration. Available polarity lexicons have applications in different areas such as psychology or health analysis, subjectivity analysis, emotion analysis, ranking based, affect based, mood based learning, content analysis and for general context analysis etc.

## 5 Open Research Problems and Future Directions

In todays' information era huge unlabelled information is available compred to minimal labelled information. Sentiment analysis research has grown lateral to many other research areas. But the vertical research growth is restricted due to lack in fundamental research which includes polarity lexicon learning. Research in polarity lexicon learning for sentiment analysis has not matured up to the mark. Considering these facts following research direction are also open research problems in learning polarity lexicon for sentiment analysis.

- Semi-supervised and unsupervised seem to be key approaches. Using them to build polarity lexicon scalable to multiple domains with minimal efforts.
- Learning some mechanism to build domain specific and generic lexicon applicable to many similar, dissimilar, and unseen domains.
- Reducing rebuilding efforts of polarity lexicon learning using some intermediate resources such as seed words etc. with key focus on utilizing learned knowledge using some technique such as transfer learning.
- Extension of available polarity lexicon in terms of size, scope, and other aspects to improve usability.
- Collaboration of available polarity lexicon in terms of usability, score/polarity etc.

## References

1. Andreevskaia A, Bergler S (2006). Mining WordNet for fuzzy sentiment: Sentiment tag extraction from WordNet glosses. In: EACL'06: Proceedings of the European Chapter of the Association for Computational Linguistics, vol. 6, pp. 209-16.
2. Anthony A, Gamon M (2005). Customizing sentiment classifiers to new domains: A case study. In: Proceedings of international conference on recent advances in natural language processing, vol. 1(3.1), Bulgaria, pp. 2-1.

3. Baccianella S, Esuli A, Sebastiani F (2010). SentiWordNet 3.0: An enhanced lexical resource for Sentiment Analysis and Opinion Mining. In: Proceedings of the seventh conference on international Language Resources and Evaluation, vol. 10, pp. 2200-4.
4. Blitzer J, Dredze M, Pereira F (2007) Biographies, Bollywood, boom-boxes and blenders: Domain adaptation for sentiment classification In: Proceedings of 45th annual meeting of the Association of Computational Linguistics, Prague, Czech Republic, June 2007, pp. 440–47.
5. Bradley MM, Lang PJ (1999). Affective norms for English words (ANEW): Instruction manual and affective ratings. Technical Report C-1, Center for Research in Psychophysiology, University of Florida.
6. Cambria E, Olsher D, Rajagopal D (2014). SenticNet 3: A common and common-sense knowledge base for cognition-driven sentiment analysis. In: Twenty-eighth AAAI conference on artificial intelligence, Quebec City 2014, pp. 1515-21.
7. Das S, Chen M (2001). Yahoo! for Amazon: Extracting market sentiment from stock message boards. In: Proceedings of the Asia pacific finance association annual conference, vol. 35, p. 43.
8. deAlbornoz JC, Plaza L, Gervas P (2012). SentiSense: An easily scalable concept-based affective lexicon for sentiment analysis. In: LREC, pp. 3562-3567.
9. Ding, X, Bing L, Yu PS (2008) A holistic lexicon-based approach to opinion mining. In: Proceedings of the conference on Web search and Web Data Mining, pp. 231-40.
10. Dodds PS, Harris KD, Kloumann IM, Bliss CA, Danforth CM (2011). Temporal Patterns of Happiness and Information in global social network: Hedonometrics and Twitter. *PLoS one,* 6(12):e26752. doi:10.1371/journal.pone.0026752
11. Esuli A, Sebastiani F (2005). Determining the semantic orientation of terms through gloss classification. In: Proceedings of the 14th ACM international conference on Information and knowledge management, pp. 617-24.
12. Esuli A, Sebastiani F (2006a). Determining term subjectivity and term orientation for opinion mining. In: Proceedings of conference of the European chapter of the Association for Computational Linguistics.
13. Esuli A, Sebastiani F (2006b). SentiWordNet: A publicly available lexical resource for opinion mining. In: Proceedings of Language Resources and Evaluation, vol. 6, pp. 417-22.
14. General Inquirer http://www.wjh.harvard.edu/~inquirer/ Accessed on 2016 Jan 20.
15. Gong B, Grauman K, Sha F (2013). Connecting the dots with landmarks: discriminatively learning domain-invariant features for unsupervised domain adaptation. In: ICML'13: Proceedings of the 30th international conference on Machine Learning (ICML), Atlanta, pp. 222-30.
16. Hassan A, Radev D (2010). Identifying text polarity using random walks. In: Proceedings of annual meeting of the Association for Computational Linguistics.
17. Hatzivassiloglou V, McKeown K (1997). Predicting the semantic orientation of adjectives. In: Proceedings of 8th conference of Association for Computational Linguistics, pp. 174-81.
18. Hearst MA (1992). Direction-based text interpretation as an information access refinement. Text-based intelligent systems: Current research and practice in information extraction and retrieval, 1:257-74.
19. Hu M, Liu B (2004). Mining and summarizing customer reviews. In: Proceedings of the tenth ACM SIGKDD international conference on Knowledge discovery and data mining, ACM, pp. 168-77.
20. Kamps J, Marx MJ, Mokken RJ, Rijke MD (2004). Using Wordnet to measure semantic orientations of adjectives. In: Proceedings.of LREC'2004, pp. 1115-18.

21. Kang H, Yoo SJ, Han D (2012). Senti-lexicon and improved Naïve Bayes algorithms for sentiment analysis of restaurant reviews, Expert Syst. Appl., 39(5):6000-10.
22. Kim SM, Hovy E (2004). Determining the sentiment of opinions. In: Proceedings of the 20th international conference on Computational Linguistics. Association for Computational Linguistics, pp. 1367.
23. Kim SM, Hovy E (2006). Identifying and analyzing judgment opinions. In: Proceedings of Human Language Technology Conference of the North American Chapter of the ACL, pp. 200-7.
24. Kiritchenko, S., Zhu, X., Mohammad, S. (2014). Sentiment Analysis of Short Informal Texts. J. Artificial Intelligence Res. 50:723-62.
25. Kiritchenko S, Zhu X, Cherry C, Mohammad S (2014). Detecting Aspects and Sentiment in Customer Reviews. In: Proceedings of the 8th international workshop on Semantic Evaluation Exercises, SemEval-2014, Dublin, Ireland, pp. 437-42.
26. Liu B (2012). Sentiment analysis and opinion mining, Morgan & Claypool Publishers.
27. Loughran T, McDonald B (2011). When is a Liability not a Liability? Textual Analysis, Dictionaries, and 10-Ks. J. of Finance, 66:1, 35-65.
28. Micro-WnOp, http://www-3.unipv.it/wnop/#Cerini07 Accessed on 2016 May 15.
29. Morinaga S, Yamanishi K, Tateishi K, Fukushima T (2002). Mining product reputations on the web. In: Proceedings of ACM SIGKDD international conference on knowledge discovery and data mining, ACM, pp. 341-49.
30. Mohammad S, Dunne C, Dorr B (2009). Generating high-coverage semantic orientation lexicons from overtly marked words and a thesaurus. In: Proceedings of the Conference on Empirical Methods in Natural Language Processing, vol. 2, pp. 599-608.
31. Mohammad SM, Kiritchenko S, Zhu X (2013). NRC-Canada: Building the state-of-the-art in sentiment analysis of tweets. In: Proceedings of seventh international workshop on Semantic evaluation exercises (SemEval-2013), Atlanta, Georgia, USA.
32. Mohammad SM, Turney PD (2010). Emotions evoked by common words and phrases: using Mechanical Turk to create an emotion lexicon. In: Proceedings of the NAACL-HLT'10 workshop on computational approaches to analysis and generation of emotion in text, California, ACL, pp. 26-34.
33. Nielsen FÅ (2011). A new ANEW: Evaluation of a word list for sentiment analysis in microblogs. arXiv preprint arXiv:1103.2903. http://arxiv.org/abs/1103.2903 Accessed on 2016 Jan 20.
34. Ohana B, Tierney B, Delany S (2011). Domain independent sentiment classification with many lexicons. In: WAINA'11: Advanced information networking and applications. IEEE workshops of international conference, pp. 632-37.
35. Pang B, Lee L, Vaithyanathan S (2002). Thumbs up? sentiment classification using machine learning techniques. In: Proceedings of conference on empirical methods in natural language processing, vol. 10, pp. 79-86.
36. Peng W, Park DH (2011). Generate adjective sentiment dictionary for social media sentiment analysis using constrained nonnegative matrix factorization. In: Proceedings of fifth international AAAI conference on weblogs and social media, pp. 273-80.
37. Qiu G, Liu B, Bu J, Chen C (2009). Expanding domain sentiment lexicon through double propagation. In: Proceedings of international joint conference on Artificial Intelligence, vol. 9, pp. 1199-04.
38. Rao D, Ravichandran D (2009). Semi-supervised polarity lexicon induction. In: Proceedings of the 12th conference of the European chapter of the ACL, pp. 675-82.
39. Rui H, Liu Y, Whinston A (2013). Whose and what chatter matters? The effect of tweets on movie sales. Decision Support Syst. 55(4):863-70.

40. Sanagar S, Gupta D (2015). Adaptation of multi-domain corpus learned seeds and polarity lexicon for sentiment analysis. In: Proceedings of international conference on computing and network communications, pp.50-58.
41. Strapparava C, Valitutti A (2004). WordNet-Affect: an affective extension of WordNet. In: LREC'04: Proceedings of 4th international conference on Language Resources and Evaluation, Lisbon, pp. 1083-86.
42. Taboada M, Brooke J, Tofiloski M, Voll K, Stede M (2011). Lexicon-based methods for sentiment analysis. Computational linguistics 37(2):267-307.
43. Turney PD (2002). Thumbs up or thumbs down? Semantic orientation applied to unsupervised classification of reviews. In: Proceedings of 40th meeting of the Association for Computational Linguistics, pp. 417–24.
44. Valitutti A, Strapparava C, Stock O (2004). Developing affective lexical resources. Psychology J. 2(1): 61-83.
45. Venugopalan M, Gupta D (2015). An enhanced polarity lexicon by learning-based method using related domain knowledge. Int. J. Information Processing & Management 6(2):61.
46. Vishnu KS, Apoorva T, Gupta D (2014). Learning domain-specific and domain independent opinion oriented lexicons using multiple domain knowledge In: Proceedings of 7th IEEE international conference on Contemporary Computing, pp. 318-23.
47. Warriner AB, Kuperman V, Brysbaert M (2013). Norms of valence, arousal, and dominance for 13,915 English lemmas. Behavior research methods, 5(4), pp. 1191-207.
48. Wiebe JM (1990). Identifying subjective characters in narrative. In: Proceedings of international conference on computational linguistics, vol. 2, pp. 401-6.
49. Wiebe J (2000). Learning subjective adjectives from corpora. In: Proceedings of national conference on artificial intelligence, pp. 735-40.
50. Wilson T, Wiebe J, Hoffmann P (2005). Recognizing Contextual Polarity in Phrase-Level Sentiment Analysis. In: Proceedings of the conference on human language technology and empirical methods in Natural Language Processing, ACL, pp. 347-54.
51. Xia R, Zong C, Hu X, Cambria E (2013). Feature ensemble plus sample selection: domain adaptation for sentiment classification. Intel. Syst. 28(3):10-8.
52. Zhang Z, Singh PM (2014). Renew: A semi-supervised framework for generating domain specific lexicons and sentiment analysis In: Proceedings of the Association for Computational Linguistics, pp. 542–51.

# MINING HIGH UTILITY ITEMSET USING GRAPHICS PROCESSOR

Maya Joshi
Department of Computer Engineering
Silver Oak College of Engineering & Technology.
Ahmedabad- 382481, India
mhjoshi11@gmail.com

Dharmesh Bhalodia
Department of Computer Engineering
Silver Oak College of Engineering & Technology.
Ahmedabad- 382481, India
dharmeshbhalodia33@gmail.com

***Abstract- In Data Mining,* Association Rule Mining is one of the most influential tasks. Several analyses and algorithm of it provides knowledge to investors or marketing manager to analysis and predict their market field and managing their records. But these procedures are not enough to originate more productive results. The traditional high utility itemset mining algorithms occupy more space, memory and time for generation of the candidate list. We presented the novel algorithm for Mining high utility itemsets using a parallel approach for transactional datasets. Therefore, the Sales Manager can use this utility itemset for their historical analysis of data, stock planning, and decision making. Our new approach is an extension of FHM algorithm, by attaching pruning method in HUIM. This utilization is improved to acquire immense efficiency on a miscellaneous platform which consists of a shared memory multiprocessor and numerous cores NVIDIA based Graphics Processing Unit (GPU) coprocessor. An empirical study and results of existing algorithm FHM are compared with the novel algorithm on NVIDIA Kepler GPUs and discovered significant improvements in computing time compare to FHM.**

***Keywords: Association Rule Mining, Frequent Pattern Mining, High Utility Itemset Mining, CUDA, GPU computing.***

## I. INTRODUCTION

Continuous and exponential furtherance of data requires being investigated in both scientific and business research. Extracting the knowledge from a substantial amount of data is mandatory for producing accurate and efficacious decisions. In this process, data mining performs a vital role in automatically extracting useful and hidden knowledge from such large databases. Different techniques have been generated to discover the characteristics of association and correlational statistics of data. Classification, Clustering, Association rule mining, Regression and Prediction are the techniques commonly used to mine the different kinds and huge amounts of data. Selection of frequently appears patterns from a transactional dataset is one of the most significant tasks of these. Frequent patterns indicate that how many numbers of times



J.M. Corchado Rodriguez et al. (eds.), *Intelligent Systems Technologies and Applications 2016*, Advances in Intelligent Systems and Computing 530,
DOI 10.1007/978-3-319-47952-1_53

an item or itemset appears in a database. Therefore, if an itemset is frequent it signifies there is a vigorous relationship between those items.

A. Association Rule Mining

Transaction data set T is given; the aim of ARM is to determine all those rules which fulfill the following criteria.[3]

(a) Association Rule: X → Y is an implication expression, where X and Y are itemsets.

(b) Rule Evaluation Metrics: X→Y is a proposition expression, where X and Y are dismembering itemsets. The efficiency of an ARM can be computed by two different terms: Support (S) and Confidence (C).

**Support (S):** $Sup \geq min\_sup$ Where, $Sup = freq(I_i, I_j)/N$
**Confidence (C):** $Conf \geq min\_conf$ Where, $Conf = freq(I_i, I_j) / freq(I_i)$

The problems arise with Frequent Pattern Mining [1] are (1) Item appears only once in the transaction (2) All items contains same weights. So, it may ignore rare item set having higher profit. It can be resolved by High Utility Itemset Mining.

In extension, current algorithms are not able to solve all frequent patterns and High Utility pattern problems in real-time. Consequently, in our novel approach, all high utility Itemset must be frequent before calculating the utility. Such information is useful for decision making and strategic planning.

B. High Utility Itemset Mining

In a transaction database, Utility of items includes subsequent two approaches: (1) The external utility (e) called as importance or value of distinct items in the dataset. The external utility represents the value associated with the item. It is independent of transactions, and which represents a profit of individual items. (2) The internal utility (i) called as the importance of items in transactions, which serves as the aggregate of the item in the transaction.

Utility of Item Set (U) = internal utility (i) * external utility (e).

HUIM algorithm extends the concept of the FPM algorithm by adding one step as follows:

Step 1: Generation of Frequent Item Set: Generate all itemsets whose support >= minimum support threshold.

Step 2: Generation of High Utility Itemset: Generate all itemsets whose utility >= minimum utility threshold.

C. Definition

Let I= {$i_1$, $i_2$, …., $i_n$} be a set of items. Now, Transactional Database D is a set of transactions D = {$T_1$, $T_2$, …., $T_n$}, where for each transaction $T_n$, $T_n \in$ I. Each transaction has a unique identifier, called its TID. Each item $i \in I$ is associated with a positive number p(i), called external utility (ex. Profit). For each transaction $T_n$ such that $i \in T_n$, a positive integer number q (i, $T_n$) is called the internal utility. The problem is to discover all frequent itemsets having higher utility or profit.

## II. Introduction To GPU

GPU [2] calculation has given a tremendous edge over the CPU, concerning computational speed. Consequently, it is a standout amongst the most interesting areas of research in the field of advanced innovative work.

The historical backdrop of Graphics processors is as long as the PC itself. The primary Graphics processor called CGA (Color Graphics Adapter) was imagined by IBM in 1981. After one and half decades of advancement, Graphics processor turns out to be all the more capable and had the capacity to support 3D increasing speed on PC desktop. At that point in 1999, the term Graphic Processing Unit was conceived when NVIDIA presented "the world's first GPU"-GeForce256 and launched it's massively parallel architecture named "CUDA" in 2006-2007. From that on, research in the fields of material science, restorative imaging thus on began to exploit GPU to quicken their applications. That is the beginning of GPGPU computing.

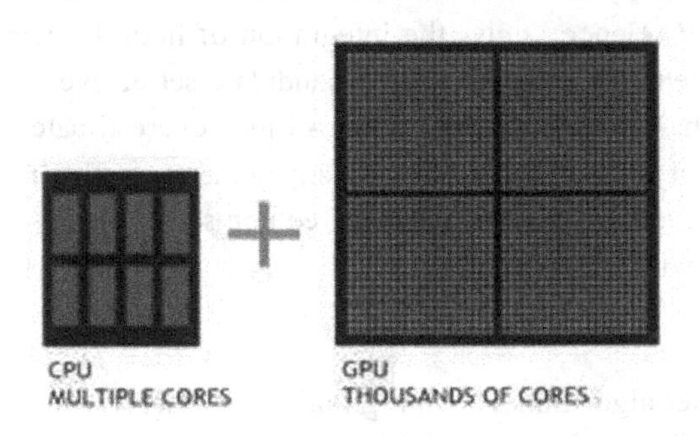


**Fig 1. Core Comparison of CPU and GPU [2]**

Parallel Programming exploits high-speed graphics cards with various processing units to perform computation speedily. Nevertheless, there are restraints and tradeoffs admitted with parallel processing that must be examined for optimal results of performance. Unlike CPU, GPU has thousands of cores. The primary function of GPU is to evaluate 3D functions. Because this kind of calculation is very heavy on CPU. Though, GPU came into existence for a graphical purpose, it has now extended computing, precision, and performance [4].

GPU computing or GPGPU (General Purpose GPU) is the usage of a GPU (graphics processing unit) to do general purpose engineering and scientific computation. The model for GPU calculation is to utilize a CPU (Host Machine) and GPU (Device Machine) together in the miscellaneous co-processing computational model. The sequential part of the function executes on the Host machine (CPU) and the computationally-intensive part is expedited by the Device Machine (GPU). From the user's perspective, the application is more expeditious because it is utilizing the better execution of the Graphics Processor to amend its own betterment.

## III. Related Works

### *A. GPAPRIORI [10]*

Yan Zhang, Fan Zhang, and J. Bakos [2011] discovered Graphics Processor Unit expedited conventional Apriori utilization which known as GPApriori. It embraces some techniques, like Candidate Generation, Pruning, and *Support Counting*. Most *techniques* either use horizontal representation or vertical representation, while GPApriori, represented incipient Bitset representation. The bitset representation needs less space, memory. It creates candidate items on CPU transfer it for support counting on GPU and it is more appropriate for designing a parallel set join operation, which is more qualified for "Bitwise-AND" operations performed on Graphics Processor for joining two transactions. The main drawback is candidate representation is not optimized.

### *B. Two-Phase [5]*

Ying Liu, Alok Choudhary and Wei-keng Liao [2005] proposed a new approach called Two-Phase Algorithm. This strategy maintains a Downward *Closure Property which is Transaction-weighted.* Hence, only the integration of high TWU (Transaction Weighted Utilization) itemsets is included into candidate set at every level throughout the level-wise search. In this algorithm, Phase - I may overestimate some itemsets that have low utility, but it never underestimates any itemsets. In phase - II, just one extra database scan is needed to filter the aggrandized itemsets. This approach is suitable for a sparse dataset with short patterns.

### *C. Hui- Miner [6]*

Liu & Qu [2012] proposed HUI-Miner algorithm. It is a high utility itemset having a utility list with list data structure. It first generates an initial utility list for itemsets of the length 1 for optimistic items. Then, HUI- Miner develops recursively a utility list for every itemset of the length k using a pair of utility lists for itemsets of the length k-1. For high utility itemset mining, every utility list for an itemset keeps the information of TIDs for each of transactions in dataset consisting the itemset, utility values of the item set in the transactions, and the sum of the utilities of the remaining items that can be included to super itemsets of the itemset in the transactions. The definite benefit of HUI-Miner is that it ignores the costly candidate generation and utility calculation.

### D. *FHM [7]*

Philippe Fournier-Viger [2014] proposed FHM algorithm. It extends the Hui-Miner Algorithm. It is a Depth- first search Algorithm. It relies on utility-lists to compute the exact utility of each itemset. In this algorithm, a novel strategy integrates called as EUCP (Estimated Utility Co-occurrence Pruning) to minimize the number of joins operations when mining high utility itemsets utilizing the utility list data structure. Estimated Utility Co- Occurrence Structure (EUCS) stores the TWU of all two-item-sets. It built during the initial database scans. EUCS illustrated as a triangular matrix or hashmap of hashmaps. The memory footprint of the EUCS structure is small. This approach is **up to 6X faster** than HUI-Miner.

### E. *CTU-Mine [8]*

Erwin et al[2007] observed that the conventional candidate-generate-and-test approach for identifying high utility itemsets is not suitable for dense datasets. Their work develops a new algorithm CTU-Mine that mines high utility itemsets using the pattern growth algorithm. A comparable contention is presented by Yu et al. Existing algorithms for high utility mining are column enumeration predicted adopting an apriori like candidate set generation-and-test approach and thus are inadequate in datasets with high dimensions. It is perplexed for evaluation due to the tree structure. This approach is felicitous for a dense dataset with the long pattern.

### F. *UP-Growth [9]*

To address issues of generating a large number of candidates, UP-Growth (V.S Tseng et al., [2010]) has been currently proposed and it utilizes PHU (Potential High Utility) model. The UP-Growth applies four strategies for reducing the number of candidate itemsets, DLU (Discarding Local Unpromising items), DGU (Discarding Global Unpromising items), DGN (Decreasing Global Node utilities), and DLN (Decreasing Local Node utilities). At this stage, DGU and DGN are applied for minimizing overestimated utilities. After that, high utility itemsets are used from the UP-Tree with DLU and DLN.

## IV. THE PROPOSED METHODOLOGY

Step1: Input: Transaction dataset with the utility of items.
Step2: Conversion from Tidset to Bit set representation.
Step3: Generating candidate sets on the host machine.
Step4: Transfer it from Host to a device for support Counting using a parallel sum reduction algorithm to increase the efficiency.
Step5: Support values results are copied back to main memory and compare it with minimum support given by the user.
Step6: Output 1: find all frequent items that have support above minimum support.
Step7: Calculate weighted Utility of Itemset and Transactions on the host machine.
Step8: Transfer it from the host machine to device machine for parallel sum reduction block-wise of Utility and Compare it with the Minimum Utility Threshold.
Step9: Output 2: List of High Utility item sets.

*a. Tidset to Bitset Representation*

| Transactions | ID |
|---|---|
| 1,2,3,4,5 | 1 |
| 2,3,4,5 | 2 |
| 3,4 | 3 |
| 1,3,4,5 | 4 |

Tidset

→

| Candidate | Tidset | Bitset |
|---|---|---|
| 1 | 1,4 | 1001 |
| 2 | 1,2 | 1100 |
| 3 | 1,2,3,4 | 1111 |
| 4 | 1,2,3,4 | 1111 |
| 5 | 1,2,4 | 1101 |

Bitset

Fig 2 Conversion of Horizontal to Vertical Transaction

Careful consideration of data elements is required as GPU operates in parallelization. The algorithm should be able to utilize the concurrent processing of multiple units supported by the GPU. Transactions are initially saved in horizontal format, that data can be easily read from disk and converted into a vertical format, using the <item, bitset> representation. Infrequent items will not appear in any frequent patterns; they can safely be abstracted from the dataset without modifying the FIM results. Therefore, vertical transaction list is designed. The Horizontal format of transactions or tidsets transactions are easy for analyzing but they arise an issue in counting operation. That makes uncertain acts and thus results in performance degradation on GPU. This is because of the GPU requires a methodology that suitable for parallel processing. Figure 2 shows the conversion from Horizontal to the vertical transaction for simplification.

*b. Bitset Access*

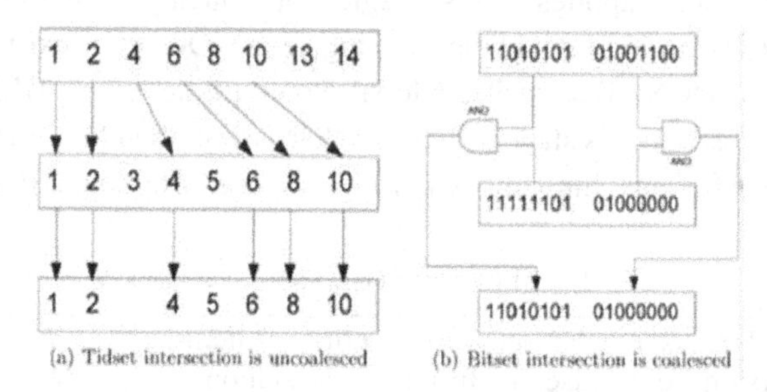

(a) Tidset intersection is uncoalesced (b) Bitset intersection is coalesced

Fig 3 Comparison of Tidset to Bitset Intersection [10]

There are two presentations Tidset and Bitset. Tidset structure is dense in nature. But, this structure makes the parallelization of mining process more difficult. While it converted into Bitset presentation it consumes more memory compare to Tidset, but it is more suited for GPU parallelization. Bitwise AND operation is used to Join two Bitset which coalesces.

*c. Support Counting On GPU*

According to figure 4. Support Counting is done parallel on GPU. For the Purpose of Parallel Sum Reduction as shown figure 5 algorithm is modified and used for counting the parallel sum of Utility (Block-wise) using GPU.

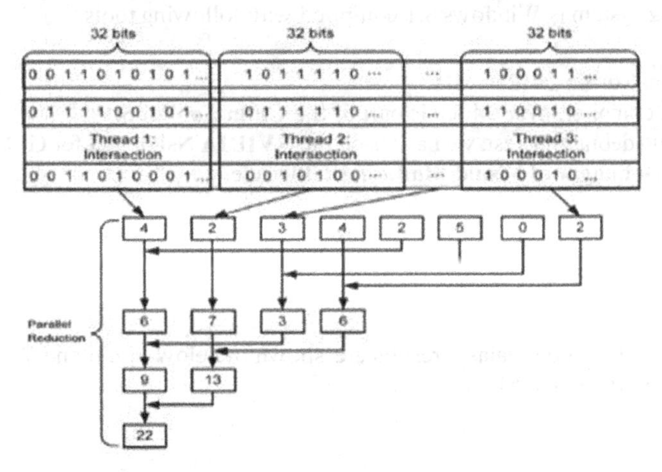


Fig 4 Support Counting Process on GPU[10]

*IV. Parallel Sum Reduction*

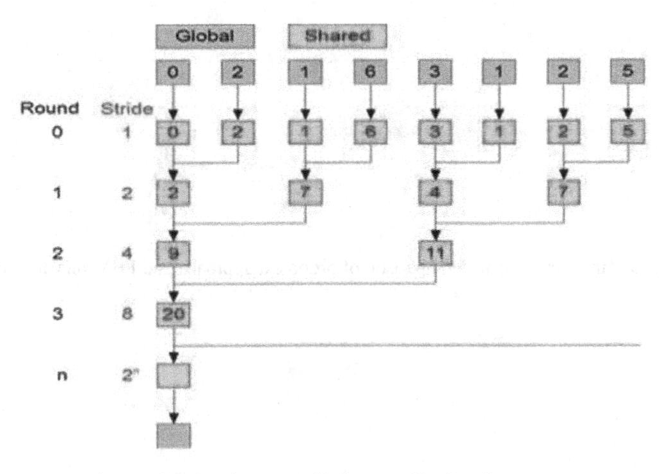


Fig 5 Block wise parallel sum Reduction

## VI. EXPERIMENTAL RESULTS

In this section, we calculate the performance of the novel algorithm. We compare it with FHM [7] algorithm, which can mine all high utility itemsets without frequent itemsets.

The experiments were executed on a workstation equipped with Intel[(R)] Core[(TM)] i3-2120 CPU @ 3.30 GHz processor, 8 GB of RAM, 64-bit architecture & a NVIDIA GTX 680 having 1530 Cuda cores.

a. *Experimental Tools*

The operating system is Windows 8.1 equipped with following tools.

- ⇨ **Visual Studio 2012** as **IDE**
- ⇨ GPU compiler, ***nvcc*** which is part of the **Cuda Toolkit 5.**
- ⇨ During debugging test, we have used the **NVIDIA Nsight 3.1** for GPU side and then **Debugger of Visual Studio** for CPU side.

Throughput is inversely proportional to support count and utility threshold. At higher support threshold and utility threshold, there is a less candidate generation and memory requirement which results in higher throughput.

On Chess and Mushroom dataset results are shown in below Fig. 6 and 7 with minimum support count 0.51.

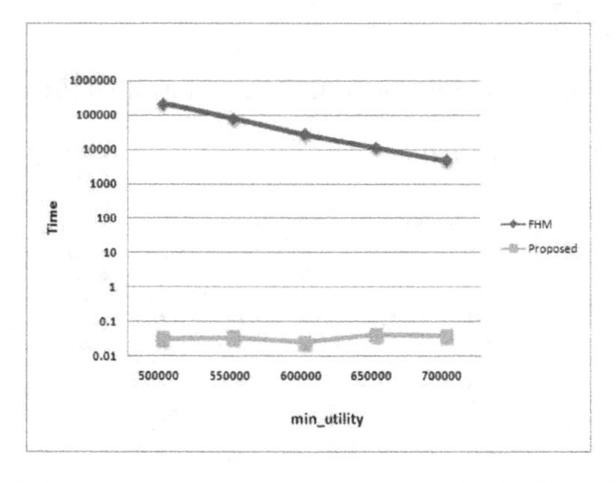


Fig. 6. Time Performance comparison of proposed approach and FHM on Chess dataset.

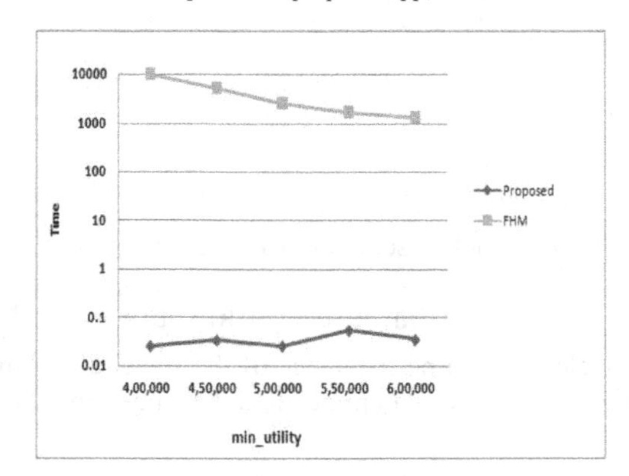


Fig. 7. Time Performance comparison of proposed approach and FHM on Mushroom dataset

Below Fig. 8 shows the time occupied by different support count on mushroom dataset with minimum utility 40,000.

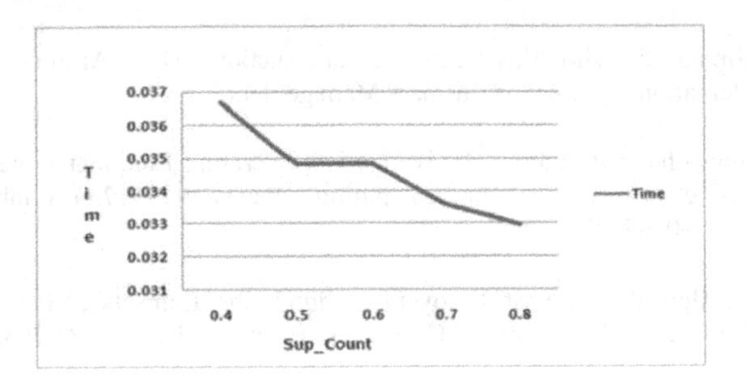


Fig. 8. Performance Analysis of proposed approach of Occupied Time for different Support Count on Mushroom dataset

## CONCLUSION

According to our survey, the evaluation of the approaches is fully relied upon the utility levels, support levels and the nature of the datasets. There are various works has been done on the High Utility Itemset Mining Algorithms Like Two-Phase, CTU-Mine, Up-Growth, Hui-Miner, FHM and for Frequent Pattern Mining using parallel programming different algorithm on GPU like GPApriori, Frontier Expansion, Tree projection, DCI, PBI & TBI and PARApriori.

In this research paper, we have introduced our newly developed and efficient HUIM approach that address the demand of Parallel approach of High utility itemsets and contribute high performance like time optimization and less candidate generation for the HUIM task on a different machine equipped with GPU.

## FUTURE WORK

Future work on this research includes how to parallelize other HUIM algorithm on GPU. We can apply different candidate representation for memory optimization. The novel algorithm which we have proposed in this research utilize only one GPU. However, we can take advantage of more parallelism with more than one GPU.

## ACKNOWLEDGMENT

We are very thankful to Mr. Vikas Tulshyan, P.G. Coordinator of SOCET, Ahmedabad, Gujarat, for providing base resources and grant for experimental work under the CUDA Teaching Center and continuous help & supports as and when needed throughout the whole research work. We would also like to thank staff members of M.E. for providing us the motivation and encouragement throughout this research work.

## REFERENCES

[1] J. Han, H. Cheng, D. Xin and X. Yan, "Frequent Pattern Mining: Current Status and Future Directions," Data Mining and Knowledge Discovery, vol. 15, no. 1, pp. 55-86, Aug 2007.

[2] Nvidia. NvidiakeplerGk110 Architecture Whitepaper .January 2013 Url : Http://Www.Nvidia.Com/Content/Pdf/Kepler/Nvidia-Kepler-Gk110-Architecture- Whitepaper.Pdf.

[3] Advance Mining of High Utility Itemsets in Transactional Data. Arumugam. P and Jose. P*. Opinion – International journal of Business Management.

[4] Yan Zhang Fan Zhang And Jason D. Bakos, "Accelerating Frequent Itemset Mining On Graphics processing Units". J Supercomputing, Pages 94–117,November2013. Doi: 10.1007/S11227-013-0887- X.

[5] "A Two-Phase Algorithm for Fast Discovery of High Utility Itemsets", Ying Liu, Wei- Keng Liao, and Alok Choudhary, Northwestern University, Evanston, IL, USA 60208,2005.

[6] Mengchi Liu Junfeng Qu, "Mining High Utility Itemsets without Candidate Generation", 2012.

[7]. "FHM: Faster High-Utility Itemset Mining using Estimated Utility Co-occurrence Pruning", Philippe Fournier-Viger1, Cheng-Wei Wu 2014.

[8] "CTU-Mine: An Efficient High Utility Itemset Mining Algorithm Using the Pattern Growth Approach" In Seventh International Conference on Computer and Information Technology (2007).

[9] "UP-Growth: An Efficient Algorithm or High Utility Itemset Mining ", Vincent S. Tseng, Cheng-Wei Wu, Bai-En Shie, and Philip S. Yu. The university of Illinois at Chicago, Chicago, Illinois, USA, 2010.

[10] "Gpapriori: GPU-Accelerated Frequent Itemset Mining", Fan Zhang, Yan Zhang, And J. Bakos. In 2011, IeeeinternationalConference On Cluster Computing(Cluster), Pages 590–594, 2011. Doi: 10.1109/Cluster. 2011.61.

# The Use of Simulation in the Management of Converter Production Logistics Processes

**Konstantin Aksyonov and Anna Antonova**

**Abstract** This paper considers an application of a simulation model to the management of converter production logistics processes. The simulation model has been developed to determine the optimal time interval of delivering the melts to the converters. The goal of optimization is to find such a time interval in which the waiting time of the service on a continuous casting machine will be minimal, and downtime of the continuous casting machine will be minimal because it influences on the resources spending and the amount of harmful emissions into the atmosphere. The simulation model has been developed in a simulation module of the metallurgical enterprise information system. The simulation module supports a multi agent simulation. Agents in the developed model are intended to describe the cutting slabs algorithm used by technologists in the metallurgical production. As a result of a series of experiments with the model the best time interval between deliveries of the melts on the converters has been found and equal to 20 minutes.

## 1 Introduction

In the steel production, the shop efficiency is determined not only by the volume of production per time, but also by the products quality, resources spending, and the amount of harmful emissions into the atmosphere. The logistics and planning of the products transportation have a direct impact on the last two characteristics. In the oxygen-converter shop (OCS), a continuous casting machine (CCM) stop leads to the need to re-start the CCM with the heating of a CCM intermediate ladle. The CCM stop may occur due to the incorrect supply of the melts to the converters. The heating of the CCM intermediate ladle is carried out up to two hours using natural gas, which on burning emits pollutants. The important tasks are the

---

Konstantin Aksyonov
Ural Federal University, Mira 19, Ekaterinburg, Russia, e-mail: wiper99@mail.ru

Anna Antonova
Ural Federal University, Mira 19, Ekaterinburg, Russia, e-mail: antonovaannas@gmail.com

 
J.M. Corchado Rodriguez et al. (eds.), *Intelligent Systems Technologies and Applications 2016*, Advances in Intelligent Systems and Computing 530,
DOI 10.1007/978-3-319-47952-1_54

analysis of the OCS shop logistics operations and choice on shop entrance the melts sequence, which would ensure continuous operation of the CCM bottling series of melts. In the study of the logistics and organizational (business) processes of OCS shop, a simulation method shows good results as applied to the optimization of production planning [3]. Development of simulation model of OCS logistics processes affecting the shop efficiency is topical.

We consider the development of the OCS shop model with the use of the simulation module of the metallurgical enterprise information system. The metallurgical enterprise information system is a web-oriented system for tracking, monitoring, modelling, analysis and improvement processes of the steel products manufacturing [2,4]. This system has been developed with the participation of Ural Federal University; the authors were involved in the development of the simulation and optimization modules. The simulation module is intended to create a simulation model with the use of the graphical notation and 'if-then' rules. The optimization module is intended to form a plan of experiments with the simulation model, perform all the experiments, and selection experiment with the best results.

## 2 Statement of the problem

Data of the processes have been collected by sensors of the automated production management systems located in the OCS divisions. Data contain information about the melt movement between the following divisions: steelmaking division (SMD), division of the secondary treatment of steel (STS), and casting division (CD). The OSC shop includes the following aggregates (see Fig. 1): three converters (K), three installations of ladle metallurgy furnace (LMF), and three continuous casting machines (CCM).

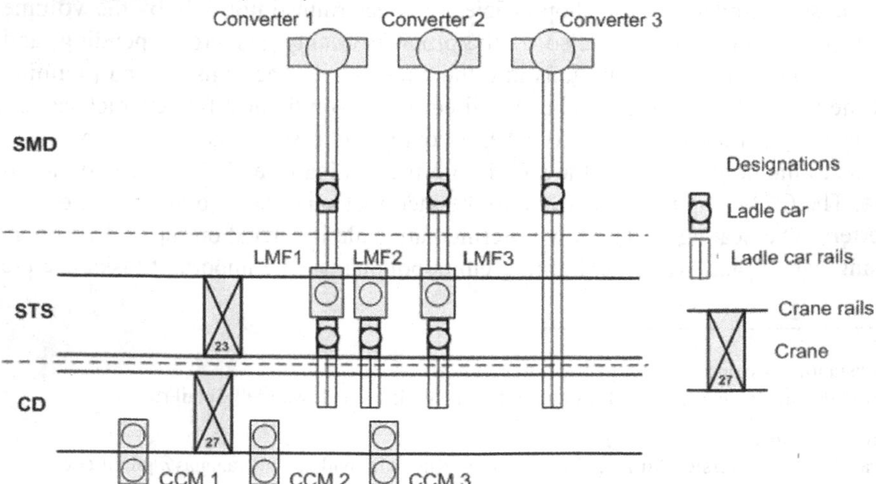


**Fig. 1** Layout of the aggregates and vehicles of the OCS shop

For example, consider a standard processing path converter – LMF – CCM. The procedure for the passage of the route from the point of view of vehicles is presented in Table 1. The objective function of the problem is to minimize the maximum downtime of the CCM waiting the ladle.

**Table 1** The OCS vehicles movement algorithm

| Vehicle | Parameter | Sequencing |
|---|---|---|
| SMD ladle car | $V_{ladle\_car} = 50$ meter per minute | 1) Post a job on the line (after the steel discharge from the converter);<br>2) move under the crane;<br>3) wait withdrawal the ladle by the crane 23;<br>4) return to the converter;<br>5) finish the work. |
| STS ladle car | $V_{ladle\_car} = 50$ meter per minute | 1) Post a job on the line (with the beginning of the ladle car 7.x work);<br>2) move to the CD division;<br>3) wait for removing the ladle from the ladle car 7.x;<br>4) back under the crane 23 to the STS division;<br>5) wait loading the ladle with melt;<br>6) move under the LMF;<br>7) wait for the end of the processing on the LMF;<br>8) move to the CD division;<br>9) wait withdrawal the ladle by the crane 23;<br>5) return to the STS division;<br>6) finish the work. |
| Crane 23 | $V_{crane} = 64$<br>$V_{crane\_trolley} = 22.7$ meter per minute | 1) Post a job (with the arrival of the ladle car 7.x to the STS division with the ladle);<br>2) come up to the ladle car;<br>3) lift the ladle by the crane;<br>4) move to the free LMF;<br>6) await the arrival of the ladle car 6.x from the CD division;<br>7) install the ladle on the ladle car 6.x by the crane;<br>8) finish the work. |
| Crane 27 | $V_{crane} = 64$<br>$V_{crane\_trolley} = 22.7$ meter per minute | 1) Post a job (with the arrival of the ladle car 6.x to the CD division with the ladle);<br>2) come up to the ladle car;<br>3) lift the ladle by the crane;<br>4) move to the desired CCM;<br>6) move the crane trolley with the ladle;<br>7) install the ladle on the CCM swivel stand;<br>8) finish the work. |

It is also necessary to take into account the following requirement: ladle can be on the CCM swivel stand in anticipation of the casting no more than 15 minutes, otherwise the steel in the ladle cools and the ladle has to be heated. It is necessary to propose options for an interval of the consistent supply of melts on converters. This interval should ensure the optimization of the objective function (1). OSC divisions work needs to analyze for 1000 minutes.

## 3 Development of the simulation model of OCS shop work

The OCS shop model has been developed using a simulation module of the metallurgical enterprise information system. In this module, the simulation model is developed via notation of multi-agent resource conversion processes (MRCP) [1]. According to the MRCP notation, model's nodes are either agents or operations. Agents in the model of OCS shop are used to implement the logic to process the orders and to management the orders attributes. Operations in the model are used to visualize the duration of the work of the shop aggregates and vehicles.

Advantages of the simulation module over existing simulation systems are: 1) support for agent-based simulation with a description of the agent's knowledge base using production rules, 2) support for the creation of models of technological, logistical and organizational (business) processes with the help of visual designer models without requiring of the users programming skills.

The developed model structure can be divided into six work units: 1) description of the order generation; 2) description of the converters work – preparation, purging, operation after purging, and draining steel operation; 3) description of the ladle car work – movement on the line, the order of the work, and interaction with cranes; 4) description of the LMF work including exception handling in the event of a new ladle car on the line 6.1.1; 5) description of the crane work – movement on the line, the order of the work, and interaction with ladle cars; 6) description of the two streams CCM elements – mold, secondary cooling zone and gas cutting. A fragment of the model structure with description of the two streams CCM is shown in Fig. 2.

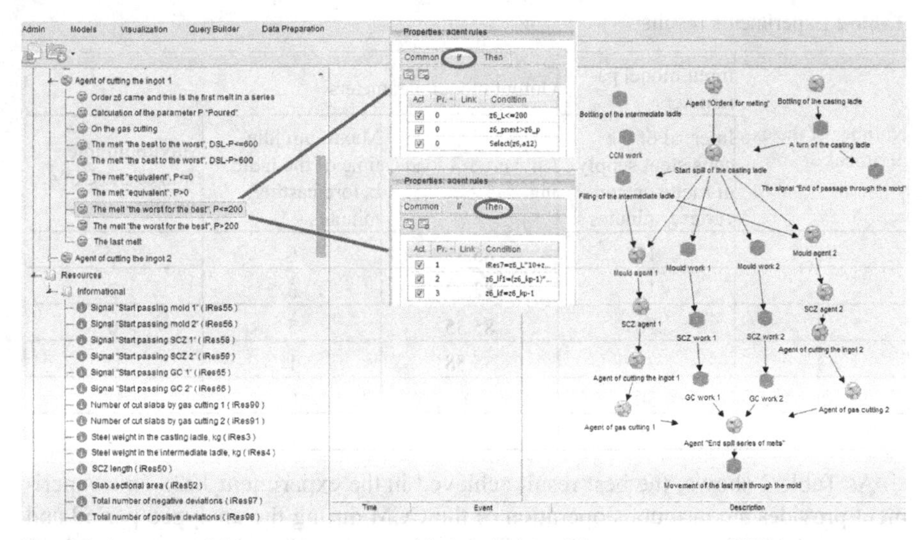


**Fig. 2** A fragment of the model structure with description of the two streams CCM

On the left below shows the elements of the model tree, this contains the rules of the agent's knowledge base. Separate situation is a rule of the form "If - Then" built using model variables (resources and orders). The "Agent of cutting the ingot 1" knowledge base comprises the expert algorithm for cutting the ingot into slabs on the melts border. The melts border appears when turning the CCM swivel stand.

## 4 Experiments results analysis

We consider the experiments with the developed model in the optimization module of the metallurgical enterprise information system. The advantages of the optimization module compared with the functionality of the existing simulation systems are: 1) use data formed as a result of the query to the data store using the query builder module of the metallurgical enterprise information system, 2) availability of access to the raw data of simulation results used for build custom graphs and reports, 3) 3D-visualization of the enterprise processes by means of the browser without the involvement of third-party tools for constructing and visualization of 3D-models, 4) support for multi-user work with the system, and 5) support the issuance of recommendations for change (optimization) of the technological, logistical and organizational (business) processes.

A series of the six experiments in which has been analysed interval of the consistent supply of melts on converters has been conducted. The results are shown in Table 2.

**Table 2** Experiments results

| Number of the experiment | Input model parameters | Output model parameters | | |
|---|---|---|---|---|
| | Interval of the consistent supply of melts on converters, minutes | Converter 3 loading, % | Maximum idle time of the ladle before casting, minutes | Maximum idle time of the CCM, minutes |
| 1 | 10 | 85,32 | 6 | 15 |
| 2 | 15 | 82,32 | 6 | 2 |
| 3 | 20 | 85,35 | 5 | 0 |
| 4 | 25 | 55,58 | 1 | 15 |
| 5 | 30 | 0 | 0 | 14 |
| 6 | 35 | 0 | 0 | 33 |

As Table 2 shows, the best result achieved in the experiment №3. This experiment provides a continuous operation of the CCM during the analyzed period and the acceptable downtime of the ladle before casting that does not exceed 15 minutes. At the end of the simulation, the report in .csv format has been built containing detailed information on changing the values of model parameters. The queue of the ladles on CCM, built with the help of the simulation report, is shown in Fig. 3. As Fig. 3 shows, the queue of the ladles on CCM not exceeds 5, according to the number of the CCM in the shop.

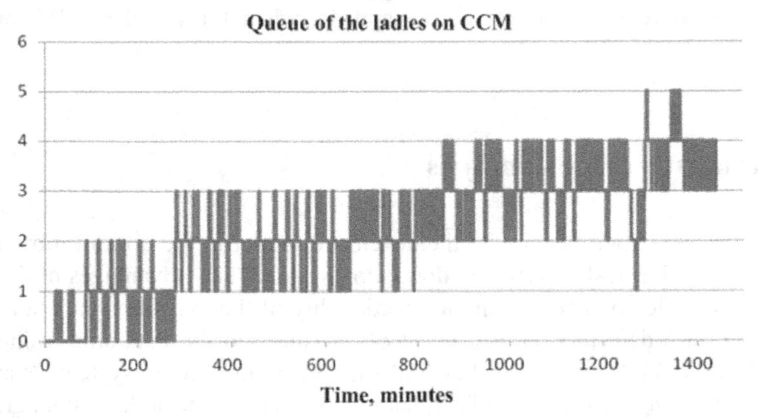


**Fig. 3** The queue of the ladles on CCM for the experiment №3

For comparison, Fig. 4 is a graph of the queue of the ladles on CCM for the experiment №2. As shown in the figure, the maximum value of the queue reaches 9 ladles. This fact causes a delay ladle before casting more than 15 minutes, and the disruption of a technological chain of steel processing.

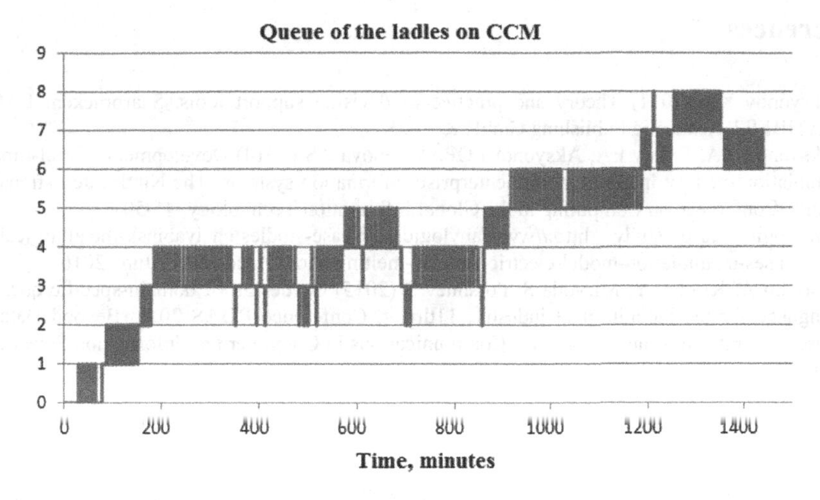


**Fig. 4** The queue of the ladles on CCM for the experiment №2

As a result of the experiments, the following inference has been obtained: the efficiency of the OCS shop associated with the continuous casting at all CCM is achieved by supplying a melt to the converters with an interval equal to 20 minutes.

# 5 Conclusion and future work

As a result of the analysis of logistics processes of the oxygen-converter shop, a simulation model of the OSC shop has been developed with the use of the simulation module of the metallurgical enterprise information system.

The model developed was used for solving the optimization of logistics management processes. As a result of the experiments, the following recommendations have been obtained. In order to prevent unplanned shutdowns in the continuous casting machine work, the interval of the consistent supply of melts on converters should be equal to 20 minutes. CCM continuity eliminates additional use of natural gas to heat the CCM intermediate ladle and reduces the amount of harmful emissions into the atmosphere.

The aim of future research is to apply the query builder and the data preparation modules of the metallurgical enterprise information system on the stage of receiving input modelling data from real production.

**Acknowledgments** The work was supported by Act 211 Government of the Russian Federation, contract № 02.A03.21.0006.

## References

1. Aksyonov KA (2011) Theory and practice of decision support tools. Saarbrucken: LAP LAMBERT Academic Publishing GmbH & Co. KG
2. Aksyonov KA, Bykov EA, Aksyonova OP, Antonova AS (2014) Development of real-time simulation models: integration with enterprise information systems. The Ninth International Multi-Conference on Computing in the Global Information Technology, 45-50
3. AnyLogic: case study. http://www.anylogic.com/case-studies/chelyabinsk-metallurgical-plant-uses-a-simulation-model-electric-furnace-melting-shop. Accessed 19 June 2016
4. Borodin A, Kiselev Y, Mirvoda S, Porshnev S (2015) On design of domain-specific query language for the metallurgical industry. 11th Int. Conference BDAS 2015: Beyond Databases, Architectures and Structures: Communications in Computer and Information Science, 505-515

# Precision Capacitance Readout Electronics for Micro sensors Using Programmable System on Chip

[1] A.Abilash, [2] S.Radha

[1] Department of ECE, SSN College of Engineering ,Chennai, India
[2] Department of ECE,SSN college of Engineering,Chennai,India
[1]abivallan1992@gmail.com ,[2] radhas@ssn.edu.in

**Abstract.** Several techniques are used for measuring the capacitance value with various range, but the need of producing an accurate, less complex and low error system is still a difficulty. In this paper a less complex capacitance measurement system with low stray capacitance and offset error is designed and implemented in Programmable system on chip (PSOC 3) development kit CY8C3866-AXI040. The principle used is linear charging and discharging of capacitor, it generates an oscillation whose frequency is inversely proportional to the capacitance value. The board is interfaced with the laboratory workstation using LabVIEW via serial communication for online data acquisition and monitoring using UART. The embedded readout system developed, measures the capacitance value ranging from nF to pF and the results are discussed. The system measures the capacitance of fabricated capacitive microsensor in its static mode, which is in pF range and it can also measure values of commercially available capacitor.

***Keywords:*** PSOC, Capacitance measurement, capacitive microsensors, LabVIEW, VISA

## 1 Introduction

Capacitance measurement has been used widely in research areas for developing new ideas in research work, evaluating the developed products and for monitoring purposes. Product development and yield engineers use capacitance measurement for analyzing the devices and for optimization purpose. Capacitive MEMS microsensors are fabricated and highly recommended because of its high sensitivity, high accuracy and resolution [1], [2]. This project describes a method to build embedded read-out for capacitive microsensor. The design involves development of embedded read out circuit and implementing the circuit in PSOC development board. This circuit can accurately measure capacitance in the range of 1-1000 nanofarad(nF), and the circuit can also measure the ranges of capacitance value available in pF , the C value of measured capacitor is converted



J.M. Corchado Rodriguez et al. (eds.), *Intelligent Systems Technologies and Applications 2016*, Advances in Intelligent Systems and Computing 530,
DOI 10.1007/978-3-319-47952-1_55

to digital readable information and is virtually monitored meanwhile the data's are stored with the help of LabVIEW(Laboratory Virtual Instrumentation Workbench).

Gautam Sarkar, et al [1] proposed the development of a measurement system with charge/discharge as its principle. Measurement system that is used to measure low value of capacitance and which minimizes many hazardous effects is described. Drift and offset errors are reduced since the measurement system does not use any integrator. The system is been configured with the help of PIC microcontroller and the data acquired are stored in PC with the help of interfacing it with a Graphical language and the information are processed, monitored based on requirements. Navid Yazdi et al [2] in his work stated that inertial force of Mass is the principle behind the sensing of acceleration. Tradeoff between Primary design parameters and the resolution of system was encountered. A charge and discharge characteristic of resistance and capacitance is used for measuring the capacitance and resistance value. This method uses a time constant principle through a semi potential rise and fall time measurement [6]. This project has a uniqueness compared to other works, which is, all the processing units used in the measurement system, both analog and digital are readily available in a single chip making it cost effective. Also, the PSoC allows the designer to customize the peripherals used based on the requirements. LabVIEW benefits this project work by virtual instrumentation which includes control measurement, data acquisition for recording and analyzing the reports. Gautam Sarkar, et al [1] work proposed the use of two input NAND gate Schmitt trigger for lowering the hysteresis problem along with PIC microcontroller for data acquisition purpose, developing the complexity in the circuit design. In [2]they have used a high impedance readout circuit and highly sensitive circuit for interfacing with bulk/ surface micro machined g- accelerometer, this helped them in reaching the higher sampling rate, but the parasitic or stray capacitance problem is also increased. Also in this paper work there are three methods described for measuring capacitance value from microsensors, all three methods includes separate components thereby increasing the complexity of the system, while in our work all the components necessary for the measurement system are in a single chip(PSOC) resolving the problem for a high complexity system. The simple system design, which uses a basic principle of linear charging and discharging is implemented in PSOC and the results are explained. To ensure the tolerance level of the measurement system a stability testing is done. Stability testing is done by assigning the threshold value as 70pF, since it is the static mode capacitance value of the microsensor and the readings are noted for a duration of 2-3 hours, and there was 0.2% error.

The paper is organized in a way as follows: a brief introduction to PSOC is explained in Section II. In Section III, the design procedures in PSOC creator are discussed. The measurement method adopted, and the implementations in PSOC are explained in Section IV and V. Section VI deals with the virtual instrumentation using VISA in LabVIEW. Finally, the result of the Capacitance measurement system along with PSOC implementation is discussed in Section VII

## 2 Programmable system on chip

Programmable system on chip(PSOC) is a microprocessor with mixture of programmable analog and digital functions in asingle chip which can be used digital electronic system design. New concepts and methodologies have been introduced such as PSOC, with the development of reconfigurable technology and system-on-chip (SoC) technology, microelectronic system level design is possible with the help of system level design and hardware\software co-design. Universities have updated their courses and syllabus to the upcoming trends and technologies [10],[11].PSOC has Programmable routing interconnects, Core CPU subsystem, memory unit, configurable analog and digital blocks in a single chip. With the help of PSOC we can code our own functions or we can use the predefined library functions and tested IPs available. PSOC has no fixed peripheral constraint, it has both Analog and digital modules embedded in it. General purpose applications are also executed in PSOC with the help of PCB attached to it. PSoC3 is a new generation of PSoC chip with Single-Cycle 8051 core up to 67MHz. Multiple pin allocation is possible in PSOC eliminating fixed functions and providing flexible routing. The PSoC architecture comprises PSoC Core, Digital System, Analog System and System Resource [4]. All the device resources are combined into a complete custom system using Configurable global bus [4].

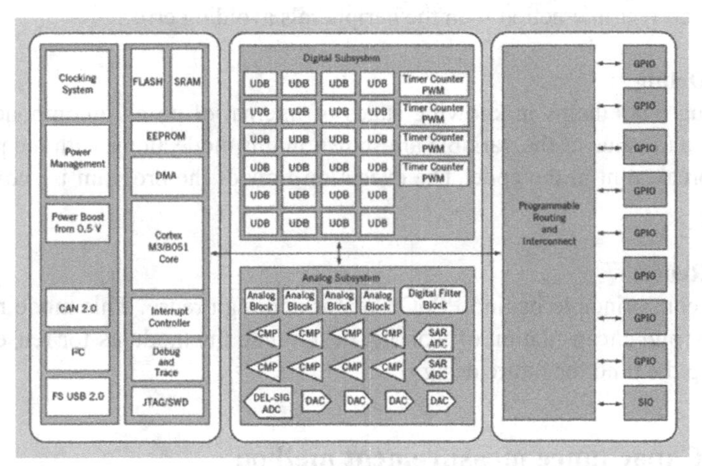


Fig 1. PSOC Block Diagram

The development kit CY8C34 which belongs to the family Flash Programmable System-on-Chip consists of eight IO ports that connect to the global digital and analog interconnect, providing access to 16 digital blocks and 12 analog blocks. The digital blocks can implement timers, counters, PWMs, UART, SPI, IDAC and the analog blocks like ADCs, DACs, DACs filters, amplifiers, and comparators are implemented [4].

# 3 Design procedure in PSoC creator

Integrated design and development environment is providing by PSOC creator. PSoC Creator and PSoC form a platform to provide uniquely flexible combination to design and implement in less time [4].

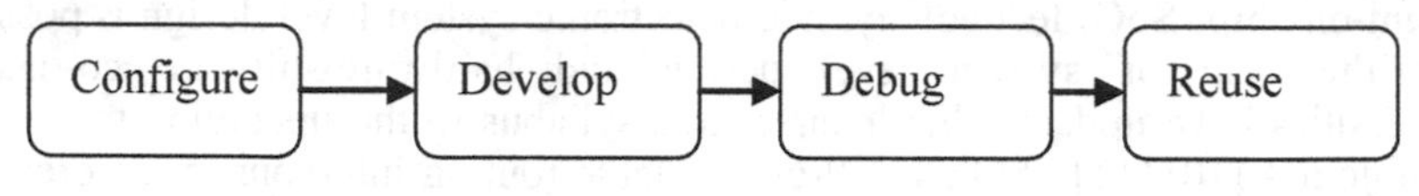


Fig 2. Design steps in PSoC

### *3.1* Configure

Configuration is done by drag and drop method, where the design schematic is formed by using the predefined, tested modules or functions available

### *3.2* Develop

In develop procedure PSoC combines C language with Application Peripheral Interface (API) which is automatically generated for each module or component in the device [4]. Error encountered in code development is reduced and it also ensures correct interaction with the peripherals avoiding error.

### *3.3* Debug

Debug mode helps in knowing states of internal of on-chip component. Type, address and value of the variable are shown in this mode along with the possibility to set breakpoint in the code. The execution flow of the program is shown in this mode.

### *3.4* Reuse

The core principle behind PSoC Creator is design reuse. This mode minimizes the error and encapsulation of working design into the modules for reuse, helps in reducing the time for future design.

# 4 Capacitance measurement method

Various measurement methods are available for measuring the capacitance value of MEMS sensor. The application in this work includes measuring the capacitance value of capacitive MEMS sensor(CMUT) in its static mode. CMUT sensor operation can be classified in to two modes, static and dynamic. Measuring the reactor wall pressure continuously, monitoring for any flaws are the primary application. Earlier methods included Non-destructive testing of the reactor wall pressure using pulse echometer, the results were limited to certain possibilities and the bandwidth was too small. Using CMUT sensor wider bandwidth and accurate results are possible[9]. The principle behind the capacitance measurement method used in this work is linear charging and discharging the external capacitor or in our case, it's a capacitive MEMS sensor [5],. The various modules available in PSOC are shown in the Figure 3. Programmable current source available in PSOC charges the external capacitor through a reset switch connected. Comparator output becomes low as the voltage reaches the threshold level (*Vth*). As the comparator uses inverted terminal logic, it enables the reset switch to discharge the capacitor, then charge cycle starts again and we get an oscillation with frequency dependent on both capacitance value and the charging current [5]. PWM is used for measurement which enables the 16 bit counter. Count value is read when the comparator's output clocks the counter and the count value it transmitted to PC using UART. The count value for the PWM time period will be proportional to the capacitance value of the external capacitor.

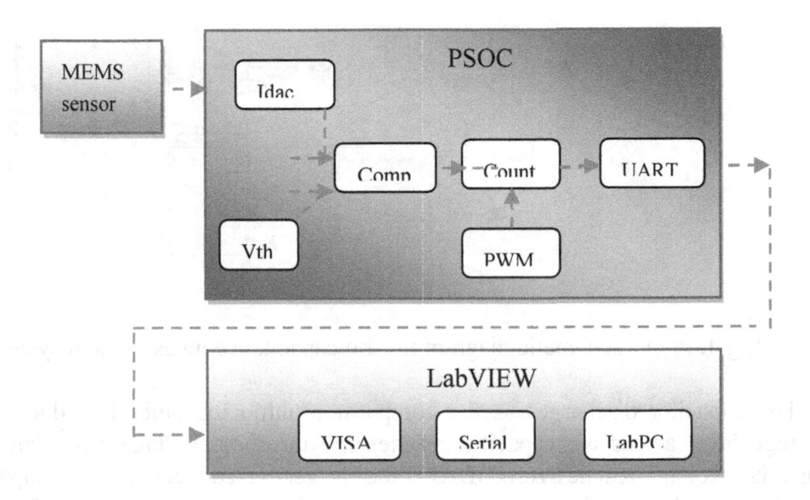


Fig 3. Architecture of Capacitance Measurement Method

## 5 Implementation in PSOC

PSOC is the most recommended device as it doesn't require any extra additional component. From block diagram, we can design the measurement system in PSOC design section using charging capacitor, comparator, Programmable current source, PWM, Counter and finally UART. IDAC module charges the capacitor through the bidirectional pin till it reaches the threshold voltage. An oscillation with frequency dependent on the capacitor value and charging current from the module is obtained. PWM is used for measuring the capacitance value by providing time widow which varies for each design [6], [7],[8]. PWM is also used for enabling the counter module. The count value from the counter is read and the value is transmitted to laboratory PC using UART. IDAC is connected to external capacitor (CINT) and also is connected to comparator which is set to inverted terminal logic. IDAC charges the capacitor as the comparator output is high and when the comparator output becomes low, IDAC gets disconnected from the external capacitor.

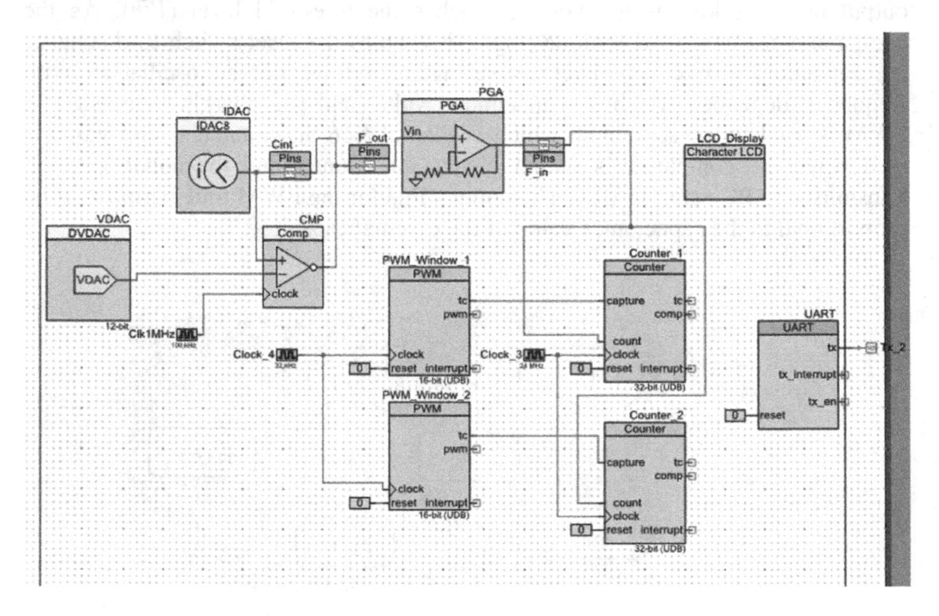


Fig 4. PSOC schematic diagram for the capacitance measurement system

The capacitor discharges as the comparator output becomes low due to the voltage level across the capacitor reaches the threshold voltage Vin. Once the capacitor voltage reaches zero, IDAC module gets re-connected to the capacitor and it starts charging it and hence we get a cycle of charge\discharge. The time period is directly proportional to the voltage, and frequency is inversely proportional [5]. Finally we can conclude by forming the following equation for measuring the capacitance value,

$$C = \frac{I}{F*V} \tag{1}$$

Where, F is the frequency of the oscillation, C is the capacitance, V is the voltage specified in VDAC, and I is the IDAC module current. Frequency is measured form the cycle either by counting the number of rising edges or falling edges. PWM specifies the time window for the measurement. Output signal [5]. FOUT is given through a PGA to the count input of the counter and the time window specified based on the requirements measures the number of rising or falling edges. The calculated capacitance value for the MEMS sensor is displayed on the LCD Display and the UART provides asynchronous communications to PC via RS 232.

## 6 Virtual instrumentation

There are many standards describing the connection between a PC and a microcontroller to exchange data (USB, RS232, GPIB...). Wires, connectors, and signals which are useful for a simple and reliable data transfer are described with the help of RS232. most of contemporary microcontrollers uses this standard communications for data transfer and data logging. This text describes the programming needed on the side of the PC running Labview to connect to the microcontroller using RS232 and exchange data[12]. The counter output value is transmitted by the UART to PC through USB interface is read by the PC[8]. This is done by using VISA configure serial port, control Vis. From the acquired counter data, calculations have been carried out to get the capacitance value using mathematical function routines in LabVIEW. Finally, the calculated output is shown in the graphical user interface (GUI) LabVIEW front panel shown in fig 5 and the block diagram representing the calculations made along with online data monitoring is shown in fig 6. LabVIEW uses VISA (virtual instrument software architecture) which is standard across the globe for configuration, measurement, automation, control, communication, data acquisition with PC through serial port and for other communications like GPIB, VXI, PXI, Ethernet and USB. Programming interface with the hardware and laboratory PC in our case is done by VISA. Interfacing hardware with other systems like Microsoft studio for visual basic, Labwindows etc. As the instrumentation hardware gets evolved, there is no need to change any configuration in VISA which makes it confident for the designer to use in any platform.NI-VISA includes various library functions and various operations can be carried simultaneously, we can change the baud rate based upon our requirements. Labview is used for virtual instrumentation, the acquired data from the measurement system is being transferred to the laboratory workstation from the remote field via serial communication i.e normal industrial standard RS232 serial communication. In the PC the programming for Labview is simulated and configured based on the user needs. Configuration includes the serial port setting which includes the baud rate,stop bits and parity bits etc. The data are continuously monitored and archived in the specified location of the laboratory PC being used.

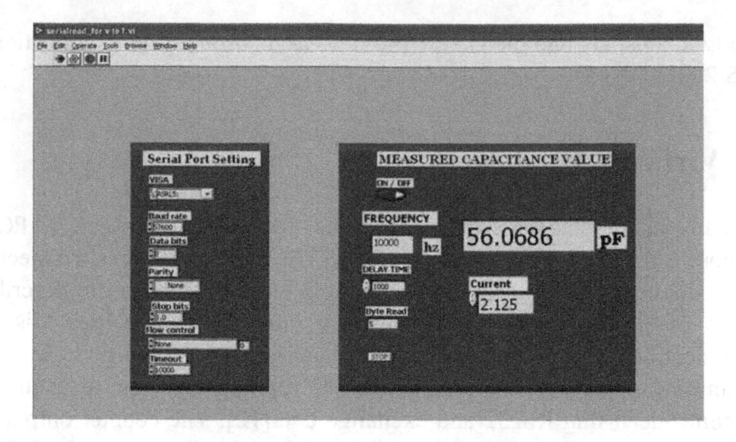


Fig 5. Capacitance value of commercial sensor is displayed in the LabVIEW

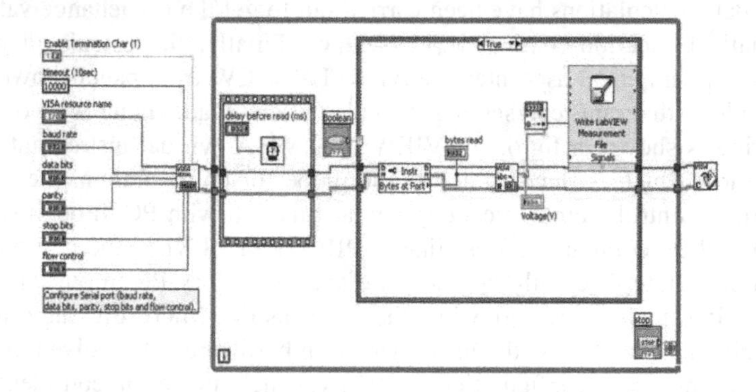


Fig 6. Online Data acquisition in LabVIEW

# 7 Results and Discussions

Capacitance embedded read out electronics has been implemented in PSOC and the following table proves the work. Table 1 involves the results of capacitance for commercially available sensors. The difference between the standard capacitance value and measured value is observed as error with 0.2%. The error is due to stray capacitance or parasitic capacitance present between the electrodes of the sensor and ground , the error is comparatively less when comparing with the work proposed by [2] which involves a parasitic capacitance of approximately 120pF. The system desgined can be used for measuring capacitance value from commercially available sensor and value from capacitive MEMS sensor.

Table 1. Measurement of capacitance in commercial sensor

| STANDARD CAPACITANCE VALUE (pf) | MEASURED CAPACITANCE VALUE(pf) |
|---|---|
| 12 | 11.9 |
| 39 | 39.57 |
| 47 | 46.17 |
| 56 | 56.06 |
| 68 | 69.26 |
| 80 | 79.15 |
| 101 | 102.2 |

Table 2. Measurement of capacitance in capacitive mems sensor

| STANDARD CAPACITANCE VALUE | MEASURED CAPACITANCE VALUE |
|---|---|
| 10 nf | 9.89 nf |
| 47 nf | 48.77 nf |
| 70 pf | 69.999 pf |

In table 2 the result of capacitance for capacitive MEMS micro sensor is observed and the value was 70pF in its static value.

## 8 Conclusion

Static value of capacitive MEMS microsensor was observed using the designed measurement system. There was ±0.2% error due to offset and stray capacitance present in the board. The repeatability during tests and the stability level of the system has been verified. The system showed a low tolerance error, while measuring the capacitance value of the Fabricated MEMS microsensor. The future work involves improvement of measurement system for better accuracy and to develop a measurement system detecting the capacitance value of MEMS sensors in dynamic mode which is expected to be in the range of femto Farad (fF).

## 9 Acknowledgment

The above mentioned experiment results are observed in Material Science Group (MSG), *Indira Gandhi Center for Atomic Research, Kalpakkam,* the author would like to thank Mrs.Usharani Ravi, scientific officer, Electronics and instrumentation section, MSG for her valuable guidance, motivation and encouragement throughout the project work and for supporting by providing valuable suggestion in this paper

## References

[1] Sarkar, G., Rakshit, A., Chatterjee, A., & Bhattacharya, K. (2013) Low Value Capacitance Measurement System with Adjustable Lead Capacitance Compensation. World Academy of Science, Engineering and Technology, International Journal of Electrical, Computer, Energetic, Electronic and Communication Engineering, 7(1), 63-70.

[2] Yazdi, N., Kulah, H., & Najafi, K. (2004, October) Precision readout circuits for capacitive microaccelerometers. In Sensors, 2004. Proceedings of IEEE (pp. 28-31). IEEE.

[3] Ye, Z., & Hua, C. (2012) An innovative method of teaching electronic system design with PSoC. Education, IEEE Transactions on, 55(3), 418-424.

[4] PSoC 3, PSoC 5 Architecture TRM (Technical Reference Manual) Document No. 001-50235 Rev. *E December 23, 2010

[5] Voltage Controlled Oscillator in PSoC 3/ PSoC 5, Application Note, http://www.cypress.com/?docID=33522.

[6] Wang, X. Z., Mo, X. Q., Yang, L. E., Zhai, Z. S., Liu, W. C., & Xiong, Z. (2014, October)A Kind of Resistance Capacitance Measurement Method Based on Time Constant. In Advanced Materials Research (Vol. 1037, pp. 156-160) Trans Tech Publications.

[7] Wang, Y., & Chodavarapu, V. P. (2015).Differential Wide Temperature Range CMOS Interface Circuit for Capacitive MEMS Pressure Sensors. Sensors, 15(2), 4253-4263.

[8] Leniček, I., Ilić, D., & Malarić, R. (2008). Determination of high-resolution digital voltmeter input parameters. Instrumentation and Measurement, IEEE Transactions on, 57(8), 1685-1688.

[9] Mu Linfeng, Zhang Wendong, He Changde, Zhang Rui, Song Jinlong, and Xue Chenyang (2015). Design and test of a capacitance detection circuit based on a transimpedance amplifier. Journal of Semiconductors, 7, 026.

[10] O. B. Adamo, P. Guturu, andM. R. Varanasi, "An innovativemethod of teaching digital system design in an undergraduate electrical and computer engineering curriculum," in *Proc. IEEE Int. Conf.Microelectron. Syst. Educ.*, 2009, pp. 25–28.

[11] Beijing, China,Dec. 15, 2011 MPW Laboratory, "Basics of application development on programmable system-on-chip," Peking University, [Online]. Available: http://www.ime.pku.edu.cn/mpw/teaching.html#jump5

[12] DušanPonikvar, September 2013 "Experiments: Labview and RS232"

# Lattice Wave Digital Filter based IIR System Identification with reduced coefficients

Akanksha Sondhi *, Richa Barsainya † and Tarun Kumar Rawat ‡

**Abstract** The purpose of this paper is to identify unknown IIR systems using a reduced order adaptive lattice wave digital filter (LWDF). For the system identification problem, LWDF structure is utilized, as it mocks-up the system with a minimal coefficient requirement, less sensitivity and robustness. The modelling technique is based on minimizing the error cost function between the higher order unknown system and reduced order identifying system. Two optimization algorithms, namely, genetic algorithm (GA) and gravitational search algorithm (GSA) are utilized for parameter estimation. By means of examples, it is shown that LWDF offers various advantages in system identification problem such as requirement of minimum number of coefficients, low mean square error (MSE), variance and standard deviation. The results demonstrate that better system identification performance is achieved by LWDF structure compared to adaptive canonic filter structure.

**Key words:** *IIR system identification, Adaptive filter, Lattice Wave digital filter, Genetic Algorithm, Gravitational search algorithm*

## 1 Introduction

Scope of adaptive filters in real-world applications like signal processing, system identification is really vast [1]. They are self-adjusting filters, whose coefficients are altered to estimate the actual parameters of an unknown system from its input and output relationship. The modelling of unknown system using infinite impulse response (IIR) system requires usage of both poles and zeroes for the aimed response

Division of Electronics and Communication Engineering
Netaji Subhas Institute of Technology, New Delhi-110045
*e-mail: sondhiakanksha14@gmail.com
†e-mail: richa.barsainya@gmail.com
‡e-mail: tarundsp@gmail.com



J.M. Corchado Rodriguez et al. (eds.), *Intelligent Systems Technologies and Applications 2016*, Advances in Intelligent Systems and Computing 530,
DOI 10.1007/978-3-319-47952-1_56

whereas, FIR uses only zeroes. However, for the appropriate level of performance, the requirement of filter coefficient is less in case of IIR in contrast to FIR. Due to this fact the adaptive IIR filters are more efficient than adaptive FIR filters [2, 3].

The problem of system identification constitutes of two sections: adaptive digital filter and adaptive algorithm. The adaptive algorithm attempts to arbitrate the adaptive filter parameters to achieve an optimal model for an unknown system, based on reducing the error function between the output of adaptive filter and unknown system, provided both systems are governed by same inputs [4]. Adaptive algorithms applied earlier for system identification were gradient search based such as least mean square, Quasi-Newton method. These traditional techniques for parameter estimation, however sometimes get trapped in local minima and slowly converge to the optimal solution. To overcome these constraints, optimization algorithms were evolved, namely GA, particle swarm optimization (PSO), GSA, etc. GA has shown potential to overcome the shortcomings of conventional methods. GA is capable of overcoming the fundamental problem of solving multidimensional problems prevalent in conventional algorithms. However, GA also exhibits a number of shortcomings.Gravitational Search Algorithm has also been utilized in this paper. It is a population based algorithm that works on the principle of Newton's law of gravitation.

Earlier the modelling of adaptive digital filter systems is done using IIR canonical filters due to their effortless implementation and analysis [5]. However, these structures have ingrained problems like sensitivity to word-length effect, stability issues and slow rate of convergence [6]. This lead to exploration and researchers came up with the new class of filters known as lattice wave digital filters (LWDF) [7, 8, 9]. Lattice wave digital filter is a specific class of wave digital filter and provides best structure for implementing IIR digital filters. LWDFs are less sensitive to the word length effect, have good dynamic range and offers enhanced stability. LWDF requires only $N$ number of coefficients for designing of $N^{th}$ order filter. In reference to that, LWDF structure models the existing system with reduced number of coefficients whereas, the IIR canonical system requires double the number of coefficients for a corresponding filter order for identification purpose. In the past few years, several efforts have been made for IIR system identification using the same order adaptive system as well as with reduced order system [10, 11]. The idea here is to focus primarily on the recursive system identification by optimizing the coefficients of the lower order adaptive system in such a way so that it imitates the higher order system. This can be achieved by minimizing the error cost function between the unknown system and adaptive system of reduced order. In this paper, two different adaptive systems, namely, canonic and LWDF are employed to test their efficiency for system identification. The parameters for comparison include the requirement of the number of coefficients, mean square error (MSE), mean, variance and standard deviation.

This paper formulates thusly: Section 2 deals with detailed explanation of LWDF. Section 3 contains the problem genesis of system identification.The brief explanation of GA and GSA is described in Section 4. Section 5 contains simulation results and quantitative analysis. The concluding remarks are mentioned under Section 6.

## 2 Lattice Wave Digital Filter

A lattice wave digital filter is an explicit class of WDF in which the realization of both the lattice branches is done by cascading of the first and second degree all-pass sections [12, 13]. These all-pass sections are implemented using symmetric 2-port adaptors and delay elements. Adaptors involves arithmetic operations viz., a multiplier and three adders. The response of each adaptor is controlled by $\gamma$ coefficient, whose value ranges from $-1 to +1$. Fig.1 shows four types of adapters classified on the basis of value of $\gamma$ with the corresponding value of $\alpha$ coefficient.

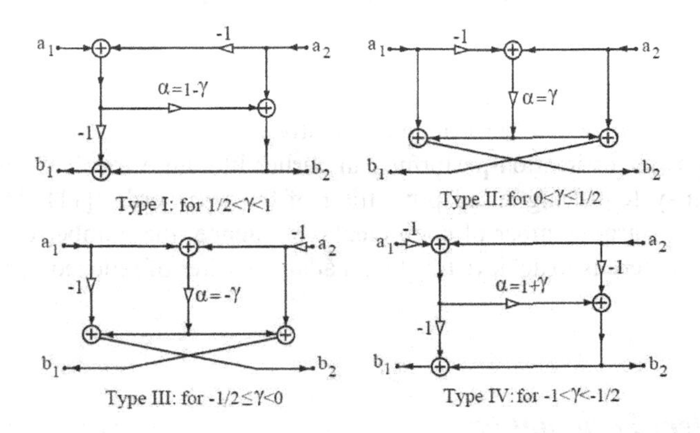


Fig. 1: Equivalents circuits of 2-port adaptors

The value of $\alpha$ coefficient which is to be implemented depends on the value of $\gamma$ and should always be greater than zero and less than or equal to 1/2, i.e. $(0 \leq \alpha \leq 1/2)$.In adaptive LWDF, even under finite arithmetic conditions, system stability is maintained, which is not in the case for adaptive IIR canonic filters.

## 3 System Identification

Contouring of system identification involves an adaptive algorithm that tries to endeavor the parameters of adaptive filter to get an optimized model for the unknown system based on minimizing an error function between the output of the adaptive filter system and the unknown system [4]. The process of identification shown in Fig.2, constitutes of sections namely, unknown system, approximating adaptive system and an adaptive algorithm.

$x(n)$ is an input signal that governs both unknown system and approximating adaptive system at the same time. The output $y(n)$ of the unknown system is mixed with a noise signal, yielding the output as $d(n)$. The output of the approximating adaptive system is $\hat{y}(n)$. Now for system identification, the error, i.e. $e(n)$ which is a

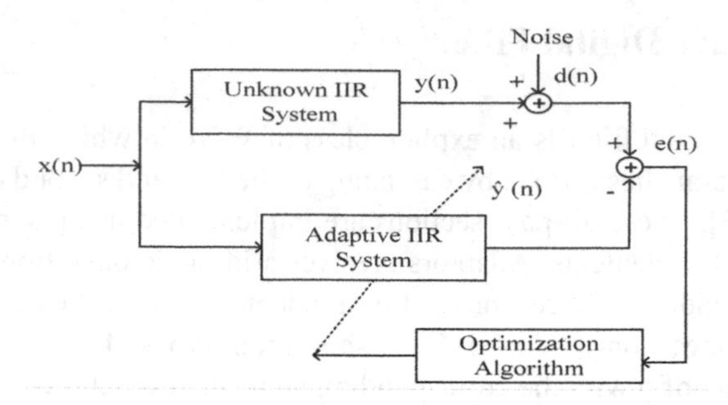


Fig. 2: Block-Diagram for system identification using adaptive IIR system.

deviation of $\hat{y}(n)$ from $y(n)$ is calculated and should be minimized, so that modelling of unknown system can be performed efficiently.

The parameter estimation performed in earlier literature involves modelling of an unknown system using an adaptive filter of the same order [11]. This leads to requirement of large number of coefficients and hence, the number of multipliers used. The idea here is to achieve this by an adaptive filter of reduced order by using optimization algorithms.

## *3.1 Problem Formulation*

The transfer function of an unknown IIR system is given by

$$H(z) = \frac{\sum_{k=0}^{m} a_k z^{-k}}{1+\sum_{k=1}^{n} b_k z^{-k}} \tag{1}$$

where $a_k$ and $b_k$ are the numerator and denominator coefficients that characterizes the unknown IIR system. IIR filter can be realized by using direct form II, where the coefficients of transfer functions are directly used as multiplier coefficients and order of the system imparts a number of delay elements involved. The transfer function of canonic direct form II structure is given by

$$G(z) = \frac{\sum_{k=0}^{m} \hat{a}_k z^{-k}}{1+\sum_{k=1}^{n} \hat{b}_k z^{-k}} \tag{2}$$

Here, $\hat{a}_k$ and $\hat{b}_k$ denotes the coefficients to be estimated by optimization. The unknown IIR system can also be modelled by using LWDF structure that is governed by a parameter $\gamma$. It's transfer function is given by

$$F(z) = \frac{-\gamma_0 + z^{-1}}{1-\gamma_0 z^{-1}} \prod_{l=1}^{m} \frac{-\gamma_{2l-1} + \gamma_{2l}(\gamma_{2l-1}-1)z^{-1} + z^{-2}}{1+\gamma_{2l}(\gamma_{2l-1}-1)z^{-1} - \gamma_{2l-1}z^{-2}} + \prod_{l=m+1}^{m+n} \frac{-\gamma_{2l-1} + \gamma_{2l}(\gamma_{2l-1}-1)z^{-1} + z^{-2}}{1+\gamma_{2l}(\gamma_{2l-1}-1)z^{-1} - \gamma_{2l-1}z^{-2}} \tag{3}$$

where $m = \frac{P-1}{2}$ and $n = \frac{Q}{2}$. Here $P$ and $Q$ are the order of two allpass filters used to formulate LWDF system. The fitness objective function, that is, sum of square error whose value is to be minimized, is given by

$$E = \sum_{n=1}^{N} [|y(n) - \hat{y}(n)|]^2 = \sum_{n=1}^{N} e^2(n) \tag{4}$$

The focus here is to evaluate the efficiency of the LWDF structure for system identification problem in comparison to canonic direct form II structure. The value of the error function is statistically calculated by employing two different optimization algorithms. The coefficients of adaptive LWDF and IIR system structures are iteratively optimized using GA and GSA until the mean square error is minimized in order to identify unknown system.

# 4 Optimization Algorithms

Heuristic algorithms, namely GA and GSA, used for IIR system modelling are briefly described under this section.

## *4.1 Genetic Algorithm (GA)*

A genetic algorithm (GA) is a search heuristic tool that mimics the process of natural selection for getting optimal solutions for optimization problems [14]. Each candidate (chromosome) with best fitted value is evolved to the next generation by the process of genetic operations and process repeats iteratively for a number of generations to get a best optimal solution. The implementation steps of GA is shown by a flowchart as given by Fig.3.

## *4.2 Gravitational Search Algorithm (GSA)*

GSA is a modern algorithm based on Newton's law of gravity and motion used to solve optimization problems [15]. The agents known as objects, interact with each other by gravitational force with the masses being their performance metric. The objects, try to occupy a position near to heavy masses and the slow movement of heavier masses assures the exploitation step of the algorithm and results in better solution represented by mass. The implementation steps of the algorithm is depicted using flowchart shown in Fig.4

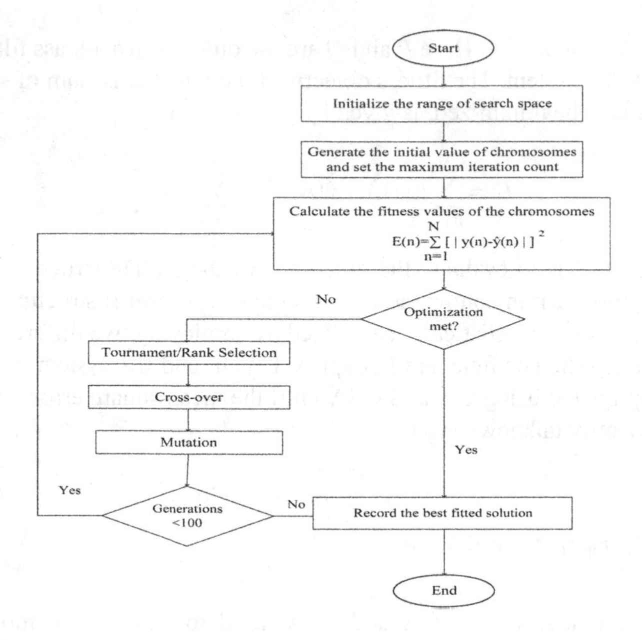


Fig. 3: Flowchart of Genetic Algorithm used for system identification

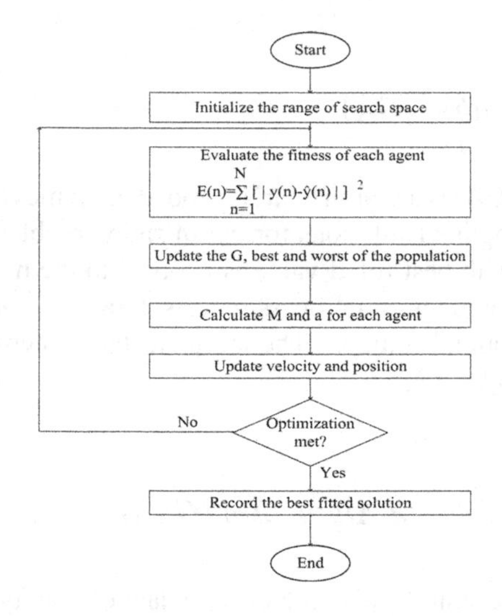


Fig. 4: Flowchart of Gravitational Search Algorithm used for system identification

# 5 Design Example

This section, by means of two different case studies compares the benefits of adaptive LWDF and canonic structures in system identification problem. We have taken two unknown systems (with symmetric and arbitrary coefficients) each for $4^{th}$ and

$6^{th}$ order. The essence here is to identify a higher order unknown system using reduced order canonic (direct form II realization) and LWDF structure. By making use of GA and GSA optimization algorithm the coefficient of lower order systems are optimized in such a way that it matches the response of higher order unknown system.The simulations are performed in MATLAB on intel core i5 windows XP professional. The optimal set of parameters for GA and GSA for solving the system identification problem are reported in Table 1.

Table 1: Control Parameters of heuristic algorithms for system identification

| Parameters | Symbol | GA | GSA |
|---|---|---|---|
| Population Size | g , gs | 25 | 25 |
| Max Iteration Cycle | N | 500 | 1500 |
| Tolerance | - | 10e-06 | 10e-06 |
| Limits of filter coefficients | Canonic | -1,+1 | -1,+1 |
| | LWDF | -1,+1 | -1,+1 |
| Selection | - | Tournament, | 3 ,5 |
| Scaling | - | Rank | - |
| Initial value of coefficients | - | 0 | 0 |
| Elite count | - | - | 2 |
| Creation function | - | Uniform | Uniform |
| Crossover fraction, function | - | - | 0.9, Scattered |
| Mutation | - | - | Gaussian |

## 5.1 *Identification of* $4^{th}$ *order unknown system using* $3^{rd}$ *order adaptive system:*

The two lowpass $4^{th}$ order Chebyshev IIR filters, one with symmetric coefficients and other with arbitrary coefficients are considered here as unknown systems [10]. The transfer functions of these systems are given as follows:

**SYSTEM 1**

$$H_1^{4th}(z) = \frac{0.1792 + 0.7168z^{-1} + 1.0752z^{-2} + 0.7168z^{-3} + 0.1792z^{-4}}{1 + 0.9036z^{-1} + 0.9648z^{-2} + 0.174z^{-3} + 0.1747z^{-4}} \tag{5}$$

**SYSTEM 2**

$$H_2^{4th}(z) = \frac{1 - 0.9z^{-1} + 0.81z^{-2} - 0.729z^{-3} + 0z^{-4}}{1 + 0.004z^{-1} + 0.2775z^{-2} - 0.2101z^{-3} + 0.14z^{-4}} \tag{6}$$

The adaptive $3^{rd}$ order IIR and LWDF structures used for identifying the above given unknown systems. The detailed analysis is given in case 1 and case 2 respectively.

*Case 1:* The transfer function of $3^{rd}$ order adaptive IIR system is given below:

$$G_{canonic}^{3rd}(z) = \frac{\hat{a}_0 + \hat{a}_1 z^{-1} + \hat{a}_2 z^{-2} + \hat{a}_3 z^{-3}}{1 + \hat{b}_1 z^{-1} + \hat{b}_2 z^{-2} + \hat{b}_3 z^{-3}} \tag{7}$$

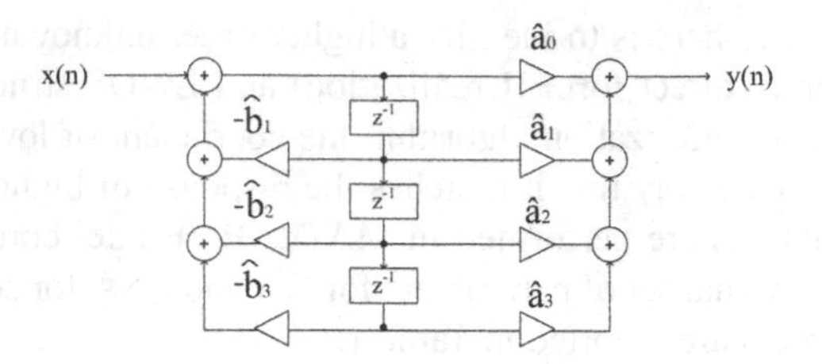


Fig. 5: Direct form II realization of adaptive $3^{rd}$ order IIR system.

where $\hat{a}_k$ and $\hat{b}_k$ are the coefficients of adaptive IIR system which are to be optimized. The direct form II structural realization of adaptive $3^{rd}$ order canonic structure is shown in Fig.5.

*Case 2:* The transfer function of $3^{rd}$ order adaptive LWDF system is given as:

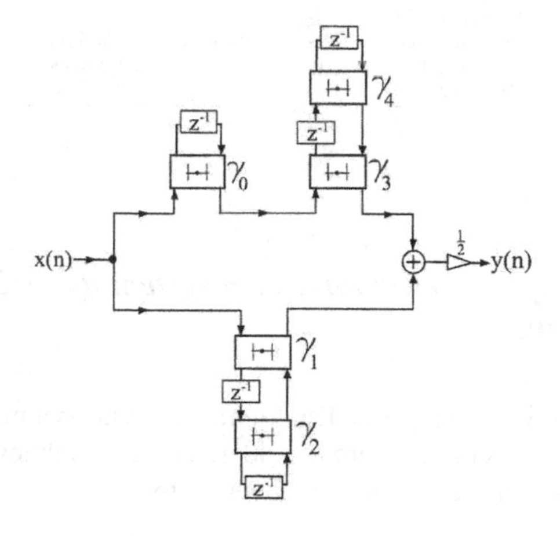


Fig. 6: LWDF realization of adaptive $3^{rd}$ order IIR system.

$$F_{lwdf}^{3rd}(z) = \frac{1}{2}\left[\frac{[(-\gamma_0-\gamma_1)+(1+\gamma_0\gamma_1-\gamma_0\gamma_1\gamma_2+\gamma_1\gamma_2+\gamma_0\gamma_2-\gamma_3)z^{-1}+(1+\gamma_0\gamma_1-\gamma_0\gamma_1\gamma_2+\gamma_1\gamma_2)+(\gamma_0\gamma_2)-\gamma_3)z^{-2}+(\gamma_0+\gamma_1)z^{-3}]}{1-(\gamma_0-\gamma_1\gamma_2+\gamma_2)z^{-1}+(-\gamma_1+\gamma_0\gamma_2-\gamma_0\gamma_2\gamma_1)z^{-2}+(\gamma_0\gamma_1)z^{-3}}\right] \tag{8}$$

where $\gamma$'s are the coefficients of LWDF system. The LWDF realization of adaptive $3^{rd}$ order IIR system is depicted in Fig.6. The fitness function provided in equation (4) is iteratively minimized to obtain the optimized coefficients of lower order adaptive LWDF and canonic structures.The mean square error, variance and standard deviation obtained for both the structures by using optimization algorithms are listed in Table 2. It is evident from the results that the LWDF approximates the symmetric unknown system (System1) more efficiently as compared to canonic, as the MSE achieved is minimum. The unknown system 2 with arbitrary coefficients

Table 2: Statiscal Analysis of $4^{th}$ order IIR system using $3^{rd}$ order canonic and LWDF employing GA and GSA

| | Canonic | | LWDF | |
|---|---|---|---|---|
| Mean square error | GA | GSA | GA | GSA |
| **SYSTEM 1** | | | | |
| Run1 | 1.5705534 | 1.3547e-15 | 0.0033298 | **5.4208e-16** |
| Run2 | 0.0079623 | 1.2411e-14 | 0.00663308 | 1.0011e-14 |
| Run3 | 3.387193 | 3.0762e-14 | 1.004967602 | 2.70354e-15 |
| Run4 | 1.2549e-04 | 1.3689e-14 | **1.007318e-04** | 1.2434e-14 |
| Run5 | 0.00808056 | 2.76854e-13 | 0.001977e-02 | 4.2752e-14 |
| **Mean** | 0.9948 | 6.7014e-14 | 0.2030 | 1.3688e-14 |
| **Var** | 2.2480 | 1.3871e-26 | 0.2010 | 2.8832e-28 |
| **Std** | 1.4993 | 1.1777e-13 | 0.4483 | 1.6980e-14 |
| **SYSTEM 2** | | | | |
| Run1 | 0.0060538 | 7.2763e-02 | 0.0055850 | 3.2000e-02 |
| Run2 | 0.0052936 | 8.29223e-02 | 0.00409870 | 4.0000e-02 |
| Run3 | 1.92516e-04 | 8.7952e-02 | **1.86710e-04** | 6.5538e-11 |
| Run4 | 1.1833970 | 8.8743e-02 | 1.05878978 | 7.4000e-02 |
| Run5 | 0.00824595 | 6.78089e-03 | 0.0070841 | **5.5733e-11** |
| **Mean** | 0.2406 | 0.0678 | 0.2151 | 0.0292 |
| **Var** | 0.2778 | 0.0012 | 0.2224 | 9.5920e-04 |
| **Std** | 0.5270 | 0.0347 | 0.4716 | 0.0310 |

as mentioned in equation (6) is also identified using adaptive LWDF and canonic structures.The same arbitrary system is previously optimized by canonic structure in [10] and the error reported in literature is $7.8212e^{-02}$ whereas, the minimum error of $5.5733e^{-11}$ is achieved in case of proposed LWDF system.

### *5.2 Identification of* $6^{th}$ *order unknown system using* $5^{th}$ *order adaptive system:*

The two lowpass $6^{th}$ order Chebyshev IIR filters, one with symmetric coefficients and other with arbitrary coefficients are considered here as unknown systems [16]. The transfer functions of these systems are given as follows:

**SYSTEM 1**

$$H_1^{6th}(z) = \frac{0.0099439 - 0.022235z^{-1} + 0.01988z^{-2} - 0.019888z^{-4} + 0.022235z^{-5} - 0.0099439z^{-6}}{1 - 3.3738z^{-1} + 6.5225z^{-2} - 7.6002z^{-3} + 6.0121z^{-4} - 2.8648z^{-5} + 0.78246z^{-6}} \tag{9}$$

**SYSTEM 2**

$$H_2^{6th}(z) = \frac{1 - 1.8z^{-2} + 1.04z^{-4} + 0.05z^{-5} + 0.192z^{-6}}{1 - 0.8z^{-2} - 0.17z^{-4} - 0.56z^{-6}} \tag{10}$$

The adaptive $5^{th}$ order IIR and LWDF structures used for identifying the above given unknown systems. The detailed analysis is given in case 1 and case 2 respectively.

*Case 1:* The transfer function of $5^{th}$ order adaptive IIR system is given below:

$$G_{canonic}^{5th}(z) = \frac{\hat{a}_0 + \hat{a}_1 z^{-1} + \hat{a}_2 z^{-2} + \hat{a}_3 z^{-3} + \hat{a}_4 z^{-4} + \hat{a}_5 z^{-5}}{1 + \hat{b}_1 z^{-1} + \hat{b}_2 z^{-2} + \hat{b}_3 z^{-3} + \hat{b}_4 z^{-4} + \hat{b}_5 z^{-5}} \tag{11}$$

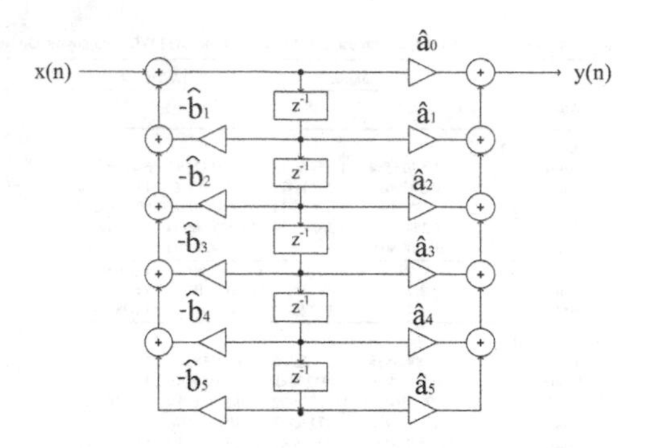


Fig. 7: Direct form II realization of adaptive $5^{th}$ order IIR system.

where $\hat{a}_k$ and $\hat{b}_k$ are the coefficients to be optimized.The direct form II structural realization of adaptive $5^{th}$ order canonic structure is shown in Fig.7.

*Case 2:* The transfer function of $5^{th}$ order adaptive LWDF system is given as:

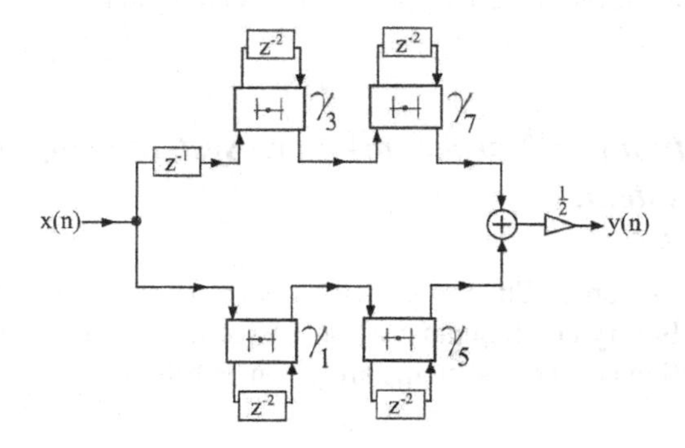


Fig. 8: LWDF realization of adaptive $5^{th}$ order IIR system.

$$F_{lwdf}^{5th}(z) = \frac{1}{2}\left[\left(\frac{-\gamma_0 + z^{-1}}{1-\gamma_0 z^{-1}}\right)\left(\frac{-\gamma_1 + \gamma_2(\gamma_1 - 1)z^{-1} + z^{-2}}{1+\gamma_2(\gamma_1 - 1)z^{-1} - \gamma_1 z^{-2}}\right) + \left(\frac{-\gamma_3 + \gamma_4(\gamma_3 - 1)z^{-1} + z^{-2}}{1+\gamma_4(\gamma_3 - 1)z^{-1} - \gamma_3 z^{-2}}\right)\right] \tag{12}$$

where, $\gamma$'s are the coefficients to be optimized.The LWDF realization of adaptive $5^{th}$ order IIR system is shown in Fig.8.

The mean square error found out by running GA and GSA for both the structures iteratively is listed in Table 3. This affirms that the adaptive LWDF outperforms the existing adaptive canonic system when applied in the system identification problem.

A comparative study of results obtained by adaptive LWDF and canonic structure for solving system identification problem is graphically shown in Fig.9. It shows the

Table 3: Statiscal Analysis of $6^{th}$ order IIR system using $5^{th}$ order canonic and LWDF employing GA and GSA

| | Canonic | | LWDF | |
|---|---|---|---|---|
| Mean square error | GA | GSA | GA | GSA |
| **SYSTEM 1** | | | | |
| Run1 | 0.00734345 | 5.7968e-06 | **3.75007549e-07** | 3.2277e-11 |
| Run2 | 8.9052370e-05 | 3.2395e-05 | 3.540030e-05 | **2.4643e-12** |
| Run3 | 4.3137e-04 | 7.0790e-07 | 4.9661185e-06 | 2.3321e-07 |
| Run4 | 9.845916e-07 | 4.6817e-12 | 2.719759e-04 | 2.1563e-09 |
| Run5 | 4.083069e-06 | 3.1456e-05 | 3.09040e-06 | 3.7651e-11 |
| **Mean** | 0.0016 | 1.4071e-05 | 6.3162e-05 | 4.7088e-08 |
| **Var** | 1.0434e-05 | 2.7076e-10 | 1.3828e-08 | 1.0826e-14 |
| **Std** | 0.0032 | 1.6455e-05 | 1.1759e-04 | 1.0405e-07 |
| **SYSTEM 2** | | | | |
| Run1 | 0.00125703 | 3.2086e-12 | 0.0011960 | 2.5902e-12 |
| Run2 | 0.005084228 | 4.0215e-12 | 0.00462308 | 2.3641e-12 |
| Run3 | 2.97680e-04 | 1.4805e-11 | 1.7433534e-05 | 4.5767e-12 |
| Run4 | 1.2443848e-04 | 3.5485e-12 | 1.889999e-05 | **1.3235e-12** |
| Run5 | 5.759743e-06 | 4.1354e-12 | **4.243410e-06** | 4.07315e-12 |
| **Mean** | 0.0014 | 5.9438e-12 | 9.7661e-04 | 2.9855e-12 |
| **Var** | 4.5923e-06 | 2.4676e-23 | 3.4163e-06 | 1.7549e-24 |
| **Std** | 0.0021 | 4.9675e-12 | 0.0018 | 1.3247e-12 |

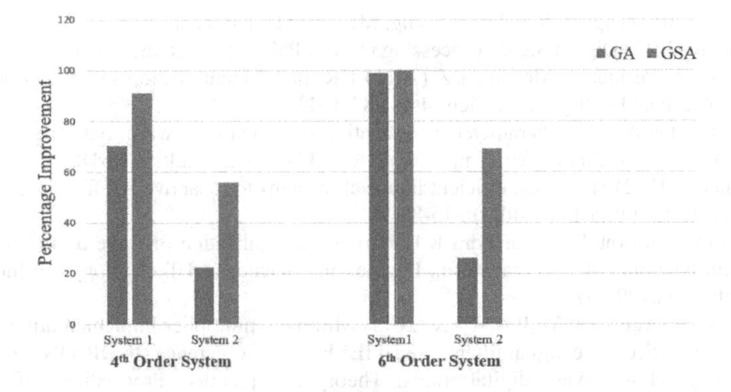


Fig. 9: Percentage improvement achieved by adaptive LWDF over canonic structure by applying GA and GSA

percentage improvement achieved in case of adaptive LWDF over adaptive canonic system, when both are utilized for identifying the same unknown system. The capability of GA and GSA in solving the identification problem is also analyzed and results shows that the optimal solution is obtained more efficiently with help of GSA.

# 6 Conclusion

The problem of system identification is solved by employing adaptive LWDF by minimizing the error function using optimization algorithm. The parameter estimation and minimization of mean square error for LWDF as well as canonic structure is done using GA and GSA. The modelling of unknown system by LWDF is found to be adept option than the canonic system, as the MSE comes out to be less. The capability of GSA in solving system identification problem is also tested over GA. The novelty lie in the fact that the higher order unknown IIR system is successfully identified by the lower order adaptive LWDF systems with requiring minimum number of coefficients for its structural realization. This leads to more computational efficiency making LWDF an adept option for system identification problem. The future research may focus on identification of nonlinear system using adaptive LWDF system.

## References


1. Hu H, Ding R (2014) Least squares based iterative identification algorithms for input nonlinear controlled autoregressive systems based on the auxiliary model. Nonlinear Dynamics. 76(1):777-84.
2. Mitra SK (2013) Digital Signal Processing. Mc Graw Hill, 4th edn.
3. Rawat TK (2015) Digital Signal Processing.Oxford Publications, India 1st edn.
4. Mostajabi T, Poshtan J, Mostajabi Z (2015) IIR model identification via evolutionary algorithms. Artificial Intelligence Review. 44(1):87-101.
5. Durmus B, Gn A (2011) Parameter identification using particle swarm optimization. International Advanced Technologies Symposium (IATS 11), Elazig, Turkey. 16-18.
6. Regalia PA (1992) Stable and efficient lattice algorithms for adaptive IIR filtering. Signal Processing, IEEE Transactions. 40(2):375-88.
7. Barsainya R, Rawat TK, Mahendra R (2016) A new realization of wave digital filters using GIC and fractional bilinear transform. Engineering Science and Technology: an International Journal. 19(1):429-437.
8. Barsainya R, Aggarwal M, Rawat TK (2015) Minimum multiplier implementation of a comb filter using lattice wave digital filter. Annual IEEE India Conference (INDICON): 1-6.
9. Fettweis A (1986) Wave digital filters: Theory and practice. Proceedings of the IEEE. 74(2):270-327.
10. Jiang S, Wang Y, Ji Z (2015) A new design method for adaptive IIR system identification using hybrid particle swarm optimization and gravitational search algorithm. Nonlinear Dynamics. 79(4):2553-76.
11. Netto SL, Agathoklis P (1998) Efficient lattice realizations of adaptive IIR algorithms. IEEE Transactions on Signal Processing. 46(1):223-7.
12. Gazsi L (1985) Explicit formulas for lattice wave digital filters. IEEE Transactions on Circuits and Systems. 32(1):68-88.
13. Yli-Kaakinen J, Saramki T (2007) A systematic algorithm for the design of lattice wave digital filters with short-coefficient wordlength. IEEE Transactions on Circuits and Systems I. 54(8):1838-51.
14. Chang WD (2006) Coefficient Estimation of IIR Filter by a Multiple Crossover Genetic Algorithm. An international Journal Computers and Mathematics with Applications. 51:1437-1444
15. Sabri NM, Puteh M, Mahmood MR (2013) A review of gravitational search algorithm. Int J Advance Soft Comput Appl. 5(3):1-39.
16. Sharifi MA, Mojallali H (2015) A modified imperialist competitive algorithm for digital IIR filter design. Optik-International Journal for Light and Electron Optics. 126(21):2979-84.

# Factors Affecting Infant Mortality Rate in India: An Analysis of Indian States

**Suriyakala Vijayakumar, Deepika M G, Amalendu Jyotishi and Deepa Gupta**

**Abstract** While there are enough efforts by the governments to reduce the infant mortality rate in developing countries, the results are not as desired. India is no exception to the case. Identifying the factors that affect the infant mortality rates would help in better targeting of the programs leading to enhanced efficiency of such programs. Earlier studies have shown the influence of socio economic factors on infant mortality rates at a global level and found that variables like fertility rate, national income, women in labour force, expenditure on health care and female literacy rates influence the infant mortality rates. The current study using the data from Indiastat.com from all states and Union Territories of India for the years 2001 and 2011 tries to establish the relationship between infant mortality rate and some of the above mentioned factors along with a few healthcare infrastructure related variables. Using a regression analysis method we not only identify the influence of the variables on infant mortality, we went a step further in identifying the performance of states and union territories in reducing IMR. The performance was measured using 'technical efficiency' analysis. We then compared the performance and growth rate of IMR to classify the states as good performers and laggards. Our results suggest that most of the major states are on track on their performance on IMR. However, a few small states and union territories like Andaman and *Nicobar* Island, Mizoram, Arunachal Pradesh as well as Jammu & Kashmir need special attention and targeting to reduce IMR.

**Keywords:** Health Infrastructure; Infant Mortality Rate (IMR); State-wise analysis; India

Suriyakala Vijayakumar[1], Deepika M G[1], Amalendu Jyotishi[1] and Deepa Gupta[2]

[1]Amrita School of Business, Amrita Vishwa Vidyapeetham, Amrita University, Bangalore India

[2]Department of Mathematics, Amrita School of Engineering, Amrita Vishwa Vidyapeetham, Amrita University, Bangalore India

Email: {surya.jan30, mgdeepika, amalendu.jyotishi, deepagupta.verma}@gmail.com



J.M. Corchado Rodriguez et al. (eds.), *Intelligent Systems Technologies and Applications 2016*, Advances in Intelligent Systems and Computing 530,
DOI 10.1007/978-3-319-47952-1_57

## 1 Introduction

Infant mortality rate (IMR) has gained attention of the academicians and policy makers as it serves as an important indicator of socio-economic development of a country. The higher infant mortality rate also enhances the uncertainty of child survival especially among low income countries and the low income communities within the countries, which in turn jeopardizes the policies to reduce fertility rate [1]. There has been several studies on identifying the causes of infant mortality. Therefore, the uncertainty associated with infant or child survival is hardly due to random factor. Once the factors are identified, the policies tend to work towards acting on those core factors in order to reduce infant mortality rates. However, the outcome of the policies in reducing infant mortality is far from desire level. Therefore, it is important to not only to revisit the factors, it is also equally important to rethink the policies based on identified factors. The divergence between the high and low income countries in terms of childhood mortality rates is staggering where the former represent only one percent of all deaths the later represents around 30 percent in low income countries [2].
Though IMRs have been consistently reduced throughout the world, divergence and disparities between the high and low income countries remain high, so as divergence among the regions within a given country [3, 4, 5]. The reasons for such variations are explained through studies examining factors affecting IMRs in different countries under different socio-economic conditions. Though such analysis is useful, international comparisons may not always suffice in arriving at factors with respect to what determines infant mortality rates and may often be subject to measurement biases. Taking into account such biases, our study aims at exploring determinants of infant mortality rates across the states of India in two time points of year 2001 and 2011. Such study may be useful to create effective policies at central and federal level to reduce IMR across the states. The paper also contributes to the literature in the area of IMR by identifying new determinants of IMR using data across the states of India in two points of time. Objectives of the study are given below:

- Identify the change in trends in infant mortality rates for different states of India for two census years of 2001 and 2011.
- Analyze factors influencing the infant mortality rates through an analysis of Indian states.

The paper is organized in the following way. We discuss about the related studies in the next section. Subsequent sections discuss the datasets, proposed approach, data description, descriptive analysis followed by the result analysis. In the penultimate section we discuss our findings and subsequently conclude the paper.

## 2 Related Work

There are two strands or approaches of studies on determinants of infant mortality namely social sciences and medical research. The focus of social science research has been from social and economic development perspective including the socio economic status of the parents, the household and community and their effects on infant mortality, assets, education and income level of the household etc. On the other hand, medical researchers have concentrated on the biological processes, endowments and conditions affecting infant health outcomes [6]. A variety of factors affecting Infant mortality are also classified as biological and socio economic [1]. The endogenous factors are biological, factors related to the formation of foetus in the womb. The age of the mother, maturity during birth and weight at birth are some of the critical biological aspects influencing infant mortality Socio-economic attributes are considered as exogenous factors. One of the causes of high infant mortality in some countries is the lack of availability of proper medical care. Most of the post neo-natal deaths especially in developing countries are due to communicable diseases and respiratory diseases such as diarrhoea, pneumonia, etc. Environmental factors including congestion, insanitation, lack of sufficient sunshine and fresh air also contribute to infant mortality. It is interesting to note that the countries with low IMR have higher proportion of infant deaths occur during the neonatal stage that are mainly genetic or biological in nature. Developed countries have mostly eliminated the environmental factors as cause of infant mortality. On the other hand, less developed and developing countries have higher proportion of infant mortality after the neonatal stage due to environmental factors [1].

[7] Study tries to test for factors affecting infant mortality rates based on a cross sectional model covering 117 countries for the year 1993. The results show that fertility rates, female participation in the labour force, per capita GNP and female literacy rates significantly affect infant mortality rates. Surprisingly government expenditure on health care as a percentage of GNP did not play a major role in determining IMRs. Of all the independent variables chosen for the study, fertility rates and female literacy rates have the strongest impact on IMRs. [8, 9] have shown that fertility rates and IMR are closely related. A high fertility rate often results in the poor health of the mother. [10] suggested that there is a strong interrelationship between IMRs and fertility rates which the author terms as 'dual causality'. When the uncertainty and risk associated with child survival is high the mother tends to have more pregnancy with a hope that a few will survive. [7] study on the contrary shows that fertility rates do indeed have an impact on IMR but not the other way. In [11] their study in Ecuador find that the declining IMR coincide with declining fertility rates. Two-way relationship between IMR and fertility rates is observed in the finding of the study. Malnutrition, water and sanitation, education of the mother, knowledge of family planning and use of contraceptives, access to health services, targeted social assistance programs, immunization, health insurance and lifestyles (like smoking and alcohol use by pregnant woman) are the important factors impacting infant mortality. [2] use a duration model to estimate the relationship between Infant mortality and socio economic status at the individual, household and community level. The study found that the full gestation

period, mother's educational attainment, marital status and the type of hospital used for delivery are the key determinants of infant survival in the first year of birth.

The incident of infant mortality is higher in rural area in comparison to urban areas [12]. Socio-economic status of the household or individual also impacts infant mortality. If the household's socio-economic status is better they tend to have lesser incident of infant mortality and vice versa [2, 13]. This relationship is based on the premise that the socioeconomic status seems to closely associated with cleaner environment, healthier nutrition and lifestyle. [12, 14] have shown mother's education has significant impact on child mortality, while [10] claim that mother's marital status is strongly associated with infant mortality. [15] found a nonlinear effect of age of mother on infant mortality having a 'U' shape relationship.

[2] in their study in Uruguay drew influence of socio-economic parameters on infant mortality and found the relationship to be negative. The policy implications drawn from the study suggest that policies directed towards improving maternal socio economic status can significantly reduce the IMR. The results based on primary data shows that low level of maternal educational attainment, delivery in public institutions and maternal unwed status are all positively related with infant mortality [2]. Gestation period between the births is also found to be significantly related to higher probabilities of infant survival [2].

In recent time urban poverty and its implications on infant mortality has drawn attention among the researchers and policy makers [16]. With the rising urban population and a substantial number of them residing in unhygienic slum areas has bearing on malnutrition and infant mortality. It is therefore not surprising that majority of infants died aged less than 4 months and are malnourished. Therefore, poverty-malnutrition-infant mortality nexus in low income urban dwelling is evident. The development of generic concept of Integrated Management of Childhood Illness (IMCI) by the World Health Organization (WHO) is a step towards recognizing comprehensive integrated assessment and treatment of sick infants [16]. A World Bank study suggests that economic growth measured as a percentage of growth of per capita Gross Domestic Product (GDP) is a key to reducing IMR [17]. Our brief review of literature suggest that several factors contribute to infant mortality. Health care and expenditure relating to that is important, at the same time, malnutrition, poverty, economic access need more comprehensive approach also address the problem of infant mortality. The debate could remain weather to divert health care expenditure to other related and important sectors or bringing those related sectors into the ambit of comprehensive community health care and thereby increasing expenditure in these important areas [18]. However, it is evident that the continuing debate on IMR warrants a critical examination of these multiple claims from a broader socio-economic development perspective.

## 3 Proposed Approach

The proposed approach comprises of four step process. Data Source, descriptive statistics, Regression Modeling and Result analysis.

**Step 1: Data Source**

Secondary data was collected for different variables which may have effect on infant mortality rate. The data was collected for all the states and union territories of India for the year 2001 and 2011. Both 2001 and 2011 being the census years, we can get the population data for all the required variables. Besides, 10-year gap provides a good understanding of the growth in the healthcare sector. Indiastat.com is the main source of all data.

**Step 2: Descriptive statistics**

Descriptive statistics provides an overview of the variables considered for analysis. Descriptive statistics also helps us in understanding the magnitude and degree of variations across the variables considered for this study.

**Step 3 Regression Modeling and Results Analysis**

Regression model is basically used to describe the influence of independent variables on dependent variable. If the independent variable is more than one, it is called as multiple linear regression model. Here, multiple linear regression method was adopted as there are many independent variables which may have an influence on the dependent variable. Linear regression model was found to be the best fit model. The general model of Regression is mentioned below

$$Y = \alpha + \Sigma\beta i x i + u \quad (1)$$

where Y is the dependent variable, $x_i$ are the independent variables, $\beta i$ are the slope coefficients of the independent variables, $\alpha$ is the intercept coefficient and $u$i is the random error term. Result analysis, this part discusses about the result obtained through descriptive statistics and linear regression analysis.

## 4 Data Description

The study is based on the secondary source of information largely gathered from the database Indiastat.com. The data is gathered on variables including infant mortality rate, female literacy rate, literacy rate, expenditure, state domestic product, female in labour force, total fertility rate, number of government hospitals, family planning, number of beds in government hospital and primary health centre for the year 2001 and 2011. The study tries to establish the relationship between the infant mortality rate as dependent variable and female literacy, state domestic profit, female labour force, total fertility rate, number of government hospital, family planning, number of government hospital beds and primary health Center as independent variables using multiple regression method.

To analyze the factors influencing infant mortality rate we initially specified a regression model with IMR as dependent variable with ten independent variables (based on theoretical and empirical literature) in our earlier attempt. However, in the final model we removed a few variables that did not show any statistically significant results or had strong multicolinearity with other variables. The definition for the variables chosen for analysis and their expected relationship with the dependent variable is explained here.

Infant Mortality Rate (IMR) – It is the dependent variable in the model. This is seen as an important indicator reflecting the country's healthcare facility and socio economic status of an economy. This is measured as number of deaths per 1000 births of children below age of one. Literacy Rate (LR) – This is measured as percentage of population. This variable explains the overall literacy rate of both male and female in the country. While female literacy rate is seen as a more prominent variable explaining IMR in the literature overall literacy rate is seen important for the reason that even male members in the family play a significant role in post child birth like taking care of all hygienic factor, regular vaccination, etc. This variable is expected to have a negative relationship with IMR. Female Literacy Rate (FLR) – It is one of the main factors which is seen to have high impact on infant mortality rate. From the literature review, it is clear that lesser the female literacy rate higher is the infant mortality rate. It is measured in terms of percentage of population and is expected to have a negative relationship with IMR. Expenditure (EXP) – This variable measures the country's overall expenditure on health care sector expressed in crores of rupees. The better the infrastructure and other facilities, mortality rate is seen to be less. Female Labour (FLR) – This is expressed as the percentage of women in Labour force. Higher this figure higher is the expected IMR since it is felt that women are posed to risk of hazardous work during their gestation period. Primary Health Centre (PHC) – PHCs are basic medical facility provided by the government across the states and Union Territories of India for providing people with basic medical facility. This variable is seen to be positively associated with IMR. State Domestic Product (SDP) – SDP is measure of total income and economic activity in a given state. Higher the SDP higher is the economic prosperity of the region and therefore this variable is expected to be negatively associated with IMR. Total Fertility Rate (TFR) – Higher fertility rate of a women is seen to increase infant deaths due to more pregnancy and negative impact on mother's health. This is measured in percentage of total population in respective states and territories. Literature has also established a two-way relation between IMR and fertility rate. Higher IMR induces the families to go for more child births with the belief that a few may survive. Human Development Index (HDI) – Human development index ranks states and Union territories based on education, life expectancy and per capita income. Higher HDI would mean high life expectancy of people and therefore this variable is expected to be negatively related with IMR. Family Planning (FP) – This measures the number of people opting for family planning which is likely to lead to less of child birth and therefore lower IMR. Government Hospital (GH) – Number of government hospitals in each state and Union territories. In India since government hospitals provide free treatment this is accessed largely by the urban and rural poor avoiding births at home which are more prone to risks. Therefore, this variable is expected to have a negative relationship with IMR. Doctor (DOC) – though there are enough physical resources available in health care sector, there should be equivalent amount of human resource (doctors) to treat them. Higher DOC may lead to reduction in IMR.

Table 1 shows the descriptive statistics of all the variable. It clearly shows the basic details like mean and standard deviation of all the variables for the year 2001 and 2011. From that it's easy to observe a few changes happened over 10 year of time. Figure 1 shows the growth rate of infant mortality from year 2001 to 2011. In that, it's clear that few states like Mizoram and Nagaland had growth rate of IMR increased over this 10 years, which is not a good sign to control infant mortality rate.

**Table 1.** Descriptive statistics of independent variable and the dependent variable

| Variables | 2001 | | 2011 | |
|---|---|---|---|---|
| | Average | Std. Dev. | Average | Std. Dev. |
| Infant Mortality Rate (per 100 birth) | 46 | 22 | 32 | 12 |
| Literacy Rate (% of Overall population) | 69 | 11 | 78 | 9 |
| Female Literacy Rate (% of Overall population) | 60 | 14 | 71 | 11 |
| Expenditure (in Crores) | 8955 | 8382 | 21052 | 24146 |
| State Domestic Profit (in Crores) | 84925 | 95686 | 158970 | 183399 |
| Female in Labour Force (in person) | 3634054 | 4398734 | 4282211 | 5237135 |
| Total Fertility Rate (in Percentage) | 3 | 1 | 2 | 1 |
| Human Development Index | 0 | 0 | 0 | 0 |
| Family Planning (in Person) | 139439 | 207945 | 130627 | 173142 |
| Total Government Hospitals (in Numbers) | 102 | 131 | 566 | 721 |
| Primary Health Centre (in Numbers) | 142 | 287 | 25 | 21 |
| Number of Doctors (in Numbers) | 1291 | 1843 | 3051 | 2993 |
| Beds in Government Hospitals (in Numbers) | 10885 | 15881 | 17963 | 21484 |

## 5 Methodology

As first step we showed the descriptive statistics of the IMR and independent variables namely, expenditure, female literacy rate, Total fertility rate and female Labour with signs of coefficients as per the theoretical expectations. In Figure 1, the growth rate of infant mortality rate from 2001 to 2011 was shown as percentage. States have shown growth in both positive and negatively. This graph will give an overview which states should be take care on higher growth rate.

First the variables are chosen based on literature review and socio economic and health infrastructure related factors which we thought will have a direct or inverse impact on infant mortality rate. As there are 11 independent variables, only few variables are used for final regression. The variables are chosen based on the correlation with the dependent variable and whether the variable is significant or not. The final

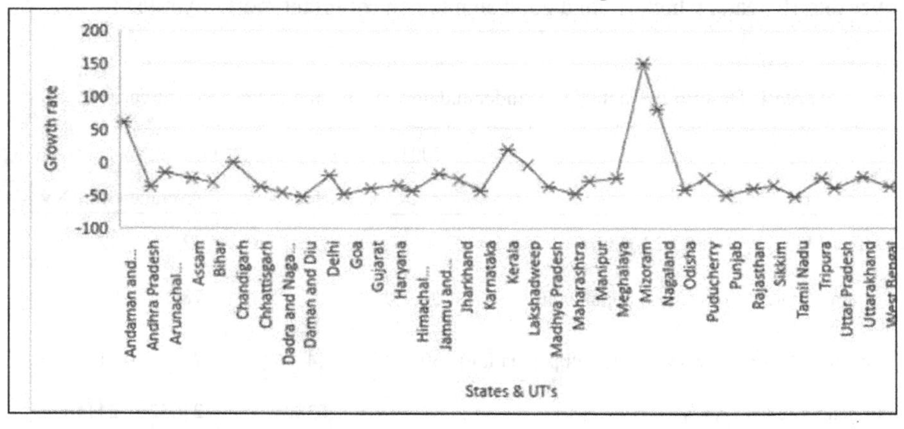


**Fig. 1.** The growth rate of Infant Mortality rate from year 2001 to 2011 across states

model is arrived by doing regression with different sets of variables both linear and log linear model. In final model we choose the four variables namely female literacy, expenditure, female labour and primary health centre which are significant when comparing to all other results with different combination of the 11 variables. The final model is linear model with the four independent variable and infant mortality rate as dependent variable.

## 6 Choice of model

To eliminate the problem of Multicollinearity, the following final linear regression model is chosen for identifying a causal relationship between IMR and a given set of independent variables

$$Y = \alpha + \beta 1x1 + \beta 2x2 + \beta 3x3 + \beta 4x4 + e_i \quad (2)$$

Where, Y = IMR, $x_1$=Female Literacy, $x_2$=Expenditure, $x_3$= Female Labour, $x_4$= PHC

, $\beta 1, \beta 2, \beta 3, \beta 4$ are respective slope coefficient of the independent variables; $e$i = random error term or stochastic variable.

## *6.1 Test for Robustness*

The R square value explains the goodness of fit of the model. Here the value of R square is reasonably high at 59 percent and Adjusted R square (adjusted with degrees of freedom) is 56.5 percent.

## *6.2 Heteroscedasticity*

This test is particularly done to find the pattern of variations in the residuals. If the variations are not uniform, then there is no heteroscedasticity. Using Breusch-Peagan test we did not find presence of hetroscadasticity in the model.

## *6.3 Auto correlation*

Similarly, we considered Durbin-Watson test for autocorrelation to identify the colinearity among the stochastic terms. The Durbin Watson value if ranges between 1.5-2.5, there is no autocorrelation. The Durbin-Watson value is 2.45022 suggesting absence of auto correlation.

## *6.4 Multicollinearity*

Multicollinearity is to find out the correlation between the independent variables. If more variables are related with each other, then it is difficult to find which independent variable is influencing the dependent variable. Multicollinearity therefore need to be removed. Variance Inflation Factor (VIF) suggest whether Multicollinearity exists or not. As a thumb rule, if VIF is <10, then there is no Multicollinearity. In this model, it suggests absence of Multicollinearity among the independent variables. Therefore, in all sense of robustness the model was identified to be a best fit model. The result for the regression is shown in Table 2. Our results show female literacy and female labour strongly explain the causal relationship with IMR with significant coefficients. This supports all earlier studies conducted with Indian data as well as other studies at global level.

**Table 2.** Regression Results for the chosen independent variable for IMR.

| Predictor | Coef | SE Coef | T | P | VIF |
|---|---|---|---|---|---|
| Constant | 92.063 | 9.478 | 9.71 | 0.000 | |
| Female Literacy | -0.8584 | 0.1344 | -6.39 | 0.000 | 1.4 |
| Expenditure | -0.0001758 | 0.0001371 | -1.28 | 0.204 | 2.9 |
| Female Labour | 0.00000119 | 0.00000060 | 1.99 | 0.051 | 3.5 |
| PHC | 0.008442 | 0.008679 | 0.97 | 0.334 | 1.4 |

**Note:** R-Sq = 59.0%, Adj R-Sq Table 2: Regression Results for the chosen independent variable for IMR.= 56.5%, Durbin – Watson Statistics = 2.45022, F stat = 23.39 (P val =0.0)

**Table 3.** Major Findings from the Study based on the growth of IMR from year 2001 to 2011 & IMR 2011

<table>
<tr><td colspan="4">Infant Mortality Rate 2011</td></tr>
<tr><td rowspan="6">Growth Rate from 2001 to 2011</td><td rowspan="3">High</td><td>High</td><td>Low</td></tr>
<tr><td>(Needs Targeting states)</td><td>(Concerned States)</td></tr>
<tr><td>Jammu and Kashmir<br>Mizoram<br>Arunachal Pradesh</td><td>Andaman and Nicobar<br>Lakshadweep<br>Delhi<br>Chandigarh<br>Nagaland</td></tr>
<tr><td rowspan="3">Low</td><td>(Acceleration States)</td><td>(Best Performing States)</td></tr>
<tr><td>Assam Chhat-tisgarh<br>Madhya Pradesh<br>Bihar<br>Odisha<br>Haryana<br>Uttar Pradesh<br>Meghalaya<br>Andhra Pradesh<br>Jharkhand<br>Rajasthan<br>Uttarakhand<br>Himachal Pradesh<br>Gujarat</td><td>West Bengal<br>Sikkim<br>Karnataka<br>Tamil Nadu<br>Dadra Nagar<br>Goa<br>Tripura<br>Puducherry<br>Punjab<br>Manipur<br>Maharashtra<br>Daman & Diu</td></tr>
</table>

Literacy rate has contributed significantly in reducing the IMR rates in developing countries. Female labour too significantly explains the causal relationship with IMR. Higher the percentage of female in labour participation higher is the infant mortality rate. However, socio economic variables like SDP, HDI (not included in the due to lack of association) and government expenditure emerge insignificant which is against the findings of some of the earlier studies [3] which explained that growth factors would automatically take care of health indicators bringing the IMR rates down.

## 7 Result and Analysis

From the analysis it is clear that female literacy rate has a huge impact on infant mortality rate. This is an overall finding. The descriptive statistics suggest that the rate of infant mortality rate has changed over these ten years, let it be increase or decrease there is an overall change in the numbers and growth rate over the ten years' gap. Cross tabulating growth rate of IMR in 2001 and 2011 shown in Table 3 and absolute IMR as two categories, we have created a matrix where states will fall in "High" or "Low" growth rate of absolute IMR. States with high IMR and growth rate are the states which needs targeting, because there is a critical issue lying in that states. Government need to look over the policy of states which are doing good and try to adopt similar policy in these states. The states which are high in growth rate and low in IMR are concerned states, where only the growth rate is the problem, which can be addressed with further better medical facilities and policy and finding out what is the root cause Female literacy can play a critical role in reduction of IMR in these states. The states with low growth rate and high IMR are the states in acceleration, where the growth rate is comparably less but IMR is high. States which are low in both growth rate and IMR are the best performing states, whose policy should be understood and replicated by other states.

## 8 Conclusion

Infant mortality rate has gained attention of academicians and policy makers as it serves as an important indicator of socio-economic development and well-being of a country. Though several factors have been effectively identified and policies have emerged at addressing those factors thereby reducing IMR, it has not yield similar results across states and regions. We have not only identified new and important variables, we also classified states into various categories that require differential attention of policies. Our study aimed at exploring determinants of infant mortality rates across the states of India in two different time points of 2001 and

2011. A comparison of IMR figures for the two census years of 2001 and 2011 for the Indian states and union territories shows that IMR figures have improved over a decade in many states of India. The average of states of IMR has reduced from 46 per 1000 in 2001 to 31 per 1000 in 2011. Year 2011 showed improved performance as compared to 2001 with some of the states showing drastic reduction in IMR figures. The highest growth in decline in 2011 from 2001 is seen for the states/union territories of Daman and Diu, Tamil Nadu, Punjab, Goa, Maharashtra, Dadra and Nagar Haveli and Karnataka. IMR in those states had declined by more than 25 percent. Mizoram, Nagaland and Andaman and Nicobar Islands shows an increase in IMR in 2011 as compared to 2001 by 60 to 150 percent. A cross sectional regression model using the ordinary least square estimation shows Female literacy and female labour strongly explain the causal relationship with IMR. This supports all earlier studies conducted with Indian data as well as cross sectional studies at global level where literacy rate seems to have contributed significantly in reducing the IMR rates in developing countries. Female labour also significantly explains the causal relationship with IMR. Higher the percentage of female in labour participation higher is the infant mortality rate. This finding needs further probing as it is contrary to the usual hypothesis. The nature of female labour participation as 'casual labourer' and in 'unorganised sector' would possibly explain why female labour participation increases IMR. However, socio economic variables like SDP, HDI (not included in the model due to lack of association) and government expenditure emerge insignificant which is against the findings of some of the earlier studies [3] which explained that growth factors would automatically take care of health indicators bringing the IMR rates down.

## References


1. Barman, Dr. Nityananda and Dipul talukdar (2014), "Socio-Demographic Factors Affecting infant mortality rate in Assam", *International Journal of Science, Environment and Technology,* 3, 1893-1900.
2. Jewell, R. Todd, Jose Martinez, Patricia Triunfo (2014), "Infant mortality in Uruguay: the effect of socioeconomic status on survival" *The Journal of Developing Areas",* 48,308-328.
3. Wagstaff A., Bustreo, F., Bryce, J., and Claeson, M., "Child Health: Reaching the Poor", *American Journal of Public Health*, 2004, 94(5), 726-736.The World Bank. 2010. World Development Indicators.
4. Riley, J.C., "Rising Life Expectancy: A Global History", Cambridge University Press, New York, 2001.
5. Houweling, T., Kunst, A., (2010), "Socio-Economic Inequalities in Childhood Mortality in Low- and Middle-Income Countries: A Review of the International Evidence", *British Medical Bulletin,* 93(1), 7-26.

6. Mosley, H.W. and Chen, L.C. (1984), "An Analytical Framework for the Study of Child Survival in Developing Countries", *Population and Development Review*, 10, 25-45.
7. Zakir, Mohammed & phanindra v. Wunnava (1999) "Factors affecting infant mortality rates: evidence from cross-sectional data" Applied Economics Letters, 6, 271-273.
8. Bhattacharya, B., Singh, K.K. and Singh, U. (1995) "Proximate determinants of fertility in Eastern Uttar Pradesh", *Human Biology*, 67, 867–86.
9. Winegarden C.R. and Bracy, P.M. (1995) "Demographic consequences of maternal-leave programs in industrial countries: evidence from fixed-effects models", *Southern Economic Journal*, 61, 1020–35.
10. Chowdhury, A.R. (1988) "The infant mortality–fertility debate: some international evidence", *Southern Economic Journal*, 54, 666–74.
11. Vos, R., J. Cuesta, M. Leon, R. Lucio and J. Rosero. (2004). "Health", *Public Expenditure Review (Ecuador).*
12. Schultz, T.P., (1993), "Mortality Decline in the Low Income World: Causes and Consequences", *Yale - Economic Growth Center*, 681.
13. Bicego, G.T. and Boerma, J.T., (1993), "Maternal Education and Child Survival: A Comparative Study of Survey Data from 17 Countries", *Social Science and Medicine*, 36(9), 1207-1227.
14. Panis, C. and Lillard, L., (1994), "Health Inputs and Child mortality: Malaysia," *Journal of Health Economics*, 13, 455-489.
15. Adebayo, S. and Fahrmeir, L., (2005), "Analysing Child Mortality in Nigeria with Geoadditive Survival Models", *Statistics in Medicine*, 24, 709-728.
16. Bhandari, Nita, Rajiv Bahl, SunitaTaneja, Jose Martines &Maharaj Bhan (2002), "Pathways to Infant Mortality in Urban Slums of Delhi, India: Implication for improving the quality of community – and Hospital-based programmes" *Centre for health and population Research*, 2, 148-155.
17. Filmer D, Pritchett LH (2001), "Estimating wealth effects without expenditure data-or tears: an application to educational enrolment's in states of India", *Demography,* 38, 115–132.
18. Henmar, L., R. Lensink, and H. White (2003), "Infant and Child Mortality in Developing Countries: Analyzing the Data for Robust Determinants," *The Journal of Development Studies*, 40(1), 101-118.

# AN AREA EFFICIENT BUILT-IN REDUNDANCY ANALYSIS FOR EMBEDDED MEMORY WITH SELECTABLE 1-D REDUNDANCY

Mr. Srirama murthy Gurugubelli[1] Mr. Darvinder Singh[2] Mr. Sadulla Shaik[3]

[1,3] Department of Electronics and Communication Engineering, VFSTR (Vignan's) University, Guntur, Andhra Pradesh, India.
**{sriramaurthy15, sadulla09}@gmail.com**
[2] INVECAS PVT.LTD, Hyderabad, Telangana, India.
**Darvinder.Singh@invecas.com**

**Abstract.** In this paper, a novel redundant mechanism for dual port embedded SRAM is presented. This work relates to 1-D (one dimensional) bit oriented redundancy algorithm to increase the reliability and yield during manufacture of memory integrated circuit chips, specifically to the efficient use of the limited space available for fuse-activated redundant circuitry on such chips, and more particularly, to replace multiple faulty memory locations using one independent redundancy element per sub-array, and reducing the number of defects-signaling fuses. In this way we double the yield and repair rate with 0.5% area penalty.

**Keywords:** Embedded SRAM, Redundancy Algorithm, 1-D bit oriented redundancy, Yield, Repair rate.

## 1 Introduction

In VLSI Industry, the memory density doubles every year by the technology scaling. Nowadays the area occupied by memory on SoC (Systems-on-chip) designs has increased to 94% [1]. For sub-micron technology and increased density of memory, these chips are more susceptible to faults. This drops the yield of the memory as well as that of SoC. From the past few years there are many redundancy techniques proposed [3]-[7] to repair embedded memories which increases yield. It is primarily done by reconfiguring the faulty cells to redundant cells [2]. The reconfiguration can be done by either bit oriented [3] or word oriented [4]-[5].

The efficiency of the redundancy analysis algorithm depends on repair rate, area overhead, and analysis time. From [3] repair rate is defined as the ratio between number of repaired memories after redundancy to the total number of faulty memories.

Repair rate depends on available spare rows or columns. There is a trade-off between repair rate and area [2]- [4]. Hence it is fixed for memory chips.



J.M. Corchado Rodriguez et al. (eds.), *Intelligent Systems Technologies and Applications 2016*, Advances in Intelligent Systems and Computing 530,
DOI 10.1007/978-3-319-47952-1_58

There are 3 possible solutions to add spare rows or columns in embedded memory.

- Spare row or spare column (1-D redundancy): The redundant elements can be either only spare rows or only spare columns [2].
- Spare row and spare columns (2-D redundancy): The redundant elements can be both spare rows and spare columns [3]-[5].
- Channel based (3-D redundancy): This is applied to memory stacks. Redundancy provided through channel, one channel can access one slice at a time [6]-[8].

The repair rates and RA (Redundancy Analysis) complexities increases from 1-D to 2-D to 3-D redundancy. From [9] redundancy algorithm proposed on DBL (Divided word line) and DWL (Divided word line) memory architecture and experimental results shows that this type of architecture has higher repair rates and improves the yield.

The rest of this paper is organized as follows. In Section II, we describe about the proposed memory block diagram and sub block level redundancy. Section III explains proposed RA algorithm. Section IV describes function of RA algorithm with example, Section V describes area over-head for implementation and in Section VI, conclusions are provided.

## 2 Proposed Memory Block Diagram

Memory core sizes increases by using DWL and DBL architecture. Here we are explaining this with an example of mXn memory array divided into 'a' column blocks and 'b' row blocks [9] is shown in Fig.1. In this each sub Array contains m/a word lines and n/b bit lines per bank. Proposed algorithm can repair maximum of '2m/a' or '2n/b' faults per bank i.e. 2 faulty cells in same column/row.

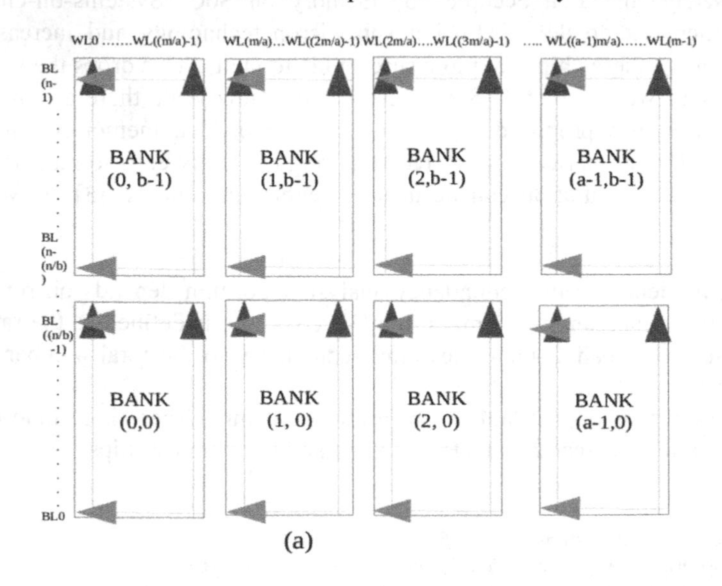


(a)

| SUB ARRAY | LOCAL IO | SUB ARRAY |
|---|---|---|
| LOCAL DECODING | LOCAL CONTROL | LOCAL DECODING |
| SUB ARRAY | LOCAL IO | SUB ARRAY |

**Fig. 1.** (a) Proposed mXn memory array. (b) Internal Blocks of each Bank.

By using power gating feature, unused banks can be powered down while reading or writing from active bank, which reduces the power consumption. Delay is significantly reduced by using two level decoding i.e. local decoding and global decoding to select the word lines and bit lines. In recent years most of papers proposed global block based redundancy (either 1-D or 2-D). In this paper sub-block level 1-D redundancy by reconfiguring the faulty bit cells to redundant bit cells by shifting the data bit by bit is proposed.

## 3 Redundancy Allocation To Sub-Array

Providing redundancy to faulty memory is a combinatorial optimization problem and complexity to find optimal solution is NP-complete, i.e. from [10] set of all instances of problem (allocation of redundancy), each instance 'I' being defined as a pair (F, c). Here 'F' is the set of feasible solutions and 'c' is cost of each element of 'F'. From past years various algorithms proposed to repair a memory by providing spare rows and columns. But physically while reconfiguring sparse faulty cells, fuses are burnt for all these fault addresses, hence RA algorithm should be cost effective and should have optimal repair rate i.e. solving a particular instance of problem $f \in F$ with minimal cost, i.e. an 'f' such that $\forall y \in F : c(f) \leq c(y)$.

Here we provide RA algorithm to cover sparse faults in the same row or column with 1-D redundancy, with minimum number of fuses and the complexity to solve the instances of a problem is degree of polynomial time, i.e. choice that will lead to a particular solution is deterministic. RA algorithm can be described with Pseudo Code in CAD for VLSI algorithm shown in Fig. 3. The list of symbols used in Pseudo code listed in Table I and i, j, k are variables.

**Table .I.** List of symbols used in Pseudo code.

| *symbol* | *description* |
|---|---|
| SFM & FM | Set of faulty memories which are repairable & Faulty memory |
| RSE<0:b-1> | Redundant segment Enable; RSE<0> for 1st segment, RSE<1> for 2nd segment.. etc. |
| RBL<0:b-1>A<br>RBL<0:b-1>B | Repaired Bit Line;<br>Store the address of IO for corresponding faulty bit cell. |

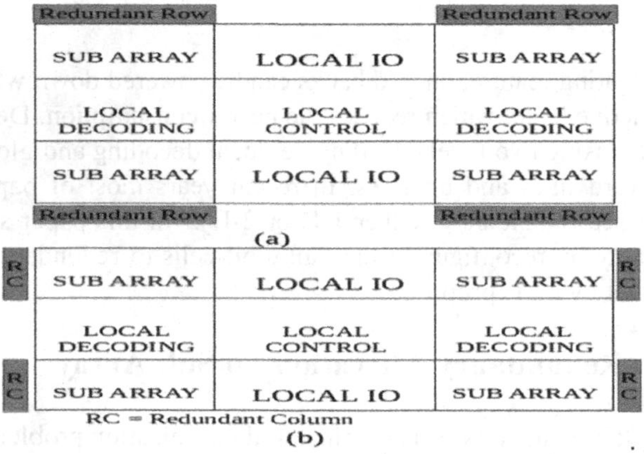

**Fig. 2.** (a) A memory array with spare Row per sub array. (b) A memory array with spare Column per sub array.

```
Foreach ( FM ∈ FSM ){

    for(i==0; i < b-1; i++)
    {
      for(j ==0; j < a-1; j++)
      {
        for(k==0; k< m/a; k++) {
       n[i][j][k]= # .of faults in bank[i][j] in column[k]
        if(n[i][j][k] != 0)
          set seg_red_enable_flag
        }
     }
     if(seg_red_enable_flag != 0)
```

```
        {
         set RSE[i]
         reconfiguration()
         }
        else
           print "there is no faulty bit cell in the bank"
       }
   }
 reconfiguration()
 {
    for(j==0; j<a-1; j++)
      {
         for(k==0; k< m/a; k++)
          {
         if (n[i][j] == 1 );i.e RBL[i]A=RBL[i]B
          {
           for(p==0; p<n/b;p++) ;each crossing kth column.
           {
             if(p< RBL[i]A) )
              allocate pth bit cell to p-1th bit ; LSB bit in the bank to Bottom spare
cell.
            else if
              allocation not require.
             }
          }
          else
          {
            for(p==0;p<n/b;p++)
            {
               if( p < RBL[i]A  )
                 allocate pth bit cell to p-1th bit cell; LSB bit in the bank to
Bottom spare cell.
                else if( p > RBL[i]B )
                  allocate pth bit cell to p+1th bit; MSB bit in the bank to Top
spare cell.
                else
                spare allocation not required.
              }
               }
           }
      }

 }
```

**Fig. 3.** Pseudo Code for Proposed 1-D Sub array level redundancy architecture.

## 4 Example

We consider a 16x16 memory and a =2, b=2 with 7 faulty bit cell in the memory shown in Fig.4(a). The corresponding spare row per sub array also shown in Fig. 4(a) and Address field notation shown Fig.4(b). The faulty Bit addresses listed in table II. The reconfiguration mechanism is shown in Fig.5(a) and (b). When 2nd WL (Word line) fires, there are 2 faulty bit cell per bank and while write/read operation, RSE[0:1] set to high, reconfiguration can be done by shifting the data in both directions. i.e. $0^{th}$ bit shifted to R0C2 position, and faulty bit "*(1,2)" allocated to $0^{th}$ position, from $2^{nd}$ to $5^{th}$ bit there is no replacement. $6^{th}$ bit to $7^{th}$ position, $7^{th}$ bit to R1C2 bit cell. Similarly for top segment also we repaired 2 faulty cells *(10,2) and *(15,2) using R2C2 and R3C3 as spare cells.

When $13^{th}$ word line fires, there is only one faulty bit cell. Reconfiguration can done by allocating the $0^{th}$ bit cells to R0C13 and from $1^{st}$ to $15^{th}$ bit there is no replacement. If $14^{th}$ WL enabled, then *(9,14) and *(14,14) repaired using R2C14 and R3C14 as spare cells.

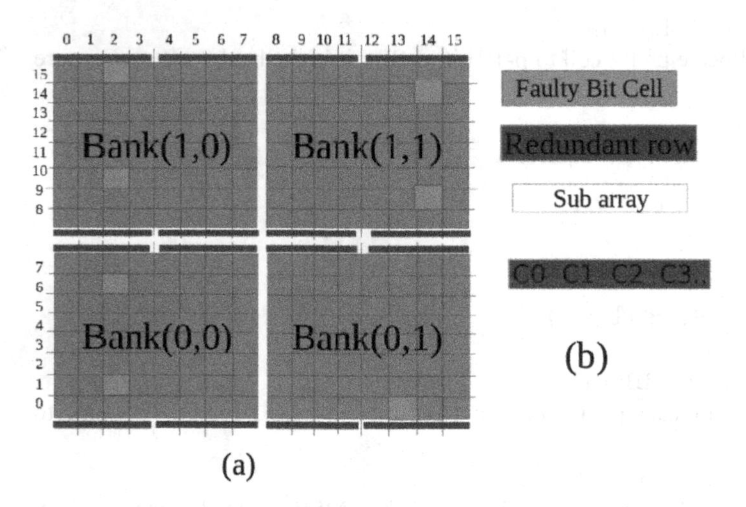


**Fig. 4.** (a) Example Faulty memory. (b) address of spare cell in redundant row.

**Table II.** Addresses of faulty bits

| S.NO | 1 | 2 | 3 | 4 | 5 | 6 | 7 |
|---|---|---|---|---|---|---|---|
| Fault Bit Address | *(1,2) | *(6,2) | *(10,2) | *(15,2) | *(0,13) | *(9,14) | *(14,14) |

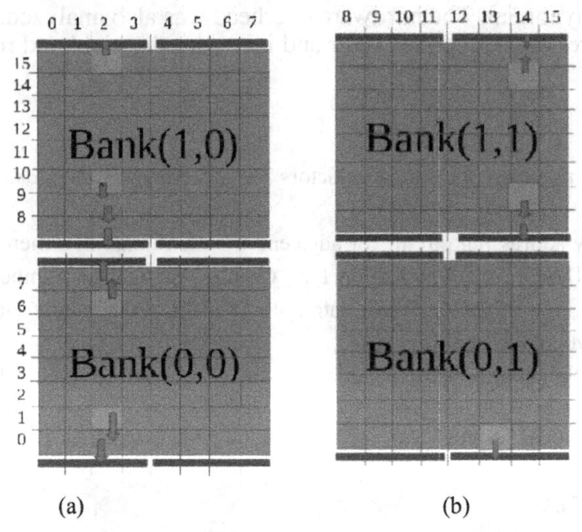


(a) (b)

**Fig. 5.** (a) Reconfiguration mechanism to repair faults S.NO 1- 4 in Table II. (b) Reconfiguration mechanism to repair faults S.NO 5-7 in Table II.

## 5 Area Overhead

The hardware overhead of memory can be hardware overhead of the bank because the bit by bit shifting needs basic one bit shifter per IO and its area is negligible. It is defined as the ratio between area of the bank with redundant rows to area of the bank. The hardware overhead for proposed redundancy architecture Fig. 2(a) and 2.(b) Formulated in equation no.1 and 2 respectively. Hardware overhead of Fig.2(a) for different sizes with various values of 'a' and 'b' with one spare row/column per sub array listed in Table III.

Area overhead per bank

For Fig.2 (a) $= 1 + \{(2*b)/n\}$.........(1)

For Fig.2 (b) $= 1 + \{(2*a)/m\}$ .........(2)

**Table III.** Area overhead per Bank

| | 1024x32 | 8192x144 | 512x1024 |
|---|---|---|---|
| a=2,b=4 | 0.25% | 0.05% | 0.0078% |
| a=4,b=4 | 0.25% | 0.05% | 0.0078% |
| A=4,b=8 | 0.5% | 0.11% | 0.015% |

## 6 CONCLUSIONS

In this paper, reconfiguration of the faulty memory (two faulty bit cells in same column per bank) cell using 1-D redundancy is proposed. Pseudo code for Redundancy Analysis algorithm also proposed. The complexity of redundancy allocation is degree of polynomial. The hardware overhead was also analyzed. The proposed algorithm ensures cost effective repair and increases the yield and repair rate.

REFERENCES

[1] "2001 technology roadmap for semiconductors" A. Allan *et al..,Computer,* Jan.2002, pages: 42 - 53.

[2] "Memory redundancy addressing circuit for adjacent columns in a memory", Publication number US5281868 A, Publication type Grant, Application number US 07/932,386, Publication date Jan 25,1994, Filing date Aug 18, 1992, Inventors Donald Morgan Original Assignee Micron Technology, Inc.

[3] "Redundancy algorithm for embedded memories with block-based architecture", Štefan Krištofík; Elena Gramatová

2013 IEEE 16th International Symposium on Design and Diagnostics of Electronic Circuits & Systems (DDECS),

Year: 2013, Pages:271-274.

[4] "A low power Built In Self Repair technique for word oriented memories", C. Banupriya; S. Chandrakala, 2014 International Conference on Electronics and Communication Systems (ICECS), Year: 2014, Pages: 1 – 5.

[4] "A BIRA algorithm for embedded memories with 2D redundancy", Shyue-Kung Lu; Yu-Cheng Tsai; Shih-Chang Huang 2005 IEEE International Workshop on Memory Technology, Design, and Testing (MTDT'05),Year: 2005, Pages: 121 – 126.

[5] "Efficient BISR Techniques for Word-Oriented Embedded Memories with Hierarchical Redundancy", Shyue-Kung Lu;

Chun-Lin Yang; Han-Wen Lin, 5th IEEE/ACIS International Conference on Computer and Information Science and 1st IEEE/ACIS International Workshop on Component-Based Software Engineering, Software Architecture and Reuse (ICIS- COMSAR'06), Year: 2006, Pages: 355 – 360.

[6] "A 3 Dimensional Built-In Self-Repair Scheme for Yield Improvement of 3 Dimensional Memories", Wooheon Kang; Changwook Lee; Hyunyul Lim; Sungho Kang, EEE Transactions on Reliability, Year: 2015, Pages: 586 – 595.

[7] "A built-in redundancy-analysis scheme for RAMs with 3D redundancy, Yi-Ju Chang; Yu-Jen Huang; Jin-Fu Li, VLSI Design, Automation and Test (VLSI-DAT), International Symposium on, Year: 2011, Pages: 1 – 4.

[8] "Redundancy Architectures for Channel-Based 3D DRAM Yield Improvement", Bing-Yang Lin; Wan-Ting Chiang; Cheng-Wen Wu; Mincent Lee; Hung-Chih Lin; hing-NenC Peng; Min-Jer Wang; 2014 International Test Conference,

Year: 2014, Pages:1-7

[9] "Efficient Built in Redundancy Analysis for Embedded Memories With 2-D Redundancy", hyue-Kung S Lu ; Yu-Chen Tsai; C. -H. Hsu; Kuo-Hua Wang; Cheng-Wen Wu, IEEE Transactions on Very Large Scale Integration (VLSI) Systems, Year: 2006, Pages:34-42.

[10] "Algorithms for VLSI Design automation by sabih H. Gerez, 3rd Edition, Wiley Publications.

# Part IV
# Applications Using Intelligent Techniques

# Performance Analysis and Implementation of Array Multiplier using various Full Adder Designs for DSP Applications: A VLSI Based Approach

Ms. Asha K A1 and Mr. Kunjan D. Shinde2

M.Tech. Student, Dept. of Electronics and Communication Engineering, PESITM, Shivamogga, India.
Assistant Professor, Dept. of Electronics and Communication Engineering, PESITM, Shivamogga, India.
Kunjan18m@gmail.com

**Abstract.** Multipliers are the significant arithmetic units which are used in various VLSI and DSP applications. Besides their crucial necessity, Multipliers are also a main source for power dissipation. Hence prior importance must be given to lessen power dissipation in order to satisfy the overall power budget for various digital circuits and systems. Multiplier performance is directly influenced by the adder cells employed, for multipliers designed using adders; therefore power dissipation problem can be solved by exploring and using better adder designs. In this paper various full adder designs are analyzed in terms of delay, power consumption and area, As the adder block is prime concern for array multiplier in order to propose an efficient Multiplier architecture. The design and implementation of full adder cells and multiplier is performed on CADENCE design suite at GPDK 180nm technology. The CMOS, GDI and Optimized full adder design is employed to implement array multiplier.

**Key words**: Array multiplier for DSP applications, Array multiplier using Full adders, Full adder, CADENCE design suite, Power, Delay and Area (Gate count).

## 1 Introduction

The fundamental units for any multiplier are the adders which are used for arithmetic functions like addition and multiplication. Computations must be performed using low-power, area-efficient adders operating at greater speed for efficient output. These energy efficient circuits are required in order to cope for the increasingly stringent demands for battery space and weight aspects in portable multimedia particularly in digital multipliers which are the basic building blocks of Digital Signal Processors. A basic multiplier can be divided into three main parts- partial product generation, partial product addition and final addition [2].



J.M. Corchado Rodriguez et al. (eds.), *Intelligent Systems Technologies and Applications 2016*, Advances in Intelligent Systems and Computing 530,
DOI 10.1007/978-3-319-47952-1_59

Binary adders are the most efficient logic elements within a digital system. Computers and processors use adder or summer in order to carry out addition operation. Any up gradation in binary adder's efficiency can boost the performance of entire system. In this paper various full adders are designed and analyzed for 4-bit and 8-bit array multiplier. The results obtained will provide required solutions by comparing the performance difference between the Array multiplier in terms of power dissipation, delay and area.

The rest of the paper is organized as section II provides literature survey carried out on various Array multipliers. Section III gives information regarding power consumption in CMOS circuits. Section IV gives knowledge about the conventional Array multiplier and its working. Section V consists of details regarding different types of 1-bit array multiplier used to analyze the behavior of Array multiplier. Section VI is about results and discussion regarding full adder and array multiplier simulations and its comparative analysis.

## 2 Literature review

The following are the somc of the papers that we have gone through and has provided the insight of current research in the relevant domain. In [1] the authors have designed and implemented 1-bit full adder, performance analysis is carried out for the design styles CMOS, GDI, TG, and GDI-PTL. CADENCE tool is used for design and implementation with GPDK 45nm which differs from GPDK 180nm technology used in this paper and some of the second order effects can be neglected here. In [2] authors have designed 4x4 array and tree multiplier using 0.18um technology, the multiplier is synthesized with 3.3v power supply. In [3] comparative study of different types of multipliers is done for low-power requirements and high speed. For comparison FPGA implemented results are considered. In [4] The conventional Array multiplier is synthesized with 16T,8T and 10T and carry bits in the final stage is not added using Ripple carry adder in final stage but it is given as input to the next left column. In [5] Conventional Array multiplier is designed in 32nm MOSFET technology using HSPICE. In [6] the power consumption equations for digital CMOS circuits are considered to cognize the power dissipation and delay in CMOS circuits. In [7] the equations for various 1-bit full adder designs are given, this gives a brief idea to understand different ways for designing 1-bit full adders. In [8] the GDI based full adder circuits is presented for low power application.

## 3 Power consumption in VLSI circuits

The primary power consumption components are switching and short circuit power are referred as dynamic power because it is consumed dynamically while the circuit is changing states i.e., when the circuit is operating. This type of power consumption contributes to the majority of the total power consumption in VLSI circuits in nanometer scale.

The three main components are:

i. **Switching power consumption**:

The power consumed while charging and discharging of the circuit capacitance during transistor switching[5].

ii. **Short circuit power consumption**:

It is the power consumed due to the short-circuit current flowing from power supply to ground during transistor switching

iii. **Static power**:

Static power consumption occurs when all the inputs are held at some logic level and the circuit is no more in the charging state. This type of dissipation becomes a problem when the circuit is in off state or is in power-down mode [4].

These power consumptions are mathematically equated as:

$$P_{total}= P_{dynamic} + P_{static} \quad (1)$$

$$P_{tr}= C_L*V_{DD}^2*(\text{clock frequency}) \quad (2)$$

Where, $P_{tr}$ = transient power consumption

$V_{DD}$ = Supply voltage

$P_{total}$ = total power consumption

$C_L$ = Load Capacitance

Leakage current $I_{leakage}$ occurs from substrate injection and sub-threshold effect. It is given by:

$$I_{leakage}= I_s*e(qv/kT-1) \quad (3)$$

Where, $I_s$= reverse saturation current

v=diode voltage

k= Boltzmann's constant($1.38*10^{-23}$ J/K)

Delay time is calculated using:

$$T_d= ((\text{clock frequency})*V_{DD})/(K*(V_{DD}-V_{th})^a \quad (4)$$

Where, K= transistors aspect ratio (W/L)

$V_{th}$= Transistor threshold voltage

a = velocity saturation index which varies between 1 and 2

The power and delay parameter are most important criteria in selection of any given design for the module and application, the tool CADENCE provides the analysis of delay and power computation with its internal algorithm for analysis, equations (1) to (4) provides the bases for understanding the performance metrics and the results are coated with reference to the values provided by the tool.

## 4 Conventional Array multiplier

Array multiplier is very regular in structure and is easier to design. Array multiplier is used for multiplication of unsigned numbers by using full adders and half adders. Full adders and half adders are connected in horizontally, vertically and diagonally to obtain sum of the partial products. If the first row of the partial products obtained is implemented using full adders, Cin will be considered as '0'. Partial products are aligned properly by simple routing and it does require any logic. Each row of adders adds a partial product to the sum, generating a new partial sum and a sequence of carriers. A nxn array multiplier requires n (n-2) full adders, n half adders and $n^2$ AND gates [3]. The delay associated with array multiplier is the time taken by the signals to propagate through each AND gates, full and half adders. Disadvantage of an array multiplier is that they are very large in size. As number of operands increases, linear array grow in size at a rate equal to the square of the operand size.

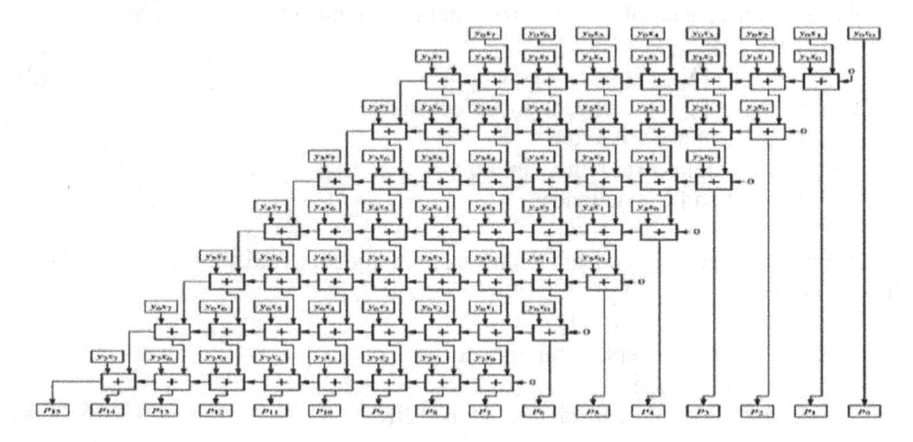

Figure 1: Block diagram representation of 8bit conventional Array multiplier

**Full adders Design**

Full adders are used to perform addition of two single bit binary numbers with a additional input signal as $C_{in}$. When an 1-bit A, 1-bit B and Cin is given as input to a full adder it performs addition to give a two bit output Sum and Cout. The input carry Cin is used for carry propagation. The expression for sum and carry as derived using its behavior is given in equation (5) and (6).

$$\text{Sum= A XOR B XOR Cin} \quad (5)$$
$$\text{Cout= (A AND B) OR (B AND Cin) OR (Cin AND A)} \quad (6)$$

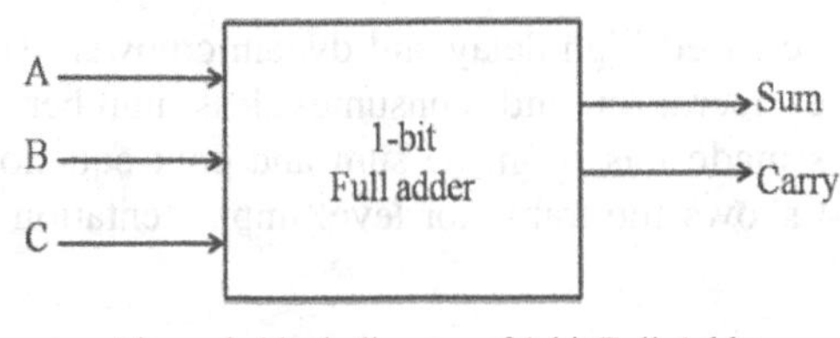


Figure 2: block diagram of 1-bit Full Adder

Table 1:Truth table of 1-bit Full adder

| A | B | Cin | Sum | Cout |
|---|---|---|---|---|
| 0 | 0 | 0 | 0 | 0 |
| 0 | 0 | 1 | 1 | 0 |
| 0 | 1 | 0 | 1 | 0 |
| 0 | 1 | 1 | 0 | 1 |
| 1 | 0 | 0 | 1 | 0 |
| 1 | 0 | 1 | 0 | 1 |
| 1 | 1 | 0 | 0 | 1 |
| 1 | 1 | 1 | 1 | 1 |

**10T full adder (GDI design style)**

In this method the 1-bit full adder design type is obtained from Sum output from the second stage of XOR circuit and the carry bit (Cout) output is calculated by multiplexing B and Cin. These structure style gives GDI based designed cell two extra input pins to use making GDI style flexible over CMOS design. GDI cell is expensive since it requires twin-well CMOS or Silicon On Insulator (SOI) process to realize. The drawback of GDI design style is degradation in the output signal which is highlighted in the simulation waveform. The number of transistor counts in this full adder 10 transistors. A,B, Cin are taken as inputs and the output of the circuit is taken from the Sum and Carry-out(Cout). From figure 3 we have transistors numbered as 1,3,5,7,9 are PMOS transistors whereas transistors numbered as 2,4,6,8,10 are NMOS transistors.

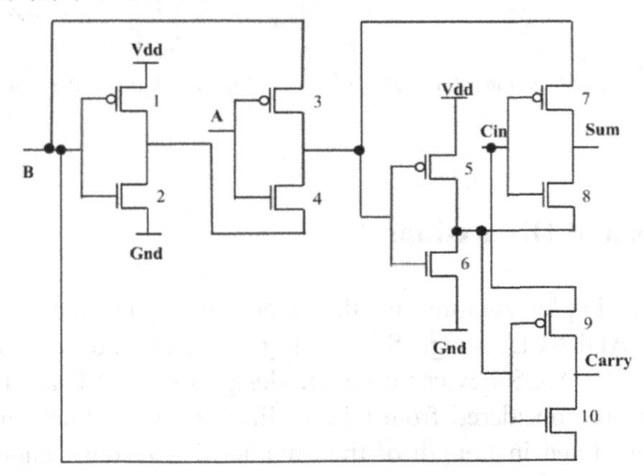


Figure 3 :Circuit diagram of 10T GDI design style

**28T full adder design (CMOS style):**

The conventional full adder based design is obtained using standard CMOS topology consists of pull-up and pull-down transistor offering full-swing output with good driving capabilities. Due to the increased number of transistors, power consumption is high and also use of PMOS transistor in pull-up network results in high input capacitances leading to increased high delay and dynamic power. This design offers reliable low voltage operation and consumes less number of transistors as this implementation is made based on the sum and cout equations provided in (1) and (2) and figure 4 shows the transistor level implementation of the same.

**54T full adder design (CMOS style):**

This is another type of realizing a full adder. CMOS design style is used for designing this 1-bit full adder. The sum output is generated in two level gate delay of XOR gate and Cout output is generated in three level gate delays of AND-OR gate logic, the delay of each gate is different with different logic used to design the given logic.

The circuit diagram is shown below:

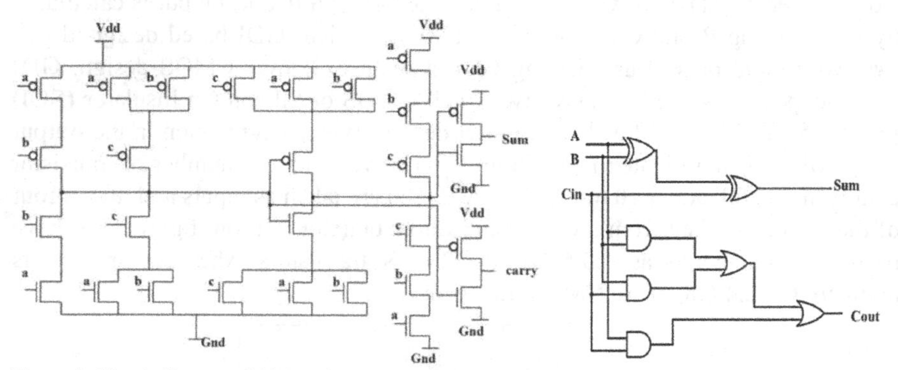


Figure 4: Circuit diagram of 28Transistor GDI

Figure 5: Circuit diagram of 54T full adder design

## 5 Results and Discussions

The design and implementation of the array multiplier using full adder is performed on CADENCE Design Suite using Virtuoso and ADE for schematic and simulations, the MOS devices used for design of AND Gate, full adder and array multiplier are considered from GPDK library with 180nm technology, the MOS devices are fixed in Length of the channel due to the technology and we have not varied the width parameter, which results in the easy of fabrication of the multiplier and helps for analysis of the same. To observer the behavior of full adder, we have applied the entire possible test patterns to obtain Sum and Cout results for each case. With the adders is used in the array multiplier architecture and the performance of multiplier is observed based on the worst case inputs applied to the multiplier, the worst case input to multiplier is one/many input cases which effectively makes all the blocks of the multiplier to function and results in the worst case delay, such a input patter is A=11111111 and B=11111111; where A and B are the inputs each of 8-bits wide and the resultant of the multiplier is product of these two signals P=1111111000000001.

A. Simulation results

The following are the schematic and simulation result of 1-bit full adder designed using CMOS, GDI and 10Transistor logic and the results of array multiplier of 4-bit and 8-bit precession.

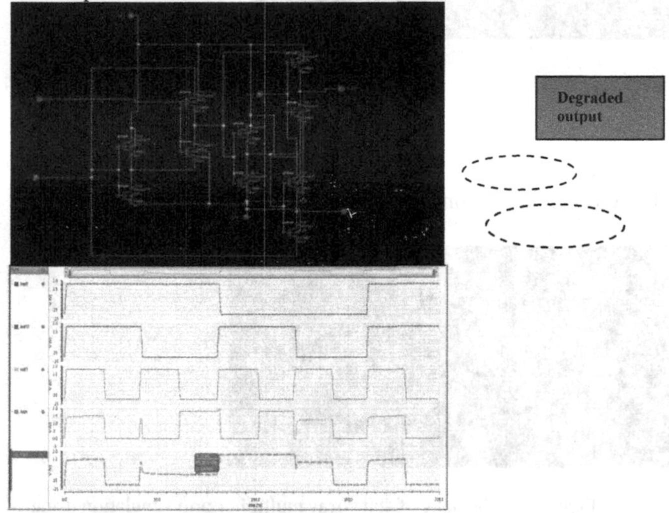


Figure 6: Schematic and simulation of 10Transistor full adder

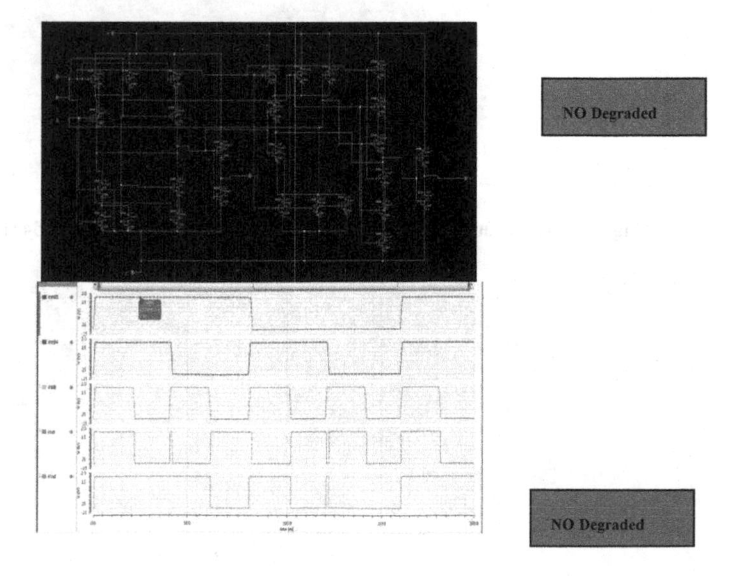

Figure 7 : Schematic and simulation of 28transistor full adder

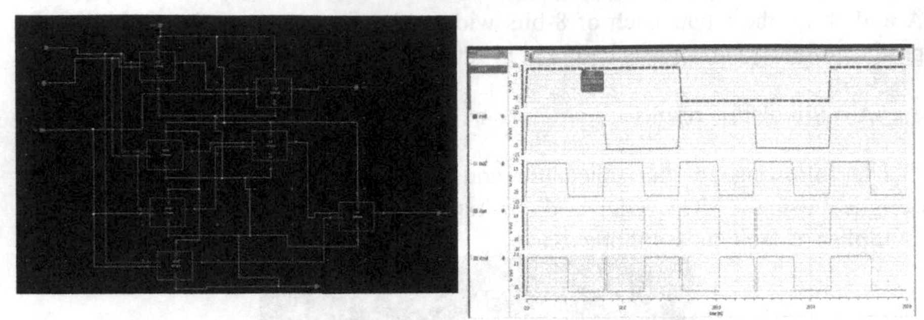

Figure 8: Schematic and simulation of 54Transistor full adder

Simulation results of Array Multiplier using various full adders mentioned above (schematic of only one type is given).

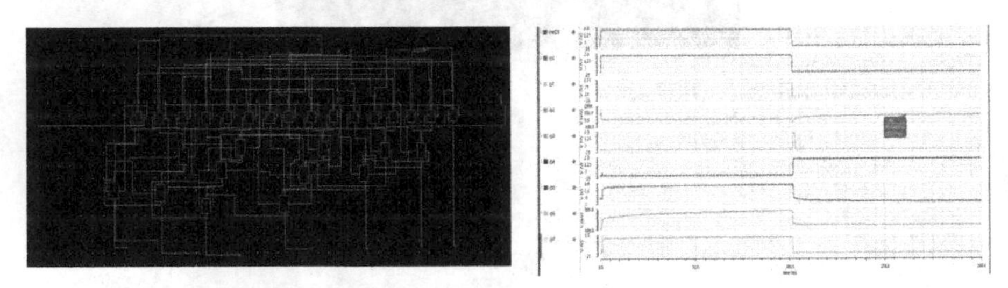

Figure 9 : Schematic of 4bit Array multiplier and simulation using 10Transistor full adder

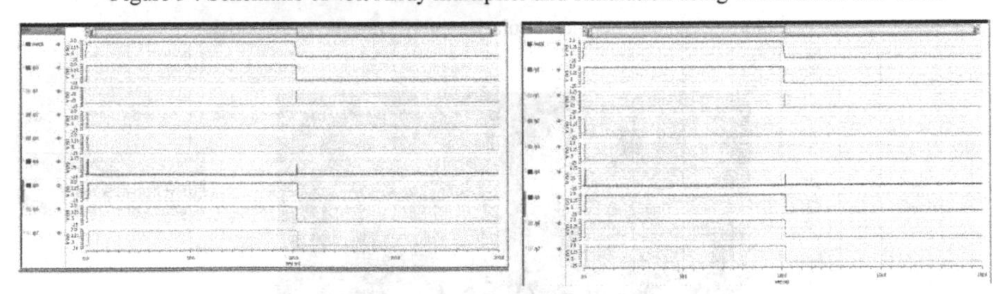

Figure 10 : Simulation of 4bit array multiplier using 28Transistor and 54Transistor full adder

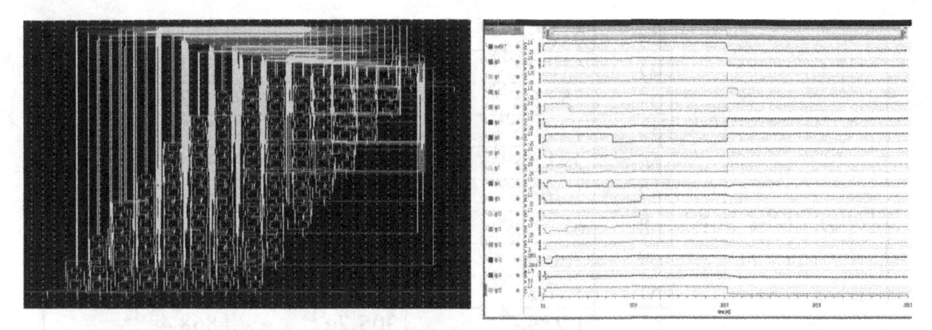

Figure 11: Schematic of 8bit Array multiplier and its simulation using 10Transistor Full adder

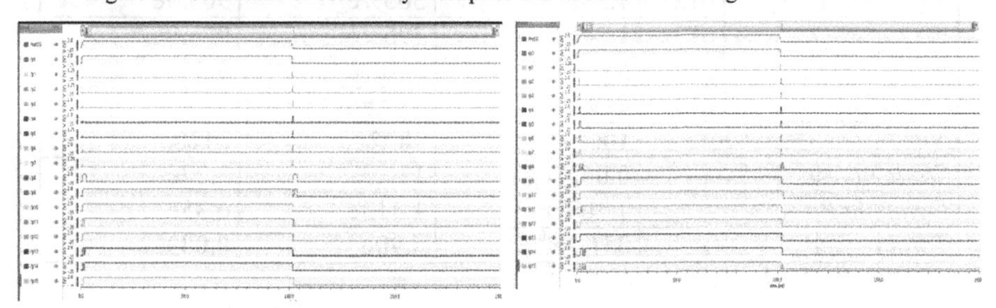

Figure 12: Simulation of 8bit Array Multiplier using 28Transistor and 54transistor Full adder

## B. Comparative Analysis

Table 2. Performance analysis of 4-bit Array Multiplier using various full adders

| *4-bit Array Multiplier* | | *10t full adders* | *28t full adders* | *54t full adders* |
|---|---|---|---|---|
| *Delay in Seconds* | p0 | $148.7e^{-12}$ | $148.7e^{-12}$ | $148.7e^{-12}$ |
| | p1 | $773.4e^{-12}$ | $305.5e^{-12}$ | $335.6e^{-12}$ |
| | p2 | $773.4e^{-12}$ | $305.5e^{-12}$ | $335.6e^{-12}$ |
| | p3 | $773.4e^{-12}$ | $309.5^{-12}$ | $572.2e^{-12}$ |
| | p4 | $773.4e^{-12}$ | $495.8^{-12}$ | $849.0e^{-12}$ |
| | p5 | $773.4e^{-12}$ | $646.2e^{-12}$ | $934.2e^{-12}$ |
| | p6 | **$884.4e^{-12}$** | $305.7e^{-12}$ | $1.081e^{-9}$ |
| | p7 | $488.5e^{-12}$ | **$891.1e^{-12}$** | **$570.2e^{-9}$** |
| *Number of Transistor* | | **248** | **392** | **600** |
| *Static power* | | $129.9e^{-6}$ | $24.04^{-6}$ | $38.71^{-6}$ |
| *Dynamic Power* | | $1.906e^{-4}$ | $2.399e^{-5}$ | $2.469e^{-5}$ |
| *Total power* | | **0.0003205** | **$4.90e^{-8}$** | **$6.34e^{-5}$** |

Table 3. Performance analysis of 8-bit Array Multiplier using various full adders

| *8-bit Array Multiplier* | | *10t full adders* | *28t full adders* | *54t full adders* |
|---|---|---|---|---|
| | p0 | $148.7e^{-12}$ | $148.7e^{-12}$ | $148.7e^{-12}$ |
| | p1 | $612.7e^{-12}$ | $305.7e^{-12}$ | $101.2e^{-9}$ |

| | | | | |
|---|---|---|---|---|
| *Delay in Seconds* | p2 | $612.7e^{-12}$ | $305.7e^{-12}$ | $347.6e^{-12}$ |
| | p3 | $612.7e^{-12}$ | $305.7e^{-12}$ | $347.6e^{-12}$ |
| | p4 | $753.9e^{-12}$ | $305.7e^{-12}$ | $347.6e^{-12}$ |
| | p5 | $1.089e^{-9}$ | $305.7e^{-12}$ | $347.6e^{-12}$ |
| | p6 | $284.5e^{-12}$ | $305.7e^{-12}$ | $334.7e^{-12}$ |
| | p7 | $1.864e^{-9}$ | $305.7e^{-12}$ | $334.7e^{-12}$ |
| | p8 | $631.5e^{-12}$ | $305.7e^{-12}$ | $800.7e^{-12}$ |
| | p9 | $760.4e^{-12}$ | $305.7e^{-12}$ | $800.7e^{-12}$ |
| | p10 | $327.5e^{-12}$ | $305.7e^{-12}$ | $1.511e^{-9}$ |
| | p11 | **$1.289e^{-9}$** | $305.7e^{-12}$ | $868.4e^{-12}$ |
| | p12 | $1.54e^{-9}$ | $305.7e^{-12}$ | $903.9e^{-12}$ |
| | p13 | $1.74e^{-9}$ | $305.7e^{-12}$ | **$975.9e^{-9}$** |
| | p14 | $553.4e^{-12}$ | $309.7e^{-12}$ | $3.225e^{-9}$ |
| | p15 | $419.8e^{-12}$ | **$815.8e^{-12}$** | $559.8e^{-12}$ |
| *Number of Transistor* | | **1008** | **1872** | **3120** |
| *Static power* | | 0.001761 | $131.3^{-6}$ | $3.27e^{-3}$ |
| *Dynamic Power* | | 1.4095e-3 | $1.313e^{-4}$ | $6.35e^{-3}$ |
| *Total power* | | **$351.5e^{-6}$** | **$4.90e^{-8}$** | **0.00962** |

It is clear from the table 2 and 3 the array multiplier consumes less number of transistors when it is designed using full adder with 10Transistors, this design might be chosen when area is the prime concern and a small amount of increase in delay and power is acceptable compared with 28Transistor full adder design based array multiplier with precision of multiplier as 8-bit and for 4-bit multiplier architecture 10Transistor based full adder is effective with Number of transistors and delay parameter.

## 6 Conclusion

This paper effectively shows the Analysis and simulation of 4-bit and 8-bit Array multiplier using 10Transistor, 28Transistor, 54Transistor full adder. The results quoted can be utilized while designing Array Multipliers. The 10T 1-bit full adder, designed with GDI style is much preferable when the area and delay is prime criteria and well suited for systolic array based applications. The degradation in output voltage clearly is one of the disadvantages which can be overcome by varying the width and length of the MOS devices. However in this paper we have not varied the width and length of the MOS devices for simplicity and to observe the response when all MOS devices have same width and length.

## 7 Acknowledgement

The Authors would like to thank the management, the Principal, and the department of Electronics and Communication Engineering of PESITM, Shivamogga for all the support and resources provided.

## References

[1] Kunjan D. Shinde, Jayashree C. Nidagundi, "Design of fast and efficient 1-bit full adder and its performance Analysis", International conference on control, Instrumentation, Communication and computational technologies (ICCICCT)
[2] Neha Maheshwari, "A design of 4x4 multiplier using 0.18um technology", International Journal of Latest Trends in Engineering and Technology (IJLTET)
[3] Sumit Vaidya, Deepak Dandekar, " Delay-power performance comparison of Multipliers in VLSI circuit design", International Journal of Computer Networks & Communications (IJCNC), Vol.2, No.4, July 2010
[4] Elancheran.J , R.Poovendran, " Implementation of 8T full adder in Array Multiplier", International Journal of Advanced Technology in Engineering and Science, Volume No.03, Issue No. 03, March 2015
[5] Kripa Mathew,S.Asha Latha,T.Ravi, E.Logashanmugam, " Design and Analysis of an Array Multiplier using and Area Efficient Full adder cell in 32nm CMOS technology", The International Journal Of Engineering And Science (Ijes) ,Volume 2 Issue 3 Pages 8-16 2013.
[6] Shaloo Yadav, Kavita Chauhan, "Study and Analysis of various types of Full adder's scheme for 250nm CMOS technology",International Journal of Electrical, Electronics and Computer Engineering, ISSN No. (Online): 2277-2626.
[7] Sardindu Panda, A.Banerjee, Dr. A.K Mukhopadhyay, "Power and delay comparison in between different types of Full Adder circuits", International Journal of Advanced Research in Electrical, Electronics and Instrumentation Engineering, Vol. 1, Issue 3,September 2012.
[8] Pankaj Kumar, Poonam Yadav, "Design and Analysis of GDI Based Full Adder Circuit for Low Power Applications", International Journal of Engineering Research and Applications, ISSN : 2248-9622, Vol. 4, Issue 3( Version 1), March 2014.

### Authors

**Ms. Asha K. A.** is pursuing M.Tech in Digital Electronics at Department of E&CE from PESITM, Shivamogga, she Received her B.E. Degree in Electrical and Electronics Engineering from

JNNCE, Shivamogga in 2014, her area of interest include VLSI and Embedded system design.

**Mr. Kunjan D. Shinde** is with PESITM Shivamogga, working as Assistant Professor in Dept. of Electronics and Communication Engineering; He received Master's Degree in Digital Electronics from SDMCET Dharwad in 2014, and received Bachelor Degree in Electronics & Communications Engineering from SDMCET Dharwad in 2012. His research interests include VLSI (Digital and Analog design), Error Control Coding, Embedded Systems, Communication Networks, Robotics and Digital system design.

# Automatic Agriculture Spraying Robot with Smart Decision Making

Sonal Sharma and Rushikesh Borse

**Abstract** The responsibility of controlling and managing the plant growth from early stage to mature harvest stage involves monitoring and identification of plant diseases, controlled irrigation and controlled use of fertilizers and pesticides. The proposed work explores the technology of wireless sensors for remote real time monitoring of vital farm parameters like humidity, environmental temperature and moisture content of the soil. We also employ the technique of image processing for vision based automatic disease detection on plant leaves. Thus this paper vigorously describes the design and construction of an autonomous mobile robot featuring plant disease detection, growth monitoring and spraying mechanism for pesticide, fertilizer and water to apply in agriculture or plant nursery. To realize this work we provide a compact, portable and a well founded platform that can survey the farmland automatically and also can identify disease and can examine the growth of the plant and accordingly spray pesticide, fertilizer and water to the plant. This approach will help farmers make right decisions by providing real-time information about the plant and it's environment using fundamental principles of Internet, Sensor's technology and Image processing.

## 1 INTRODUCTION

The main business of Indian people is agriculture and the economy of the nation is decided by agriculture. The essential nutrients for plant growth are commonly generates in its surroundings. The plant development process depends on the conditions of the environment, where plant grows.



J.M. Corchado Rodriguez et al. (eds.), *Intelligent Systems Technologies and Applications 2016*, Advances in Intelligent Systems and Computing 530,
DOI 10.1007/978-3-319-47952-1_60

The plant development process depends on the conditions of the environment, where plant grows. The necessary parameters like, humidity, light, moisture, ambient temperature and $CO_2$ etc. are consists in the environment. Deep understanding of all these factors and their relationships can help the farmer to get much familiar with any of the potential problems that will affect the health of the plants and thereby more appropriate and accurate measures can be taken to get rid of these problems [5].

Temperature is one of the factor that influences the plant development process the most. Each plant species presents a disparate temperature range within which they can grow normally. Above of this range, the processes required for plant's life stop as the enzymes become inactive. The loss of moisture from the plant is controlled by the humidity. As humidity increases the plant's development process will also get affected because fungal diseases will spread rapidly and air will also become saturated with vapour which will than restrict transpiration [5]. The productivity of the crop is also affected by other major biological parameters such as pests, disease and soil. Soil moisture is the key parameter which can be used to determine the right time to irrigate and right amount of water to supply. These parameters can be controlled by human beings for improvising the production of crop [14]. Robots are increasingly taking over many operations from humans where repeatability in routine task and precision in work are required and also the situation where human operators are exposed to risk [10]. One of these tasks involves the raising of crops in greenhouse, where most of the operations on the crop are still performed manually by the human operators even though they are often highly perpetual [10].The water content information of soil can be used to automate the irrigation systems by using dielectric moisture sensors and accordingly control the actuators for water, rather then having a fixed irrigation schedule with specific duration [3]. Autonomous mobile robots are now relegated as a part Precision Agriculture for optimizing field management task of agriculture in order to enhance environment supervision, crop perception and economics [11]. The proposed work explores the technology of wireless sensors for remote real time monitoring of vital farm parameters like humidity, environmental temperature and moisture content of the soil. We also employ the technique of image processing for vision based automatic disease detection on plant leaves. To realize this work we provide a compact, portable and a well founded platform that can survey the farmland automatically and also can identify disease and can examine the growth of the plant and accordingly spray pesticide, fertilizer and water to the plant. This approach will help farmers make right decisions.

The plant development process depends on the conditions of the environment, where plant grows. The necessary parameters like, humidity, light, moisture, ambient temperature and $CO_2$ etc. are consists in the environment. Deep understanding of all these factors and their relationships can help the farmer to get much familiar with any of the potential problems that will affect the health of the plants and thereby more appropriate and accurate measures can be taken to get rid of these problems [5].

---

Sonal Sharma,
Sinhgad Academy of Engineering, Pune, India, e-mail:sonalsharma7381@gmail.com

Rushikesh Borse
Sinhgad Academy of Engineering, Pune, India, e-mail: rpborse.sae@sinhgad.edu

Temperature is one of the factor that influences the plant development process the most. Each plant species presents a disparate temperature range within which they can grow normally. Above of this range, the processes required for plant's life stop as the enzymes become inactive. The loss of moisture from the plant is controlled by the humidity. As humidity increases the plant's development process will also get affected because fungal diseases will spread rapidly and air will also become saturated with vapour which will than restrict transpiration [5]. The productivity of the crop is also affected by other major biological parameters such as pests, disease and soil. Soil moisture is the key parameter which can be used to determine the right time to irrigate and right amount of water to supply. These parameters can be controlled by human beings for improvising the production of crop [14]. Robots are increasingly taking over many operations from humans where repeatability in routine task and precision in work are required and also the situation where human operators are exposed to risk [10]. One of these tasks involves the raising of crops in greenhouse, where most of the operations on the crop are still performed manually by the human operators even though they are often highly perpetual [10].The water content information of soil can be used to automate the irrigation systems by using dielectric moisture sensors and accordingly control the actuators for water, rather then having a fixed irrigation schedule with specific duration [3]. Autonomous mobile robots are now relegated as a part Precision Agriculture for optimizing field management task of agriculture in order to enhance environment supervision, crop perception and economics [11]. The proposed work explores the technology of wireless sensors for remote real time monitoring of vital farm parameters like humidity, environmental temperature and moisture content of the soil. We also employ the technique of image processing for vision based automatic disease detection on plant leaves. To realize this work we provide a compact, portable and a well founded platform that can survey the farmland automatically and also can identify disease and can examine the growth of the plant and accordingly spray pesticide, fertilizer and water to the plant. This approach will help farmers make right decisions.

Further, the paper has been described as given here; In section 2, we have discussed about the previous work in the similar domain. Section 3 describes block diagram of the system that has been proposed. Section 4 describes the overall construction of agriculture vision system. This section includes the hardware portion and software constrains of the system along with their respective results and at last in section 5 we have drawn the conclusion.

## 2 Related Works

For the crop quality management the detection of plant diseases and their extent have always been a major concern in agriculture field. Till date, many researches have already been done in the same problem area. Many disease detection systems based on image processing techniques have been proposed earlier. Some already developed systems area is explained below:

Van Li et al. [12] have also provided a measurement method for pest detection on leaves. This system enables automatic spraying of pesticide to the pest's position on the leaves by using binocular stereo vision system. Sai Kirthi Pilli et al.[9] has given AI based embedded algorithms that pick regions of interest and perform color transformations to examine type and extend of diseases, the sprayer mechanism that sprays the requisite pesticide. A Camaro et al. [1] has shown the processing algorithm that involves conversion of the RGB image of diseased leaf of the plant, into another color transformations H, I3a and I3b. R. Pydipati et al. [8] has shown research that used the color co-occurrence method (CCM) to check whether texture based hue, saturation, and intensity (HSI) color features together with statistical classification algorithms could be used to identify diseased and normal citrus leaves under laboratory conditions. Tao Liu et al. [11] offered a Navigation system mainly includes controllers, electromagnetic sensors for spraying pesticide in green house environment. An autonomous pesticide spraying device would be of the utmost importance in avoidance of human exposure to the hazardous chemicals and to ensure that the optimal amount of spray is applied. In the same context irrigation systems are also required for precise use of water; one such solution has given by Joaqun Gutirrez [3]. Some other systems and work regarding smart farming have also been seen which are listed in the table 1.

Table 1 List of existing systems and techniques they employ

| Ref no. | Year | Author name | Subject | Proposed concept/innovation |
|---|---|---|---|---|
| [9]. . | 2015 | SaiKirthi Pilli et al. | Crop disease detection using image processing | It provides a small platform to detect diseases and spray pesticide automatically. Testing has been done with cotton and groundnut plantations. |
| [7]. | 2014 | PengJian-sheng | An Intelligent Robot System for Spraying Pesticides | It proposes the design of an intelligent wireless controlled robot for spraying pesticides. For monitoring the core micro controller, a wireless router for the network connection point, and a camera for capturing video is employed. |
| [3]. | 2014 | Joaquin Gutierrez et al. | Automatic Irrigation System Using a WSN and GPRS Module | An automated irrigation system was proposed to optimize use of water for agricultural crops. The system has a Distributed wireless network of temperature and soil moisture sensors. Gateway unit handles sensor information, triggers the actuators. |

| | | | | |
|---|---|---|---|---|
| [4]. | 2012 | K. Prema et al | Fuzzy Controller and Virtual Instrumentation based remotely operated agricultural Robot | An Agricultural Robot is treated as server system with internet connection and assigned static IP address. The client system can be a PC with internet connection |
| [1]. | 2008-2009 | A. Camaro et al. | Automatic identification of plant disease based on image processing algorithm | This study narrates an image processing based method that identifies all visual symptoms of plant diseases, from an analysis of the colored images |
| [12]. | 2009 | Van Li et al. | Automatic Pest Detection and Spray in Greenhouse | A depth measurement method for pests identification on leaves with a binocular stereo vision system. |
| [8]. | 2002 | R. Pydipati et al. | Citrus disease identification with the help of color texture features analysis | The leaf sample discriminant analysis by using CCM textural features was proposed. It achieved classification accuracies of over 95 percent for all classes while using hue and saturation texture features. |
| [6]. | 2002 | Naiqian Zhang et al. | Precision agriculture -a worldwide overview | This paper provides an overview of worldwide development and current status of precision agriculture Technique |

## 3 Proposed System

Figure 1 displays the block diagram of the proposed system. Each block shows an important element of the system. As can be seen ARM7 is used as the control unit. Various sensors also have incorporated for environment monitoring. Three water motors are included for spraying purpose of each liquid.

Web cam has been included which enables the vision capability of the system. Zigbee and GSM are used for wireless communication purpose.

# 4 Design Considerations of the System

In the present situation any system design is incomplete without a software and hardware combination. The system proposed in section 3 is detailed in further two subsections.

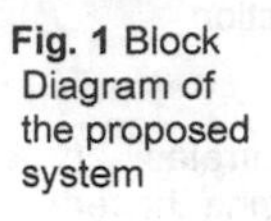
**Fig. 1** Block Diagram of the proposed system

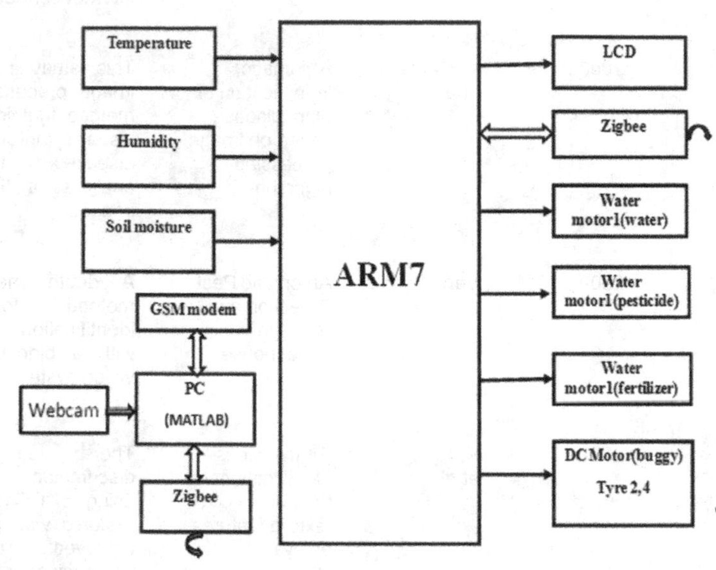


## *4.1 Hardware Details*

The hardware part consists of a microcontroller ARM7, three sensors (Humidity sensor, temperature sensor and a soil moisture sensor), as shown in figure it has three container of 500 ml for spraying of pesticide, fertilizer and water. Two dc motor are required to physically run the robotic buggy. A webcam is used for capturing the images of plant through MATLAB Command, in WI-FI environment. Along with it system has GSM modem for sending SMS to farmer and personal computer with MATLAB software

### 4.1.1 System control section

The system's controlling part is composed by LPC2138 microcontroller as the core controller. The LPC2138 microcontrollers are based on a 32/16 bit ARM7TDMI-S CPU with real-time emulation and embedded trace support. It has 64 pins.

### 4.1.2 Wireless Sensor Unit

The sensor unit is made up of a RF transceiver, sensors, a microcontroller, and power sources. Here the wireless sensor unit is based on the microcontroller LPC2138 that controls radio modem ZigBee and processes information from the soil moisture sensor IC LM358, Temperature sensor LM35 and humidity sensor (Precision Centigrade Sensors).

### 4.1.3 Driver Modules

DC motors are used to physically run the application as per the requirements configured in software. The dc motor operates on 12v. To drive a dc motor, it requires a dc motor driver called L293D.

### 4.1.4 Trolley Modules

Trolley module is a carrier trolley that carries the liquid tanks and the entire hardware system. Our main challenge was to design an adjustable chassis which could carry a load of 5-10 Kgs. Hence chassis of the vehicle is made up of wooden and it can step in a variety of composite cement road pavement, littlie mud, gravel, grass etc. Paving is not that much necessary with big tires and higher power batteries it can move over grass and fields too but yes care should be taken of there must not be any pothole or pit.

### 4.1.5 Spray Modules

Spray module composed of a spray head, pumps, relays, Servos, and DC machine. The spray system consists of three tanks. The valves are controlled by the on-board microprocessor electronically. As the robot passes over reflective places on the ground as per the plants distance, the pump is turned on and off to enable selective spraying of the diseased. To spray from the spray head, the microcontroller I/O port can control the working status of the pump.

### 4.1.6 Implementation

Figure 2 shows the hardware assembly of robot that has been implemented. Figure 4 shows the working algorithm of control section. Microcontroller LPC2318 is used as the main controller. For serial communication with the computer, RS232 module has been used on board. Figure 3 shows the agriculture robot field set-up.

## *4.2 Software Details*

In order to build the disease and growth identification capability of ROBOT, image processing technique has used. The basic steps, involved in this technique are acquisition of an image, image pre-processing followed by segmentation, feature extraction and then statistical analysis and classification MATLAB Software is used for disease detection and growth measurement based on various algorithms that configure the software part of the system. We have included two different flows one for disease detection and one for finding out the growth of the plant as shown in figure 5. For this purpose we will mount a wireless camera on the robot buggy and then connecting web cam to PC which has the MATLAB software through Wi-Fi connectivity.

**Fig. 2** Agriculture Robot hardware assembly

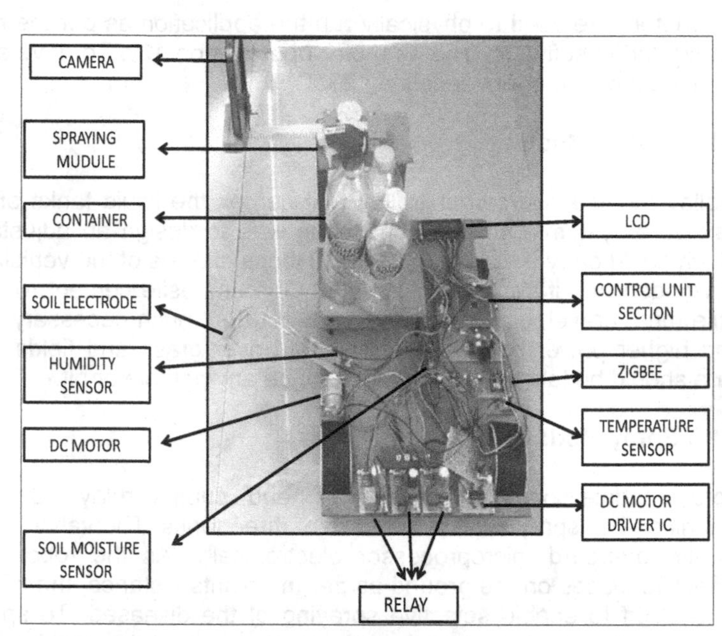


**Fig. 3** Experimental Agriculture robot field setup

### *4.2.1 Image Acquisition*

The initial stage of a vision system is the acquisition of an image. Image capture module comprises of a camera (web-cam) or a android mobile phone camera and wireless fidelity (Wi-Fi) internet connection. Today with the help of Android based OS and its applications an image can be directly fed from the mobile to MATLAB. IP web-cam app and Wi-Fi are required for image acquisition.

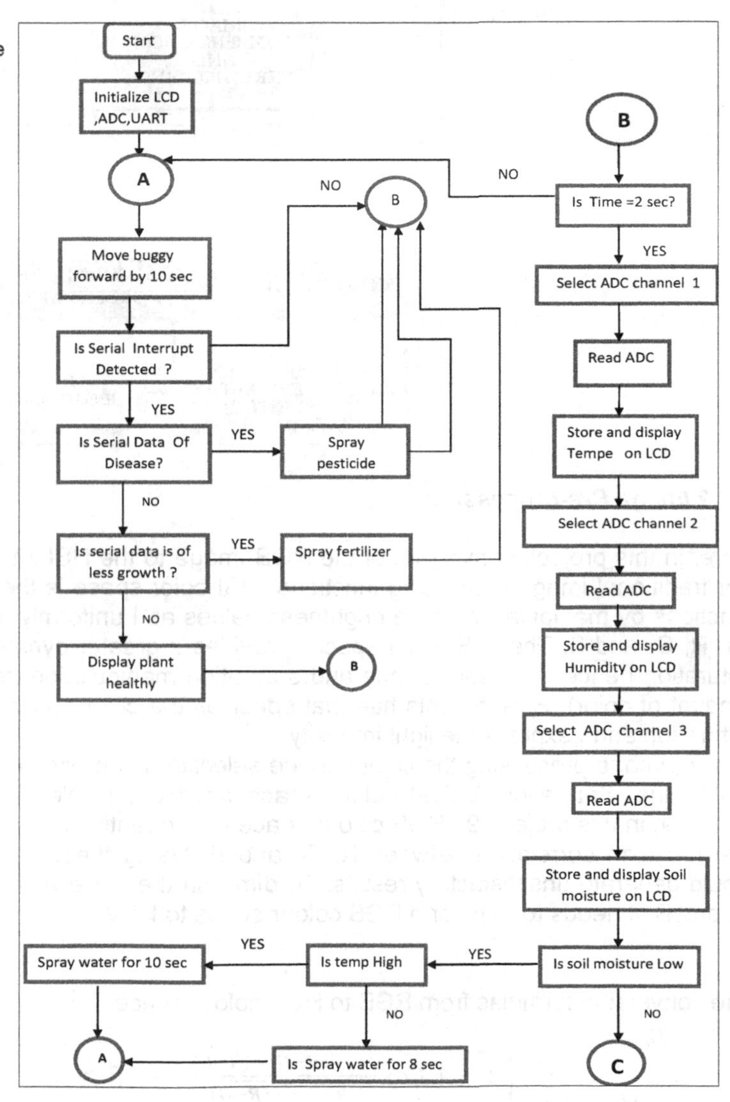


**Fig. 4** Working algorithm for the control section

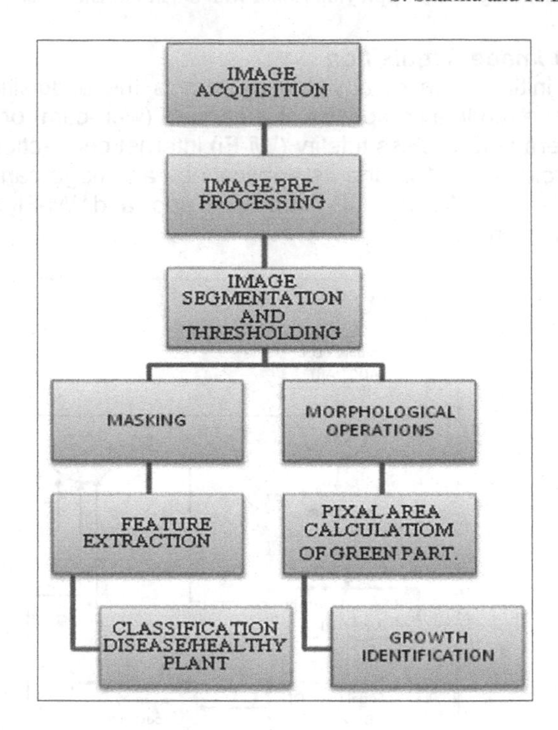


**Fig. 5** Flowchart describes the flow of the System for disease detection and growth identification

### *4.2.2 Image Pre-processing*

Here, in this project, conversion of the RGB image to the HSI image is done. For traditional image processing functions, HSI color space is the best which functions by manipulation of the brightness values as I uniformly dependants on R, G, and B. The HSI color space provides a greater dynamic range of saturation hence it is used for hue and saturation manipulation (to adjust the amount of color). H represents hue that specifies the color purity, S indicates saturation and I express the light intensity.

As for image processing the colour space selection is an essential step. The HSV (Hue, Saturation, Value) colour space have been selected for growth detection in this project [9]. RGB colour space is frequently used colour space, there is high correlation between R, G, and B. Using these values directly would generate unsatisfactory results. To diminish the correlation between R, G, and B, it needs to transform RGB colour space to HSV.

The conversion formulas from RGB to HSV colour space.

$$Hue\ (H) = \begin{cases} 2 - A.Cos\frac{[(R-G)+(R-B)]}{\sqrt[2]{(R-G)^2+(R-G)(G-B)}},\ B > G \\ A.Cos\frac{[(R-G)+(R-B)]}{\sqrt[2]{(R-G)^2+(R-G)(G-B)}},\ B \leq G \end{cases} \quad (4.1)$$

$$Saturation(S) = \frac{\max(R,G,B) - \min(R,G,B)}{max(R+G+B)} \quad (4.2)$$

$$Value(V) = \frac{R+G+B}{255} \quad (4.3)$$

### *4.2.3 Image Segmentation*

To separate different regions in an image that do not intersect each other, image segmentation is used [13].

For the purpose of disease detection on plant, image segmentation is done using K-means clustering [2]. The K-means clustering algorithm attempts to classify objects (pixels in our case) rested on a set of features into K number of classes. The classification is done by minimizing the sum of squares of distances between the objects and corresponding cluster or class centroid. Three clusters proved to be sufficient in this project.

After that the masking of the green pixels are done on the boundaries. Two steps are involved in this: At first most of the pixels of green colour are recognized afterwards to obtain varying threshold value Global image threshold has been applied. The green pixels are then asked as- If pixel intensities have green component less than the calculated threshold value then this pixel's R, G and B components are vanished. Second step requires deleting pixels with zero components, and those at the boundaries of the infected clusters.

In order to find the growth of the plant a different segmentation techniques have been applied here in order to determine the total leaf and the lesion area pixels. This involves converting an input image into grayscale image by taking the image of the diseased plant with white background would produce large difference in gray values of background and object. After the holes are eliminated and region filling is done in white region in order to have binary image that contains leaf region. At last image is scanned from top to bottom and from left to right to count pixels of the total leaf area. If the calculated pixel count of the green part is above a particular set point (pixel count) then the growth of the plant will be resulted as normal otherwise abnormal.

### *4.2.4 Computation of Integrated Feature Vector*

In this project, the Gray Level Co-Matrix (GLCM) is used to extract the statistical data from the texture of an image[2].

glcm = graycomatrix(I) produce a gray-level co-occurrence matrix (GLCM) from input image . Graycomatrix creates the GLCM by calculating how frequently a pixel with gray-level (grayscale intensity) value i occurs horizontally adjacent to a pixel with the value j. (You can also specify other pixel spatial relationships using the 'Offsets' parameter – see Parameters.) Each of the element (i,j) in glcm specifies how many times the pixel with value i occurred horizontally adjacent to a pixel with value j.

The Gray level Co-Matrix computes various Statistical Data some of them are:

1. Entropy (E)

$$E = -\sum_{i,j} p(i,j) \log\big(p(i,j)\big) \tag{4.4}$$

In a system entropy signifies the disorder. In case of random distribution system would present a high entropy. Images with solid tone would have a 0 value for entropy measure. Features are used to tell about the entropy for heavy and smooth textures Texture analysis computes the spatial disorder.

2. Correlation (r)

$$r = \frac{\sum_{i,j}(ij)\mathrm{p(i,j)} - \mu_x\mu_y}{\sigma_x\sigma_y} \tag{4.5}$$

Where, $\mathrm{p}_x$ and $\mathrm{p}_y$ are the partial probability density functions. $\mu_x\mu_y$ are the means and $\sigma_x\sigma_y$ are the Standard deviations of $\mathrm{p}_x$ and $\mathrm{p}_y$.
$\psi$

3. Energy (e)

$$e = \sum_{i,j} p(i,j)^2 \tag{4.6}$$

Energy is exactly opposite to the entropy measurement. It is the estimation of homogeneity. Texture with larger homogeneity would result higher energy.

4. Homogeneity (h)

$$h = \sum_{i.j} \frac{1}{1-(i-j)^2} p(i,j) \tag{4.7}$$

Uniformity estimation of non zero entries in the gray level co-matrix represents the homogeneity. GLCM homogeneity would be high if there are lot of pixels with similar grey level and when there is varying gray level values, GLCM homogeneity is low. An image having no variation, homogeneity value is 1 and an image having little changes would have high homogeneity value.

#### *4.2.5 Support Vector Machine (SVM)*

Support Vector Machine is one of the machine learning technique which is basically used for classification. It's a kernel based classifier. At first it was developed for linear separation which was able to classify data into two classes only. In our project SVM has been used for classifying diseased leaf and healthy leaf. For this data set has been prepared by our manually. Plant's leaf samples were acquired from mobile and these leaves are divided into normal and diseased subsets. The diseased subset contains samples of two types of diseases: Brown and yellow Spots.

### *4.2.6 Image processing results and analysis*

For disease detection on plant leaf at first image is captured and segmentation is performed with k-mean clustering. After that region of interest (ROI) is selected by choosing a particular cluster containing part of diseased area. Figure 6 shows the algorithm implementation steps for disease detection. Figure 7 shows the image processing result of disease detection.

To find growth of the plant at first the images of the plant are taken by the android mobile camera through MATLAB command. After acquiring the image firstly, the conversion of RGB to HSV was made to as HSV. Then some morphological operation were performed to eliminate irrelevant part and to have more focus over the green leaf part. At last the growth of the plant is observed by calculating the area of pixel of the green part of the total leaf area. Whether, this count is above a particular pixel count. Figure 8 shows the result of image processing for growth detection of plant and Figure 9 shows the result of both the algorithm (Disease and growth) together on MATLAB. Similarly results for healthy plant with normal growth, healthy plant with abnormal growth and diseased plant with abnormal growth were also obtained.

**Fig.6** Flowchart of Algorithm Implementation steps for disease detection Experimental

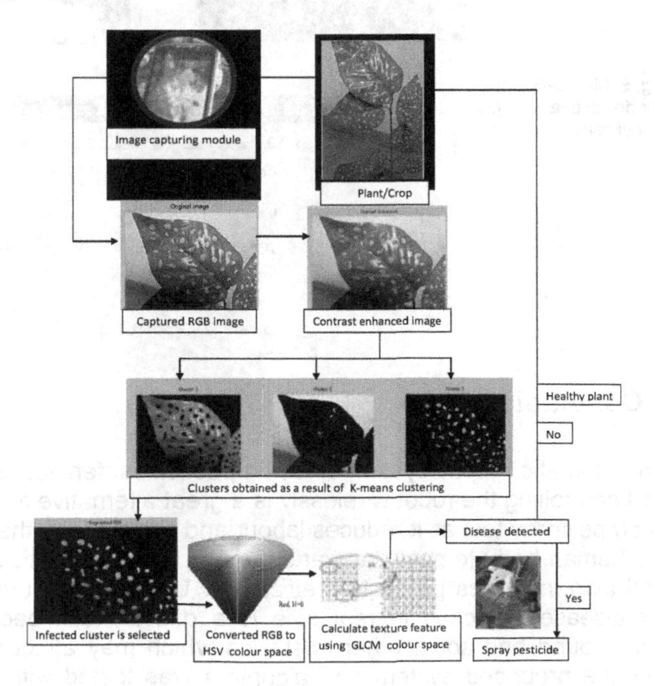

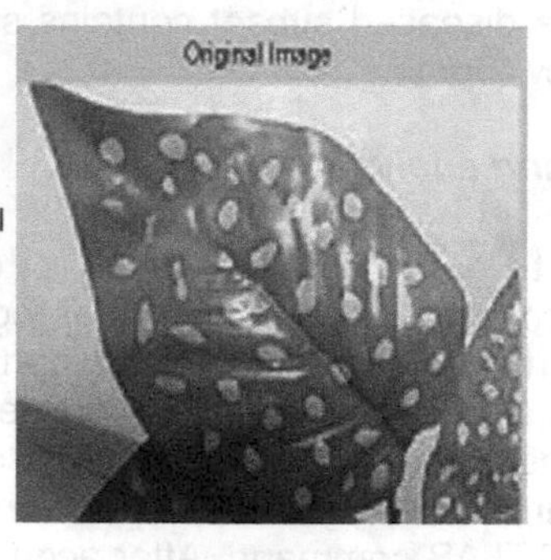


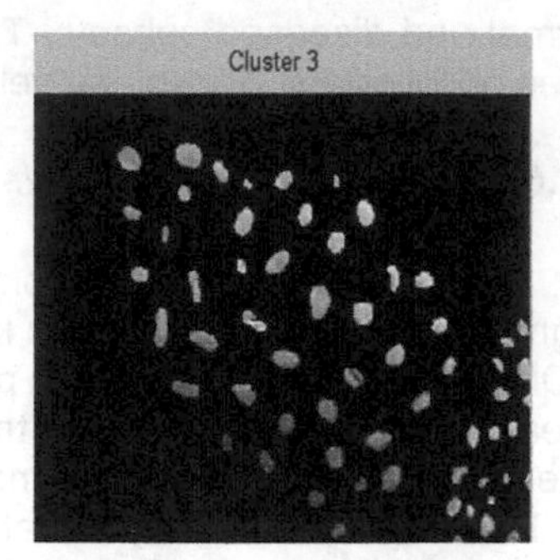


**Fig. 7** Disease Detection after segmentation

(a) Original image
(b) Segmented ROI of disease

**Fig. 8** Growth Identification of plant

(a) Original image
(b) Leaf area pixels Calculation

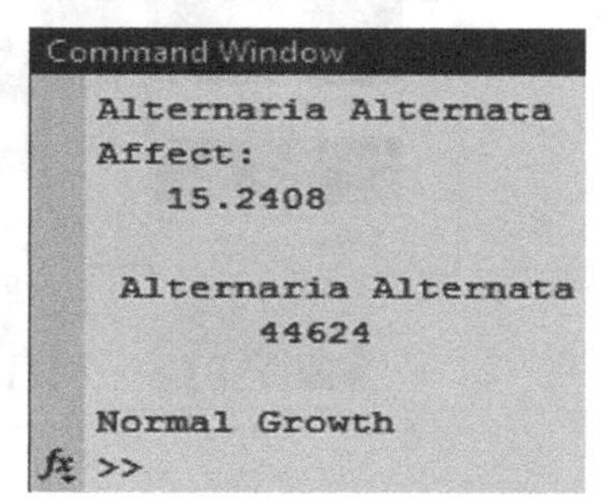


**Fig. 9** MATLAB output For disease and Growth Identification

## 5 Conclusions

An automatic robotic system spraying pesticides, fertilizer and water to plants and controlling the robot wirelessly is a great alternative to manual completion of crops spray test as it reduces labour and protect from the direct exposure of the human body to pesticides and thereby reduces pesticide harm to people, and also improves production efficiency. Using image processing technique for disease detection in real time is a difficult task because of sun light, background and some other obstacles which may affect results sometimes. Still the proposed system and algorithm was tested with plants and results were found to be satisfactory. The motive of identifying disease on the leaves and the growth of the plant was successfully achieved. The proposed system is a prototype which shows how one system can be used to overcome many problems in a farm land. Proposed system has been constructed as an autonomous robot for agricultural that could provide real time disease detection of plants along

with the controlled spraying of pesticide, fertilizer and water. There are certain limitation with this proposed system like communication range, camera quality, capacity of containers which can be successfully overcome with the increased range of resources, budget, knowledge and time. In future, this work could be further enhanced to a full stand-alone form.

## References


1. A. Camargoa, J.S. Smith. (2009). An image-processing based algorithm to automatically identify plant disease visual symptoms. *elsevier journal of computer and electronics in agriculture.* doi:10.1016/j.biosystemseng.2008.09.030
2. H. Al-Hiary, S. Bani-Ahmad, M. Reyalat, M. Braik and Z, AL Rahamneh. (2011). Fast and Accurate Detection and Classification of Plant Diseases. *International Journal of Computer Applications.* Volume 17. doi:10.5120/2183-2754.
3. Joaquin Gutierrez, Juan Francisco, Villa-Medina, Alejandra Nieto- Garibay, and Miguel Angel Porta-Gandara. (2014). Automated Irrigation System Using a Wireless Sensor Network and GPRS Module. *IEEE Transactions On Instrumentation And Measurement*, Vol. 63.doi:10.1109/TIM.2013.2276487.
4. K. Prema N. Senthil Kumar, S.S. Dash, Sudhakar Chowdar. (2012). Online control of remote operated agricultural Robot using Fuzzy Controller and Virtual Instrumentation. *IEEE-International Conference On Advances In Engineering, Science And Management* (ICAESM):196 – 201.
5. Mohamed Rawidean Mohd Kassim, Ibrahim Mat, Ahmad Nizar Harun. Wireless Sensor Network in Precision Agriculture Application. (2014). *IEEE conference MIMOS,* Malaysia. doi:10.1109/CITS.2014.6878963.
6. Naiqian Zhang, Maohua Wang, Ning Wang. (2002). Precision agriculture-a worldwide overview. *Computers and Electronics in Agriculture elsevier*, volume 36:113-132, doi:10.1016/S0168-1699(02)00096-0.
7. Peng Jian-sheng. (2014). An Intelligent Robot System for Spraying Pesticides. *The open electrical and electronic Engineering Journal,8* : 435-444.
8. R. Pydipati, T.F. Burks, W.S. Lee. (2006). Identification of citrus disease using color texture features and discriminant analysis. *Computers and Electronics in Agriculture.* doi:10.1016/j.compag.2006.01.004.
9. Sai Kirthi Pilli, Bharathiraja Nallathambi, Smith Jessy George, Vivek Diwanji. (2015). eAGROBOT- A robot for early crop disease detection using image processing. *IEEE Sponsored 2nd International Conference On Electronics And Communication System (ICECS).* doi:10.1109/ECS.2014.7090754.
10. Snehal M. Deshmukh, Dr. S.R.Gengaje. (2015). ARM- based pesticide spraying robot. *International Journal of Engineering Research and General Science.* Volume 3: ISSN 2091-2730
11. Tao Liu, Bin Zhang, Jixing Jia. (2011). Electromagnetic navigation system design of the green house spraying robot. *IEEE second international conference (MACE),* Hohhot.doi:10.1109/MACE.2011.5987400.

12. Van Li Chunlei, Xia Jangmyung Lee. (2015). Vision-based pest detection and automatic spray of greenhouse plant. *International conference on (ISIE), Korea*.doi: 10.1109/ISIE.2009.5218251.
13. Weizheng, S., Yachun, W., Zhanliang, C., and Hongda (2008).Grading method of leaf spot disease based on image processing. *International Conference on Computer Science and Software Engineering*:12–14.doi:10.1109/CSSE.2008.1649
14. Zhang Junxiong, Cao Zhengyong, Geng Chuangxing (2009). Research on Precision Target spray robot in greenhouse. *Transactions of the Chinese Society of Agricultural Engineering*. Volume 25:70-73.

# Intelligent System for Wayfinding through Unknown Complex Indoor Environment

Sobika S and Rajathilagam B

**Abstract** For a complex indoor environment, Wayfinding is knowing the environment and navigating within it. It is only possible if the person knows the place very well. If a person goes into an unknown environment, he/she may need assistance for wayfinding. GPS technology which works very well and is popularly used in outdoor navigation cannot be relied for indoor wayfinding as the signal strength is weak. This paper proposes a system which can be used as an assistance for navigating through an unknown environment with the aid of Wi-Fi, visual landmarks including corridors, staircase and others. This work aims at creating a custom route map for an unfamiliar complex indoor setting through visual perception and graphic information.

## 1 Introduction

Wayfinding generally means identifying users current position, identifying the location to which the user desires to go and identifying how to reach the destination from the initial position. Wayfinding in outdoor can be easily done with GPS technology. But the same in an indoor environment is a difficult task as the GPS satellite links are blocked and is unreliable inside a building. Wayfinding became important as the building structures of urban centers, hospitals, shopping malls etc became huge and complex. In such environment the graphic systems in the building such as the

Sobika S
Sobika S, Department of Computer Science and Engineering, Amrita School of Engineering, Coimbatore, Amrita Vishwa Vidyapeetham, Amrita University, India
e-mail: sobika.apr93@gmail.com

Rajathilagam B
Rajathilagam B, Department of Computer Science and Engineering, Amrita School of Engineering, Coimbatore, Amrita Vishwa Vidyapeetham, Amrita University, India
e-mail: b_rajathilagam@cb.amrita.edu



J.M. Corchado Rodriguez et al. (eds.), *Intelligent Systems Technologies and Applications 2016*, Advances in Intelligent Systems and Computing 530,
DOI 10.1007/978-3-319-47952-1_61

maps, direction boards, sign boards and other signage systems became insufficient for proper wayfinding. Different approaches for wayfinding have been proposed. Smart phones are used in wayfinding in blend with other technologies like Wi-Fi, Bluetooth, RFID tags, fiduciary markers, sensors and other wireless technologies.

Here a system which uses the direction board information available in the environment for wayfinding is proposed. This include three phases: rough location estimation of the user, direction symbol identification from the direction board and identifying the movement of the user.

## 2 Related Works

Certain methods create the floor plan of the building which can be used later and certain others identify the position of the user and guide them to the destination by showing the path in the map. Sankar A. et al proposed a smartphone application that will capture, visualize and reconstruct the indoor environment with Manhattan world elements [10]. The application use the sensors within the smartphone camera, accelerometer, gyroscope and magnetometer for indoor reconstruction. But the system is susceptible to shake and the image may become blurred. Giovanni Pintore et al, proposed a method similar to the previous one which doesnt restrict to Manhattan world elements alone [9]. But narrow corridors with different shapes is difficult to reconstruct. Hile et al. proposed an indoor positioning method where the camera image of a landmark is compared with the previously saved images and based on the match camera pose is identified [4]. This combined with Wi-Fi signal strength is used to identify the position of the user and then arrows are augmented and users are guided to destination. This method fails if the floor have complex pattern and if the floor and wall have same color. Another approach proposed by Mulloni et al. uses camera phone to determine the user location by detecting frame marker, a new type of fiduciary marker [2]. RSS feed is used to update the details. This system was deployed in a conference which helped users to reach the destination from the information represented as 2D as well as 3D map. The usage of camera resulted in quick drainage of battery. Alberto Serra et al proposed a navigation system based on inertial navigation system which uses the sensors with in the smartphone [12]. Once the initial position is found and map is downloaded by scanning a 2D barcode, with the help of dead reckoning method movement of the user is marked in the map. Coughlan et al. have proposed a wayfinding system using camera phone for visually impaired persons [6], [3]. This paper proposes the use of color markers as beacons, near any text or barcode. This system may have difficulty in working in crowded environment. Mitsuaki Nozawa et al. have proposed an indoor navigation system in smartphone using view based navigation [8]. In this method the image features comparison and matching is done to identify the user position. Hyunho Lee et al. proposed a method for recognizing objects like hallway entrance, hallway and pillar in indoor environment for vision-based navigation using smartphone camera without using image database [5]. Images are captured using a smartphone and features are extracted from each image and compared with the already defined features to recognize the indoor objects. The system cannot work properly if the indoor objects are hidden by any obstacles. Adheen Ajay and D. Venkataraman discussed different feature extraction algorithm for SLAM problems [1]. Nevetha M P and Baskar A discussed text extraction from scene images which helps in automated assistance [7]. Sara Paiva discusses about a dead reckoning system for pedestrian tracking using inbuilt sensors in the smartphone [11].

The focus of this paper is on identifying the structural components of the indoor environment as part of the other landmarks that may be visually identified in the place to assist the wayfinding algorithm.

## 3 Proposed System

For wayfinding, the initial position of the user is required. This can be obtained by using the Wi-Fi connection in the building and by identifying to which access point the smartphone is connected. Generally in a building for proper wayfinding graphic systems like direction board, sign board etc. will be there. Our system makes use of these existing infrastructure in the building. From the direction board the text and direction symbols will be extracted and user will be guided accordingly. Throughout the way the information from the direction board is used to get to the destination. The user reads the details from the direction boards available in the path until the user reaches the destination. Using this information and certain landmarks in the way a map is generated, which can be used to find the way back as well as for future use. Fig. 1 shows the complete system architecture. Generally the aim of this system is to observe and comprehend the graphic system in the environment as the user make their way through the building.

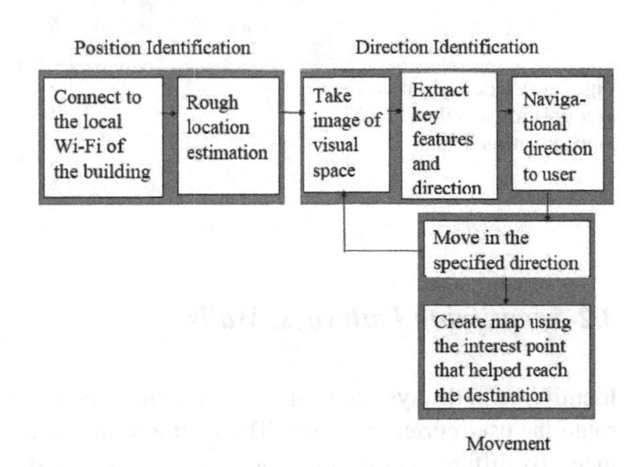


**Fig. 1** System Architecture

## *3.1 Indoor Positioning*

Indoor positioning of the user can be found with the help of Wi-Fi connection in the smartphone. The smartphone in the users hand is connected to the Wi-Fi in the building. At a time the smartphone will be connected to only one access point. The signal strength of Wi-Fi from an access point connected to a smartphone will be different at each position. The details obtained from the access points are: SSID, BSSID, IP address, MAC address, frequency, link speed, Rssi, Rssi level. These details obtained from each of the location in the building is saved in a database. Once the smartphone is connected to a Wi-Fi, according to which access point the smartphone is connected and the signal strength of the Wi-Fi, rough location estimation can be done by comparing these details with those in the database. The Fig. 2(a) shows a representation of access point and user position in a floor of a building. The numbers 1, 2, 3, 4, 5 in the figure represents rooms and the blue oval with A and B marked is the access points. The signal strength received to the user from the access points A and B will be different. From the figure it is understood that the user is standing near to access point A. So the strongest signal will be from access point A and thus the smartphone in users hand will be connected to access point A. Fig. 2(b) shows the details of the access point obtained when the smartphone is connected to Wi-Fi in the building. This details is compared with database and the rough location information is retrieved : near to access point A which is near to room no 1, 2 and 5.

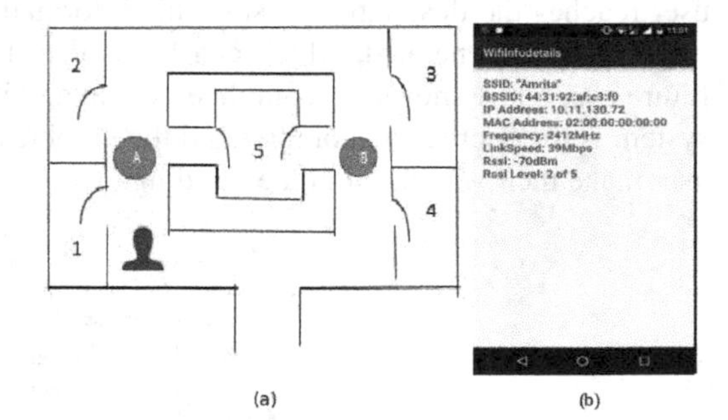


**Fig. 2** (a)Access point and user position in a floor, (b)Wi-Fi information details

## *3.2 Identifying Pathways, Walls*

Identifying pathways and walls in the visual space gives a clear understanding of the route the user currently taking. These structural elements better define the environment. By differentiating floor from wall the corridor space can be distinguished. The pathways may be identified using edge detection and feature comparison to standard elements of a pathway. In the case of a long straight corridor, a straight edge can be obtained at the point where the wall meets the floor. If there is a proper color difference between the wall and the floor, using the color value the floor can be identified. Identification of floor is important if the user is blind. In this case by differentiating wall and floor as well as identifying a clear pathway without obstruction will be useful for guiding the blind user in the environment.

### *3.3 Identifying Direction Boards*

The wayfinding can be done by making use of the objects and features in the visual space. As the proposed system dont use any markers or QR codes demonstrated by other papers, graphic systems available in the environment can be used as the information for wayfinding. Assume that the building have proper direction boards. From the direction board text and direction symbols are extracted.

In order to reach a destination, the user will type in the destination as input and this will be compared with the direction boards to give a proper direction. The user takes the image of the direction board he sees first. While taking the image sometimes the image will be rotated. Instead of portrait, the image will be obtained as landscape. This can be rotated back to portrait by using the ExifInterface in android. Using Tesseract, an open source OCR engine, optical character recognition can be done. The trained data; English language for printed text is saved to the sdcard of the smartphone. The input will be compared with the OCR result. If a match is found the corresponding direction symbol is extracted and the navigation instruction is given to user. This process is continued till the user reaches the destination. Tesseract gives a reasonably good results if the font, color and background color is proper. If the user is standing facing the direction board and taking the image the result will be good.

For direction symbol extraction template matching is used. For four directions straight, left, right, back; four template of direction symbol is used as shown in Fig. 3. The direction board image is given as input. This will be the source image and the images in Fig. 3 is the template image. By sliding each of the template one pixel at a time from right to left and top to bottom over the source image the similarity metric is measured. The source image is compared with all the templates and did the matching and normalizing. For each of the source-template pair a maximum value of similarity is obtained. By analyzing different inputs and the similarity measure a threshold value is set, such that when a source image is given as input direction symbol in that is correctly obtained.

**Fig. 3** Template images for right, left, back and straight

#### 3.3.1 Challenges

When the image of the direction board is taken, the orientation of the image may be different as shown in Fig. 4. The user taking the image may not be standing directly facing the direction board. In this case text extraction and direction symbol extraction is difficult. The OCR result will give wrong output. So first, the perspective distortion of the image have to be corrected. By doing the morphological operations background subtraction is done and only direction board is left. Now the four corners of the direction board is detected using Harris corner method. These four corners are properly ordered either in clockwise or anti clockwise direction. The perspective distortion correction can be done by using plane homography and projective transformation.

**Fig. 4** Direction boards with perspective distortion

### *3.4 Landmarks*

Something which is unique and which catch users eye easily and which stand out in the environment can be considered as a landmark. How people remember the path travelled is usually by creating a mental map of the surrounding by identifying certain points as landmarks and linking the navigational instructions with those landmarks. So including certain landmarks in the wayfinding system will increase the understandability of the route map. People will find it easy to navigate themselves along the path by viewing this landmarks along with the navigational instructions. Fig. 5 shows example of landmarks in shopping mall. The escalator and the board shown in the figure is easy to find in the environment. Navigational instructions can be clubbed to this landmark and it will make the user easy to remember the route. If these kinds of landmark information is included in the route map, wayfinding in an unfamiliar environment will be easy.

### *3.5 Building Route Map*

The path from the initial position of the user to the destination is to be represented as a model which the user understands. For that the entire path have to track. The details obtained from the direction board as well as the details from the mobile sen-

**Fig. 5** Landmark based navigation

sors is used to give a meaningful representation of the path. Building of the route map includes initial position estimation, identifying direction boards and landmarks and using mobile sensors to track the entire path. In the case of an unfamiliar environment, the user might not be knowing his current position and how to go from there to the desired location. So identifying the initial position is important. By identifying the Wi-Fi signal strength and the access point to which the smartphone is connected, initial position is identified. This will be marked in the map. Whenever the user passes a landmark, it can be included in the route map for giving more clarity of the path. When a direction board is seen in the environment, the required destination is searched in that and the user gets the navigational instructions. Throughout the path identification of the landmark and direction board is done and the user reaches the destination. This whole path is tracked using the sensors in the smartphone for generating the route map. The path ones generated may be reused. The detailed explanation of the sensors is mentioned in the following section.

### 3.5.1 Mobile Sensors

By using the sensors within the smartphone itself, the movement of the user can be found. How much distance the user travelled and the direction in which the user is walking can be found out with the help of these sensors.

Accelerometer

For the smartphone 3 axis X, Y, Z is there. All the sensor will give values in these three axes. The accelerometer value is influenced by gravity. So these have to be eliminated. From the raw accelerometer value gravity is subtracted by passing the accelerometer value through low pass filter and the linear acceleration is obtained.

$$Linear \quad acceleration = Total \quad acceleration - gravity \tag{1}$$

We assume that the user is holding the phone in his hand horizontal to the floor, with the screen facing up. So the acceleration value in the Y axis will give the acceleration of movement of the user. The peak in the acceleration value represents a step. The accelerometer value above a particular threshold is considered as a step. The number of steps multiplied with the step length will give the total distance travelled. The average step length is taken as 2.4 feet. Fig. 6 shows the accelerometer reading in the y axis during walking.

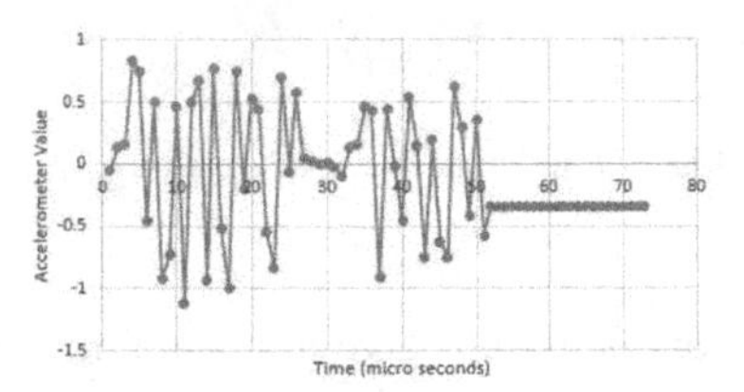


**Fig. 6** Accelerometer reading of Y axis

Forward Direction

The forward direction of movement of the user can be found from magnetometer sensor in the smartphone. The magnetometer sensor will give the orientation in relation to the Earths magnetic field. So for identifying the direction change, the azimuth value is obtained. The azimuth value before a direction change or turn is kept as reference. This is considered as a straight walk. During this straight walk there will not be much change in the azimuth value. All the azimuth value will be almost similar. When the user take a turn, the direction at which the smartphone in his hand was initially facing also changes. As the azimuth value is obtained at the rate of microseconds, when the user take a turn there will be a drastic change in the azimuth value. The azimuth value will be either increasing or decreasing depending upon the direction at which the user is turning. This is represented in Fig. 9.

## 4 Results and Discussion

The prototype of the proposed system, which is still under development uses the camera, accelerometer sensor and magnetometer sensor in the smartphone. The direction board details gives the user which direction to move. The Fig. 7(a) shows the result of OCR on a direction board : WAY TO CENTRAL LIBRARY. The Fig. 7(b) shows the template matching done on the arrow and that image is given to do OCR and the text; FACULTY ROOM is correctly obtained as result.

The perspective distortion on the direction board is corrected by identifying the four corners and then doing the distortion correctness algorithm. The result of perspective distortion is shown in Fig. 8.

The sensors in the smartphone will calculate the distance travelled and the turn made by the user. The two turns made by the user is marked in the plot of azimuth value in Fig. 9 . In the graph the straight line shows a straight path and a slope or a variation of value shows a turn made by the user. A path usually contains a straight

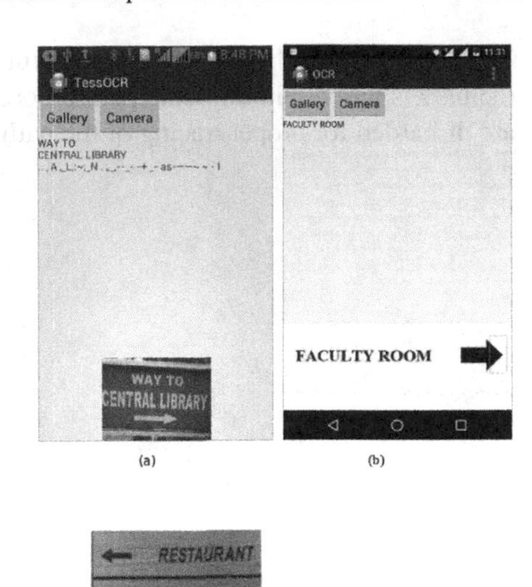


**Fig. 7** (a)OCR,(b)OCR on template matched image

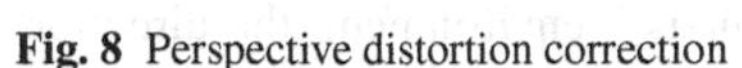

**Fig. 8** Perspective distortion correction

path followed by turns; if any turn is there in the path. The turn can be identified by analyzing the azimuth values obtained. Initial direction of walk is considered as straight path. Azimuth value will be almost same at this point. When user turns, increasing or decreasing of azimuth value occurs. While comparing the difference of value during turn and previous straight path, if the result is positive then right turn and if result is negative it is left turn. In the Fig. 9, the first turn is left and second turn is right.

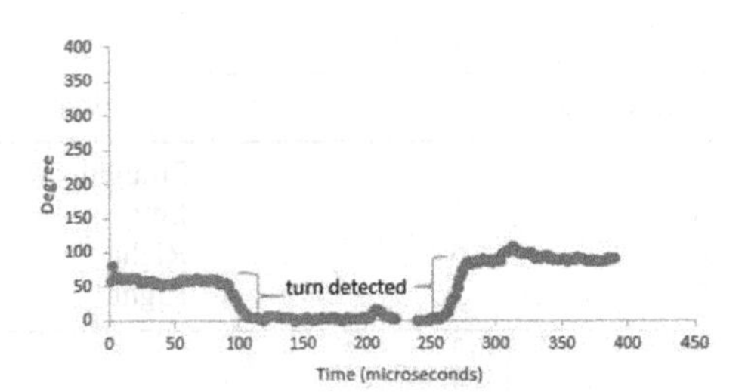


**Fig. 9** Magnetometer values showing the turn detection

A turn can occur while the user is moving or the user can turn while standing at the same place and might continue the previous path. A challenge in turn detection is identifying the second category. Fig. 10 shows the azimuth value plotted when the user walk then stop and turn and continue the previous path. From the figure it is understood that two turns occurs simultaneously without any straight path in between.

The azimuth value before and after the turn is almost similar which represents that the user is moving in the same path before and after the turn. These kinds of turns are discarded for proper tracing of the path.

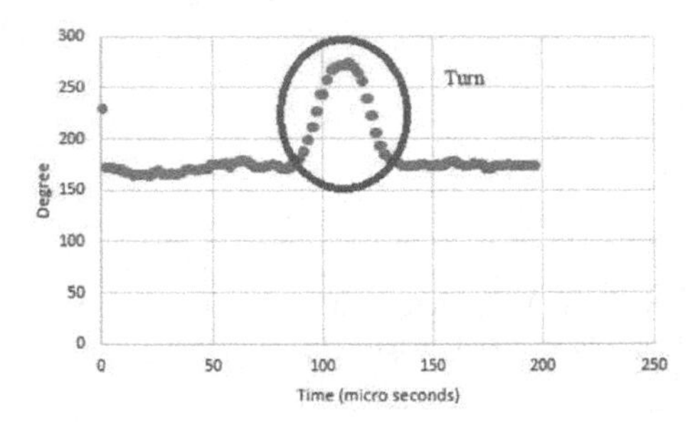


**Fig. 10** Turning while standing at the same point

Another challenge is when there is a path at 90 degree right direction and another at 45 degree right direction. In this case direction of turn cannot be generalized as right. It might cause confusion for which path to take. The change in azimuth value will be difficult to differentiate for both the case. The user may not be moving in the same way always. If a direction board is there in such environment, the direction symbol in that can be used to represent the turn in the map. The distance of the straight path and direction of turn is saved as shown in Table 1. This is the case where the user is moving from the entrance to room no 5 in the Fig. 1. Table 1 represents move 6 steps straight, 8 steps left, 8 steps right and 7 steps right. Here a landmark is marked which is room no 1. A direction board was available in the turn to room no 5. These details are represented in a graphical representation in Fig. 11(a). The star represents a landmark and the clipart of direction board represents that a direction board is available in that position. Fig. 11(b) shows the representation of the path in the floor plan.

**Table 1** Example of a route

| Step Count | Distance(m) | Direction |
|---|---|---|
| 6 | 4.5 | Straight |
| 8 | 6 | Left |
| 8 | 6 | Right |
| 7 | 5.3 | Right |

The proposed system have been experimented on a shopping mall shown in Fig. 12. In a crowded environment the step count measurement were a little disturbed. The turns made by the user was properly detected. The lack of proper direction boards in the environment made a little confusions. Excluding that the path travelled by the user were traced correctly. Using the information from the step count and

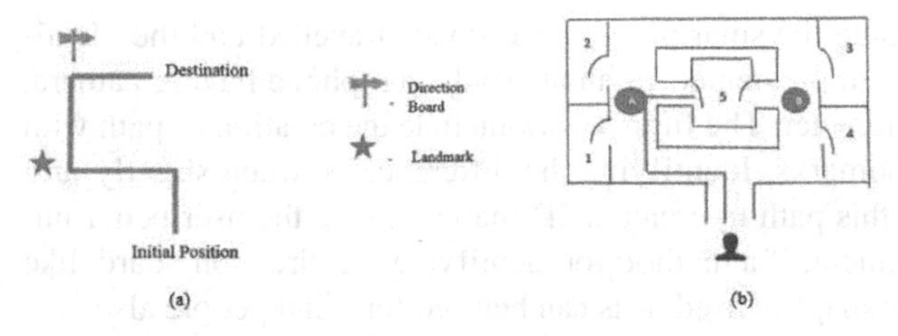


**Fig. 11** (a)Example of Graphical Representation of route, (b)The route represented on floor plan

**Fig. 12** Indoor images of a shopping mall

turns ones obtained, and also observing the environment another person will be able to travel through the same path without much difficulty.

In an actual map or a floor plan of the building, it gives the whole picture of the building. The user may be interested in only a particular part of the building. So giving such a user the floor plan is meaningless. Also identifying the position of the user in the floor plan is difficult. The user have to interpret the map and navigate himself along the map which is difficult and time consuming. Usually the map of the building will be fixed at a particular point in the building. So the user have to memorize the path which is not so appealing. The orientation of the map may not be in relation to the user which poses more difficulty.

In the proposed system only the required path will be there from source to destination. Positioning of the user and navigation along the path is not difficult. The path is generated in the smartphone, so no need to memorize. The user can look and move along the path.

The components of the proposed work have been tested individually and the overall route building is being experimented across a few complex indoor environments. The work presented here is part of an ongoing work to develop a full-fledged wayfinding system for unknown indoor buildings.

## 5 Conclusion and Future Enhancement

In this paper a system for wayfinding which uses visual and inertial wayfinding approach in an indoor environment is proposed. The visual approach is the identification of text and directions from the direction board. The inertial navigation approach is the usage of sensors inside the smartphone for distance travelled and the identifying turns. The system is implemented in an android smartphone having camera, accelerometer and magnetometer. The future work include the creation of path with more information and landmarks. Identifying the difference between slightly turn and fully turn. Extruding this path to create a 3D map will give the user better understanding of the environment. If a method for identifying the direction board, like by identifying the color or shape is used, this can be used for blind people also.

## References


1. Ajay A, Venkataraman D (2013) A Survey on Sensing Methods and Features Extraction Algorithms for SLAM Problem. International Journal of Computer Science, Engineering and Applications (IJCSEA). 3(1): 59-63. DOI: 10.5121/ijcsea.2013.3105.
2. Alessandro Mulloni, Daniel Wagner, Istvan Barakonyi, Dieter Schmalstieg (2009) Indoor Positioning and Navigation with Camera Phones, IEEE Pervasive Computing, vol.8, no. 2, pp. 22-31, doi:10.1109/MPRV.2009.30.
3. Coughlan J, Manduchi R, Shen H (2006) Cell phone-based wayfinding for the visually impaired. In: 1st International Workshop on Mobile Vision, Graz, Austria.
4. Hile H, Borriello G (2008) Positioning and Orientation in Indoor Environments Using Camera Phones. IEEE Computer Graphics and Applications 28, 4, ACM Press, 32-39.
5. Hyunho Lee, Jaehun Kim, Chulki Kim, Minah Seo, Seok Lee, Soojung Hur, Taikjin Lee (2014) Object recognition for vision based navigation in indoor environment without using image database, In 18th International symposium on consumer electronics (ISCE 2014).
6. Manduchi R, Coughlan J, Ivanchenko V (2008) Search Strategies of Visually Impaired Persons using a Camera Phone Wayfinding System. 11th Intl. Conf. on Computers Helping People with Special Needs (ICCHP '08); Linz, Austria.
7. Nevetha M P, Baskar A (2015) Applications of Text Detection and its Challenges: A Review, Proceedings of the Third International Symposium on Women in Computing and Informatics, Kochi, India [doi:10.1145/2791405.2791555].
8. Nozawa M, Hagiwara Y, Choi Y (2012) Indoor human navigation system on smartphones using view-based navigation. In: 12th international conference on control, automation and systems (ICCAS), 2012, pp 1916-1919.
9. Pintore G, Gobbetti E (2014) Effective mobile mapping of multi-room indoor structures. Vis. Comput., 30, 707-716.
10. Sankar A, Seitz S (2012) Capturing indoor scenes with smartphones. In Proceedings of the 25th Annual ACM Symposium on User Interface Software and Technology (UIST '12), Cambridge, MA, USA, pp. 403-412.
11. Sara Paiva (2015) Domain independent pedestrain dead reckoning system for tracking and localization. In 18th International Conference on Computational Science and Engineering.
12. Serra A, Dess T, Carboni D, Popescu V, Atzori L (2010) Inertial Navigation Systems for User - Centric Indoor Applications, Networked and Electronic Media Summit, Barcelona.

# Genetic Algorithm Based Suggestion Approach for State Wise Crop Production in India

**Saakshi Gusain, Kunal Kansal, Tribikram Pradhan**

**Abstract.** Agriculture is considered as the backbone of Indian economy. Despite the tremendous increase in industrialization and advancement in technology, a majority of India's population is dependent on agriculture. Agriculture in India is not uniform throughout. In this paper, rule based classification has been applied to classify the states based on the amount of rainfall. The accuracy of Classification results is validated using Random Forest algorithm. Also we have found the major crops that can be grown in a state using a proposed algorithm by analyzing the soil, temperature and rainfall data. Due to time constraint and non-optimal usage of land resources, cultivation of all the suggested crops is not possible. Therefore, Genetic Algorithm is applied to give the best possible suggestion for crop cultivation across various states in India. So our proposed method will help to maximize the overall agriculture production in India.
***Keywords:*** Correlation, Prediction, Classification, Random Forest, Genetic Algorithm.

## 1. Introduction

Agriculture plays a significant role in Indian economy. Majority of India's population is dependent on agriculture. The agriculture sector accounts for about 19% share, of the total value of country's export and also contributes about 27.4% to the gross domestic product. The agricultural production has kept pace with the popular growth rate of 21% per annum but from past year's record there is slight declination in production because of factors like industrialization and increasing tertiary sectors.

Agriculture is not uniform in India, all sorts of different topography exist such as snowy mountain peaks of Himalayas in North, dry and sandy Thar Desert in Rajasthan, plains in UP, Bihar and West Bengal, Deccan Plateau in central India, peninsular in south, and islands. There are four major seasons in India Jan-Feb is Winter, Mar-May is Summer, Jun-Sept is Monsoon, and Oct-Dec is Post Monsoon. Based on the season crops are classified as Kharif crops, Rabi crops and Zaid crops. Kharif crops are sown in Jul-Sept i.e. monsoon season and are harvested in September and October. Major Kharif crops are rice, jowar, maize, cotton, ragi, bajra, sugarcane and jute. Rabi crops are sown in Oct-Dec i.e. post monsoon season andrvested in April and May. Major Rabi crops are wheat, peas, pulses, mustard and rapeseed. Zaid crops are sown in summer season and harvested during monsoon season. Major Zaid

Tribikram Pradhan, Dept.of Information and Communication Technology, Manipal Institute of Technology, Manipal University, tribikram.pradhan@manipal.edu



J.M. Corchado Rodriguez et al. (eds.), *Intelligent Systems Technologies and Applications 2016*, Advances in Intelligent Systems and Computing 530,
DOI 10.1007/978-3-319-47952-1_62

crops are rice, maize, sunflower, vegetables and groundnut. Agriculture in India is also dependent on the rainfall i.e. precipitation caused by the monsoon winds. Due to lack of irrigation facilities majority of farmers depend on rainfall for the irrigation of crops. Production amount of crop varies as Rainfall varies, if good monsoon then production will increase but if inadequate or extreme monsoon then production decreases immensely thereby shortage of crops and hence inflation in price.Not only climatic conditions and rainfall Indian agriculture also depends on the soil type. Varied variety of soil is found across varied topography of India. Different major types of soil found in India are Mountain soil, Red soil, Alluvial soil, Laterite soil, Loamy and Marshy soil, Black soil, Dry Arid soil, Forest soil, Saline soil. Soils have nutrients which are classified as primary nutrients and secondary nutrients. Primary nutrients include Sodium (Na), Phosphorous (Ph), Potash (K). Secondary nutrients include Iron (Fe), Magnesium (Mg), Aluminum (Al), Salts etc. Different crops require different nutrients i.e. Soil on which it can be grown.

The paper consists of the following: Section 2 comprises of the Literature Survey followed by Methodology section. Section 5 shows the case studies and obtaining results using the above mentioned methodology. Section 6 finally concludes the paper along with suggestions.

## 2. Related Work

70% of India's population is indirectly or directly dependent on the agriculture due to which rainfall prediction plays a major role and so the prediction of rainfall analysis has always been a very highly researched field in India [1].The weather forecast is performed by usingNeural network technique. For performing weather forecasting, neural network techniques such as, back-propagation algorithm and supervised learning are used. The total food grain yield gets influenced by the variations in rainfall over India [2]. An increase in rainfall intensity and frequent occurrence of heavy rains may cause flood, while a reduction can cause a drought, both will severely affect crop production. Therefore, the effect of wet and dry rainfall spells on the food grain yields must be studied. By using the method of linear trend the de-trended data is done and backward differences are used. Piecewise linear trends in the data are significantly reduced using the method of backward difference. Various techniques can be used for prediction of rainfall. Another such technique used by Goutami [3] was multiple linear regressions. She pointed out that choosing the input matrix arbitrarily is a better way than Artificial Neural Network, as it did not analyze the autocorrelation structure of rainfall time series.

A.Kulkarni [4] mentioned another technique under classification i.e. fuzzy c-Means. This method does not force a pattern to get classified into only one cluster, unlike the hard clustering methods, such as kmeans clustering technique and Map-to-Map correlation technique, rather it assigns varyingmembership to each and every cluster. Monsoon rainfall features can also be evaluated by the merged rainfall observations and the estimates done by satellites as performed by S. K. Roy Bhowmik [5]. At a resolution of 1° grid, the objective analysis for daily rainfall of the Indian monsoon region has been carried by INSAT derived precipitation estimates and merging dense land rainfall observations. This daily analysis reproduced detailed characteristics of Indian summer monsoon which was based on observations from the high dense rain gauge. Another paper by Abhishek [6] shows Geo-spatial analysis of the temporal trends of kharif crop phenology metrics over India and its relationships with rainfall parameters. Using Kendall's correlation test, the results showed significant similarity between the number of rainy days and SGS also the number of rainy days and amount of rainfall during the rainy season were found to have significant effect over the SiNDVI in 25–30 % of the study area.

## 3. Methodology

In this paper, firstly we predict the state wise amount of rainfall using previous 50 years of data using different parameters associated with rainfall and selecting only those which are strongly positively correlated using Kendall's correlation analysis. To determine the upcoming amount of rainfall in future we have used the Random forest classification techniques to assign labels to each state based on certain ranges of rainfall. Secondly, using the Crop, Rainfall, Soil, and Climate data we will predict the possible states for a particular crop to be cultivated using the algorithm mentioned below. After getting various crops for individual states we need to find the similarity among crops with the help of Pearson correlation coefficient technique. We have to consider only those pairs of crops which are positively correlated. Based on the result obtained from above mentioned steps, we will suggest suitable crops for states. Thirdly, using the Genetic Algorithm, we will suggest the set of crops that should be cultivated in a state over a year. The main purpose of our paper is to maximize and increase the agriculture and optimally utilize the resources so that the economy of India increases and farmers will gain profit thereby improving their condition.

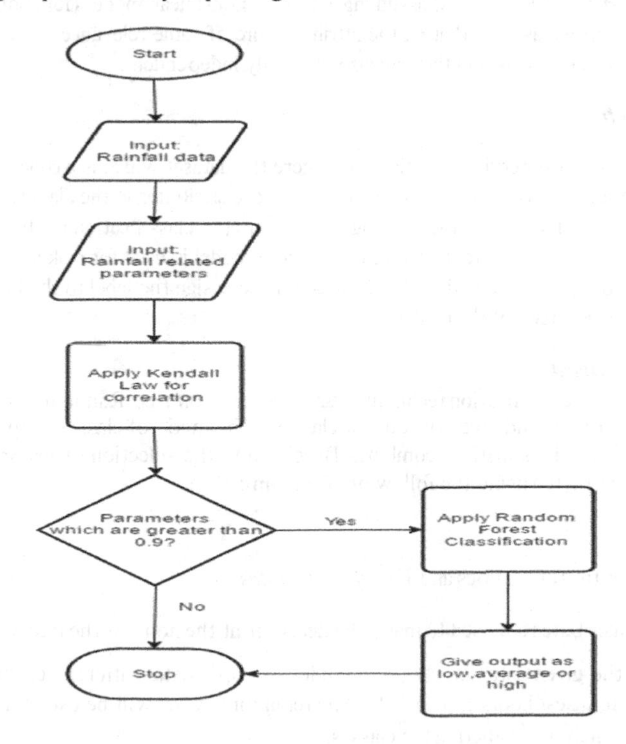


Figure 1. Flowchart of Rainfall classification

### *3.1 Correlation*

Correlation is a statistical technique that will show whether and how strongly the random pairs of variables are related to each other. Many measures are used to find the correlation analysis. For our paper we will be using Kendall's Tau coefficient correlation analysis to find the primary factors, through which we will predict the amount of rainfall in India.

### *3.2 Kendall's Tau coefficient correlation analysis*

Kendall's tau analysis method is one of the correlation analysis technique which is defined below.

The Kendall $\tau$ coefficient is defined as: $\tau = n^{*}(n-1)/2$

The coefficient Tau must lie between $-1 \leq \tau \leq 1$. If coefficient value is 1 then it is perfect agreement between two rankings and if -1 then the disagreement is perfect for the same. If both the two variables are independent, then the coefficient value will be zero.

### *3.3 Prediction*

Prediction is a supervised learning task where the data is used directly to predict the value of a class of a new instance. In this we predict the value of a given continuous-valued variable which is based on other variable's values, assuming linear or non-linear model (function) of dependency. In Prediction we assume that all the attributes are of same relevance i.e. are equally important for classification task and they are also statically independent.

### *3.4 Classification*

Classification is also a supervised learning task where the dataset which is a collection of records contains a large set of attributes, out of which, one of the attributes is the class (label). Classification is divided into two steps- the learning step, where the classification model is constructed and the classification step where the above constructed model is used for making predictions for the label (class) of the given test data. Here the goal is to assign the label to the testing data and also calculate the accuracy of the model.

### *3.5 Random Forest*

Random Forest is a classification technique that uses decision tree. Random Forest comprises of many decision trees and gives the output class i.e. the mode of classes output by each of the individual trees. This method combines Bagging and the selection of features randomly. Each tree is constructed using the following algorithm:

**Algorithm:**

**Step 1-**Let N be the total tuples and *V be the attributes*.

**Step 2-** The $v$ attributes are used to make the decision at the node of the tree, $v < V$.

**Step 3-** From the given data set, choose n random sample data with replacement from all $N$ available training cases/ bootstrap sample. The remaining cases will be used for error estimation in the tree, thereby predicting the classes.

**Step 4-** Randomly choose $v$ variables, for each given node of the tree and among these the best split attribute is decided for that particular node.

**Step 5-**Each tree is fully grown and not pruned.

For prediction of label, the testing data is pulled down the tree. This procedure is iterated over all the random trees generated, and the majority vote of all the trees is reported as the prediction of random forest.

### *3.6 Genetic Algorithm*

Genetic Algorithms are based on ideas of natural selection and study of genetic. They are the adaptive heuristic search algorithms. An intelligent usage of a random search is represented by the genetic algorithm which is used to solve problems related to optimization. The basic methods applied in the GAs are designed for process simulation in the natural systems that are necessary for necessary for evolution, especially those that follow the Charles Darwin's principle "survival of the fittest". In nature, the competition among the various species for scanty resources, results in the fittest individuals that adapt to the changes and dominate over the weaker ones.

**Basic Genetic Algorithm**

**Procedure Genetic Algorithm**

1. x = 0
2. Initialize population P(x)
3. Evaluate P(x)
4. **Do Until** (done)
5. x = x + 1
6. Parent_selection P(x)
7. Mutate P(x)
8. Evaluate P(x)
9. Survive P(x)
10. **End do**
11. Recombine P(x)
12. **End**

An encoding schema or genetic representation for the potential solutions. The creation of initial population for potential solutions. An evaluation function which acts as environment, rating the solutions in the terms of "fitness". Genetic operators that alters composition of the offspring. The values used for various parameters in the GA.

**Chromosome Encoding**: A chromosome must contain information about the solution that it represents. A binary string is used for encoding. Example:

Chromosome 1: 1000100101010100    Chromosome 2: 1000111001111100

**Initial Population:** Creation of population of chromosomes randomly i.e. binary codes of given lengths.

**Evaluation:** This function is equivalent to the initial function f(x) in which the chromosome represents binary code for the real value, x.

**Alteration:** In this, based on population evaluated in previous iteration, the new population is selected. Depending on the fitness value or objective function value an individual may be selected.

**Crossover:** It selects genes from the parent chromosomes and thus creates new offspring's. To do this, some crossover point is randomly chosen and everything before this point is copied from first parent and everything after this point is copied from the second parent. To all pairs of selected individuals, crossover is not applied. The choice is made depending on the specified probability termed as the Crossover probability (PC), whose value lies between 0.5 and 1.
Example:

| | |
|---|---|
| Chromosome 1 | 10001 \| 00101010100 |
| Chromosome 2 | 10001 \| 11001111100 |
| Offspring 1 | 10001 \| 11001111100 |
| Offspring 2 | 10001 \| 00101010100 |

**Mutation:** After the crossover, mutation will be performed. This will prevent, falling of all the solutions inside the population into the local optimum. The new offspring's are changed by mutation, randomly. For binary encoding, few of the randomly chosen bits can be switched from 0 → 1 or from 1 → 0.Example

| | |
|---|---|
| Original Offspring 1 | 10001 11001111100 |
| Original Offspring 2 | 10001 00101010100 |
| Mutated offspring 1 | 10001 01001111100 |
| Mutated offspring 2 | 10001 10101010100 |

### *3.7 Algorithm*

**Algorithm 1:** Labelling each state with the season-wise rainfall

1. **do for** each state
2. Input the data set for rainfall (winter, pre-monsoon, monsoon, post-monsoon).
3. Label1 ← calculate mean value for winter Jan-Feb (year-wise)
4. Label2 ← calculate mean value for pre-monsoon Mar-May (year-wise)
5. Label3 ← calculate mean value for monsoon Jun-Sep (year-wise)
6. Label4 ← calculate mean value for post-monsoon Oct-Dec (year-wise)
7. Determine the lower and higher values in each column season wise and also calculate their mean value.
8. LV← lower value
9. HV← higher value
10. MV← mean value
11. Determine the range for the value and assign the label as follows
12. **For** all s in Season **do**
13. **For** all v in year **do**
14. Low← (Value <LV)
15. Average← (LV<Value<HV)
16. High← (Value>HV)
17. **End**

**Algorithm 2:** Determining the possible states for cultivation for crops

1. procedure main ()
2. **begin:**
3. **for** all C in Crops do
4. **for** all S in States do
5. flag1←checkRain-Range (C_rain,S)
6. flag2←checkSoil (C.soil,S)
7. flag3←checkClimate (C.climate,S)
8. **if** (flag1==1&flag2==1&flag3==1)
9. C.Label←suitable
10. **else**
11. C.Label← not suitable
12. **End**

**Algorithm 3:** Determining whether crop rainfall lies in range of state rainfall

```
1. procedure checkRain-Range (rain,State)
2. begin:
3. if rainfall is in range State.rainfall
4.     flag 1=1
5. else
6.     flag 1=0
7. Return flag 1
8. End
```

**Algorithm 4:** Determining soil required by crop is available in state or not

```
1. procedure checkSoil (soil,State)
2. begin:
3. for all S in State.soil do
4.     if soil == S
5.         flag2 = 0
6. else
7.     flag2 = 0
8. Return flag2
9. End
```

**Algorithm 5:** Determining climate required by crop is suitable in state

```
1. procedure checkClimate (climate,State)
2. begin:
3. if climate is in range State.climate
4.     flag3=1;
5. else
6.     flag3=0;
7. Return flag3
8. End
```

## 4. Case Study/Result & Discussion

In this section the result of the application of Kendall's Tau coefficient correlation analysis is shown to calculate the correlation among various predictors such as Precipitation (PPT), Minimum Temperature (MNT), Average Temperature (AT), reference-crop-Evapotranspiration (RCE), Cloud Cover(CC), Wet Day Frequency (WDF), Vapour-Pressure(VP), Diurnal-Temperature Range (DTR), Ground-Frost-Frequency (GFF), Potential Evapotranspiration(PE), Maximum Temperature(MXT). This data is obtained from India Water Portal [6], for the years 1951 to 2014. The average value of the various factors has been considered for a better prediction result. Table1 shows the data used in the Correlation analysis.

**Table 1.** Data set containing factors of precipitation for district Bangalore

| Factors | Jan | Feb | Mar | Apr | May | Jun | Jul | Aug | Sept | Oct | Nov | Dec |
|---|---|---|---|---|---|---|---|---|---|---|---|---|
| PPT | 3.33 | 4.32 | 9.84 | 51.2 | 127. | 80.7 | 94.7 | 97.1 | 122. | 164. | 66.3 | 16.5 |
| Mnt | 16.0 | 17.4 | 19.5 | 21.4 | 21.2 | 20.4 | 19.9 | 20.0 | 19.7 | 19.5 | 18.0 | 16.4 |
| Mxt | 28.2 | 30.6 | 33.1 | 34.2 | 32.9 | 29.4 | 28.0 | 28.2 | 29.0 | 28.6 | 27.7 | 27.1 |
| At | 22.1 | 24.0 | 26.3 | 27.8 | 27.1 | 24.9 | 23.9 | 24.1 | 24.3 | 24.1 | 22.8 | 21.7 |
| Cc | 25.8 | 22.8 | 24.1 | 35.0 | 46.4 | 70.5 | 80.1 | 72.8 | 63.7 | 58.0 | 45.8 | 36.1 |

| | | | | | | | | | | | | |
|---|---|---|---|---|---|---|---|---|---|---|---|---|
| Vp | 16.9 | 17.2 | 18.5 | 21.6 | 23.1 | 23.5 | 23.3 | 23.3 | 23.2 | 22.7 | 20.4 | 18.2 |
| Wdf | 0.45 | 0.47 | 0.99 | 3.18 | 5.86 | 4.56 | 6.69 | 6.04 | 6.98 | 7.97 | 3.81 | 1.48 |
| Dtr | 12.2 | 13.1 | 13.5 | 12.8 | 11.7 | 9.00 | 8.14 | 8.27 | 9.29 | 9.15 | 9.69 | 10.7 |
| Gff | 0.43 | 0.02 | 0.00 | 0.00 | 0.00 | 0.00 | 0.00 | 0.00 | 0.00 | 0.00 | 0.00 | 0.10 |
| Rce | 3.89 | 4.67 | 5.53 | 5.88 | 5.62 | 4.68 | 4.35 | 4.38 | 4.49 | 4.07 | 3.66 | 3.50 |
| Pe | 5.85 | 6.64 | 7.42 | 7.56 | 7.22 | 5.88 | 5.33 | 5.45 | 5.75 | 5.43 | 5.28 | 5.30 |

All of the above factors are correlated by using the method as shown in Section A.1.The final data set obtained after applying Kendall's Tau coefficient correlation analysis is given in Table2.

**Table 2.** Correlation Analysis Output

| Factors | Ppt | Mnt | Mxt | At | Cc | Vp | Wdf | Dtr | Gff | Rce | Pe |
|---|---|---|---|---|---|---|---|---|---|---|---|
| PPT | 0.99 | 0.36 | 0.06 | 0.24 | 0.54 | 0.48 | 0.87 | -0.4 | -0.5 | 0.09 | -0.0 |
| Mnt | 0.36 | 0.99 | 0.45 | 0.69 | 0.33 | 0.57 | 0.24 | -0.1 | 0.48 | 0.60 | 0.36 |
| Mxt | 0.06 | 0.45 | 0.99 | 0.75 | -0.2 | 0.03 | -0.0 | 0.36 | 0.06 | 0.84 | 0.84 |
| At | 0.24 | 0.69 | 0.75 | 0.99 | 0.03 | 0.27 | 0.12 | 0.12 | -0.1 | 0.84 | 0.66 |
| Cc | 0.54 | 0.33 | -0.2 | 0.03 | 0.99 | 0.75 | 0.66 | -0.8 | -0.6 | 0.6 | -0.3 |
| Vp | 0.48 | 0.57 | 0.03 | 0.27 | 0.75 | 0.99 | 0.60 | -0.6 | -0.7 | 0.21 | -0.6 |
| Wdf | 0.87 | 0.24 | -0.6 | 0.12 | 0.66 | 0.60 | 0.99 | 0.57 | -0.6 | -0.3 | -0.2 |
| Dtr | -0.4 | -0.1 | 0.36 | 0.12 | -0.8 | -0.6 | -0.5 | 0.99 | 0.57 | 0.21 | 0.45 |
| Gff | -0.5 | -0.4 | 0.06 | -0.1 | -0.6 | -0.7 | -0.6 | 0.57 | 0.99 | -0.0 | 0.15 |
| Rce | 0.09 | 0.60 | 0.84 | 0.84 | -0.6 | 0.18 | -0.0 | 0.21 | -0.1 | 0.99 | 0.75 |
| Pe | -0.0 | 0.36 | 0.84 | 0.66 | -0.3 | -0.0 | -0.1 | 0.45 | -0.0 | 0.75 | 0.99 |

By the above analysis it is found that only the four predictor factors Reference Crop Evapotranspiration (RCE), Maximum Temperature (MXT), Average Temperature (AT) and Potential Evapotranspiration (PE) are sufficient to determine the rainfall prediction. After obtaining the highly positively correlated variables RCE, MXT, AT, PE we will use the values of these variables (60 years) and predict the amount of rainfall in future state wise. Using Algorithm 1 we will assign the label. Using Algorithm 1 we will find the label (Average, Low, High) for each state season-wise. After obtaining this result we will apply the Random Forest method to obtain the classification rules and will also Cross-Validate the results obtained by Algorithm 1 using Rapid Miner Tool for authentication.

**Table 3. Algorithm 1 Output**

| State | Winter | Pre-Monsoon | Monsoon | Post-Monsoon | Prediction |
|---|---|---|---|---|---|
| Andhra Pradesh | Low | Average | Average | Average | Average |
| Arunachal Pradesh | Average | Average | Average | Average | Average |
| Assam | Average | Average | Average | Average | Average |
| Bihar | Average | Average | Average | Average | Average |

| | | | | | |
|---|---|---|---|---|---|
| Chhattisgarh | Average | Average | Average | Average | Average |
| West Bengal | Average | Average | Average | Average | Average |
| Gujarat Region | Low | Low | Average | Low | Low |
| Haryana | Average | Average | Average | Average | Average |
| Himachal Pradesh | Average | Average | Average | Average | Average |
| Jharkhand | Average | Average | Average | Average | Average |
| Karnataka | Low | Average | Average | Average | Average |
| Kerala | Average | Average | Average | Average | Average |
| Konkan | Low | Low | Average | Average | Average |
| Madhya Pradesh | Average | Average | Average | Average | Average |
| Maharashtra | Average | Average | Average | Average | Average |
| Eastern Region | Average | Average | Average | Average | Average |
| Orissa | Average | Average | Average | Average | Average |
| Punjab | Average | Average | Average | Average | Average |
| Rajasthan | Low | Average | Average | Low | Average |
| Sikkim | Average | Average | Average | Average | Average |
| Tamil Nadu | Average | Average | Average | Average | Average |
| Telangana | Average | Average | Average | Average | Average |
| Uttar Pradesh | Average | Average | Average | Average | Average |
| Uttaranchal | Average | Average | Average | Average | Average |

After running through Algorithm 1 we will cross validate the result using Random Forest in Rapid Miner tool.

**Figure 2. Snapshot of cross validation in random forest**

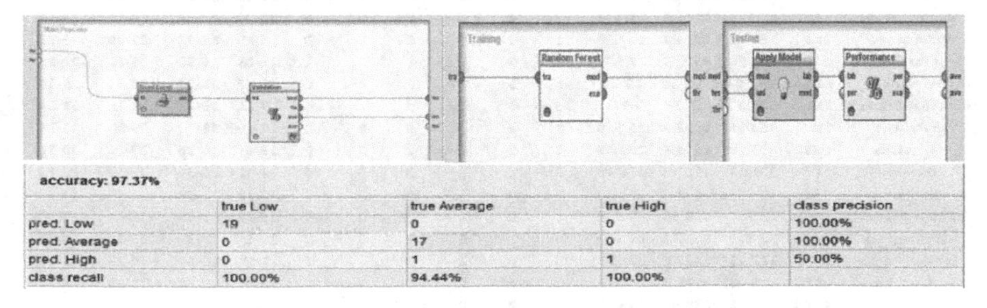

accuracy: 97.37%

| | true Low | true Average | true High | class precision |
|---|---|---|---|---|
| pred. Low | 19 | 0 | 0 | 100.00% |
| pred. Average | 0 | 17 | 0 | 100.00% |
| pred. High | 0 | 1 | 1 | 50.00% |
| class recall | 100.00% | 94.44% | 100.00% | |

After getting the rainfall analysis, we suggest the possible states in India for the growth of a particular crop. For this purpose, we take into consideration only the major crops of India. With the help of Algorithm 2, we will find the possible states for crop production. Based on the various parameters like rainfall, soil and climate of the various crops and states we have analyzed the result. Table 4 shows the data of rainfall, soil and temperature of 15 major crops of India. This data will be used as the input for Algorithm 2.

**Table 4.** Data set containing rainfall, soil & temperature of major crops.

| Crop | Rain-fall(mm) | Temp (°C) | Alluvial | Red | Laterite | Black | Forest | Desert |
|---|---|---|---|---|---|---|---|---|
| Rice | 1150-1250 | 20-30 | 1 | 1 | 1 | 1 | 0 | 0 |

| | | | | | | | | |
|---|---|---|---|---|---|---|---|---|
| Wheat | 300-1000 | 15-25 | 1 | 1 | 1 | 1 | 1 | 1 |
| Jowar | 1000 | 26-33 | 1 | 0 | 0 | 1 | 0 | 0 |
| Bajra | 400-600 | 20-30 | 0 | 1 | 0 | 1 | 0 | 1 |
| Maize | 500-1000 | 20-30 | 1 | 1 | 0 | 1 | 1 | 1 |
| Arhar | 60-65 | 20-35 | 0 | 1 | 1 | 0 | 0 | 1 |
| Urad | 50-65 | 25-31 | 0 | 1 | 1 | 0 | 0 | 0 |
| Moong | 850-1000 | 25-35 | 0 | 1 | 1 | 0 | 0 | 1 |
| Tea | 1500-2500 | 21-29 | 0 | 0 | 1 | 0 | 1 | 0 |
| Coffee | 1000-2000 | 16-24 | 0 | 0 | 1 | 0 | 1 | 0 |
| Cotton | 600-1000 | 22-30 | 1 | 1 | 1 | 1 | 0 | 0 |
| Jute | 1500-2000 | 33030 | 1 | 0 | 0 | 0 | 0 | 0 |
| GNut | 400-600 | 20-35 | 1 | 1 | 0 | 1 | 0 | 1 |
| Rubber | 2000-3000 | 20-35 | 0 | 0 | 1 | 0 | 0 | 0 |
| Sugar-cane | 1000-1750 | 20-35 | 1 | 1 | 1 | 1 | 0 | 0 |

Table 5 below shows the data of rainfall, soil and temperature for the states of India. This data will also be used as the input for Algorithm 2.

**Table 5.** Data set containing rainfall, soil & temperature of the states.

| States | Winter (mm) | Pre Monsoon (mm) | Monsoon (mm) | Post Monsoon (mm) | Alluvial | Red | Laterite | Black | Forest | Desert | Winter (°C) | Pre Monsoon (°C) | Monsoon (°C) | Post Monsoor (°C) |
|---|---|---|---|---|---|---|---|---|---|---|---|---|---|---|
| Andhra Pradesh | 13.36 | 83.794 | 483.4 | 276.56 | 1 | 1 | 0 | 1 | 0 | 0 | 20-25 | 27-33 | 25-28 | 20-25 |
| Arunachal Pradesh | 124.8 | 654.55 | 1872.4 | 217.99 | 0 | 1 | 0 | 0 | 1 | 0 | 15-21 | upto 40 | 22-30 | 22-30 |
| Assam | 42.99 | 617.52 | 1825.4 | 200.49 | 1 | 0 | 1 | 0 | 0 | 0 | 10-15. | 21-25 | 24-28 | 22-26 |
| Bihar | 23.39 | 80.061 | 992.85 | 76.153 | 1 | 0 | 0 | 0 | 0 | 0 | 15-20 | 26-32 | 26-32 | 20-25 |
| Chhatisgarh | 22.96 | 47.473 | 1151.6 | 75.345 | 0 | 1 | 0 | 0 | 0 | 0 | 20-24 | 27-34 | 26-28 | 20-25 |
| West Bengal | 31.51 | 178.12 | 1154.4 | 147.98 | 1 | 1 | 0 | 0 | 0 | 0 | 19-23 | 26-32 | 27-30 | 19-25 |
| Gujarat Region | 1.297 | 5.6773 | 698.83 | 27.602 | 1 | 0 | 0 | 1 | 0 | 1 | 19-23 | 26-32 | 28-32 | 22-27 |
| Haryana | 32.54 | 38.147 | 450.24 | 25.13 | 1 | 0 | 0 | 0 | 0 | 0 | 14-17 | 23-32 | 29-33 | 15-24 |
| Himachal Pradesh | 178.7 | 229.21 | 738.89 | 94 | 0 | 0 | 0 | 0 | 1 | 0 | 10-15. | 19-29 | 23-26 | 13-20 |
| Jharkhand | 31.71 | 82.684 | 1052.3 | 95.928 | 0 | 1 | 0 | 0 | 0 | 0 | 16-19 | 23-32 | 25-30. | 16-25 |
| Karnataka | 3.662 | 133.59 | 1443.6 | 202.45 | 0 | 1 | 1 | 1 | 0 | 0 | 19-24 | 20-25 | 20-25 | 20-25 |
| Kerala | 24.26 | 375.35 | 1967.3 | 478.19 | 0 | 1 | 1 | 0 | 0 | 0 | 22-30 | 25-35 | 23-30 | 22-32 |
| Konkan | 0.569 | 36.645 | 2903.2 | 142.05 | 0 | 0 | 1 | 0 | 0 | 0 | 20-32 | 23-34 | 23-30 | 22-32 |
| Madhya Pradesh | 16.99 | 23.807 | 893.58 | 67.953 | 0 | 1 | 0 | 1 | 0 | 0 | 16-20 | 25-34 | 25-32 | 17-25 |
| Maharashtra | 12.32 | 30.312 | 813.14 | 84.422 | 0 | 1 | 1 | 1 | 0 | 0 | 20-25 | 27-34 | 25-30 | 20-26 |
| Eastern region | 44.5 | 522.45 | 1477.6 | 221.97 | 0 | 0 | 1 | 0 | 0 | 0 | 12-15. | 17-23. | 20-25 | 15-18 |
| Orissa | 28.87 | 120.95 | 1148.9 | 146.25 | 0 | 1 | 1 | 0 | 0 | 0 | 18-23 | 25-31 | 25-30 | 18-24 |
| Punjab | 48.64 | 51.308 | 475.62 | 33.625 | 1 | 0 | 0 | 0 | 0 | 1 | 12-17. | 21-32 | 30-35 | 15-26 |
| Rajasthan | 9.283 | 17.775 | 431.83 | 17.691 | 1 | 0 | 1 | 1 | 0 | 1 | 15-20 | 24-35 | 28-35 | 17-27 |
| Sikkim | 49.17 | 465.59 | 2064.7 | 174.95 | 1 | 0 | 0 | 0 | 1 | 0 | 12-16. | 19-24 | 23-26 | 15-24 |
| Tamil Nadu | 29.08 | 131.45 | 338.79 | 448.51 | 1 | 1 | 1 | 0 | 0 | 0 | 22-25 | 25-30 | 25-30 | 22-26 |
| Telangana | 12.16 | 57.345 | 751.87 | 115.56 | 0 | 1 | 0 | 1 | 0 | 0 | 20-25 | 27-33 | 25-28 | 20-25 |
| Uttar Pradesh | 29.7 | 32.419 | 853.65 | 53.08 | 1 | 0 | 0 | 0 | 0 | 0 | 14-17 | 22-32 | 29-35 | 15-26 |
| Uttaranchal | 113.5 | 151.26 | 1100.8 | 70.961 | 1 | 0 | 0 | 0 | 1 | 0 | 10-15. | 19-29 | 25-30 | 13-24 |

Possible crops that can be cultivated in various states, which we got after applying algorithm 2 has been shown in Table 6. For representation we have used "Y"-crop can be produced and "N"-crop cannot be produced. For example In state Karnataka the crops like rice, jowar, maize, tea, coffee and sugarcane having the value Y shows that the above crops can be produced with the available growing condition of Karnataka. Eg. Karnataka= {rice, jowar, maize, tea, coffee, sugarcane}

**Table 6. Algorithm 2 Output.**

| State | Rice | Wheat | Jowar | Bajra | Maize | Arhar | Urad | Moong | Tea | Coffee | Cotton | Jute | Nuts | Rubber | Sugarcane |
|---|---|---|---|---|---|---|---|---|---|---|---|---|---|---|---|
| Andhra Pradesh | Y | Y | Y | Y | Y | N | N | Y | N | N | Y | N | Y | N | Y |
| Arunachal Pradesh | Y | Y | Y | N | Y | N | N | Y | Y | Y | Y | N | N | N | Y |
| Assam | Y | Y | N | N | Y | N | N | N | Y | Y | Y | Y | Y | Y | Y |
| Bihar | Y | N | Y | N | Y | N | N | N | N | N | Y | Y | N | N | Y |
| Chhattisgarh | Y | N | N | N | Y | Y | Y | N | N | N | N | N | N | N | Y |
| West Bengal | Y | N | N | N | Y | N | N | N | N | N | Y | N | N | N | Y |
| Gujarat Region | N | N | N | Y | Y | N | N | N | N | N | Y | N | Y | N | N |
| Haryana & Delhi | Y | N | N | N | N | N | N | N | N | N | N | N | Y | N | N |
| Himachal Pradesh | Y | N | N | N | Y | N | N | N | N | N | N | N | N | N | N |
| Jharkhand | Y | N | N | N | Y | Y | Y | Y | N | N | Y | N | N | N | Y |
| Karnataka | Y | N | Y | N | Y | N | N | N | Y | Y | N | N | N | N | Y |
| Kerala | Y | Y | N | Y | Y | N | N | N | Y | Y | N | N | Y | Y | N |
| Konkan& Goa | Y | N | N | N | N | N | N | N | N | N | N | N | N | Y | N |
| Madhya Pradesh | N | Y | Y | N | Y | Y | Y | Y | N | N | Y | N | N | N | N |
| Maharashtra | N | Y | Y | N | Y | Y | Y | Y | N | N | Y | N | N | N | N |
| North Region | N | Y | N | N | N | N | N | N | Y | Y | N | N | N | N | Y |
| Orissa | Y | N | N | N | N | N | N | N | N | Y | N | N | N | N | Y |
| Punjab | N | N | N | Y | N | Y | Y | N | N | N | N | N | Y | N | N |
| Rajasthan | N | N | N | Y | N | N | N | N | N | N | N | N | Y | N | N |
| Sikkim | Y | Y | Y | N | N | N | N | N | Y | N | N | Y | Y | N | N |
| Tamil Nadu | Y | Y | N | Y | N | N | N | N | N | N | N | N | N | N | N |
| Telangana | Y | Y | N | N | Y | Y | Y | N | N | N | Y | N | N | N | N |
| Uttar Pradesh | N | N | N | N | Y | N | N | N | N | N | Y | N | N | N | N |
| Uttaranchal | N | Y | Y | N | N | N | N | N | N | Y | N | N | N | N | Y |

After suggesting the possible states for the growth of major crops, now we will find the correlation between the different crops to know the similarity amongthem. Applyingthe Pearson Correlation analysis, a strongly positive correlation coefficient with a value of 0.99 has been taken into consideration. Based on the above calculation we have suggested the highly correlated crops with high similarity. The correlation results has been shown in Table 7By using the dataset given in Table 4 we have computed the Correlation among crops.

**Table 7.Table showing output for Crop Correlation:**

| | Rice | Wheat | Jowar | Bajra | Maize | Arhar | urad | moong | tea | coffee | cotton | jute | nut | Rubber | sugarcane |
|---|---|---|---|---|---|---|---|---|---|---|---|---|---|---|---|
| Rice | 1 | | | | | | | | | | | | | | |
| Wheat | 1 | 1 | | | | | | | | | | | | | |
| Jowar | 1 | 0.993 | 1 | | | | | | | | | | | | |
| Bajra | 1 | 0.982 | 0.997 | 1 | | | | | | | | | | | |
| Maize | 1 | 1 | 0.996 | 0.99 | 1 | | | | | | | | | | |
| Arhar | 0.3 | 0.351 | 0.461 | 0.52 | 0.377 | 1 | | | | | | | | | |
| urad | 0.2 | 0.334 | 0.445 | 0.5 | 0.36 | 1 | 1 | | | | | | | | |
| moong | 1 | 1 | 0.995 | 0.99 | 1 | 0.372 | 0.35 | 1 | | | | | | | |
| tea | 1 | 0.985 | 0.957 | 0.94 | 0.979 | 0.184 | 0.17 | 0.9808 | 1 | | | | | | |
| coffee | 1 | 0.986 | 0.959 | 0.94 | 0.981 | 0.192 | 0.17 | 0.9823 | 1 | 1 | | | | | |
| cotton | 1 | 1 | 0.995 | 0.99 | 1 | 0.372 | 0.35 | 0.9998 | 1 | 0.982 | 1 | | | | |
| jute | 1 | 0.989 | 0.963 | 0.94 | 0.984 | 0.208 | 0.19 | 0.9853 | 1 | 1 | 0.9852 | 1 | | | |
| groundnu | 0.9 | 0.973 | 0.994 | 1 | 0.979 | 0.555 | 0.54 | 0.9779 | 0.9 | 0.922 | 0.978 | 0.9 | 1 | | |
| Rubber | 1 | 0.982 | 0.952 | 0.93 | 0.976 | 0.17 | 0.15 | 0.978 | 1 | 1 | 0.9779 | 1 | 0.9 | 1 | |
| Sugarcane | 1 | 0.997 | 0.979 | 0.96 | 0.994 | 0.273 | 0.26 | 0.9945 | 1 | 0.996 | 0.9945 | 1 | 1 | 0.9944 | 1 |

Finally the highly correlated crops are shown in Table 8 based upon the result obtained from previous Table 7.We have listed the correlated crops which have almost same agricultural needs and may be cultivated together with the same growing condition.

**Table 8: Highly Correlated Crops:**

| CROP | CORRELATED CROPS |
|---|---|
| RICE | Wheat, Maize, Moong, Tea, Coffee, Cotton, Jute, Rubber, Sugarcane |
| WHEAT | Rice, Jowar, Maize, Moong, Cotton, Sugarcane |
| JOWAR | Wheat, Bajra, Maize, Moong, Cotton, Groundnut |
| BAJRA | Groundnut, Jowar |
| MAIZE | Rice, Wheat, Jowar, Moong, Cotton, Sugarcane |
| ARHAR | Urad |
| URAD | Arhar |
| MOONG | Cotton, Sugarcane, Rice, Wheat, Jowar, Maize |
| TEA | Coffee, Jute, Rubber, Sugarcane, Rice |
| COFFEE | Tea, Jute, Rubber, Sugarcane, Rice |
| COTTON | Sugarcane, Rice, Wheat, Jowar, Maize, Moong |
| JUTE | Rubber, Sugarcane, Rice, Tea, Coffee |
| GNUT | Jowar, Bajra |
| RUBBER | Sugarcane, Rice, Tea, Coffee, Jute |
| SUGARCANE | Rice, Wheat, Maize, Moong, Tea, Coffee, Jute, Rubber |

By using the result obtained in Table 6-the Algorithm 2 Output and combining the result from Table 8-the correlation output, we will suggest all possible crops which may be produced in the different states of India. For example in Karnataka the various crops like wheat, moong, cotton, jute, groundnut, and rubber will be added to the previous result.

Karnataka= {rice, jowar, maize, tea, coffee, sugarcane} + {wheat, moong, cotton, jute, ground-nut, rubber}

Karnataka= {rice, jowar, maize, tea, coffee, sugarcane, wheat, moong, cotton, jute, groundnut, rubber}

**Table 9**: State-wise Crops:

| State | Rice | Wheat | Jowar | Bajra | Maize | Arhar | Urad | Moong | Tea | Coffee | Cotton | Jute | Groundnut | Rubber | Sugarcane |
|---|---|---|---|---|---|---|---|---|---|---|---|---|---|---|---|
| Andhra Pradesh | Y | Y | Y | Y | Y | N | N | Y | N | N | Y | N | Y | N | Y |
| Arunachal Pradesh | Y | Y | Y | N | Y | N | N | Y | Y | Y | Y | Y | N | Y | Y |
| Assam & Meghalaya | Y | Y | Y | Y | Y | N | N | Y | Y | Y | Y | Y | Y | Y | Y |
| Bihar | Y | Y | Y | N | Y | N | N | Y | Y | Y | Y | Y | N | Y | Y |
| Chhatisgarh | Y | Y | Y | N | Y | Y | Y | Y | Y | Y | Y | Y | N | Y | Y |
| Gangetic West Bengal | Y | Y | Y | N | Y | N | N | Y | Y | Y | Y | Y | N | Y | Y |
| Gujarat Region | Y | Y | Y | Y | Y | N | N | Y | N | N | Y | N | Y | N | Y |
| Haryana & Delhi | Y | Y | Y | Y | Y | N | N | Y | Y | Y | Y | Y | Y | Y | Y |
| Himachal Pradesh | Y | Y | Y | N | Y | N | N | Y | Y | Y | Y | Y | N | Y | Y |
| Jharkhand | Y | Y | Y | Y | Y | Y | Y | Y | Y | Y | Y | Y | Y | Y | Y |
| Karnataka | Y | Y | Y | N | Y | N | N | Y | Y | Y | Y | Y | Y | Y | Y |
| Kerala | Y | Y | Y | Y | Y | N | N | Y | Y | Y | Y | Y | Y | Y | Y |
| Konkan & Goa | Y | Y | N | N | Y | N | N | Y | Y | Y | Y | Y | N | Y | Y |

| | | | | | | | | | | | | | | | |
|---|---|---|---|---|---|---|---|---|---|---|---|---|---|---|---|
| Himachal Pradesh | Y | Y | Y | N | Y | N | N | Y | Y | Y | Y | Y | N | Y | Y |
| Jharkhand | Y | Y | Y | Y | Y | Y | Y | Y | Y | Y | Y | Y | Y | Y | Y |
| Karnataka | Y | Y | Y | N | Y | N | N | Y | Y | Y | Y | Y | Y | Y | Y |
| Kerala | Y | Y | Y | Y | Y | N | N | Y | Y | Y | Y | Y | Y | Y | Y |
| Konkan & Goa | Y | Y | N | N | Y | N | N | Y | Y | Y | Y | Y | N | Y | Y |
| Madhya Pradesh | Y | Y | Y | Y | Y | Y | Y | Y | N | N | Y | N | Y | N | Y |
| Maharashtra | Y | Y | Y | N | Y | Y | Y | Y | N | N | Y | N | Y | N | Y |
| North Eastern region | Y | Y | Y | N | Y | N | N | Y | Y | Y | Y | Y | N | Y | Y |
| Orissa | Y | Y | N | N | Y | N | N | Y | Y | Y | Y | Y | N | Y | Y |
| Punjab | Y | Y | Y | Y | Y | Y | Y | Y | N | N | N | Y | Y | Y | Y |
| Rajasthan | N | N | Y | Y | N | N | N | N | N | N | N | N | Y | N | N |
| Sikkim | Y | Y | Y | N | Y | N | N | Y | Y | Y | Y | Y | Y | Y | Y |
| Tamil Nadu | Y | Y | Y | Y | Y | N | N | Y | Y | Y | Y | Y | Y | Y | Y |
| Telangana | Y | Y | Y | N | Y | Y | Y | Y | Y | Y | Y | Y | N | Y | Y |
| Uttar Pradesh | Y | Y | Y | N | Y | N | N | Y | N | N | Y | N | N | N | Y |
| Uttaranchal | Y | Y | Y | Y | Y | N | N | Y | Y | Y | Y | Y | Y | Y | Y |

By verifying the result obtained above with the actual crop production statistics, we suggest the crops whose cultivation is not being practiced presently. The suggestion is shown in Table 10.For example in state Karnataka the following crops moong, jute and rubber may also be produced which was not being produced. Karnataka= {moong, jute and rubber}.

**Table 10: Suggested Crops**

| State | Suggestion |
|---|---|
| Andhra Pradesh | No suggestion |
| Arunachal Pradesh | Wheat, Maize, Moong, Coffee, Cotton, Jute, Rubber, Sugarcane |
| Assam | Wheat, Maize, Moong, Coffee, Cotton, Groundnut, Rubber, Sugarcane |
| Bihar | Coffee, Cotton, Rubber |
| Chhattisgarh | Tea, Coffee, Cotton, Rubber, jute |
| West Bengal | Coffee, Rubber |
| Gujarat Region | No suggestion |
| Haryana | Tea, Coffee |
| Himachal Pradesh | Jowar, Moong, Coffee, Cotton, Jute, Rubber, Sugarcane |
| Jharkhand | Maize, Arhar, Urad, Moong, Tea, Coffee, Cotton, Groundnut, Jute, Rubber, Sugarcane |
| Karnataka | Moong, Jute, Rubber |
| Kerala | Wheat, Maize, Moong, Cotton, Jute, Groundnut, Sugarcane |
| Konkan | Wheat, Maize, Moong, Tea, Coffee, Cotton, Jute, Rubber, Sugarcane |
| Madhya Pradesh | Jowar, Bajra, Groundnut |
| Maharashtra | No suggestion |
| Eastern Region | Wheat, Jowar, Maize, Moong, Coffee, Cotton, Jute, Groundnut, Rubber, Sugarcane |
| Orissa | Coffee, Cotton, Rubber |
| Punjab | Jowar, Groundnut, Rubber |
| Rajasthan | No suggestion |
| Sikkim | Wheat, Jowar, Maize, Moong, Coffee, Cotton, Jute, Groundnut, Rubber, Sugarcane |

| | |
|---|---|
| Tamil Nadu | Wheat, Moong, Jute |
| Telangana | Arhar, Urad |
| Uttar Pradesh | Tea, Coffee, Jute, Rubber |
| Uttaranchal | Jowar, Bajra, Maize, Moong, Coffee, Cotton, Jute, Groundnut, Rubber |

## 5. Algorithm: Selection of Crops for production using Genetic Algorithm

Based on the suggested crops, we observed that it's not possible to cultivate all crops in a particular state even if it's satisfying all criteria of agricultural needs for growing crops. We found some suggestions for the cultivation of crops that are not currently being grown in different states in India. But due to factors such as seasonal constraints and land availability it is not possible to cultivate all the suggested crops at the same time or on the same land. Almost all states have possibility of producing around 5 to 9 new crops. So we have given a randomized approach for the selection of the types of crops for productivity of corresponding states throughout the year. For this randomized approach we have used genetic algorithm. The selection of best combination of crops can be achieved using the following strategy. An encoding scheme has been devised to achieve the identification of selected crops. The allocation of the encoded bits is in the following manner:

1. First bit is allocated to the selection of Important Crops like rice and Wheat.
2. Second and third bits are allocated to the selection of little important Crops like jowar, bajra, maize and moong.
3. Fourth, fifth and sixth bits are allocated to the selection of less Important Crops like arhar, urad, tea, coffee, cotton.
4. Last two bits are allocated to the selection of Crops like sugarcane, ground nuts, rubber, jute

**Table 11: Initial encoding scheme various crops selection**:

| Bit Number | Type of crop based on season wise climate | Crop Name | Genetic code |
|---|---|---|---|
| 1 | Season I | Rice | 0 |
| 1 | Season I | Wheat | 1 |
| 2 and 3 | Season II | Jowar | 0 0 |
| 2 and 3 | Season II | Bajra | 01 |
| 2 and 3 | Season II | Maize | 10 |
| 2 and 3 | Season II | Moong | 11 |
| 4,5 and 6 | Season III | Arhar | 000 |
| 4,5 and 6 | Season III | Urad | 010 |
| 4,5 and 6 | Season III | Tea | 011 |
| 4,5 and 6 | Season III | Cotton | 100 |
| 4,5 and 6 | Season III | Coffee | 111 |
| 7 and 8 | Season IV | Sugarcane | 00 |
| 7 and 8 | Season IV | Jute | 01 |
| 7 and 8 | Season IV | Rubber | 10 |
| 7 and 8 | Season IV | Groundnut | 11 |

Let us take an example of Karnataka. So for the selection purpose, now we select few random chromosomes (8 bits) for the selection purpose of the crops as described in the encoding

scheme. For every single chromosome, we count the occurrence of 1's in the particular chromosome and that count is known as fitness of the chromosome. After the calculation of fitness, for every chromosome we have determined the average fitness i.e. x in this case.
We find the chromosomes that have a value of fitness greater than or equal to the average fitness value. Next, from 12 chromosomes, 8 of them are selected. Now we consider a pair of any two chromosomes and perform crossover. We apply crossover on chromosomes pair wise. After the results obtained we decide the random mutation sites for the resultant chromosomes.

**Table 12: Initial chromosomes selection and fitness evaluation:**

| Bit 1 | Bit 2 | Bit 3 | Bit 4 | Bit 5 | Bit 6 | Bit 7 | Bit 8 | Fitness | Selected Y or N |
|---|---|---|---|---|---|---|---|---|---|
| 1 | 0 | 1 | 1 | 1 | 0 | 0 | 0 | 4 | Y |
| 0 | 0 | 1 | 0 | 0 | 0 | 1 | 0 | 2 | N |
| 0 | 1 | 0 | 0 | 0 | 1 | 1 | 1 | 4 | Y |
| 1 | 0 | 1 | 1 | 1 | 1 | 0 | 0 | 5 | Y |
| 0 | 1 | 0 | 1 | 0 | 1 | 1 | 1 | 5 | Y |
| 1 | 0 | 0 | 1 | 0 | 1 | 1 | 1 | 5 | Y |
| 0 | 1 | 0 | 0 | 0 | 0 | 1 | 1 | 3 | N |
| 1 | 0 | 1 | 1 | 1 | 0 | 1 | 0 | 5 | Y |
| 0 | 1 | 1 | 0 | 0 | 0 | 0 | 0 | 2 | N |
| 1 | 0 | 1 | 0 | 1 | 1 | 0 | 1 | 5 | Y |
| 1 | 1 | 0 | 1 | 0 | 1 | 0 | 0 | 4 | N |
| 1 | 1 | 0 | 1 | 0 | 1 | 0 | 0 | 4 | N |

**Average Fitness Value: 48/12=4**

After mutation, we obtained the set of offspring's. We select the valid offspring's among the resultant offspring's to get the selected crops for a corresponding state.

**Table 13: Intermediate steps obtained by performing mutation and crossover for the selection of crop:**

| SI no. | Selected Bits | Crossover sites | Result of cross over | Mutation Sites | Result |
|---|---|---|---|---|---|
| 1 | 11010111 | 4 | 11011000 | 7 | 11011010 |
| 2 | 10111000 | 6 | 10110111 | 6 | 10110011 |
| 3 | 10111100 | 6 | 10111111 | 6 | 10111011 |
| 4 | 1010111 | 6 | 01010100 | 4 | 01000100 |
| 5 | 10011111 | 3 | 10011010 | 3 | 10111010 |

| 6 | 10111010 | 3 | 10111111 | 5 | 10110111 |
|---|---|---|---|---|---|
| 7 | 11101101 | 5 | 11101100 | 5 | 11100100 |
| 8 | 11010100 | 5 | 11010101 | 3 | 11110101 |

All combination obtained from the genetic algorithm fails to give the desired selection of crops (especially 4, 5 and 6 bits should be either 000,010,011,100 or 111). Except the second bit no such results has the above parameters correct. So the answer is the second bit i.e. 1 01 100 11. So from the above results we can select the crops as follows:
First bit is 1 so it's Wheat.
Second and third bit are 01 so it's Bajra.
Fourth, fifth and sixth bits are 100 so it's Cotton.
Seventh and eighth bits are 11 so it's Groundnut.
Karnataka= {Wheat, Bajra, Cotton, Groundnut}.

## 6. Conclusion & Future Work

In this paper we have analyzed the rainfall data of various states over last 65 years with the help of highly correlated rainfall parameters. Next, by making use of soil, temperature and rainfall data we have suggested the major crops that can be produced in the states of India. Also we computed the similarity among crops with the help of Pearson correlation analysis method. After getting the possible crops along with the similar crops we have suggested few more crops which are currently not being produced in a given state. As per the previous results obtained we get many crops that can be grown in a state which is not possible in a year. So finally, Genetic Algorithm has been used to randomly recommend four crops in a year in a particular state. The research can be further extended by considering the factors such as land use pattern, natural sources of irrigation like Rivers, lakes etc. and the usage of modern tools and equipment to increase agricultural production.

## References

1. Time Series Analysis of Forecasting Indian Rainfall by Akashdeep , Gupta, Anjali Gautam http://www.ijies.org/attachments/File/v1i6/F0222051613.pdf

2. Impact of monsoon rainfall on the total foodgrain yield over India http://www.ias.ac.in/jess/jul2014/1129.pdf

3. The Prediction of Indian Monsoon Rainfall: A Regression Approach http://www-personal.umich.edu/~copyrght/image/solstice/sum07/Solstice_GoutamiED.pdf

4. Rainfall Patterns over India: Classification with Fuzzy c-Means Method http://link.springer.com/article/10.1007/s00704005019

5. Rainfall analysis for Indian monsoon region using the merged rain gauge observations and satellite estimates: Evaluation of monsoon rainfall features .http://link.springer.com/article/10.1007%2Fs12040-007-0019-1

6. Mishra, K. K., Brajesh Kumar Singh, Akash Punhani, and Uma Singh. "Combining neural networks and genetic algorithms to predict and to maximize lemon grass oil production." In 2009 International Joint Conference on Computational Sciences and Optimization, pp. 297-299. IEEE, 2009.

7. Singh, B.K., Punhani, A. and Mishra, K.K., 2009, December. Hybrid computational method for optimizing isabgol production. In 2009 World Congress on Nature & Biologically Inspired Computing (NaBIC).

8. Singh, B. K., A. Punhani, and R. Nigam. "Notice of Retraction Performance evaluation of sigmoid functions with hybrid computational method for Optimizing Lemon Grass Oil Production." In Computer Science and Information Technology (ICCSIT), 2010 3rd IEEE International Conference on, vol. 9, pp. 379-383. IEEE, 2010.

9. Geo-spatial analysis of the temporal trends of *kharif* crop phenology metrics over India and its relationships with rainfall parameters

10. http://www.yourarticlelibrary.com/soil/soil-groups-8-major-soil-groups-available-in-india/13902/

11. http://www.mapsofindia.com/maps/schoolchildrens/major-soil-types-map.html

12. http://listz.in/top-10-rice-producing-states-in-india.html

13. http://listz.in/top-10-wheat-producing-states-in-india.html

14. http://www.preservearticles.com/2012020422630/complete-information-on-area-and-productionofjowar-in-india.html

15. http://listz.in/top-10-sugarcane-producing-states-in-india.html

16. [http://listz.in/top-5-largest-coffee-producing-states-in-india.html

17. http://listz.in/top-10-cotton-producing-states-in-india.html

18. http://www.onlinegk.com/geography/crops-and-leading-producers-states/

19. http://www.commoditiescontrol.com/eagritrader/staticpages/index.php?id=89

20. https://data.gov.in/catalog/area-weighted-monthly-seasonal-and-annual-rainfall-mm-36meteorological-subdivisions

21. https://data.gov.in/catalog/gross-area-under-irrigation-crops

22. https://data.gov.in/resources/production-major-crops/download

# Fuzzy based Autonomous Parallel Parking Challenges in Real time Scenario

[1]Naitik Nakrani and [2]Maulin Joshi

**Abstract** Fuzzy based automation in automobile industries has attracted many researcher in recent years for their ability to adapt human like expertise. In this paper, combination of fuzzy based navigation and parallel parking problem is discussed. Navigation algorithm makes a vehicle able to reach to destination avoiding nearby obstacles. Main objective of this paper is to highlight and explore challenges present in parking system. Multipurpose parallel parking system is mainly focused. Different forward and reverse parking situations are considered in parallel parking. Matlab simulations are also provided to aid and validate claim of challenges.

## 1 Introduction

Increasing interest in automation of automobiles attracts many researchers in recent years. Many automobile industries are working in developing autonomous vehicle by putting artificial intelligence in machine. Autonomous vehicle parking [2, 5, 7, 8, 9, 10, 12] is also one part of complete autonomous vehicle where vehicle parks itself by sensing nearby conditions. It comes with many challenges out of which few is being introduced in this paper.

---

[1] Naitik Nakrani
Assistant Professor, CGPIT,UTU, Bardoli, naitik.nakrani@utu.ac.in
[2] Maulin Joshi
Professor, SCET, Surat, maulin.joshi@scet.ac.in

 
J.M. Corchado Rodriguez et al. (eds.), *Intelligent Systems Technologies and Applications 2016*, Advances in Intelligent Systems and Computing 530,
DOI 10.1007/978-3-319-47952-1_63

Autonomous Parking is ability of vehicle to start from its initial location and posture and reach to its final location and posture in a parking area. Typically parking problem is classified into: Garage parking, Parallel parking and Diagonal parking. This paper describes difficulties present into parallel parking where parking is being done in parallel to curb of the road. Parallel parking has many advantages over the other stage of parking like less space required, no need for specified parking slots and it can be used in streets, markets, society anywhere. Although choice of parking totally depends on actual environment conditions.

Many literatures [2, 5, 8, 9, 10, 12] have discussed and proposed techniques for autonomous parallel parking. In most of methods, steering control mechanism is described and experienced. Also different trajectories are also described with taking care of non- holonomic constraints. Basically controlling methods of vehicle to achieve parking are categorized into two approaches. One is path planning approach [2, 3, 4] where feasible path is pre-planned taking account of environment and non-holonomic constraints. Control strategy generates command to manoeuvre vehicle on that path. Many trajectories [1, 2, 4, 12] have been used for such path generation but among them all fifth order polynomial [5, 12] provides better steer control fulfilling differential drive mechanism. Second is soft computing based approach [5, 6, 7, 8, 9, 10], which uses feedback to decide each and every step size and motion of vehicle. Fuzzy, neural or genetics based approaches have been widely popular since last many years. Advantage of second is it does not require pre knowledge of environment and it can take decision on its own during runtime in any difficult circumstances. Thus it makes them online path planning.

Generally Vehicle is having nonlinear and time varying kinematics equations with non-holonomic constraints. It is hard to control them automatically with any traditional control strategies. Any human driver has that skill set to manoeuvre vehicle in perfect manner by sensing nearby situations without even knowing kinematics equations. Such intelligence can be implemented in vehicle using some linguistic logic theories like Fuzzy control. Since last two decades fuzzy logic is arising as a promising tool to formulate and translate human skills to machine language.

This paper is organized as follows. In section 2 well designed and suitable parking system is described. It generally has idea to distribute and divide entire parking problem into three steps: Navigation, via point location and parking. In section 3 navigation is adapted [11, 13] which used 4 input 2 output fuzzy system. In section 4 choice of via points are described. One of this via point will act as start location for parking. Also parallel parking mechanism is adapted [5] which is 2 input 1 output fuzzy system. Challenges present in this domain are explained in section 5. Simulations are provided to give support at each stages. Section 6 concludes this paper.

## 2 Parking System

### *2.1 Architecture*

In this section overall idea of parking scenario is described. Parking problem can be generalized as extended version of navigation. Parking including navigation can make high level of autonomy in vehicle. Fig. 1 shows a parking environment with parking slot, few points (called as via points for further reference) angles at 22.5° and road map.

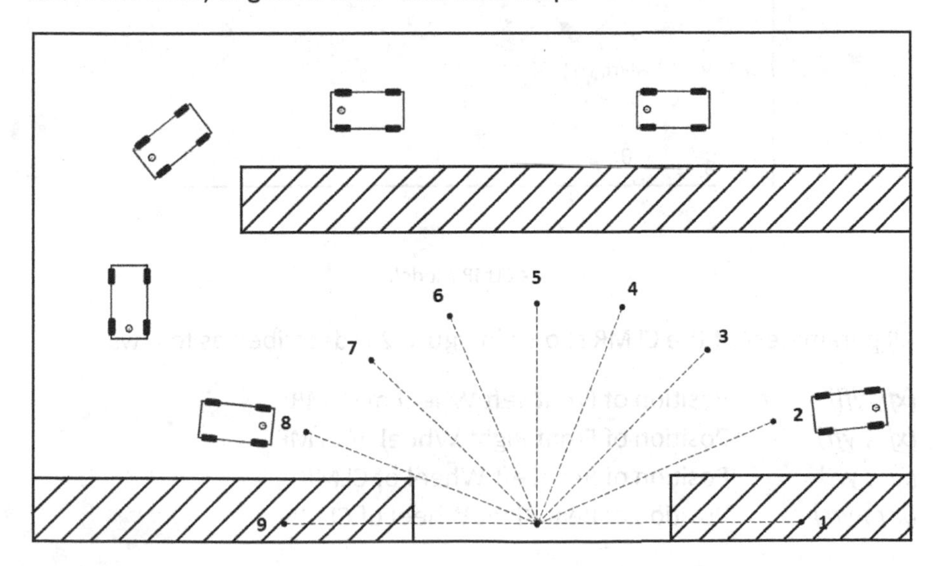


**Fig. 1** A Parking System Overview.

Before actual parking problem starts, vehicle has to reach to nearby via points through navigation. For Navigation task, destination location will be any one of this via points. Selection of via points is depends on environment. I.e. point numbers 1 and 9 will not be used as destination of navigation because of their inaccessibility. Once vehicle reach to any one of the via points algorithm must be shift to parking control. After this, either path planning or soft computing based approach can be used.

## *2.2 Car like mobile robot model (CLMR)*

Consider a car like mobile robot model as shown in figure 2. Vehicle is able to turn its front wheels into left and right but they remain parallel to each other. Rear wheels are allowed to roll but not slip and they are parallel to car body.

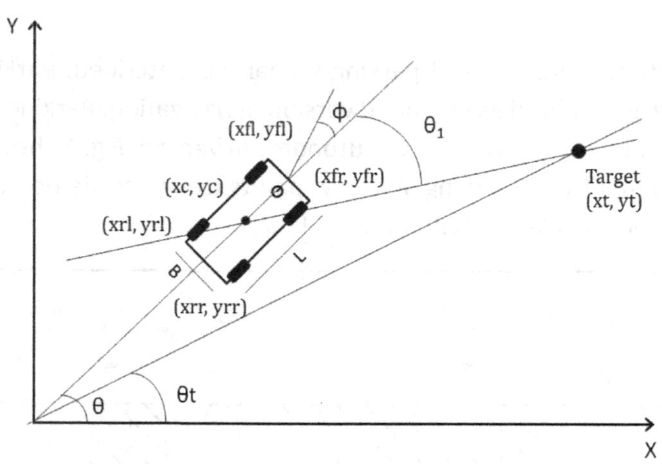


**Fig. 2** CLMR Model.

All parameters of the CLMR shown in figure 2 is described as follows:

| | |
|---|---|
| *(xfl, yfl)* | Position of Front Left Wheel of CLMR |
| *(xfr , yfr)* | Position of Front Right Wheel of CLMR |
| *(xrl , yrl)* | Position of Rear Left Wheel of CLMR |
| *(xrr , yrr)* | Position of Rear Right Wheel of CLMR |

| | |
|---|---|
| *(xc , yc)* | Position of Center of CLMR with respect to its length and width |
| *(xt , yt)* | Position of Target (if any) |
| θ | Orientation of Vehicle Body with respect to X axis |
| θt | Orientation of Target with respect to X axis |
| θ1 | Angle between CLMR body and Target |
| ϕ | Orientation of the steering-wheels with respect to the CLMR body i.e. steering angle |
| L | Length of CLMR |
| B | Width of CLMR |

The kinematics equations of CLMR model is as follows:

$$\theta_{new} = \theta_{old} + \dot{\theta} * dt \tag{1}$$

$$x_{new} = x_{old} + v * \cos(\theta_{new}) * dt \tag{2}$$

$$y_{new} = y_{old} + v * \sin(\theta_{new}) * dt \tag{3}$$

Equations 1, 2 and 3 are used to obtain new position of vehicle at each instances. $\dot{\theta}$ is normally output of fuzzy system. Velocity is assumed to be constant during parking.

## 3 Navigation

### *3.1 System Design*

Navigation problem discussed earlier problem is adapted from our earlier work [11, 13]. It discussed neuro fuzzy based Autonomous Mobile robot navigation using 4 input 2 output fuzzy system with unique range measurement algorithm. Total nine sensors are used to take distance measurements from nearby obstacles, which then grouped intro 3, left, middle and right and min ( each group sensor ) is given as 3 input to fuzzy system. Fourth input is heading angle which is derived from target position and current position of CLMR. Front Left and Front right wheel velocities are 2 output of fuzzy system.

Membership functions of all input and output used in [11, 13] is shown in figure 3. Among four inputs all left, front and right sensor inputs membership function is given in fig. 3(a). Heading angle is fuzzified into Negative, Zero and Positive membership functions as shown in fig. 3(b). Output of fuzzy system is individual velocities of front left and front right wheels, which are fuzzified into same slow, med and fast membership function as per fig. 3(c).

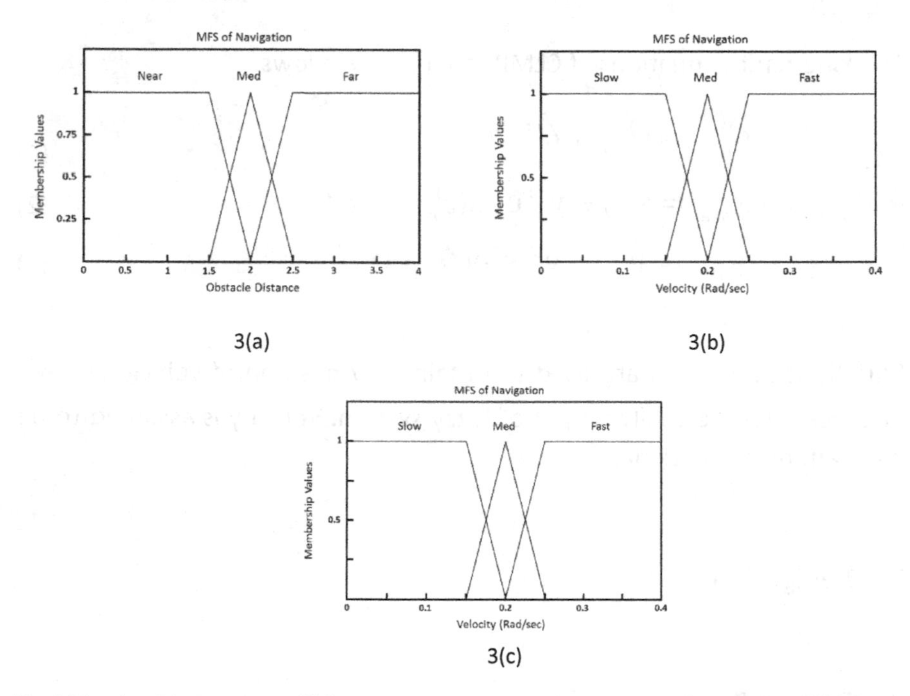


**Fig.** 3 Membership Functions of (a) Input: Sensor Distance (b) Input: Heading Angle and (c) Output: Left and Right Velocity.

**Table 1.** Fuzzy If – then rules for Navigation

| | If | | | | | Then | |
|---|---|---|---|---|---|---|---|
| Rule no. | Fuzzy Behaviour | Left Obs | Front Obs | Right Obs | Head Ang | Left Vel | Right Vel |
| 1 | Target Steer | Far | Far | Far | N | Low | Fast |
| 2 | Target Steer | Far | Far | Far | Z | Fast | Fast |
| 3 | Target Steer | Far | Far | Far | P | Fast | Low |
| 4 | Obstacle Avoid-ance | Near | Near | Far | N | Fast | Low |
| 5 | Edge Following | Far | Far | Near | P | Med | Med |

Based on this input information, fuzzy rule base is defined to realize various behaviour like target steer, obstacle avoidance, edge following etc. few samples of these rules is given in table 1. These fuzzy rules show that the robot mainly adjusts its motion direction and quickly moves to the target if there are no obstacles around the robot. When sensors data indicates that there exist obstacles nearby robot; it must try to change its path in order to avoid those obstacles.

### *3.2 Simulation results*

In this section simulation results are provided to support navigation system. A Matlab environment is designed to validate fig. 1. A Car like mobile robot with Length =1.5 m and width =0.75 m is assumed. Total 9 different sensors equally separated at $\pi/8$ are used to detect nearby obstacle distance in radial direction upto 3 meter. Heading angle is assumed with respect to Y axis. Total 81 Fuzzy rules are used to calculate updated turning angle. Defuzzification method used is centroid.

Fig. 4(a) indicates path that CLMR takes while navigating through any one of the via points. For forward parking via points can be opted from any 4 points which has x location less then target x location. If vehicle is to be parked in reverse manner it has to overcome the parking slot and need to reach any one of 4 points whose x location is grater then target x location. Fig. 4 (b) shows path than CLMR takes to go ahead of parking. Decision of Forward and Reverse parking is taken manually here for only demonstrating purpose. Red dots indicated Front side of CLMR

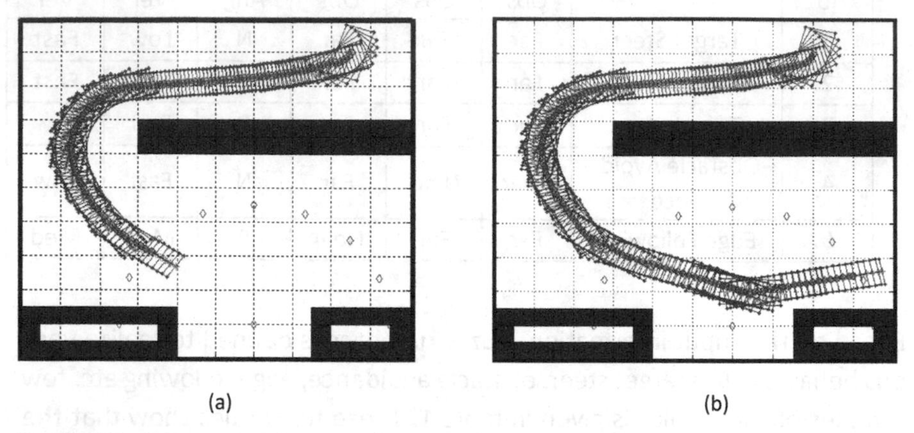

(a) (b)

**Fig. 4.**(a) Navigation to Initial Location for Forward Parking.(b) Navigation to Initial Location for Reverse Parking.

# 4 Parallel Parking

## 4.1 System Description

Once vehicle reach to any one of the via points mentioned in fig. 1 through navigation, it has to be switched to parking problem. Based on availability of parking space and environment constraint decision needs to be taken for forward or reverse parking. As per fig.1 , point8 can act as starting location for forward parking and point9 can act as starting location for reverse parking.

Parking algorithm used here is based on Li and Chang [5]. They proposed Fuzzy parallel parking control (FPPC) with aim to follow vehicle onto fifth order reference trajectories. Fifth order polynomial path is mentioned as most suitable path considering kinematic constraints [1, 5, 12]. The parameters used to construct FPPC are starting and final location of vehicle, orientation of vehicle θ, orientation of target θt and angle between vehicle body and target ϕ. This parameters are used to determine change on steering angle i.e. steering rate using linguistic fuzzy rule base system.

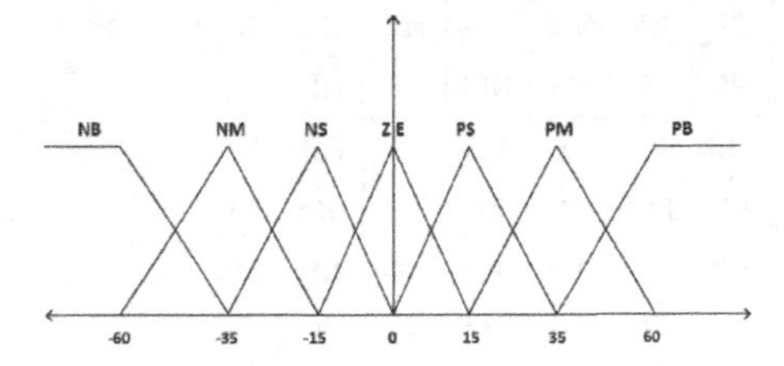


**Fig. 5 (a)** Membership function of Input 1.

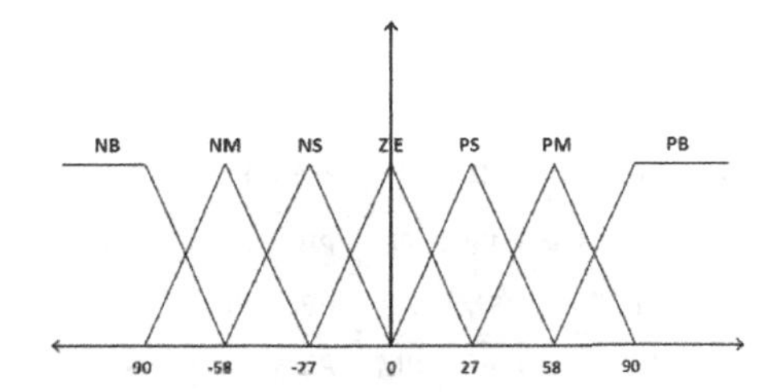


**Fig. 5 (b)** Membership function of Input 2.

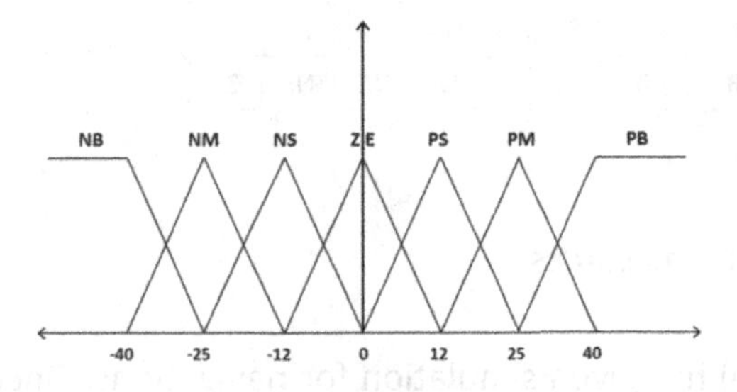


**Fig. 5 (c)** Membership function of Output.

Input and output variables used in [5] are fuzzyfied into 7 various membership function shown in fig.5 (a), 5 (b) and 5(c). Based on 7 different fuzzy partition 49 Fuzzy rules has been designed for parallel forward and parallel reverse parking control. These rules are based on linguistic human driver expertise converted into machine learning. All set of rules are shown in Table 2 and Table 3. Input1 is angle difference between θ1 and θt, input2 is angle difference between θ and θt and output is steering angle ϕ. All angles are measured with respect to X – axis.

**Table 2.** Fuzzy Rule for Reverse FPPC [5]

| i1 \ i2 Φ | **NB** | **NM** | **NS** | **ZE** | **PS** | **PM** | **PB** |
|---|---|---|---|---|---|---|---|
| **NB** | ZE | NS | NM | NB | NB | NB | NB |
| **NM** | PS | ZE | NS | NM | NB | NB | NB |
| **NS** | PM | PS | ZE | NS | NM | NB | NB |
| **ZE** | PB | PM | PS | ZE | NS | NM | NB |
| **PS** | PB | PB | PM | PS | ZE | NS | NM |
| **PM** | PB | PB | PB | PM | PS | ZE | NS |
| **PB** | PB | PB | PB | PB | PM | PS | ZE |

**Table 3.** Fuzzy Rule for Forward FPPC [5]

| i1 \ i2 Φ | **NB** | **NM** | **NS** | **ZE** | **PS** | **PM** | **PB** |
|---|---|---|---|---|---|---|---|
| **NB** | ZE | PS | PM | PB | PB | PB | PB |
| **NM** | NS | ZE | PS | PM | PB | PB | PB |
| **NS** | NM | NS | ZE | PS | PM | PB | PB |
| **ZE** | NB | NM | NS | ZE | PS | PM | PB |
| **PS** | NB | NB | NM | NS | ZE | PS | PM |
| **PM** | NB | NB | NB | NM | NS | ZE | PS |
| **PB** | NB | NB | NB | NB | NM | NS | ZE |

## 4.2 Simulation results

Section 3 (b) has given simulation for navigations. Once CLMR reached to any of appropriate via points control is switched to Parking control. For parking above mentioned system is used. Dimension of Environment and CLMR is unchanged. Angles used in parking system are measured with respect to X axis. Total 49 Rules different for Forward and Reverse FPPC is used to calculate turning angle as per Table 2 and Table 3. For parking problem velocity is assumed to constant as 1 m/s. Differential drive equation given in section2 (b) are used to calculate updated turning angle at each step. Here also centroid method is used as defuzzification. Forward Parking is shown in fig. 6 (a) and Reverse parking is shown in fig. 6(b).

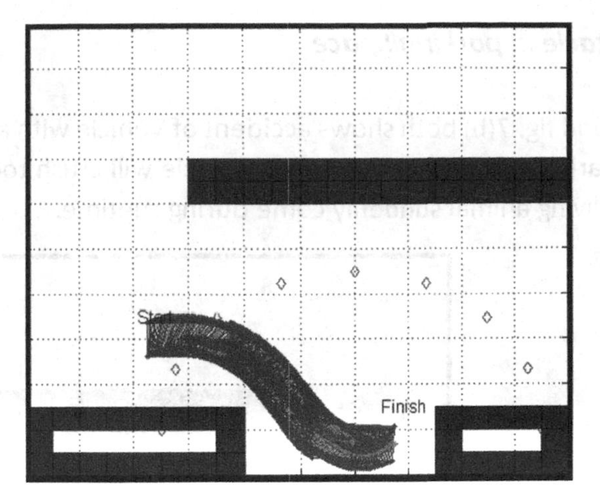


**Fig. 6 (a)** Forward Parking from Via Points.

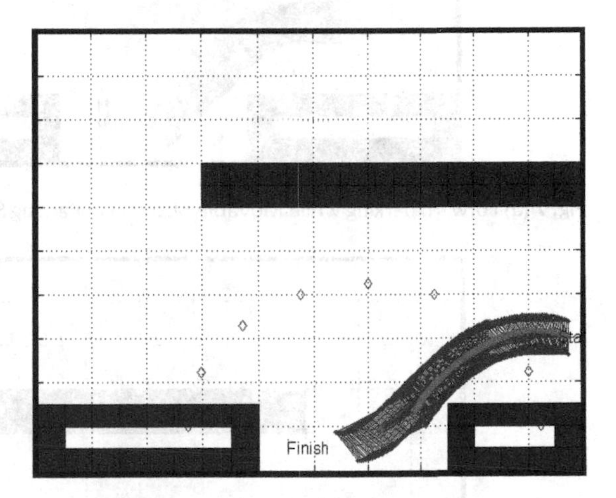


**Fig. 6 (b)**Reverse Parking From Via Points.

## 5 Challenges

All the earlier approaches used to solve parking problem have many limitations. The control system described early does not consider the avoidance of unforeseen obstacles in the parking space. Parking environment and Obstacles are not considered as dynamic in any cases. Following two challenges are discussed where popular parking problem fails.

## *5.1 Obstacle in parking Space*

Fig. 7(a) and fig. 7(b) both shows accident of Vehicle with any obstacle present in parking during manoeuvring. Vehicle will crush too if there is present any living animal suddenly come during runtime.

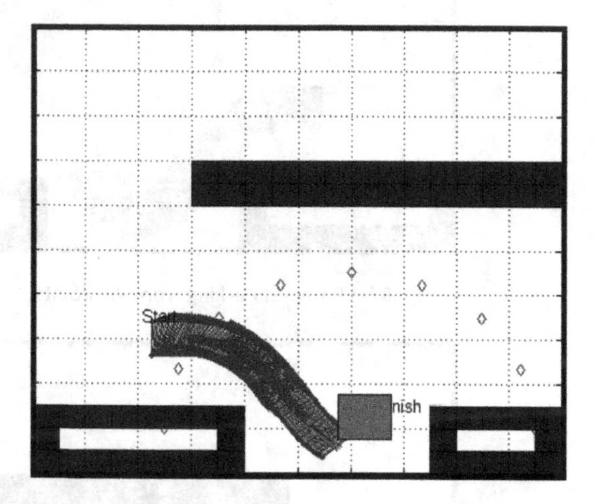

**Fig. 7 (a)** Forward parking while Movable obstacle in Parking Space.

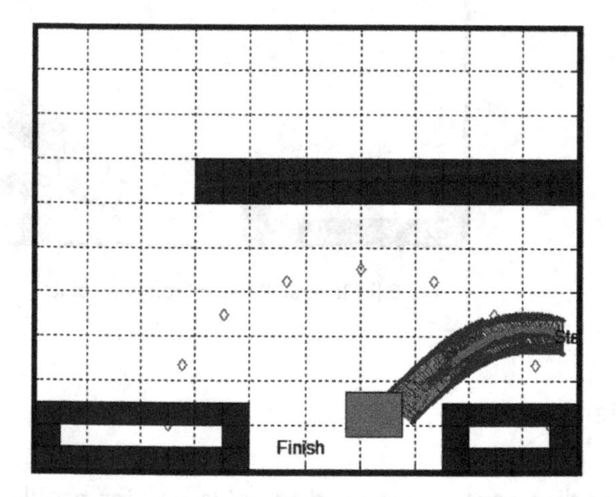


**Fig. 7 (b)** Reverse parking while Movable obstacle in Parking Space.

### *5.2 Size of Vehicle*

For this challenge dimension of vehicle is assumed double as compared to previous simulation results. Fig. 8(a) and 8(b) shows failure of parking because in both cases CLMR touched corner of parking and also it goes deep onto curb. No such adaptive fuzzy mechanism is provided to solve such problems where vehicle dimensions are taken into considerations.

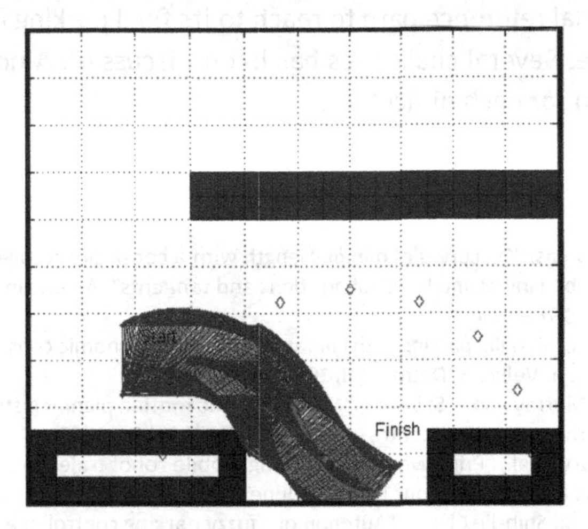


**Fig. 8 (a)** Forward parking while considering larger dimension of CLMR.

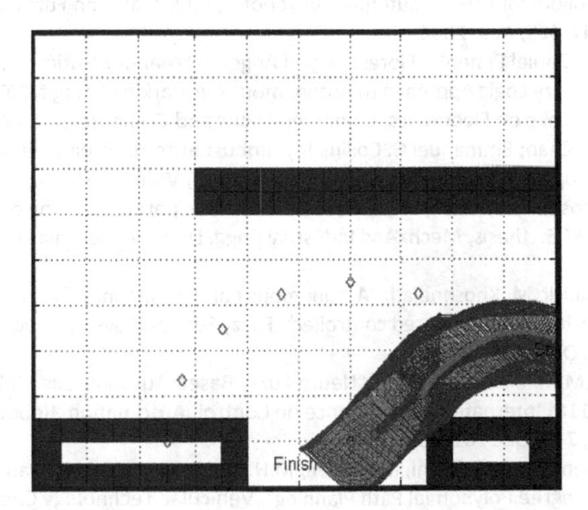


**Fig. 8 (b)** Reverse parking while considering larger dimension of CLMR.

## 6 Conclusion

In this paper futuristic parking system combining navigation and core parking control is perceived. Navigation for CLMR works perfectly in even dynamic or static environment, which makes it suitable for any real time system. Selection of via points is depends on system requirements. Parallel parking generated fuzzy control decisions so that CLMR follows fifth order polynomial reference path to reach to its final parking location inside parking space. Several challenges has been discussed. A novel solution can be proposed for each difficulties.

# References

1 L.E. Dubins, "On curves of minimal length with a constraint on average curvature and with prescribed initial and terminal positions and tangents", American Journal of Math 79 (1957) 497 – 516
2 D. Lyon, "Parallel parking with curvature and nonholonomic constraints," in Proc. Symp. Intelligent Vehicles, Detroit, MI, 1992, pp.341–346
3 R. M. Murray and S. S. Sastry, "Nonholonomic motion planning:steering using sinusoids," IEEE Trans. Automat. Contr., vol. 38, pp.700–716, May 1993
4 S. Fleury et al., "Primitives for smoothing mobile robot trajectories", IEEE transactions on robotics and automation, 11, No. 3, June 1995.
5 Li T.-H.S., Shih-Jie Chang, "Autonomous fuzzy parking control of a car-like mobile robot", IEEE Trans. on Syst. Man Cybern. A, Syst. Humans, vol.33, no.4, pp.451-465, July 2003.
6 Baturone I., Moreno-Velo F.J., Sánchez-Solano S., Ollero A., "Automatic design of fuzzy controllers for car-like autonomous robots", IEEE Trans. on Fuzzy Syst., vol.12, no.4, pp.447-465, Aug. 2004.
7 Carlos Daniel Pimentel Flores; Miguel Ángel; Hernández Gutiérrez; Rubén Alejos Palomares, "Fuzzy Logic Approach to Autonomous Car Parking Using MATLAB", IEEE International Conference on Electronics, Communications and Computers, Jan 2005
8 Yanan Zhao; Emmanuel G. Collins Jr., "Robust automatic parallel parking in tight spaces via fuzzy logic", Robotics and Autonomous System, Volume 51, Issues 2–3, 31 May 2005
9 M. khoshnejad, "Fuzzy logic based autonomous parallel parking of a car-like mobile robot", M.S. thesis, Mech. And Industrial Engg. Dept., Concordia University, Montrealo, CA, 2006
10 K. Demirli, M. Khoshnejad, "Autonomous parallel parking of a car-like mobile robot by a neuro-fuzzy sensor-based controller", Fuzzy Sets and Systems, vol. 160, Issue 19, pp. 2876-2891, Oct. 2009.
11 Joshi M.M. and Zaveri M.A., "Neuro-Fuzzy Based Autonomous Mobile Robot Navigation", IEEE 11th International Conference on Control, Automation, Robotics and Vision , ICARCV 2010, 7-10 Dec 2010, Singapore
12 Shuwen Zhang; Simkani, M.; Zadeh, M.H., "Automatic Vehicle Parallel Parking Design Using Fifth Degree Polynomial Path Planning", Vehicular Technology Conference (VTC Fall), 2011 IEEE, pp.1-4, 5-8 Sept 2011.
13 Joshi M.M., Ph.d thesis "Fuzzy Logic Based Autonomous Mobile Robot Navigation", SardarVallabhbhai National Institute of Technology, July 2012

# Application of a Hybrid Relation Extraction Framework for Intelligent Natural Language Processing

Lavika Goel, Rashi Khandelwal, Eloy Retamino, Suraj Nair, Alois Knoll

**Abstract** When an intelligent system needs to carry out a task, it needs to understand the instructions given by the user. But natural language instructions are unstructured and cannot be resolved by a machine without processing. Hence Natural Language Processing (NLP) needs to be done by extracting relations between the words in the input sentences. As a result of this, the input gets structured in the form of relations which are then stored in the system's knowledge base. In this domain, majorly two kinds of extraction techniques have been discovered and exploited - rule based and machine learning based. These approaches have been separately used for text classification, data mining, etc. However progress still needs to be made in the field of information extraction from human instructions. The work done here, takes both the approaches, combines them to form a hybrid algorithm and applies this to the domain of human robot interactions. The approach first uses rules and patterns to extract candidate relations. It then uses a machine learning classifier called Support Vector Machine (SVM) to learn and identify the correct relations. The algorithm is then validated against a standard text corpus taken from the RoCKIn transcriptions and the accuracy achieved is shown to be around 91%.

Lavika Goel
Birla Institute of Technology and Science, Pilani, India e-mail: lavika.goel@pilani.bits-pilani.ac.in

Rashi Khandelwal
Birla Institute of Technology and Science, Pilani, India e-mail: rashi.khandelwal2011@gmail.com

Eloy Retamino
TUM CREATE, Singapore e-mail: eloy.retamino@tum-create.edu.sg

Suraj Nair
TUM CREATE, Singapore e-mail: suraj.nair@tum-create.edu.sg

Alois Knoll
Technical University of Munich (TUM), Germany e-mail: knoll@in.tum.de



J.M. Corchado Rodriguez et al. (eds.), *Intelligent Systems Technologies and Applications 2016*, Advances in Intelligent Systems and Computing 530,
DOI 10.1007/978-3-319-47952-1_64

# 1 Introduction

Natural Language Processing (NLP) comprises of two parts :

1. Extracting the meaning of natural language input and converting it to a relational form.
2. Converting the system generated output back to a human readable form.

The former is more challenging because of the intrinsic ambiguities that arise in human languages. Hence an automated and efficient tool is needed to unambiguously extract the meaning and relations from human inputs and store them in an intelligent system's knowledge base. The meaning of the input is understood by extracting relations occurring between the words in the sentence. Additional information is acquired by analyzing the sentence structure. Once the important words and the relations between them are identified, the sentence is transformed into a relational representation. This representation is finally integrated into the bigger knowledge graph, which the system can later access.

The following section talks about the earlier works done in this area and explains how the work done here combines and applies these methods to new application domains, like human robot interactions. The section on the proposed framework discusses the different parts of the algorithm. At first the text corpus and its use is described. The pre-processing phase is then discussed, where the natural language sentences are transformed into a set of relations. Next, the section on matching and identification of the potential candidates, based on rules and patterns, is explained. These relation triples are converted to feature vectors in the following step, and are passed to the machine learning classifier. The classifier either classifies them into meaningful relations or discards them. Thereafter, the obtained results are illustrated and discussed, concluding with the future scope of this work.

# 2 Related Work

Relation extraction from text has been an important area of research for a long time. People have proposed methodologies which have been used by different applications like search engines, text classifiers, query answering systems, etc. However in the domain of human robot interactions there is still a lack of automated mechanisms that can understand natural language instructions, resolve ambiguities and integrate new information into the system's knowledge. There is also an absence of publicly available corpora made up of human instructions, that are required for validating new approaches.

For relation extraction from text, initially rule based approaches were implemented, where inputs were matched against set rules and patterns. Manually defined patterns needed an extensive knowledge of the domain, and even when this domain knowledge was available, the algorithms had low recall values. Results for specific queries were good but extending the patterns each time new queries were encountered, required tremendous efforts.

As a result machine learning was deployed, in order that the system could learn new relation types. Different kinds of machine learning classifiers were modeled and used [1]. At first, kernel methods were explored [2], [3] for classification. An improvement to this, was to adopt semi-supervised learning [4] with feature based methods. For such classifiers, feature sets had to be generated. The type and number of lexical and syntactic features that were used, varied according to the different domains and applications. Eventually Support Vector Machines (SVM) were observed to give the best results.

Although these different techniques had already been discovered, they were not readily used for processing human inputs given to robotic systems. There was also a lack of available data sets that captured such human robot interactions, making it difficult to validate and compare the approaches.

The work done here is different from earlier attempts and algorithms, in that it proposes the use of a hybrid approach integrating both rule based and machine learning based methods. This significantly reduces the shortcomings that were encountered while using the methods separately. It also validates the algorithm against a recently released data set that has transcriptions comprising of instructions given by a human to a robot. The results obtained after applying the algorithm to the data set are quite promising.

## 3 Proposed Framework

The proposed approach is a hybrid algorithm which combines both rule based and machine learning approaches for extracting relations from natural language input. The overall outline of the approach is as shown in Figure 1. Each of the steps is explained in the upcoming sections.

### *3.1 Text Corpus*

For the relation extraction task, inputs in the form of natural language instructions given by a human to a robot are needed. For this purpose the corpus made available by RoCKIn is used. During RoCKIn (Robot Competitions Kick Innovation in Cognitive Systems and Robotics) events, several data sets have been collected to be redistributed to the robotics community. These include transcriptions of monologues, wherein a human gives simple commands and instructions to a robot, to perform certain tasks.

## 3.2 Pre-Processing Phase

The natural language input in its unstructured form cannot be understood by a machine and thus it has to be processed into a relational representation.

This phase extracts the different lexical units (nouns, verbs, etc) from the sentence. First the sentence is disintegrated and broken into individual words and tokens by a tokenizer. This step removes the less important words. Next, the tokens are tagged with their Part of Speech (POS) tags, using the Part of Speech (POS) Tagger. At the same time the verbs are converted into their root forms so as to simplify the process. For example a verb like "seeing" is converted to its root form "see". If there are proper nouns, they are further labeled as names of people, places or organizations. This tagging process is called Named Entity Recognition (NER). These pre-processing functionalities are implemented using Python's Natural Language Tool Kit (NLTK). At the end of this phase, a parse tree depicting the structure of the sentence and the relations between the words is obtained.

Figure 2 shows the parse tree obtained after tagging the different words in the sample input sentence, "John is seeing Anna". Along with the parts of speech tags, John and Anna are also tagged as names of persons. When a user gives this input, the system needs to identify the relation triple [John, see, Anna] where John is the subject, see is the predicate and Anna is the object.

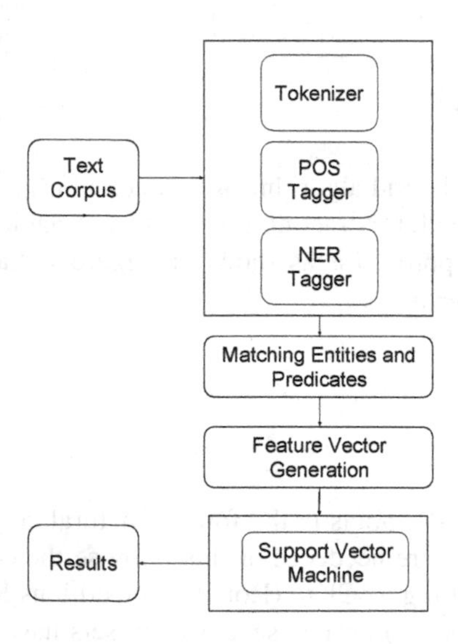


**Fig. 1** Outline of the algorithm

**Table 1** Count of different tokens

| | Entities | Predicates | Others | Total |
|---|---|---|---|---|
| Transcription 1 | 266 | 242 | 783 | 1291 |
| Transcription 2 | 313 | 226 | 912 | 1451 |
| Transcription 3 | 534 | 470 | 1765 | 2769 |

## *3.3 Matching Entities and Predicates*

The proposed algorithm combines the rule based and machine learning approaches. The rule based approach states that all relations in the input sentence should be in the form of triples : [Subject, Predicate, Object]. The subject and object are called entities. The general trend in English sentences shows that the subjects and objects are mostly nouns, hence the same restriction is applied here. The predicate is the relation connecting the subject and object, and it can belong to the following parts of speech: verb, preposition, adverb and adjective. Once these rules are applied, the sets of potential subjects, objects and predicates for the relations is obtained. Table 1 shows the count of the different kinds of tokens that were extracted from the actual input sentences, taken from three different transcriptions of the RoCKIn data set.

Since the rule based approaches have low recall values, incorrect relations might be produced if these sets are directly used. Hence a further step of using classifiers is carried out. The triple as it is cannot be passed to the classifier. And along with the triple, information about the context of the sentence also needs to be passed for better classification Hence, these triple combinations are converted to suitable feature vectors. Feature vectors are a set of numerical values that indicate different aspects of the relation. In order to get these numerical values WordNet is used. WordNet is made up of an exhaustive collection of all the important lexical types that are present in the English language. As a pre-processing step the entire graph of WordNet is captured and each of the entities present are mapped to a certain value. This hash map is stored and is then used to obtain numerical values for words in order to form the feature vector. Sometimes the exact word may not be present in

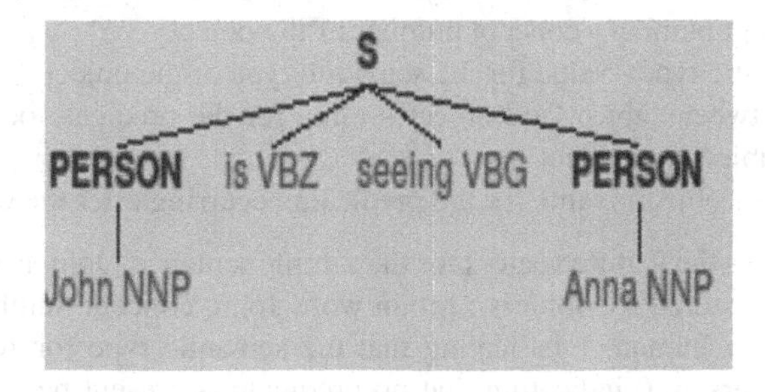


**Fig. 2** Parse tree of the sentence "John is seeing Anna"

the WordNet graph, but related words might be there. Hence, while looking up for values of words, WordNet synsets and hypernyms are useful. Synsets are the set of synonyms for a word. Hypernyms are words or entities in the WordNet graph that are more general in context and are ancestors of the word in the hierarchical structure. Hence a thorough search for the suitable synset or hypernym is also done, to get the word that is closest in meaning to our main word.

## 3.4 Feature Vector Generation

As mentioned previously, nouns make up the subject and object types of the relations. Thus it is useful to store the information for all the nouns, and then generate feature vectors between all combinations of subject and object, in order to capture every possible relation in the input. These feature vectors are then either classified as meaningful relations or are discarded. For every noun, the following information in the form of a set of values is stored :

1. Word - entity word
2. Concept number - the value for the noun in the WordNet graph that exactly matches to, or is closest in meaning to this word
3. Semantic type - value for the category (name of person, organization, location) to which this entity belongs
4. Predicate before - value for the predicate (verb, adjective, preposition, adverb) in the WordNet graph, occurring before the entity in the sentence
5. Distance - distance of the word from the root in the parse tree
6. Predicate after - value for the predicate (verb, adjective, preposition, adverb) in the WordNet graph, occurring after the entity in the sentence

The vectors for the subject and object are then combined to get the feature vector for the entire triple: [Subject, Predicate, Object] having the following attributes :

1. Subject concept number - concept number of the subject
2. Subject semantic type - value for the semantic type of the subject
3. Predicate before subject - value for the predicate occurring before the subject
4. Distance - distance between the subject and object in the parse tree
5. Object concept number - concept number of the object
6. Object semantic type - value for the semantic type of the object
7. Predicate between subject and object - value for the predicate occurring in between the subject and object
8. Predicate after object - value for the predicate occurring after the object

Figure 3 shows the feature vectors for the sample sentence "John is seeing Anna". The information stored for John is : [noun word John, concept number for human because John is a human, 1 indicating that the semantic type for John is that its the name of a person, 0 indicating that no predicate is present before John, value showing the distance of the word John from the root of the parse tree, value from

WordNet for "see" which is the predicate occurring after John in the sentence]. Similarly the vector for Anna is formed, with the difference being that the predicate occurs before Anna in the sentence. Finally the two vectors are combined to form the feature vector for the triple [John, see, Anna]. The values represent : [concept number for John, semantic type value for John, value of predicate occurring before John, distance between John and Anna in the parse tree, concept number for Anna, semantic type value for Anna, predicate occurring between John and Anna in the sentence, value of predicate occurring after Anna]. This is the vector which is passed to the classifier.

## 3.5 Relation Classification

This phase does the actual classification of relations that have been extracted from the sentence. The classifier takes the feature vector and classifies it into one of the relations. Here relations refers to the different predicates that can occur in the input sentence and can form a part of the triple [Subject, Predicate, Object ]. For example in the sentence "John is seeing Anna", the predicate is "see", which is the root form of the verb "seeing". Hence "see" is the required relation that connects John and Anna. If this predicate has been encountered before and the classifier is already trained for it then it would be classified correctly. If a new predicate is encountered, incorrect classification may result and the user can then correctly annotate the triple and add it to the training data set for further learning. Support Vector Machines (SVM) have been used in earlier works relating to text classification and are shown to have higher accuracies. Hence in this work, a multiclass Support Vector Machine has been used where the classes are the different relations that can occur in the sentence. Each sentence from the data set is taken as the input and the feature vectors are then automatically generated and fed into the classifier.

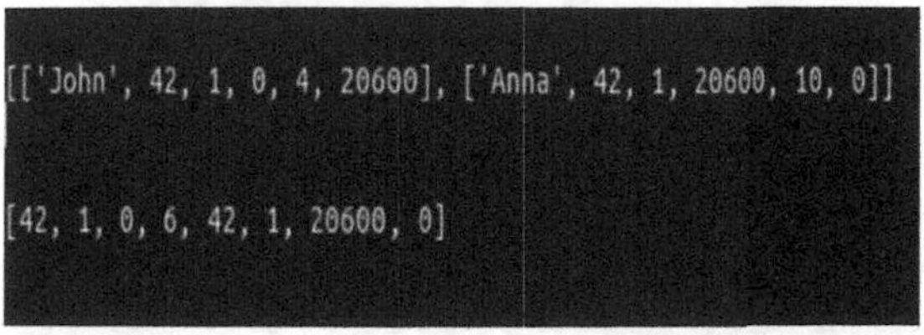

```
[['John', 42, 1, 0, 4, 20600], ['Anna', 42, 1, 20600, 10, 0]]

[42, 1, 0, 6, 42, 1, 20600, 0]
```

**Fig. 3** Feature vectors generated for the input sentence "John is seeing Anna"

**Table 2** Precision and Accuracy values

| | Precision | Accuracy |
|---|---|---|
| Transcription 1 | 71% | 89% |
| Transcription 2 | 70% | 90% |
| Transcription 3 | 72% | 93% |

## 4 Results

The approach proposed in this work falls into the category of semi supervised learning because the entire data corpus is not annotated initially. A few training samples are annotated after which the classifier is trained and tested. While testing, the annotator is asked to specify any incorrect classifications or missing relations. This information is used to further train the classifier. Hence a continuous process of training and testing goes on.

The data set comprises of three transcriptions obtained from the RoCKIn corpus, and each one is further divided into five equal clusters of sentences, for the purpose of continuous training. Table 2 shows the accuracy and precision values, calculated by taking the average of the values for the five clusters in each of the transcriptions. Within each individual transcription, as we move from one cluster to another, the precision and accuracy values increase. This trend is also depicted in Figures 4 and 5. Figure 4 shows the increasing accuracy curves for each of the transcriptions. The y axis shows the accuracy values whereas the x axis shows the cluster number. For each of the transcriptions, it is observed that the accuracy value improves with more training. The training data set is appended after every cluster is tested.

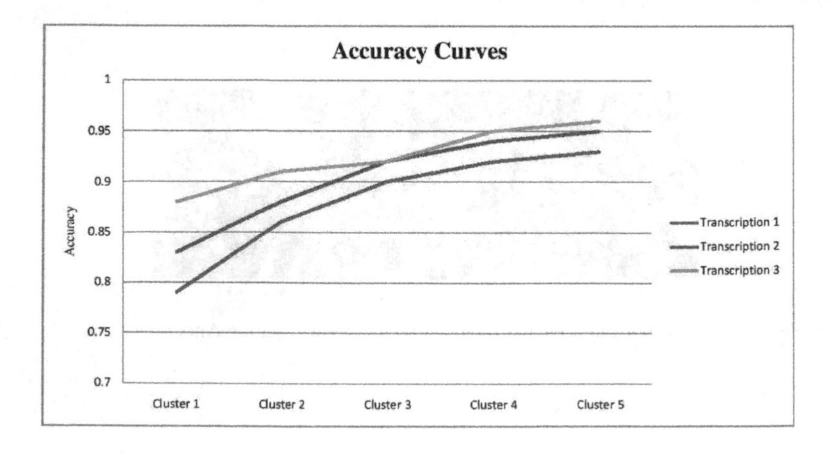


**Fig. 4** Accuracy curves for the three transcriptions

The final accuracy achieved is quite satisfactory. Figure 5 shows a similar trend for precision values. The average accuracy achieved by the classifier over all the three transcriptions is around 91%.

## 5 Conclusions and Future Work

For extracting relations from natural language inputs, rule based and machine learning techniques have been individually used since a long time. The rule based methods have low recall values. The machine learning methods, on the other hand show low values initially, but the accuracy and precision values increase with training. The proposed algorithm thus combines both these techniques, in order to obtain accurate relations. The results are obtained by validating the algorithm against a standard data set for the human robot interaction domain. The inputs are taken from the RoCKIn competition transcriptions, which capture the typical instructions given by a human to a robot.

For improving the obtained results, in addition to using global repositories, local knowledge bases like ontologies and databases can be built. Machine learning classifiers can be used with different kinds of lexical and syntactic feature attributes, depending on the domain of the inputs. In future, more techniques can be combined, and data sets for different domains can be made available to the scientific community, in order to compare and contrast these new approaches as well as help in the advancement of the state of the art.

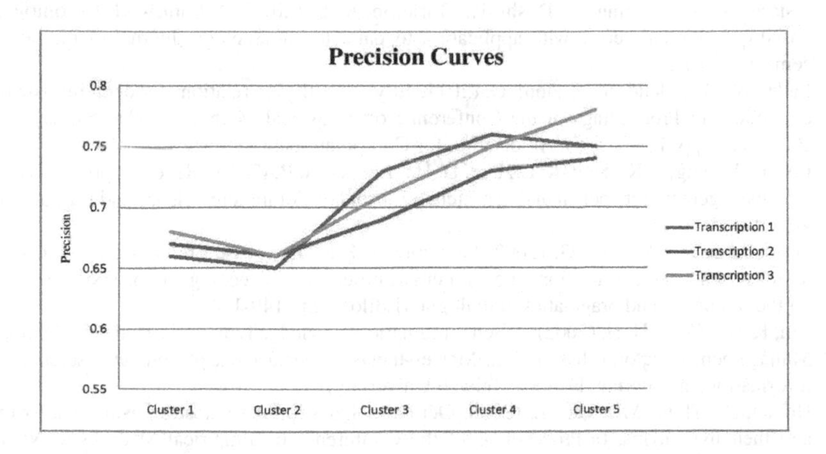


**Fig. 5** Precision curves for the three transcriptions

**Acknowledgements** This work was financially supported by the Singapore National Research Foundation under its Campus for Research Excellence And Technological Enterprise (CREATE) programme.

## References

1. Grau, B., Ligozat, A., Minard, A.. (2011). Multi-class SVM for Relation Extraction from Clinical Reports. RANLP.
2. Culotta, A., Sorensen, J. (2004, July). Dependency tree kernels for relation extraction. In Proceedings of the 42nd Annual Meeting on Association for Computational Linguistics (p. 423). Association for Computational Linguistics.
3. Dmitry Zelenko, Chinatsu Aone, and Anthony Richardella. 2003. Kernel methods for relation extraction. J. Mach. Learn. Res. 3 (March 2003), 1083-1106.
4. Zhang, Z. (2004, November). Weakly-supervised relation classification for information extraction. In Proceedings of the thirteenth ACM international conference on Information and knowledge management (pp. 581-588). ACM.
5. RoCKIn File Repository - http://rockinrobotchallenge.eu/pydio/data/public/9fc13f
6. Wang, T., Li, Y., Bontcheva, K., Cunningham, H., Wang, J. (2006). Automatic extraction of hierarchical relations from text (pp. 215-229). Springer Berlin Heidelberg.
7. Jang, H., Lim, J., Lim, J. H., Park, S. J., Lee, K. C., Park, S. H. (2006). Finding the evidence for protein-protein interactions from PubMed abstracts. Bioinformatics, 22(14), e220-e226.
8. Vargas-Vera, M., Motta, E. (2004). AQUAontology-based question answering system. In MICAI 2004: Advances in Artificial Intelligence (pp. 468-477). Springer Berlin Heidelberg.
9. Anantharangachar, R., Ramani, S., Rajagopalan, S. (2013). Ontology guided information extraction from unstructured text. arXiv preprint arXiv:1302.1335.
10. Kordjamshidi, P., Frasconi, P., Van Otterlo, M., Moens, M. F., De Raedt, L. (2011). Relational learning for spatial relation extraction from natural language. In Inductive Logic Programming (pp. 204-220). Springer Berlin Heidelberg.
11. Asiaee, A. H., Minning, T., Doshi, P., Tarleton, R. L. (2015). A framework for ontology-based question answering with application to parasite immunology. Journal of biomedical semantics, 6(1), 1.
12. Fader, A., Soderland, S., Etzioni, O. (2011, July). Identifying relations for open information extraction. In Proceedings of the Conference on Empirical Methods in Natural Language Processing (pp. 1535-1545). Association for Computational Linguistics.
13. Lee, C. J., Jung, S. K., Kim, K. D., Lee, D. H., Lee, G. G. B. (2010). Recent approaches to dialog management for spoken dialog systems. Journal of Computing Science and Engineering, 4(1), 1-22.
14. Quesada, J. F., Amores, G. (2002, September). Knowledgebased reference resolution for dialogue management in a home domain environment. In Proceedings of the sixth workshop on the semantics and pragmatics of dialogue (Edilog) (pp. 149-154).
15. Lin, K. K., Chen, H. H. (2002). A Semiautomatic Knowledge Extraction Model for Dialogue Management (Doctoral dissertation, Master thesis of department of computer science and information engineering, National Taiwan University).
16. Huang, Z., Thint, M., Qin, Z. (2008, October). Question classification using head words and their hypernyms. In Proceedings of the Conference on Empirical Methods in Natural Language Processing (pp. 927-936). Association for Computational Linguistics.
17. Ramezani, M., Witschel, H. F., Braun, S., Zacharias, V. (2010). Using machine learning to support continuous ontology development. In Knowledge Engineering and Management by the Masses (pp. 381-390). Springer Berlin Heidelberg.
18. Lee, W. N., Shah, N., Sundlass, K., Musen, M. A. (2008, November). Comparison of ontology-based semantic-similarity measures. In AMIA.

19. Williams, S. (2000). Anaphoric reference and ellipsis resolution in a telephone-based spoken language system for accessing email. Corpus-based and Computational Approaches to Discourse Anaphora, 3, 171.
20. Gieselmann, P. (2004, July). Reference resolution mechanisms in dialogue management. In Proceedings of the 8th workshop on the semantics and pragmatics of dialogue (CATALOG) (pp. 28-34).
21. Soon, W. M., Ng, H. T., Lim, D. C. Y. (2001). A machine learning approach to coreference resolution of noun phrases. Computational linguistics, 27(4), 521-544.
22. Orasan, C., Evans, R. J. (2007). NP animacy identification for anaphora resolution. Journal of Artificial Intelligence Research, 79-103.
23. Trausan-Matu, S., Rebedea, T. (2009). Ontology-based analyze of chat conversations. An urban development case. In Proceedings of Towntology Conference, Liege.
24. Muzaffar, A. W., Azam, F., Qamar, U., Mir, S. R., Latif, M. (2015, January). A Hybrid Approach to Extract and Classify Relation from Biomedical Text. In Proceedings of the International Conference on Information and Knowledge Engineering (IKE) (p. 17). The Steering Committee of The World Congress in Computer Science, Computer Engineering and Applied Computing (WorldComp).
25. Giuliano, C., Lavelli, A., Romano, L. (2006, April). Exploiting shallow linguistic information for relation extraction from biomedical literature. In EACL (Vol. 18, pp. 401-408).
26. Lindes, P., Lonsdale, D. W., Embley, D. W. (2015, February). Ontology-Based Information Extraction with a Cognitive Agent. In Twenty-Ninth AAAI Conference on Artificial Intelligence.

# Anaphora Resolution in Hindi: A Hybrid Approach

A. Ashima , B, Sukhnandan kaur, C. Rajni Mohana*[*]

Jaypee University of Information Technology,, Waknaghat, P.O. Waknaghat Teh Kandaghat, Distt. Solan, 173 234, (H.P.), India;
A. Email: ashi.chd92@gmail.com, B. Email : sukh136215@gmail.com, C. Email: rajni.mohana@juit.ac.in,

**Abstract.** Now-a-days machines have the ability of upturning the large matrices with speed and grace but they still fail to master the basics of the written and spoken languages to the machine. It is difficult to address computer as though addressing another person. So, there are various methods to resolve the problem of refereeing and this process is called anaphora resolution. Anaphora is very challenging job in Hindi language because the style of writing is changed with respect to the expressions. In this paper, we have used hybrid approach which is the combination of rule based and learning based to resolve gender, number agreement, co-reference and animistic knowledge in Hindi domain. The results are computed by using various globally accepted evaluation metrics like MUC, B3, CEAF, F- score on three different data sets. The accuracy of the system is evaluated by kappa statistics.

**Keywords:** Rule based approach, Learning based approach, CPG rules, Gender agreement, Number agreement, Co-reference, Animistic Knowledge.

## 1 Introduction

NLP is the natural language processing is referred with the intercommunication between computers and human languages [1]. It is associated with area of artificial intelligence and computer science. The basic aim of NLP is to develop such software which can understand various human languages as well as analyze and generate the languages that human used naturally. So, a machine can interact with the human by using simple human language. It is easy for a human to understand, learn and use the human languages, symbol system but it is hard for a computer to master it. It uses the general knowledge for representing and reasoning. It has many challenges in representation like word sense disambiguation, sentiment analysis, summarization from paragraph to sentences, correction of the search query, language detection, missing apostrophes, etc. Anaphora is one of the difficult task to address the computer as though human addressing another person [2]. We are focusing the pronoun resolution in Hindi. The resolution of the anaphora is tough job in Hindi language.

---

[*] Corresponding author.

 
J.M. Corchado Rodriguez et al. (eds.), *Intelligent Systems Technologies and Applications 2016*, Advances in Intelligent Systems and Computing 530,
DOI 10.1007/978-3-319-47952-1_65

For example:

बासप्पा ने सिद्धेश्वर हाई स्कूल बीजापुर से मैट्रिक की परीक्षा उत्तीर्ण की। इन्हें फुटबॉल एवं तैराकी के अलावा कुछ भारतीय परम्परागत खेलों में भी रुचि थी ।

Now in the above example 'बासप्पा' is the noun and 'इन्हें' is the pronoun. 'इन्हें' pronoun is referring to the entity 'बासप्पा'. The process of identification of the referent is known Anaphora Resolution.

Hindi language is not depending on the semantic and syntactic structure. It is independent from the word order. It is purely verb-based language. Many researchers are examined the problem of pronouns resolution in Hindi domain. In this paper, we resolve the pronouns that are based on the gender, number agreement, co-reference and animistic knowledge that are explained below [3]:

1.1 ***Animistic Knowledge***: The referents are filtered out on the basis of living and non living things. For example:

"रवि *हर दिन* सलाद *खाता था और अपनी* मां *को भी खिलाता था* |"

In the above example pronoun "अपनी" cite to "रवि" as referent "अपनी" is an animistic (living) pronoun that cite to animistic (living) noun.

1.2 ***Co-reference resolution:*** It is the job of referring all expressions to the same entity in the discourse. For example:

"दीपक *ने* बाघ *देखा। वह बहुत सुंदर था।*"

In this example "*वह*" *cites either to the* "बाघ" or "दीपक" but "*वह*" cites *to the* "बाघ", it is close to the "बाघ".

1.3 **Number Agreement**: The referents are filtered out on the basis of singular and plural. For example:

"दीपक *और* रवि भाई *हैं और वे बहुत* बुद्धिमान *हैं*|"

Now, in this example pronoun "वे" refers to "दीपक *और* रवि" that is plural noun.

1.4 ***Gender Agreement:*** The referents are filtered out on the basis of masculine and feminine.

For example:

"दीपक *ने* मॉल *से* बल्ला *खरीदा। वह उसे* खेलता है|"

"आशिमा *ने* मॉल *से* पर्स *खरीदा। वह उसे पसंद करती है*|"

Now, Hindi is a verb-based language. The verbs are used to resolve pronouns based on gender agreement. The verbs "करता है" and "करती है", it can be understandable that "उसे" mentions to masculine and feminine respectively.

In this paper, we attempt to resolve the all the above issues by using the hybrid approach which is the combination of rule based and learning based approach. In this, we have designed heuristic rules and used existing CPG rule (Computational Paninian Grammar Dependency Rules) to train our system to resolve the pronouns.

### *1.5 Motivation*

Many researchers like (Hobb, 1978) [4] , (Carter,1987) [5] , (Carbonell and Brown, 1988) [6] [7] , (Lappin and Leass,1994) [8] , (Mitkov, 1996) [9][10][11], (CogNIAC, 1997) [12] used rule based and learning based approaches to resolve the referents. The rule based approach is based on handwritten rules in the form of regular expression and context -pattern which helps to find the accurate entity whenever there is ambiguity. It depends on the lexical and dictionary knowledge for tagging each word. The advantage of rule based approach is that it examines at semantic and syntax extent. The drawback is that a large number of rules and linguistic knowledge is required to blanket the all the features of language. The learning based approach extract the linguistic knowledge automatically to tag every word in the given text from the training corpus. The advantage of rule based approach is that the system can be trained with sufficient small set of rules for tagging and the disadvantage is that it takes large time to train the system. The combination of both approaches reduces the complexity in tagging and uses appropriate cues to disambiguate the words (resolving pronouns) in less time.

Further the paper follows as: In section 2, the summary of the literature survey on the anaphora resolution is presented. In section 3, the proposed work is explained. Experimental results are described in section 4. Finally, the whole paper is concluding is section 5.

## 2 Related Work

In anaphora resolution a lot of the work is done in other European Languages and English but less amount of work is done in Hindi language. In English Language most of the work is divided into two fundamental approaches that are Rule based and Learning based approaches.

The issues of the anaphora resolution were first examined by the Jesperson in 1954. [Hobbs, 1978] [4] noticed the anaphora resolution problem in the two nouns that mentions to the pronoun. In earlier the anaphora resolution algorithms are based on the heuristics. These algorithms did not use the syntactic constraints. The anaphora resolution algorithm is shifted from the traditional syntax and semantic to the simple semantic and syntax. Some researchers like [Bernnar, Friedman and Polard, 1987], [Sinder, 1981, 1983], [Grosz, Joshi and Weinstin, l983, 1986], and [Webber, 1988] represent the different versions of the discourse based algorithms. The features like coherence and focusing are used to identify the pronouns. After 1986 , A system is characterized into two approaches that are Knowledge based and other alternative approaches like [Corbonell and Brown ,1988] [6] Rich and [Luperfiy,1988] and [Asher and Wada ,1988],. presents a variety of syntactic , discourse and semantic factors which combine into the multi-dimensional metric for ordering the pronouns. [Lappin and Lease, 1994] [18] algorithm work on the syntactic representation that is generated by the grammar parser. In this the weights are assigned using the linguistics based ideas. [Sobha L and R.N.Patnaik, 1999] presented the algorithm which assigned the weight to the nouns by using the threshold value and linguists rules. The threshold value is given to the correct items. These items may be non pronominal or pronominal. There are many other algorithms which used the machine learning approach. [Mitkov, 1997] [10][13] drafted algorithm which is based on the semi-automatic annotation of pronoun phrases that are pairs in discourse . This algorithm is based on knowledge-poor and robust anaphora resolution method that is followed by post-editing. There are many systems like CogNIAC [12] which gave high rate of pronouns resolution by using the information of the part of speech tagging, noun phrase recognition, parse tree , basis salience features like number , gender and recency etc, sentence detection etc but it did not resolve the pronouns in the case of ambiguity. In the previous work some approaches use the dependency relations for the anaphora resolution. First one is [Uppalapu and Sharma, 2009] [14] discuss the dependency structure for the Hindi language. This algorithm is the improved version of the S-list algorithm of [Prasad and Strube, 2008] [15]. It used the traditional grammatical relations that is (subject >object >Indirect object >others) for ranking the items in the list and used the Paninian grammatical relations (CPG) i.e. (k1 >k2 >k3 >k4>others). However this algorithm only used the dependency relations as attribute to order the item in the centering based approach. The dependency structure and relation has the syntactic attributes and properties. The dependency structure must be explored to describe the different nouns for different pronouns that are used for resolving the anaphora. [Bjrkelund and Kuhn, 2012] [16] describe the dependency structure for English language. They are using the dependency relations as an attribute for the co-reference resolution in a fully learning based approach. The combination of dependency structure and phrase – structure gives better result than using the phrase structure alone. [Praveen Dhwake, 2013] [17] drafted approach to resolve Entity-pronoun references in Hindi called hybrid approach and used dependency structures as a source of syntactic information. In this approach the dependency structure were used a rule-based caliber for resolving the simple pronouns and decision tree classifier was used to solved the more puzzling sentences, using grammatical and semantic features. The results show that by using the dependency structures that gives syntactic knowledge which helps to resolve particular pronouns. Semantic information such as animacy and Named Entity categories further helped to improve the resolution accuracy.

## 3 Proposed Work

The proposed System is shown in Fig.1. It implements a proposed hybrid algorithm which is a combination of both rules based and learning based approach [19]. In rule based approach, we use heuristics rule and CPG rules (Anncora) [20] for finding the gender, number, co-reference, animistic based pronouns. These rules act as the filter for the candidates. In Learning based approach, we use learning techniques to train our system so that it can calculate POS (part of speech) that is based on the available knowledge for Hindi in the WordNet library. The system uses limited rules of the Hindi dependency tree bank for explaining the dependency and animistic knowledge.

### 3.1 System Design

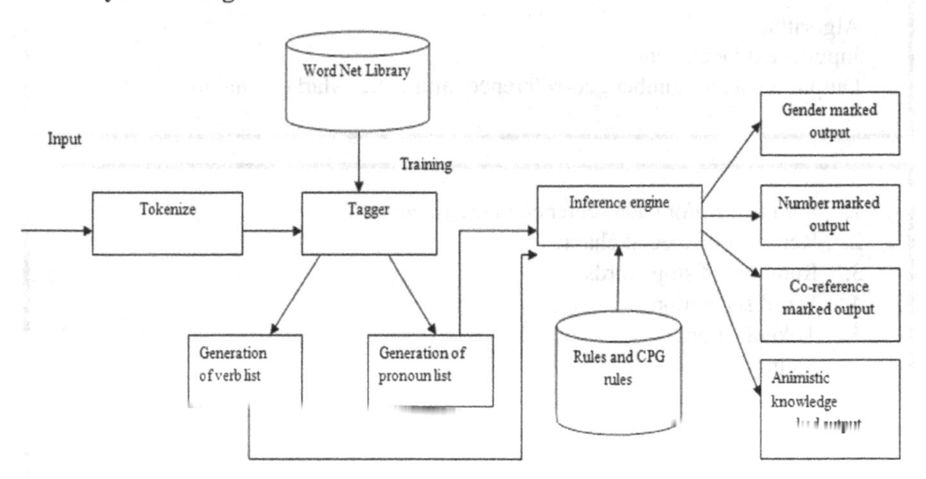


**Fig. 1:** System Design

Initially, the text is auto identified by machine. Then it is broken into pieces by removing the special characters, stop words called token. These tokens are added to the word list. The POS tag is calculated by using the WordNet library according to part of speech two lists are generated one is verb list and second is pronoun list. The rules are applied to the both list and after the filtration the output is generated on the bases of gender, number, co-reference and animistic knowledge pronouns list. The rules are applied to both lists and after the filtration the output is generated on the bases of gender, number, co-reference and animistic knowledge.

### 3.2 Algorithm

We have implemented the algorithm of the system described in Fig. 2. This algorithm is applied on the part of tagger and inference engine of system design in Fig. 1. In our algorithm, the data set is taken as the input. The data is preprocessed which include the Removal of special character: like '.' ; '<'; '>' ; ',' ; '_' ; '-' ; '\\' ; '(' ; ')' ; ':' ; ';' ; '[' ; ']' ; '{' ; '}' or '/':, removal of stop words that refer to the most common words in a language, word separation, tokenization that is the task of chopping it up into pieces, called tokens, stemming that refers to a crude heuristic process that chops off the ends of words in the hope of achieving the goal that is accurate most of the time, and often includes the removal of derivational affixes, case conversion: change lower case to upper case or vice-versa. After that word list is created and POS tag is calculated. According to the part of speech, the verb and pronoun list is generated and rules are applied to these lists, after the filtration of the lists the gender, number, co-reference and animacy marked output is generated.

Algorithm:
Input: Text Document
Output: Gender, number , co-reference, animistic Marked Output

1. Set i to 1 // For each sentence in the document
2. Removal of special character.
3. Removal of stop words
4. Word separation
5. Tokenization
6. Stemming
7. Case conversion
8. Create wordlist
9. Selection of possible candidate
10. POS tagging
11. Apply linguistic rules for gender agreement
    i. If the gender of the verb contains ‘a’/ ‘◌ा ‘at the last then the noun is marked as masculine gender.
    ii. If the gender of the verb contains ‘e’/ ‘◌ी ‘at the last then the noun is marked as feminine gender.
    iii. If the gender of the verb contains ‘a’/ ‘◌ा ‘at the last then the pronoun is marked as masculine gender.
    iv. If the gender of the verb contains ‘e’/ ‘◌ी ‘at the last then the pronoun is marked as feminine gender.
12. Apply linguistic rules for Number agreement
    i. If the word of the verb contains ‘ae’/ ‘◌े ‘ and ‘aae’/ ‘◌ै’at the last then the noun/ pronoun is plural .
    ii. If the word of the verb contains ‘au’/ ‘‘◌ु “ , ‘e’/ ‘◌ी ‘ ‘and ‘a’; ‘◌ा’ at the last then the noun/
13. Apply co-reference rule
    i. Calculate the term frequency.
    ii. Assign the weight to nearest entity
14. Apply CPG rules
    i. (k1 >k2 >k3 >k4>others)
15. Marked out put
16. i ++
17. If (i< n ) go to step 2 else end.

**Fig. 2:** Proposed Algorithm of the System

## 3.3 CPG Rules

The CPG rules are known as computational Paninian grammatical relations or rule [20] like (k1 >k2 >k3 >k4>others). The some rules are explained below in Table 1:

**Table 1:** Explanation of Paninian grammatical relations

| Sr No. | Tag Name | Tag description | Example |
|---|---|---|---|
| 1 | k1 | karta (doer/agent/subject) *Karta* is defined as the 'most independent' of all the *karakas* (participants). | (1) **rAma** bETA hE (2) **sIwA** KIra sAwI hE |
| 2 | k2 | karma (object/patient) *Karma* is the locus of the result implied by the verb root. | (1) *rAma rojZa* ***eka seba*** *KAwA hE* |
| 3 | k3 | karana (instrument) *karana* karaka denotes the instrument of an action expressed by a verb root. The activity of *karana* helps in achieving the activity of the main action. | (1) *rAma ne* ***cAkU se*** *seba kAtA* (2) *sIwA ne* ***pAnI se*** *GadZe koBarA* (1) |
| 4 | k4 | sampradaana (recipient) Sampradana karaka is the recipient/beneficiary of an action. It is the person/object for whom the karma is intended. | *(1) rAma ne mohana ko Kira xI* *(2) rAma ne hari se yaha kahA* |
| 5 | k5 | apaadaana (source) apadana karaka indicates the source of the activity, i.e. the point of departure. A noun indicates the point of detachment for a verb expressing an activity which involves movement away from is apadana. | *(1) rAma ne cammaca se katorI se Kira KAyI* *(2) cora pulisa se BagawA hE* |
| 6 | ccof | Conjunct of relation Co-ordination and sub-ordination. | *(1) rAma seba KAwA hE Ora sIwA xUXa pIwI hE* *(2) rAma ne SyAma se kahA ki vaha kala nahIM AyegA* |

## 3.4 Pseudo Code for Co-reference

The algorithm for calculating the co-reference is explained in Fig. 3. In this higher weight is assigned to noun which is near to pronoun. . In this process first the training is given to the system by library Hindi keywords and dependency tagger. The POS is calculated by using POS tagger. After calculating the part of speech of the words verb and pronoun lists are generated then check the list if the word is pronoun than the higher weight is assigned to the nearest noun.

```
I: Input file fi containing Hindi text

Li: Library of Hindi Keywords

POS: Part of Speech Tagger for each word

Di: Dependency tagger for each word in each sentence containing tag name Tn and tag
    description  Td
Wj : Weight of the word
ccof : Conjunct of relation
```

```
for each line in I
        Store line Lj
        Obtain wj by tokenizing line Lj
end for
Train the algorithm using library Li
Train Dependency tagger Di with file I
for each word wj in I
        Find POSw using POS tagger
        Use Di for obtaining Tnw and Tiw
        if Tnw equals k1
                Assign weight wk to word wj
                Obtain POS of word wj
                if word wj is pronoun
                        Check for weight of nearby noun with Tnw equals k1
                        Find verb in vicinity and assign weight
                        flag=1
                end if
                if flag is not equal to 1
                        Check for word wj equals pk1 or k7t
                        Assign weight lower than previous
                        Find POS equals ccof
                        Add all weights
                end if
        end if
        Obtain final weight and normalize
        Assign Noun to word wj with tag name Tnw
end for
Compare obtained response chain with key chain
```

**Fig. 3:** Pseudo Code for Co-reference

# 4 Experiment and Result

For evaluating the performance of the anaphora, we have used three different data sets as shown in table 2. Total data consists of 4 news articles with 45 sentences, 581 words, 30 pronouns in data set 1. In data set 2, data consists of children story with 34 sentences, 444 words and 28 pronouns. In data set 3, data consists of 2 bibliographies with 71 sentences, 824 words and 67 pronouns. The sentences are complex and have all types of pronouns like first, second, third, reflexive, etc.

**Table 2:** Dataset by Category

| Data set | No of sentences | Total of words | No of pronouns | Resolved Pronouns by system | Correctly resolved pronouns |
|---|---|---|---|---|---|
| News | 45 | 581 | 30 | 37 | 23 |
| Story | 34 | 444 | 28 | 33 | 25 |
| Bibliography | 71 | 824 | 67 | 62 | 52 |

### 4.1 Notations

- There are two systems one is called GOLD system in which the pronouns are detected and resolved by the human and second one is called SYS system in which the pronouns are detected and resolved by the machine. Singleton mention refers to the noun which occurs only single time after resolving the pronouns by the system.
- Doubleton mention refers to the nouns which occur two times after pronoun resolution in the system.
- Multiple mentions refer to the frequency of the nouns after the pronoun resolution in the system.
- After resolving the pronouns by the machine is called response chain and these predicted pronouns pointing to the noun called key chain
- False positive (FP) is refer to the pronoun mentions to entity is false in response chain
- False negative (FP) means the entity according to key chain but are something else in response chain.
- S (d) is defined as the records detected by the system.
- S1 (d) is the correct or relevant records detected by the system.
- K (d) is defined as the relevant records in the document.
- N is the total no of mentions, specifically,
  - $S(d) = \{S_j : j = 1, 2, \cdots, |S(d)|\}$, Equation (1)
  - $S1(d) = \{S1_x : x = 1, 2, \cdots, |S1(d)|\}$, Equation (2)
  - $K(d) = \{Kk : i = 1, 2, \cdots, |K(d)|\}$, Equation (3)

Where $K_k$ is series of mentions in K(d), $S1_x$ is series of mentions in S1(d) and $S_j$ is series of mentions in S(d) respectively.

There are many evolutions metric. Some are described below:

i. **MUC**

It is the link based evolution matrix called message understanding co-reference [21] that calculates the minimum link between the mention (GOLD and SYS). To calculate the recall (R), the total no of links between GOLD and SYS is divided by the minimum number of links that are required to specify GOLD. To calculate precision

(P), this number is divided by the minimum number of links that are required to specify SYS [22]. The process of calculation the MUC is given below:

- Find the number of singleton mentions in the document (d).
- Find the number of doubleton mentions in the document.
- Find the number of multiple mentions in the document.
- Find the precision (P) = $\sum_1^N (S1 \setminus S) \setminus N$ Equation(4)
- Find the recall (R) = $\sum_1^N (S1 \setminus K) \setminus N$ Equation(5)
- Find the MUC F-score = $\frac{2 P * R}{(P+R)}$ Equation(6)

ii. **B3**

In this, to obtain recall all the intersecting mentions between the SYS and GOLD is calculated and divided by the total number of mention in the GOLD [23]. To obtain the precision the number of joining mentions between the SYS and GOLD is calculated and divided by the total no of mentions in the SYS. The process of calculating the B3 is given below:

- Calculate the B3 precision

  Singletons are ignored for false positive while consider for false negative. Calculate the B3(P) by using equation 4.
- Calculate the B3 recall

  Singletons are considered for FN. B3 (R) is calculated by using equation 5.
- B3 F-score is calculated by using the equation 6.

iii. **CEAF**

It is the best method to compute the one to one mapping between the entity in the SYS and GOLD which means each SYS entity is mapping to at most single GOLD entity [24]. It maximizes the similarity of the best mapping. The score of recall and precision are same when true mentions are involved. In both the common mentions between every two mapped entities is divided by the no mentions. The process of calculating the CEAF is given below:

- In response discard all singletons and doubletons

  In original

  if Precision

  Keep doubletons

  If recall

  Discard doubletons
- Find the precision (P) using equation 4.
- Find the recall (R) using equation 5.
- Find the CEAF F-score using equation 6.

iv. **F – score**

The F- score combines precision and recall [25]. RECALL is the ratio of the number of applicable entities retrieved to the total number of applicable entities in the database. PRECISION is the ratio of the number of applicable entities retrieved to the total number of inapplicable and applicable entities retrieved. It is usually expressed as a percentage. The process of finding the F-score is given below:

- Find the recall ( R) = $S1\backslash S$ Equation(7)
- Find the precision (P) = $S1\backslash K$ Equation(8)
- Find the F-score using the equation 6.

## 4 Fleiss's Kappa

It is the statistical evaluation matrix which is used to check the reliability of the agreement over the multiple raters [26]. P(A) is the proportion of the time the judges agreed. P(E) is the proportion of the time they would be calculated to concurred by chance. The strength of the agreement is described in table 3.

- For number of judges that assign category j to pronouns i = $x_{ij}$
- The Fleiss's kappa is calculated as :

$$k = \frac{P(A) - P(E)}{1 - P(E)} \quad \text{Equation (9)}$$

- The numerator of equation indicates the degree of agreement that is attainable above chance.
- The denominator indicates the degree of agreement actually achieved above chance.

**Table 3:** Strength of agreement

| Kappa statistic | Strength of agreement |
|---|---|
| <0.00 | Poor |
| 0.0 to 0.20 | Slight |
| 0.21 to 0.40 | Fair |
| 0.41 to 0.60 | Moderate |
| 0.61 to 0.80 | Substantial |
| 0.81 to 1.0 | Almost perfect |

We have computed the result by using various metrics are shown in Table 4. The minimum link between the GOLD and SYS is evaluated by using the MUC metric. It gave 75.85% result on news article, 91.52% on story and 86.15% on bibliography. It gave higher results because singletons are considered in it. Singletons are ignored for false positive while consider for false negative in the B3 evaluation and it gave 65.67% result on the news, 90.32% on the story and 80.99% on the bibliography. To compute, one to one mapping between the entity in the SYS and GOLD the CEAF metric is used which gave 75.45% result on news, 85.89% on story and 72.75%on bibliography. The F- score is also computed and it gave 68.62% on news, 81.96% on story and 80.61% on bibliography.

**Table 4:** Results of hybrid approach on data sets

| Dataset | MUC | B3 | CEAF | F-score |
|---|---|---|---|---|
| News | 75.85% | 65.67% | 75.45% | 68.62% |
| Story | 91.52% | 90.32% | 85.89% | 81.96% |
| Bibliography | 86.15% | 80.99% | 72.75% | 80.61% |

The results are represented in graph in Fig. 4. We can see, the MUC of the news articles is less than other domain because the news article contains complex pronouns and less no of singletons. The B3, CEAF, F-score of the story is high as comparative to other domain. Generally in the previous work other researches computed the result on the bases of F-score metric. Other researchers have 61-65% F-score on the news domain and our system gave 68.62%. They have 60-64% and 76-81% F-score on story and bibliography domain and our system gave 81.96% and 80.61%. The overall performance is calculated in Table 5.

In this we can see that the story domain has higher percentage rather than other because it is a straightforward narrative style with extremely low sentence structure complexity. It is observed that success rate of solving the pronoun varies with the structure of sentences. Hindi has no proper structure. So, the success rate depends on the style of writing. The different article domain has different way of writing that affects the performance of the system. The overall performance of the system according to the different metrics is MUC gave 84.50, B3 is 78.9%, CEAF is 78.03% and F-score is 77.06%. As compared to the previous work, the overall performance is calculated by using F-score that is 60 to70% and our system gave 77.06 % which means it is higher accurate than others system.

**Table 5:** Average result of overall system

| Evaluation metrics | Overall performance of system |
|---|---|
| MUC | 84.50% |
| B3 | 78.9% |
| CEAF | 78.03% |
| F-score | 77.06% |

We have computed the overall system result with other evaluation metrics. MUC gave the higher results rather than other metrics as shown in Fig. 5 and B3, F-score and CEAF scores are usually lower than MUC on datasets Because in B3. F-score and CEAF singletons are annotated because a great percentage of the score is simply due to the resolution of singletons.

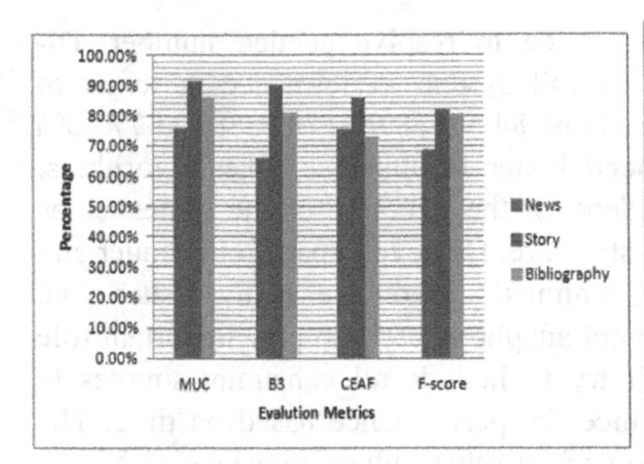


Fig. 4: Result of evaluation metrics on datasets

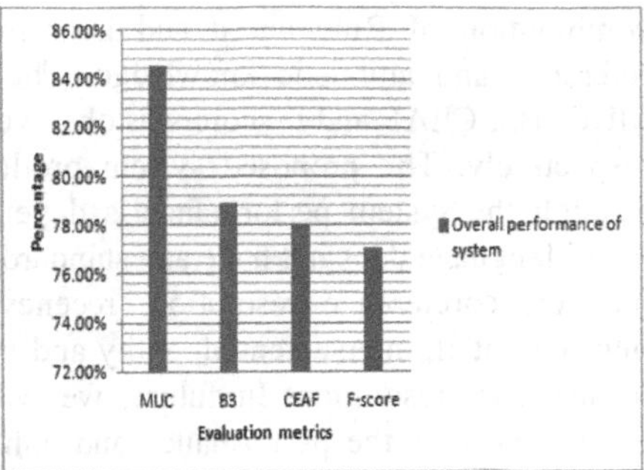


Fig. 5: Overall System Performance

### 4.2 Result of the agreement

The agreement on behave of judges the accuracy of the system is calculated with the evaluation metric kappa that is designed for categorical judgments and corrects a simple agreement rate for the rate of chance agreement. We have conducted the experiment over 3 data sets annotated by 2 raters. Annotators were asked to assign categories like relevant, non relevant etc as stated, according to the type of entity it refers to. The result is shown in Table 6.

**Table 6:** Kappa statics for Dataset

| Data set | P(A) | P(E) | Kappa |
|---|---|---|---|
| News | 0.9729 | 0.7662 | 0.6532 |
| Story | 0.9090 | 0.6836 | 0.7124 |
| Bibliography | 0.9354 | 0.7997 | 0.6775 |

The overall kappa for our system is 0.681. According to Table 3, the strength of our agreement is substantial.

## 5 Conclusion

Finally, to summarize, this paper analyzed the benefaction of various researchers who worked on various research issues. Through literature survey, we were able to identify various research gaps in anaphora resolution like recency factor, Animistic knowledge, Gender, Number agreement, NER, Pronoun resolution, etc. It can be seen that the problem of anaphora resolution is challenging but not uncontrollable. From last few years, anaphor resolution has gain a large attention. A large amount of work has demonstrated which have shown good results but in Hindi language. On the basis of related work (in Hindi), this manuscript has proposed a hybrid approach which is a combination of Rule based and Learning based to resolve gender, number, co-reference and animistic knowledge. The overall system performance in terms of MUC, B3, CEAF and F-score was observed to be 84.50, 78.9%, 78.03% and 77.06% respectively. The proposed system produced better results than other algorithms. Though the system performance is dependent on the structure of the sentences as Hindi language does not have any standard structure. However, apart from gender and number, coreference resolution, recency, animistic there are many issues like intrasentential, intersentential, entity and event anaphora, etc also play important role in anaphora resolution. In future, we will try to include all constraint sources to further increase the performance and enhance the performance based on time. The proposed approach will be further applied to various other Indian languages.

## References

1. J. L. Vicedo and A. Ferr´andez. Importance of pronominal anaphora resolution in question answering systems.In Proceedings of the 38th Annual Meeting on Association for Computational Linguistics, page, 555–562. Association for Computational Linguistics, 2000.
2. Wikipedia. Anaphora (linguistics )— Wikipedia, the free encyclopedia, 2014. [Online; accessed 17-January-2014].
3. Smita Singh, Priya Lakhmani,Dr.Pratistha Mathur and Dr.Sudha Morwal "Anaphora Resolution In HINDI Language Using Gazetteer Method",*International Journal on Computational Sciences & Applications IJCSA,* vol.4, pp.567-569, Ju. 2014
4. Hobbs, Jerry, "Resolving pronoun references",*Lingua*, vol.44, pp. 311-338, Jan.1978.
5. Carter, David M "A shallow processing approach to anaphor resolution", PhD thesis, Univ. of Cambridge, 1987.
6. Carbonell, J.G. and Brown, R.D., 1988, August. Anaphora resolution: a multi-strategy approach. In *Proceedings of the 12th conference on Computational linguistics-Volume 1* (pp. 96-101). Association for Computational Linguistics.
7. Dagan, I. and Itai, A., 1990, August. Automatic processing of large corpora for the resolution of anaphora references. In *Proceedings of the 13th conference on Computational linguistics-Volume 3* (pp. 330-332). Association for Computational Linguistics.
8. Mitkov, R., 1998, August. Robust pronoun resolution with limited knowledge. In *Proceedings of the 17th international conference on Computational linguistics-Volume 2* (pp. 869-875). Association for Computational Linguistics.
9. Mitkov, R., 1996. Anaphora resolution: a combination of linguistic and statistical approaches. In *Proceedings of the Discourse Anaphora And Resolution Colloquium, DAARC96.*

10. Connolly, D., Burger, J.D. and Day, D.S., 1997. A machine learning approach to anaphoric reference. In *New Methods in Language Processing* (pp. 133-144).
11. Boguraev, Branimir, Christopher Kennedy,"Salience based content characterization of documents", *ACL'97/EACL'97 workshop on Intelligent scalable text summarization*, 3-9, Madrid, Spain, 1997
12. Baldwin, Breck, "CogNIAC: high precision core- ference with limited knowledge and linguistic resources" *ACL'97 workshop on Operational factors in practical, robust anaphora resolution*, 38-45, Madrid, Spain, 1997.
13. Kameyama, Megumi, "Recognizing referential links: an information extraction perspective",*ACL'97/EACL'97 workshop on Intelligent scalable text summarization*, 3-9, Madrid, Spain, 1997.
14. Bhargav Uppalapu, Dipti Misra Sharma "Pronoun Resolution For Hindi" in Proc. *DAARC2009*,vol. 5847, April 22, 2009
15. R. Prasad and M. Strube. Discourse salience and pronoun resolution in hindi. U. Penn Working Papers inLinguistics, 6:189–208, 2000.
16. A. Bjrkelund and J. Kuhn. Phrase structures and dependencies for end-to-end coreference resolution. In Proceedings of COLING 2012: Posters, pages 145–154. The COLING 2012 Organizing Committee, 2012.
17. Dakwale, P., Mujadia, V. and Sharma, D.M., 2013. A Hybrid Approach for Anaphora Resolution in Hindi. in Proc *6th International Joint Conference on Natural Language Processing, IJCNLP* , Nagoya, Japan, Oct. 14-18, 2013 (pp. 977-981).
18. Mitkov, R., 1998, August. Robust pronoun resolution with limited knowledge. In *Proceedings of the 17th international conference on Computational linguistics-Volume 2* (pp. 869-875). Association for Computational Linguistics.
19. Kamlesh Dutta , Saroj Kaushik "Anaphor Resolution Approaches :, *Web Journal of Formal Computation and Cognitive Linguistics*, vol.10, pp. 71-76, Jan. 2008.
20. Sharma, D.M., Sangal, R., Bai, L., Begam, R. and Ramakrishnamacharyulu, K.V., 2007. AnnCorra: TreeBanks for Indian Languages. *Annotation Guidelines (manuscript), IIIT, Hyderabad, India.*
21. Aberdeen, J., Burger, J., Day, D., Hirschman, L., Robinson, P. and Vilain, M., 1995, November. MITRE: description of the Alembic system used for MUC-6. In *Proceedings of the 6th conference on Message understanding*(pp. 141-155). Association for Computational Linguistics.
22. Recasens, M. and Hovy, E., 2011. BLANC: Implementing the Rand index for coreference evaluation. *Natural Language Engineering*, *17*(04), pp.485-510.
23. Amit Bagga and Breck Baldwin. 1998. Algorithms for scoring coreference chains. In *Proceedings of the LRECWorkshop on Linguistic Coreference*, page563–566.

24. Xiaoqiang Luo. 2005. On coreference resolution performance metrics. In *Proceedings of Human LanguageTechnology Conference and Conference on Empirical Methods in Natural Language Processing*,pages 25-27.
25. Kaur, Sukhnandan, and Rajni Mohana. "A roadmap of sentiment analysis and its research directions." International Journal of Knowledge and Learning 10, no. 3 (2015): 296-323
26. J. Fleiss. Measuring nominal scale agreement among many raters. Psychological bulletin, 76(5):378, 1971.

# A Personalized Social Network Based Cross Domain Recommender System

**Sharu Vinayak[1] ,Richa Sharma[2] Rahul Singh[3]**

[1,2,3]Chandigarh University,Mohali,140413,India

Email: [1]sharuvinayak@gmail.com, [2]richas30@yahoo.com, [3]rahulsinghcse25@gmail.com

*Abstract*— In the last few years recommender systems has become one of the most popular research field. Although with time various new algorithms have been introduced for improving recommendations but there are some areas in this research field that still need to be concentrated on. Cross domain recommendations and recommender systems and social networks are two of the research challenges that need to be explored more. In this paper we have proposed a novel idea for making recommendations in one domain using information from the other domain. The information has been extracted from popular social networking site Facebook.com. The proposed approach has been successful in providing good recommendations.

*Index Terms*—Content Based Recommender Systems, Collaborative Filtering Systems, Hybrid Recommender Systems, Cross Domain Recommender Systems .

## 1 INTRODUCTION

Recommender system is one of the trending research fields in the present era. Many researches have been done and are still being done in this field. In literature maximum of the recommender systems make recommendations for a single domain [1][2][3][4][5][6]. For example, MovieLens compute recommendation for movies only [2] i.e. movies recommended to the user are closely related to those that he/she already likes. On the other hand , the aim of Konstan *et al.*[4] and Billsus and Pazzan [7] is to recommend news, recommending music albums is the goal of Shardanand and Maes [9] and Burke[8] recommends restaurants etc. Proceeding with this paper a domain refers to set of those items that are similar to one another in characteristics that can be easily distinguished for instance, movies, TV programs, music , games etc. In actual we do not stick to a fixed definition and would use it in a more supple form, as a domain can be split up into more precise ones like books into textbooks and novels.

Despite being useful a single domain recommendation is not enough in some of the cases for instance, Amit likes romantic movies and enjoys watching the movie The Fault In Our Stars. Single domain recommender systems will recommend him with the movies that are similar to the one liked by him. However , for an online shopping site



J.M. Corchado Rodriguez et al. (eds.), *Intelligent Systems Technologies and Applications 2016*, Advances in Intelligent Systems and Computing 530,
DOI 10.1007/978-3-319-47952-1_65

it may constitute of items like movie DVDs , Books , Music CDs etc. Therefore, as per our belief the recommender system would be more effective if it can recommend Amit with other items apart from movies. For example, book with the same title can be recommended to Amit or some other related books too can be recommended to him. The biggest dearth of current recommender systems is that they focus on getting input from user about their interests or tastes in an area inorder to produce recommendations in the same area.

The core perception that inspires our study i.e. the cross domain recommendation is highlighted in the above example. Inspite of being applied in the real world this concept remains a rarely studied research issue. On one hand where census scrutiny of popular items that abandon the personalization of recommendation are counted upon by current recommender systems , this paper tries to fix this gap through the study of cross domain and personalized recommendation. The remainder of this paper is as follows: section 2 discusses about the field of recommender systems, section 3 constitutes related work. In section 4 the proposed system has been explained. Section 5 gives an overview of recommender system evaluation and section 6 at last concludes the paper along with some future goals.

# 2 RECOMMENDER SYSTEMS

Recommendations always have been a part and parcel of human lives in one way or the other. Earlier recommendations or as they are commonly known suggestions were made mouth to mouth like one person suggesting his fellow mates to a read that he like and that according to him they may also like. In the recent years there has been a slight change in this whole process and that is it has taken the shape of an information filtering system i.e. tools and techniques that refer you with the items you may like based on various parameters may be your likes , your past search history etc.

## 2.1 Recommender System Components

A recommender system constitutes of Five basic components as shown in figure 1.

### 2.1.1 Data Collection and Processing

Recommender systems rely completely on the data to analyse both the user likings and to make predictions accordingly. So, investing more time on the collection and processing of data is all worth it. The main aim of this component is to gather data and then cleanse it and form the most generalised form of data possible because the quality of recommendations produced is directly proportional to the quality of input data.

### 2.1.2 Recommendation Model

Undeniably, recommendation model is the core component of any RS that produces recommendations for the users. It not only gathers user data like his likings and disliking and the details about the items but also predicts the list of items that might interest the user. By far, a huge amount of work in this field has been done on this crucial component of RS.

### 2.1.3 Data Post –Processing

Before presenting recommendations to the users some post processing needs to be done. At this moment recommendations need to be filtered and rearranged accordingly. The main motive of this component is to make recommendations look more sensible.

### 2.1.4 Online Modules

A set of online modules is further used for serving and tracking the use of the post-processed recommendations. At this stage decisions like what should be stored in the logs are made for both reporting the performance of the system as well as learning from its usage and interactions.

### 2.1.5 User Interface

The last component of a RS plays a crucial role as it is the medium through which recommendations produced are made visible to the users. This component defines not only what a user can see but also how the user can impart with the recommender. The usefulness of a RS is highly effected by the type of user interface being designed. For instance, making the users understand how and why is an item recommended to them can be a good practice for a RS.

The above mentioned components can be both developed parallel or sequentially depending upon the type of structure that might suit the developer, his team and the system goals.

## 3 CROSS DOMAIN RECOMMENDER SYSTEMS

Cross domain recommender systems with no exact definition can be simply stated as transferring knowledge from one module to produce recommendations in another module. Here the first module is called the source domain and the other module is known as target domain.

Although in the field of recommendation systems there are various research challenges like Group Recommendations, Impact Of Recommendations,

Recommendations and Social Networks etc. cross domain is one of these and is popularly trending amongst the research community. Initial studies on cross domain were presented by Winoto and Tang [11] , in their studies the authors hypothesized that the precision of recommendations of a cross domain recommender systems may be less as compared to that of a single domain recommender system.

The topic has become oh so trending and vastly popular these days due to the arising sparsity issue.

Although numerous papers and researches have been carried out on single domain recommender systems yet they lack to alleviate the problem of sparsity. This problem arises due to lack of user ratings for an item that further result in a sparse matrix.

Bin Li *et al.* [12] in their work authors have generated a new algorithm namely the Code Book Transfer(CBT) algorithm to user movie ratings given by the user from source domain for recommending books to the users in the target domain. The focus is not to have users or items in the two domains to be identical or even overlap. Empirical tests are performed inorder to explore how accurate recommendations are produced by CBT as compared to the existing algorithms.

As not much work and papers have been written related to cross domain recommendations there is lack of information about this issue but as much as the data is available it highlights that the main goal of cross domain recommendations is to resolve the sparsity problem in collaborative filtering approach.

Manuel Enrich *et al.*[13] in their work have presented an approach in which they recommend items in one domain based on the user ratings in other domain, where the domains are completely disjoint and auxiliary domains. The main purpose of authors in this paper is to deal with the cold start issue by exploiting cross domain recommendation.

There are four domain levels in cross domain recommender systems namely,

- *Attribute Level:* The attributes and type of the recommended items are same.. If two items vary in the value of certain attribute then they are contemplated to be of different domains. For example, if two movies belong to different genres say, romantic and thriller then they are said to be belonging to different domains. 12% of work in cross domain recommendation has been done in this domain level.
- *Type Level:* items are of similar type and share some attributes. For two items that have different attribute sets are said to be a part of different domains. 9% of work in cross domain recommendation has been done in this domain level.
- *Item Level:* items are neither similar in type and also most of the attributes of the recommended items are different from one another. 55% of work in cross domain recommendation has been done in this domain level.

- *System Level:* Items belonging to different systems that are examined as different domains are recommended. 24% of work in cross domain recommendation has been done in this domain level.

### *3.1 Goals Of Cross Domain Recommender Systems*

Following are the goals of cross domain recommender systems:

- improving accuracy
- addressing the cold start problem
- offering added value to recommendations
- enhancing user models.

### *3.2 Cross Domain Recommender Systems Research Issues*

A few research issues related to cross domain recommendations are as follows:

- Alliance of cross domain and contextual recommendations.
- Cross domain recommendations for reducing the user model abstraction effort.
- Collecting new and real datasets for cross domain recommendations.

As not much work and papers have been written related to cross domain recommendations there is lack of information about this issue but as much as the data is available it highlights that the main goal of cross domain recommendations is to resolve the sparsity problem in collaborative filtering approach.

Figure 1 Recommender System Components

## 4 Proposed Methodology

In this dissertation a novel approach that deals with research issues such as cross domain recommendations and integration of social networks with RS has been proposed. The proposed methodology is as shown in figure 2.

As the approach suggests a list of books will be recommended to the user based on his tastes in movies. APIs of Social Networking site Facebook.com will be used to fetch user information from his social media profile , this information includes username, facebookID , profile picture and list of movies liked by him.

To fetch information about movies IMDB APIs will be used. At last, four algorithms available in the apache mahout framework will be evaluated on the basis of lowest *root mean square value (RMSE).*

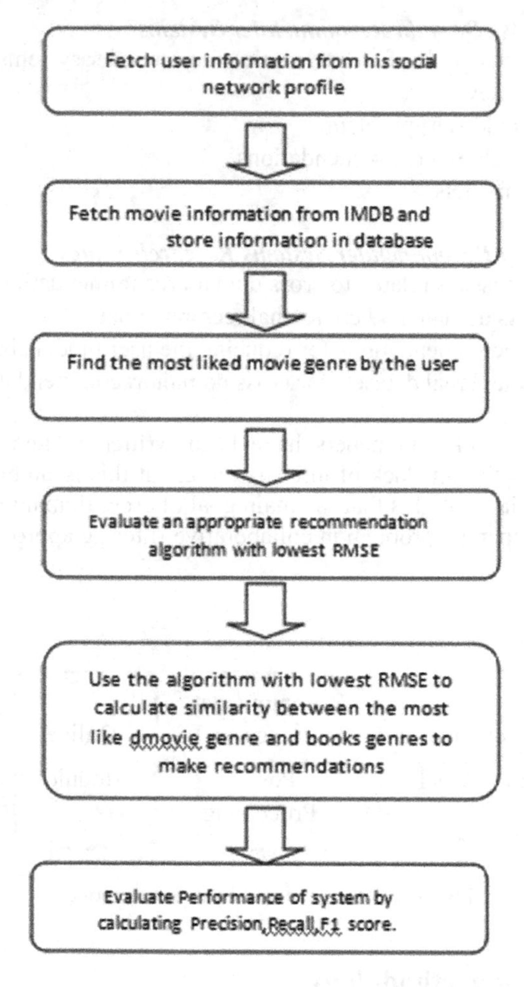


Figure 2 Flow Chart Of Proposed Methodology

The algorithm with lowest RMSE will be used to find similarity between the most liked genre by the user in terms of movies with the genre of the books available in the database and books with highest similarity will be recommended to the user.

### 4.1 System Architecture

Basically our system is dependent on both Facebook APIs and IMDB API for successfully providing recommendations.

The architecture of our system is as shown in Figure 3

- **Data Gathering:**

APIs of the popular sites *Facebook* and *Imdb* are used inorder to gather information about the user and movies.

- **Data Storage:**

Data gathered using APIs and information about the books is stored in database. Data for books has been gathered using WECA tool.

- **Connectivity:**

Fast internet connection is required so that APIs of both sites i.e. *Facebook* and *Imdb* can fetch data at a faster rate.

- **Recommendation:**

For producing précised recommendations four algorithms that are given in the apache mahout library have been evaluated on the basis of *Root Mean Square Parameter(RMSE).*

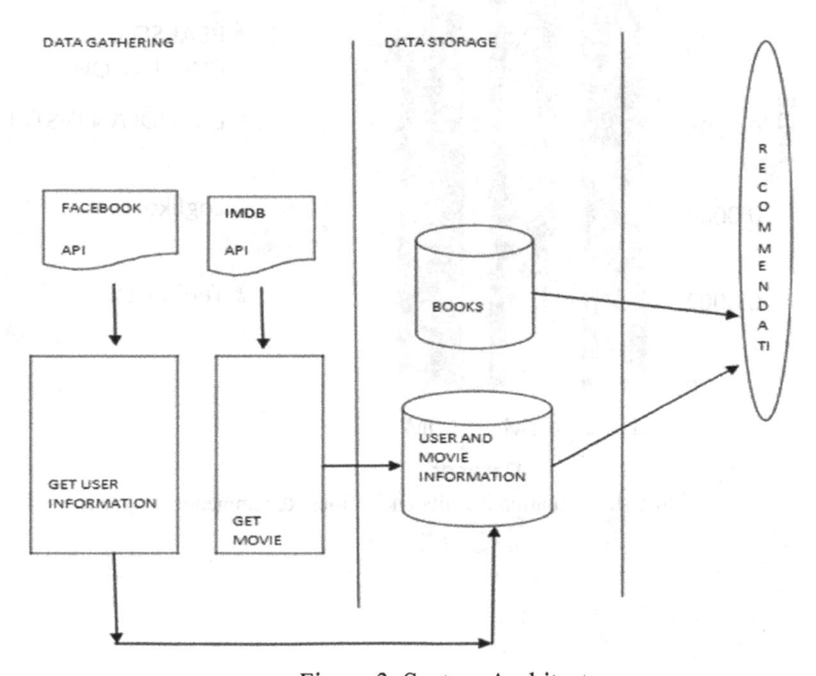


Figure 3. System Architecture

# 5 RESULTS

## 5.1 Evaluation Of Recommender Algorithms

Four recommender algorithms available in the popular apache mahout framework i.e. Pearson correlation coefficient, LogLikeLiHood, Tanimoto and Euclidean Distance have been evaluated on test sets i.e. 100K, 1M and 10M data sets of MovieLens.com( consists of movie ratings data) available on the GroupLens website.

The veracity of these algorithms has been calculated using *root mean square error (RMSE). RMSE* is one of the most popular evaluation metrics used to evaluate the veracity of the protended ratings. Both MAE (*mean absolute error)* and RMSE are similar in nature except for the fact that RMSE puts more emphasis on deviation.

Figure 4 shows evaluation results of the four algorithms on different datasets.

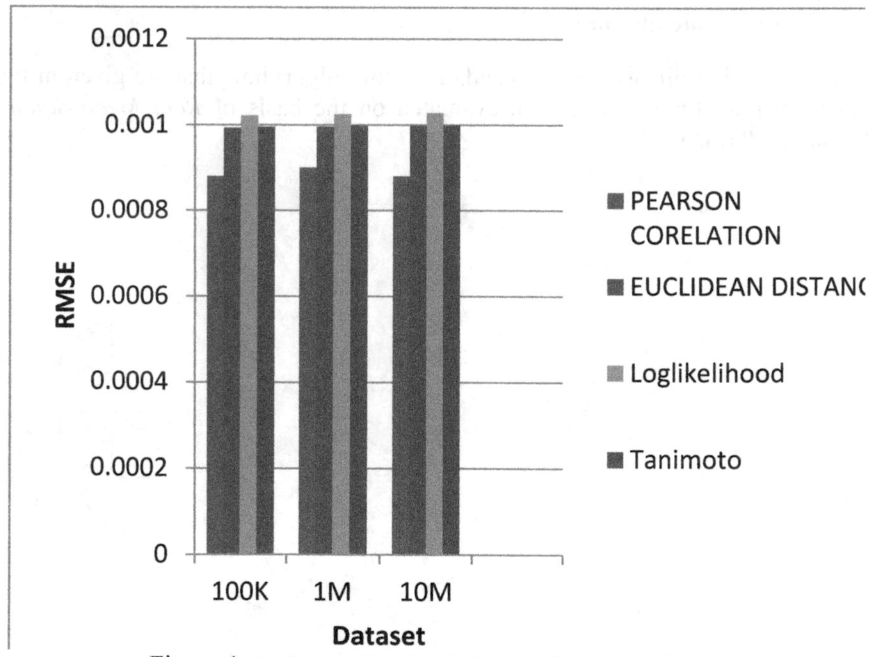


Figure 4. Evaluation Results Of Various Recommender Algorithms

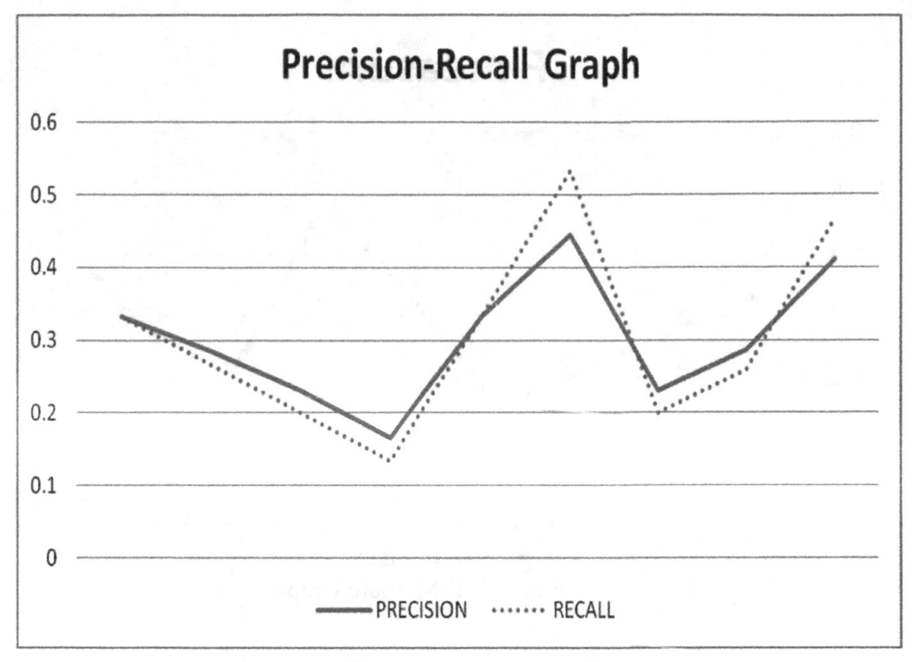


Figure 5. Precision-Recall Graph

## 5.2 Proposed System Evaluation

The results of the recommender system are evaluated further using four parameters namely,

- Precision: is the measure of how many correctly items have been recommended.
- Recall: is the measure of number of correct positive results divided by the number of positive results that should have been returned.
- F-measure: is the test's accuracy.

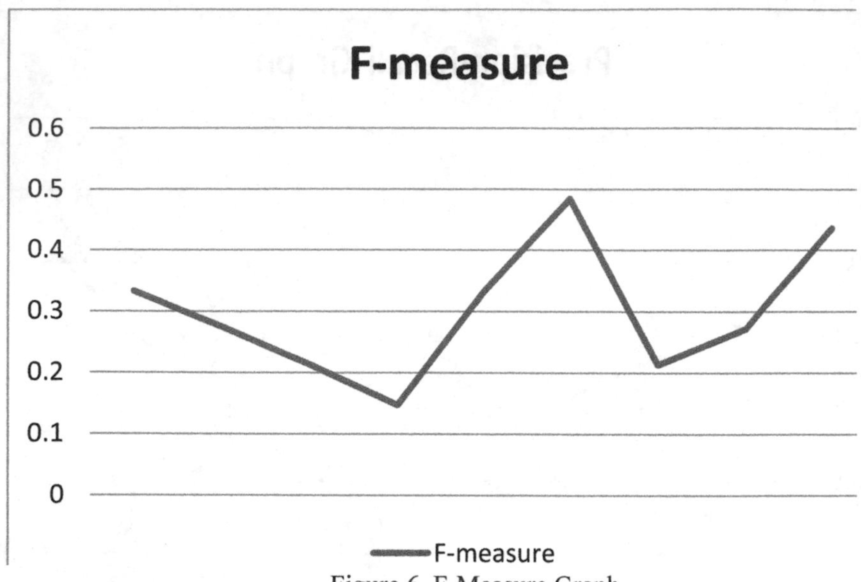


Figure 6. F-Measure Graph

## 6 CONCLUSION AND FUTURE SCOPE

We have proposed a novel idea that works on two research issues related to RS and also tries to solve the common problems faced in terms of RS i.e. cold start problem and sparsity. Our system has done justice to the problem statement and works efficiently; since all the data about user and movies is fetched from online portals it reduces the cold start issue and sparsity problem. Also as the results show the proposed system has tried working its best on the cross domain recommendations and integration of social networks with RS.

Due to security permissions on various user accounts on Facebook we were able to gather only a few users in order to evaluate our system. In future the number of users can be increased; also the length of the book database shall be increased in order to get better results.

## REFERENCES

[1] Berkovsky, S., Kuflik, T and Ricci, F., "Entertainment Personalization Mechanism through Cross-domain User Modeling," in Proc. 1st Int'l Conf. Intelligent Technologies for Interactive Entertainment, LNAI 3814, pp. 215-219, 2005.

[2] Herlocker, J., Konstan, J., Borchers, A. and Riedl, J., "An Algorithmic Framework for Performing Collaborative Filtering," in Proc. 22nd Annual Int'l ACM

SIGIR Conf. Research and Development in Information Retrieval (SIGIR'99), pp. 230-237, 1999.

[3] Joachims, T., Freitag, D. and Mitchell, T., "WebWatcher: A Tour Guide for the World Wide Web," in Proc. 15th Int'l Joint Conf. on AI (IJCAI'97), pp. 770-775, 1997.

[4] Konstan, J.A.,Miller,B.N., Maltz, D.,Herlocker, J.L.,Gordon, L. R. and Riedl, J., "GroupLens: Applying Collaborative Filtering to Usenet News," CACM 40, 3, pp. 77-87, 1997.

[5] Li, Y., Liu, L., and Li, X., "A Hybrid Collaborative Filtering Method for Multiple-interests and Multiple-content Recommendation in E-Commerce," Expert Sys. with Applications 28, pp. 67-77, 2005.

[6] Tang, T.Y., Winoto, P. and Chan, K.C.C., "Scaling Down Candidate Sets Based on the Temporal Feature of Items for Improved Hybrid Recommendations," in Intelligent Techniques in Web Personalization (B. Mobasher and S. S. Anand

[7] Eds.), LNAI 3169, pp. 169-185, 2004. Billsus, D. and Pazzani, M., "User Modeling for Adaptive News Access," User Modeling and User-Adaptive Interaction 10, 2-3, pp. 147-180, 2000.

[8] Burke, R., "Hybrid Recommender Systems: Survey and Experiments," User Modeling and User Adaptive Interaction 12,

4, pp. 331-370, 2002.

[9] Shardanand, U. and Maes, P., "Social Information Filtering: Algorithms for Automating 'Word of Mouth'," in Proc. ACM SIGCHI Conf. Human Factors in Computing Sys., (ACM CHI'1995), pp. 210-217, 1995.

[10] B. Sarwar, G. Karypis, J. Konstan, and J. Riedl. Item-based collaborative filtering recommendation algorithms. Proceedings of the 10th International Conference on World Wide Web (WWW), I:285–295, 2001.

[11] Winoto, P., Tang, T.: If You Like the Devil Wears Prada

the Book, Will You also Enjoy the Devil Wears Prada the Movie? A Study of Cross-Domain Recommendations. New Generation Computing 26, pp.

209–225 (2008)

[12] Li, B., Yang, Q., Xue, X.: Transfer learning for collaborative filtering via a rating-matrix generative model. In: Proc. of ICML 2009, pp. 617-624 (2009)

[13]Enrich, Manuel, Matthias Braunhofer, and Francesco Ricci. "Cold-start management with cross-domain collaborative filtering and tags." In *E-Commerce and Web Technologies*, pp. 101-112. Springer Berlin Heidelberg, 2013.

[14]Shani, Guy, and Asela Gunawardana. "Evaluating recommendation systems." In *Recommender systems handbook*, pp. 257-297. Springer US, 2011.

[15] Umanets, Artem, Artur Ferreira, and Nuno Leite. "GuideMe–A tourist guide with a recommender system and social interaction." *Procedia Technology* 17 (2014): 407-414.

[16]Zhang, Y., Cao, B., Yeung, D.Y.: Multi-domain collaborative filtering. In: Proc. of UAI 2010, pp. 725–732 (2010) [17]Carmagnola, F., Cena, F., Gena, C.: User Model Interoperability: A Survey. User Modeling and User-Adapted Interaction 21(3), pp. 285–331(2011)

[18]Driskill, R., Riedl, J.: Recommender Systems for E-Commerce: Challenges and Opportunities. AAAI'99 Workshop on Artificial Intelligence for Electronic Commerce, pp. 73–76 (1999)

[19]Kitts, B., Freed, D., Vrieze, M.: Cross-sell: A Fast Promotion-tunable Customer-item Recommendation Method based on Conditionally Independent Probabilities. 6th ACM SIGKDD Conference on Knowledge Discovery and Data Mining, pp. 437–446 (2000)

[20]Cantador, Iván, and Paolo Cremonesi. "Tutorial on cross-domain recommender systems." In *Proceedings of the 8th ACM Conference on Recommender systems*, pp. 401-402. ACM (2014)

[21]Cremonesi, Paolo, Antonio Tripodi, and Roberto Turrin. "Cross-domain recommender systems." In *Data Mining Workshops (ICDMW), 2011 IEEE 11th International Conference on*, pp. 496-503. Ieee, (2011)

[22] Dutta, Pallab, and A. Kumaravel. "A Novel Approach to Trust based Identification of Leaders in Social Networks." *Indian Journal of Science and Technology* 9, no. 10 (2016).

[23] Mamlouk, Lamia, and Olivier Segard. "Big Data and Intrusiveness: Marketing Issues." *Indian Journal of Science and Technology* 8, no. S4 (2015): 189-193.

[24] Reddy, C. Abilash, and V. Subramaniyaswamy. "An Enhanced Travel Package Recommendation System based on Location Dependent Social Data." *Indian Journal of Science and Technology* 8, no. 16 (2015): 1.

## Authors' Profiles

Sharu Vinayak belongs to Ludhiana,Punjab,India.Born on 8th of October, 1992. The author is currently pursuing M.E. in Computer Science and Engineering from Chandigarh University and has done B.Tech in Information Technology from Ludhiana College Of Engineering and Technology. Programming and learning about new technologies interests the author.

**Richa Sharma**: Belongs to Bilaspur, Himachal Pradesh and born on March 6, 1992. She is a Research student for Masters Degree for Computer Science Engineering in Chandigarh University. She received her Bachelor's Degree from Punjab Technical University in 2014. The field of her thesis is Recommender Systems. Her areas of interests for research are R language for building Recommender Systems, Data Privacy and Security.

Rahul Singh born on $15^{th}$ of april 1989 is an assistant professor at Chandigarh University, Mohali. He has done his B.Tech in computer science and engineering from Jhansi and M.E. in computer science and engineering from Thapar University,Patiala. Theory of computation and Software engineering are his areas of interes

# Diagnosis of Liver Disease Using Correlation Distance Metric Based K-Nearest Neighbor Approach

Aman Singh[1,*] and Babita Pandey

[1]Department of Computer Science and Engineering, Lovely Professional University, Jalandhar, Punjab – 144411, India
amansingh.x@gmail.com
[2]Department of Computer Applications, Lovely Professional University, Jalandhar, Punjab-144411, India,
shukla.babita@yahoo.co.in

**Abstract.** Mining meaningful information from huge medical datasets is a key aspect of automated disease diagnosis. In recent years, liver disease has emerged as one of the commonly occurring disease worldwide. In this study, a correlation distance metric and nearest rule based k-nearest neighbor approach is presented as an effective prediction model for liver disease. Intelligent classification algorithms employed on liver patient dataset are linear discriminant analysis (LDA), diagonal linear discriminant analysis (DLDA), quadratic discriminant analysis (QDA), diagonal quadratic discriminant analysis (DQDA), least squares support vector machine (LSSVM) and k-nearest neighbor (KNN) based approaches. K-fold cross validation method is used to validate the performance of mentioned classifiers. It is observed that KNN based approaches are superior to all classifiers in terms of attained accuracy, sensitivity, specificity, positive predictive value (PPV) and negative predictive (NPV) value rates. Furthermore, KNN with correlation distance metric and nearest rule based machine learning approach emerged as the best predictive model with highest diagnostic accuracy. Especially, the proposed model attained remarkable sensitivity by reducing the false negative rates.

**Keywords:** Liver disease diagnosis; k nearest neighbor; data mining; classification

## 1 Introduction

The wide availability of computer aided systems for medicine is evident but an accurate and well-organized diagnosis is still considered to be an art. To look inside a patient for finding principal causes of a disease is impossible as human body functions in an intricate way. Generally, diagnosis is being conducted on the basis of symptoms present and by analyzing the history of patient lifestyle. Assessment of patients needs proficient and experienced physicians in dealing their multifaceted cases. Solving these cases is not easy as the dimensions of medical science have been extremely expanding. Development of powerful diagnostic frameworks using classifi-



J.M. Corchado Rodriguez et al. (eds.), *Intelligent Systems Technologies and Applications 2016*, Advances in Intelligent Systems and Computing 530,
DOI 10.1007/978-3-319-47952-1_67

cation methods is becoming a key factor of improvement in medical assessment. Similarly, a range of expert systems have also been designed for liver disease diagnosis [1].

A healthy liver leads to healthy life. Liver performs various metabolic functions include filtration of blood, detoxification of chemicals, metabolizing drugs and bile secretion [2]. It also assists in digestion, absorption and processing of food. Improper working of any of the liver function leads to liver disease. Factors that can cause liver damage are heavy long term alcohol consumption, accumulation of excess amount of body fat, high salt intake, and overuse of medications. General symptoms of liver disease are queasiness, appetite loss, sleepiness, and chronic weakness. While the disease progresses, severe symptoms may include jaundice, hemorrhage, inflamed abdomen and decline in mental abilities. The most common category of liver disease are alcoholic liver disease and nonalcoholic fatty liver disease [1–4].

Literature study proves the wide usage of classification techniques and methods in liver disease diagnosis. Artificial neural network (ANN), decision trees, fuzzy logic (FL), rule-based reasoning (RBR), case-based reasoning (CBR), support vector machine (SVM), genetic algorithm (GA), artificial immune system (AIS) and particle swarm optimization (PSO) are deployed individually or in integration. General liver disorder was classified using ANN based approach [5,6], hepatitis disease was diagnosed using simulated annealing and SVM [7], and using feed-forward neural network [8]. SVM was also implemented to classify primary biliary cirrhosis [9]. ANN was also being used to predict liver fibrosis severity in subjects with HCV [10] and to assess complexity level in hepatitis examination samples [11]. CMAC neural network was deployed for classifying hepatitis B, hepatitis C and cirrhosis [12]. C5.0 decision tree and AdaBoost were used to categorize chronic hepatitis B and C [13]. C4.5 decision tree was used to examine liver cirrhosis [14]. Fuzzy logics were used for hepatitis diagnosis [15], general liver disorder diagnosis [16–18], and for classifying liver disease as alcoholic liver damage, primary hepatoma, liver cirrhosis and cholelithiasis [19]. ANN-CBR integration was used to test hepatitis in patients [2]. AN-FL was used to deal with class imbalance problem and to enhance classification accuracy of liver patient data [20]. AIS-FL was used to evaluate prediction accuracy of liver disorder in patients [21]. ANN-GA was used to classify liver fibrosis [22]. ANN-PSO and ANN-CBR-RBR hybridization were proposed for hepatitis diagnosis [23,24].

Mortality rates are rapidly increasing in liver disease cases. This indicates the need of computer-aided systems for assessing liver patients. The study accordingly proposed a correlation distance metric and nearest rule based KNN prediction model for learning, analysis and diagnosis of liver disease. The model is more liable and accurate in comparison with other traditional classifiers. Attained results of presented model are compared with other classifiers include LDA, DLDA, QDA, DQDA and LSSVM using statistical parameters. These parameters are accuracy, sensitivity, specificity, PPV and NPV rates. It also shows the capability to act as specialist assistant in training clinicians and to perform in time prediction of disease by minimizing the false positives.

The rest of the paper is organized as follows. Section 2 presents the proposed intelligent system and the other classifiers implemented. Section 3 describes with achieved simulation results. Finally, section 4 concludes the study.

## 2 Methodologies

Primarily physicians play a vital role in taking final decision on a patient assessment and treatment. Nevertheless, applicability of classification algorithms boost the prediction rate accuracy and also acts as a second opinion on substantiating the sickness. Consequently, this section presents the description of classification algorithms deployed in the study for liver disease diagnosis. These classifiers include LDA, DLDA, QDA, DQDA, LSSVM and correlation distance metric and nearest rule based KNN approach which are introduced as follows.

LDA is widely used for categorization and dimensionality reduction of data. It performs efficiently for unequal frequencies in within-class matrices. It works on the concept of maximum separability by capitalize on the ratio of between-class variance and within-class variance. It draws decision region between available classes in data which actually helps in understanding the feature data distribution [25,26]. For example, a given dataset have A classes; $\mu_b$ is mean vector of class b where b=1, 2, . . A; $X_b$ is total samples within class b where b=1, 2, .. A.

$$X = \sum_{b=0}^{A} X_b \quad (1)$$

$$M_p = \sum_{b=1}^{A} \sum_{c=1}^{X_c} (y_c - \mu_b)(y_c - \mu_b)^T \quad (2)$$

$$M_q = \sum_{b=1}^{A} (\mu_b - \mu)(\mu_b - \mu)^T \quad (3)$$

$$\mu = 1/A \sum_{b=1}^{A} \mu_b \quad (4)$$

where X is total number of samples, Mp , Mq and μ is the within-class scatter metric, between-class scatter metric and mean of entire dataset respectively. On contrary, DLDA assumes each class with same diagonal covariance.

QDA classifier performs same calculations as LDA and is also considered as a more generalized version of the later. The difference lies between both is the computation of discriminant functions [27]. Unlike LDA which draw an average from all three classes, QDA uses unique covariance matrices. It takes mean and sigma as a parameter for the available class. The covariance metric which is unique for each class is represented by sigma. On contrary, DQDA is a modification to QDA in which off diagonal elements are positioned to zero for each class covariance metric. It fits in the family of naive bayes classifier that follows multivariate normality. The given class should have more than two observations as the variance of a feature cannot be approximated with less than two.

SVM is a linear learning machine in which the structural risk of misclassifying is minimized choosing a hyperplane between two classes. On margins of an optimal hyperplane the training data is called as support vectors which help in classifying the features to their respective classes. These support vectors are being determined during the learning process. If data is nonlinearly separable then kernel functions draw the data in elevated dimensional attribute space. Specific selection of a hyperplane and kernel function is required to design an efficient support vector machine. In this study, least square hyperplane and radial basis kernel function were used to classify the patient data [28,29].

KNN is a semi-supervised and competitive learning method that belongs to the family of instance based algorithms. It creates its model based on training dataset and predicts a new data case by searching training data for the k-most similar cases. It strongly retains all observations selected at the time of training. This prediction data case of k-most similar cases is recapitulated and returned as the forecast for a new case. The selection of distance metric functions for finding similarity measure depends on structure of data. Available functions are euclidean, cityblock, cosine, correlation and hamming out of which correlation performed best for this study and hamming was not supportive as it can only be used for categorical or binary data [30, 31]. Let's assume a pa-by-q data of metric A that can be represented as pa (1-by-q) row vectors a1 , a2 , . . . . apa , and pb-by-q data of metric B that can be represented as pb (1-by-q) row vectors b1, b2, . . . . bpb. The correlation distance is the statistical difference between vector au and bv are defined in Eq. (5).

$$d_{uv} = \left(1 - \frac{(a_u - \bar{a}_u)(b_v - \bar{b}_v)'}{\sqrt{(a_u - \bar{a}_u)(a_u - \bar{a}_u)'}\sqrt{(b_v - \bar{b}_v)(b_v - \bar{b}_v)'}}\right) \quad (5)$$

$$where\ \bar{a}_u = \frac{1}{q}\sum\nolimits_j a_{uj} \quad (6)$$

$$\bar{b}_v = \frac{1}{q}\sum\nolimits_j a_{vj} \quad (7)$$

## 3 Simulations and Results

The liver health examination data used in this study has the objective of improving the ability of diagnosing liver disease based on features collected. This dataset is available with university of california machine learning repository. Samples in this collection are 583 and each sample consists of 11 features as entrance parameters out which 10 are contributing as inputs and one is acting as a target for determining a person as sick or healthy individual. These features include age (patient's age), gender (patient's gender), TB (total bilirubin), DB (direct bilirubin), ALP (alkaline phosphatase), SGPT (alamine aminotransferase), SGOT (aspartate aminotransferase), TP (total proteins), ALB (albumin), A/G ratio (albumin and globulin ratio) and Selector (field used to split data into two sets - sick or healthy. Data was divided into training and testing part using 10-fold cross validation method. Diagnostic results of classification algorithms were compared using statistical parameters which are defined in Eq. (8), (9), (10), (11) and (12) respectively.

$$Accuracy = \frac{TP + TN}{TP + TN + FP + FN} \quad (8)$$

$$Sensitivity = \frac{TP}{TP + FN} \quad (9)$$

$$Specificity = \frac{TN}{TN + FP} \quad (10)$$

$$PPV = \frac{TP}{TP + FP} \quad (11)$$

$$NPV = \frac{TN}{TN + FN} \quad (12)$$

where TN designates true negative (normal people rightly identified as normal), TP is true positive (diseased people rightly identified as diseased), FN is false negative (diseased people wrongly identified as normal), and FP expresses false positive (normal people incorrectly identified as diseased).

Classification algorithms deployed were LDA, DLDA, QDA, DQDA and KNN based approaches. These KNN approaches include KNN with euclidean, cityblock, cosine, and correlation distance matrices. Each mentioned KNN variant was implemented using nearest, random and consensus rules for classifying the cases. Fig. 1, 2, 3, 4, and 5 illustrates the comparison among KNN based approaches using statistical parameters mentioned in eq. (8), (9), (10), (11) and (12). KNN with euclidean distance metric and nearest rule had 100% (training) and 96.05% (testing) accuracy, KNN with euclidean distance metric and random rule had 100% (training) and 95.54% (testing) accuracy, KNN with euclidean distance metric and consensus rule had 100% (training) and 95.37% (testing) accuracy, KNN with cityblock distance metric and nearest rule had 100% (training) and 95.88% (testing) accuracy, KNN with cityblock distance metric and random rule had 100% (training) and 96.23% (testing) accuracy, KNN with cityblock distance metric and consensus rule had 100% (training) and 95.88% (testing) accuracy, KNN with cosine distance metric and nearest rule had 100% (training) and 96.23% (testing) accuracy, KNN with cosine distance metric and random rule had 100% (training) and 95.57% (testing) accuracy, KNN with cosine distance metric and consensus rule had 100% (training) and 95.88% (testing) accuracy, KNN with correlation distance metric and nearest rule had 100% (training) and 96.74% (testing) accuracy, KNN with correlation distance metric and random rule had 100% (training) and 96.23% (testing) accuracy and KNN with correlation distance metric and consensus rule had 100% (training) and 96.05% (testing) accuracy.

KNN with euclidean distance metric and nearest rule had 100% (training) and 91.62% (testing) sensitivity, KNN with euclidean distance metric and random rule had 100% (training) and 93.41% (testing) sensitivity, KNN with euclidean distance metric and consensus rule had 100% (training) and 93.41% (testing) sensitivity, KNN with cityblock distance metric and nearest rule had 100% (training) and 94.01% (testing) sensitivity, KNN with cityblock distance metric and random rule had 100% (training) and 95.81% (testing) sensitivity, KNN with cityblock distance metric and consensus rule had 100% (training) and 94.01% (testing) sensitivity, KNN with cosine distance metric and nearest rule had 100% (training) and 92.81% (testing) sensitivity, KNN with cosine distance metric and random rule had 100% (training) and 90.79% (testing) sensitivity, KNN with cosine distance metric and consensus rule had 100% (training) and 90.42% (testing) sensitivity, KNN with correlation distance metric and nearest rule had 100% (training) and 95.81% (testing) sensitivity, KNN with correlation distance metric and nearest rule had 100% (training) and 94.61% (testing) sensitivity and KNN with correlation distance metric and nearest rule had 100% (training) and 92.22% (testing) sensitivity.

KNN with euclidean distance metric and nearest rule had 100% (training) and 97.84% (testing) specificity, KNN with euclidean distance metric and random rule had 100% (training) and 96.39% (testing) specificity, KNN with euclidean distance metric and consensus rule had 100% (training) and 96.15% (testing) specificity, KNN with cityblock distance metric and nearest rule had 100% (training) and 96.63% (testing) specificity, KNN with cityblock distance metric and random rule had 100% (training) and 96.39% (testing) specificity, KNN with cityblock distance metric and consensus rule had 100% (training) and 96.63% (testing) specificity, KNN with cosine distance metric and nearest rule had 100% (training) and 97.6% (testing) specificity, KNN with cosine distance metric and random rule had 100% (training) and 97.51% (testing) specificity, KNN with cosine distance metric and consensus rule had 100% (training) and 98.08% (testing) specificity, KNN with correlation distance metric and nearest rule had 100% (training) and 97.12% (testing) specificity, KNN with correlation distance metric and random rule had 100% (training) and 96.88% (testing) specificity and KNN with correlation distance metric and consensus rule had 100% (training) and 97.6% (testing) specificity.

KNN with euclidean distance metric and nearest rule had 100% (training) and 94.44% (testing) PPV, KNN with euclidean distance metric and random rule had 100% (training) and 91.23% (testing) PPV, KNN with euclidean distance metric and consensus rule had 100% (training) and 90.07% (testing) PPV, KNN with cityblock distance metric and nearest rule had 100% (training) and 91.81% (testing) PPV, KNN with cityblock distance metric and random rule had 100% (training) and 91.43% (testing) PPV, KNN with cityblock distance metric and consensus rule had 100% (training) and 91.81% (testing) PPV, KNN with cosine distance metric and nearest rule had 100% (training) and 93.94% (testing) PPV, KNN with cosine distance metric and random rule had 100% (training) and 93.66% (testing) PPV, KNN with cosine distance metric and consensus rule had 100% (training) and 94.97% (testing) PPV, KNN with correlation distance metric and nearest rule had 100% (training) and 93.02% (testing) PPV, KNN with correlation distance metric and random rule had 100% (training) and 92.4% (testing) PPV, and KNN with correlation distance metric and consensus rule had 100% (training) and 93.9% (testing) PPV.

KNN with euclidean distance metric and nearest rule had 100% (training) and 96.67% (testing) NPV, KNN with euclidean distance metric and random rule had 100% (training) and 97.33% (testing) NPV, KNN with euclidean distance metric and consensus rule had 100% (training) and 97.32% (testing) NPV, KNN with cityblock distance metric and nearest rule had 100% (training) and 97.57% (testing) NPV, KNN with cityblock distance metric and random rule had 100% (training) and 98.28% (testing) NPV, KNN with cityblock distance metric and consensus rule had 100% (training) and 97.57% (testing) NPV, KNN with cosine distance metric and nearest rule had 100% (training) and 97.13% (testing) NPV, KNN with cosine distance metric and random rule had 100% (training) and 96.31% (testing) NPV, KNN with cosine distance metric and consensus rule had 100% (training) and 96.23% (testing) NPV, KNN with correlation distance metric and nearest rule had 100% (training) and 98.3% (testing) NPV, KNN with correlation distance metric and random rule had 100% (training) and 97.82% (testing) NPV, and KNN with correlation distance metric and consensus rule had 100% (training) and 96.9% (testing) NPV. Finally, KNN with correlation distance metric and nearest rule was found superior to other KNN based approaches in diagnosing liver disease.

To select the best predictive model for liver disease diagnosis, achieved results of the finalized KNN approach were compared with obtained results of LDA, DLDA, QDA, DQDA and SVM classifiers. LDA had an accuracy of 63.81% (training) and 63.98% (testing), sensitivity of 56.76% (training) and 56.49% (testing), specificity of 81.76% (training) and 82.63% (testing), PPV of 88.8% (training) and 89.02% (testing), NPV of 42.61% (training) and 43.26% (testing), DLDA had an accuracy of 62.1% (training) and 61.92% (testing), sensitivity of 55.44% (training) and 55.05% (testing), specificity of 79.05% (training) and 79.04% (testing), PPV of 87.8% (training) and 86.74% (testing), NPV of 63.81% (training) and 63.98 (testing), QDA had an accuracy of 56.19% (training) and 55.23% (testing), sensitivity of 40.91% (training) and 39.9% (testing), specificity of 94.04% (training) and 93.41% (testing), PPV of

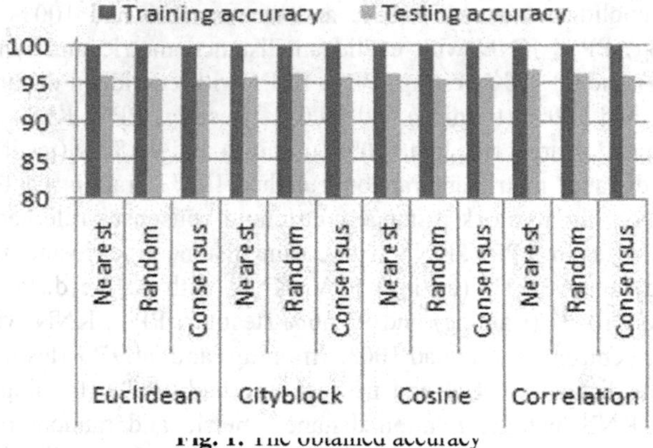


Fig. 1. The obtained accuracy

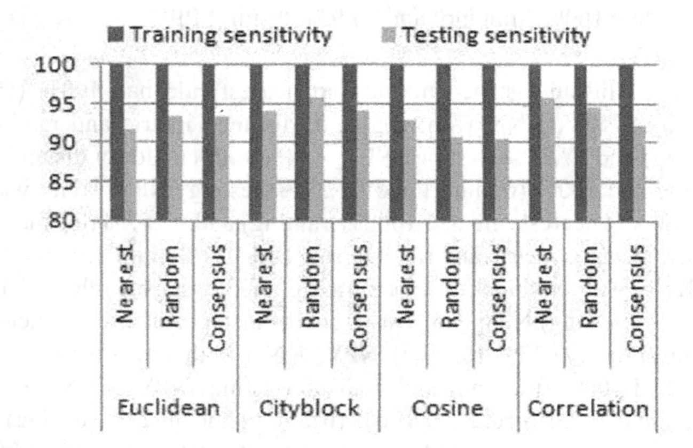


**Fig. 2.** The obtained sensitivity

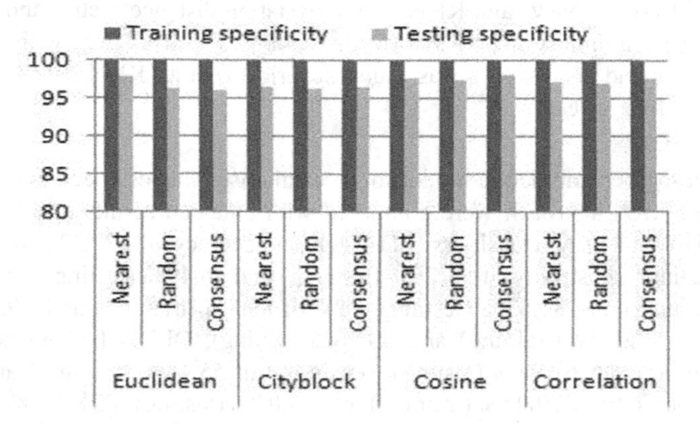


**Fig. 3.** The obtained specificity

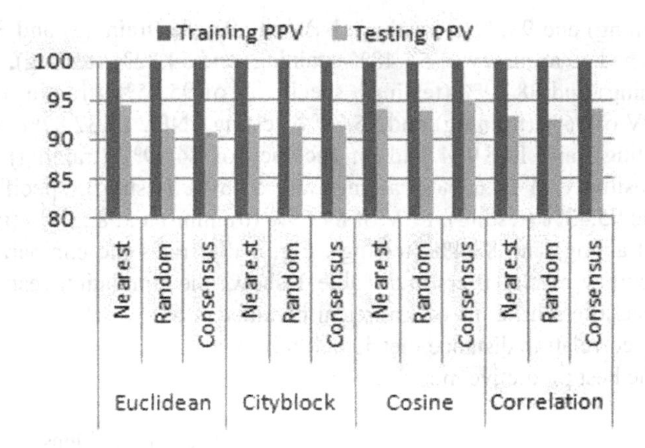


**Fig. 4.** The obtained PPV

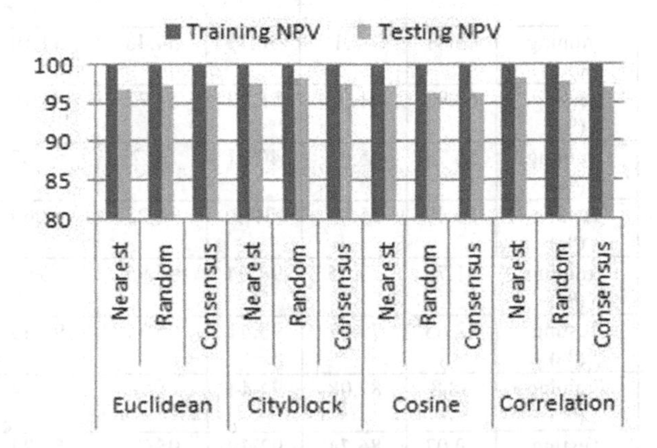


**Fig. 5.** The obtained NPV

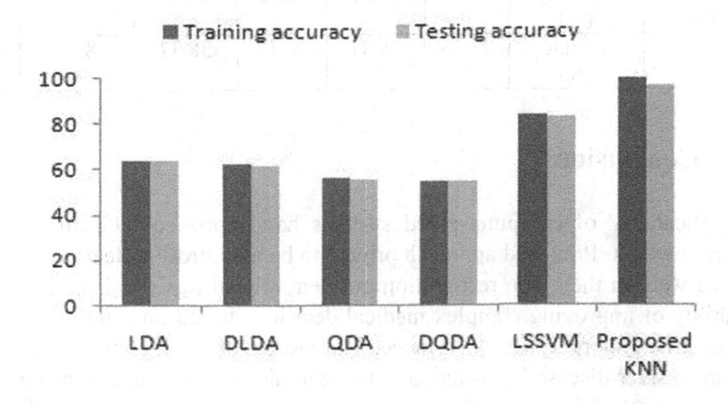


**Fig. 6.** The comparative view of achieved accuracy of classifiers

94.44% (training) and 93.79% (testing), NPV of 39.12% (training) and 38.42% (testing), DQDA had an accuracy of 54.48% (training) and 54.72% (testing), sensitivity of 38.2% (training) and 38.22% (testing), specificity of 95.95% (training) and 95.81% (testing), PPV of 96% (training) and 95.78% (testing), NPV of 37.87% (training) and 38.37% (testing), and LSSVM had an accuracy of 84.19% (training) and 83.19% (testing), sensitivity of 54.67% (training) and 52.69% (testing), specificity of 96% (training) and 95.43% (testing), PPV of 84.54% (training) and 82.24% (testing), NPV of 84.11% (training) and 83.4% (testing). Fig. 6 illustrates the comparative view of obtained accuracy of classifiers and Table 1 shows the simulation results of all implemented classifiers in terms of statistical parameters out which proposed KNN approach (with correlation distance metric and nearest rule) appears to take the lead and is selected the best predictive model for liver disease diagnosis.

**Table 1.** The prediction results of classification algorithms

| Classification method | | LDA | DLDA | QDA | DQDA | LSSVM | Proposed KNN |
|---|---|---|---|---|---|---|---|
| Accuracy | Training (%) | 63.81 | 62.1 | 56.19 | 54.48 | 84.19 | 100 |
| | Testing (%) | 63.98 | 61.92 | 55.23 | 54.72 | 83.19 | 96.74 |
| Sensitivity | Training (%) | 56.76 | 55.44 | 40.91 | 38.2 | 54.67 | 100 |
| | Testing (%) | 56.49 | 55.05 | 39.9 | 38.22 | 52.69 | 95.81 |
| Specificity | Training (%) | 81.76 | 79.05 | 94.04 | 95.95 | 96 | 100 |
| | Testing (%) | 82.63 | 79.04 | 93.41 | 95.81 | 95.43 | 97.12 |
| PPV | Training (%) | 88.8 | 87.08 | 94.44 | 96 | 84.54 | 100 |
| | Testing (%) | 89.02 | 86.74 | 93.79 | 95.78 | 82.24 | 93.02 |
| NPV | Training (%) | 42.61 | 56.37 | 39.12 | 37.87 | 84.11 | 100 |
| | Testing (%) | 43.26 | 56.87 | 38.42 | 38.37 | 83.4 | 98.3 |

## 4 Conclusion

Applicability of computer-aided systems has improved medical practices to an immense extent. Proposed approach proved to be a controlling learning technique that worked well on the given recognition problem related to liver disease. It showed the capability of improving complex medical decisions based on training data by finding k closest neighbors to a new instance. The presented intelligent system significantly diagnoses liver disease by means of euclidean distance metric and nearest rule based KNN approach. Disease prediction was carried out using a vast data of five hundred and eighty three samples from diverse patients. False negative rates were reduced by dividing health examination data into training and testing. Experimentation confirmed that the results of proposed KNN approach were superior to LDA, DLDA, QDA, DQDA and LSSVM classifiers. Thousands people lose their lives because of erroneous evaluation and inappropriate treatment as medical cases are still largely influenced by the subjectivity of clinicians. Therefore, development of computer aided systems in medicine seems to be of great use in assisting physicians and in providing training to novice researchers.

## References

[1] A. Singh, B. Pandey, Intelligent techniques and applications in liver disorders: a survey, Int. J. Biomed. Eng. Technol. 16 (2014) 27–70.

[2] C.-L. Chuang, Case-based reasoning support for liver disease diagnosis., Artif. Intell. Med. 53 (2011) 15–23.

[3] R.H. Lin, C.L. Chuang, A hybrid diagnosis model for determining the types of the liver disease, Comput. Biol. Med. 40 (2010) 665–670.

[4] E.L. Yu, J.B. Schwimmer, J.E. Lavine, Non-alcoholic fatty liver disease: epidemiology, pathophysiology, diagnosis and treatment, Paediatr. Child Health (Oxford). 20 (2010) 26–29.

[5] G.S. Babu, S. Suresh, Meta-cognitive RBF Network and its Projection Based Learning algorithm for classification problems, Appl. Soft Comput. 13 (2013) 654–666.

[6] M. Aldape-Perez, C. Yanez-Marquez, O. Camacho-Nieto, A. J Arguelles-Cruz, An associative memory approach to medical decision support systems., Comput. Methods Programs Biomed. 106 (2012) 287–307.

[7] J.S. Sartakhti, M.H. Zangooei, K. Mozafari, Hepatitis disease diagnosis using a novel hybrid method based on support vector machine and simulated annealing (SVM-SA), Comput. Methods Programs Biomed. 108 (2015) 570–579.

[8] S. Ansari, I. Shafi, A. Ansari, Diagnosis of liver disease induced by hepatitis virus using Artificial Neural Networks, Multitopic Conf. (INMIC), 2011 IEEE .... (2011) 8–12.

[9] P. Revesz, T. Triplet, Classification integration and reclassification using constraint databases, Artif. Intell. Med. 49 (2010) 79–91.

[10] A.M. Hashem, M.E.M. Rasmy, K.M. Wahba, O.G. Shaker, Prediction of the degree of liver fibrosis using different pattern recognition techniques, in: 2010 5th Cairo Int. Biomed. Eng. Conf. CIBEC 2010, 2010: pp. 210–214.

[11] D. a. Elizondo, R. Birkenhead, M. Gamez, N. Garcia, E. Alfaro, Linear separability and classification complexity, Expert Syst. Appl. 39 (2012) 7796–7807.

[12] İ.Ö. Bucak, S. Baki, Diagnosis of liver disease by using CMAC neural network approach, Expert Syst. Appl. 37 (2010) 6157–6164.

[13] A.G. Floares, Intelligent clinical decision supports for interferon treatment in chronic hepatitis C and B based on i-biopsy™, in: 2009 Int. Jt. Conf. Neural Networks, 2009: pp. 855–860.

[14] W. Yan, M. Lizhuang, L. Xiaowei, L. Ping, Correlation between Child-Pugh Degree and the Four Examinations of Traditional Chinese Medicine (TCM) with Liver Cirrhosis, 2008 Int. Conf. Biomed. Eng. Informatics. (2008) 858–862.
[15] O.U. Obot, S.S. Udoh, A framework for fuzzy diagnosis of hepatitis, 2011 World Congr. Inf. Commun. Technol. (2011) 439–443.
[16] P. Luukka, Fuzzy beans in classification, Expert Syst. Appl. 38 (2011) 4798–4801.
[17] I. Gadaras, L. Mikhailov, An interpretable fuzzy rule-based classification methodology for medical diagnosis, Artif. Intell. Med. 47 (2009) 25–41.
[18] M. Neshat, M. Yaghobi, M.B. Naghibi, A. Esmaelzadeh, Fuzzy expert system design for diagnosis of liver disorders, in: Proc. - 2008 Int. Symp. Knowl. Acquis. Model. KAM 2008, 2008: pp. 252–256.
[19] L.K. Ming, L.C. Kiong, L.W. Soong, Autonomous and deterministic supervised fuzzy clustering with data imputation capabilities, Appl. Soft Comput. 11 (2011) 1117–1125.
[20] D.-C. Li, C.-W. Liu, S.C. Hu, A learning method for the class imbalance problem with medical data sets., Comput. Biol. Med. 40 (2010) 509–518.
[21] E. Mężyk, O. Unold, Mining fuzzy rules using an Artificial Immune System with fuzzy partition learning, Appl. Soft Comput. 11 (2011) 1965–1974.
[22] F. Gorunescu, S. Belciug, M. Gorunescu, R. Badea, Intelligent decision-making for liver fibrosis stadialization based on tandem feature selection and evolutionary-driven neural network, Expert Syst. Appl. 39 (2012) 12824–12832.
[23] O.U. Obot, F.M.E. Uzoka, A framework for application of neuro-case-rule base hybridization in medical diagnosis, Appl. Soft Comput. J. 9 (2009) 245–253.
[24] S.N. Qasem, S.M. Shamsuddin, Radial basis function network based on time variant multi-objective particle swarm optimization for medical diseases diagnosis, Appl. Soft Comput. 11 (2011) 1427–1438.
[25] J. Ye, Q. Li, A two-stage linear discriminant analysis via QR-decomposition., IEEE Trans. Pattern Anal. Mach. Intell. 27 (2005) 929–41.
[26] Y. Guo, T. Hastie, R. Tibshirani, Regularized linear discriminant analysis and its application in microarrays., Biostatistics. 8 (2007) 86–100.
[27] S. Srivastava, M.R. Gupta, B.A. Frigyik, Bayesian Quadratic Discriminant Analysis, J. Mach. Learn. Res. 8 (2007) 1277–1305.
[28] C. Cortes, V. Vapnik, Support-Vector Networks, Mach. Learn. 20 (1995) 273–297. doi:10.1023/A:1022627411411.
[29] D. Tsujinishi, S. Abe, Fuzzy least squares support vector machines for multiclass problems, in: Neural Networks, 2003: pp. 785–792.
[30] S. Sun, R. Huang, An adaptive k-nearest neighbor algorithm, in: Proc. - 2010 7th Int. Conf. Fuzzy Syst. Knowl. Discov. FSKD 2010, 2010: pp. 91–94.
[31] H. Samet, K-nearest neighbor finding using MaxNearestDist, IEEE Trans. Pattern Anal. Mach. Intell. 30 (2008) 243–252.

# Sensorless Control of PMSM Drive with Neural Network Observer using a Modified SVPWM strategy

**Shoeb Hussain, Mohammad Abid Bazaz**

**Abstract** In this paper sensorless controlled PMSM drive is presented with neural network designed for speed and position estimation of the motor. Multi-level inverter (MLI) is operated using a modified space vector modulation (SVPWM) strategy for generation of 5-level and 7-level voltages. The sensorless control estimates the value of speed and position from calculations based on measured current and voltage. The issue with sensorless control strategy in state estimation arises with motor parameter variation and with distortions in current and voltage. The use of neural network observer deals with the issue of motor parameter variation. The use of MLI improves estimation with reduction in distortion in current further improved with the use of proposed SVPWM. The proposed scheme uses lesser switching states thus reducing the power loss compared to conventional scheme. Simulation is carried out in MATLAB on PMSM drive in order to test the physical performance of the drive.

## 1 Introduction

The PMSM drives offer a number of advantages over induction motor drives [1]. With lesser rotor power losses and higher power density, these drives are finding increased utilization in commercial and industrial applications. Vector control strategy improves the dynamic operation of these drives. Speed sensors and position encoders are a part of strategy in vector controlled drives. In order to improve reliability and lower costs in the system, sensors are eliminated in the sensorless scheme.

Sensorless scheme uses calculations based on measured values of current and voltage in order to estimate speed and position in the drive. Observers have been designed for speed estimation in AC drives [2]. Most important among these are the Luenberger observer and Kalman filter types [3, 4, 5].

Shoeb Hussain
Department of Electrical Engineering, NIT Srinagar, shoeb_15phd13@nitsri.net

Mohammad Abid Bazaz
Department of Electrical Engineering, NIT Srinagar, abid@nitsri.net



J.M. Corchado Rodriguez et al. (eds.), *Intelligent Systems Technologies and Applications 2016*, Advances in Intelligent Systems and Computing 530,
DOI 10.1007/978-3-319-47952-1_68

The problem arises in sensorless control scheme with motor parameter variation. Most of the observers depend on the motor model for estimation of speed and position. Since while operation of motors, parameters like resistances and inductances vary resulting in improper estimation of speed and position. However the use of Artificial Neural Network (ANN) based observers are very effective in motor parameter estimation. These intelligent controllers don't require the exact mathematical model of the system [6]. As such they offer a solution for design of motor model independent observers. This paper thus utilises an ANN based speed and position estimator for PMSM drive. Besides with Kalman filter and its types, mathematics involved is complex along with hectic tuning of the filter parameters. Again with ANN based observers, the numerical complexity is reduced.

Besides the motor parameter variation, load and speed variation can result in wrong estimation. In [7], the authors have shown that with these variations, Kalman filter suffers. The proposed model however is quite robust and immune to such variations. Also distortion in current and voltage can lead to inaccurate estimation. As such in this paper, MLI is used for operation of the drive [8, 9]. MLI offers nearly sinusoidal voltage waveform as a result of which the THD in current is reduced. The paper presents a modified SVPWM strategy for drive operation. The proposed model uses an octadecagonal and triacontagonal structures for operation of 5-level and 7-level MLI. The modulation strategy offers lesser switching states, reducing distortion in current and lower power losses.

The paper has been framed in 7 sections. Section 2 presents the design of neural network observer, section 3 discusses the sliding mode controller for speed control of the drive, section 4 proposes the modified SVPWM strategy for the MLI, section 5 presents the simulation results and finally the conclusions in section 6.

## 2 Neural Network Observer

Just like the use of ANN as controllers, these can also be designed as observers. The use of ANN for speed estimation in AC drives reduces the mathematical complexity involved with Kalman filter and sliding mode observer. Besides, neural network also does not require the exact knowledge of mathematical model of the system making the speed observer parameter independent.

Initially the motor is run with arbitrary weights, thus, the output is much deviated from the target for a given information design. The output produced from the ANN is compared with target continuously and weights are balanced likewise in a feed-forward way utilizing back- propagation calculation until the error turns out to be little. The back propagation calculation depends on Levenberg-Marquardt calculation. The weights are balanced as

$$w(k+1) = w(k) - (j(k)^T j(k) + \mu I)^{-1} j(k)e(k) \tag{01}$$

where, j is jacobian matrix, I is idenity matrix, e is error between output and desired speed, μ is combination coefficient.

The ANN utilized as a part of the plan has three layers, the input, hidden and output layers as shown in Fig. 1. The input layer has five neurons and the output layer has one neuron, while the hidden layer is optimised for fifteen neurons. The five inputs have been chosen as $i_d$, $i_q$, $v_d$, $v_q$ and $\omega_r$ *(k-1)* in case of the PMSM drive; output is speed $\omega_r$ (k). The output of every neuron is associated with every one of the neurons in the forward layer through a weight. The bias signal is additionally coupled to every one of the neurons of the hidden layer through a weight.

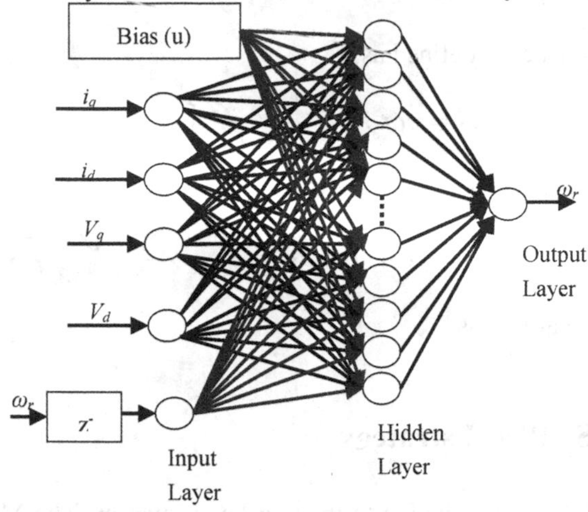


**Fig. 1.** Block diagram of neural network structure designed as an Observer for PMSM drive

## 3 Sliding Mode Controller

The SMC is essentially a versatile controller intended for variable control structures [1]. The utilization of SMC is feasible for both linear and non-linear frameworks. The controller results in vigorous execution of the drive for parameter variations and load changes. The SMC powers the framework to track or slide along a reference direction characterized in state plane by the utilization of switching calculation. The use of SMC for PMSM drive improves the operation of the drive. With load and speed variations, estimation suffers, which demands robust controller for execution of the drive. SMC is highly robust and capable of handling load and speed variations.

The SMC can be connected to PMSM drive in order to draw superior performance independent of variations in moment of inertia *'J'*, damping coefficient *'B'* and load torque *'$T_L$'*. The control system is reliant on the sliding surface. For the vector controlled PMSM drive, error is computed as

$$e(t) = \omega_r^* - \omega_r \tag{02}$$

where, ωr* and ωr are the reference and measured rotor speed

With classical SMC, derivative term is introduced. Hence, integral based SMC is used to define the sliding trajectory as a result of which the control is continuous and independent of derivative [10, 11].

From the PMSM model, the torque equation is

$$\dot{\omega}_r(t) = \frac{1.5 * P * \lambda * i_q - B * \omega - T_L}{J} \tag{03}$$

The sliding surface is defined as:

$$S = e(t) + \frac{3 * P * \lambda}{2 * J} \int e(t)dt \tag{04}$$

Combining (02), (03) and (04), we get

$$i_q^* = qe + f * sgn(S) + \frac{2 * J}{3 * P * \lambda} \dot{\omega}^* \tag{05}$$

where, $q$ and $f$ are gains

## 4 Modified SVPWM strategy

MLIs offer a superior alternative to the two level inverter. The MLIs offer higher voltage and power, taking care of ability enhancing the execution of AC drives [1]. As already has been stated, speed estimation is dependent on calculations based on measured currents. Hence an efficient estimation strategy will make sure the current distortion is minimised. The MLIs likewise brings about reduction in harmonic content of current, which enhances the estimation of speed in sensorless controlled drives. MLIs can be driven by space vector control modulators for enhanced execution [12]. In this paper, a modified space vector scheme has been employed for a MLI. The space vector scheme is presented for a 5-level and a 7-level MLI.

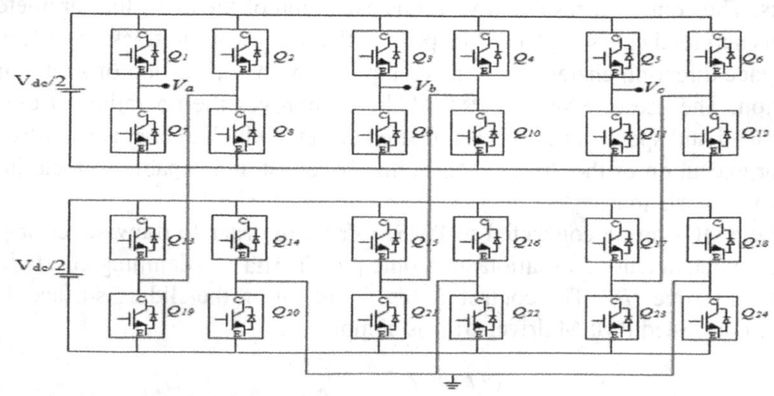


**Fig. 2.** Cascaded 5 level H-bridge Multi-level Inverter

Herein a cascaded H-bridge inverter setup is executed keeping in mind the end goal to produce multi-level voltage. Sensorless control of cutting edge drives using MLIs is being looked into for improving the execution of system. The MLI used as a part of the arrangement for a 5-level inverter is a 24 switch based cascaded H-bridge inverter (involving two H-bridges in parallel). IGBTs are used in the MLI as they offer high current dealing with limit at reasonably high switching frequency. The inverter drives the vector controlled AC drive. The MLI is shown in Fig. 2 where in two DC voltage sources are used for inverter operation.

SVPWM improves the operation of a variable frequency drive. The SVPWM plan involves exceptional figuring as a progressed PWM methodology [13, 14]. The SVPWM method offers advantage over current modulation with better DC voltage usage and more straightforward execution. The proposed strategy shows an octadecagonal space vector modulation structure for the MLI of a 5-level voltage and triacontagonal structure for MLI of 7-level voltage as shown in Fig. 3 and Fig. 4 respectively.

The fundamental thought about the modulation plan is to utilize the DC bus voltage from one H-bridge inverter absolutely and the DC voltage from the other H-bridge is utilized as part at uniform intervals. This results in multi-level operation of the cascaded inverter setup. The space vector structure so got is octadecagonal for 5-level operation that involves 18 divisions using 12 switching states and triacontagonal for 7-level operation with 30 divisions and 18 switching states.

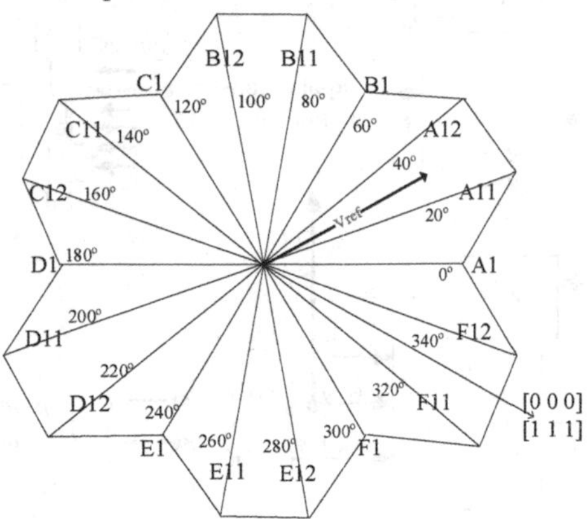


**Fig. 3.** Octadecagonal structure of SVM scheme used for switching 5-level inverter

## 5 Simulation Results

The simulation of PMSM motor carried out as shown in Fig. 5 is first carried out in MATLAB Simulink. The simulation is completed on a PMSM motor with motor parameters indicated in Table 2. The simulation is contemplated with speed/position observer as implemented through neural network. The speed control has been carried out using sliding mode controller. The multi-level inverter is utilized for drive operation for both 5-level and 7-level inverter and compared with a 2-level inverter. The altered SVPWM plan is utilized for inverter operation.

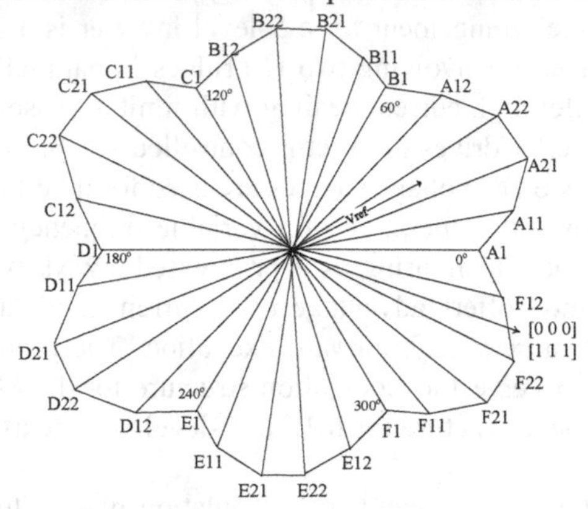


**Fig. 4.** Triacontagon structure of SVM scheme used for switching 7-level inverter

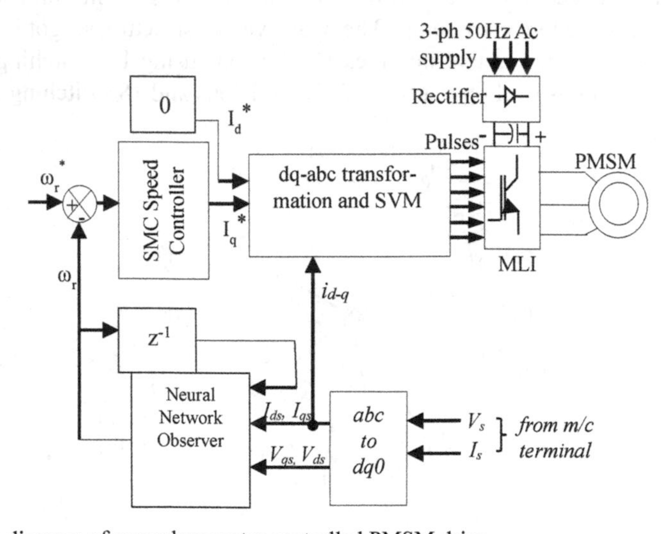


**Fig. 5.** Block diagram of sensorless vector controlled PMSM drive

## *5.1 Simulation with 2-level inverter*

Fig. 6 illustrates the three phase stator current, rotor speed and electromagnetic torque of the PMSM drive when subject to no load condition. The neural network observer evaluates the speed and position of the drive and SMC is utilized for speed control. The motor is driven through a 2-level based VSI. At first the speed is set at ω=180 rad/s and after time t=0.1 sec, the speed is changed to ω = - 180 rad/s. It can be seen that the estimation is accurate despite speed variation.

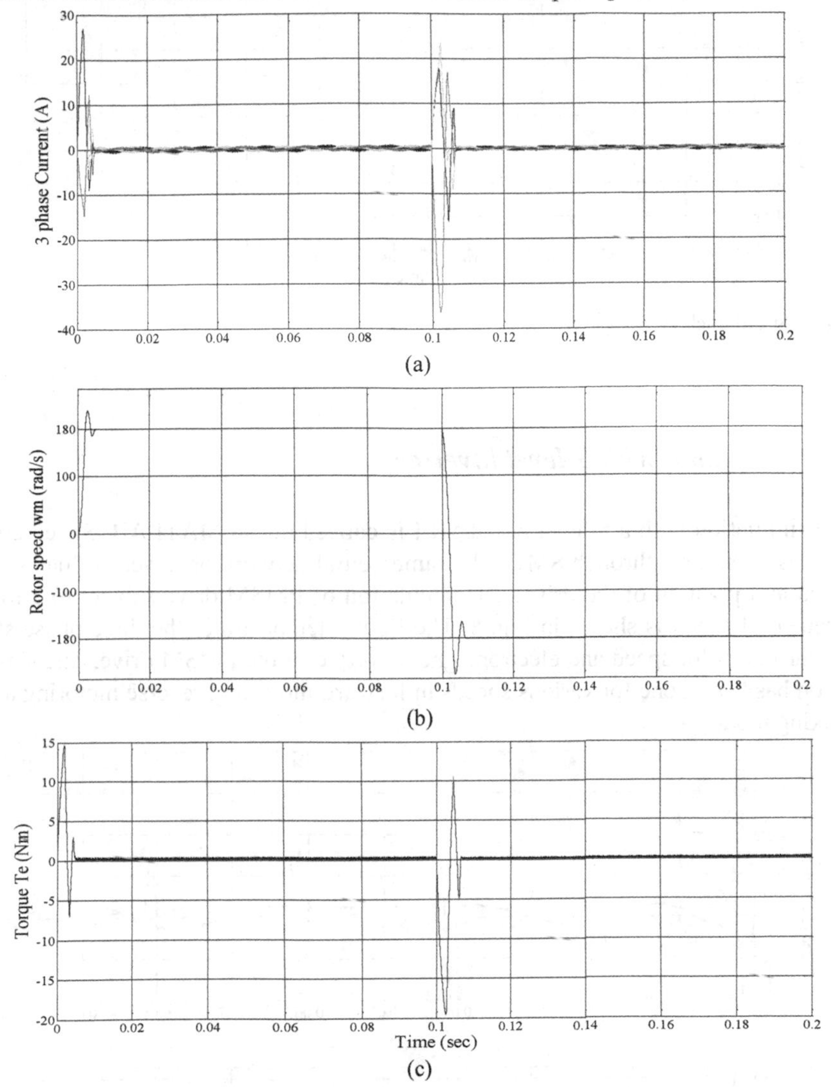


**Fig. 6.** Performance of the PMSM drive for speed variation (a) Stator Current (b) Rotor Speed (c) Torque at no load using a 2-level inverter

In Fig. 7 speed estimation utilizing ANN can be observed for all three motor working modes, i.e. forward motoring mode, reverse motoring mode and stopping mode. Here the motor is kept running at no load. The controller makes it conceivable to track speed with no error. Likewise the advantage with ANN observer is the estimation of speed closer to zero speed area.

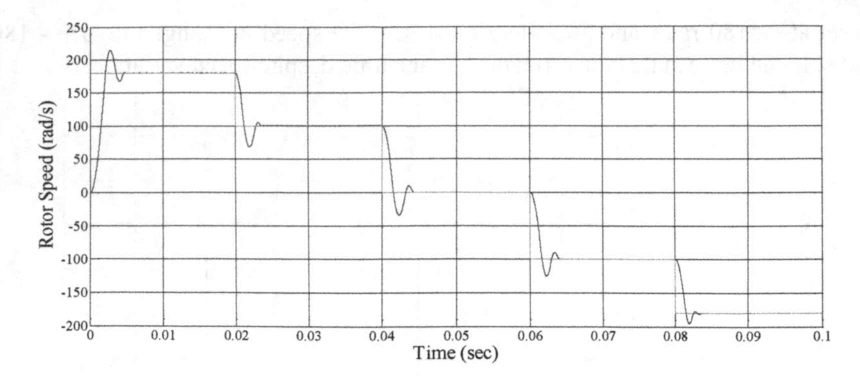


**Fig. 7.** Speed tracking using ANN observer at no load

## *5.2 Simulation with 5-level inverter*

The simulation with a 5-level based MLI is carried out in MATLAB. Speed control has been done through SMC. The same neural network observer evaluates the speed and position of the drive. The simulation of PMSM drive subject to a load torque of 1.5 Nm is shown in Fig. 8. The figure demonstrates the three phase stator current, rotor speed and electromagnetic torque of the PMSM drive. The simulation has been done for various speeds in forward motoring, reverse motoring and braking mode.

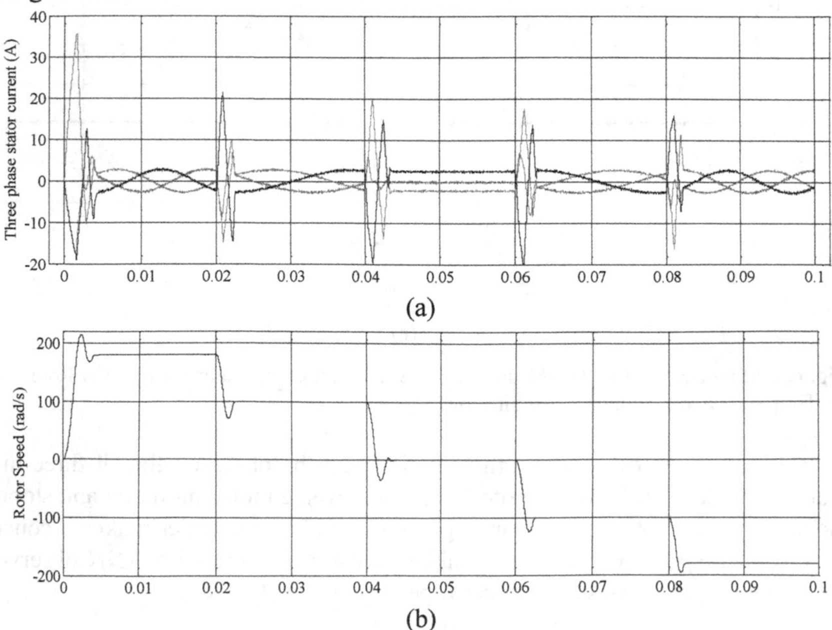

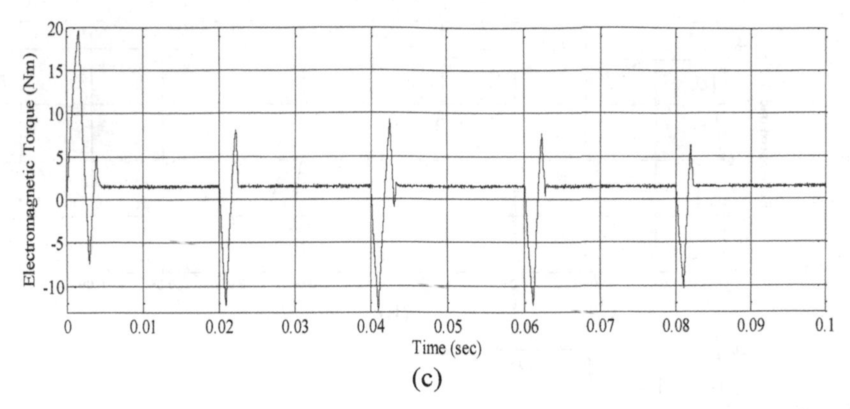


(c)

**Fig. 8.** Performance of the PMSM drive for speed variation (a) Stator Current (b) Rotor Speed (c) Torque at $T_L$=1.5 Nm with 5-level inverter

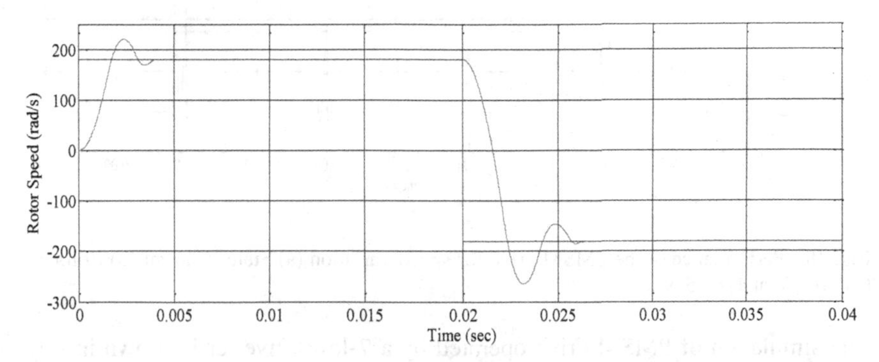


**Fig. 9.** Reference and Estimated Rotor speed at No load with 5-level inverter

In Fig. 9, the reference and estimated speed are compared. The estimated speed tracks reference speed exactly. The drive in this case has been subject to no load.

## *5.3 Simulation with 7-level inverter*

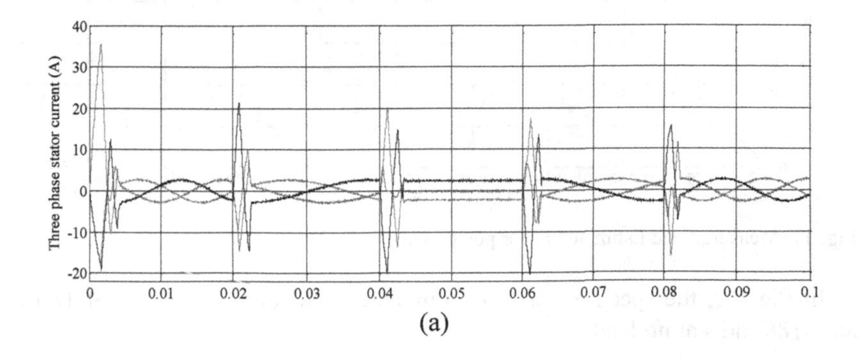


(a)

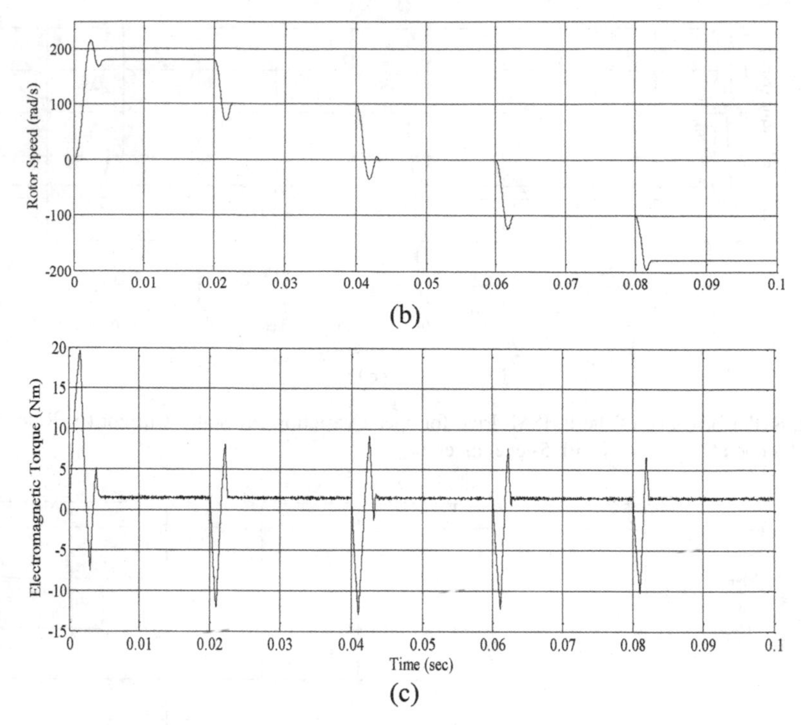


**Fig. 10.** Performance of the PMSM drive for speed variation (a) Stator Current (b) Rotor Speed (c) Torque at $T_L$=1.5 Nm

The simulation of PMSM drive operated by a 7-level inverter is shown in Fig. 10 through Fig 12. The three phase stator current, rotor speed and electromagnetic torque are observed in Fig. 10 for various speed conditions in motoring and braking modes. Observation drawn show superior performance of the drive.

Rotor position comparison between measured and estimated values is drawn in Fig. 11 for different speed conditions at load.

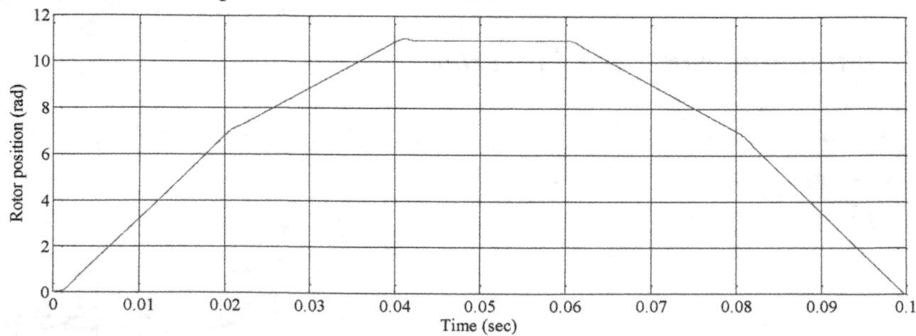


**Fig. 11.** Measured and Estimated Rotor position at $T_L$=1.5 Nm

In Fig. 12, the speed estimate is compared with reference speed of 180 rad/s and -180 rad/s at no load.

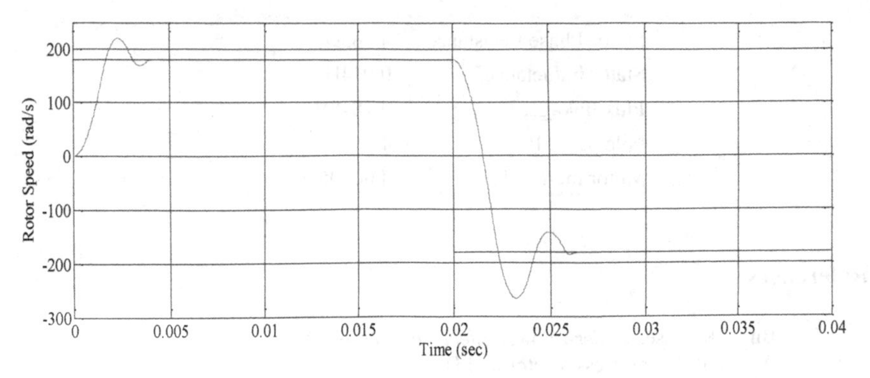


**Fig. 12.** Reference and Estimated Rotor speed at No load

# 6 Conclusion

As the motor is in operation, motor parameters vary. Observers modelled as such that depend on the motor parameters, thus result in inaccurate readings. The paper thus presents the use of neural network observer for speed/position estimation in PMSM drives. The neural network observer offers a model independent design solution for the problem. Furthermore the neural network observer does not involve numerical computational complexities as with other observers like extended Kalman filter, unscented Kalman filter etc. The neural network observer can track speed in low speed regions also with high accuracy. Looking at the torque profiles, the observer results in lesser torque throbs. As a consequence of this, the stator current profile is likewise enhanced with estimation more exact.

The drive has been operated using a multi-level inverter. The use of MLI for drive operation improves the nature of current by reducing the harmonic distortion in the current. The utilization of SVM for MLI based drives thus results in enhanced current profile as an aftereffect of which the speed/position observer turns out to be more powerful in nature. The use of SMC also enhances the execution of the drive by providing stability against speed and load variations. Simulation results show the superior performance of the proposed scheme.

## Appendix

**Table 1.** PMSM motor specifications

| | |
|---|---|
| DC link voltage | 300 V |
| Load torque | 1.7 Nm |
| Rated speed | 392.7 rps |

| | |
|---|---|
| Stator Phase Resistance | 4.765Ω |
| Stator inductance | 0.014H |
| Flux linkage, λ | 0.1848V.s |
| Pole pairs, P | 4 |
| Motor inertia, J | .0001051Kgm$^2$ |

## References


1. Bimal K. Bose, *Modern Power Electronics & AC Drives*, Prentice Hall India, 2003
2. A Vas, P., "Sensorless Vector and Direct Torque Control", New York: Oxford University Press, 1998
3. A Huang M C, Moses A J, Anayi F, Yao X G. "Linear Kalman filter (LKF) sensorless control for permanent magnet synchronous motor based on orthogonal output linear model," *International Symposium on Power Electronics, Electrical Drives, Automation and Motion*, 2006. May, 23rd - 26th, 2006, pp. 1381-1386.
4. M. Pacas, "Sensorless drives in industrial applications", *IEEE Ind. Electron. Mag.*, vol. 5, no. 2, pp. 16-23, 2011
5. N. K. Quang, N. T. Hieu, and Q. P. Ha, "FPGA-Based Sensorless PMSM Speed Control Using Reduced-Order Extended Kalman Filters", *IEEE transactions on Industrial Electronics*, Vol. 61, No. 12, December 2014, pp. 6574-6582
6. M.N. Cirstea, A. Dinu, J.G. Khor, M. McCormick, *Neural and Fuzzy Logic Control of Drives and Power Systems*, Newnes publications, 2002
7. Tze-Fun Chan, Pieter Borsje and Weimin Wang, "Application of Unscented Kalman Filter to Sensorless Permanent-Magnet Synchronous Motor Drive", *Electric Machines and Drives Conference, IEMDC 09*, pp. 631-638
8. A S. Kamel S. Mark A. Greg, "Sensorless control of induction motors using multilevel converters", *IET Power Electron.*, 2012, Vol. 5, Iss. 2, pp. 269–279
9. Schroedl, M.; 'Sensorless control of AC machines at low speed and standstill based on the INFORM method', *IEEE IAS Annual Meeting*, 1996, vol. 4, pp. 270–277
10. Rui Guo, Xiuping Wang, Junyou Zhao and Wenbo Yu (2011), "Fuzzy Sliding Mode Direct Torque Control for PMSM", Eighth International Conference on Fuzzy Systems and Knowledge Discovery (FSKD), pp. 511-514.
11. Bhim Singh, B.P. Singh, Sanjeet Dwivedi, (2006) "DSP Based Implementation of Sliding Mode Speed Controller for Vector Controlled Permanent Magnet Synchronous Motor Drive", Proceedings of India International Conference on Power Electronics 2006, pp. 22-27.
12. Seo, J. H., Choi, C.H., Hyun, D.S.: "A new simplified space-vector PWM method for three level inverters", *IEEE Trans. Power Electron.*, 2001, Vol. 16, no. 4, pp.545 -550
13. Dujic, D., Grandi, G., Jones, M., Levi, E., "A space vector PWM scheme for multifrequency output voltage generation with multiphase voltage-source inverters", IEEE Trans. Ind. Electron. , vol. 55, no. 5, 2008, pp.1943 -1955
14. Tekwani, P. N., Kanchan, R. S., Gopakumar, K., "Current-error space-vector-based hysteresis PWM controller for three-level voltage source inverter fed drives", Proc. Inst. Elect. Engg, Elect. Power Appl., Vol. 152, No. 5, 2005, pp.1283 -1295

# A Multimodel Approach for Schizophrenia Diagnosis using fMRI and sMRI Dataset

A. Varshney, C. Prakash, N. Mittal and P. Singh

**Abstract** Schizophrenia is an acute psychotic disorder, reflected as unusual social conduct. The exact reason for the Schizophrenia is still unknown. At Present it is deprived of any established clinical diagnostic test. Study reveals that unbalanced brain chemicals, cells, environment, and genetics contribute toward this disease. Its diagnosis is through the external observation of behavioral symptoms. Healthcare specialists take the help of Functional magnetic resonance imaging (fMRI) to identify schizophrenia patients by comparing the brain activation patterns with the healthy subjects. This paper presents a novel approach for the cognitive state, classifier for Schizophrenia. A multivariate fusion model by combining Functional Network Connectivity (FNC) and Source Based Morphometry (SBM) features obtained from fMRI and sMRI techniques, is used for the classification of Schizophreniac patients and Healthy subjects.

## 1 Introduction

Schizophrenia is an acute psychotic disorder, characterized by unusual delusions, perceptions, and abnormal social conduct. At present, there is no established clinical diagnostic test to identify this diseases exactly. Earlier, Healthcare specialists used to depend purely on behavioral symptoms of the patients. But now with the advances in the brain mapping technologies such as magnetic resonance imaging (MRI), experts can identify the deviation in subject's brain activity pattern using

Achin Varshney, Chandra Prakash, Namita Mittal, Pushpendra Singh
Malaviya National Institute of Technology, Jaipur, India-302017
e-mail: achinvarshney1@gmail.com

Chandra Prakash e-mail: cse.cprakash@gmail.com
· Namita Mittal e-mail: mittalnamita@gmail.com ·
Pushpendra Singh e-mail: mertiya.pushpendra@gmail.com



J.M. Corchado Rodriguez et al. (eds.), *Intelligent Systems Technologies and Applications 2016*, Advances in Intelligent Systems and Computing 530,
DOI 10.1007/978-3-319-47952-1_69

imaging techniques. Computer-aided diagnosis (CAD) can offer insights for better analytical and treatment purposes by identifying abnormal brain regions.

Functional magnetic resonance imaging (fMRI) is a neuroimaging technique that captures the metabolic functions within the brain. It identifies brain activation regions by gathering information about oxygen consumption by the brain tissues. It utilises the dependency of the magnetic properties of hemoglobin on the amount of oxygen it carries which is called the Blood Oxygen Level Dependent (BOLD) effect [1]. On the other hand Structural magnetic resonance imaging (sMRI) is used for the examination of the anatomy and pathology of the brain. It is used to analyse neurodegenerative diseases such as schizophrenia and Alzheimers disease. Most of the work done till now is based on features that were derived from fMRI data without combining the features of the sMRI data. Healthcare specialists take the help of Functional magnetic resonance imaging (fMRI) to identify schizophrenia patients by comparing the brain activation patterns with the normal subjects.

## *1.1 Functional Network Connectivity (FNC)*

Functional magnetic resonance imaging (fMRI) is a neuroimaging technique that captures the metabolic functions within the brain. It identifies brain activation regions by gathering information about oxygen consumption by the brain tissues [2]. It utilises the dependency of the magnetic properties of haemoglobin on the amount of oxygen it carries which is called the Blood Oxygen Level Dependent (BOLD) effect. On the other hand Structural magnetic resonance imaging (sMRI) is used for the examination of the anatomy and pathology of the brain. It is used to analyse neurodegenerative diseases such as schizophrenia and Alzheimer disease. Most of the work done till now is based on features that were derived from fMRI data without combining the features of the sMRI data. Healthcare specialists take the help of Functional magnetic resonance imaging (fMRI) to identify schizophrenia patients by comparing the brain activation patterns with the normal subjects.

The earlier proposed technique of classification of schizophrenia using fMRI data [3] involved the preprocessing of the raw fMRI data and the analysis of training data with group ICA. The Time course and Spatial map feature of the affected subject were evaluated using Back reconstruction. Functional network connectivity value were calculated for selected components, to analyse time course of the subject. Linear Discriminant Classifier, Fisher Linear, Logistic, KNN, Naive Bayes, Decision tree, Quadratic, Neural Network, SVM etc classifier were used to evaluate the accuracy of schizophrenia on the dataset. Result of Non-linear discriminative method was better than linear classifiers. For analyzing the performance of the used classifiers, Cross validation with Leave-one-out technique was used.

## 1.2 Source-Based Morphometry (SBM)

Source-Based Morphometry (SBM) [4] approach is an advancement over voxel-based morphometry (VBM) technique [5]. In SBM, Independent component analysis (ICA) is used to find the features of the brain imaging for statistical analysis. It is used to identify and analyse the gray area in the brain of Schizophrenia patients. SBM consist of three steps: Image preprocessing, ICA and statistical analyses of the image. It has been found that the foremost significant gray matter lies within bilateral lobe in schizophrenia subjects.A near-zero loading for a given ICA-derived brain map indicates that the brain regions indicated in that map are lowly gifted within the subject (i.e., the gray-matter concentration in those regions are terribly low in that subject). Because this data is derived from structural MRI scans, SBM values are thought of to be the structural modality options describing patterns of the brain structure.

In [6], discriminatory features were obtained by using a 3 phase dimensional reduction technique. ICA based features were extracted by taking both spatial and structural information of the voxels. Then clusters in spatially contiguous voxels were identified. Singular value decomposition (SVD) [7] was used to collect the information features by considering variance. Then a hybrid multivariate forward feature selection by considering Fisher's discriminant ratio and Pearsons correlation is used for the selection of best features.

Literature suggests that most of the brain mapping methods focus either on structural or functional connectome modality [8]. For better accuracy in the classification of an abnormal pattern in brain, there is need of multimodel brain mapping methods with enhanced feature extraction techniques.

The first section is introduction with literature survey. Our proposed methodology is discussed in section 2. Machine Learning algorithm classifier are presented in section 3 followed by result in section 4. Section 5 presents the conclusion.

# 2 METHODOLOGY

The proposed methodology is shown in Fig.1. Data preprocessing is the first and initial step in the Schizophrenia cognitive state classification.Creation of Spatial maps using Group ICA is finally followed by the combination of reduced FNC and SBM features from fMRI and sMRI datasets.

## 2.1 Data Preprocessing

Preprocessing of both the Datasets, fMRI and sMRI is done using statistical parametric mapping toolbox version 12 (SPM12) as discussed in [9].

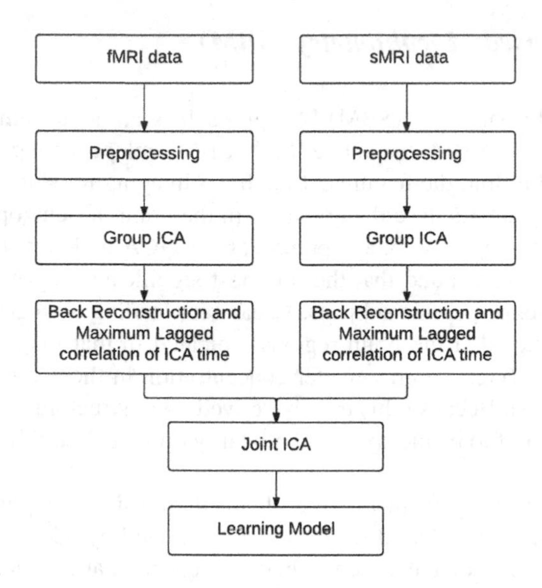


Fig. 1: Methodology of the Proposed System

## *2.2 Creation of spatial maps using Group ICA*

3D Spatial Maps can be extracted by using Independent component analysis (ICA) from 4D MRI data. Number of independent components (ICs) is decided on the basis of Modified minimum description length (MDL) criterion [11]. Group spatial ICA [12] occurs in two levels after dimension reduction using Principal Component Analysis at the individual level and concatenation of reduced individual subjects to form aggregated data. This aggregated data is further reduced to the desired optimal number of components using Group-PCA. Infomax algorithm [13] is used with ICA to develop 3-D ICA spatial map along with ICA timecourse with z-score. The value of z-score is used as 3-D spatial maps for the representation of individual components. Similarily for the gray matter images, spatial ICA is done as implemented in the GIFT ToolBox using Infomax Algorithm to obtain features.

## *2.3 Calculation of FNC and SBM Loadings*

The FNC toolbox in Matlab was used for the FNC analysis. The FNC toolbox computes maximum lagged correlation among the components as computed in [14]. ICA components's timecouse is calculated followed by the pearson correlation coefficient between the time courses. Similarily, SBM correlation loadings are determined from their spatial maps.

### 2.4 Combination of Reduced Features from fMRI and sMRI Data

The FNC and SBM Loadings are fused together in an input matrix using multivariate fusion model based on ICA. Sui et al. discuss the Joint ICA [15] as a technique to analyze multiple modalities based on the same set of subjects. It is implemented in the fusion ICA toolbox(FIT).

## 3 Machine Learning Algorithms

After the feature extraction from the group ICA from both fMRI and sMRI dataset, Naive Bayes and Support Vector Machines are used as learning models.

### 3.1 Naive Bayes Classification (NBC)

It is a probabilistic approach used in case of availability of prior knowledge and is thus considered as a supervised learning approach [16]. It assumes Gaussian distribution of values for each class. From a given set of training data, it learns the conditional probability of each feature $x_i \in$ X having class name c $\in$ class label set C.

Equation (1) estimates the probability of a given test data point belonging to a class c when its set of features $x_i$ is given. It is called Naive because it assumes that all the data sets are independent of each other and this is helpful in calculating the probabilities from a large set of training data.

$$P(C=c|X=x) = \frac{P(X=x|C=c)P(C=c)}{P(X=x)} \tag{1}$$

### 3.2 Support Vector Machines (SVM)

It is based on the concept of optimally separating hyperplanes and part of supervised learning [17]. It is a non-probabilistic method that projects the input space onto a high-dimensional feature space non-linearly. The essential action inside an SVM classifier is the minimization of the following objective function:

$$\begin{gathered} \min \frac{1}{2} \alpha^T Q_\alpha - e^T \alpha \\ \text{subject to} \quad y^T \alpha = 0 \\ 0 \leq \alpha_i \leq C, \quad i = 1....n \end{gathered} \tag{2}$$

In Eqn (2), Q is an $n \times n$ positive semidefinite matrix and $Q_{ij} \equiv y_i y_j$. e represents the vector of all ones, $C > 0$ is the upper bound. The decision function that defines the separating hyperplane for an SVM is given by:

$$\text{sgn}\left(\sum_{i=1}^{n} y_i \alpha_i K(x_i, x) + \rho\right) \tag{3}$$

The SVM is a robust classifier that essentially uses a kernel function to define the distribution of data points inside the feature space. In Eqn (3), $\rho$ stands for the bias term and K($x_i$, $x$) represents the kernel function that relates the subset of training data vectors $x_i$ to the testing data vectors $x$. The different possible kernels are linear, Radial Basis Function (RBF), polynomial etc.

## 4 Results and Discussion

The dataset used in this work is acquired through Center of Biomedical Research Excellence in Brain Function and Mental Illness(CORBE).

Naive Bayes and Support Vector Machines are used as classifiers for learning a decision model. For cross validation K-fold scheme is used. The results over 5 fold validation is shown in Table 1.

When only the Functional Network Connectivity (FNC) Loadings were used in Naive Bayes Technique, the average and mean accuracy was found to be 81.94% and 82% respectively. When the FNC Loading were combined with the Source Based Morphometry (SBM) Features, the average and mean accuracy got improved to 90.82% and 91 % respectively in NBC classifier.

When only the Functional Network Connectivity (FNC) Loadings were used in SVM, the average and mean accuracy was 82.4%. When the FNC Loading were

Table 1: k-Fold(k=5) Cross validation Results ( in terms of Accuracy (%))

| | 1st Fold | 2nd Fold | 3rd Fold | 4th Fold | 5th Fold | Average Accuracy |
|---|---|---|---|---|---|---|
| **NBC- FNC** | 0.8208 | 0.8218 | 0.8218 | 0.8208 | 0.8209 | 81.94 |
| **NBC-(FNC+SBM)** | 0.9109 | 0.9100 | 0.9063 | 0.9071 | 0.9090 | 90.82 |
| **SVM- FNC** | 0.8274 | 0.8248 | 0.8242 | 0.8275 | 0.8194 | 82.4 |
| **SVM-FNC+SBM** | 0.9749 | 0.9745 | 0.9781 | 0.9780 | 0.9770 | 97.65 |

Table 2: Tuning hyper-parameters

| **Kernel** | **C** | **Gamma** | **Accuracy** |
|---|---|---|---|
| Linear | 1 | - | 0.981 |
| Linear | 10 | - | 0.982 |
| Linear | 100 | - | 0.981 |
| Linear | 1000 | - | 0.981 |
| RBF | 1 | 0.0001 | 0.895 |
| RBF | 10 | 0.0001 | 0.97 |
| RBF | 100 | 0.0001 | 0.988 |
| RBF | 1000 | 0.0001 | 0.988 |
| RBF | 1 | 0.001 | 0.967 |
| RBF | 10 | 0.001 | 0.984 |
| RBF | 100 | 0.001 | 0.988 |
| RBF | 1000 | 0.001 | 0.988 |

combined with the Source Based Morphometry (SBM) Features in SVM, the average and mean accuracy got improved to 97.65% and 98% respectively.

For evaluating the SVM performance score, Grid Search is used in which we try to get the optimal values for parameters C and gamma by iterating through the parameter space for the best Cross-validation. The parameter C, which is the regularization parameter common to all the SVM kernels determines the cost for the miss-classification.Higher the value of C, higher is the miss-classification cost and thereby the decision boundary tries to fit the maximum training examples correctly. A low C value makes the decision surface as smooth whereas a very high value of C can make the classifier to overfit the model. Gamma defines how much influence a single training example has.The larger Gamma is, the closer other examples must be to be affected. So a small gamma will give a low bias and high variance while a large gamma will give higher bias and low variance.

In this work we use linear and Radial Basis Function (RBF) kernel for Support Vector Machine (SVM). Linear kernel is considered as degenerate version of RBF. RBF consider *C* and *gamma* parameter in training phase. C control the misclassification of training examples with decision surface. *gamma* specifies the influence of a single training example. Table 2 list the different values used for C, kernel and gamma for Support Vector Classifier and the best parameters found on the set.

Table 3 lists the score obtained on evaluation set. Using the proposed system healthy subject were identified 100 percent and the F1 score is 1. F1 score for Schizophrenia's subject is 0.98.

Table 3: Scores obtained on Evaluation Set

| Category | Precision | Recall | F1 score |
|---|---|---|---|
| Healthy Control | 1.00 | 1.00 | 1.00 |
| Schizophreniac | 0.98 | 0.98 | 0.98 |

## 5 Conclusion

It is not possible to suggest the exact reason for Schizophrenia which is a pyschotic disorder. Health care professionals use fMRI for mapping the activities of certain brain functions as potential cause for the disease. In this paper for better accuracy in the classification of an abnormal pattern in the brain, we propose multimodel brain mapping methods using both fMRI and sMRI dataset with ICA feature extraction techniques. The classifier prediction accuracy got improved upto 91% using the Multimodal fusion of both FNC and SBM values as compared to single FNC values 82% in case of Naive Bayes Classifier . Similarly, the prediction accuracy got improved upto 98% using the Multimodal fusion of both FNC and SBM values as compared to single FNC values 82% for the Support Vector Machine Classifier. As a future scope one can improve the features selection process and analyse the result on this multimodel approach.

**Acknowledgements** The authors would like to acknowledge Biomedical Informatics Research Network, for providing access to the dataset, under the following support: U24-RR021992, Function BIRN (for functional data).

## References

1. S. Ogawa, T.M. Lee, A.R. Kay, D.W. Tank, Brain magnetic resonance imaging with contrast dependent on blood oxygenation, Proc. Natl. Acad. Sci. U.S.A. 87, 9868-9872, 1990.
2. B. Rashid, E. Damaraju, G. D. Pearlson, V. D. Calhoun,Dynamic connectivity states estimated from resting fMRI Identify differences among Schizophrenia, bipolar disorder, and healthy control subjects. Frontiers in Human Neuroscience. 8:897, 2014.
3. M. R. Arbabshirani, K. A. Kiehl, G. D. Pearlson, V. D. Calhoun,Classification of schizophrenia patients based on resting-state functional network connectivity, Frontiers in Neurosci.,Volume 7, 2013.
4. J.M Segall, E.A. Allen, R.E. Jung, E.B. Erhardt, S.K. Arja, K. Kiehl, V. D. Calhoun, Correspondence between structure and function in the human brain at rest,Front. Hum. Neurosci. ,2012.
5. Xu L, Groth KM, Pearlson G, Schretlen DJ, Calhoun VD. Source-Based Morphometry: The Use of Independent Component Analysis to Identify Gray Matter Differences With Application to Schizophrenia. Human brain mapping. 30(3):711-724, 2009.

6. A. Juneja and B. Rana and R.K. Agrawal,A combination of singular value decomposition and multivariate feature selection method for diagnosis of schizophrenia using fMRI,Biomedical Signal Processing and Control,Volume 27, 2016.
7. G.H. Golub, C. Reinsch, Singular value decomposition and least squares solutions, Numer. Math. 14 (5), 403-420, 1970.
8. Hilgetag, Claus, Thomas R. Knosche, and Max Planck. "Correspondence between structure and function in the human brain at rest." Mapping the connectome: Multi-level analysis of brain connectivity, 2012.
9. Wellcome-Trust-Centre-for-Neuroimaging, SPM12 Statistical Parametric Mapping, University College London, 2009, From http://www.fil.ion.ucl.ac.uk/spm/software/spm12/.
10. A. Juneja, B. Rana and R. K. Agrawal, "A Novel Approach for Classification of Schizophrenia Patients and Healthy Subjects Using Auditory Oddball Functional MRI," Artificial Intelligence (MICAI), 2014 13th Mexican International Conference on, Tuxtla Gutierrez, 2014, pp. 75-81.
11. V.D. Calhoun, T. Adali, G.D. Pearlson, A method for making group inferences from functional MRI data using independent component analysis, Hum. Brain Mapp. 14 (3) (2001) 140-151.
12. Y.O. Li, T. Adali, V.D. Calhoun, Estimating the number of independent components for functional magnetic resonance imaging data, Hum. Brain Mapp. 28 (11) (2007) 1251-1266.
13. A.J. Bell, T.J. Sejnowski, An information-maximization approach to blind separation and blind deconvolution, Neural Comput. 7 (6) (1995) 1129-1159.
14. M.J. Jafri, G. D. Pearlson,M. Stevens, and V. D. Calhoun, (2008). A method for functional network connectivity among spatially independent resting-state components in schizophrenia. Neuroimage 39, 1666-1681.
15. J. Sui, T. Adali, Q. Yu, V. D. Calhoun, A Review of Multivariate Methods for Multimodal Fusion of Brain Imaging Data. Journal of neuroscience methods. 2012;204(1):68-81. doi:10.1016/j.jneumeth.2011.10.031.
16. Leung, K. M. Naive bayesian classifier. Polytechnic University Department of Computer Science/Finance and Risk Engineering, 2007.
17. Kohavi, Ron. "Scaling Up the Accuracy of Naive-Bayes Classifiers: A Decision-Tree Hybrid." In KDD, vol. 96, pp. 202-207. 1996.

# An Innovative Solution for effective enhancement of Total Technical Life (TTL) of an Aircraft

Balachandran A[1], PR Suresh[2], Shriram K Vasudevan[3], Akshay Balachandran[4].
[1]Research Scholar, PRIST University, India.
[2]Professor, Mechanical Engineering, NSS College of engineering, Palakkad, India.
[3]Dept. of Computer Science and Engineering, Amrita School of Engineering, Amrita Vishwa Vidyapeetham, Amrita University, India.
[4]Dept. of Electrical and Electronics Engineering, Amrita School of Engineering, Amrita Vishwa Vidyapeetham, Amrita University, India.

ABSTRACT

All the countries in the globe eventually happen to spend a lot of money towards strengthening their respective Air Force every year by spending a lot of money. Based on the observation made on the budgets framed, it will be clearly visible for anyone to understand that about 60 to 70 percentage of total money is spent towards technology up-grade and maintenance. Coming to the civilian aircraft maintenance and upgrade, the cost and budget is no inferior to the former class of military aviation budgets. We have taken this as challenge and we have come up with a solution for enhancing TTL (Total Technical Life) and reduced maintenance cost. We have built an embedded system based portable solution for measuring the usage of the aircraft and based on the usage and utilization, the maintenance can be made and this will greatly reduce the cost. The stress and strain measurements has lead us to calculate the cyclic stress that the aircraft undergoes and based on which the maintenance can be carried out. The complete implementation has been done and results are analyzed along with challenges which were resolved later.

**Keywords**: TTL (Total Technical Life), Maintenance, stress and strain measurement, aircraft, cyclic stress, Arduino computing board



J.M. Corchado Rodriguez et al. (eds.), *Intelligent Systems Technologies and Applications 2016*, Advances in Intelligent Systems and Computing 530,
DOI 10.1007/978-3-319-47952-1_70

## 1. Introduction

The area that has been chosen for research is highly challenging and demanding. One can understand the importance of this research through the results which will definitely reduce the total maintenance cost for the aircraft. Be it civilian or military, the proposed model would be suited well with carrying out minor changes based on the structure of the aircraft.

No major technological or design modifications need to be carried out except onboard monitoring of the strain level for a preferred area of the selected aircraft. To define the TTL, it is the effective flying time for an aircraft which manufacturers have cleared based on their design life keeping factor of safety in mind and is same across the all the planes for the entire fleet [1].

Considering the effective flying time of an aircraft, say any fighter plane Mirage 2000, being 7,000 hours, the calendar life will be somewhere around 25 years. The flight as one could understand needs intermediate and periodical servicing to be carried out at regular intervals as per guidelines suggested by manufacturers. So, if an aircraft is flown in harsh conditions throughout the lifespan, it deserves a definite and careful maintenance compared with an aircraft which did not undergo the same difficult flying envelopes [4]. Also, there could be aircraft which might not have come out of hanger for a longer duration due to various reasons, but, time of 25 years would have been completed. Ideally, all these aircraft should not be given same treatment with respect to maintenance. This research is focusing on this burning and need of the hour issue of underutilization problems without compromising the aircraft safety.

To be clear, the design of the set up used embedded system based solution with various parameters being monitored at specific areas of aircraft structures. The research has been concentrated only over few parts of aircraft, rather than plunging into the entire aircraft system but into the very critical and important parts of the system, from which the enhancement of the Total Technical Life of the aircraft is made possible.

## 2. Implementation Strategy

This section deals about the components and the way they have been interconnected to achieve the target. The following components serving as the major stake holders in the research.

- Model Aircraft (UAV)
- Arduino computing machine with glass shield
- Sensors (Strain Gauges)
- Instrumentation Amplifier
- A Rigid camera to be housed in the aircraft.
- MicroSD card and house for MircoSD card.

Before getting in depth into the component analysis, it would be important to look into the architectural / construction details. The following figure. 1 represents the architectural model.

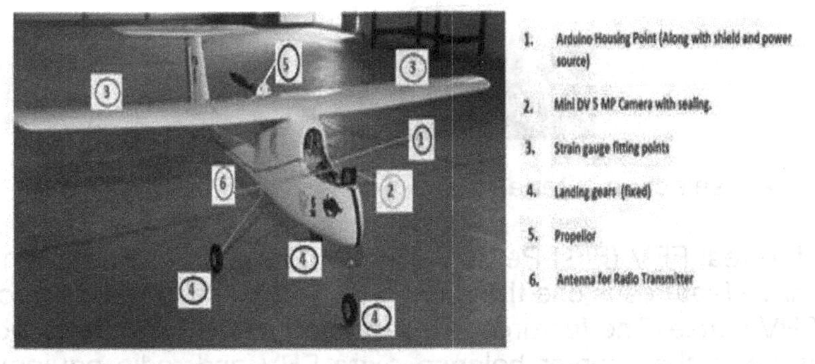


Fig. 1 UAV aircraft and its components.

## *2.1 Component usage analysis*

### a) Eagle UAV Model Aircraft

The model aircraft preferred was Eagle UAV (Unmanned Aerial Vehicle). There are many aircraft models available these days and selecting the right aircraft was challenge. Through lot of studies and analysis, it is understood that the following three points are very mandatory for picking the aircraft for the research:

1. Large wing span
2. Space for housing any auxiliary units (computing unit/camera/power source)
3. Sleek and rigid aerodynamic body shape to withstand stress.

The Eagle UAV model appeared to have all the above mentioned features and so hence is the selection made. The following representation shown in figure 2 is presented in support of the selection of this UAV.

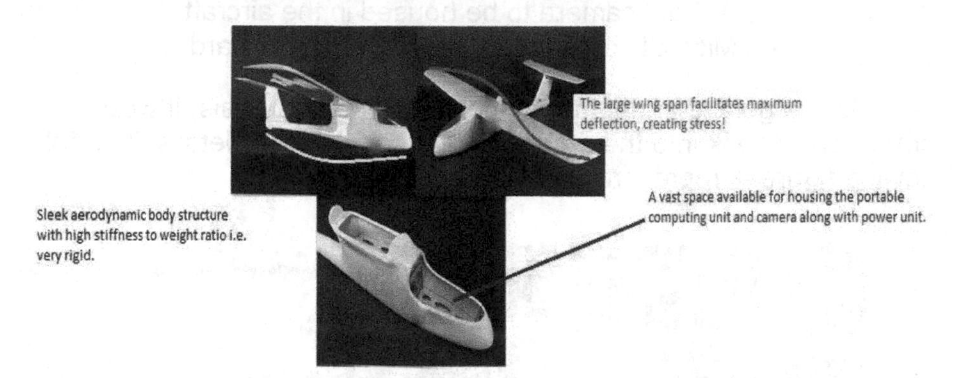

The large wing span facilitates maximum deflection, creating stress!
A vast space available for housing the portable computing unit and camera along with power unit.
Sleek aerodynamic body structure with high stiffness to weight ratio i.e. very rigid.


Fig. 2 UAV selection criteria.

Also, the real FPV (First Person View) model which comes with the must have features is one the easy ones to build, this comes packed with FPV sense. The features like rear mounted elevator and rudder servos helps the counter balance extra FPV and radio equipment. There are large internal ply mounting trays for installing batteries, transmitters, which fall into the practical features of the FPV, allowing easy access and configuration of components used for the research. The large canopy area is present for allowing an array of camera equipment which is not mounted to the frame allowing the user to configure the model as per the user requirement.

This is a hand launch model can be launched in any terrain. There would be plenty of power and lift for the initial launch only because of the combination of a large wing and the recommended 4s power train. With the flat bottom and rear skid, the fiberglass fuselage is ultra-tough and ideally shaped for landing on grass. However, an electric engine is used to provide positive power to the model and the electric engine is powered by battery. The aircraft model is made of balsa wood which has got high stiffness to weight ratio [7]. Also The landing gear system was specially designed to suit the model so that the model can be tested in any terrain.

Other than the above mentioned points, the UAV has the following specifications:

- Wing Area: 34.3" $dm^2$
- Wing Span: 1669mm (65.7")
- Flying Weight: 2300g
- Motor Mount Diameter: 58mm
- Material: Fiberglass & Ply/Balsa
- Fuselage: 1181mm (46.5")
- Minimum 4 Channel, 4 x Mini servo
- Distance From Centre of Mount to Fuse: 114mm (max prop diameter 8")

### b) Arduino Processing Kit (Computing system)

There were a lot of challenges in the selection of computing platform. There are numerous boards available like Raspberry-Pi, Arduino and Beagleboard. Arduino was preferred as support is available through the online communities is a huge plus point. Arduino is an open source electronics prototyping microcontroller based development board. It can also be described as a programmable logic controller. For the further enhancements to be carried out with the same implementation, one can go ahead with FPGA (Field Programmable Gate Array) or any advance computing platform with appropriate software support.

### c) Sensors – Strain Gauges

The strain gauges play a major role in the complete experiment. Strain gauge is placed at the location where the aircraft will face the maximum load condition. As the definition says, a strain gauge is a device for indication of the strain of the material or structure or in our case is an aircraft wing and closer to the point of attachment [6]. For our research, we deploy 4 strain gauges connected in series and implanted at the wings of the model aircraft. The strain gauge preferred out of a lot of available in the market is Omega 120 Ω Strain Gauge.

The following were the features enabling us to use this particular strain gauge from Omega. The proven features are:

- Can be used for both static and dynamic applications, we belong to latter.
- The body of the strain gauge is mechanically very strong. So, it can withstand heavy loading conditions and also could give correct measurement.
- It has a very small bending radius.
- Broad Temperature Range enabling us to use it at different altitudes.
- Encapsulated for Added Durability getting more flexibility as a feature for the strain gauge.

The following figure. 3 reveals the exact structure of the strain gauge used.

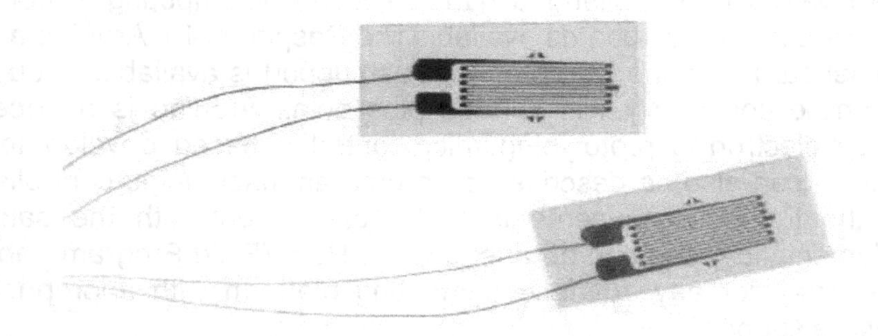

Fig. 3 Strain gauges used in the project – Omega 120 Ohms.

The strain gauges are connected in Full bridge format. Wheatstone bridge over the Kelvin Bridge was preferred as more number of resistors are required in the Kelvin Bridge. A Wheatstone bridge can be configured in three ways. It can be full bridge, half bridge or quarter bridge configuration. A full bridge is the one which will have all of the strain gauges connected to it active. It is easily understandable that half bridge rectifier will have two of its connected gauge active. Quarter Bridge is the one which will have only one gauge. We prefer the full bridge and we have all the four strain gauges connected, but point of focus is given at one particular strain gauge where the strain measurement is found to be perfect. Figure 4 reveals the full bridge configuration used.

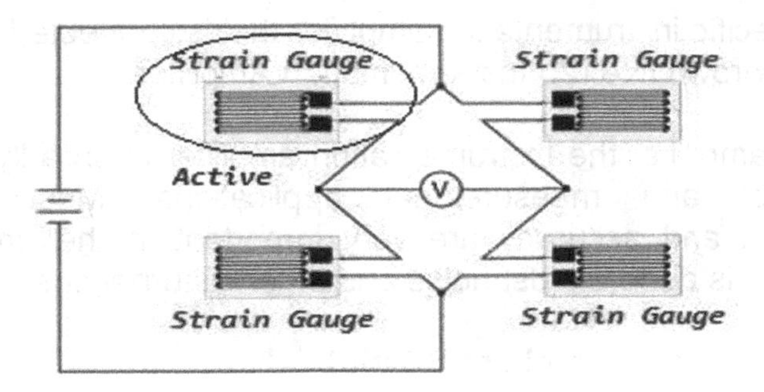


Fig. 4 Strain Gauge – Full Bridge

Strain gauge placement in the aircraft wing is presented below in the figure 5 shown below.

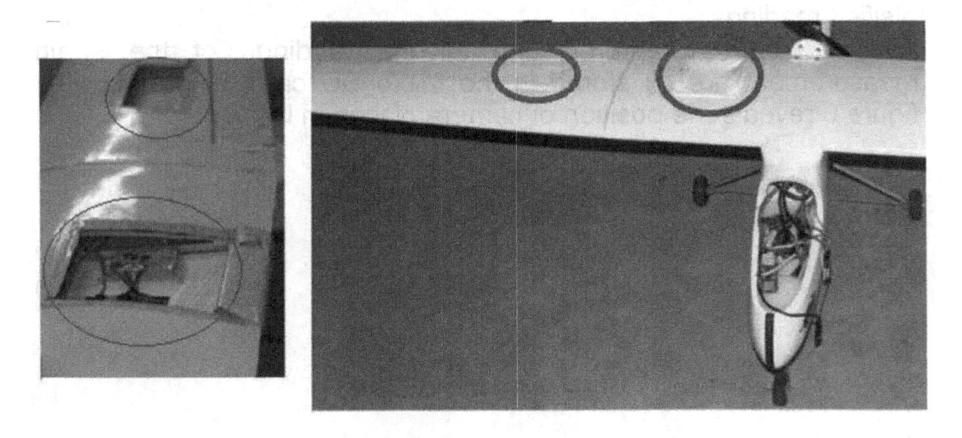

Fig. 5 Strain gauge placement.

Though the wing has a large span, we have plotted the strain gauge which reads the strain at spar [3]. The reason being simple, spar has high bending loads when the aircraft wing undergoes high deflection. The next component to be discussed is instrumentation amplifier. This component carries a lot of importance as it keeps the data valid through amplification.

**d) Instrumentation amplifier**

Load cells (series of 4 strain gauges) only make a very small change in voltage and hence an instrumentation amplifier is used to increase the voltage to something we can use.

The specific instrumentation amplifier that is connected to the model is Burr-Brown INA125 Instrumentation Amplifier.

The in-amp, i.e. the instrumentation amplifier is broadly used in the industrial and measurement applications. Most importantly, precision and accuracy are very important in the project as the research is done amidst noise and other disturbances.

**e) Camera and Micro SD slot.**

There is a camera fitted on the front portion of the cockpit which captures the maneuver of the aircraft. The camera is used in order to co- relate the reading obtained with the planned maneuver. The camera can be used in understanding the flying pattern which generates strain reports and can help in eliminating/filtering the falsified readings.
All the pictures captured and all the readings of the strain measurement are all stored in the Micro SD card. The following figure 6 reveals the position of camera placed in UAV.

Fig.6 Camera position.

A traditional Avionics UAV Radio Transmitter set is used for controlling the aircraft, remotely. Internal wires used in the UAV are all traditional copper wires.

## 3. Components Connection Diagram (Circuit)

Since all the components which are used in the project have been discussed, it is inevitable to know the way they are interconnected to each other.

The following figure represents the circuit which connects all parts together to get the UAV flying. Load cells (an arrangement of strain gauges) only make a very small change in voltage and therefore an instrumentation amplifier is used to increase the voltage output. Figure 7 reveals the complete connections and the ready to fly setup of the UAV, which has been flown to take the results. Figure 8 represents the complete aircraft with all connections up and ready to fly.

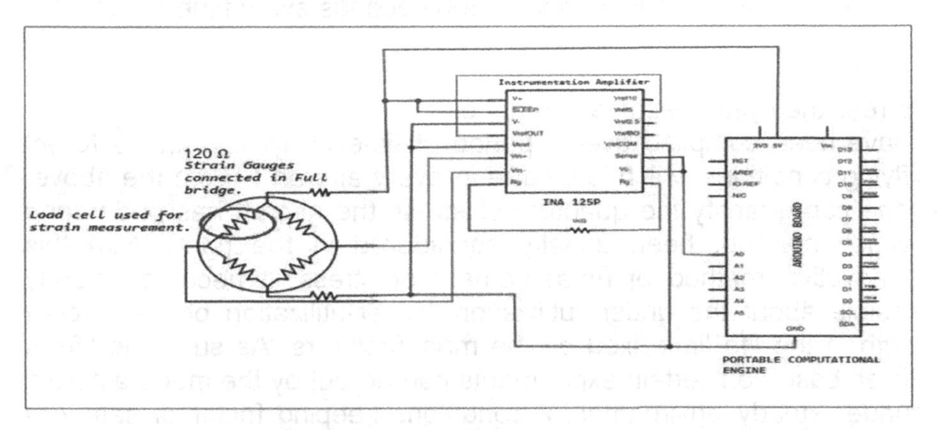


Fig.7 Complete circuit diagram – Strain measurement.

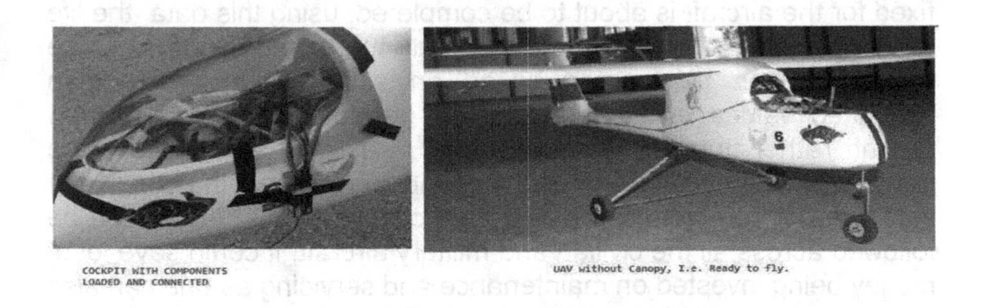

COCKPIT WITH COMPONENTS LOADED AND CONNECTED
UAV without Canopy, I.e. Ready to fly.


Fig. 8 UAV with components loaded and connected.

## 4. Results and Analysis

The readings obtained from one of the strain gauges are in the form of voltage. The variation in the voltage measurement is an instrument to understand how much strain the UAV has undergone at a particular instance. The UAV was flown for many times and the readings at different flying conditions were also taken. Weather conditions played a very major role and were also one of the influencing factors for the strain measurement.

UAV had been flown in conditions which were very smooth without any jitter and readings were measured. Also, the UAV had been made to fly in rude conditions making it to undergo heavier load conditions. The readings were taken into note and it was very informative to understand the strain pattern. We also measured the strain during takeoff and landing conditions under different wind speed conditions. All the measurements are summarized below in the graph. The flying time was even for all the conditions and we flew the UAV for a minimum period of 300 seconds every time.

With the available above results, one can come know how much stress the flying machine has undergone during flying. These results have been compiled over a period of time of flying under different flying conditions which is unique to every aircraft. From the above, one can quantify the quantum of stress the aircraft has undergone which has not been usually considered in the past. With this innovative method of measurement of stress, it becomes clearly visible about the under- utilization / over-utilization of the aircraft with in the life limit fixed by the manufacturers. As such the life is fixed based on certain experiments carried out by the manufacturers under steady environmental conditions keeping factor of safety in mind. Now with this experiment the data on stress pattern can be captured for each aircraft of the fleet. When the Total Technical life fixed for the aircraft is about to be completed, using this data, the life extension study can be carried out more meaningfully. The extensions can be granted based on actual flying rather than estimated values and the extension period will differ for each aircraft rather than for the entire fleet. This will also remove the inhibitions in the minds of users of the aero plane that the flight safety is not compromised at the cost of utilization. If this can be followed across all the civilian and military aircraft, it could save lot of money being invested on maintenance and servicing as one can also give extensions to the servicing periodicity based on the physical utilization of the aero plane. In future, new maintenance policy will emerge by undertaking a study of the data analytics of various aircraft of the same fleet. To conclude, the work is of national interest to save the avoidable expenditure which is spent on aircraft maintenance and to avoid under-utilization of aircraft.

## 5. Challenges Faced

The research itself was a very big challenge to begin with as there were no enough literature available. To collect the relevant material and selection of components of the UAV was another challenge.

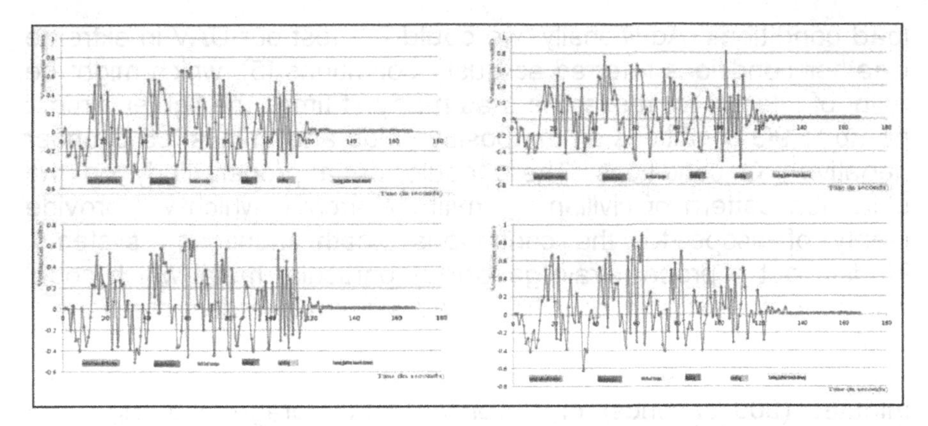

Fig. 9 Strain measured in Voltage at different flying conditions.

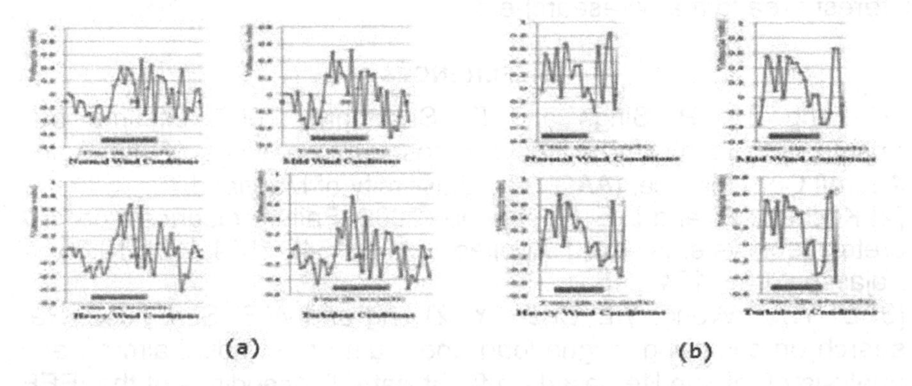


Fig. 10 (a) shows Take-off parameters Fig 10. (b) Shows Climbing parameters

Housing the electronic components including fixing of the Arduino board without compromising the equilibrium of the aircraft structure was one of the challenges. During the first flight, extracting the actual data by removing the noise is important challenge for which noise removal circuit was used. Selection of appropriate load cells (strain gauge) was another major area of challenge. Major challenge was in the area of voltage measurement from strain gauge. Finally, to protect the UAV from birds was a big trouble as they tried consistently to follow the UAV since it was generating sound.

## 6. FUTURE WORK

This is a beginning of continuous health monitoring especially on a fixed part of the aircraft [2]. Nevertheless, there are lot more things that can be done in the same area and can lead to fruitful research future. UAV was taken and tested for the strain under different wind load conditions. Additionally, we could not test our UAV in extreme weather conditions referred as gusty conditions [5], which might be point of interest for researchers as many at times the fighter aircraft go up to 9G conditions on the positive side and minus 2.5 G under negative G conditions. There could be a proper study on the utilization pattern of civilian vs. military aircraft, which will provide plenty of scope for the continuous heath monitoring systems. Finding out a proper strain gauge for particular metal which could withstand the extreme conditions and still could collect results is a challenge, which can be carried out as a research by itself. Over and above all these, collection of data has been done for about five minutes (300 seconds) of fly only. Data storage and Big Data Analytics during the aircraft flying for a long period would be of interest area to many researchers.

## References

[1] Feng, D., P. Singh and P. Sandborn, 2007. Lifetime buy optimization to minimize lifecycle cost. Proceedings of the Aging Aircraft Con- ference, (AAC' 07), University of Maryland.

[2] Kechida, S. and N.E. Debbache, 2005. Failure diagnosis on discrete event systems. Am. J. Applied Sci., 2: 1547-1551. DOI: 10.3844 / ajassp .2005 .1547. 1551

[3] Li, Y., Z. Wang, Y.L. Chen, Y. Zhang and W.S. Sun, 2012. Research on compiling fatigue load spectrum of individual aircraft and analysis of fatigue life based on flight data. Proceedings of the IEEE Conference on Prognostics and System Health Management, May 23-25, IEEE Xplore Press, Beijing, pp: 1-5. DOI: 10.1109/PHM.2012.6228937

[4] Meseroll, R.J., C.J. Kirkos and R.A. Shannon, 2007. Data mining Navy Fight and maintenance data to affect repair. Proceedings of the IEEE Autotestcon, Sept. 17-20, IEEE Xplore Press, Baltimore, MD., pp: 476-481. DOI: 10.1109/AUTEST.2007.4374256

[5] Sausen, R., C. Fichter and G. Amanatidis, 2004. Aviation, Atmosphere and Climate, number 83 in Air pollution research report. Proceedings of an International Conference Aviation, Atmosphere and Climate (AAC), (AAC' 04).

[7] Stress, Strain and Strain gauges by John.M. Cimbala, Penn State University (Latest Revision 24 Oct 2013).

[8] Elastic properties of wood, The Young's Moduli, Moduli of Rigidity and Poission's Ratios of Balso and Quipo No 1528 Reviewed 1962 by United States Department of Agriculture in cooperation with University of Wisconsin-Madison.

# ANFIS Based Speed Controller for a Direct Torque Controlled Induction Motor Drive

Hadhiq Khan, Shoeb Hussain, Mohammad Abid Bazaz

**Abstract** This paper presents a Neuro-Fuzzy adaptive controller for speed control of a three phase direct torque controlled induction motor drive. The Direct Torque Control (DTC) scheme is one of the most advanced methods for controlling the flux and electromagnetic torque of machines. Control of electromagnetic torque/speed in these drives for high performance applications requires a highly robust and adaptive controller. Adaptive Neural-Fuzzy Inference System (ANFIS) is a hybrid between Artificial Neural Networks (ANN) and Fuzzy Logic Control (FLC) that enhances the execution of direct torque controlled drives and overcomes the difficulties in the physical implementation of high performance drives.

MATLAB/SIMULINK implementation of 15 hp, 50 Hz, 4 pole squirrel cage induction motor controlled with the DTC scheme is presented in this paper. The PI controller used for speed control in conventional DTC drives is substituted by the ANFIS based controller. Simulation results show the use of ANFIS decreases the response time along with reduction in torque ripples.

## 1 Introduction

Squirrel cage induction motors find applications in motor driven pumps, washing machines, air conditioning and heating systems, servo drives, hybrid electric vehicles, domestic appliances etc. Low price, small size and weight, rugged and robust

Hadhiq Khan
Department of Electrical Engineering, NIT Srinagar, Email: hadhiqkhan@gmail.com
Shoeb Hussain
Department of Electrical Engineering, NIT Srinagar, Email: shoeb_ 15phd13@nitsri.net
Mohammad Abid Bazaz
Department of Electrical Engineering, NIT Srinagar, Email: abid@nitsri.net



J.M. Corchado Rodriguez et al. (eds.), *Intelligent Systems Technologies and Applications 2016*, Advances in Intelligent Systems and Computing 530,
DOI 10.1007/978-3-319-47952-1_71

contruction, the absence of commutators and brushes are some of the distinct advantages that make induction motors the most preferred motors in electric drive systems.

Induction motors happen to be constant speed motors. With the use of power electronic converters, induction motors can be employed in variable speed applications. To interface between the fixed voltage and frequency utility supply with motors, power electronic converters are used. High frequency and low loss power semiconductor devices are being increasingly used for manufacturing efficient converters.

Compared to the control of dc motors, the control of induction motors is difficult owing to the fact that the induction motor is a dynamic, nonlinear system. Unpredictable disturbances such as noise and load changes and the uncertainties in the machine parameters further complicate the control problem.

To reduce the complex nonlinear structure, advanced control techniques such as field oriented control (FOC) [1], [2] and direct torque control (DTC) [3], [4] have been developed offering fast and dynamic torque response. The DTC scheme, as the name indicates is a direct control of the torque and flux of the machine. The electromagnetic torque and the flux generated in the machine is compared to the reference values of the torque and flux in hysteresis comparators. The outputs of the two comparators along with the stator flux position determine the voltage vector from a lookup table.

For controlling the various states of the inverter, control pulses are generated. The output of the inverter is eventually fed to the induction motor.

Even though FOC and DTC methods have been effective, a number of drawbacks are observed ( distortions in currents and torque due to change in the flux sector, the need for high sampling frequency for digital implementation of the comparators, sensitivity to variations in the machine parameters etc).

To mitigate the shortcomings in the DTC scheme, intelligent control techniques using human motivated techniques, pattern recognition and decision making are being adopted [5], [6]. Some of the methods based on the concept of Artificial intelligence (AI) are Artificial Neural Network (ANN), Expert System (ES), Fuzzy Logic Control (FLC) to mention a few. AI based techniques when implemented in motor control have shown improved performance [7].

Generally, motor control drives rely on the use of PI controllers. These controllers, however, are sensitive to system non-linearities, variations in parameters and any unwanted disturbances. These disadvantages can be overcome by the use of AI based intelligent controllers. Fuzzy logic and artificial neural networks are two of the most popular systems for use as intelligent controllers and have shown an improved performance over conventional controllers [8], [9], [10]. ANFIS (adaptive neural network based fuzzy inference system), a hybrid of ANN and FLC is another popular control scheme being used in high performance drives, having superior design and performance characteristics [6].

The use of an ANFIS based speed controller for a direct torque controlled induction motor drive is presented in this paper. The PI controller used in the conventional DTC scheme is replaced with the ANFIS based speed controller.

The paper is arranged in the following sections: Section 2 presents the dynamic dynamic-quadrature axis model of the motor. The concept of direct torque control is introduced in Section 3 followed by the control strategy in Section 4. The ANFIS speed controller is presented in Section 5 followed by the simulation results and conclusion in Sections 6 and 7 respectively.

## 2 D - Q Model of Induction Motor

At the core of the induction motor DTC scheme lies the dynamic direct-quadrature axis model of the induction motor. The steady state model of the induction motor is useful only if the steady state characteristics of the motor are to be studied. For high performance drives, the transient as well as the steady state response needs to be taken into account which is not possible without the use of the d-q model. A detailed description of the model is illustrated in [11].

The generalised state equations governing the dynamics of the motor rotating with a mechanical speed $\omega_r$ are:

$$\frac{dF_{qs}}{dt} = \omega_b \left[ v_{qs} - \frac{\omega_e}{\omega_b} F_{ds} - \frac{R_s}{X_{ls}} \left( F_{qs} - F_{qm} \right) \right] \tag{1}$$

$$\frac{dF_{ds}}{dt} = \omega_b \left[ v_{ds} + \frac{\omega_e}{\omega_b} F_{qs} - \frac{R_s}{X_{ls}} \left( F_{ds} - F_{dm} \right) \right] \tag{2}$$

$$\frac{dF_{qr}}{dt} = -\omega_b \left[ \frac{(\omega_e - \omega_r)}{\omega_b} F_{dr} + \frac{R_r}{X_{lr}} \left( F_{qr} - F_{qm} \right) \right] \tag{3}$$

$$\frac{dF_{dr}}{dt} = -\omega_b \left[ \frac{-(\omega_e - \omega_r)}{\omega_b} F_{qr} + \frac{R_r}{X_{lr}} \left( F_{dr} - F_{dm} \right) \right] \tag{4}$$

$$F_{qm} = \frac{X_{ml}}{X_{ls}} F_{qs} + \frac{X_{ml}}{X_{lr}} F_{qr} \tag{5}$$

$$F_{dm} = \frac{X_{ml}}{X_{ls}} F_{ds} + \frac{X_{ml}}{X_{lr}} F_{dr} \tag{6}$$

$$X_{ml} = \frac{1}{\frac{1}{X_m} + \frac{1}{X_{ls}} + \frac{1}{X_{lr}}} \tag{7}$$

where the flux linkage state variable is denoted by $F$. $ds$ and $qs$ in subscripts correspond to the dynamic and quadrature axis stator variables. $dr$ and $qr$ in subscripts refer to the rotor variables. $\omega_b$ is the base frequency of the machine. $\omega_e$ is the frequency of an arbitrarily rotating reference frame. All calculations in the DTC scheme discussed in this paper are performed in the stationary reference frame, with

$\omega_e$ set to 0. The motor currents are obtained from the flux linkages as:

$$i_{qs} = \frac{F_{qs} - F_{qm}}{X_{ls}}, \quad i_{qr} = \frac{F_{qr} - F_{qm}}{X_{lr}} \tag{8}$$

$$i_{ds} = \frac{F_{ds} - F_{dm}}{X_{ls}}, \quad i_{dr} = \frac{F_{dr} - F_{dm}}{X_{lr}} \tag{9}$$

Detailed simulations of the model presented above is explained in [12].

# 3 Direct Torque Control

The direct torque control scheme was introduced by Depenbrock and Takahashi [3], [4] in the 1980s. The performance of the DTC scheme is comparable to that of other field oriented control methods and is simpler to implement. The need for repeated transformations from one reference frame to another, as in case of FOC drives is eliminated as all the calculations are performed in a single reference frame.

The electromagnetic torque and the stator flux developed in the machine are compared to their respective reference values which are input to hysteresis comparators. The outputs of the two hysteresis comparators along with the position of the vector of the stator flux generates control signals which are fed to a voltage source inverter (VSI). The VSI in turn, drives the induction motor mode [13].

The fundamental assumption in DTC is that if the stator resistance voltage can be taken to be equal to zero, the flux in the stator of the induction motor can be assumed to be equal to the integration of the voltage. Alternatively, the applied voltage can be thought to be equal to the time derivative of the stator flux. The relationship between the stator flux and the applied voltage can be thus expressed as:

$$\bar{V}_s = \frac{d}{dt}(\bar{\psi}_s) \tag{10}$$

or

$$\Delta\bar{\psi}_s = \bar{V}_s.\Delta t \tag{11}$$

A closer look at these equations reveals that the we can change $\bar{\psi}_s$, the stator flux, by applying an appropriate the stator voltage $\bar{V}_s$ for time $\Delta t$.

The output voltage of the inverter, therefore, is impressed upon the stator flux vector. To obtain a desired value of the flux of the stator, appropriate voltage vectors have to be applied [14]. For short transients, the rotor flux is assumed to be constant since the rotor time constant is significantly large.

Flux and torque references are the command inputs of the control scheme. When the system is operated in the torque control mode, a torque input is applied, whereas in the speed control mode, the reference speed is input to a PI controller which translates the signal to an equivalent torque reference to be input to the system. Magnitudes of the stator flux and torque are estimated from the d-q model. Comparison between the calculated and the reference values is done. Flux and torque hysteresis bands, between which the two quantities are allowed to vary, are defined. If the errors fall outside the hysteresis bands, appropriate control signals to increase/decrease the torque and flux are required.

The location of the stator flux vector in space is also to be calculated. A two dimensional complex plane is divided into six sectors which span 60 degrees each. A total of eight voltage vectors $V_0$ through $V_7$ are defined. The vectors $V_0$ and $V_7$ maintain the torque and flux constant and hence are called zero vectors. Vectors $V_1$ through $V_6$ are active vectors responsible for increasing/decreasing flux and torque.

After each sampling interval, appropriate voltage vectors are selected to keep the stator flux and electromagnetic torque within limit [15]. The width of the hysteresis band has a direct influence on the inverter switching frequency. With a decrease in width of the band, the switching frequency increases. The switching frequency of the VSI is low when a wide hysteresis band is set. Consequently, the system response is poor.

# 4 Control Scheme

The basic dircect torque control scheme of an induction motor has two parallel modes: for controlling flux and torque respectively. The flux reference is directly input to the system in the first mode while as the reference torque (taken as the output of the speed controller) is the input in the second mode. A signal computation block is used to compute the values of the electromagnetic torque, the stator flux magnitude and the sector of the stator flux vector. Hysteresis controllers, whose input are the estimated and the reference quantities, produce digital outputs. The outputs of these controllers along with the sector location are used to generate control signals for the VSI. The inverter output voltages are in turn fed to the induction motor.

## *4.1 Signal Estimation Block*

The stator d - axis and q - axis flux magnitudes can be estimated using the relations:

$$\psi_{ds} = \int \left(v_{ds} - R_s i_{ds}\right) dt, \quad \psi_{qs} = \int \left(v_{qs} - R_s i_{qs}\right) dt \tag{12}$$

$$|\psi_s| = \sqrt{\psi_{ds}^2 + \psi_{qs}^2} \tag{13}$$

The electromagnetic torque can be estimated using the relation :

$$T_e = \frac{3}{2}\frac{P}{2}\frac{1}{\omega_b}\left(\psi_{ds} i_{qs} - \psi_{qs} i_{ds}\right) \tag{14}$$

Six sectors, each spanning an angle $\frac{\pi}{3}$ are defined in a two dimensional complex plane. The location of the flux space vector in the complex plane can be calculated by:

$$\phi_s = \tan^{-1}\left(\frac{\psi_{qs}}{\psi_{ds}}\right) \tag{15}$$

The value of the sector angle is used to select in the voltage vector.

## 4.2 Hysteresis Controllers

In this block, the estimated and reference values of torque and stator flux are compared. Appropriate commands to increasing or decreasing torque and flux are generated to keep both of them within their respective hysteresis bands. In case an increment in stator flux is required, $\Delta\psi_s = 1$ and for a decrement $\Delta\psi_s = -1$

$$\Delta\psi_s = 1 \quad if \quad |\psi_s| \leq |\psi_s^*| - |hysteresis\ band| \tag{16}$$

$$\Delta\psi_s = -1 \quad if \quad |\psi_s| \geq |\psi_s^*| + |hysteresis\ band| \tag{17}$$

To increase the electromagnetic torque $\Delta T_e = 1$, to decrease torque $\Delta T_e = -1$. To keep torque constant, $\Delta T_e = 0$. A three level control is used for controlling torque and a two level control for controlling flux [11]:

$$\Delta T_e = 1 \quad if \quad T_e \leq T_e^* - |hysteresis\ band| \tag{18}$$

$$\Delta T_e = 0 \quad if \quad T_e = T_e^* \tag{19}$$

$$\Delta T_e = -1 \quad if \quad T_e \geq T_e^* + |hysteresis\ band| \tag{20}$$

## 4.3 Inverter Switching Table

Table 1 shows the outputs of the hysteresis comparators with the sector number. The voltage vectors corresponding to the hysteresis outputs and the sector numbers are shown. The switch positions to obtain the desired vector are given in the following table:

Table 1: Voltage Vector Switching

| Output of Hysteresis Controller | | Number of the Sector | | | | | |
|---|---|---|---|---|---|---|---|
| | | 1 | 2 | 3 | 4 | 5 | 6 |
| $\Delta\psi_s = +1$ | $\Delta T_e = +1$ | $\mathbf{V_2}$ | $\mathbf{V_3}$ | $\mathbf{V_4}$ | $\mathbf{V_5}$ | $\mathbf{V_6}$ | $\mathbf{V_1}$ |
| | $\Delta T_e = 0$ | $\mathbf{V_0}$ | $\mathbf{V_7}$ | $\mathbf{V_0}$ | $\mathbf{V_7}$ | $\mathbf{V_0}$ | $\mathbf{V_7}$ |
| | $\Delta T_e = -1$ | $\mathbf{V_6}$ | $\mathbf{V_1}$ | $\mathbf{V_2}$ | $\mathbf{V_3}$ | $\mathbf{V_4}$ | $\mathbf{V_5}$ |
| $\Delta\psi_s = -1$ | $\Delta_{Te} = +1$ | $\mathbf{V_3}$ | $\mathbf{V_4}$ | $\mathbf{V_5}$ | $\mathbf{V_6}$ | $\mathbf{V_1}$ | $\mathbf{V_2}$ |
| | $\Delta T_e = 0$ | $\mathbf{V_7}$ | $\mathbf{V_0}$ | $\mathbf{V_7}$ | $\mathbf{V_0}$ | $\mathbf{V_7}$ | $\mathbf{V_0}$ |
| | $\Delta T_e = -1$ | $\mathbf{V_5}$ | $\mathbf{V_6}$ | $\mathbf{V_1}$ | $\mathbf{V_2}$ | $\mathbf{V_3}$ | $\mathbf{V_4}$ |

Table 2: Switch States and Voltage Vectors

| | $\mathbf{V}_0$ | $\mathbf{V}_1$ | $\mathbf{V}_2$ | $\mathbf{V}_3$ | $\mathbf{V}_4$ | $\mathbf{V}_5$ | $\mathbf{V}_6$ | $\mathbf{V}_7$ |
|---|---|---|---|---|---|---|---|---|
| $S_A$ | 0 | 1 | 1 | 0 | 0 | 0 | 1 | 1 |
| $S_B$ | 0 | 0 | 1 | 1 | 1 | 0 | 0 | 1 |
| $S_C$ | 0 | 0 | 0 | 0 | 1 | 1 | 1 | 1 |

The DTC scheme uses a VSI to drive the induction motor model. The primary function of the inverter is the conversion of control signals to voltage signals which are fed to the motor. Signals $S_{A,B,C}$ control the states of the inverter. For the three phases, a total of six switches are used. For each phase, two swiches are provided. When signal $S_A$ is 1, the switch $A_h$ is swicthed on and $A_l$ is switched off. Similarly, when the signal $S_A$ is 0, $A_h$ is switched off and $A_l$ is switched on. The same procedure fashion, the other switches can be controlled. The voltage vector which is the output of the VSI is [16] :

$$\mathbf{V}_s = \frac{2}{3}V_{dc}\left(S_A + e^{j\frac{2\pi}{3}}S_B + e^{j\frac{4\pi}{3}}S_C\right) \tag{21}$$

## 5 ANFIS Speed Controller

An induction motor drive controlled by the DTC has two modes of operation: The *Torque Control* mode in which the input to the system is a reference torque. The second mode is the *Speed Control* mode where a PI controller is used whose input is the difference between the reference speed and the actual speed of the motor [17].

In high performance drives, a number of control strategies are available for use as speed controllers. Fuzzy logic, neural networks and other intelligent control techniques are increasingly being implemented in control applications demanding high precision.

ANFIS is a hybrid scheme which combines fuzzy logic and artificial neural network [18]. The principles of fuzzy logic and neural networks are combined to create a better system with and enhanced design and superior performance.

In an ANFIS, a neural network is used to design a fuzzy inference system. If the input/output mappings are available for a fuzzy system, the neural network training method can be used to design the membership functions and the rule table of the fuzzy model.

The ANN is based on the concept of input/output mapping. It is trained to associate patterns between the inputs and the output analogous to how the brain works. Given a set of input data, the network is trained to produce a corresponding output pattern.

Proper mapping between the input and output can be made possible by appropriately adjusting the weights, which relate the input and the output. The training is initiated with random weights.A comparison between the output pattern for a given input with the desired output pattern is made. The weights are accordingly adjusted with each iteration till the error between the computed and the desired values falls within acceptable limits. At the termination of the training process, the network must be in a position to recall all the input and output patterns and if required, should be able to perform interpolation and extrapolation by little amounts.

A fuzzy inference system maps input characteristics to input membership functions, input membership functions to rules. The rules are mapped to a set of output characteristics which in turn are mapped to output characteristics. The output characteristics are mapped to output membership functions which ultimately terminated as output membership functions, and the output membership functions to the output. Two of the most popular fuzzy inference systems are the Mamdani [19] and the Sugeno [20] systems.

In this paper, the speed controller is based on ANFIS. Depending on a certain training scenario, the ANFIS controller is provided with a data set. It comprises complete information of the process under various operating conditions. A proper value of torque is provided based on the measured and the reference speed [21]. Both the input as well as the target data are contained in the data set. The ANFIS controller trains the fuzzy system on the based on this data set. To train the fuzzy inference system, a neural network is used [8].

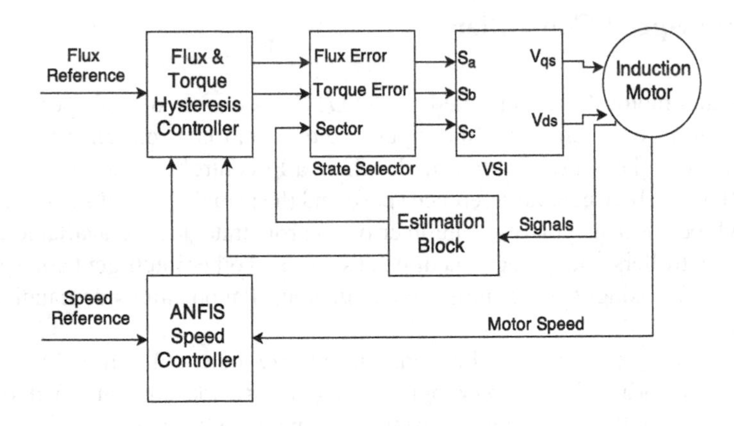


Fig. 1: ANFIS Based DTC Scheme

Figure 2 shows the ANFIS structure. It consists of five layers. The input layer contains the information about the reference and measured speeds and the error between the two. The input membership functions of the fuzzy system occupies the second layer of the system. A 7*7 gaussian membership function type system is selected. The third layer is used to generate the rules of the fuzzy system. The fourth layer is the output membership function layer. The output membership functions are constant values in a Sugeno based fuzzy system. Finally, in the output layer, the activated control signal is generated by processing the weighted signal through an activation function [18].

The back-propagation algorithm is used to compare the output pattern of the neural network with the desired output pattern. Supervised learning technique is used to train the controller. The control signal is fed forwards and the error is fed backwards. A reduction in error is possible if the number of iterations is increased. The fuzzy system is to be trained. This is done by adjusting the weights to reduce the error. After adjusting the final weights, rules governing the fuzzy system, the input membership function and the output membership function are obtained. Figure 3 shows the input membership functions. The ANFIS thus derived is used for controlling the speed of the induction motor drive.

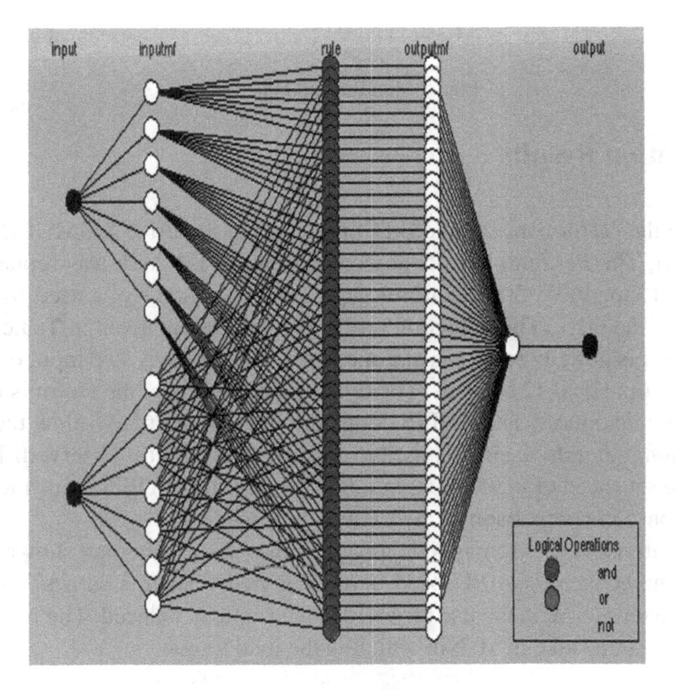


Fig. 2: ANFIS Structure

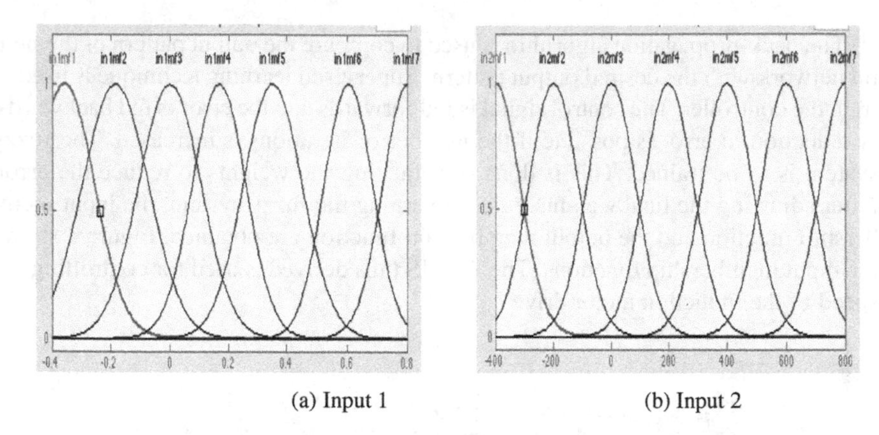


(a) Input 1 (b) Input 2

Fig. 3: Membership Functions

## 6 Simulation Results

To analyze the performance of the DTC induction motor drive a SIMULINK model was created. The PI controller of the conventional DTC drive was replaced by an ANFIS. A 15hp, 400V, 50Hz squirrel cage induction motor was used to study the dynamics of the drive. The induction motor parameters are given in Table 3.

The motor is initially set to run at a speed of −200 rad/s; A step input of 250 rad/s is applied at time t = 0.12 seconds. The dynamic behaviour of the motor is illustrated in the following figures. The ANFIS controller has the ability to follow the speed of the induction motor drive effectively. No considerable offset is observed. The motor reaches the set speed of −200 rad/s in 0.05 seconds. The ANFIS controller shows a faster response in comparison to the PI speed controller.

The variations of electromagnetic torque in the two schemes are shown in figures 7 and 8. The conventional PI speed controller based scheme shows considerable ripple in the torque. With the use of ANFIS, the ripple is reduced. The torque settles at the steady state value of 10 Nm which is the load torque.

Table 3: Motor Parameters

| | |
|---|---|
| Output Power(hp) | 15 |
| Resistance (Stator) $R_s$, $\Omega$ | 0.371 |
| Resistance (Rotor) $R_r$, $\Omega$ | 0.415 |
| Leakage Inductance (Stator) $L_{ls}$, mH | 2.72 |
| Leakage Inductance (Rotor) $L_{lr}$, mH | 3.3 |
| Inductance (Magnetizing) $L_m$, mH | 84.33 |
| Frequency (base) $\omega_b$, rad/s | 314 |
| Inertia $J$, kgm$^2$ | 0.02 |
| DC Voltage $V_{dc}$, V | 400 |
| Poles $P$ | 4 |

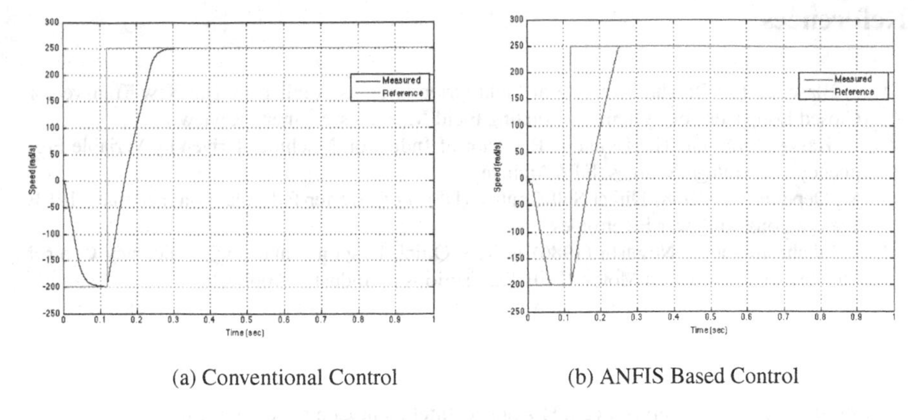


(a) Conventional Control (b) ANFIS Based Control

Fig. 4: Mechanical Speed v/s Time

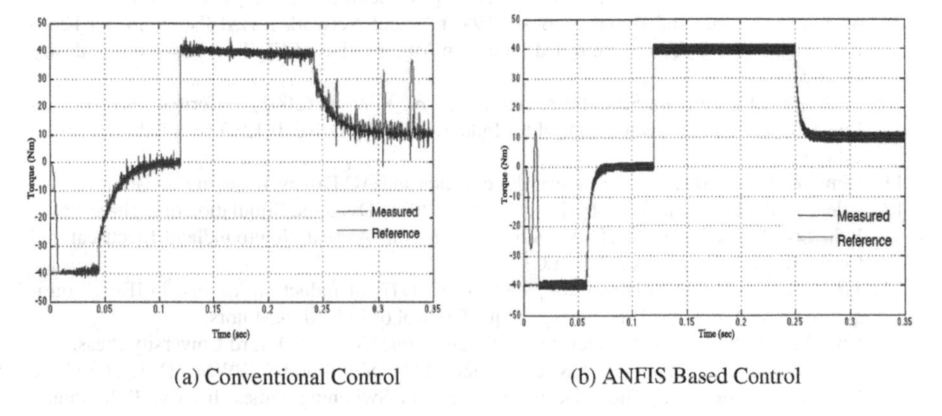


(a) Conventional Control (b) ANFIS Based Control

Fig. 5: Electromagnetic Torque v/s Time

# 7 Conclusion

ANFIS implementation for an induction motor drive has been presented in this paper. Different operating conditions have been studied through simulations.

When compared with conventional controllers, it is seen that with the use of ANFIS for speed control, the speed of the drive can be regulated more efficiently. Forward and reverse motoring mode speeds have been tracked and the drive system shows a smooth transition between the two modes. The ripple in the electromagnetic torque is considerably reduced. Also, the response time of the system is reduced.

## References

1. F. Blaschke (1972) The Principle of Field Orientation as Applied to The New Transvector Closed Loop Control System for Rotating Field Machines. Siemens Review
2. K. Hasse (1968) On The Dynamic Behavior of Induction Machines Driven by Variable Frequency and Voltage Sources. ETZ Archive.
3. M. Depenbrock (1988) Direct Self Control (DSC) of inverter-fed induction machines. IEEE Transactions on Power Electronics
4. I. Takahashi and T. Nogushi (1986) A New Quick Response and High Efficiency Control Strategy of an Induction Motor. IEEE Transactions on Industry Applications
5. K. S. Narendra and S. Mukhopadhyay (1996) Intelligent Control Using Neural Networks. IEEE Press, New York.
6. Bimal K. Bose (1997) Expert System, Fuzzy Logic and Neural Networks in Power Electronics and Drives. IEEE Press, New Jersey
7. Tze-Fun Chan and Keli Shi (2011) Applied Intelligent Control of Induction Motor Drives. John Wiley and Sons.
8. Shoeb Hussain and Mohammad Abid Bazaz (2014) ANFIS Implementation on a Three Phase Vector Controlled Induction Motor with Efficiency Optimisation. In : International Conference on Ciruits, Systems, Communication and Information Technology (CSCITA)
9. M. Godoy Simces and Bimal K. Bose (1995) Neural Network Based Estimation of Feedback Signals for a Vector Controlled Induction Motor Drive. IEEE Transactions on Industry Applications
10. M. Nasir Uddin, Tawfik S. Radwan, and M. Azizur Rahman (2002) Performances of Fuzzy-Logic-Based Indirect Vector Control for Induction Motor Drive. IEEE Trans. Industry Applications
11. Bimal K. Bose (2002) Modern Power Electronics and AC Drives. Pearson Education Inc.
12. Adel Aktaib, Daw Ghanim and M. A. Rahman (2011) Dynamic Simulation of a Three-Phase Induction Motor Using MATLAB Simulink. In 20th Annual Newfoundland Electrical and Computer Eng. Conference (NECEC).
13. J.R.G. Schofield (1995) Direct Torque Control - DTC of Induction Motors. In IEEE Colloquium on Vector Control and Direct Torque Control of Induction Motors.
14. Peter Vas (1998) Sensorless Vector and Direct Torque Control. Oxford University Press.
15. A. Kumar, B.G. Fernandes, and K. Chatterjee (2004) Simplified SVPWM - DTC of 3 phase Induction Motor Using The Concept of Imaginary Switching Times. In: The 30th Annual Conference of the IEEE Industrial Electronics Society, Korea.
16. H.F. Abdul Wahab and H. Sanusi (2008) Simulink Model of Direct Torque Control of Induction Machine. American Journal of Applied Sciences
17. Haitham Abu-Rub, Atif Iqbal and Jaroslaw Guzinski (2012) High Performance Control of AC Drives. John Wiley and Sons.
18. J.S.R. Jang (1993) ANFIS: Adaptive-Network-Based Fuzzy Inference System. IEEE Transactions on Systems, Man and Cybernetics
19. E. H. Mamdani and S. Assilian (1975) An Experiment in Linguistic Synthesis with a Fuzzy Logic Controller. International Journal of Man-Machine Studies
20. M. Sugeno (1985) Industrial Applications of Fuzzy Control. Elsevier Science Pub. Co.
21. C. T. Lin and C. S. George Lee (1996) Neural Fuzzy Systems: A Neuro-Fuzzy Synergism to Intelligent Systems. Prentice Hall.

# Design of a Multi-Priority Triage Chair for Crowded Remote Healthcare Centers in Sub-Saharan Africa

Santhi Kumaran[1] and Jimmy NSENGA[2]

[1]University of Rwanda, Kigali, Rwanda
santhikr@yahoo.com
[2]CETIC, Gosselies, Belgium
jimmy.nsenga@cetic.be

**Abstract.** Remote healthcare centers located in sub-Saharan of Africa are still facing a shortage of healthcare practitioners, yielding long queue of patients in waiting rooms during several hours. This situation increases the risk that critical patients are consulted too late, motivating the importance of priority-based patient scheduling systems. This paper presents the design of a Multi-Priority Triage Chair (MPTC) to be installed at the entrance of patient's waiting rooms such that each new patient first sits in the MPTC to measure its vital signs and register his/her other priority parameters such as arrival time, travel time or distance between the concerned center and its home, and so on. The proposed MPTC will then update in real-time the consultation planning in order to statistically minimize (i) the number of critical patients not treated within a pre-defined waiting time, (ii) the number of patients waiting for more than a pre-defined period and (iii) the number of patients living relatively far who gets their consultation postponed to another day.

**Keywords:** Triage, Priority Patient Scheduling; vital signs; scarce resource management.

## 1 Introduction

The very basic patient scheduling technique is known as first-come-first-serve (FCFS) [2]. However, this strategy is inherently sub-optimal considering that for instance a patient in a critical condition may get access too late to a consultation or a patient living very far from the healthcare center would get his/her consultation postponed, while the doctor or the nurse has spend most of its scarce time consulting non-priority patients. The process of prioritizing patients when resources are insufficient is commonly known as Triage.

One important parameter to take into account when prioritizing patient consultation is the criticality of a patient health state. This information may be estimated by



J.M. Corchado Rodriguez et al. (eds.), *Intelligent Systems Technologies and Applications 2016*, Advances in Intelligent Systems and Computing 530,
DOI 10.1007/978-3-319-47952-1_72

measuring vital signs such as temperature, pulse, blood pressure, respiration rate, oxygen saturation and so on, and calculating a score commonly known as emergency safety index (ESI) [3]. Triage-based on vital signs have been considered in different scenarios. In [4], an electronic triage tag (E-triage) have been used to determine a transportation order of patients that maximizes the life-saving ratio in the case of a disaster, thus considering the latest vital signs and temporal variation in the survival probability of each patient, the time for an ambulance to transport the patient to an appropriate hospital, and other factors. In [5], vital signs have been considered to decompress overcrowded emergency departments (EDs) by identifying a subset of elderly patients for whom the likelihood of admission is very high, allowing the ordering of an inpatient bed at triage and avoiding delays in admission. On top of the criticality of a patient health state, in [6] the authors have also considered the waiting time as an additional priority parameter for patient scheduling in EDs. However, considering rural and remote healthcare facilities in Africa, the travel time or distance between the center and patient's home turns out to be an important parameter to take into account while scheduling patients, thus for instance to avoid that patients living very far from the healthcare center get their consultation postponed to another day.

This paper presents the design of a Multi-Priority Triage Chair (MPTC) to be installed at the entrance of patient's waiting rooms, such that each new patient first sits in to measure its vital signs and register other priority parameters such as its arrival time, travel time between the concerned center and its home, and so on. The proposed MPTC will then update in real-time the consultation planning in order to statistically minimize (i) the number of critical patients not treated within a predefined time after their arrival, (ii) the number of patients waiting for more than a pre-defined period and (iii) the number of patients living relatively far who gets their consultation postponed to another day.

This paper is organized as follows. Section 2 presents the operating context model of the proposed MPTC. Section 3 describes the main components of the MPTC with a special focus on the multi-priority scheduling strategy. A conclusion is presented in Section 4.

## 2 Operating Context Model for the proposed MPTC

Fig. 1 presents the operating context model of the proposed Multi-Priority Triage Chair (MPTC). When a new patient arrives at the healthcare center, he/she will be immediately undergone an identification and preliminary check-up process as follows: (i) Patient sits on the vacant MPTC, (ii) Authenticate him/her self by entering its electronic identity card (eID) and providing his/her home address. A patient unique identification number (Pid) is then generated. (iii) By following the instructions displayed on the interactive chair screen, patient vital signs are sequentially measured, eventually by the help of a support staff who may be for instance an intern with some basic skills to use the system, (iv).

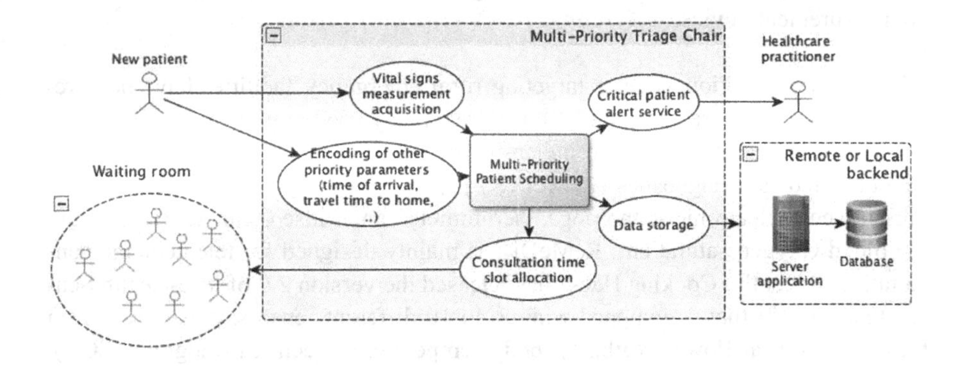


**Fig. 1.** Operating Context Model of the Multi-Priority Triage Chair.

At the end of the measurement process, vital signs along with the patient id, patient home address and arrival time are then transmitted to the inner embedded processor, which then locally executes an adaptive multi-priority patient-scheduling algorithm.

Once the patient is done with the preliminary check, he/she may now get access to the waiting room, where the consultation order of queuing patients is displayed on a flat screen. Furthermore, both raw vital signs data and analytic results for each patient are then sent to the health center web server for further analysis and storage. Note that if the measured vital signs indicate a very critical health anomaly of the patient, then an immediate alert may be generated to inform available healthcare practitioners.

## 3 Design of the proposed MPTC

This section presents the main components of the MPTC with a special focus on the multi-priority patient scheduling strategy.

### 3.1 Hardware for patient priority parameters acquisition

As stated above, the considered patient priority parameters are vital signs, waiting time and distance or time travel from home to the concerned healthcare center. In the following, we focus mainly on vital signs measurements that require a dedicated measurement device.

Vital signs acquisition systems targeting rural emergency facilities have been reported in the literature and some of them are already used in practice. Neurosynaptic Communications Pvt Ltd, an indian-based ICT company, develops a kit called Remote Medical Diagnostics (ReMeDi®) [7] composed by ECG machine, blood pressure cuff, electronic stethoscope, thermometer, and Pulse Oximeter (to measure the blood Oxygen saturation). ReMeDi® is mainly designed for telemedicine consultation. Recently, Cooking Hacks has released the version 2.0 of its e-Health Sensor Platform [8] that is equipped with up to 10 different sensors: pulse, oxygen in blood (SPO2), airflow (breathing), body temperature, electrocar-diogram (ECG), glucometer, galvanic skin response (GSR - sweating), blood pres-sure (sphygmomanometer), patient position (accelerometer) and mus-cle/eletromyography sensor (EMG).

This platform will be integrated with an open embedded hardware such Arduino and Raspberry Pi, which features a powerful embedded process, to implement the core hardware of the MPTC.

### 3.2 Multi-priority patient scheduling strategy

This section starts by a mathematical formulation of the multi-priority patient-scheduling problem. Next, the targeted objective functions are described before explaining the process to compute the optimum configuration parameters.

**Mathematical Formulation of the multi-priority scheduling problem**

Fig. 2 shows a model of the proposed scheduling algorithm process. Let us define $P(t)$ to be the number of queuing patients at a given instant $t$.

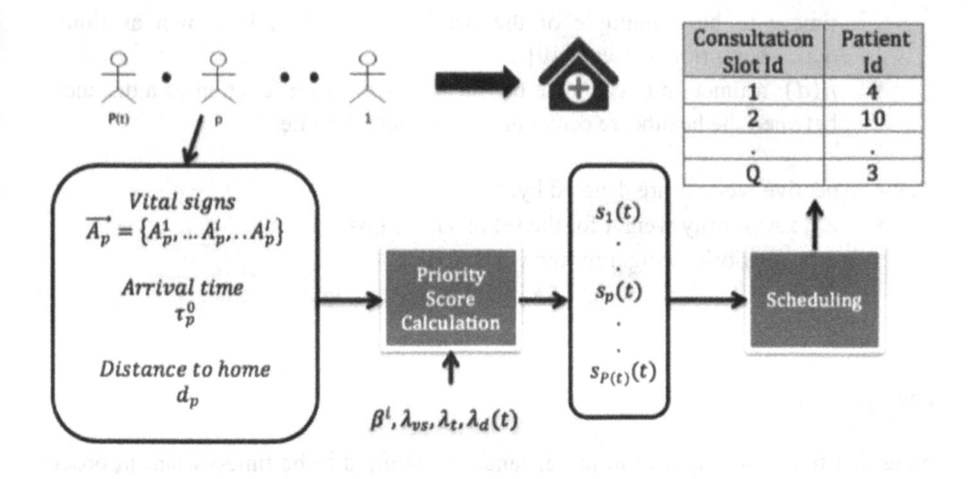


**Fig. 2.** Model of the proposed MPTC algorithm process.

Each patient $p$ sitting in the waiting room has the following parameters:

- $A_p^i$ : Measurement of the $i^{th}$ vital sign. The set of all vital signs measurement for this patient is represented by $\overrightarrow{A_p} = \{A_p^1, \dots A_p^i \dots A_p^l\}$.
- $\tau_p^0$ : Time of arrival at the health center
- $d_p$: Distance between his/her home and the health center

At a given instant $t$, a priority score $s_p(t)$ of a patient $p$ is given by the following weighted sum function (1):

$$s_p(t) = \beta\, S_p + (1 - \beta)\left[\lambda_{vs} * f(\bar{A}_p) + \lambda_t * g(t - \tau_p^0) + \lambda_d(t) * h(d_p)\right] \quad (1)$$

with

$$\beta = \begin{cases} 1, & \text{if the system operator intervene to provide manually a score } S_p \\ 0, & \text{if the score is automatically calculated} \end{cases}$$

The calculated score is a weighted sum function with:

- $f(\overrightarrow{A_p})$: a function that produces a single-parameter representation of the patient status by combining different vital signs measurement. An example of this kind of function is the Visensia Safety Index (VSI) [9],
- $g(\Delta t)$: a function that calculates a priority score as function of the waiting time $t - \tau_p^0$, meaning the difference between the current time $t$ and the time of arrival $\tau_p^0$. For simplicity, we will consider the patient's priority at

time $t$ to be a multiple of the waiting time. This is known as time-dependent priority queue [10],

- $h(d)$: a function to calculate the priority score as a function of a distance between the healthcare center and the patient's home.

Their respective weights are denoted by:

- $\lambda_{vs}$ : A priority weight for the set of vital signs
- $\lambda_t$ : A priority weight for the waiting time
- $\lambda_d\,(t)$ : A priority weight for the distance to home

such that $\lambda_{vs}^2+\lambda_t^2 + \lambda_d^2(t) = 1$. The FCFS strategy is instantiated by setting $\lambda_t = 1$ and $\lambda_{vs} = \lambda_d(t) = 0$.

Note that the weight related to the distance is assumed to be time-variant in order for instance to increase the priority score of patients living far away from the center as the closing time of the center is approaching.

**Targeted objective functions**

The patient-scheduling algorithm will be implemented to achieve the following objectives:

- Minimize the number of patients in critical health status are not treated within a pre-defined waiting period.
- Minimize the number of patients who wait more than a pre-defined maximum waiting period.
- Minimize the risk that patients living far from the center are not treated the same day (postponing consultations due to lack time).

Once the weighted mathematical functions are defined, the optimal configuration of the MPTC is obtained by calculating their corresponding optimum weights such that the above objective functions are achieved.

**Solving the multi-objective multi-priority scheduling problem**

With the processing power of the selected embedded platform (Arduino or Raspberry PI), the first setup of the MPTC will use an exhaustive search strategy to find the optimum weights $\widetilde{\lambda_{vs}}, \widetilde{\lambda_t}$ and $\widetilde{\lambda_d\,(t)}$. The input data required for an exhaustive search will be generated as follows:

- *Time of arrival of each patient* is modeled as a random process based on a Poisson distribution [10],

- *Safety Index of each patient:* According to [11], test data for each vital sign can be generated by considering a normal distribution. The mean and standard deviation for each vital sign are presented in Table 1. Therefore, a safety index, which is a kind of weighted sum of vital signs measurement, has an exponential modified Gaussian (EMG) distribution.
- *Home distance:* we consider a normal distribution to generate the distance between the healthcare center and each patient's home.

It is very important to note that the calculation should be carried for each healthcare center. In fact, the different healthcare centers feature different levels of patient frequentation, shortage deficit, remote location, and so on. The algorithm parameters can be progressively personalized for each healthcare center by using machine learning techniques.

**Table. 1.** Mean and Standard Deviation of Each Vital Sign [11]

| Vital Sign | Mean | Standard Deviation |
|---|---|---|
| Heart Rate | 80 BPM | 1.5 |
| Blood Pressure | 112 mmHg | 2 |
| Bod Temperature | 36,8° | 0.2 |
| SpO2 | 97% | 0.5 |
| Respiration Rate | 12 breaths per min | 1 |

**Scheduling policy**

At this point, we assume that we have been able to calculate the optimum set of priority weights $\widetilde{\lambda_{vs}}$, $\widetilde{\lambda_t}$ and $\widetilde{\lambda_d(t)}$ via exhaustive simulations. Then, the scheduling can follow the following rules:

*Step 1: Classify patient in priority categories*

Based on priority scores $\vec{S}(t) = \{s_1(t), \ldots s_p(t), \ldots s_{P(t)}(t)\}$, the waiting patients are classified in *N* priority clusters $\vec{C} = \{C(1), \ldots C(i) \ldots C(N)\}$ such that the i[th] cluster contains patients whose priority score is $S_i < s_p(t) \leq S_{i+1}$, with $S_i$ and $S_{i+1}$ being respectively the minimum and the maximum threshold score to belong to the i[th] cluster. The number of priority clusters with their associated thresholds will be calculated via exhaustive simulation.

*Step 2: Consultation time slot assignment*

Once the priority clusters are created, patients who are assigned in clusters with the highest priority patients will be treated first. Inside a cluster, the FCFS will be applied.

### 3.3 Additional functionalities

**Critical patient alert**

If the vital signs of a patient indicate a very critical health state, the MPTC will generate an emergency alert to inform available healthcare practitioners without considering other priority parameters.

**Data backup**

Once the vital signs are measured and other priority parameters acquired, the MPTC will then send the collected data to a secured web server, which is ideally located in the same health center. For local web server, wireless local area network (WLAN) technologies such as WiFi can be used while for remote server, one may rely on 3G/4G cellular networks. To ensure patient data privacy and security, patient data will be encrypted and then send via a HTTPS (secure) channel to ensure a point-to-point security tunnel between the MPTC and the web server.

## 4 Conclusions

Optimizing the productivity of the scarce healthcare resources working in rural and remote area of Sub-Saharan Africa is a very important challenge. With the growing availability of low cost vital signs sensor kits and low-cost open hardware embedded processors, the development of triage system that prioritizes the scheduling of patients is becoming more and more affordable for rural emergency facilities.

In this paper, we have proposed a design of a Multi-Priority Triage Chair (MPTC) that enables to adapt in real-time the consultation time slots among queuing patients, by considering the criticality of vital signs measurements, waiting time and distance between the patient's home and the concerned health center. The proposed system is expected to minimize (i) the number of critical patients not treated within a pre-defined time after their arrival, (ii) the number of patients waiting for more than a pre-defined period and (iii) the number of patients living relatively far who gets their consultation postponed to another day.

The next step in our research is to evaluate the performances of the proposed system via monte-carlo simulations. The prototyping of the MPTC will be carried out under the African Center of Excellence in Internet of Things (ACEIoT), a five-year project to be funded by the World Bank. The first pilot experimentations will be conducted in remote healthcare centers located in Rwanda. Note that proposed system can be used in other developing regions than Sub-Saharan Africa, which are experiencing similar challenges in terms of "shortage of practitioners, remote hospitals, and so on".

## References

[1] World Health Organization (WHO): World Health Statistics 2015. Available from http://www.who.int/gho/publications/world_health_statistics/2015/en/
[2] Obamiro, J.K.: Queuing theory and patient satisfaction - An overview of terminology and application in ante-natal care unit, Bulletin Vol. LXII No. 1/2010, Economic Sciences Series, Petroleum-Gas University of Ploiesti, Romania.
[3] Mirhaghi A, Kooshiar H, Esmaeili H, Ebrahimi M.: Outcomes for Emergency Severity Index Triage Implementation in the Emergency Department. Journal of Clinical and Diagnostic Research : JCDR. 2015;9(4):OC04-OC07. doi:10.7860/JCDR/2015/11791.5737.
[4] Mizumoto, T., Sun, W., Yasumoto, K., Ito, M.: Transportation Scheduling Method for Patients in MCI using Electronic Triage Tag, eTELEMED, pages 1–7, 2011
[5] LaMantia MA1 et al.: Predicting hospital admission and returns to the emergency department for elderly patients. 2010 Mar;17(3):252-9. doi: 10.1111/j.1553-2712.2009.00675.x.
[6] Sharif, A. B., Stanford, D. A., Taylor, P. & Ziedins, I. : A multi-class multi-server accumulating priority queue with application to health care. Operations Research for HealthCare 3(2), 73–79
[7] ReMeDi® Solution: http://www.neurosynaptic.com/remedi-solution/
[8] e-Health Sensor Platform V2.0 for Arduino and Raspberry Pi https://www.cooking-hacks.com/documentation/tutorials/ehealth-biometric-sensor-platform-arduino-raspberry-pi-medical
[9] Rui Duarte , Zulian Liu: "Visensia for early detection of deteriorating vital signs in adults in hospital", August 2015, https://www.nice.org.uk/advice/mib36/chapter/technology-overview
[10] L. Kleinrock: A delay dependent queue discipline. Naval Research Logistics Quarterly, 11(3-4):329–341, 1964
[11] Lo, Owen et al.: Patient simulator: towards testing and validation. In: Pervasive Health 2011. ICST, Dublin.
[12] Lin D et al.: Estimating the waiting time of multi-priority emergency patients with downstream blocking. Health Care Management Science. 2014;17(1):88-99. doi:10.1007/s10729-013-9241-3

# Inter-Emotion Conversion using Dynamic Time Warping and Prosody Imposition

Susmitha Vekkot and Shikha Tripathi

**Abstract** The objective of this work is to explore the importance of parameters contributing to synthesis of expression in vocal communication. The algorithm discussed in this paper uses a combination of Dynamic Time Warping (DTW) and prosody manipulation to inter-convert emotions among one another and compares with neutral to emotion conversion using objective and subjective performance indices. Existing explicit control methods are based on prosody modification using neutral speech as starting point and have not explored the possibility of conversion between inter-related emotions. Also, most of the previous work relies entirely on perception tests for evaluation of speech quality post synthesis. In this paper, the objective comparison in terms of error percentage is verified with forced choice perception test results. Both indicate the effectiveness of inter-emotion conversion by speech with better quality. The same is also depicted by synthesis results and spectrograms.

## 1 Introduction

Emotion is the essence of speech. Human expression of emotion can be through gestures, facial expressions or speech tone. Vocal emotions convey the state of speaker through linguistic and para-linguistic expressions. Linguistic part of speech consists of semantic cues categorizing various emotions while emotional activation states can be grouped in para-linguistic representation. In cases where text-level interpretation of emotion is not sufficient, the subtle variations of non-linguistic struc-

Susmitha Vekkot
Dept of Electronics & Communication Engineering, Amrita School of Engineering, Bengaluru, Amrita Vishwa Vidyapeetham, Amrita University, India, e-mail: v_susmitha@blr.amrita.edu

Shikha Tripathi
Dept of Electronics & Communication Engineering, Amrita School of Engineering, Bengaluru, Amrita Vishwa Vidyapeetham, Amrita University, India, e-mail: t_shikha@blr.amrita.edu



J.M. Corchado Rodriguez et al. (eds.), *Intelligent Systems Technologies and Applications 2016*, Advances in Intelligent Systems and Computing 530,
DOI 10.1007/978-3-319-47952-1_73

ture of vocal flow are more helpful in judging the underlying emotional state of the speaker. Technological developments during the last decade have made human-machine interaction systems more robust and natural. Good quality speech synthesis systems which can incorporate the appropriate emotions at right places are needed for making dialogue systems effective and human-like [1]. Thus, the importance of analysing, synthesising and incorporating required emotion in speech is the motivation for the present work. Different approaches can be used for generating expressive speech. In first approach, known as implicit control, the vocoder is designed so that the text-to-speech synthesis system itself incorporates the necessary emotion and generates speech which is intelligible. But such systems require large training dataset and complex computation [2]. The second approach deals with incorporating necessary emotions in TTS system generated neutral speech, known as explicit control. Here, the main challenge lies in accurately determining the minute changes in emotion-dictating parameters and manipulating these in appropriate manner to generate emotional speech which is natural and human-like. The whole process of generating emotional speech by explicit control involves analysis, estimation and incorporation of parameters specific to each target emotion. Most research in explicit control focusses on conversion by prosody modification on LP residual and LP co-efficients [3].

Several researchers have contributed towards analysing prosodic parameters responsible for various emotions and most of them deal with converting neutral speech to expressive speech. Some of the earlier work in this direction are given in [4], [5], [6]. Earliest development in explicit control was through formant synthesis and diphone concatenation [7], [8]. Later work focussed on modification of key prosodic features like pitch, duration and intensity contour according to predefined set of rules developed for generation of target emotion. As emotional traits are distributed non-uniformly across the entire utterance, uniform modification of prosodic parameters will not give best results. Therefore, most researchers incorporated sentence, phrase, word or phoneme level non-uniform modification factors for generating speech loaded with target expression [9], [10], [11]. A vowel-based duration modification is proposed in [12] which operates by detecting vowel onset points as the anchor points for prosody modification.

Literature suggests that major contribution of researchers have been in conversion from neutral to expressive speech by modifying prosodic parameters. A large quantum of research is directed towards LP based analysis and synthesis wherein the parameter modifications are done on LP residual and co-efficients. The methods developed for analysis are computationally extensive and operates on various transforms rather than the actual speech itself.

This work is directed towards emotional inter-conversion which is important in cases where dynamic variations across emotions is required or there is non-availability of neutral emotion in source database. The paper is organised as follows. Section 2 discusses about emotional speech database used in this work while section 3 deals with Dynamic Time Warping and proposed methodology. Section 4 depicts the results and evaluation. Finally, the paper is concluded in section 5 with insights into future challenges.

## 2 Emotional Speech Database

The speech database used in this work is IITKGP-SESC (Indian Institute of Technology Kharagpur Simulated Emotion Speech Corpus). Speech is recorded using 10 (5 male and 5 female) experienced professional artists from All India Radio (AIR) Vijayawada, India. The Telugu (an Indian language) version of database is used for experiments. Fifteen emotionally unbiased Telugu sentences are recorded in which each artist speaks with 8 emotions viz. neutral, anger, happy, fear, compassion, sarcasm, surprise and disgust. Each emotion has 1500 utterances, thus making the total utterances to 12000, The entire recording lasts for around 7 hours [13]. For the current study, we have considered 5 basic emotions in IIT-KGP database viz neutral, anger, happy, fear and surprise from male and female artists. Inter-emotion conversion among the above emotions is carried out and results are verified and tabulated.

## 3 Proposed Method

The algorithm proposed in this paper uses a combination of Dynamic Time Warping (DTW), a dynamic programming approach for aligning the utterances and further modification of prosody according to the target expression contours. The detailed description of DTW and the proposed algorithm is given in subsections 3.1 and 3.2.

### *3.1 Dynamic Time Warping (DTW)*

Inter-emotion conversion requires that the source and target utterance be time aligned. The DTW algorithm accomplishes the alignment by matching source and target speech patterns, represented by the sequences of vectors $(X_1, X_2, \ldots, X_m)$ and $(Y_1, Y_2, \ldots, Y_n)$, where each vector $X_i$ represents source pattern and $Y_j$ the target pattern corresponding to the shape of vocal tract for $i^{th}$ and $j^{th}$ frames, respectively [14]. The two patterns are compared using inter-vector distance and dynamic programming with warping path constraints. The DTW algorithm relies on the statement that optimal path to any target point (i, j) in two dimensional space must pass through (i-1, j),(i-1, j-1) or (i, j-1) [14]. Thus, the minimum distance to any (i, j) is given by equation 1 as:

$$D_m(i,j) = d(i,j) + min\{d(i-1,j), d(i-1,j-1), d(i,j-1)\} \quad (1)$$

where d(i, j) is inter-pattern vector distance between source and target. The algorithm determines minimum accumulated distance to point (m,n) recursively, thus giving rise to warping path. Warping path determines which features need to be changed for inter-emotion conversion, their co-ordinate location or duration (with respect to time) and the extent to which feature vector should be transformed. The amount of transformation in this case is dictated by the pitch contour, intensity contour and formants of target expressive speech.

## *3.2 Proposed Algorithm*

The proposed algorithm operates in two stages. In the first stage, a dynamic time warping algorithm is used to determine the optimum distance separating the source and target utterances. The warped output is subjected to prosody modification to generate target speech. The steps involved are given below:

1. Read the source X(n) and target Y(n) expressive speech separately.
2. Select and pre-process the utterances for dynamic time warping.
3. Find the minimum accumulated distance between each frame of source and target utterance by using equation 2:

$$D_m(i,j) = d(i,j) + minimumof\{d(i-1,j), d(i-1,j-1), d(i,j-1)\} \quad (2)$$

4. Decision regarding the extent of parameter change and duration modification are taken by warping path constraints and target features.
5. Here, the target features considered are prosody parameters pitch contour, intensity contour and formant frequencies corresponding to each emotional utterance.
6. The warped utterance is further subjected to prosody modification as per the sequence below:
   - The formants existing in target speech i.e.$F_{1t}$,$F_{2t}$, $F_{3t}$ and $F_{4t}$ are copied to dynamic time warped source.
   - The intensity and pitch contour ($F_{0t}$) of the target emotional speech are then copied to the manipulated warped version.

- Prosody imposition on warped utterance is carried out by target pitch contour, intensity contour, frequency spectrum imposition using Equalization curves and formants.
- Using the new pitch contour $F0_{new}$, intensity and formants $F_{inew}$, the expressive speech is synthesised by concatenation with overlap add.
- The resulting speech after dynamic time warping and prosody imposition is compared with target.

## 4 Results and Validation

The results obtained for inter-emotion conversion is compared with neutral to emotion conversion. It is observed that in emotions with overlapping features/characteristics, inter-emotion conversion gives better subjective and objective results. The actual speech waveforms and spectrograms of target and synthesised emotions are plotted in Figures 1, 2, 3 and 4. In each of the conversion cases, the results obtained are analysed and the best results corresponding to particular source emotion are considered and represented.

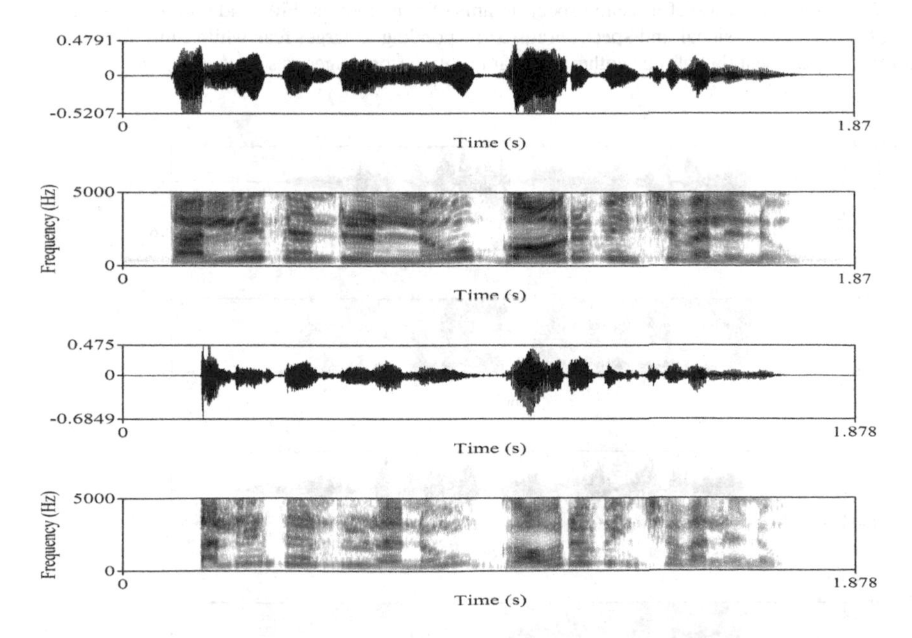


**Fig. 1** Comparison of waveforms and spectrograms of anger emotion: First and second sub-figures represent the speech signal and spectrogram corresponding to target anger while third and fourth sub-figures represent that of the synthesised anger from neutral source.

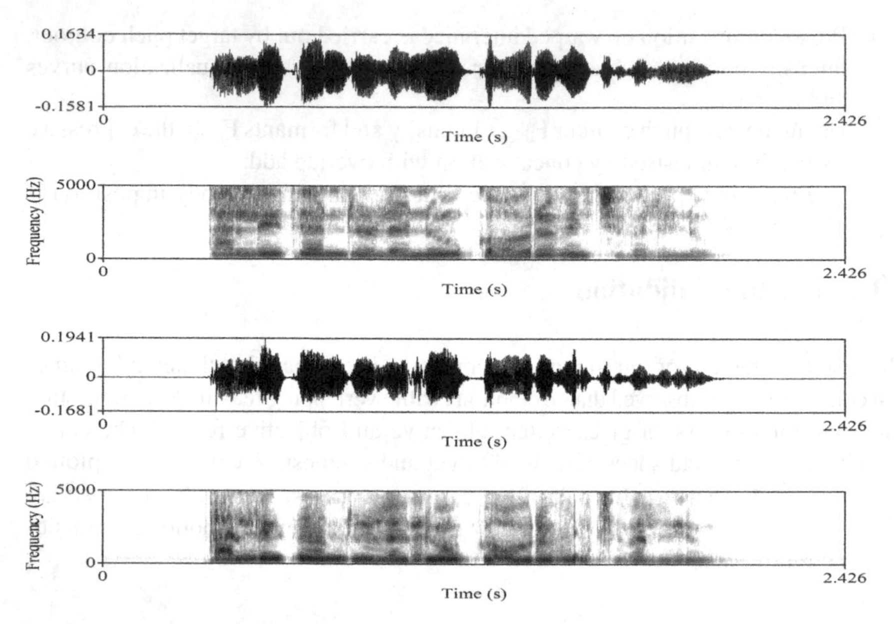


**Fig. 2** Comparison of waveforms and spectrograms of fear emotion: First and second sub-figures represent the speech signal and spectrogram corresponding to target fear while third and fourth sub-figures represent that of the synthesised fear emotion from anger as source emotion.

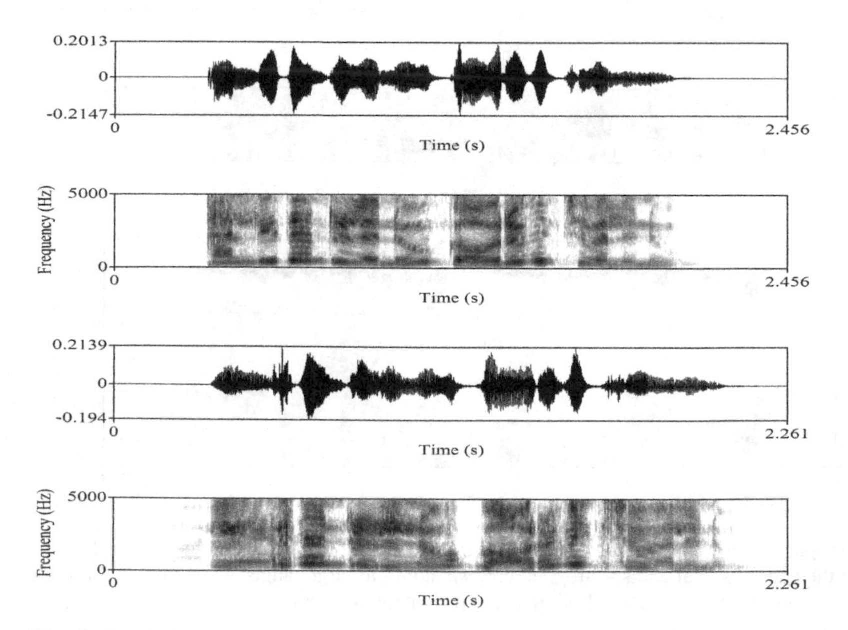


**Fig. 3** Comparison of waveforms and spectrograms of happiness: First and second sub-figures represent the speech signal and spectrogram corresponding to target happiness while third and fourth sub-figures represent that of the happiness emotion synthesised from fear source.

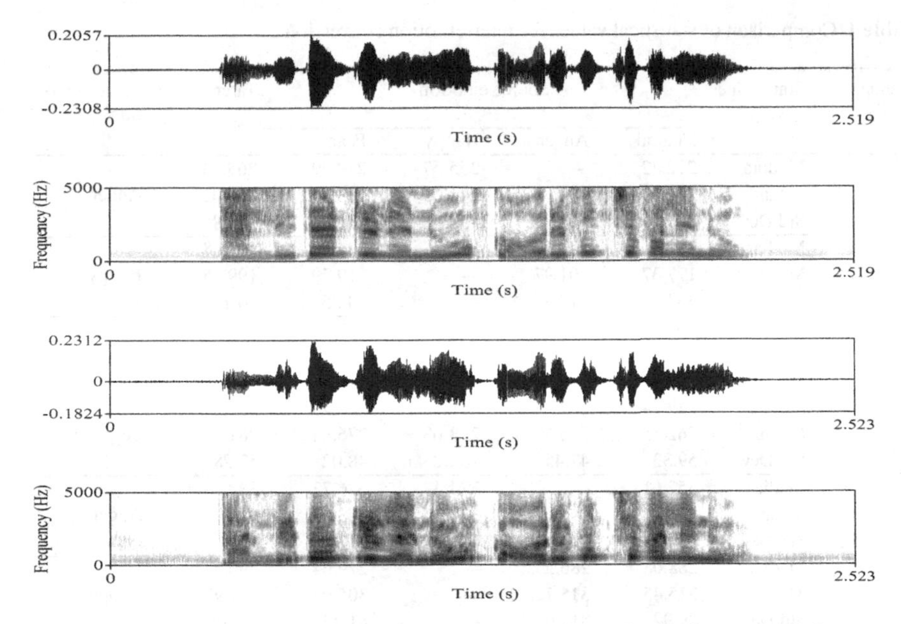


**Fig. 4** Comparison of waveforms and spectrograms of surprise: First and second sub-figures represent the speech signal and spectrogram corresponding to target surprise while third and fourth sub-figures represent that of the surprise emotion synthesised from happiness.

The synthesised speech samples and spectrograms are very much similar to the target emotion in each case. This proves the effectiveness of inter-emotion conversion by the proposed algorithm in the paper. As the utterances are time aligned before prosodic changes, the difference in duration between the two utterances are compensated properly.

## *4.1 Objective Comparison*

Objective comparison involved calculation of statistical parameters of synthesised emotion after conversion and analysing the similarity of pitch contours of target and original emotion. The table indicates that conversion is successful as the statistical values obtained in each case differ from actual by less than 5%. Table 1 lists the results of statistical comparison. Also, from table 1, it can be inferred that in most of the cases, conversion from a related emotion yields better results than neutral to emotion conversion.

The objective comparison in table 1 is obtained using one sample each of male and female target utterances of each emotion considered.

**Table 1** Comparison of statistical values for inter-emotion conversion

| Gender | Stat. Values | Source emotion | | | | Target value | Target emotion |
|---|---|---|---|---|---|---|---|
| | | Neutral | Anger | Happy | Fear | | |
| Male | Median | 272.32 | — | 255.57 | 270.29 | 268.63 | Anger |
| | Mean | 261.35 | — | 241.09 | 263.12 | 260.02 | |
| | Std Dev. | 47.91 | — | 49.23 | 48.28 | 48.28 | |
| | Median | 170.4 | 203.65 | — | 224.29 | 193.68 | Happy |
| | Mean | 177.37 | 201.97 | — | 219.79 | 199.55 | |
| | Std Dev. | 35.96 | 31.84 | — | 33.85 | 40.19 | |
| | Median | 243.85 | 246.56 | 247.58 | — | 244.14 | Fear |
| | Mean | 239.73 | 247.86 | 248.44 | — | 245.24 | |
| | Std Dev. | 35.67 | 21.8 | 20.98 | — | 20.28 | |
| | Median | 279.74 | 289.42 | 288.52 | 288.45 | 288.81 | Surprise |
| | Mean | 262.15 | 276.66 | 278.63 | 276.84 | 281.76 | |
| | Std Dev. | 59.33 | 47.45 | 46.32 | 48.02 | 57.28 | |
| Female | Median | 357.61 | — | 374.38 | 376.72 | 338.34 | Anger |
| | Mean | 364.46 | — | 376.45 | 373.07 | 350.97 | |
| | Std Dev. | 83.07 | — | 78.59 | 76.96 | 86.92 | |
| | Median | 288.08 | 286.53 | — | 279.64 | 290.52 | Happy |
| | Mean | 315.45 | 315.72 | — | 305.09 | 317.97 | |
| | Std Dev. | 79.92 | 81.76 | — | 81.43 | 81.61 | |
| | Median | 252.83 | 255.06 | 257.47 | — | 256.29 | Fear |
| | Mean | 247.69 | 260.04 | 257.15 | — | 260.38 | |
| | Std Dev. | 45.99 | 33.09 | 45.74 | — | 32.59 | |
| | Median | 337.21 | 309.45 | 337.96 | 330.03 | 334.57 | Surprise |
| | Mean | 330.85 | 319.09 | 338.64 | 331.48 | 332.99 | |
| | Std Dev. | 85.82 | 74.38 | 78.34 | 75.62 | 78.1 | |

## 4.2 Subjective Test

Subjective evaluation of the effectiveness of comparison is also determined by a listening test. The test is carried out by 11 subjects, with a mix of male/female and age groups in the range of 20-40. The subjects consisted of students, faculty and research scholars working in Robotic Research Center. Each subject was made to listen to the target emotion in the database. 34 utterances, of both male and female speakers were played to the group and they were asked to rate each utterance on a scale of 5 (with rank of perception ranging from 1 to 5) depending on the perceptive quality. In the perception test, samples converted from neutral and across emotions were tested separately and compared. Table 2 gives the judgement scale used for quality check of synthesised emotion.
Based on these, Comparison Mean-Opinion Scores (CMOS) were calculated for each emotion and the scores obtained for each case is listed in table 3.

**Table 2** Ranking used to judge the effectiveness of conversion by comparing similarity to target

| Ranking | Speech Quality | Perceptual distortion |
|---|---|---|
| 1 | Bad | Annoying and objectionable |
| 2 | Poor | Sounds different from target |
| 3 | Fair | Slightly annoying but perceivable as target emotion |
| 4 | Good | Almost like target emotion |
| 5 | Excellent | Exactly like target emotion |

**Table 3** Comparison Mean Opinion Scores (CMOS) for neutral to emotion conversion versus inter-emotion conversion: N-Neutral, A-Anger, H-Happy, F-Fear, S-Surprise

| Comparison Mean Opinion Scores(CMOS) | | | | | | | | | | | | |
|---|---|---|---|---|---|---|---|---|---|---|---|---|
| Male | | | | | | | | | | | | |
| Anger | | | Happy | | | Fear | | | Surprise | | | |
| N-A | F-A | H-A | N-H | A-H | F-H | N-F | A-F | H-F | N-S | A-S | F-S | H-S |
| 3.63 | 3.0 | 3.18 | 2.82 | 3.0 | 2.64 | 2.1 | 3.0 | 2.36 | 3.1 | 3.1 | 3.0 | 3.1 |
| Female | | | | | | | | | | | | |
| Anger | | | Happy | | | Fear | | | Surprise | | | |
| N-A | F-A | H-A | N-H | A-H | F-H | N-F | A-F | H-F | N-S | A-S | F-S | H-S |
| 2.8 | 2.72 | 2.9 | 3.0 | 3.1 | 3.1 | 3.27 | 1.8 | 3.36 | 3.1 | 1.6 | 3.27 | 3.18 |

From the table, it can be seen that the maximum value of CMOS is obtained for conversion to/from anger emotion. This is expected since anger is easiest to synthesise with clear intensity variations and higher pitch values. For other emotions like fear and surprise, an interesting trend is observed. The characteristics of fear, surprise and happiness can be perceived better when they are synthesised from emotions with similar characteristics than from neutral speech. For eg. conversion to surprise gives better results when source emotion is happiness than from neutral. This is because the inflexions associated with happy, surprise and fear have some common harmonics in them which aids in synthesis. It is difficult to incorporate the finer level of details required in synthesis of complex emotions like fear and surprise for conversion from neutral source.

In [3], the conversion is effected by Dynamic Time Warping of source and target LP co-efficients wherein LP co-coefficients of neutral source are replaced with that of target while prosody modification is done in LP residual, obtained after LP analysis. Here the speech is re-synthesised by inverse filtering using modified LP residual. In our work, we have further explored the possibility of inter-conversion among similar emotions from different sources and compared objectively and subjectively with neutral to emotion conversion. The approach used here can be used for dynamic emotion conversion task and is computationally very efficient since the algorithm operates directly on speech signal.

The subjective results have been graphically represented in Figure 5. The objective and subjective comparison results are similar in the sense that for conversion to surprise, the maximum CMOS values are obtained in most cases which is correlated by the large reduction in error percentage. This means that the maximum recognition rates are obtained for surprise emotion. Also, conversion to fear gives good results in perception. This is validated objectively in figure 6 which gives a comparatively lesser value of deviation between actual and target values. The error values and CMOS relation is inversely proportional i.e the better the CMOS value, lesser should be the deviation from target. The error percentage values are calculated by comparing the median values across emotions. Median is chosen since it is less prone to irregularities in database [15]. An interesting fact is that, while neutral-anger conversion gives best perception by virtue of CMOS values, surprise gives best results objectivcly. Also, the inflexions associated with surprise are correctly depicted when conversion is effected from a related emotion like happiness.

The subjective results are compared with objective evaluations. In each case, the error (%) between the synthesised and target emotions have been calculated and plotted as a column chart in figure 6.

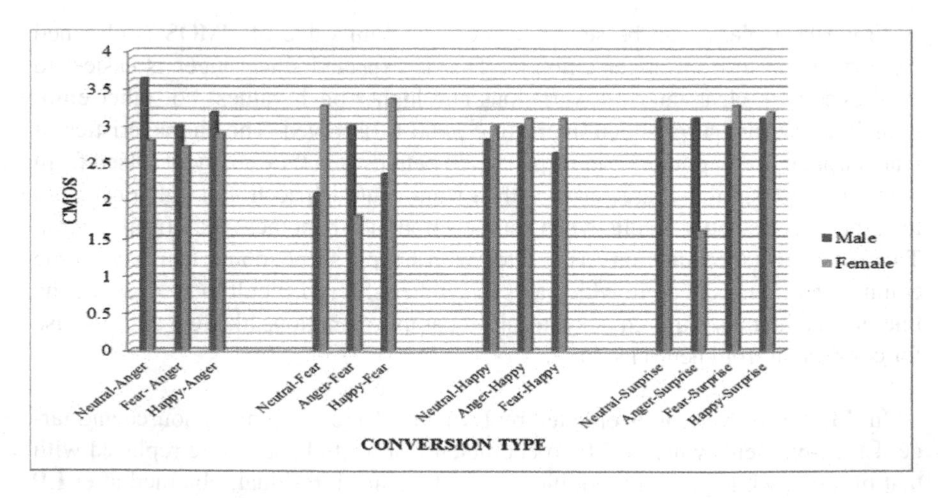


**Fig. 5** Comparison of CMOS values for conversion across emotions and from neutral to emotion. Maximum CMOS value is obtained for conversion from neutral to anger

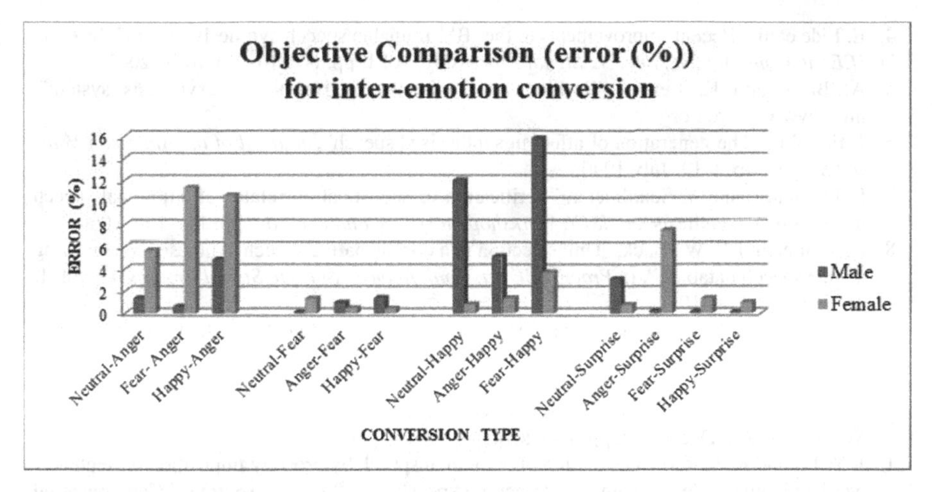


**Fig. 6** Comparison of error (%) in inter-emotion conversion. The percentage error has been calculated by using median value of pitch in each case. Minimum error is obtained for conversion to surprise from happiness and fear emotions.

## 5 Conclusion

Inter-emotion conversion between basic emotions viz. anger, happy, fear and surprise has been accomplished using the principles of dynamic time warping and prosody imposition. The effectiveness of proposed algorithm is verified using objective and subjective evaluations. The results yielded through comparisons prove that inter-conversion of vocal emotions is very effectively obtained with less time and computational complexity. However, the dynamic time warping algorithm used here is text-dependent ie in cases where warping path (co-efficients) of target emotion is not available, the algorithm cannot be used. Also, in most of the cases, the lack of expressiveness of some emotions in database posed a problem while doing perception tests as the listeners found it difficult to relate even the original file to the target emotion. The challenge lies in developing a more robust technique for emotion conversion which is both speaker and database independent thereby, dictating the requirement of a more emotive database.

## References


1. M. Theune, K. Meijis and D. Heylen, "Generating expressive speech for storytelling applications", *IEEE Transactions on Audio, Speech and Language Processing*, pp. 1137-1144,2006.
2. M. Schroder, "Expressing degree of activation in synthetic speech", *IEEE Transactions on Audio, Speech and Language Processing*, vol.14,no.4, pp. 1128-1136, 2006.
3. D. Govind and S. R. M. Prasanna. "Expressive Speech Synthesis Using Prosodic Modification and Dynamic Time Warping", In *Proceedings of NCC*, Jan 16-18, IIT Guwahati, pp.290-293, 2009.

4. E. Eide et al. "Recent improvements to the IBM trainable speech synthesis system", In:*Proc. IEE Int. Conf. Acoust., Speech, Signal Processing*, vol 1, pp. I-708-I-711 vol.1, 2003.
5. A. Black and K. Lenzo, "Building voices in the festival speech synthesis system", http://www.festvox.org.
6. J. E. Cahn, "The generation of affect in synthesised speech", *Journal of the American Voice I/O Society*, pp. 1-19, July 1990.
7. F. Burkhardt and W. Sendlmeier, "Verification of acoustical correlates of emotional speech using formant synthesis", in *ISCA Workshop on speech and emotion*, pp. 151-156, 2000.
8. A. J. Hunt and A. W. Black, "Unit selection in a concatenative speech synthesis system using a large speech database", in *Proc. IEEE Int. Conf. Acoust., Speech, Signal Processing*, vol. 1, pp. 373-376, May 1996.
9. D. Govind and S. R. M. Prasanna. "Dynamic prosody modification using zero frequency filtered signal", *Circuits, Syst. Signal Process.*, doi. 10.1007/s10772-012-9155-3.
10. A. K. Vuppala and S. R. Kaidiri. "Neutral to Anger speech conversion using non-uniform duration modification", in:*Proc. 9th International Conference on Industrial and Information Systems (ICIIS)*, Dec 15-17, pp. 1-4, 2014.
11. J. Yadav and K. S. Rao. "Generation of emotional speech by prosody imposition on sentence, word and syllable level fragments of neutral speech", in *Proc. International Conference on Cognitive Computing and Information Processing (CCIP)*, March 3-4, pp. 1-5, 2015.
12. K. S. Rao and A. K. Vuppala. "Non-uniform time scale modification using instants of significant excitation and vowel onset points", *Speech Communication*, 55, pp. 745-756, 2013.
13. Shashidhar G. Koolagudi, Sudhamay Maity, Vuppala Anil Kumar, Saswat Chakrabarti, K S Rao. "IITKGP-SESC: Speech Database for Emotion Analysis", in *Proc. of IC3*, Noida, pp.485–492, 2009.
14. P. Gangamohan, V. K. Mittal and B. Yegnanarayana. "A Flexible Analysis Synthesis Tool (FAST) for studying the characteristic features of emotion in speech", in *Proc. 9th Annual IEEE Consumer Communications & Networking Conference*, pp.266-270, 2012.
15. B. Akanksh, S. Vekkot and S. Tripathi. "Inter-conversion of emotions in speech using TD-PSOLA", in *Proc. SIRS 2015: Advances in Signal Processing and Intelligent Recognition Systems*, Trivandrum, pp.367–378, 2015.

# Implementing and Deploying Magnetic Material Testing as an Online Laboratory

Rakhi Radhamani[1], Dhanush Kumar[1], Krishnashree Achuthan[2], Bipin Nair[1], and Shyam Diwakar[1,*]

[1] Amrita School of Biotechnology, Amrita Vishwa Vidyapeetham(Amrita University), Kollam, India

[2] Amrita School of Engineering, Amrita Vishwa Vidyapeetham (Amrita University),Kollam, India

*shyam@amrita.edu

**Abstract.** Hysteresis loop tracing (HLT) experiment is an undergraduate experiment for physics and engineering students to demonstrate magnetic properties of ferrite materials. In this paper, we explore a new approach of setting- up triggered testing of magnetic hysteresis via a remotely controlled loop tracer. To aid student learners, through an experimental design, we focused on factors such as analytical expression of mathematical model and modeling of reversible changes, which were crucial for learning hysterisis components. The goal was to study the phenomena of magnetic hysteresis and to calculate the retentivity, coercivity and saturation magnetization of a material using a hybrid model including simulation and remotely controlled hysteresis loop tracer. The remotely controlled equipment allowed recording the applied magnetic field (H) from an internet-enabled computer. To analyze learning experiences using online laboratories, we evaluated usage of online experiment among engineering students (N=200) by organized hands-on workshops and direct feedback collection. We found students adapted to use simualtions and remotely controlled lab equipment augmenting laboratory skills, equipment accessibility and blended learning experiences.

**Keywords:** virtual labs, Ferromagnetism, simulation, hysteresis loop, blended learning, remote labs.

## 1 Introduction

Laboratory based courses together with hands-on experimentation have a vital role in engineering and science education [1]. In recent years, educational researchers consider the significance of adding ICT enabled learning in science education to facilitate experimental studies beyond limits of a classroom [2]. The prominent usage of virtual laboratories in education has been reported to provide new insights to support education [3], [4]. Blended learning approch of implemeting virtual lab courses together with physical experimentations have became a most popular scenario in most of the university sectors [5]. Government is also taking up initiatives to bring virtual lab



J.M. Corchado Rodriguez et al. (eds.), *Intelligent Systems Technologies and Applications 2016*, Advances in Intelligent Systems and Computing 530,
DOI 10.1007/978-3-319-47952-1_74

based education as an e-learning platform for augmenting the current education strategies [6], [7].

Many universities and research institutes have already launched virtual and remote laboratories on the web, which are accessible to users around the world. 'Netlabs' developed by University of South Australia (http://netlab.unisa.edu.au/), 'iCampus iLabs' from Massachusetts Institute of Technology (http://icampus.mit.edu/projects/ilabs/),and 'Remotely controlled lab' from the Palacký University of Olomouc (http://ictphysics.upol.cz/remotelab/) are good examples of remote laboratories that allow users to conduct experiments remotely on real laboratory equipment. Virtual Amrita Laboratories Universilizing Education (VALUE) have been implemented as part of Ministry of Human Resource Department's (MHRD) ICT initiative, in collaboration with several Indian Institutes of Technology (IITs) and other universities in India. The project focused on reproducing real lab experiments to a virtual environment using techniques such as animation, simulation and remotely controlled experiments (http://vlab.co.in/) . Animations graphically deliver the wet lab procedures with a close semblance to physical experimentations. Simulations were designed with mathematical models for reconstructing real datasets. Remote labs provides users with access to real lab equipment, from a distant location, through internet [8]. Access to experiments were based on user registration that have been employed to track statisctics of experiment usage. VLCAP platform was used for implementing and deploying virtual labs [9]. Such online classroom instructional materials were used as supplementary laboratory courses and as interactive textbooks apart from scheduled classroom hours [10]. Several pedagogical studies were carried among university students and professors to analyze the role of virtual and remote laboratories in augmenting laboratory education[11]. Studies suggested the use of remote laboratories in augmenting laboratory education of undergraduate and postgraduate students in universities . Role of remotely controlled experiments as a flexible learning and teaching platform were analyzed using remotely controlled light microscope experiment as a learning exercise[12]. Studies also showed the implementation strategies of remote triggered photovoltaic solar cell experiment and the fundamental characteristics of photovoltaic solar cells, a basic undergraduate experiment in science and engineering course [13].

In this paper, we have discussed about the remotely controlled hysteresis loop tracer, a physical experiment used to determine various magnetic parameters commonly performed in physics laboratories of most engineering courses that was designed and deployed as a virtual laboratory experiment. We also implemented a hysteresis loop tracer simulator for improving pedagogical experience in learning and teaching. The objective was to virtualize magnetic tracer and establish a remote tool for teaching and learning magnetic hysteresis process, later analyzing the impact of such online educational tools amongst student users.

### 1.2 Overview of Hysteresis Loop Tracer Experiment

When a magnetic material is subjected to a cycle of magnetization, the graph intensity of magnetization (M) vs. magnetizing field (H) gives a closed curve called M-H loop. Magnetization intensity M does not become zero when magnetizing field H is reduced to zero. Thus the intensity of magnetization M at every stage lags behind the applied field H. This property is called magnetic hysteresis. The M-H loop is called hysteresis loop. The shape and area of the loop are different for different materials [14]. The phenomenon of magnetic hysteresis includes identifying the magnetic properties such as coercivity, retentivity, saturation magnetization, and thereby establishing the identity of the material itself (Fig.1). The presence of multiple components in the same sample can be recognized by distinctive changes in the slope of the hysteresis curves as a function of applied field [15], [16].

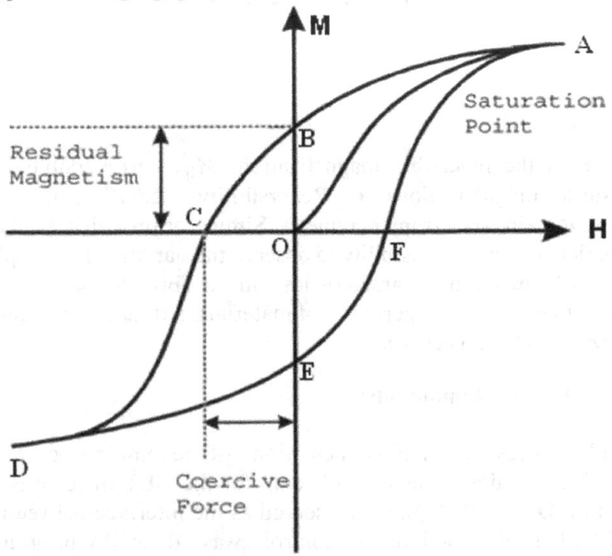


Fig.1. Magnetic hysteresis loop showing coercive force OC, residual magnetism OB, saturation point A

Even though equipment is commonly available in laboratories, access to laboratory usage hours restricts student's practice on such equipment. Also the equipment was costly and needs to be maintained. In situations, where universities were not well localized to suit requirements of laboratory education, virtual and remotye experimentations were implemented as a supplementary learning environment to train students "anytime-anywhere" and for repeating trials in order to facilitate learning based on trial-and-error, which has been known to increase learning interest and motivation of students [6].

## 2 Materials and Methods

### 2.1 Mathematical Modeling of Hysteresis loop tracer

As a part of modeling hysteresis loop tracer, a HLT Simulator was developed in MATLAB (Mathworks, USA) that used mathematical models with a series of instructions, which facilitates user interaction. The Jiles - Atherton Model [17] was chosen for simulating the magnetic nature of the material. Magnetic hysteresis was recorded from the following equations (1), (2) and (3).

$$\frac{dM_{irr}}{dH} = \frac{M_{an}-M_{irr}}{d\ \times k-\alpha\ \times(M_{an}-M_{irr})} \quad (1)$$

$$\frac{dM_{rev}}{dH} = c \times \left(\frac{dM_{an}}{dH} - \frac{dM_{irr}}{dH}\right) \quad (2)$$

$$M = \mathrm{M}_{rev} + M_{irr}$$

(3)

Where, $M_{rev}$ is the reversible magnetization, $M_{irr}$ - irreversible magnetization, $M_{an}$ - anhysteric magnetizations, c- Reversibility coefficient, k - Coercivity parameter, α - Domain Anisotropy parameter. Simulator provided the user a better user interface experience and the possibility to observe the variations in the plot as a level of interpretation of material characteristics in a broader sense. To observe changes in magnetic properties of materials, soft and hard magnets were used for characterizing hysteresis curve.

### 2.2 Architecture of Remote labs

To provide online access, actual implementation of the remotely controlled hardware setup of the HLT was done. The control setup for the HLT tracer was connected to a Data Acquisition Device (DAQ), which served as the interface between the server and the device [11]. Effective modeling of control was done by programming access to the DAQ through a computer. Server hardware was optimized to handle a large number of requests. All experimental setup[18] remained connected to the server. The server received the requests from online users and processed commands to equipment hardware through DAQ and the data from the equipment was sent back to the user. The server communicated with the client through a service-broker. On completion of an experiment, the server notifies the user through a service broker when the results became available. The setup used a web server to communicate between a data acquisition module and the device which was to be remotely triggered. Data traffic and connection to the clients were handled by an apache server[12]. In order to avoid hardware damage, client access was delimited to pre-selected controls necessary to perform the experiment (Fig.2).

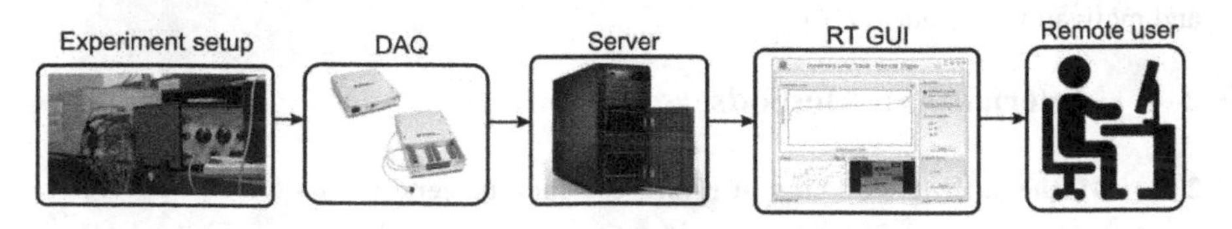


Fig.2. Block Diagram of Remotely Controlled HLT Setup.

### 2.3 Remote Controlled Hysteresis Loop Tracer - Hardware setup

The hardware circuit consists of a sweep unit, integrator and differentiator unit for obtaining the voltages proportional to the input magnetic field (H), magnetization (M) and its first and second-order derivatives corresponding to the speed and phase-change of magnetization . Magnetic field was obtained with a solenoid driven on AC. Magnetic field was calibrated with a Hall Probe for uniformity and the field was calculated via AC current passing through the solenoid. A small resistance in series with the solenoid serves the purpose of taking a signal $e_1$ corresponding to H. The signal $e_2$(corresponding to $\frac{dM}{dt}$)was taken from the pick-up coil placed at the centre of the solenoid containing the sample andwas integrated and phase-corrected. This signal was then subtracted from the reference signal $e_1$ and amplified to give the signal corresponding to M. The $e_1$signal was also subtracted from $e_3$ maintaining the correct ratio (to account for demagnetization and area ratio) and amplified to give signal corresponding to H. $e_2$ was also passed through the differentiator for getting signal corresponding to $\frac{d^2M}{dt^2}$ which was used for phase identification. The solenoid and pickup coil arrangement had a maximum current specification of 4A. Also, 10-turn helipots were used as variable resistors for initial calibration of the experiment. Varying amplitudes for the input magnetic field were obtained from a 10-tap transformer with output voltages from 9.6 V to 126 V and a maximum current rating of 4A. The operational amplifier (op-amp) used were standard 741 op-amp integrated chips (ICs).

### 2.4 Theory of operation of remotely controlled Hysteresis Loop Tracer

Students could access experiment Graphical User Interface (GUI) through a web browser. The GUI consisted of controls for performing the remote triggered hysteresis loop which included a slider to choose the input magnetic field values and a set of radio buttons to choose the plot type to be displayed. User sent data and access information to the remote server while triggering the experiment remotely. The control signals from the client machine were transmitted to the remote server in XML format. For interfacing the hardware to the server, DAQ Module 6221 USB M-Series was used. Inputs to the hardware circuit were obtained from the server from the DAQ Virtual Instrument [16].The digital inputs were employed for selecting the amplitude of the applied magnetic field. The user was able to switch the input amplitude of the applied magnetic field. Switching the applied magnetic field was done by establishing switching between the taps of the transformer in a remote manner. Using this structure the remote user could change the input magnetic field by using the GUI-based slider. The relay structure consisted of a set of 10 relays driven by the relay driver ULN2803A. The analog voltage outputs coming from the hardware corresponding to the applied magnetic field (H) and the magnetization of the sample (M) was relayed to the server and then sent to the user. User was able to see the M-H curve on the GUI (Fig.3). Additionally, user could also switch between three outputs, M, $\frac{dM}{dt}$ and $\frac{d^2M}{dt^2}$. User could see the M-H plot, $\frac{d\ M}{dt}$-H plot or the $\frac{d^2M}{dt^2}$-H plot according to their choice on the graph plot element placed on the GUI. The remotely-controlled magnetic hysteris loop experiment is openly available via http://vlab.amrita.edu/index.php?sub=1&brch=195&sim=800&cnt=1.

### 2.5 Usage and Usability analysis via Field trials

To estimate user behavior and usability, several field trials were conducted in different universities in India to undergraduate engineering students. After providing basic instructions on how to use virtual and remote labs, each of the 200 participants were allowed to perform remote experiments independently. Among 200 students, 50 students were specifically allowed to study Magnetic Material Characterization via Hystersis. After completing the lab exercise, a questionnaire-based direct feedback was collected from participants to analyze their learning experiences using virtual lab. The questions for analysis were as follows:

1. Virtual lab techniques such as animation, simulation and remote labs helped in understanding experiment concepts necessary for using it in a physical laboratory.
2. Would you prefer including simulations to augment laboratory access in your studies?
3. Do you agree that remote lab helps to provide students with a feel of real equipment?
4. Would you support blended learning approach of using simulation and remote labs in their learning?

User's responses (yes/no) were used for assessment

## 3 Implementation

### 3.1 HLT Simulator Implementation

The coding of the Jiles - Atherton model HLT Simulator was done in two parts: coding of the hystplot function and linking of the function with a Graphical user interface (GUI) using MATLAB. A "hystplot" function accepted the inputs namely, relative permeability and five Jiles –Atherton parameters, which defined the nature of sample under observation. The amplitude of the alternating magnetic field applied was also given as an input to the function. Euler's Method was used for calculating the value of M for each value of input H. Hysteresis plot created in the simulator GUI (Fig.3) We also recreated the simulator on Adobe Action Script for the web-based virtual lab.

### 3.2 Remotely Controlled Hysteresis Loop Tracer

The hardware circuit modeled in PSpice was implemented on hardware and interfaced to the server-side computer using the DAQ 6221 Module. 9 digital inputs and two analog inputs of the DAQ Module were used. In this design, applied magnetic field was provided using a rotating "Dial" whereas output plot choice was enabled by "Knob". The hysteresis loop and the plots for the rate and phase of hysteresis were obtained in the server-side computer on the interface (Fig 4A,4B and 4C). The results, although in agreement with the actual experimental setup have showed minor differences due to errors and noise introduced during interfacing.

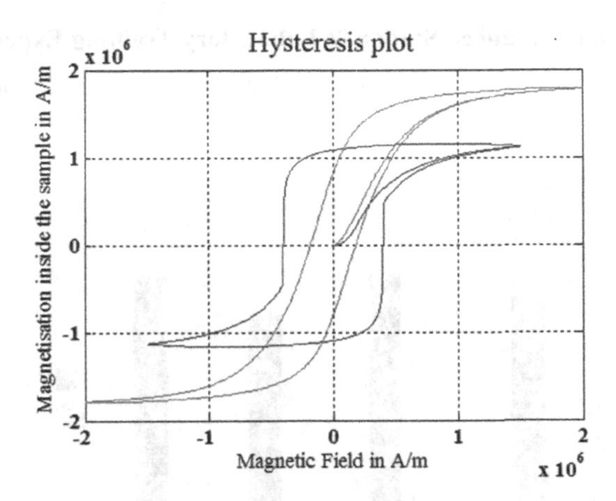


Fig 3. Simulated Hysteresis plots of two materials

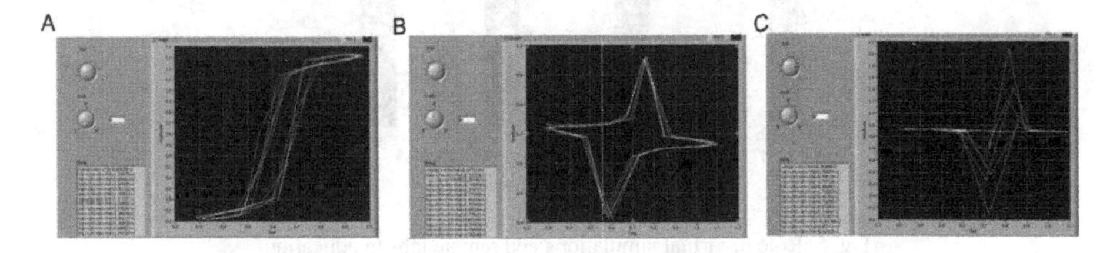


Fig. 4 Snapshots showing the plots for A) Hysteresis Curve B) Speed and C) Phase of hysteresis

## 4 Results

### 4.1 Virtual Lab Simulator Augmented Student Access

Among the student participants, 84% of them indicated that virtual lab techniques such as animation, simulation and remote labs helped them in understanding experiment concepts necessary for using it in a physical laboratory. 90% of them preffered to include simulations in their learning to augment laboratory access (Fig.5, responses for feedback questions Q1 and Q2). Specific feedback from 50 undergraduate students who have performed HLT experiment indicated all of them (100%) were able to reproduce hysteresis plots and tesed their understanding about the basics of hysteresis and the impact of varying different Jiles-Atherton parameters on the hysteresis loop of two magnetic materials A and B,via the simulator platform without the help of an instructor. Students were able to interpret the graphical analysis, as the thinner loop represents material B and the thicker loop corresponds to material A. From the graph obtained in GUI, students were able to calculate saturation magnetization, reversibility coefficient, coercivity parameter and domain anisotropy parameter.

### 4.2 Remote lab Enhanced Student's Laboratory Training Experience

Among the student participants, 82% agreed that remote lab helped them to provide with a feel of real laboratory equipment. 88% indicated using blended learning approach with simulation and remote labs in their studies (Fig.5, responses for feedback questions Q3 and Q4).

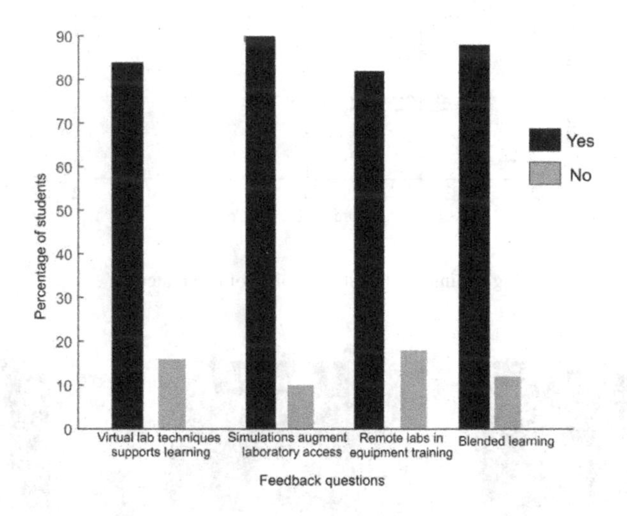


Fig. 5. Role of virtual simulations and remote labs in education

Specific feedback collected from the study participants of HLT experiment were used to determine the adaptation of remotely controlled labs as a lab exercise in their learning process.

User responses (86%) indicated that the materials provided by the online labs are easily adaptable tools for students to improve their skills in laboratory practices.Some of theparticipant commented (as is, from our feedback forms):"In our college laboratory, we are sharing lab equipments and performing the experiment as a group. Virtual and remote lab setup helped me practicing the experiment individually", “Interesting and valuable platform!! If this is successfully implemented in all the colleges, there will be definitely a change in the quality of students passing out.” "Although the remote lab didn’t feel as real as the actual lab, remote labs lets students practice the experiment many times and compare the results in order to have a more fundamental idea. "Overall analysis suggested the usage of online laboratories enhanced laboratory education amongst student groups.

## 5 Discussion

The implementation of HLT devices indicates a possibility of interfacing expensive technologies for standard classroom usage. Mathematical simulation allowed augmenting conceptual knowledge related to laboratory equipment and supported student practice time on devices beyond classroom hours. Having both simulators and remotely controlled devices allowed students to compare between noises generated during recordings to ideal mathematical reconstructions. Remote triggering of the HLT experiment showed promising results towards extending such setups for the study of ferrite materials. Student feedback indicated choices of including remote experimentation as part of curriculum. Studies on virtual simulation and remote labs implicate the roles of online labs in reducing instructor-dependence thus promoting student-centered learning process. The implementation of the HLT experiment facilitated the pedagogical use of online equipment as a learning platform for students with time or cost limited access to real world experiments. Common problems with remote setups were connectivity and slot availability problems. ICT enabled virtual and remote labs in education shows promising outlook given the trend of student and teacher usage as teaching and learning tools.

## 5 Conclusion

The implementation of a remote triggering of the HLT experiment opens up new approaches of online education and experimentation. The implementation of a remote triggered, portable HLT device opens up a new approach in ferrite material testing in which experiments can be done onsite. Usage of remote labs as an additional classroom material for teaching skills also represents novel changes in engineering research. The emerging achievements of virtual and remote labs in creating online courses, is an area which is extending and thus needs a further research. Large-scaletests will be needed to analyze and provide the assessment. The virtual lab is free for use and can be accessed via http://vlab.amrita.edu

**Acknowledgements.** This work derives direction and ideas from the chancellor of Amrita University, Sri Mata Amritanandamayi Devi. The authors would like to thank Devidayal Soman, Jessin P.A, Sreenath P.S, Srikanth S, Rahul Srikumar, Hareesh Singanamala for their contributions in this work. This work was funded under the Sakshat project (Phase 1 &Phase 2) of NME-ICT, Ministry of HRD, Government of India.

## References


1. Lyle, F.D., Rosa J Albert: The role of the laboratory in Undergraduate Engineering Education. J. Eng. Educ. 43, 121–130 (2005).
2. Majumdar, S.: Emerging Trends in ICT for Education and Training. Dir. Gen. Asia Pacific Reg. IVETA. (2015).
3. Nair, B., Krishnan, R., Nizar, N., Radhamani, R., Rajan, K., Yoosef, A., Sujatha, G., Radhamony, V., Achuthan, K., Diwakar, S.: Role of ICT visualization-oriented virtual laboratories in Universities for enhancing biotechnology education - VALUE initiative : Case study and impacts. FormaMente. Vol. VII, 209–226 (2012).
4. Diwakar, S., Radhamani, R., Sujatha, G., Sasidharakurup, H., Shekhar, A., Achuthan, K., Nedungadi, P., Raman, R., Nair, B.: Usage and Diffusion of Biotechnology Virtual Labs for Enhancing University Education in India's Urban and Rural Areas. In: E-Learning as a Socio-Cultural System: A Multidimensional Analysis. pp. 63–83 (2014).
5. Bijlani, K., Manoj, P., Venkat Rangan, P.: VIEW: A framework for interactive elearning in a virtual world. CEUR Workshop Proc. 333, 177–187 (2008).
6. Diwakar, S., Achuthan, K., Nedungadi, P., Nair, B.: Biotechnology Virtual Labs: Facilitating Laboratory Access Anytime-Anywhere for Classroom Education. InTech (2012).
7. Dykes, M., Parchoma, G.: Engaging learners in online discussions. TechTrends. 48, 16–23 (2004).
8. Radhamani, R., Sasidarakurup, H., Sujatha, G., Nair, B., Achuthan, K., Diwakar, S.: Virtual Labs Improve Student's Performance in a Classroom. In: Vincenti, G., Bucciero, A., and Vaz de Carvalho, C. (eds.) E-Learning, E-Education, and Online Training. Springer International Publishing, Cham (2014).
9. Raman, R., Nedungadi, P., Achuthan, K., Diwakar, S.: Integrating Collaboration and Accessibility for Deploying Virtual Labs Using VLCAP, http://www.i-scholar.in/index.php/ITJEMASTTUE/article/view/59247, (2011).
10. Huang, C.: Changing learning with new interactive and media-rich instruction environments: virtual labs case study report. Comput. Med. Imaging Graph. 27, 157–64 (2003).
11. Auer, M., Pester, A., Ursutiu, D., Samoila, C.: Distributed virtual and remote labs in engineering. In: IEEE International Conference on Industrial Technology, 2003. pp. 1208–1213. IEEE (2003).
12. Kumar, D., Singanamala, H., Achuthan, K., Srivastava, S., Nair, B., Diwakar, S.: Implementing a Remote-Triggered Light Microscope: Enabling Lab Access via VALUE Virtual labs. In: Proceedings of the 2014 International Conference on Interdisciplinary Advances in Applied Computing - ICONIAAC '14. pp. 1–6 (2014).
13. Freeman, J., Nagarajan, A., Parangan, M., Kumar, D., Diwakar, S., Achuthan, K.: Remote Triggered Photovoltaic Solar Cell Lab : Effective Implementation Strategies for Virtual Labs. In: IEEE International Conference on Technology Enhanced Education. pp. 1–7 (2012).
14. Sung, H.W.F., Rudowicz, C.: Physics behind the magnetic hysteresis loop - A survey of misconceptions in magnetism literature. J. Magn. Magn. Mater. 260, 250–260 (2003).
15. Mooney, S.: The Use of Mineral Magnetic Parameters to Characterize Archaeological Ochres. J. Archaeol. Sci. 30, 511–523 (2002).
16. Hejda, P., Zelinka, T.: Modelling of hysteresis processes in magnetic rock samples using the Preisach diagram. Phys. Earth Planet. Inter. 63, 32–40 (1990).
17. Jiles, D., Atherton, D.: Ferromagnetic hysteresis. IEEE Trans. Magn. 19, 2183–2185 (1983).
18. Gustavsson, I., Olsson, T., Åkesson, H., Zackrisson, J.: A Remote Electronics Laboratory for Physical Experiments using Virtual Breadboards. Proc. 2005 Am. Soc. Eng. Educ. Annu. Conf. Expo. (2005).

# Hybrid Associative Classification Model for Mild Steel Defect Analysis

Veena N. Jokhakar[1], S. V. Patel[2]

[1]M.Sc.(I.T.) Programme, Veer Narmad South Gujarat University, Surat, India
veena.jokhakar@gmail.com
[2]Sarvajanik College of Engineering and Technology, Gujarat Technological University, India
patelsv@gmail.com

**ABSTRACT.** Quality of the steel coil manufactured in a steel plant is influenced by several parameters during the manufacturing process. Coiling temperature deviation defect is one of the major issues. This defect causes steels metallurgical properties to diverge in the final product. In order to find the cause of this defect, various parameter values sensed by sensors are stored in database. Many approaches exist to analyze these data in order to find the cause of the defect. This paper presents a novel model HACDC (Hybrid Associative Classification with Distance Correlation) to analyze causality for coiling temperature deviation. Due to the combination of association rule, distance correlation and ensemble techniques we achieve an accuracy of 95% which is quite better than other approaches. Moreover, to the best of our knowledge, this is the first implementation of random forest algorithm in analyzing steel coil defects.

**Keywords:** Apriori; Random forest; steel defect diagnosis; distance correlation

## 1 INTRODUCTION

Manufacturing industries like steel basically focus on production of defect free products to remain competitive in the market. But as the steel manufacturing process is very complex in nature, defect diagnosis has been an area of interest in research since last few years. Basically a variation in the characteristics in the final product is called as a defect. These are two types of defects, surface and dimensional defect. Surface defects are like holes, patches, scratches etc, while dimensional defects are deviation in dimensional characteristics from the anticipated value like thickness, width, profile, coiling temperature, finishing temperature, etc.



J.M. Corchado Rodriguez et al. (eds.), *Intelligent Systems Technologies and Applications 2016*, Advances in Intelligent Systems and Computing 530,
DOI 10.1007/978-3-319-47952-1_75

We in this work are concentrating on coiling temperature deviation defect of steel coil.

A typical coil manufacturing process, involves large number of processing steps, each step requires firm statistical process control and quality control mechanisms and keep track of all minutiae of the processes that go out-of-control or could not maintain required steel coil parameters. Though the entire process is tracked in detail, it many times leads to unknown issues or defects in the production. In order to control such integrated process, a system which is intelligent enough to do a great amount of data crunching and analysis on the data to dig out the cause of the defect is needed. It needs to gather, systematize and infer the wide variety of information available during each sub process.

Our work specifically concentrates on the dimensional defect called coiling temperature deviation. All these parameters are non-linear in nature. The exact reason for the deviation either high or low of target values are always a big issue to be solved that takes long time sometimes in several days and sometimes not even found. If an intelligent system generates all the possible rules for the generation of this problem, then the future problems may be avoided and save time, money and hard work. This work is to solve problem by generating rules that cause the coiling temperature that go high or low.

The current approach followed in the company (for whom this data analysis is carried out) is visualization technique. Using the graphs plotted for each process variable, a manual comparison of graphs is done for the pattern followed by the graph of the process parameters. If the process parameters follow the same pattern as coiling temperature graph, the resultant parameter is depicted as the causal parameter. This parameter is then controlled for the further processing of coils. This evaluation usually takes hours, or some time days together and many times the problem is left unsolved. Statistical process and quality control is also used for quality analysis of products during the manufacturing process like x-bar and r-chart, as well as post analysis of defects like p-charts etc. However, manufacturing process is usually so complex that traditional statistical techniques or data management tools are not sufficient to extract learning patterns and knowledge for the quality improvement.

Other industries have been using different approaches for fault diagnosis too. Authors in [1] have used linear regression and MLPCA based regression coefficient for quality diagnosis. However, statistics alone cannot be used for generating rules for taking decision. Data mining is being applied for quality diagnosis and improvement in complicated manufacturing process like semiconductor and steel making in recent years [2]. Many recent researches in fault detection and prediction have used data mining in their work such as neural networks, genetic algorithms, support vector machines etc.

This research proposes a new model named HACDC (Hybrid Associative Classification with Distance Correlation), composed of combination of association, statistics and classification techniques. It proves to be an efficient model composed of well known apriori algorithm and random forest (a powerful classifier) and outperforms other existing methodologies in terms of performance and accuracy. We also find that no work has yet been performed using this powerful technique in defect diagnosis in steel industry. The proposed model produces results that are easy to understand.

The paper is organized as follows:

## 2 BACKGROUND

Steel manufacturing is a multifarious process that consists of a variety of sub processes. During the process various sensors measure various related parameters. As the slab comes out of tunnel furnace (TF), it gets in contact with external atmosphere that leads to formation of a layer. This is separated by the descaler(DS) by spraying water from top and bottom at a specified temperature. Pyrometer 1(P1) and Pyrometer 2(P2) read the temperature of the strip. This reading gives the entry temperature. The number of measurements taken depends upon the length of the slab. The longer the slab larger the readings are. Average number of reading taken at entry side is about 240. This reading are taken every one ms. At this time, all the entry details are sent to the Mill Setup Computer to generate initial values of strands in MSC and CSC for laminar and Coiler section. Rolling Stands reduce thickness of the strip. Thickness is reduced maximum at S1 and gradually reduced in rest. S5, S6 and S7 majorly work on main aspects of providing Finished and final touch to Strip related to its shape and Profile. These stands consist of two sets of Work Rolls known as Top WR and Bottom WR and Two sets of Backup Rolls i.e. Top BUR and Bottom BUR. Rolls are cooled by water continuously and also inter-strand cooling is done. The number of reading again in each of these stands also vary, in stand 1-3 the reading are taken every 1 or 2 ms, and from 5 to 7 stand the readings are taken every 4-8 ms. This number of readings taken are called as segments. These leads to totally imbalance number of record reading or each sector of the mill. Thickness Gauge is for measuring the thickness of the strip. Multifunctional gauge measures Width, Profile, flatness, Thickness of the strip. And an integrated Pyrometer reads FT. It is used to get the readings of many Properties of coil and hence named so. Again the number of reading taken of finishing mill temperature is about 250 and above readings depending upon the length.

Laminar cooling is located between the Finishing Mill Stand F7 and Down Coiler. Its Main task is to cool down the strip in a uniform (laminar) fashion to Coiling Temperature (CT). Each of two zones consists of 4 banks. Each bank consists of 4 sets of 2 sided nozzles that spray water. A last Zone Known as Trimming Zone is used for trimming purpose. In front of the laminar cooling line and behind each cooling section, cross spraying units are installed. A set of rolls driven by motors on which Strip passes called as roller table is situated after laminar. There are two pyrometers that measure the Coiling temperature. Once the Strip leaves the laminar zone these pyro meter read the temperature known as Coiling Temperature. Down Coiler, coils the strip. It has set of Pinch Rolls, Wrapper Rolls and Mandrel which wraps and coils the strip. Figure 1 shows the structure of a CSP mill from tunnel furnace to down coiler.

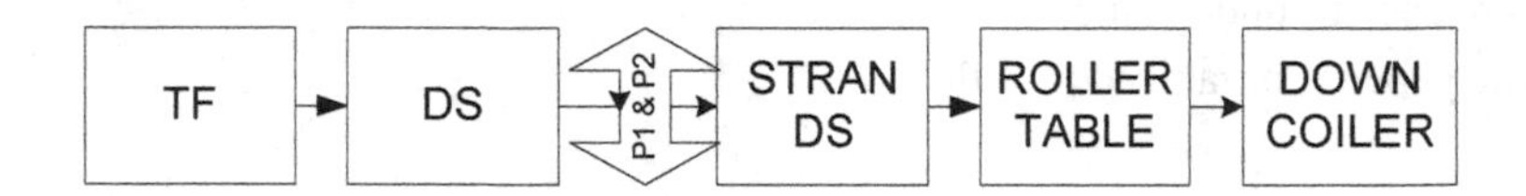


Figure 1: CSP Mill

# 3 RELATED WORK

Intelligent systems that implement machine learning make the purposeful task more precise. Listed below are various machine learning approaches implemented to address problem fault or defect diagnosis. Figure 2: shows the current approaches applied to the problem.

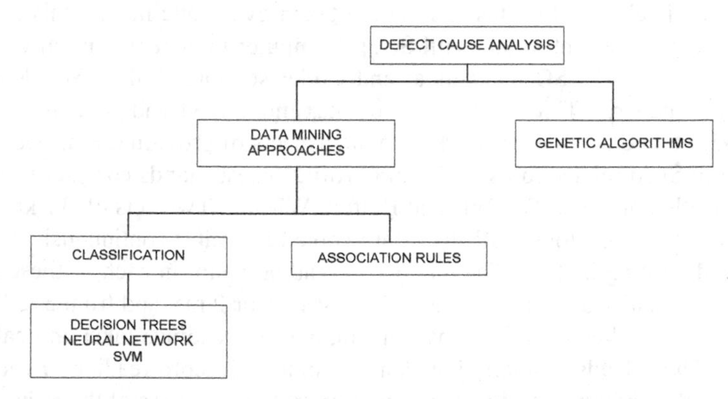


Figure 2: Defect Cause Analysis Tree

Use of Decision Tree for quality and defect analysis has been addressed by BernaBakir et al. [4] ,Shu-guang He et al.[5] and Martin Vlado et al. [6] .The experiments by authors in [4] show the use of C5.0 algorithm, which is an improved version of C4.5. They used Clementine 10.1 to execute this algorithm. After constructing the tree, global pruning was performed with 75% severity to avoid over fitting. Accuracy estimated for the model was 92.15% for the training set. The decision tree model examined nine process variables to be important on the response, defect types, and extracted ten rules associated with these significant input variables. They also show that logistic regression when used for defect analysis, gave unsatisfactory results. Instead decision trees; CART I resulted in 64 % accuracy and CART II resulted in 92% accuracy. The experiments by authors in [5] proposed a knowledge based continuous quality improvement in manufacturing quality. They proposed a unusual model from DMAIC(Define-Measure-Analyse-Imporove-Control) as a goal-driven approach. They collected 1000 records at random from the process. Each record consisted of 4 factors. The records were classified on defect rates. Defect rate > 3.0% were set as H and others as L. They used Decision tree C5.0 for the analysis and generated 5 rules. Authors in [6] recognized an algorithmic decision tree, for the forecast of cracks development in hot rolling. They assessed prediction of V-shaped and U-shaped cracks in steel slab during hot rolling using C4.5 program. The use of Neural Networks for the stated purpose has been addressed by the [8], Shankar Kumar Choudhari[9], Bulsari [11] , M. Li. S. Sethi[10] and FahmiArif et al.[7]. Authors in [8] gathered average values and process measured with different frequencies. They collected 1326 steel strips and 50 variables from indicative measurements and 127 averages values. They used SOM neural networks to tackle this task, as it is able to visualize nonlinear dependencies. The mean accuracy of the model was perhaps lower than 90% for scale defect prediction. Experts in [9] also demonstrate the applicability of regression and neural network in defect detection at Tata steel. Authors in [10] offered the groundwork results of a data mining study of a production line involving hundreds of variables related to various engineering processes involved in manufacturing coated glass. The study was carried out using two nonlinear, nonparametric approaches, namely neural network and CART, to model the relationship between the qualities of the coating and machine readings. Authors in [11] observed excellent results when using multi-layered feed-forward artificial neural network (ANN) models to forecast the silicon content of hot metal from a blast furnace. Time-dependencies (time lags) between each of the different inputs used and the output (silicon content) were found to be important. Experts in [7] proposed a structured model based on neural network with radial basis function for defect component analysis along with PCA. Advantage of this approach is that this can be applied to both linear and non linear data. uses less statistical training, has the ability to detect all possible interactions between predictor variables. Disadvantages of the approach are that it is difficult to interpret, "black box" nature, greater computational burden and is prone to over fitting. They got accuracy of 90%.

Golriz Amooee et al. [12], Kai-Ying Chen et al. [13] and Sankar Mahadevan et al.[14] show SVM based approaches to address the issue. Authors in [12] showed comparison between various mining algorithms for fault detection. They conclude Support Vector Machines(SVM) have the best processing time and also overall high accuracy. Then they applied C5 model to generate predictions rules. Experts in [13] proposed a SVM based model which integrates a dimension reduction scheme to analyze the failures of turbines in thermal power facilities. They operate on thermal power plant data to assess the efficiency of the SVM based model. They conclude that that SVM outperforms linear discriminant analysis (LDA) and back-propagation neural networks (BPN) in classification performance. Authors in [14] proposed an approach for fault detection and diagnosis based on One-Class Support Vector Machines (1-class SVM). The approach is based on a non-linear distance metric measured in a feature space. Drawback of this technique is its high algorithmic complexity and extensive memory requirements in large-scale tasks.

Shun Gang He et al. [5], Sayed Mehran Sharafi et al. [15] and Wei Chau Chen et al.[16] propose use of Association rule in defect diagnosis. Experts in [5] state SPC and association rule based model during manufacturing the process variable that undergo variation. For that purpose they have used SPC to identify the variations and then used association rule for diagnosis of cause of the problem. Here apriori algorithm minimum antecedent support was set to 0.65% and confidence was set to 80%. They considered 62,592 products, 16,905 products failed the test. They concluded that 95% of the products that failed in the test were due to assembly line 4. Authors in [15] used Association rules, Decision Tree and neural networks for recognition of steel defects on surface. After performing preprocessing the number of variables was condensed from 186 to 36 at the modeling phase. The study also tried using these algorithms to reduce product defects with Pits & Blister defect.

They show that the accuracy percentage of each model out performs other depending on the carbon content and other elements.

Experts in [16] depicted Root-cause Machine Identifier (RMI) method using the method of association rule mining to resolve the problem of defect detection competently and effectively. This analysis was defect case based.

Genetic algorithm based approaches were address by JarnoHaapamaki et al.[17] and Danny La [18]. For prediction of scale defect prediction, experts in [17] showed Genetic algorithm (GA) based method. They show Self Organizing Map based model for identifying the most promising variables and then used these variables in GA for the prediction. Average error was 0,0957[1/m2]. Authors of [18] used genetic algorithm to create a controller for the HMT. Given current conditions (specified by the current HMT, which is referred to as the "operating point," and current values of the input variables), the solution should determine what changes need to be made to each variable in order to achieve a desired HMT at a desired time (some hours into the future).

As per the study, it was found that none have considered the deviation percentage of process parameters and similarity of data movement that could depict the cause of the problem. Moreover, we find that defect causality analysis needs research due to non-linear data property and complexity. Finally, we appreciate the current approaches that have all addressed the issue, but still needs performance improvements; hence we propose a new model further.

# 4 RANDOM FOREST

A Random Forest is a classifier consisting of a collection of tree. It grows many trees as an alternative of a single tree. To classify a new entity from an input vector, it puts the input vector in each tree in the forest. Each tree classifies the data that are called as tree "votes" for that class. The forest chooses the classification having the the majority votes.

Increasing the correlation between the trees increases the forest error rate vice a versa a tree with a lower error rate is becomes strong classifier. Reducing number of attributes reduces both the correlation and the strength. Increasing it increases both. Somewhere in between is an ideal range of m. Figure 3 shows the way random forest are grown.

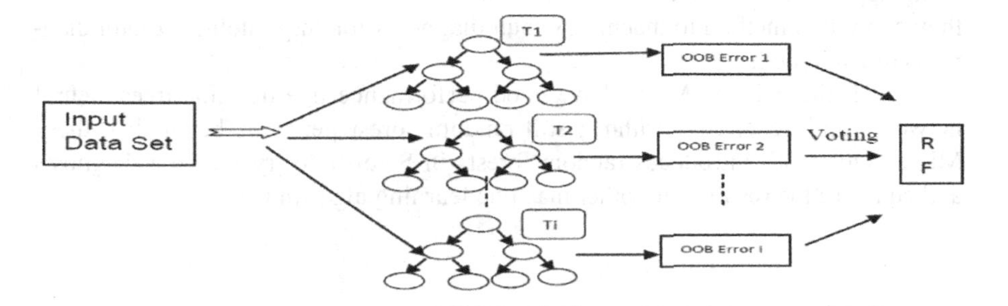


Figure 3: Random forest working methodology

The training set for each tree is drawn by boost strap sampling with replacement, about one-third of the cases are left out of the sample. This left out data is called as out of bag data (OOB).This data is then used to calculate an unbiased estimate of the classification error as trees are grown. After each tree is built, all of the data are run down the tree, and proximities are computed for each pair of cases. If two cases occupy the same terminal node, their proximity is increased by one. At the end of the run, the proximities are normalized by dividing by the number of trees. Proximities are used in replacing missing data, locating outliers, and producing illuminating low-dimensional views of the data. As the OOB data is used or error estimation, the division of data as test set is also not necessary to be done. As these trees are not pruned, these fully grown trees are fairly unbiased but are highly variable. This variance is reduced via bagging and randomization. As in CART pruning is performed to reduce variation. Random forest also provides insights of data by providing variable importance measures, partial dependence, proximities, prototypes, etc.

Byron P. et. al.[19] state that for a less number of variables, the ANN classification was competitive, but as the number of variables was increased, the boosting results proved more competent and superior to the ANN technique.Lidia Auret et.

al. [20] found that random forest feature extraction depicted similar fault analysis performance for The Tennessee Eastman process and improved fault detection performance for the nonlinear model, and better fault detection performance for the calcium carbide process; as compared to principal component analysis. The efficiency of random forests on fault diagnosis of power electronic circuit experimented by determining best possible configurations of random forest is shown by Jin-Ding Cai et.al. [21].Manolis Maragoudakis et.al[22] applied two different types of randomness, inserted into the individual trees of a random forest. They evaluated results obtained by other classical Machine Learning algorithms, such as neural networks, Classification and Regression Trees (CART), Naive Bayes, and K-nearest neighbor (KNN). They used a dataset of Gas turbine for fault diagnosis.

Zhiyuan Yanget. al. in [23] state that random forest achieved good results when being applied to medicine, biology machine learning and other areas but however they apply this method to machinery fault diagnosis for ship intelligent fault diagnosis of chain box.

Though the existing MLAs have good performance like decision trees, neural networks and genetic algorithms; still random forest performs better than these MLAs. Our work introduces random forest in Steel industry for fault diagnosis and compare the results with other machine learning algorithms.

# 5 PROPOSED HYBRID MODEL

We propose a new model that is broadly divided as data preprocessing, relation identification (association), statistical derivation and classification. The purpose is, to find the most closely related parameters then to look at the similarity of the pattern followed and then classifying the defect called coiling temperature. It thereby turns to represent the causal parameter for the cause of the defect.

We used association rule mining as the first step to find the most appropriate columns having relation. Distance correlation is calculated for relevant and high confidence factor associations. Finally, we perform classification. The model we obtain can then be used to classify new examples to achieve a better perception of the defect causing parameters.

As we had a parameter CT to be classified, with lots of predictor columns, we select random forest classifier and well known apriori algorithm for association finding as per the study looking at the advantages.

The dataset we were provided was the real time data gathered from sensors of process related measurements. These measurements of process may be called sample readings taken at a moment of time for a particular location of the steel sheet being processed. Length of the slab that comes out of tunnel furnace is of 30-40 m. As this slab enters the finishing mill, in stands, the length is increased and the thickness is decreased, to achieve the target dimensions. The measurements taken at every location of mill; is hence different, that results in imbalanced set of readings. Moreover, the measurements taken for each location of slab; or the thinner product; sheet; is called segments. This results in multiple segments reading. This hence leads to many missing entries in the readings. All these data are totally non-linear in nature. The best possible way of addressing issue is to visualize graphs and compare, which is already being done. But this does not guarantee finding the non linear dependency in totality. It needs human intelligence and experience also to identify the daily raising new issues. Hence we develop a new intelligent layered approach for addressing this. We address all the issues of data structuring as well as non–linear dependency using dCor. This works exactly the way a human compares a visual graph.

Steps involved are:

## 5.1 Data preprocessing

Steel data is extremely imbalanced dataset that is only 20-25 % of the production is found to be defective. The steps performed were Filtration, Aggregation, Missing Values imputation, Equalizing by Proportionate sampling.

After the earlier steps performed on data, it is left with missing values in stand 5-7 as the reading in these stands were taken every 4-8 ms. From literature survey, we found number of techniques for missing value replacement like, median, average, min, max etc. From our study and experiments, we found harmonic mean as best suitable for our purpose. So, we impute these missing values with harmonic mean of last non zero value and next non zero value after study for achieving higher accuracy.

Further, the number of records in datasets are mismatched; ie. Not equal; hence we take proportionate samples of finishing temperature, keeping the entry and stand data to 100% records. We hence go for minimum loss of data. This still again leaves us with 300-500 records per coil depending upon the length of the coil.

## 5.2 Association identification

We perform association rule mining to find the close association with high support and confidence factor.

### 5.2.1 Data Reduction

It is very important to reduce the number of attributes fed for mining. Hence to achieve this, after rigorous study and to have higher accuracy, we use Association rules for Data Reduction. Columns that are related to Coiling temperature are considered, leaving the thickness based columns aside, with the help of expert advice. Further, these associated variables are used for calculating similarity measure.

### 5.2.2 Binning

Apriori work well for factor data but not for numeric data, hence we performed unsupervised binning by quantiles and converted the data into factor.

Algorithm was then executed with min support as 0.01 and confidence factor as 0.75, with 493719 transactions. This generated 625 purned rule(s).

Table 1 shows the few associations with confidence value greater than 0.75.

**Table 1: Association rules**

| Rules | Support | Confidence | Lift |
|---|---|---|---|
| {CC_WR_CF_XS7=j} => {CC_WR_CF_ES7=j} | 0.09439175 | 0.9439156 | 9.439137 |
| {CC_WR_CL_IN_TMP5=g} => {CC_WR_CL_IN_TMP6=g} | 0.09422364 | 0.9422345 | 9.422326 |
| {CC_WR_CL_IN_TMP6=g} => {CC_WR_CL_IN_TMP5=g} | 0.09422364 | 0.9422345 | 9.422326 |
| {CC_WR_CF_ES6=c} => {CC_WR_CF_XS6=c} | 0.09416287 | 0.9416268 | 9.416249 |
| {SLIP_FORWARD5=j} => {CC_WR_CF_XS6=j} | 0.09098293 | 0.9098274 | 9.098256 |
| {WR_LIN_SPD5=g} => {CC_WR_CF_XS6=g} | 0.08811490 | 0.8811472 | 8.811454 |
| {MEA_TEM_FRONT_FM_AVG=h} => {SLAB_TRANSFER_SPD=h} | 0.07579413 | 0.7579397 | 7.579382 |

## 5.3 Statistical layer

Association rules depict the correlation, but not as positive or negative correlation. Hence to be accurate, new measure called distance correlation is calculated. We performed rescaling, calculated coefficient of variation, normalizing by z- score technique, distance correlation calculation.

The attributes found to have strong association (correlation) by association rule mining are initially normalized by calculating z-score and then DCOR is calculated, which allows us to find whether the pattern of two the column measurement follow same pattern with negative or positive correlation.

Finally the last step to Factor target variable.

After performing the relevant data processing, data reduction and distance correlation calculation, the dataset fed for further processing was as shown in table 2.

| Feature Name | Type | Meaning |
|---|---|---|
| CoilID | Varchar | Unique Id of Coil (9 character) |
| steel_gradeid | Nchar | Steel grade based on the chemical properties |
| Ischild | Derived | ischild based on coilid |
| cvET | Derived | Variation in Entry temperature |
| ref_hx 5 - 7 | Float | Reference thickness of stand 5 to 7 |
| Cvslbspd | Derived | Variation in slab entry speed |
| dcorETSLBSPD | Derived | Distance correlation between entry temperature and entry speed |
| Dcorsspdft | Derived | Distance correlation between entry speed and finishing temperature |
| dcorETFT | Derived | Distance correlation between entry temperature and finshing temperature |
| dstdevwr_spd5 - 7 | Float | Variation in work roll speed of stand 5 -7 |
| dcorwrspd7_ft | Derived | Distance correlation of stand 7 work roll speed and finishing temperature |
| dstdevloo_act_tension5 - 7 | Derived | Variation of looper actual tension of stand 5- 7 |
| dstddevagc_act_gauge_corr 5 - 7 | Derived | Variation of actual gauge control stand 5 -7 |
| dstdevslip_forward5- 7 | Derived | Variation of slip forward speed in stand 5 - 7 |
| dCC_WR_CL_IN_TMP 5- 7FT | Derived | Distance correlation between work roll cooling interstand temperature and finishing temperature 5 -7 |
| dAvgCC_WR_CL_IN_TMP5 - 7FT | Derived | Variation of work roll cooling interstand temperature and finishing temperature 5 -7 |
| dCC_WR_CF_ES5- 7FT | Derived | Distance correlation between work roll cooling fan entry side temperature and finishing temperature 5 -7 |
| dCC_WR_CF_XS5- 7FT | Derived | Distance correlation between work roll cooling fan exit side temperature and finishing temperature 5 -7 |
| ftStdev | Derived | Variation in finishing temperature |
| WATER_FLOW_TOP1-44 | Derived | Water flow top 1 -44 |
| WATER_FLOW_BOT1-44 | Derived | Water flow bottom 1- 44 |
| Dslipccwrxs | Derived | Distance correlation between slip forward and cooling fan work roll exit |
| dwrintmp56 | Derived | Distance correlation between work roll interstand temperature 5,6 |
| dwrlinspd56 | Derived | Distance correlation between work roll linear speed |
| CROSS_SPRAY_ON_FLAG1-12 | Derived | Cross spray 1- 12 |
| CTFlag | Derived and factored | Coiling temperature flag (target) |

**Table 2: Data fed for classification**

## 5.4 Classification layer

We finally perform a black box machine learning with random forest algorithm to these values to generate trees. We generated 1000 trees using regularized random forest (RRF) library. As this technique is black box, we use inTrees framework to visualize the rules. Classification rules generated can help engineers to identify the problem area like variation% of control parameters and the similarity of control parameter movement that could be the cause of the defect.

# 6 RESULTS

We created 1000 tree ensemble. Random forest is black box; we used inTrees to see the non redundant rules in readable form.Table 3 shows 5 rules generated.

We observed that a miss classification rate of 20% in the test set and OOB error of 23%

The rules generated are explained with measures as “len”, ”freq”, ”pred”, ”condition”, ”err”, that is the number of attribute-value pairs in a condition, percentage of data satisfying a condition, outcome of a rule and the error rate of a rule respectively. “ImpRRF” shows the most important rule, taking into consideration both error and frequency. Hence we see that even though rule 3 has high frequency, but due to error rate; its importance decreases. Rule with highest importance show that when the steel grade is 'SRCCRM06','SRCCRM28','SRCCRMB2', 'SRCCRP01','SRCDRWC1','SRCDRWG2','SRCLNC32' and the distance correlation factor between slab speed and finishing mill temperature is greater than 1.8 that is it is less significantly familial and the variation in 6th stand work roll speed is less than 5.37 then the coiling temperature tends to go high, hence here stand 6 speed is more responsible to cause the defect.

Greater distance correlation depicts stronger resemblance, depicts the cause of defect.

Fig 3 shows the OOB error rate, that depicts the error rate calculated with the out of bag data, ie. 1/3rd set of data from left from boot strap sampling during creation of random forest. With this we achieved accuracy of the model to be 95%

Table 4 shows comparisons made with existing techniques and solution with HACDC model, concluding HACDC achieves higher performance in terms of accuracy and accurately determining the cause of the defect.

| Sr. no. | Length | Frequency | Error | Condition | Prediction | ImpR RF |
|---|---|---|---|---|---|---|
| 1 | 3 | 0.52 | 0.18 | steel_grade in c('SRCCRM06','SRCCRM28','SRCCRMB2','SRCCRP01','SRCDRWC1','SRCDRWG2','SRCLNC32') & dcorsspdft>0.185236220573201 & dstdevwr_spd6<=5.3779382339644 | H | 1 |
| 2 | 4 | 0.34 | 0.07 | steel_gradeid in c('SRCCRM06','SRCCRM28','SRCCRP01','SRCDRWC1','SRCDRWG2','SRCLNC32','SRCTRN33') & dcorsspdft>0.18 & dstdevwr_spd6<=4.09 & dstdevwr_spd6>1.00 | H | 0.37 |
| 3 | 5 | 0.08 | 0.15 | steel_gradeid in c('SRCCRM06','SRCCRMB2') & dCC_WR_CF_XS7FT>0.11 & dslipccwrxs<=0.001 & dwrlinspd56<=0.95 & dwrlinspd56>0.82 | L | 0.21 |
| 4 | 4 | 0.06 | 0.18 | steel_gradeid in c('SRCCRM06','SRCCRMB2') & dslipccwrxs<=0.0007 & dwrlinspd56<=0.94 & dwrlinspd56>0.82 | L | 0.19 |
| 5 | 3 | 0.16 | 0.07 | dstdevwr_spd7<=10.20 & dstdevwr_spd7>6.15 & dwrlinspd56>0.93 | H | 0.17 |

**Table 3: CT Defect causing rules**

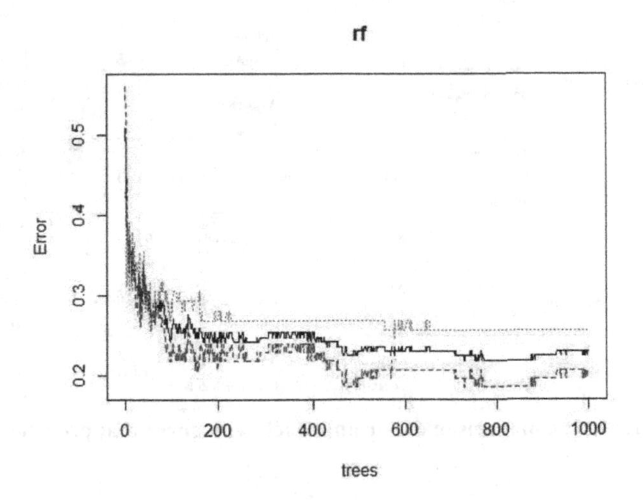


**Fig 3: shows the OOB error rate**

| Ref No | Model Used | Problem addressed | Input Variables And Instances | Method Used | Statistically Significant(S),Error Rates(E),Accuracy(A) | Comments |
|---|---|---|---|---|---|---|
| 6 | Decision Tree | prediction of cracks evolution | 5,* | C4.5 | * | Evaluated U,V shaped cracks. |
| 7 | " | Faultless manufacturing | 22,1115 | PCA and ID3 in multistage | 90(A) | Probability of misclassification of negative class is high with combination of PCA+ID3 than other techniques. |
| 4 | " | Defect analysis | * | CART I<br>CART II | 64(A)<br>92(A) | CART II was better than CART I and Logistic Regression |
| 5 | " | Continuous quality improvement | *,1000 | C5.0 | * | Generated 5 Rules and used these for Parameter optimization |
| 15 | " | Prediction of defects on steel surface | *,DS1-2300<br>DS2-350 | Decision tree | 83(A) | For low carbon products(0.0%-0.15%) |
| 15 | Neural Network | Prediction of defects on steel surface | *,DS1-2300,DS2-350 | Neural Network | 76(A) | products with molybdenum |
| 8 | " | Scale defect prediction | 50 + 127 avg. val.,1326 | SOM neural networks | <90(A) | Visualization of the rolling and defect modeling |
| 5 | Association Mining | Identify variations and diagnose cause | *,62592 | SPC for variation identification and Association Rules Mining | * | 95% of the products failed due to assembly line 4 |
| 15 | " | Prediction of Pits and Blister Defect on steel surface | *,DS1-2300,DS2-350 | Association rules | 64(A) | For low carbon products(0.15% - 0.25%) |
| 12 | SVM | Fault detection | 8,* | SVM and others | 90(A) | SVM proved to be better than other |
| 17 | Genetic Algorithm | Scale defect prediction | 50 + 127 avg. val,1326 | Genetic algorithm | 0.0640(E) | Performance not as good as neural network. |
| HACDC | Random Forest | Coiling Temperature Defect Cause | 1116 coils, instances 56 attributes | Apriori, Dcor,Random Forest | 95(A) | Outperforms all the other work done for Steel Defect Cause |

**Table 4: Comparison of our approach with current approaches**

# 7 CONCLUSION

The issue of defect cause diagnosis problem of steel industry has been addressed in our work. To deal with the problem we propose a new hybrid model HACDC, that uses association rule, distance correlation and classification techniques. The dataset used is highly non-linear, imbalanced with varied number of measurements. We investigate distance correlation that is a new distance based correlation measurement that can deal both non linear and linear data. We have introduced random forest, a classification algorithm, on steel coil defect dataset, though some work exists using this machine earning algorithm in fields like medicine, bioscience, stock market and software engineering. We find novel unseen rules with accuracy of 95% that proves to overcome the current existing solutions. Which can help the engineers to control the production according to the rules generated, or if they have seen any of the rules being applied to current ongoing process, the next production can be taken care.

# REFERENCES


[1] Veena Jokhakar, S.V.Patel, A Review of Business Intelligence Techniques for Mild Steel Defect Diagnosis, International Journal of Computer Applications (0975 – 8887) Volume 113 – No. 10, March 2015

[2] Dharminder Kumar, Suman, Performance Analysis Of Various Data Mining Algorithms: A Review, International Journal Of Computer Applications (0975 – 8887), Volume 32– No.6, October 2011

[3] B. Bakır, Bakır, I. Batmaz, F. A. Güntürkün, I. A. Ipekçi, G. Köksal, And N. E. Özdemirel , Defect Cause Modeling With Decision Tree And Regression Analysis, World Academy Of Science, Engineering And Technology International Journal Of Mechanical, Aerospace, Industrial And Mechatronics Engineering Vol:2 No:12, 2008

[4]Berna Bakir, Defect Cause Modeling With Decision Tree And Regression Analysis: A Case Study In Casting Industry, A Thesis Submitted To The Graduate School Of Informatics Of Middle East Technical University, May 2007

[5]Shu-Gauge He,ZhenHe,G. Alan Wang And Lili, Quality Improvement Using Data Mining In Manufacturing G Processes, Data Mining And Knowledge Discovery In Real Life Applications, ISBN 978-3-902613-53-0,Pp.438,Feb 2009

[6]MartinVlado,RóbertBidulský;LuciaGulová,KristínaMachová,JanaBidulská, JánValíˇCek And JánSas, The Production Of Cracks Evolution In Continuously Cast Steel Slab, High Temp. Mater. Proc., Vol. 1–2 (2011), Pp. 105–111, Copyright © 2011 De Gruyter. Doi 10.1515/Htmp.2011.014

[7]FahmiArif,NannaSuryana,BurairahHussin, A Data Mining Approach For Developing Quality Prediction Model In Multi-Stage Manufacturin,© 2013 By Ijca Journal, Volume 69 - Number 22

[8]Jarno J. Haapamaki, Satu M. Tamminen And JuhaJRoning,Data Mining Methods In Hot Steel Rolling For Scale Defect Prediction, Proceedings of the IASTED International Conference on Artificial Intelligence and Applications, 2005

[9]Shakar Kumar Choudhari, Sunil Kumar, Vinit K Mathur, Application Of Datamining Technology At Tata Steel,2004

[10]M. Li. S. .Sethi, J. Luciow,K. Wangner, Mining Production Data With Neural Network & Cart, In Conf. Rec. Ieee Int. Conf. Data Mining, 2003.

[11] Bulsari, AbhayAndSaxen, Henrik. "Classification Of Blast Furnace Probe Temperatures Using Neural Networks." Steel Research.
Vol. 66. 1995

[12] Golriz Amooee, Behrouz Minaei-Bidgoli, Malihe Bagheri-Devnavi, A Comparison Between Data Mining Production Algorithms For Fault Detection,IJSCI ,Vol.8,Issue 6,No 3, Nov 2011, Issn :1694-0814

[13]Kai-Ying Chen, Long-Sheng Chen, Mu-Chen Chen, Chia-Lung Lee , Using SVM Based Method For Equipment Fault Detection In A Thermal Power Plant, Computers In Industry 62, 42–50, Journal Homepage: Www.Elsevier.Com/Locate/Compind ,2011

[14]Sankar Mahadevan And Sirish L. Shah, Fault Detection And Diagnosis In Process Data Using Support Vector Machines, Jounal Of Process Control 19, Science Direct, 1627-1639,2009

[15]Sayed Mehran Sharafi, Hamid Reza Esmaeily, Applying Data Mining Methods To Predict Defects On Steel Surface, Jounal Of Theoretical And Applied Information Technology,©2005-2010

[16]Wei-Chou Chen, Shian-Shyong Tseng, Ching-Yao Wang, A novel manufacturing defect detection method using association rule mining techniques, ELSEVIER,Expert Systems with Applications 29 (2005) 807–815

[17]JarnoIIaapamaki And JuhaRoning,Genetic Algorithms In Hot Steel Rolling For Scale Defect Prediction,World Academy Of Science ,Engineering And Technology 5 2007

[18]By Danny Lai, Supervised By Dr. Amar Gupta, Using Genetic Algorithms As A Controller For Hot Metal Temperature In Blast Furnace Processes, 2000

[19] Byron P. Roea, Hai-Jun Yanga,_, JiZhub, Yong Liuc, Ion Stancuc, Gordon Mcgregord, Boosted Decision Trees As An Alternative To Artificial Neural Networks For Particle Identification, Nuclear Instruments And Methods In Physics Research A 543 (2005) 577–584,Science Direct, Available Online 25 January 2005

[20]Lidia Auret , Process Monitoring and Fault Diagnosis using Random Forests , Dissertation presented for the Degree of DOCTOR OF PHILOSOPHY (Extractive Metallurgical Engineering) , December 2010

[21] Jin-Ding Cai, Ren-Wu Yan, Fault Diagnosis of Power Electronic Circuit Based on Random Forests Algorithm, 2009 Fifth International Conference on Natural Computation,IEEE, 2009

[22] ManolisMaragoudakis, Euripides Loukis and Panagiotis-ProdromosPantelides, Random Forests Identification of Gas Turbine Faults, 19th International Conference on Systems Engineering,IEEE

[23]Zhiyuan Yang, QinmingTan ,The Application of Random Forest and Morphology Analysis to Fault Diagnosis on the Chain box of ships, Third International Symposium on Intelligent Information Technology and Security Informatics,IEEE

# Improving the performance of Wavelet based Machine Fault Diagnosis System using Locality Constrained Linear Coding

Vinay Krishna, Piruthvi Chendur P, Abhilash P P, Reuben Thomas Abraham, Gopinath R, Santhosh Kumar C

**Abstract** Support Vector Machine (SVM) is a popular machine learning algorithm used widely in the field of machine fault diagnosis. In this paper, we experiment with SVM kernels to diagnose the inter turn short circuit faults in a 3kVA synchronous generator. We extract wavelet features from the current signals captured from the synchronous generator. From the experiments, it is observed that the performance of baseline system is not satisfactory because of the inherent non linear characteristic of the features. Feature transformation techniques such as Principal Component Analysis (PCA) and Locality-constrained Linear Coding (LLC) are experimented to improve the performance of the baseline system. Although PCA allows for choosing dimensions with maximum variance, the dimension reduction always contributes to underperformance. On the other hand, LLC uses a codebook of basis vectors to map the features onto higher dimensional space where a computationally efficient linear kernel can be used. Experiments and results reveal that LLC outperforms PCA by improving the baseline system with an overall accuracy of 25.87 %, 21.47 %, and 21.79 % for the R, Y, and B phase faults respectively.

## 1 Introduction

Periodic maintenance is a typical fault diagnosis method in industries. It is performed at regular time periods irrespective of the machine condition. Thus, periodic

---

Vinay Krishna, Piruthvi Chendur P, Abhilash P P, Reuben Thomas Abraham, Gopinath R, Santhosh Kumar C
Machine Intelligence Research Lab,
Dept. of Electronics and Communication Engineering,
Amrita School of Engineering, Coimbatore.
Amrita Vishwa Vidyapeetham,
Amrita University,
India.
e-mail: vinay_krishna35007@yahoo.com

 
J.M. Corchado Rodriguez et al. (eds.), *Intelligent Systems Technologies and Applications 2016*, Advances in Intelligent Systems and Computing 530,
DOI 10.1007/978-3-319-47952-1_76

maintenance requires high maintenance cost. Later, Condition Based Maintenance (CBM) was introduced inorder to reduce the maintenance costs. It is performed only when an abnormal behavior is suspected in the machine. CBM reduces the downtime and improves the productivity and safety operations [1].

The steps involved in CBM technique are data acquisition, feature extraction, feature selection, feature transformation and classification. Data acquisition involves extraction of signals under fault and no fault conditions. The features extracted are used for capturing fault discriminant information from the raw data [2]. Time domain [3], frequency domain [4], wavelet domain [5] and empirical mode decomposition [6] are the signal processing based feature extraction methods used widely in machine health monitoring systems.

Feature selection algorithm is used to improve the performance of machine health monitoring system by removing the redundancy in features [7–9]. Feature transformation techniques are used for better fault discrimination [10]. Classification involves training the fault models for the different conditions of the machine and testing the reliability of the model. The learning algorithm used predominantly in many applications including fault diagnosis is the Support Vector Machine (SVM) [11].

In this work, we diagnose the inter-turn faults in the stator windings of the 3 kVA synchronous generator. The current signals are used for the feature extraction. First, the baseline system is developed using wavelet features, since the time and frequency domain features are not capable of extracting inherent information from non-stationary signals. For revealing fault related information, the multi resolution ability of discrete wavelet transform (DWT) makes it more suitable for fault diagnosis [12]. We experimented with linear SVM kernel and other standard non-linear kernels polynomial, sigmoid, and RBF. The experiments and results show that the performance of the learning algorithm is not satisfactory since the kernels used does not match the features or vice versa.

The capability of the learning algorithm can be enhanced by using either of the following approaches: 1) The features that fit the kernel can be selected 2) Kernels can be singled out to match the features 3) Features can be mapped onto a higher dimensional space where a linear kernel can be used for classification. In this work, experiments are conducted on the wavelet based features that are mapped onto unique dimensional spaces with the use of Principal Component Analysis (PCA) and Locality Constrained linear Coding (LLC) [13–15] which can enhance the performance of fault diagnosis. PCA is a dimension reduction technique that translates the initial variables into a smaller set of uncorrelated variables with a negligible loss of information. LLC is a feature transformation technique that maps the features to a higher dimensional space where they are linearly separable. LLC is used to represent the features as a linear combination of k-nearest basis vectors or neighbors (kNN). Previous experiments conducted on various feature transformation techniques revealed that LLC offered a superior performance [16]. LLC offers a better performance and minimal computational complexity against the Sparse Coding [17] as a result of the constraint on locality. This suggests that the feature transformation technique LLC can be relied on to improve the performance of the fault diagnosis system [16]. Experiments conducted by Sreekumar et al. using LLC for effective fault diagnosis of rotating machine using vibration signals also revealed that LLC offered a better performance incomparison with baseline SVM system [18]. Gopinath et al. experimented LLC for scalable fault models by capturing the intelligence from the 3 kVA synchronous generator and scaled up the fault models to monitor the condition of the 5 kVA synchronous generator. PCA was also used to remove the system dependent features after feature transformation using LLC [19]. Gopinath et al. experimented nuisance attribute projection algorithm to remove the system dependent features effectively across the 3 kVA and 5 kVA synchronous generators [20]. Cost sensitive classification and regression tree algorithm was used to improve the performance of fault diagnosis of the 3 kVA synchronous generators for mission critical applications. Sensitivity (alarm accuracy) was improved at the expense of the specificity (no-alarm accuracy) [21]. In this paper, the performance of PCA and LLC using wavelet statistical features is compared to improve the fault diagnosis in 3 kVA synchronous generators.

# 2 System Description

## 2.1 *Experimental Setup and Data Collection*

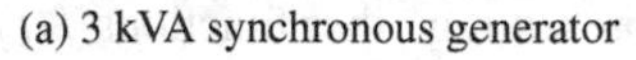

(a) 3 kVA synchronous generator (b) Experimental testbed

Fig. 1: Experimental setup

The experimental setup consists of a synchronous generator, three phase resistive load, current and voltage sensors, and data acquisition system. The current through and the voltage across the load are measured by the sensors. The Data acquisition system uses NI PXI 6221 for interfacing the sensors. Short circuit faults can occur in either of the winding coils, stator or field, in the case of generators. These short circuit faults are injected under a controlled environment.

In generators, tappings are normally present at 0% and 100% of the coil windings in the case of stator and field winding coils. But in order to inject faults at different magnitudes, the 3 kVA synchronous generator is customized. Of the total turns (27 turns), Inter-turn faults are injected at about 30% (8 turns), 60% (16 turns) and 82% (22 turns) [22]. Fig. 1 presents the photograph of the customized synchronous generator and the experimental setup. Each experiment is conducted for about 10 seconds. Since the signals are sampled at a rate of 1 kHz, the number of samples obtained becomes 10,000 samples / trial. The data set is then obtained by repeating the process for different fault conditions. Fig. 2 represents the current signatures obtained from the 3 kVA generator.

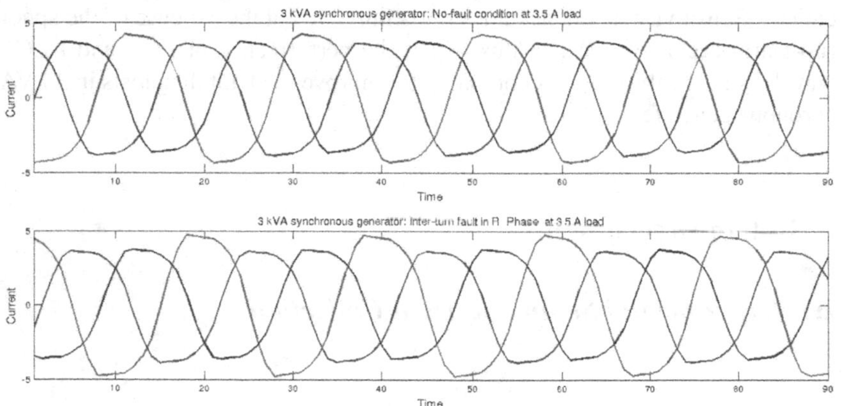


Fig. 2: Captured current signature under fault and no fault conditions

## 2.2 Feature Extraction using Wavelet Transform

Wavelets are used as a mathematical tool for extracting information from the signals. There are a number of wavelets such as Morlet, Mexican Hat, Haar Daubechies, Symlets, Coiflets, and Biorthogonal spline wavelets. These wavelets can be used for analyzing continuous and discrete signals using continuous and discrete transforms. Wavelets form the basis functions $\psi_k(t)$ in representing other functions $f(t)$.

$$f(t) = \sum_k \alpha_k \psi_k(t) \tag{1}$$

The basis functions are representated by the mother wavelet and the father wavelet or the scaling function , $\phi(t)$. The father wavelet signifies the smooth low frequency components of the signal while the mother wavelet is a representation of the high frequency details by noting the amount of stretching of the wavelet.

### 2.2.1 Wavelet Transform

Wavelet transforms are localized both in time domain and as well as in frequency domain, unlike fourier transform. Wavelet series, Continuous Wavelet Transform (CWT) and Discrete Wavelet Transform (DWT) are the classical methods of performing Wavelet Transforms. Wavelet series maps a continuous function into a series of coefficients [23]. In CWT, the wavelet coefficients are obtained by scaling and translating basis function across the length of the signal. The CWT of a continuous time signal $f(t)$ is obtained as

$$CWT_{\psi}f(b,s) = \int_{-\infty}^{+\infty} f(t)\psi_{b,s}^{*}\,dt \tag{2}$$

where

$$\psi_{b,s}^{*}(t) = \frac{1}{\sqrt{b}}\psi^{*}\left(\frac{t-s}{b}\right), \text{ and } b,s \in R,\ a \neq 0$$

$s$ the scaling parameter and a measure of frequency and $b$ the translation parameter and a measure of time. The DWT is the discrete counterpart to the CWT, which is used in this work. In DWT, the wavelets are discretely sampled. As a result, DWT serves as a fast and a lossy transformation. The wavelet coefficients of the signal $f(K)$ is given by

$$DWT_{\psi}f(s_1,s_2) = \sum_{k} f(k)\psi_{s_1,s_2}^{*}(k) \tag{3}$$

where

$$\psi_{s_1,s_2}^{*}(k) = \frac{1}{\sqrt{c^m}}\psi^{*}\left(\frac{k-s_2dc^{s_1}}{c^{s_1}}\right)$$

and $s_1, s_2$ are the scaling and sampling parameters.

In this paper, experiments were conducted on daubechies and symlets for appropriate decomposition levels. These wavelets are used to decompose the current signals into approximate coefficients and detail multi resolute coefficients. This is referred to as multi resolution analysis. These coefficients are not directly fed into the fault diagnosis system. The dimensions of these patterns are kept as low as possible to avoid a large computational time and poor performance, by choosing a suitable feature extraction methodology. Table 1 lists the different statistical features extracted from the wavelet coefficients and their formulations.

## *2.3 Principal Component Analysis(PCA)*

Principal Component Analysis (PCA) is a feature transformation technique used mainly for the purpose of dimensionality reduction. PCA linearly transforms original set of variables into substantially smaller set of uncorrelated variables that retains most of the information from the original data set. Let $X$ be the set of all mean normalized features $x_1, x_2, \ldots, x_n$ and $\Sigma_X = \frac{1}{n-1}XX^T$ be the covariance matrix of the set of features, $X$. The aim of this method is to determine new basis set of vectors, $P$ such that

Table 1: Wavelet domain Statistical Features [23]

| Feature | Detail Coefficients | Approximate Coefficients |
|---|---|---|
| Mean | $feat_i = \frac{1}{N}\sum_{j=1}^{N} d_{ij}$ | $feat_{11} = \frac{1}{N}\sum_{j=1}^{N} c_{lj}$ |
| Standard Deviation | $feat_{i+11} = \frac{1}{N}\sum_{j=1}^{N}(d_{ij}-\mu_1)^2$ | $feat_{22} = \frac{1}{N}\sum_{j=1}^{N}(c_{lj}-\mu_2)^2$ |
| Skewness | $feat_{i+22} = \sqrt{\frac{N}{24}}\left\{\frac{1}{N}\sum_{j=1}^{N}\left(\frac{d_{ij}-\mu_1}{\sigma_1}\right)^4-3\right\}$ | $feat_{33} = \sqrt{\frac{N}{24}}\left\{\frac{1}{N}\sum_{j=1}^{N}\left(\frac{c_{lj}-\mu_2}{\sigma_2}\right)^4-3\right\}$ |
| Kurtosis | $feat_{i+33} = \sqrt{\frac{1}{6N}\sum_{j=1}^{N}\left(\frac{d_{ij}-\mu_1}{\sigma_1}\right)^3}$ | $feat_{44} = \sqrt{\frac{1}{6N}\sum_{j=1}^{N}\left(\frac{c_{lj}-\mu_2}{\sigma_2}\right)^3}$ |
| RMS | $feat_{i+44} = \sqrt{\frac{1}{N}\sum_{j=1}^{N} d_{ij}^2}$ | $feat_{55} = \sqrt{\frac{1}{N}\sum_{j=1}^{N} c_{lj}^2}$ |
| Form Factor | $feat_{i+55} = \frac{\mu_1}{rms_1}$ | $feat_{66} = \frac{\mu_2}{rms_2}$ |
| Crest Factor | $feat_{i+66} = \frac{peak}{rms_1}$ | $feat_{77} = \frac{peak}{rms_2}$ |
| Energy | $feat_{i+77} = \sum_{j=1}^{N}\left\|d_{ij}^2\right\|$ | $feat_{88} = \sum_{j=1}^{N}\left\|c_{lj}^2\right\|$ |
| Shannon Entropy | $feat_{i+88} = -\sum_{j=1}^{N} d_{ij}^2 log(d_{ij}^2)$ | $feat_{99} = -\sum_{j=1}^{N} c_{lj}^2 log(c_{lj}^2)$ |
| Log energy Entropy | $feat_{i+99} = \sum_{j=1}^{N} log(d_{ij}^2)$ | $feat_{110} = \sum_{j=1}^{N} log(c_{lj}^2)$ |
| Inter-quartile Range | $feat_{i+110} = d_i(75) - d_i(25)$ | $feat_{121} = c_j(75) - c_j(25)$ |

$$Y = PX \tag{4}$$

and the corresponding covariance matrix,

$$\Sigma_Y = \frac{1}{n-1}YY^T = \frac{1}{n-1}P(XX^T)P^T \tag{5}$$

is diagonalized. Thus, the redundancy in features is eliminated and the new set of features $Y = [y_1, y_2, \ldots, y_k]$ is uncorrelated.

## *2.4 Locality-constrained Linear Coding (LLC)*

LLC is a feature mapping technique for representing non-linear features which improve the performance of linear classifiers. Let $X = [x_1, x_2, ......, x_N] \in R^{D\times N}$ be the set of all $D$-dimensional input vectors. Each input feature vector is represented by the k-nearest basis vectors (kNN) as a linear combination. These basis vectors is also sometimes referred to as the codebooks. The codebooks can be generated by means of k-means clustering where centroid of each cluster forms the basis vectors.

Given the codebook $B = [b_1, b_2, ......, b_M] \in R^{D \times M}$, each input data is converted into an $M$-dimensional code by the algorithm. These LLC codes are sparse, high dimensional and used for capturing the patterns present in the input data. The criterion for LLC can be written as:

$$\min_{C} \sum_{i=1}^{N} \| x_i - Bc_i \|^2 + \lambda \| d_i \odot c_i \| \tag{6}$$

$$s.t. 1^T c_i = 1, \forall_i$$

where $\odot$ represents element-wise multiplication. Depending on its similarity to $x_i$, Locality adaptor $d_i$ allows unique freedom for each basis vectors. Locality descriptors are expressed as [13]

$$d_i = exp\left(\frac{dist(x_i, B)}{\sigma}\right) \tag{7}$$

where

$$dist(x_i, B) = [dist(x_i, b_1), ..., dist(x_i, b_M)]^T$$

defines the Euclidean distance between $x_i$ and $b_j$. The locality descriptor $d_i$ is further normalized to lie between (0,1].

The weight decay speed for the locality adaptor $d_i$ is adjusted by varying the parameter $\sigma$. The constraint $1^T c_i = 1$ establishes the requirements of shift invariance for the LLC code. Locality is given higher importance than sparsity as the locality must lead to sparsity [15]. Thus the LLC codes are sparse in the sense that it has only few significant values and not in the sense of $l^0$ norm. Practically, sparsity is attained by simply thresholding small coefficients to zero. LLC selects the local basis from basis set of vectors to form the local coordinate system. Instead of using equation 11, encoding process can be sped up by using $k\,(k < M)$ nearest neighbours alone instead of considering all of the bases. The following is the criteria for the above explained methodology of fast approximation LLC [13].

$$\min_{\widetilde{C}} \sum_{i=1}^{N} \| x_i - \widetilde{c}_i B_i \|^2 \tag{8}$$

$$s.t. 1^T \widetilde{c}_i = 1, \forall_i$$

Fast approximation LLC method is used in this paper which reduces the computational complexity from $O(M^2)$ to $O(M + k^2)$ and thus improves the speed of the encoding process. Fig. 3 describes the methodology for the overall system.

## *2.5 Support Vector Machine(SVM)*

Statistical learning theory proposed by Vapnik and Chervonenkis [11] formed the basis for the learning algorithm. SVM constructs a separating hyperplane for clas-

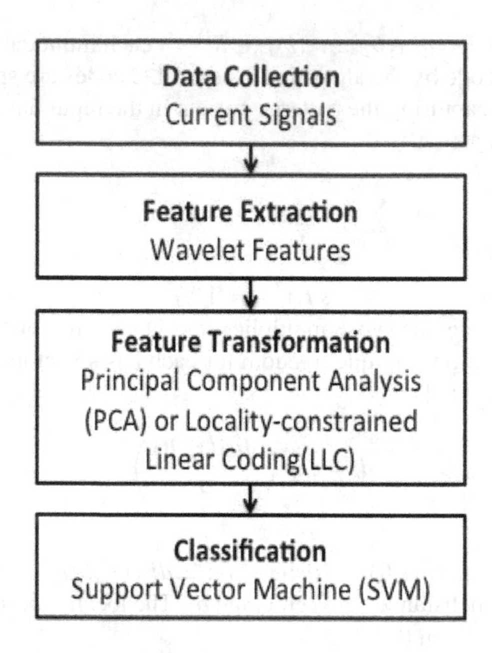


Fig. 3: Overall System Description

sification in a high dimensional space. The hyperplane provides a good separation when it is at a maximal distance from the nearest training data point or the support vector of any class. For a linearly separable data set, two parallel hyperplanes are selected that bounds the two classes of data. These hyperplanes are defined by the equations $x.w+b=-1$ and $x.w+b=1$, where $w$ is the normal vector to the plane, $x$ the set of all training data and $b$ the bias. The maximum margin hyperplane is the one that lies exactly at the centre of the bounding planes. Fig. 4 describes the approach in which the SVM determines the optimal hyperplane. Mathematically, the margin can be expressed as

$$\frac{2}{||w||^2} \tag{9}$$

In order to obtain this maximum margin, the above equation is rewritten as a minimization problem as:

$$\min_{w,b} \frac{||w||^2}{2} \tag{10}$$

subject to the constraint

$$y_i(w\,.\,x_i - b) \geq 1, \forall_i.$$

where $y_i = \{-1, 1\}$ forms the class label. The $w$ and $b$ that solve this optimization problem determine the classifier. Using Lagrange dual minimization problem:

$$L(w,b,\alpha) = \min_{w,b} \max_{\alpha \geq 0} \left\{ \frac{1}{2} \|w\|^2 - \sum_{i=1}^{m} \alpha_i [y_i(w^T x_i - b) - 1] \right\} \tag{11}$$

and using Karush Kuhn Tucker (KKT) condition [24], solution can be obtained as:

$$w = \sum_{i=1}^{m} \alpha_i y_i x_i \tag{12}$$

$$b = \frac{1}{N_{SV}} \sum_{i=1}^{N} (w^T x_i - y_i) \tag{13}$$

where $\alpha_k$ is the Lagrange multiplier. The problem written in its dual form by using the solution obtained for $w$ as:

$$L(w,b,\alpha) = \max_{\alpha_i} \left\{ \sum_{i=1}^{m} \alpha_i - \frac{1}{2} \sum_{i,j} \alpha_i \alpha_j y_i y_j k(x_i, x_j) \right\} \tag{14}$$

subject to

$$\alpha_i \geq 0 \text{ and } \sum_{i=1}^{m} \alpha_i y_i = 0$$

where $k(x_i, x_j)$ is the kernel function. The decision function obtained by SVM learning is given as [11]

$$d(x) = sign(w^T x + b) = sign(\sum_{i=1}^{n} \alpha_i y_i x^T x - b) \tag{15}$$

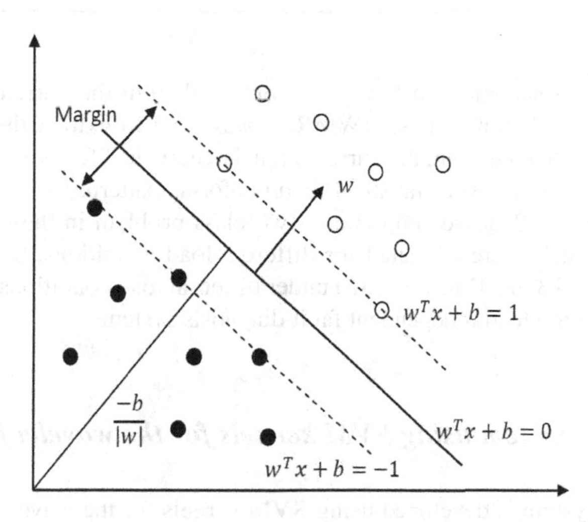


Fig. 4: Optimal Hyperplane for Linearly separable case

Non-linear kernels are typically used to transform the features onto a higher dimensional space, when the data is linearly inseparable. In this higher dimensional space, the features can be classified by using a linear hyperplane. The experiments uses both SVM linear kernel and standard non-linear kernels for fault diagnosis.

## 3 Experiments and Results

The following were the experiments conducted on the current signals obtained from the 3kVA synchronous generator:

1. Developing baseline system using SVM kernels for the wavelet features
2. Improving the performance of the baseline system using feature transformation techniques, PCA and LLC

In this work, current signals are acquired from the three phase generator under fault and no fault condition. Table 2 presents the details of data set used for training and testing purposes.

Table 2: Details of the acquired data

| Data | No Fault | Fault | Total |
|---|---|---|---|
| **Train data** | 13300 | 10640 | 23940 |
| **Test data** | 5852 | 4683 | 10535 |
| **Total** | 19152 | 15323 | 34475 |

Wavelet based statistical features were obtained from the current signals. The wavelet multi resolution analysis (WMRA) was used to extract the approximate and detailed coefficients from the current signals. Statistical features were extracted from these coefficients. The statistical features forms patterns of fault and no fault signatures in R, Y, B phases implying a two class problem in these three phases. The current signals were collected for different load conditions, 0.5A, 1A, 1.5A, 2A, 2.5A, 3A and 3.5A. Data collected under different load conditions are combined together to develop load independent fault diagnosis system.

### *3.1 Baseline system using SVM kernels for the wavelet features*

The baseline system is developed using SVM kernels for the wavelets Daubechies (DBn) and Symlet (SYMn) where *n* represents the order of the respective wavelet. The different levels of decomposition are experimented to better the performance of the health monitoring system. The performance of the baseline system is listed in

Table 3. The experiments revealed that linear kernel performed better with wavelet DB8 level 5, DB8 level 10 for polynomial kernel, SYM2 level 5 for RBF kernel. For sigmoid kernel, the performance does not improve for the different wavelets and their decomposition levels.

Table 3: Performance of the baseline system

| Kernel | Wavelet | R Phase | Y Phase | B Phase |
|---|---|---|---|---|
| **Linear** | **DB8 Level 5** | **72.52** | **77.82** | **77.17** |
| **Polynomial** | DB8 Level 10 | 53.48 | 55.80 | 72.10 |
| **RBF** | SYM2 Level 5 | 57.90 | 57.10 | 56.29 |
| **Sigmoid** | DBn and SYMn | 55.55 | 55.55 | 55.55 |

It can be observed that the performance of baseline system is poor attributed by the use of non-linear features that does not match the kernels used. Further, the fault signatures differ widely with different load conditions. The performance of the baseline system can be improved by mapping the features onto a unique dimensional space while removing the load dependency of the features using the feature transformation techniques, LLC and PCA [13].

### *3.2 Improving the Fault Diagnosis System using PCA*

The covariance matrix is calculated from the mean centred data of the training dataset. The eigen vectors and the eigen values for the corresponding covariance matrix can be used to reconstruct the basis for the feature vectors. PCA allows for feature selection by varying the number of basis(principal components) for the fault diagnosis system. The number of principal components is varied to identify the optimal number for a better performance. Table 4 presents the performance of the PCA based fault diagnosis system using SVM for different kernels. PCA improves the performance of the best baseline system (linear kernel) by 19.07%, 6.60% and 20.86% respectively for R, Y, and B phases respectively. It can inferred that the elimination of redundant load dependent dimensions contributes to the improvement in performance of the health monitoring system. Inorder to further improve the performance of fault diagnosis, LLC is experimented on the linear baseline system to enchance the fault diagnosis system.

### *3.3 Improving the Fault Diagnosis System using LLC*

The codebook, set of basis vectors is first computed from the training data set by means of Lind-Buzo-Grey algorithm [25]. Then, the feature vectors from test and

Table 4: Performance of PCA-SVM kernel based fault diagnosis system

| Kernel | R Phase | Y Phase | B Phase |
|---|---|---|---|
| **Linear** | 76.03 | 84.42 | 86.69 |
| **Polynomial** | **91.59** | **81.93** | **98.03** |
| **RBF** | 48.57 | 46.95 | 53.33 |
| **Sigmoid** | 55.55 | 55.55 | 55.55 |

training data sets are represented by the $k$- nearest basis vectors as a linear combination by means of codes. Training and testing the classifier is done by using the same codebook. Multiple trials are conducted by varying size of codebooks and the number of neighbors for identifying the best performance. The LLC codes are used by SVM for training and testing. The experiments show that DB8 level 5 (Refer table 4) improves the baseline system performance. The improvement in performance is calculated as 25.87%, 21.47% and 21.78% for R, Y, and B phases respectively. It can be observed that the load dependency factors are eliminated with the help of load independent basis vectors (codebooks) for the representation of the input feature vector. This results in the LLC-SVM fault diagnosis system being more consistent and offers a better performance in comparison with the PCA based system.

Table 5: Improved Performance of DB8 Level 5 using LLC

| Phase | Codebook Size | 64 | 128 | 256 | 512 | 1024 | 2048 | 4096 |
|---|---|---|---|---|---|---|---|---|
| **R Phase** | **10** | 88.82 | 92.06 | 94.40 | 95.62 | 95.86 | 96.27 | 95.98 |
| | **20** | 90.86 | 93.66 | 96.21 | 97.10 | 97.58 | 97.47 | 97.13 |
| | **30** | 90.86 | 93.66 | 96.21 | 97.61 | 98.05 | 98.09 | 97.79 |
| | **40** | 90.86 | 93.66 | 96.21 | 97.61 | 98.18 | **98.39** | 98.20 |
| **Y Phase** | **10** | 84.45 | 91.93 | 94.57 | 96.26 | 96.96 | 96.85 | 95.12 |
| | **20** | 88.02 | 94.47 | 97.59 | 97.99 | 98.69 | 98.43 | 97.93 |
| | **30** | 89.91 | 94.47 | 97.59 | 98.79 | 99.11 | 98.95 | 98.60 |
| | **40** | 90.62 | 94.47 | 97.59 | 98.79 | **99.29** | 99.18 | 98.90 |
| **B Phase** | **10** | 85.40 | 93.83 | 93.23 | 93.44 | 94.59 | 94.75 | 95.30 |
| | **20** | 90.1 | 94.89 | 97.05 | 97.86 | 98.14 | 98.51 | 98.64 |
| | **30** | 92.89 | 94.89 | 97.57 | 98.50 | 98.55 | 98.91 | 98.94 |
| | **40** | 92.89 | 94.89 | 97.57 | 98.5 0 | 98.89 | **98.96** | 98.94 |

## 4 Conclusion

In this work, we developed a system to diagnose the inter-turn faults in a 3 kVA synchronous generator. Wavelet based statistical features were extracted from the current signals. The baseline system was developed using the Support Vector Machine (SVM) linear, polynomial, sigmoid and RBF kernels. From the experiments, it was observed that the baseline system performance was not encouraging since the features were non-linear. The performance of the baseline system was improved by feature transformation techniques such as Principal Component Analysis (PCA) and Locality-constrained Linear Coding (LLC). Experiments and results show that, PCA helps to improve the performance of the baseline system by 19.07%, 4.11% and 20.86% for the R, Y, and B phases respectively. We further experimented with LLC to map the features onto the higher dimensional space and then used linear SVM. LLC improves the performance of the baseline system by 25.87%, 21.47% and 21.78% for the R, Y, and B phases respectively. It was noted that LLC outperforms the PCA in improving the fault diagnosis of 3 kVA synchronous generator effectively using wavelet features.

## References


1. Andrew KS Jardine, Daming Lin, and Dragan Banjevic. A review on machinery diagnostics and prognostics implementing condition-based maintenance. *Mechanical systems and signal processing*, 20(7):1483–1510, 2006.
2. Haining Liu, Chengliang Liu, and Yixiang Huang. Adaptive feature extraction using sparse coding for machinery fault diagnosis. *Mechanical Systems and Signal Processing*, 25(2):558–574, 2011.
3. B Samanta and KR Al-Balushi. Artificial neural network based fault diagnostics of rolling element bearings using time-domain features. *Mechanical systems and signal processing*, 17(2):317–328, 2003.
4. YD Chen, R Du, and LS Qu. Fault features of large rotating machinery and diagnosis using sensor fusion. *Journal of Sound and Vibration*, 188(2):227–242, 1995.
5. Ruqiang Yan, Robert X Gao, and Changting Wang. Experimental evaluation of a unified time-scale-frequency technique for bearing defect feature extraction. *Journal of Vibration and Acoustics*, 131(4):041012, 2009.
6. Ruqiang Yan and Robert X Gao. Rotary machine health diagnosis based on empirical mode decomposition. *Journal of Vibration and Acoustics*, 130(2):21007, 2008.
7. Sylvain Verron, Teodor Tiplica, and Abdessamad Kobi. Fault detection and identification with a new feature selection based on mutual information. *Journal of Process Control*, 18(5):479–490, 2008.
8. Leo H Chiang, Mark E Kotanchek, and Arthur K Kordon. Fault diagnosis based on fisher discriminant analysis and support vector machines. *Computers & chemical engineering*, 28(8):1389–1401, 2004.
9. Kui Zhang, Yuhua Li, Philip Scarf, and Andrew Ball. Feature selection for high-dimensional machinery fault diagnosis data using multiple models and radial basis function networks. *Neurocomputing*, 74(17):2941–2952, 2011.
10. Achmad Widodo, Bo-Suk Yang, and Tian Han. Combination of independent component analysis and support vector machines for intelligent faults diagnosis of induction motors. *Expert Systems with Applications*, 32(2):299–312, 2007.

11. Vladimir Vapnik. *The nature of statistical learning theory*. Springer Science & Business Media, 2013.
12. BS Kim, SH Lee, MG Lee, J Ni, JY Song, and CW Lee. A comparative study on damage detection in speed-up and coast-down process of grinding spindle-typed rotor-bearing system. *Journal of materials processing technology*, 187:30–36, 2007.
13. Jinjun Wang, Jianchao Yang, Kai Yu, Fengjun Lv, Thomas Huang, and Yihong Gong. Locality-constrained linear coding for image classification. In *Computer Vision and Pattern Recognition (CVPR), 2010 IEEE Conference on*, pages 3360–3367. IEEE, 2010.
14. Jianchao Yang, Kai Yu, Yihong Gong, and Tingwen Huang. Linear spatial pyramid matching using sparse coding for image classification. In *Computer Vision and Pattern Recognition, 2009. CVPR 2009. IEEE Conference on*, pages 1794–1801. IEEE, 2009.
15. Kai Yu, Tong Zhang, and Yihong Gong. Nonlinear learning using local coordinate coding. In *Advances in neural information processing systems*, pages 2223–2231, 2009.
16. R Gopinath, C Santhosh Kumar, K Vishnuprasad, and KI Ramachandran. Feature mapping techniques for improving the performance of fault diagnosis of synchronous generator. In *International Journal of Prognostics and Health Management*, 6(2), pages 12p, 2015.
17. Gopinath R. Santhosh Kumar C. Vaisakh, B.K. and M. Ganesan. Condition monitoring of synchronous generator using sparse coding. *International Journal of Applied Engineering Research*, 10:26689–26697, 2015.
18. KT Sreekumar, R Gopinath, M Pushparajan, Aparna S Raghunath, C Santhosh Kumar, KI Ramachandran, and M Saimurugan. Locality constrained linear coding for fault diagnosis of rotating machines using vibration analysis. In *2015 Annual IEEE India Conference (INDICON)*, pages 1–6. IEEE, 2015.
19. Santhosh Kumar .C., Gopinath, R., and K.I. Ramachandran. Scalable fault models for diagnosis of synchronous generators. *Int. J. Intelligent Systems Technologies and Applications*, 15(1):35–51, 2016.
20. R Gopinath, C Santhosh Kumar, KI Ramachandran, V Upendranath, and PVR Sai Kiran. Intelligent fault diagnosis of synchronous generators. *Expert Systems with Applications*, 45:142–149, 2016.
21. R Gopinath, C Santhosh Kumar, V Vaijeyanthi, and KI Ramachandran. Fine tuning machine fault diagnosis system towards mission critical applications. In *Intelligent Systems Technologies and Applications*, pages 217–226. Springer, 2016.
22. R Gopinath, TNP Nambiar, S Abhishek, S Manoj Pramodh, M Pushparajan, KI Ramachandran, C Senthil Kumar, and R Thirugnanam. Fault injection capable synchronous generator for condition based maintenance. In *Intelligent Systems and Control (ISCO), 2013 7th International Conference on*, pages 60–64. IEEE, 2013.
23. Hüseyin Erişti, Ayşegül Uçar, and Yakup Demir. Wavelet-based feature extraction and selection for classification of power system disturbances using support vector machines. *Electric power systems research*, 80(7):743–752, 2010.
24. H. W. Kuhn and A. W. Tucker. Nonlinear programming. In *Proceedings of the Second Berkeley Symposium on Mathematical Statistics and Probability*, pages 481–492, Berkeley, Calif., 1951. University of California Press.
25. Yoseph Linde, Andres Buzo, and Robert M Gray. An algorithm for vector quantizer design. *Communications, IEEE Transactions on*, 28(1):84–95, 1980.

# Investigation of Effect of Butanol Addition on Cyclic Variability in a Diesel Engine Using Wavelets

Rakesh kumar Maurya; Mohit Raj Saxena

**Abstract** This study focuses on the experimental investigation of the cyclic variations of maximum cylinder pressure ($P_{max}$) in a stationary diesel engine using continuous wavelet transform. Experiments were performed on a stationary diesel engine at a constant speed (1500 rpm) for low, medium and high engine load conditions with neat diesel and butanol/diesel blends (10%, 20%, and 30% butanol by volume). In-cylinder pressure history data was recorded for 2000 consecutive engine operating cycles for the investigation of cyclic variability. Cyclic variations were analyzed for maximum cylinder pressure. The results indicated that variations in the $P_{max}$ is highest at lower load condition and decreases with an increase in the engine load. Global wavelet spectrum (GWS) power decreases with an increase in the engine operating load indicating decrease in cyclic variability with the engine load. The results also revealed that lower cyclic variations obtained with butanol/diesel blends in comparison to neat diesel.

**Keywords:** Cyclic variations; Wavelets; Diesel engine; Combustion; Cylinder pressure

## 1 Introduction

Combustion in a diesel engine is dependent on a number of factors like air-fuel ratio of combustible mixture, the flow characteristics of the mixture inside the cylinder, in-cylinder temperature, injection parameters of fuel, engine operating conditions, etc. These factors have a direct impact on the performance, combustion and emission characteristics of an internal combustion engine. Worldwide engine manufacturers have been trying to develop an engine with higher thermal efficien-

1



J.M. Corchado Rodriguez et al. (eds.), *Intelligent Systems Technologies and Applications 2016*, Advances in Intelligent Systems and Computing 530,
DOI 10.1007/978-3-319-47952-1_77

cy and lower emissions. Compression ignition (CI) engines using diesel, are commonly used in heavy- duty vehicles due to their higher power output and thermal efficiency. In order to meet stringent legislation limits, after-treatment devices have been commonly used [1]. Alternative fuel is one of the ways to achieve higher thermal efficiency with lower emissions. To achieve the stable combustion with alternative fuels, cyclic variations in combustion need to be minimized. The cyclic variation in combustion cycles may lead to increase in the emissions, reduce the engine power output, increases the engine noise and degrade the engine life [2, 3].Thus, it is required to diagnose and control the cyclic variations in CI engines fuelled with either neat diesel or any alternative fuel. Several studies [4, 5] indicated that biodiesel could be used directly in the CI engine to reduce the emissions. Conversely, biodiesel is not used significantly in CI engine due to the high nitrogen oxides ($NO_x$) emissions and durability issues such as deposition formation, fuel filter plugging, carbonization of the injector tip, etc. Presently researchers focused their attention towards butanol as an alternative fuel for CI engines due to its inimitable properties.

Several techniques (such as statistical methods, stochastic methods, wavelet transform etc.) were used for the analysis of cyclic variations in engines of different combustion strategies [6-10]. Wavelet transform is prominently used to investigate the cyclic variations due to its capability of analyzing signals in frequency as well as time domain simultaneously, which is the advantage of wavelet transform over the Fourier transform. A study experimentally investigated the effect of compression ratio and engine operating load on the cyclic variations in the total heat release (THR) and indicated mean effective pressure (IMEP) using wavelet transformation in CI engine [10]. Their results revealed that highest cyclic variation was observed at lower load condition while the engine was operated at lowest CR. Sen et al. [12] investigated the cyclic variations in mean indicated pressure at various rotational speeds of the CI engine. Their results indicated that the strong periodicities come out in lower frequency bands and has repeatability in several combustion engine cycles. Pan et al. [13] diagnosed the impact of CR, EGR and boost pressure on cyclic variations of port injected gasoline engine. Their results indicated that cyclic variations reduce with an increase in compression ratio and intake air pressure for a given EGR while higher cyclic variations were observed with increase in EGR ratio. Few studies [14-17, 22-24] conducted the analysis of cyclic variability in engine fuelled with alternative fuels.

Sen et al. [14] experimentally investigated the effect of direct injection of natural gas on the cyclic variability in the SI-engine. They used the CWT to diagnose the variations in IMEP at different compression ratios with two different engine operating speeds. Moreover, cross wavelet analysis was used to find out the interrelationship between the total combustion duration and IMEP, main combustion and flame development duration. Their results indicate that a strong interconnection of IMEP with total combustion and main combustion duration. In another study, Sen et al. [15] determined the impact of hydrogen addition with natural gas on the cyclic variations of natural gas in spark ignition engine. Discrete wavelet trans-form

(DWT) was applied for the investigation of variations in the time series of IMEP. They used root mean square (RMS) value to determine the cyclic variability at each discrete level. Their results indicated that in addition of hydrogen, RMS value of signal reduces at all discrete levels as compared to 100% natural gas. Baggio et al. [22] simulated the cyclic variations using quasi-dimensional simulations in spark ignition engine fuelled by gasoline-hydrogen blends. Their results show that amplitude of the oscillations decrease with increase in the hydrogen percentage for leaner operating condition. Ali et al. [23] investigated the cyclic variation in an engine fuelled with palm-biodiesel and diesel blends. They re-corded and analyzed the cylinder pressure data of 200 consecutive engine cycles. They found cyclic variations decreases with increase in the fraction of the bio-diesel in the blended fuel. The lowest cyclic variations were observed with 30% biodiesel in the blended fuel as compared to neat diesel, 10% and 20% biodiesel blends. In another study, Ali et al. [24] experimentally investigated the effect of the addition of diethyl ether in biodiesel-diesel blend on cyclic variations of combustion parameters in diesel engine. Experiments were conducted at 2500 rpm and cylinder pressure data recorded for 200 engine cycles. Results show that cyclic variation in IMEP increases with increase in the fraction of diethyl ether. Moreover, their results depicted intermediate and long periodicities in data series using diesel, while short periodicity obtained intermittently with biodiesel-diesel blend. A recent study experimentally investigated the cyclic variability of IMEP and THR in conventional diesel engine using symbol sequence analysis technique [25]. Results revealed that cyclic variations of IMEP and THR shifts from stochastic to deterministic behavior at different engine load conditions and higher number of cycles required for closed loop combustion control than just previous engine cycle.
However, it is necessary to fully understand the characteristics of cyclic variations in diesel engines. Several studies conducted on cyclic variations in diesel engine jointly cover fairly good ranges for some of the operating conditions. However, very few studies are conducted on conventional stationary diesel engine. The present study investigates the cyclic variability in a stationary diesel engine using wavelet transform at different engine operating load conditions and fuel blends. Cyclic variability in the maximum cylinder pressure ($P_{max}$) is investigated for neat diesel and 10%, 20% and 30% butanol/diesel blends.

## 2 Experimental Setup and Methodology

A single cylinder, stationary genset diesel engine was coupled with an eddy current dynamometer. The schematic of experimental test rig is shown in Fig.1. The neat diesel or butanol/diesel blended fuel was injected at 23° bTDC into the cylinder by a mechanically operated fuel injector. Experiments were conducted at constant speed of 1500 rpm at 0%, 50% and 100% (low, medium and high) engine operating load conditions. In-cylinder pressure history data were recorded for

2000cycles. In-cylinder pressure history data was measured by a piezoelectric pressure transducer which is mounted in the cylinder head. The crank angle encoder of resolution 1 CAD used for measuring crank angle position.

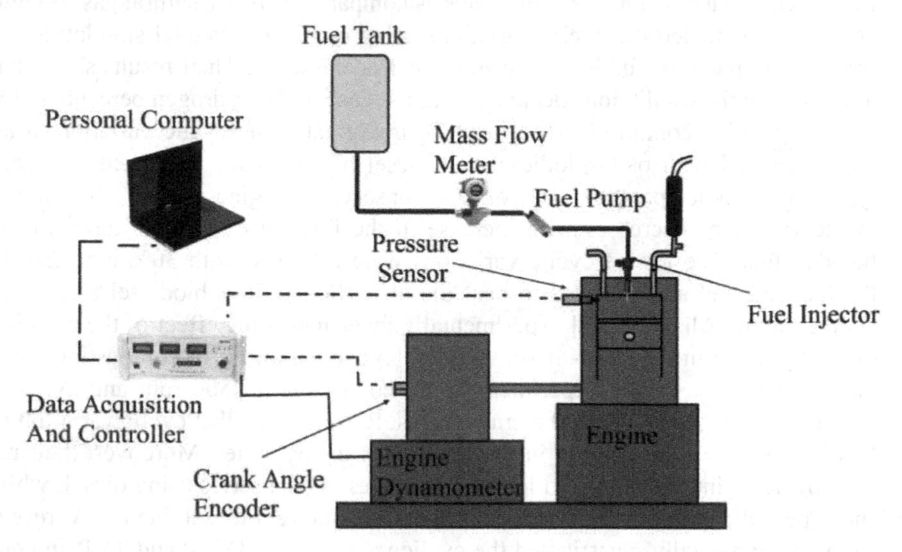


**Fig. 1.** Schematic of experimental test rig.

The continuous wavelet transform (CWT) with respect to a wavelet ψ(t) is given by equation (1).

$$CWT(a,b) = \frac{1}{\sqrt{|a|}} \int_{-\infty}^{\infty} x(t)\psi^*\left(\frac{t-b}{a}\right), a,b \in R, a \neq 0 \tag{1}$$

Where x (t) is continuous signal, ψ(t) is a mother wavelet with unit energy, ψ*(t) indicates its conjugate and 'a' and 'b' are scaling and translating parameter respectively.

The Wavelet Power Spectrum (WPS) provides information regarding the fluctuations of variances at different scales or frequencies. The magnitude of signal energy at a certain scale 'a' and particular location 'n' is obtained by the squared modulus of CWT. This is referred as wavelet power spectrum and is also presented as scalogram. WPS is normalized by dividing with  so that power relative to white noise is obtained and is shown in equation (2).

$$WPS = |CWT_n(a)|^2 \tag{2}$$

The normalized WPS is depicted in equation (3)

$$WPS_n = \frac{|CWT_n(a)|^2}{\sigma^2} \tag{3}$$

Where  is the standard deviation. A CWT may generally be complex function (real and imaginary part), so a modulus in that case would actually mean the amplitude of CWT. WPS depends on time as well as scale (frequency) represented by a surface. The contours of the surface can be plotted on a plane to obtain a time scale representation of WPS. Through WPS important information can be obtained, such as events with higher variances and their frequency of occurrences, and the duration of time for which they persist. This information can be used to modify and control the system.

Global wavelet spectrum (GWS) is the time average of the WPS and calculated by the equation (4)

$$GWS = W_s = \frac{1}{N}\sum_{n=1}^{N}\left|CWT_n(a)\right|^2 \quad (4)$$

The Global wavelet spectrum is represented by $W_s$. The peak locations in global wavelet give an indication about the dominant periodicities in the time data series.

## 3 Results and Discussion

This section presents the analysis of the cyclic variations of maximum cylinder pressure ($P_{max}$) in stationary diesel engine operated on diesel butanol blends. Continuous wavelet transform (CWT) is used to analyze the variations in engine combustion cycles. Fig. 2 shows the time series of $P_{max}$ for neat diesel and butanol/diesel blends (10%, 20%, and 30% butanol) at low, medium and higher load conditions at a constant engine speed of 1500 rpm. Fig.2. indicates that mean value for $P_{max}$ increases with the increase in engine load and highest for full (100%) load condition because of the higher amount of fuel burned in the engine cycle at higher load condition. Higher cycle-to-cycle variations appear at lower load conditions (fig. 2) in comparison to higher engine load. At lower engine load, combustion temperature is lower which may lead to higher variation as diesel combustion is initiated by auto-ignition. Wavelet transform can be used to analyze the cyclic variations as it gives the frequency and time information simultaneously [16, 18, 19]. By using CWT, WPS and GPS are generated from the time series plots of $P_{max}$. Fig.3-6 shows the spectrograms (including WPS and GPS) for neat diesel and butanol/diesel blends. The squared modulus of CWT indicates the energy of the signal, which is typically known as WPS [16].

In WPS, the horizontal axis shows the number of engine cycles, while the vertical axis represents the periodicity of time series (inverse frequency) and sequential variations. In GPS, horizontal axis denotes the power while the vertical represents the period. Spikes in the GWS curve represent the prevailing periodicities in the time series. Colors in the spectrogram represent the energy of the signal, which denotes the magnitude of wavelet transforms [20]. Colors in the WPS indicate the frequency of the signal. Red color denotes the maximum energy of the signal while the blue color represents the minimum energy of the signal. Cone of Influence (COI) is the region where edge effects become important and was defined by Torrence et al. [18] as e-folding time for autocorrelation of WPS at each scale. The e-folding time is taken in such a way that at the edges where discontinuities exist, the magnitude of wavelet power reduces by a factor $e^{-2}$. The region inside the COI is considered and the region outside this is neglected.

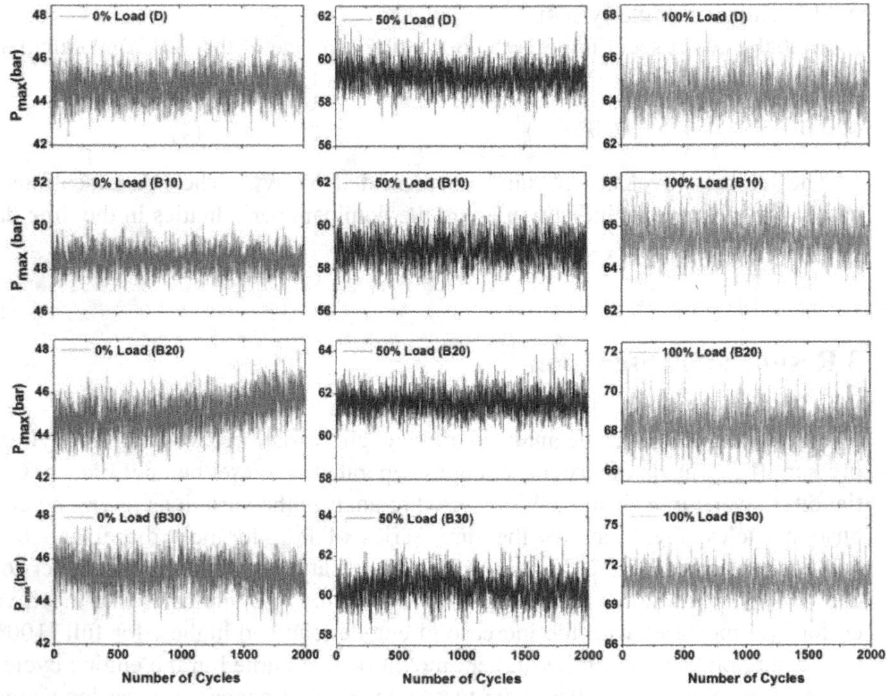


**Fig. 2** Time series plots of $P_{max}$ for neat diesel and butanol/diesel blends at different load conditions.

GWS and WPS for the lower load condition is shown in fig.3(a).It can be depicted from the fig.3(a) that the periodic band of 256-520, have highest cyclic variability during cycles of 117-1328 (highest color intensity). It can also be depicted from fig.3(a) that moderated intensity periodic bands of 4-8 occur intermittently throughout the cycles. The occurrence of strong periodicity bands in the spectrogram indicates the higher cyclic variations. WPS and GWS for medium load condition is shown in fig.3(b).Fig.3(b) indicates the moderated and strong periodicity of 11-20, 16-40 and 20-64 periods occur in the cycle range of 138-251, 734-835 and 1174-1280 respectively.WPS and GWS for higher load condition is shown in fig.3(c). Fig.3(c) indicates that the moderated periodicity band of 16-32 and 26-40 occur between the cycle range of 963-1025 and 1705-1801 respectively. A moderate periodicity band of 4-8 period was observed intermittently during the cycles at all operating load conditions. It is very interesting to note that in all figures of GWS (fig. 3(a), 3(b) and 3(c)), peaks or spikes of power reduces with an increase in engine operating load which indicates a reduction in cyclic variations with an increase in engine operating load. At lower load condition, comparatively lower amount of fuel is burnt in the combustion chamber leading to the lower combustion temperature. At lower load condition, lower cylinder wall temperature could increase the ignition delay which may lead to higher cyclic variability.

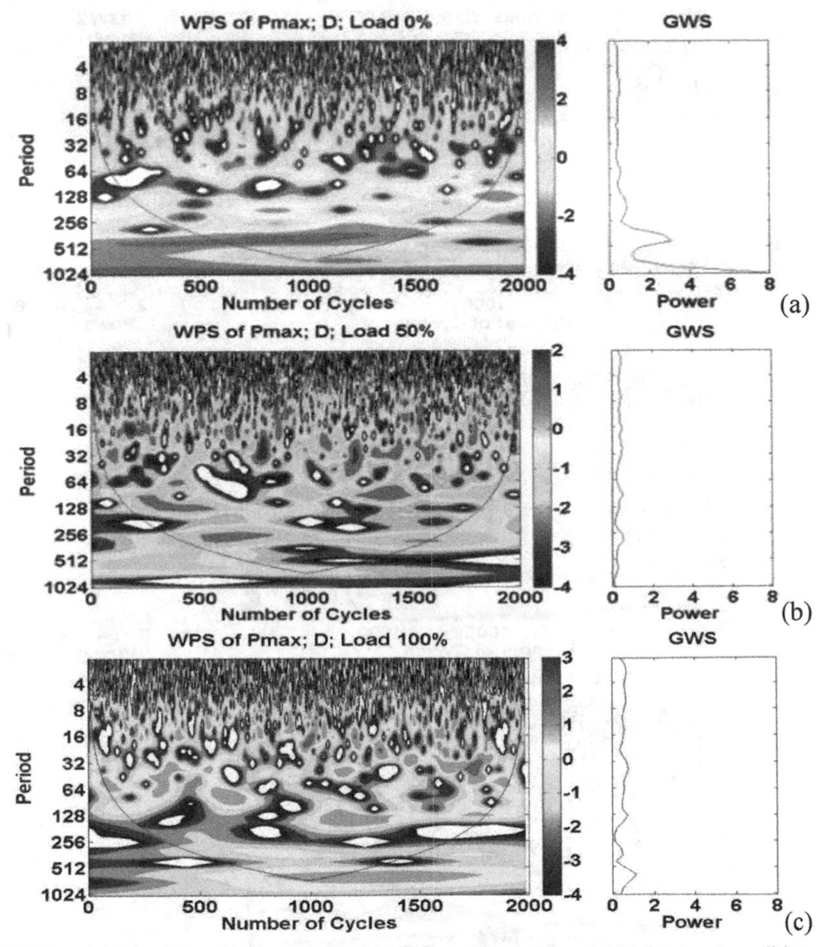


**Fig.3** WPS and GPS of $P_{max}$ for neat diesel at (a) lower; (b) medium; (c) higher load conditions.

Fig. 4(a) shows the WPS and GWS for 10% butanol and diesel blends at lower load condition. It can be depicted from the figure that the moderated periodicity of 55-125 period occurs in the cycle ranges from 298-559. It can also be noticed from fig.4(a) that moderated periodicity occurs in a lower periodic zone as compared to neat diesel (as shown in fig.3(a)). From fig.4(a) it can also be observed that lower peak power achieved in GWS (as compared to neat diesel in fig.3(a)) indicates lower cyclic variations as compared to neat diesel at lower load condition. It might be due to the higher oxygen content in the fuel blend, which enhance the atomization of fuel causes more fuel burnt in the cylinder, which reduces the cyclic variations. Similarly, on increasing the load cyclic variability reduces as shown in fig. 4(b) and 4(c).In fig.4, GWS power slightly increases with load. It is because of the amplification of the GWS power at higher period [21].

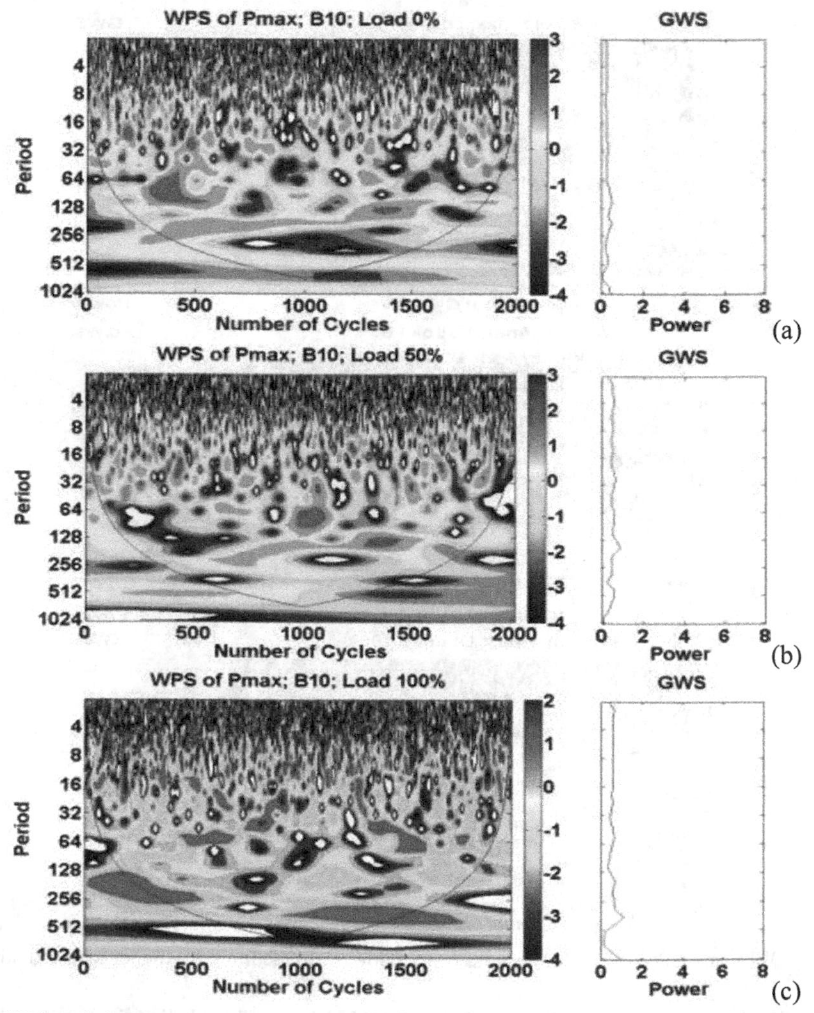


**Fig.4** WPS and GPS of $P_{max}$ for neat 10% butanol/diesel blend at (a) lower; (b) medium; (c) higher load conditions.

Fig.5 shows the WPS and GWS for 20% butanol and diesel blends at low, medium and higher load conditions. It can be observed from fig.5(a) that the moderate and strong periodicity of 120-240 and 512-1024 period occurs in the cycle ranges of 274-488 and 1177-2000 respectively. Since the strong periodicity of

512-1024 period occurs below the cone of influence, this result cannot be considered for analysis.

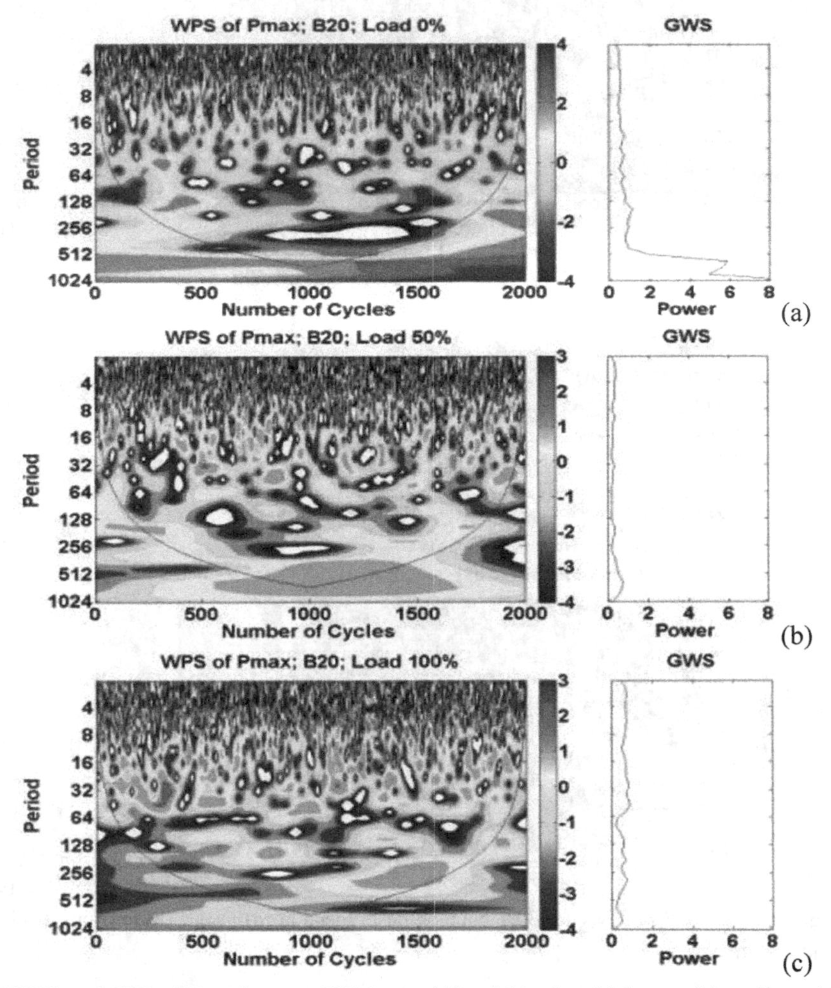


**Fig.5** WPS and GPS of $P_{max}$ for neat 20% butanol/diesel blend at (a) lower; (b) medium; (c) higher load conditions.

The large peak in GWS (in fig.5(a)) might be because of the amplification of the GWS power at higher period [21]. At moderated and higher load conditions, the moderated periodicity observed in lower period zone and GWS power reduces with increases with an increase in engine operating load (as shown in fig.5).

WPS and GWS for 30% butanol and diesel blends at low, medium and higher load condition is shown in fig.6. Similar results were obtained with 30% butanol/diesel blend as with 10% and 20% butanol/diesel blends. In the fig.6(a), moderated periodicity of 320-512 period occurs in the cycle range of 117-1325.

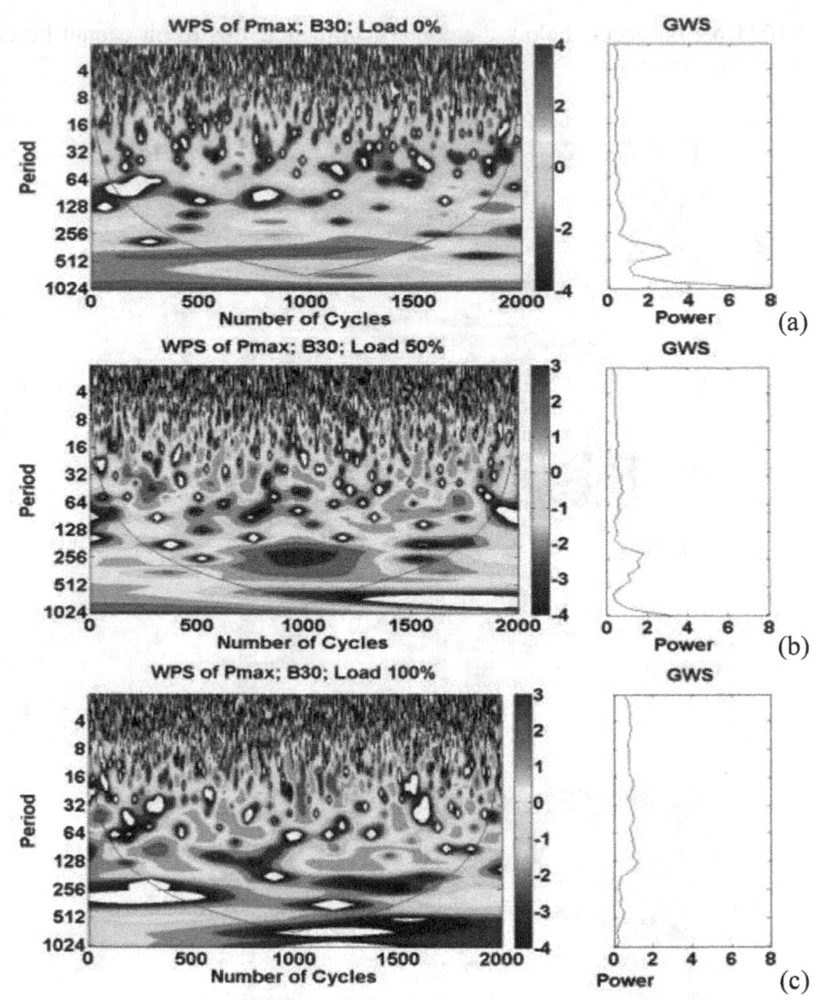


**Fig.6** WPS and GPS of $P_{max}$ for 30% butanol/diesel blend at (a) lower; (b) medium; (c) higher load conditions

Fig.6(b) shows the WPS and GWS for 30% butanol and diesel blends at medium load conditions. The strongest periodicity of 32-64 and 130-265 period occur between the cycle ranges of 269-310 and 823-1165 respectively. WPS and GWS for 30% butanol and diesel blends at higher load conditions is shown in fig.6(c). It is observed from the fig.6 that the GWS power is reduced with an increase in engine operating load which indicates that reduction in cyclic variability with engine load. Furthermore, from fig.4-6 it is also observed that peak GWS power decrease with increases in the fraction of butanol/diesel blends up to the 20% butanol in the blend. On further increasing the fraction of butanol in the blend (30% butanol/diesel blend) the GWS power slightly increases, which suggests an increase in the cyclic variability. This might be because of lower cetane number of fuel blend which leads to unstable combustion.

# 4 Conclusions

Present study experimentally investigated the cyclic variations in peak cylinder pressure for stationary diesel engine with neat diesel and three butanol/diesel blends at different engine load conditions using wavelet transform. It was found that lower load condition has a higher GWS power and hence cyclic variability is also higher. Higher engine load cyclic variability decreases as GWS power decreases. It was also found that at lower load conditions, cyclic variations observed at higher periods while at higher load condition, cyclic variability shifted toward the lower periodic zone. Results revealed that an increase in the fraction of butanol (up to 20% butanol) in the blend cyclic variability reduces as compared to neat diesel fuel. Further increasing the fraction of butanol in the blend (30% butanol) cyclic variability slightly increases.

# References


1. Johnson, T.V.: Review of Diesel Emissions and Control 2010-01-0301(2010).
2. Atkins, R.D.:An introduction to engine testing and development. Warrendale, SAE International (2009).
3. Pundir, B.P., Zvonow, V.A., Gupta, C.P.: Effect of charge non-homogeneity on cycle-by-cycle variations in combustion in SI engines. No. 810774. SAE Technical Paper (1981).
4. Raheman, H., Jena, P.C., Jadav, S.S.: Performance of a diesel engine with blends of biodiesel (from a mixture of oils) and high-speed die-sel. International Journal of Energy and Environmental Engineering. 4(1), 1-9(2013).
5. Dhar, A., Agarwal, A.K.: Experimental investigations of effect of Karanja biodiesel on tribological properties of lubricating oil in a compression ignition engine. Fuel 130, 112-119(2014).
6. Sen, A.K., Litak, G., Kaminski, T.,Wendeker, M.: Multifractal and statistical analyses of heat release fluctuations in a spark ignition engine. Chaos: An In-terdisciplinary Journal of Nonlinear Science 18(3), 033115:1-6(2008).
7. Rakopoulos, D.C., Rakopoulos, C.D., Giakoumis, E.G., Papagiannakis, R.G., Kyritsis, D.C.: Experimental-stochastic investigation of the combustion cyclic variability in HSDI diesel engine using ethanol–diesel fuelblends. Fuel, 87(8), 1478-1491 (2008).
8. Maurya, R.K., Agarwal, A.K.: Statistical analysis of the cyclic variations of heat release parameters in HCCI combustion of methanol and gaso-line. Appl.Energ. 89(1), 228-236 (2012).
9. Daw, C.S., Finney, C.E.A., Green, J.B.,Kennel, M.B., Thomas, J.F., Connolly, F.T.: A simple model for cyclic variations in a spark-ignition engine." SAE transactions 105, 2297-2306(1996).

10. Maurya, R.K., Saxena, M.R., Nekkanti, A.: Experimental Investigation of Cyclic Variation in a Diesel Engine Using Wavelets.Intelligent Systems Technologies and Applications, Advances in Intelligent Systems and Computing, Springer International Publishing 384, 247-257(2016).
11. Maurya, R.K., Agarwal, A.K.: Experimental investigation of the effect of the intake air temperature and mixture quality on the combustion of a methanol-and gasoline-fuelled homogeneous charge compression ignition engine.Proc. Inst. Mech. Eng. D J. Automob. Eng.223(11), 1445-1458(2009).
12. Sen, A.K., Longwic, R.,Litak, G., Gorski, K.: Analysis of cycle-to-cycle pres-sure oscillations in a diesel engine. Mech. Syst. Signal Pr. 22(2), 362-373(2008).
13. Pan, M., Shu, G., Wei, H., Zhu, T., Liang, Y., Liu, C.: Effects of EGR, com-pression ratio and boost pressure on cyclic variation of PFI gasoline engine at WOT operation. Appl. Therm. Eng.64(1), 491-498(2014).
14. Sen, A.K., Zheng, J., Huang, Z.: Dynamics of cycle-to-cycle variations in a natural gas direct-injection spark-ignition engine. Appl. Energ. 88(7), 2324-2334(2011).
15. Sen, A.K., Wang, J., Huang,Z.: Investigating the effect of hydrogen addition on cyclic variability in a natural gas spark ignition engine: Wavelet multireso-lution analysis. Appl. Energ. 88(12), 4860-4866(2011).
16. Sen, A.K., Ash, S.K., Huang, B., Huang, Z.: Effect of exhaust gas recircula-tion on the cycle-to-cycle variations in a natural gas spark ignition en-gine. Appl. Therm. Eng. 31(14), 2247-2253(2011).
17. Selim, M.Y.E.: Effect of engine parameters and gaseous fuel type on the cy-clic variability of dual fuel engines. Fuel 84(7), 961-971(2005).
18. Torrence, C., Compo, G.P.: A practical guide to wavelet analysis. Bull. Amer. Meteor. Soc. 79(1), 61-78(1998).
19. Wu, J.D., Chen, J.C.: Continuous wavelet transform technique for fault signal diagnosis of internal combustion engines. NDT & E International 39(4), 304-311(2006).
20. Tily, R., Brace,C.J.: Analysis of cyclic variability in combustion in internal combustion engines using wavelets. Proc. Inst. Mech. Eng. D J. Automob. Eng. 225(3), 341–353(2011).
21. Wu, S., Liu, Q.: Some problems on the global wavelet spectrum.J. Ocean Univ. China, 4(4), 398-402(2005).
22. Martínez-Boggio, S.D., Curto-Risso, P.L., Medina, A., Hernández, A.C.: Simulation of cycle-to-cycle variations on spark ignition engines fueled with gasoline-hydrogen blends. Int. J. Hydrogen Energ.,41(21), 9087-9099 (2016).
23. Ali, O.M., Mamat, R., Abdullah, N.R., Abdullah, A.A.: Analysis of blended fuel properties and engine performance with palm biodiesel–diesel blended fuel. Renew.Energ., 86, 59-67 (2016).
24. Ali, O.M., Mamat, R., Masjuki, H.H.,Abdullah, A.A.: Analysis of blended fuel properties and cycle-to-cycle variation in a diesel engine with a diethyl ether additive. Energ. Convers. Manage., 108, 511-519 (2016).
25. Maurya, R.K.: Investigation of Deterministic and Random Cyclic Patterns in a Conventional Diesel Engine Using Symbol Sequence Analysis. In Advanced Computing and Communication Technologies, Springer Singa-pore, 549-556 (2016).

# Intelligent Energy Conservation: Indoor Temperature Forecasting with Extreme Learning Machine

Sachin Kumar, Saibal K Pal and Ram Pal Singh

**Abstract** At present, most of the buildings are using process of heating, ventilation and air conditioning(HVAC)-systems. HVAC systems are also responsible for consumption of huge amount of energy. Home automation techniques are being used to reduce the waste of resources especially energy that is available to us in the form of temperature, electricity, water, sunlight, etc. Forecasting and predicting the future demand of the energy can help us to maintain and to reduce the cost of energy in the buildings. In this paper, we use the experiments the small medium large system (SMLsystem) which is the house built at the university of CEU cardinal Herrera (CEU-UCH) for competition named Solar Decathlon 2013. With the data available from this experiments, we try to predict and forecast the future temperature condition intelligently for energy conservation with the model based on Extreme Learning Machine(ELM). This will help in determining the energy needs of the buildings and further will help in efficient utilization and conservation of energy.

**Key words:** ELM . OSELM . Energy Conservation . Indoor Temperature . Prediction . Intelligent System

---

Sachin Kumar
Department of Computer Science, University of Delhi, Delhi, India
e-mail: sachin.blessed@gmail.com

Saibal K Pal
SAG Lab, Matcalfe House, Defence Research Development Organisation, Delhi, India
e-mail: skptech@yahoo.com

Ram Pal Singh
Department of Computer Science, Deen Dayal Upadhaya College, University of Delhi, Delhi, India.
e-mail: rprana@gmail.com



J.M. Corchado Rodriguez et al. (eds.), *Intelligent Systems Technologies and Applications 2016*, Advances in Intelligent Systems and Computing 530,
DOI 10.1007/978-3-319-47952-1_78

# 1 Introduction

Across the world, the building sector consumes more energy in the form of electricity as around 40 per cent [2] [16]. Recent research and development across the globe has stated that energy consumed in buildings represents a 40 per cent rate of total energy consumed. In this more than a 50 percent is consumed by the process of Heating, Ventilation and Air Conditioning. All these processed combined together are called HVAC systems [9]. In addition to this, the Spanish Institute for Diversification and Saving of Energy (IDAE) [11] supported by the Spanish Government gave details that Spanish households consumption is 30 per cent of total energy expenditure of the country. Such details and figures compel the governments to deliberate on how to improve the energy consumption for the future [2]. Building construction and maintenance and energy use in it is dependent on the type of building, location and material used and other factors. HVAC systems could exemplify up to 40 per cent of the total energy intake of a building [9]. The activation and deactivation of HVAC systems depends on the comfort parameters that have been established. The most important factor is the indoor temperature which is directly related with the notion of comfort. Important research is going and state of art can be found in [9]. In order to know more accurately, to do research and develop and suggest policy related answers, The School of Technical Sciences of Universidad CEU-UCH built a solar-powered house that is known as Small Medium Large System (SMLsystem) [2] in order to participate in the Solar Decathlon Europe competition 2012 [1] [2] [11]. The house integrates a whole range of different technologies to improve energy efficiency. The objective of such construction was to build a near zero-energy house. The aim is to design and build houses that consume as few natural resources as possible while minimize the production of waste products during their life cycle. Special emphasis is placed on reducing energy consumption and obtaining all the needed energy from the sun [11]. While ensuring energy efficiency it must be taken into account that a person could much easier recognize its activities if its comfort is ensured and there are no negative factors involved such as cold, heat, low light, noise, low air quality, etc. which could disturb the person. The concept of a smart or intelligent environment highlights technical solutions based on sensor networks, artificial intelligence, machine learning and agent-based software [12].To implement energy related intelligent forecasting and then taking decision based on insights of such intelligent systems can help in energy conservation. In fact, machine learning techniques have been widely used for a range of applications devoted to model of energy systems [15] [16] [18], through the estimation of artificial neural network (ANN) models by using historical records. This paper focuses on multivariate forecasting using as input feature's different weather indicators, based on previous work [18]. From energy efficiency perspective, Indoor temperature forecasting is an attractive problem that has been widely studied in the literature in [2] [3] [15]. A forecasting model developed in one particular location can give high performance but in some other location could be totally useless. It may be because of geographical weather conditions, using different constructive materials, energy transfer and HVAC systems [2]. Algorithms and technologies which are dynamic and can adapt to such changes needs to be developed. On-line learning algorithms [4] are needed to learn a model from scratch or to adapt a pre-trained model in such scenarios. In prediction, Artificial neural networks (ANNs) have been applied extensively [9] [16]. More developments in ANNs have helped in improvement over previous basic ANN architectures which have introduced Deep architectures with many levels of non-linearity and the ability of representing complex features from its inputs [17]. All this advances helped in development of advanced models that have theoretical ability to learn very complex features and in achieving better generalization performance. Empirical studies such as [4] have demonstrated that training deep networks that have more than 2 or 3 layers using random weights initialization and supervised training provide worse results than training shallow architectures. Machine Learning Techniques such as Extreme Learning Machine (ELM) [5], Online Sequential Extreme Learning Machine (OS-ELM) [14] could be used in such type of learning. This paper presents a study of ELM [8]and OS-ELM [14], and their ability to learn a real indoor temperature forecasting task. In the future it is expected that deep ANNs could learn trend and seasonality of time series improving the overall efficiency of the experimental framework.

## 2 Extreme Learning Machine

In our research paper, we used ELM to analyse the given time series as ELM training process is a simple linear regression and predict the future values by training ELM using $M$ training samples $X = (x_{i1}, x_{i2}, ..., x_{iM})\ \varepsilon\Re^M$ and producing the $N$ output values $Y = (y_{i1}, y_{i2}, ..., y_{iN})$. In this section we shall discuss ELM algorithm proposed by Huang et al. [8] [7] for (SLFNs) for solving classification and regression problems. However, in ELM selection of optimal number of hidden layer neuron is time consuming and may be removed by using kernel based ELM. For $M$ random samples $(x_i, y_i)\varepsilon\Re^M \times \Re^N$ where $x_i$ is $M \times 1$ input vector and $y_i$ is $N \times 1$ output vector. Let us assume SLFNs has $M$ input samples and $N$ output layers with $k$ neurons and $P$ hidden neurons. Let $G$ be a real value function such that $G(w_i, b_i, x)$ will be the output of the $i^{th}$ hidden neuron with respect to the input vector $X$ and $b_i\varepsilon\Re$ is the bias corresponding to the input vector $X$ and weight vector $w_i = (w_{i1}, w_{i2}, ..., w_{iM})$ where $w_{ij}$ is weight of the connection between the $i^{th}$ input neuron and the $j^{th}$ hidden neuron of the input layer. The output function $Y$ will be given by

$$Y_P(x_j) = \sum_{i=1}^{P} \eta_i G(w_i, b_i, x_j), \quad j = 1, \cdots, M \tag{1}$$

where,$\eta$, is output weight matrix. Equation (1) can be rewritten as

$$\mathbb{Z}\eta = Y \tag{2}$$

where

$$\mathbb{Z}(\hat{w},\hat{b},\hat{x}) = \begin{bmatrix} G(w_1,b_1,x_1) & \cdots & G(w_P,b_P,x_1) \\ \vdots & \ddots & \vdots \\ G(w_1,b_1,x_M) & \cdots & G(w_P,b_P,x_M) \end{bmatrix}_{M\times P} \tag{3}$$

where

$$\hat{w} = \{w_1,w_2,\cdots,w_P\}$$
$$\hat{b} = \{b_1,b_2,\cdots,b_P\}$$
$$\hat{x} = \{x_1,x_2,\cdots,x_M\}$$

$$\eta = \begin{bmatrix} \eta_1^T \\ \eta_2^T \\ \vdots \\ \eta_P^T \end{bmatrix}_{P\times N} \quad and \quad Y = \begin{bmatrix} y_1^T \\ y_2^T \\ \vdots \\ y_M^T \end{bmatrix}_{M\times N} \tag{4}$$

where $\mathbb{Z}$ is hidden layer output matrix of (SLFNs) with $i^{th}$ column of $\mathbb{Z}$ being the $i^{th}$ hidden neuron output with respect to the input $\hat{x}$. As ELM requires no tuning of its parameters $w_i$ and $b_i$ during the training period, as they are randomly assigned and Equation (2) becomes a linear equation whose output weights can be estimated as follows

$$\eta = \mathbb{Z}^{\dagger} Y \tag{5}$$

where $\mathbb{Z}^{\dagger}$ is Moore-Penrose generalised inverse [13] of hidden layer output matrix $\mathbb{Z}$. It can be calculated when $\mathbb{Z}^T\mathbb{Z}$ is non singular $\mathbb{Z}^{\dagger} = (\mathbb{Z}^T\mathbb{Z})^{-1}\mathbb{Z}^T$ For the additive hidden neurons the activation functions $g(x) : \Re \to \Re$ such as radial bias function (RBF), sigmoid function (Sig), sinusoid function (sin), $G(w_i,b_i,x)$ is given by

$$G(w_i,b_i,x) = g(w_i \cdot x + b_i), b_i \varepsilon \Re \tag{6}$$

where $w_i$ is the weight vector connecting the input layer to the $i^{th}$ hidden neuron and $b_i$ is the bias of the $i^{th}$ hidden neuron , $w_i \cdot x$ denotes the inner product of the vector $w_i$ and $x$ in $\Re^n$.

## 3 Online Sequential ELM (OSELM)

Online sequential ELM introduced by Liang et al. [14] was an improvement over ELM as it can handle the real time data more efficiently. In certain situations the entire data set may not be available initially but keeps arriving at different intervals of time which OSELM can handle such data effectively. OSELM can learn from the training samples by taking one element at a time or by grouping the samples into varying size chunks [14]. In OSELM the input weights and biases are randomly generated and fixed and based on this the output weight matrix $\eta$ is determined using the recursive least squares algorithms.

$$\tilde{\eta} = \mathbb{Z}^{\dagger} Y \tag{7}$$

where $rank(\mathbb{Z}) = P$, $P$ is the number of hidden neurons. Sequential solution of Equation (7) results in implementing OSELM.

**Learning Algorithm of OSELM:** Given training sample $\mathbb{N} = \{(x_i, y_i) \| x_i \varepsilon \Re^M, y_i \varepsilon \Re^N, i = 1, \cdots, N\}$ where $N$ training data samples are available initially [5] [6] [8].

1. **Initialization Phase:**
   Take a small chunk of training data $\mathbb{N}_0 = \{(x_i, y_i) \| i = 1, \cdots, N_0\}$ from the given training set $\mathbb{N}$ where $N_0 \geq P$

   a. Arbitrarily generate the hidden neuron parameters$(w_i, b_i), i = 1, .P$
   b. Compute initial hidden layer output matrix $\mathbb{Z}_0$

$$\mathbb{Z}_0 = \begin{bmatrix} G(w_1 b_1 x_1) & \cdots & G(w_P b_P x_1) \\ \vdots & \ddots & \vdots \\ G(w_1 b_1 x_{N_0}) & \cdots & G(w_P b_P x_{N_0}) \end{bmatrix}_{N_0 \times P} \tag{8}$$

   c. Approximate the initial output weight matrix $\eta^{(0)}$ where it is given

$$\eta^{(0)} = \mathbb{Z}_0^{\dagger} Y_0 \tag{9}$$

   where $\mathbb{Z}_0$ and $Y_0$ is defined as $\mathbb{Z}_0^{\dagger} = (\mathbb{Z}_0^T \mathbb{Z}_0)^{-1} \mathbb{Z}_0^T$, $\mathbb{G}_0 = \mathbb{Z}_0^T \mathbb{Z}_0$ and $Y_0 = \{y_1, \cdots, y_{N_0}\}^T$
   d. Set $k = 0$

2. **Sequential Learning Phase:**
   a. Take $(k+1)^{th}$ chunk of training data

$$\mathbb{N}_{k+1} = \{(x_i, y_i) | i = (\sum_{j=0}^{k} N_j) + 1, \cdots, (\sum_{j=0}^{k+1} N_j) \tag{10}$$

   where each chunk is of varying sizes $N_0, N_1, \cdots, N_{k+1}$.
   b. Set $Y_{(k+1)}$ as given by array

$$Y_{(k+1)} = \begin{bmatrix} y^T_{(\Sigma_{j=0}^{k} N_j + 1)} \\ \vdots \\ y^T_{(\Sigma_{j=0}^{(k+1)} N_j)} \end{bmatrix}_{N_{(k+1)} \times N} \tag{11}$$

   c. Compute $\mathbb{Z}_{(k+1)}$ as given at bottom of the page[1]

1

d. Compute $(k+1)^{th}$ output weight matrix $\eta^{(k+1)}$ sequentially using the following equation

$$\eta^{(k+1)} = \eta^{(k)} + \mathbb{G}^{-1}_{(k+1)} \mathbb{Z}^T_{(k+1)} (Y_{(k+1)} - \mathbb{Z}_{(k+1)} \eta^{(k)}) \tag{13}$$

where $\mathbb{G}_{k+1} = \mathbb{G}_k + \mathbb{Z}^T_{k+1} \mathbb{Z}_{k+1}$

e. Set k=k+1. Goto step 2(a)

## 4 Time Series Forecasting

Time series is defined as the sequences of data points from any process continuously over the period of time which shows trends and patterns. Formalization can be done as a sequence of scalars from a variable/input $x$ obtained as output of the observed process: $(x) = r_0(x), r_1(x), ..., r_{i-1}(x), r_i(x), r_{i+1}(x), ....$ In the time series, we are interested in the value of one variable whose value might be dependent on itself only or on some other attributes. This variable in which we are interested is called forecasted variable and other variables which contribute to the changing value of variable are called covariates. If value of the forecasted variable $x$ is depended on its previous values only such as $r(x) = r_0(x), r_1(x), r_{i1}(x), r_i(x), r_{i+1}(x), ...$ where $i$ is up to $N$ then it is called Univariate time series forecasting. Opposite to this, when $r(x) = r(x), r(y_i), r(y_2)....r(y_M)$, where $y_i$ are additional variables and $M$ is no of additional variables, it is called multivariate forecasting. In any forecast model function $F$, $F$ receives some inputs as some past values of some variables. These variables are selected by some information gains approaches to ensure that they contribute to that variables. Such contributing variables can be called covariates and denotes as $(y_1, y_2, ....., y_M)$where $M$ is total no of such variables. The forecast function model produces as output in one values or window of values. If it is window of the values, then Window can be represented as Z. In mathematical form, it is like $\hat{s}_{t+1}(x0), \hat{s}_{t+2}(x0), ...., \hat{s}_{t+Z}(x0) = F((y_0), (y_1), ...., (y_C))$ where F is some functional combination of values of covariates. In this equation, $(y_i) = s^t_{t-I(x)+1}$ where $I(x)$ represents past values of variable $x$, and $t$ the current time instant. The number of past values $I(x)$ and no of covariates are important to ensure good performance of the model but these things are not easy as it all depends on estimation. Different type of approaches are available in the literature, estimation of $I(x)$ by using linear or non-linear auto-correlation statistics, estimation of $I(x)$ by trial-and-error, and a combination of models varying the value of $I(x)$ [18].

---

$$\mathbb{Z}_{(k+1)} = \begin{bmatrix} G(w_1, b_1, x_{(\sum_{j=0}^{k} N_j+1)}) & \cdots & G(w_P, b_P, x_{(\sum_{j=0}^{k} N_j+1)}) \\ \vdots & \ddots & \vdots \\ G(w_1, b_1, x_{(\sum_{j=0}^{k} N_j+1)}) & \cdots & G(w_P, b_P, x_{(\sum_{j=0}^{k} N_j+1)}) \end{bmatrix}_{N_0 \times P} \tag{12}$$

## 4.1 Evaluation Determinants

Now after the implementation of the model, we have two values of forecasted variable: forecasted and actual values. The performance of Model is analyzed by using the determinants like the root mean square error (RMSE) function. Mathematical Formulation for RMSE

$$RMSE(t) = sqrt(1/W) \sum_{p=1}^{W} (s_{t+w}(x_0) - \hat{s}_{t+z}(x_0))^2 \tag{14}$$

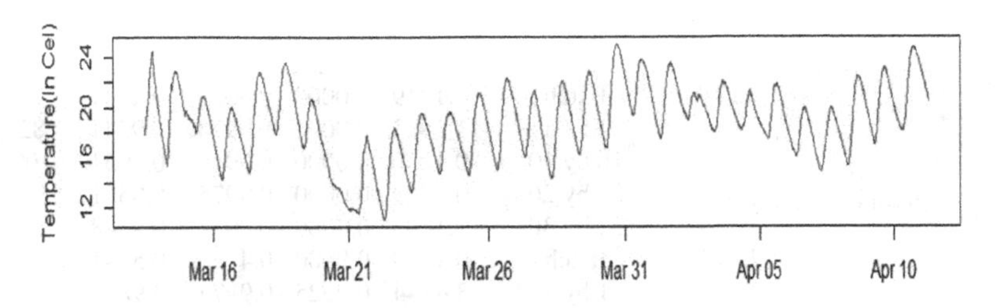


Fig. 1: SMLsystem Indoor Temperature Time Series for Year 2011 Line wise

## 4.2 Data Description and Preprocessing

The data is collected from the experiment of the SML2010 [1] which was responsible for generating time series data sets using sensors in house. It corresponds to around 40 days of monitoring data. The data was sampled every minute and it was computed and uploaded with mean values of 15 minute. For each hour computed mean is at 0 min, 15 min, 30 min, and 45 min. Data have total no of 24 attributes and a total training and testing samples 2764(28 days) and 1373(14 days) testing samples respectively. One time feature and five sensor signals were taken into formulation by [18] but the three Indoor Temperature, Hour feature and Sun irradiance were found to be the most important. All the above described variables have high correlation coefficient with temperature. The data set can be downloaded from the UCI machine learning repository [10]. In order to prepare data for prediction task, there is need for the preprocessing of data. Both the time series are generated to predicted with some step ahead prediction. This prediction window is denoted by $V$ and is having value 10 in experiments is determined on the basis of the four past samples: $[S_{n-v}, S_{n-v-6}, S_{n-v-12}, S_{n-v-18}]$. From this nth input and output instance is

$$Xn = [S_{n-v}, S_{n-v-6}, S_{n-v-12}, S_{n-v-18}] \tag{15}$$

$$Y_n = S_n \tag{16}$$

Here, integer values like 0, 6, 12 and 18 used in equation are called shift values in our experiment and $M$ is number of covariates after converting univariate time series to multivariate arrangement. In this simulation the division of data into training and testing data is 75 and 25 per cent respectively.

## 5 Experimental Results

| Activation Function | Algorithm | Learning Mode | Time | | RMSE(Min Error) | | Nodes |
|---|---|---|---|---|---|---|---|
| | | | Training | Testing | Training | Testing | |
| Sigmoid | ELM | Batch | 0.1719 | 0.0000 | 0.4083 | 0.4041 | 12 |
| | OS-ELM | 1 by 1 | 1.2813 | 0.0000 | 0.9974 | 0.9241 | 82 |
| | | 10 by 10 | 0.1250 | 0.0000 | 0.9976 | 0.9951 | 100 |
| | | 20 by 20 | 0.1719 | 0.0000 | 0.9975 | 0.9591 | 91 |
| | | 30 by 30 | 0.1406 | 0.0000 | 0.9977 | 0.9548 | 48 |
| RBF | ELM | Batch | 0.1718 | 0.0000 | 0.4796 | 0.5841 | 95 |
| | OS-ELM | 1 by 1 | 3.4844 | 0.0625 | 0.9974 | 0.9840 | 73 |
| | | 10 by 10 | 0.5000 | 0.1250 | 0.9978 | 0.9821 | 89 |
| | | 20 by 20 | 0.3594 | 0.1250 | 0.9977 | 0.9841 | 86 |
| | | 30 by 30 | 0.2500 | 0.1094 | 0.9978 | 0.9891 | 87 |
| HardLim | ELM | Batch | 0.1250 | 0.0000 | 0.4713 | 0.4721 | 37 |
| | OS-ELM | 1 by 1 | 1.7813 | 0.0000 | 0.992 | 0.9361 | 90 |
| | | 10 by 10 | 0.2969 | 0.0000 | 0.9909 | 0.9561 | 91 |
| | | 20 by 20 | 0.2344 | 0.0000 | 0.9916 | 0.9621 | 99 |
| | | 30 by 30 | 0.1875 | 0.0469 | 0.9901 | 0.9575 | 66 |

Table 1: SMLSystem Indoor Temperature Forecasting with $N = 8$ variables

In Experiment process, after data preprocessing and normalization phase, two three combinations of covariates were applied and results were obtained. Apart from this, different combination were run on the ELM and OS-ELM models for indoor temperature. In some cases Multi step ahead value $V$ was performed and in some cases it was indicated value prediction. Tables I, II and III shows the obtained results in normal circumstances when only criteria is to get maximum accuracy in the form of least RMSE. Results shown are promising and obtained different accuracy based on RMSE and other parameters. Next three tables IV, V and VI shows the result with ensemble approach with iterative and repetitive execution on the same parameter to obtain robust and generalised result and average is taken. Here value $P$ denotes the non of iteration performed with the same and values of parameters. Bothy type of results reflects the that experiment conducted gives good optimal results in terms of accuracy with different combination of covariates. Experiments also give balanced results in case of generalisation of the approach without compromising on accuracy front. In both approaches, Results shown are for training and testing component of the data. Among combinations of co variates and data show better performance in

| Activation Function | Algorithm | Learning Mode | Time | | RMSE(Min-Error) | | Nodes |
|---|---|---|---|---|---|---|---|
| | | | Training | Testing | Training | Testing | |
| Sigmoid | ELM | Batch | 0.0625 | 0.1563 | 0.0664 | 0.0642 | 46 |
| | OS-ELM | 1 by 1 | 0.7969 | 0.0000 | 0.0170 | 0.0162 | 40 |
| | | 10 by 10 | 0.1563 | 0.0000 | 0.0174 | 0.0157 | 35 |
| | | 20 by 20 | 0.1094 | 0.0000 | 0.0171 | 0.0159 | 43 |
| | | 30 by 30 | 0.125 | 0.0000 | 0.0166 | 0.0165 | 37 |
| RBF | ELM | Batch | 0.1719 | 0.0000 | 0.0674 | 0.0690 | 98 |
| | OS-ELM | 1 by 1 | 2.6719 | 0.0625 | 0.0164 | 0.0194 | 66 |
| | | 10 by 10 | 0.3438 | 0.0625 | 0.0164 | 0.0192 | 83 |
| | | 20 by 20 | 0.3438 | 0.0313 | 0.0166 | 0.0202 | 58 |
| | | 30 by 30 | 0.2344 | 0.0625 | 0.0163 | 0.0192 | 81 |
| HardLim | ELM | Batch | 0.1406 | 0.0000 | 0.1644 | 0.1721 | 95 |
| | OS-ELM | 1 by 1 | 1.7969 | 0.0000 | 0.0627 | 0.0712 | 86 |
| | | 10 by 10 | 0.2656 | 0.0000 | 0.0557 | 0.0692 | 88 |
| | | 20 by 20 | 0.2031 | 0.0000 | 0.0616 | 0.0678 | 94 |
| | | 30 by 30 | 0.1250 | 0.0000 | 0.0633 | 0.0738 | 83 |

Table 2: SMLSystem Indoor Temperature Forecasting Only with Indoor Temperature, $M = 6$, $Shift = 6$ and $V = 10$

| Activation Function | Algorithm | Learning Mode | Time | | RMSE(Min Error) | | Nodes |
|---|---|---|---|---|---|---|---|
| | | | Training | Testing | Training | Testing | |
| Sigmoid | ELM | Batch | 0.1250 | 0.0000 | 0.9808 | 0.9841 | 6 |
| | OS-ELM | 1 by 1 | 0.2969 | 0.0000 | 0.9995 | 0.9981 | 7 |
| | | 10 by 10 | 0.0625 | 0.0000 | 0.9995 | 0.9989 | 5 |
| | | 20 by 20 | 0.0625 | 0.0000 | 0.9995 | 0.9990 | 4 |
| | | 30 by 30 | 0.0622 | 0.0000 | 0.9995 | 0.9991 | 2 |
| RBF | ELM | Batch | 0.0469 | 0.0000 | 0.9761 | 0.9851 | 4 |
| | OS-ELM | 1 by 1 | 0.1563 | 0.0000 | 0.9995 | 0.9989 | 2 |
| | | 10 by 10 | 0.0469 | 0.0000 | 0.9995 | 0.9990 | 4 |
| | | 20 by 20 | 0.0469 | 0.0313 | 0.9995 | 0.9991 | 1 |
| | | 30 by 30 | 0.0625 | 0.0000 | 0.9995 | 0.9991 | 4 |
| HardLim | ELM | Batch | 0.1875 | 0.0000 | 0.9015 | 0.9920 | 8 |
| | OS-ELM | 1 by 1 | 1.8438 | 0.0000 | 0.9731 | 0.9931 | 27 |
| | | 10 by 10 | 0.2188 | 0.0000 | 0.9695 | 0.9951 | 9 |
| | | 20 by 20 | 0.1563 | 0.0000 | 0.9738 | 0.9962 | 4 |
| | | 30 by 30 | 0.1719 | 0.0000 | 0.9696 | 0.9970 | 9 |

Table 3: SMLSystem Time Series for Indoor Time Series Forecasting with Three Variables Indoor Temperature, Hour and Sun Irradiance

| Activation Function | Algorithm | Learning Mode | P | Average Time | | Ave-RMSE | | Nodes |
|---|---|---|---|---|---|---|---|---|
| | | | | Training | Testing | Training | Testing | |
| Sigmoid | ELM | Batch | 6 | 0.0120 | 0.0009 | 0.5590 | 0.4531 | 12 |
| | OS-ELM | 1 by 1 | 10 | 0.7871 | 0.0094 | 0.9984 | 1.0032 | 82 |
| | | 10 by 10 | 6 | 0.2110 | 0.0072 | 0.9978 | 0.9977 | 100 |
| | | 20 by 20 | 6 | 0.1420 | 0.0085 | 0.9979 | 1.0121 | 91 |
| | | 30 by 30 | 6 | 0.6480 | 0.0052 | 0.9991 | 0.9994 | 48 |
| RBF | ELM | Batch | 6 | 0.1500 | 0.0289 | 0.4994 | 0.7337 | 95 |
| | OS-ELM | 1 by 1 | 6 | 1.9300 | 0.0415 | 1.0125 | 0.9988 | 73 |
| | | 10 by 10 | 10 | 0.3901 | 0.0594 | 0.9983 | 1.0122 | 89 |
| | | 20 by 20 | 10 | 0.2501 | 0.0592 | 0.9984 | 1.0171 | 86 |
| | | 30 by 30 | 10 | 0.2161 | 0.0819 | 1.0194 | 0.9983 | 87 |
| HardLim | ELM | Batch | 6 | 0.0487 | 0.0033 | 0.5601 | 0.5301 | 37 |
| | OS-ELM | 1 by 1 | 6 | 1.2600 | 0.0097 | 0.9938 | 0.9996 | 90 |
| | | 10 by 10 | 6 | 0.2161 | 0.0102 | 0.9938 | 0.9986 | 91 |
| | | 20 by 20 | 10 | 0.1870 | 0.0108 | 0.9928 | 0.9983 | 99 |
| | | 30 by 30 | 10 | 0.0913 | 0.0071 | 0.9958 | 0.9986 | 66 |

Table 4: SMLSystem Time Series for Indoor Time Series Forecasting with $N = 8$ variables Ensemble

| Activation Function | Algorithm | Learning Mode | P | Ave-Time | | Ave-RMSE | | Nodes |
|---|---|---|---|---|---|---|---|---|
| | | | | Training | Testing | Training | Testing | |
| Sigmoid | ELM | Batch | 10 | 0.0541 | 0.0187 | 0.0766 | 0.0694 | 46 |
| | OS-ELM | 1 by 1 | 5 | 0.4109 | 0.0109 | 0.0205 | 0.0171 | 40 |
| | | 10 by 10 | 20 | 0.0994 | 0.0025 | 0.0218 | 0.0176 | 35 |
| | | 20 by 20 | 10 | 0.0906 | 0.0031 | 0.0201 | 0.0171 | 43 |
| | | 30 by 30 | 20 | 0.0703 | 0.0062 | 0.0213 | 0.0176 | 37 |
| RBF | ELM | Batch | 2 | 0.1421 | 0.0171 | 0.0690 | 0.0795 | 98 |
| | OS-ELM | 1 by 1 | 10 | 2.4406 | 0.0507 | 0.0203 | 0.0241 | 66 |
| | | 10 by 10 | 10 | 0.2604 | 0.0208 | 0.0250 | 0.0409 | 83 |
| | | 20 by 20 | 10 | 0.1731 | 0.0187 | 0.0221 | 0.0262 | 58 |
| | | 30 by 30 | 5 | 0.2218 | 0.0500 | 0.0224 | 0.0302 | 81 |
| HardLim | ELM | Batch | 10 | 0.1572 | 0.0031 | 0.1891 | 0.0694 | 95 |
| | OS-ELM | 1 by 1 | 15 | 1.4475 | 0.0010 | 0.0782 | 0.0893 | 86 |
| | | 10 by 10 | 15 | 0.2179 | 0.0031 | 0.0743 | 0.0860 | 88 |
| | | 20 by 20 | 15 | 0.1677 | 0.0093 | 0.0741 | 0.0880 | 94 |
| | | 30 by 30 | 20 | 0.1256 | 0.0025 | 0.0854 | 0.0965 | 83 |

Table 5: SMLSystem Indoor Temperature Forecasting Only with Indoor Temperature, $M = 6$, $Shift = 6$ and $V = 10$ Ensemble

| Activation Function | Algorithm | Learning Mode | P | Time | | Ave-RMSE | | Nodes |
|---|---|---|---|---|---|---|---|---|
| | | | | Training | Testing | Training | Testing | |
| Sigmoid | ELM | Batch | | 0.2156 | 0.0046 | 0.9996 | 0.9991 | 6 |
| | OS-ELM | 1 by 1 | 5 | 0.2156 | 0.0000 | 0.9995 | 1.0360 | 7 |
| | | 10 by 10 | 5 | 0.2156 | 0.0000 | 0.9996 | 0.9991 | 5 |
| | | 20 by 20 | 5 | 0.2156 | 0.0000 | 0.9995 | 0.9988 | 4 |
| | | 30 by 30 | 5 | 0.2156 | 0.0000 | 0.9996 | 0.9991 | 2 |
| RBF | ELM | Batch | 5 | 0.2156 | 0.0000 | 0.9996 | 0.9991 | 4 |
| | OS-ELM | 1 by 1 | 5 | 0.0000 | 2.4406 | 0.9996 | 0.9995 | 2 |
| | | 10 by 10 | 5 | 0.0328 | 0.0031 | 0.9996 | 0.9991 | 4 |
| | | 20 by 20 | 5 | 0.0375 | 0.0000 | 0.9996 | 0.9992 | 1 |
| | | 30 by 30 | 5 | 0.0156 | 0.0020 | 0.9996 | 0.9994 | 4 |
| HardLim | ELM | Batch | 5 | 0.2156 | 0.0000 | 0.9996 | 0.9991 | 8 |
| | OS-ELM | 1 by 1 | 20 | 0.2918 | 0.0000 | 0.9936 | 0.9998 | 27 |
| | | 10 by 10 | 20 | 0.0890 | 0.0000 | 0.9985 | 0.9998 | 9 |
| | | 20 by 20 | 30 | 0.0468 | 0.0000 | 0.9994 | 0.9989 | 4 |
| | | 30 by 30 | 30 | 0.0468 | 0.0000 | 0.9988 | 0.9993 | 9 |

Table 6: SMLSystem Time Series for Indoor Time Series Forecasting with Three Variables Indoor Temperature, Hour and Sun Irradiance

the test partition. In this experiments three sets of covariates were taken. One set is with the eight potential co variates. Other with only one as indoor temperature and last one is potential three covariates as mentioned in table description. With these combination and ELM and OS ELM, experiments were conducted and results are displayed. Results in both the approaches are good and better in case of OS ELM approach with activation functions. The most significant combination of covariates is time hour and sun irradiance which achieves least RMSE and optimal accuracy both the cases and generally for all covariates combinations.

## 6 Conclusion

An accurate forecasting of indoor temperature could yield an energy-efficient control with HVAC systems. A multivariate forecasting methods for prediction of indoor temperature has been performed using Extreme Learning Machines and its variants. Several combinations of covariates, forecasting model combination and their comparison on the basis of standard Root Mean Square Error methods have been performed. study found that using only indoor temperature with the hour that is categorical variable and sun irradiance together, gives a good level of accuracy without going for all the covariates. The data used for this study includes to one month and a week of a Southern Europe house. Another development and further extension could be to perform experiments using several months of data in other houses where weather conditions may change among seasons and locations and during day. As future work, could be implementation of different techniques such as deep learning with potential feature selection tools. A deeper study to understand the relationship between covariates would also be interesting.

## References


1. Solar decathlon europe competition, 2012. *United States Department of Energy*, 2011.
2. On-line learning of indoor temperature forecasting models towards energy efficiency. *Energy and Buildings*, 83:162 – 172, 2014.
3. Guillermo Escriv-Escriv, Carlos lvarez Bel, Carlos Roldn-Blay, and Manuel Alczar-Ortega. New artificial neural network prediction method for electrical consumption forecasting based on building end-uses. *Energy and Buildings*, 43(11):3112 – 3119, 2011.
4. Robert Hecht-Nielsen. Theory of the backpropagation neural network. In *Neural Networks, 1989. IJCNN., International Joint Conference on*, pages 593–605. IEEE, 1989.
5. Gao Huang, Guang-Bin Huang, Shiji Song, and Keyou You. Trends in extreme learning machines: A review. *Neural Networks*, 61:32 – 48, 2015.
6. Guang-Bin Huang. An insight into extreme learning machines: random neurons, random features and kernels. *Cognitive Computation*, 6(3):376–390, 2014.
7. Guang-Bin Huang, Hongming Zhou, Xiaojian Ding, and Rui Zhang. Extreme learning machine for regression and multiclass classification. *Systems, Man, and Cybernetics, Part B: Cybernetics, IEEE Transactions on*, 42(2):513–529, 2012.
8. Guang-Bin Huang, Qin-Yu Zhu, and Chee-Kheong Siew. Extreme learning machine: theory and applications. *Neurocomputing*, 70(1):489–501, 2006.
9. E. Camponogara J. Normey-Rico M. Berenguel P. Ortigosa J. lvarez, J. Redondo. Optimizing building comfort temperature regulation via model predictive control. *Energy and Buildings*, 57:361372, 2012.
10. M. Lichman K. Bache. Uci machine learning repository. *University of California*, 2013.
11. F. Scapolo J. Leijten-J. Burgelman K. Ducatel, M. Bogdanowicz. Istag scenarios for ambient intelligence, 2010.
12. Eija Kaasinen, Tiina Kymlinen, Marketta Niemel, Thomas Olsson, Minni Kanerva, and Veikko Ikonen. A user-centric view of intelligent environments: User expectations, user experience and user role in building intelligent environments. *Computers*, 2(1):1, 2013.
13. JJ Koliha, Dragan Djordjević, and Dragana Cvetković. Moore–penrose inverse in rings with involution. *Linear Algebra and its Applications*, 426(2):371–381, 2007.
14. Nan-Ying Liang, Guang-Bin Huang, Paramasivan Saratchandran, and Narasimhan Sundararajan. A fast and accurate online sequential learning algorithm for feedforward networks. *Neural Networks, IEEE Transactions on*, 17(6):1411–1423, 2006.
15. Fernando Mateo, Juan Jos Carrasco, Abderrahim Sellami, Mnica Milln-Giraldo, Manuel Domnguez, and Emilio Soria-Olivas. "machine learning methods to forecast temperature in buildings. *Expert Systems with Applications*, 40(4):1061 – 1068, 2013.
16. S. Silva E. Conceic P. Ferreira, A. Ruano. Neural networks based predictive control for thermal comfort and energy savings in public buildings, energy and buildings. *Energy and Buildings*, 55:238251, 2007.
17. Mark Weiser. Some computer science issues in ubiquitous computing. *Commun. ACM*, 36(7):75–84, July 1993.
18. Francisco Zamora-Martnez, Pablo Romeu, Paloma Botella-Rocamora, and Juan Pardo. Towards energy efficiency: Forecasting indoor temperature via multivariate analysis. *Energies*, 6(9):4639, 2013.

# Development of Real Time Helmet based Authentication with Smart Dashboard for Two Wheelers

**Ashish Kumar Pardeshi, Hitesh Pahuja, Balwinder Singh**

Abstract—**Nowadays, it is mandatory to wear helmet in all countries for safety purpose. At the time of accident helmet may save a rider's life. Usually most of the people avoid wearing helmet. Thus to encourage rider to wear helmet and to provide smart features to two wheelers, a work is proposed here which will authenticate the engine ignition only when the rider will wear a helmet and provide smart dashboard on the two wheeler for making communication with helmet when worn by the rider. The communication between helmet and smart dashboard is established using Bluetooth, thus providing secure and confident link between the two. Network of limit switch with proper mechanical assembly installed inside the helmet is used to detect the helmet wearing phenomenon. Smart dashboard has graphical touch screen, thus enabling Graphical User Interface (GUI) for the features like maintaining the bike service records, automatic head light control mechanism through LDR (Light Dependent Resistor) and One Time Password (OTP) input for two wheeler access in the absent of helmet. Smart dashboard is also equipped with GPS and GSM to locate the Vehicle position, in case of vehicle lost or theft through a missed call. During emergency rider can also contact to the contacts list in the dashboard via specific button on GUI.**

## 1 Introduction

Two wheeler riders constitute the largest proportion (71 %) of vehicles users in India and, compared with other vehicle users, this group has a higher proportion of road traffic injuries [1]. In India 136834 person reportedly dead due to road traffic injuries in 2011 and an estimated 2 million people have disabilities as a result of road traffic crashes [1]. Among two wheelers, injury to head and neck is often the

Ashish Kumar Pardeshi
Centre for Development of Advance Computing, A Scientific Society of Ministry of Communication & Information Technology, Mohali, Punjab, India.
Email: ashishkumarpardeshi@gmail.com

Hitesh Pahuja
Email: hitesh@cdac.in
Balwinder Singh
Email: balwinder@cdac.in



J.M. Corchado Rodriguez et al. (eds.), *Intelligent Systems Technologies and Applications 2016*, Advances in Intelligent Systems and Computing 530,
DOI 10.1007/978-3-319-47952-1_79

main cause of death and disabilities, and helmet can reduce this risk substantially [3]. Non-use of helmet is associated with injuries and disabilities that result in higher treatment cost [3]. Also, enforcement of helmet laws has demonstrated a decrease in rates of head injuries deaths, while repealing these laws has shown an increase in these rates [3].

The objective of this presented work is to develop a system which will make mandatory for the riders to wear helmet in order to access their two wheelers. These systems can be installed to the bike at the time of its manufacturing and these advance helmets can be delivered to the customer at time of vehicle purchasing. The Engine will only ignite when rider will wear the helmet. The other aim of this proposed work is also to provide smart dashboard with the help of graphical touch screen TFT LCD with Simple User Interface to access the features for the vehicle and its maintenance.

The presented work deals with two parts: 1) Smart Dashboard and 2) Helmet Circuitry. Smart dashboard is equipped with graphical TFT touch screen, Bluetooth, Real time Clock, LDR, GPS and GSM. All these modules being interfaced to Arduino Mega Board, which is the main controller board for the Smart Dashboard. GPS and GSM is used for tracking the vehicle location in case of vehicle theft or lost and rider's emergency. Real Time Clock will provide clock data and keep the smart dashboard update thus maintaining the service records of vehicle. LDR is used for Automatic Setting for Head Light control [4]. All the settings and features can be easily access by the rider through a simple and easy user interface on graphical TFT touch screen.

Helmet circuitry is installed inside the helmet and equipped with network of limit switches, Bluetooth, 3.7v lithium polymer battery, battery charger circuit and ATMEGA 328 controller board. Helmet circuitry is powered with 3.7v lithium polymer battery can charged through on helmet charging circuit. Network of limit Switches will help the controller board inside the helmet to detect the wearing of helmet by the rider and thus ATMEGA 328 controller board will enable the helmet Bluetooth (in Slave Mode) to be paired with Smart Dashboard Bluetooth (in Master Mode) and establish a link between the two in order to provide the vehicle engine ignition [5].

## 2 Technical Studies

The development of the proposed work will be described in detail in this section. This proposed work has two modules, one is bike unit in the form of smart dashboard to be attached on the two wheeler and other is helmet circuitry to be embedded into helmet. The block diagram for Smart Dashboard and Helmet circuitry are shown in Fig. 1 and Fig. 2 respectively. Both of this modules have their own processing units and power supplies.

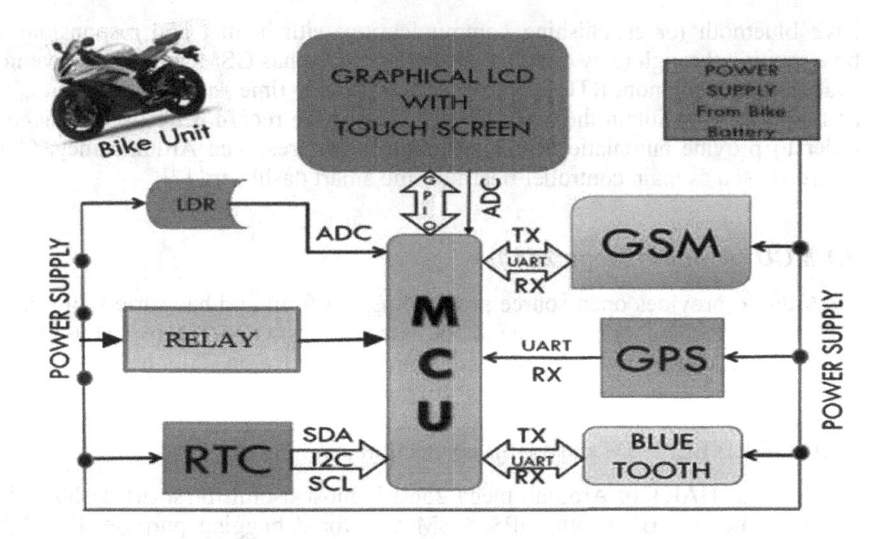


**Fig. 1** Bike Unit (Smart Dashboard)

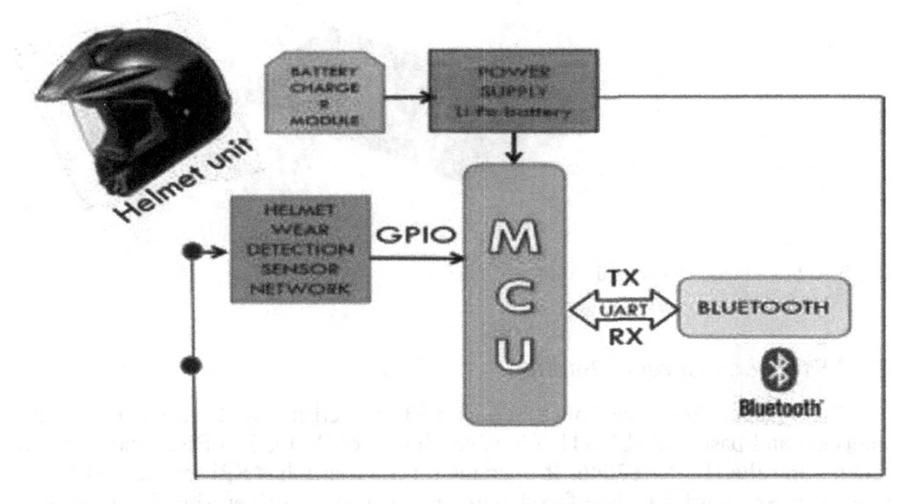


**Fig.2** Helmet Unit (Helmet Circuitry)

The bike unit is basically a smart dashboard equipped with TFT Touch Screen Display which provide graphical user Interface with touch input access for enjoying the bike features. Normally this kind of display are most common to four wheelers, where driver can access the various functionality through the user interface in the form of application. Similar system is tried to put on the two wheeler with functionality and required features for the two wheelers [6]. The Smart Dashboard also have bluetooth for establishing communication with helmet and responsible for bike ignition through relay circuitry. With this it also has GSM and GPS for vehicle Tracking phenomenon. RTC is also added to provide time and date display feature and helping to maintain the real time vehicle service record. LDR is interfaced in order to provide automatic head light control features. The Arduino mega 2560 Board is used as main controller board for the smart dashboard [7].

## 2.1 MCU (Arduino Mega 2560)

Arduino provides open source prototyping platform and has variety of boards, In Smart Dashboard Arduino mega 2560 board as shown in Figure 3 is used. This board has ATmega 2560 8-bit microcontroller as System on Chip. It operates on 16 MHz clock frequency with operating voltage of 5 volts. It provides 54 digital input/output pins, 16 analog input pins, 4 UART, 256 KB of flash memory, 8 KB SRAM, 4 KB EEPROM and a USB connection.

The four UART of Arduino mega 2560 is most useful for smart dashboard in order to interface Bluetooth, GPS, GSM and for debugging purpose. EEPROM (Electrical erasable programmable read only memory) is used for retaining the status of bike head light Control data and for one-time password. The board can be easily programmed with on board USB to UART converter [8].

**Fig. 3** Arduino mega 2560 board

## 2.2 TFT Touch Screen Display

The 3.2-inch Resistive Touch Screen TFT is used in smart dashboard for user interface and based on ILI9341, it has resolution of 240 x 320 Pixel and on board touch controller IC XPT2046. It operates on 3.3V and has SPI interface with microcontroller board and interfaced with the help TFT driver shield for Arduino mega board. Figure 4 shows below is the TFT touch screen with driver [15].

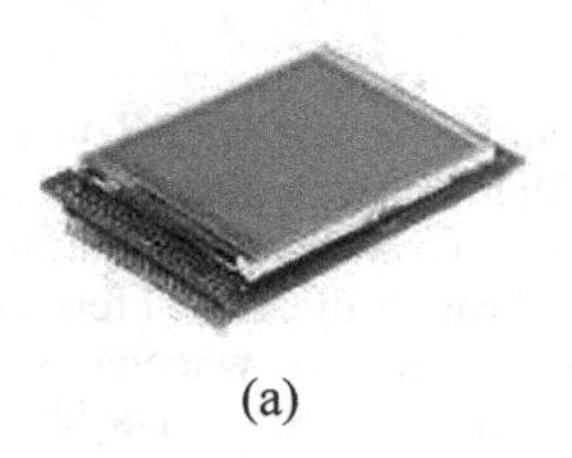

(a)

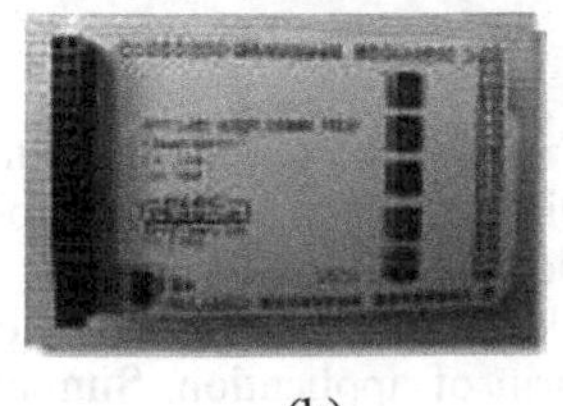

(b)

**Fig. 4** (a) 3.2" TFT Touch Screen and (b) Driver Shield

### 2.3 Bluetooth HC-05 (Master Mode)

Bluetooth HC-05 as shown in Figure 5 module is based on serial port protocol (SPP) which is basically designed for transparent wireless serial connection setup. The module has enhanced data rates that is 3 Mbps modulation with complete 2.4 GHz radio transceiver and baseband. This module can be configured as both Master and Slave mode at a time with the help of AT commands and provide UART (Rx and Tx) interface with programmable baud rate. In Smart Dashboard Bluetooth module HC-05 is configured as Master mode and programmed for Auto pairing to the Helmet Bluetooth for establishing wireless link between Smart Dashboard and helmet, when helmet is worn by the rider [9].

**Fig. 5** Bluetooth HC-05

### 2.4 LDR Circuit

Light Dependent Resistor (LDR) is special type of resistor whose resistance decreases with increasing the number of Light packet on its sensing surface and vice-versa. Thus the more the Luminous, less will be resistance. Figure 6 (a) shows the LDR with small graph explaining its name. LDR are always used in voltage divider circuit and generally provide analog output thus can be interfaced to internal ADC of microcontroller. Figure 6 (b) shows the voltage divider circuit for the LDR. And the output voltage can be derived from equation 1. LDR is interfaced to smart dashboard for the feature of Automatic Headlight Control of the vehicle where the rider can calibrate the LDR value for head light to be ON or OFF [10].

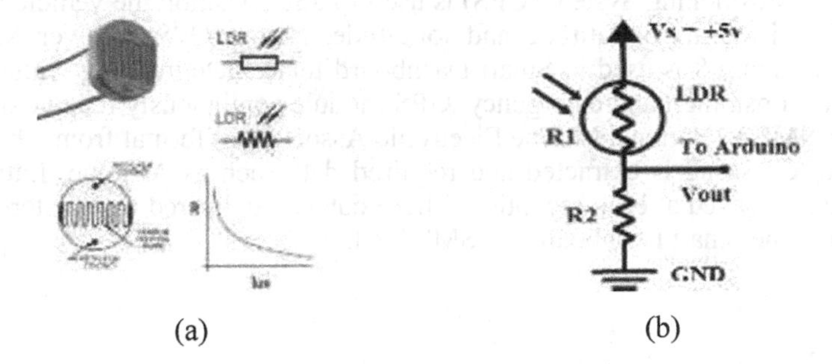


**Fig. 6** (a) LDR with Graph and (b) Voltage divider Circuit

Vout = (R2 * Vs) / (R1 + R2) (1)

Where: R1 = Resistance across LDR

R2 = Parallel resistor across LDR

Vs = Source Voltage

Vout = Analog Output to Arduino ADC

### 2.5 GSM Module

Global System for Mobile Communication (GSM) module as shown in Figure 7 SIM900 feature the full quad-band operation with SMS, Calling and GPRS Data functionality. AT commands are used to communicate and configure the GSM module by the microcontroller board. GSM Module interface through UART (Rx and Tx) to Microcontroller board. Smart Dashboard is able to generate SMS for Emergency, OTP, Vehicle Tracking and Unauthorized Access via GSM Module through the AT commands used in Code Development [11].

**Fig.7** GSM SIM900 Module

### 2.6 GPS Module

Global Positioning System (GPS) is used to track/position the vehicle using satellite data in terms of latitude and longitude. Neo-m GPS receiver Module as shown in Figure 8 is used in Smart Dashboard for vehicle tracking feature in case of vehicle Lost, theft and emergency. GPS module continuously receive data in the form of NMEA (National Marine Electronic Association) format from which String "$GGRMC" string is extracted and required data such as Altitude, Latitude and Longitude is saved at each reception. These data are delivered to registered mobile number in the Smart Dashboard via SMS [10].

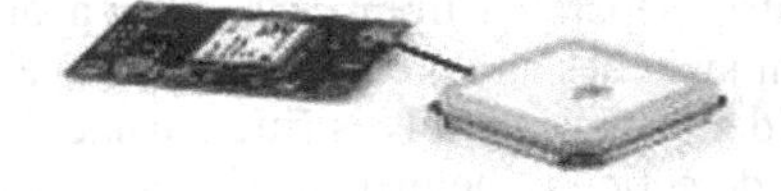

**Fig. 8** NEO-M GPS Receiver Module

### *2.7 Relay Circuitry*

Relay is an electromechanical switch controlled electronically and switches the higher voltage device. Relay has a coil, on energizing the coil the COM terminal of relay get connected to NO (normally open) terminal of relay otherwise COM is only connected to NC (normally closed) terminal [12].

Two 12v single pole and single throw relay modules are interfaced with smart dashboard for engine Ignition and headlight control separately. Relay module interfaces relay with microcontroller board with the help of optical isolators thus providing safety to the microcontroller circuitry. Figure 9 shows the relay circuitry.

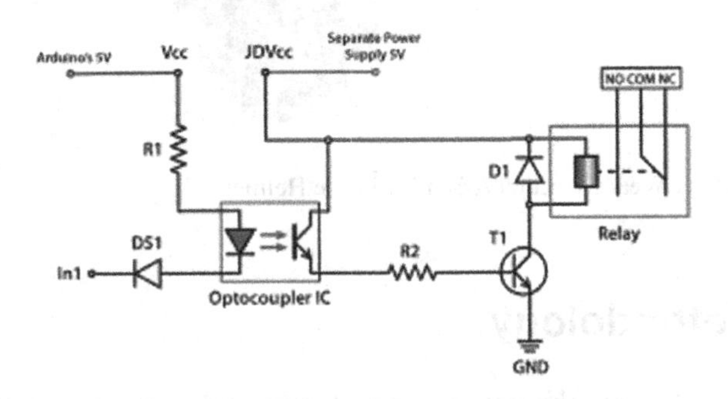


**Fig. 9** Relay Module using Optocoupler

### *2.8 Helmet Wear Detection Network*

Figure 10 shows the helmet detection assembly with Limit switches are used to detect the helmet worn by the rider. Limit switch is a kind of small mechanical momentary switch which provide continuity between its common (COM) and normally open (NO) terminal when the lever of switch is pressed otherwise it shows continuity between common (COM) and normally closed (NC) terminal. In prototype three limit switches are used to make a network of switches with proper mechanical assembly get fitted inside the helmet. All limit switches are connected in series thus making mandatory for all limit switches to be pressed during the helmet worn by the rider in order to access the two wheeler's ignition [13].

Limit switches are placed between two strip separated from each other apart through spring. Holding strip holds the all limit switches and pressing strip press all the limit switch when helmet is worn by the rider. This assembly is fitted inside the helmet with proper installation and allow easy detection of helmet wearing and safe around the rider's head [14].

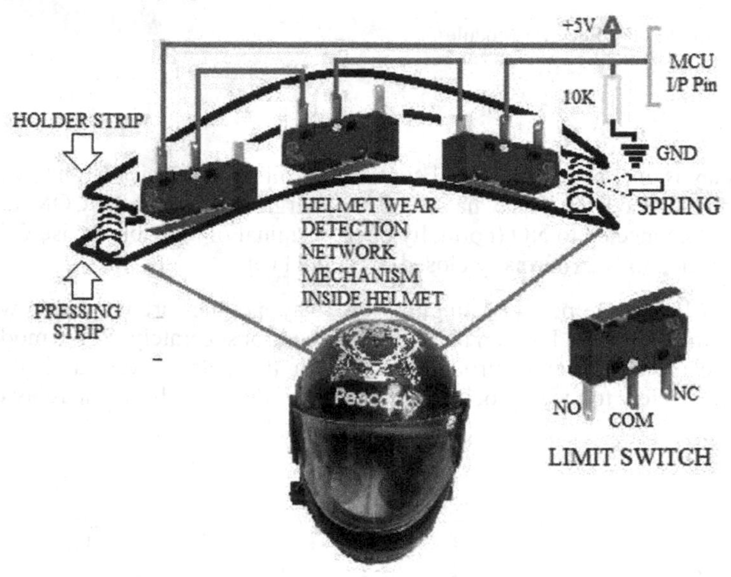


**Fig. 10** Helmet Wear Detection Assembly inside Helmet

# 3 Methodology

Figure 11 shows the methodology of proposed work in the form of Flow Chart. The description of flow chart is as follows:

Initially the rider will insert the key and turning on key will provide power to smart dashboard. Smart dashboard will initialize all interfaced modules that is RTC, Bluetooth, GSM, GPS, TFT Touch screen and cutoff the Ignition relay (Engine will not ignite).

Now the Dashboard will display the Date and Time and applications icon with helmet worn message, that it will ask for rider to wear helmet for riding the vehicle and dashboard's Master Bluetooth will continuously search for helmets Slave Bluetooth for pairing. For accessing the vehicle rider needs to wear the helmet, now when rider wears the helmet, then helmet Worn detection network inside the helmet gets activated and power the helmet's slave Bluetooth and get paired with dashboard's master Bluetooth module exchange the authentication code to establish communication link between them and thereby smart dashboard will switch ON the ignition relay to ignite the vehicle ignition on kick or self-start.

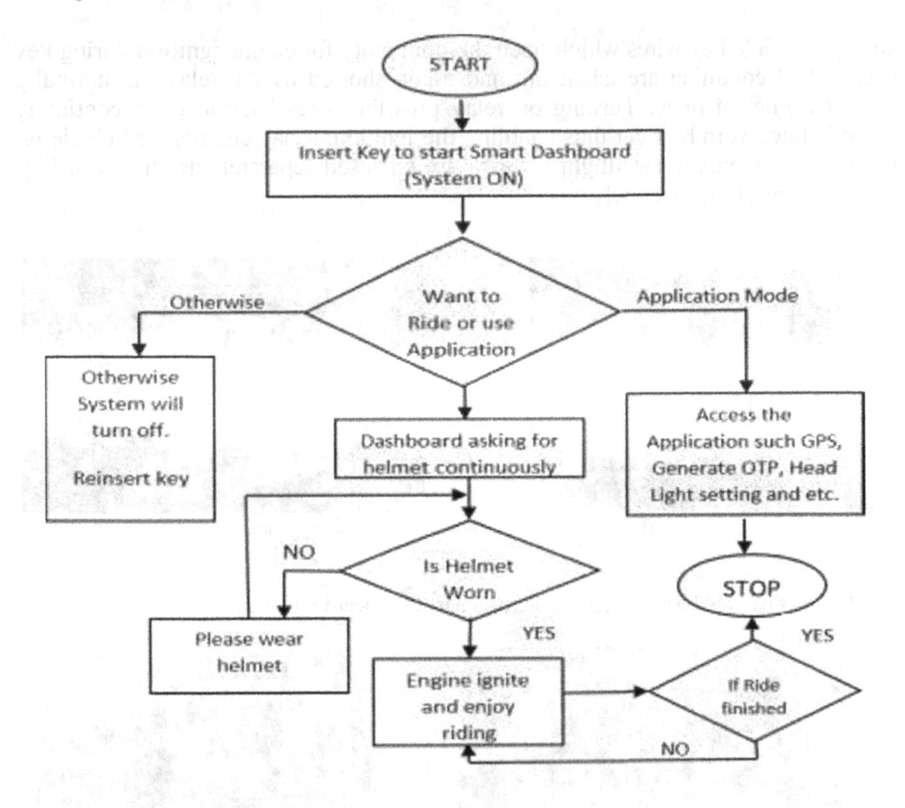


**Fig. 11** Methodology of proposed work

Other than riding the vehicle, rider can use the different applications on smart dashboards in the form of simple and easy to use Graphical User Interface (GUI) for accessing the features like vehicle service records- where rider can check and notified for the next service as well as have the last three service records update, Vehicle Automatic Headlight control- here rider can set the headlight control in either automatic mode (based on set LDR calibrated value, headlight will turn on and off) or Manual Mode (control by the switches on vehicle), Positioning of vehicle-using GPS vehicle can be track during lost, theft and emergency. Generating OTP-using GSM rider can be notified for OTP and tracking location in the form of SMS.

## 4 Result and Discussion

This Product was developed into two parts: Smart dashboard (Vehicle unit) and Helmet circuitry (Helmet unit). For both the units Single sided PCB was prototyped using the Eagle PCB designing software with proper layout considerations.

Smart Dashboard was prototyped into three Single Sided PCB, structured in the form of Shields as shown in Figure 12. Smart Dashboard structure was mounted with the Arduino Mega 2560 Boards, RTC module, GPS Module, GSM module, Bluetooth, TFT touch Screen, LDR with basic Components soldered for the Power supply. Vehicle key wires which open the continuity for engine ignition during key was in ON condition are taken out and again shorted by the relay via normally close terminal of relay. Turning on relay puts the wires back to open continuity when helmet worn by rider thus enabling the ignition. Relay circuits for vehicle ignition and Automatic Headlight Control are enclosed separately as shown in Figure13 (a) and (b) respectively.

(a)

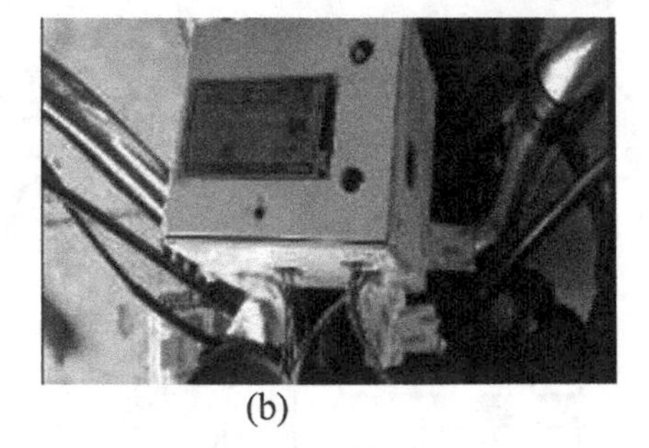

(b)

**Fig. 12** (a) Smart dashboard Circuitry and (b) Smart Dashboard Prototype

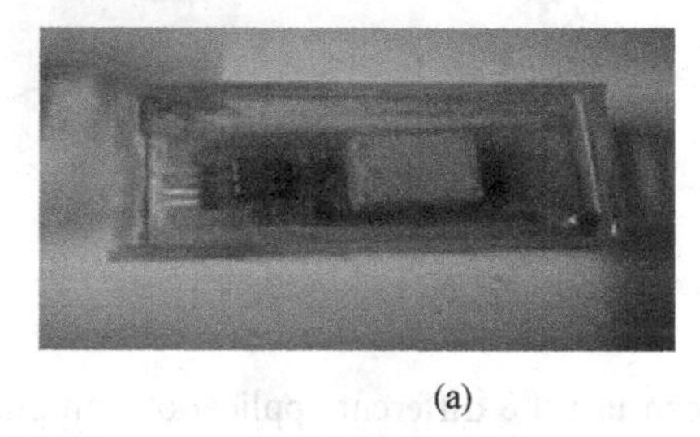

(a)

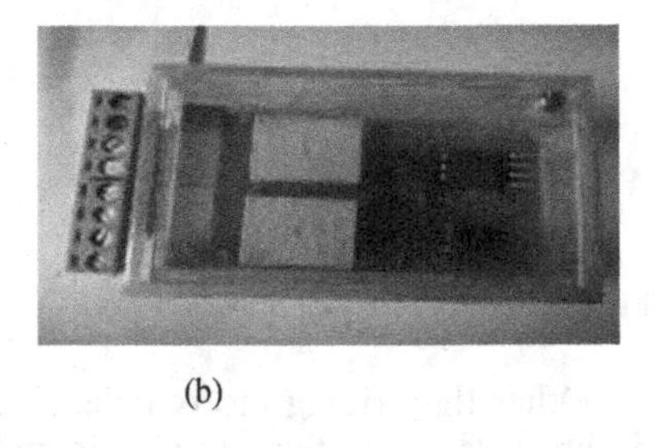

(b)

**Fig. 13** (a) Ignition Relay Prototype and (b) Headlight Relay prototype

For Helmet circuitry the single PCB was fabricated and mounted with Arduino Pro Mini, Bluetooth, Battery charger, battery with BSS138 MOSFET soldered for switching the power for Bluetooth. Helmet prototypes shown in Fig.14 (a) and (b).

(a)

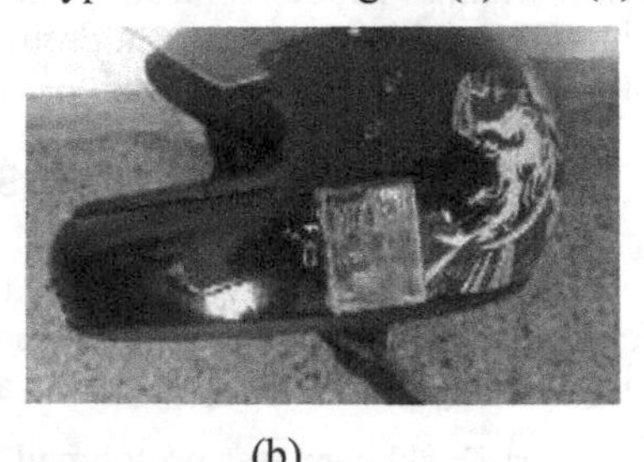

(b)

**Fig. 14** (a) Helmet Circuitry and (b) Helmet Prototype

The Code for both the unit was developed under the Arduino IDE and written in C language with the help of few libraries. The Proposed work was successfully completed with real time testing of five objectives as under follows:

- Simple Graphical User Interface (GUI); Easy to use user interface was developed for accessing the system.
- Helmet Authentication; The Bike Engine was only igniting when the rider worn the helmet. GUI for Helmet Authentication are shown in Figure15. There are several situations that promoted the Engine ignitions are shown in Table I.
- Vehicle Tracking Facility; GPS was providing the Latitude and Longitude accurately for which field test was performed around the campus of CDAC and results are shown in Table II. Thus able to locate the vehicle (GUI for GPS and Google map are shown in Figure16 and GSM Module was sending SMS to the owner for GPS data and OTP as shown in Fig .17.
- Automatic Head Light; LDR analog data was digitally converted and monitored continuously for switching the Head Light of vehicle if set in Auto Mode as GUI is shown in Figure18. The graph for calibrated LDR decimal value along morning, noon, evening and night is shown in Figure 19.
- Service Records; maintained with the help of RTC module and Database is created in internal EEPROM of Arduino Mega Board GUI as shown in Figure 20.

**TABLE 1** Situation for Engine Ignition

| HELMET TECHNICAL ISSUE | HELMET WORN | SMART DASHBOARD BLUETOOTH | HELMET BLUETOOTH | BLUETOOTH CONNECTIVITY STATUS | VEHICLE ENGINE ACCESS |
|---|---|---|---|---|---|
| No Issues | Not Worn | Enabled | Disabled | Not Connected | Not Granted |
| No Issues | Worn | Enabled | Enabled | Paired and Connected | Granted |
| Battery Not Charged or technical Fault | Not Worn | Enabled + OTP Input Via Touchpad | Disabled | Not Connected | Not Granted |

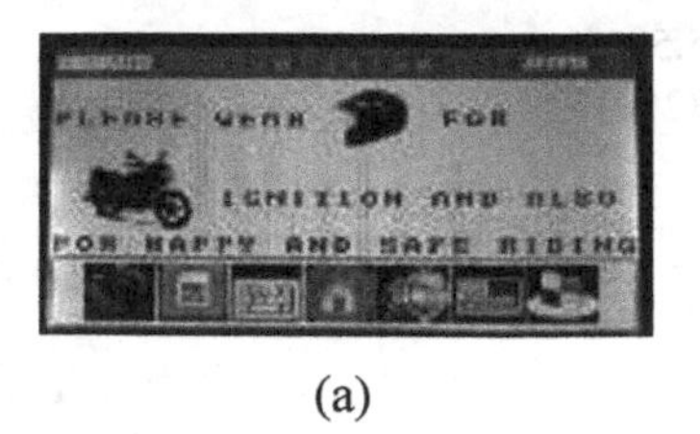

(a)

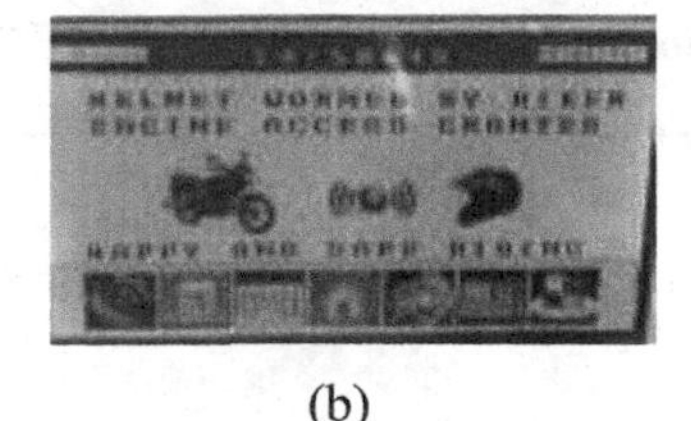

(b)

**Fig. 15** (a) Before Helmet Authentication Display and (b) After Helmet Authentication Display

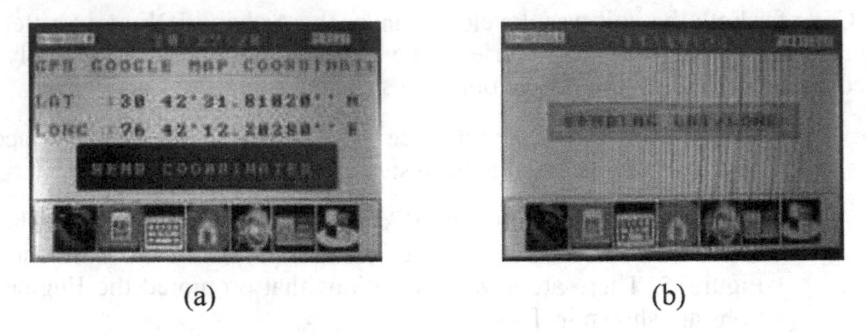


**Fig. 16** (a) GPS Data Display and (b) Sending SMS

**TABLE 2** Field Test of GPS in Proposed System

| Location | GPS Coordinates | Verification on Google Map |
|---|---|---|
| **CDAC Mohali<br>A-34, Phase 8,<br>Industrial Area<br>Sector 71, SAS Nagar<br>Mohali, Punjab 160071** | N 30˚42′29.05″<br>(30.7080)<br>E 76˚42′14.70″<br>(76.7040) | Verified |
| **IVY Hospital<br>Sector 71, SAS Nagar<br>Mohali, Punjab 160071** | N 30˚42′24.41″<br>(30.7067)<br>E 76˚42′29.50″<br>(76.7081) | Verified |
| **PCL Chowk<br>Phase 3B-2,<br>Sector 60, SAS Nagar<br>Mohali, Punjab 160059** | N 30˚42′37.59″<br>(30.7104)<br>E 76˚42′45.724″<br>(76.7127) | Verified |
| **3B-2 Chowk<br>Phase 3B-1,<br>Sector 60, SAS Nagar<br>Mohali, Punjab 160059** | N 30˚42′49.74″<br>(30.7138)<br>E 76˚43′05.596″<br>(76.7182) | Verified |

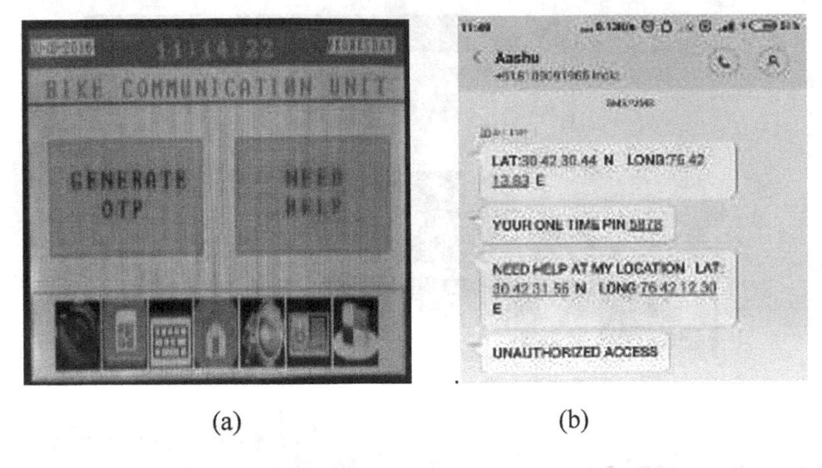


(a) (b)

**Fig. 17** (a) GUI for SMS communication and (b) Received SMS

**Fig. 18** GUI for Head Light Controls

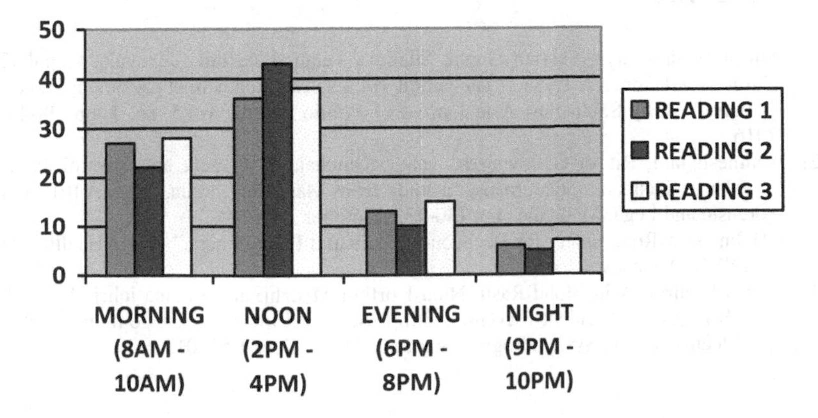


**Fig. 19** Graph LDR value versus Time

**Fig. 20** GUI for Service Records

## 5 Conclusion

The system is able to make wearing helmet mandatory for the rider for those who driving two wheeler vehicle. Otherwise the vehicle will not start (Engine will not Ignite). Thus the safety of rider is achieved to some extent by this proposed work. Moreover, violating of traffic rules of not wearing helmet will be reduced dramatically. Smart dashboard is also providing the rider to maintain the vehicle, by providing service records and enabling the rider to access the vehicle smartly using Touch GUIs. The Proposed system also provide SMS notification to the user and emergency contacts in Smart Dashboard for OTP, GPS Tracking Data and Emergency need. The Smart Dashboard and helmet circuitry was installed into the Splendor motor bike and Helmet respectively. Which gave successful real Time testing of proposed work effortlessly. Therefore, Government should make this system mandatory for every two wheeler owners. The proposed work is newly researched and innovative product for two wheeler automobile industry which soon will be implemented by companies after approval from transport ministry.

## References


[1] Shirin Wadhwaniya, Shivam Gupta, Shailaja Tetali, Lakshmi K josyula, Gopal Krishna Gururaj, and Adnan A Hyder, "The validity of self-reported helmet use among motorcyclist in India," WHO South East Asia Journal of Public Health., vol.5, no. 1, pp. 38-44, April 2016.

[2] Animesh jain, Ritesh G. Menezes, Tanuj Kanchan, S. Gagan, and Rashmi Jain, "Two wheeler accidents on Indian roads – a study from Mangalore, India," ELSEVIER Journal of Forensic and Legal Medicine., pp. 130-133, October 2008.

[3] "Helmets: A Road Safety for Decision-Makers and Practitioners," World Health Organization (WHO), Geneva: 2006.

[4] Mohd Khairul, Afiq Mohd Rasli, Nina Korlina Madzhi, and Juliana johri "Smart Helmet with Sensors for Accident Prevention," in Proc. IEEE International Conference on Electrical, Electronics and System Engineering., pp. 21-26, Dec. 4-5, 2013.

[5] Manasi Pental, Monali Jadhav, and Priyanka Girme, "Biker rider's safety using helmet," International Journal of electrical and electronics Engineering and Telecommunication (IJEETC)., vol. 4, no. 2, pp. 37-44, April 2015.
[6] A Viral Vijay, Ajay Singh, Bhanwar Veer Singh, Abhimanyu Yadav, Blessy Varghese, and Ankit, "Hi-tech helmet and accidental free transportation system," International Journal Advanced Technology and Engineering., vol. 2, no. 6, pp. 67-70, May 2015.
[7] Rattapoom Waranusast, Nannaphat Bundo, Vasan Timtong, and Chainarong Tangnoi, "Machine vision techniques for motorcycle safety helmet detection," in proc. IEEE 28th International Conference on Image and Computer Vision New Zealand., pp. 35-40, Nov. 29, 2013.
[8] K Sudarsan and P. Kumaraguru Diderot, "Helmet for road hazard warning with wireless bike authentication and traffic adaptive Mp3 playback," International Journal of Science and Research (IJSR)., vol. 3, no. 3, pp.340-345, March 2014.
[9] M.S. joshi, Deepali, and V. Mahajan, "Arm 7 based theft control, accident detection & vehicle positioning system," International Journal of Innovative and Exploring Engineering (IJITEE)., vol. 4, no. 2, pp. 29-31 July 2014.
[10] Sawant Supriya C, Dr. Bombale U.L., and Patil T.B, "An intelligent vehicle control and monitoring using Arm," International Journal of Engineering and Innovative Technology (IJEIT)., vol. 2, no. 4, pp. 56-59, Oct. 2012.
[11] Pham Hoang dat, Micheal Drieberg, and Nguyen Chi Cuong "Development of vehicle tracking system using GPS and GSM modem," in Proc. IEEE International Conference on Open Systems., pp.89-94, Dec. 2-4, 2013.
[12] R. Prudhvi Raj, Ch. Shri Krishna Kanth, A. Bhargav, and K. Bharath, "Smart-tec helmet" Advance in Electronic and Electric Engineering., vol.4, no. 5 pp.493-498, 2014.
[13] Muhammad Ridhwan Ahmad Fuad and Michael Drieberg, "Remote Vehicle Tracking System using GSM Modem and Google Map," in Proc. IEEE International Conference on Sustainable Utilization and Development in Engineering and Technology., pp. 16-19,2013
[14] Aneesh Alocious and Thomas George, "Embedded system Controlled Smart Helmet," International Journal of Advanced Research in Electrical & Electronics and Instrumentation Engineering (IJAREEIE)., vol. 4, special issue 1, pp. 165-171, March 2015.
[15] Mangesh Jadhawar, Gauri, Kandepalli, Ashlesa Kohade, and Rajkumar Komati, "Smart Helmet Safety System using Atmega 32," International Journal of Research in Engineering and Technology (IJRET)., vol. 5, no. 5, pp. 287-289, May 2016.

[15] [illegible]

# A Simplified Exposition of Sparsity Inducing Penalty Functions for Denoising

Shivkaran Singh and Sachin Kumar S and Soman K P

**Abstract** This paper attempts to provide a pedagogical view to the approach of denoising using non-convex regularization developed by Ankit Parekh et al. The present paper proposes a simplified signal denoising approach by explicitly using sub-band matrices of decimated wavelet transform matrix. The objective function involves convex and non-convex terms in which the convexity of the overall function is restrained by parameterized non-convex term. The solution to this convex-optimization problem is obtained by employing Majorization-Minimization iterative algorithm. For the experimentation purpose, different wavelet filters such as daubechies, coiflets and reverse biorthogonal were used.

## 1 Introduction

Signal denoising is a task where we estimate our original signal $x \in \mathbb{R}^n$ from its noisy observation, say $z \in \mathbb{R}^n$, represented as

$$z = x + n, \tag{1}$$

where $n$ represents Additive White Gaussian Noise (AWGN) with standard deviation $\sigma$. Sparsity based solutions have a clear advantage over other solutions because of low space and time complexity. A successful method of convex denoising using tight frame regularization was derived by [1]. This paper aims to provide a pedagogical explanation to the theory proposed by [1]. In this work, a denoising method using Non-Convex Penalty (NCP) functions and decimated wavelet transform matrix $W$ [2] is proposed. Wavelet transform matrix $(W)$ contains sub-band wavelet

Shivkaran Singh · Sachin Kumar S · Soman K P
Center for Computational Engineering and Networking (CEN)
Amrita School of Engineering, Coimbatore
Amrita Vishwa Vidyapeetham, Amrita University, India 641112
e-mail: shvkrn.s@email.address



J.M. Corchado Rodriguez et al. (eds.), *Intelligent Systems Technologies and Applications 2016*, Advances in Intelligent Systems and Computing 530,
DOI 10.1007/978-3-319-47952-1_80

matrices corresponding to different levels of transformation. The problem formulation is derived using $W$ because it is less complicated and easy to manipulate. The problem formulation for the signal denoising problem using $W$ is given as

$$\underset{x}{\text{argmin}} \left\{ F(x) := \frac{1}{2}||z-x||_2^2 + \lambda_1 \sum_{i_1} \psi(W_1 x; c_1)_{i_1} + .... \right.$$

$$\left. ..... + \lambda_{L+1} \sum_{i_{L+1}} \psi(W_{L+1} x; c_{L+1})_{i_{L+1}} \right\} \tag{2}$$

where $\lambda_1, \lambda_2 .... \lambda_{L+1}$ are the regularization parameters corresponding to different levels of transformation ($\lambda > 0$ always), $W_1, W_2 ..... W_{L+1}$ are the sub-band matrices of W corresponding to different levels of transformations, $\psi(.;c) : \mathbb{R} \rightarrow \mathbb{R}$ is a non-convex sparsity inducing function parametrized by $c$ and $L$ represents levels of transformation of W. In Section 3, more clarity about (2) and different sub-band matrices of W is provided. Parameter $c$ is used to adjust the convexity of overall objective function $F(x)$. The formulation comprises of convex and non-convex (non-smooth) terms. For sparse processing of the signal, penalty function, $\psi(x;c)$, should be chosen such that it improves the sparsity of $x$ [3]. In this work, NCP functions such as *logarithmic*, *arctangent* and *rational penalty function* were used. The use of NCP functions leads to better sparsity. Among all mentioned NCP functions,*arctangent penalty function* delivered better results as it converges more rapidly to the identity function, see Fig. 3 in [4].

To solve the denoising problem, Majorization-Minimization (M-M) algorithm instead of Alternating Direction Multiplier Method (ADMM) [1] was used, which provided computational advantage and improved Root Mean Squared Error (RMSE) value (see Fig. 3). Further, we tried different filters associated with different wavelets namely — Biorthogonal, Coiflets, Daubechies, Symlets and Reverse Biorthogonal Wavelets. Reverse biorthogonal filters provided the competing results. The entire experimentation was performed using MATLAB. In Section 2, the properties of sparsity inducing non-smooth penalty functions are accentuated. Later we described the M-M class of algorithms [5], highlighting majorization-minimization. In Section 3, first the notations involved are addressed and then the M-M algorithm used in our formulation is explained. Later in Section **??**, 1-D example and corresponding experimental results are provided.

# 2 Methodology

## 2.1 Sparsity Inducing Functions

The convex proxy for sparsity i.e. $l_1 norm$ has a special importance in sparse signal processing. Nevertheless, sparsity inducing NCP functions provide better realization in several signal estimation problems. In the formulation (2), parameter $c$ will ensure that the addition of a non-smooth penalty function doesn't alter the overall convexity of cost function $F$. Parameter $c$ can be chosen to make NCP function maximally non convex [4]. In order to maintain convexity of the cost function $F$, range derived in [3] is utilized, which is given by

$$0 < c_j < \frac{1}{\lambda_j} \tag{3}$$

where $j = 0, 1, 2....L$ corresponds to the different levels of transformation. If parameter $c$ fails to comply with (3), then global optimal solution (minima) can't be ensured, means the geometry of the NCP function affects the solution of the overall cost function. Later we will see in section 2 that $c_j = 1/\lambda_j$ provided competing results. Alongwith the aforementioned condition, any non-convex penalty function $\psi(.;c) : \mathbb{R} \rightarrow \mathbb{R}$ must adheres to the following conditions [1] to provide global minima

1. $\psi$ is continuous function on $\mathbb{R}$
2. $\psi$ is differentiable twice on $\mathbb{R} \setminus \{0\}$
3. Slope of $\psi(x)$ is unity at immediate right of origin, and on immediate left its negative unity

$$\psi'(0+) = 1 \; and \; \psi'(0-) = -1$$

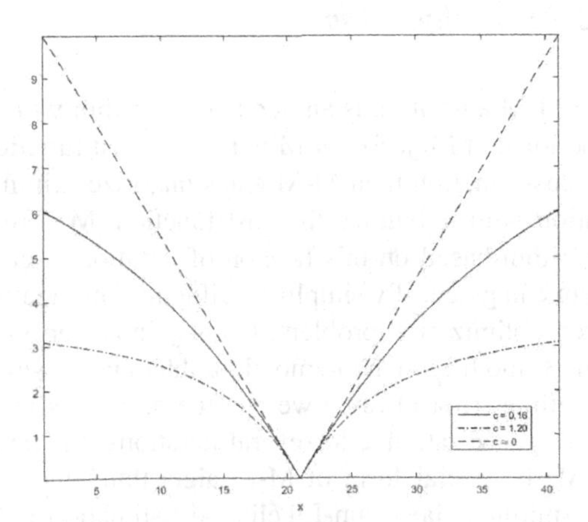


**Fig. 1** Logarithmic NCP function

4. For $x > 0$, $\psi'(x)$ is decreasing from 1 to 0 and for $x < 0$, $\psi'(x)$ is increasing from $-1$ to 0 (Refer Figure 1 to visualize)
5. $\psi(x)$ is symmetric i.e. function is unchanged by any permutation of variable x.

$$\psi(-x;c) = \psi(x;c)$$

6. $l_1 norm$ can be retrieved as a special case of $\psi(x;c)$, with $c = 0$
7. The greatest lower bound to the value of $\psi''(x)$ is $-c$

Examples of several NCP functions used in experimentation are listed in Table 1. NCP function is not differentiable at origin, because of a discontinuity of $\psi(x)$ at $x = 0$, which is illustrated in Fig.1. In Fig.1, the variation in the function with different values of parameter $c$ could be observed. It turns out that, mathematically we can generalize the derivative of a non-differentiable function using sub-gradient. In convex analysis, sub-gradients are used when convex function is not differentiable [6].

Table 1: Examples of NCP Functions

| NC Penalty Function | Expression |
|---|---|
| Rational | $\psi(x) = \frac{\lvert x\rvert}{1+c\lvert x\rvert/2}$ |
| Logarithmic | $\psi(x) = \frac{1}{c}\log(1+c\lvert x\rvert)$ |
| Arctangent | $\psi(x) = \frac{2}{c\sqrt{3}}\left(\tan^{-1}\left(\frac{1+2c\lvert x\rvert}{\sqrt{3}}\right) - \frac{\pi}{6}\right)$ |

## 2.2 M-M algorithm

The M-M algorithm is an iterative algorithm which operates by generating a proxy function that Majorize or Minorize the cost function. If the intention is to minimize the cost function then M-M reads majorize-minimize. On the other hand if the intention is to maximize the cost function, MM reads minorize-minimize [7]. Any algorithm based on this fashion of iteration is called M-M algorithm. M-M algorithm can gracefully simplify a difficult optimization problem by (1) linearizing the given optimization problem, (2) disjoining the variables in the optimization problem (3) modifiying non-smooth problem into a smooth problem [5]. It turns out that sometimes (not always) we pay the price of simplifying the problem with slower convergence rate due to several iterations. In statistics, Expectation-Maximization (E-M) is special form of M-M algorithm. It is a widely applicable approach for determining Maximum-Likelihood-Estimation (MLE) [8]. The M-M algorithm is usually easier to grasp than E-M algorithm. In this paper, we used M-M algorithm for minimizing the cost function, hence it reads majorize-minimize. Further explanation will consider the case of minimizing the cost function. The initial part of M-M algorithm's implementation is to define the majorizer ($G_m(x)$), $m = 0, 1, 2...$, where m denotes the current majorizer. The idea behind using a majorizer is that it is easier to solve than $F(x)$ ($G_m(x)$ must be convex function). A function $G_m(x)$ is called a majorizer of another function $F(x)$ at $x = x_m$ iff

$$F(x) \leq G_m(x)\ for\ all\ x \tag{4}$$

$$F(x) = G_m(x)\ at\ x = x_m \tag{5}$$

In other words, (4) signifies that the surface of $G_m(x)$ always lies above the surface of $F(x)$ and (5) signifies that $G_m(x)$ is tangent to $F(x)$ at point $x = x_m$. The basic intuition regarding equation (4) and (5) could be developed from Fig. 2. In M-M algorithm, for each iteration we must find a majorizer first with upper bound and then minimize it. Hence the name Majorization-Minimization. If $x_{m+1}$ is minima of current proxy function ($G_m(x)$), M-M algorithm derives the cost function $F(x)$ downwards. It is demonstrated in [5] that numerical stability of M-M approach depends on decreasing the proxy function rather than minimizing it. In a similar way, M-M approach works for multi-dimensional function where it is more effective.

As a general practice, quadratic functions are preferred as majorizers because derivative provides a linear system. A polynomial of higher order could also be used, however it will make the solution difficult (non-linear system). To minimize $F(x)$ using M-M approach, we can either majorize $\frac{1}{2}||y - x||^2$ [9] or NCP function [10] or both. In our experimentation, we majorized the NCP function.

## 3 Problem Formulation

In this paper the importance of NCP functions for 1-D signal denoising is addressed as NCP functions promotes sparsity. The NCP function should be chosen to assure

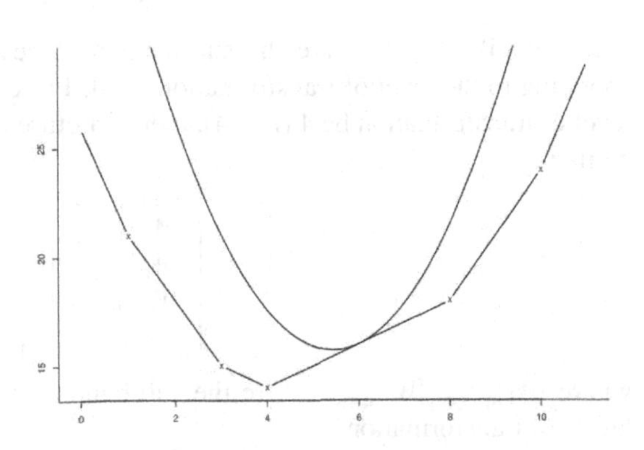


**Fig. 2** A Majorizing Function for Piecewise Defined Linear Function

the convexity of overall cost function. A benefit of this approach is that we can arrive at the solution by using convex optimization methods.

### 3.1 Notations

The signal $x$ to be estimated is represented by an N-Point vector

$$x = [x_1, x_2, x_3, ........., x(N)]$$

The NCP function $\psi(x;c)$ used in our experimentation parametrized by $c$ is represented as [4]

$$\psi(x) = \frac{2}{a\sqrt{3}}\left(\tan^{-1}\left(\frac{1+2a|x|}{\sqrt{3}}\right) - \frac{\pi}{6}\right)$$

The wavelet transform matrix is represented by a square matrix [2] W. The Matlab syntax to generate the wavelet matrix is as follows $W = wavmat(N, F_1, F_2, L)$ where N denotes the length of the signal, $F_1$ and $F_2$ are decomposition/reconstruction filters which can be obtained by Matlab function $wfilters('wavelet-name')$; for different wavelets, L denotes the transformation level (we used $L = 4$).

### 3.2 Sub-band matrices of wavelet transform matrix

Structure of a wavelet transform matrix W could be interpreted as

$$W = \begin{bmatrix} W_1 \\ W_2 \\ \vdots \\ \vdots \\ W_{L+1} \end{bmatrix}$$

where $W_1, W_2, .....W_{L+1}$ are the sub-band matrices of W with dimensions corresponding to the level of transformation used. For e.g. Let signal length be $N = 64$, level of transformation be 4 ($L = 4$), then structure of wavelet transform matrix will be like

$$W = \begin{bmatrix} [W_1]_{4\times64} \\ [W_2]_{4\times64} \\ [W_3]_{8\times64} \\ [W_4]_{16\times64} \\ [W_5]_{32\times64} \end{bmatrix}_{64\times64}$$

where $[W_1]_{4\times64}$, $[W_2]_{4\times64}$ ... are the sub-band matrices corresponding to different levels of transformation.

### 3.3 Algorithm

Consider the cost function in (2), M-M algorithm generates a sequence of simpler optimization problems as

$$x_{m+1} = \underset{x}{\operatorname{argmin}}\, G_m(x_m) \tag{6}$$

i.e. in each iteration we are solving a smaller convex optimization problem. The expression in (6) is used to update $x_m$ in each iteration with $x_0 = z$ as initialization. Each iteration in M-M algorithm will have different majorizer which must fit an upper bound for NCP function i.e. it should follow (4) & (5). As mentioned in Section 2, we used second order polynomial as majorizer for simpler solution to our optimization problem.

As mentioned in [11], scalar case majorizer for NCP function could be given by

$$g(x;s) = \frac{\psi'(s)}{2s}x^2 + \left(\psi(s) - \frac{s}{2}\psi'(s)\right) \tag{7}$$

It can be modified as

$$g(x;s) = \frac{\psi'(s)}{2s}x^2 + \kappa \tag{8}$$

where $\kappa$ is

$$\psi(s) - \frac{s}{2}\psi'(s) \tag{9}$$

the term in parenthesis (i.e. $\kappa$) in (7) is merely a constant which can be avoided for solving optimization problem.
Equivalently, the corresponding vector case can be derived as

$$\begin{aligned} g([Wx],[Ws]) = [Wx]_1^T \frac{\psi'([Ws]_1)}{2[Ws]_1}[Wx]_1 + \ldots \\ \ldots + [Wx]_n^T \frac{\psi'([Ws]_n)}{2[Ws]_n}[Wx]_n + \kappa \end{aligned}$$

where W is Wavelet matrix, $[Wx]_n$ is the $n_{th}$ component of vector $Wx$, $[Ws]_n$ is the $n_{th}$ component of vector $Ws$. Above equation can also be written as

$$g([Wx],[Ws]) = \sum_{i=1}^{n} [Wx]_i^T \frac{\psi'([Ws]_i)}{2[Ws]_i}[Wx]_i + \kappa \tag{10}$$

The more compact form of (10) is given by

$$g([Wx],[Ws]) = \frac{1}{2}(Wx)^T \Lambda (Wx) + \kappa \tag{11}$$

Where

$$\Lambda = \begin{bmatrix} \frac{\psi'([Ws]_1)}{[Ws]_1} & & \\ & \ddots & \\ & & \frac{\psi'([Ws]_n)}{[Ws]_n} \end{bmatrix}$$

Therefore, using (4) we can say that

$$g\left([Wx],[Ws]\right) \geq \psi([Wx]) \tag{12}$$

Hence $g([Wx],[Ws])$ is a majorizer to $\psi([Wx])$. Therefore, using (11) & (12), we can directly give majorizer for (2) by

$$G(x,s) = \frac{1}{2}||z-x||^2 + \frac{1}{2}\lambda(Wx)^T\Lambda(Wx) + \kappa \tag{13}$$

To avoid any further confusion, $\Lambda$ will consume all the different $\lambda$'s. Finally. It will appear as

$$\Lambda^\lambda = \begin{bmatrix} (\lambda_1)\frac{\psi'([Ws]_1)}{[Ws]_1} & & \\ & \ddots & \\ & & (\lambda_{L+1})\frac{\psi'([Ws]_n)}{[Ws]_n} \end{bmatrix} \tag{14}$$

expression in (13) becomes

$$G(x,s) = \frac{1}{2}||z-x||^2 + \frac{1}{2}(Wx)^T\Lambda^\lambda(Wx) + \kappa \tag{15}$$

Using M-M algorithm, $x_m$ can be calculated by minimizing (16) as

$$x_{m+1} = \underset{x}{\text{argmin}}\, G_m(x,s) \tag{16}$$

$$x_{m+1} = \underset{x}{\text{argmin}}\, \frac{1}{2}||z-x||^2 + \frac{1}{2}(Wx)^T\Lambda^\lambda(Wx) + \kappa \tag{17}$$

$$x_{m+1} = \underset{x}{\text{argmin}}\, \frac{1}{2}||z-x||^2 + \lambda\frac{1}{2}x^TW^T\Lambda^\lambda Wx + \kappa \tag{18}$$

minimizing (18), we can easily explicitly arrive at

$$x_{m+1} = \left(I + W^T\Lambda^\lambda W\right)^{-1} z \tag{19}$$

The only problem with the update equation arrives when the term $[Ws]_n$ goes to zero, the corresponding entries in $\Lambda$ becomes infinite. Therefore, expression (19) may become infinite. However, To avoid this unstable state, *Woodbury Matrix Identity* (more commonly called as the matrix inversion lemma) [12] could be used, which is given in the form

$$(A+XBY)^{-1} = A^{-1} - A^{-1}X(B^{-1}+YA^{-1}X)^{-1}YA^{-1} \tag{20}$$

using (20), our update equation (19) becomes

$$x_{m+1} = z - W^T \left( \Lambda^{-\lambda} + WW^T \right)^{-1} \tag{21}$$

Clearly (21) solves for $\Lambda^{-\lambda}$ instead of $\Lambda$. Therefore, entries of $\Lambda^{-\lambda}$ will become zero, instead of becoming infinity. Above algorithm can be easily implemented using MATLAB. In depth analysis of M-M algorithm can be found in Chapter 12 of [5].

The M-M approach explained above is summarized as
*Step 1*: Set $m = 0$. Initialize by setting $x_0 = z$
*Step 2*: Choose $G_m(x)$ for $F(x)$ such that

$$F(x) \leq G_m(x) \; for \; all \; x$$

$$F(x) = G_m(x) \; at \; x = x_m$$

*Step 3*: Find $x_{m+1}$ by minimizing

$$x_{m+1} = \underset{x}{\operatorname{argmin}} \, G_m(x, s)$$

*Step 4*: Increment $m$ using $m = m + 1$, jump to *Step 2*

## 4 Example

During experimentation, $1-D$ signal denoising problemwas considered. The synthetic noisy signal used for experimentation purpose was generated using $MakeSignal()$ function from Wavelab tool with Additive White Gaussian Noise of $\sigma = 4$. The wavelab tool is available at: `http://statweb.stanford.edu/~wavelab/` The value of parameter $c$ which maintains the convexity could be is calculated using (3). Maximal sparse solution was noted with $c = 1/\lambda$. To compute the values associated with $\lambda_j$, expression given by [1] was further modified as

$$\lambda_j = \sigma \times \beta \times 2^{-j/2} \times \frac{N}{2^{(\log(N/2)-j)}}, 1 \leq j \leq L \tag{22}$$

Where $N$ denotes the length of the signal and L denotes transformation level. Note that, we used $\lambda_0 = 1$ for first sub-band matrix $W_1$. Further, different $(\lambda_j)$ for different band of wavelet transform matrix was used. As mentioned in Section 3, arc tangent NCP function is used. Using trial and error, 4-Level transformation in Wavelet Matrix (i.e. $L = 4$) gave better results. Further, reconstruction and decomposition high pass filters of reverse biorthogonal wavelet ($rbio2.2$) were employed, which

Table 2: RMSE values for different wavelet filters

| Wavelet filter | Root Mean Square Error |
|---|---|
| Biorthogonal ('bior1.3') | 1.6881 |
| Coiflets ('coif1') | 1.5902 |
| Daubechies ('db2') | 1.4844 |
| Biorthogonal ('bior2.2') | 1.4811 |
| Reverse Biorthogonal ('rbio2.2') | 1.4565 |

provided improved results. However, experiment with biorthogonal, daubechies, coiflets and Symlets wavelets was also conducted. Symlets family of wavelets is merely a modified version of daubechies with similar properties and improved symmetry. Symlets provided the identical result as of daubechies. Table 2 lists the different RMSE values obtained using different wavelet filters with $\beta = 0.98$. The outcomes obtained (Table 2) were compared with outcomes obtained in [1] and $l_1$ regularization. It can be observed in Figure 3 that Non-convex regularization with decimated wavelet gave better results. In Fig 3, it can be observed that non-convex regularization using decimated wavelets preserved the peaks of the given signal. Therefore, same approach could be used for signals pertaining sharp peaks. Example of one such signal is Electrocardiogram (ECG) signal.

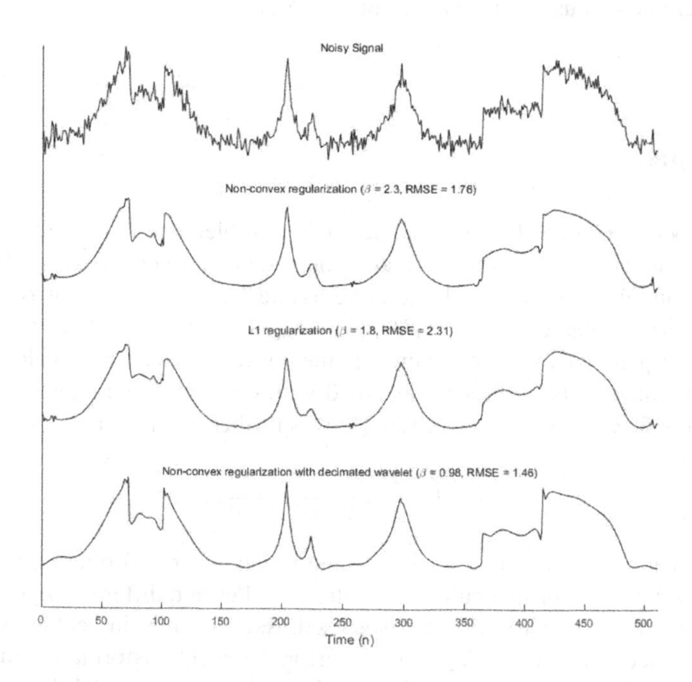


Fig. 3: Example of 1-D Denoising

## 5 Conclusion

The attempt in this paper is to provide a pedagogical approach to the ingenious methodology proposed by Ankit parekh et al [1]. An approach for 1-D signal denoising using decimated wavelet transform is proposed. The sub-band nature of wavelet transform matrix (W) is exploited to get easier and better understanding and solution for the given denoising problem. The problem formulation comprised a smooth and a non-smooth term with a parameter $c$ which controls the overall convexity. The solution to this formulation is obtained using M-M iterative algorithm. The proposed approach offered better experimental results than non-convex regularization [1] and $l_1$ regularization. The same functional procedure could be extended for denoising a noisy image.

## References

1. Ankit Parekh and Ivan W Selesnick. Convex denoising using non-convex tight frame regularization. *Signal Processing Letters, IEEE*, 22(10):1786–1790, 2015.
2. Jie Yan. Wavelet matrix. *Dept. of Electrical and Computer Engineering, University of Victoria, Victoria, BC, Canada*, 2009.
3. Ivan Selesnick. Penalty and shrinkage functions for sparse signal processing. *Connexions*, 2012.
4. Ivan W Selesnick and Ilker Bayram. Sparse signal estimation by maximally sparse convex optimization. *Signal Processing, IEEE Transactions on*, 62(5):1078–1092, 2014.
5. Kenneth Lange. *Numerical analysis for statisticians*. Springer Science & Business Media, 2010.
6. Jan van Tiel. *Convex analysis*. John Wiley, 1984.
7. David R Hunter and Kenneth Lange. A tutorial on mm algorithms. *The American Statistician*, 58(1):30–37, 2004.
8. Geoffrey McLachlan and Thriyambakam Krishnan. *The EM algorithm and extensions*, volume 382. John Wiley & Sons, 2007.
9. Ingrid Daubechies, Michel Defrise, and Christine De Mol. An iterative thresholding algorithm for linear inverse problems with a sparsity constraint. *Communications on pure and applied mathematics*, 57(11):1413–1457, 2004.
10. Ivan Selesnick. Total variation denoising (an mm algorithm). *NYU Polytechnic School of Engineering Lecture Notes*, 2012.
11. Ivan W Selesnick, Ankit Parekh, and Ilker Bayram. Convex 1-d total variation denoising with non-convex regularization. *Signal Processing Letters, IEEE*, 22(2):141–144, 2015.
12. Mario AT Figueiredo, J Bioucas Dias, Joao P Oliveira, and Robert D Nowak. On total variation denoising: A new majorization-minimization algorithm and an experimental comparisonwith wavalet denoising. In *Image Processing, 2006 IEEE International Conference on*, pages 2633–2636. IEEE, 2006.

# Author Index





J.M. Corchado Rodriguez et al. (eds.), *Intelligent Systems Technologies and Applications 2016*, Advances in Intelligent Systems and Computing 530,
DOI 10.1007/978-3-319-47952-1